HANDBOOK OF
IMAGE AND
VIDEO PROCESSING

Academic Press Series in Communications, Networking, and Multimedia

EDITOR-IN-CHIEF

Jerry D. Gibson
Southern Methodist University

This series has been established to bring together a variety of publications that represent the latest in cutting-edge research, theory, and applications of modern communication systems. All traditional and modern aspects of communications as well as all methods of computer communications are to be included. The series will include professional handbooks, books on communication methods and standards, and research books for engineers and managers in the world-wide communications industry.

HANDBOOK OF
IMAGE AND
VIDEO PROCESSING

EDITOR

AL BOVIK

DEPARTMENT OF ELECTRICAL AND COMPUTER ENGINEERING
THE UNIVERSITY OF TEXAS AT AUSTIN
AUSTIN, TEXAS

ACADEMIC PRESS

A Harcourt Science and Technology Company

SAN DIEGO / SAN FRANCISCO / NEW YORK / BOSTON / LONDON / SYDNEY / TOKYO

ACADEMIC PRESS
A Harcourt Science and Technology Company
525 B Street, Suite 1900, San Diego, CA 92101-4495, USA
http://www.academicpress.com

Academic Press
Harcourt Place, 32 Jamestown Road, London, NW1 7BY, UK
http://www.hbuk.co.uk/ap/

Library of Congress Catalog Number: 99-69120

ISBN: 0-12-119790-5

Printed in Canada

00 01 02 03 04 05 FR 9 8 7 6 5 4 3 2 1

Preface

This *Handbook* represents contributions from most of the world's leading educators and active research experts working in the area of *Digital Image and Video Processing*. Such a volume comes at a very appropriate time, since finding and applying improved methods for the acquisition, compression, analysis, and manipulation of visual information in digital format has become a focal point of the ongoing revolution in information, communication and computing. Moreover, with the advent of the world-wide web and digital wireless technology, digital image and video processing will continue to capture a significant share of "high technology" research and development in the future. This *Handbook* is intended to serve as the basic reference point on image and video processing, both for those just entering the field as well as seasoned engineers, computer scientists, and applied scientists that are developing tomorrow's image and video products and services.

The goal of producing a truly comprehensive, in-depth volume on *Digital Image and Video Processing* is a daunting one, since the field is now quite large and multidisciplinary. Textbooks, which are usually intended for a specific classroom audience, either cover only a relatively small portion of the material, or fail to do more than scratch the surface of many topics. Moreover, any textbook must represent the specific point of view of its author, which, in this era of specialization, can be incomplete. The advantage of the current *Handbook* format is that every topic is presented in detail by a distinguished expert who is involved in teaching or researching it on a daily basis.

This volume has the ambitious intention of providing a resource that covers introductory, intermediate and advanced topics with equal clarity. Because of this, the *Handbook* can serve equally well as reference resource and as classroom textbook. As a reference, the *Handbook* offers essentially all of the material that is likely to be needed by most practitioners. Those needing further details will likely need to refer to the academic literature, such as the *IEEE Transactions on Image Processing*. As a textbook, the *Handbook* offers easy-to-read material at different levels of presentation, including several introductory and tutorial chapters and the most basic image processing techniques. The *Handbook* therefore can be used as a basic text in introductory, junior- and senior-level undergraduate, and graduate-level courses in digital image and/or video processing. Moreover, the *Handbook* is ideally suited for short courses taught in industry forums at any or all of these levels. Feel free to contact the Editor of this volume for one such set of computer-based lectures (representing 40 hours of material).

The *Handbook* is divided into ten major sections covering more than 50 Chapters. Following an Introduction, Section 2 of the *Handbook* introduces the reader to the most basic methods of gray-level and binary image processing, and to the essential tools of image Fourier analysis and linear convolution systems. Section 3 covers basic methods for image and video recovery, including enhancement, restoration, and reconstruction. Basic Chapters on Enhancement and Restoration serve the novice. Section 4 deals with the basic modeling and analysis of digital images and video, and includes Chapters on wavelets, color, human visual modeling, segmentation, and edge detection. A valuable Chapter on currently available software resources is given at the end. Sections 5 and 6 deal with the major topics of image and video compression, respectively, including the JPEG and MPEG standards. Sections 7 and 8 discuss the practical aspects of image and video acquisition, sampling, printing, and assessment. Section 9 is devoted to the multimedia topics of image and video databases, storage, retrieval, and networking. And finally, the *Handbook* concludes with eight exciting Chapters dealing with applications. These have been selected for their timely interest, as well as their illustrative power of how image processing and analysis can be effectively applied to problems of significant practical interest.

As Editor and Co-Author of this *Handbook*, I am very happy that it has been selected to lead off a major new series of handbooks on Communications, Networking, and Multimedia to be published by Academic Press. I believe that this is a real testament to the current and growing importance of digital image and video processing. For this opportunity I would like to thank Jerry Gibson, the series Editor, and Joel Claypool, the Executive Editor, for their faith and encouragement along the way.

Last, and far from least, I'd like to thank the many co-authors who have contributed such a fine collection of articles to this *Handbook*. They have been a model of professionalism, timeliness, and responsiveness. Because of this, it was my pleasure to carefully read and comment on every single word of every Chapter, and it has been very enjoyable to see the project unfold. I feel that this *Handbook of Image and Video Processing* will serve as an essential and indispensable resource for many years to come.

Al Bovik
Austin, Texas
1999

Editor

Al Bovik is the General Dynamics Endowed Fellow and Professor in the Department of Electrical and Computer Engineering at the University of Texas at Austin, where he is the Associate Director of the Center for Vision and Image Sciences. He has published nearly 300 technical articles in the general area of image and video processing areas and holds two U.S. patents.

Dr. Bovik is a recipient of the IEEE Signal Processing Society Meritorious Service Award (1998), and is a two-time Honorable Mention winner of the international Pattern Recognition Society Award. He is a Fellow of the IEEE, is the Editor-in-Chief of the *IEEE Transactions on Image Processing*, serves on many other boards and panels, and was the Founding General Chairman of the IEEE International Conference on Image Processing, which was first held in Austin, Texas in 1994.

Contributors

Scott T. Acton
Oklahoma State University
Stillwater, Oklahoma

Jake K. Aggarwal
The University of Texas at Austin
Austin, Texas

Jan P. Allebach
Purdue University
West Lafayette, Indiana

Rashid Ansari
University of Illinois at Chicago
Chicago, Illinois

Supavadee Aramvith
University of Washington
Seattle, Washington

Gonzalo Arce
University of Delaware
Newark, Delaware

Barry Barnett
The University of Texas at Austin
Austin, Texas 78759

Keith A. Bartels
Southwest Research Institute
San Antonio, Texas

Jan Biemond
Delft University of Technology
Delft, The Netherlands

Charles G. Boncelet, Jr.
University of Delaware
Newark, Delaware

Charles A. Bouman
Purdue University
West Lafayette, Indiana

Alan C. Bovik
The University of Texas at Austin
Austin, Texas

Kevin W. Bowyer
University of South Florida
Tampa, Florida

Walter Carrara
Nonlinear Dynamics, Inc.
Ann Arbor, Michigan

Rama Chellappa
University of Maryland
College Park, Maryland

Tsuhan Chen
Carnegie Mellon University
Pittsburgh, Pennsylvania

Rolf Clackdoyle
Medical Imaging Research Laboratory
University of Utah

Lawrence K. Cormack
The University of Texas at Austin
Austin, Texas

Edward J. Delp
Purdue University
West Lafayette, Indiana

Mita D. Desai
The University of Texas at San Antonio
San Antonio, Texas

Kenneth R. Diller
The University of Texas at Austin
Austin, Texas

Eric Dubois
University of Ottawa
Ottawa, Ontario, Canada

Adriana Dumitras
University of British Columbia
Vancouver, British Columbia, Canada

Touradj Ebrahimi
EPFL
Lausanne, Switzerland

Berna Erol
University of British Columbia
Vancouver, British Columbia, Canada

Brian L. Evans
The University of Texas at Austin
Austin, Texas

P. Fieguth
University of Waterloo
Ontario, Canada

Nikolas P. Galatsanos
Illinois Institute of Technology
Chicago, Illinois

Joydeep Ghosh
The University of Texas at Austin
Austin, Texas

Ron Goodman
ERIM International, Inc.
Ann Arbor, Michigan

Ulf Grendander
Brown University
Providence, Rhode Island

G. M. Haley
Ameritech
Hoffman Estates, Illinois

Soo-Chul Han
Lucent Technologies
Murray Hill, New Jersey

Joe Havlicek
University of Oklahoma
Norman, Oklahoma

Michael D. Heath
University of South Florida
Tampa, Florida

William E. Higgins
Pennsylvania State University
University Park, Pennsylvania

Shih-Ta Hsiang
Rensselaer Polytechnic Institute
Troy, New York

Thomas S. Huang
University of Illinois at Urbana–Champaign
Urbana, Illinois

Anil Jain
Michigan State University
East Lansing, Michigan

Lina J. Karam
Arizona State University
Tempe, Arizona

William C. Karl
Boston University
Boston, Massachusetts

Aggelos K. Katsaggelos
Northwestern University
Evanston, Illinois

Mohammad A. Khan
Georgia Institute of Technology
Atlanta, Georgia

Janusz Konrad
INRS Télécommunications
Verdun, Quebec, Canada

Faouzi Kossentini
University of British Columbia
Vancouver, British Columbia, Canada

Murat Kunt
Signal Processing Laboratory, EPFL
Lausanne, Switzerland

Reginald L. Lagendijk
Delft University of Technology
Delft, The Netherlands

Sridhar Lakshmanan
University of Michigan – Dearborn
Dearborn, Michigan

Richard M. Leahy
University of Southern California
Los Angeles, California

Wei-Ying Ma
Hewlett-Packard Laboratories
Palo Alto, California

Chhandomay Mandal
The University of Texas at Austin
Austin, Texas

B. S. Manjunath
University of California
Santa Barbara, California

Petros Maragos
National Technical University of Athens
Athens, Greece

Nasir Memon
Polytechnic University
Brooklyn, New York

Fatima A. Merchant
Perceptive Scientific Instruments, Inc.
League City, Texas

Michael I. Miller
Johns Hopkins University
Baltimore, Maryland

Phillip A. Mlsna
Northern Arizona University
Flagstaff, Arizona

Baback Moghaddam
Mitsubishi Electric Research Laboratory
  (MERL)
Cambridge, Massachusetts

Pierre Moulin
University of Illinois
Urbana, Illinois

John Mullan
University of Delaware
Newark, Delaware

T. Naveen
Tektronix
Beaverton, Oregon

Sharath Pankanti
IBM T. J. Watson Research Center
Yorktown Heights, New York

Thrasyvoulos N. Pappas
Northwestern University
Evanston, Illinois

Jose Luis Paredes
University of Delaware
Newark, Delaware

Alex Pentland
Massachusetts Institute of Technology
Cambridge, Massachusetts

Lucio F. C. Pessoa
Motorola, Inc.
Austin, Texas

Ioannis Pitas
University of Thessaloniki
Thessaloniki, Greece

Kannan Ramchandran
University of California, Berkeley
Berkeley, California

Joseph M. Reinhardt
University of Iowa
Iowa City, Iowa

Jeffrey J. Rodriguez
The University of Arizona
Tucson, Arizona

Peter M. B. van Roosmalen
Delft University of Technology
Delft, The Netherlands

Yong Rui
Microsoft Research
Redmond, Washington

Martha Saenz
Purdue University
West Lafayette, Indiana

Robert J. Safranek
Lucent Technologies
Murray Hill, New Jersey

Paul Salama
Purdue University
West Lafayette, Indiana

Dan Schonfeld
University of Illinois at Chicago
Chicago, Illinois

Timothy J. Schulz
Michigan Technological University
Houghton, Michigan

K. Clint Slatton
The University of Texas at Austin
Austin, Texas

Mark J. T. Smith
Georgia Institute of Technology
Atlanta, Georgia

Michael A. Smith
Carnegie Mellon University
Pittsburgh, Pennsylvania

Shridhar Srinivasan
Sensar Corporation
Princeton, New Jersey

Anuj Srivastava
Florida State University
Tallahassee, Florida

Ming-Ting Sun
University of Washington
Seattle, Washington

A. Murat Tekalp
University of Rochester
Rochester, New York

Daniel Tretter
Hewlett Packard Laboratories
Palo Alto, California

H. Joel Trussell
North Carolina State University
Raleigh, North Carolina

Chun-Jen Tsai
Northwestern University
Evanston, Illinois

Baba C. Vemuri
University of Florida
Gainesville, Florida

George Voyatzis
University of Thessaloniki
Thessaloniki, Greece

D. Wang
Samsung Electronics
San Jose, California

Dong Wei
Drexel University
Philadelphia, Pennsylvania

Miles N. Wernick
Illinois Institute of Technology
Chicago, Illinois

Ping Wah Wong
Gainwise Limited
Cupertino, California

John W. Woods
Rensselaer Polytechnic Institute
Troy, New York

Zixiang Xiong
Texas A&M University
College Station, Texas

Jun Zhang
University of Wisconsin at Milwaukee
Milwaukee, Wisconsin

Huaibin Zhao
The University of Texas at Austin
Austin, Texas

Contents

SECTION I Introduction

SECTION II Basic Image Processing Techniques

SECTION III Image and Video Processing

Image and Video Enhancement and Restoration

SECTION IV   Image and Video Analysis

SECTION V   Image Compression

SECTION VI   Video Compression

SECTION VII   Image and Video Acquisition

SECTION VIII   Image and Video Rendering and Assessment

SECTION IX   Image and Video Storage, Retrieval and Communication

SECTION X   Applications of Image Processing

Introduction

I

1.1

Introduction to Digital Image and Video Processing

Alan C. Bovik
The University of Texas at Austin

As we enter the new millennium, scarcely a week passes where we do not hear an announcement of some new technological breakthrough in the areas of digital computation and telecommunication. Particularly exciting has been the participation of the general public in these developments, as affordable computers and the incredible explosion of the World Wide Web have brought a flood of instant information into a large and increasing percentage of homes and businesses. Most of this information is designed for *visual* consumption in the form of text, graphics, and pictures, or integrated *multimedia* presentations. *Digital images* and *digital video* are, respectively, pictures and movies that have been converted into a computer-readable binary format consisting of logical 0s and 1s. Usually, by an image we mean a still picture that does not change with time, whereas a video evolves with time and generally contains moving and/or changing objects. Digital images or video are usually obtained by converting continuous signals into digital format, although "direct digital" systems are becoming more prevalent. Likewise, digital visual signals are viewed by using diverse display media, included digital printers, computer monitors, and digital projection devices. The frequency with which information is transmitted, stored, processed, and displayed in a digital visual format is increasing rapidly, and thus the design of engineering methods for efficiently transmitting, maintaining, and even improving the visual integrity of this information is of heightened interest.

One aspect of image processing that makes it such an interesting topic of study is the amazing diversity of applications that use image processing or analysis techniques. Virtually every branch of science has subdisciplines that use recording devices or sensors to collect image data from the universe around us, as depicted in Fig. 1. These data are often multidimensional and can be arranged in a format that is suitable for human viewing. Viewable datasets like this can be regarded as images, and they can be processed by using established techniques for image processing, even if the information has not been derived from visible-light sources. Moreover, the data may be recorded as they change over time, and with faster sensors and recording devices, it is becoming easier to acquire and analyze digital video datasets. By mining the rich spatiotemporal information that is available in video, one can often analyze the growth or evolutionary properties of dynamic physical phenomena or of living specimens.

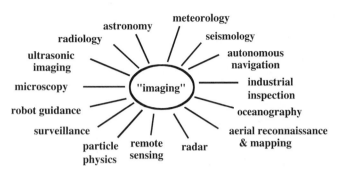

FIGURE 1   Part of the universe of image processing applications.

Types of Images

Another rich aspect of digital imaging is the diversity of image types that arise, and that can derive from nearly every type of radiation. Indeed, some of the most exciting developments in medical imaging have arisen from new sensors that record image data from previously little-used sources of radiation, such as PET (positron emission tomography) and MRI (magnetic resonance imaging), or that sense radiation in new ways, as in CAT (computer-aided tomography), where X-ray data are collected from multiple angles to form a rich aggregate image.

There is an amazing availability of radiation to be sensed, recorded as images or video, and viewed, analyzed, transmitted, or stored. In our daily experience we think of "what we see" as being "what is there," but in truth, our eyes record very little of the information that is available at any given moment. As with any sensor, the human eye has a limited bandwidth. The band of electromagnetic (EM) radiation that we are able to see, or "visible light," is quite small, as can be seen from the plot of the EM band in Fig. 2. Note that the horizontal axis is logarithmic! At any given moment, we see very little of the available radiation that is going on around us, although certainly enough to get around. From an evolutionary perspective, the band of EM wavelengths that the human eye perceives is perhaps optimal, since the volume of data is reduced, and the data that are used are highly reliable and abundantly available (the Sun emits strongly in the visible bands, and the Earth's atmosphere is also largely transparent in the visible wavelengths). Nevertheless, radiation from other bands can be quite useful as we attempt to glean the fullest possible amount of information from the world around us. Indeed,

certain branches of science sense and record images from nearly all of the EM spectrum, and they use the information to give a better picture of physical reality. For example, astronomers are often identified according to the type of data that they specialize in, e.g., radio astronomers, X-ray astronomers, and so on. Non-EM radiation is also useful for imaging. A good example are the high-frequency sound waves (ultrasound) that are used to create images of the human body, and the low-frequency sound waves that are used by prospecting companies to create images of the Earth's subsurface.

One commonality that can be made regarding nearly all images is that radiation is emitted from some source, then interacts with some material, and then is sensed and ultimately transduced into an electrical signal, which may then be digitized. The resulting images can then be used to extract information about the radiation source, and/or about the objects with which the radiation interacts.

We may loosely classify images according to the way in which the interaction occurs, understanding that the division is sometimes unclear, and that images may be of multiple types. Figure 3 depicts these various image types.

Reflection images sense radiation that has been reflected from the surfaces of objects. The radiation itself may be ambient or artificial, and it may be from a localized source, or from multiple or extended sources. Most of our daily experience of optical imaging through the eye is of reflection images. Common non-visible examples include radar images, sonar images, and some types of electron microscope images. The type of information that can be extracted from reflection images is primarily about object surfaces, that is, their shapes, texture, color, reflectivity, and so on.

Emission images are even simpler, since in this case the objects being imaged are self-luminous. Examples include thermal or infrared images, which are commonly encountered in medical,

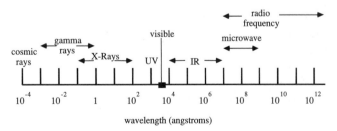

FIGURE 2   The electromagnetic spectrum.

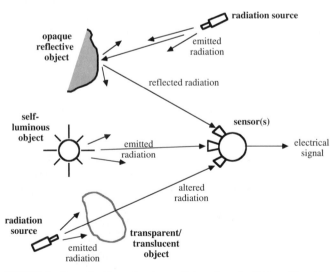

FIGURE 3   Recording the various types of interaction of radiation with matter.

astronomical, and military applications, self-luminous visible-light objects, such as light bulbs and stars, and MRI images, which sense particle emissions. In images of this type, the information to be had is often primarily internal to the object; the image may reveal how the object creates radiation, and thence something of the internal structure of the object being imaged. However, it may also be external; for example, a thermal camera can be used in low-light situations to produce useful images of a scene containing warm objects, such as people.

Finally, *absorption images* yield information about the internal structure of objects. In this case, the radiation passes through objects and is partially absorbed or attenuated by the material composing them. The degree of absorption dictates the level of the sensed radiation in the recorded image. Examples include X-ray images, transmission microscopic images, and certain types of sonic images.

Of course, the preceding classification into types is informal, and a given image may contain objects that interact with radiation in different ways. More important is to realize that images come from many different radiation sources and objects, and that the purpose of imaging is usually to extract information about either the source and/or the objects, by sensing the reflected or transmitted radiation, and examining the way in which it has interacted with the objects, which can reveal physical information about both source and objects.

Figure 4 depicts some representative examples of each of the preceding categories of images. Figures 4(a) and 4(b) depict reflection images arising in the visible-light band and in the microwave band, respectively. The former is quite recognizable; the latter is a synthetic aperture radar image of DFW airport. Figs. 4(c) and 4(d) are emission images, and depict, respectively, a forward-looking infrared (FLIR) image, and a visible-light image of the globular star cluster Omega Centauri. Perhaps the reader can probably guess the type of object that is of interest in Fig. 4(c). The object in Fig. 4(d), which consists of over a million stars, is visible with the unaided eye at lower northern latitudes. Lastly, Figs. 4(e) and 4(f), which are absorption images, are of a digital (radiographic) mammogram and a conventional light micrograph, respectively.

Scale of Images

Examining the pictures in Fig. 4 reveals another image diversity: *scale*. In our daily experience we ordinarily encounter and visualize objects that are within 3 or 4 orders of magnitude of 1 m. However, devices for image magnification and amplification have made it possible to extend the realm of "vision" into the cosmos, where it has become possible to image extended structures extending over as much as 10^{30} m, and into the microcosmos, where it has become possible to acquire images of objects as small as 10^{-10} m. Hence we are able to image from the grandest scale to the minutest scales, over a range of 40 orders

of magnitude, and as we will find, the techniques of image and video processing are generally applicable to images taken at any of these scales.

Scale has another important interpretation, in the sense that any given image can contain objects that exist at scales different from other objects in the same image, or that even exist at multiple scales simultaneously. In fact, this is the rule rather than the exception. For example, in Fig. 4(a), at a small scale of observation, the image contains the bas-relief patterns cast onto the coins. At a slightly larger scale, strong circular structures arose. However, at a yet larger scale, the coins can be seen to be organized into a highly coherent spiral pattern. Similarly, examination of Fig. 4(d) at a small scale reveals small bright objects corresponding to stars; at a larger scale, it is found that the stars are nonuniformly distributed over the image, with a tight cluster having a density that sharply increases toward the center of the image. This concept of multiscale is a powerful one, and it is the basis for many of the algorithms that will be described in the chapters of this *Handbook*.

Dimension of Images

An important feature of digital images and video is that they are *multidimensional signals*, meaning that they are functions of more than a single variable. In the classic study of *digital signal processing*, the signals are usually one-dimensional functions of time. Images, however, are functions of two, and perhaps three space dimensions, whereas digital video as a function includes a third (or fourth) time dimension as well. The dimension of a signal is the number of coordinates that are required to index a given point in the image, as depicted in Fig. 5. A consequence of this is that digital image processing, and especially digital video processing, is quite data intensive, meaning that significant computational and storage resources are often required.

Digitization of Images

The environment around us exists, at any reasonable scale of observation, in a space/time continuum. Likewise, the signals and images that are abundantly available in the environment (before being sensed) are naturally *analog*. By analog, we mean two things: that the signal exists on a continuous (space/time) domain, and that also takes values that come from a continuum of possibilities. However, this *Handbook* is about processing *digital* image and video signals, which means that once the image or video signal is sensed, it must be converted into a computer-readable, digital format. By digital, we also mean two things: that the signal is defined on a discrete (space/time) domain, and that it takes values from a discrete set of possibilities. Before digital processing can commence, a process of *analog-to-digital conversion*

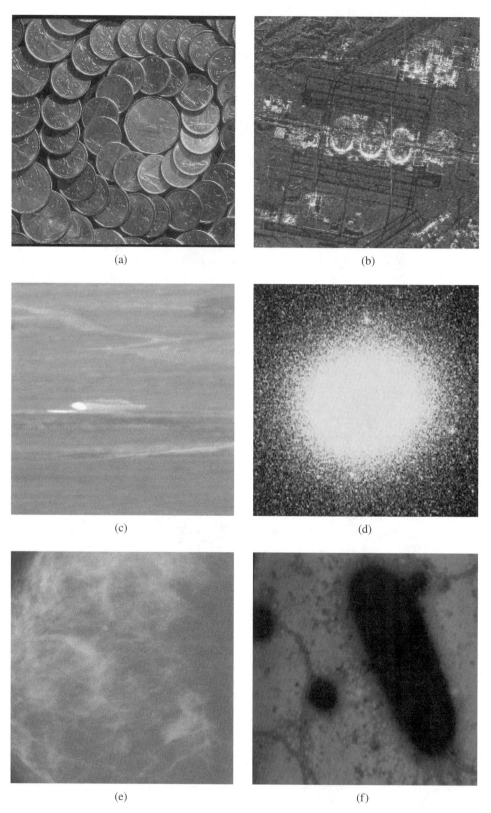

FIGURE 4   Examples of (a), (b), reflection; (c), (d), emission; and (e), (f) absorption image types.

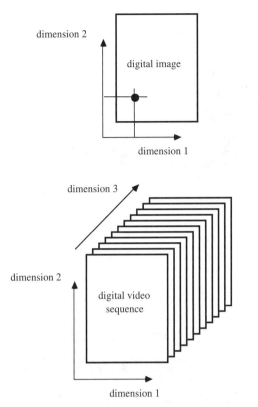

FIGURE 5   The dimensionality of images and video.

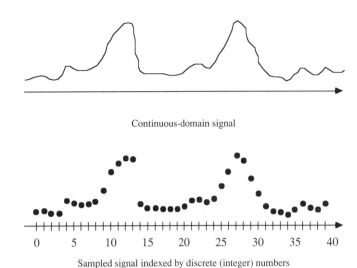

Continuous-domain signal

Sampled signal indexed by discrete (integer) numbers

FIGURE 6   Sampling a continuous-domain one-dimensional signal.

(A/D conversion) must occur. A/D conversion consists of two distinct subprocesses: *sampling* and *quantization*.

Sampled Images

Sampling is the process of converting a continuous-space (or continuous-space/time) signal into a discrete-space (or discrete-space/time) signal. The sampling of continuous signals is a rich topic that is effectively approached with the tools of linear systems theory. The mathematics of sampling, along with practical implementations, are addressed elsewhere in this *Handbook*. In this Introductory Chapter, however, it is worth giving the reader a feel for the process of sampling and the need to sample a signal sufficiently densely. For a continuous signal of given space/time dimensions, there are mathematical reasons why there is a lower bound on the space/time sampling frequency (which determines the minimum possible number of samples) required to retain the information in the signal. However, image processing is a visual discipline, and it is more fundamental to realize that what is usually important is that the process of sampling does not lose *visual information*. Simply stated, the sampled image or video signal must "look good," meaning that it does not suffer too much from a loss of visual resolution, or from artifacts that can arise from the process of sampling.

Figure 6 illustrates the result of sampling a one-dimensional continuous-domain signal. It is easy to see that the samples collectively describe the gross shape of the original signal very nicely,

but that smaller variations and structures are harder to discern or may be lost. Mathematically, information may have been lost, meaning that it might not be possible to reconstruct the original continuous signal from the samples (as determined by the Sampling Theorem; see Chapters 2.3 and 7.1). Supposing that the signal is part of an image, e.g., is a single scan line of an image displayed on a monitor, then the visual quality may or may not be reduced in the sampled version. Of course, the concept of visual quality varies from person to person, and it also depends on the conditions under which the image is viewed, such as the viewing distance.

Note that in Fig. 6, the samples are indexed by integer numbers. In fact, the sampled signal can be viewed as a vector of numbers. If the signal is finite in extent, then the signal vector can be stored and digitally processed as an array; hence the integer indexing becomes quite natural and useful. Likewise, image and video signals that are space/time sampled are generally indexed by integers along each sampled dimension, allowing them to be easily processed as multidimensional arrays of numbers. As shown in Fig. 7, a sampled image is an array of sampled image values that are usually arranged in a row–column format. Each of the indexed array elements is often called a *picture element*, or *pixel* for short. The term *pel* has also been used, but has faded in usage probably because it is less descriptive and not as

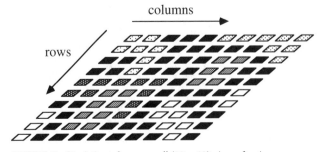

columns

rows

FIGURE 7   Depiction of a very small (10×10) piece of an image array.

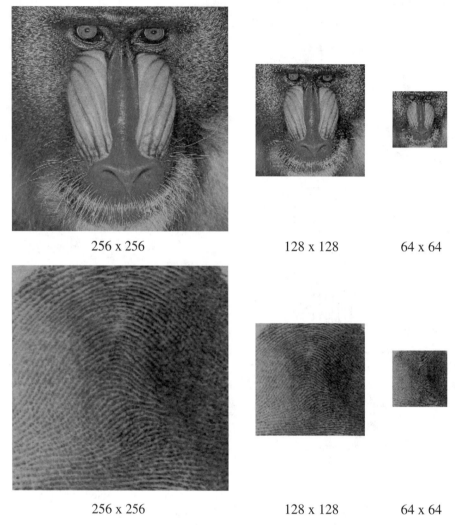

| 256 x 256 | 128 x 128 | 64 x 64 |

| 256 x 256 | 128 x 128 | 64 x 64 |

FIGURE 8   Examples of the visual effect of different image sampling densities.

catchy. The number of rows and columns in a sampled image is also often selected to be a power of 2, because this simplifies computer addressing of the samples, and also because certain algorithms, such as discrete Fourier transforms, are particularly efficient when operating on signals that have dimensions that are powers of 2. Images are nearly always rectangular (hence indexed on a Cartesian grid), and they are often square, although the horizontal dimension is often longer, especially in video signals, where an aspect ratio of 4 : 3 is common.

As mentioned in the preceding text, the effects of insufficient sampling ("undersampling") can be visually obvious. Figure 8 shows two very illustrative examples of image sampling. The two images, which we call "mandrill" and "fingerprint," both contain a significant amount of interesting visual detail that substantially defines the content of the images. Each image is shown at three different sampling densities: 256×256 (or $2^8 \times 2^8 = 65,536$ samples), 128×128 (or $2^7 \times 2^7 = 16,384$ samples), and 64×64 (or $2^6 \times 2^6 = 4,096$ samples). Of course, in both cases, all three scales of images are digital, and so there is potential loss of in-

formation relative to the original analog image. However, the perceptual quality of the images can easily be seen to degrade rather rapidly; note the whiskers on the mandrill's face, which lose all coherency in the 64×64 image. The 64×64 fingerprint is very interesting, since the pattern has completely changed! It almost appears as a different fingerprint. This results from an undersampling effect know as *aliasing*, in which image frequencies appear that have no physical meaning (in this case, creating a false pattern). Aliasing, and its mathematical interpretation, will be discussed further in Chapter 2.3 in the context of the Sampling Theorem.

Quantized Images

The other part of image digitization is *quantization*. The values that a (single-valued) image takes are usually *intensities*, since they are a record of the intensity of the signal incident on the sensor, e.g., the photon count or the amplitude of a measured

a pixel   8-bit representation

FIGURE 9   Illustration of an 8-bit representation of a quantized pixel.

wave function. Intensity is a positive quantity. If the image is represented visually, using shades of gray (like a black-and-white photograph), then the pixel values are referred to as *gray levels*. Of course, broadly speaking, an image may be multivalued at each pixel (such as a color image), or an image may have negative pixel values, in which case it is not an intensity function. In any case, the image values must be quantized for digital processing.

Quantization is the process of converting a *continuous-valued image*, which has a continuous range (set of values that it can take), into a *discrete-valued image*, which has a discrete range. This is ordinarily done by a process of rounding, truncation, or some other irreversible, nonlinear process of information destruction. Quantization is a necessary precursor to digital processing, since the image intensities must be represented with a finite precision (limited by word length) in any digital processor.

When the gray level of an image pixel is quantized, it is assigned to be one of a finite set of numbers, which is the *gray-level range*. Once the discrete set of values defining the gray-level range is known or decided, then a simple and efficient method of quantization is simply to round the image pixel values to the respective nearest members of the intensity range. These rounded values can be any numbers, but for conceptual convenience and ease of digital formatting, they are then usually mapped by a linear transformation into a finite set of nonnegative integers $\{0, \ldots, K - 1\}$, where K is a power of 2: $K = 2^B$. Hence the number of allowable gray levels is K, and the number of bits allocated to each pixel's gray level is B. Usually $1 \le B \le 8$ with $B = 1$ (for binary images) and $B = 8$ (where each gray level conveniently occupies a byte) being the most common bit depths (see Fig. 9). Multivalued images, such as color images, require

quantization of the components either individually or collectively ("vector quantization"); for example, a three-component color image is frequently represented with 24 bits per pixel of color precision.

Unlike sampling, quantization is a difficult topic to analyze, because it is nonlinear. Moreover, most theoretical treatments of signal processing assume that the signals under study are *not* quantized, because this tends to greatly complicate the analysis. In contrast, quantization is an essential ingredient of any (lossy) signal compression algorithm, where the goal can be thought of as finding an optimal quantization strategy that simultaneously minimizes the volume of data contained in the signal, while disturbing the fidelity of the signal as little as possible. With simple quantization, such as gray-level rounding, the main concern is that the pixel intensities or gray levels must be quantized with sufficient precision that excessive information is not lost. Unlike sampling, there is no simple mathematical measurement of information loss from quantization. However, while the effects of quantization are difficult to express mathematically, the effects are visually obvious.

Each of the images depicted in Figs. 4 and 8 is represented with 8 bits of gray-level resolution — meaning that bits less significant than the eighth bit have been rounded or truncated. This number of bits is quite common for two reasons. First, using more bits will generally *not* improve the visual appearance of the image — the adapted human eye usually is unable to see improvements beyond 6 bits (although the total range that can be seen under different conditions can exceed 10 bits) — hence using more bits would be wasteful. Second, each pixel is then conveniently represented by a byte. There are exceptions: in certain scientific or medical applications, 12, 16, or even more bits may be retained for more exhaustive examination by human or by machine.

Figures 10 and 11 depict two images at various levels of gray-level resolution. A reduced resolution (from 8 bits) was obtained by simply truncating the appropriate number of less-significant bits from each pixel's gray level. Figure 10 depicts the 256×256 digital image "fingerprint" represented at 4, 2, and 1 bit of gray-level resolution. At 4 bits, the fingerprint is nearly indistinguishable from the 8-bit representation of Fig. 8. At 2 bits, the image

FIGURE 10   Quantization of the 256×256 image "fingerprint." Clockwise from left: 4, 2, and 1 bits per pixel.

FIGURE 11   Quantization of the 256 × 256 image "eggs." Clockwise from upper left: 8, 4, 2, and 1 bits per pixel.

has lost a significant amount of information, making the print difficult to read. At 1 bit, the *binary* image that results is likewise hard to read. In practice, binarization of fingerprints is often used to make the print more distinctive. With the use of simple truncation–quantization, most of the print is lost because it was inked insufficiently on the left, and to excess on the right. Generally, bit truncation is a poor method for creating a binary image from a gray-level image. See Chapter 2.2 for better methods of image binarization.

Figure 11 shows another example of gray-level quantization. The image "eggs" is quantized at 8, 4, 2, and 1 bit of gray-level resolution. At 8 bits, the image is very agreeable. At 4 bits, the eggs take on the appearance of being striped or painted like Easter eggs. This effect is known as "false contouring," and re-sults when inadequate gray-scale resolution is used to represent smoothly varying regions of an image. In such places, the effects of a (quantized) gray level can be visually exaggerated, leading to an appearance of false structures. At 2 bits and 1 bit, significant information has been lost from the image, making it difficult to recognize.

A quantized image can be thought of as a stacked set of single-bit images (known as *bit planes*) corresponding to the gray-level resolution depths. The most significant bits of every pixel com-prise the top bit plane, and so on. Figure 12 depicts a 10 × 10 digital image as a stack of B bit planes. Special-purpose image processing algorithms are occasionally applied to the individual bit planes.

Color Images

Of course, the visual experience of the normal human eye is not limited to gray scales — *color* is an extremely important aspect of images. It is also an important aspect of digital images. In a very general sense, color conveys a variety of rich information that describes the *quality* of objects, and as such, it has much to do with visual *impression*. For example, it is known that different colors have the potential to evoke different emotional responses. The perception of color is allowed by the color-sensitive neurons known as *cones* that are located in the retina of the eye. The cones

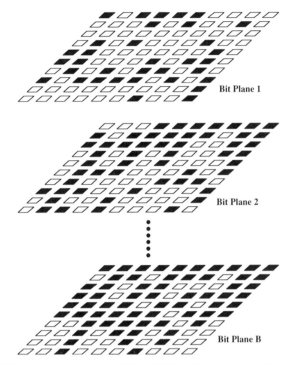

FIGURE 12 Depiction of a small (10×10) digital image as a stack of bit planes ranging from most significant (top) to least significant (bottom).

are responsive to normal light levels and are distributed with greatest density near the center of the retina, known as *fovea* (along the direct line of sight). The *rods* are neurons that are sensitive at low-light levels and are not capable of distinguishing color wavelengths. They are distributed with greatest density around the periphery of the fovea, with very low density near the line of sight. Indeed, one may experience this phenomenon by observing a dim point target (such as a star) under dark conditions. If one's gaze is shifted slightly off center, then the dim object suddenly becomes easier to see.

In the normal human eye, colors are sensed as near-linear combinations of long, medium, and short wavelengths, which roughly correspond to the three *primary colors* that are used in standard video camera systems: Red (R), Green (G), and Blue (B). The way in which visible-light wavelengths map to RGB camera color coordinates is a complicated topic, although standard tables have been devised based on extensive experiments. A number of other color coordinate systems are also used in image processing, printing, and display systems, such as the YIQ (luminance, in-phase chromatic, quadratic chromatic) color coordinate system. Loosely speaking, the YIQ coordinate system attempts to separate the perceived image *brightness* (luminance) from the chromatic components of the image by means of an invertible linear transformation:

$$\begin{bmatrix} Y \\ I \\ Q \end{bmatrix} = \begin{bmatrix} 0.299 & 0.587 & 0.114 \\ 0.596 & -0.275 & -0.321 \\ 0.212 & -0.523 & 0.311 \end{bmatrix} \begin{bmatrix} R \\ G \\ B \end{bmatrix}. \quad (1)$$

The RGB system is used by color cameras and video display systems, whereas the YIQ is the standard color representation used in broadcast television. Both representations are used in practical image and video processing systems, along with several other representations.

Most of the theory and algorithms for digital image and video processing have been developed for single-valued, monochromatic (gray level), or intensity-only images, whereas color images are vector-valued signals. Indeed, many of the approaches described in this *Handbook* are developed for single-valued images. However, these techniques are often applied (suboptimally) to color image data by regarding each color component as a separate image to be processed and by recombining the results afterward. As seen in Fig. 13, the R, G, and B components contain a considerable amount of overlapping information. Each of them is a valid image in the same sense as the image seen through colored spectacles, and can be processed as such. Conversely, however, if the color components are collectively available, then vector image processing algorithms can often be designed that achieve optimal results by taking this information into account. For example, a vector-based image enhancement algorithm applied to the "cherries" image in Fig. 13 might adapt by giving less importance to enhancing the blue component, since the image signal is weaker in that band.

Chromanance is usually associated with slower amplitude variations than is luminance, since it usually is associated with fewer image details or rapid changes in value. The human eye has a greater spatial bandwidth allocated for luminance perception than for chromatic perception. This is exploited by compression algorithms that use alternate color representations, such as YIQ, and store, transmit, or process the chromatic components using a lower bandwidth (fewer bits) than the luminance component. Image and video compression algorithms achieve increased efficiencies through this strategy.

Size of Image Data

The amount of data in visual signals is usually quite large, and it increases geometrically with the dimensionality of the data. This impacts nearly every aspect of image and video processing; data volume is a major issue in the processing, storage, transmission, and display of image and video information. The storage required for a single monochromatic digital still image that has (row $\times$ column) dimensions $N \times M$ and B bits of gray-level resolution is NMB bits. For the purpose of discussion we will assume that the image is square ($N = M$), although images of any aspect ratio are common. Most commonly, $B = 8$ (1 byte/pixel) unless the image is binary or is special purpose. If the image is vector valued, e.g., color, then the data volume is multiplied by the vector dimension. Digital images that are delivered by commercially available image digitizers are typically of an approximate size of 512×512 pixels, which is large enough to fill much of a monitor screen. Images both larger (ranging up to 4096×4096

FIGURE 13   Color image of "cherries" (top left), and (clockwise) its red, green, and blue components. (See color section, p. C–1.)

or more) and smaller (as small as 16 × 16) are commonly encountered. Table 1 depicts the required storage for a variety of

TABLE 1   Data-volume requirements for digital still images of various sizes, bit depths, and vector dimension

Spatial Dimensions	Pixel Resolution (bits)	Image Type	Data Volume (bytes)
128 × 128	1	Monochromatic	2,048
256 × 256	1	Monochromatic	8,192
512 × 512	1	Monochromatic	32,768
1024 × 1024	1	Monochromatic	131,072
128 × 128	8	Monochromatic	16,384
256 × 256	8	Monochromatic	65,536
512 × 512	8	Monochromatic	262,144
1024 × 1024	8	Monochromatic	1,048,576
128 × 128	3	Trichromatic	6,144
256 × 256	3	Trichromatic	24,576
512 × 512	3	Trichromatic	98,304
1024 × 1024	3	Trichromatic	393,216
128 × 128	24	Trichromatic	49,152
256 × 256	24	Trichromatic	196,608
512 × 512	24	Trichromatic	786,432
1024 × 1024	24	Trichromatic	3,145,728

image resolution parameters, assuming that there has been no compression of the data. Of course, the spatial extent (area) of the image exerts the greatest effect on the data volume. A single 512 × 512 × 8 color image requires nearly a megabyte of digital storage space, which only a few years ago was a lot. More recently, even large images are suitable for viewing and manipulation on home personal computers (PCs), although they are somewhat inconvenient for transmission over existing telephone networks.

However, when the additional time dimension is introduced, the picture changes completely. Digital video is extremely storage intensive. Standard video systems display visual information at a rate of 30 images/s for reasons related to human visual latency (at slower rates, there is a perceivable "flicker"). A 512 × 512 × 24 color video sequence thus occupies 23.6 megabytes for *each* second of viewing. A 2-hour digital film at the same resolution levels would thus require ∼*85 gigabytes* of storage at nowhere near theatre quality. That is alot of data, even for today's computer systems. Fortunately, images and video generally contain a significant degree of redundancy along each dimension. Taking this into account along with measurements of human visual response, it is possible to significantly compress digital images and video streams to acceptable levels. Sections 5 and 6 of this *Handbook* contain a number of chapters devoted to these topics. Moreover, the pace of information delivery is expected to significantly increase in the future, as significant additional bandwidths become available in the form of gigabit and terabit Ethernet networks, digital subscriber lines that use existing telephone networks, and public cable systems. These developments in telecommunications technology, along with improved algorithms for digital image and video transmission, promise a future that will be rich in visual information content in nearly every medium.

Digital Video

A significant portion of this *Handbook* is devoted to the topic of *digital video processing*. In recent years, hardware technologies and standards activities have matured to the point that it is becoming feasible to transmit, store, process, and view video signals that are stored in digital formats, and to share video signals between different platforms and application areas. This is a natural evolution, since temporal change, which is usually associated with motion of some type, is often the most important property of a visual signal.

Beyond this, there is a wealth of applications that stand to benefit from digital video technologies, and it is no exaggeration to say that the blossoming digital video industry represents many billions of dollars in research investments. The payoff from this research will be new advances in digital video processing theory, algorithms, and hardware that are expected to result in many billions more in revenues and profits. It is safe to say that digital video is very much the current frontier and the future of image processing research and development. The existing and expected applications of digital video are either growing rapidly or are expected to explode once the requisite technologies become available.

Some of the notable emerging digital video applications are as follows:

- video teleconferencing
- video telephony
- digital TV, including high-definition television (HDTV)
- internet video
- medical video
- dynamic scientific visualization
- multimedia video
- video instruction
- digital cinema

Sampled Video

Of course, the digital processing of video requires that the video stream be in a digital format, meaning that it must be sampled and quantized. Video quantization is essentially the same as image quantization. However, video sampling involves taking samples along a new and different (time) dimension. As such, it involves some different concepts and techniques.

First and foremost, the time dimension has a direction associated with it, unlike the space dimensions, which are ordinarily regarded as directionless until a coordinate system is artificially imposed upon it. Time proceeds from the past toward the future, with an origin that exists only in the current moment. Video is often processed in "real time," which (loosely) means that the result of processing appears effectively "instantaneously" (usually in a perceptual sense) once the input becomes available. Such a processing system cannot depend on more than a few future video samples. Moreover, it must process the video data quickly enough that the result appears instantaneous. Because of the vast data volume involved, the design of fast algorithms and hardware devices is a major priority.

In principle, an analog video signal $I(x, y, t)$, where (x, y) denote continuous space coordinates and t denotes continuous time, is continuous in both the space and time dimensions, since the radiation flux that is incident on a video sensor is continuous at normal scales of observation. However, the analog video that is viewed on display monitors is *not* truly analog, since it is sampled along one space dimension and along the time dimension. Practical so-called analog video systems, such as television and monitors, represent video as a one-dimensional electrical signal $V(t)$. Prior to display, a one-dimensional signal is obtained by sampling $I(x, y, t)$ along the vertical (y) space direction and along the time (t) direction. This is called scanning, and the result is a series of time samples, which are complete pictures or *frames*, each of which is composed of space samples, or *scan lines*.

Two types of video scanning are commonly used: *progressive scanning* and *interlaced scanning*. A progressive scan traces a complete frame, line by line from top to bottom, at a scan rate of Δt s/frame. High-resolution computer monitors are a good example, with a scan rate of $\Delta t = 1/72$ s. Figure 14 depicts progressive scanning on a standard monitor.

A description of interlaced scanning requires that some other definitions be made. For both types of scanning, the *refresh rate* is the frame rate at which information is displayed on a monitor. It is important that the frame rate be high enough, since otherwise the displayed video will appear to "flicker." The human eye detects flicker if the refresh rate is less than ∼50 frames/s. Clearly, computer monitors (72 frames/s) exceed this rate by almost 50%. However, in many other systems, notably television, such fast refresh rates are not possible unless spatial resolution is severely compromised because of bandwidth limitations. Interlaced scanning is a solution to this. In $P : 1$ interlacing, every Pth line is refreshed at each frame refresh. The subframes in interlaced video are called *fields*; hence P fields constitute a frame. The most common is 2 : 1 interlacing, which is used in standard television systems, as depicted in Fig. 14. In 2 : 1 interlacing, the two fields are usually referred to as the top and bottom fields. In this way, flicker is effectively eliminated provided that the field refresh rate is above the visual limit of ∼50 Hz. Broadcast television in the U.S. uses a frame rate of 30 Hz; hence the field rate

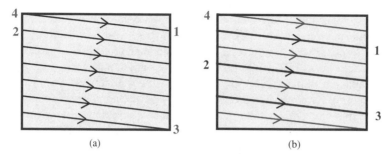

FIGURE 14 Video scanning: (a) Progressive video scanning. At the end of a scan (**1**), the electron gun spot snaps back to (**2**). A blank signal is sent in the interim. After reaching the end of a frame (**3**), the spot snaps back to (**4**). A synchronization pulse then signals the start of another frame. (b) Interlaced video scanning. Red and blue fields are alternately scanned left to right and top to bottom. At the end of scan (**1**), the spot snaps to (**2**). At the end of the blue field (**3**), the spot snaps to (**4**) (new field).

is 60 Hz, which is well above the limit. The reader may wonder if there is a loss of visual information, since the video is being effectively subsampled by a factor of 2 in the vertical space dimension in order to increase the apparent frame rate. In fact there is, since image motion may change the picture between fields. However, the effect is ameliorated to a significant degree by standard monitors and TV screens, which have screen phosphors with a *persistence* (glow time) that just matches the frame rate; hence each field persists until the matching field is sent.

Digital video is obtained either by sampling an analog video signal $V(t)$, or by directly sampling the three-dimensional space–time intensity distribution that is incident on a sensor. In either case, what results is a time sequence of two-dimensional spatial intensity arrays, or equivalently, a three-dimensional space–time array. If a progressive analog video is sampled, then the sampling is rectangular and properly indexed in an obvious manner, as illustrated in Fig. 15. If an interlaced analog video is sampled, then the digital video is interlaced also as shown in Fig. 16. Of course, if an interlaced video stream is sent to a system that processes or displays noninterlaced video, then the video data must first be converted or *deinterlaced* to obtain a standard progressive video stream before the accepting system will be able to handle it.

Video Transmission

The data volume of digital video is usually described in terms of bandwidth or bit rate. As described in Chapter 6.1, the bandwidth

of digital video streams (without compression) that match the current visual resolution of current television systems exceeds 100 megabits/s (mbps). Proposed digital television formats such as HDTV promise to multiply this by a factor of at least 4. By contrast, the networks that are currently available to handle digital data are quite limited. Conventional telephone lines (POTS) delivers only 56 kilobits/s (kbps), although digital subscriber lines (DSLs) promise to multiply this by a factor of 30 or more. Similarly, ISDN (Integrated Services Digital Network) lines that are currently available allow for data bandwidths equal to 64 p kbps, where $1 \leq p \leq 30$, which falls far short of the necessary data rate to handle full digital video. Dedicated T1 lines (1.5 mbps) also handle only a small fraction of the necessary bandwidth. Ethernet and cable systems, which currently can handle as much as 1 gigabit/s (gbps) are capable of handling raw digital video, but they have problems delivering multiple streams over the same network. The problem is similar to that of delivering large amounts of water through small pipelines. Either the data rate (water pressure) must be increased, or the data volume must be reduced.

Fortunately, unlike water, digital video can be compressed very effectively because of the redundancy inherent in the data, and because of an increased understanding of what components in the video stream are actually visible. Because of many years of research into image and video compression, it is now possible to

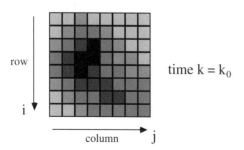

FIGURE 15 A single frame from a sampled progressive video sequence.

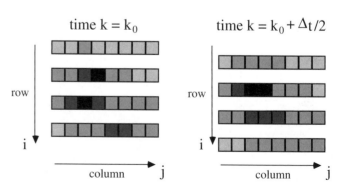

FIGURE 16 A single frame (two fields) from a sampled 2 : 1 interlaced video sequence.

transmit digital video data over a broad spectrum of networks, and we may expect that digital video will arrive in a majority of homes in the near future. Based on research developments along these lines, a number of world standards have recently emerged, or are under discussion, for video compression, video syntax, and video formatting. The use of standards allows for a common protocol for video and ensures that the consumer will be able to accept the same video inputs with products from different manufacturers. The current and emerging video standards broadly extend standards for still images that have been in use for a number of years. Several chapters are devoted to describing these standards, while others deal with emerging techniques that may effect future standards. It is certain, in any case, that we have entered a new era in which digital visual data will play an important role in education, entertainment, personal communications, broadcast, the Internet, and many other aspects of daily life.

Objectives of this Handbook

The goals of this *Handbook* are ambitious, since it is intended to reach a broad audience that is interested in a wide variety of image and video processing applications. Moreover, it is intended to be accessible to readers that have a diverse background, and that represent a wide spectrum of levels of preparation and engineering or computer education. However, a *Handbook* format is ideally suited for this multiuser purpose, since it allows for a presentation that adapts to the reader's needs. In the early part of the *Handbook* we present very basic material that is easily accessible even for novices to the image processing field. These chapters are also useful for review, for basic reference, and as support for later chapters. In every major section of the *Handbook*, basic introductory material is presented, as well as more advanced chapters that take the reader deeper into the subject.

Unlike textbooks on image processing, the *Handbook* is therefore not geared toward a specified level of presentation, nor does it uniformly assume a specific educational background. There is material that is available for the beginning image processing user, as well as for the expert. The *Handbook* is also unlike a textbook in that it is not limited to a specific point of view given by a single author. Instead, leaders from image and video processing education, industry, and research have been called upon to explain the topical material from their own daily experience. By calling upon most of the leading experts in the field, we have been able to provide a complete coverage of the image and video processing area without sacrificing any level of understanding of any particular area.

Because of its broad spectrum of coverage, we expect that the *Handbook of Image and Video Processing* will serve as an excellent textbook as well as reference. It has been our objective to keep the student's needs in mind, and we believe that the material contained herein is appropriate to be used for classroom presentations ranging from the introductory undergraduate level, to the upper-division undergraduate, to the graduate level. Although the *Handbook* does not include "problems in the back,"

this is not a drawback since the many examples provided in every chapter are sufficient to give the student a deep understanding of the function of the various image and video processing algorithms. This field is very much a visual science, and the principles underlying it are best taught with visual examples. Of course, we also foresee the *Handbook* as providing easy reference, background, and guidance for image and video processing professionals working in industry and research.

Our specific objectives are to

- provide the practicing engineer and the student with a highly accessible resource for learning and using image/video processing algorithms and theory
- provide the essential understanding of the various image and video processing standards that exist or are emerging, and that are driving today's explosive industry
- provide an understanding of what images are, how they are modeled, and give an introduction to how they are perceived
- provide the necessary practical background to allow the engineer student to acquire and process his or her own digital image or video data
- provide a diverse set of example applications, as separate complete chapters, that are explained in sufficient depth to serve as extensible models to the reader's own potential applications

The *Handbook* succeeds in achieving these goals, primarily because of the many years of broad educational and practical experience that the many contributing authors bring to bear in explaining the topics contained herein.

Organization of the Handbook

Since this *Handbook* is emphatically about *processing* images and video, the next section is immediately devoted to basic algorithms for image processing, instead of surveying methods and devices for image acquisition at the outset, as many textbooks do. Section 2 is divided into three chapters, which respectively introduce the reader to the most fundamental two-dimensional image processing techniques. Chapter 2.1 lays out basic methods for gray-level image processing, which includes point operations, the image histogram, and simple image algebra. The methods described there stand alone as algorithms that can be applied to most images, but they also set the stage and the notation for the more involved methods discussed in later chapters. Chapter 2.2 describes basic methods for image binarization and for binary image processing, with emphasis on morphological binary image processing. The algorithms described there are among the most widely used in applications, especially in the biomedical area. Chapter 2.3 explains the basics of the Fourier transform and frequency-domain analysis, including discretization of the Fourier transform and discrete convolution. Special emphasis is placed on explaining frequency-domain concepts through visual examples. Fourier image analysis provides a unique opportunity for visualizing the meaning of frequencies as components of signals. This approach reveals insights that are difficult to capture in one-dimensional, graphical discussions.

Section 3 of the *Handbook* deals with methods for correcting distortions or uncertainties in images and for improving image information by combining images taken from multiple views. Quite frequently the visual data that are acquired have been in some way corrupted. Acknowledging this and developing algorithms for dealing with it is especially critical since the human capacity for detecting errors, degradations, and delays in digitally delivered visual data is quite high. Image and video signals are derived from imperfect sensors, and the processes of digitally converting and transmitting these signals are subject to errors. There are many types of errors that can occur in image or video data, including, for example, blur from motion or defocus; noise that is added as part of a sensing or transmission process; bit, pixel, or frame loss as the data are copied or read; or artifacts that are introduced by an image or video compression algorithm. As such, it is important to be able to model these errors, so that numerical algorithms can be developed to ameliorate them in such a way as to improve the data for visual consumption. Section 3 contains three broad categories of topics. The first is image/video enhancement, in which the goal is to remove noise from an image while retaining the perceptual fidelity of the visual information; these are seen to be conflicting goals. Chapters are included that describe very basic linear methods; highly efficient nonlinear methods; and recently developed and very powerful wavelet methods; and also extensions to video enhancement. The second broad category is image/video restoration, in which it is assumed that the visual information has been degraded by a distortion function, such as defocus, motion blur, or atmospheric distortion, and more than likely, by noise as well. The goal is to remove the distortion and attenuate the noise, while again preserving the perceptual fidelity of the information contained within. And again, it is found that a balanced attack on conflicting requirements is required in solving these difficult, ill-posed problems. The treatment again begins with a basic, introductory chapter; ensuing chapters build on this basis and discuss methods for restoring multichannel images (such as color images); multiframe images (i.e., using information from multiple images taken of the same scene); iterative methods for restoration; and extensions to video restoration. Related topics that are considered are motion detection and estimation, which is essential for handling many problems in video processing, and a general framework for regularizing ill-posed restoration problems. Finally, the third category involves the extraction of enriched information about the environment by combining images taken from multiple views of the same scene. This includes chapters on methods for computed stereopsis and for image stabilization and mosaicking.

Section 4 of the *Handbook* deals with methods for image and video analysis. Not all images or videos are intended for direct human visual consumption. Instead, in many situations it is of interest to automate the process of repetitively interpreting the content of multiple images or video data through the use of an *image* or *video analysis algorithm*. For example, it may be desired to *classify* parts of images or videos as being of some type, or it may be desired to *detect* or *recognize* objects contained in the data sets. If one is able to develop a reliable computer algorithm that consistently achieves success in the desired task, and if one has access to a computer that is fast enough, then a tremendous savings in man hours can be attained. The advantage of such a system increases with the number of times that the task must be done and with the speed with which it can be automatically accomplished. Of course, problems of this type are typically quite difficult, and in many situations it is not possible to approach, or even come close to, the efficiency of the human visual system. However, if the application is specific enough, and if the process of image acquisition can be sufficiently controlled (to limit the variability of the image data), then tremendous efficiencies can be achieved. With some exceptions, image/video analysis systems are quite complex, but they are often composed at least in part of subalgorithms that are common to other image/video analysis applications. Section 4 of this *Handbook* outlines some of the basic models and algorithms that are encountered in practical systems. The first set of chapters deals with image models and representations that are commonly used in every aspect of image/video processing. This starts with a chapter on models of the human visual system. Much progress has been made in recent years in modeling the brain and the functions of the optics and the neurons along the visual pathway (although much remains to be learned as well). Because images and videos that are processed are nearly always intended for eventual visual consumption by humans, in the design of these algorithms it is imperative that the receiver be taken into account, as with any communication system. After all, vision is very much a form of dense communication, and images are the medium of information. The human eye–brain system is the receiver. This is followed by chapters on wavelet image representations, random field image models, image modulation models, image noise models, and image color models, which are referred to in many other places in the *Handbook*. These chapters may be thought of as a core reference section of the *Handbook* that supports the entire presentation. Methods for image/video classification and segmentation are described next; these basic tools are used in a wide diversity of analysis applications. Complementary to these are two chapters on edge and boundary detection, in which the goal is finding the boundaries of regions, namely, sudden changes in image intensities, rather than finding (segmenting out) and classifying regions directly. The approach taken depends on the application. Finally, a chapter is given that reviews currently available software for image and video processing.

As described earlier in this introductory chapter, image and video information is highly data intensive. Sections 5 and 6 of the *Handbook* deal with methods for compressing this data. Section 5 deals with still image compression, beginning with several basic chapters of lossless compression, and on several useful general approaches for image compression. In some realms, these approaches compete, but each has its advantages and subsequent appropriate applications. The existing JPEG standards for both lossy and lossless compression are described next. Although these standards are quite complex, they are described in sufficient detail to allow for the practical design of systems that accept and transmit JPEG data sets.

Section 6 extends these ideas to video compression, beginning with an introductory chapter that discusses the basic ideas and that uses the H.261 standard as an example. The H.261 standard, which is used for video teleconferencing systems, is the starting point for later video compression standards, such as MPEG. The following two chapters are on especially promising methods for future and emerging video compression systems: wavelet-based methods, in which the video data are decomposed into multiple subimages (scales or subbands), and object-based methods, in which objects in the video stream are identified and coded separately across frames, even (or especially) in the presence of motion. Finally, chapters on the existing MPEG-I and MPEG-II and emerging MPEG-IV and MPEG-VII standards for video compression are given, again in sufficient detail to enable the practicing engineer to put the concepts to use.

Section 7 deals with image and video scanning, sampling, and interpolation. These important topics give the basics for understanding image acquisition, converting images and video into digital format, and for resizing or spatially manipulating images. Section 8 deals with the visualization of image and video information. One chapter focuses on the halftoning and display of images, and another on methods for assessing the quality of images, especially compressed images.

With the recent significant activity in *multimedia*, of which image and video is the most significant component, methods for databasing, access/retrieval, archiving, indexing, networking, and securing image and video information are of high interest. These topics are dealt with in detail in Section 9 of the *Handbook*.

Finally, Section 10 includes eight chapters on a diverse set of image processing applications that are quite representative of the universe of applications that exist. Many of the chapters in this section have *analysis, classification,* or *recognition* as a main goal, but reaching these goals inevitably requires the use of a broad spectrum of image/video processing subalgorithms for enhancement, restoration, detection, motion, and so on. The work that is reported in these chapters is likely to have significant impact on science, industry, and even on daily life. It is hoped that readers are able to translate the lessons learned in these chapters, and in the preceding material, into their own research or product development work in image and/or video processing. For students, it is hoped that they now possess the required reference material that will allow them to acquire the basic knowledge to be able to begin a research or development career in this fast-moving and rapidly growing field.

Acknowledgment

Many thanks to Prof. Joel Trussell for carefully reading and commenting on this introductory chapter.

Basic Image Processing Techniques

2.1

Basic Gray-Level Image Processing

Alan C. Bovik
*The University of Texas
at Austin*

1 Introduction

This Chapter, and the two that follow, describe the most commonly used and most basic tools for digital image processing. For many simple image analysis tasks, such as contrast enhancement, noise removal, object location, and frequency analysis, much of the necessary collection of instruments can be found in Chapters 2.1–2.3. Moreover, these chapters supply the basic groundwork that is needed for the more extensive developments that are given in the subsequent chapters of the *Handbook*.

In this chapter, we study basic gray-level digital image processing operations. The types of operations studied fall into three classes.

The first are *point operations*, or image processing operations that are applied to individual pixels only. Thus, interactions and dependencies between neighboring pixels are not considered, nor are operations that consider multiple pixels simultaneously to determine an output. Since spatial information, such as a pixel's location and the values of its neighbors, are not considered, point operations are defined as functions of pixel intensity only. The basic tool for understanding, analyzing, and designing image point operations is the *image histogram*, which will be introduced below.

The second class includes *arithmetic operations* between images of the same spatial dimensions. These are also point operations in the sense that spatial information is not considered, although information is shared between images on a pointwise basis. Generally, these have special purposes, e.g., for noise reduction and change or motion detection.

The third class of operations are *geometric image operations*. These are complementary to point operations in the sense that they are not defined as functions of image intensity. Instead, they are functions of spatial position only. Operations of this type change the appearance of images by changing the coordinates of the intensities. This can be as simple as image translation or rotation, or it may include more complex operations that distort or bend an image, or "morph" a video sequence. Since our goal, however, is to concentrate on digital image processing of real-world images, rather than the production of special effects, only the most basic geometric transformations will be considered. More complex and time-varying geometric effects are more properly considered within the science of *computer graphics*.

2 Notation

Point operations, algebraic operations, and geometric operations are easily defined on images of any dimensionality, including digital video data. For simplicity of presentation, we will restrict our discussion to two-dimensional images only. The extensions to three or higher dimensions are not difficult, especially in the case of point operations, which are independent of dimensionality. In fact, spatial/temporal information is not considered in their definition or application.

We will also only consider monochromatic images, since extensions to color or other multispectral images is either trivial, in that the same operations are applied identically to each band (e.g., R, G, B), or they are defined as more complex color space operations, which goes beyond what we want to cover in this basic chapter.

Suppose then that the single-valued image $f(\mathbf{n})$ to be considered is defined on a two-dimensional discrete-space coordinate system $\mathbf{n} = (n_1, n_2)$. The image is assumed to be of finite support, with image domain $[0, N-1] \times [0, M-1]$. Hence the nonzero image data can be contained in a matrix or array of dimensions $N \times M$ (rows, columns). This *discrete-space* image will have originated by sampling a continuous image $f(x, y)$ (see Chapter 7.1). Furthermore, the image $f(\mathbf{n})$ is assumed to be *quantized* to K levels $\{0, \ldots, K-1\}$; hence each pixel value takes one of these integer values (Chapter 1.1). For simplicity, we will refer to these values as *gray levels*, reflecting the way in which monochromatic images are usually displayed. Since $f(\mathbf{n})$ is both discrete-space and quantized, it is *digital*.

3 Image Histogram

The basic tool that is used in designing point operations on digital images (and many other operations as well) is the *image histogram*. The histogram H_f of the digital image f is a plot or graph of the *frequency of occurrence* of each gray level in f. Hence, H_f is a one-dimensional function with domain $\{0, \ldots, K-1\}$ and possible range extending from 0 to the number of pixels in the image, NM.

The histogram is given explicitly by

$$H_f(k) = J \tag{1}$$

if f contains *exactly J* occurrences of gray level k, for each $k = 0, \ldots, K-1$. Thus, an algorithm to compute the image histogram involves a simple counting of gray levels, which can be

accomplished even as the image is scanned. Every image processing development environment and software library contains basic histogram computation, manipulation, and display routines (Chapter 4.12).

Since the histogram represents a reduction of dimensionality relative to the original image f, information is lost — the image f cannot be deduced from the histogram H_f except in trivial cases (when the image is constant valued). In fact, the number of images that share the same arbitrary histogram H_f is astronomical. Given an image f with a particular histogram H_f, every image that is a spatial shuffling of the gray levels of f has the same histogram H_f.

The histogram H_f contains no spatial information about f — it describes the frequency of the gray levels in f and nothing more. However, this information is still very rich, and many useful image processing operations can be derived from the image histogram. Indeed, a simple visual display of H_f reveals much about the image. By examining the appearance of a histogram, it is possible to ascertain whether the gray levels are distributed primarily at lower (darker) gray levels, or vice versa. Although this can be ascertained to some degree by visual examination of the image itself, the human eye has a tremendous ability to adapt to overall changes in luminance, which may obscure shifts in the gray-level distribution. The histogram supplies an absolute method of determining an image's gray-level distribution.

For example, the *average optical density*, or *AOD*, is the basic measure of an image's overall average brightness or gray level. It can be computed directly from the image:

$$\text{AOD}(f) = \frac{1}{NM} \sum_{n_1=0}^{N-1} \sum_{n_2=0}^{M-1} f(n_1, n_2) \tag{2}$$

or it can be computed from the image histogram:

$$\text{AOD}(f) = \frac{1}{NM} \sum_{k=0}^{K-1} k H_f(k). \tag{3}$$

The AOD is a useful and simple meter for estimating the center of an image's gray-level distribution. A target value for the AOD might be specified when designing a point operation to change the overall gray-level distribution of an image.

Figure 1 depicts two hypothetical image histograms. The one on the left has a heavier distribution of gray levels close to zero

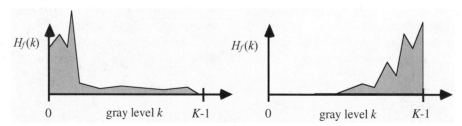

FIGURE 1 Histograms of images with gray-level distribution skewed toward darker (left) and brighter (right) gray levels. It is possible that these images are underexposed and overexposed, respectively.

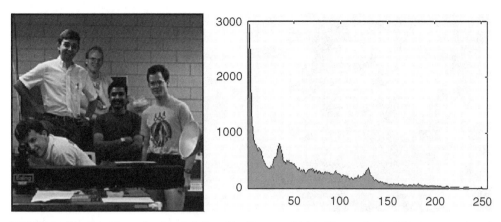

FIGURE 2 The digital image "students" (left) and its histogram (right). The gray levels of this image are skewed toward the left, and the image appears slightly underexposed.

(and a low AOD), while the one on the right is skewed toward the right (a high AOD). Since image gray levels are usually displayed with lower numbers' indicating darker pixels, the image on the left corresponds to a predominantly dark image. This may occur if the image f was originally underexposed prior to digitization, or if it was taken under poor lighting levels, or perhaps the process of digitization was performed improperly. A skewed histogram often indicates a problem in gray-level allocation. The image on the right may have been overexposed or taken in very bright light.

Figure 2 depicts the 256×256 ($N = M = 256$) gray-level digital image "students" with a gray-scale range $\{0, \ldots, 255\}$, and its computed histogram. Although the image contains a broad distribution of gray levels, the histogram is heavily skewed toward the dark end, and the image appears to be poorly exposed. It is of interest to consider techniques that attempt to "equalize" this distribution of gray levels. One of the important applications of image point operations is to correct for poor exposures like the one in Fig. 2. Of course, there may be limitations to the effectiveness of any attempt to recover an image from poor exposure, since information may be lost. For example, in Fig. 2, the gray levels saturate at the low end of the scale, making it difficult or impossible to distinguish features at low brightness levels.

More generally, an image may have a histogram that reveals a poor usage of the available gray-scale range. An image with a compact histogram, as depicted in Fig. 3, will often have a poor visual contrast or a washed-out appearance. If the gray-scale range is filled out, also depicted in Fig. 3, then the image tends to have a higher contrast and a more distinctive appearance. As will be shown, there are specific point operations that effectively expand the gray-scale distribution of an image.

Figure 4 depicts the 256×256 gray-level image "books" and its histogram. The histogram clearly reveals that nearly all of the gray levels that occur in the image fall within a small range of gray scales, and the image is of correspondingly poor contrast.

It is possible that an image may be taken under correct lighting and exposure conditions, but that there is still a skewing of the gray-level distribution toward one end of the gray-scale or that the histogram is unusually compressed. An example would be an image of the night sky, which is dark nearly everywhere. In such a case, the appearance of the image may be normal but the histogram will be very skewed. In some situations, it may still be of interest to attempt to enhance or reveal otherwise difficult-to-see details in the image by the application of an appropriate point operation.

4 Linear Point Operations on Images

A *point operation* on a digital image $f(\mathbf{n})$ is a function h of a single variable applied identically to every pixel in the

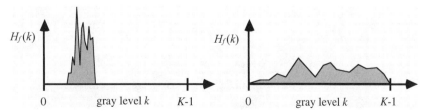

FIGURE 3 Histograms of images that make poor (left) and good (right) use of the available gray-scale range. A compressed histogram often indicates an image with a poor visual contrast. A well-distributed histogram often has a higher contrast and better visibility of detail.

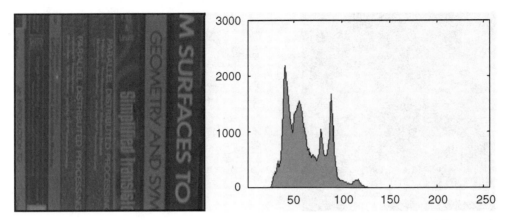

FIGURE 4   Digital image "books" (left) and its histogram (right). The image makes poor use of the available gray-scale range.

image, thus creating a new, modified image $g(\mathbf{n})$. Hence at each coordinate $\mathbf{n}$,

$$g(\mathbf{n}) = h[f(\mathbf{n})]. \qquad (4)$$

The form of the function h is determined by the task at hand. However, since each output $g(\mathbf{n})$ is a function of a single pixel value only, the effects that can be obtained by a point operation are somewhat limited. Specifically, no spatial information is utilized in Eq. (4), and there is no change made in the spatial relationships between pixels in the transformed image. Thus, point operations do not effect the spatial positions of objects in an image, nor their shapes. Instead, each pixel value or gray level is increased or decreased (or unchanged) according to the relation in Eq. (4). Therefore, a point operation h does change the gray-level distribution or histogram of an image, and hence the overall appearance of the image.

Of course, there is an unlimited variety of possible effects that can be produced by selection of the function h that defines the point operation of Eq. (4). Of these, the simplest are the *linear point operations*, where h is taken to be a simple linear function of gray level:

$$g(\mathbf{n}) = P f(\mathbf{n}) + L. \qquad (5)$$

Linear point operations can be viewed as providing a gray-level additive offset L and a gray-level multiplicative scaling P of the image f. Offset and scaling provide different effects, and so we will consider them separately before examining the overall linear point operation of Eq. (5).

The *saturation* conditions $|g(\mathbf{n})| < 0$ and $|g(\mathbf{n})| > K - 1$ are to be avoided if possible, since the gray levels are then not properly defined, which can lead to severe errors in processing or display of the result. The designer needs to be aware of this so steps can be taken to ensure that the image is not distorted by values falling outside the range. If a specific wordlength has been allocated to represent the gray level, then saturation may result in an overflow or underflow condition, leading to very large errors. A simple way to handle this is to simply clip those values falling outside of the allowable gray-scale range to the endpoint values. Hence, if $|g(\mathbf{n}_0)| < 0$ at some coordinate $\mathbf{n}_0$, then set $|g(\mathbf{n}_0)| = 0$ instead. Likewise, if $|g(\mathbf{n}_0)| > K - 1$, then fix $|g(\mathbf{n}_0)| = K - 1$.

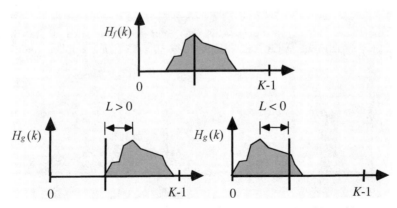

FIGURE 5   Effect of additive offset on the image histogram. Top: original image histogram; bottom: positive (left) and negative (right) offsets shift the histogram to the right and to the left, respectively.

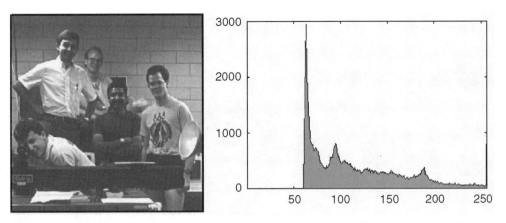

FIGURE 6 Left: Additive offset of the image of students in Fig. 2 by amount 60. Observe the clipping spike in the histogram to the right at gray level 255.

Of course, the result is no longer strictly a linear point operation. Care must be taken, since information is lost in the clipping operation, and the image may appear artificially flat in some areas if whole regions become clipped.

4.1 Additive Image Offset

Suppose $P = 1$ and L is an integer satisfying $|L| \leq K - 1$. An *additive image offset* has the form

$$g(\mathbf{n}) = f(\mathbf{n}) + L. \tag{6}$$

Here we have prescribed a range of values that L can take. We have taken L to be an integer, since we are assuming that images are quantized into integers in the range $\{0, \ldots, K - 1\}$. We have also assumed that $|L|$ falls in this range, since otherwise all of the values of $g(\mathbf{n})$ will fall outside the allowable gray-scale range.

In Eq. (6), if $L > 0$, then $g(\mathbf{n})$ will be a brightened version of the image $f(\mathbf{n})$. Since spatial relationships between pixels are unaffected, the appearance of the image will otherwise be essentially the same. Likewise, if $L < 0$, then $g(\mathbf{n})$ will be a dimmed version of the $f(\mathbf{n})$. The histograms of the two images have a

simple relationship:

$$H_g(k) = H_f(k - L). \tag{7}$$

Thus, an offset L corresponds to a shift of the histogram by amount L to the left or to the right, as depicted in Fig. 5.

Figures 6 and 7 show the result of applying an additive offset to the images of students and books in Figs. 2 and 4, respectively. In both cases, the overall visibility of the images has been somewhat increased, but there has not been an improvement in the contrast. Hence, while each image as a whole is easier to see, the details in the image are no more visible than they were in the original. Figure 6 is a good example of saturation; a large number of gray levels were clipped at the high end (gray-level 255). In this case, clipping did not result in much loss of information.

Additive image offsets can be used to calibrate images to a given average brightness level. For example, suppose we desire to compare multiple images $f_1, f_2, \ldots, f_n$ of the same scene, taken at different times. These might be surveillance images taken of a secure area that experiences changes in overall ambient illumination. These variations could occur because the area is exposed to daylight.

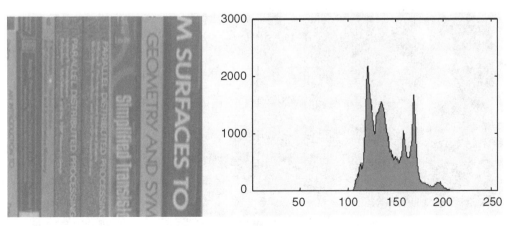

FIGURE 7 Left: Additive offset of the image of books in Fig. 4 by amount 80.

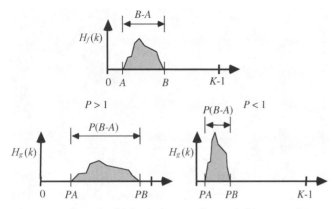

FIGURE 8 Effects of multiplicative image scaling on the histogram. If $P > 1$, the histogram is expanded, leading to more complete use of the gray-scale range. If $P < 1$, the histogram is contracted, leading to possible information loss and (usually) a less striking image.

A simple approach to counteract these effects is to equalize the AODs of the images. A reasonable AOD is the gray-scale center $K/2$, although other values may be used depending on the application. Letting $L_m = \text{AOD}(f_m)$, for $m = 1, \ldots, n$, the "AOD-equalized" images $g_1, g_2, \ldots, g_n$ are given by

$$g_m(\mathbf{n}) = f_m(\mathbf{n}) - L_m + K/2. \qquad (8)$$

The resulting images then have identical AOD $K/2$.

4.2 Multiplicative Image Scaling

Next we consider the scaling aspect of linear point operations. Suppose that $L = 0$ and $P > 0$. Then, a *multiplicative image scaling* by factor P is given by

$$g(\mathbf{n}) = P f(\mathbf{n}). \qquad (9)$$

Here, P is assumed positive since $g(\mathbf{n})$ must be positive. Note that we have not constrained P to be an integer, since this would usually leave few useful values of P; for example, even taking $P = 2$ will severely saturate most images. If an integer result is

required, then a practical definition for the output is to *round* the result in Eq. (9):

$$g(\mathbf{n}) = \text{INT}[P f(\mathbf{n}) + 0.5], \qquad (10)$$

where $\text{INT}[R]$ denotes the nearest integer that is less than or equal to R.

The effect that multiplicative scaling has on an image depends on whether P is larger or smaller than one. If $P > 1$, then the gray levels of g will cover a broader range than those of f. Conversely, if $P < 1$, then g will have a narrower gray-level distribution than f. In terms of the image histogram,

$$H_g\{\text{INT}[P k + 0.5]\} = H_f(k). \qquad (11)$$

Hence, multiplicative scaling by a factor P either stretches or compresses the image histogram. Note that for quantized images, it is not proper to assume that Eq. (11) implies $H_g(k) = H_f(k/P)$, since the argument of $H_f(k/P)$ may not be an integer.

Figure 8 depicts the effect of multiplicative scaling on a hypothetical histogram. For $P > 1$, the histogram is expanded (and hence, saturation is quite possible), while for $P < 1$, the histogram is contracted. If the histogram is contracted, then multiple gray levels in f may map to single gray levels in g, since the number of gray levels is finite. This implies a possible loss of information. If the histogram is expanded, then spaces may appear between the histogram bins where gray levels are not being mapped. This, however, does not represent a loss of information and usually will not lead to visual information loss.

As a rule of thumb, histogram expansion often leads to a more distinctive image that makes better use of the gray-scale range, provided that saturation effects are not visually noticeable. Histogram contraction usually leads to the opposite: an image with reduced visibility of detail that is less striking. However, these are only rules of thumb, and there are exceptions. An image may have a gray-scale spread that is too extensive, and it may benefit from scaling with $P < 1$.

Figure 9 shows the image of students following a multiplicative scaling with $P = 0.75$, resulting in compression of the

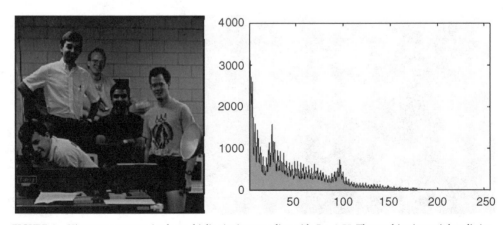

FIGURE 9 Histogram compression by multiplicative image scaling with $P = 0.75$. The resulting image is less distinctive. Note also the regularly spaced tall spikes in the histogram; these are gray levels that are being "stacked," resulting in a loss of information, since they can no longer be distinguished.

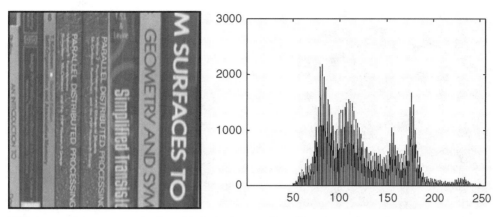

FIGURE 10 Histogram expansion by multiplicative image scaling with $P = 2.0$. The resulting image is much more visually appealing. Note the regularly spaced gaps in the histogram that appear when the discrete histogram values are spread out. This does not imply a loss of information or visual fidelity.

histogram. The resulting image is darker and less contrasted. Figure 10 shows the image of books following scaling with $P = 2$. In this case, the resulting image is much brighter and has a better visual resolution of gray levels. Note that most of the high end of the gray-scale range is now used, although the low end is not.

4.3 Image Negative

The first example of a linear point operation that uses both scaling and offset is the *image negative*, which is given by $P = -1$ and $L = K - 1$. Hence

$$g(\mathbf{n}) = -f(\mathbf{n}) + (K - 1) \qquad (12)$$

and

$$H_g(k) = H_f(K - 1 - k). \qquad (13)$$

Scaling by $P = -1$ reverses (flips) the histogram; the additive offset $L = K - 1$ is required so that all values of the result are positive and fall in the allowable gray-scale range. This operation creates a *digital negative image*, unless the image is already a negative, in which case a positive is created. It should be mentioned that unless the digital negative of Eq. (12) is being computed, $P > 0$ in nearly every application of linear point operations.

An important application of Eq. (12) occurs when a negative is scanned (digitized), and it is desired to view the positive image. Figure 11 depicts the negative image associated with "students." Sometimes, the negative image is viewed intentionally, when the positive image itself is very dark. A common example of this is for the examination of telescopic images of star fields and faint galaxies. In the negative image, faint bright objects appear as dark objects against a bright background, which can be easier to see.

4.4 Full-Scale Histogram Stretch

We have already mentioned that an image that has a broadly distributed histogram tends to be more visually distinctive. The *full-scale histogram stretch*, which is also often called a *contrast*

 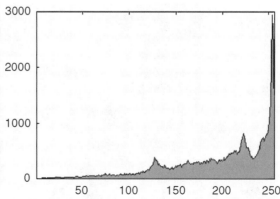

FIGURE 11 Example of an image negative with the resulting reversed histogram.

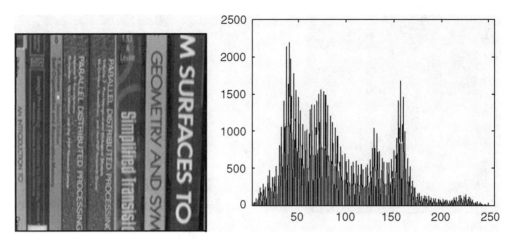

FIGURE 12   Full-scale histogram stretch of the image of books.

stretch, is a simple linear point operation that expands the image histogram to fill the entire available gray-scale range. This is such a desirable operation that the full-scale histogram stretch is easily the most common linear point operation. Every image processing programming environment and library contains it as a basic tool. Many image display routines incorporate it as a basic feature. Indeed, commercially available digital video cameras for home and professional use generally apply a full-scale histogram stretch to the acquired image before being stored in camera memory. It is called automatic gain control (AGC) on these devices.

The definition of the multiplicative scaling and additive offset factors in the full-scale histogram stretch depend on the image f. Suppose that f has a compressed histogram with maximum gray-level value B and minimum value A, as shown in Fig. 8 (top):

$$A = \min_{\mathbf{n}}\{f(\mathbf{n})\}, \qquad B = \max_{\mathbf{n}}\{f(\mathbf{n})\}. \tag{14}$$

The goal is to find a linear point operation of the form of Eq. (5) that maps gray levels A and B in the original image to gray levels 0 and $K - 1$ in the transformed image. This can be expressed in two linear equations:

$$PA + L = 0 \tag{15}$$

and

$$PB + L = K - 1 \tag{16}$$

in the two unknowns (P, L), with solutions

$$P = \left(\frac{K - 1}{B - A}\right) \tag{17}$$

and

$$L = -A\left(\frac{K - 1}{B - A}\right). \tag{18}$$

Hence, the overall full-scale histogram stretch is given by

$$g(\mathbf{n}) = \text{FSHS}[f(\mathbf{n})] = \left(\frac{K - 1}{B - A}\right)[f(\mathbf{n}) - A]. \tag{19}$$

We make the shorthand notation FSHS, since Eq. (19) will prove to be commonly useful as an addendum to other algorithms. The operation in Eq. (19) can produce dramatic improvements in the visual quality of an image suffering from a poor (narrow) gray-scale distribution. Figure 12 shows the result of applying the FSHS to the images of books. The contrast and visibility of the image was, as expected, greatly improved. The accompanying histogram, which now fills the available range, also shows the characteristics gaps of an expanded discrete histogram.

If the image f already has a broad gray-level range, then the histogram stretch may produce little or no effect. For example, the image of students (Fig. 2) has gray scales covering the entire available range, as seen in the histogram accompanying the image. Therefore, Eq. (19) has no effect on "students." This is unfortunate, since we have already commented that "students" might benefit from a histogram manipulation that would redistribute the gray level densities. Such a transformation would have to nonlinearly reallocate the image's gray-level values. Such nonlinear point operations are described next.

5 Nonlinear Point Operations on Images

We now consider *nonlinear point* operations of the form

$$g(\mathbf{n}) = h[f(\mathbf{n})], \tag{20}$$

where the function h is nonlinear. Obviously, this encompasses a wide range of possibilities. However, there are only a few functions h that are used with any great degree of regularity. Some of these are functional tools that are used as part of larger, multistep algorithms, such as absolute value, square, and square-root functions. One such simple nonlinear function that is very

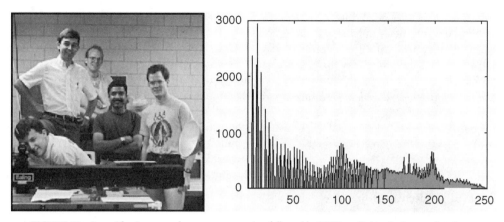

FIGURE 13   Logarithmic gray-scale range compression followed by FSHS applied to the image of students.

commonly used is the logarithmic point operation, which we describe in detail.

5.1 Logarithmic Point Operations

Assuming that the image $f(\mathbf{n})$ is positive valued, the *logarithmic point operation* is defined by a composition of two operations: a point logarithmic operation, followed by a full-scale histogram stretch:

$$g(\mathbf{n}) = \text{FSHS}\{\log[1 + f(\mathbf{n})]\}. \tag{21}$$

Adding unity to the image avoids the possibility of taking the logarithm of zero. The logarithm itself acts to nonlinearly compress the gray-level range. All of the gray level is compressed to the range $[0, \log(K)]$. However, larger (brighter) gray levels are compressed much more severely than are smaller gray levels. The subsequent FSHS operation then acts to linearly expand the log-compressed gray levels to fill the gray-scale range. In the transformed image, dim objects in the original are now allocated a much larger percentage of the gray-scale range, hence improving their visibility.

The logarithmic point operation is an excellent choice for improving the appearance of the image of students, as shown in Fig. 13. The original image (Fig. 2) was not a candidate for FSHS because of its broad histogram. The appearance of the original suffers because many of the important features of the image are obscured by darkness. The histogram is significantly spread at these low brightness levels, as can be seen by comparing it to Fig. 2, and also by the gaps that appear in the low end of the histogram. This does not occur at brighter gray levels.

Certain applications quite commonly use logarithmic point operations. For example, in astronomical imaging, a relatively few bright pixels (stars and bright galaxies, etc.) tend to dominate the visual perception of the image, while much of the interesting information lies at low bright levels (e.g., large, faint nebulae). By compressing the bright intensities much more heavily, then applying FSHS, the faint, interesting details visually emerge.

Later, in Chapter 2.3, the Fourier transforms of images will be studied. The Fourier transform magnitudes, which are of the same dimensionalities as images, will be displayed as intensity arrays for visual consumption. However, the Fourier transforms of most images are dominated visually by the Fourier coefficients of a relatively few low frequencies, so the coefficients of important high frequencies are usually difficult or impossible to see. However, a point logarithmic operation usually suffices to ameliorate this problem, and so image Fourier transforms are usually displayed following the application of Eq. (21), both in this *Handbook* and elsewhere.

5.2 Histogram Equalization

One of the most important nonlinear point operations is *histogram equalization*, also called *histogram flattening*. The idea behind it extends that of FSHS: not only should an image fill the available gray-scale range, but it should be uniformly distributed over that range. Hence, an idealized goal is a flat histogram. Although care must be taken in applying a powerful nonlinear transformation that actually changes the shape of the image histogram, rather than just stretching it, there are good mathematical reasons for regarding a flat histogram as a desirable goal. In a certain sense,[1] an image with a perfectly flat histogram contains the largest possible amount of *information* or complexity.

In order to explain histogram equalization, it will be necessary to make some refined definitions of the image histogram. For an image containing NM pixels, the *normalized image histogram* is given by

$$p_f(k) = \frac{1}{NM} H_f(k) \tag{22}$$

for $k = 0, \ldots, K - 1$. This function has the property that

$$\sum_{k=0}^{K-1} p_f(k) = 1. \tag{23}$$

---

[1]In the sense of maximum entropy; see Chapter 5.1.

The normalized histogram $p_f(k)$ has a valid interpretation as the empirical probability density (mass function) of the gray-level values of image f. In other words, if a pixel coordinate **n** is chosen at random, then $p_f(k)$ is the probability that $f(\mathbf{n}) = k$: $p_f(k) = \Pr\{f(\mathbf{n}) = k\}$.

We also define the *cumulative normalized image histogram* to be

$$P_f(r) = \sum_{k=0}^{r} p_f(k); \quad r = 0, \ldots, K - 1. \tag{24}$$

The function $P_f(r)$ is an empirical probability distribution function; hence it is a nondecreasing function, and also $P_f(K - 1) = 1$. It has the probabilistic interpretation that for a randomly selected image coordinate **n**, $P_f(r) = \Pr\{f(\mathbf{n}) \leq r\}$. From Eq. (24) it is also true that

$$p_f(k) = P_f(k) - P_f(k - 1); \quad k = 0, \ldots, K - 1, \tag{25}$$

so $P_f(k)$ and $p_f(k)$ can be obtained from each other. Both are complete descriptions of the gray-level distribution of the image f.

To understand the process of *digital histogram equalization*, we first explain the process by supposing that the normalized and cumulative histograms are functions of continuous variables. We will then formulate the digital case of an approximation of the continuous process. Hence, suppose that $p_f(x)$ and $P_f(x)$ are functions of a continuous variable x. They may be regarded as image probability density function (pdf) and cumulative distribution function (cdf), with relationship $p_f(x) = \mathrm{d}P_f(x)/\mathrm{d}x$. We will also assume that P_f^{-1} exists. Since P_f is nondecreasing, this is either true or P_f^{-1} can be defined by a convention. In this hypothetical continuous case, we claim that the image

$$\mathrm{FSHS}(g) \tag{26}$$

where

$$g = P_f(f) \tag{27}$$

has a uniform (flat) histogram. In Eq. (26), $P_f(f)$ denotes that

P_f is applied on a pixelwise basis to f:

$$g(\mathbf{n}) = P_f[f(\mathbf{n})] \tag{28}$$

for all **n**. Since P_f is a continuous function, Eqs. (26)–(28) represent a smooth mapping of the histogram of image f to an image with a smooth histogram. At first, Eq. (27) may seem confusing since the function P_f that is computed from f is then applied to f. To see that a flat histogram is obtained, we use the probabilistic interpretation of the histogram. The cumulative histogram of the resulting image g is

$$\begin{aligned} P_g(x) &= \Pr\{g \leq x\} = \Pr\{P_f(f) \leq x\} \\ &= \Pr\{f \leq P_f^{-1}(x)\} = P_f\{P_f^{-1}(x)\} = x \end{aligned} \tag{29}$$

for $0 \leq x \leq 1$. Finally, the normalized histogram of g is

$$p_g(x) = \mathrm{d}P_g(x)/\mathrm{d}x = 1 \tag{30}$$

for $0 \leq x \leq 1$. Since $p_g(x)$ is defined only for $0 \leq x \leq 1$, the FSHS in Eq. (26) is required to stretch the flattened histogram to fill the gray-scale range.

To flatten the histogram of a digital image f, first compute the discrete cumulative normalized histogram $P_f(k)$, apply Eq. (28) at each **n**, and then Eq. (26) to the result. However, while an image with a perfectly flat histogram is the result in the ideal continuous case outlined herein, in the digital case the output histogram is only approximately flat, or more accurately, more flat than the input histogram. This follows since Eqs. (26)–(28) collectively are a point operation on the image f, so every occurrence of gray level k maps to $P_f(k)$ in g. Hence, histogram bins are never reduced in amplitude by Eqs. (26)–(28), although they may increase if multiple gray levels map to the same value (thus destroying information). Hence, the histogram cannot be truly equalized by this procedure.

Figures 14 and 15 show histogram equalization applied to our ongoing example images of students and books, respectively. Both images are much more striking and viewable than the original. As can be seen, the resulting histograms are not really flat;

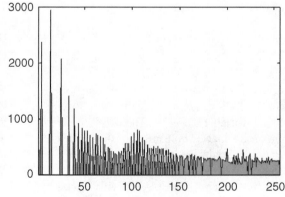

FIGURE 14   Histogram equalization applied to the image of students.

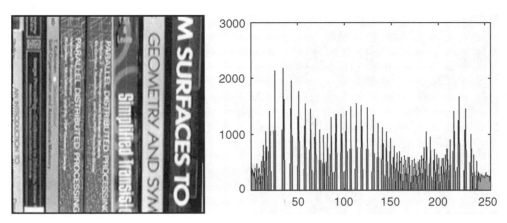

FIGURE 15 Histogram equalization applied to the image of books.

it is flatter in the sense that the histograms are spread as much as possible. However, the heights of peaks are *not* reduced. As is often the case with expansive point operations, gaps or spaces appear in the output histogram. These are not a problem unless the gaps become large and some of the histogram bins become isolated. This amounts to an excess of quantization in that range of gray levels, which may result in false contouring (Chapter 1.1).

5.3 Histogram Shaping

In some applications, it is desired to transform the image into one that has a histogram of a specific shape. The process of histogram shaping generalizes histogram equalization, which is the special case in which the target shape is flat. Histogram shaping can be applied when multiple images of the same scene, but taken under mildly different lighting conditions, are to be compared. This extends the idea of AOD equalization described earlier in this chapter. When the histograms are shaped to match, the comparison may exclude minor lighting effects. Alternately, it may be that the histogram of one image is shaped to match that of another, again usually for the purpose of comparison. Or it might simply be that a certain histogram shape, such as a Gaussian, produces visually agreeable results for a certain class of images.

Histogram shaping is also accomplished by a nonlinear point operation defined in terms of the empirical image probabilities or histogram functions. Again, exact results are obtained in the hypothetical continuous-scale case. Suppose that the target (continuous) cumulative histogram function is $Q(x)$, and that Q^{-1} exists. Then let

$$g = Q^{-1}[P_f(f)], \tag{31}$$

where both functions in the composition are applied on a pixel-wise basis. The cumulative histogram of g is then

$$
\begin{aligned}
P_g(x) &= \Pr\{g \le x\} = \Pr\{Q^{-1}[P_f(f)] \le x\} \\
&= \Pr\{P_f(f) \le Q(x)\} = \Pr\{f \le P_f^{-1}[Q(x)]\} \\
&= P_f\{P_f^{-1}[Q(x)]\} = Q(x), \tag{32}
\end{aligned}
$$

as desired. Note that the FSHS is not required in this instance. Of course, Eq. (32) can only be approximated when the image f is digital. In such cases, the specified target cumulative histogram function $Q(k)$ is discrete, and some convention for defining Q^{-1} should be adopted, particularly if Q is computed from a target image and is unknown in advance. One common convention is to define

$$Q^{-1}(k) = \min_s\{s: Q(s) \ge k\}. \tag{33}$$

As an example, Fig. 16 depicts the result of shaping the histogram of "books" to match the shape of an inverted "V" centered at the middle gray level and extending across the entire gray scale. Again, a perfect V is not produced, although an image of very high contrast is still produced. Instead, the histogram shape that results is a crude approximation to the target.

6 Arithmetic Operations between Images

We now consider *arithmetic operations* defined on multiple images. The basic operations are pointwise image addition/subtraction and pointwise image multiplication/division. Since digital images are defined as arrays of numbers, these operations have to be defined carefully.

Suppose we have n images of dimensions $N \times M f_1, f_2, \ldots, f_n$. It is important that they be of the same dimensions since we will be defining operations between corresponding array elements (having the same indices).

The *sum* of n images is given by

$$f_1 + f_2 + \cdots + f_n = \sum_{m=1}^{n} f_m, \tag{34}$$

while for any two images f_r, f_s the *image difference* is

$$f_r - f_s. \tag{35}$$

The *pointwise product* of the n images $f_1, \ldots, f_n$ is denoted by

$$f_1 \otimes f_2 \otimes \cdots \otimes f_n = \prod_{m=1}^{n} f_m, \tag{36}$$

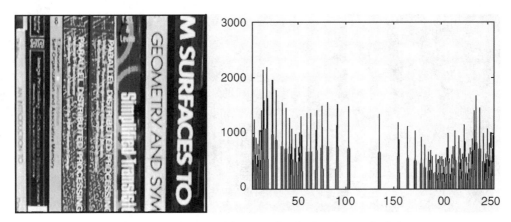

FIGURE 16   Histogram of the image of books shaped to match a "V".

where in Eq. (36) we do *not* infer that the matrix product is being taken. Instead, the product is defined on a pointwise basis. Hence $g = f_1 \otimes f_2 \otimes \cdots \otimes f_n$ if and only if

$$g(\mathbf{n}) = f_1(\mathbf{n}) f_2(\mathbf{n}) \cdots f_n(\mathbf{n}) \qquad (37)$$

for every $\mathbf{n}$. In order to clarify the distinction between matrix product and pointwise array product, we introduce the special notation $\otimes$ to denote the pointwise product. Given two images f_r, f_s the *pointwise image quotient* is denoted

$$g = f_r \Delta f_s \qquad (38)$$

if for every $\mathbf{n}$ it is true that $f_s(\mathbf{n}) \neq 0$ and

$$g(\mathbf{n}) = f_r(\mathbf{n})/f_s(\mathbf{n}). \qquad (39)$$

The pointwise matrix product and quotient are mainly useful when Fourier transforms of images are manipulated, as will be seen in Chapter 2.3. However, the pointwise image sum and difference, despite their simplicity, have important applications that we will examine next.

6.1 Image Averaging for Noise Reduction

Images that occur in practical applications invariably suffer from random degradations that are collectively referred to as *noise*. These degradations arise from numerous sources, including radiation scatter from the surface before the image is sensed; electrical noise in the sensor or camera; channel noise as the image is transmitted over a communication channel; bit errors after the image is digitized, and so on. A good review of various image noise models is given in Chapter 4.4 of this *Handbook*.

The most common generic noise model is *additive noise*, where a noisy observed image is taken to be the sum of an original, uncorrupted image g and a noise image q:

$$f = g + q, \qquad (40)$$

where q is an two-dimensional $N \times M$ random matrix, with

elements $q(\mathbf{n})$ that are random variables. Chapter 4.4 develops the requisite mathematics for understanding random quantities and provides the basis for noise filtering. In this basic chapter we will not require this more advanced development. Instead, we make the simple assumption that the noise is *zero mean*. If the noise is zero mean, then the average (or sample mean) of n independently occurring noise matrices $q_1, q_2, \ldots, q_n$ tends toward zero as n grows large:[2]

$$\left(\frac{1}{n}\right) \sum_{m=1}^{n} q_m \approx \mathbf{0}, \qquad (41)$$

where $\mathbf{0}$ denotes the $N \times M$ matrix of zeros.

Now suppose that we are able to obtain n images $f_1, f_2, \ldots, f_n$ of the same scene. The images are assumed to be noisy versions of an original image g, where the noise is zero mean and additive:

$$f_m = g + q_m \qquad (42)$$

for $m = 1, \ldots, n$. Hence, the images are assumed either to be taken in rapid succession, so that there is no motion between frames, or under conditions where there is no motion in the scene. In this way only the noise contribution varies from image to image.

By averaging the multiple noisy images of Eq. (42), we find

$$\begin{aligned}
\left(\frac{1}{n}\right) \sum_{m=1}^{n} f_m &= \left(\frac{1}{n}\right) \sum_{m=1}^{n} (g + q_m) \\
&= \left(\frac{1}{n}\right) \sum_{m=1}^{n} g + \left(\frac{1}{n}\right) \sum_{m=1}^{n} q_m \\
&= g + \left(\frac{1}{n}\right) \sum_{m=1}^{n} q_m \\
&\approx g, \qquad (43)
\end{aligned}$$

[2]More accurately, the noise must be assumed *mean ergodic*, which means that the sample mean approaches the statistical mean over large sample sizes. This assumption is usually quite reasonable. The statistical mean is defined in Section 4.4.

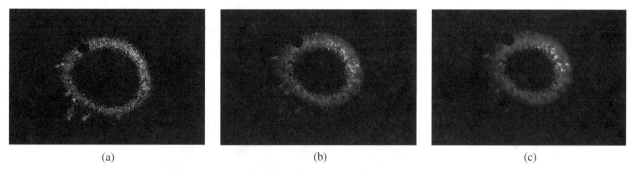

FIGURE 17 Example of image averaging for noise reduction. (a) Single noisy image; (b) average of four frames; (c) average of 16 frames. (Courtesy of Chris Neils of The University of Texas at Austin.)

using Eq. (41). If a large enough number of frames are averaged together, then the resulting image should be nearly noise free, and hence should approximate the original image. The amount of noise reduction can be quite significant; one can expect a reduction in the noise variance by a factor n. Of course, this is subject to inaccuracies in the model, e.g., if there is any change in the scene itself, or if there are any dependencies between the noise images (in an extreme case, the noise images might be identical), then the reduction in the noise will be limited.

Figure 17 depicts the process of noise reduction by frame averaging in an actual example of confocal microscope imaging (Chapter 10.7). The image(s) are of *Macroalga Valonia microphysa*, imaged with a laser scanning confocal microscope (LSCM). The dark ring is chlorophyll fluorescing under Ar laser excitation. As can be seen, in this case the process of image averaging is quite effective in reducing the apparent noise content and in improving the visual resolution of the object being imaged.

6.2 Image Differencing for Change Detection

Often it is of interest to detect changes that occur in images taken of the same scene but at different times. If the time instants are closely placed, e.g., adjacent frames in a video sequence, then the goal of change detection amounts to image motion detection (Chapter 3.8). There are many applications of motion detection and analysis. For example, in video compression algorithms, compression performance is improved by exploiting redundancies that are tracked along the motion trajectories of image objects that are in motion. Detected motion is also useful for tracking targets, for recognizing objects by their motion, and for computing three-dimensional scene information from two-dimensional motion.

If the time separation between frames is not small, then change detection can involve the discovery of gross scene changes. This can be useful for security or surveillance cameras, or in automated visual inspection systems, for example. In either case, the basic technique for change detection is the *image difference*. Suppose that f_1 and f_2 are images to be compared. Then the absolute difference image

$$g = |f_1 - f_2| \qquad (44)$$

will embody those changes or differences that have occurred between the images. At coordinates **n** where there has been little change, $g(\mathbf{n})$ will be small. Where change has occurred, $g(\mathbf{n})$ can be quite large. Figure 18 depicts image differencing. In the difference image, large changes are displayed as brighter intensity values. Since significant change has occurred, there are many bright intensity values. This difference image could be processed by an automatic change detection algorithm. A simple series of steps that might be taken would be to binarize the difference image, thus separating change from nonchange, using a threshold (Chapter 2.2), counting the number of high-change pixels, and finally, deciding whether the change is significant enough to take some action. Sophisticated variations of this theme are currently in practical use. The histogram in Fig. 18(d) is instructive, since it is characteristic of differenced images; many zero or small gray-level changes occur, with the incidence of larger changes falling off rapidly.

7 Geometric Image Operations

We conclude this chapter with a brief discussion of *geometric image operations*. Geometric image operations are, in a sense, the opposite of point operations: they modify the spatial positions and spatial relationships of pixels, but they do not modify gray-level values. Generally, these operations can be quite complex and computationally intensive, especially when applied to video sequences. However, the more complex geometric operations are not much used in engineering image processing, although they are heavily used in the computer graphics field. The reason for this is that image processing is primarily concerned with correcting or improving images of the real world; hence complex geometric operations, which distort images, are less frequently used. Computer graphics, however, is primarily concerned with creating images of an unreal world, or at least a visually modified

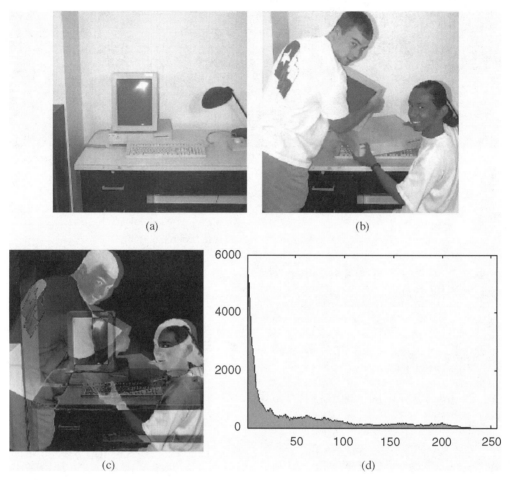

FIGURE 18   Image differencing example. (a) Original placid scene; (b) a theft is occurring! (c) the difference image with brighter points' signifying larger changes; (d) the histogram of (c).

reality, and subsequently geometric distortions are commonly used in that discipline.

A geometric image operation generally requires two steps. The first is a spatial mapping of the coordinates of an original image f to define a new image g:

$$g(\mathbf{n}) = f(\mathbf{n}') = f[\mathbf{a}(\mathbf{n})]. \qquad (45)$$

Thus, geometric image operations are defined as functions of position rather than intensity. The two-dimensional, two-valued mapping function $\mathbf{a}(\mathbf{n}) = [a_1(n_1, n_2), a_2(n_1, n_2)]$ is usually defined to be continuous and smoothly changing, but the coordinates $\mathbf{a}(\mathbf{n})$ that are delivered are not generally integers. For example, if $\mathbf{a}(\mathbf{n}) = (n_1/3, n_2/4)$, then $g(\mathbf{n}) = f(n_1/3, n_2/4)$, which is not defined for most values of (n_1, n_2). The question then is, which value(s) of f are used to define $g(\mathbf{n})$, when the mapping does not fall on the standard discrete lattice?

Thus implies the need for the second operation: *interpolation* of noninteger coordinates $a_1(n_1, n_2)$ and $a_2(n_1, n_2)$ to integer values, so that g can be expressed in a standard row–column format. There are many possible approaches for accomplishing interpolation; we will look at two of the simplest: *nearest neighbor interpolation*, and *bilinear interpolation*. The first of these is too simplistic for many tasks, whereas the second is effective for most.

7.1 Nearest-Neighbor Interpolation

Here, the geometrically transformed coordinates are mapped to the nearest integer coordinates of f:

$$g(\mathbf{n}) = f\{\text{INT}[a_1(n_1, n_2) + 0.5], \text{INT}[a_2(n_1, n_2) + 0.5]\}, \quad (46)$$

where $\text{INT}[R]$ denotes the nearest integer that is less than or equal to R. Hence, the coordinates are *rounded* prior to assigning them to g. This certainly solves the problem of finding integer coordinates of the input image, but it is quite simplistic, and, in practice, it may deliver less than impressive results. For example, several coordinates to be mapped may round to the same values, creating a block of pixels in the output image of the same value. This may give an impression of "blocking," or of structure that is not physically meaningful. The effect is particularly noticeable along sudden changes in intensity, or "edges," which may appear jagged following nearest neighbor interpolation.

7.2 Bilinear Interpolation

Bilinear interpolation produces a smoother interpolation than does the nearest-neighbor approach. Given four neighboring image coordinates $f(n_{10}, n_{20})$, $f(n_{11}, n_{21})$, $f(n_{12}, n_{22})$, and $f(n_{13}, n_{23})$ — these can be the four nearest neighbors of $f[\mathbf{a}(\mathbf{n})]$ — then the geometrically transformed image $g(n_1, n_2)$ is computed as

$$g(n_1, n_2) = A_0 + A_1 n_1 + A_2 n_2 + A_3 n_1 n_2, \qquad (47)$$

which is a bilinear function in the coordinates (n_1, n_2). The bilinear weights A_0, A_1, A_2, and A_3 are found by solving

$$\begin{bmatrix} A_0 \\ A_1 \\ A_2 \\ A_3 \end{bmatrix} = \begin{bmatrix} 1 & n_{10} & n_{20} & n_{10}n_{20} \\ 1 & n_{11} & n_{21} & n_{11}n_{21} \\ 1 & n_{12} & n_{22} & n_{12}n_{22} \\ 1 & n_{13} & n_{23} & n_{13}n_{23} \end{bmatrix}^{-1} \begin{bmatrix} f(n_{10}, n_{20}) \\ f(n_{11}, n_{21}) \\ f(n_{12}, n_{22}) \\ f(n_{13}, n_{23}) \end{bmatrix}. \qquad (48)$$

Thus, $g(n_1, n_2)$ is defined to be a linear combination of the gray levels of its four nearest neighbors. The linear combination defined by Eq. (48) is in fact the value assigned to $g(n_1, n_2)$ when the best (least-squares) planar fit is made to these four neighbors. This process of optimal averaging produces a visually smoother result.

Regardless of the interpolation approach that is used, it is possible that the mapping coordinates $a_1(n_1, n_2)$, $a_2(n_1, n_2)$ do not fall within the pixel ranges

$$0 \le a_1(n_1, n_2) \le N - 1$$

and/or $\qquad\qquad\qquad\qquad\qquad\qquad\qquad (49)$

$$0 \le a_2(n_1, n_2) \le M - 1,$$

in which case it is not possible to define the geometrically transformed image at these coordinates. Usually a nominal value is assigned, such as $g(\mathbf{n}) = 0$, at these locations.

7.3 Image Translation

The most basic geometric transformation is the *image translation*, where

$$a_1(n_1, n_2) = n_1 - b_1, \qquad a_2(n_1, n_2) = n_2 - b_2, \qquad (50)$$

where (b_1, b_2) are integer constants. In this case $g(n_1, n_2) = f(n_1 - b_1, n_2 - b_2)$, which is a simple shift or translation of g by an amount b_1 in the vertical (row) direction and an amount b_2 in the horizontal direction. This operation is used in image display systems, when it is desired to move an image about, and

it is also used in algorithms, such as image convolution (Chapter 2.3), where images are shifted relative to a reference. Since integer shifts can be defined in either direction, there is usually no need for the interpolation step.

7.4 Image Rotation

Rotation of the image g by an angle θ relative to the horizontal (n_1) axis is accomplished by the following transformations:

$$a_1(n_1, n_2) = n_1 \cos \theta - n_2 \sin \theta,$$

$$a_2(n_1, n_2) = n_1 \sin \theta + n_2 \cos \theta. \qquad (51)$$

The simplest cases are: $\theta = 90°$, where $[a_1(n_1, n_2), a_2(n_1, n_2)] = (-n_2, n_1)$; $\theta = 180°$, where $[a_1(n_1, n_2), a_2(n_1, n_2)] = (-n_1, -n_2)$; and $\theta = -90°$, where $[a_1(n_1, n_2), a_2(n_1, n_2)] = (n_2, -n_1)$. Since the rotation point is not defined here as the center of the image, the arguments of Eq. (51) may fall outside of the image domain. This may be ameliorated by applying an image translation either before or after the rotation to obtain coordinate values in the nominal range.

7.5 Image Zoom

The *image zoom* either magnifies or minifies the input image according to the mapping functions

$$a_1(n_1, n_2) = n_1/c, \qquad a_2(n_1, n_2) = n_2/d, \qquad (52)$$

where $c \ge 1$ and $d \ge 1$ to achieve magnification, and $c < 1$ and $d < 1$ to achieve minification. If applied to the entire image, then the image size is also changed by a factor $c(d)$ along the vertical (horizontal) direction. If only a small part of an image is to be zoomed, then a translation may be made to the corner of that region, the zoom applied, and then the image cropped.

The image zoom is a good example of a geometric operation for which the type of interpolation is important, particularly at high magnifications. With nearest neighbor interpolation, many values in the zoomed image may be assigned the same gray scale, resulting in a severe "blotching" or "blocking" effect. The bilinear interpolation usually supplies a much more viable alternative.

Figure 19 depicts a $4\times$ zoom operation applied to the image in Fig. 13 (logarithmically transformed "students"). The image was first zoomed, creating a much larger image (16 times as many pixels). The image was then translated to a point of interest (selected, e.g., by a mouse), and then it was cropped to size 256×256 pixels around this point. Both nearest-neighbor and bilinear interpolation were applied for the purpose of comparison. Both provide a nice close-up of the original, making the faces much more identifiable. However, the bilinear result is much smoother, and it does not contain the blocking artifacts that can make recognition of the image difficult.

(a)   (b)

FIGURE 19   Example of (4×) image zoom followed by interpolation. (a) Nearest-neighbor interpolation; (b) bilinear interpolation.

It is important to understand that image zoom followed by interpolation does not inject *any* new information into the image, although the magnified image may appear easier to see and interpret. The image zoom is only an interpolation of known information.

Acknowledgment

Many thanks to Dr. Scott Acton for carefully reading and commenting on this chapter.

2.2

Basic Binary Image Processing

Alan C. Bovik
The University of Texas at Austin

Mita D. Desai
The University of Texas at San Antonio

1 Introduction

In this second chapter on basic methods, we explain and demonstrate fundamental tools for the processing of *binary* digital images. Binary image processing is of special interest, since an image in binary format can be processed with very fast logical (Boolean) operators. Often, a binary image has been obtained by abstracting essential information from a gray-level image, such as object location, object boundaries, or the presence or absence of some image property.

As seen in the previous two chapters, a digital image is an array of numbers or sampled image intensities. Each gray level is quantized or assigned one of a finite set of numbers represented by B bits. In a binary image, only one bit is assigned to each pixel: $B = 1$, implying two possible gray-level values, 0 and 1. These two values are usually interpreted as Boolean; hence each pixel can take on the logical values 0 or 1, or equivalently, "true" or "false." For example, these values might indicate the absence or presence of some image property in an associated gray-level image of the same size, where 1 at a given coordinate indicates the presence of the property at that coordinate in the gray-level image, and 0 otherwise. This image property is quite commonly a sufficiently high or low intensity (brightness), although more abstract properties, such as the presence or absence of certain objects, or smoothness or nonsmoothness, etc., might be indicated.

Since most image display systems and software assume images of eight or more bits per pixel, the question arises as to how binary images are displayed. Usually, they are displayed using the two extreme gray tones, black and white, which are ordinarily represented by 0 and 255, respectively, in a gray-scale display environment, as depicted in Fig. 1. There is no established convention for the Boolean values that are assigned to "black" and to "white." In this chapter we will uniformly use 1 to represent black (displayed as gray-level 0) and 0 to represent white (displayed as gray-level 255). However, the assignments are quite commonly reversed, and it is important to note that the Boolean values 0 and 1 have no physical significance other than what the user assigns to them.

Binary images arise in a number of ways. Usually, they are created from gray-level images for simplified processing or for printing (see Chapter 8.1 on image halftoning). However, certain types of sensors directly deliver a binary image output. Such devices are usually associated with printed, handwritten, or line drawing images, with the input signal being entered by hand on a pressure sensitive tablet, a resistive pad, or a light pen.

In such a device, the (binary) image is first initialized prior to image acquisition:

$$g(\mathbf{n}) = 0 \tag{1}$$

at all coordinates $\mathbf{n}$. When pressure, a change of resistance, or light is sensed at some image coordinate $\mathbf{n}_0$, then the image is assigned the value 1:

$$g(\mathbf{n}_0) = 1 \tag{2}$$

This continues until the user completes the drawing, as depicted in Fig. 2. These simple devices are quite useful for entering engineering drawings, handprinted characters, or other binary graphics in a binary image format.

FIGURE 1   A 10 × 10 binary image.

2 Image Thresholding

Usually, a binary image is obtained from a gray-level image by some process of information abstraction. The advantage of the *B*-fold reduction in the required image storage space is offset by what can be a significant loss of information in the resulting binary image. However, if the process is accomplished with care, then a simple abstraction of information can be obtained that can enhance subsequent processing, analysis, or interpretation of the image.

The simplest such abstraction is the process of *image thresholding*, which can be thought of as an extreme form of gray-level quantization. Suppose that a gray-level image *f* can take *K* possible gray levels 0, 1, 2, . . . , *K* − 1. Define an integer threshold, *T*, that lies in the gray-scale range of $T \in \{0, 1, 2, \ldots, K - 1\}$. The process of thresholding is a process of simple comparison: each pixel value in *f* is compared to *T*. Based on this comparison, a binary decision is made that defines the value of the corresponding pixel in an output binary image *g*:

$$g(\mathbf{n}) = \begin{cases} 0 & \text{if } f(\mathbf{n}) \geq T \\ 1 & \text{if } f(\mathbf{n}) < T \end{cases} \tag{3}$$

Of course, the threshold *T* that is used is of critical importance, since it controls the particular abstraction of information that is obtained. Indeed, different thresholds can produce different valuable abstractions of the image. Other thresholds may produce little valuable information at all. It is instructive to observe the result of thresholding an image at many different levels in sequence. Figure 3 depicts the image "mandrill" (Fig. 8 of Chapter 1.1) thresholded at four different levels. Each produces different information, or in the case of Figs. 3(a) and 3(d), very little useful information. Among these, Fig. 3(c) probably contains the most visual information, although it is far from ideal. The four threshold values (50, 100, 150, and 200) were chosen without the use of any visual criterion.

As will be seen, image thresholding can often produce a binary image result that is quite useful for simplified processing, interpretation, or display. However, some gray-level images do not lead to any interesting binary result regardless of the chosen threshold *T*.

Several questions arise: Given a gray-level image, how does one decide whether binarization of the image by gray-level thresholding will produce a useful result? Can this be decided automatically by a computer algorithm? Assuming that thresholding is likely to be successful, how does one decide on a threshold level *T*? These are apparently simple questions pertaining to a very simple operation. However, these questions turn out to be quite difficult to answer in the general case. In other cases, the answer is simpler. In all cases, however, the basic tool for understanding the process of image thresholding is the image histogram, which was defined and studied in Chapter 2.1.

Thresholding is most commonly and effectively applied to images that can be characterized as having *bimodal histograms*. Figure 4 depicts two hypothetical image histograms. The one on the left has two clear modes; the one at the right either has a single mode, or two heavily overlapping, poorly separated modes.

Bimodal histograms are often (but not always) associated with images that contain objects and backgrounds having a significantly different average brightness. This may imply bright objects on a dark background, or dark objects on a bright background. The goal, in many applications, is to separate the objects from the background, and to label them as object or as background. If the image histogram contains well-separated modes associated with an object and with a background, then thresholding can be the means for achieving this separation. Practical examples of gray-level images with well-separated bimodal histograms are not hard to find. For example, an image of machine-printed type (like that being currently read), or of handprinted characters, will have a very distinctive separation between object and background. Examples abound in biomedical applications, where it is often possible to control the lighting of objects and background. Standard bright-field microscope images of single or multiple cells (micrographs) typically contain bright objects against a darker background. In many industry applications, it is also possible to control the relative brightness of objects of interest and the backgrounds they are set against. For example, machine parts that are being imaged (perhaps in an automated inspection application) may be placed on a mechanical conveyor that has substantially different reflectance properties than the objects.

Given an image with a bimodal histogram, a general strategy for thresholding is to place the threshold *T* between the image modes, as depicted in Fig. 4(a). Many "optimal" strategies have been suggested for deciding the exact placement of the threshold between the peaks. Most of these are based on an assumed

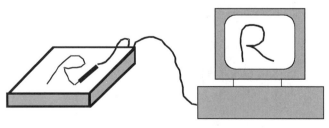

FIGURE 2   Simple binary image device.

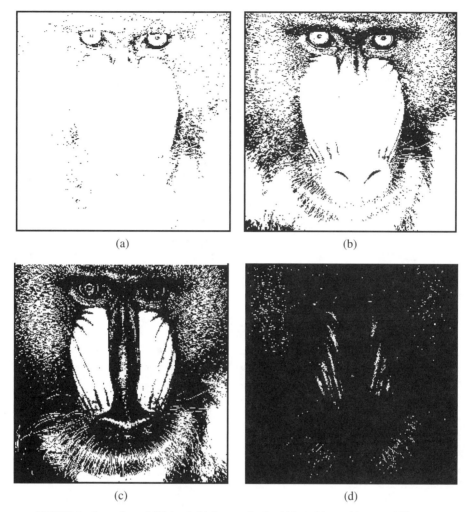

FIGURE 3   Image "mandrill" thresholded at gray levels of (a) 50, (b) 100, (c) 150, and (d) 150.

statistical model for the histogram, and by posing the decision of labeling a given pixel as "object" versus "background" as a statistical inference problem. In the simplest version, two hypotheses are posed:

H_0:   The pixel belongs to gray level Population 0.
H_1:   The pixel belongs to gray level Population 1.

Here pixels from population 0 and 1 have conditional probability density functions (pdf's) $p_f(a \,|\, H_0)$ and $p_f(a \,|\, H_1)$, respectively,

under the two hypotheses. If it is also known (or estimated) that H_0 is true with probability p_0 and that H_1 is true with probability $p_1 (p_0 + p_1 = 1)$, then the decision may be cast as a likelihood ratio test. If an observed pixel has gray level $f(\mathbf{n}) = k$, then the decision may be rendered according to

$$\frac{p_f(k \,|\, H_1)}{p_f(k \,|\, H_0)} \underset{H_0}{\overset{H_1}{\gtrless}} \frac{p_0}{p_1}. \tag{4}$$

The decision whether to assign logical 0 or 1 to a pixel can thus

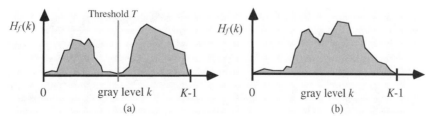

FIGURE 4   Hypothetical histograms: (a) well-separated modes and (b) poorly separated or indistinct modes.)

be regarded as applying a simple statistical test to each pixel. In relation (4), the conditional pdf's may be taken as the modes of a bimodal histogram. Algorithmically, this means that they must be fit to the histogram by using some criterion, such as least squares. This is usually quite difficult, since it must be decided that there are indeed two separate modes, the locations (centers) and widths of the modes must be estimated, and a model for the shape of the modes must be assumed. Depending on the assumed shape of the modes (in a given application, the shape might be predictable), specific probability models might be applied, e.g., the modes might be taken to have the shape of Gaussian pdf's (Chapter 4.5). The prior probabilities p_0 and p_1 are often easier to model, since in many applications the relative areas of object and background can be estimated or given reasonable values based on empirical observations.

A likelihood ratio test such as relation (4) will place the image threshold T somewhere between the two modes of the image histogram. Unfortunately, any simple statistical model of the image does not account for such important factors as object/background continuity, visual appearance to a human observer, non-uniform illumination or surface reflectance effects, and so on. Hence, with rare exceptions, a statistical approach such as relation (4) will not produce as good a result as would a human decision maker making a manual threshold selection.

Placing the threshold T between two obvious modes of a histogram may yield acceptable results, as depicted in Fig. 4(a). The problem is significantly complicated, however, if the image contains multiple distinct modes or if the image is nonmodal or level. Multimodal histograms can occur when the image contains multiple objects of different average brightness on a uniform background. In such cases, simple thresholding will exclude some objects (Fig. 5). Nonmodal or flat histograms usually imply more complex images, containing significant gray-level variation, detail, non-uniform lighting or reflection, etc. (Fig. 5). Such images are often not amenable to a simple thresholding process, especially if the goal is to achieve figure–ground separation. However, all of these comments are, at best, rules of thumb. An image with a bimodal histogram might not yield good results when thresholded at any level, while an image with a perfectly flat histogram might yield an ideal result. It is a good mental exercise to consider when these latter cases might occur.

Figures 6–8 shows several images, their histograms, and the thresholded image results. In Fig. 6, a good threshold level for the micrograph of the cellular specimens was taken to be $T = 180$. This falls between the two large modes of the histogram (there are many smaller modes) and was deemed to be visually optimal by one user. In the binarized image, the individual cells are not perfectly separated from the background. The reason for this is that the illuminated cells have non-uniform brightness profiles, being much brighter toward the centers. Taking the threshold higher ($T = 200$), however, does not lead to improved results, since the bright background then begins to fall below threshold.

Figure 7 depicts a *negative* (for better visualization) of a digitized mammogram. Mammography is the key diagnostic tool for the detection of breast cancer, and in the future, digital tools for mammographic imaging and analysis will be used. The image again shows two strong modes, with several smaller modes. The first threshold chosen ($T = 190$) was selected at the minimum point between the large modes. The resulting binary image has the nice result of separating the region of the breast from the background. However, radiologists are often interested in the detailed structure of the breast and in the brightest (darkest in the negative) areas, which might indicate tumors or microcalcifications. Figure 7(d) shows the result of thresholding at the lower level of 125 (higher level in the positive image), successfully isolating much of the interesting structure.

Generally, the best binarization results by means of thresholding are obtained by direct human operator intervention. Indeed, most general-purpose image processing environments have thresholding routines that allow user interaction. However, even with a human picking a visually "optimal" value of T, thresholding rarely gives perfect results. There is nearly always some misclassification of object as background, and vice versa. For example, in the image "micrograph," no value of T is able to successfully extract the objects from the background; instead, most of the objects have "holes" in them, and there is a sprinkling of black pixels in the background as well.

Because of these limitations of the thresholding process, it is usually necessary to apply some kind of *region correction* algorithms to the binarized image. The goal of such algorithms is to correct the misclassification errors that occur. This requires identifying misclassified background points as object points,

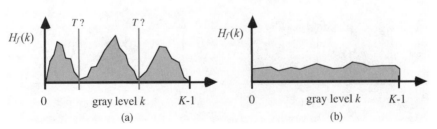

FIGURE 5 Hypothetical histograms: (a) Multimodal, showing the difficulty of threshold selection; (b) nonmodal, for which the threshold selection is quite difficult or impossible.

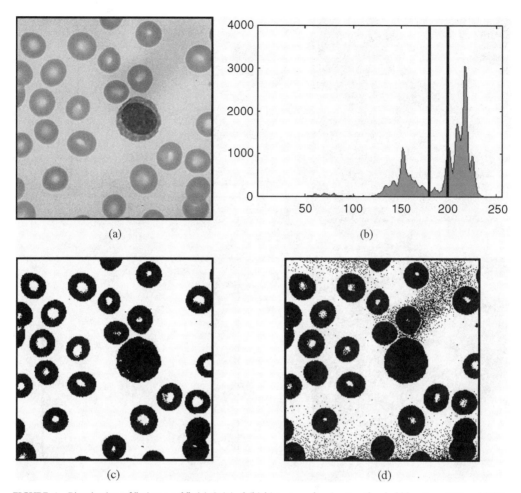

FIGURE 6 Binarization of "micrograph": (a) Original (b) histogram showing two threshold locations (180 and 200), and (c) and (d) resulting binarized images.

and vice versa. These operations are usually applied directly to the binary images, although it is possible to augment the process by also incorporating information from the original gray-scale image. Much of the remainder of this chapter will be devoted to algorithms for region correction of thresholded binary images.

3 Region Labeling

A simple but powerful tool for identifying and labeling the various objects in a binary image is a process called *region labeling*, *blob coloring*, or *connected component identification*. It is useful since once they are individually labeled, the objects can be separately manipulated, displayed, or modified. For example, the term "blob coloring" refers to the possibility of displaying each object with a different identifying color, once labeled.

Region labeling seeks to identify connected groups of pixels in a binary image f that all have the same binary value. The simplest such algorithm accomplishes this by scanning the entire image (left to right, top to bottom), searching for occurrences

of pixels of the same binary value and connected along the horizontal or vertical directions. The algorithm can be made slightly more complex by also searching for diagonal connections, but this is usually unnecessary. A record of connected pixel groups is maintained in a separate label array r having the same dimensions as f, as the image is scanned. The following algorithm steps explain the process, in which the region labels used are positive integers.

3.1 Region Labeling Algorithm

1. Given an $N \times M$ binary image f, initialize an associated $N \times M$ region label array: $r(\mathbf{n}) = 0$ for all $\mathbf{n}$. Also initialize a region number counter: $k = 1$.

   Then, scanning the image from left to right and top to bottom, for every $\mathbf{n}$ do the following:
2. If $f(\mathbf{n}) = 0$ then do nothing.
3. If $f(\mathbf{n}) = 1$ and also $f(\mathbf{n} - (1, 0)) = f(\mathbf{n} - (0, 1)) = 0$, as depicted in Fig. 8(a), then set $r(\mathbf{n}) = 0$ and $k = k + 1$. In this case the left and upper neighbors of $f(\mathbf{n})$ do not belong to objects.

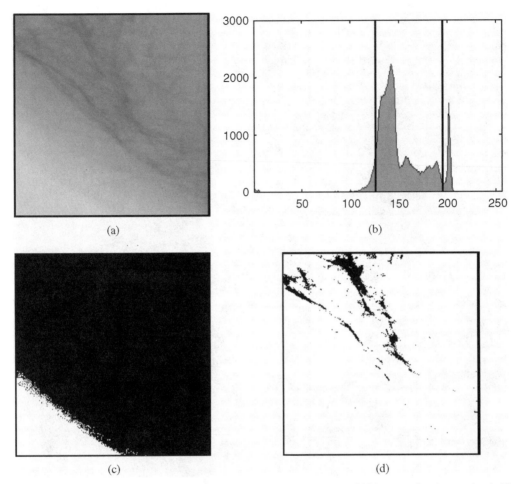

FIGURE 7 Binarization of "mammogram": (a) Original *negative* mammogram; (b) histogram showing two threshold locations (190 and 125), and (c) and (d) resulting binarized images.

4. If $f(\mathbf{n}) = 1$, $f(\mathbf{n} - (1, 0)) = 1$, and $f(\mathbf{n} - (0, 1)) = 0$, Fig. 8(b), then set $r(\mathbf{n}) = r(\mathbf{n} - (1, 0))$. In this case the upper neighbor $f(\mathbf{n} - (1, 0))$ belongs to the same object as $f(\mathbf{n})$.

5. If $f(\mathbf{n}) = 1$, $f(\mathbf{n} - (1, 0)) = 0$, and $f(\mathbf{n} - (0, 1)) = 1$, Fig. 8(c), then set $r(\mathbf{n}) = r(\mathbf{n} - (0, 1))$. In this case the left neighbor $f(\mathbf{n} - (0, 1))$ belongs to the same object as $f(\mathbf{n})$.

6. If $f(\mathbf{n}) = 1$, and $f(\mathbf{n} - (1, 0)) = f(\mathbf{n} - (0, 1)) = 1$, Fig. 8(d), then set $r(\mathbf{n}) = r(\mathbf{n} - (0, 1))$. If $r(\mathbf{n} - (0, 1)) \neq r(\mathbf{n} - (1, 0))$, then record the labels $r(\mathbf{n} - (0, 1))$ and $r(\mathbf{n} - (1, 0))$ as equivalent. In this case both the left and upper neighbors belong to the same object as $f(\mathbf{n})$, although they may have been labeled differently.

A simple application of region labeling is the measurement of object area. This can be accomplished by defining a vector **c** with elements $c(k)$ that are the pixel area (pixel count) of region k.

3.2 Region Counting Algorithm

Initialize $\mathbf{c} = \mathbf{0}$. For every **n** do the following:

1. If $f(\mathbf{n}) = 0$ then do nothing.
2. If $f(\mathbf{n}) = 1$, then $c[r(\mathbf{n})] = c[r(\mathbf{n})] + 1$.

Another simple but powerful application of region labeling is the removal of minor regions or objects from a binary image. The ways in which this is done depends on the application. It may be desired that only a single object should remain (generally, the largest object), or it may be desired that any object with a pixel area less than some minimum value should be deleted. A variation is that the minimum value is computed as a percentage of the largest object in the image. The following algorithm depicts the second possibility.

FIGURE 8 Pixel neighbor relationships used in a region labeling algorithm. In each of (a)–(d), $f(\mathbf{n})$ is the lower right pixel.

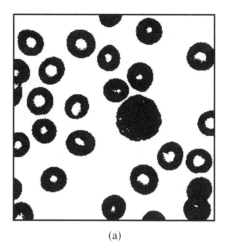 

(a)   (b)

FIGURE 9   Result of applying the region labeling -— counting–removal algorithms to (a) the binarized image in Fig. 6(c) and (b) then to the image in (a), but in the polarity-reversed mode.

3.3 Minor Region Removal Algorithm

Assume a minimum allowable object size of S pixels. For every $\mathbf{n}$ do the following.

1. If $f(\mathbf{n}) = 0$ then do nothing.
2. If $f(\mathbf{n}) = 1$ and $c[r(\mathbf{n})] < S$, then set $f(\mathbf{n}) = 0$.

Of course, all of the above algorithms can be operated in reverse polarity, by interchanging 0 for 1 and 1 for 0 everywhere.

An important application of region labeling/region counting/minor region removal is in the correction of thresholded binary images. The application of a binarizing threshold to a gray-level image inevitably produces an imperfect binary image, with such errors as extraneous objects or holes in objects. These can arise from noise, unexpected objects (such as dust on a lens), and generally, non-uniformities in the surface reflectances and illuminations of the objects and background.

Figure 9 depicts the result of sequentially applying the region labeling/region counting/minor region removal algorithms to the binarized micrograph image in Fig. 6(c). The series of algorithms was first applied to Fig. 6(c) as above to remove extraneous small black objects, using a size threshold of 500 pixels as shown in Fig. 9(a). It was then applied again to this modified image, but in the polarity-reversed mode, to remove the many object holes, this time using a threshold of 1000 pixels. The result shown in Fig. 9(b) is a dramatic improvement over the original binarized result, given that the goal was to achieve a clean separation of the objects in the image from the background.

4 Binary Image Morphology

We next turn to a much broader and more powerful class of binary image processing operations that collectively fall under the name *binary image morphology*. These are closely related to (in fact, are the same as in a mathematical sense) the gray-level morphological operations described in Chapter 3.3. As the name indicates, these operators modify the *shapes* of the objects in an image.

4.1 Logical Operations

The morphological operators are defined in terms of simple logical operations on local groups of pixels. The logical operators that are used are the simple NOT, AND, OR, and MAJ (majority) operators. Given a binary variable x, $\mathrm{NOT}(x)$ is its logical complement. Given a set of binary variables $x_1, \ldots, x_n$, the operation $\mathrm{AND}(x_1, \ldots, x_n)$ returns value 1 if and only if $x_1 = \cdots = x_n = 1$ and 0 otherwise. The operation $\mathrm{OR}(x_1, \ldots, x_n)$ returns value 0 if and only if $x_1 = \cdots = x_n = 0$ and 1 otherwise. Finally, if n is odd, the operation $\mathrm{MAJ}(x_1, \ldots, x_n)$ returns value 1 if and only if a majority of $(x_1, \ldots, x_n)$ equal 1 and 0 otherwise.

We observe in passing the DeMorgan's Laws for binary arithmetic, specifically

$$\mathrm{NOT}[\mathrm{AND}(x_1, \ldots, x_n)] = \mathrm{OR}[\mathrm{NOT}(x_1), \ldots, \mathrm{NOT}(x_n)], \quad (5)$$

$$\mathrm{NOT}[\mathrm{OR}(x_1, \ldots, x_n)] = \mathrm{AND}[\mathrm{NOT}(x_1), \ldots, \mathrm{NOT}(x_n)], \quad (6)$$

which characterizes the duality of the basic logical operators AND and OR under complementation. However, note that

$$\mathrm{NOT}[\mathrm{MAJ}(x_1, \ldots, x_n)] = \mathrm{MAJ}[\mathrm{NOT}(x_1), \ldots, \mathrm{NOT}(x_n)]. \quad (7)$$

Hence MAJ is its own dual under complementation.

4.2 Windows

As mentioned, morphological operators change the shapes of objects by using *local logical operations*. Since they are local operators, a formal methodology must be defined for making the operations occur on a local basis. The mechanism for doing this is the *window*.

A window defines a geometric rule according to which gray levels are collected from the vicinity of a given pixel coordinate. It is called a window since it is often visualized as a moving collection of empty pixels that is passed over the image. A morphological operation is (conceptually) defined by moving a window over the binary image to be modified, in such a way that it is eventually centered over every image pixel, where a local logical operation is performed. Usually this is done row by row, column by column, although it can be accomplished at every pixel simultaneously, if a massively parallel-processing computer is used.

Usually, a window is defined to have an approximate circular shape (a digital circle cannot be exactly realized) since it is desired that the window, and hence, the morphological operator, be rotation invariant. This means that if an object in the image is rotated through some angle, then the response of the morphological operator will be unchanged other than also being rotated. While rotational symmetry cannot be exactly obtained, symmetry across two axes can be obtained, guaranteeing that the response be at least reflection invariant. Window size also significantly effects the results, as will be seen.

A formal definition of windowing is needed in order to define the various morphological operators. A window **B** is a set of $2P + 1$ coordinate shifts $b_i = (n_i, m_i)$ centered around $(0, 0)$:

$$\mathbf{B} = \{b_1, \ldots, b_{2P+1}\} = \{(n_1, m_1), \ldots, (n_{2P+1}, m_{2P+1})\}$$

Some examples of common one-dimensional (row and column) windows are

$$\mathbf{B} = \text{ROW}[2P + 1] = \{(0, m); m = -P, \ldots, P\} \quad (8)$$

$$\mathbf{B} = \text{COL}[2P + 1] = \{(n, 0); n = -P, \ldots, P\} \quad (9)$$

and some common two-dimensional windows are

$$\mathbf{B} = \text{SQUARE}[(2P + 1)^2] = \{(n, m); n, m = -P, \ldots, P\} \quad (10)$$

$$\mathbf{B} = \text{CROSS}[4P + 1] = \text{ROW} (2P + 1) \cup \text{COL}(2P + 1) \quad (11)$$

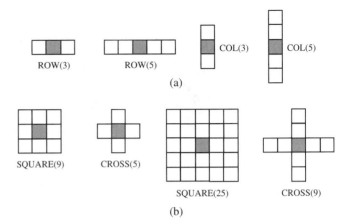

FIGURE 10 Examples of windows: (a) one-dimensional, $\text{ROW}(2P + 1)$ and $\text{COL}(2P + 1)$ for $P = 1, 2$; (b) two-dimensional, $\text{SQUARE}[(2P + 1)^2]$ and $\text{CROSS}[4P + 1]$ for $P = 1, 2$. The window is centered over the shaded pixel.

with obvious shape-descriptive names. In each of Eqs. (8)–(11), the quantity in brackets is the number of coordinates shifts in the window, hence also the number of local gray levels that will be collected by the window at each image coordinate. Note that the windows of Eqs. (8)–(11) are each defined with an odd number $2P + 1$ coordinate shifts. This is because the operators are symmetrical: pixels are collected in pairs from opposite sides of the center pixel or $(0, 0)$ coordinate shift, plus the $(0, 0)$ coordinate shift is always included. Examples of each the windows of Eqs. (8)–(11) are shown in Fig. 10. The example window shapes in Eqs. (8)–(11) and in Fig. 10 are by no means the only possibilities, but they are (by far) the most common implementations because of the simple row–column indexing of the coordinate shifts.

The action of gray-level collection by a moving window creates the *windowed set*. Given a binary image f and a window **B**, the windowed set at image coordinate **n** is given by

$$\mathbf{B} f(\mathbf{n}) = \{f(\mathbf{n} - \mathbf{m}); \mathbf{m} \in \mathbf{B}\}, \quad (12)$$

which, conceptually, is the set of image pixels covered by **B** when it is centered at coordinate **n**. Examples of windowed sets associated with some of the windows in Eqs. (8)–(11) and Fig. 10 are as follows:

$$\mathbf{B} = \text{ROW}(3): \quad \mathbf{B} f(n_1, n_2) = \{f(n_1, n_2 - 1), f(n_1, n_2), f(n_1, n_2 + 1)\} \quad (13)$$

$$\mathbf{B} = \text{COL}(3): \quad \mathbf{B} f(n_1, n_2) = \{f(n_1 - 1, n_2), f(n_1, n_2), f(n_1 + 1, n_2)\} \quad (14)$$

$$\mathbf{B} = \text{SQUARE}(9): \quad \mathbf{B} f(n_1, n_2) = \{f(n_1 - 1, n_2 - 1), f(n_1 - 1, n_2), f(n_1 - 1, n_2 + 1),$$
$$f(n_1, n_2 - 1), f(n_1, n_2), f(n_1, n_2 + 1),$$
$$f(n_1 + 1, n_2 - 1), f(n_1 + 1, n_2), f(n_1 + 1, n_2 + 1)\} \quad (15)$$

$$\mathbf{B} = \text{CROSS}(5): \quad \mathbf{B} f(n_1, n_2) = \{f(n_1 - 1, n_2), f(n_1, n_2 - 1), f(n_1, n_2), f(n_1, n_2 + 1), f(n_1 + 1, n_2)\} \quad (16)$$

where the elements of Eqs. (13)–(16) have been arranged to show the geometry of the windowed sets when centered over coordinate $\mathbf{n} = (n_1, n_2)$. Conceptually, the window may be thought of as capturing a series of miniature images as it is passed over the image, row by row, column by column.

One last note regarding windows involves the definition of the windowed set when the window is centered near the boundary edge of the image. In this case, some of the elements of the windowed set will be undefined, since the window will overlap "empty space" beyond the image boundary. The simplest and most common approach is to use *pixel replication*: set each undefined windowed set value equal to the gray level of the nearest known pixel. This has the advantage of simplicity, and also the intuitive value that the world just beyond the borders of the image probably does not change very much. Figure 11 depicts the process of pixel replication.

4.3 Morphological Filters

Morphological filters are Boolean filters. Given an image f, a many-to-one binary or Boolean function h, and a window $\mathbf{B}$, the Boolean-filtered image $g = h(f)$ is given by

$$g(\mathbf{n}) = h[\mathbf{B} f(\mathbf{n})] \qquad (17)$$

at every $\mathbf{n}$ over the image domain. Thus, at each $\mathbf{n}$, the filter collects local pixels according to a geometrical rule into a windowed set, performs a Boolean operation on them, and returns the single Boolean result $g(\mathbf{n})$.

The most common Boolean operations that are used are AND, OR, and MAJ. They are used to create the following simple, yet powerful *morphological filters*. These filters act on the objects in the image by shaping them: expanding or shrinking them, smoothing them, and eliminating too-small features.

The binary *dilation filter* is defined by

$$g(\mathbf{n}) = OR[\mathbf{B} f(\mathbf{n})] \qquad (18)$$

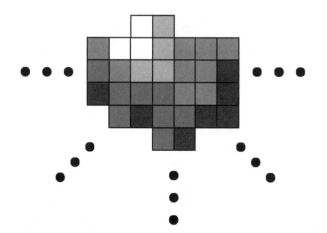

FIGURE 11 Depiction of pixel replication for a window centered near the (top) image boundary.

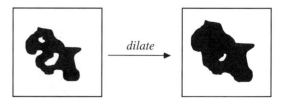

FIGURE 12 Illustration of dilation of a binary 1-valued object; the smallest hole and gap were filled.

and is denoted $g = \text{dilate}(f, \mathbf{B})$. The binary *erosion filter* is defined by

$$g(\mathbf{n}) = AND[\mathbf{B} f(\mathbf{n})] \qquad (19)$$

and is denoted $g = \text{erode}(f, \mathbf{B})$. Finally, the binary *majority filter* is defined by

$$g(\mathbf{n}) = MAJ[\mathbf{B} f(\mathbf{n})] \qquad (20)$$

and is denoted $g = \text{majority}(f, \mathbf{B})$. Next we explain the response behavior of these filters.

The *dilate* filter expands the size of the foreground, object, or one-valued regions in the binary image f. Here the 1-valued pixels are assumed to be black because of the convention we have assumed, but this is not necessary. The process of dilation also smoothes the boundaries of objects, removing gaps or bays of too-narrow width and also removing object holes of too-small size. Generally, a hole or gap will be filled if the dilation window cannot fit into it. These actions are depicted in Fig. 12, while Fig. 13 shows the result of dilating an actual binary image. Note that dilation using $\mathbf{B} = \text{SQUARE}(9)$ removed most of the small holes and gaps, while using $\mathbf{B} = \text{SQUARE}(25)$ removed nearly all of them. It is also interesting to observe that dilation with the larger window nearly completed a bridge between two of the large masses. Dilation with CROSS(9) highlights an interesting effect: individual, isolated 1-valued or BLACK pixels were dilated into larger objects having the same shape as the window. This can also be seen with the results using the SQUARE windows. This effect underlines the importance of using symmetric windows, preferably with near rotational symmetry, since then smoother results are obtained.

The *erode* filter shrinks the size of the foreground, object, or 1-valued regions in the binary image f. Alternately, it expands the size of the background or 0-valued regions. The process of erosion smoothes the boundaries of objects, but in a different way than dilation: it removes peninsulas or fingers of too-narrow width, and also it removes 1-valued objects of too-small size. Generally, an isolated object will be eliminated if the dilation window cannot fit into it. The effects of erode are depicted in Fig. 14. Figure 15 shows the result of applying the *erode* filter to the binary image "cell." Erosion using $\mathbf{B} = \text{SQUARE}(9)$ removed many of the small objects and fingers, while using $\mathbf{B} = \text{SQUARE}(25)$ removed most of them. As an example of intense smoothing, $\mathbf{B} = \text{SQUARE}(81)$, a 9×9 square window, was also applied.

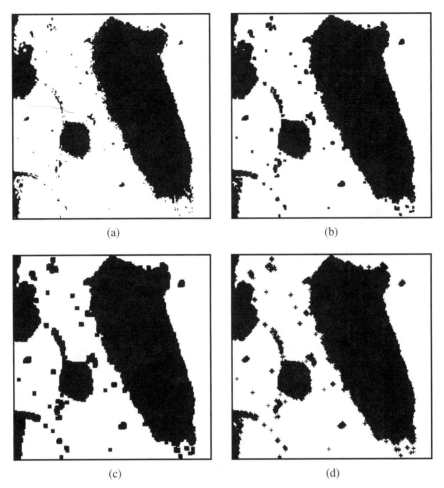

FIGURE 13   Dilation of a binary image. (a) Binarized image "cells." Dilate with (b) **B** = SQUARE(9), (c) **B** = SQUARE(25), and (d) **B** = CROSS(9).

Erosion with CROSS(9) again produced a good result, except at a few isolated points where isolated 0-valued or WHITE pixels were expanded into larger +-shaped objects.

An important property of the erode and dilate filters is the relationship that exists between them. In fact, in reality they are the same operation, in the dual (complementary) sense. Indeed, given a binary image f and an arbitrary window **B**, it is true that

$$\text{dilate}(f, \mathbf{B}) = \text{NOT}\{\text{erode}[\text{NOT}(f), \mathbf{B}]\} \qquad (21)$$

$$\text{erode}(f, \mathbf{B}) = \text{NOT}\{\text{dilate}[\text{NOT}(f), \mathbf{B}]\}. \qquad (22)$$

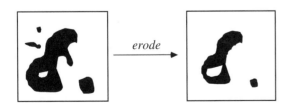

FIGURE 14   Illustration of erosion of a binary 1-valued object. The smallest objects and peninsula were eliminated.

Equations (21) and (22) are a simple consequence of the De-Morgan's Laws (5) and (6). A correct interpretation of this is that erosion of the 1-valued or BLACK regions of an image is the same as dilation of the 0-valued or WHITE regions — and vice versa.

An important and common misconception must be mentioned. Erode and dilate filters shrink and expand the sizes of 1-valued objects in a binary image. However, they are *not* inverse operations of one another. Dilating an eroded image (or eroding a dilated image) very rarely yields the original image. In particular, dilation cannot recreate peninsulas, fingers, or small objects that have been eliminated by erosion. Likewise, erosion cannot unfill holes filled by dilation or recreate gaps or bays filled by dilation. Even without these effects, erosion generally will not exactly recreate the same shapes that have been modified by dilation, and vice versa.

Before discussing the third common Boolean filter, the majority, we will consider further the idea of sequentially applying erode and dilate filters to an image. One reason for doing this is that the erode and dilate filters have the effect of changing the

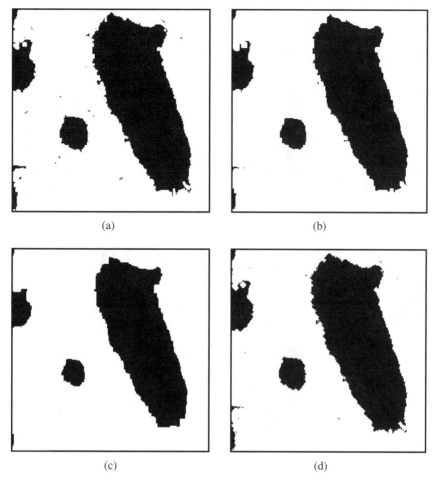

FIGURE 15 Erosion of the binary image "cells." Erode with (a) **B** = SQUARE(9), (b) **B** = SQUARE(25), (c) **B** = SQUARE(81), and (d) **B** = CROSS(9).

sizes of objects, as well as smoothing them. For some objects this is desirable, e.g., when an extraneous object is shrunk to the point of disappearing; however, often it is undesirable, since it may be desired to further process or analyze the image. For example, it may be of interest to label the objects and compute their sizes, as in Section 3 of this chapter.

Although erode and dilate are not inverse operations of one another, they are approximate inverses in the sense that if they are performed in sequence on the same image with the same window **B**, then object and holes that are not eliminated will be returned to their approximate sizes. We thus define the size-preserving smoothing morphological operators termed *open filter* and *close filter*, as follows:

$$\text{open}(f, \mathbf{B}) = \text{dilate}[\text{erode}(f, \mathbf{B}), \mathbf{B}] \qquad (23)$$

$$\text{close}(f, \mathbf{B}) = \text{erode}[\text{dilate}(f, \mathbf{B}), \mathbf{B}]. \qquad (24)$$

Hence, the opening (closing) of image f is the erosion (dilation) with window **B** followed by dilation (erosion) with window **B**. The morphological filters open and close have the same smooth-

ing properties as erode and dilate, respectively, but they do not generally effect the sizes of sufficiently large objects much (other than pixel loss from pruned holes, gaps or bays, or pixel gain from eliminated peninsulas).

Figure 16 depicts the results of applying the open and close operations to the binary image "cell," using the windows **B** = SQUARE(25) and **B** = SQUARE(81). Large windows were used to illustrate the powerful smoothing effect of these morphological smoothers. As can be seen, the open filters did an excellent job of eliminating what might be referred to as "black noise" — the extraneous 1-valued objects and other features — leaving smooth, connected, and appropriately sized large objects. By comparison, the close filters smoothed the image intensely as well, but without removing the undesirable black noise. In this particular example, the result of the open filter is probably preferable to that of close, since the extraneous BLACK structures present more of a problem in the image.

It is important to understand that the open and close filters are *unidirectional* or *biased* filters in the sense that they remove one type of "noise" (either extraneous WHITE or BLACK features), but not both. Hence, open and close are somewhat

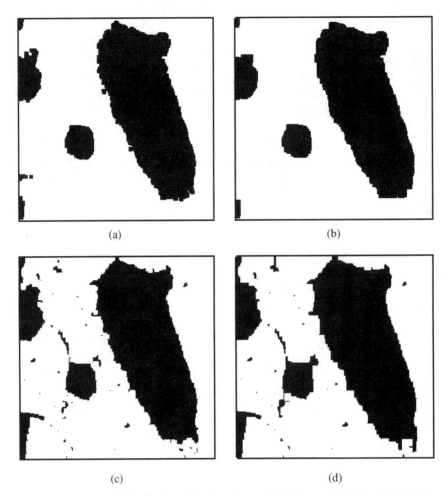

FIGURE 16   Open and close filtering of the binary image "cells." Open with (a) **B** = SQUARE(25), (b) **B** = SQUARE(81); close with (c) **B** = SQUARE(25), (d) **B** = SQUARE(81).

special-purpose binary image smoothers that are used when too-small BLACK and WHITE objects (respectively) are to be removed.

It is worth noting that the close and open filters are again in fact the same filters, in the dual sense. Given a binary image f and an arbitrary window **B**,

$$\text{close}(f, \mathbf{B}) = \text{NOT}\{\text{open}[\text{NOT}(f), \mathbf{B}]\} \quad (25)$$

$$\text{open}(f, \mathbf{B}) = \text{NOT}\{\text{close}[\text{NOT}(f), \mathbf{B}]\}. \quad (26)$$

In most binary smoothing applications, it is desired to create an unbiased smoothing of the image. This can be accomplished by a further concatenation of filtering operations, applying open and close operations in sequence on the same image with the same window **B**. The resulting images will then be smoothed *bidirectionally*. We thus define the unbiased smoothing morphological operators *close–open filter* and *open–close filter*, as follows:

$$\text{close–open}(f, \mathbf{B}) = \text{close}[\text{open}(f, \mathbf{B}), \mathbf{B}] \quad (27)$$

$$\text{open–close}(f, \mathbf{B}) = \text{open}[\text{close}(f, \mathbf{B}), \mathbf{B}]. \quad (28)$$

Hence, the close–open (open–close) of image f is the open

(close) of f with window **B** followed by the close (open) of the result with window **B**. The morphological filters close–open and open–close in Eqs. (27) and (28) are general-purpose, bidirectional, size-preserving smoothers. Of course, they may each be interpreted as a sequence of four basic morphological operations (erosions and dilations).

The close–open and open–close filters are quite similar but are not mathematically identical. Both remove too-small structures without affecting size much. Both are powerful shape smoothers. However, differences between the processing results can be easily seen. These mainly manifest as a function of the first operation performed in the processing sequence. One notable difference between close–open and open–close is that close–open often links together neighboring holes (since erode is the first step), while open–close often links neighboring objects together (since dilate is the first step). The differences are usually somewhat subtle, yet often visible upon close inspection.

Figure 17 shows the result of applying the close–open and the open–close filters to the ongoing binary image example. As can be seen the results (for **B** fixed) are very similar, although the close–open filtered results are somewhat cleaner, as expected.

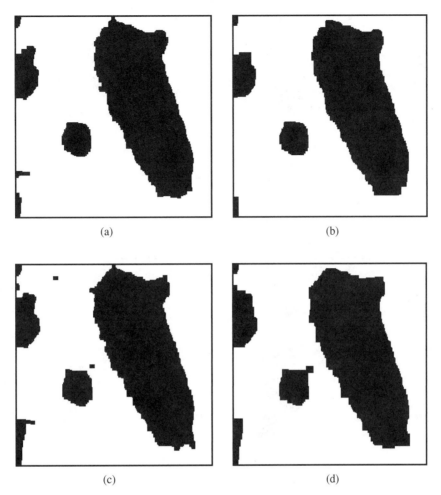

(a)   (b)

(c)   (d)

FIGURE 17 Close–open and open–close filtering of the binary image "cells." Close–open with (a) **B** = SQUARE(25), (b) **B** = SQUARE(81); Open–close with (c) **B** = SQUARE(25), (d) **B** = SQUARE(81).

There are also only small differences between the results obtained using the medium and larger windows, because of the intense smoothing that is occurring. To fully appreciate the power of these smoothers, compare it to the original binarized image "cells" in Fig. 13(a).

The reader may wonder whether further sequencing of the filtered responses will produce different results. If the filters are properly alternated as in the construction of the close–open and open–close filters, then the dual filters become increasingly similar. However, the smoothing power can most easily be increased by simply taking the window size to be larger.

Once again, the close–open and open–close filters are dual filters under complementation.

We now return to the final binary smoothing filter, the *majority filter*. The majority filter is also known as the binary *median filter*, since it may be regarded as a special case (the binary case) of the gray-level median filter (Chapter 3.2).

The majority filter has similar attributes as the close–open and open–close filters: it removes too-small objects, holes, gaps, bays and peninsulas (both 1-valued and 0-valued small features),

and it also does not generally change the size of objects or of background, as depicted in Fig. 18. It is less biased than any of the other morphological filters, since it does not have an initial erode or dilate operation to set the bias. In fact, majority is its own dual under complementation, since

$$\text{majority}(f, \mathbf{B}) = \text{NOT}\{\text{majority}[\text{NOT}(f), \mathbf{B}]\} \qquad (29)$$

The majority filter is a powerful, unbiased shape smoother. However, for a given filter size, it does not have the same degree of smoothing power as close–open or open–close.

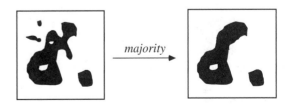

FIGURE 18 Effect of majority filtering. The smallest holes, gaps, fingers, and extraneous objects are eliminated.

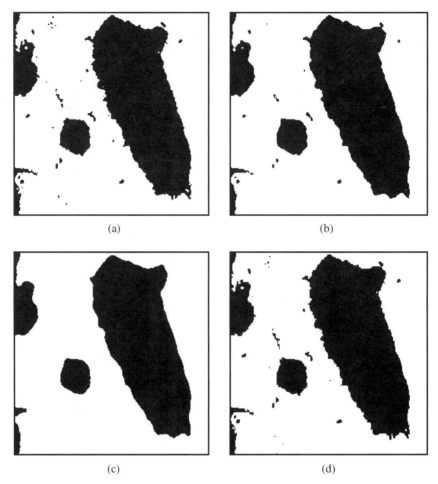

(a)

(b)

(c)

(d)

FIGURE 19   Majority or median filtering of the binary image "cells." Majority with (a) **B** = SQUARE(9), (b) **B** = SQUARE(25); Majority with (c) **B** = SQUARE(81), (d) **B** = CROSS(9).

Figure 19 shows the result of applying the majority or *binary median* filter to the image "cell." As can be seen, the results obtained are very smooth. Comparison with the results of open–close and close–open are favorable, since the boundaries of the major smoothed objects are much smoother in the case of the median filter, for both window shapes used and for each size. The majority filter is quite commonly used for smoothing noisy binary images of this type because of these nice properties. The more general gray-level median filter (Chapter 3.2) is also among the most-used image processing filters.

4.4 Morphological Boundary Detection

The morphological filters are quite effective for smoothing binary images, but they have other important applications as well. One such application is *boundary detection*, which is the binary case of the more general edge detectors studied in Chapters 4.11 and 4.12.

At first glance, boundary detection may seem trivial, since the boundary points can be simply defined as the transitions from 1 to 0 (and vice versa). However, when there is noise present,

boundary detection becomes quite sensitive to small noise artifacts, leading to many useless detected edges. Another approach which allows for smoothing of the object boundaries involves the use of morphological operators.

The "difference" between a binary image and a dilated (or eroded) version of it is one effective way of detecting the object boundaries. Usually it is best that the window **B** that is used be small, so that the difference between image and dilation is not too large (leading to thick, ambiguous detected edges). A simple and effective "difference" measure is the two-input exclusive-OR operator, XOR. The XOR takes logical value 1 only if its two inputs are different. The boundary detector then becomes simply

$$\text{boundary}(f, \mathbf{B}) = \text{XOR}[f, \text{dilate}(f, \mathbf{B})]. \quad (30)$$

The result of this operation as applied to the binary image "cells" is shown in Fig. 20(a), using **B** = SQUARE(9). As can be seen, essentially all of the BLACK–WHITE transitions are marked as boundary points. Often, this is the desired result. However, in other instances, it is desired to detect only the major object

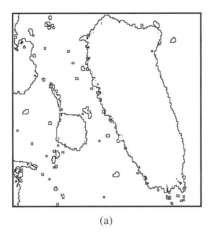 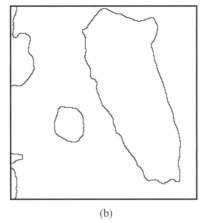

(a)   (b)

FIGURE 20 Object boundary detection. Application of boundary(f, **B**) to (a) the image "cells"; (b) the majority-filtered image in Fig. 19(c).

boundary points. This can be accomplished by first smoothing the image with a close–open, open–close, or majority filter. The result of this smoothed boundary detection process is shown in Fig. 20(b). In this case, the result is much cleaner, as only the major boundary points are discovered.

5 Binary Image Representation and Compression

In several later chapters, methods for compressing gray-level images are studied in detail. Compressed images are representations that require less storage than the nominal storage. This is generally accomplished by coding of the data based on measured statistics, rearrangement of the data to exploit patterns and redundancies in the data, and (in the case of lossy compression), quantization of information. The goal is that the image, when decompressed, either looks very much like the original despite a loss of some information (lossy compression), or is not different from the original (lossless compression).

Methods for lossless compression of images are discussed in Chapter 5.1. Those methods can generally be adapted to both gray-level and binary images. Here, we will look at two methods for lossless binary image representation that exploit an assumed structure for the images. In both methods the image data are represented in a new format that exploits the structure. The first method is *run-length coding*, which is so called because it seeks to exploit the redundancy of long run lengths or runs of constant value 1 or 0 in the binary data. It is thus appropriate for the coding/compression of binary images containing large areas of constant value 1 and 0. The second method, *chain coding*, is appropriate for binary images containing binary contours, such as the boundary images shown in Fig. 20. Chain coding achieves compression by exploiting this assumption. The chain code is also an information-rich, highly manipulable representation that can be used for shape analysis.

5.1 Run-Length Coding

The number of bits required to naively store an $N \times M$ binary image is NM. This can be significantly reduced if it is known that the binary image is smooth in the sense that it is composed primarily of large areas of constant 1 and/or 0 value.

The basic method of run-length coding is quite simple. Assume that the binary image f is to be stored or transmitted on a row-by-row basis. Then for each image row numbered m, the following algorithm steps are used.

1. Store the first pixel value (0 or 1) in row m in a 1-bit buffer as a reference.
2. Set the run counter $c = 1$.
3. For each pixel in the row,
   - Examine the next pixel to the right.
   - If it is the same as the current pixel, set $c = c + 1$.
   - If different from the current pixel, store c in a buffer of length b and set $c = 1$.
   - Continue until end of row is reached.

Thus, each run length is stored by using b bits. This requires that an overall buffer with segments of lengths b be reserved to store the run lengths. Run-length coding yields excellent lossless compressions, provided that the image contains lots of constant runs. Caution is necessary, since if the image contains only very short runs, then run-length coding can actually increase the required storage.

Figure 21 depicts two hypothetical image rows. In each case, the first symbol stored in a 1-bit buffer will be logical 1. The run-length code for Fig. 21(a) would be "1," 7, 5, 8, 3, 1 ..., with symbols after the "1" stored with b bits. The first five runs in this sequence have average length $24/5 = 4.8$; hence if $b \leq 4$, then compression will occur. Of course, the compression can be much higher, since there may be runs of lengths in the dozens or hundreds, leading to very high compressions.

In the worst-case example of Fig. 21(b), however, the storage actually increases b-fold! Hence, care is needed when applying

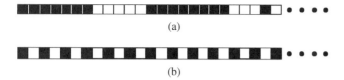

(a)

(b)

FIGURE 21   Example rows of a binary image, depicting (a) reasonable and (b) unreasonable scenarios for run-length coding.

this method. The apparent rule of thumb, if it can be applied *a priori*, is that the average run length L of the image should satisfy $L > b$ if compression is to occur. In fact, the compression ratio will be approximately L/b.

Run-length coding is also used in scenarios other than binary image coding. It can also be adapted to situations in which there are run lengths of any value. For example, in the JPEG lossy image compression standard for gray-level images (see Chapter 5.5), a form of run-length coding is used to code runs of zero-valued frequency-domain coefficients. This run-length coding is an important factor in the good compression performance of JPEG. A more abstract form of run-length coding is also responsible for some of the excellent compression performance of recently developed wavelet image compression algorithms (Chapter 5.4).

5.2 Chain Coding

Chain coding is an efficient representation of binary images composed of contours. We will refer to these as "contour images." We assume that contour images are composed only of single-pixel width, connected contours (straight or curved). These arise from processes of edge detection or boundary detection, such as the morphological boundary detection method just described, or the results of some of the edge detectors described in Chapters 4.11 and 4.12 when applied to gray-scale images.

The basic idea of chain coding is to code contour directions instead of naïve bit-by-bit binary image coding or even coordinate representations of the contours. Chain coding is based on identifying and storing the directions from each pixel to its neighbor pixel on each contour. Before this process is defined, it is necessary to clarify the various types of neighbors that are associated with a given pixel in a binary image. Figure 22 depicts two neighborhood systems around a pixel (shaded). To the left are depicted the *4-neighbors* of the pixel, which are connected along the horizontal and vertical directions. The set of 4-neighbors of

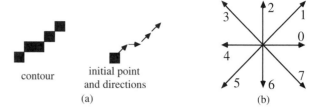

FIGURE 23   Representation of a binary contour by direction codes. (a) A connected contour can be represented exactly by an initial point and the subsequent directions. (b) Only 8 direction codes are required.

a pixel located at coordinate $\mathbf{n}$ will be denoted $N_4(\mathbf{n})$. To the right are the *8-neighbors* of the shaded pixel in the center of the grouping. These include the pixels connected along the diagonal directions. The set of 8-neighbors of a pixel located at coordinate $\mathbf{n}$ will be denoted $N_8(\mathbf{n})$.

If the initial coordinate $\mathbf{n}_0$ of an 8-connected contour is known, then the rest of the contour can be represented without loss of information by the directions along which the contour propagates, as depicted in Fig. 23(a). The initial coordinate can be an endpoint, if the contour is open, or an arbitrary point, if the contour is closed. The contour can be reconstructed from the directions, if the initial coordinate is known. Since there are only eight directions that are possible, then a simple 8-neighbor direction code may be used. The integers $\{0, \ldots, 7\}$ suffice for this, as shown in Fig. 23(b). Of course, the direction codes 0, 1, 2, 3, 4, 5, 6, 7 can be represented by their 3-bit binary equivalents: 000, 001, 010, 011, 100, 101, 110, 111. Hence, each point on the contour *after* the initial point can be coded by 3 bits. The initial point of each contour requires $\lceil \log_2(MN) \rceil$ bits, where $\lceil \cdot \rceil$ denotes the ceiling function: $\lceil x \rceil$ = the smallest integer that is greater than or equal to x. For long contours, storage of the initial coordinates is incidental.

Figure 24 shows an example of chain coding of a short contour. After the initial coordinate $\mathbf{n}_0 = (n_0, m_0)$ is stored, the chain code for the remainder of the contour is: 1, 0, 1, 1, 1, 1, 3, 3, 3, 4, 4, 5, 4 in integer format, or 001, 000, 001, 001, 001, 001, 011, 011, 011, 100, 100, 101, 100 in binary format. Chain coding is an efficient representation. For example, if the image dimensions $N = M = 512$, then representing the contour by storing the coordinates of each contour point requires six times as much storage as the chain code.

FIGURE 22   Depiction of the 4-neighbors and the 8-neighbors of a pixel (shaded).

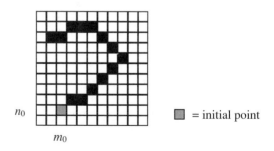

FIGURE 24   Depiction of chain coding.

2.3

Basic Tools for Image Fourier Analysis

Alan C. Bovik

*The University of Texas
at Austin*

1 Introduction

In this third chapter on basic methods, the basic mathematical and algorithmic tools for the frequency-domain analysis of digital images are explained. Also introduced is the two-dimensional discrete-space convolution. Convolution is the basis for linear filtering, which plays a central role in many places in this *Handbook*. An understanding of frequency-domain and linear filtering concepts is essential to be able to comprehend such significant topics as image and video enhancement, restoration, compression, segmentation, and wavelet-based methods. Exploring these ideas in a two-dimensional setting has the advantage that frequency-domain concepts and transforms can be visualized as images, often enhancing the accessibility of ideas.

2 Discrete-Space Sinusoids

Before defining any frequency-based transforms, first we shall explore the concept of *image frequency,* or more generally, of *two-dimensional frequency.* Many readers may have a basic background in the frequency-domain analysis of one-dimensional signals and systems. The basic theories in two dimensions are founded on the same principles. However, there are some exten-

sions. For example, a two-dimensional frequency component, or sinusoidal function, is characterized not only by its location (phase shift) and its frequency of oscillation, but also by its direction of oscillation.

Sinusoidal functions will play an essential role in all of the developments in this chapter. A *two-dimensional discrete-space sinusoid* is a function of the form

$$\sin[2\pi(Um + Vn)]. \tag{1}$$

Unlike a one-dimensional sinusoid, function (1) has two frequencies, U and V (with units of cycles/pixel), which represent the frequency of oscillation along the vertical (m) and horizontal (n) spatial image dimensions. Generally, a two-dimensional sinusoid oscillates (is nonconstant) along every direction except for the direction orthogonal to the direction of fastest oscillation. The frequency of this fastest oscillation is the *radial frequency,* i.e.,

$$\Omega = \sqrt{U^2 + V^2}, \tag{2}$$

which has the same units as U and V, and the direction of this fastest oscillation is the angle, i.e.,

$$\theta = \tan^{-1}\left(\frac{V}{U}\right), \tag{3}$$

with units of radians. Associated with function (1) is the complex exponential function

$$\exp[j2\pi(Um + Vn)] = \cos[2\pi(Um + Vn)]$$
$$+ j\sin[2\pi(Um + Vn)], \qquad (4)$$

where $j = \sqrt{-1}$ is the pure imaginary number.

In general, sinusoidal functions can be defined on discrete integer grids; hence functions (1) and (4) hold for all integers $-\infty < m, n < \infty$. However, sinusoidal functions of infinite duration are not encountered in practice, although they are useful for image modeling and in certain image decompositions that we will explore.

In practice, discrete-space images are confined to finite $M \times N$ sampling grids, and we will also find it convenient to utilize *finite-extent* ($M \times N$) *two-dimensional discrete-space sinusoids*, which are defined only for integers

$$0 \le m \le M - 1, \quad 0 \le n \le N - 1, \qquad (5)$$

and undefined elsewhere. A sinusoidal function that is confined to domain (5) can be contained within an image matrix of dimensions $M \times N$, and is thus easily manipulated digitally.

In the case of finite sinusoids defined on finite grids (5), it will often be convenient to use the scaled frequencies

$$(u, v) = (MU, NV) \qquad (6)$$

which have the visually intuitive units of cycles/image. With this, two-dimensional sinusoid (1) defined on finite grid (5) can be re-expressed as

$$\sin\left[2\pi\left(\frac{u}{M}m + \frac{v}{N}n\right)\right], \qquad (7)$$

with similar redefinition of complex exponential (4).

Figure 1 depicts several discrete-space sinusoids of dimensions 256×256 displayed as intensity images after linear mapping the gray scale of each to the range 0–255. Because of the nonlinear

(a)

(b)

(c)

(d)

FIGURE 1 Examples of finite two-dimensional discrete-space sinusoidal functions. The scaled frequencies of Eq. (6) measured in cycles/image are (a) $u = 1$, $v = 4$; (b) $u = 10$, $v = 5$; (c) $u = 15$, $v = 35$; and (d) $u = 65$, $v = 35$.

response of the eye, the functions in Fig. 1 look somewhat more like square waves than smoothly varying sinusoids, particularly at higher frequencies. However, if any of the images in Fig. 1 is sampled along a straight line of arbitrary orientation, the result is an ideal (sampled) sinusoid.

A peculiarity of discrete-space (or discrete-time) sinusoids is that they have a maximum possible *physical* frequency at which they can oscillate. Although the frequency variables (u, v) or (U, V) may be taken arbitrarily large, these large values do not correspond to arbitrarily large physical oscillation frequencies. The ramifications of this are quite deep and significant, and they relate to the restrictions placed on sampling of continuous-space images (the Sampling Theorem) and the Nyquist frequency. The sampling of images and video is covered in Chapters 7.1 and 7.2.

As an example of this principle, we will study a one-dimensional example of a discrete sinusoid. Consider the finite cosine function, $\cos\{2\pi[(u/M)m + (v/N)n]\} = \cos[2\pi(u/16)m]$, which results by taking $M = N = 16$, and $v = 0$. This is a cosine wave propagating in the m direction only (all columns are the same) at frequency u (cycles/image).

Figure 2 depicts the one-dimensional cosine for various values of u. As can be seen, the physical oscillation frequency increases until $u = 8$; for incrementally larger values of u, however, the physical frequency diminishes. In fact, the function is period-16 in the frequency index u:

$$\cos\left(2\pi \frac{u}{16} m\right) = \cos\left[2\pi \frac{(u + 16k)}{16} m\right] \tag{8}$$

for all integers k. Indeed, the highest physical frequency of $\cos[2\pi(u/M)m]$ occurs at $u = M/2 + kM$, (for M even) for all integers k. At these periodically placed frequencies, Eq. (8) is equal to $(-1)^m$; the fastest discrete-index oscillation is the

alternating signal. This observation will be important next as we define the various frequency-domain image transforms.

3 Discrete-Space Fourier Transform

The *discrete-space Fourier transform*, or *DSFT*, of a given discrete-space image f is given by

$$F(U, V) = \sum_{m=-\infty}^{\infty} \sum_{n=-\infty}^{\infty} f(m, n) e^{-j2\pi(Um+Vn)}, \tag{9}$$

with the *inverse discrete-space Fourier transform* (IDSFT),

$$f(m, n) = \int_{-0.5}^{0.5} \int_{-0.5}^{0.5} F(U, V) e^{j2\pi(Um+Vn)} dU \, dV. \tag{10}$$

When Eqs. (9) and (10) hold, we will often make the notation $f \overset{\Im}{\leftrightarrow} F$ and say that f, F form a *DSFT pair*. The units of the frequencies (U, V) in Eqs. (9) and (10) are cycles/pixel. It should be noted that, unlike continuous Fourier transforms, the DSFT is asymmetrical in that the forward transform F is continuous in the frequency variables (U, V), while the image or inverse transform is discrete. Thus, the DSFT is defined as a summation, while the IDSFT is defined as an integral.

There are several ways of interpreting the DSFT in Eqs. (9) and (10). The most usual mathematical interpretation of Eq. (10) is as a decomposition of $f(m, n)$ into *orthonormal* complex exponential basis functions $e^{j2\pi(Um+Vn)}$ that satisfy

$$\int_{-0.5}^{0.5} \int_{-0.5}^{0.5} e^{j2\pi(Um+Vn)} e^{-j2\pi(Up+Vq)} dU \, dV$$

$$= \begin{cases} 1; & m = p, \quad n = q \\ 0; & \text{otherwise} \end{cases}. \tag{11}$$

Another (somewhat less precise) interpretation is the engineering concept of the transformation, without loss, of space-domain image information into frequency-domain image information. Representing the image information in the frequency domain has significant conceptual and algorithmic advantages, as will be seen. A third interpretation is a physical one, in which the image is viewed as the result of a sophisticated constructive-destructive interference wave pattern. By assigning each of the infinite number of complex exponential wave functions $e^{j2\pi(Um+Vn)}$ the appropriate complex weights $F(U, V)$, one can recreate the intricate structure of any discrete-space image exactly as an interference sum.

The DSFT possesses a number of important properties that will be useful in defining applications. In the following, assume that $f \overset{\Im}{\leftrightarrow} F$, $g \overset{\Im}{\leftrightarrow} G$, and $h \overset{\Im}{\leftrightarrow} H$.

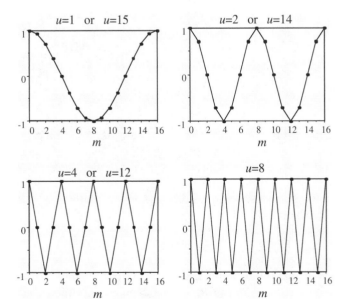

FIGURE 2 Illustration of physical versus numerical frequencies of discrete-space sinusoids.

3.1 Linearity of DSFT

Given images f, g and arbitrary complex constants a, b, the following holds:

$$af + bg \overset{\mathfrak{F}}{\leftrightarrow} aF + bG. \tag{12}$$

This property of linearity follows directly from Eq. (9), and it can be extended to a weighted sum of any countable number of images. It is fundamental to many of the properties of, and operations involving, the DSFT.

3.2 Inversion of DSFT

The two-dimensional function $F(U, V)$ uniquely satisfies relationships (9) and (10). That the inversion holds can be easily shown by substituting Eq. (9) into Eq. (10), reversing the order of sum and integral, and then applying Eq. (11).

3.3 Magnitude and Phase of DSFT

The DSFT F of an image f is generally complex valued. As such it can be written in the form

$$F(U, V) = R(U, V) + jI(U, V), \tag{13}$$

where

$$R(U, V) = \sum_{m=-\infty}^{\infty} \sum_{n=-\infty}^{\infty} f(m, n) \cos[2\pi(Um + Vn)] \tag{14}$$

and

$$I(U, V) = -\sum_{m=-\infty}^{\infty} \sum_{n=-\infty}^{\infty} f(m, n) \sin[2\pi(Um + Vn)] \tag{15}$$

are the real and imaginary parts of $F(U, V)$, respectively.

The DSFT can also be written in the often-convenient phasor form

$$F(U, V) = |F(U, V)|e^{j\angle F(U,V)}, \tag{16}$$

where the *magnitude spectrum* of image f is

$$|F(U, V)| = \sqrt{R^2(U, V) + I^2(U, V)} \tag{17}$$
$$= \sqrt{F(U, V)F^*(U, V)} \tag{18}$$

where the asterisk denotes the complex conjugation. The *phase spectrum* of image f is

$$\angle F(U, V) = \tan^{-1}\left[\frac{I(U, V)}{R(U, V)}\right]. \tag{19}$$

3.4 Symmetry of DSFT

If the image f is real, which is usually the case, then the DSFT is *conjugate symmetric*:

$$F(U, V) = F^*(-U, -V), \tag{20}$$

which means that the DSFT is completely specified by its values over any half-plane. Hence, if f is real, the DSFT is redundant. From Eq. (20), it follows that the magnitude spectrum is *even symmetric*:

$$|F(U, V)| = |F(-U, -V)|, \tag{21}$$

while the phase spectrum is *odd symmetric*:

$$\angle F(U, V) = -\angle F(-U, -V), \tag{22}$$

3.5 Translation of DSFT

Multiplying (or modulating) the discrete-space image $f(m, n)$ by a two-dimensional complex exponential wave function, $\exp[j2\pi(U_0 m + V_0 n)]$, results in a translation of the DSFT:

$$f(m, n) \exp[j2\pi(U_0 m + V_0 n)] \overset{\mathfrak{F}}{\leftrightarrow} F(U - U_0, V - V_0). \tag{23}$$

Likewise, translating the image f by amounts m_0, n_0 produces a modulated DSFT:

$$f(m - m_0, n - n_0) \overset{\mathfrak{F}}{\leftrightarrow} F(U, V) \exp[-j2\pi(U m_0 + V n_0)] \tag{24}$$

3.6 Convolution and the DSFT

Given two images or two-dimensional functions f and h, their *two-dimensional discrete-space linear convolution* is given by

$$g(m, n) = f(m, n) * h(m, n) = h(m, n) * f(m, n)$$
$$= \sum_{p=-\infty}^{\infty} \sum_{q=-\infty}^{\infty} f(p, q)h(m - p, n - q). \tag{25}$$

The linear convolution expresses the result of passing an image signal f through a two-dimensional linear convolution system h (or vice versa). The commutativity of the convolution is easily seen by making a substitution of variables in the double sum in Eq. (25).

If g, f, and h satisfy spatial convolution relationship (25), then their DSFTs satisfy

$$G(U, V) = F(U, V)H(U, V); \tag{26}$$

hence convolution in the space domain corresponds directly to multiplication in the spatial frequency domain. This important property is significant both conceptually, as a simple and direct means for effecting the frequency content of an image, and computationally, since the linear convolution has such a simple expression in the frequency domain.

The two-dimensional DSFT is the basic mathematical tool for analyzing the frequency-domain content of two-dimensional discrete-space images. However, it has a major drawback for digital image processing applications: the DSFT $F(U, V)$ of a discrete-space image $f(m, n)$ is continuous in the frequency coordinates (U, V); there are an uncountably infinite number of values to compute. As such, discrete (digital) processing or display in the frequency domain is not possible using the DSFT unless it is modified in some way. Fortunately, this is possible when the image f is of finite dimensions. In fact, by sampling the DSFT in the frequency domain we are able to create a computable Fourier domain transform.

4 Two-Dimensional Discrete Fourier Transform (DFT)

Now we restrict our attention to the practical case of discrete-space images that are of finite extent. Hence assume that image $f(m, n)$; can be expressed as a matrix $\mathbf{f} = [f(m, n) 0 \le m \le M - 1, 0 \le n \le N - 1]$. As we will show, a finite-extent image matrix $\mathbf{f}$ can be represented exactly as a *finite* weighted sum of two-dimensional frequency components, instead of an infinite number. This leads to computable and numerically manipulable frequency-domain representations. Before showing how this is done, we shall introduce a special notation for the complex exponential that will simplify much of the ensuing development.

We will use

$$W_K = \exp\left[-j\frac{2\pi}{K}\right] \qquad (27)$$

as a shorthand for the basic complex exponential, where K is the dimension along one of the image axes ($K = N$ or $K = M$). The notation of Eq. (27) makes it possible to index the various elementary frequency components at arbitrary spatial and frequency coordinates by simple exponentiation:

$$W_M^{um} W_N^{vn} = \cos\left[2\pi\left(\frac{u}{M}m + \frac{v}{N}n\right)\right]$$
$$- j\sin\left[2\pi\left(\frac{u}{M}m + \frac{v}{N}n\right)\right]. \qquad (28)$$

This process of space and frequency indexing by exponentiation greatly simplifies the manipulation of frequency components and the definition of the DFT. Indeed, it is possible to develop frequency-domain concepts and frequency transforms without the use of complex numbers (and in fact some of these, such as the discrete cosine transform, or DCT, are widely used, especially in image/video compression; see Chapters 5.5, 5.6, 6.4, and 6.5 of this *Handbook*).

For the purpose of analysis and basic theory, it is much simpler to use W_M^{um} and W_N^{vn} to represent finite-extent (of dimensions

M and N) frequency components oscillating at u (cycles/image) and v (cycles/image) in the m and n directions, respectively. Clearly,

$$|W_M^{um} W_N^{vn}| = 1 \qquad (29)$$

and

$$\angle W_M^{um} W_N^{vn} = -2\pi\left(\frac{u}{M}m + \frac{v}{N}n\right). \qquad (30)$$

Observe that the minimum physical frequency of W_M^{um} periodically occurs at the indices $u = kM$ for all integers k:

$$W_M^{kMm} = 1 \qquad (31)$$

for any integer m; the minimum oscillation is no oscillation. If M is even, the maximum physical frequency periodically occurs at the indices $u = kM + M/2$:

$$W_M^{(kM+M/2)m} = 1e^{-j\pi m} = (-1)^m, \qquad (32)$$

which is the discrete period-2 (alternating) function, the highest possible discrete oscillation frequency.

The *two-dimensional DFT* of the finite-extent ($M \times N$) image $\mathbf{f}$ is given by

$$\tilde{F}(u, v) = \sum_{m=0}^{M-1} \sum_{n=0}^{N-1} f(m, n) W_M^{um} W_N^{vn} \qquad (33)$$

for integer frequencies $0 \le u \le M - 1, 0 \le v \le N - 1$. Hence, the DFT is also of finite extent $M \times N$, and can be expressed as a (generally complex-valued) matrix $\tilde{\mathbf{F}} = [\tilde{F}(u, v); 0 \le u \le M - 1, 0 \le v \le N - 1]$. It has a unique *inverse discrete Fourier transform*, or *IDFT*:

$$f(m, n) = \frac{1}{MN} \sum_{u=0}^{M-1} \sum_{v=0}^{N-1} \tilde{F}(u, v) W_M^{-um} W_N^{-vn} \qquad (34)$$

for $0 \le m \le M - 1, 0 \le n \le N - 1$. When Eqs. (33) and (34) hold, it is often denoted $\mathbf{f} \overset{\text{DFT}}{\longleftrightarrow} \tilde{\mathbf{F}}$, and we say that $\mathbf{f}, \tilde{\mathbf{F}}$ form a *DFT pair*.

A number of observations regarding the DFT and its relationship to the DSFT are necessary. First, the DFT and IDFT are symmetrical, since both forward and inverse transforms are defined as sums. In fact they have the same form, except for the polarity of the exponents and a scaling factor. Secondly, both forward and inverse transforms are finite sums; both $\tilde{\mathbf{F}}$ and $\mathbf{f}$ can be represented uniquely as *finite* weighted sums of *finite-extent* complex exponentials with integer-indexed frequencies. Thus, for example, any 256×256 digital image can be expressed as the weighted sum of $256^2 = 65,536$ complex exponential (sinusoid) functions, including those with real parts shown in Fig. 1. Note that the frequencies (u, v) are scaled so that their units are in cycles/image, as in Eq. (6) and Fig. 1.

Most importantly, the DFT has a direct relationship to the DSFT. In fact, the DFT of an $M \times N$ image $\mathbf{f}$ is a uniformly *sampled* version of the DSFT of $\mathbf{f}$:

$$\tilde{F}(u, v) = F(U, V)\big|_{U=\frac{u}{M}, V=\frac{v}{N}} \qquad (35)$$

for integer frequency indices $0 \leq u \leq M - 1, 0 \leq v \leq N - 1$. Since $\mathbf{f}$ is of finite extent and contains MN elements, the DFT $\tilde{\mathbf{F}}$ is conservative in that it also requires only MN elements to contain complete information about $\mathbf{f}$ (to be exactly invertible). Also, since $\tilde{\mathbf{F}}$ is simply evenly spaced samples of F, many of the properties of the DSFT translate directly with little or no modification to the DFT.

4.1 Linearity and Invertibility of DFT

The DFT is linear in the sense of formula (12). It is uniquely invertible, as can be established by substituting Eq. (33) into Eq. (34), reversing the order of summation, and using the fact that the discrete complex exponentials are also orthogonal:

$$\sum_{u=0}^{M-1} \sum_{v=0}^{N-1} \left(W_M^{um} W_M^{-up} \right) \left(W_N^{vn} W_N^{-vq} \right)$$
$$= \begin{cases} MN; & m = p, \quad n = q \\ 0; & \text{otherwise} \end{cases}. \qquad (36)$$

The DFT matrix $\tilde{\mathbf{F}}$ is generally complex; hence it has an associated magnitude spectrum matrix, denoted

$$|\tilde{\mathbf{F}}| = [|\tilde{F}(u, v)|; 0 \leq u \leq M - 1, 0 \leq v \leq N - 1], \qquad (37)$$

and phase spectrum matrix, denoted

$$\angle\tilde{\mathbf{F}} = [\angle\tilde{F}(u, v); 0 \leq u \leq M - 1, 0 \leq v \leq N - 1]. \qquad (38)$$

The elements of $|\tilde{\mathbf{F}}|$ and $\angle\tilde{\mathbf{F}}$ are computed in the same way as the DSFT magnitude and phase of Eqs. (16)–(19).

4.2 Symmetry of DFT

Like the DSFT, if $\mathbf{f}$ is real valued, then the DFT matrix is conjugate symmetric, but in the matrix sense:

$$\tilde{F}(u, v) = \tilde{F}^*(M - u, N - v) \qquad (39)$$

for $0 \leq u \leq M - 1, 0 \leq v \leq N - 1$. This follows easily by substitution of the reversed and translated frequency indices $(M - u, N - v)$ into forward DFT Eq. (33). An apparent repercussion of Eq. (39) is that the DFT $\tilde{\mathbf{F}}$ matrix is redundant and hence can represent the $M \times N$ image with only approximately $MN/2$ DFT coefficients. This mystery is resolved by realizing that $\tilde{\mathbf{F}}$ is complex valued and hence requires twice the storage for real and imaginary components. If $\mathbf{f}$ is not real valued, then Eq. (39) does not hold.

Of course, Eq. (39) implies symmetries of the magnitude and phase spectra:

$$|\tilde{F}(u, v)| = |\tilde{F}(M - u, N - v)| \qquad (40)$$

and

$$\angle\tilde{F}(u, v) = -\angle\tilde{F}(M - u, N - v) \qquad (41)$$

for $0 \leq u \leq M - 1, 0 \leq v \leq N - 1$.

4.3 Periodicity of DFT

Another property of the DSFT that carries over to the DFT is frequency periodicity. Recall that the DSFT $F(U, V)$ has unit period in U and V. The DFT matrix $\tilde{\mathbf{F}}$ was defined to be of finite extent $M \times N$. However, forward DFT Eq. (33) admits the possibility of evaluating $\tilde{F}(u, v)$ outside of the range $0 \leq u \leq M - 1, 0 \leq v \leq N - 1$. It turns out that $\tilde{F}(u, v)$ is period-M and period-N along the u and v dimensions, respectively. For any integers k, l,

$$\tilde{F}(u + kM, v + lN) = \tilde{F}(u, v) \qquad (42)$$

for every $0 \leq u \leq M - 1, 0 \leq v \leq N - 1$. This follows easily by substitution of the periodically extended frequency indices $(u + kM, v + lN)$ into forward DFT Eq. (33). Interpretation (42) of the DFT is called the *periodic extension* of the DFT. It is defined for all integer frequencies u, v.

Although many properties of the DFT are the same, or similar to those of the DSFT, certain important properties are different. These effects arise from sampling the DSFT to create the DFT.

4.4 Image Periodicity Implied by DFT

A seemingly innocuous yet extremely important consequence of sampling the DSFT is that the resulting DFT equations imply that the image $\mathbf{f}$ is itself periodic. In fact, IDFT Eq. (34) implies that for any integers k, l,

$$f(m + kM, n + lN) = f(m, n) \qquad (43)$$

for every $0 \leq m \leq M - 1, 0 \leq n \leq N - 1$. This follows easily by substitution of the periodically extended space indices $(m + kM, n + lN)$ into inverse DFT Eq. (34).

Clearly, finite-extent digital images arise from imaging the real world through finite field-of-view (FOV) devices, such as cameras, and outside that FOV, the world does not repeat itself periodically, *ad infinitum*. The implied periodicity of $\mathbf{f}$ is purely a synthetic effect that derives from sampling the DSFT. Nevertheless, it is of paramount importance, since any algorithm that is developed, and that uses the DFT, will operate as though the DFT-transformed image were spatially periodic in the sense of Eq. (43). One important property and application of the DFT that is effected by this spatial periodicity is the frequency-domain convolution property.

4.5 Cyclic Convolution Property of the DFT

One of this most significant properties of the DSFT is the linear convolution property, Eqs. (25) and (26), which says that space-domain convolution corresponds to frequency-domain multiplication:

$$f * h \overset{\mathfrak{F}}{\leftrightarrow} F H. \tag{44}$$

This useful property makes it possible to analyze and design linear convolution-based systems in the frequency domain. Unfortunately, property (44) does not hold for the DFT; a product of DFTs does not correspond (inverse transform) to the linear convolution of the original DFT-transformed functions or images. However, it does correspond to another type of convolution, variously known as *cyclic convolution, circular convolution,* or *wraparound convolution.*

We will demonstrate the form of the cyclic convolution by deriving it. Consider the two $M \times N$ image functions $\mathbf{f} \overset{\text{DFT}}{\longleftrightarrow} \tilde{\mathbf{F}}$ and $\mathbf{h} \overset{\text{DFT}}{\longleftrightarrow} \tilde{\mathbf{H}}$. Define the *pointwise* matrix product[1]

$$\tilde{\mathbf{G}} = \tilde{\mathbf{F}} \otimes \tilde{\mathbf{H}} \tag{45}$$

according to

$$\tilde{G}(u, v) = \tilde{F}(u, v) \tilde{H}(u, v) \tag{46}$$

for $0 \leq u \leq M - 1, 0 \leq v \leq N - 1$. Thus we are interested in the form of $\mathbf{g}$. For each $0 \leq m \leq M - 1, 0 \leq n \leq N - 1$, we have that

$$
\begin{aligned}
g(m, n) &= \frac{1}{MN} \sum_{u=0}^{M-1} \sum_{v=0}^{N-1} \tilde{G}(u, v) W_M^{-um} W_N^{-vn} \\
&= \frac{1}{MN} \sum_{u=0}^{M-1} \sum_{v=0}^{N-1} \tilde{F}(u, v) \tilde{H}(u, v) W_M^{-um} W_N^{-vn} \\
&= \frac{1}{MN} \sum_{u=0}^{M-1} \sum_{v=0}^{N-1} \left\{ \sum_{p=0}^{M-1} \sum_{q=0}^{N-1} f(p, q) W_M^{up} W_N^{vq} \right\} \\
&\quad \times \left\{ \sum_{r=0}^{M-1} \sum_{s=0}^{N-1} h(r, s) W_M^{-ur} W_N^{vs} \right\} W_M^{-um} W_N^{-vn}
\end{aligned}
\tag{47}
$$

by substitution of the definitions of $\tilde{F}(u, v)$ and $\tilde{H}(u, v)$. Rearranging the order of the summations to collect all of the complex exponentials inside the innermost summation reveals that

$$
\begin{aligned}
g(m, n) = \frac{1}{MN} \sum_{p=0}^{M-1} \sum_{q=0}^{N-1} f(p, q) \sum_{r=0}^{M-1} \sum_{s=0}^{N-1} h(r, s) \\
\times \sum_{u=0}^{M-1} \sum_{v=0}^{N-1} W_M^{u(p+r-m)} W_N^{v(q+s-n)}.
\end{aligned}
\tag{48}
$$

Now, from Eq. (36), the innermost summation

$$
\begin{aligned}
\sum_{u=0}^{M-1} \sum_{v=0}^{N-1} W_M^{u(p+r-m)} W_N^{v(q+s-n)} \\
= \begin{cases} MN; & r = m - p, \quad s = n - q \\ 0; & \text{otherwise} \end{cases}.
\end{aligned}
\tag{49}
$$

hence

$$g(m, n) = \sum_{p=0}^{M-1} \sum_{q=0}^{N-1} f(p, q) h\left[(m - p)_M, (n - q)_N\right] \tag{50}$$

$$= f(m, n) \circledast h(m, n) = h(m, n) \circledast f(m, n) \tag{51}$$

where $(x)_N = x \bmod N$ and the symbol $\circledast$ denotes the two-dimensional cyclic convolution.[2] The final step of obtaining Eq. (50) from Eq. (49) follows since the argument of the shifted and twice-reversed (along each axis) function $h(m - p, n - q)$ finds no meaning whenever $(m - p) \notin \{0, \ldots, M - 1\}$ or $(n - q) \notin \{0, \ldots, N - 1\}$, since h is undefined outside of those coordinates. However, because the DFT was used to compute $g(m, n)$, then the periodic extension of $h(m - p, n - q)$ is implied, which can be expressed as $h[(m - p)_M, (n - q)_N]$. Hence Eq. (50) follows. That $\circledast$ is commutative is easily established by a substitution of variables in Eq. (50). It can also be seen that cyclic convolution is a form of linear convolution, but with one (either, but not both) of the two functions being periodically extended. Hence

$$
\begin{aligned}
f(m, n) \circledast h(m, n) &= f(m, n) * h[(m)_M, (n)_N] \\
&= f[(m)_M, (n)_N] * h(m, n).
\end{aligned}
\tag{52}
$$

This *cyclic convolution property* of the DFT is unfortunate, since in the majority of applications it is not desired to compute the cyclic convolution of two image functions. Instead, what is frequently desired is the linear convolution of two functions, as in the case of linear filtering. In both linear and cyclic convolution, the two functions are superimposed, with one function reversed along both axes and shifted to the point at which the convolution is being computed. The product of the functions is computed at every point of overlap, with the sum of products being the convolution. In the case of the cyclic convolution, one (not both) of the functions is periodically extended, hence the overlap is much larger and wraps around the image boundaries. This produces a significant error with respect to the correct linear convolution result. This error is called *spatial aliasing*, since the wraparound error contributes false information to the convolution sum.

Figure 3 depicts the linear and cyclic convolutions of two hypothetical $M \times N$ images f and h at a point (m_0, n_0). From the figure, it can be seen that the wraparound error can overwhelm

---

[1] As opposed to the standard matrix product.

[2] Modular arithmetic is remaindering. Hence $(x)_N$ is the integer remainder of (x/N).

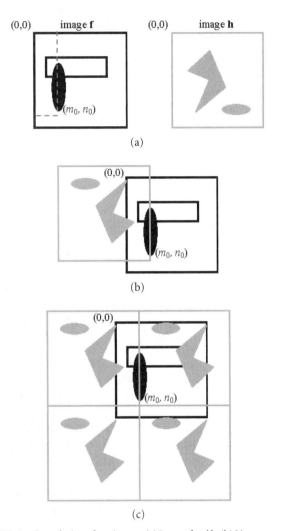

FIGURE 3 Convolution of two images. (a) Images **f** and **h**. (b) Linear convolution result at (m_0, n_0) is computed as the sum of products where **f** and **h** overlap. (c) Cyclic convolution result at (m_0, n_0) is computed as the sum of products where **f** and the periodically extended **h** overlap.

the linear convolution contribution. Note in Fig. 3(b) that although linear convolution sum (25) extends over the indices $0 \le m \le M - 1$ and $0 \le n \le N - 1$, the overlap is restricted to the indices.

4.6 Linear Convolution by Using the DFT

Fortunately, it turns out that it is possible to compute the linear convolution of two arbitrary finite-extent two-dimensional discrete-space functions or images by using the DFT. The process requires modifying the functions to be convolved prior to taking the product of their DFTs. The modification acts to cancel the effects of spatial aliasing. Suppose more generally that f and h are two arbitrary finite-extent images of dimensions $M \times N$ and $P \times Q$, respectively. We are interested in computing the linear convolution $g = f * h$ using the DFT. We assume the general case where the images, f, h do not have the same dimensions, since

in most applications an image is convolved with a filter function of different (usually much smaller) extent.

Clearly,

$$
g(m, n) = f(m, n) * h(m, n)
$$
$$
= \sum_{p=0}^{M-1} \sum_{q=0}^{N-1} f(p, q) h(m - p, n - q). \quad (53)
$$

Inverting the pointwise products of the DFTs $\tilde{\mathbf{F}} \otimes \tilde{\mathbf{H}}$ will not lead to Eq. (53), since wraparound error will occur. To cancel the wraparound error, the functions **f** and **h** are modified by increasing their size by *zero padding* them. Zero padding means that the arrays **f** and **h** are expanded into larger arrays, denoted $\hat{\mathbf{f}}$ and $\hat{\mathbf{h}}$, by filling the empty spaces with zeros. To compute the linear convolution, the pointwise product $\tilde{\hat{\mathbf{G}}} = \hat{\mathbf{F}} \otimes \hat{\mathbf{H}}$ of the DFTs of the zero-padded functions $\hat{\mathbf{f}}$ and $\hat{\mathbf{h}}$ is computed. The inverse DFT $\hat{\mathbf{g}}$ of $\tilde{\hat{\mathbf{G}}}$ then contains the correct linear convolution result.

The question remains as to how many zeros are used to pad the functions **f** and **h**. The answer to this lies in understanding how zero padding works and how large the linear convolution result should be. Zero padding acts to cancel the spatial aliasing error (wraparound) of the DFT by supplying zeros where the wraparound products occur. Hence the wraparound products are all zero and contribute nothing to the convolution sum. This leaves only the linear convolution contribution to the result. To understand how many zeros are needed, one must realize that the resulting product DFT $\tilde{\hat{\mathbf{G}}}$ corresponds to a periodic function $\hat{\mathbf{g}}$. If the horizontal or vertical periods are too small (not enough zero padding), the periodic replicas will overlap (spatial aliasing). If the periods are just large enough, then the periodic replicas will be contiguous instead of overlapping; hence spatial aliasing will be canceled. Padding with more zeros than this results in excess computation. Figure 4 depicts the successful result of zero padding to eliminate wraparound error.

The correct period lengths are equal to the lengths of the correct linear convolution result. The linear convolution result of two arbitrary $M \times N$ and $P \times Q$ image functions will generally be $(M + P - 1) \times (N + Q - 1)$, hence we would like the DFT $\tilde{\hat{\mathbf{G}}}$ to have these dimensions. Therefore, the $M \times N$ function **f** and the $P \times Q$ function **h** must both be zero padded to size $(M + P - 1) \times (N + Q - 1)$. This yields the correct linear convolution result:

$$
\hat{\mathbf{g}} = \hat{\mathbf{f}} \circledast \hat{\mathbf{h}} = \mathbf{f} * \mathbf{h}. \quad (54)
$$

In most cases, linear convolution is performed between an image and a filter function much smaller than the image: $M \gg P$ and $N \gg Q$. In such cases the result is not much larger than the image, and often only the $M \times N$ portion indexed $0 \le m \le M - 1$, $0 \le n \le N - 1$ is retained. The reasoning behind this is first, that it may be desirable to retain images of size MN only, and second, that the linear convolution result beyond the

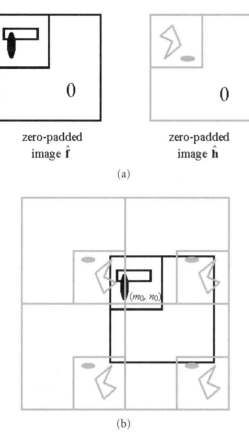

zero-padded
image $\hat{\mathbf{f}}$

zero-padded
image $\hat{\mathbf{h}}$

(a)

(b)

FIGURE 4 Linear convolution of the same two images as Fig. 2 by zero padding and cyclic convolution (via the DFT). (a) Zero-padded images $\hat{\mathbf{f}}$ and $\hat{\mathbf{h}}$. (b) Cyclic convolution at (m_0, n_0) computed as the sum of products where $\hat{\mathbf{f}}$ and the periodically extended $\hat{h}$ overlap. These products are zero except over the range $0 \leq p \leq m_0$ and $0 \leq q \leq n_0$.

borders of the original image may be of little interest, since the original image was zero there anyway.

4.7 Computation of the DFT

Inspection of the DFT, relation (33) reveals that computation of each of the MN DFT coefficients requires on the order of MN complex multiplies/additions. Hence, of the order of $M^2 N^2$ complex multiplies and additions are needed to compute the overall DFT of an $M \times N$ image $\mathbf{f}$. For example, if $M = N = 512$, then of the order of $2^{36} = 6.9 \times 10^{10}$ complex multiplies/additions are needed, which is a very large number. Of course, these numbers assume a naïve implementation without any optimization. Fortunately, fast algorithms for DFT computation, collectively referred to as *fast Fourier transform* (FFT) algorithms, have been intensively studied for many years. We will not delve into the design of these, since it goes beyond what we want to accomplish in a *Handbook* and also since they are available in any image processing programming library or development environment (Chapter 4.13 reviews these) and most math library programs.

The FFT offers a computational complexity of order not exceeding $MN \log_2(MN)$, which represents a considerable speedup. For example, if $M = N = 512$, then the complexity is of the order of $9 \times 2^{19} = 4.7 \times 10^6$. This represents a very typical speedup of more than 14,500:1 !

Analysis of the complexity of cyclic convolution is similar. If two images of the same size $M \times N$ are convolved, then again, the naïve complexity is on the order of $M^2 N^2$ complex multiplies and additions. If the DFT of each image is computed, the resulting DFTs pointwise multiplied, and the inverse DFT of this product calculated, then the overall complexity is of the order of $MN \log_2(2M^3 N^3)$. For the common case $M = N = 512$, the speedup still exceeds 4700:1.

If linear convolution is computed with the DFT, the computation is increased somewhat since the images are increased in size by zero padding. Hence the speedup of DFT-based linear convolution is somewhat reduced (although in a fixed hardware realization, the known existence of these zeros can be used to effect a speedup). However, if the functions being linearly convolved are both not small, then the DFT approach will always be faster. If one of the functions is very small, say covering fewer than 32 samples (such as a small linear filter template), then it is possible that direct space-domain computation of the linear convolution may be faster than DFT-based computation. However, there is no strict rule of thumb to determine this lower cutoff size, since it depends on the filter shape, the algorithms used to compute DFTs and convolutions, any special-purpose hardware, and so on.

4.8 Displaying the DFT

It is often of interest to visualize the DFT of an image. This is possible since the DFT is a sampled function of finite (periodic) extent. Displaying one period of the DFT of image $\mathbf{f}$ reveals a picture of the frequency content of the image. Since the DFT is complex, one can display either the magnitude spectrum $|\tilde{\mathbf{F}}|$ or the phase spectrum $\angle \tilde{\mathbf{F}}$ as a single two-dimensional intensity image.

However, the phase spectrum $\angle \tilde{\mathbf{F}}$ is usually not visually revealing when displayed. Generally it appears quite random, and so usually the magnitude spectrum $|\tilde{\mathbf{F}}|$ only is absorbed visually. This is not intended to imply that image phase information is not important; in fact, it is exquisitely important, since it determines the relative shifts of the component complex exponential functions that make up the DFT decomposition. Modifying or ignoring image phase will destroy the delicate constructive-destructive interference pattern of the sinusoids that make up the image.

As briefly noted in Chapter 2.1, displays of the Fourier transform magnitude will tend to be visually dominated by the low-frequency and zero-frequency coefficients, often to such an extent that the DFT magnitude appears as a single spot. This is highly undesirable, since most of the interesting information usually occurs at frequencies away from the lowest frequencies. An effective way to bring out the higher-frequency coefficients

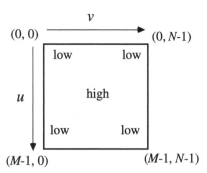

FIGURE 5   Distribution of high- and low-frequency DFT coefficients.

for visual display is by means of a point logarithmic operation: instead of displaying $|\tilde{\mathbf{F}}|$, display

$$\log_2[1 + |\tilde{F}(u, v)|] \qquad (55)$$

for $0 \leq u \leq M - 1$, $0 \leq v \leq N - 1$. This has the effect of compressing all of the DFT magnitudes, but larger magnitudes much more so. Of course, since all of the logarithmic magnitudes will be quite small, a full-scale histogram stretch should then be applied to fill the gray-scale range.

Another consideration when displaying the DFT of a discrete-space image is illustrated in Fig. 5. In the DFT formulation, a single $M \times N$ period of the DFT is sufficient to represent the image information, and also for display. However, the DFT matrix is even symmetric across both diagonals. More importantly, the center of symmetry occurs in the image center, where the high-frequency coefficients are clustered near $(u, v) = (M/2, N/2)$. This is contrary to conventional intuition, since in most engineering applications, Fourier transform magnitudes are displayed with zero and low-frequency coefficients at the center. This is particularly true of one-dimensional continuous Fourier transform magnitudes, which are plotted as graphs with the zero frequency at the origin. This is also visually convenient, since the dominant lower frequency coefficients then are clustered together at the center, instead of being scattered about the display.

A natural way of remedying this is to instead display the shifted DFT magnitude

$$|\tilde{F}(u - M/2, v - N/2)| \qquad (56)$$

for $0 \leq u \leq M - 1$, $0 \leq v \leq N - 1$. This can be accomplished in a simple way by taking the DFT of

$$(-1)^{m+n} f(m, n) \overset{\text{DFT}}{\longleftrightarrow} \tilde{F}(u - M/2, v - N/2). \qquad (57)$$

Relation (57) follows since $(-1)^{m+n} = e^{j\pi(m+n)}$; hence from translation property (23) the DSFT is shifted by amount 1/2 cycles/pixel along both dimensions; since the DFT uses the scaled frequencies of Eq. (6), the DFT is shifted by $M/2$ and $N/2$ cycles/image in the u and v directions, respectively.

Figure 6 illustrates the display of the DFT of the image "fingerprint" image, which is Fig. 8 of Chapter 1.1. As can be seen, the DFT phase is visually unrevealing, while the DFT magnitude

is most visually revealing when it is centered and logarithmically compressed.

5 Understanding Image Frequencies and the DFT

It is sometimes easy to lose track of the meaning of the DFT and of the frequency content of an image in all of the (necessary!) mathematics. When using the DFT, it is important to remember that the DFT is a detailed map of the frequency content of the image, which can be visually digested as well as digitally processed. It is a useful exercise to examine the DFT of images, particularly the DFT magnitudes, since it reveals much about the distribution and meaning of image frequencies. It is also useful to consider what happens when the image frequencies are modified in certain simple ways, since this reveals further insights into spatial frequencies, and it also moves toward understanding how image frequencies can be systematically modified to produce useful results.

In the following paragraphs we will present and discuss a number of interesting digital images along with their DFT magnitudes represented as intensity images. When examining these, recall that bright regions in the DFT magnitude "image" correspond to frequencies that have large magnitudes in the real image. Also, in some cases, the DFT magnitudes have been logarithmically compressed and centered by means of relations (55) and (57), respectively, for improved visual interpretation.

Most engineers and scientists are introduced to Fourier domain concepts in a one-dimensional setting. One-dimensional signal frequencies have a single attribute—that of being either high or low frequency. Two-dimensional (and higher-dimensional) signal frequencies have richer descriptions characterized by both magnitude and direction,[3] which lend themselves well to visualization. We will seek intuition into these attributes as we separately consider the *granularity* of image frequencies, corresponding to the radial frequency of Eq. (2), and the *orientation* of image frequencies, corresponding to the frequency angle of Eq. (3).

5.1 Frequency Granularity

The granularity of an image frequency refers to its radial frequency. Granularity describes the appearance of an image that is strongly characterized by the radial frequency portrait of the DFT. An abundance of large coefficients near the DFT origin corresponds to the existence large, smooth, image components, often of smooth image surfaces or background. Note that nearly every image will have a significant peak at the DFT origin (unless it is very dark), since from Eq. (33) it is the summed intensity of

---

[3]Strictly speaking, one-dimensional frequencies can be positive or negative going. This polarity may be regarded as a directional attribute, without much meaning for real-valued one-dimensional signals.

FIGURE 6 Display of DFT of the image "fingerprint" from Chapter 1.1. (a) DFT magnitude (logarithmically compressed and histogram stretched); (b) DFT phase; (c) Centered DFT (logarithmically compressed and histogram stretched); (d) Centered DFT (without logarithmic compression).

the image (integrated optical density):

$$\tilde{F}(0, 0) = \sum_{m=0}^{M-1} \sum_{n=0}^{N-1} f(m, n). \tag{58}$$

The image "fingerprint" (Fig. 8 of Chapter 1.1) with the DFT magnitude shown in Fig. 6(c) just above is an excellent example of image granularity. The image contains relatively little low-frequency or very high-frequency energy, but does contain an abundance of midfrequency energy, as can be seen in the symmetrically placed half-arcs above and below the frequency origin. The "fingerprint" image is a good example of an image that is primarily bandpass.

Figure 7 depicts the image "peppers" and its DFT magnitude. The image contains primarily smooth intensity surfaces separated by abrupt intensity changes. The smooth surfaces contribute to the heavy distribution of low-frequency DFT coefficients, while the intensity transitions ("edges") contribute a noticeable amount of midfrequencies to higher frequencies over a broad range of orientations.

Finally, Fig. 8, the image "cane," depicts an image of a repetitive weave pattern that exhibits a number of repetitive peaks in the DFT magnitude image. These are *harmonics* that naturally appear in signals (such as music signals) or images that contain periodic or nearly periodic structures.

FIGURE 7   Image of peppers (left) and its DFT magnitude (right).

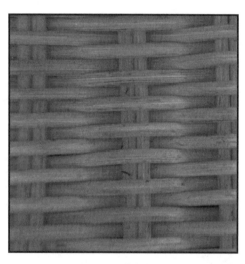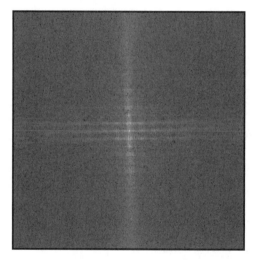

FIGURE 8   Image cane (left) and its DFT magnitude (right).

As an experiment toward understanding frequency content, suppose that we define several zero–one image frequency masks, as depicted in Fig. 9. By masking (multiplying) the DFT $\tilde{F}$ of an image $\mathbf{f}$ with each of these, one will produce, following an inverse DFT, a resulting image containing only low, middle, or high frequencies. In the following, we show examples of this operation. The astute reader may have observed that the zero–one

low-frequency   mid-frequency   high-frequency
mask   mask   mask

FIGURE 9   Image radial frequency masks. Black pixels take value 1, and white pixels take value 0.

frequency masks, which are defined in the DFT domain, may be regarded as DFTs with IDFTs defined in the space domain. Since we are taking the products of functions in the DFT domain, it has the interpretation of cyclic convolution of Eqs. (46)–(51) in the space domain. Therefore the following examples should not be thought of as low-pass, bandpass, or high-pass linear filtering operations in the proper sense. Instead, these are instructive examples in which image frequencies are being directly removed. The approach is not a substitute for a proper linear filtering of the image by using a space-domain filter that has been DFT transformed with proper zero padding. In particular, the naïve demonstration here does dictate how the frequencies between the DFT frequencies (frequency samples) are effected, as a properly designed linear filter does.

In all of the examples, the image DFT was computed, multiplied by a zero–one frequency mask, and inverse discrete Fourier transformed. Finally, a full-scale histogram stretch was applied to map the result to the gray-level range (0, 255), since otherwise the resulting image is not guaranteed to be positive.

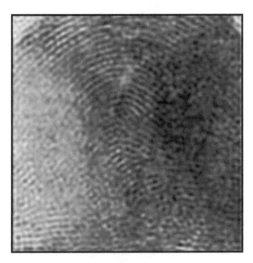

FIGURE 10   Image of fingerprint processed with the (left) low-frequency and the (right) midfrequency DFT masks.

In the first example, shown in Fig. 10, the image "fingerprint" is shown following treatment with the low-frequency mask and the midfrequency mask. The low-frequency result looks much blurred, and there is an apparent loss of information. However, the midfrequency result seems to enhance and isolate much of the interesting ridge information about the fingerprint.

In the second example (Fig. 11), the image "peppers" was treated with the midfrequency DFT mask and the high-frequency DFT mask. The midfrequency image is visually quite interesting since it is apparent that the sharp intensity changes were significantly enhanced. A similar effect was produced with the higher-frequency mask, but with greater emphasis on sharp details.

5.2 Frequency Orientation

The orientation of an image frequency refers to its angle. The term orientation applied to an image or image component describes those aspects of the image that contribute to an appearance that is strongly characterized by the frequency orientation portrait of the DFT. If the DFT is brighter along a specific orientation, then the image contains highly oriented components along that direction.

The image of the fingerprint, with DFT magnitude in Fig. 6(c), is also an excellent example of image orientation. The DFT contains significant midfrequency energy between the approximate orientations 45–135° from the horizontal axis. This corresponds perfectly to the orientations of the ridge patterns in the fingerprint image.

Figure 12 shows the image "planks," which contains a strong directional component. This manifests as a very strong extended peak extending from lower left to upper right in the DFT magnitude. Figure 13 ("escher") exhibits several such extended peaks, corresponding to strongly oriented structures in the horizontal and slightly off-diagonal directions.

Again, an instructive experiment can be developed by defining zero–one image frequency masks, this time tuned to different

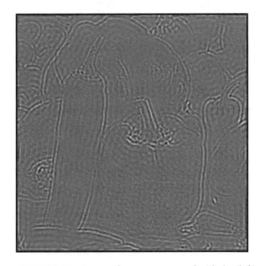

FIGURE 11   Image of peppers processed with the (left) midfrequency and the (right) high-frequency DFT masks.

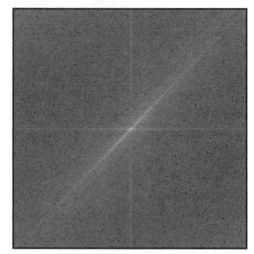

FIGURE 12   Image of planks (left) and DFT magnitude (right).

FIGURE 13   Image of escher (left) and DFT magnitude (right).

orientation frequency bands instead of radial frequency bands. Several such oriented frequency masks are depicted in Fig. 14A.

As a first example, the DFT of the image "planks" was modified by two orientation masks. In Figure 14B (left), an orientation mask that allows the frequencies in the range 40–50° only (as well as the symmetrically placed frequencies 220–230°) was applied. This was designed to capture the bright ridge of DFT coefficients easily seen in Fig. 12. As can be seen, the strong oriented informa-

tion describing the cracks in the planks and some of the oriented grain is all that remains. Possibly, this information could be used by some automated process. Then, in Fig. 14B (right), the frequencies in the much larger ranges 50–220° (and −130–40°) were admitted. These are the complementary frequencies to the first range chosen, and they contain all the other information other than the strongly oriented component. As can be seen, this residual image contains little oriented structure.

As a first example, the DFT of the image "escher" was also modified by two orientation masks. In Fig. 15 (left), an orientation mask that allows the frequencies in the range −25–25° (and 155–205°) only was applied. This captured the strong horizontal frequency ridge in the image, corresponding primarily to the strong vertical (building) structures. Then, in Fig. 15 (right), frequencies in the vertically oriented ranges 45–135° (and 225–315°) were admitted. This time completely different structures were highlighted, including the diagonal waterways, the background steps, and the paddlewheel.

FIGURE 14A   Examples of image frequency orientation masks.

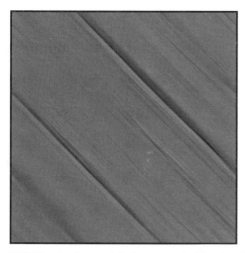 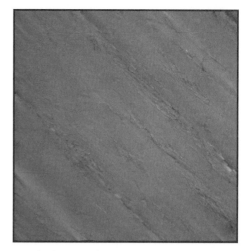

FIGURE 14B Image of planks processed with oriented DFT masks that allow frequencies in the range (measured from the horizontal axis) of (left) 40–50° (and 220–230°), and (right) 50–220° (and −130–40°).

6 Related Topics in this *Handbook*

The Fourier transform is one of the most basic tools for image processing, or for that matter, the processing of any kind of signal. It appears throughout this *Handbook* in various contexts.

One topic that was not touched on in this basic chapter is the frequency-domain analysis of sampling continuous images/video to create discrete-space images/video. Understanding the relationship between the DSFT and the DFT (spectrum of digital image signals) and the continuous Fourier transform of the original, unsampled image is basic to understanding the information content, and possible losses of information, in digital images. These topics are ably handled in Chapters 7.1 and 7.2 of this *Handbook*. Sampling issues were not covered in this chapter, since it was felt that most users deal with digital images that have been already created. Hence, the emphasis is on the immediate processing, and sampling issues are offered as a background understanding.

Fourier domain concepts and linear convolution pervade most of the chapters in Section 3 of the *Handbook*, since linear filtering, restoration, enhancement, and reconstruction all depend on these concepts. Most of the mathematical models for images and video in Section 4 of the *Handbook* have strong connections to Fourier analysis, especially the wavelet models, which extend the ideas of Fourier techniques in very powerful ways. Extended frequency-domain concepts are also heavily utilized in Sections 5 and 6 (image and video compression) of the *Handbook*, although the transforms used differ somewhat from the DFT.

Acknowledgment

My thanks to Dr. Hung-Ta Pai for carefully reading and commenting on this chapter.

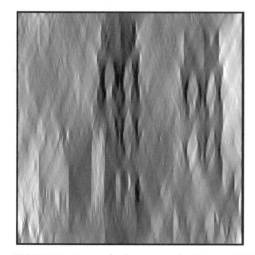 

FIGURE 15 Image of escher processed with oriented DFT masks that allow frequencies in the range (measured from the horizontal axis) of (left) −25–25° (and 155–205°), and (right) 45–135° (and 225–315°).

Image and Video Processing

Image and Video Enhancement and Restoration

Reconstruction from Multiple Images

3.1

Basic Linear Filtering with Application to Image Enhancement

Alan C. Bovik
The University of Texas at Austin

Scott T. Acton
Oklahoma State University

1 Introduction

Linear system theory and linear filtering play a central role in digital image and video processing. Many of the most potent techniques for modifying, improving, or representing digital visual data are expressed in terms of linear systems concepts. Linear filters are used for generic tasks such as image/video contrast improvement, denoising, and sharpening, as well as for more object- or feature-specific tasks such as target matching and feature enhancement.

Much of this *Handbook* deals with the application of linear filters to image and video enhancement, restoration, reconstruction, detection, segmentation, compression, and transmission. The goal of this chapter is to introduce some of the basic supporting ideas of linear systems theory as they apply to digital image filtering, and to outline some of the applications. Special emphasis is given to the topic of linear image enhancement.

We will require some basic concepts and definitions in order to proceed. The basic two-dimensional discrete-space signal is the *two-dimensional impulse function*, defined by

$$\delta(m - p, n - q) = \begin{cases} 1, & m = p \text{ and } n = q \\ 0, & \text{otherwise} \end{cases}. \quad (1)$$

Thus, Eq. (1) takes unit value at coordinate (p, q) and is everywhere else zero. The function in Eq. (1) is often termed the *Kronecker delta function* or the *unit sample sequence* [1]. It plays the same role and has the same significance as the so-called

Dirac delta function of continuous system theory. Specifically, the response of linear systems to Eq. (1) will be used to characterize the general responses of such systems.

Any discrete-space image f may be expressed in terms of the impulse function in Eq. (1):

$$f(m, n) = \sum_{p=-\infty}^{\infty} \sum_{q=-\infty}^{\infty} f(m - p, n - q)\delta(p, q)$$

$$= \sum_{p=-\infty}^{\infty} \sum_{q=-\infty}^{\infty} f(p, q)\delta(m - p, n - q). \quad (2)$$

Expression (2), called the *sifting property*, has two meaningful interpretations here. First, any discrete-space image can be written as a sum of weighted, shifted unit impulses. Each weighted impulse comprises one of the pixels of the image. Second, the sum in Eq. (2) is in fact a discrete-space linear convolution. As is apparent, the linear convolution of any image f with the impulse function δ returns the function unchanged.

The impulse function effectively describes certain systems known as *linear space-invariant systems*. We explain these terms next.

A two-dimensional system **L** is a process of image transformation, as shown in Fig. 1.

We can write

$$g(m, n) = \mathbf{L}[f(m, n)]. \quad (3)$$

The system **L** is *linear* if and only if for any $f_1(m, n)$, $f_2(m, n)$

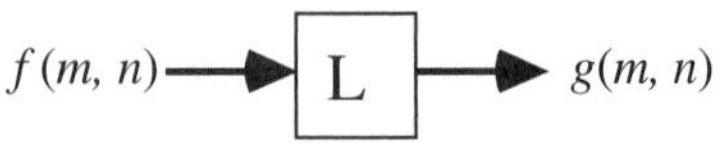

$$f(m, n) \longrightarrow \boxed{\text{L}} \longrightarrow g(m, n)$$

FIGURE 1   Two-dimensional input–output system.

such that

$$g_1(m, n) = \mathbf{L}[\, f_1(m, n)], \qquad g_2(m, n) = \mathbf{L}[\, f_2(m, n)] \quad (4)$$

and any two constants a, b, then

$$ag_1(m, n) + bg_2(m, n) = \mathbf{L}[af_1(m, n) + bf_2(m, n)] \quad (5)$$

for every (m, n). This is often called the *superposition property* of linear systems.

The system $\mathbf{L}$ is *shift invariant* if for every $f(m, n)$ such that Eq. (3) holds, then also

$$g(m - p, n - q) = \mathbf{L}[\, f(m - p, n - q)] \quad (6)$$

for any (p, q). Thus, a spatial shift in the input to $\mathbf{L}$ produces no change in the output, except for an identical shift.

The rest of this chapter will be devoted to studying systems that are linear and shift invariant (LSI). In this and other chapters, it will be found that LSI systems can be used for many powerful image and video processing tasks. In yet other chapters, non-linearity or space variance will be shown to afford certain advantages, particularly in surmounting the inherent limitations of LSI systems.

2 Impulse Response, Linear Convolution, and Frequency Response

The *unit impulse response* of a two-dimensional input–output system $\mathbf{L}$ is

$$\mathbf{L}[\delta(m - p, n - q)] = h(m, n; p, q). \quad (7)$$

This is the response of system $\mathbf{L}$, at spatial position (m, n), to an impulse located at spatial position (p, q). Generally, the impulse response is a function of these four spatial variables. However, if the system $\mathbf{L}$ is space invariant, then if

$$\mathbf{L}[\delta(m, n)] = h(m, n) \quad (8)$$

is the response to an impulse applied at the spatial origin, then also

$$\mathbf{L}[\delta(m - p, n - q)] = h(m - p, n - q), \quad (9)$$

which means that the response to an impulse applied at any spatial position can be found from the impulse response in Eq. (8).

As already mentioned, the discrete-space impulse response $h(m, n)$ completely characterizes the input–output response of

LSI input–output systems. This means that if the impulse response is known, then an expression can be found for the response to any input. The form of the expression is two-dimensional discrete-space linear convolution.

Consider the generic system $\mathbf{L}$ shown in Fig. 1, with input $f(m, n)$ and output $g(m, n)$. Assume that the response is due to the input f only (the system would be at rest without the input). Then, from Eq. (2):

$$g(m, n) = \mathbf{L}[\, f(m, n)]$$

$$= \mathbf{L}\left[\sum_{p=-\infty}^{\infty} \sum_{q=-\infty}^{\infty} f(p, q)\delta(m - p, n - q) \right] \quad (10)$$

If the system is known to be linear, then

$$g(m, n) = \sum_{p=-\infty}^{\infty} \sum_{q=-\infty}^{\infty} f(p, q)\mathbf{L}[\delta(m - p, n - q)] \quad (11)$$

$$= \sum_{p=-\infty}^{\infty} \sum_{q=-\infty}^{\infty} f(p, q)h(m, n; p, q), \quad (12)$$

which is all that generally can be said without further knowledge of the system and the input. If it is known that the system is space invariant (hence LSI), then Eq. (12) becomes

$$g(m, n) = \sum_{p=-\infty}^{\infty} \sum_{q=-\infty}^{\infty} f(p, q)h(m - p, n - q) \quad (13)$$

$$= f(m, n) * h(m, n), \quad (14)$$

which is the two-dimensional discrete space linear convolution of input f with impulse response h.

The linear convolution expresses the output of a wide variety of electrical and mechanical systems. In continuous systems, the convolution is expressed as an integral. For example, with lumped electrical circuits, the convolution integral is computed in terms of the passive circuit elements (resistors, inductors, and capacitors). In optical systems, the integral utilizes the point-spread functions of the optics. The operations occur effectively instantaneously, with the computational speed limited only by the speed of the electrons or photons through the system elements.

However, in discrete signal and image processing systems, the discrete convolutions are calculated sums of products. This convolution can be directly evaluated at each coordinate (m, n) by a digital processor, or, as discussed in Chapter 2.3, it can be computed by using the discrete cosine transform (DFT) using a fast Fourier transform (FFT) algorithm. Of course, if the exact linear convolution is desired, this means that the involved functions must be appropriately *zero padded* prior to using the DFT, as discussed in Chapter 2.3. The DFT/FFT approach is usually, but not always faster. If an image is being convolved with a very small spatial filter, then direct computation of Eq. (14) can be faster.

Suppose that the input to a discrete LSI system with impulse response $h(m, n)$ is a complex exponential function:

$$f(m, n) = e^{2\pi j(Um+Vn)}$$
$$= \cos[2\pi(Um+Vn)] + j\sin[2\pi(Um+Vn)]. \quad (15)$$

Then the system response is the linear convolution

$$g(m, n) = \sum_{p=-\infty}^{\infty} \sum_{q=-\infty}^{\infty} h(p, q) f(m-p, n-q)$$

$$= \sum_{p=-\infty}^{\infty} \sum_{q=-\infty}^{\infty} h(p, q) e^{2\pi j[U(m-p)+V(n-q)]} \quad (16)$$

$$= e^{2\pi j(Um+Vn)} \sum_{p=-\infty}^{\infty} \sum_{q=-\infty}^{\infty} h(p, q) e^{-2\pi j(Up+Vq)}, \quad (17)$$

which is exactly the input $f(m, n) = e^{2\pi j(Um+Vn)}$ multiplied by a function of (U, V) only:

$$H(U, V) = \sum_{p=-\infty}^{\infty} \sum_{q=-\infty}^{\infty} h(p, q) e^{-2\pi j(Up+Vq)}$$

$$= |H(U, V)|e^{j\angle H(U,V)}. \quad (18)$$

The function $H(U, V)$, which is immediately identified as the discrete-space Fourier transform (or DSFT, discussed extensively in Chapter 2.3) of the system impulse response, is called the *frequency response* of the system.

From Eq. (17) it may be seen that that the response to any complex exponential sinusoid function, with frequencies (U, V), is the same sinusoid, but with its amplitude scaled by the system *magnitude response* $|H(U, V)|$ evaluated at (U, V) and with a shift equal to the system *phase response* $\angle H(U, V)$ at (U, V). The complex sinusoids are the unique functions that have this invariance property in LSI systems.

As mentioned, the impulse response $h(m, n)$ of a LSI system is sufficient to express the response of the system to any input.[1] The frequency response $H(U, V)$ is uniquely obtainable from the impulse response (and vice versa) and so contains sufficient information to compute the response to any input that has a DSFT. In fact, the output can be expressed in terms of the frequency response by $G(U, V) = F(U, V)H(U, V)$ and by the DFT/FFT with appropriate zero padding. In fact, throughout this chapter and elsewhere, it may be assumed that whenever a DFT is being used to compute linear convolution, the appropriate zero padding has been applied to avoid the wrap-around effect of the cyclic convolution.

Usually, linear image processing filters are characterized in terms of their frequency responses, specifically by their spectrum shaping properties. Coarse common descriptions that apply to two-dimensional image processing include low-pass, band-pass, or high-pass. In such cases the frequency response is primarily a function of radial frequency and may even be circularly symmetric, viz., a function of $U^2 + V^2$ only. In other cases the filter may be strongly directional or oriented, with response strongly depending on the frequency angle of the input. Of course, the terms low pass, bandpass, high pass, and oriented are only rough qualitative descriptions of a system frequency response. Each broad class of filters has some generalized applications. For example, low-pass filters strongly attenuate all but the "lower" radial image frequencies (as determined by some bandwidth or cutoff frequency), and so are primarily smoothing filters. They are commonly used to reduce high-frequency noise, or to eliminate all but coarse image features, or to reduce the bandwidth of an image prior to transmission through a low-bandwidth communication channel or before subsampling the image (see Chapter 7.1).

A (radial frequency) bandpass filter attenuates all but an intermediate range of "middle" radial frequencies. This is commonly used for the enhancement of certain image features, such as edges (sudden transitions in intensity) or the ridges in a fingerprint. A high-pass filter attenuates all but the "higher" radial frequencies, or commonly, significantly amplifies high frequencies without attenuating lower frequencies. This approach is often used for correcting images that have suffered unwanted low-frequency attenuation (blurring); see Chapter 3.5.

Oriented filters, which either attenuate frequencies falling outside of a narrow range of orientations, or amplify a narrow range of angular frequencies, tend to be more specialized. For example, it may be desirable to enhance vertical image features as a prelude to detecting vertical structures, such as buildings.

Of course, filters may be a combination of types, such as bandpass and oriented. In fact, such filters are the most common types of basis functions used in the powerful wavelet image decompositions (Chapters 4.2) that have recently found so many applications in image analysis (Chapter 4.4), human visual modeling (Chapter 4.1), and image and video compression (Chapters 5.4 and 6.2).

In the remainder of this chapter, we introduce the simple but important application of linear filtering for *linear image enhancement*, which specifically means attempting to smooth image noise while not disturbing the original image structure.[2]

---

[1] Strictly speaking, for any *bounded* input, and provided that the system is stable. In practical image processing systems, the inputs are invariably bounded. Also, almost all image processing filters do not involve feedback and hence are naturally stable.

[2] The term "image enhancement" has been widely used in the past to describe any operation that improves image quality by some criteria. However, in recent years, the meaning of the term has evolved to denote image-preserving noise smoothing. This primarily serves to distinguish it from similar-sounding terms, such as "image restoration" and "image reconstruction," which also have taken specific meanings.

3 Linear Image Enhancement

The term "enhancement" implies a process whereby the visual quality of the image is improved. However, the term "image enhancement" has come to specifically mean a process of smoothing irregularities or noise that has somehow corrupted the image, while modifying the original image information as little as possible. The noise is usually modeled as an additive noise or as a multiplicative noise. We will consider additive noise now. As noted in Chapter 4.5, multiplicative noise, which is the other common type, can be converted into additive noise in a homomorphic filtering approach.

Before considering methods for image enhancement, we will make a simple model for additive noise. Chapter 4.5 of this *Handbook* greatly elaborates image noise models, which prove particularly useful for studying image enhancement filters that are nonlinear.

We will make the practical assumption that an observed noisy image is of finite extent $M \times N$: $\mathbf{f} = [f(m, n); 0 \le m \le M - 1, 0 \le n \le N - 1]$. We model $\mathbf{f}$ as a sum of an original image $\mathbf{o}$ and a *noise image* $\mathbf{q}$:

$$\mathbf{f} = \mathbf{o} + \mathbf{q}, \tag{19}$$

where $\mathbf{n} = (m, n)$. The additive noise image $\mathbf{q}$ models an undesirable, unpredictable corruption of $\mathbf{o}$. The process $\mathbf{q}$ is called a *two-dimensional random process* or a *random field*. Random additive noise can occur as thermal circuit noise, communication channel noise, sensor noise, and so on. Quite commonly, the noise is present in the image signal before it is sampled, so the noise is also sampled coincident with the image.

In Eq. (19), both the original image and noise image are unknown. The goal of enhancement is to recover an image $\mathbf{g}$ that resembles $\mathbf{o}$ as closely as possible by reducing $\mathbf{q}$. If there is an adequate model for the noise, then the problem of finding $\mathbf{g}$ can be posed as an *image estimation* problem, where $\mathbf{g}$ is found as the solution to a statistical optimization problem. Basic methods for image estimation are also discussed in Chapter 4.5, and in some of the following chapters on image enhancement using nonlinear filters.

With the tools of Fourier analysis and linear convolution in hand, we will now outline the basic approach of image enhancement by linear filtering. More often than not, the detailed statistics of the noise process $\mathbf{q}$ are unknown. In such cases, a simple linear filter approach can yield acceptable results, if the noise satisfies certain simple assumptions.

We will assume a *zero-mean additive white noise* model. The zero-mean model is used in Chapter 2.1, in the context of frame averaging. The process $\mathbf{q}$ is *zero mean* if the average or *sample mean* of R arbitrary noise samples

$$\left(\frac{1}{R}\right) \sum_{r=1}^{R} q(m_r, n_r) \to 0 \tag{20}$$

as R grows large (provided that the noise process is mean ergodic, which means that the sample mean approaches the statistical mean for large samples).

The term *white noise* is an idealized model for noise that has, on the average, a broad spectrum. It is a simplified model for *wideband noise*. More precisely, if $Q(U, V)$ is the DSFT of the noise process $\mathbf{q}$, then Q is also a random process. It is called the *energy spectrum* of the random process $\mathbf{q}$. If the noise process is white, then the average squared magnitude of $Q(U, V)$ is constant over all frequencies in the range $[-\pi, \pi]$. In the ensemble sense, this means that the sample average of the magnitude spectra of R noise images generated from the same source becomes constant for large R:

$$\left(\frac{1}{R}\right) \sum_{r=1}^{R} |Q_r(U, V)| \to \eta \tag{21}$$

for all (U, V) as R grows large. The square η^2 of the constant level is called the *noise power*. Since $\mathbf{q}$ has finite extent $M \times N$, it has a DFT $\tilde{\mathbf{Q}} = [\tilde{Q}(u, v): 0 \le u \le M-1, 0 \le v \le N-1]$. On average, the magnitude of the noise DFT $\tilde{\mathbf{Q}}$ will also be flat. Of course, it is highly unlikely that a given noise DSFT or DFT will actually have a flat magnitude spectrum. However, it is an effective simplified model for unknown, unpredictable broadband noise.

Images are also generally thought of as relatively broadband signals. Significant visual information may reside at mid-to-high spatial frequencies, since visually significant image details such as edges, lines, and textures typically contain higher frequencies. However, the magnitude spectrum of the image at higher image frequencies is usually low; most of the image power resides in the low frequencies contributed by the dominant luminance effects. Nevertheless, the higher image frequencies are visually significant.

The basic approach to linear image enhancement is low-pass filtering. There are different types of low-pass filters that can be used; several will be studied in the following. For a given filter type, different degrees of smoothing can be obtained by adjusting the filter bandwidth. A narrower bandwidth low-pass filter will reject more of the high-frequency content of a white or broadband noise, but it may also degrade the image content by attenuating important high-frequency image details. This is a tradeoff that is difficult to balance.

Next we describe and compare several smoothing low-pass filters that are commonly used for linear image enhancement.

3.1 Moving Average Filter

The moving average filter can be described in several equivalent ways. First, with the use of the notion of *windowing* introduced in Chapter 2.2, the moving average can be defined as an algebraic operation performed on local image neighborhoods according to a geometric law defined by the window. Given an image $\mathbf{f}$ to be filtered and a window **B** that collects gray-level pixels according to a geometric rule (defined by the window shape), then the moving average-filtered image **g** is given by

$$g(\mathbf{n}) = \text{AVE}[\mathbf{B}\,f(\mathbf{n})]. \qquad (22)$$

where the operation AVE computes the sample average of its arguments. Thus, the local average is computed over each local neighborhood of the image, producing a powerful smoothing effect. The windows are usually selected to be symmetric, as with those used for binary morphological image filtering (Chapter 2.2).

Since the average is a linear operation, it is also true that

$$g(\mathbf{n}) = \text{AVE}[\mathbf{B}o(\mathbf{n})] + \text{AVE}[\mathbf{B}q(\mathbf{n})]. \qquad (23)$$

Because the noise process **q** is assumed to be zero mean in the sense of Eq. (20), then the last term in Eq. (23) will tend to zero as the filter window is increased. Thus, the moving average filter has the desirable effect of reducing zero-mean image noise toward zero. However, the filter also affects the original image information. It is desirable that $\text{AVE}[\mathbf{B}o(\mathbf{n})] \approx o(\mathbf{n})$ at each **n**, but this will not be the case everywhere in the image if the filter window is too large. The moving average filter, which is low pass, will blur the image, especially as the window span is increased. Balancing this tradeoff is often a difficult task.

The moving average filter operation, Eq. (22), is actually a linear convolution. In fact, the impulse response of the filter is defined as having value $1/R$ over the span covered by the window when centered at the spatial origin $(0, 0)$, and zero elsewhere, where R is the number of elements in the window.

For example, if the window is SQUARE $[(2P + 1)^2]$, which is the most common configuration (it is defined in Chapter 2.2), then the average filter impulse response is given by

$$h(m, n) = \begin{cases} 1/(2P + 1)^2, & -P \leq m, n \leq P \\ 0, & \text{otherwise} \end{cases}. \qquad (24)$$

The frequency response of the moving average filter, Eq. (24), is

$$H(U, V) = \frac{\sin[(2P + 1)\pi U]}{(2P + 1)\sin(\pi U)} \frac{\sin[(2P + 1)\pi V]}{(2P + 1)\sin(\pi V)}. \qquad (25)$$

The *half-peak bandwidth* is often used for image processing filters. The half-peak (or 3 dB) cutoff frequencies occur on the locus of points (U, V) where $|H(U, V)|$ falls to $1/2$. For filter (25), this locus intersects the U axis and V axis at the cutoffs $U_{\text{half-peak}}, V_{\text{half-peak}} \approx 0.6/(2P + 1)$ cycles/pixel.

As depicted in Fig. 2, the magnitude response $|H(U, V)|$ of filter (25) exhibits considerable *sidelobes*. In fact, the number of sidelobes in the range $[0, \pi]$ is P. As P is increased, the filter bandwidth naturally decreases (more high-frequency at-

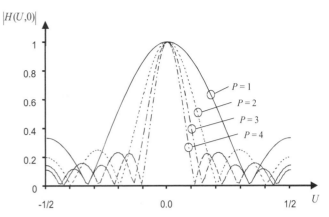

FIGURE 2 Plots of $|H(U, V)|$ given in Eq. (25) along $V = 0$, for $P = 1, 2, 3, 4$. As the filter span is increased, the bandwidth decreases. The number of sidelobes in the range $[0, \pi]$ is P.

tenuation or smoothing), but the overall sidelobe energy does not. The sidelobes are in fact a significant drawback, since there is considerable *noise leakage* at high noise frequencies. These residual noise frequencies remain to degrade the image. Nevertheless, the moving average filter has been commonly used because of its general effectiveness in the sense of Eq. (23) and because of its simplicity (ease of programming).

The moving average filter can be implemented either as a direct two-dimensional convolution in the space domain, or by use of DFTs to compute the linear convolution (see Chapter 2.3). It should be noted that the impulse response of the moving average filter is defined here as centered at the frequency origin. If the DFT is to be used, then the impulse response must be periodically extended, with the repetition period equal to the span of the DFT. This will result in impulse response coefficients' being distributed at the corners of the impulse response image, rather than being defined on negative space coordinates.

Since application of the moving average filter balances a tradeoff between noise smoothing and image smoothing, the filter span is usually taken to be an intermediate value. For images of the most common sizes, e.g., 256×256 or 512×512, typical (SQUARE) average filter sizes range from 3×3 to 15×15. The upper end provides significant (and probably excessive) smoothing, since 225 image samples are being averaged to produce each new image value. Of course, if an image suffers from severe noise, then a larger window might be warranted. A large window might also be acceptable if it is known that the original image is very smooth everywhere.

Figure 3 depicts the application of the moving average filter to an image that has had zero-mean white Gaussian noise added to it. In the current context, the distribution (Gaussian) of the noise is not relevant, although the meaning can be found in Chapter 4.5. The original image is included for comparison. The image was filtered with SQUARE-shaped moving average filters of spans 5×5 and 9×9, producing images with significantly

FIGURE 3 Examples of applications of a moving average filter: (a) original image "eggs"; (b) image with additive Gaussian white noise; moving average filtered image, using (c) SQUARE(25) window (5 × 5) and (d) SQUARE(81) window (9 × 9).

different appearances from each other as well as the noisy image. With the 5 × 5 filter, the noise is inadequately smoothed, yet the image has been blurred noticeably. The result of the 9 × 9 moving average filter is much smoother, although the noise influence is still visible, with some higher noise frequency components managing to leak through the filter, resulting in a mottled appearance.

3.2 Ideal Low-Pass Filter

As an alternative to the average filter, a filter may designed explicitly with no sidelobes by forcing the frequency response to be zero outside of a given radial cutoff frequency Ω_c,

$$H(U, V) = \begin{cases} 1, & \sqrt{U^2 + V^2} \leq \Omega_c \\ 0, & \text{otherwise} \end{cases}, \quad (26)$$

or outside of a rectangle defined by cutoff frequencies along the U and V axes,

$$H(U, V) = \begin{cases} 1, & |U| \leq U_c, \quad |V| \leq V_c \\ 0, & \text{otherwise} \end{cases}. \quad (27)$$

Such a filter is called *ideal low-pass filter* (ideal LPF) because of its idealized characteristic. We will study Eq. (27) rather than Eq. (26) since it is easier to describe the impulse response of the filter. If the region of frequencies passed by Eq. (26) is square, then there is little practical difference in the two filters if $U_c = V_c = \Omega_c$.

The impulse response of the ideal low-pass filter of Eq. (27) is given explicitly by

$$h(m, n) = U_c V_c \, \text{sinc}(2\pi U_c m) \, \text{sinc}(2\pi V_c n), \quad (28)$$

where $\text{sinc}(x) = (\sin x/x)$. Despite the seemingly "ideal" nature of this filter, it has some major drawbacks. First, it cannot be implemented exactly as a linear convolution, since impulse response (28) is infinite in extent (it never decays to zero). Therefore it must be approximated. One way is to simply truncate the impulse response, which in image processing applications is often satisfactory. However, this has the effect of introducing *ripple* near the frequency discontinuity, producing unwanted noise leakage. The introduced ripple is a manifestation of the well-known *Gibbs phenomena* studied in standard signal processing texts [1]. The ripple can be reduced by using a tapered truncation of the impulse response, e.g., by multiplying Eq. (28) with a Hamming window [1]. If the response is truncated to image size $M \times N$, then the ripple will be restricted to the vicinity of the locus of cutoff frequencies, which may make little difference in the filter performance. Alternately, the ideal LPF can be approximated by a Butterworth filter or other ideal LPF approximating function. The Butterworth filter has frequency response [2]

$$H(U, V) = \frac{1}{1 + ((\sqrt{U^2 + V^2})/(\Omega_c))^{2K}}. \tag{29}$$

and, in principle, can be made to agree with the ideal LPF with arbitrary precision by taking the filter order K large enough. However, Eq. (29) also has an infinite-extent impulse response with no known closed-form solution. Hence, to be implemented it must also be spatially truncated (approximated), which reduces the approximation effectiveness of the filter [2].

It should be noted that if a filter impulse response is truncated, then it should also be slightly modified by adding a constant level to each coefficient. The constant should be selected such that the filter coefficients sum to unity. This is commonly done since it is generally desirable that the response of the filter to the (0, 0) spatial frequency be unity, and since for any filter

$$H(0, 0) = \sum_{p=-\infty}^{\infty} \sum_{q=-\infty}^{\infty} h(p, q). \tag{30}$$

The second major drawback of the ideal LPF is the phenomena known as *ringing*. This term arises from the characteristic response of the ideal LPF to highly concentrated bright spots in an image. Such spots are impulselike, and so the local response has the appearance of the impulse response of the filter. For the circularly symmetric ideal LPF in Eq. (26), the response consists of a blurred version of the impulse surrounded by sinclike spatial sidelobes, which have the appearances of rings surrounding the main lobe.

In practical application, the ringing phenomenon creates more of a problem because of the *edge response* of the ideal LPF. In the simplistic case, the image consists of a single one-dimensional step edge: $s(m, n) = s(n) = 1$ for $n \geq 0$ and $s(n) = 0$, otherwise. Figure 4 depicts the response of the ideal LPF with impulse

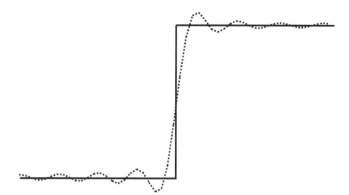

FIGURE 4 Depiction of edge ringing. The step edge is shown as a continuous curve; the linear convolution response of ideal LPF (28) is shown as a dotted curve.

response (28) to the step edge. The step response of the ideal LPF oscillates (rings) because the sinc function oscillates about the zero level. In the convolution sum, the impulse response alternately makes positive and negative contribution, creating overshoots and undershoots in the vicinity of the edge profile. Most digital images contain numerous steplike light-to-dark or dark-to-light image transitions; hence, application of the ideal LPF will tend to contribute considerable ringing artifacts to images. Since edges contain much of the significant information about the image, and since the eye tends to be sensitive to ringing artifacts, often the ideal LPF and its derivatives are not a good choice for image smoothing. However, if it is desired to strictly bandlimit the image as closely as possible, then the ideal LPF is a necessary choice.

Once an impulse response for an approximation to the ideal LPF has been decided, then the usual approach to implementation again entails zero padding both the image and the impulse response, using the periodic extension, taking the product of their DFTs (using an FFT algorithm), and defining the result as the inverse DFT. This was done in the example of Fig. 5, which depicts application of the ideal LPF using two cutoff frequencies. This was implemented by using a truncated ideal LPF without any special windowing. The dominant characteristic of the filtered images is the ringing, manifested as a strong mottling in both images. A very strong oriented ringing can be easily seen near the upper and lower borders of the image.

3.3 Gaussian Filter

As we have seen, *filter sidelobes* in either the space or spatial frequency domain contribute a negative effect to the responses of noise-smoothing linear image enhancement filters. Frequency-domain sidelobes lead to noise leakage, and space-domain sidelobes lead to ringing artifacts. A filter with sidelobes in neither domain is the *Gaussian filter*, with impulse response

$$h(m, n) = \frac{1}{2\pi\sigma^2} e^{-(m^2 + n^2)/2\sigma^2}. \tag{31}$$

(a)   (b)

FIGURE 5 Example of application of ideal low-pass filter to the noisy image in Fig. 3(b). The image is filtered with the radial frequency cutoff of (a) 30.72 cycles/image and (b) 17.07 cycles/image. These cutoff frequencies are the same as the half-peak cutoff frequencies used in Fig. 3.

Impulse response (31) is also infinite in extent, but it falls off rapidly away from the origin. In this case, the frequency response is closely approximated by

$$H(U, V) \approx e^{-2\pi^2\sigma^2(U^2+V^2)} \quad \text{for } |U|, |V| < 1/2, \quad (32)$$

which is also a Gaussian function. Neither Eq. (31) nor Eq. (32) shows any sidelobes; instead, both impulse and frequency response decay smoothly. The Gaussian filter is noted for the absence of ringing and noise leakage artifacts. The half-

peak radial frequency bandwidth of Eq. (32) is easily found to be

$$\Omega_c = \frac{1}{\pi\sigma}\sqrt{\ln\sqrt{2}} \approx \frac{0.187}{\sigma}. \quad (33)$$

If it is possible to decide an appropriate cutoff frequency Ω_c, then the cutoff frequency may be fixed by setting $\sigma = 0.187/\Omega_c$ pixels. The filter may then be implemented by truncating Eq. (31) using this value of σ, adjusting the coefficients to sum to one, zero padding both impulse response and image (taking care

(a)   (b)

FIGURE 6 Example of the application of a Gaussian filter to the noisy image in Fig. 3(b). The image is filtered with the radial frequency cutoff of (a) 30.72 cycles/image ($\sigma \approx 1.56$ pixels) and (b) 17.07 cycles/image ($\sigma \approx 2.80$ pixels). These cutoff frequencies are the same as the half-peak cutoff frequencies used in Figs. 3 and 5.

 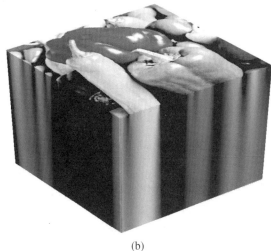

(a) (b)

FIGURE 7 Depiction of the scale-space property of a Gaussian low-pass filter. In (b), the image in (a) is Gaussian filtered with progressively larger values of σ (narrower bandwidths), producing successively smoother and more diffuse versions of the original. These are "stacked" to produce a data cube with the original image on top to produce the representation shown in (b).

to use the periodic extension of the impulse response implied by the DFT), multiplying DFTs, and taking the inverse DFT to be the result. The results obtained (see Fig. 6) are much better than those computed by using the ideal LPF, and they are slightly better than those obtained with the moving average filter, because of the reduced noise leakage.

Figure 7 shows the result of filtering an image with a Gaussian filter of successively larger σ values. As the value of σ is increased, small-scale structures such as noise and details are reduced to a greater degree. The sequence of images shown in Fig. 7(b) is a *Gaussian scale space*, where each scaled image is calculated by convolving the original image with a Gaussian filter of increasing σ value [3].

The Gaussian scale space may be thought of as evolving over time t. At time t, the scale space image $\mathbf{g}_t$ is given by

$$\mathbf{g}_t = h_\sigma * \mathbf{f}, \tag{34}$$

where $\mathbf{h}_\sigma$ is a Gaussian filter with scale factor σ, and $\mathbf{f}$ is the initial image. The time-scale relationship is defined by $\sigma = \sqrt{t}$. As σ is increased, less significant image features and noise begin to disappear, leaving only large-scale image features.

The Gaussian scale space may also be viewed as the evolving solution of a partial differential equation [3, 4]:

$$\frac{\partial \mathbf{g}_t}{\partial t} = \nabla^2 \mathbf{g}_t, \tag{35}$$

where $\nabla^2 \mathbf{g}_t$ is the Laplacian of $\mathbf{g}_t$. For an extended discussion of scale-space and partial differential equation methods, see Chapter 4.12 of this *Handbook*.

4 Discussion

Linear filters are omnipresent in image and video processing. Firmly established in the theory of linear systems, linear filters are the basis of processing signals of arbitrary dimensions. Since the advent of the fast Fourier transform in the 1960's, the linear filter has also been an attractive device in terms of computational expense. However, it must be noted that linear filters are performance limited for image enhancement applications. From the several experiments performed in this chapter, it can be seen that the removal of broadband noise from most images by means of linear filtering is impossible without some degradation (blurring) of the image information content. This limitation is due to the fact that complete frequency separation between signal and broadband noise is rarely practicable. Alternative solutions that remedy the deficiencies of linear filtering have been devised, resulting in a variety of powerful nonlinear image enhancement alternatives. These are discussed in Chapters 3.2–3.4 of this *Handbook*.

References

[1] A. V. Oppenheim and R. W. Schafer, *Discrete-Time Signal Processing* (Prentice-Hall, Englewood Cliffs, NJ, 1989).

[2] R. C. Gonzalez and R. E. Woods, *Digital Image Processing* (Addison-Wesley, Reading, MA, 1993).

[3] A. P. Witkin, "Scale-space filtering," in *Proceedings of the International Joint Conference on Artificial Intelligence* (IJCAI, Inc., Karlsruhe, Germany, 1983), pp. 1019–1022.

[4] J. J. Koenderink, "The structure of images," *Biol. Cybern.* **50**, 363–370 (1984).

3.2

Nonlinear Filtering for Image Analysis and Enhancement

Gonzalo R. Arce,
José L. Paredes,
and John Mullan
University of Delaware

1 Introduction

Digital image enhancement and analysis have played, and will continue to play, an important role in scientific, industrial, and military applications. In addition to these applications, image enhancement and analysis are increasingly being used in consumer electronics. Internet Web users, for instance, not only rely on built-in image processing protocols such as JPEG and interpolation, but they also have become image processing users equipped with powerful yet inexpensive software such as PhotoShop. Users not only retrieve digital images from the Web but are now able to acquire their own by use of digital cameras or through digitization services of standard 35-mm analog film. The end result is that consumers are beginning to use home computers to enhance and manipulate their own digital pictures. Image enhancement refers to processes seeking to improve the visual appearance of an image. As an example, image enhancement might be used to emphasize the edges within the image. This edge-enhanced image would be more visually pleasing to the naked eye, or perhaps could serve as an input to a machine that would detect the edges and perhaps make measurements of shape and size of the detected edges. Image enhancement is important because of its usefulness in virtually all image processing applications.

José L. Paredes is also with the University of Los Andes, Mérida-Venezuela.

Image enhancement tools are often classified into (a) point operations, and (b) spatial operators. Point operations include contrast stretching, noise clipping, histogram modification, and pseudo-coloring. Point operations are, in general, simple nonlinear operations that are well known in the image processing literature and are covered elsewhere in this *Handbook*. Spatial operations used in image processing today are, in contrast, typically linear operations. The reason for this is that spatial linear operations are simple and easily implemented. Although linear image enhancement tools are often adequate in many applications, significant advantages in image enhancement can be attained if nonlinear techniques are applied [1]. Nonlinear methods effectively preserve edges and details of images, whereas methods using linear operators tend to blur and distort them. Additionally, nonlinear image enhancement tools are less susceptible to noise. Noise is always present because of the physical randomness of image acquisition systems. For example, underexposure and low-light conditions in analog photography lead to images with film-grain noise, which, together with the image signal itself, are captured during the digitization process.

This article focuses on nonlinear and spatial image enhancement and analysis. The nonlinear tools described in this article are easily implemented on currently available computers. Rather than using linear combinations of pixel values within a local window, these tools use the local weighted median. In Section 2, the principles of weighted medians (WMs) are presented.

Weighted medians have striking analogies with traditional linear FIR filters, yet their behavior is often markedly different. In Section 3, we show how WM filters can be easily used for noise removal. In particular, the center WM filter is described as a tunable filter highly effective in impulsive noise. Section 4 focuses on image enlargement, or zooming, using WM filter structures that, unlike standard linear interpolation methods, provide little edge degradation. Section 5 describes image sharpening algorithms based on WM filters. These methods offer significant advantages over traditional linear sharpening tools whenever noise is present in the underlying images. Section 6 goes beyond image enhancement and focuses on the analysis of images. In particular, edge-detection methods based on WM filters are described as well as their advantages over traditional edge-detection algorithms.

2 Weighted Median Smoothers and Filters

2.1 Running Median Smoothers

The running median was first suggested as a nonlinear smoother for time series data by Tukey in 1974 [2]. To define the running median smoother, let $\{x(\cdot)\}$ be a discrete time sequence. The running median passes a window over the sequence $\{x(\cdot)\}$ that selects, at each instant n, a set of samples to comprise the observation vector $\mathbf{x}(n)$. The observation window is centered at n, resulting in

$$\mathbf{x}(n) = [x(n - N_L), \ldots, x(n), \ldots, x(n + N_R)]^T, \quad (1)$$

where N_L and N_R may range in value over the nonnegative integers and $N = N_L + N_R + 1$ is the window size. The median smoother operating on the input sequence $\{x(\cdot)\}$ produces the output sequence $\{y\}$, where at time index n

$$y(n) = \text{MEDIAN}[x(n - N_L), \ldots, x(n), \ldots, x(n + N_R)] \quad (2)$$

$$= \text{MEDIAN}[x_1(n), \ldots, x_N(n)], \quad (3)$$

where $x_i(n) = x(n - N_L + 1 - i)$ for $i = 1, 2, \ldots, N$. That is, the samples in the observation window are sorted and the middle, or median, value is taken as the output. If $x_{(1)}, x_{(2)}, \ldots, x_{(N)}$ are the sorted samples in the observation window, the median smoother outputs

$$y(n) = \begin{cases} x_{\left(\frac{N+1}{2}\right)} & \text{if } N \text{ is odd} \\ \frac{x_{\left(\frac{N}{2}\right)} + x_{\left(\frac{N}{2}+1\right)}}{2} & \text{otherwise} \end{cases}. \quad (4)$$

In most cases, the window is symmetric about $x(n)$ and $N_L = N_R$.

The input sequence $\{x(\cdot)\}$ may be either finite or infinite in extent. For the finite case, the samples of $\{x(\cdot)\}$ can be indexed as $x(1), x(2), \ldots, x(L)$, where L is the length of the sequence. Because of the symmetric nature of the observation window, the window extends beyond a finite extent input sequence at both

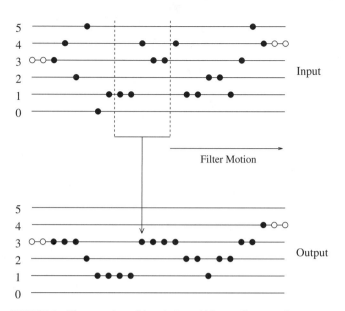

FIGURE 1 The operation of the window width 5 median smoother: ○, appended points.

the beginning and end. These end effects are generally accounted for by appending N_L samples at the beginning and N_R samples at the end of $\{x(\cdot)\}$. Although the appended samples can be arbitrarily chosen, typically these are selected so that the points appended at the beginning of the sequence have the same value as the first signal point, and the points appended at the end of the sequence all have the value of the last signal point.

To illustrate the appending of input sequence and the median smoother operation, consider the input signal $\{x(\cdot)\}$ of Fig. 1. In this example, $\{x(\cdot)\}$ consists of 20 observations from a six-level process, $\{x : x(n) \in \{0, 1, \ldots, 5\}, n = 1, 2, \ldots, 20\}$. The figure shows the input sequence and the resulting output sequence for a window size 5 median smoother. Note that to account for edge effects, two samples have been appended to both the beginning and end of the sequence. The median smoother output at the window location shown in the figure is

$$y(9) = \text{MEDIAN}[x(7), x(8), x(9), x(10), x(11)]$$

$$= \text{MEDIAN}[1, 1, 4, 3, 3] = 3.$$

Running medians can be extended to a recursive mode by replacing the "causal" input samples in the median smoother by previously derived output samples [3]. The output of the recursive median smoother is given by

$$y(n) = \text{MEDIAN}[y(n - N_L), \ldots, y(n - 1),$$

$$x(n), \ldots, x(n + N_R)]. \quad (5)$$

In recursive median smoothing, the center sample in the observation window is modified before the window is moved to the next position. In this manner, the output at each window location replaces the old input value at the center of the window. With the same amount of operations, recursive median smoothers have better noise attenuation capabilities than their nonrecursive counterparts [4, 5]. Alternatively, recursive median smoothers require smaller window lengths than their nonrecursive counterparts in order to attain a desired level of noise attenuation. Consequently, for the same level of noise attenuation, recursive median smoothers often yield less signal distortion. In image processing applications, the running median window spans a local two-dimensional (2-D) area. Typically, an $N \times N$ area is included in the observation window. The processing, however, is identical to the one-dimensional (1-D) case in the sense that the samples in the observation window are sorted and the middle value is taken as the output.

The running 1-D or 2-D median, at each instant in time, computes the sample median. The sample median, in many respects, resemble the sample mean. Given N samples $x_1, \ldots, x_N$, the sample mean, $\bar{x}$, and sample median, $\tilde{x}$, minimize the expression

$$G(\beta) = \sum_{i=1}^{N} |x_i - \beta|^p \tag{6}$$

for $p = 2$ and $p = 1$, respectively. Thus, the median of an odd number of samples emerges as the sample whose sum of absolute distances to all other samples in the set is the smallest. Likewise, the sample mean is given by the value β whose square distance to all samples in the set is the smallest possible. The analogy between the sample mean and median extends into the statistical domain of parameter estimation, where it can be shown that the sample median is the maximum likelihood (ML) estimator of location of a constant parameter in Laplacian noise. Likewise, the sample mean is the ML estimator of location of a constant parameter in Gaussian noise [6]. This result has profound implications in signal processing, as most tasks where non-Gaussian noise is present will benefit from signal processing structures using medians, particularly when the noise statistics can be characterized by probability densities having tails heavier than Gaussian tails (which leads to noise with impulsive characteristics) [7–9].

2.2 Weighted Median Smoothers

Although the median is a robust estimator that possesses many optimality properties, the performance of running medians is limited by the fact that it is temporally blind. That is, all observation samples are treated equally regardless of their location within the observation window. Much like weights can be incorporated into the sample mean to form a weighted mean, a weighted median can be defined as the sample that minimizes the weighted cost function

$$G_p(\beta) = \sum_{i=1}^{N} W_i |x_i - \beta|^p, \tag{7}$$

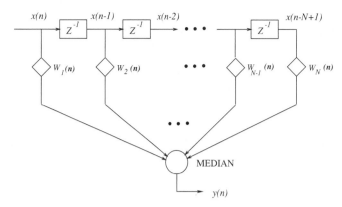

FIGURE 2 The weighted median smoothing operation.

for $p = 1$. For $p = 2$, the cost function of Eq. (7) is quadratic and the value β minimizing it is the normalized weighted mean

$$\hat{\beta} = \arg \min_{\beta} \sum_{i=1}^{N} W_i (x_i - \beta)^2 = \frac{\sum_{i=1}^{N} W_i x_i}{\sum_{i=1}^{N} W_i}, \tag{8}$$

with $W_i > 0$. For $p = 1$, $G_1(\beta)$ is piecewise linear and convex for $W_i \geq 0$. The value β minimizing Eq. (7) is thus guaranteed to be one of the samples $x_1, x_2, \ldots, x_N$ and is referred to as the weighted median, originally introduced over a hundred years ago by Edgemore [10]. After some algebraic manipulations, it can be shown that the running weighted median output is computed as

$$y(n) = \text{MEDIAN}[W_1 \diamond x_1(n), W_2 \diamond x_2(n), \ldots, W_N \diamond x_N(n)], \tag{9}$$

where $W_i > 0$ and $\diamond$ is the replication operator defined as $W_i \diamond x_i = \overbrace{x_i, x_i, \ldots, x_i}^{W_i \text{ times}}$. Weighted median smoothers were introduced in the signal processing literature by Brownigg in 1984 and have since received considerable attention [11–13]. The WM smoothing operation can be schematically described as in Fig. 2.

Weighted Median Smoothing Computation

Consider the window size 5 WM smoother defined by the symmetric weight vector $\mathbf{W} = [1, 2, 3, 2, 1]$. For the observation $\mathbf{x}(n) = [12, 6, 4, 1, 9]$, the weighted median smoother output is found as

$$
\begin{aligned}
y(n) &= \text{MEDIAN}[12 \diamond 1, 6 \diamond 2, 4 \diamond 3, 1 \diamond 2, 9 \diamond 1] \\
&= \text{MEDIAN}[12, 6, 6, 4, 4, 4, 1, 1, 9] \tag{10} \\
&= \text{MEDIAN}[1, 1, 4, 4, \underline{4}, 6, 6, 9, 12] = 4
\end{aligned}
$$

where the median value is underlined in Eq. (10). The large weighting on the center input sample results in this sample being taken as the output. As a comparison, the standard median output for the given input is $y(n) = 6$.

Although the smoother weights in the above example are integer valued, the standard WM smoother definition clearly allows for positive real-valued weights. The WM smoother output for this case is as follows.

1. Calculate the threshold $T_0 = \frac{1}{2} \sum_{i=1}^{N} W_i$.
2. Sort the samples in the observation vector $\mathbf{x}(n)$.
3. Sum the weights corresponding to the sorted samples, beginning with the maximum sample and continuing down in order.
4. The output is the sample whose weight causes the sum to become greater than or equal to T_0.

To illustrate the WM smoother operation for positive real-valued weights, consider the WM smoother defined by $\mathbf{W} = [0.1, 0.1, 0.2, 0.2, 0.1]$. The output for this smoother operating on $\mathbf{x}(n) = [12, 6, 4, 1, 9]$ is found as follows. Summing the weights gives the threshold $T_0 = \frac{1}{2} \sum_{i=1}^{5} W_i = 0.35$. The observation samples, sorted observation samples, their corresponding weight, and the partial sum of weights (from each ordered sample to the maximum) are

observation samples	12,	6,	4,	1,	9
corresponding weights	0.1,	0.1,	0.2,	0.2,	0.1
sorted observation samples	1,	4,	6,	9,	12
corresponding weights	0.2,	0.2,	0.1,	0.1,	0.1
partial weight sums	0.7,	<u>0.5</u>,	0.3,	0.2,	0.1

$$(11)$$

Thus, the output is 4 since when starting from the right (maximum sample) and summing the weights, the threshold $T_0 = 0.35$ is not reached until the weight associated with 4 is added.

An interesting characteristic of WM smoothers is that the nature of a WM smoother is not modified if its weights are multiplied by a positive constant. Thus, the same filter characteristics can be synthesized by different sets of weights. Although the WM smoother admits real-valued positive weights, it turns out that any WM smoother based on real-valued positive weights has an equivalent integer-valued weight representation [14]. Consequently, there are only a finite number of WM smoothers for a given window size. The number of WM smoothers, however, grows rapidly with window size [13].

Weighted median smoothers can also operate on a recursive mode. The output of a recursive WM smoother is given by

$$y(n) = \text{MEDIAN}[W_{-N_1} \diamond y(n - N_1), \ldots, W_{-1} \diamond y(n - 1),$$
$$W_0 \diamond x(n), \ldots, W_{N_1} \diamond x(n + N_1)], \quad (12)$$

where the weights W_i are as before constrained to be positive valued. Recursive WM smoothers offer advantages over WM smoothers in the same way that recursive medians have advantages over their nonrecursive counterparts. In fact, recursive WM smoothers can synthesize nonrecursive WM smoothers of much longer window sizes [14].

2.2.1 The Center Weighted Median Smoother

The weighting mechanism of WM smoothers allows for great flexibility in emphasizing or deemphasizing specific input samples. In most applications, not all samples are equally important. Because of the symmetric nature of the observation window, the sample most correlated with the desired estimate is, in general, the center observation sample. This observation leads to the center weighted median (CWM) smoother, which is a relatively simple subset of WM smoother that has proven useful in many applications [12].

The CWM smoother is realized by allowing only the center observation sample to be weighted. Thus, the output of the CWM smoother is given by

$$y(n) = \text{MEDIAN}[x_1, \ldots, x_{c-1}, W_c \diamond x_c, x_{c+1}, \ldots, x_N], \quad (13)$$

where W_c is an odd positive integer and $c = (N+1)/2 = N_1 + 1$ is the index of the center sample. When $W_c = 1$, the operator is a median smoother, and for $W_c \geq N$, the CWM reduces to an identity operation.

The effect of varying the center sample weight is perhaps best seen by way of an example. Consider a segment of recorded speech. The voiced waveform "a" noise is shown at the top of Fig. 3. This speech signal is taken as the input of a CWM smoother of window size 9. The outputs of the CWM, as the weight parameter $W_c = 2w + 1$ for $w = 0, \ldots, 3$, are shown in the figure. Clearly, as W_c is increased less smoothing occurs. This response of the CWM smoother is explained by relating the weight W_c and the CWM smoother output to select order statistics (OS).

The CWM smoother has an intuitive interpretation. It turns out that the output of a CWM smoother is equivalent to computing

$$y(n) = \text{MEDIAN}[x_{(k)}, x_c, x_{(N-k+1)}], \quad (14)$$

where $k = (N + 2 - W_c)/2$ for $1 \leq W_c \leq N$, and $k = 1$ for $W_c > N$ [12]. Since $x(n)$ is the center sample in the observation window, i.e., $x_c = x(n)$, the output of the smoother is identical to the input as long as the $x(n)$ lies in the interval $[x_{(k)}, x_{(N+1-k)}]$. If the center input sample is greater than $x_{(N+1-k)}$ then the smoothing outputs $x_{(N+1-k)}$, guarding against a high rank order (large) aberrant data point being taken as the output. Similarly, the smoother's output is $x_{(k)}$ if the sample $x(n)$ is smaller than this order statistic. This CWM smoother performance characteristic is illustrated in Figs. 4 and 5. Figure 4 shows how the input sample is left unaltered if it is between the trimming statistics $x_{(k)}$ and $x_{(N+1-k)}$ and mapped to one of these statistics if it is outside this range. Figure 5 shows an example of the CWM smoother operating on a constant-valued sequence in additive Laplacian noise. Along with the input and output, the trimming statistics are shown as an upper and lower bound on the filtered signal. It is easily seen how increasing k will tighten the range in which the input is passed directly to the output.

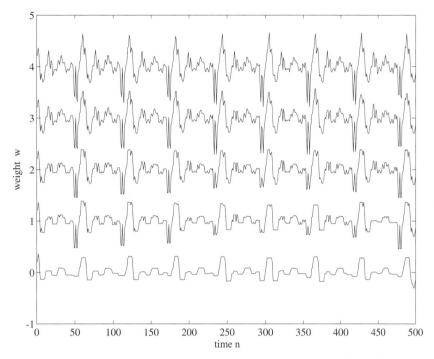

FIGURE 3 Effects of increasing the center weight of a CWM smoother of window size $N = 9$ operating on the voiced speech "a". The CWM smoother output is shown for $W_c = 2w + 1$, with $w = 0, 1, 2, 3$. Note that for $W_c = 1$ the CWM reduces to median smoothing, and for $W_c = 9$ it becomes the identity operator.

2.2.2 Permutation Weighted Median Smoothers

The principle behind the CWM smoother lies in the ability to emphasize, or deemphasize, the center sample of the window by tuning the center weight, while keeping the weight values of all other samples at unity. In essence, the value given to the center weight indicates the "reliability" of the center sample. If the center sample does not contain an impulse (high reliability), it would be desirable to make the center weight large such that no smoothing takes place (identity filter). In contrast, if an impulse was present in the center of the window (low reliability), no emphasis should be given to the center sample (impulse), and the center weight should be given the smallest possible weight, i.e. $W_c = 1$, reducing the CWM smoother structure to a simple median. Notably, this adaptation of the center weight can be easily achieved by considering the center sample's rank among all pixels in the window [15, 16]. More precisely, denoting the rank of the center sample of the window at a given location as $R_c(n)$, then the simplest *permutation* WM smoother is defined by the following modification of the CWM smoothing operation:

$$W_c(n) = \begin{cases} N & \text{if } T_L \leq R_c(n) \leq T_U, \\ 1 & \text{otherwise} \end{cases} \tag{15}$$

where N is the window size and $1 \leq T_L \leq T_U \leq N$ are two adjustable threshold parameters that determine the degree of smoothing. Note that the weight in Eq. (15) is data adaptive and may change between two values with n. The smaller (larger) the threshold parameter T_L (T_U) is set to, the better the detail preservation. Generally, T_L and T_U are set symmetrically around the median. If the underlying noise distribution was not symmetric about the origin, a nonsymmetric assignment of the thresholds would be appropriate.

The data-adaptive structure of the smoother in Eq. (15) can be extended so that the center weight is not only switched between two possible values, but can take on N different values:

$$W_c(n) = \begin{cases} W_{c(j)}(n) & \text{if } R_c(n) = j, \quad j \in \{1, 2, \ldots, N\} \\ 0 & \text{otherwise} \end{cases} \tag{16}$$

Thus, the weight assigned to x_c is drawn from the center weight set $\{W_{c(1)}, W_{c(2)}, \ldots, W_{c(N)}\}$. With an increased number of weights, the smoother in Eq. (16) can perform better although the design of the weights is no longer trivial and optimization algorithms are needed [15, 16]. A further generalization

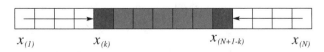

$x_{(1)}$ $\qquad$ $x_{(k)}$ $\qquad\qquad$ $x_{(N+1-k)}$ $\qquad$ $x_{(N)}$

FIGURE 4 CWM smoothing operation. The center observation sample is mapped to the order statistic $x_{(k)}$ ($x_{(N+1-k)}$) if the center sample is less (greater) than $x_{(k)}$ ($x_{(N+1-k)}$) and left unaltered otherwise.

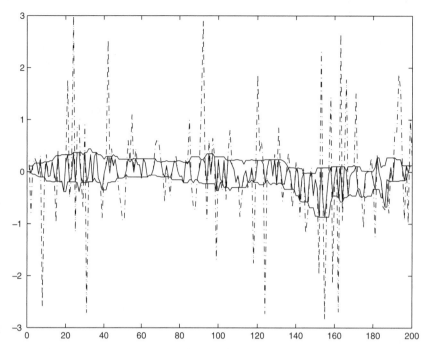

FIGURE 5 Example of the CWM smoother operating on a Laplacian distributed sequence with unit variance. Shown are the input ($- \cdot - \cdot -$) and output (—) sequences as well as the trimming statistics $x_{(k)}$ and $x_{(N+1-k)}$. The window size is 25 and $k = 7$.

of Eq. (16) is feasible when weights are given to all samples in the window, but when the value of each weight is data dependent and determined by the rank of the corresponding sample. In this case, the output of the permutation WM smoother is found as

$$y(n) = \text{MEDIAN}\big[W_{1(R_1)} \diamond x_1(n),$$
$$W_{2(R_2)} \diamond x_2(n), \ldots, W_{N(R_N)} \diamond x_N(n) \big], \qquad (17)$$

where $W_{i(R_i)}$ is the weight assigned to $x_i(n)$ and selected according to the sample's rank R_i. The weight assigned to x_i is drawn from the weight set $\{ W_{i(1)}, W_{i(2)}, \ldots, W_{i(N)} \}$. Having N weights per sample, a total of N^2 samples need to be stored in the computation of Eq. (17). In general, optimization algorithms are needed to design the set of weights although in some cases the design is simple, as with the smoother in Eq. (15). Permutation WM smoothers can provide significant improvement in performance at the higher cost of memory cells [15].

2.2.3 Threshold Decomposition and Stack Smoothers

An important tool for the analysis and design of weighted median smoothers is the threshold decomposition property [17]. Given an integer-valued set of samples $x_1, x_2, \ldots, x_N$ forming the vector $\mathbf{x} = [x_1, x_2, \ldots, x_N]^T$, where $x_i \in \{-M, \ldots, -1, 0, \ldots, M\}$, the threshold decomposition of $\mathbf{x}$ amounts to decomposing this vector into $2M$ binary vectors $\mathbf{x}^{-M+1}, \ldots, \mathbf{x}^0, \ldots, \mathbf{x}^M$, where the ith element of $\mathbf{x}^m$ is defined

by

$$x_i^m = T^m(x_i) = \begin{cases} 1 & \text{if } x_i \geq m \\ -1 & \text{if } x_i < m \end{cases}, \qquad (18)$$

where $T^m(\cdot)$ is referred to as the thresholding operator. With the use of the sgn function, the above can be written as $x_i^m = \text{sgn}(x_i - m^-)$, where m^- represents a real number approaching the integer m from the left. Although defined for integer-valued signals, the thresholding operation in Eq. (18) can be extended to noninteger signals with a finite number of quantization levels. The threshold decomposition of the vector $\mathbf{x} = [0, 0, 2, -2, 1, 1, 0, -1, -1]^T$ with $M = 2$, for instance, leads to the 4 binary vectors

$$\begin{aligned}
\mathbf{x}^2 &= [-1, -1, \quad 1, -1, -1, -1, -1, -1, -1]^T, \\
\mathbf{x}^1 &= [-1, -1, \quad 1, -1, \quad 1, \quad 1, -1, -1, -1]^T, \\
\mathbf{x}^0 &= [\;\; 1, \quad 1, \quad 1, -1, \quad 1, \quad 1, \quad 1, -1, -1]^T, \\
\mathbf{x}^{-1} &= [\;\; 1, \quad 1, \quad 1, -1, \quad 1, \quad 1, \quad 1, \quad 1, \quad 1]^T.
\end{aligned} \qquad (19)$$

Threshold decomposition has several important properties. First, threshold decomposition is reversible. Given a set of thresholded signals, each of the samples in $\mathbf{x}$ can be exactly reconstructed as

$$x_i = \frac{1}{2} \sum_{m=-M+1}^{M} x_i^m. \qquad (20)$$

Thus, an integer-valued discrete-time signal has a unique threshold signal representation, and vice versa:

$$x_i \overset{T.D.}{\longleftrightarrow} \{x_i^m\},$$

where $\overset{T.D.}{\longleftrightarrow}$ denotes the one-to-one mapping provided by the threshold decomposition operation.

The set of threshold decomposed variables obey the following set of partial ordering rules. For all thresholding levels $m > \ell$, it can be shown that $x_i^m \le x_i^\ell$. In particular, if $x_i^m = 1$ then $x_i^\ell = 1$ for all $\ell < m$. Similarly, if $x_i^\ell = -1$ then $x_i^m = -1$, for all $m > \ell$. The partial order relationships among samples across the various thresholded levels emerge naturally in thresholding and are referred to as *stacking constraints* [18].

Threshold decomposition is of particular importance in weighted median smoothing since they are commutable operations. That is, applying a weighted median smoother to a $2M + 1$ valued signal is equivalent to decomposing the signal to $2M$ binary thresholded signals, processing each binary signal separately with the corresponding WM smoother, and then adding the binary outputs together to obtain the integer-valued output. Thus, the weighted median smoothing of a set of samples $x_1, x_2, \ldots, x_N$ is related to the set of the thresholded weighted median smoothed signals as [14, 17]

Weighted MEDIAN$(x_1, \ldots, x_N)$

$$= \frac{1}{2} \sum_{m=-M+1}^{M} \text{Weighted MEDIAN}\big(x_1^m, \ldots, x_N^m\big). \quad (21)$$

Since $x_i \overset{T.D.}{\longleftrightarrow} \{x_i^m\}$ and Weighted MEDIAN $(x_i|_{i=1}^N) \overset{T.D.}{\longleftrightarrow}$ {Weigthed MEDIAN$(x_i^m|_{i=1}^N)$}, the relationship in Eq. (21) establishes a *weak* superposition property satisfied by the nonlinear median operator, which is important because the effects of median smoothing on binary signals are much easier to analyze than those on multilevel signals. In fact, the weighted median operation on binary samples reduces to a simple Boolean operation. The median of three binary samples x_1, x_2, x_3, for example, is equivalent to $x_1 x_2 + x_2 x_3 + x_1 x_3$, where the $+$ (OR) and $x_i x_j$ (AND) Boolean operators in the $\{-1, 1\}$ domain are defined as

$$x_i + x_j = \max(x_i, x_j),$$
$$x_i x_j = \min(x_i, x_j). \quad (22)$$

Note that the operations in Eq. (22) are also valid for the standard Boolean operations in the $\{0, 1\}$ domain.

The framework of threshold decomposition and Boolean operations has led to the general class of nonlinear smoothers referred here to as stack smoothers [18], whose output is defined by

$$S(x_1, \ldots, x_N) = \frac{1}{2} \sum_{m=-M+1}^{M} f\big(x_1^m, \ldots, x_N^m\big) \quad (23)$$

where $f(\cdot)$ is a Boolean operation satisfying Eq. (22) and

the stacking property. More precisely, if two binary vectors $\mathbf{u} \in \{-1, 1\}^N$ and $\mathbf{v} \in \{-1, 1\}^N$ stack, i.e., $u_i \ge v_i$ for all $i \in \{1, \ldots, N\}$, then their respective outputs stack, i.e., $f(\mathbf{u}) \ge f(\mathbf{v})$. A necessary and sufficient condition for a function to possess the stacking property is that it can be expressed as a Boolean function that contains no complements of input variables [19]. Such functions are known as *positive Boolean* functions (PBFs).

Given a positive Boolean function $f(x_1^m, \ldots, x_N^m)$ that characterizes a stack smoother, it is possible to find the equivalent smoother in the integer domain by replacing the binary AND and OR Boolean functions acting on the x_i's with max and min operations acting on the multilevel x_i samples. A more intuitive class of smoothers is obtained, however, if the positive Boolean functions are further restricted [14]. When self-duality and separability is imposed, for instance, the equivalent integer domain stack smoothers reduce to the well-known class of weighted median smoothers with positive weights. For example, if the Boolean function in the stack smoother representation is selected as $f(x_1, x_2, x_3, x_4) = x_1 x_3 x_4 + x_2 x_4 + x_2 x_3 + x_1 x_2$, the equivalent WM smoother takes on the positive weights $(W_1, W_2, W_3, W_4) = (1, 2, 1, 1)$. The procedure of how to obtain the weights W_i from the PBF is described in [14].

2.3 Weighted Median Filters

Admitting only positive weights, WM smoothers are severely constrained as they are, in essence, smoothers having low-pass type filtering characteristics. A large number of engineering applications require bandpass or high-pass frequency filtering characteristics. Linear FIR equalizers admitting only positive filter weights, for instance, would lead to completely unacceptable results. Thus, it is not surprising that weighted median smoothers admitting only positive weights lead to unacceptable results in a number of applications.

Much like the sample mean can be generalized to the rich class of linear FIR filters, there is a logical way to generalize the median to an equivalently rich class of weighted median filters that admit both positive and negative weights [20]. It turns out that the extension is not only natural, leading to a significantly richer filter class, but is simple as well. Perhaps the simplest approach to derive the class of weighted median filters with real-valued weights is by analogy. The sample mean $\bar{\beta} = \text{MEAN}(x_1, x_2, \ldots, x_N)$ can be generalized to the class of linear FIR filters as

$$\beta = \text{MEAN}(W_1 x_1, W_2 x_2, \ldots, W_N x_N), \quad (24)$$

where $x_i \in R$. In order for the analogy to be applied to the median filter structure, Eq. (24) must be written as

$$\bar{\beta} = \text{MEAN}(|W_1| \operatorname{sgn}(W_1) x_1, |W_2| \operatorname{sgn}(W_2) x_2, \ldots,$$
$$|W_N| \operatorname{sgn}(W_N) x_N) \quad (25)$$

where the sgn of the weight affects the corresponding input sample and the weighting is constrained to be nonnegative.

By analogy, the class of weighted median filters admitting real-valued weights emerges as [20]

$$\tilde{\beta} = \text{MEDIAN}[|W_1| \diamond \text{sgn}(W_1)x_1, |W_2| \diamond \text{sgn}(W_2)x_2,$$

$$\ldots, |W_N| \diamond \text{sgn}(W_N)x_N], \tag{26}$$

with $W_i \in R$ for $i = 1, 2, \ldots, N$. Again, the weight sgns are uncoupled from the weight magnitude values and are merged with the observation samples. The weight magnitudes play the equivalent role of positive weights in the framework of weighted median smoothers. It is simple to show that the weighted mean (normalized) and the weighted median operations shown in Eqs. (25) and (26) respectively minimize

$$G_2(\beta) = \sum_{i=1}^{N} |W_i|(\text{sgn}(W_i)x_i - \beta)^2,$$

$$\tag{27}$$

$$G_1(\beta) = \sum_{i=1}^{N} |W_i||\text{sgn}(W_i)x_i - \beta|.$$

While $G_2(\beta)$ is a convex continuous function, $G_1(\beta)$ is a convex but piecewise linear function whose minimum point is guaranteed to be one of the sgned input samples, i.e., $\text{sgn}(W_i)x_i$.

Weighted Median Filter Computation

The WM filter output for noninteger weights can be determined as follows [20].

1. Calculate the threshold $T_0 = \frac{1}{2}\sum_{i=1}^{N} |W_i|$.
2. Sort the sgned observation samples $\text{sgn}(W_i)x_i$.
3. Sum the magnitude of the weights corresponding to the sorted sgned samples beginning with the maximum and continuing down in order.
4. The output is the sgned sample whose magnitude weight causes the sum to become greater than or equal to T_0.

The following example illustrates this procedure. Consider the window size 5 WM filter defined by the real valued weights $[W_1, W_2, W_3, W_4, W_5]^T = [0.1, 0.2, 0.3, -0.2, 0.1]^T$. The output for this filter operating on the observation set $[x_1, x_2, x_3, x_4, x_5]^T = [-2, 2, -1, 3, 6]^T$ is found as follows. Summing the absolute weights gives the threshold $T_0 = \frac{1}{2}\sum_{i=1}^{5} |W_i| = 0.45$. The sgned observation samples, sorted observation samples, their corresponding weight, and the partial sum of weights (from each ordered sample to the maximum) are.

observation samples	-2,	2,	-1,	3,	6
corresponding weights	0.1,	0.2,	0.3,	-0.2,	0.1
sorted signed observation samples	-3,	-2,	-1,	2,	6
corresponding absolute weights	0.2,	0.1,	0.3,	0.2,	0.1
partial weight sums	0.9,	0.7,	<u>0.6</u>,	0.3,	0.1

Thus, the output is -1 since when starting from the right (maximum sample) and summing the weights, the threshold $T_0 = 0.45$ is not reached until the weight associated with -1 is added. The underlined sum value above indicates that this is the first sum which meets or exceeds the threshold.

The effect that negative weights have on the weighted median operation is similar to the effect that negative weights have on linear FIR filter outputs. Figure 6 illustrates this concept, where $G_2(\beta)$ and $G_1(\beta)$, the cost functions associated with linear FIR and weighted median filters, respectively, are plotted as a function of β. Recall that the output of each filter is the value minimizing the cost function. The input samples are again selected as $[x_1, x_2, x_3, x_4, x_5] = [-2, 2, -1, 3, 6]$, and two sets of weights are used. The first set is $[W_1, W_2, W_3, W_4, W_5] = [0.1, 0.2, 0.3, 0.2, 0.1]$, where all the coefficients are positive, and the second set is $[0.1, 0.2, 0.3, -0.2, 0.1]$, where W_4 has been changed, with respect to the first set of weights, from

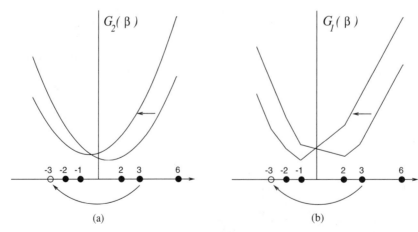

(a)   (b)

FIGURE 6   Effects of negative weighting on the cost functions $G_2(\beta)$ and $G_1(\beta)$. The input samples are $[x_1, x_2, x_3, x_4, x_5]^T = [-2, 2, -1, 3, 6]^T$, which are filtered by the two set of weights $[0.1, 0.2, 0.3, 0.2, 0.1]^T$ and $[0.1, 0.2, 0.3, -0.2, 0.1]^T$, respectively.

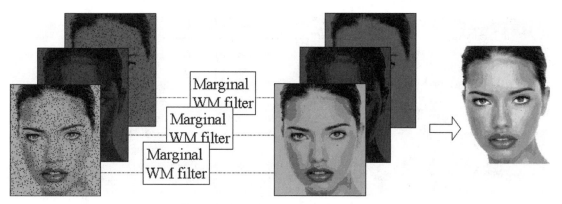

FIGURE 7 Center WM filter applied to each component independently. (See color section, p. C–2.)

0.2 to -0.2. Figure 6(a) shows the cost functions $G_2(\beta)$ of the linear FIR filter for the two sets of filter weights. Notice that by changing the sgn of W_4, we are effectively moving x_4 to its new location $\text{sgn}(W_4)x_4 = -3$. This, in turn, pulls the minimum of the cost function toward the relocated sample $\text{sgn}(W_4)x_4$. Negatively weighting x_4 on $G_1(\beta)$ has a similar effect, as shown in Fig. 6(b). In this case, the minimum is pulled toward the new location of $\text{sgn}(W_4)x_4$. The minimum, however, occurs at one of the samples $\text{sgn}(W_i)x_i$. More details on WM filtering can be found in [20, 21].

2.4 Vector Weighted Median Filters

The extension of the weighted median for use with color images is straightforward. Although sorting multicomponent (vector) pixel values and selecting the middle value is not well defined [22–26], as in the scalar case, the WM operation acting on multicomponent pixels resorts to the least absolute deviation sum definition of the WM operation [27]. Thus, the filter output is defined as the vector-valued sample that minimizes a weighted cost function.

Although we concentrate on the filtering of color images, the concepts defined in this section can also be applied to the filtering of N-component imagery [28]. Color images are represented by three components: red, green, and blue, with combinations of

these to produce the entire color spectrum. The weighted median filtering operation of a color image can be achieved in a number of ways [22–26, 29], two of which we summarize below.

2.4.1 Marginal WM filter

The simplest approach to WM filtering a color image is to process each component independently by a scalar WM filter. This operation is depicted in Fig. 7, where the green, blue, and red components of a color image are filtered independently and then combined to produce the filtered color image. A drawback associated with this method is that different components can be strongly correlated and, if each component is processed separately, this correlation is not exploited. In addition, since each component is filtered independently, the filter outputs can combine to produce colors not present in the original image. The advantage of marginal processing is the computational simplicity.

2.4.2 Vector WM filter

A more logical extension, yet significantly more computationally expensive approach, is found through the minimization of a weighted cost function that takes into account the multicomponent nature of the data. Here, the filtering operation processes all components jointly such that the cross-correlations

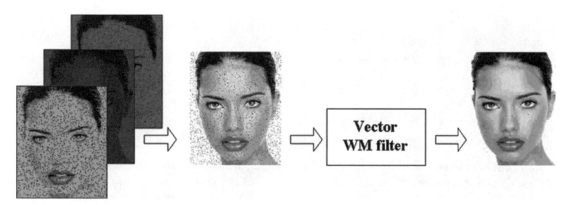

FIGURE 8 Center vector WM filter applied in the three-dimensional space. (See color section, p. C–2.)

between components are exploited. As is shown in Fig. 8, the three components are jointly filtered by a vector WM filter leading to a filtered color image. Vector WM filtering requires the extension of the original WM filter definition as follows. Define $\underline{x}_i = [x_i^1, x_i^2, x_i^3]^T$ as a three-dimensional vector, where x_i^1, x_i^2 and x_i^3 are respectively the red, green, and blue components of the ith pixel in a color image, and recall that the weighted median of a set of one-dimensional samples x_i $i = 1, \ldots, N$ is given by

$$\hat{\beta} = \arg \min_{\beta} \sum_{i=1}^{N} |W_i| |\mathrm{sgn}(W_i)x_i - \beta|. \qquad (28)$$

Extending this definition to a set of three-dimensional vectors $\underline{x}_i$ for $i = 1, \ldots, N$ leads to

$$\hat{\underline{\beta}} = \arg \min_{\underline{\beta}} \sum_{i=1}^{N} |W_i| \|\underline{s}_i - \underline{\beta}\| \qquad (29)$$

where $\hat{\underline{\beta}} = [\hat{\beta}^1, \hat{\beta}^2, \hat{\beta}^3]^T$, $\underline{s}_i = \mathrm{sgn}(W_i)\underline{x}_i$, and $\| \cdot \|$ is the L_2 norm defined as

$$\|\underline{s}_i - \underline{\beta}\| = \left[(s_i^1 - \beta^1)^2 + (s_i^2 - \beta^2)^2 + (s_i^3 - \beta^3)^2 \right]^{1/2}. \qquad (30)$$

The definition of the vector weighted median filter is very similar to that of the vector WM smoother introduced in [27]. Unlike the one-dimensional case, $\hat{\underline{\beta}}$ is not generally one of the $\underline{s}_i$; indeed, there is no closed form solution for $\hat{\underline{\beta}}$. Moreover, solving Eq. (29) involves a minimization problem in a three-dimensional space that can be computationally expensive. To overcome these shortcomings, a suboptimal solution for Eq. (29) is found if $\hat{\underline{\beta}}$ is restricted to be one of the sgned samples $\underline{s}_i$. This leads to the following definition:

The vector WM filter output of $\underline{x}_1, \ldots, \underline{x}_N$ is the value of $\hat{\underline{\beta}}$, with $\hat{\underline{\beta}} \in \{\underline{s}_1, \ldots, \underline{s}_N\}$ such that

$$\sum_{i=1}^{N} |W_i| \|\hat{\underline{\beta}} - \underline{s}_i\| \leq \sum_{i=1}^{N} |W_i| \|\underline{s}_j - \underline{s}_i\|, \quad \text{for all } j = 1, \ldots, N. \qquad (31)$$

This definition can be implemented as follows.

- For each signed sample $\underline{s}_j$, compute the distances to all the other sgned samples ($\|\underline{s}_j - \underline{s}_i\|$) for $i = 1, \ldots, N$ using Eq. (30).
- Compute the sum of the weighted distances given by the right side of Eq. (31).
- Choose as filter output the sample $\underline{s}_j$ that produces the minimum sum of the weighted distances.

Although vector WM filter, as presented above, is defined for color images, these definitions are readily adapted to filter any N-component imagery.

3 Image Noise Cleaning

Median smoothers are widely used in image processing to clean images corrupted by noise. Median filters are particularly effective at removing outliers. Often referred to as "salt-and-pepper" noise, outliers are often present because of bit errors in transmission, or they are introduced during the signal acquisition stage. Impulsive noise in images can also occur as a result to damage to analog film. Although a weighted median smoother can be designed to "best" remove the noise, CWM smoothers often provide similar results at a much lower complexity [12]. By simply tuning the center weight, a user can obtain the desired level of smoothing. Of course, as the center weight is decreased to attain the desired level of impulse suppresion, the output image will suffer increased distortion, particularly around the image's fine details. Nonetheless, CWM smoothers can be highly effective in removing salt-and-pepper noise while preserving the fine image details. Figures 9(a) and 9(b) depict a noise-free gray-scale image and the corresponding image with salt-and-pepper noise. Each pixel in the image has a 10% probability of being contaminated with an impulse. The impulses occur randomly and were generated by MATLAB's imnoise function. Figures 9(c) and 9(d) depict the noisy image processed with a 5×5 window CWM smoother with center weights 15 and 5, respectively. The impulse-rejection and detail-preservation tradeoff in CWM smoothing is clearly illustrated in Figs. 9(c) and 9(d). A color version of the "portrait" image was also corrupted by salt-and-pepper noise and filtered using CWM. Marginal CWM smoothing was performed in Fig. 10. The differences between marginal and vector WM processing will be illustrated shortly.

At the extreme, for $W_c = 1$, the CWM smoother reduces to the median smoother, which is effective at removing impulsive noise. It is, however, unable to preserve the image's fine details [30]. Figure 11 shows enlarged sections of the noise-free image (left), and of the noisy image after the median smoother has been applied (center). Severe blurring is introduced by the median smoother and readily apparent in Fig. 11. As a reference, the output of a running mean of the same size is also shown in Fig. 11 (right). The image is severely degraded as each impulse is smeared to neighboring pixels by the averaging operation.

Figures 9 and 10 show that CWM smoothers can be effective at removing impulsive noise. If increased detail preservation is sought and the center weight is increased, CWM smoothers begin to breakdown and impulses appear on the output. One simple way to ameliorate this limitation is to employ a recursive mode of operation. In essence, past inputs are replaced by previous outputs as described in Eq. (12), with the only difference being that only the center sample is weighted. All the other samples in the window are weighted by one. Figure 12 shows enlarged sections of the nonrecursive CWM filter (left) and of the corresponding recursive CWM smoother, both with the same center weight ($W_c = 15$). This figure illustrates the increased noise attenuation provided by recursion without the loss of image resolution.

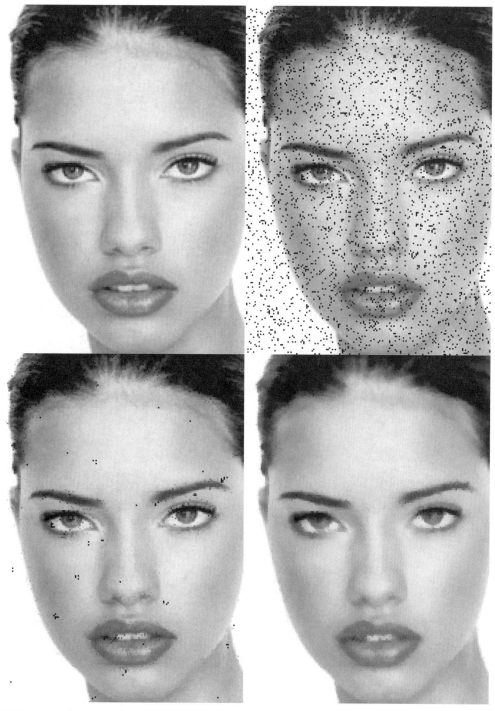

FIGURE 9 Impulse noise cleaning with a 5 × 5 CWM smoother: (a) original gray-scale "portrait" image, (b) image with salt-and-pepper noise, (c) CWM smoother with $W_c = 15$, (d) CWM smoother with $W_c = 5$.

Both recursive and nonrecursive CWM smoothers, can produced outputs with disturbing artifacts, particularly when the center weights are increased in order to improve the detail-preservation characteristics of the smoothers. The artifacts are most apparent around the image's edges and details. Edges at the output appear jagged, and impulsive noise can break through next to the image detail features. The distinct response of the CWM smoother in different regions of the image is due to the fact that images are nonstationary in nature. Abrupt changes in the image's local mean and texture carry most of the visual information content. CWM smoothers process the entire image with fixed weights and are inherently limited in this sense by their

FIGURE 10   Impulse noise cleaning with a 5 × 5 CWM smoother: (a) original "portrait" image, (b) image with salt- and-pepper noise, (c) CWM smoother with $W_c = 16$, (d) CWM smoother with $W_c = 5$. (See color section, p. C–3.)

static nature. Although some improvement is attained by introducing recursion or by using more weights in a properly designed WM smoother structure, these approaches are also static and do not properly address the nonstationarity nature of images.

Significant improvement in noise attenuation and detail preservation can be attained if permutation WM filter structures are used. Figure 12 (right) shows the output of the permutation CWM filter in Eq. (15) when the salt-and-pepper degraded "por-trait" image is inputted. The parameters were given the values $T_L = 6$ and $T_U = 20$. The improvement achieved by switching W_c between just two different values is significant. The impulses are deleted without exception, the details are preserved, and the jagged artifacts typical of CWM smoothers are not present in the output.

Figures 10–12 depict the results of marginal component filtering. Figure 13 illustrates the differences between marginal and

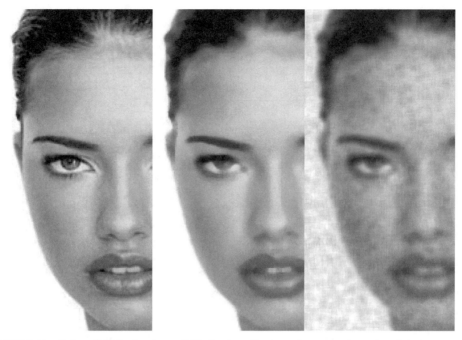

FIGURE 11 (Enlarged) Noise-free image (left), 5 × 5 median smoother output (center), and 5 × 5 mean smoother (right). (See color section, p. C–4.)

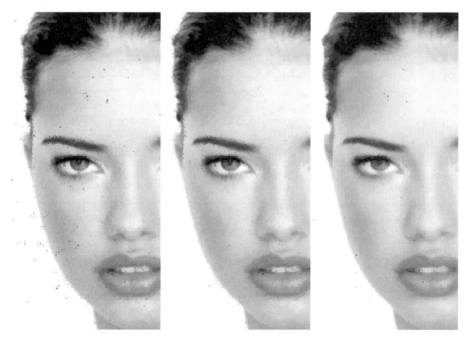

FIGURE 12 (Enlarged) CWM smoother output (left), recursive CWM smoother output (center), and permutation CWM smoother output (right). Window size is 5 × 5. (See color section, p. C-4.)

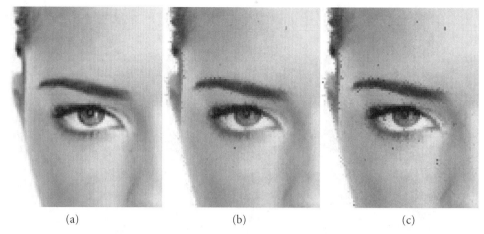

FIGURE 13   (a) Original image, (b) filtered image using a marginal WM filter, (c) filtered image using a vector WM filter. (See color section, p. C–4.)

vector processing. Figure 13(a) shows the original image, Fig. 13(b) shows the filtered image using marginal filtering, and Fig. 13(c) shows the filtered image using vector filtering. As Fig. 13 shows, the marginal processing of color images removes more noise than in the vector approach; however, it can introduce new color artifacts to the image.

4 Image Zooming

Zooming an image is an important task used in many applications, including the World Wide Web, digital video, DVDs, and scientific imaging. When zooming, pixels are inserted into the image in order to expand the size of the image, and the major task is the interpolation of the new pixels from the surrounding original pixels. Weighted medians have been applied to similar problems requiring interpolation, such as interlace to progressive video conversion for television systems [13]. The advantage of using the weighted median in interpolation over traditional linear methods is better edge preservation and less of a "blocky" look to edges.

To introduce the idea of interpolation, suppose that a small matrix must be zoomed by a factor of 2, and the median of the closest two (or four) original pixels is used to interpolate each new pixel:

$$
\begin{bmatrix} 7 & 8 & 5 \\ 6 & 10 & 9 \end{bmatrix}
\xrightarrow{\boxed{\text{Zero Interlace}}}
\begin{bmatrix} 7 & 0 & 8 & 0 & 5 & 0 \\ 0 & 0 & 0 & 0 & 0 & 0 \\ 6 & 0 & 10 & 0 & 9 & 0 \\ 0 & 0 & 0 & 0 & 0 & 0 \end{bmatrix}
$$

$$
\xrightarrow{\boxed{\text{Median Interpolation}}}
\begin{bmatrix} 7 & 7.5 & 8 & 6.5 & 5 & 5 \\ 6.5 & 7.5 & 9 & 8.5 & 7 & 7 \\ 6 & 8 & 10 & 9.5 & 9 & 9 \\ 6 & 8 & 10 & 9.5 & 9 & 9 \end{bmatrix}
$$

Zooming commonly requires a change in the image dimensions by a noninteger factor, such as a 50% zoom where the dimensions must be 1.5 times the original. Also, a change in the

length-to-width ratio might be needed if the horizontal and vertical zoom factors are different. The simplest way to accomplish zooming of arbitrary scale is to double the size of the original as many times as needed to obtain an image larger than the target size in all dimensions, interpolating new pixels on each expansion. Then the desired image can be attained by subsampling the larger image, or taking pixels at regular intervals from the larger image in order to obtain an image with the correct length and width. The subsampling of images and the possible filtering needed are topics well known in traditional image processing; thus we will focus on the problem of doubling the size of an image.

A digital image is represented by an array of values, each value defining the color of a pixel of the image. Whether the color is constrained to be a shade of gray, in which case only one value is needed to define the brightness of each pixel, or whether three values are needed to define the red, green, and blue components of each pixel does not affect the definition of the technique of weighted median interpolation. The only difference between gray-scale and color images is that an ordinary weighted median is used in gray-scale images whereas color requires a vector weighted median.

To double the size of an image, first an empty array is constructed with twice the number of rows and columns as the original [Fig. 14(a)], and the original pixels are placed into alternating rows and columns [the "00" pixels in Fig. 14(a)]. To interpolate the remaining pixels, the method known as polyphase interpolation is used. In the method, each new pixel with four original pixels at its four corners [the "11" pixels in Fig. 14(b)] is interpolated first by using the weighted median of the four nearest original pixels as the value for that pixel. Since all original pixels are equally trustworthy and the same distance from the pixel being interpolated, a weight of 1 is used for the four nearest original pixels. The resulting array is shown in Fig. 14(c). The remaining pixels are determined by taking a weighted median of the four closest pixels. Thus each of the "01" pixels in Fig. 14(c) is interpolated by using two original pixels to the left and right and

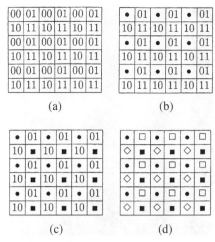

FIGURE 14 The steps of polyphase interpolation.

two previously interpolated pixels above and below. Similarly, the "10" pixels are interpolated with original pixels above and below and interpolated pixels ("11" pixels) to the right and left.

Since the "11" pixels were interpolated, they are less reliable than the original pixels and should be given lower weights in determining the "01" and "10" pixels. Therefore the "11" pixels are given weights of 0.5 in the median to determine the "01" and "10" pixels, while the "00" original pixels have weights of 1 associated with them. The weight of 0.5 is used because it implies that when both "11" pixels have values that are not between the two "00" pixel values then one of the "00" pixels or their average will be used. Thus "11" pixels differing from the "00" pixels do not greatly affect the result of the weighted median. Only when the "11" pixels lie between the two "00" pixels, they have a direct effect on the interpolation. The choice of 0.5 for the weight is arbitrary, since any weight greater than 0 and less than 1 will produce the same result. When the polyphase method is implemented, the "01" and "10" pixels must be treated differently because the orientation of the two closest original pixels is different for the two types of pixels. Figure 14(d) shows the final result of doubling the size of the original array.

To illustrate the process, consider an expansion of the gray-scale image represented by an array of pixels, the pixel in the ith row and jth column having brightness $a_{i,j}$. The array $a_{i,j}$ will be interpolated into the array $x_{i,j}^{pq}$, with p and q taking values 0 or 1, indicating in the same way as above the type of interpolation required:

$$
\begin{bmatrix} a_{1,1} & a_{1,2} & a_{1,3} \\ a_{2,1} & a_{2,2} & a_{2,3} \\ a_{3,1} & a_{3,2} & a_{3,3} \end{bmatrix}
$$

$$
\Longrightarrow
\begin{bmatrix}
x_{1,1}^{00} & x_{1,1}^{01} & x_{1,2}^{00} & x_{1,2}^{01} & x_{1,3}^{00} & x_{1,3}^{01} \\
x_{1,1}^{10} & x_{1,1}^{11} & x_{1,2}^{10} & x_{1,2}^{11} & x_{1,3}^{10} & x_{1,3}^{11} \\
x_{2,1}^{00} & x_{2,1}^{01} & x_{2,2}^{00} & x_{2,2}^{01} & x_{2,3}^{00} & x_{2,3}^{01} \\
x_{2,1}^{10} & x_{2,1}^{11} & x_{2,2}^{10} & x_{2,2}^{11} & x_{2,3}^{10} & x_{2,3}^{01} \\
x_{3,1}^{00} & x_{3,1}^{01} & x_{3,2}^{00} & x_{3,2}^{01} & x_{3,3}^{00} & x_{3,3}^{11} \\
x_{3,1}^{10} & x_{3,1}^{11} & x_{3,2}^{10} & x_{3,2}^{11} & x_{3,3}^{10} & x_{3,3}^{11}
\end{bmatrix}
$$

The pixels are interpolated as follows:

$$
x_{i,j}^{00} = a_{i,j}
$$

$$
x_{i,j}^{11} = \text{MEDIAN}[a_{i,j}, a_{i+1,j}, a_{i,j+1}, a_{i+1,j+1}]
$$

$$
x_{i,j}^{01} = \text{MEDIAN}[a_{i,j}, a_{i,j+1}, 0.5 \diamond x_{i-1,j}^{11}, 0.5 \diamond x_{i+1,j}^{11}]
$$

$$
x_{i,j}^{10} = \text{MEDIAN}[a_{i,j}, a_{i+1,j}, 0.5 \diamond x_{i,j-1}^{11}, 0.5 \diamond x_{i,j+1}^{11}].
$$

An example of median interpolation compared with bilinear interpolation is given in Fig. 15. Bilinear interpolation uses the average of the nearest two original pixels to interpolate the "01" and "10" pixels in Fig. 14(b) and the average of the nearest four original pixels for the "11" pixels. The edge-preserving advantage of the weighted median interpolation is readily seen in the figure.

5 Image Sharpening

Human perception is highly sensitive to edges and fine details of an image, and since they are composed primarily by high-frequency components, the visual quality of an image can be enormously degraded if the high frequencies are attenuated or completely removed. In contrast, enhancing the high-frequency components of an image leads to an improvement in the visual quality. *Image sharpening refers to any enhancement technique that highlights edges and fine details in an image.* Image sharpening is widely used in printing and photographic industries for increasing the local contrast and sharpening the images. In principle, image sharpening consists of adding to the original image a signal that is proportional to a high-pass filtered version of the original image. Figure 16 illustrates this procedure, often referred to as unsharp masking [31, 32], on a one-dimensional signal. As shown in Fig. 16, the original image is first filtered by a high-pass filter that extracts the high-frequency components, and then a scaled version of the high-pass filter output is added to the original image, thus producing a sharpened image of the original. Note that the homogeneous regions of the signal, i.e., where the signal is constant, remain unchanged. The sharpening operation can be represented by

$$
s_{i,j} = x_{i,j} + \lambda \mathcal{F}(x_{i,j}), \tag{32}
$$

where $x_{i,j}$ is the original pixel value at the coordinate (i, j), $\mathcal{F}(\cdot)$ is the high-pass filter, λ is a tuning parameter greater than or equal to zero, and $s_{i,j}$ is the sharpened pixel at the coordinate (i, j). The value taken by λ depends on the grade of sharpness desired. Increasing λ yields a more sharpened image.

If color images are used $x_{i,j}$, $s_{i,j}$, and λ are three-component vectors, whereas if gray-scale images are used $x_{i,j}$, $s_{i,j}$, and λ are single-component vectors. Thus the process described here can be applied to either gray-scale or color images, with the only difference being that vector filters have to be used in sharpening color images whereas single-component filters are used with gray-scale images.

FIGURE 15 Example of zooming. Original is at the top with the area of interest outlined in white. On the lower left is the bilinear interpolation of the area, and on the lower right the WM interpolation.

The key point in the effective sharpening process lies in the choice of the high-pass filtering operation. Traditionally, linear filters have been used to implement the high-pass filter; however, linear techniques can lead to unacceptable results if the original image is corrupted with noise. A tradeoff between noise attenuation and edge highlighting can be obtained if a weighted median filter with appropriated weights is used. To illustrate this, consider a WM filter applied to a gray-scale image where the following filter mask is used:

$$W = \frac{1}{3} \begin{bmatrix} -1 & -1 & -1 \\ -1 & 8 & -1 \\ -1 & -1 & -1. \end{bmatrix}. \qquad (33)$$

Because of the weight coefficients in Eq. (33), for each position of the moving window, the output is proportional to the difference between the center pixel and the smallest pixel around the center pixel. Thus, the filter output takes relatively large values for prominent edges in an image, and small values in regions that are fairly smooth, being zero only in regions that have a constant gray level.

Although this filter can effectively extract the edges contained in an image, the effect that this filtering operation has over negative-slope edges is different from that obtained for positive-slope edges.[1] Since the filter output is proportional to the

---

[1] A change from a gray level to a lower gray level is referred to as a negative-slope edge, whereas a change from a gray level to a higher gray level is referred to as a positive-slope edge.

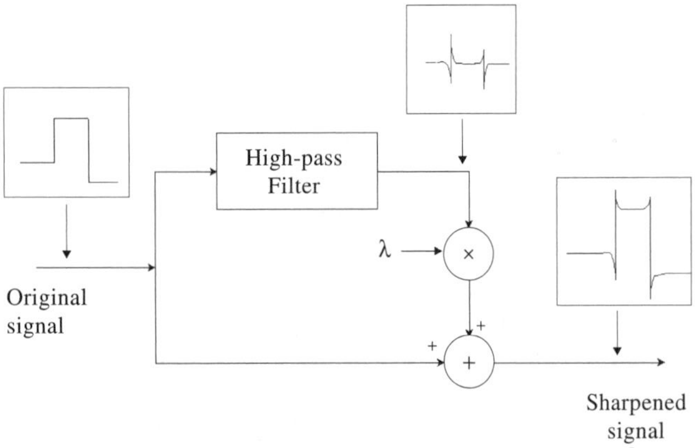

FIGURE 16 Image sharpening by high-frequency emphasis.

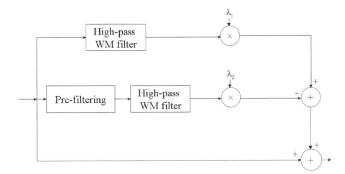

FIGURE 17 Image sharpening based on the weighted median filter.

difference between the center pixel and the smallest pixel around the center, for negative-slope edges, the center pixel takes small values producing small values at the filter output. Moreover, the filter output is zero if the smallest pixel around the center pixel and the center pixel have the same values. This implies that negative-slope edges are not extracted in the same way as positive-slope edges. To overcome this limitation the basic image sharpening structure shown in Fig. 16 must be modified such that positive-slope edges as well as negative-slope edges are highlighted in the same proportion. A simple way to accomplish that is: (a) extract the positive-slope edges by filtering the original image with the filter mask described above; (b) extract the negative-slope edges by first preprocessing the original image such that the negative-slope edges become positive-slope edges, and then filter the preprocessed image with the filter described above; (c) combine appropriately the original image, the filtered version of the original image, and the filtered version of the preprocessed image to form the sharpened image.

Thus both positive-slope edges and negative-slope edges are equally highlighted. This procedure is illustrated in Fig. 17, where the top branch extracts the positive-slope edges and the middle branch extracts the negative-slope edges. In order to understand the effects of edge sharpening, a row of a test image is plotted in Fig. 18 together with a row of the sharpened image when only the positive-slope edges are highlighted, Fig. 18(a), only

the negative-slope edges are highlighted, Fig. 18(b), and both positive-slope and negative-slope edges are jointly highlighted, Fig. 18(c).

In Fig. 17, λ_1 and λ_2 are tuning parameters that control the amount of sharpness desired in the positive-slope direction and in the negative-slope direction, respectively. The values of λ_1 and λ_2 are generally selected to be equal. The output of the prefiltering operation is defined as

$$x'_{i,j} = M - x_{i,j}, \qquad (34)$$

with M equal to the maximum pixel value of the original image. This prefiltering operation can be thought of as a flipping and a shifting operation of the values of the original image such that the negative-slope edges are converted in positive-slope edges. Since the original image and the pre-filtered image are filtered by the same WM filter, the positive-slope edges and negative-slopes edges are sharpened in the same way.

In Fig. 19, the performance of the WM filter image sharpening is compared with that of traditional image sharpening based on linear FIR filters. For the linear sharpener, the scheme shown in Fig. 16 was used. The parameter λ was set to 1 for the clean image and to 0.75 for the noise image. For the WM sharpener, the scheme of Fig. 17 was used with $\lambda_1 = \lambda_2 = 2$ for the clean image, and $\lambda_1 = \lambda_2 = 1.5$ for the noise image. The filter mask given by Eq. (33) was used in both linear and median image sharpening. As before, each component of the color image was processed separately.

6 Edge Detection

Edge detection is an important tool in image analysis, and it is necessary for applications of computer vision in which objects have to be recognized by their outlines. An edge-detection algorithm should show the locations of major edges in the image while ignoring false edges caused by noise. The most common approach used for edge detection is illustrated in Fig. 20. A

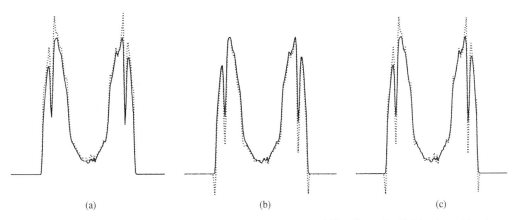

    (a)   (b)   (c)

FIGURE 18 Original row of a test image (solid curve) and row sharpened (dotted curve) with (a) only positive-slope edges, (b) only negative-slope edges, and (c) both positive and negative-slope edges.

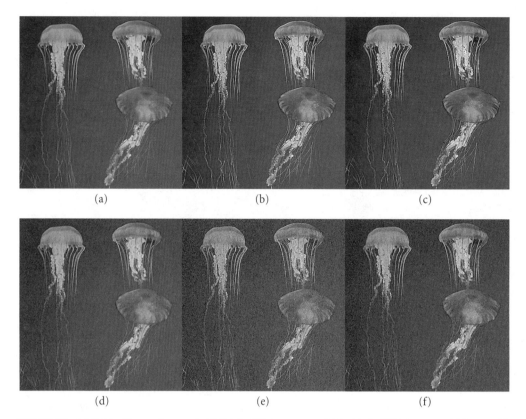

FIGURE 19   (a) Original image sharpened with (b) the FIR sharpener, and (c) with the WM sharpener. (d) Image with added Gaussian noise sharpened with (e) the FIR sharpener, and (f) the WM sharpener. (See color section, p. C–5.)

high-pass filter is applied to the image to obtain the amount of change present in the image at every pixel. The output of the filter is thresholded to determine those pixels that have a high enough rate of change to be considered lying on an edge; i.e., all pixels with filter output greater than some value T are taken as edge pixels. The value of T is a tunable parameter that can be adjusted to give the best visual results. High thresholds lose some of the real edges, while low values result in many false edges; thus a tradeoff has to be made to get the best results. Other techniques such as edge thinning can be applied to further pinpoint the location of the edges in an image.

The most common linear filter used for the initial high-pass filtering is the Sobel operator, which uses the following 3×3 masks:

$$\begin{bmatrix} -1 & -2 & -1 \\ 0 & 0 & 0 \\ 1 & 2 & 1 \end{bmatrix} \begin{bmatrix} -1 & 0 & 1 \\ -2 & 0 & 2 \\ -1 & 0 & 1 \end{bmatrix}$$

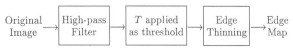

FIGURE 20   The process of edge detection.

These two masks, called Sobel masks, are convolved with the image separately to measure the strength of horizontal edges and vertical edges, respectively, present at each pixel. Thus if the amount to which a horizontal edge is present at the pixel in the ith row and jth column is represented as $E_{i,j}^{h}$, and if the vertical edge indicator is $E_{i,j}^{v}$, then the values are:

$$E_{i,j}^{h} = -x_{i-1,j-1} - 2x_{i-1,j} - x_{i-1,j+1} + x_{i+1,j-1}$$
$$+ 2x_{i+1,j} + x_{i+1,j+1}$$
$$E_{i,j}^{v} = -x_{i-1,j-1} - 2x_{i,j-1} - x_{i+1,j-1} + x_{i-1,j+1}$$
$$+ 2x_{i,j+1} + x_{i+1,j+1}$$

The two strengths are combined to find the total amount to which any edge exists at the pixel: $E_{i,s}^{total} = \sqrt{(E_{i,s}^{h})^2 + (E_{i,s}^{v})^2}$. This value is then compared to the threshold T to determine the existence of an edge.

In place of the use of linear high-pass filters, weighted median filters can be used. To apply weighted medians to the high-pass filtering, the weights from the Sobel masks can be used. The Sobel linear high-pass filters take a weighted difference between the pixels on either side of $x_{i,j}$. In contrast, if the same weights are used in a weighted median filter, the value returned is the difference between the lowest-valued pixels on either side of $x_{i,j}$.

FIGURE 21 (a) Original image, (b) edge detector using linear method, and (c) median method.

If the pixel values are then flipped about some middle value, the difference between the *highest* pixels on either side can also be obtained. The flipping can be achieved by finding some maximum pixel value M and using $x'_{i,j} = M - x_{i,j}$ as the "flipped" value of $x_{i,j}$, thus causing the highest values to become the lowest. The lower of the two differences across the pixel can then be used as the indicator of the presence of an edge. If there is a true edge present, then both differences should be high in magnitude, while if noise causes one of the differences to be too high, the other difference is not necessarily affected. Thus the horizontal and vertical edge indicators are:

$$E_{i,j}^{h} =$$
$$\min\left(\text{MEDIAN}\begin{bmatrix} -1 \diamond x_{i-1,j-1}, & -2 \diamond x_{i-1,j}, & -1 \diamond x_{i-1,j+1}, \\ 1 \diamond x_{i+1,j-1}, & 2 \diamond x_{i+1,j}, & 1 \diamond x_{i+1,j+1} \end{bmatrix},\right.$$
$$\left.\text{MEDIAN}\begin{bmatrix} -1 \diamond x'_{i-1,j-1}, & -2 \diamond x'_{i-1,j}, & -1 \diamond x'_{i-1,j+1} \\ 1 \diamond x'_{i+1,j-1}, & 2 \diamond x'_{i+1,j}, & 1 \diamond x'_{i+1,j+1} \end{bmatrix}\right)$$

$$E_{i,j}^{v} = \min\left(\text{MEDIAN}\begin{bmatrix} -1 \diamond x_{i-1,j-1}, & 1 \diamond x_{i-1,j+1}, \\ -2 \diamond x_{i,j-1}, & 2 \diamond x_{i,j+1}, \\ -1 \diamond x_{i+1,j-1}, & 1 \diamond x_{i+1,j+1}, \end{bmatrix},\right.$$
$$\left.\text{MEDIAN}\begin{bmatrix} -1 \diamond x'_{i-1,j-1}, & 1 \diamond x'_{i-1,j+1}, \\ -2 \diamond x'_{i,j-1}, & 2 \diamond x'_{i,j+1}, \\ -1 \diamond x'_{i+1,j-1}, & 1 \diamond x'_{i+1,j+1} \end{bmatrix}\right)$$

and the strength of horizontal and vertical edges $E_{(i,j)}^{h,v}$ is determined in the same way as the linear case: $E_{i,j}^{h,v} = \sqrt{(E_{i,j}^{h})^2 + (E_{i,j}^{v})^2}$.

Another addition to the weighted median method is necessary in order to detect diagonal edges. Horizontal and vertical indicators are not sufficient to register diagonal edges, so the following two masks must also be used:

$$\begin{bmatrix} -2 & -1 & 0 \\ -1 & 0 & 1 \\ 0 & 1 & 2 \end{bmatrix} \begin{bmatrix} 0 & 1 & 2 \\ -1 & 0 & 1 \\ -2 & -1 & 0 \end{bmatrix}.$$

These masks can be applied to the image just as the Sobel masks above. Thus the strengths of the two types of diagonal edges are

$E_{i,j}^{d1}$ for diagonal edges going from the bottom left of the image to the top right (using the mask on the left above) and $E_{i,j}^{d2}$ for diagonal edges from top left to bottom right (the mask on the right), and the values are given by

$$E_{i,j}^{d1} = \min\left(\text{MEDIAN}\begin{bmatrix} -2 \diamond x_{i-1,j-1}, & -1 \diamond x_{i-1,j}, \\ -1 \diamond x_{i,j-1}, & 1 \diamond x_{i,j+1}, \\ 1 \diamond x_{i+1,j}, & 2 \diamond x_{i+1,j+1}, \end{bmatrix},\right.$$
$$\left.\text{MEDIAN}\begin{bmatrix} -2 \diamond x'_{i-1,j-1}, & -1 \diamond x'_{i-1,j}, \\ -1 \diamond x'_{i,j-1}, & 1 \diamond x'_{i,j+1}, \\ 1 \diamond x'_{i+1,j}, & 2 \diamond x'_{i+1,j+1} \end{bmatrix}\right)$$

$$E_{i,j}^{d2} = \min\left(\text{MEDIAN}\begin{bmatrix} 1 \diamond x_{i-1,j}, & 2 \diamond x_{i-1,j+1}, \\ -1 \diamond x_{i,j-1}, & 1 \diamond x_{i,j+1}, \\ -2 \diamond x_{i+1,j-1}, & -1 \diamond x_{i+1,j} \end{bmatrix},\right.$$
$$\left.\text{MEDIAN}\begin{bmatrix} 1 \diamond x'_{i-1,j}, & 2 \diamond x'_{i-1,j+1}, \\ -1 \diamond x'_{i,j-1}, & 1 \diamond x'_{i,j+1}, \\ -2 \diamond x'_{i+1,j-1}, & -1 \diamond x'_{i+1,j} \end{bmatrix}\right).$$

A diagonal edge strength is determined in the same way as the horizontal and vertical edge strength above: $E_{i,j}^{d1,d2} = \sqrt{(E_{i,j}^{d1})^2 + (E_{i,j}^{d2})^2}$. The indicator of all edges in any direction is the maximum of the two strengths $E_{i,j}^{h,v}$ and $E_{i,j}^{d1,d2}$: $E_{i,j}^{total} = \max(E_{i,j}^{h,v}, E_{i,j}^{d1,d2})$. As in the linear case, this value is compared to the threshold T to determine whether a pixel lies on an edge. Figure 21 shows the results of calculating $E_{i,j}^{total}$ for an image. The results of the median edge detection are similar to the results of using the Sobel linear operator. Other approaches for edge detector based on median filter can be found in [33–36].

7 Conclusion

The principles behind WM smoothers and WM filters have been presented in this article, as well as some of the applications of these nonlinear signal processing structures in image enhancement. It should be apparent to the reader that many similarities exist between linear and median filters. As illustrated in this article, there are several applications in image enhancement were

WM filters provide significant advantages over traditional image enhancement methods using linear filters. The methods presented here, and other image enhancement methods that can be easily developed using WM filters, are computationally simple and provide significant advantages, and consequently can be used in emerging consumer electronic products, PC and internet imaging tools, medical and biomedical imaging systems, and of course in military applications.

Acknowledgment

This work was supported in part by the National Science Foundation under grant MIP-9530923.

References

[1] Y. Lee and S. Kassam, "Generalized median filtering and related nonlinear filtering techniques," *IEEE Trans. on Acoustics, Speech and Signal Processing*, ASSP-**33**, June 1985.

[2] J. W. Tukey, "Nonlinear (nonsuperimposable) methods for smoothing data," in *Conf. Rec.*, (Eascon), 1974.

[3] T. A. Nodes and N. C. Gallagher, Jr., "Median filters: some modifications and their properties," *IEEE Transactions on Acoustics, Speech, and Signal Processing* **29**, Oct. 1982.

[4] G. R. Arce and N. C. Gallagher, " Stochastic analysis of the recursive median filter process," *IEEE Transactions on Information Theory*, IT-**34**, July 1988.

[5] G. R. Arce, N. C. Gallagher, Jr., and T. A. Nodes, "Median filters: theory and aplications," in *Advances in Computer Vision and Image Processing* (T. S. Huang, ed.) **2**, JAI Press, 1986.

[6] E. Lehmann, *Theory of Point Estimation* (New York, J. Wiley) 1983.

[7] A. Bovik, T. Huang, and D. Munson, "A generalization of median filtering using linear combinations of order statistics," *IEEE Trans. on Acoustics, Speech and Signal Processing*, ASSP-**31**, Dec. 1983.

[8] H. A. David, *Order Statistics* (New York, J. Wiley, 1981).

[9] B. C. Arnold, N. Balakrishnan, and H. Nagaraja, *A First Course in Order Statistics* (New York, J. Wiley, 1992).

[10] F. Y. Edgeworth, "A new method of reducing observations relating to several quantities," *Phil. Mag. (Fifth Series)* **24**, 1887.

[11] D. R. K. Brownrigg, "The weighted median filter," *Commun. Assoc. Comput. Mach.* **27**, Aug. 1984.

[12] S.-J. Ko and Y. H. Lee, "Center weighted median filters and their applications to image enhancement," *IEEE Transactions on Circuits and Systems* **38**, Sept. 1991.

[13] L. Yin, R. Yang, M. Gabbouj, and Y. Neuvo, "Weighted median filters: a tutorial," *Trans. on circuits and Systems II* **41**, May 1996.

[14] O. Yli-Harja, J. Astola, and Y. Neuvo, "Analysis of the properties of median and weighted median filters using threshold logic and stack filter representation," *IEEE Transactions on Acoustics, Speech, and Signal Processing* **39**, Feb. 1991.

[15] G. R. Arce, T. A. Hall, and K. E. Barner, "Permutation weighted order statistic filters," *IEEE Transactions on Image Processing* **4**, Aug. 1995.

[16] R. C. Hardie and K. E. Barner, "Rank conditioned rank selection filters for signal restoration," *IEEE Transactions on Image Processing* **3**, Mar. 1994.

[17] J. Fitch, E. Coyle, and N. Gallagher, "Median filtering by threshold decomposition," *IEEE Trans. Acoustics, Speech and Signal Processing*, ASSP-**32**, Dec. 1984.

[18] P. Wendt, E. J. Coyle, and N. C. Gallagher, Jr., "Stack filters," *IEEE Transactions on Acoustics, Speech, and Signal Processing* **34**, Aug. 1986.

[19] E. N. Gilbert, "Lattice-theoretic properties of frontal switching functions," *J. Math. Phys.* **33**, Apr. 1954.

[20] G. R. Arce, "A general weighted median filter structure admitting real-valued weights," *IEEE Transactions on Signal Processing*, SP-**46**, DEc. 1998.

[21] J. L. Paredes and G. R. Arce, "Stack filters, stack smoothers, and mirror threshold decomposition," *IEEE Transactions on Signal Processing* **47**(10), Oct. 1999.

[22] V. Barnett, "The ordering of multivariate data," *J. Royal Statistical Society* **139**, Part 3, 1976.

[23] P. E. Trahanias, D. Karakos, and A. N. Venetsanopoulos, "Directional Processing of Color Images: Theory and Experimental Result," *IEEE Transactions on Image Processing* **5**, June 1996.

[24] R. Hardie and G. R, Arce, "Ranking in R^p and its use in multivariate image estimation," *IEEE Transactions on Circuits Syst. for Video Technol.* **1**, June 1991.

[25] I. Pitas and P. Tsakalides, "Multivariate ordering in color image filtering," *IEEE Transactions on Circuits Syst. for Video Technol.* **1**, Feb. 1991.

[26] V. Koivunen, "Nonlinear Filtering of Multivariate Images Under Robust Error Criterion," *IEEE Transactions on Image Processing* **5**, June 1996.

[27] J. Astola, P. H. L. Yin, and Y. Neuvo, "Vector median filters," *Proceedings of the IEEE* **78**, April 1990.

[28] V. Koivunen, N. Himayat, and S. Kassam, "Nonlinear filtering techniques for multivariate images — Design and robustness characterization," *Signal Processing* **57**, Feb. 1997.

[29] C. Small, "A survey of multidimensional medians," *Int. Stat. Rev.* **58**, no. 3, 1990.

[30] A. Bovik, "Streaking in median filtered images," *IEEE Trans. on Acoustics, Speech and Signal Processing*, ASSP-**35**, Apr 1987.

[31] A. K. Jain, *Fundamentals of Digital Image Processing* (Englewood Cliffs, New Jersey, Prentice Hall, 1989).

[32] J. S. Lim, *Two-dimensional Signal and Image Processing* (Englewood Cliffs, New Jersey, Prentice Hall, 1990).

[33] I. Pitas and A. Venetsanopoulos, "Edge detectors based on order statistics," *IEEE Trans. Pattern Analysis and Machine Intelligence*, PAMI-**8**, July 1986.

[34] A. Bovik and D. Munson, "Edge detection using median comparisons," *Computer Vision Graphics and Image Processing* **33**, July 1986.

[35] D. Lau and G. R. Arce, "Edge detector using weighted median filter," Tech. Rep. 97-05-15, Department of Computer and Electrical Engineering, University of Delaware, 1997.

[36] I. Pitas and A. Venetsanopoulos, "Nonlinear order statistic filters for image and edge detection," *Signal Processing* **10**, no. 4, 1986.

3.3

Morphological Filtering for Image Enhancement and Detection

Petros Maragos
National Technical University of Athens

Lúcio F. C. Pessoa
Motorola, Inc.

1 Introduction

The goals of image enhancement include the improvement of the visibility and perceptibility of the various regions into which an image can be partitioned and of the detectability of the image features inside these regions. These goals include tasks such as cleaning the image from various types of noise; enhancing the contrast among adjacent regions or features; simplifying the image by means of selective smoothing or elimination of features at certain scales; and retaining only features at certain desirable scales. While traditional approaches for solving these above tasks have used mainly tools of linear systems, there is a growing understanding that linear approaches are not well suitable or even fail to solve problems involving geometrical aspects of the image. Thus there is a need for nonlinear approaches. A powerful nonlinear methodology that can successfully solve these problems is mathematical morphology.

Mathematical morphology is a set- and lattice-theoretic methodology for image analysis, which aims at quantitatively describing the geometrical structure of image objects. It was initiated [17] in the late 1960's to analyze binary images from geological and biomedical data as well as to formalize and extend earlier or parallel work [12, 13] on binary pattern recognition based on cellular automata and Boolean/threshold logic. In the late 1970's it was extended to gray-level images [17]. In the mid-1980's it was brought to the mainstream of image/signal processing and related to other nonlinear filtering approaches [7, 8]. Finally, in the late 1980's and 1990's it was generalized to arbitrary lattices [2, 18]. The above evolution of ideas has formed what we call nowadays the field of *morphological image processing*, which is a broad and coherent collection of theoretical concepts, nonlinear filters, design methodologies, and applications systems. Its rich theoretical framework, algorithmic efficiency, easy implementability on special hardware, and suitability for many shape-oriented problems have propelled its widespread usage and further advancement by many academic and industry groups working on various problems in image processing, computer vision, and pattern recognition.

This chapter provides a brief introduction to the application of morphological image processing to image enhancement and detection. There are several motivations for using morphological filters for such problems. First, it is of paramount importance to preserve, uncover, or detect the geometric structure of image objects. Thus, morphological filters, which are more suitable than linear filters for shape analysis, play a major role for geometry-based enhancement and detection. Further, they offer efficient solutions to other nonlinear tasks such as non-Gaussian noise suppression. This task can also be accomplished (with similar performance) by a closely related class of nonlinear systems, the median, rank, and stack filters, which also outperform linear filters in non-Gaussian noise suppression. Finally, the elementary morphological operators[1] are the building blocks for large classes of nonlinear image processing systems, which include rank and stack filters.

2 Morphological Image Operators

2.1 Morphological Filters for Binary Images

Given a sampled[2] binary image signal $f[x]$ with values 1 for the image object and 0 for the background, typical image transformations involving a moving window set $W = \{y_1, y_2, \ldots, y_n\}$ of n sample indexes would be

$$\psi_b(f)[x] = b(f[x - y_1], \ldots, f[x - y_n]), \quad (1)$$

where $b(v_1, \ldots, v_n)$ is a Boolean function of n variables. The mapping $f \mapsto \psi_b(f)$ is called a *Boolean filter*. When the Boolean function b is varied, a large variety of Boolean filters can be obtained. For example, choosing a Boolean AND for b would *shrink* the input image object, whereas a Boolean OR would *expand* it. Numerous other Boolean filters are possible, since there are 2^{2^n} possible Boolean functions of n variables. The main applications of such Boolean image operations have been in biomedical image processing, character recognition, object detection, and general two-dimensional (2-D) shape analysis [12, 13].

Among the important concepts offered by mathematical morphology was to use *sets* to represent binary images and set operations to represent binary image transformations. Specifically, given a binary image, let the object be represented by the set X and its background by the set complement X^c. The Boolean OR transformation of X by a (window) set B is equivalent to the Minkowski set addition $\oplus$, also called **dilation**, of X by B:

$$X \oplus B \equiv \{x + y : x \in X, y \in B\} = \bigcup_{y \in B} X_{+y} \quad (2)$$

where $X_{+y} \equiv \{x + y : x \in X\}$ is the *translation* of X along the

vector y. Likewise, if $B^r \equiv \{x : -x \in B\}$ is the *reflection* of B with respect to the origin, the Boolean AND transformation of X by B^r is equivalent to the Minkowski set subtraction $\ominus$, also called **erosion**, of X by B:

$$X \ominus B \equiv \{x : B_{+x} \subseteq X\} = \bigcap_{y \in B} X_{-y}. \quad (3)$$

Cascading erosion and dilation creates two other operations, the *opening*, $X \circ B \equiv (X \ominus B) \oplus B$, and the *closing*, $X \bullet B \equiv (X \oplus B) \ominus B$, of X by B. In applications, B is usually called a *structuring element* and has a simple geometrical shape and a size smaller than the image X. If B has a regular shape, e.g., a small disk, then both opening and closing act as nonlinear filters that smooth the contours of the input image. Namely, if X is viewed as a flat island, the opening suppresses the sharp capes and cuts the narrow isthmuses of X, whereas the closing fills in the thin gulfs and small holes.

There is a *duality* between dilation and erosion since $X \oplus B = (X^c \ominus B^r)^c$; i.e., dilation of an image object by B is equivalent to eroding its background by B^r and complementing the result. A similar duality exists between closing and opening.

2.2 Morphological Filters for Gray-Level Images

Extending morphological operators from binary to gray-level images can be done by using set representations of signals and transforming these input sets by means of morphological set operations. Thus, consider an image signal $f(x)$ defined on the continuous or discrete plane $\mathbb{ID} = \mathbb{R}^2$ or $\mathbb{Z}^2$ and assuming values in $\overline{\mathbb{R}} = \mathbb{R} \cup \{-\infty, \infty\}$. Thresholding f at all amplitude levels v produces an ensemble of binary images represented by the *threshold sets*,

$$\Theta_v(f) \equiv \{x \in \mathbb{ID} : f(x) \geq v\}, \quad -\infty < v < +\infty. \quad (4)$$

The image can be exactly reconstructed from all its threshold sets since

$$f(x) = \sup\{v \in \mathbb{R} : x \in \Theta_v(f)\}, \quad (5)$$

where "sup" denotes supremum.[3] Transforming each threshold set of the input signal f by a set operator Ψ and viewing the transformed sets as threshold sets of a new image creates [7, 17] a *flat* image operator ψ, whose output signal is

$$\psi(f)(x) = \sup\{v \in \mathbb{R} : x \in \Psi[\Theta_v(f)]\}. \quad (6)$$

For example, if Ψ is the set dilation and erosion by B, the above procedure creates the two most elementary morphological image

---

[1]The term "morphological operator," which means a morphological signal transformation, shall be used interchangeably with "morphological filter," in analogy to the terminology "rank or linear filter."

[2]Signals of a continuous variable $x \in \mathbb{R}^d$ are usually denoted by $f(x)$, whereas for signals with discrete variable $x \in \mathbb{Z}$ we write $f[x]$. $\mathbb{Z}$ and $\mathbb{R}$ denote, respectively, the set of reals and integers.

[3]Given a set X of real numbers, the *supremum* of X is its lowest upper bound. If X is finite (or infinite but closed from above), its supremum coincides with its maximum.

operators: the dilation and erosion of $f(x)$ by a set B:

$$(f \oplus B)(x) \equiv \bigvee_{y \in B} f(x - y) \tag{7}$$

$$(f \ominus B)(x) \equiv \bigwedge_{y \in B} f(x + y) \tag{8}$$

where $\bigvee$ denotes supremum (or maximum for finite B) and $\bigwedge$ denotes infimum (or minimum for finite B). Flat erosion (dilation) of a function f by a small convex set B reduces (increases) the peaks (valleys) and enlarges the minima (maxima) of the function. The flat opening $f \circ B = (f \ominus B) \oplus B$ of f by B smooths the graph of f from below by cutting down its peaks, whereas the closing $f \bullet B = (f \oplus B) \ominus B$ smooths it from above by filling up its valleys.

The most general translation-invariant morphological dilation and erosion of a gray-level image signal $f(x)$ by another signal g are:

$$(f \oplus g)(x) \equiv \bigvee_{y \in \mathbb{D}} f(x - y) + g(y), \tag{9}$$

$$(f \ominus g)(x) \equiv \bigwedge_{y \in \mathbb{D}} f(x + y) - g(y). \tag{10}$$

Note that signal dilation is a nonlinear convolution where the sum of products in the standard linear convolution is replaced by a max of sums.

2.3 Universality of Morphological Operators[4]

Dilations or erosions can be combined in many ways to create more complex morphological operators that can solve a broad variety of problems in image analysis and nonlinear filtering. Their versatility is further strengthened by a theory outlined in [7, 8] that represents a broad class of nonlinear and linear operators as a minimal combination of erosions or dilations. Here we summarize the main results of this theory, restricting our discussion only to discrete 2-D image signals.

Any *translation-invariant* set operator Ψ is uniquely characterized by its *kernel*, $\mathrm{Ker}(\Psi) \equiv \{X \in \mathbb{Z}^2 : 0 \in \Psi(X)\}$. The kernel representation requires an infinite number of erosions or dilations. A more efficient (requiring less erosions) representation uses only a substructure of the kernel, its *basis*, $\mathrm{Bas}(\Psi)$, defined as the collection of kernel elements that are *minimal* with respect to the partial ordering $\subseteq$. If Ψ is also *increasing* (i.e., $X \subseteq Y \implies \Psi(X) \subseteq \Psi(Y)$) and *upper semicontinuous* (i.e., $\Psi(\cap_n X_n) = \cap_n \Psi(X_n)$ for any decreasing set sequence X_n), then Ψ has a nonempty basis and can be represented exactly as a union of erosions by its basis sets:

$$\Psi(X) = \bigcup_{A \in \mathrm{Bas}(\Psi)} X \ominus A. \tag{11}$$

The morphological basis representation has also been extended to gray-level signal operators. As a special case, if ϕ is a flat signal operator as in Eq. (6) that is translation invariant and commutes with thresholding, then ϕ can be represented as a supremum of erosions by the basis sets of its corresponding set operator Φ:

$$\phi(f) = \bigwedge_{A \in \mathrm{Bas}(\Phi)} f \ominus A. \tag{12}$$

By duality, there is also an alternative representation where a set operator Ψ satisfying the above three assumptions can be realized exactly as the intersection of dilations by the reflected basis sets of its *dual operator* $\Psi^d(X) \equiv [\Psi(X^c)]^c$. There is also a similar dual representation of signal operators as an infimum of dilations.

Given the wide applicability of erosions/dilations, their parallelism, and their simple implementations, the morphological representation theory supports a general purpose image processing (software or hardware) module that can perform erosions/dilations, based on which numerous other complex image operations can be built.

2.4 Median, Rank, and Stack Filters

Flat erosion and dilation of a discrete image signal $f[x]$ by a finite window $W = \{y_1, \ldots, y_n\} \subseteq \mathbb{Z}^2$ is a moving local minimum or maximum. Replacing min/max with a more general rank leads to rank filters. At each location $x \in \mathbb{Z}^2$, sorting the signal values within the reflected and shifted n-point window $(W^r)_{+x}$ in decreasing order and picking the pth largest value, $p = 1, 2, \ldots, n$, yields the output signal from the pth *rank filter*:

$$(f \square_p W)[x] \equiv p\text{th rank of } (f[x - y_1], \ldots, f[x - y_n]). \tag{13}$$

For odd n and $p = (n + 1)/2$ we obtain the *median* filter. Rank filters and especially medians have been applied mainly to suppress impulse noise or noise whose probability density has heavier tails than the Gaussian for enhancement of image and other signals, since they can remove this type of noise without blurring edges, as would be the case for linear filtering. A discussion of median-type filters can be found in Chapter 3.2.

If the input image is binary, the rank filter output is also binary since sorting preserves a signal's range. Rank filtering of binary images involves only counting of points and no sorting. Namely, if the set $S \subseteq \mathbb{Z}^2$ represents an input binary image, the output set produced by the pth *rank set filter* is

$$S \square_p W \equiv \{x : \mathrm{card}((W^r)_{+x} \cap S) \geq p\}, \tag{14}$$

where $\mathrm{card}(X)$ denotes the cardinality (i.e., number of points) of a set X.

All rank operators *commute with thresholding*; i.e.,

$$\Theta_v[f \square_p W] = [\Theta_v(f)] \square_p W, \qquad \forall v, \forall p, \tag{15}$$

where $\Theta_v(f)$ is the binary image resulting from thresholding f at level v. This property is also shared by all morphological operators that are finite compositions or maxima/minima of flat dilations and erosions by finite structuring elements. All such signal operators ψ that have a corresponding set operator Ψ and commute with thresholding can be alternatively implemented by means of *threshold superposition* as in Eq. (6). Further, since the binary version of all the above discrete translation-invariant finite-window operators can be described by their generating Boolean function as in Eq. (1), all that is needed in synthesizing their corresponding gray-level image filters is knowledge of this Boolean function. Specifically, let $f_v[x]$ be the binary images represented by the threshold sets $\Theta_v(f)$ of an input gray-level image $f[x]$. Transforming all f_v with an increasing (i.e., containing no complemented variables) Boolean function $b(u_1, \ldots, u_n)$ in place of the set operator Ψ in Eq. (6) creates a class of nonlinear signal operators by means of threshold superposition, called *stack filters* [1, 7]:

$$\phi_b(f)[x] \equiv \sup\{v \in \mathbb{R} : b(f_v[x - y_1], \ldots, f_v[x - y_n]) = 1\}.$$
(16)

The use of Boolean functions facilitates the design of such discrete flat operators with determinable structural properties. Since each increasing Boolean function can be uniquely represented by an irreducible sum (product) of product (sum) terms, and each product (sum) term corresponds to an erosion (dilation), each stack filter can be represented as a finite maximum (minimum) of flat erosions (dilations) [7]. Because of their representation by means of erosions/dilations (which have a geometric interpretation) and Boolean functions (which are related to mathematical logic), stack filters can be analyzed or designed not only in terms of their statistical properties for image denoising but also in terms of their geometric and logic properties for preserving selected image structures.

2.5 Morphological Operators and Lattice Theory

A more general formalization [2, 18] of morphological operators views them as operators on complete lattices. A *complete lattice* is a set $\mathcal{L}$ equipped with a partial ordering $\leq$ such that $(\mathcal{L}, \leq)$ has the algebraic structure of a *partially ordered set* where the supremum and infimum of any of its subsets exist in $\mathcal{L}$. For any subset $\mathcal{K} \subseteq \mathcal{L}$, its *supremum* $\bigvee \mathcal{K}$ and *infimum* $\bigwedge \mathcal{K}$ are defined as the lowest (with respect to $\leq$) upper bound and greatest lower bound of $\mathcal{K}$, respectively. The two main examples of complete lattices used in morphological image processing are: (i) the space of all binary images represented by subsets of the plane $\mathbb{D}$ where the $\bigvee/\bigwedge$ lattice operations are the set union/intersection, and (ii) the space of all gray-level image signals $f : \mathbb{D} \to \overline{\mathbb{R}}$, where the $\bigvee/\bigwedge$ lattice operations are the supremum/infimum of sets of real numbers. An operator ψ on $\mathcal{L}$ is called **increasing** if it preserves the partial ordering, i.e., $f \leq g$ implies $\psi(f) \geq \psi(g)$. Increasing operators are of great importance, and among them

four fundamental examples are as follows:

$$\delta \text{ is } \mathbf{dilation} \Longleftrightarrow \delta\left(\bigvee_{i \in I} f_i\right) = \bigvee_{i \in I} \delta(f_i), \tag{17}$$

$$\epsilon \text{ is } \mathbf{errosion} \Longleftrightarrow \epsilon\left(\bigwedge_{i \in I} f_i\right) = \bigwedge_{i \in I} \epsilon(f_i), \tag{18}$$

$$\alpha \text{ is } \mathbf{opening} \Longleftrightarrow \alpha \text{ is increasing, idempotent,} \\ \text{and antiextensive,} \tag{19}$$

$$\beta \text{ is } \mathbf{closing} \Longleftrightarrow \beta \text{ is increasing, idempotent,} \\ \text{and extensive,} \tag{20}$$

where I is an arbitrary index set, idempotence means that $\alpha(\alpha(f)) = \alpha(f)$, and (anti-)extensivity of $(\alpha)\beta$ means that $\alpha(f) \leq f \leq \beta(f)$ for all f.

These definitions allow broad classes of signal operators to be grouped as lattice dilations, or erosions, or openings, or closings and their common properties to be studied under the unifying lattice framework. Thus, the translation-invariant morphological dilations $\oplus$, erosions $\ominus$, openings $\circ$, and closings $\bullet$ are simple special cases of their lattice counterparts.

3 Morphological Filters for Enhancement

3.1 Image Smoothing or Simplification

3.1.1 Lattice Opening Filters

The three types of nonlinear filters defined below are lattice openings in the sense of operation (19) and have proven to be very useful for image enhancement.

If a 2-D image f contains one-dimensional (1-D) objects, e.g. lines, and B is a 2-D disklike structuring element, then the simple opening or closing of f by B will eliminate these 1-D objects. Another problem arises when f contains large-scale objects with sharp corners that have to be preserved; in such cases opening or closing f by a disk B will round these corners. These two problems could be avoided in some cases if we replace the conventional opening with a *radial opening*,

$$\alpha(f) = \bigvee_\theta f \circ L_\theta, \tag{21}$$

where the sets L_θ are rotated versions of a line segment L at various angles $\theta \in [0, 2\pi)$. This has the effect of preserving an object in f if this object is left unchanged after the opening by L_θ in at least one of the possible orientations θ. See Fig. 1 for examples.

There are numerous image enhancement problems in which what is needed is suppression of arbitrarily shaped connected components in the input image whose areas (number of pixels) are smaller than a certain threshold n. This can be accomplished by the *area opening* of size n, which, for binary images, keeps

Original

Gaussian Linear Convolutions

Morphological Clos-Openings

Morphological Radial Clos-Openings

Morphological Clos-Openings by Reconstruction

| scale=4 | scale=8 | scale=16 | scale=32 |

FIGURE 1 Linear and morphological multiscale image smoothers. (The scale parameter was defined as the variance of the Gaussians for linear convolutions; the radius of the structuring element for clos-openings; and the scale of the marker for the reconstruction filters.)

only the connected components whose area is $\geq n$ and eliminates the rest. The area opening can also be extended to gray-level images.

Consider now a set $X = \cup_i X_i$ as a union of disjoint connected components X_i and let $M \subseteq X_j$ be a *marker* in the jth component; i.e., M could be a single point or some feature set in X that lies only in X_j. Let us define the *opening by reconstruction* as the operator

$$\mathrm{MR}_X(M) \equiv \text{connected component of } X \text{ containing } M. \quad (22)$$

This is a lattice opening that from the input set M yields as output exactly the component X_j containing the marker. Its output is called the *morphological reconstruction* of the component from the marker. It can extract large-scale components of the image from knowledge only of a smaller marker inside them. An algorithm to implement the opening by reconstruction is based on the *conditional dilation* of M by B within X:

$$\delta_{B|X}(M) \equiv (M \oplus B) \cap X. \quad (23)$$

If B is a disk with a radius smaller than the distance between X_j and any of the other components, then by iterating this conditional dilation we can obtain in the limit

$$\lim_{n \to \infty} \underbrace{(\delta_{B|X} \dots (\delta_{B|X}(\delta_{B|X}(M))))}_{n \text{ times}} = \mathrm{MR}_X(M)$$

the whole component X_j. Replacing the binary with gray-level images, the set dilation with function dilation, and $\cap$ with $\wedge$ yields the gray-level opening by reconstruction. Openings (and closings) by reconstruction have proven to be extremely useful for image simplification because they can suppress small features and keep only large-scale objects without any smoothing of their boundaries. Examples are shown in Fig. 1.

3.1.2 Multiscale Morphological Smoothers

Multiscale image analysis has recently emerged as a useful framework for many computer vision and image processing tasks, including (i) noise suppression at various scales and (ii) feature detection at large scales followed by refinement of their location or value at smaller scales. Most of the previous work in this area was based on a *linear* multiscale smoothing, i.e., convolutions with a Gaussian with a variance proportional to scale. However, these linear smoothers blur or shift image edges, as shown in Fig. 1. In contrast, there is a variety of *nonlinear* smoothing filters, including the morphological openings and closings that can provide a multiscale image ensemble [8, 17] and avoid the above shortcomings of linear smoothers. For example, Fig. 1 shows three types of *clos-openings* (i.e., cascades of openings followed by closings): (1) flat clos-openings by a 2-D disklike structuring element that preserve the vertical image edges but may distort horizontal edges by fitting the shape of the

structuring element; (2) radial flat clos-openings that preserve both the vertical edges as well as any line features along the directions ($0°, 45°, 90°, 135°$) of the four line segments used as structuring elements; (3) gray-level clos-openings by reconstruction, which are especially useful because they can extract the exact outline of a certain object by locking on it while smoothing out all its surroundings. The marker for the opening (closing) by reconstruction was an erosion (dilation) of the original image by a disk of radius equal to scale.

The required building blocks for the above morphological smoothers are the multiscale dilations and erosions. The simplest multiscale dilation and erosion of an image $f(x)$ at scales $t > 0$ are the flat dilations/erosions of f by scaled versions $tB = \{tz : z \in B\}$ of a unit-scale planar compact convex set B (e.g., a disk, a rhombus, and a square),

$$\delta(x, t) \equiv (f \oplus tB)(x), \qquad \varepsilon(x, t) \equiv (f \ominus tB)(x), \quad (24)$$

which apply both to gray-level and binary images. One discrete approach to implement multiscale dilations and erosions is to use scale recursion, i.e., $f \oplus (n+1)B = (f \oplus nB) \oplus B$, where $n = 0, 1, 2, \dots$, and nB denotes the n-fold dilation of B with itself. An alternative and more recent approach that uses continuous models for multiscale smoothing is based on *partial differential equations (PDEs)*. This was inspired by the modeling of linear multiscale image smoothing by means of the isotropic heat diffusion PDE $\partial U/\partial t = \nabla^2 U$, where $U(x, t)$ is the convolution of the initial image $f(x) = U(x, 0)$ with a Gaussian at scale t. Similarly, the multiscale dilation $\delta(x, t)$ of f by a disk of radious (scale) t can be generated as a weak solution of the following nonlinear PDE:

$$\frac{\partial \delta}{\partial t} = \|\nabla \delta\|, \quad (25)$$

with initial condition $\delta(x, 0) = f(x)$, where ∇ denotes the spatial gradient operator and $\| \cdot \|$ is the Euclidean norm. The generating PDE for the erosion is $\partial \varepsilon/\partial t = -\|\nabla \varepsilon\|$. A review and references of the PDE approach to multiscale morphology can be found in [6]. In general, the PDE approach yields very close approximations to Euclidean multiscale morphology with arbitrary subpixel accuracy.

3.1.3 Noise Suppression by Median and Alternating Sequential Filters

In their behavior as nonlinear smoothers, as shown in Fig. 2, the medians act similarly to an *open-closing* $(f \circ B) \bullet B$ by a convex set B of diameter approximately half the diameter of the median window [7]. The open-closing has the advantages over the median that it requires less computation and decomposes the noise suppression task into two independent steps, i.e., suppressing positive spikes via the opening and negative spikes via the closing.

FIGURE 2 (a) Original clean image. (b) Noisy image obtained by corrupting the original with two-level salt and pepper noise occuring with probability 0.1 (peak signal-to-noise ratio or PSNR = 18.9 dB). (c) Open-closing of noisy image by a 2 × 2-pel square (PSNR = 25.4 dB). (d) Median of noisy image by a 3 × 3-pel square (PSNR = 25.4 dB).

The popularity and efficiency of the simple morphological openings and closings to suppress impulse noise is supported by the following theoretical development [19]. Assume a class of sufficiently smooth random input images that is the collection of all subsets of a finite mask W that are open(or closed) with respect to a set B and assign a uniform probability distribution on this collection. Then, a discrete binary input image X is a random realization from this collection; i.e., use ideas from random sets [17] to model X. Further, X is corrupted by a union (or intersection) noise N, which is a 2-D sequence of independent identically distributed (i.i.d.) binary Bernoulli random variables with probability $p \in [0, 1)$ of occurrence at each pixel. The observed image is the noisy version $Y = X \cup N$ (or $Y = X \cap N$).

Then, the maximum *a posteriori* estimate [19] of the original X given the noisy image Y is the opening (or closing) of the observed Y by B.

Another useful generalization of openings and closings involves cascading open-closings $\beta_t \alpha_t$ at multiple scales $t = 1, \ldots, r$, where $\alpha_t(f) = f \circ tB$ and $\beta_t(f) = f \bullet tB$. This generates a class of efficient nonlinear smoothing filters

$$\psi_{asf}(f) = \beta_r \alpha_r \ldots \beta_2 \alpha_2 \beta_1 \alpha_1(f), \qquad (26)$$

called *alternating sequential filters*, which smooth progressively from the smallest scale possible up to a maximum scale r and have a broad range of applications [18]. Their optimal design is addressed in [16].

3.2 Edge or Contrast Enhancement

3.2.1 Morphological Gradients

Consider the difference between the flat dilation and erosion of an image f by a symmetric disklike set B containing the origin whose diameter diam (B) is very small:

$$\text{edge}(f) = \frac{(f \oplus B) - (f \ominus B)}{\text{diam}(B)}. \tag{27}$$

If f is binary, edge(f) extracts its boundary. If f is gray level, the above residual enhances its edges [9, 17] by yielding an approximation to $\|\nabla f\|$, which is obtained in the limit of Eq. (27) as diam$(B) \to 0$ (see Fig. 3). Further, thresholding this morphological gradient leads to binary edge detection.

The symmetric morphological gradient (27) is the average of two asymmetric ones: the erosion gradient $f - (f \ominus B)$ and the dilation gradient $(f \oplus B) - f$. The symmetric or asymmetric morphological edge-enhancing gradients can be made more robust for edge detection by first smoothing the input image with a linear blur [4]. These hybrid edge-detection schemes that largely contain morphological gradients are computationally more efficient and perform comparably or in some cases better than several conventional schemes based only on linear filters.

3.2.2 Toggle Contrast Filter

Consider a gray-level image $f[x]$ and a small-size symmetric disklike structuring element B containing the origin. The following discrete nonlinear filter [3] can enhance the local contrast of f by sharpening its edges:

$$\psi(f)[x] =$$
$$\begin{cases} (f \oplus B)[x] & \text{if } (f \oplus B)[x] - f[x] \le f[x] - (f \ominus B)[x] \\ (f \ominus B)[x] & \text{if } (f \oplus B)[x] - f[x] > f[x] - (f \ominus B)[x] \end{cases}. \tag{28}$$

At each pixel x, the output value of this filter *toggles* between the value of the dilation of f by B (i.e., the maximum of f inside the moving window B centered) at x and the value of its erosion by B (i.e., the minimum of f within the same window) according to which is closer to the input value $f[x]$. The toggle filter is usually applied not only once but is *iterated*. The more iterations, the more contrast enhancement. Further, the iterations converge to a *limit (fixed point)* [3] reached after a finite number of iterations. Examples are shown in Figs. 4 and 5.

As discussed in [6, 15], the above discrete toggle filter is closely related to the operation and numerical algorithm behind a

nonlinear (shock-wave) PDE proposed in [10] to deblur images and/or enhance their contrast by edge sharpening. For 1-D images such a PDE is

$$\frac{\partial u}{\partial t} = -\left| \frac{\partial u}{\partial x} \right| \text{sign}\left(\frac{\partial^2 u}{\partial x^2} \right). \tag{29}$$

Starting at $t = 0$, with the blurred image $u(x, 0) = f(x)$ as the initial data, and running the numerical algorithm implementing this PDE until some time t yields a filtered image $u(x, t)$. Its goal is to restore blurred edges sharply, accurately, and in a nonoscillatory way by propagating shocks (i.e., discontinuities in the signal derivatives). Steady state is reached as $t \to \infty$. Over convex regions ($\partial^2 u / \partial x^2 > 0$) this PDE acts as a 1-D erosion PDE $\partial u / \partial t = -|\partial u / \partial x|$, which models multiscale erosion of $f(x)$ by the horizontal line segment $[-t, t]$ and shifts parts of the graph of $u(x, t)$ with positive (negative) slope to the right (left) but does not move the extrema or inflection points. Over concave regions ($\partial^2 u / \partial x^2 < 0$) it acts as a 1-D dilation PDE $\partial u / \partial t = |\partial u / \partial x|$, which models multiscale dilation of $f(x)$ by the same segment and reverses the direction of propagation. For certain piecewise-constant signals blurred by means of linear convolution with finite-window smooth tapered symmetric kernels, the shock filtering $u(x, 0) \mapsto u(x, \infty)$ can recover the original signal and thus achieve an exact deconvolution [10]; an example of such a case is shown in Fig. 4.

4 Morphological Filters for Detection

4.1 Morphological Correlation

Consider two real-valued discrete image signals $f[x]$ and $g[x]$. Assume that g is a signal pattern to be found in f. To find which shifted version of g "best" matches f, a standard approach has been to search for the shift lag y that minimizes the *mean squared error* $E_2[y] = \sum_{x \in W}(f[x + y] - g[x])^2$ over some subset W of $\mathbb{Z}^2$. Under certain assumptions, this matching criterion is equivalent to maximizing the *linear cross-correlation* $L_{fg}[y] \equiv \sum_{x \in W} f[x + y]g[x]$ between f and g. A discussion of linear template matching can be found in Chapter 3.1.

Although less mathematical tractable than the mean squared error criterion, a statistically more robust criterion is to minimize the *mean absolute error*

$$E_1[y] \equiv \sum_{x \in W} |f[x + y] - g[x]|. $$

This mean absolute error criterion corresponds to a nonlinear signal correlation used for signal matching; see [8] for a review. Specifically, since $|a - b| = a + b - 2\min(a, b)$, under certain assumptions (e.g., if the error norm and the correlation is

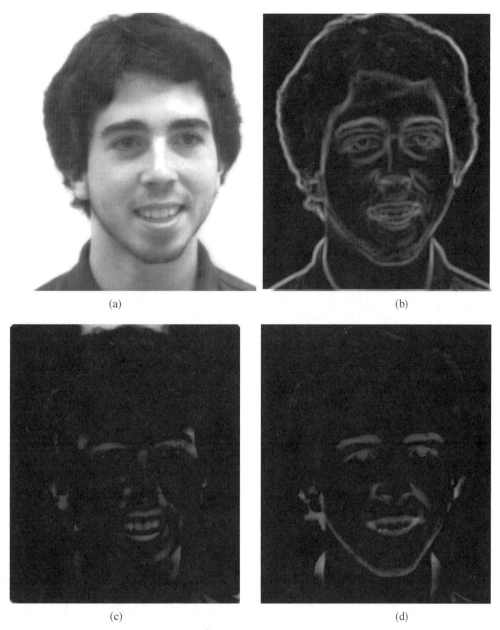

FIGURE 3 Morphological edge and blob detectors. (a) Image f. (b) Edges: morphological gradient $f \oplus B - f \ominus B$, where B is a small discrete disklike set. (c) Peaks: $f - f \circ B$. (d) Valleys: $f \bullet B - f$.

normalized by dividing it with the average area under the signals f and g), minimizing $E_1[y]$ is equivalent to maximizing the *morphological* cross-correlation

$$M_{fg}[y] \equiv \sum_{x \in W} \min(f[x + y], g[x]). \qquad (30)$$

It can be shown experimentally and theoretically that the detection of g in f is indicated by a sharper matching peak in $M_{fg}[y]$ than in $L_{fg}[y]$. In addition, the morphological (sum of minima) correlation is faster than the linear (sum of

products) correlation. These two advantages of the morphological correlation coupled with the relative robustness of the mean absolute error criterion make it promising for general signal matching.

4.2 Binary Object Detection and Rank Filtering

Let us approach the problem of binary image object detection in the presence of noise from the viewpoint of statistical hypothesis testing and rank filtering. Assume that the observed discrete

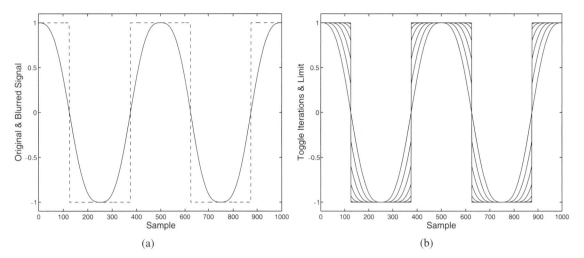

FIGURE 4 (a) Original signal (dashed curve) and its blurring (solid curve) by means of convolution with a finite positive symmetric tapered impulse response. (b) Filtered versions of the blurred signal in (a) produced by iterating the 1-D toggle filter, with $B = \{-1, 0, 1\}$, until convergence to the limit signal, reached at 125 iterations; the displayed filtered signals correspond to iteration indexes that are multiples of 20.

binary image $f[x]$ within a mask W has been generated under one of the following two probabilistic hypotheses:

$$H_0: \quad f[x] = e[x], \quad x \in W.$$

$$H_1: \quad f[x] = |g[x - y] - e[x]|, \quad x \in W.$$

Hypothesis $H_1(H_0)$ stands for "object present" (object not present) at pixel location y. The *object* $g[x]$ is a deterministic binary template. The *noise* $e[x]$ is a stationary binary random field, which is a 2-D sequence of i.i.d. random variables taking value 1 with probability p and 0 with probability $1 - p$, where $0 < p < 0.5$. The *mask* $W = G_{+y}$ is a finite set of pixels equal to the region G of support of g shifted to location y at which the decision is taken.(For notational simplicity, G is assumed to be symmetric, i.e., $G = G^r$.) The absolute-difference superposition between g and e under H_1 forces f to always have values 0 or 1. Intuitively, such a signal-to-noise superposition means that the noise e toggles the value of g from 1 to 0 and from 0 to 1 with probability p at each pixel. This noise model can be viewed either as the common binary symmetric channel noise in signal transmission or as a binary version of the salt and pepper noise. To decide whether the object g occurs at y we use a Bayes decision rule that minimizes the total probability of error and hence leads to the *likelihood ratio test*:

$$\frac{Pr(f/H_1)}{Pr(f/H_0)} \underset{H_0}{\overset{H_1}{\gtrless}} \frac{Pr(H_0)}{Pr(H_1)}, \tag{31}$$

where $Pr(f/H_i)$ are the likelihoods of H_i with respect to the observed image f, and $Pr(H_i)$ are the *a priori* probabilities.

This is equivalent to

$$M_{fg}[y] = \sum_{x \in W} \min(f[x], g[x - y]) \underset{H_0}{\overset{H_1}{\gtrless}} \theta$$

$$\theta = \frac{1}{2} \left(\frac{\log[Pr(H_0)/Pr(H_1)]}{\log[(1-p)/p]} + \mathrm{card}(G) \right) \tag{32}$$

Thus, the selected statistical criterion and noise model lead to compute the morphological (or equivalently linear) binary correlation between a noisy image and a known image object and compare it to a threshold for deciding whether the object is present.

Thus, optimum detection in a binary image f of the presence of a binary object g requires comparing the binary correlation between f and g to a threshold θ. This is equivalent[5] to performing a rth rank filtering on f by a set G equal to the support of g, where $1 \le r \le \mathrm{card}(G)$ and r is related to θ. Thus, the rank r reflects the area portion of (or a probabilistic confidence score for) the shifted template existing around pixel y. For example, if $Pr(H_0) = Pr(H_1)$, then $r = \theta = \mathrm{card}\,(G)/2$ and hence the binary median filter by G becomes the optimum detector.

4.3 Hit–Miss Filter

The set erosion (3) can also be viewed as Boolean template matching since it gives the center points at which the shifted

---

[5]An alternative implementation and view of binary rank filtering is by means of *thresholded convolutions*, in which a binary image is linearly convolved with the indicator function of a set G with $n = \mathrm{card}(G)$ pixels and then the result is thresholded at an integer level r between 1 and n; this yields the output of the rth rank filter by G acting on the input image.

FIGURE 5 (a) Original image f. (b) Blurred image g obtained by an out-of-focus camera digitizing f. (c) Output of the 2-D toggle filter acting on g (B was a small symmetric disklike set containing the origin). (d) Limit of iterations of the toggle filter on g (reached at 150 iterations).

structuring element fits inside the image object. If we now consider a set A probing the image object X and another set B probing the background X^c, the set of points at which the shifted pair (A, B) fits inside the image X is the *hit–miss transformation* of X by (A, B):

$$X \otimes (A, B) \equiv \{x : A_{+x} \subseteq X, B_{+x} \subseteq X^c\}. \qquad (33)$$

In the discrete case, this can be represented by a Boolean product function whose uncomplemented (complemented) variables correspond to points of $A(B)$. It has been used extensively for binary feature detection [17]. It can actually model all binary template matching schemes in binary pattern recognition that use a pair of a positive and a negative template [13].

In the presence of noise, the hit–miss filter can be made more robust by replacing the erosions in its definitions with rank filters

that do not require an exact fitting of the whole template pair (A, B) inside the image but only a part of it.

4.4 Morphological Peak/Valley Feature Detection

Residuals between openings or closings and the original image offer an intuitively simple and mathematically formal way for peak or valley detection. Specifically, subtracting from an input image f its opening by a compact convex set B yields an output consisting of the image peaks whose support cannot contain B. This is the *top-hat transformation* [9],

$$\text{peak}(f) = f - (f \circ B), \qquad (34)$$

which has found numerous applications in geometric feature

detection [17]. It can detect *bright blobs*, i.e., regions with significantly brighter intensities relative to the surroundings. The shape of the detected peak's support is controlled by the shape of B, whereas the scale of the peak is controlled by the size of B. Similarly, to detect *dark blobs*, modeled as image intensity valleys, we can use the valley detector

$$\text{valley}(f) = (f \bullet B) - f. \qquad (35)$$

See Fig. 3 for examples.

The morphological peak/valley detectors are simple, efficient, and have some advantages over curvature-based approaches. Their applicability in situations in which the peaks or valleys are not clearly separated from their surroundings is further strengthened by generalizing them in the following way. The conventional opening in Eq. (34) is replaced by a general lattice opening such as an area opening or opening by reconstruction. This generalization allows a more effective estimation of the image background surroundings around the peak and hence a better detection of the peak.

5 Optimal Design of Morphological Filters for Enhancement

5.1 Brief Survey of Existing Design Approaches

Morphological and rank/stack filters are useful for image enhancement and are closely related since they can all be represented as maxima of morphological erosions [7]. Despite the wide application of these nonlinear filters, very few ideas exist for their optimal design. The current four main approaches are (a) designing morphological filters as a finite union of erosions [5] based on the morphological basis representation theory (outlined in Section 2.3); (b) designing stack filters by means of threshold decomposition and linear programming [1]; (c) designing morphological networks, using either voting logic and rank tracing learning or simulated annealing [20]; and (d) designing morphological/rank filters by means of a gradient-based adaptive optimization [14]. Approach (a) is limited to binary increasing filters. Approach (b) is limited to increasing filters processing nonnegative quantized signals. Approach (c) requires a long time to train and convergence is complex. In contrast, approach (d) is more general since it applies to both increasing and nonincreasing filters and to both binary and real-valued signals. The major difficulty involved is that rank functions are *not* differentiable, which imposes a deadlock on how to adapt the coefficients of morphological/rank filters using a gradient-based algorithm. The methodology described in this section is an extension and improvement to the design methodology (d), leading to a new approach that is simpler, more intuitive, and numerically more robust.

For various signal processing applications it is sometimes useful to mix in the same system both nonlinear and linear filtering strategies. Thus, hybrid systems, composed by linear and nonlinear (rank-type) subsystems, have frequently been proposed in the research literature. A typical example is the class of L filters that are linear combinations of rank filters. Several adaptive algorithms have also been developed for their design, which illustrated the potential of adaptive hybrid filters for image processing applications, especially in the presence of non-Gaussian noise.

Given the applicability of hybrid systems and the relatively few existing ideas to design their nonlinear part, in this section we present a general class of nonlinear systems, called *morphological/rank/linear* (**MRL**) filters [11], that contains as special cases morphological, rank, and linear filters, and we develop an efficient method for their adaptive optimal design. MRL filters consist of a linear combination between a morphological/rank filter and a linear FIR filter. Their nonlinear component is based on a rank function, from which the basic morphological operators of erosion and dilation can be obtained as special cases.

5.2 MRL Filters

We shall use a vector notation to represent the values of the 1-D or 2-D sampled signal (after some enumeration of the signal samples) inside an n-point moving window. Let $\underline{x} = (x_1, x_2, \ldots, x_n)$ in $\mathbb{R}^n$ represent the input signal segment and y be the output value from the filter. The MRL filter is defined as the shift-invariant system whose local signal transformation rule $\underline{x} \mapsto y$ is given by

$$y \equiv \lambda\alpha + (1 - \lambda)\beta,$$
$$\alpha = \mathcal{R}_r(\underline{x} + \underline{a}) = \mathcal{R}_r(x_1 + a_1, x_2 + a_2, \ldots, x_n + a_n), \qquad (36)$$
$$\beta = \underline{x} \cdot \underline{b}^T = x_1 b_1 + x_2 b_2 + \cdots + x_n b_n,$$

where $\lambda \in \mathbb{R}$, $\underline{a}, \underline{b} \in \mathbb{R}^n$, and $(\cdot)^T$ denotes transposition. $\mathcal{R}_r(\underline{t})$ is the rth rank function of $t \in \mathbb{R}^n$. It is evaluated by sorting the components of $\underline{t} = (t_1, t_2, \ldots, t_n)$ in decreasing order, $t_{(1)} \geq t_{(2)} \cdots \geq t_{(n)}$, and picking the rth element of the sorted list; i.e., $\mathcal{R}_r(\underline{t}) \equiv t_{(r)}, r = 1, 2, \ldots, n$. The vector $\underline{b} = (b_1, b_2, \ldots, b_n)$ corresponds to the coefficients of the linear FIR filter, and the vector $\underline{a} = (a_1, a_2, \ldots, a_n)$ represents the coefficients of the morphological/rank filter. We call $\underline{a}$ the "structuring element" because for $r = 1$ and $r = n$ the rank filter becomes the morphological dilation and erosion by a structuring function equal to $\pm\underline{a}$ within its support. For $1 < r < n$, we use $\underline{a}$ to generalize the standard unweighted rank operations to filters with weights. The median is obtained when $r = \lfloor n/2 + 1 \rfloor$. Besides these two sets of weights, the rank r and the mixing parameter λ will also be included in the training process for the filter design. If $\lambda \in [0, 1]$, the MRL filter becomes a convex combination of its components, so that when we increase the contribution of one component, the other one decreases. From Eq. (36) it follows that computing each output sample requires $2n + 1$ additions, $n + 2$ multiplications, and an n-point sorting operation.

Because of the use of a gradient-based adaptive algorithm, derivatives of rank functions will be needed. Since these functions are not differentiable in the common sense, we will propose a simple design alternative using "rank indicator vectors" and "smoothed impulses." We define the unit sample function $q(v)$, $v \in \mathbb{R}$, as

$$q(v) \equiv \begin{cases} 1, & \text{if } v \equiv 0 \\ 0, & \text{otherwise} \end{cases}. \qquad (37)$$

Applying q to all components of a vector $\underline{v} \in \mathbb{R}^n$ yields a vector unit sample function

$$\mathcal{Q}(\underline{v}) \equiv (q(v_1), q(v_2), \ldots, q(v_n)).$$

Given a vector $\underline{t} = (t_1, t_2, \ldots, t_n)$ in $\mathbb{R}^n$, and a rank $r \in \{1, 2, \ldots, n\}$, the rth *rank indicator vector* $\underline{c}$ of $\underline{t}$ is defined by

$$\underline{c}(\underline{t}, r) \equiv \frac{\mathcal{Q}(z\underline{1} - \underline{t})}{\mathcal{Q}(z\underline{1} - \underline{t}) \cdot \underline{1}^T}, \quad z = \mathcal{R}_r(\underline{t}), \qquad (38)$$

where $\underline{1} = (1, 1, \ldots, 1)$. Thus, the rank indicator vector marks the locations in $\underline{t}$ where the z value occurs. It has many interesting properties [11], which include the following. It has unit area:

$$\underline{c} \cdot \underline{1}^T = 1.$$

It yields an inner-product representation of the rank function:

$$\underline{c} \cdot \underline{t}^T = \mathcal{R}_r(\underline{t}).$$

Further, for r fixed, if $\underline{c}$ is constant in a neighborhood of some $\underline{t}_0$, then the rth rank function $\mathcal{R}_r(\underline{t})$ is differentiable at $\underline{t}_0$ and

$$\left. \frac{\partial \mathcal{R}_r(\underline{t})}{\partial \underline{t}} \right|_{\underline{t}=\underline{t}_0} = \underline{c}(\underline{t}_0, r). \qquad (39)$$

At points in whose neighborhood $\underline{c}$ is not constant, the rank function is not differentiable.

At points where the function $z = \mathcal{R}_r(\underline{t})$ is not differentiable, a possible *design choice* is to assign the vector $\underline{c}$ as a one-sided value of the discontinuous $\partial z / \partial \underline{t}$. Further, since the rank indicator vector will be used to estimate derivatives and it is based on the discontinuous unit sample function, a simple approach to avoid abrupt changes and achieve numerical robustness is to replace the unit sample function by a *smoothed impulse* $q_\sigma(v)$ that depends on a scale parameter $\sigma \geq 0$ and has at least the following required properties:

$$q_\sigma(v) = q_\sigma(-v) \quad \text{(symmetry)},$$
$$q_\sigma(v) \rightarrow q(v) \forall v \quad \text{as } \sigma \rightarrow 0, \qquad (40)$$
$$q_\sigma(v) \rightarrow 1 \forall v \quad \text{as } \sigma \rightarrow \infty.$$

Functions like $\exp\left[-\frac{1}{2}(v/\sigma)^2\right]$ or $\operatorname{sech}^2(v/\sigma)$ are natural choices for $q_\sigma(v)$.

From the filter definition (36), we see that our design goal is to specify a set of parameters $\underline{a}$, $\underline{b}$, r, and λ in such away that some design requirement is met. However, instead of using the integer rank parameter r directly in the training equations, we work with a real variable ρ implicitly defined by the following rescaling:

$$r \equiv \left\lfloor n - \frac{n-1}{1 + \exp(-\rho)} + 0.5 \right\rfloor, \quad \rho \in \mathbb{R}, \qquad (41)$$

where $\lfloor \cdot + 0.5 \rfloor$ denotes the usual rounding operation and n is the dimension of the input signal vector $\underline{x}$ inside the moving window. Thus, the weight vector to be used in the filter design task is defined by

$$\underline{w} \equiv (\underline{a}, \rho, \underline{b}, \lambda), \qquad (42)$$

but any of its components may be fixed during the process.

5.3 Least-Mean-Square Approach to Designing Optimal MRL Filters

Our framework for adaptive design is related to adaptive filtering, in which the design is viewed as a learning process and the filter parameters are iteratively adapted until convergence is achieved. The usual approach to adaptively adjust the vector $\underline{w}$, and therefore design the filter, is to define a cost function $J(\underline{w})$, estimate its gradient $\nabla J(\underline{w})$, and update $\underline{w}$ by the iterative (recursive) formula

$$\underline{w}(i+1) = \underline{w}(i) - \mu_0 \nabla J(\underline{w})|_{\underline{w}=\underline{w}(i)}, \qquad (43)$$

so that the value of the cost function tends to decrease at each step. The positive constant μ_0 is usually called the *step size* and regulates the tradeoff between stability and speed of convergence of the iterative procedure. Iteration (43) starts with an initial guess $\underline{w}(0)$ and is terminated when some desired condition is reached. This approach is commonly known as the *method of steepest descent*.

As cost function J, for the ith update $\underline{w}(i)$ of the weight vector, we use

$$J(\underline{w}(i)) = \frac{1}{M} \sum_{k=i-M+1}^{i} e^2(k), \qquad (44)$$

where $M = 1, 2, \ldots$ is a memory parameter, and the *instantaneous error*

$$e(k) = d(k) - y(k) \qquad (45)$$

is the difference between the desired output signal $d(k)$ and the actual filter output $y(k)$ for the training sample k. The memory parameter M controls the smoothness of the updating process. If we are processing noiseless signals, it is sometimes better to simply set $M = 1$ (minimum computational complexity). In contrast, if we are processing noisy signals, we should use $M > 1$ and sufficiently large to reduce the noise influence during the training process. Further, it is possible to make a training process convergent by using a larger value of M.

Hence, the resulting adaptation algorithm, called the *averaged least-mean-square* (*LMS*) algorithm, is

$$\underline{w}(i+1) = \underline{w}(i) + \frac{\mu}{M} \sum_{k=i-M+1}^{i} e(k) \frac{\partial y(k)}{\partial \underline{w}} \Bigg|_{\underline{w}=\underline{w}(i)},$$

$$i = 0, 1, 2, \ldots, \quad (46)$$

where $\mu = 2\mu_0$. From Eqs. (42) and (36),

$$\frac{\partial y}{\partial \underline{w}} = \left(\frac{\partial y}{\partial \underline{a}}, \frac{\partial y}{\partial \rho}, \frac{\partial y}{\partial \underline{b}}, \frac{\partial y}{\partial \lambda} \right) = \left[\lambda \frac{\partial \alpha}{\partial \underline{a}}, \lambda \frac{\partial \alpha}{\partial \rho}, (1 - \lambda)\underline{x}, \alpha - \beta \right].$$

$$(47)$$

According to Eq. (39) and our design choice, we set

$$\frac{\partial \alpha}{\partial \underline{a}} = \underline{c} = \frac{\mathcal{Q}(\alpha \underline{1} - \underline{x} - \underline{a})}{\mathcal{Q}(\alpha \underline{1} - \underline{x} - \underline{a}) \cdot \underline{1}^T}, \quad \alpha = \mathcal{R}_r(\underline{x} + \underline{a}). \quad (48)$$

The final unknown is $s = \partial \alpha / \partial \rho$, which will be one more design choice. Notice from Eqs. (41) and (36) that $s \geq 0$. If all the elements of $\underline{t} = \underline{x} + \underline{a}$ are identical, then the rank r does not play any role, so that $s = 0$ whenever this happens. In contrast, if only one element of $\underline{t}$ is equal to α, then variations in the rank r can drastically modify the output α; in this case s should assume a maximum value. Thus, a possible simple choice for s is

$$\frac{\partial \alpha}{\partial \rho} = s \equiv 1 - \frac{1}{n} \mathcal{Q}(\alpha \underline{1} - \underline{x} - \underline{a}) \cdot \underline{1}^T, \quad \alpha = \mathcal{R}_r(\underline{x} + \underline{a}), \quad (49)$$

where n is the dimension of $\underline{x}$.

Finally, to improve the numerical robustness of the training algorithm, we will frequently replace the unit sample function by smoothed impulses, obeying Eq. (40), in which case an appropriate smoothing parameter σ should be selected. A natural choice of a smoothed impulse is $q_\sigma(v) = \exp[-\frac{1}{2}(v/\sigma)^2]$, $\sigma > 0$. The choice of this nonlinearity will affect only the gradient estimation step in design procedure (46). We should use small values of σ such that $q_\sigma(v)$ is close enough to $q(v)$. A possible systematic way to select the smoothing parameter σ could be to set $|q_\sigma(v)| \leq \epsilon$ for $|v| \geq \delta$, so that, for some desired ϵ and δ, $\sigma = \delta/\sqrt{\ln(1/\epsilon^2)}$.

Theoretical conditions for convergence of training process (46) can be derived under the following considerations. The goal is to find upper bounds μ_w to the step size μ, such that Eq. (46) can converge if $0 < \mu < \mu_w$. We assume the framework of system identification with noiseless signals, and we consider the training process of only one element of $\underline{w}$ at a time, while the others are optimally fixed. This means that given the original and

transformed signals, and three parameters (sets) of the original $\underline{w}^* = (\underline{a}^*, \rho^*, \underline{b}^*, \lambda^*)$ used to transform the input signal, we will use Eq. (46) to track only the fourth unknown parameter (set) of $\underline{w}^*$ in a noiseless environment. If training process (46) is convergent, then $\lim_{i \to \infty} \| \underline{w}(i) - \underline{w}^* \| = 0$, where $\| \cdot \|$ is some error norm. By analyzing the behavior of $\| \underline{w}(i) - \underline{w}^* \|$, under the above assumptions, conditions for convergence have been found in [11].

5.4 Application of Optimal MRL Filters to Enhancement

The proper operation of training process (46) has been verified in [11] through experiments confirming that, if the conditions for convergence are met, our design algorithm converges fast to the real parameters of the MRL filter within small error distances.

We illustrate its applicability to image enhancement by an experiment.[6] The goal here it to restore an image corrupted by non-Gaussian noise. Hence, the input signal is a noisy image, and the desired signal is the original (noiseless) image. The noisy image for training the filter was generated by first corrupting the original image with a 47-dB additive Gaussian white noise, and then with a 10% multivalued impulse noise. After the MRL filter is designed, another noisy image (with similar type of perturbation) is used for testing. The optimal filter parameters were estimated after scanning the image twice during the training process. We used training algorithm (46) with $M = 1$ and $\mu = 0.1$, and we started the process with an unbiased combination between a flat median and the identity, i.e.,

$$\underline{a}_0 = \begin{bmatrix} 0 & 0 & 0 \\ 0 & 0 & 0 \\ 0 & 0 & 0 \end{bmatrix}, \quad \underline{b}_0 = \begin{bmatrix} 0 & 0 & 0 \\ 0 & 1 & 0 \\ 0 & 0 & 0 \end{bmatrix}, \quad \rho_0 = 0, \quad \lambda_0 = 0.5.$$

The final trained parameters of the filter were

$$\underline{a} = \begin{bmatrix} 0.75 & 0.00 & 0.05 \\ -0.46 & -0.01 & 0.71 \\ -0.09 & -0.02 & -0.51 \end{bmatrix}, \quad \underline{b} = \begin{bmatrix} 0.01 & 0.19 & -0.01 \\ 0.13 & 0.86 & 0.07 \\ 0.00 & 0.13 & -0.02 \end{bmatrix},$$

$$r = 5, \quad \lambda = 0.98,$$

which represents a biased combination between a nonflat median filter and a linear FIR filter, where some elements of $\underline{a}$ and $\underline{b}$ present more influence in the filtering process.

Figure 6 shows the results of using the designed MRL filter with a test image, and its comparison with a flat median filter of the same window size. The noisy image used for training is not

---

[6] Implementation details: The images are scanned twice during the training process, following a zig-zag path from top to bottom, and then from bottom to top. The local input vector $\underline{x}$ is obtained at each pixel by column-by-column indexing of the image values inside an n-point square window centered around the pixel. The vectors $\underline{a}$ and $\underline{b}$ are indexed the same way. The unit sample function $q(v)$ is approximated by $q_\sigma(v) = \exp[-\frac{1}{2}(v/\sigma)^2]$, with $\sigma = 0.001$. The image values are normalized to be in the range $[0, 1]$.

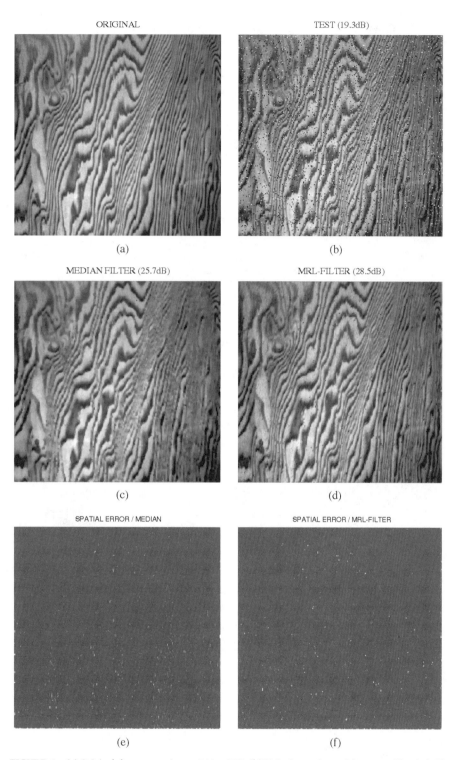

FIGURE 6   (a) Original clean texture image (240 × 250). (b) Noisy image: image (a) corrupted by a hybrid 47-dB additive Gaussian white noise and 10% multivalued impulse noise (PSNR = 19.3 dB). (c) Noisy image restored by a flat 3 × 3 median filter (PSNR = 25.7 dB). (d) Noisy image restored by the designed 3 × 3 MRL filter (PSNR = 28.5 dB). (e) Spatial error map of the flat median filter; lighter areas indicate higher errors. (f) Spatial error map of the MRL filter.

included there because the (noisy) images used for training and testing are simply different realizations of the same perturbation process. Observe that the MRL filter outperformed the median filter by approximately 3 dB. Spatial error plots are also included, which show that the optimal MRL filter preserves better the image structure since its corresponding spatial error is more uncorrelated than the error of the median filter.

For the type of noise used in this experiment, we must have at least part of the original (noiseless) image; otherwise, we would not be able to provide a good estimate to the optimal filter parameters during training process (46). In order to validate this point, we repeated the above experiment with 100×100 subimages of the training image (only 17% of the pixels), and the resulting MRL filter still outperformed the median filter by approximately 2.3 dB. There are situations, however, in which we can use only the noisy image together with some filter constraints and design the filter that is closest to the identity [14]. But this approach is only appropriate for certain types of impulse noise.

An exhaustive comparison of different filter structures for noise cancellation is beyond the scope of this chapter. Nevertheless, this experiment was extended with the adaptive design of a 3×3 L filter under the same conditions. Starting the L filter with a flat median, even after scanning the image four times during the training process, we found the resulting L filter was just 0.2 dB better than the (flat) median filter.

Acknowledgment

Part of this chapter dealt with the authors' research work, which was supported by the U.S. National Science Foundation under grants MIPS-86-58150 and MIP-94-21677.

References

[1] E. J. Coyle and J. H. Lin, "Stack filters and the mean absolute error criterion," *IEEE Trans. Acoust. Speech Signal Process.* **36**, 1244–1254, (1988).

[2] H. J. A. M. Heijmans, *Morphological Image Operators* (Academic, Boston, 1994).

[3] H. P. Kramer and J. B. Bruckner, "Iterations of a nonlinear transformation for enhancement of digital images," *Pattern Recog.* **7**, 53–58 (1975).

[4] J. S. J. Lee, R. M. Haralick, and L. G. Shapiro, "Morphologic edge detection," *IEEE Trans. Rob. Autom.* **RA-3**, 142–156 (1987).

[5] R. P. Loce and E. R. Dougherty, "Facilitation of optimal binary morphological filter design via structuring element libraries and design constraints," *Opt. Eng.* **31**, 1008–1025 (1992).

[6] P. Maragos, "Partial differential equations in image analysis: continuous modeling, discrete processing," in *Signal Processing IX: Theories and Applications*, (EURASIP Press, 1998), Vol. II, pp. 527–536.

[7] P. Maragos and R. W. Schafer, "Morphological filters. Part I: Their set-theoretic analysis and relations to linear shift-invariant filters. Part II: Their relations to median, order-statistic, and stack filters," *IEEE Trans. Acoust. Speech Signal Process.* **35**, 1153–1184 (1987); *ibid*, **37**, 597 (1989).

[8] P. Maragos and R. W. Schafer, "Morphological systems for multi-dimensional signal processing," *Proc. IEEE* **78**, 690–710 (1990).

[9] F. Meyer, "Contrast feature extraction," in *Special Issues of Practical Metallography*, J. L. Chermant, ed. (Riederer-Verlag, Stuttgart, 1978), pp. 374–380.

[10] S. Osher and L. I. Rudin, "Feature-oriented image enhancement using Schock filters," *SIAM J. Numer. Anal.* **27**, 919–940 (1990).

[11] L. F. C. Pessoa and P. Maragos, "MRL-Filters: A general class of nonlinear systems and their optimal design for image processing," *IEEE Trans. Image Process.* **7**, 966–978 (1998).

[12] K. Preston, Jr., and M. J. B. Duff, *Modern Cellular Automata* (Plenum, New York, 1984).

[13] A. Rosenfeld and A. C. Kak, *Digital Picture Processing* (Academic, New York, 1982), Vols. 1 and 2.

[14] P. Salembier, "Adaptive rank order based filters," *Signal Process.* **27**, 1–25 (1992).

[15] J. G. M. Schavemaker, M. J. T. Reinders, and R. Van den Boomgaard, "Image sharpening by morphological filtering," Presented at the IEEE Workshop on Nonlinear Signal & Image Processing, MacKinac Island, Michigan, Sept. 1997.

[16] D. Schonfeld and J. Goutsias, "Optimal morphological pattern restoration from noisy binary images," *IEEE Trans. Pattern Anal. Machine Intell.* **13**, 14–29 (1991).

[17] J. Serra, *Image Analysis and Mathematical Morphology* (Academic, New York, 1982).

[18] J. Serra, ed., *Image Analysis and Mathematical Morphology, Vol. 2: Theoretical Advances* (Academic, New York, 1988).

[19] N. D. Sidiropoulos, J. S. Baras, and C. A. Berenstein, "Optimal filtering of digital binary images corrupted by union/intersection noise," *IEEE Trans. Image Process.* **3**, 382–403 (1994).

[20] S. S. Wilson, "Training structuring elements in morphological networks," in *Mathematical Morphology in Image Processing*, E. R. Dougherty, ed. (Marcel Dekker, New York, 1993).

3.4

Wavelet Denoising for Image Enhancement

Dong Wei
Drexel University

Alan C. Bovik
The University of Texas at Austin

1 Introduction

Image processing is a science that uncovers information about images. Enhancement of an image is necessary to improve appearance or to highlight some aspect of the information contained in the image. Whenever an image is converted from one form to another, e.g., acquired, copied, scanned, digitized, transmitted, displayed, printed, or compressed, many types of noise or noiselike degradations can be present in the image. For instance, when an analog image is digitized, the resulting digital image contains quantization noise; when an image is halftoned for printing, the resulting binary image contains halftoning noise; when an image is transmitted through a communication channel, the received image contains channel noise; when an image is compressed, the decompressed image contains compression errors. Hence, an important subject is the development of image enhancement algorithms that remove (smooth) noise artifacts while retaining image structure.

Digital images can be conveniently represented and manipulated as matrices containing the light intensity or color information at each spatially sampled points. The term *monochrome digital image*, or simply *digital image*, refers to a two-dimensional light intensity function $f(n_1, n_2)$, where n_1 and n_2 denote spatial coordinates, the value of $f(n_1, n_2)$ is proportional to the brightness (or gray level) of the image at that point, and n_1, n_2, and $f(n_1, n_2)$ are integers.

The problem of image denoising is to recover an image $f(n_1, n_2)$ from the observation $g(n_1, n_2)$, which is distorted by noise (or noiselike degradation) $q(n_1, n_2)$; i.e.,

$$g(n_1, n_2) = f(n_1, n_2) + q(n_1, n_2). \tag{1}$$

Chapter 3.1 considers methods for linear image restoration. The classical image denoising techniques are based on *filtering*, which can be classified into two categories: *linear filtering* and *nonlinear filtering*. Linear filtering-based denoising is based on low-pass filtering to suppress high-frequency noise. The simplest low-pass filter is spatial averaging. Linear filtering can be implemented in either the spatial domain or the frequency domain (usually by means of fast Fourier transforms). Nonlinear filters used in denoising include order statistic filters and morphological filters. The most popular nonlinear filter is the median filter, which is a special type of order statistic filter. For detailed discussions of these nonlinear filters, see Chapters 3.2 (median filters), 3.3 (morphological filters), and 4.4 (order statistic filters).

The basic difficulty with these filtering-based denoising techniques is that, if applied indiscriminately, they tend to blur the image, which is usually objectionable. In particular, one usually wants to avoid blurring sharp edges or lines that occur in the image.

Recently, wavelet-based denoising techniques have been recognized as powerful tools for denoising. Different from those filtering-based classical methods, wavelet-based methods can be viewed as transform-domain point processing.

2 Wavelet Shrinkage Denoising

2.1 The Discrete Wavelet Transform

Before introducing wavelet-based denoising techniques, we first briefly review relevant basics of the discrete wavelet transform (See Chapter 4.1 for a fuller introduction to wavelets).

The *discrete wavelet transform* (DWT) is a multiresolution (or multiscale) representation. The DWT is implemented by means of multirate filterbanks.

Figure 1 shows an implementation of a three-level forward DWT based on a two-channel recursive filterbank, where $h_0(n)$ and $h_1(n)$ are low-pass and high-pass analysis filters, respectively, and the block $\downarrow 2$ represents the downsampling operator by a factor 2. The input signal $x(n)$ is recursively decomposed into a total of four subband signals: a coarse signal, $c_3(n)$, and three detail signals, $d_1(n)$, $d_2(n)$, and $d_3(n)$, of three resolutions.

Figure 2 plots an implementation of a three-level inverse DWT based on a two-channel recursive filterbank, where $\tilde{h}_0(n)$ and $\tilde{h}_1(n)$ are low-pass and high-pass synthesis filters, respectively, and the block $\uparrow 2$ represents the upsampling operator by a factor 2. The four subband signals $c_3(n)$, $d_3(n)$, $d_2(n)$, and $d_1(n)$ are recursively combined to reconstruct the output signal $x(n)$. The four finite impulse response filters satisfy

$$h_1(n) = (-1)^n h_0(n), \tag{2}$$

$$\tilde{h}_0(n) = h_0(1 - n), \tag{3}$$

$$\tilde{h}_1(n) = (-1)^n h_0(1 - n), \tag{4}$$

so that the output of the inverse DWT is identical to the input of the forward DWT and the resulting DWT is an orthonormal transform.

For a signal of length N, the computational complexity of its DWT is $O(N)$, provided that the length of the filter $h_0(n)$ is negligible compared to N.

The two-dimensional (2-D) DWT of a 2-D signal can be implemented by using the one-dimensional (1-D) DWT in a separable fashion. At each level of decomposition (or reconstruction), the 1-D forward DWT (or inverse DWT) is first applied to every row of the signal and then applied to every column of the

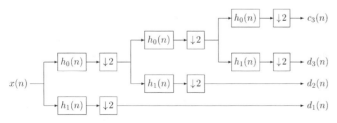

FIGURE 1 A three-level forward DWT based on a two-channel iterative filterbank.

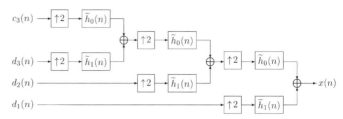

FIGURE 2 A three-level inverse DWT based on a two-channel iterative filterbank.

resulting data. For an image of size $N \times M$, the computational complexity of its 2-D DWT is $O(NM)$, provided that the length of the filter $h_0(n)$ is negligible compared to both N and M.

2.2 The Donoho–Johnstone Method

The method of wavelet shrinkage denoising was developed principally by Donoho and Johnstone [1–3]. Suppose we want to recover a one-dimensional signal f from a noisy observation g; i.e.,

$$g(n) = f(n) + q(n) \tag{5}$$

for $n = 0, 1, \ldots, N - 1$, where q is additive noise. The method attempts to reject noise by damping or thresholding in the wavelet domain. The estimate of the signal f is given by

$$\hat{f} = \mathcal{W}^{-1} \mathcal{T}_\lambda \mathcal{W} g \tag{6}$$

where the operators $\mathcal{W}$ and $\mathcal{W}^{-1}$ stand for the forward and inverse discrete wavelet transforms, respectively, and $\mathcal{T}_\lambda$ is a wavelet-domain pointwise thresholding operator with a threshold λ.

The key idea of wavelet shrinkage is that the wavelet representation can separate the signal and the noise. The DWT compacts the energy of the signal into a small number of DWT coefficients having large amplitudes, and it spreads the energy of the noise over a large number of DWT coefficients having small amplitudes. Hence, a thresholding operation attenuates noise energy by removing those small coefficients while maintaining signal energy by keeping these large coefficients unchanged.

There are two types of basic thresholding rules. For a given function $p(y)$, the *hard thresholding* operator is defined as

$$(\mathcal{T}_\lambda p)(y) = \begin{cases} p(y), & \text{if } |p(y)| > \lambda \\ 0, & \text{otherwise} \end{cases}, \tag{7}$$

and the *soft thresholding* operator is defined as

$$(\mathcal{T}_\lambda p)(y) = \begin{cases} p(y) - \lambda, & \text{if } p(y) > \lambda \\ p(y) + \lambda, & \text{if } p(y) < -\lambda \\ 0, & \text{otherwise.} \end{cases} \tag{8}$$

Since both hard thresholding and soft thresholding are nonlinear operators, wavelet shrinkage is a type of nonlinear processing. Since $\mathcal{T}_\lambda$ is a point processing operator, its computational complexity is $O(1)$. Hence, the complexity of the DWT-based wavelet shrinkage is $O(N)$ for a length-N signal.

If the parameter λ is too large, then the thresholding operation will remove a significant amount of signal energy, i.e., it causes oversmoothing. If λ is too small, then a significant amount of noise will not be suppressed. Several approaches have been proposed for selecting the threshold λ. The simplest are *VisuShrink* [1] and *SureShrink*, which is based on *Stein's Unbiased Risk Estimate* (SURE).

Both soft thresholding and hard thresholding require that the energy of the reconstructed signal $\hat{f}$ is lower than the energy of the noisy observation g. If an appropriate threshold is chosen, then the energy suppressed in wavelet shrinkage is mostly corresponding to the noise q. Therefore, the true signal f is not weakened after denoising.

2.3 Shift-Invariant Wavelet Shrinkage

One disadvantage of the DWT is that it is not a shift-invariant[1] transform. For instance, the DWT of $x(n-1)$ is not a shifted version of the DWT of $x(n)$. Such a shift variance is caused by the downsampling and upsampling operations. It has been argued [4] that DWT-based wavelet shrinkage sometimes produces visual artifacts such as the "pseudo-Gibbs phenomena" in the neighborhood of discontinuities in the signal due to the lack of shift invariance of the DWT. In order to avoid the problem caused by the DWT, Coifman and Donoho and Lang *et al.* independently proposed to use the *undecimated DWT* (UDWT) in wavelet shrinkage to achieve shift invariance [4, 5].

Figure 3 illustrates an implementation of a two-level forward UDWT based on a two-channel recursive filterbank. At each level of decomposition, both odd-indexed and even-indexed samples at the outputs of the filters $h_0(n)$ and $h_1(n)$ are maintained without decimation. Since there is no downsampler in the forward UDWT, the transform is a shift-invariant representation.

Since the number of UDWT coefficients is larger than the signal length, the inverse UDWT is not unique. In Fig. 3, if the filterbank satisfies Eqs. (2), (3), and (4), then the signal $x(n)$ can be exactly reconstructed from each of the four sets of UDWT coefficients: $\{c_2^{oo}(n), d_2^{oo}(n), d_1^{o}(n)\}$, $\{c_2^{oe}(n), d_2^{oe}(n), d_1^{o}(n)\}$, $\{c_2^{eo}(n), d_2^{eo}(n), d_1^{e}(n)\}$, and $\{c_2^{ee}(n), d_2^{ee}(n), d_1^{e}(n)\}$. For denoising applications, it is appropriate to reconstruct $x(n)$ by averaging all possible reconstructions.

It has been demonstrated in [4] and [5] that the UDWT-based denoising achieves considerably better performance than the DWT-based denoising. The cost of such an improvement in performance is the increase in computational complexity. For a length-N signal, if the length of the filter $h_0(n)$ is negligible

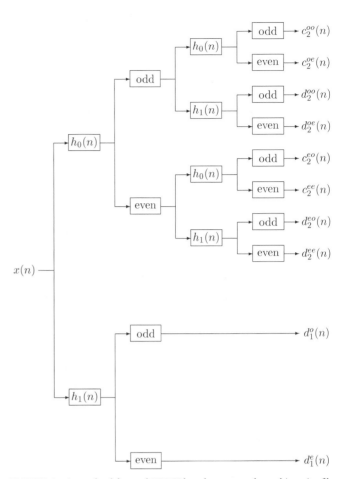

FIGURE 3 A two-level forward UDWT based on a two-channel iterative filterbank (the "odd" and "even" blocks stand for the downsamplers that sample odd-indexed and even-indexed outputs from the preceding filter, respectively).

compared to N, then the computational complexity of the UDWT is $O(N \log_2 N)$, which is higher than that of the DWT.

3 Image Enhancement by Means of Wavelet Shrinkage

3.1 Suppression of Additive Noise

Although wavelet shrinkage was originally proposed for removing noise in 1-D signals, it can be straightforwardly extended to images and other 2-D signals. Replacing the 1-D DWT by the 2-D DWT, we can apply directly the thresholding operation on the 2-D DWT coefficients. Hence, the computational complexity of the 2-D DWT-based wavelet shrinkage is $O(NM)$ for an image of size $N \times M$.

The 2-D version of the Donoho–Johnstone method has been extended to more sophisticated variations. Xu *et al.* proposed a wavelet-domain adaptive thresholding scheme to better preserve significant image features, which were identified by the spatial correlation of the wavelet coefficients at different scales [6].

Thresholding was performed only on the wavelet coefficients that do not correspond to any image features. A similar method was proposed by Hilton and Ogden in [7], where the significant wavelet coefficients were determined by recursive hypothesis tests. Malfait and Roose combined the wavelet representation and a Markov random field image model to incorporate a Bayesian statistical description for manipulating the wavelet coefficients of the noisy image [8]. Weyrich and Warhola applied the method of generalized cross validation to determine shrinkage parameters [9]. In [10], Chambolle *et al.* provided sharp estimates of the best wavelet shrinkage parameter in removing Gaussian noise from images.

Successful applications of denoising by wavelet shrinkage include the reduction of speckles in radar images [11] and the removal of noise in magnetic resonance imaging (MRI) data [6, 12, 13].

3.2 Removal of Blocking Artifacts in DCT-Coded Images

Lossy image coding is essential in many visual communications applications because a limited transmission bandwidth or storage space often does not permit lossless image coding, where compression ratios are typically low. However, the quality of lossy-coded images can be severely degraded and unacceptable, especially at low bit rates. The distortion caused by compression usually manifests itself as various perceptually annoying artifacts. This problem calls for postprocessing or enhancement of compressed images [14].

Most current image and video compression standards, such as JPEG (Chapter 5.5), H.261 (Chapter 6.1), MPEG-1, and MPEG-2 (Chapter 6.4), adopt the block discrete cosine transform (DCT). At the encoder, an image, a video frame, or a motion-compensated residual image is first partitioned into 8×8 nonoverlapping blocks of pixels. Then, an 8×8 DCT is performed on each block and the resulting transform coefficients are quantized and entropy coded. This independent processing of blocks does not take into account the between-block pixel correlations. Therefore, at low bit rates, such an encoding scheme typically leads to blocking artifacts, which manifest themselves as artificial discontinuities between adjacent blocks. In general, blocking artifacts are the most perceptually annoying distortion in images and video compressed by the various standards. The suppression of blocking artifacts has been studied as an image enhancement problem and as an image restoration problem. An overview of various approaches can be found in [14].

Though wavelet shrinkage techniques were originally proposed for the attenuation of signal-independent Gaussian noise, they work as well for the suppression of other types of distortion. In particular, wavelet shrinkage has been successful in removing coding artifacts in compressed images. Gopinath *et al.* first applied the Donoho–Johnstone method to attenuate blocking artifacts and obtained considerable improvement in terms

of both the objective and subjective image quality [15]. The success of wavelet shrinkage in the enhancement of compressed images is a result of the *compression property* of wavelet bases [16]. In a compressed image, the remaining important features (e.g., edges) after compression are typically dominant and global, and the coding artifacts are subdominant and local (e.g., the blocking artifacts in block DCT-coded images). The wavelet transform compacts the energy of those features into a small number of wavelet coefficients having large magnitude, and spreads the energy of the coding error into a large number of wavelet coefficients having small magnitude; i.e., the image features and the coding artifacts are well separated in the wavelet domain. Therefore, among the wavelet coefficients of a compressed image, those large coefficients very likely correspond to the original image, and those small ones very likely correspond to the coding artifacts. Naturally, keeping large coefficients and eliminating small ones (i.e., setting them to zero), or thresholding, will reduce the energy of the coding error.

Better enhancement performance can be achieved by using the UDWT-based shrinkage [17, 18] at the expense of increasing the postprocessing complexity from $O(NM)$ to $O(NM \log_2(NM))$ for an $N \times M$ image. For image coding applications in which fast decoding is desired, it is appropriate to use low-complexity postprocessing methods. In [19], the optimal shift-invariant wavelet packet basis is searched at the encoder and the basis is used at the decoder to attenuate the coding artifacts. Such a scheme achieves comparable enhancement performance with the UDWT-based method and possesses a low post-processing complexity $O(NM)$. The expenses are twofold: increase of encoding complexity, which is tolerable in many applications, and overhead bits required to code the optimal basis, which have a negligible effect on compression ratio.

4 Examples

In our simulations, we choose 512×512 8-bit gray-scale test images. We apply two wavelet shrinkage methods based on the DWT and the UDWT, respectively, to the distorted images and compare their enhancement performance in terms of both objective and subjective quality.

We use *peak signal-to-noise ratio* (PSNR) as the metric for objective image quality. The PSNR is defined as

$$PSNR = 10 \log_{10} \left\{ \frac{255^2}{\frac{1}{NM} \sum_{n_1=1}^{N} \sum_{n_2=1}^{M} [f(n_1, n_2) - \hat{f}(n_1, n_2)]^2} \right\}$$
(9)

where $f(n_1, n_2)$ and $\hat{f}(n_1, n_2)$, $1 \leq n_1 \leq N$, $1 \leq n_2 \leq M$, are the original image and the noisy image (or the enhanced image) with size $N \times M$, respectively.

We choose Daubechies' eight-tap orthonormal wavelet filterbank for both the DWT and the UDWT [20]. We perform five-level wavelet decomposition and reconstruction. We apply soft

FIGURE 4 Enhancement of a noisy "Barbara" image: (a) the original Barbara image; (b) image corrupted by Gaussian noise; (c) image enhanced with the DWT-based method; (d) image enhanced with the UDWT-based method.

thresholding and hard thresholding for the DWT-based shrinkage and the UDWT-based shrinkage, respectively.

4.1 Gaussian Noise

Figure 4 illustrates an example of removing additive white Gaussian noise by means of wavelet shrinkage. Figures 4(a) and 4(b) display the original "Barbara" image and a noisy version, respectively. The PSNR of the noisy image is 24.6 dB. Figures 4(c) and 4(d) show the images enhanced by means of wavelet shrinkage based on the DWT and the UDWT, respectively. The PSNRs of the two enhanced images are 28.3 and 30.1 dB, respectively. Comparing the four images, we conclude that the perceptual quality of the enhanced images are significantly better than the noisy image: noise is greatly removed while sharp image features are well preserved without noticeable blurring. Although both methods improve the objective and subjective quality of the distorted

image, the UDWT-based method achieves better performance, i.e., higher PSNR and better subjective quality, than the DWT-based method.

4.2 Blocking Artifacts

Figure 5 illustrates an example of suppressing blocking artifacts in JPEG-compressed images. Figure 5(a) is a part of the original "Lena" image. Figure 5(b) is the same part of a JPEG-compressed version at 0.25 bit per pixel (bpp), where blocking artifacts are clearly visible. The PSNR of the compressed image is 30.4 dB. Figures 5(c) and 5(d) are the corresponding parts in the images enhanced by means of DWT-based shrinkage and UDWT-based shrinkage, respectively. The PSNRs of the two enhanced images are 31.1 and 31.4 dB, respectively; i.e., the UDWT-based shrinkage achieves better objective quality. Although both of them have better visual quality than the JPEG-compressed one, the artifacts

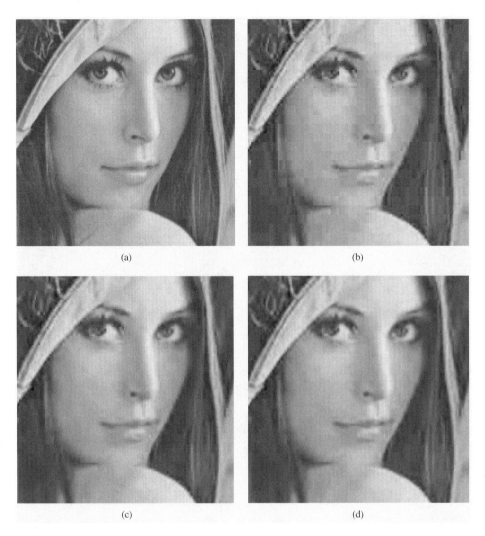

FIGURE 5   Enhancement of a JPEG-compressed "Lena" image: (a) a region of the original Lena image; (b) JPEG-compressed image; (c) image postprocessed by the DWT-based method; (d) image postprocessed by the UDWT-based method.

are more completely removed in Fig. 5(d) than in Fig. 5(c); i.e., the UDWT-based method achieves a better tradeoff between the suppression of coding artifacts and the preservation of image features.

5 Summary

We have presented an overview of image enhancement by means of wavelet denoising. Compared to many classical filtering-based methods, wavelet-based methods can achieve a better tradeoff between noise reduction and feature preservation. Another advantage of wavelet denoising is its low computational complexity. Waveket denoising is a powerful tool for image enhancement. The success of wavelet image denoising derives from the same property as does the success of wavelet image compression algorithms (Chapter 5.4): the compact image representation provided by the discrete wavelet transform.

References

[1] D. L. Donoho and I. M. Johnstone, "Ideal Spatial Adaptation via Wavelet Shrinkage," *Biometrika* **81**, 425–455 (1994).

[2] D. L. Donoho. "De-noising by soft-thresholding," *IEEE Trans. Inform. Theory* **41**, 613–627 (1995).

[3] D. L. Donoho and I. M. Johnstone, "Adaptation to unknown smoothness via wavelet shrinkage," *J. Amer. Stat. Assoc.* **90**, 1200–1224 (1995).

[4] R. R. Coifman and D. L. Donoho, "Translation-invariant denoising," in *Wavelets and Statistics*, A. Antoniadis and G. Oppenheim, eds., (Springer, Berlin, 1995), pp. 125–150.

[5] M. Lang, H. Guo, J. E. Odegard, C. S. Burrus, and R. O. Wells, Jr., "Noise reduction using an undecimated discrete wavelet transform," *IEEE Signal Process. Lett.* **3**, 10–12 (1996).

[6] Y. Xu, J. B. Weaver, D. M. Healy, Jr., and J. Lu, "Wavelet transform domain filters: A spatially selective noise filtration technique," *IEEE Trans. Image Process.* **3**, 747–758 (1994).

[7] M. L. Hilton and R. T. Ogden, "Data analytic wavelet threshold

selection in 2-D signal denoising," *IEEE Trans. Signal Process.* **45**, 496–500 (1997).

[8] M. Malfait and D. Roose, "Wavelet-based image denoising using a Markov random field *a priori* model," *IEEE Trans. Image Process.* **6**, 549–565 (1997).

[9] N. Weyrich and G. T. Warhola, "Wavelet shrinkage and generalized validation for image denoising," *IEEE Trans. Image Process.* **7**, 82–90 (1998).

[10] A. Chambolle, R. A. DeVore, N. Lee, and B. J. Lucier, "Nonlinear wavelet image processing: Variational problems, compression, and noise removal through wavelet shrinkage," *IEEE Trans. Image Process.* **7**, 319–335 (1998).

[11] P. Moulin, "A wavelet regularization method for diffuse radar-target imaging and speckle-noise reduction," *J. Math. Imaging Vision* **3**, 123–134 (1993).

[12] J. B. Weaver, Y. Xu, D. M. Healy, Jr., and L. D. Cromwell, "Filtering noise from images with wavelet transforms," *Magn. Reson. Med.* **21**, 288–295 (1991).

[13] M. L. Hilton, T. Ogden, D. Hattery, G. Eden, and B. Jawerth, "Wavelet denoising of functional MRI data," in *Wavelets in Medicine and Biology*, A. Aldroubi and M. Unser, eds., (CRC Press, Boca Raton, FL, 1996), pp. 93–114.

[14] M.-Y. Shen and C.-C. J. Kuo, "Review of postprocessing techniques for compression artifact removal," *J. Visual Commun. Image Rep.* **9**, 2–14 (1998). Special Issue on High-Fidelity Media Processing.

[15] R. A. Gopinath, M. Lang, H. Guo, and J. E. Odegard, "Wavelet-based post-processing of low bit rate transform coded images," in *Proc. IEEE Int. Conf. Image Processing* (IEEE, New York, 1994), Vol II, pp. 913–917.

[16] D. L. Donoho, "Unconditional bases are optimal bases for data compression and for statistical estimation," *Appl. Comput. Harmon. Anal.* **1**, 100–115 (1993).

[17] D. Wei and C. S. Burrus, "Optimal wavelet thresholding for various coding schemes," in *Proc. IEEE Int. Conf. Image Processing* (IEEE, New York, 1995), Vol. I, pp. 610–613.

[18] Z. Xiong, M. T. Orchard, and Y.-Q. Zhang, "A deblocking algorithm for JPEG compressed images using overcomplete wavelet representations," *IEEE Trans. Circuits Syst. Video Technol.* **7**, 433–437 (1997).

[19] D. Wei and A. C. Bovik, "Enhancement of compressed images by optimal shift-invariant wavelet packet basis," *J. Visual Commun. Image Rep.* **9**, 15–24 (1998). Special Issue on High-Fidelity Media Processing.

[20] I. Daubechies, *Ten Lectures on Wavelets* (Soc. Indus. Appl. Math., Philadelphia, PA, 1992).

3.5

Basic Methods for Image Restoration and Identification

Reginald L. Lagendijk
and Jan Biemond
Delft University of Technology

1 Introduction

Images are produced to record or display useful information. Because of imperfections in the imaging and capturing process, however, the recorded image invariably represents a degraded version of the original scene. The undoing of these imperfections is crucial to many of the subsequent image processing tasks. There exists a wide range of different degradations that have to be taken into account, covering for instance noise, geometrical degradations (pin-cushion distortion), illumination and color imperfections (under- or overexposure, saturation), and blur. This chapter concentrates on basic methods for removing blur from recorded sampled (spatially discrete) images. There are many excellent overview articles, journal papers, and textbooks on the subject of image restoration and identification. Readers interested in more details than given in this chapter are referred to [2, 3, 9, 11, 14].

Blurring is a form of bandwidth reduction of an ideal image caused by the imperfect image formation process. It can be caused by relative motion between the camera and the original scene, or by an optical system that is out of focus. When aerial photographs are produced for remote sensing purposes, blurs are introduced by atmospheric turbulence, aberrations in the optical system, and relative motion between the camera and the ground. Such blurring is not confined to optical images; for example, electron micrographs are corrupted by spherical aberrations of the electron lenses, and CT scans suffer from X-ray scatter.

In addition to these blurring effects, noise always corrupts any recorded image. Noise may be introduced by the medium through which the image is created (random absorption or scatter effects), by the recording medium (sensor noise), by measurement errors due to the limited accuracy of the recording system, and by quantization of the data for digital storage.

The field of *image restoration* (sometimes referred to as image deblurring or image deconvolution) is concerned with the reconstruction or estimation of the uncorrupted image from a blurred and noisy one. Essentially, it tries to perform an operation on the image that is the inverse of the imperfections in the image formation system. In the use of image restoration methods, the characteristics of the degrading system and the noise are assumed to be known *a priori*. In practical situations, however, one may not be able to obtain this information directly from the image formation process. The goal of *blur identification* is to estimate the attributes of the imperfect imaging system from the observed degraded image itself prior to the restoration process. The combination of image restoration and blur identification is often referred to as *blind image deconvolution* [11].

Image restoration algorithms distinguish themselves from image *enhancement* methods in that they are based on models for the degrading process and for the ideal image. For those cases in which a fairly accurate blur model is available, powerful restoration algorithms can be arrived at. Unfortunately, in numerous practical cases of interest the modeling of the blur is unfeasible, rendering restoration impossible. The limited validity of blur models is often a factor of disappointment, but one should realize that if none of the blur models described in this chapter are applicable, the corrupted image may well be beyond restoration. Therefore, no matter how powerful blur identification and restoration algorithms are, the objective *when capturing* an image undeniably is to avoid the need for restoring the image.

The image restoration methods that are described in this chapter fall under the class of *linear spatially invariant restoration* filters. We assume that the blurring function acts as a convolution kernel or *point-spread function* $d(n_1, n_2)$ that does not vary spatially. It is also assumed that the statistical properties (mean and correlation function) of the image and noise do not change spatially. Under these conditions the restoration process can be carried out by means of a linear filter of which the point-spread function is spatially invariant, i.e., is constant throughout the image. These modeling assumptions can be mathematically formulated as follows. If we denote by $f(n_1, n_2)$ the desired ideal spatially discrete image that does not contain any blur or noise, then the recorded image $g(n_1, n_2)$ is modeled as [see also Fig. 1(a)] [1]

$$g(n_1, n_2) = d(n_1, n_2) * f(n_1, n_2) + w(n_1, n_2)$$
$$= \sum_{k_1=0}^{N-1} \sum_{k_2=0}^{M-1} d(k_1, k_2) f(n_1 - k_1, n_2 - k_2) + w(n_1, n_2)$$
(1)

Here $w(n_1, n_2)$ is the noise that corrupts the blurred image. Clearly the objective of image restoration is to make an estimate $\hat{f}(n_1, n_2)$ of the ideal image $f(n_1, n_2)$, given only the degraded image $g(n_1, n_2)$, the blurring function $d(n_1, n_2)$ and some information about the statistical properties of the ideal image and the noise.

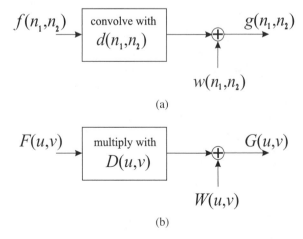

(a)

(b)

FIGURE 1 Image formation model in the (a) spatial domain and (b) Fourier domain.

An alternative way of describing Eq. (1) is through its spectral equivalence. By applying discrete Fourier transforms to Eq. (1), we obtain the following representation [see also Fig. 1(b)]:

$$G(u, v) = D(u, v) F(u, v) + W(u, v),$$
(2)

where (u, v) are the spatial frequency coordinates, and capitals represent Fourier transforms. Either Eq. (1) or (2) can be used for developing restoration algorithms. In practice the spectral representation is more often used since it leads to efficient implementations of restoration filters in the (discrete) Fourier domain.

In Eqs. (1) and (2), the noise $w(n_1, n_2)$ is modeled as an additive term. Typically the noise is considered to have a zero mean and to be white, i.e., spatially uncorrelated. In statistical terms this can be expressed as follows [15]:

$$E[w(n_1, n_2)] \approx \sum_{k_1=0}^{N-1} \sum_{k_2=0}^{M-1} w(k_1, k_2) = 0$$
(3a)

$$R_w(k_1, k_2) = E[w(n_1, n_2)w(n_1 - k_1, n_2 - k_2)]$$
$$\approx \sum_{n_1=0}^{N-1} \sum_{n_2=0}^{M-1} w(n_1, n_2)w(n_1 - k_1, n_2 - k_2)$$
$$= \begin{cases} \sigma_w^2 & \text{if } k_1 = k_2 = 0 \\ 0 & \text{elsewhere} \end{cases}.$$
(3b)

Here σ_w^2 is the variance or power of the noise and $E[\ ]$ refers to the expected value operator. The approximate equality indicates that on the average Eq. (3) should hold, but that for a given image Eq. (3) holds only approximately as a result of replacing the expectation by a pixelwise summation over the image. Sometimes the noise is assumed to have a Gaussian probability density function, but for none of the restoration algorithms described in this chapter is this a necessary condition.

In general the noise $w(n_1, n_2)$ may not be independent of the ideal image $f(n_1, n_2)$. This may happen, for instance, if the image formation process contains nonlinear components, or if the noise is multiplicative instead of additive. Unfortunately, this dependency is often difficult to model or to estimate. Therefore, noise and ideal image are usually assumed to be orthogonal, which is — in this case — equivalent to being uncorrelated because the noise has zero mean. In statistical terms expressed, the following condition holds:

$$R_{fw}(k_1, k_2) = E[f(n_1, n_2)w(n_1 - k_1, n_2 - k_2)]$$
$$\approx \sum_{n_1=0}^{N-1} \sum_{n_2=0}^{M-1} f(n_1, n_2)w(n_1 - k_1, n_2 - k_2) = 0.$$
(4)

Models (1)–(4) form the foundations for the class of linear spatially invariant image restoration and accompanying blur identification algorithms. In particular these models apply to monochromatic images. For color images, two approaches can be taken. In the first place one can extend Eqs. (1)–(4) to incorporate multiple color components. In many practical cases of interest this is indeed the proper way of modeling the problem of color image restoration, since the degradations of the different color components (such as the tristimulus signals red–green–blue, luminance–hue–saturation, or luminance–chrominance) are not independent. This leads to a class of algorithms known as "multiframe filters" [5, 9]. A second, more pragmatic way of dealing with color images is to assume that the noises and blurs in each of the color components are independent. The restoration of the color components can then be carried out independently as well, meaning that one simply regards each color component as a monochromatic image by itself, forgetting about the other color components. Though obviously this model might be in error, acceptable results have been achieved in this way.

The outline of this chapter is as follows. In Section 2, we first describe several important models for linear blurs, namely motion blur, out-of-focus blur, and blur due to atmospheric turbulence. In Section 3, three classes of restoration algorithms are introduced and described in detail, namely the inverse filter, the Wiener and constrained least-squares filter, and the iterative restoration filters. In Section 4, two basic approaches to blur identification will be described briefly.

2 Blur Models

The blurring of images is modeled in Eq. (1) as the convolution of an ideal image with a two-dimensional (2-D) point-spread function (PSF), $d(n_1, n_2)$. The interpretation of Eq. (1) is that *if* the ideal image $f(n_1, n_2)$ would consist of a single intensity point or point source, this point would be recorded as a spread-out intensity pattern[1] $d(n_1, n_2)$; hence the name *point-spread* function.

It is worth noticing that point-spread functions in this chapter are not a function of the spatial location under consideration, i.e., they are spatially invariant. Essentially this means that the image is blurred in exactly the *same* way at *every* spatial location. Point-spread functions that do not follow this assumption are, for instance, due to rotational blurs (turning wheels) or local blurs (a person out of focus while the background is in focus). The modeling, restoration, and identification of images degraded by spatially varying blurs is outside the scope of this chapter and is actually still a largely unsolved problem.

In most cases the blurring of images is a spatially continuous process. Since identification and restoration algorithms are always based on spatially discrete images, we present the blur models in their continuous forms, followed by their discrete (sampled) counterparts. We assume that the sampling rate of the images has been chosen high enough to minimize the

(aliasing) errors involved in going from the continuous to discrete models.

The spatially continuous PSF $d(x, y)$ of any blur satisfies three constraints, namely:

- $d(x, y)$ takes on nonnegative values only, because of the physics of the underlying image formation process;
- when real-valued images are dealt with the point-spread function $d(x, y)$ is real-valued too;
- the imperfections in the image formation process are modeled as passive operations on the data, i.e., no "energy" is absorbed or generated. Consequently, for spatially continuous blurs and for spatially discrete blurs the PSF is constrained to satisfy

$$\int_{-\infty}^{\infty} \int_{-\infty}^{\infty} d(x, y)\, \mathrm{d}x\, \mathrm{d}y = 1, \qquad (5\mathrm{a})$$

$$\sum_{n_1=0}^{N-1} \sum_{n_2=0}^{M-1} d(n_1, n_2) = 1, \qquad (5\mathrm{b})$$

respectively. In the following paragraphs we present four common point-spread functions, which are encountered regularly in practical situations of interest.

2.1 No Blur

In case in which the recorded image is imaged perfectly, no blur will be apparent in the discrete image. The spatially continuous PSF can then be modeled as a Dirac delta function:

$$d(x, y) = \delta(x, y) \qquad (6\mathrm{a})$$

and the spatially discrete PSF as a unit pulse:

$$d(n_1, n_2) = \delta(n_1, n_2) = \begin{cases} 1 & \text{if } n_1 = n_2 = 0 \\ 0 & \text{elsewhere} \end{cases}. \qquad (6\mathrm{b})$$

Theoretically, Eq. (6a) can never be satisfied. However, as long as the amount of "spreading" in the continuous image is smaller than the sampling grid applied to obtain the discrete image, Eq. (6b) will be arrived at.

2.2 Linear Motion Blur

Many types of motion blur can be distinguished, all of which are due to relative motion between the recording device and the scene. This can be in the form of a translation, a rotation, a sudden change of scale, or some combinations of these. Here only the important case of a global translation will be considered.

When the scene to be recorded translates relative to the camera at a constant velocity v_{relative} under an angle of ϕ radians with the horizontal axis during the exposure interval $[0, t_{\text{exposure}}]$, the distortion is one dimensional. Defining the "length of motion"

---

[1]Ignoring the noise for a moment.

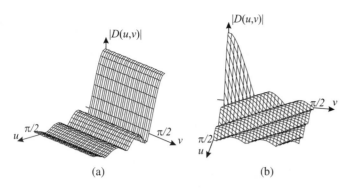

(a)　　　　　　　　　　　　　　　　　(b)

FIGURE 2　PSF of motion blur in the Fourier domain, showing $|D(u, v)|$, for (a) $L = 7.5$ and $\phi = 0$; (b) $L = 7.5$ and $\phi = \pi/4$

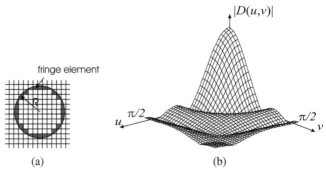

(a)　　　　　　　　　　　　　　　　　(b)

FIGURE 3　(a) Fringe elements of discrete out-of-focus blur that are calculated by integration; (b) PSF in the Fourier domain, showing $|D(u, v)|$, for $R = 2.5$.

by $L = v_{\text{relative}} \, t_{\text{exposure}}$, we find the PSF is given by

$$d(x, y; L, \phi) = \begin{cases} \frac{1}{L} & \text{if } \sqrt{x^2 + y^2} \le \frac{L}{2}, \; \frac{x}{y} = -\tan\phi \\ 0 & \text{elsewhere} \end{cases}. \quad (7a)$$

The discrete version of Eq. (7a) is not easily captured in a closed form expression in general. For the special case that $\phi = 0$, an appropriate approximation is

$$d(n_1, n_2; L)$$
$$= \begin{cases} \frac{1}{L} & \text{if } n_1 = 0, |n_2| \le \left\lfloor \frac{L-1}{2} \right\rfloor \\ \frac{1}{2L}\left\{ (L-1) - 2\left\lfloor \frac{L-1}{2} \right\rfloor \right\} & \text{if } n_1 = 0, |n_2| = \left\lceil \frac{L-1}{2} \right\rceil \\ 0 & \text{elsewhere} \end{cases}. \quad (7b)$$

Figure 2(a) shows the modulus of the Fourier transform of the PSF of motion blur with $L = 7.5$ and $\phi = 0$. This figure illustrates that the blur is effectively a horizontal low-pass filtering operation and that the blur has spectral zeros along characteristic lines. The interline spacing of these characteristic zero pattern is (for the case that $N = M$) approximately equal to N/L. Figure 2(b) shows the modulus of the Fourier transform for the case of $L = 7.5$ and $\phi = \pi/4$.

2.3 Uniform Out-of-Focus Blur

When a camera images a three-dimensional (3-D) scene onto a 2-D imaging plane, some parts of the scene are in focus while other parts are not. If the aperture of the camera is circular, the image of any point source is a small disk, known as the circle of confusion (COC). The degree of defocus (diameter of the COC) depends on the focal length and the aperture number of the lens, and the distance between camera and object. An accurate model not only describes the diameter of the COC, but also the intensity distribution within the COC. However, if the degree of defocusing is large relative to the wavelengths considered, a geometrical approach can be followed resulting in a uniform intensity distribution within the COC. The spatially continuous

PSF of this uniform out-of-focus blur with radius R is given by

$$d(x, y; R) = \begin{cases} \frac{1}{\pi R^2} & \text{if } \sqrt{x^2 + y^2} \le R^2 \\ 0 & \text{elsewhere} \end{cases}. \quad (8a)$$

Also for this PSF the discrete version $d(n_1, n_2)$ is not easily arrived at. A coarse approximation is the following spatially discrete PSF:

$$d(n_1, n_2; R) = \begin{cases} \frac{1}{C} & \text{if } \sqrt{n_1^2 + n_2^2} \le R^2 \\ 0 & \text{elsewhere} \end{cases}, \quad (8b)$$

where C is a constant that must be chosen so that Eq. (5b) is satisfied. Approximation (8b) is incorrect for the fringe elements of the point-spread function. A more accurate model for the fringe elements would involve the integration of the area covered by the spatially continuous PSF, as illustrated in Fig. 3. Figure 3(a) shows the fringe elements that have to be calculated by integration. Figure 3(b) shows the modulus of the Fourier transform of the PSF for $R = 2.5$. Again, a low-pass behavior can be observed (in this case both horizontally and vertically), as well as a characteristic pattern of spectral zeros.

2.4 Atmospheric Turbulence Blur

Atmospheric turbulence is a severe limitation in remote sensing. Although the blur introduced by atmospheric turbulence depends on a variety of factors (such as temperature, wind speed, and exposure time), for long-term exposures the point-spread function can be described reasonably well by a Gaussian function:

$$d(x, y; \sigma_G) = C \exp\left(-\frac{x^2 + y^2}{2\sigma_G^2} \right). \quad (9a)$$

Here σ_G determines the amount of spread of the blur, and the constant C is to be chosen so that Eq. (5a) is satisfied. Since Eq. (9a) constitutes a PSF that is separable in a horizontal and a vertical component, the discrete version of Eq. (9a) is usually obtained by first computing a one-dimensional (1-D) discrete

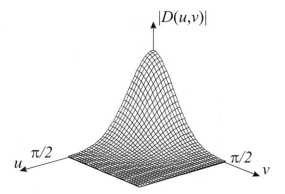

FIGURE 4 Gaussian PSF in the Fourier domain ($\sigma_G = 1.2$).

Gaussian PSF $\tilde{d}(n)$. This 1-D PSF is found by a numerical discretization of the continuous PSF. For each PSF element $\tilde{d}(n)$, the 1-D continuous PSF is integrated over the area covered by the 1-D sampling grid, namely $[n - 1/2, n + 1/2]$:

$$\tilde{d}(n; \sigma_G) = C \int_{n-1/2}^{n+1/2} \exp\left(-\frac{x^2}{2\sigma_G^2}\right) dx. \quad (9b)$$

Since the spatially continuous PSF does not have a finite support, it has to be truncated properly. The spatially discrete approximation of Eq. (9a) is then given by

$$d(n_1, n_2; \sigma_G) = \tilde{d}(n_1; \sigma_G)\tilde{d}(n_2; \sigma_G). \quad (9c)$$

Figure 4 shows this PSF in the spectral domain ($\sigma_G = 1.2$). Observe that Gaussian blurs do not have exact spectral zeros.

3 Image Restoration Algorithms

In this section we will assume that the PSF of the blur is satisfactorily known. A number of methods will be introduced for removing the blur from the recorded image $g(n_1, n_2)$ using a linear filter. If the point-spread function of the linear restoration filter, denoted by $h(n_1, n_2)$, has been designed, the restored image is given by

$$\hat{f}(n_1, n_2) = h(n_1, n_2) * g(n_1, n_2)$$
$$= \sum_{k_1=0}^{N-1} \sum_{k_2=0}^{M-1} h(k_1, k_2) g(n_1 - k_1, n_2 - k_2) \quad (10a)$$

or in the spectral domain by

$$\hat{F}(u, v) = H(u, v) G(u, v). \quad (10b)$$

The objective of this section is to design appropriate restoration filters $h(n_1, n_2)$ or $H(u, v)$ for use in Eq. (10).

In image restoration the improvement in quality of the restored image over the recorded blurred one is measured by the signal-to-noise ratio improvement. The signal-to-noise-ratio of the recorded (blurred and noisy) image is defined as follows in decibels:

$$\begin{aligned} &\mathrm{SNR}_g \\ &= 10 \log_{10}\left(\frac{\text{variance of the ideal image } f(n_1, n_2)}{\text{variance of the difference image } g(n_1, n_2) - f(n_1, n_2)}\right) \text{(dB).} \end{aligned}$$
$$(11a)$$

The signal-to-noise ratio of the restored image is similarly defined as

$$\begin{aligned} &\mathrm{SNR}_{\hat{f}} \\ &= 10 \log_{10}\left(\frac{\text{variance of the ideal image } f(n_1, n_2)}{\text{variance of the difference image } \hat{f}(n_1, n_2) - f(n_1, n_2)}\right) \text{(dB).} \end{aligned}$$
$$(11b)$$

Then, the improvement in signal-to-noise ratio (SNR) is given by

$$\begin{aligned} &\Delta\mathrm{SNR} = \mathrm{SNR}_{\hat{f}} - \mathrm{SNR}_g \\ &= 10 \log_{10}\left(\frac{\text{variance of the difference image } g(n_1, n_2) - f(n_1, n_2)}{\text{variance of the difference image } \hat{f}(n_1, n_2) - f(n_1, n_2)}\right) \text{(dB).} \end{aligned}$$
$$(11c)$$

The improvement in SNR is basically a measure that expresses the reduction of disagreement with the ideal image when comparing the distorted and restored image. Note that all of the above signal-to-noise measures can only be computed in the case in which the ideal image $f(n_1, n_2)$ is available, i.e., in an experimental setup or in a design phase of the restoration algorithm. When restoration filters are applied to real images for which the ideal image is not available, often only the visual judgment of the restored image can be relied upon. For this reason it is desirable for a restoration filter to be somewhat "tunable" to the liking of the user.

3.1 Inverse Filter

An inverse filter is a linear filter whose point-spread function $h_{\mathrm{inv}}(n_1, n_2)$ is the inverse of the blurring function $d(n_1, n_2)$, in the sense that

$$\begin{aligned} &h_{\mathrm{inv}}(n_1, n_2) * d(n_1, n_2) \\ &= \sum_{k_1=0}^{N-1} \sum_{k_2=0}^{M-1} h_{\mathrm{inv}}(k_1, k_2) d(n_1 - k_1, n_2 - k_2) \\ &= \delta(n_1, n_2). \end{aligned} \quad (12)$$

When formulated as in Eq. (12), inverse filters seem difficult to

FIGURE 5 (a) Image out-of-focus with $\text{SNR}_g = 10.3$ dB (noise variance $= 0.35$). (b) Inverse filtered image. (c) Magnitude of the Fourier transform of the restored image. The DC component lies in the center of the image. The oriented white lines are spectral components of the image with large energy. (d) Magnitude of the Fourier transform of the inverse filter response.

design. However, the spectral counterpart of Eq. (12) immediately shows the solution to this design problem [1]:

$$H_{\text{inv}}(u, v) D(u, v) = 1 \Rightarrow H_{\text{inv}}(u, v) = \frac{1}{D(u, v)}. \quad (13)$$

The advantage of the *inverse filter* is that it requires only the blur PSF as *a priori* knowledge, and that it allows for perfect restoration in the case that noise is absent, as one can easily see by substituting Eq. (13) into Eq. (10b):

$$\begin{aligned}
\hat{F}_{\text{inv}}(u, v) &= H_{\text{inv}}(u, v) G(u, v) \\
&= \frac{1}{D(u, v)} (D(u, v) F(u, v) + W(u, v)) \\
&= F(u, v) + \frac{W(u, v)}{D(u, v)}. \quad (14)
\end{aligned}$$

If the noise is absent, the second term in Eq. (14) disappears so that the restored image is identical to the ideal image. Unfortunately, several problems exist with Eq. (14). In the first place the inverse filter may not exist because $D(u, v)$ is zero at selected

frequencies (u, v). This happens for both the linear motion blur and the out-of-focus blur described in the previous section. Second, even if the blurring function's spectral representation $D(u, v)$ does not actually go to zero but becomes small, the second term in Eq. (14) — known as the inverse filtered noise — will become very large. Inverse filtered images are therefore often dominated by excessively amplified noise.[2]

Figure 5(a) shows an image degraded by out-of-focus blur ($R = 2.5$) and noise. The inverse filtered version is shown in Fig. 5(b), clearly illustrating its uselessness. The Fourier transforms of the restored image and of $H_{\text{inv}}(u, v)$ are shown in Figs. 5(c) and 5(d), respectively, demonstrating that indeed the spectral zeros of the PSF cause problems.

3.2 Least-Squares Filters

For the noise sensitivity of the inverse filter to be overcome, a number of restoration filters have been developed; these are collectively called least-squares filters. We describe the two most

---

[2]In the literature, this effect is commonly referred to as the ill-conditionedness or ill-posedness of the restoration problem.

commonly used filters from this collection, namely the Wiener filter and the constrained least-squares filter.

The Wiener filter is a linear spatially invariant filter of the form of Eq. (10a), in which the point-spread function $h(n_1, n_2)$ is chosen such that it minimizes the mean-squared error (MSE) between the ideal and the restored image. This criterion attempts to make the difference between the ideal image and the restored one — i.e., the remaining restoration error — as small as possible *on the average*:

$$\text{MSE} = E(f(n_1, n_2) - \hat{f}(n_1, n_2))^2$$
$$\approx \sum_{n_1=0}^{N-1} \sum_{n_2=0}^{M-1} (f(n_1, n_2) - \hat{f}(n_1, n_2))^2, \quad (15)$$

where $\hat{f}(n_1, n_2)$ is given by Eq. (10a). The solution of this minimization problem is known as the *Wiener filter*, and it is easiest defined in the spectral domain:

$$H_{\text{wiener}}(u, v) = \frac{D^*(u, v)}{D^*(u, v)D(u, v) + [S_w(u, v)/S_f(u, v)]}. \quad (16)$$

Here $D^*(u, v)$ is the complex conjugate of $D(u, v)$, and $S_f(u, v)$ and $S_w(u, v)$ are the power spectrum of the ideal image and the noise, respectively. The power spectrum is a measure for the average signal power per spatial frequency (u, v) carried by the image. In the noiseless case we have $S_w(u, v) = 0$, so that the Wiener filter approximates the inverse filter:

$$H_{\text{wiener}}(u, v)\big|_{S_w(u,v) \to 0} = \begin{cases} \frac{1}{D(u, v)} & \text{for } D(u, v) \neq 0 \\ 0 & \text{for } D(u, v) = 0 \end{cases} \quad (17)$$

For the more typical situation in which the recorded image is noisy, the Wiener filter trades off the restoration by inverse filtering and suppression of noise for those frequencies where $D(u, v) \to 0$. The important factors in this tradeoff are the power spectra of the ideal image and the noise. For spatial frequencies where $S_w(u, v) \ll S_f(u, v)$, the Wiener filter approaches the inverse filter, while for spatial frequencies where $S_w(u, v) \gg S_f(u, v)$ the Wiener filter acts as a frequency rejection filter, i.e., $H_{\text{wiener}}(u, v) \to 0$.

If we assume that the noise is uncorrelated (white noise), its power spectrum is determined by the noise variance only:

$$S_w(u, v) = \sigma_w^2 \quad \text{for all } (u, v). \quad (18)$$

Thus, it is sufficient to estimate the noise variance from the recorded image to get an estimate of $S_w(u, v)$. The estimation of the noise variance can also be left to the user of the Wiener filter as if it were a tunable parameter. Small values of σ_w^2 will yield a result close to the inverse filter, whereas large values will oversmooth the restored image.

The estimation of $S_f(u, v)$ is somewhat more problematic since the ideal image is obviously not available. There are three

possible approaches to take. In the first place, one can replace $S_f(u, v)$ by an estimate of the power spectrum of the blurred image and compensate for the variance of the noise σ_w^2:

$$S_f(u, v) \approx S_g(u, v) - \sigma_w^2 \approx \frac{1}{NM} G^*(u, v)G(u, v) - \sigma_w^2. \quad (19)$$

The above used estimator for the power spectrum $S_g(u, v)$ of $g(n_1, n_2)$ is known as the periodogram. This estimator requires little *a priori* knowledge, but it is known to have several shortcomings. More elaborate estimators for the power spectrum exist, but these require much more *a priori* knowledge.

A second approach is to estimate the power spectrum $S_f(u, v)$ from a set of representative images. These representative images are to be taken from a collection of images that have a content "similar" to the image that has to be restored. Of course, one still needs an appropriate estimator to obtain the power spectrum from the set of representative images.

The third and final approach is to use a statistical model for the ideal image. Often these models incorporate parameters that can be tuned to the actual image being used. A widely used image model — not only popular in image restoration but also in image compression — is the following 2-D causal autoregressive model [8]:

$$f(n_1, n_2) = a_{0,1} f(n_1, n_2 - 1) + a_{1,1} f(n_1 - 1, n_2 - 1)$$
$$+ a_{1,0} f(n_1 - 1, n_2) + v(n_1, n_2). \quad (20a)$$

In this model the intensities at the spatial location (n_1, n_2) are described as the sum of weighted intensities at neighboring spatial locations and a small unpredictable component $v(n_1, n_2)$. The unpredictable component is often modeled as white noise with variance σ_v^2. Table 1 gives numerical examples for mean-square error estimates of the *prediction coefficients* $a_{i,j}$ for some images. For the mean-square error estimation of these parameters, first the 2-D autocorrelation function has been estimated, which is then used in the Yule–Walker equations [8]. Once the model parameters for Eq. (20a) have been chosen, the power spectrum can be calculated to be equal to

$$S_f(u, v) = \frac{\sigma_v^2}{|1 - a_{0,1} e^{-ju} - a_{1,1} e^{-ju-jv} - a_{1,0} e^{-jv}|^2}. \quad (20b)$$

TABLE 1 Prediction coefficients and variance of $v(n_1, n_2)$ for four images[a]

Image	$a_{0,1}$	$a_{1,1}$	$a_{1,0}$	σ_v^2
Cameraman	0.709	−0.467	0.739	231.8
Lena	0.511	−0.343	0.812	132.7
Trevor White	0.759	−0.525	0.764	33.0
White noise	−0.008	−0.003	−0.002	5470.1

[a] These are computed in the MSE optimal sense by the Yule–Walker equations.

FIGURE 6 (a) Wiener restoration of image in Fig. 5(a) with assumed noise variance equal to 35.0 (ΔSNR = 3.7 dB). (b) Restoration using the correct noise variance of 0.35 (ΔSNR = 8.8 dB). (c) Restoration assuming the noise variance is 0.0035 (ΔSNR = 1.1 dB). (d) Magnitude of the Fourier transform of the restored image in (b).

The tradeoff between noise smoothing and deblurring that is made by the Wiener filter is illustrated in Fig. 6. Going from 6(a) to 6(c), the variance of the noise in the degraded image, i.e., σ_w^2, has been estimated too large, optimally, and too small, respectively. The visual differences, as well as the differences in improvement in SNR (ΔSNR) are substantial. The power spectrum of the original image has been calculated from model (20a). From the results it is clear that the excessive noise amplification of the earlier example is no longer present because of the masking of the spectral zeros [see Fig. 6(d)]. Typical artifacts of the Wiener restoration — and actually of most restoration filters — are the residual blur in the image and the "ringing" or "halo" artifacts present near edges in the restored image.

The *constrained least-squares filter* [7] is another approach for overcoming some of the difficulties of the inverse filter (excessive noise amplification) and of the Wiener filter (estimation of the power spectrum of the ideal image), while still retaining the simplicity of a spatially invariant linear filter. If the restoration is a good one, the blurred version of the restored image should be approximately equal to the recorded distorted image. That is,

$$d(n_1, n_2) * \hat{f}(n_1, n_2) \approx g(n_1, n_2). \tag{21}$$

With the inverse filter the approximation is made exact, which leads to problems because a match is made to noisy data. A more reasonable expectation for the restored image is that it satisfies

$$\begin{aligned}
&\| g(n_1, n_2) - d(n_1, n_2) * \hat{f}(n_1, n_2) \|^2 \\
&= \sum_{k_1=0}^{N-1} \sum_{k_2=0}^{M-1} (g(k_1, k_2) - d(k_1, k_2) * \hat{f}(k_1, k_2))^2 \approx \sigma_w^2.
\end{aligned} \tag{22}$$

There are potentially many solutions that satisfy relation (22). A second criterion must be used to choose among them. A common criterion, acknowledging the fact that the inverse filter tends to amplify the noise $w(n_1, n_2)$, is to select the solution that is as "smooth" as possible. If we let $c(n_1, n_2)$ represent the point-spread function of a 2-D high-pass filter, then among

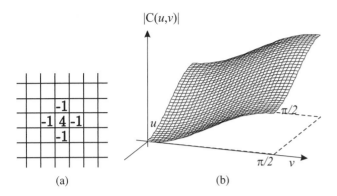

FIGURE 7 Two-dimensional discrete approximation of the second derivative operation: (a) PSF $c(n_1, n_2)$, and (b) spectral representation.

the solutions satisfying relation (22) the solution is chosen that minimizes

$$\Omega(\hat{f}(n_1, n_2)) = \|c(n_1, n_2) * \hat{f}(n_1, n_2)\|^2$$
$$= \sum_{k_1=0}^{N-1} \sum_{k_2=0}^{M-1} (c(k_1, k_2) * \hat{f}(k_1, k_2))^2. \quad (23)$$

The interpretation of $\Omega(\hat{f}(k_1, k_2))$ is that it gives a measure for the high-frequency content of the restored image. Minimizing this measure subject to the constraint of Eq. (22) will give a solution that is both within the collection of potential solutions of Eq. (22) and has as little high-frequency content as possible at the same time. A typical choice for $c(n_1, n_2)$ is the discrete approximation of the second derivative shown in Fig. 7, also known as the 2-D Laplacian operator. For more details on the subject of discrete derivative operators, refer to Chapter 4.10 of this *Handbook*.

The solution to the above minimization problem is the constrained least-squares filter $H_{cls}(u, v)$ that is easiest formulated

in the discrete Fourier domain:

$$H_{cls}(u, v) = \frac{D^*(u, v)}{D^*(u, v) D(u, v) + \alpha C^*(u, v) C(u, v)}. \quad (24)$$

Here α is a tuning or *regularization* parameter that should be chosen such that Eq. (22) is satisfied. Though analytical approaches exist to estimate α [9], the regularization parameter is usually considered user tunable.

It should be noted that although their motivations are quite different, the formulation of the Wiener filter, Eq. (16), and constrained least-squares filter, Eq. (24), are quite similar. Indeed these filters perform equally well, and they behave similarly in the case that the variance of the noise, σ_w^2, approaches zero. Figure 8 shows restoration results obtained by the constrained least-squares filter using three different values of α. A final remark about $\Omega(\hat{f}(n_1, n_2))$ is that the inclusion of this criterion is strongly related to the use of an image model. A vast amount of literature exists on the usage of more complicated image models, especially the ones inspired by 2-D autoregressive processes [17] and the Markov random field theory [6].

3.3 Iterative Filters

The filters formulated in the previous two sections are usually implemented in the Fourier domain using Eq. (10b). Compared to the spatial domain implementation in Eq. (10a), the direct convolution with the 2-D point-spread function $h(n_1, n_2)$ can be avoided. This is a great advantage because $h(n_1, n_2)$ has a very large support, and typically contains NM nonzero filter coefficients even if the PSF of the blur has a small support that contains only a few nonzero coefficients. There are, however, two situations in which spatial domain convolutions are preferred over the Fourier domain implementation, namely:

- in situations in which the dimensions of the image to be restored are very large, and

FIGURE 8 (a) Constrained least-squares restoration of image in Fig. 5(a) with $\alpha = 2 \times 10^{-2}$ (ΔSNR $= 1.7$ dB), (b) $\alpha = 2 \times 10^{-4}$ (ΔSNR $= 6.9$ dB), (c) $\alpha = 2 \times 10^{-6}$ (ΔSNR $= 0.8$ dB).

- in cases in which additional knowledge is available about the restored image, especially if this knowledge cannot be cast in the form of Eq. (23). An example is the *a priori* knowledge that image intensities are always positive. Both in the Wiener and the constrained least-squares filter the restored image may come out with negative intensities, simply because negative restored signal values are not explicitly prohibited in the design of the restoration filter.

Iterative restoration filters provide a means to handle the above situations elegantly [3, 10, 14]. The basic form of iterative restoration filters is the one that iteratively approaches the solution of the inverse filter, and it is given by the following spatial domain iteration:

$$\hat{f}_{i+1}(n_1, n_2) = \hat{f}_i(n_1, n_2)$$
$$+ \beta[g(n_1, n_2) - d(n_1, n_2) * \hat{f}_i(n_1, n_2)]. \quad (25)$$

Here $\hat{f}_i(n_1, n_2)$ is the restoration result after i iterations. Usually in the first iteration $\hat{f}_0(n_1, n_2)$ is chosen to be identical to zero or identical to $g(n_1, n_2)$. Iteration (25) has been independently discovered many times, and is referred to as the van Cittert, Bially,

or Landweber iteration. As can be seen from Eq. (25), during the iterations the blurred version of the current restoration result $\hat{f}_i(n_1, n_2)$ is compared to the recorded image $g(n_1, n_2)$. The difference between the two is scaled and added to the current restoration result to give the next restoration result.

With iterative algorithms, there are two important concerns — does it converge and, if so, to what limiting solution? Analyzing Eq. (25) shows that convergence occurs if the convergence parameter β satisfies

$$|1 - \beta D(u, v)| < 1 \quad \text{for all } (u, v). \quad (26a)$$

Using the fact that $|D(u, v)| \leq 1$, this condition simplifies to

$$0 < \beta < 2, \qquad D(u, v) > 0. \quad (26b)$$

If the number of iterations becomes very large, then $f_i(n_1, n_2)$ approaches the solution of the inverse filter:

$$\lim_{i \to \infty} \hat{f}_i(n_1, n_2) = h_{\text{inv}}(n_1, n_2) * g(n_1, n_2). \quad (27)$$

Figure 9 shows four restored images obtained by iteration (25).

(a)

(b)

(c)

(d)

FIGURE 9 Iterative restoration ($\beta = 1.9$) of the image in Fig. 5(a) after (a) 10 iterations (ΔSNR = 1.6 dB), (b) 100 iterations (ΔSNR = 5.0 dB), (c) 500 iterations (ΔSNR = 6.6 dB), (d) 5000 iterations (ΔSNR = −2.6 dB).

Clearly, as the iteration progresses, the restored image is dominated more and more by inverse filtered noise.

Iterative scheme (25) has several advantages and disadvantages that we discuss next. The first advantage is that Eq. (25) does not require the convolution of images with 2-D PSFs containing many coefficients. The only convolution is that of the restored image with the PSF of the blur, which has relatively few coefficients.

The second advantage is that no Fourier transforms are required, making Eq. (25) applicable to images of arbitrary size. The third advantage is that although the iteration produces the inverse filtered image as a result if the iteration is continued indefinitely, the iteration can be terminated whenever an acceptable restoration result has been achieved. Starting off with a blurred image, the iteration progressively deblurs the image. At the same time the noise will be amplified more and more as the iteration continues. It is now usually left to the user to trade off the degree of restoration against the noise amplification, and to stop the iteration when an acceptable partially deblurred result has been achieved.

The fourth advantage is that the basic form of Eq. (25) can be extended to include all types of *a priori* knowledge. First all knowledge is formulated in the form of projective operations on the image [4]. After applying a projective operation the (restored) image satisfies the *a priori* knowledge reflected by that operator. For instance, the fact that image intensities are always positive can be formulated as the following projective operation P:

$$P[\hat{f}(n_1, n_2)] = \begin{cases} \hat{f}(n_1, n_2) & \text{if } \hat{f}(n_1, n_2) \geq 0 \\ 0 & \text{if } \hat{f}(n_1, n_2) < 0 \end{cases}. \quad (28)$$

By inclusion of this projection P in the iteration, the final image after convergence of the iteration and all of the intermediate images will not contain negative intensities. The resulting iterative restoration algorithm now becomes

$$\hat{f}_{i+1}(n_1, n_2) = P[\hat{f}_i(n_1, n_2) \\ + \beta(g(n_1, n_2) - d(n_1, n_2) * \hat{f}_i(n_1, n_2))]. \quad (29)$$

The requirements on β for convergence as well as the properties of the final image after convergence are difficult to analyze and fall outside the scope of this chapter. Practical values for β are typically around 1. Further, not all projections P can be used in iteration (29), but only *convex* projections. A loose definition of a convex projection is the following. If two images $f^{(1)}(n_1, n_2)$ and $f^{(2)}(n_1, n_2)$ both satisfy the *a priori* information described by the projection P, then also the combined image

$$f^{(c)}(n_1, n_2) = \varepsilon f^{(1)}(n_1, n_2) + (1 - \varepsilon) f^{(2)}(n_1, n_2) \quad (30)$$

must satisfy this *a priori* information for all values of ε between 0 and 1.

A final advantage of iterative schemes is that they are easily extended for spatially variant restoration, i.e., restoration where either the PSF of the blur or the model of the ideal image (for instance, the prediction coefficients in Eq. (20)) vary locally [9, 14].

On the negative side, the iterative scheme of Eq. (25) has two disadvantages. In the first place the second requirement in Eq. (26b), namely that $D(u, v) > 0$, is not satisfied by many blurs, like motion blur and out-of-focus blur. This causes Eq. (25) to diverge for these types of blur. In the second place, unlike the Wiener and constrained least-squares filter the basic scheme does not include any knowledge about the spectral behavior of the noise and the ideal image. Both disadvantages can be corrected by modifying the basic iterative scheme as follows:

$$\hat{f}_{i+1}(n_1, n_2) \\ = (\delta(n_1, n_2) - \alpha\beta c(-n_1, -n_2) * c(n_1, n_2)) * \hat{f}_i(n_1, n_2) \\ + \beta d(-n_1, -n_2) * (g(n_1, n_2) - d(n_1, n_2) * \hat{f}_i(n_1, n_2)). \quad (31)$$

Here α and $c(n_1, n_2)$ have the same meaning as in the constrained least-squares filter. Though the convergence requirements are more difficult to analyze, it is no longer necessary for $D(u, v)$ to be positive for all spatial frequencies. If the iteration is continued indefinitely, Eq. (31) will produce the constrained least-squares filtered image as result. In practice the iteration is terminated long before convergence. The precise termination point of the iterative scheme gives the user an additional degree of freedom over the direct implementation of the constrained least-squares filter. It is noteworthy that although Eq. (31) seems to involve many more convolutions than Eq. (25), a reorganization of terms is possible, revealing that many of those convolutions can be carried out once and off line, and that only one convolution is needed per iteration:

$$\hat{f}_{i+1}(n_1, n_2) = g^d(n_1, n_2) + k(n_1, n_2) * \hat{f}_i(n_1, n_2), \quad (32a)$$

where the image $g^d(n_1, n_2)$ and the fixed convolution kernel $k(n_1, n_2)$ are given by

$$g^d(n_1, n_2) = \beta d(-n_1, -n_2) * g(n_1, n_2)$$
$$k(n_1, n_2) = \delta(n_1, n_2) - \alpha\beta c(-n_1, -n_2) * c(n_1, n_2) \quad (32b)$$
$$- \beta d(-n_1, -n_2) * d(n_1, n_2).$$

A second — and very significant — disadvantage of iterations (25) and (29)–(32) is the slow convergence. Per iteration the restored image $\hat{f}_i(n_1, n_2)$ changes only a little. Many iteration steps are therefore required before an acceptable point for termination of the iteration is reached. The reason is that the above iteration is essentially a *steepest descent* optimization algorithms, which are known to be slow in convergence. It is possible to reformulate the iterations in the form of, for instance, a *conjugate*

(a)   (b)

FIGURE 10   (a) Restored image illustrating the effect of the boundary value problem. The image was blurred by the motion blur shown in Fig. 2(a) and restored by using the constrained least-squares filter. (b) Preprocessed blurred image at its borders such that the boundary value problem is solved.

gradient algorithm, which exhibits a much higher convergence rate [14].

3.4 Boundary Value Problem

Images are always recorded by sensors of finite spatial extent. Since the convolution of the ideal image with the PSF of the blur extends beyond the borders of the observed degraded image, part of the information that is necessary to restore the border pixels is not available to the restoration process. This problem is known as the *boundary value problem*, and it poses a severe problem to restoration filters. Although at first glance the boundary value problem seems to have a negligible effect because it affects only border pixels, this is not true at all. The point-spread function of the restoration filter has a very large support, typically as large as the image itself. Consequently, the effect of missing information at the borders of the image propagates throughout the image, in this way deteriorating the entire image. Figure 10(a) shows an example of a case in which the missing information immediately outside the borders of the image is assumed to be equal to the mean value of the image, yielding dominant horizontal oscillation patterns caused by the restoration of the horizontal motion blur.

Two solutions to the boundary value problem are used in practice. The choice depends on whether a spatial domain or a Fourier domain restoration filter is used. In a spatial domain filter, missing image information outside the observed image can be estimated by extrapolating the available image data. In the extrapolation, a model for the observed image can be used, such as the one in Eq. (20), or more simple procedures can be used, such as mirroring the image data with respect to the image border. For instance, image data missing on the left-hand side of the image could be estimated as follows:

$$g(n_1, n_2 - k) = g(n_1, n_2 + k) \quad \text{for } k = 1, 2, 3, \ldots . \quad (33)$$

In case Fourier domain restoration filters are used, such as the ones in Eqs. (16) or (24), one should realize that discrete Fourier transforms assume periodicity of the data to be transformed. Effectively in 2-D Fourier transforms this means that the left- and right-hand sides of the image are implicitly assumed to be connected, as well as the top and bottom part of the image. A consequence of this property — implicit to discrete Fourier transforms — is that missing image information at the left-hand side of the image will be taken from the right-hand side, and vice versa. Clearly in practice this image data may not correspond to the actual (but missing data) at all. A common way to fix this problem is to interpolate the image data at the borders such that the intensities at the left- and right-hand side as well as the top and bottom of the image transit smoothly. Figure 10(b) shows what the restored image looks like if a border of five columns or rows is used for linearly interpolating between the image boundaries. Other forms of interpolation could be used, but in practice mostly linear interpolation suffices. All restored images shown in this chapter have been preprocessed in this way to solve the boundary value problem.

4 Blur Identification Algorithms

In the previous section it was assumed that the point-spread function $d(n_1, n_2)$ of the blur was known. In many practical cases the actual restoration process has to be preceded by the identification of this point-spread function. If the camera misadjustment, object distances, object motion, and camera motion are known, we could, in theory, determine the PSF analytically. Such situations are, however, rare. A more common situation is that the blur is estimated from the observed image itself.

The blur identification procedure starts out with the choice of a parametric model for the point-spread function. One category of parametric blur models has been given in Section 2. As an example, if the blur were known to be caused by motion, the blur

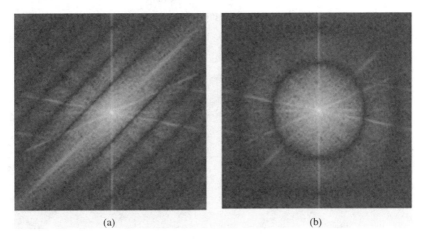

FIGURE 11 $|G(u, v)|$ of 2 blurred images.

identification procedure would estimate the length and direction of the motion.

A second category of parametric blur models is the one that describes the point-spread function $d(n_1, n_2)$ as a (small) set of coefficients within a given finite support. Within this support the value of the PSF coefficients has to be estimated. For instance, if an initial analysis shows that the blur in the image resembles out-of-focus blur, which, however, cannot be described parametrically by Eq. (8b), the blur PSF can be modeled as a square matrix of, say, size 3 by 3, or 5 by 5. The blur identification then requires the estimation of 9 or 25 PSF coefficients, respectively. This section describes the basics of the above two categories of blur estimation.

4.1 Spectral Blur Estimation

In Figs. 2 and 3 we have seen that two important classes of blurs, namely motion and out-of-focus blur, have spectral zeros. The structure of the zero patterns characterizes the type and degree of blur within these two classes. Since the degraded image is described by Eq. (2), the spectral zeros of the PSF should also be visible in the Fourier transform $G(u, v)$, albeit that the zero pattern might be slightly masked by the presence of the noise.

Figure 11 shows the modulus of the Fourier transform of two images, one subjected to motion blur and one to out-of-focus blur. From these images, the structure and location of the zero patterns can be estimated. In the case in which the pattern contains dominant parallel lines of zeros, an estimate of the length and angle of motion can be made. In case dominant circular patterns occur, out-of-focus blur can be inferred and the degree of out of focus (the parameter R in Eq. (8)) can be estimated.

An alternative to the above method for identifying motion blur involves the computation of the two-dimensional cepstrum of $g(n_1, n_2)$. The ceptrum is the inverse Fourier transform of the logarithm of $|G(u, v)|$. Thus,

$$\tilde{g}(n_1, n_2) = -\mathsf{F}^{-1}\{\log |G(u, v)|\}, \tag{34}$$

where F^{-1} is the inverse Fourier transform operator. If the noise can be neglected, $\tilde{g}(n_1, n_2)$ has a large spike at a distance L from the origin. Its position indicates the direction and extent of the motion blur. Figure 12 illustrates this effect for an image with the motion blur from Fig. 2(b).

4.2 Maximum-Likelihood Blur Estimation

In case the point-spread function does not have characteristic spectral zeros or in case a parametric blur model such as motion or out-of-focus blur cannot be assumed, the individual coefficients of the point-spread function have to be estimated. To this end maximum-likelihood estimation procedures for the unknown coefficients have been developed [9, 12, 13, 18]. Maximum-likelihood estimation is a well-known technique for parameter estimation in situations in which no stochastic knowledge is available about the parameters to be estimated [15].

Most maximum-likelihood identification techniques begin by assuming that the ideal image can be described with the 2-D autoregressive model, Eq. (20a). The parameters of this image model — that is, the prediction coefficients $a_{i, j}$ and the variance σ_v^2 of the white noise $v(n_1, n_2)$ — are not necessarily assumed to be known.

If we can assume that both the observation noise $w(n_1, n_2)$ and the image model noise $v(n_1, n_2)$ are Gaussian distributed, the log-likelihood function of the observed image, given the image model and blur parameters, can be formulated. Although the log-likelihood function can be formulated in the spatial domain, its spectral version is slightly easier to compute [13]:

$$L(\theta) = -\sum_u \sum_v \left(\log P(u, v) + \frac{|G(u, v)|^2}{P(u, v)} \right), \tag{35a}$$

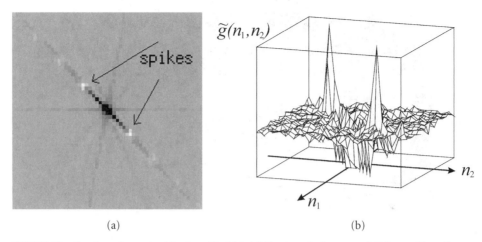

(a)   (b)

FIGURE 12   Cepstrum for motion blur from Fig. 2(c). (a) Cepstrum is shown as a 2-D image. The spikes
appear as bright spots around the center of the image. (b) Cepstrum is shown as a surface plot.

where θ symbolizes the set of parameters to be estimated, i.e., $\theta = \{a_{i,j}, \sigma_v^2, d(n_1, n_2), \sigma_w^2\}$, and $P(u, v)$ is defined as

$$P(u, v) = \sigma_v^2 \frac{|D(u, v)|^2}{|1 - A(u, v)|^2} + \sigma_w^2. \qquad (35b)$$

Here $A(u, v)$ is the discrete 2-D Fourier transform of $a_{i,j}$.

The objective of maximum-likelihood blur estimation is now to find those values for the parameters $a_{i,j}, \sigma_v^2, d(n, n_2)$, and σ_w^2 that maximize the log-likelihood function $L(\theta)$. From the perspective of parameter estimation, the optimal parameter values best explain the observed degraded image. A careful analysis of Eq. (35) shows that the maximum likelihood blur estimation problem is closely related to the identification of 2-D autoregressive moving-average (ARMA) stochastic processes [13, 16].

The maximum likelihood estimation approach has several problems that require nontrivial solutions. Actually the differentiation between state-of-the-art blur identification procedures is mostly in the way they handle these problems [11]. In the first place, some constraints must be enforced in order to obtain a unique estimate for the point-spread function. Typical

constraints are:

- the energy conservation principle, as described by Eq. (5b),
- symmetry of the point-spread function of the blur, i.e., $d(-n_1, -n_2) = d(n_1, n_2)$.

Second, log-likelihood function (35) is highly nonlinear and has many local maxima. This makes the optimization of Eq. (35) difficult, no matter what optimization procedure is used. In general, maximum-likelihood blur identification procedures require good initializations of the parameters to be estimated in order to ensure converge to the global optimum. Alternatively, multiscale techniques could be used, but no ready-to-go or best approach has been agreed upon so far.

Given reasonable initial estimates for θ, various approaches exist for the optimization of $L(\theta)$. They share the property of being iterative. Besides standard gradient-based searches, an attractive alternative exists in the form of the expectation-minimization (EM) algorithm. The EM algorithm is a general procedure for finding maximum-likelihood parameter estimates. When applied to the blur identification procedure, an iterative scheme results that consists of two steps [12, 18] (see Fig. 13).

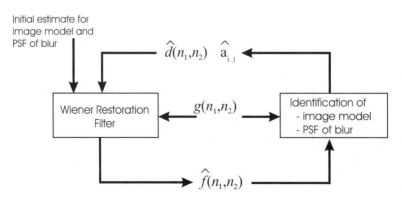

FIGURE 13   Maximum-likelihood blur estimation by the EM procedure.

Expectation step

Given an estimate of the parameters θ, a restored image $\hat{f}_E(n_1, n_2)$ is computed by the Wiener restoration filter, Eq. (16). The power spectrum is computed by Eq. (20b), using the given image model parameter $a_{i,j}$ and σ_v^2.

Maximization step

Given the image restored during the expectation step, a new estimate of θ can be computed. First, from the restored image $\hat{f}_E(n_1, n_2)$ the image model parameters $a_{i,j}$, σ_v^2 can be estimated directly. Second, from the approximate relation

$$g(n_1, n_2) \approx d(n_1, n_2) * \hat{f}_E(n_1, n_2) \qquad (36)$$

and the constraints imposed on $d(n_1, n_2)$, the coefficients of the point-spread function can be estimated by standard system identification procedures [14].

By alternating the E step and the M step, one achieves convergence to a (local) optimum of the log-likelihood function. A particular attractive property of this iteration is that although the overall optimization is nonlinear in the parameters θ, the individual steps in the EM algorithm are entirely linear. Furthermore, as the iteration progresses, intermediate restoration results are obtained that allow for monitoring of the identification process.

As conclusion we observe that the field of blur identification has been significantly less thoroughly studied and developed than the classical problem of image restoration. Research in image restoration continues with a focus on blur identification, using for instance cumulants and generalized cross-validation [11].

References

[1] H. C. Andrews and B. R. Hunt, *Digital Image Restoration* (Prentice-Hall, Englewood Cliffs, NJ, 1977).

[2] M. R. Banha and A. K. Katsaggelos, "Digital image restoration," *IEEE Signal Process. Mag.* **14**, 24–41 (1997).

[3] J. Biemond, R. L. Lagendijk, and R. M. Mersereau, "Iterative methods for image deblurring," *Proc. IEEE* **78**, 856–883 (1990).

[4] P. L. Combettes, "The foundation of set theoretic estimation," *Proc. IEEE* **81**, 182–208 (1993).

[5] N. P. Galatsanos and R. Chin, "Digital restoration of multichannel images," *IEEE Trans. Signal Process.* **37**, 415–421 (1989).

[6] F. Jeng and J. W. Woods, "Compound Gauss-Markov random fields for image estimation," *IEEE Trans. Signal Process.* **39**, 683–697 (1991).

[7] B. R. Hunt, "The application of constrained least squares estimation to image restoration by digital computer," *IEEE Trans. Comput.* **2**, 805–812 (1973).

[8] A. K. Jain, "Advances in mathematical models for image processing," *Proc. IEEE* **69**, 502–528 (1981).

[9] A. K. Katsaggelos, ed., *Digital Image Restoration* (Springer-Verlag, New York, 1991).

[10] A. K. Katsaggelos, "Iterative image restoration algorithm," *Opt. Eng.* **28**, 735–748 (1989).

[11] D. Kundur and D. Hatzinakos, "Blind image deconvolution," *IEEE Signal Process. Mag.* **13**, 43–64 (1996).

[12] R. L. Lagendijk, J. Biemond, and D. E. Boekee, "Identification and restoration of noisy blurred images using the expectation-maximization algorithm," *IEEE Trans. Acoust. Speech Signal Process.* **38**, 1180–1191 (1990).

[13] R. L. Lagendijk, A. M. Tekalp, and J. Biemond, "Maximum likelihood image and blur identification: a unifying approach," *J. Opt. Eng.* **29**, 422–435 (1990).

[14] R. L. Lagendijk and J. Biemond, *Iterative Identification and Restoration of Images* (Kluwer, Boston, MA, 1991).

[15] H. Stark and J. W. Woods, *Probability, Random Processes, and Estimation Theory for Engineers* (Prentice-Hall, Upper Saddle River, NJ, 1986).

[16] A. M. Tekalp, H. Kaufman, and J. W. Woods, "Identification of image and blur parameters for the restoration of non-causal blurs," *IEEE Trans. Acoust. Speech Signal Process.* **34**, 963–972 (1986).

[17] J. W. Woods and V. K. Ingle, "Kalman filtering in two-dimensions — further results," *IEEE Trans. Acoust. Speech Signal Process.* **29**, 188–197 (1981).

[18] Y. L. You and M. Kaveh, "A regularization approach to joint blur identification and image restoration," *IEEE Trans. Image Process.* **5**, 416–428 (1996).

3.6

Regularization in Image Restoration and Reconstruction

William C. Karl
Boston University

1 Introduction

This chapter focuses on the need for and use of regularization methods in the solution of image restoration and reconstruction problems. The methods discussed here are applicable to a variety of such problems. These applications to specific problems, including implementation specifics, are discussed in greater detail in the other chapters of the *Handbook*. Our aim here is to provide a unifying view of the difficulties that arise, and the tools that are used, in the analysis and solution of these problems. In the remainder of this section a general model for common image restoration and reconstruction problems is presented, together with the standard least-squares approach taken for solving these problems. A discussion of the issues leading to the need for regularization is provided. In Section 2, so-called direct regularization methods are treated; in Section 3, iterative methods of regularization are discussed. In Section 4 an overview is given of the important problem of parameter choice in regularization. Section 5 concludes the chapter.

1.1 Image Restoration and Reconstruction Problems

Image restoration and reconstruction problems have as their goal the recovery of a desired, unknown, image of interest $f(x, y)$ based on the observation of a related set of distorted data $g(x, y)$. These problems are generally distinguished from image enhancement problems by their assumed knowledge and use of a *distortion model* relating the unknown $f(x, y)$ to the observed $g(x, y)$. In particular, we focus here on distortion models captured by a linear integral equation of the following form:

$$g(x, y) = \int_{-\infty}^{\infty} \int_{-\infty}^{\infty} h(x, y; x', y') f(x', y') \, dx' \, dy', \quad (1)$$

where $h(x, y; x', y')$ is the kernel or response function of the distorting system, often termed the point-spread function (PSF). Such a relationship is called a Fredholm integral equation of the first kind [1] and captures most situations of engineering interest. Note that Eq. (1), while linear, allows for the possibility of shift-variant system functions.

Examples of image *restoration* problems that can be captured by distortion model (1) are discussed in Chapter 3.5 and include compensation for incorrect camera focus, removal of uniform motion blur, and correction of blur caused by atmospheric turbulence. All these examples involve a spatially invariant PSF; that is, $h(x, y; x', y')$ is only a function of $x - x'$ and $y - y'$. One of the most famous examples of image restoration involving a *spatially varying* point-spread function is provided by the Hubble Space Telescope, where a flaw in the main mirror resulted in a spatially varying distortion of the acquired images.

Examples of image *reconstruction* problems fitting into the framework of Eq. (1) include those involving reconstruction based on projection data. Many physical problems can be cast into this or a very similar tomographic type of framework, including the following: medical computer-aided tomography, single photon emission tomography, atmospheric tomography, geophysical tomography, radio astronomy, and synthetic aperture radar imaging. The simplest model for these types of problems relates the observed projection $g(t, \theta)$ at angle θ and offset t to the underlying field $f(x, y)$ through a spatially varying PSF given by [2]:

$$h(t, \theta; x', y') = \delta(t - x' \cos(\theta) - y' \sin(\theta)) \qquad (2)$$

The set of projected data $g(t, \theta)$ is often called a sinogram. See Chapter 10.2 for more detail.

Even when the unknown field is modeled as continuous, the data, of necessity, are often discrete because of the nature of the sensor. Assuming there are N_g observations, Eq. (1) can be written as

$$g_i = g(x_i, y_i) = \int_{-\infty}^{\infty} \int_{-\infty}^{\infty} h_i(x', y') f(x', y') \, dx' \, dy',$$
$$1 \le i \le N_g, \quad (3)$$

where $h_i(x', y') = h(x_i, y_i; x', y')$ denotes the kernel corresponding to the ith observation. In Eq. (3) the discrete data have been described as simple *samples* of the continuous observations. This can always be done, since any averaging induced by the response function of the instrument can be included in the specification of $h_i(x', y')$.

Finally, the unknown image $f(x, y)$ itself is commonly described in terms of a discrete and finite set of parameters. In particular, assume that the image can be adequately represented by a weighted sum of N_f basis functions $\phi_j(x, y)$, $j = 1, \ldots, N_f$ as follows:

$$f(x, y) = \sum_{j=1}^{N_f} f_j \, \phi_j(x, y). \qquad (4)$$

For example, the basis functions $\phi_j(x, y)$ are commonly chosen

to be the set of unit height boxes corresponding to array of square pixels, though other bases (e.g., wavelets; see Chapter 4.1) have also found favor. Given the expansion in Eq. (4), the image is then represented by the collection of N_f coefficients f_j. For example, if a square $N \times M$ pixel array is used, then $N_f = NM$ and the f_j simply become the values of the pixels themselves.

Substituting Eq. (4) into Eq. (3) and simplifying yields the following completely discrete relationship between the set of observations g_i and the collection of unknown image coefficients f_j:

$$g_i = \sum_{j=1}^{N_f} H_{ij} \, f_j, \quad 1 \le i \le N_g, \qquad (5)$$

where H_{ij} is given by

$$H_{ij} = \int_{-\infty}^{\infty} \int_{-\infty}^{\infty} h_i(x', y') \, \phi_j(x', y') \, dx' \, dy',$$
$$1 \le i \le N_g, \quad 1 \le j \le N_f \quad (6)$$

and represents the inner product of the ith observation kernel $h_i(x, y)$ with the jth basis function $\phi_j(x, y)$. Collecting all the observations g_i and image unknowns f_j into a single matrix equation yields a matrix equation capturing the observation process:

$$g = Hf, \qquad (7)$$

where the length N_g vector g, the length N_f vector f, and the $N_g \times N_f$ matrix H follow naturally from Eqs. (4), (5), and (6). When a rectangular pixel basis is used for $\phi_j(x, y)$ and the system is shift invariant, so that $h(x, y; x', y') = h(x - x', y - y')$, the resulting matrix H will exhibit a banded block-Toeplitz structure with Toeplitz blocks — that is, the structure of H is of the form

$$H = \begin{bmatrix} H_0 & H_1 & H_2 & \cdots & H_M \\ H_{-1} & H_0 & H_1 & \cdots & H_{M-1} \\ H_{-2} & H_{-1} & H_0 & \ddots & \vdots \\ \vdots & \ddots & \ddots & \ddots & H_1 \\ H_{-N} & \cdots & H_{-2} & H_{-1} & H_0 \end{bmatrix}, \qquad (8)$$

where the blocks H_i themselves internally exhibit the same banded Toeplitz structure. This structure is just a reflection of the linear convolution operation underlying the problem. Such linear convolutional problems can be represented by equivalent circular convolutional problems through appropriate zero padding. When this circulant embedding is done, the corresponding matrix H will then possess a block-circulant structure with circulant blocks [3]. In this case the structure of the associated H will be

(a)   (b)

FIGURE 1   (a) Original cameraman image (256×256). (b) Image distorted by 7-pixel horizontal motion blur and 30-dB SNR additive noise.

of the form

$$H = \begin{bmatrix} H_0 & H_1 & H_2 & \cdots & H_M \\ H_M & H_0 & H_1 & \cdots & H_{M-1} \\ H_{M-1} & H_M & H_0 & \ddots & \vdots \\ \vdots & \ddots & \ddots & \ddots & H_1 \\ H_1 & \cdots & H_{M-2} & H_{M-1} & H_0 \end{bmatrix}, \quad (9)$$

where the block rows are simple circular shifts of each other and each block element itself possesses this structure. The significance of the block-circulant form is that there exist efficient computational approaches for problems with such structure in H, corresponding to the application of Fourier-based, frequency domain techniques.

In practice, our measured data are corrupted by inevitable perturbations or noise, so that our unknown image f is actually related to our data through

$$g = Hf + q, \quad (10)$$

where q is a vector of perturbation or noise samples. In what follows, the focus will be on the discrete or sampled data case as represented by Eq. (7) or (10).

For purposes of illustration throughout the coming discussion, two example problems will be considered. The first example is an image restoration problem involving restoration of an image distorted by spatially invariant horizontal motion blur. The original 256×256 image is shown in Fig. 1(a). The distorted data, corresponding to application of a length 7-pixel horizontal motion blur followed by the addition of white Gaussian noise for an SNR[1] of 30 dB, is shown in Fig. 1(b).

[1] SNR (dB) $\equiv 10 \log_{10}[\text{Var}(Hf)/\text{Var}(q)]$, where Var($z$) denotes the variance of z.

The second example problem is an image reconstruction problem, involving reconstruction of an image from noisy tomographic data. The original 50×50 phantom image is shown in Fig. 2(a). The noisy projection data are shown in Fig. 2(b), with the horizontal axis corresponding to angle θ and the vertical axis corresponding to projection offset t. These data corresponds to application of Eq. (2) with 20 angles evenly spaced between $\theta = 0°$ and $\theta = 180°$ and 125 samples in t per angle followed by addition of white Gaussian noise for an SNR of 30 dB. This example represents a challenging inversion problem that might arise in nondestructive testing.

1.2 Least-Squares and Generalized Solutions

The image restoration or reconstruction problem can be seen to be equivalent to one of solving for the unknown vector f given knowledge of the data vector g and the distorting system matrix H. At first sight, a simple matrix inverse would seem to provide the solution, but this approach does not lead to usable solutions. There are four basic issues that must be dealt with in inverting the effects of H to find an estimate $\hat{f}$ of f. First, there is the problem of solution existence. Given an observation g in Eq. (7) there may not exist any f that solves this equation with equality, because of the presence of noise. Second, there is the problem of solution uniqueness. If the null space of H is nonempty, then there are objects or images that are "unobservable" in the data. The null space of H is the collection of all input images that produce zero output. An example would be the set of DC or constant images when H is a high-pass filter. Such components may exist in the true scene, but they do not appear in the observations. In these cases, there will be many choices of f that produce the same set of observations, and it must be decided which is the "right one." Such a situation arises, for example, when H represents a filter whose magnitude goes to zero for some range of frequencies, in

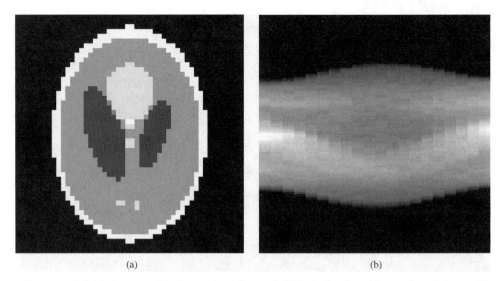

FIGURE 2   (a) Original tomographic phantom image (50×50). (b) Projection data with 20 angles and 125 samples per angle corrupted by additive noise, 30-dB SNR.

which case images differing in these bands will produce identical observations. Third, there is the problem of solution stability. It is desired that the estimate of f remain relatively the same in the face of perturbations to the observations (either caused by noise, uncertainty, or numerical roundoff). These first three elements are the basis of the classic Hadamard definition of an ill-posed problem [4,5]. In addition to these problems, a final issue exists. Equation (7) only represents the observations and says nothing about any prior knowledge about the solution. In general, more information will be available, and a way to include it in the solution is needed. Regularization will prove to be the means to deal with all these problems.

The standard approach taken to inverting Eq. (7) will now be examined and its weaknesses explained in the context of the above discussion. The typical (and reasonable) solution to the first problem of solution existence is to seek a *least-squares solution* to the set of inconsistent equations represented by Eq. (7). That is, the estimate is defined as the least-squares fit to the observed data:

$$\hat{f}_{\text{LS}} = \arg\min_{f} \|g - Hf\|_2^2, \tag{11}$$

where $\|z\|_2^2 = \sum_i z_i^2$ denotes the ℓ_2 norm and arg denotes the argument producing the minimum (as opposed to the value of the minimum itself). A weighted error norm is also sometimes used in the specification of Eq. (11) to give certain observations increased importance in the solution: $\|g - Hf\|_W^2 = \sum_i w_i [g - Hf]_i^2$. If H has full column rank, the null space of H is empty, and the estimate is unique and is obtained as the solution to the following set of normal equations [4]:

$$(H^T H)\hat{f}_{\text{LS}} = H^T g. \tag{12}$$

When the null space of H is not empty, the second inversion difficulty of non-uniqueness, caused by the presence of unobservable images, must also be dealt with. What is typically done in these situations is to seek the unique solution of minimum energy or norm among the collection of least-squares solutions. This *generalized solution* is usually denoted by $\hat{f}^+$ and defined as:

$$\hat{f}^+ = \arg\min_{f} \|f\|_2 \text{ subject to } \min\|g - Hf\|_2. \tag{13}$$

The generalized solution is often expressed as $\hat{f}^+ = H^+ g$, where H^+ is called the generalized inverse of H (note that H^+ is defined implicitly through Eq. (13)). Thus generalized solutions are least-squares solutions of minimum size or energy. Since components of the solution f that are unobservable do not improve the fit to the data, but only serve to increase the solution energy, the generalized solution corresponds to the least-squares solution with no unobservable components, i.e., with no component in the null space of H. Note that when the null space of H is empty (for example, when we have at least as many independent observations g_i as unknowns f_j), the generalized and least-squares solutions are the same.

To understand how the generalized solution functions, consider a simple filtering situation in which the underlying PSF is shift invariant (such as our deblurring problem) and the corresponding H is a circulant matrix. In this shift-invariant, filtering context the generalized solution method is sometimes referred to as "inverse filtering" (see Chapter 3.5). In this case, H can be diagonalized by the matrix F, which performs the two-dimensional (2-D) discrete Fourier transform (DFT) on an image (represented as a vector) [3]. In particular, letting tildes denote transform quantities, we have

$$H = F^{-1}\tilde{H}F, \qquad H^T = F^{-1}\tilde{H}^* F, \tag{14}$$

where $\tilde{H}$ is a diagonal matrix and $\tilde{H}^*$ denotes the complex conjugate of $\tilde{H}$. The diagonal elements of $\tilde{H}$ are just the 2-D DFT coefficients $\tilde{h}_i$ of the PSF of this circulant problem; $\text{diag}[\tilde{H}] = \tilde{h} = Fh$, where h is given by, for example, the first column of H. Applying these relationships to Eq. (12), we obtain the following frequency domain characterization of the generalized solution:

$$\tilde{H}^* \tilde{H} \tilde{f}^+ = \tilde{H}^* \tilde{g}, \qquad (15)$$

where $\tilde{f}^+$ is a vector of the 2-D DFT coefficients of the generalized solution, and $\tilde{g}$ is a vector of the 2-D DFT coefficients of the data. This set of equations is diagonal, so each component of the solution may be solved for separately:

$$\tilde{f}_i^+ = \begin{cases} \left(\frac{1}{\tilde{h}_i}\right) \tilde{g}_i, & |\tilde{h}_i| \neq 0 \\ 0 & \text{otherwise} \end{cases}. \qquad (16)$$

Thus, the generalized solution performs simple inverse filtering where the frequency response magnitude is nonzero, and sets the solution to zero otherwise.

For general, nonconvolutional problems (e.g., for tomographic problems) the 2-D DFT matrix F does not provide a diagonalizing decomposition of H as in Eq. (14). There is, however, a generalization of this idea to arbitrary, shift-varying PSF system matrices called the singular value decomposition (SVD) [6]. The SVD is an important tool for understanding and analyzing inverse problems. The SVD of an $N_g \times N_f$ matrix H is a decomposition of the matrix H of the following form:

$$H = USV^T = \sum_{k=1}^{p} \sigma_k u_k v_k^T, \qquad (17)$$

where U is an $N_g \times N_g$ matrix, V is an $N_f \times N_f$ matrix, and S is an $N_g \times N_f$ diagonal matrix with the values $\sigma_1, \sigma_2, \ldots, \sigma_p$ arranged on its main diagonal and zeros elsewhere, where $p = \min(N_g, N_f)$. The orthonormal columns u_i of U are called the left singular vectors, the orthonormal columns v_i of V are called the right singular vectors, the σ_i are called the singular values, and the set of triples $\{\sigma_i, u_i, v_i\}, 1 \leq i \leq p$ is called the singular system of H. Further, if r is the rank of H, then the σ_i satisfy:

$$\sigma_1 \geq \cdots \geq \sigma_r > \sigma_{r+1} = \cdots = \sigma_p = 0. \qquad (18)$$

The calculation of the entire SVD is too computationally expensive for general problems larger than modest size, though the insight it provides makes it a useful conceptual tool nonetheless. It is possible, however, to efficiently calculate the SVD for certain structured problems (such as for problems with a separable PSF [3]) or to calculate only parts of the SVD for general problems. Such calculations of the SVD can be done in a numerically robust fashion, and many software tools exist for this purpose.

The SVD allows the development of an analysis similar to Eq. (16) for general problems. In particular, the generalized solution can be expressed in terms of the elements of the SVD as follows:

$$\hat{f}^+ = \sum_{i=1}^{r} \frac{u_i^T g}{\sigma_i} v_i. \qquad (19)$$

This expression, valid for any H, whether convolutional or not, may be interpreted as follows. The observed data g are decomposed with respect to the set of basis images $\{u_i\}$ (yielding the coefficients $u_i^T g$). The coefficients of this representation are scaled by $1/\sigma_i$ and then used as the weights of an expansion of $\hat{f}^+$ with respect to the new set of basis images $\{v_i\}$. Note, in particular, that the sum only runs up to r. The components v_i of the reconstruction for $i > r$ correspond to $\sigma_i = 0$ and are omitted from the solution. These components correspond precisely to images that will be unobserved in the data. For example, if H were a low-pass filter, DC image components would be omitted from the solution. Note that for a linear shift-invariant problem, where frequency domain techniques are applicable, solution (19) is equivalent to inverting the system frequency response at those frequencies where it is nonzero, and setting the solution to zero at those frequencies where it is zero, as previously discussed.

1.3 The Need for Regularization

A number of observations about the drawbacks of generalized solution (13) or (19) may be made. First, the generalized solution makes no attempt to reconstruct components of the image that are unobservable in the data (i.e. in the null space of H). For example, if a given pixel is obscured from view, the generalized solution will set its value to zero despite the fact that all values near it might be visible (and hence, despite the fact that a good estimate as to its value may be made). Second, and perhaps more seriously, the generalized solution is "unstable" in the face of perturbations to the data — that is, small changes in the data — lead to large changes to the solution. To understand why this is so, note that most physical PSF system matrices H have the property that their singular values σ_i tend gradually to zero as i increases and, further, the singular image vectors u_i and v_i corresponding to these small σ_i are high-frequency in nature. The consequences of this behavior are substantial. In particular, the impact of the data on the ith coefficient of generalized solution (19) can be expressed as

$$\frac{u_i^T g}{\sigma_i} = v_i^T f + \frac{u_i^T q}{\sigma_i}. \qquad (20)$$

There are two terms on the right-hand side of Eq. (20): the first term is due to the true image and the second term is due to the noise. For large values of the index i, these terms are like high-frequency Fourier coefficients of the respective elements (since u_i and v_i are typically high frequency). The high-frequency contribution from the true image $v_i^T f$ will generally be much smaller

FIGURE 3   Generalized solutions $\hat{f}^+$ corresponding to data in Figs. 1(b) (left) and 2(b) (right).

than that due to the noise $u_i^T q$, since images tend to be lower frequency than noise. Further, the contribution from the noise is then *amplified* by the large factor $1/\sigma_i$. Overall then, the solution will be dominated by very large, oscillatory terms that are due to the noise. Another way of understanding this undesirable behavior follows from the generalized solution's insistence on reducing the data fit error above all else. If the data have noise, the solution $\hat{f}^+$ will be distorted in an attempt to fit to the noise components. Figure 3 shows the generalized solutions corresponding to the motion-blur restoration example of Fig. 1 and the tomographic reconstruction example of Fig. 2. The solutions have been truncated to the original range of the images in each case, either [0, 255] for the motion-blur example or [0, 1] for the tomographic example. Clearly these solutions are unacceptable.

The above insight provides not only a way of understanding why it is likely that the generalized inverse solution will have difficulties, but also a way of analyzing specific problems. In particular, since the generalized solution fails because of the explosion of the coefficients $u_i^T g/\sigma_i$ in the sum of Eq. (19), potential inversion difficulties can be seen by plotting the quantities $|u_i^T g|$, σ_i and the ratio $|u_i^T g|/\sigma_i$ versus i [7]. Demonstrations of such plots for the two example problems are shown in Fig. 4. In both cases, for large values of the index i the coefficients $|u_i^T g|$ level off because of noise while the associated σ_i continue to decrease, and thus the corresponding reconstruction coefficients in this range become very large.

It is for these reasons that the generalized solution is an unsatisfactory approach to the problems of image restoration and

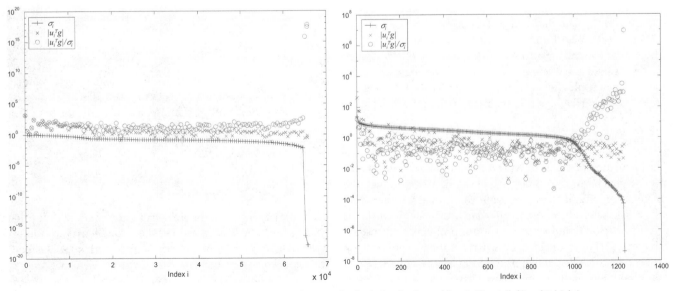

FIGURE 4   Plots of the components comprising the generalized solution for the problem in Figs. 1 (left) and 2 (right).

reconstruction in all but the lowest noise situations. These difficulties are generally a reflection of the ill-posed nature of the underlying continuous problem, as reflected in ill-conditioning of the system PSF matrix H. The answer to these difficulties is found through what is known as *regularization*. The purpose of regularization is to allow the inclusion of prior knowledge to stabilize the solution in the face of noise and allow the identification of physically meaningful and reasonable estimates. The basic idea is to constrain the solution in some way so as to avoid the oscillatory nature of the noise dominated solution observed in Fig. 3 [4].

A *regularization method* is often formally defined as an inversion method depending on a single real parameter $\alpha \geq 0$, which yields a family of approximate solutions $\hat{f}(\alpha)$ with the following two properties: first, for large enough α the regularized solution $\hat{f}(\alpha)$ is stable in the face of perturbations or noise in the data (unlike the generalized solution) and, second, as α goes to zero the unregularized generalized solution is recovered: $\hat{f}(\alpha) \rightarrow \hat{f}^+$ as $\alpha \rightarrow 0$. The parameter α is called the "regularization parameter" and controls the tradeoff between solution stability (i.e., noise propagation) and nearness of the regularized solution $\hat{f}(\alpha)$ to the unregularized solution $\hat{f}^+$ (i.e., approximation error in the absence of noise). Since the generalized solution represents the highest possible fidelity to the data, another way of viewing the role of α is in controlling the tradeoff between the impact of data and the impact of prior knowledge on the solution. There are a wide array of regularization methods, and an exhaustive treatment is beyond the scope of this chapter. The aim of this chapter is to provide a summary of the main approaches and ideas.

2 Direct Regularization Methods

In this section what are known as "direct" regularization methods are examined. These methods are conceptually defined by a direct computation, though they may utilize, for example, iterative methods in the computation of a practical solution. In Section 3 the use of iterative methods as a regularization approach in their own right is discussed.

2.1 Truncated SVD Regularization

From the discussion in Section 1.3, it can be seen that the stability problems of the generalized solution are associated with the large gain given the noise that is due to the smallest singular values σ_i. A logical remedy is to simply truncate small singular values to zero. This approach to regularization is called *truncated SVD* (TSVD) or numerical filtering [4,7]. Indeed, such truncation is almost always done to some extent in the definition of the numerical rank of a problem, so TSVD simply does this to a greater extent. In fact, one interpretation of TSVD is as defining the rank of H *relative* to the noise in the problem. The TSVD regularized solution can be usefully defined based on Eq. (19) in

the following way:

$$\hat{f}_{\text{TSVD}}(\alpha) = \sum_{i=1}^{r} w_{i,\alpha} \frac{u_i^T g}{\sigma_i} v_i, \quad (21)$$

where $w_{i,\alpha}$ is a set of weights or filter factors given by

$$w_{i,\alpha} = \begin{cases} 1 & i \leq k(\alpha) \\ 0 & i > k(\alpha) \end{cases}, \quad (22)$$

with the positive integer $k(\alpha) = \lfloor \alpha^{-1} \rfloor$, where $\lfloor x \rfloor$ denotes x rounded to the next smaller integer. Defined in this way, TSVD has the properties of a formal regularization method. TSVD simply throws the offending components of the solution out, but it does not introduce any new components. As a result, TSVD solutions, although stabilized against noise, make no attempt to include image components that are unobservable in the data (like the original generalized solution).

Another way of understanding the TSVD solution is as follows. If H_k denotes the closest rank-k approximation to H, then, by analogy to Eq. (13), the TSVD solution $\hat{f}_{\text{TSVD}}(\alpha)$ in Eq. (21) is also given by

$$\hat{f}_{\text{TSVD}}(\alpha) = \arg\min\|f\|_2 \text{ subject to } \min\|g - H_k f\|_2, \quad (23)$$

which shows that the TSVD method can be thought of as directly approximating the original problem H by a nearby H_k that is better conditioned and less sensitive. In terms of its impact on reconstruction coefficients, the TSVD method corresponds to the choice of an ideal step weighting function $w_{i,\alpha}$ applied to the coefficients of the generalized solution. Certainly other weighting functions could and have been applied [4]. Indeed, some regularization methods are precisely interpretable in this way, as will be discussed.

Figure 5 shows truncated SVD solutions corresponding to the motion-blur restoration example of Fig. 1 and the tomographic reconstruction problem of Fig. 2. For the motion-blur restoration problem, the solution used only approximately 40,000 of the over 65,000 singular values of the complete generalized reconstruction, whereas for the tomographic reconstruction problem the solution used only ~800 of the full 2500 singular values. As can be seen, the noise amplification of the generalized reconstruction has indeed been controlled in both cases. In the motion-blur example some vertical ringing caused by edge effects can be seen.

2.2 Tikhonov Regularization

Perhaps the most widely referenced regularization method is the Tikhonov method. The key idea behind the Tikhonov method is to directly incorporate prior information about the image f through the inclusion of an additional term to the original least-squares cost function. In particular, the Tikhonov regularized estimate is defined as the solution to the following minimization

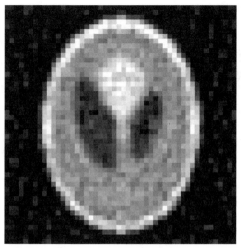

FIGURE 5   Truncated SVD solutions corresponding to data in Figs. 1(b) (left) and 2(b) (right).

problem:

$$\hat{f}_{\text{Tik}}(\alpha) = \arg\min_f \|g - Hf\|_2^2 + \alpha^2 \|Lf\|_2^2. \qquad (24)$$

The first term in Eq. (24) is the same ℓ_2 residual norm appearing in the least-squares approach and ensures fidelity to data. The second term in Eq. (24) is called the "regularizer" or "side constraint" and captures prior knowledge about the expected behavior of f through an additional ℓ_2 penalty term involving just the image. The regularization parameter α controls the tradeoff between the two terms. The minimizer of Eq. (24) is the solution to the following set of normal equations:

$$(H^T H + \alpha^2 L^T L)\,\hat{f}_{\text{Tik}}(\alpha) = H^T g. \qquad (25)$$

This set of linear equations can be compared to the equivalent set of Eq. (12) obtained for the unregularized least-squares solution. A solution to Eq. (25) exists and will be unique if the null spaces of H and L are distinct. There are a number of ways to obtain the Tikhonov solution from Eq. (25), including matrix inversion, iterative methods, and the use of factorizations like the SVD (or its generalizations) to diagonalize the system of equations.

To gain a deeper appreciation of the functioning of Tikhonov regularization, first consider the case when $L = I$, a diagonal matrix of ones. The corresponding side constraint term in Eq. (24) then simply measures the "size" or energy of f and thus, by inclusion in the overall cost function, directly prevents the pixel values of f from becoming too large (as happened in the unregularized generalized solution). The effect of α in this case is to trade off the fidelity to the data with the energy in the solution. With the use of the definition of the SVD combined with Eq. (25), the Tikhonov solution when $L = I$ can be

expressed as

$$\hat{f}_{\text{Tik}}(\alpha) = \sum_{i=1}^{r} \left(\frac{\sigma_i^2}{\sigma_i^2 + \alpha^2} \right) \frac{u_i^T g}{\sigma_i}\, v_i. \qquad (26)$$

Comparing this expression to Eq. (19), one can define an associated set of weight or filter factors $w_{i,\alpha}$ for Tikhonov regularization with $L = I$ as follows:

$$w_{i,\alpha} = \frac{\sigma_i^2}{\sigma_i^2 + \alpha^2}. \qquad (27)$$

In contrast to the ideal step behavior of the TSVD weights in Eq. (21), the Tikhonov weights decay like a "double-pole" lowpass filter, where the "pole" occurs at $\sigma_i = \alpha$. Thus, Tikhonov regularization with $L = I$ can be seen to function similarly to TSVD, in that the impact of the higher index singular values on the solution is attenuated. Another consequence of this similarity is that when $L = I$, the Tikhonov solution again makes no attempt to reconstruct image components that are unobservable in the data.

The case when $L \neq I$ is more interesting. Usually L is chosen as a derivative or gradient operator so that $\|Lf\|$ is a measure of the variability or roughness of the estimate. Common choices for L include discrete approximations to the 2-D gradient or Laplacian operators, resulting in measures of image slope and curvature, respectively. Such operators are described in Chapter 4.10. Inclusion of such terms in Eq. (24) forces solutions with limited high-frequency energy and thus captures a prior belief that solution images should be smooth. An expression for the Tikhonov solution when $L \neq I$ that is similar in spirit to Eq. (26) can be derived in terms of the generalized SVD of the pair (H, L) [6, 7], but this is beyond the scope of this chapter. Interestingly, it can be shown that the Tikhonov solution when $L \neq I$ does contain image components that are unobservable in the data, and thus allows for extrapolation from the data. Note, it

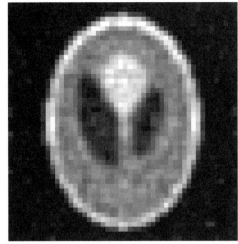

FIGURE 6 Tikhonov regularized solutions when L is a gradient operator corresponding to the data in Figs. 1(b) (left) and (b) (right).

is also possible to consider the addition of multiple terms of the form $\|L_i f\|_2$, to create weighted derivative penalties of multiple orders, such as arise in Sobolev norms.

Figure 6 shows Tikhonov regularized solutions for both the motion-blur restoration example of Fig. 1 and the tomographic reconstruction example of Fig. 2 when $L = D$ is chosen as a discrete approximation of the gradient operator, so that the elements of Df are just the brightness changes in the image. The additional smoothing introduced through the use of a gradient-based L in the Tikhonov solutions can be seen in the reduced oscillation or variability of the reconstructed images.

Before leaving Tikhonov regularization it is worth noting that the following two inequality constrained least-squares problems are essentially the same as the Tikhonov method:

$$\hat{f} = \arg\min\|g - Hf\|_2 \text{ subject to } \|Lf\|_2 \leq 1/\lambda_1, \quad (28)$$

$$\hat{f} = \arg\min\|Lf\|_2 \text{ subject to } \|g - Hf\|_2 \leq \lambda_2. \quad (29)$$

The nonnegative scalars λ_1 and λ_2 play the roles of regularization parameters. The solution to each of these problems is the same as that obtained from Eq. (24) for a suitably chosen value of α that depends in a nonlinear way on λ_1 or λ_2. The latter approach is also related to a method for choosing the regularization parameter called the "discrepancy principle," which we discuss in Section 4.

While Eq. (26), and its generalization when $L \neq I$, gives an explicit expression for the Tikhonov solution in terms of the SVD, for large problems computation of the SVD may not be practical and other means must be sought to solve Eq. (25). When H and L have circulant structure (corresponding to a shift-invariant filter), these equations are diagonalized by the DFT matrix and the problem can be easily solved in the frequency domain. Often,

even when this is not the case, the set of equations in (25) possess a sparse and banded structure and may be efficiently solved by using iterative schemes, such as preconditioned conjugate gradient.

2.3 Nonquadratic Regularization

The basic Tikhonov method is based on the addition of a quadratic penalty $\|Lf\|_2$ to the standard least-squares (and hence quadratic) data fidelity criterion, as shown in Eq. (24). The motivation for this addition was the stabilization of the generalized solution through the inclusion of prior knowledge in the form of a side constraint. The use of such quadratic, ℓ_2-based criteria for the data and regularizer leads to linear problem (25) for the Tikhonov solution, and thus results in an inverse filter that is a linear function of the data. While such linear processing is desirable, since it leads to straightforward and reasonably efficient computation methods, it is also limiting, in that far more powerful results are possible if nonlinear methods are allowed. In particular, when used for suppressing the effect of high-frequency noise, such linear filters, by their nature, also reduce high-frequency energy in the true image and hence blur edges. For this reason, the generalization of the Tikhonov approach through the inclusion of certain nonquadratic criteria is now considered. To this end, consider estimates obtained as the solution of the following generalized formulation:

$$\hat{f}(\alpha) = \arg\min_f J_1(f, g) + \alpha^2 J_2(f), \quad (30)$$

where $J_1(f, g)$ represents a general distance measure between the data and its prediction based on the estimated f and $J_2(f)$ is a general regularizing penalty. Both costs may be a nonquadratic function of the elements of f. Next, a number of popular and interesting choices for $J_1(f, g)$ and $J_2(f)$ are examined.

Maximum Entropy Regularization

Perhaps the most widely used nonquadratic regularization approach is the maximum entropy method. The entropy of a positive valued image in the discrete case may be defined as

$$-J_2(f) = -\sum_{i=1}^{N_f} f_i \log(f_i) \qquad (31)$$

and can be taken as a measure of the uncertainty in the image. This interpretation follows from information theoretic considerations when the image is normalized so that $\sum_{i=1}^{N_f} f_i = 1$, and may thus be interpreted as a probability density function [5]. In this case, it can be argued that the maximum entropy solution is the most noncommittal with respect to missing information. A simpler motivation for the use of the entropy criterion is that it ensures positive solutions. Combining entropy cost (31) with a standard quadratic data fidelity term for $J_1(f, g)$ yields the maximum entropy estimate as the solution of

$$\hat{f}_{\mathrm{me}}(\alpha) = \arg\min_f \|g - Hf\|_2^2 + \alpha^2 \sum_{i=1}^{N_f} f_i \log(f_i). \qquad (32)$$

There are a number of variants on this idea involving related definitions of entropy, cross-entropy, and divergence [5]. Experience has shown that this method provides image reconstructions with greater energy concentration (i.e., most coefficients are small and a few are very large) relative to quadratic Tikhonov approaches. For example, when the f_i represent pixel values, the approach has resulted in sharper reconstructions of point objects, such as star fields in astronomical images. The difficulty with formulation (32) is that it leads to a nonlinear optimization problem for the solution, which must be solved iteratively.

Figure 7 shows maximum entropy solutions corresponding to both the motion-blur restoration example of Fig. 1 and the tomographic reconstruction example of Fig. 2. Note that these two examples are not particularly well matched to the maximum entropy approach, since in both cases the true image is not composed of pointlike objects. Still, the maximum entropy side constraint has again succeeded in controlling the noise amplification observed in the generalized reconstruction. For the tomography example, note that small variations in the large background region have been supressed and the energy in the reconstruction has been concentrated within the reconstructed object. In the motion-blur example, the central portion of the reconstruction is sharp, but the edges again show some vertical ringing caused by boundary effects.

Total Variation Regularization

Another nonquadratic side constraint that has achieved popularity in recent years is the total variation measure:

$$J_2(f) = \|Df\|_1 = \sum_{i=1}^{N_f} |[Df]_i|, \qquad (33)$$

where $\|z\|_1$ denotes the ℓ_1 norm (i.e., the sum of the absolute value of the elements), and D is a discrete approximation to the gradient operator described in Chapter 4.10, so that the elements of Df are just the brightness changes in the image. The total variation estimate is obtained by combining Eq. (33) with the standard quadratic data fidelity term for $J_1(f, g)$ to yield

$$\hat{f}_{\mathrm{TV}}(\alpha) = \arg\min_f \|g - Hf\|_2^2 + \alpha^2 \sum_{i=1}^{N_f} |[Df]_i|. \qquad (34)$$

The total variation of a signal is just the total amount of change the signal goes through and can be thought of as a measure of signal variability. Thus, it is well suited to use as a side constraint and seems similar to standard Tikhonov

 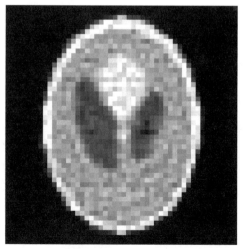

FIGURE 7   Maximum entropy solutions corresponding to the data in Figs. 1(b) (left) and 2(b) (right).

regularization with a derivative constraint. But, unlike standard quadratic Tikhonov solutions, total variation regularized answers can contain localized steep gradients, so that edges are preserved in the reconstructions. For these reasons, total variation has been suggested as the "right" regularizer for image reconstruction problems [8].

The difficulty with formulation (34) is that it again leads to a challenging nonlinear optimization problem that is caused by the nondifferentiability of the total variation cost. One approach to overcoming this challenge leads to an interesting formulation of the total variation problem. It has been shown that the total variation estimate is the solution of the following set of equations in the limit as $\beta \to 0$:

$$(H^T H + \alpha^2 D^T W_\beta(\hat{f}_{TV}) D) \hat{f}_{TV} = H^T g, \qquad (35)$$

where the diagonal weight matrix $W_\beta(f)$ depends on f and β and is given by

$$W_\beta(f) = \frac{1}{2} \text{diag}\left[\frac{1}{\sqrt{|[Df]_i|^2 + \beta}}\right], \qquad (36)$$

with $\beta > 0$ a constant. Equation (35) is obtained by smoothly approximating the ℓ_1 norm of the derivative: $\|Df\|_1 \approx \sum_{i=1}^n \sqrt{|[Df]_i|^2 + \beta}$.

Formulation (36) is interesting in that it gives insight into the difference between total variation regularization and standard quadratic Tikhonov regularization with $L = D$. Note that the latter case would result in a set of equations similar to Eq. (35) but with $W = I$. Thus, the effect of the change to a total variation cost is the incorporation of a spatially varying weighting of each derivative penalty term by $1/\sqrt{|[Df]_i|^2 + \beta}$. When the local derivative $|[Df]_i|^2$ is small, the weight goes to a large value, imposing greater smoothness to the solution in these regions. When the local derivative $|[Df]_i|^2$ is large, the weight goes to a

small value, allowing large gradients in the solution coefficients at these points.

Computationally, Eq. (35) is still nonlinear, since the weight matrix depends on f. However, it suggests a simple fixed point iteration for f, only requiring the solution of a standard linear problem at each step:

$$(H^T H + \alpha^2 D^T W_\beta(f^{(k)}) D) f^{(k+1)} = H^T g, \qquad (37)$$

where β is typically set to a small value.

With the use of the iterative approach of Eq. (37), total variation solutions to the motion-blur restoration example of Fig. 1 and the tomographic reconstruction example of Fig. 2 were generated. Figure 8 shows these total variation solutions. In addition to suppressing excessive noise growth, total variation achieves excellent edge preservation and structure recovery in both cases. These results are typical of total variation reconstructions and have led to the great popularity of this and similar methods in recent years.

Other Nonquadratic Regularization

More generally, a variety of nonquadratic choices for $J_1(f, g)$ and $J_2(f)$ have been considered. In general, these measures have the characteristic that they do not penalize large values of their argument as much as the standard quadratic ℓ_2 penalty does. Indeed, maximum entropy and total variation can both be viewed in this context, in that both are simple changes of the side constraint to a size or energy measure that is less drastic than squared energy. Another choice with these properties is the general family of ℓ_p norms:

$$J(z) = \|z\|_p = \left(\sum_{i=1}^{N_z} |z_i|^p\right)^{1/p}, \qquad (38)$$

with $1 \le p \le 2$. With p chosen in this range, these norms are

FIGURE 8   Total variation solutions corresponding to the data in Figs. 1(b) (left) and 2(b) (right).

less severe in penalizing large values than the ℓ_2 norm, yet they are still convex functions of the argument (and so still result in tractable algorithms).

Measures with even more drastic penalty "attenuation" based on nonconvex functions have been considered. An example is the so called weak-membrane cost:

$$J(z) = \min\left(\|z\|_2^2, \alpha\right). \tag{39}$$

When used in the data fidelity term $J_1(f, g)$, these measures are related to notions of robust estimation [9] and provide robustness to outliers in the data and also to model uncertainty. When used in a side constraint $J_2(f)$ on the gradient of the image Df, these measures preserve edges in regularized reconstructions and produce results similar in quality to the total variation solution discussed previously. The difficulty with the use of such nonconvex costs is computational, though the search for efficient approaches to such problems has been the subject of active study [10].

2.4 Statistical Methods

We now discuss a statistical view of regularization. If the noise q and the unknown image f are viewed as random fields, then we may seek the maximum *a posteriori* (MAP) estimate of f as that value which maximizes the posterior density $p(f \mid g)$. Using Bayes rule and the monotonicity properties of the logarithm, we obtain

$$\hat{f}_{\text{MAP}} = \arg\max_f p(f \mid g) = \arg\max_f \ln p(g \mid f) + \ln p(f). \tag{40}$$

Notice that this cost function has two terms: a data-dependent term $\ln p(g \mid f)$, called the log-likelihood function, and a term $\ln p(f)$, dependent only on f, termed the prior model. These two terms are similar to the two terms in the Tikhonov functional, Eq. (24). The likelihood function captures the dependence of the data on the field and enforces fidelity to data in Eq. (40). The prior model term captures our *a priori* knowledge about f in the absence of data, and it allows incorporation of this information into the estimate.

To be concrete, consider the widely used case of Gaussian statistics:

$$g = Hf + q, \qquad q \sim N(0, \Lambda_q), \tag{41}$$

$$f \sim N(0, \Lambda_f), \tag{42}$$

where $f \sim N(m, \Lambda)$ denotes that f is a Gaussian random vector with mean m and covariance matrix Λ, as described in Chapter 4.4. Under these assumptions $\ln p(g \mid f) \propto -\frac{1}{2}\|g - Hf\|_{\Lambda_q^{-1}}^2$ and $\ln p(f) \propto -\frac{1}{2}\|f\|_{\Lambda_f^{-1}}^2$, and upon substitu-

tion into Eq. (40) we obtain

$$\hat{f}_{\text{MAP}} = \arg\min_f \|g - Hf\|_{\Lambda_q^{-1}}^2 + \|f\|_{\Lambda_f^{-1}}^2. \tag{43}$$

The corresponding set of normal equations defining the MAP estimate are given by

$$\left(H^T \Lambda_q^{-1} H + \Lambda_f^{-1}\right) \hat{f}_{\text{MAP}} = H^T \Lambda_q^{-1} g \tag{44}$$

The solution of Eq. (44) is also the linear minimum mean square error (MMSE) estimate in the general (i.e., non-Gaussian) case. The MMSE estimate minimizes $E[\|f - \hat{f}\|_2^2]$, where $E[z]$ denotes the expected value of z.

A particularly interesting prior model for f corresponds to

$$Df = r, \qquad r \sim N(0, \lambda_r I), \tag{45}$$

where D is a discrete approximation of the gradient operator described in Chapter 4.10. Equation (45) implies a Gaussian prior model for f with covariance $\Lambda_f = \lambda_r (D^T D)^{-1}$ (assuming, for convenience, that D is invertible). Prior model (45) essentially says that the *increments* of f are uncorrelated with variance λ_r — that is, that f itself corresponds to a Brownian motion type of model. Clearly, real images are not Brownian motions, yet the continuity of Brownian motion models suggest this model may be reasonable for image restoration.

To demonstrate this insight, the Brownian motion prior image model of Eq. (45) is combined with an uncorrelated observation noise model $\Lambda_q = \lambda_q I$ in Eq. (41) to obtain MAP estimates for both the motion-blur restoration example of Fig. 1 and the tomographic reconstruction example of Fig. 2. In each case, the variance λ_r is set to the variance of the derivative image Df and the variance λ_q is set to the additive noise variance. In Fig. 9 the resulting MAP-based solutions for the two examples are shown.

In addition to an estimate, the statistical framework also provides an expression for an associated measure of estimate uncertainty through the error covariance matrix $\Lambda_e = E[ee^T]$, where $e = f - \hat{f}$. For the MAP estimate in Eq. (43), the error covariance is given by

$$\Lambda_e = \left(H^T \Lambda_q^{-1} H + \Lambda_f^{-1}\right)^{-1}. \tag{46}$$

The diagonal entries of Λ_e are the variances of the individual estimation errors and have a natural use for estimate evaluation and data fusion. Note also that the trace of Λ_e is the mean square error of the estimate. In practice, Λ_e is usually a large, full matrix, and the calculation of all its elements is impractical. There are methods, however, to estimate its diagonal elements [11].

In the stationary case, the matrices in Eq. (44) possess a block-circulant structure and the entire set of equations can be solved on an element-by-element basis in the discrete frequency

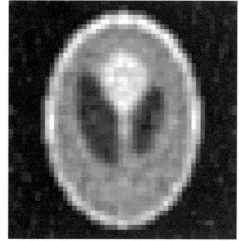

FIGURE 9 Brownian motion prior-based MAP solutions corresponding to the data in Figs. 1(b) (left) and 2(b) (right).

domain, as was done, c.f. Eq. (16). The MAP estimator then reduces to the Wiener filter:

$$\tilde{f}_i = \left(\frac{\tilde{h}_i^* S_{f_i}}{|\tilde{h}_i|^2 S_{f_i} + S_{q_i}} \right) \tilde{g}_i, \tag{47}$$

where $*$ denotes complex conjugate, $\tilde{z}_i$ denotes the ith coefficient of the 2-D DFT of the corresponding image z, and S_{z_i} denotes the ith element of the power spectral density of the random image z. The power spectral density is the 2-D DFT of the corresponding covariance matrix. The Wiener filter is discussed in Chapter 3.9.

Before proceeding, it is useful to consider the relationship between Bayesian MAP estimates and Tikhonov regularization. From Eqs. (43) and (44) we can see that in the Gaussian case (or linear MMSE case) the MAP estimate is essentially the same as a general Tikhonov estimate for a particular choice of weighting matrices. For example, suppose that both the noise model and the prior model correspond to uncorrelated random variables so that $\Lambda_q = \lambda_q I$ and $\Lambda_f = \lambda_f I$. Then Eq. (43) is equivalent to

$$\hat{f}_{\text{MAP}} = \arg\min_f \|g - Hf\|_2^2 + \frac{\lambda_q}{\lambda_f} \|f\|_2^2, \tag{48}$$

which is precisely the Tikhonov estimate when $L = I$ and the regularization parameter $\alpha^2 = \lambda_q/\lambda_f$. This association provides a natural interpretation to the regularization parameter as a measure of the relative uncertainty between the data and the prior. As another example, note that the MAP estimate corresponding to prior model (45) coupled with observation model (42) with $\Lambda_q = \lambda_q I$ will be the same as a standard Tikhonov estimate with $L = D$ and $\alpha^2 = \lambda_q/\lambda_r$.

While so far the MAP estimate has been interpreted in the Tikhonov context, it is also possible to interpret particular cost choices in the Tikhonov formulation as statistical models for the underlying field and noise. For example, consider the total variation formulation in Eq. (34). Comparing this cost function to Eq. (40), we find it reasonable to make the following probabilistic associations:

$$\ln p(g \mid f) \propto -\|g - Hf\|_2^2, \quad \ln p(f) \propto -\alpha^2 \sum_{i=1}^{N_f} |[Df]_i|, \tag{49}$$

which is consistent with the following statistical observation and prior models for the situation:

$$g = Hf + q, \qquad q \sim N(0, I), \tag{50}$$

$$p(f) \sim \prod_{i=1}^{N_f} e^{-\alpha^2 |[Df]_i|}. \tag{51}$$

The statistical prior model for f has *increments* $[Df]_i$ that are independent identically distributed (IID) according to a Laplacian density. In contrast, standard Tikhonov regularization with $L = D$ corresponds to the Brownian motion-type prior model, Eq. (45), with increments that are also IID but Gaussian distributed.

Finally, while we have emphasized the similarity of the MAP estimate to Tikhonov methods, there are, of course, situations particularly well matched to a statistical perspective. A very important example arises in applications such as in low-light-level imaging on CCD arrays, compensating for film grain noise, and certain types of tomographic imaging (e.g., PET and SPECT). In these applications, the discrete, counting nature of the imaging device is important and the likelihood term $p(g \mid f)$ is well modeled by a Poisson density function, leading to a signal-dependent noise model. The estimate resulting from use of such a model will be different from a standard Tikhonov solution, and in some instances can be significantly better. More generally, if a statistical description of the observation process and prior knowledge is available through physical modeling or first principle arguments, then the MAP formulation provides a rational way to combine this information together with measures of uncertainty to generate an estimate.

2.5 Parametric Methods

Another direct method for focusing the information in data and turning an ill-conditioned or poorly posed problem into a well-conditioned one is based on changing the parameterization of the problem. The most common representational choice for the unknown image, as mentioned previously, is to parameterize the problem in terms of the values of a regular grid of rectangular pixels. The difficulty with such a parameterization is that there are usually a large number of pixel values, which must then be estimated from the available data. For example, an image represented on a 512×512 square pixel array has over 250,000 unknowns to estimate!

Perhaps the simplest change in parameterization is a change in basis. An obvious example is provided by the SVD. By parameterizing the solution in terms of a reduced number of singular vectors, we saw that regularization of the resulting solution could be obtained (for example, through TSVD). The SVD is based completely on the distorting operator H, and hence contains no information about the underlying image f or the observed data g. By using parameterizations better matched to these other pieces of the problem it is reasonable to expect better results. This insight has resulted in the use of other decompositions and expansions [12] in the solution of image restoration and reconstruction problems, including the wavelet representations discussed in Chapter 4.1. These methods share the aim of using bases with parsimonious representations of the unknown f, which thus serve to focus the information in the data into a few, robustly estimated coefficients.

Generalizing such changes of basis, prior information can be used to construct a representation directly capturing knowledge of the structure or geometry of the objects in the underlying image f. For example, the scene might be represented as being composed of a number of simple geometric shapes, with the parameters of these shapes taken as the unknowns. For example, such approaches have been taken in connection with problems of tomographic reconstruction. The advantage of such representations is that the number of unknowns can be dramatically reduced to, say, tens or hundreds rather than hundreds of thousands, thus offering the possibility of better estimation of these fewer unknowns. The disadvantage is that the resulting optimization problems are generally nonlinear, can be expensive, and require good initializations to avoid converging to local minima of the cost.

3 Iterative Regularization Methods

One reason for an interest in iterative methods in association with regularization is purely computational. These methods provide efficient solutions of the Tikhonov or MAP normal equations, Eq. (25) or (44). Their attractions in this regard are several. First, reasonable approximate solutions can often be obtained with few iterations, and thus with far less computation than required for exact solution of these equations. Second, iterative approaches avoid the memory intensive factorizations or explicit inverses required for exact calculation of a solution, which is critical for very large problems. Finally, many iterative schemes are naturally parallelizable, and thus can be easily implemented on parallel hardware for additional speed.

Interestingly, however, when applied to the *unregularized* problem, Eq. (12), and terminated long *before convergence*, iterative methods provide a smoothing effect to the corresponding solution. As a result, they can be viewed as a regularization method in their own right [13]. Such use is examined in this section. More detail on various iterative algorithms for restoration can be found in Chapter 3.10. The reason the regularization behavior of iterative algorithms occurs is that the low-frequency (i.e., smooth) components of the solution tend to converge faster than the high-frequency (i.e., rough) components. Hence, for such iterative schemes, the number of iterations plays the role of the inverse of the regularization parameter α, so fewer iterations corresponds to greater regularization (and larger α).

To gain insight into the regularizing behavior of iterative methods, consider a simple Landweber fixed point iteration for the solution of Eq. (12). This basic iterative scheme appears under a variety of names in different disciplines (e.g., the Van Cittert iteration in image reconstruction or the Gerchberg–Papoulis algorithm for bandwidth extrapolation). The iteration is given by

$$
\begin{aligned}
\hat{f}_{\text{Lw}}^{(k+1)} &= \hat{f}_{\text{Lw}}^{(k)} + \gamma H^T \big(g - H \hat{f}_{\text{Lw}}^{(k)} \big) \\
&= \gamma H^T g + \big(I - \gamma H^T H \big) \hat{f}_{\text{Lw}}^{(k)},
\end{aligned} \tag{52}
$$

where γ is a real relaxation parameter satisfying $0 < \gamma < 2/\sigma_{\max}^2$ and $\sigma_{\max}$ is the maximum singular value of H. If the iteration is started with $\hat{f}_{\text{Lw}}^{(0)} = 0$, then the estimate after k steps is given by [5, 13]:

$$
\hat{f}_{\text{Lw}}^{(k)} = \sum_{i=1}^{p} \Big[1 - \big(1 - \gamma \sigma_i^2 \big)^k \Big] \frac{u_i^T g}{\sigma_i} \, v_i, \tag{53}
$$

where $\{\sigma_i, u_i, v_i\}$ is the singular system of H. Comparing this expression with Eq. (21) or (26), we find the effect of the iterative scheme is again to weight or filter the coefficients of unregularized generalized solution (19), where the weight or filter function is now given by

$$
w_{i,k} = 1 - \big(1 - \gamma \sigma_i^2 \big)^k. \tag{54}
$$

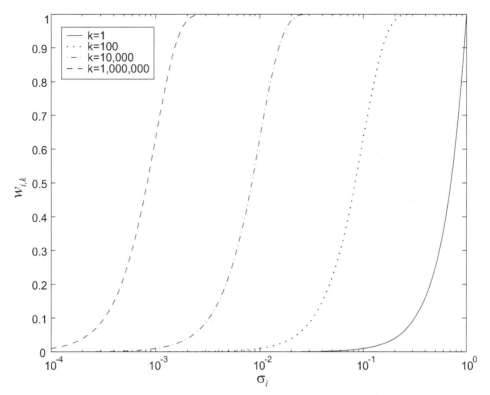

FIGURE 10 Plots of the Landweber weight function $w_{i,k}$ of Eq. (54) versus singular value σ_i for various numbers of iterations k.

This function is plotted in Fig. 10 for $\gamma = 1$ for a variety of values of iteration count k. As can be seen, it has a steplike behavior as a function of the size of σ_i, where the location of the transition depends on the number of iterations [7], so the iteration count of the iterative method does indeed play the role of the (inverse of the) regularization parameter.

An implication of this behavior is that, to obtain reasonable estimates from such an iterative method applied to the unregularized normal equations, a stopping rule is needed, or the generalized inverse will ultimately be obtained. This phenomenon is known as "semiconvergence" [5]. Figure 11 shows Landweber iterative solutions corresponding to the motion-blur restoration example of Fig. 1 after various numbers of iterations, and Fig. 12 shows the corresponding solutions for the tomographic reconstruction problem of Fig. 2. In both examples, when too few iterations are performed the resulting solution is overregularized, blurred, and missing significant image structure. Conversely, when too many iterations are performed, the corresponding solutions are underregularized and begin to display the excessive noise amplification characteristic of the generalized solution.

While Landweber iteration (52) is simple to understand and analyze, its convergence rate is slow, which motivates the use of other iterative methods in many problems. One example is the conjugate gradient (CG) method, which is one of the most powerful and widely used methods for the solution of symmetric, sparse linear systems of equations [14]. It has been shown that when CG is applied to the unregularized normal equations, Eq. (12), the corresponding estimate $\hat{f}_{\mathrm{CG}}^{(k)}$ after k iterations is given by the solution to the following problem:

$$\hat{f}_{\mathrm{CG}}^{(k)} = \arg\min \|g - Hf\|_2 \quad \text{subject to}$$
$$f \in \mathcal{K}_k(H^T H, H^T g), \quad (55)$$

where $\mathcal{K}_k(H^T H, H^T g) = \mathrm{span}\{H^T g, (H^T H) H^T g, \ldots, (H^T H)^{k-1} H^T g\}$ is called the Krylov subspace associated to the normal equations. Thus, k iterations of CG again regularize the problem, this time through the use of a Krylov subspace constraint on the solution [instead of a quadratic side constraint as in Eq. (28)]. The regularizing effect arises from the property that the Krylov subspace $\mathcal{K}_k(H^T H, H^T g)$ approximates the subspace $\mathrm{span}\{v_1, \ldots, v_k\}$ spanned by the first k right singular vectors. While the situation is more complicated than in the Landweber iteration case, weight or filter factors $w_{i,k}$ that depend on k can also be defined for the CG method. These weight factors are observed to have a similar attenuating behavior for the large singular values as for the Landweber case, with a rolloff that is also dependent on the number of iterations.

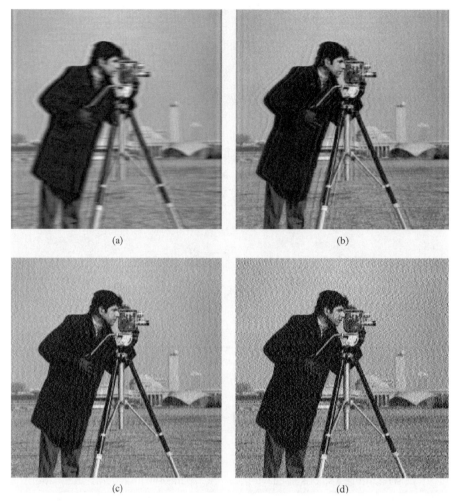

(a)

(b)

(c)

(d)

FIGURE 11   Iterative Landweber solution of the unregularized normal equations for the example of Fig. 1, with (a) five, (b) 50, (c) 500, and (d) 5,000 iterations.

4 Regularization Parameter Choice

Regularization, by stabilizing the estimate in the face of noise amplification, inherently involves a tradeoff between fidelity to the data and fidelity to some set of prior information. These two components are generally measured through the residual norm $\|g - H\hat{f}\|$, or more generally $J_1(\hat{f}, g)$, and the side constraint norm $\|L\hat{f}\|$, or more generally $J_2(\hat{f})$. The regularization parameter α controls this tradeoff, and an important part of the solution of any problem is finding a reasonable value for α.

In this section, five methods for choosing the regularization parameter will be discussed: choice based on visual criterion; the discrepancy principle, based on some knowledge of the noise; the L-curve criterion based on a plot of the residual norm versus the side constraint norm; generalized cross-validation, based on minimizing prediction errors; and statistical parameter choice, based on modeling the underlying processes.

4.1 Visual Inspection

Often the main tradeoff dealt with in regularization is between the excessive noise amplification that occurs in the absence of regularization and oversmoothing of the solution if too much is used. Further, there may be considerable prior knowledge on the part of the viewer about the characteristics of the underlying scene — as arises in the restoration of images of natural scenes. In such cases, it may be entirely reasonable to choose the regularization parameter through simple visual inspection of regularized images as the regularization parameter is varied. This approach is well suited, for example, to iterative methods, in which the number of iterations effectively sets the regularization parameter. Since iterative methods are terminated long before convergence is achieved when they are used as a form of regularization, the intermediate estimates are simply monitored as the iteration proceeds and the iteration is stopped when noise distortions are observed to be entering the solution. This process can be seen in the examples of Figs. 11 and 12; when few

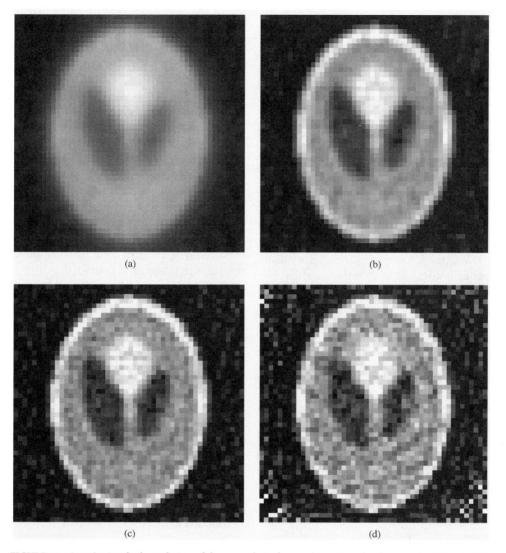

FIGURE 12 Iterative Landweber solution of the unregularized normal equations for the example of Fig. 2, with (a) five, (b) 50, (c) 500, and (d) 50,000 iterations.

iterations have been performed, the solution appears overregularized. As more iterations are done the detail in the solution is recovered. Finally, as too many iterations are performed the solution becomes corrupted by noise effects. This visual approach to choosing α is clearly problematic in cases in which the viewer has little prior understanding of the structure of the scene being imaged or in cases in which the reconstructed field itself is very smooth, making it difficult to visually evaluate over- from underregularized solutions.

4.2 The Discrepancy Principle

If there is knowledge about the perturbation or noise q in Eq. (10), then it makes sense to use it in choosing α. When viewed deterministically, this information is often in the form of knowledge about the size or energy of the perturbation:

$$\|q\|_2 \le \delta_q. \qquad (56)$$

This knowledge provides a bound on the residual norm $\|g - Hf\|_2 \le \delta_q$. In a stochastic setting, such information can take the form of knowledge of the noise variance λ_q.

Since the price for overfitting the solution to the data (i.e., for underregularizing) is excessive noise amplification (as seen in the generalized solution), it makes sense to choose the regularization parameter large enough that the data fit error achieves this bound, but no larger (to avoid overregularizing). This idea is behind the discrepancy principal approach to choosing the regularization parameter, attributed to Morozov. Formally, the regularization parameter α is chosen as that value for which the residual norm achieves the equality

$$\|g - H\hat{f}(\alpha)\|^2 = \delta_q^2, \tag{57}$$

or, in the stochastic setting, where the residual norm equals λ_q. There also exist generalized versions of the discrepancy principle that incorporate knowledge of perturbations to the model H as well. Finally, note that in the deterministic case the value of α provided by the discrepancy principle generally leads to some overregularization, since the actual perturbation may be smaller than the given bound. Conversely, specification of a bound in Eq. (56) that is too small can lead to undesirable noise growth in the solution.

Use of the discrepancy principle requires knowledge of the perturbation bound δ_q or noise variance λ_q. Sometimes this quantity may be obtained from physical considerations, prior knowledge, or direct estimation from the data. When this is not the case, parameter choice methods are required which avoid the need for such knowledge. Two such approaches are examined next.

4.3 The L-Curve

Since all regularization methods involve a tradeoff between fidelity to the data, as measured by the residual norm, and the fidelity to some prior information, as measured by the side constraint norm, it would seem natural to choose a regularization parameter based on the behavior of these two terms as α is varied. Indeed, a graphical plot of $\|L\hat{f}(\alpha)\|_2$ versus $\|g - H\hat{f}(\alpha)\|_2$ on a log–log scale as α is varied is called the L-curve and has been proposed as a means to choose the regularization parameter [7]. Note, especially, that α is a parameter along this curve. The L-curve, shown schematically in Fig. 13 (c.f. [7]) has a characteristic "L" shape (hence its name), which consists of a vertical part and a horizontal part. The vertical part corresponds to underregularized estimates, where the solution is dominated by the amplified noise. In this region, small changes to α have a large effect on the size or energy of $\hat{f}$, but a relatively small impact on the data fit. The horizontal part of the L-curve corresponds

to oversmoothed estimates, where the solution is dominated by residual fit errors. In this region changes to α affect the size of $\hat{f}$ weakly, but produce a large change in the fit error.

The idea behind the L-curve approach for choosing the regularization parameter is that the corner between the horizontal and vertical portions of the curve defines the transition between over- and underregularization, and thus represents a balance between these two extremes and the best choice of α. The point on the curve corresponding to this α is shown as α^* in Fig. 13. While the notion of choosing α to correspond to the corner of the L-curve is natural and intuitive, there exists the issue of defining exactly what is meant by the "corner" of this curve. A number of definitions have been proposed, including the point of maximum curvature, the point closest to a reference location, and the point of tangency with a line of slope -1. The last definition is especially interesting, since it can be shown that the optimal α for this criterion must satisfy

$$\alpha^2 = \frac{\|g - H\hat{f}(\alpha)\|_2^2}{\|L\hat{f}(\alpha)\|_2^2}. \tag{58}$$

The right hand side of Eq. (58) can be loosely interpreted as the ratio of an estimated noise variance to an estimated signal variance for zero-mean images with $L = I$, and thus appears similar in spirit to Eq. (48).

4.4 Generalized Cross-Validation

Another popular method for choosing the regularization parameter that does not require knowledge of the noise properties is generalized cross-validation (GCV). The basic idea behind cross-validation is to minimize the set of prediction errors — that is, to choose α so that the regularized solution obtained with a data point removed predicts this missing point well when averaged over all ways of removing a point. This viewpoint leads to minimization with respect to α of the following GCV function:

$$\mathcal{V}(\alpha) = \frac{\|g - H\hat{f}(\alpha)\|_2^2}{[\text{trace}\,(I - HH^\#)]^2}, \tag{59}$$

where $H^\#$ denotes the linear operator that generates the regularized solution when applied to data, so that $\hat{f}(\alpha) = H^\# g$. The value of α that minimizes the cost $\mathcal{V}(\alpha)$ in Eq. (59) is also an estimate of the value of α that minimizes the mean square error $E[\|Hf - H\hat{f}(\alpha)\|_2^2]$ [15].

Note that only the data are used in the calculation of $\mathcal{V}(\alpha)$ and no prior knowledge of, e.g., the noise amplitude, is required. However, there are also a number of difficulties related to the computation of the GCV cost in Eq. (59). First, the operator $HH^\#$ must be found. While specifying this quantity is straightforward for, e.g., Tikhonov regularization [where it may be written completely in terms of the filter weights $w_{i,\alpha}$ of Eq. (27)], it may prove inconvenient for other regularization methods (e.g.,

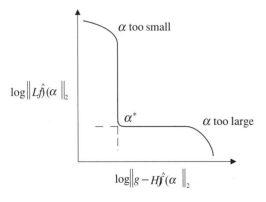

FIGURE 13   Structure of the L-curve.

for iterative methods; though see [7], Sec. 7.4). Finally, in some cases the GCV cost curve is quite flat, leading to numerical problems in finding the minimum of $\mathcal{V}(\alpha)$, which can result in overly small values of α.

4.5 Statistical Approaches

Our last method of parameter choice is not really a parameter choice technique per say, but rather an estimation approach. As discussed in Section 2.4, given a statistical model of the observation process through $p(g \mid f)$ and of the prior information about f through $p(f)$, the MAP estimate is obtained by solving optimization problem (40). Note that there are no undetermined parameters to set in this formulation. In the statistical view, the problem of regularization parameter determination is exchanged for a problem of *statistical modeling* through the specification of $p(g \mid f)$ and $p(f)$. The tradeoff between data and prior inherent in the choice of the regularization parameter α is captured in the modeling of the relative uncertainties in the processes g and f. Sometimes the densities $p(g \mid f)$ and $p(f)$ follow from physical considerations or direct experimental investigation. Such is the case in the Poisson observation model for $p(g \mid f)$ often used in tomographic and film-based imaging problems [2, 3]. In such cases, the Bayesian point of view provides a natural and rational way of balancing data and prior.

For many problems, however, the specification of the densities $p(g \mid f)$ and $p(f)$ may appear to be a daunting task. For example, what is the "right" prior density $p(f)$ for the pixels in an image of a natural scene? Identifying such a density at first seems a much more difficult undertaking than finding a good value of the single scalar parameter α. Fortunately, from an engineering standpoint, the goal is usually not to most accurately model the field f or observation g, but rather to find a *reasonable* statistical model that leads to tractable computation of a good estimate. In this regard, relatively simple statistical models may suffice for the purposes of image restoration and reconstruction. Further, the statistical nature of these models may suggest rational choices of their parameters not obvious from the Tikhonov point of view. For example, as discussed c.f. Eq. (48), under a white Gaussian assumption for both the observation noise and the prior, the regularization parameter α^2 can be identified with the variance ratio λ_q/λ_f, where λ_f corresponds to the variance of the underlying image and λ_q is the variance of the noise. Another example is provided by the Brownian motion image model of Eq. (45). This case corresponded to Tikhonov regularization with $L = D$ and $\alpha^2 = \lambda_q/\lambda_r$, where now λ_r is the variance of the *derivative* image.

5 Summary

In this chapter we have discussed the need for regularization in problems of image restoration and reconstruction. We have given an overview of the issues that arise in these problems and

the means to deal with them. The two driving forces in the need for regularization are noise amplification and lack of data. The primary idea behind regularization is the inclusion of prior knowledge to counteract these effects. Though there are a large variety of ways to view both these problems and their solution, there is also a great amount of commonality in their essence. In applying such methods to image processing problems (as opposed to one-dimensional signals), all these approaches lead to optimization problems requiring considerable computation. Fortunately, powerful computational resources are becoming available on the desktop of nearly all engineers, and there is a wealth of complementary software tools to aid in their application; see, e.g., [16]. The goal of this chapter has been to provide a unifying view of this area.

Throughout the chapter, we have assumed that the distortion model, as captured by H, is perfectly known and the only uncertainty is due to noise q in the observations g. Often, however, the knowledge of H is not perfect, and in such cases the uncertainty in this model must be dealt with as well. Sometimes it is sufficient to simply treat such model uncertainty as a larger effective observation noise. Alternatively, the uncertainty in H may be explicitly included in the formulation of the inversion problem. Such an approach leads naturally to a method known as total least squares (TLS) [6], which is simply the extension of the least-squares idea to include minimization of the square error in both the model and data. Regularized versions of TLS also exist [17] and have shown improved results over the basic least-squares methods.

Further Reading

Chapter 3.5 discusses the basics of image restoration. Chapter 3.6 presents problems arising in multichannel image restoration. Chapter 3.7 treats multiframe image restoration, and Chapter 3.9 is focused on video restoration. In Chapter 3.10 there is a more in-depth discussion of iterative methods of image restoration. Chapter 10.2 examines image reconstruction from projections and its application. There are also a number of accessible, yet more extensive, treatments of this material in the general literature. A readable engineering treatment of discrete inverse problems is given in [7], and an associated package of numerical tools is presented in [16]. A deeper theoretical treatment of the topic of data inversion can be found in [4,5]. The iterative approach to image restoration is studied in [13]. Iterative solution methods in general are discussed in depth in [14].

Acknowledgments

I thank D. Castanon, M. Cetin, and R. Weisenseel for their help in generating the examples of this chapter and J. O'Neill for proofreading the text.

References

[1] R. Kress, *Linear Integral Equations* (Springer-Verlag, New York, 1989).

[2] A. C. Kak and M. Slaley, *Principles of Computerized Tomographic Imaging* (IEEE, Piscataway, NJ, 1987).

[3] H. C. Andrews and B. R. Hunt, *Digital Image Restoration* (Prentice-Hall, Englewood Cliffs, NJ, 1977).

[4] M. Bertero, "Linear inverse and ill-posed problems," in *Advances in Electronics and Electron Physics* (Academic, New York, 1989), Vol. 75.

[5] H. W. Engl, M. Hanke, and A. Neubauer, *Regularization of Inverse Problems* (Kluwer, Dordrecht, The Netherlands, 1996).

[6] G. H. Golub and C. F. Van Loan, *Matrix Computations.* Johns Hopkins studies in the Mathematical Sciences (Johns Hopkins U. Press, Baltimore, MD, 1989).

[7] P. C. Hansen, *Rank-Deficient and Discrete Ill-Posed Problems* (SIAM, Philadelphia, 1998).

[8] L. I. Rudin, S. Osher, and E. Fatemi, "Nonlinear total variation based noise removal algorithms," *Physica D* **60**, 259–268 (1992).

[9] P. J. Huber, *Robust Statistics.* Wiley Series in Probability and Mathematical Statistics (Wiley, New York, 1981).

[10] D. Geman and C. Yang, "Nonlinear image recovery with half-quadratic regularization," *IEEE Trans. Image Proc.* **4**, 932–945 (1995).

[11] A. M. Erisman and W. Tinney, "On computing certain elements of the inverse of a sparse matrix," *Commun. ACM* **18**, 177–179 (1975).

[12] D. L. Donoho, "Nonlinear solution of linear inverse problems by wavelet-vagulette decomposition," *Appl. Comput. Harmonic Anal.* **2**, 101–126 (1995).

[13] R. L. Lagendijk and J. Biemond, *Iterative Identification and Restoration of Images* (Kluwer, Boston, 1991).

[14] O. Axelsson, *Iterative Solution Methods* (Cambridge U. Press, Cambridge, UK, 1994).

[15] G. Wahba, *Spline Models for Observational Data* (SIAM, Philadelphia, 1990).

[16] P. C. Hansen, "Regularization tools: a Matlab package for analysis and solution of discrete ill-posed problems," *Numer. Algorithms* **6**, 1–35 (1994).

[17] R. D. Fierro, G. H. Golub, P. C. Hansen, and D. P. O'Leary, "Regularization by truncated total least-squares," *SIAM J. Sci. Comp.* **18**, 1223–1241 (1997).

3.7

Multichannel Image Recovery

Nikolas P. Galatsanos
and Miles N. Wernick
Illinois Institute of Technology

Aggelos K. Katsaggelos
Northwestern University

1 Introduction

Color images, video images, medical images obtained by multiple scanners, and multispectral satellite images consist of multiple image frames or *channels* (Fig. 1). These image channels depict the same scene or object observed either by different sensors or at different times, and thus have substantial commonality among them. We use the term *multichannel image* to refer to any collection of image channels that are not identical but that exhibit strong between-channel correlations.

In this chapter we focus on the problem of image recovery as it applies specifically to multichannel images. *Image recovery* refers to the computation of an image from observed data that alone do not uniquely define the desired image. Important examples are image denoising, image deblurring, decoding of compressed images, and medical image reconstruction.

In image recovery, ambiguities in inferring the desired image from the observations usually arise from uncertainty produced by noise. These ambiguities can only be reduced if, in addition to information provided by the observed data, one also has prior knowledge about the desired image. In many applications the most powerful piece of prior information is that the desired image is smooth (spatially correlated), whereas the noise is not. Multichannel images offer the possibility of exploiting

correlations *between* the channels in addition to those within each channel. By utilizing this extra information, multichannel image recovery can yield tremendous benefits over separate recovery of the component channels.

In a broad category of image recovery techniques, the image is computed by optimizing an objective function that quantifies correspondence of the image to the observed data as well as prior knowledge about the true image. Two frameworks have been developed to describe the use of prior information as an aid in image recovery: the deterministic formulation with regularization [7] and the statistical formulation [4]. Conceptually, these frameworks are quite different, but in many applications they lead to identical algorithms.

In the deterministic formulation, the problem is posed in terms of inversion of the observation model, with regularization used to improve the stability of the solution (for more details see Chapter 3.11). In the statistical formulation, the statistics of the noise and the desired image are incorporated explicitly, and they are used to describe the desired characteristics of the image. In principle, these formulations apply equally well to multichannel images and single-channel images, but in practice two significant problems must be addressed.

First, an appropriate model must be developed to express the relationships among the image channels. Regularization

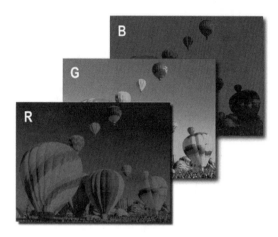

FIGURE 1 Example of a multichannel image. A color image consists of three color components (channels) that are highly correlated with one another. Similarly, a video image sequence consists of a collection of closely related images. (See color section, p. C–5.)

and statistical methods reflecting within-channel relationships are much better developed than those describing between-channel relationships, and they are often easier to work with. For example, while the power spectrum (the Fourier transform of the autocorrelation) is a useful statistical descriptor for one channel, the cross power spectrum (the Fourier transform of the cross-correlation) describing multiple channels is less tractable because it is complex. Second, a multichannel image has many more pixels than each of its channels, so approaches that minimize the computations are typically sought. Several such approaches are described in the following sections.

The goal of this chapter is to provide a concise summary of the theory of multichannel image recovery. We classify multichannel recovery methods into two broad approaches, each of which is illustrated through a practical application. In the first, which we term the *explicit* approach, all the channels of the multichannel image are processed collectively, and regularization operators or prior distributions are used to express the between- and within-channel relationships. In the second, which we term the *implicit* approach, the same effect is obtained indirectly by (1) applying a Karhunen–Loève (KL) transform that decorrelates the channels, (2) recovering the channels separately in a single-channel fashion, and (3) inverting the KL transform. This approach has a substantial computational advantage over the explicit approach, as we explain later, but it can only be applied in certain situations.

The rest of this chapter is organized as follows. In Section 2 we present the multichannel observation model, and we review basic image recovery approaches in Section 3. In Section 4 we describe the explicit approach and illustrate it by using the example of restoration of video image sequences. In Section 5, we explain the implicit approach and illustrate it by using the example of reconstruction of time-varying medical images. We conclude with a summary in Section 6.

2 Imaging Model

Throughout this chapter we use symbols in boldface type to denote multichannel quantities. We assume the following discrete model for multichannel imaging:

$$\mathbf{g} = \mathbf{H}\mathbf{f} + \mathbf{n}, \tag{1}$$

where $\mathbf{g}$, $\mathbf{f}$, and $\mathbf{n}$ are vectors representing the observed (degraded) multichannel data, the true multichannel image, and random noise, respectively, and matrix $\mathbf{H}$ denotes the linear multichannel degradation operator. Lexicographic ordering is used to represent the images as vectors by stacking all their rows or columns in one long vector. Then, each multichannel image is a concatenation of its K component channels, i.e.,

$$\mathbf{g} = \begin{bmatrix} g_1 \\ g_2 \\ \vdots \\ g_K \end{bmatrix}, \quad \mathbf{f} = \begin{bmatrix} f_1 \\ f_2 \\ \vdots \\ f_K \end{bmatrix}, \quad \mathbf{n} = \begin{bmatrix} n_1 \\ n_2 \\ \vdots \\ n_K \end{bmatrix}, \tag{2}$$

where g_k, f_k, and n_k denote individual channels of the observations, the true image, and noise, respectively. In its most general form, the linear multichannel degradation operator can be written as

$$\mathbf{H} = \begin{bmatrix} H_{11} & H_{12} & \cdots & H_{1K} \\ H_{21} & H_{22} & \cdots & H_{2K} \\ \cdots & \cdots & \cdots & \cdots \\ H_{K1} & H_{N2} & \cdots & H_{KK} \end{bmatrix}, \tag{3}$$

where the diagonal blocks represent within-channel degradations, and the off-diagonal blocks represent between-channel degradations. We will assume that each source channel f_i has $N \times M$ pixels and that each channel of the observations g_i has $L \times P$ pixels. Therefore, f_i is a $MN \times 1$ vector and g_i is a $LP \times 1$ vector; consequently, H_{ij} is a $LP \times MN$ matrix and $\mathbf{H}$ is a $KLP \times KMN$ matrix. If the degradation is shift invariant the product $H_{ij} f_j$ should in principle represent an ordinary linear convolution operation. However, we would like to use the discrete Fourier transform (DFT) to compute the convolutions rapidly, and this can be done only for circular convolutions. Fortunately, circular convolution and linear convolution produce the same result if we first embed each image in a larger array of zeros (see [8], p. 145). A matrix H_{ij} that, when multiplied by an image f_j, produces the effect of circular convolution has what is known as a *block-circulant* structure. Block-circulant matrices are diagonalized by the DFT [8], thus leading to simplified calculations.

For purposes of notational simplicity, we assume throughout that the multichannel source image $\mathbf{f}$ and noise $\mathbf{n}$ have zero mean. The source image usually does not obey this assumption in reality, but the equations can easily be modified to accommodate a nonzero mean by the introduction of appropriate corrections.

3 Multichannel Image Estimation Approaches

Image recovery is most often achieved by constructing an objective function to quantify the quality of an image estimate, then optimizing that function to obtain the desired result. Some important objective functions and solutions from estimation theory are reviewed in this section.

3.1 Linear Minimum Mean Square Error Estimation

The mean square error (MSE), defined as

$$\text{MSE}(\hat{\mathbf{f}}) = E\{\|\mathbf{f} - \hat{\mathbf{f}}\|^2\}, \quad (4)$$

is a measure of the quality of the image estimate $\hat{\mathbf{f}}$. Here, $E\{\cdot\}$ denotes the expectation operator. Among the images that are a linear function of the data (i.e., $\hat{\mathbf{f}} = \mathbf{Ag}$ where $\mathbf{A}$ is a matrix), the image that minimizes the MSE is known as the linear minimum mean square error (LMMSE) estimate, $\mathbf{f}_{\text{LMMSE}}$. Assuming $\mathbf{f}$ and $\mathbf{n}$ are uncorrelated [11], the LMMSE solution is found by finding the matrix $\mathbf{A}$ that minimizes the MSE. The resulting LMMSE solution is

$$\mathbf{f}_{\text{LMMSE}} = \mathbf{C}_f \mathbf{H}^T (\mathbf{H}\mathbf{C}_f \mathbf{H}^T + \mathbf{C}_n)^{-1}\mathbf{g}, \quad (5)$$

where $\mathbf{C}_n$ is the covariance matrix of the multichannel noise vector $\mathbf{n}$, and $\mathbf{C}_f$ is the $KNM \times KNM$ covariance matrix of the multichannel image vector $\mathbf{f}$, defined as

$$\mathbf{C}_f = E\{\mathbf{f}\mathbf{f}^T\} = \begin{bmatrix} C_{11} & C_{12} & \cdots & C_{1K} \\ C_{21} & C_{22} & \cdots & C_{2K} \\ \cdots & \cdots & \cdots & \cdots \\ C_{K1} & C_{K2} & \cdots & C_{KK} \end{bmatrix}, \quad (6)$$

where $C_{ij} = E\{f_j f_i^T\}$.

With use of the matrix inversion lemma [11] it is easy to show that

$$\mathbf{C}_f \mathbf{H}^T (\mathbf{H}\mathbf{C}_f \mathbf{H} + \mathbf{C}_n)^{-1} = (\mathbf{H}^T \mathbf{C}_n^{-1} \mathbf{H} + \mathbf{C}_f^{-1})^{-1} \mathbf{H}^T \mathbf{C}_n^{-1}. \quad (7)$$

Thus, we can also write

$$\mathbf{f}_{\text{LMMSE}} = (\mathbf{H}^T \mathbf{C}_n^{-1} \mathbf{H} + \mathbf{C}_f^{-1})^{-1} \mathbf{H}^T \mathbf{C}_n^{-1} \mathbf{g}. \quad (8)$$

3.2 Regularized Weighted Least-Squares Estimation

The weighted least-squares (WLS) estimate of $\mathbf{f}$ is

$$\mathbf{f}_{\text{WLS}} = \arg\min_{\mathbf{f}} \{(\mathbf{Hf} - \mathbf{g})^T \mathbf{C}_n^{-1} (\mathbf{Hf} - \mathbf{g})\}, \quad (9)$$

where $\arg\min_{\mathbf{f}}\{J(\mathbf{f})\}$ denotes the vector $\mathbf{f}$ that minimizes $J(\mathbf{f})$. Here, $\mathbf{C}_n$ is the covariance matrix of the noise, which in this context is assumed to be diagonal. In the presence of noise, the WLS solution will usually be very noisy. This occurs because the matrix $\mathbf{H}$ is often ill conditioned or singular; therefore some of its singular values are close to or equal to zero. In this case, the solution is unstable and highly sensitive to noise. Regularization is a well-known solution to this instability, in which the ill-posed problem is replaced by a well-posed problem for which the solution is an acceptable approximation (see [4, 5] and Chapter 3.11).

In the WLS formulation, regularization can be achieved by adding to the WLS functional a term that takes on large values if the image is noisy. A term that achieves this is $\|\mathbf{Qf}\|^2$, where $\mathbf{Q}$ is a high-pass filter operator. Incorporating this term, the regularized WLS (RWLS) estimate is obtained as follows:

$$\mathbf{f}_{\text{RWLS}} = \arg\min_{\mathbf{f}} \{(\mathbf{Hf} - \mathbf{g})^T \mathbf{C}_n^{-1} (\mathbf{Hf} - \mathbf{g}) + \lambda\|\mathbf{Qf}\|^2\}, \quad (10)$$

in which λ, known as the regularization parameter, controls the tradeoff between fidelity to the data (reflected by the first term of the objective function) and smoothness of the estimate (reflected by the second term) [7]. Solving for $\mathbf{f}_{\text{RWLS}}$ we obtain

$$\mathbf{f}_{\text{RWLS}} = (\mathbf{H}^T \mathbf{C}_n^{-1} \mathbf{H} + \lambda \mathbf{Q}^T \mathbf{Q})^{-1} \mathbf{H}^T \mathbf{C}_n^{-1} \mathbf{g}. \quad (11)$$

Note that the RWLS and the LMMSE estimates are equivalent when $\mathbf{C}_f^{-1} = \lambda \mathbf{Q}^T \mathbf{Q}$.

A special case of RWLS estimation occurs if the noise is white, i.e., if $\mathbf{C}_n = \sigma^2 \mathbf{I}$. In this case the RWLS functional reduces to the regularized least-squares (RLS) functional:

$$\mathbf{f}_{\text{RLS}} = \arg\min_{\mathbf{f}}\{J(\mathbf{f})\} = \arg\min_{\mathbf{f}}\{(\mathbf{Hf} - \mathbf{g})^T (\mathbf{Hf} - \mathbf{g}) + \lambda\|\mathbf{Qf}\|^2\}. \quad (12)$$

3.3 Multichannel Regularization

In multichannel image recovery, the operator $\mathbf{Q}$ enforces smoothness not only within each image channel, but also between the channels, thus achieving an additional measure of noise suppression, and often producing dramatically better images. In ordinary single-channel recovery of two-dimensional (2-D) images, one can define Q as a discrete two-dimensional Laplacian operator, which represents 2-D convolution with the following mask:

$$\begin{pmatrix} 0 & 1 & 0 \\ 1 & -4 & 1 \\ 0 & 1 & 0 \end{pmatrix}. \quad (13)$$

For multichannel recovery, one can use the three-dimensional (3-D) Laplacian $\mathbf{Q}$, which implies correlations between channels. The 3-D Laplacian [7] can be performed as a 3-D convolution, which can be written in equation form for the *l*th channel as

$$[\mathbf{Qf}]_l(i, j) = -6 f_l(i, j)(i, j) + f_l(i - 1, j) + f_l(i + 1, j)$$
$$+ f_l(i, j - 1) + f_l(i, j + 1) f_{l-1}(i, j)$$
$$+ f_{l+1}(i, j). \tag{14}$$

Though they would yield excellent results, the closed-form solutions given in the previous section usually cannot be computed directly because of the large number of dimensions of multichannel images. For example, to restore a three-channel color image having 512×512 pixels per channel, direct computation of $\mathbf{f}_{\mathrm{LMMSE}}$ using Eq. (5) would require the inversion of matrices of dimension $786,432 \times 786,432$. To sidestep this dimensionality problem, computationally efficient algorithms must be designed. Next, we present two such approaches that lead to practical multichannel image recovery techniques.

4 Explicit Multichannel Recovery Approaches

4.1 Space-Invariant Multichannel Recovery

The difficulty of directly implementing the solutions reviewed in the previous section lies in the complexity of inverting large matrices. In this section we show that, if the multichannel imaging system is space invariant and the noise and signal are stationary, then the required inversion is easily performed by using the fact that block-circulant matrices are diagonalized by the discrete Fourier transform [8]. To simplify our discussion of space-invariant multichannel imaging, we assume in this section that the observed and true image channels are square and have the same number of pixels, i.e., $M = N = L = P$.

Let us assume that the imaging system is space invariant and that the channels of the source image **f** are jointly stationary, i.e.,

$$E\{f_i(x, y) f_j(x', y')\} = [\mathbf{C}_f]_{ij}(x - x', y - y'), \tag{15}$$

where $f_i(x, y)$ denotes pixel (x, y) of image channel f_i, and $[\mathbf{C}_f]_{ij}$ is the covariance matrix of f_i and f_j. Let us make the same assumption about the noise channels n_i. In this case the covariance matrices $[\mathbf{C}_f]_{ij}$ and $[\mathbf{C}_n]_{ij}$ can be approximated by $M^2 \times M^2$ block-circulant matrices. The matrices $\mathbf{H}, \mathbf{C}_f$, and $\mathbf{C}_n$ are composed of $M^2 \times M^2$ block-circulant blocks, but *they are not themselves block circulant*.

Now let us define the multichannel DFT as $\mathbf{W} = \mathrm{diag}$ $\{W, W, \ldots, W\}$, where W is a $M^2 \times M^2$ matrix representing the two-dimensional DFT, and $\mathbf{W}$ has K blocks. Using $\mathbf{W}$ to

transform the LMMSE solution in Eq. (5), we obtain

$$\mathbf{Wf}_{\mathrm{LMMSE}} = \mathbf{WC}_f \mathbf{W}^{-1} \mathbf{WH}^T \mathbf{W}^{-1} (\mathbf{WHW}^{-1} \mathbf{WC}_f \mathbf{W}^{-1} \mathbf{WHW}^{-1}$$
$$+ \mathbf{WC}_n \mathbf{W}^{-1})^{-1} \mathbf{Wg}, \tag{16}$$

or, equivalently,

$$\mathbf{F}_{\mathrm{LMMSE}} = \mathbf{D}_{C_f} (\mathbf{D}_H)^h \mathbf{\Delta}^{-1} \mathbf{G} \tag{17}$$

where

$$\mathbf{\Delta} = \mathbf{D}_H \mathbf{D}_{C_f} (\mathbf{D}_H)^h + \mathbf{D}_{C_n} \tag{18}$$

and h denotes the Hermitian of a matrix. The $KM^2 \times 1$ vectors $\mathbf{F}_{\mathrm{LMMSE}}$ and $\mathbf{G}$ are the DFTs of the multichannel vectors $\mathbf{f}_{\mathrm{LMMSE}}$ and $\mathbf{g}$, respectively. The matrices $\mathbf{D}_{C_f}, \mathbf{D}_{C_n}$, and $\mathbf{D}_H$ are obtained by using $\mathbf{W}$ to transform $\mathbf{C}_f, \mathbf{C}_n$, and $\mathbf{H}$, respectively (e.g., $\mathbf{D}_{C_f} = \mathbf{W}^{-1} \mathbf{C}_f \mathbf{W}$). The matrix $\mathbf{\Delta}$ has a special form that allows the inversion to be readily performed; thus the difficulty of computing $\mathbf{f}_{\mathrm{LMMSE}}$ has been eliminated.

Any matrix **C** having block-circulant blocks C_{ij} can be transformed with the multichannel DFT into a matrix **D** having diagonal blocks as follows:

$$\mathbf{D} = \mathbf{WCW}^{-1}$$
$$= \begin{bmatrix} WC_{11}W^{-1} & WC_{12}W^{-1} & \cdots & WC_{1K}W^{-1} \\ WC_{21}W^{-1} & WC_{22}W^{-1} & \cdots & WC_{2K}W^{-1} \\ \cdots & \cdots & \cdots & \cdots \\ WC_{K1}W^{-1} & WC_{K2}W^{-1} & \cdots & WC_{KK}W^{-1} \end{bmatrix}. \tag{19}$$

Although the blocks of **D** are diagonal, **D** is not itself diagonal. Any matrix having this property is termed a *nondiagonal block-diagonal* (NDBD) matrix. Matrices $\mathbf{D}_{C_f}, \mathbf{D}_{C_n}$, and $\mathbf{D}_H$ are also NDBD [5,6].

NDBD matrices have two useful properties that lead to a tractable method for inverting $\mathbf{\Delta}$ and thus obtaining the LMMSE solution in Eq. (17). First, the set of NDBD matrices is closed under addition, multiplication, and inversion [5,6]. Therefore, because $\mathbf{D}_H$ and $\mathbf{D}_{C_f}$ are NDBD matrices, so is $\mathbf{\Delta}$. Second, a $KM \times KM$ NDBD matrix such as $\mathbf{\Delta}$ can be rearranged into a matrix having M nonzero $K \times K$ blocks along its diagonal by applying a row operation transformation **T** [6, 10] to obtain $\mathbf{T\Delta T}^T = \mathrm{diag}\{R_1, R_2, \ldots, R_M\}$, where each R_j is a general $K \times K$ matrix.

Once transformed, the originally intractable problem of inverting the $KM \times KM$ matrix $(\mathbf{HC}_f \mathbf{H}^T + \mathbf{C}_n)$ in Eq. (5) is reduced to one of separately inverting the $K \times K$ blocks R_j, $j = 1, \ldots, M$, of $\mathbf{T\Delta T}^T$. Because the number of channels K is usually much smaller than the number pixels M in each channel, the inversion problem is greatly simplified.

FIGURE 2 Example of a multichannel LMMSE restoration: original (upper left), degraded (upper right), restored single-channel statistics obtained from original (middle left), restored single-channel statistics obtained from degraded original (middle right), restored multichannel statistics obtained from original (lower left), restored multichannel statistics obtained from degraded (lower right). (See color section, p. C–6.)

4.2 Numerical Experiment

A numerical experiment with color images is shown to demonstrate the improvement that results from the application of multichannel as compared to single-channel LMMSE restoration. For this experiment different distortions were applied to each of the red (R), green (G), and blue (B) channels of the original image, which is shown at the upper-left side of Fig. 2. The red channel was blurred by vertical motion blur over 7 pixels, the green channel by horizontal blur over 9 pixels, and the blue channel by a 7×7 pill-box blur. In all cases the blurs were symmetric around the origin.

The variance of the noise added to the blurred data is defined by using the blurred signal-to-noise ratio (BSNR) metric. These metrics are given per channel i,

$$\text{BSNR} = 10 \log_{10} \frac{\| H_i f_i \|^2}{M \sigma^2} \text{ dB}, \qquad (20)$$

where M is the total number of pixels in f_i and σ^2 is the variance of the additive noise.

Noise was added to all three channels with corresponding BSNRs of 20, 30, and 40 dB. The degraded image is shown in the upper-right side of Fig. 2. The degraded image was restored by using a single-channel LMMSE filter to restore each channel independently, and the multichannel LMMSE filter. In both cases the required power spectra and cross-power spectra were evaluated by using the original image, to establish the upper bound of performance, as well as the available noisy-blurred image, as a more realistic scenario. They were computed in all cases with Daniell's Periodogram (the regular periodogram was spatially averaged using a 5×5 window).

The results of the single-channel LMMSE restoration are shown in the middle of Fig. 2, left (use of original image for power spectra estimation) and right (use of degraded image for power spectra estimation).

The results of the multichannel LMMSE restoration are shown in Fig. 2 at the bottom, left (use of original image for cross-power spectra estimation) and left (use of degraded image for cross-power spectra estimation). From these experiments it is clear that multichannel restoration produces visually more pleasing results than single-channel restoration.

4.3 Space-Variant Multichannel Recovery Approaches

In many cases the degradation and/or the regularization/covariance matrix may not be space invariant. In such cases the frequency-domain approach described in the previous section cannot be applied because the matrices involved are not NDBD and thus direction inversion of $\boldsymbol{\Delta}$ in Eq. (17) is not possible. Instead, the RLS solution $\hat{\mathbf{f}}$ that minimizes Eq. (12) must be computed iteratively. Taking the gradient of $J(\mathbf{f})$ in Eq. (12) yields

$$(\mathbf{H}^T\mathbf{H} + \lambda\mathbf{Q}^T\mathbf{Q})\hat{\mathbf{f}} = \mathbf{H}^T\mathbf{g}. \tag{21}$$

This equation can be solved by using the method of successive approximations [14], which yields the following iteration:

$$\hat{\mathbf{f}}^{(0)} = \mathbf{0}$$
$$\hat{\mathbf{f}}^{(k+1)} = \hat{\mathbf{f}}^{(k)} + \alpha\big[\mathbf{H}^T\mathbf{g} - (\mathbf{H}^T\mathbf{H} + \lambda\mathbf{Q}^T\mathbf{Q})\hat{\mathbf{f}}^{(k)}\big], \tag{22}$$

where $\hat{\mathbf{f}}^k$ is the image estimate at iteration k, and α, known as the *relaxation parameter*, is a scalar that controls the convergence properties of the iteration. It is easy to verify that a stationary point of this iteration satisfies Eq. (21).

4.4 Application to Restoration of Moving Image Sequences

With the recent explosion of multimedia applications, the restoration of image sequences is becoming an increasingly important problem. The purpose of image-sequence restoration is to recover information lost during image sensing, recording, transmission, and storage. Usually, image sequences consist of image frames of the same object or scene taken at closely spaced time intervals; therefore, they often exhibit a high degree of between-frame correlation. In the context of multichannel image recovery, we refer to the image frames as *channels*.

The correlation structure in an image sequence is often much more complicated than in a still color image because there is motion between frames. The *displacement vector* (DV) represents the motion of an image patch from one frame to the next, and the *displacement vector field* (DVF), which describes the motion of various pixels, is indispensable for describing the between-frame correlations; for details on DVF estimation for this application see Ref. [1]. For example, if the image patch occupying pixel (i, j) in frame l has displacement vector (m, n), it appears in pixel $(i + m, j + n)$ in frame $l + 1$. Thus, there will be strong between-frame correlation between $f_l(i, j)$ and $f_{l+1}(i + m, j + n)$. The correlation structure described by the DVF is not space invariant in most situations; therefore, in these cases, the frequency-domain approach described previously cannot be applied.

To accommodate motion in the RLS formulation, the regularization operator must be modified to reflect the fact that a pixel in frame l is not necessarily correlated with the same pixel in frames $l + 1$ and $l - 1$, but rather with pixels that are offset by the corresponding displacement vectors. To express this, we modify the 3-D Laplacian operator $\mathbf{Q}$ defined in Eq. (14) to obtain the 3-D motion-compensated Laplacian (3DMCL), defined by

$$\begin{aligned}
[\mathbf{Q}_{3DMCL}\mathbf{f}]_l(i, j) = & -6f_l(i, j)(i, j) + f_l(i - 1, j) \\
& + f_l(i + 1, j) + f_l(i, j - 1) + f_l(i, j + 1) \\
& + f_{l-1}\big(i + m^{(i,j)}_{(l-1,l)}, j + n^{(i,j)}_{(l-1,l)}\big) \\
& + f_{l+1}\big(i + m^{(i,j)}_{(l+1,l)}, j + n^{(i,j)}_{(l+1,l)}\big),
\end{aligned} \tag{23}$$

where $(m^{(i,j)}_{(l+k,l)}, n^{(i,j)}_{(l+k,l)})$, $k = -1, 1$ represents the DV between frames l and $(l + k)$ for pixel (i, j). It is easy to generalize this operator to capture the temporal correlation between more than three channels.

The iterative algorithm in Eq. (22) is easily implemented by using the regularization operator $\mathbf{Q}_{3DMCL}$ because it is assumed symmetric, i.e., $\mathbf{Q}_{3DMCL} = \mathbf{Q}^T_{3DMCL}$. Therefore, in Eq. (22) $\mathbf{Q}^T_{3DMCL}\mathbf{Q}_{3DMCL}\mathbf{f}$ is computed by applying $\mathbf{Q}_{3DMCL}$ twice; for more details, see [1]. Integer DVs are used in the example shown in this section. The generalization of this approach to noninteger DVs for recovery of compressed video is presented in [16].

The application of multichannel image restoration to image sequences with motion is demonstrated by the following experimental example. Ten frames (each of size 256×256) from the "Trevor White" sequence were used as test images. The results obtained by multichannel restoration are compared with the results obtained by restoring each frame separately (henceforth referred to as Model 0), using an independent-channel version of Eq. (22) in which the regularization operator is $\mathbf{Q} = \text{diag}\{Q, Q, \ldots, Q\}$, where Q represents the convolution with the 2-D Laplacian kernel defined in Eq. (13).

To apply the iteration in Eq. (22) the DVF must first be estimated. Four approaches were used for this task. We refer to these approaches, combined with the iteration in Eq. (22), as Models 1–4, which are defined as follows. In Model 1 the DVF is estimated directly from the degraded images. In Model 2 the DVF is estimated from the images restored by Model 0. In Model 3 the DVF is estimated from the images restored by Model 2. In Model 4 the original image sequence is used to obtain the DVFs. Model 4 is used to test the upper bound of performance of the proposed multichannel restoration algorithm.

In Models 1–4 the DVF is computed from either the degraded, the restored, or the source image. A block-search algorithm (BSA) was used to estimate the between channel DVFs. The motion vector at pixel (i, j) between frames l and k was found by matching a 5×5 window centered at pixel (i, j) of frame l to a 5×5 window in frame k. An exhaustive search over a 31×31 area centered at pixel (i, j) of frame k was used, and the matching metric was the sum of the squared errors.

Two experiments are summarized here (more are described in [1]), in which all five models were tested and compared. The variance of the noise added to the blurred data is defined by using the blurred BSNR metric that was defined in Eq. (20). As an objective measure of performance of the restoration algorithms, the improvement signal-to-noise ratio (ISNR) metric was used. This metric is given by

$$\text{ISNR} = 10 \log_{10} \frac{\| f_i - g_i \|^2}{\| f_i - \hat{f}_i \|^2} \text{ dB}, \qquad (24)$$

where the vectors f_i, g_i, and $\hat{f}_i$ are the ith channel of the original image, the degraded image, and the restored image, respectively.

In both experiments the relaxation parameter α was obtained numerically by using a method, based on the Rayleigh quotient, described in [1]. The value of the regularization parameter λ was chosen to be equal to $(10^{\frac{\text{BSNR}}{10}})^{-1}$ [6]. To restore all 10 frames of the image sequence, six five-channel multichannel filters were used in which a five-channel multichannel regularization operator similar to the one in Eq. (23) was used. Except for the first two and last two frames of the sequence, a five-channel noncausal filter was used to restore each frame. This filter used both the two previous and the two following frames of the frame being restored.

Ten frames (frames 41–50) of the Trevor White sequence were blurred by an 11×11 uniform blur. The point-spread function of this blur is given by

$$h(i, j) = \begin{cases} \frac{1}{121} & \text{if } -5 \leq i, j \leq 5 \\ 0 & \text{otherwise.} \end{cases} \qquad (25)$$

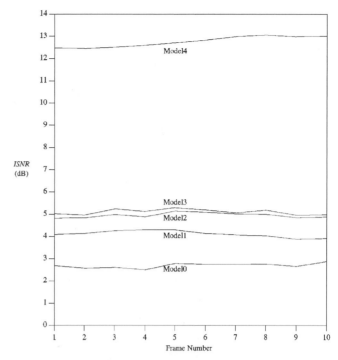

FIGURE 3 ISNR plots for Experiment I, case (i): BSNR = 10 dB, 11×11 blur, $\alpha = 0.1$, and $\lambda = 0.1$.

Cases (i) and (ii) corresponding, respectively, to 10 and 30 dB BSNR of additive white Gaussian noise were examined. Plots of the ISNR are shown in Figs. 3 and 4. In Figs. 5, 6, and 7 the 8th frame of this experiment is shown for cases (i) and (ii).

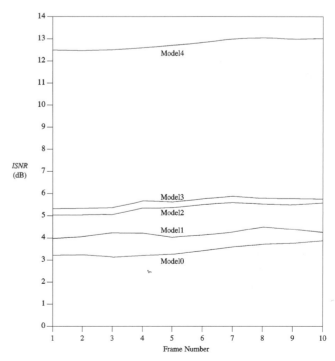

FIGURE 4 ISNR plots for Experiment I, case (ii): BSNR = 30 dB, 11×11 blur, $\alpha = 2.0$, and $\lambda = 0.001$.

FIGURE 5 Original Image twy048 (top). Experiment I case (i) (bottom left): degraded image, with 11×11 blur and 10 dB of BSNR additive noise. Experiment I case (ii) (bottom right): degraded image, with 11×11 blur and 30 dB of BSNR additive noise.

The original and the degraded images are shown in Fig. 5. In Figs. 6 and 7 the restored images from this experiment are shown.

Both the visual and the PSNR results of this experiment demonstrate that (1) the multichannel regularization greatly improves the restored images, and (2) the accuracy of the between-channel knowledge that is incorporated is crucial to the quality of the results.

5 Implicit Approach to Multichannel Image Recovery

The purpose of multichannel image recovery is to make use of the correlations between channels of the source image for purposes of noise suppression. Unfortunately, the between-channel smoothing required to exploit this information can greatly increase the computational cost. In the implicit approach, the computational burden is dramatically reduced by applying a KL

transformation to the observed data prior to processing. For a review of the KL transform see [9], p. 163.

The computational savings result from two important functions of the KL transform: decorrelation and compression. Because the KL transform decorrelates the source channels, it eliminates the need for cross-channel smoothing. In addition, because the KL transform compresses the significant signal information, it effectively reduces the number of channels that must be processed. For example, a 50-channel source image with highly correlated channels might be described almost perfectly by only five KL channels, with the remaining 45 channels dominated by noise. In such an example, only five channels of data would have to be processed instead of the original 50.

The basic steps of the implicit approach to multichannel image recovery are as follows: (1) apply a KL transformation to the data; (2) discard the channels of the KL-transformed data that are dominated by noise; (3) recover an image channel from each of the remaining KL-domain data channels separately; and (4) apply an inverse KL transform to the recovered channels to convert them back to the original domain.

FIGURE 6 Experiment I case (i): restored images using Model 0 (upper left), Model 1 (upper right), Model 3 (lower left), and Model 4 (lower right).

Having outlined the steps of the basic algorithm, let us now explain and justify it. In this section, N denotes the total number of pixels in each image, M denotes the total number of elements in each observation, and K represents the number of channels. The implicit approach is based on the assumption that the multichannel covariance matrix $\mathbf{C}_f$ in Eq. (6) is separable into a spatial part $C_f^{(s)}$ of dimension $K \times K$ and a temporal part $C_f^{(t)}$ of dimension $N \times N$ as follows:

$$\mathbf{C}_f = C_f^{(t)} \otimes C_f^{(s)}, \tag{26}$$

where $\otimes$ denotes the Kronecker product. In the recovery algorithm this calls for separate temporal and spatial regularization operations. This separability assumption is best suited for imaging of motion-free objects or scenes; however, it has been shown in [13] to work extremely well in reconstructing image sequences of the beating heart, as we will show later.

Decorrelation of the channels of the source image is achieved by application of a KL transformation Φ, which is the transpose of the eigenvector matrix of $C_f^{(t)}$, i.e.,

$$C_f^{(t)} \Phi^t = \Phi^t D, \tag{27}$$

where $D = \text{diag}\{d_1, \ldots, d_K\}$ and d_l is the lth eigenvalue of $C_f^{(t)}$.

The limitation of the implicit approach is that it involves, in addition to the separability condition in Eq. (26), the following assumptions, which may not hold in some applications.

1. *The system matrix must be of the form* $\mathbf{H} = \text{diag}\{H, H, \ldots, H\} = I \otimes H$. Each block H in the multichannel system matrix $\mathbf{H}$ denotes the system matrix describing the degradation of one image channel. This form for $\mathbf{H}$ represents the situation in which every image channel is degraded in the same way, and the channels are degraded independently of one another.

2. *The multichannel noise covariance must be of the form* $\mathbf{C}_n = \text{diag}\{C_n, C_n, \ldots, C_n\} = I \otimes C_n$. This means that every channel of the multichannel observations must be have the same noise covariance matrix and that the noise channels must be uncorrelated.

FIGURE 7 Experiment I case (ii): restored images using Model 0 (upper left), Model 1 (upper right), Model 3 (lower left), and Model 4 (lower right).

5.1 KL Transformation of the Multichannel Imaging Model

The values of pixel i in the multichannel image and pixel m in the multichannel observations form, respectively, the $K \times 1$ vectors

$$
\begin{aligned}
\mathbf{f}(j) &= [f_1(i), f_2(i), \dots, f_K(i)]^T, \\
\mathbf{g}(i) &= [g_1(m), g_2(m), \dots, g_K(m)]^T.
\end{aligned}
\tag{28}
$$

In terms of these vectors, the form of $\mathbf{C}_f$ in Eq. (26) can be written as

$$
E\{\mathbf{f}(i)\mathbf{f}^T(j)\} = \left[C_f^{(s)} \right]_{ij} C_f^{(t)}, \quad i, j = 1, \dots, N. \tag{29}
$$

If we define the KL-domain quantity $\tilde{\mathbf{f}}(i) = \Phi \mathbf{f}(i)$, then

$$
E\{\tilde{\mathbf{f}}(i)\tilde{\mathbf{f}}^T(j)\} = \left[C_f^{(s)} \right]_{ij} D. \tag{30}
$$

Equation (30) indicates that the transformed vector $\tilde{\mathbf{f}}(i)$ does not exhibit any between-channel correlations. Thus, if recovery is performed in the KL domain, the need for between-channel smoothing is eliminated. As applied to the multichannel vector $\mathbf{f}$, the KL transform is represented by a multichannel transformation matrix $\mathbf{A}_M$, defined as

$$
\mathbf{A}_M = \Phi \otimes I_M, \tag{31}
$$

where I_M denotes the $M \times M$ identity matrix. Applying $\mathbf{A}_M$ to both sides of the multichannel imaging model in Eq. (1), we obtain

$$
\mathbf{A}_M E\{\mathbf{g}\} = \mathbf{A}_M \mathbf{H} \mathbf{f}, \tag{32}
$$

where $E\{\cdot\}$ denotes expectation with respect to the noise $\mathbf{n}$. Using properties of the Kronecker product, we rewrite $\mathbf{A}_M \mathbf{H}$

as follows:

$$\mathbf{A}_M \mathbf{H} = (\Phi \otimes I_M)(I_K \otimes H) = (\Phi I_K) \otimes (I_M H)$$

$$= (I_K \Phi) \otimes (H I_N) = (I_K \otimes H)(\Phi \otimes I_N) = \mathbf{H} \mathbf{A}_N. \tag{33}$$

Interchanging the transformation matrix and the expectation operator in Eq. (32), using the result in Eq. (33), and defining the transformed quantities $\tilde{\mathbf{g}} = \mathbf{A}_M \mathbf{g}$ and $\tilde{\mathbf{f}} = \mathbf{A}_N \mathbf{f}$ yields the following transformed imaging model:

$$E\{\tilde{\mathbf{g}}\} = \mathbf{H} \tilde{\mathbf{f}}. \tag{34}$$

Note that Eq. (34) has precisely the same form as the original linear imaging model; thus a solution in the KL domain can be accomplished by use of existing recovery approaches.

5.2 KL Transformation of the RWLS Cost Functional

We define a more general version of the multichannel RWLS functional introduced in Eq. (12) as follows:

$$J(\mathbf{f}) = (\mathbf{g} - \mathbf{H}\mathbf{f})^T \mathbf{C}_n^{-1} (\mathbf{g} - \mathbf{H}\mathbf{f}) + \lambda \mathbf{f}^T \mathbf{C}_f^{-1} \mathbf{f}. \tag{35}$$

In this section we show that this RWLS functional is simplified greatly by the KL transformation under the conditions described previously. In a statistical interpretation of this functional, $\mathbf{C}_f$ should be chosen to be the covariance matrix of the multichannel image $\mathbf{f}$; therefore, $C_f^{(t)}$ and $C_f^{(s)}$ should be chosen to be the covariance matrices expressing the between-channel and within-channel covariances, respectively. As described earlier, we choose $C_f^{(s)} = (\lambda Q^T Q)^{-1}$, where Q is a matrix representing a discrete approximation of the 2-D Laplacian operator.

To transform the multichannel RWLS cost functional, we begin by writing the quantities of interest in terms of their KL-domain counterparts (identified with a tilde) as follows:

$$\mathbf{g} = \mathbf{A}_M^T \tilde{\mathbf{g}}, \qquad \mathbf{f} = \mathbf{A}_N^T \tilde{\mathbf{f}}. \tag{36}$$

Note that $\mathbf{A}_N$ and $\mathbf{A}_M$ are orthogonal matrices, so $\mathbf{A}_N^T = \mathbf{A}_N^{-1}$ and $\mathbf{A}_M^T = \mathbf{A}_M^{-1}$. Substituting for these quantities in Eq. (35), we obtain

$$J(\tilde{\mathbf{f}}) = \left(\mathbf{A}_M^T \tilde{\mathbf{g}} - \mathbf{H} \mathbf{A}_N^T \tilde{\mathbf{f}} \right)^T \mathbf{C}_n^{-1} \left(\mathbf{A}_M^T \tilde{\mathbf{g}} - \mathbf{H} \mathbf{A}_N^T \tilde{\mathbf{f}} \right)$$

$$+ \lambda \mathbf{A}_N^T \tilde{\mathbf{f}}^T \left[C_f^{(t)} \otimes C_f^{(s)} \right]^{-1} \mathbf{A}_N^T \tilde{\mathbf{f}}. \tag{37}$$

In a manner similar to that used to derive Eq. (33), it can be shown that

$$\mathbf{H} \mathbf{A}_N^T \tilde{\mathbf{f}} = \mathbf{A}_M^T \mathbf{H} \tilde{\mathbf{f}}. \tag{38}$$

Using Eq. (38), distributing the transpose operations, and factoring out the transformation matrices, we obtain

$$J(\tilde{\mathbf{f}}) = (\tilde{\mathbf{g}} - \mathbf{H}\tilde{\mathbf{f}})^T \mathbf{A}_M \mathbf{C}_n^{-1} \mathbf{A}_M^T (\tilde{\mathbf{g}} - \mathbf{H}\tilde{\mathbf{f}})$$

$$+ \lambda \tilde{\mathbf{f}}^T \left\{ \mathbf{A}_N \left[C_f^{(t)} \otimes C_f^{(s)} \right]^{-1} \mathbf{A}_N^T \right\} \tilde{\mathbf{f}}. \tag{39}$$

Using Eq. (31), we rewrite the term in curly braces as follows:

$$\mathbf{A}_N \left[C_f^{(t)} \otimes C_f^{(s)} \right]^{-1} \mathbf{A}_N^T = D^{-1} \otimes \left[C_f^{(s)} \right]^{-1}, \tag{40}$$

where D is the eigenvalue matrix defined in Eq. (27). It is easy to show that

$$\mathbf{A}_M \mathbf{C}_n^{-1} \mathbf{A}_M^T = \mathbf{C}_{\tilde{n}}^{-1}, \tag{41}$$

where $\mathbf{C}_{\tilde{n}}$ is the covariance matrix of the observations in the KL domain. With use of Eqs. (40) and (41), Eq. (39) becomes

$$J(\tilde{\mathbf{f}}) = (\tilde{\mathbf{g}} - \mathbf{H}\tilde{\mathbf{f}})^T \mathbf{C}_{\tilde{n}}^{-1} (\tilde{\mathbf{g}} - \mathbf{H}\tilde{\mathbf{f}}) + \lambda \tilde{\mathbf{f}}^T \left\{ D^{-1} \otimes \left[C_f^{(s)} \right]^{-1} \right\} \tilde{\mathbf{f}}. \tag{42}$$

Using the assumption $\mathbf{C}_{\tilde{n}}^{-1} = I \otimes C_n^{(s)^{-1}}$, since $\mathbf{H}$ and $\mathbf{C}_{\tilde{n}}^{-1}$ are block diagonal, and D and I are diagonal, Eq. (42) reduces to

$$J(\tilde{\mathbf{f}}) = \sum_{l=1}^{K} J_l(\tilde{f}_l), \tag{43}$$

in which

$$J_l(\tilde{f}_l) = (\tilde{g}_l - H\tilde{f}_l)^T C_{\tilde{n}_l}^{-1} (\tilde{g}_l - H\tilde{f}_l) + \left(\frac{\lambda}{d_l} \right) \tilde{f}_l^T \left[C_f^{(s)} \right]^{-1} \tilde{f}_l, \tag{44}$$

where $\tilde{f}_l$ and $\tilde{g}_l$ are the lth KL components of $\mathbf{f}$, and $\mathbf{g}$, respectively, and d_l is the eigenvalue associated with the lth KL basis vector.

5.3 Space-Invariant Image Restoration by the Implicit Approach

Image restoration (deblurring) problems can be solved especially easily when the degradation operator H and the covariance matrices $C_f^{(s)}$ and C_{n_l} are circulant. As in the general case, application of the implicit approach begins with computation of the covariance matrix $C_f^{(t)}$ and its eigenvectors in Φ, which are used to transform the observed multichannel image to the KL domain. Then conventional Wiener filters [8] can be implemented in the DFT domain in closed form, and applied one by one to the significant KL-domain channels to restore them. Finally, the multichannel image is obtained by inverting the KL transform.

FIGURE 8   Example frames (numbers 1, 2, 3, 10, 20, and 40) from a sequence of 44 frames of dynamic PET data. In dynamic PET the object typically is stationary, but is changing in time (data courtesy of Jogeshwar Mukherjee).

FIGURE 9   First six Karhunen-Loève components of the dynamic PET data in Fig. 8. The remaining 38 components look similar to the sixth and are dominated by noise. Only the first three contain significant signal information.

5.4 Space-Variant Image Recovery by the Implicit Approach

When the degradation is not shift invariant and/or the statistics are not stationary, the recovered KL-domain channels must be computed iteratively. The RWLS functional $J(\mathbf{f})$ can be minimized by minimizing $J_l(\tilde{f}_l)$ separately in Eq. (44). Since $J_l(\tilde{f}_l)$ is quadratic with respect to $\tilde{f}_l$, a number of iterative minimization methods, including the conjugate gradient algorithm [3], can be used to find $\tilde{f}_l$. Theoretically, the conjugate gradient method is guaranteed to converge in N steps (the dimension of each image channel), but a much smaller number of iterations is sufficient for good results in practice.

5.5 Multichannel Reconstruction of Medical Image Sequences

In this section we describe an application of the implicit multichannel recovery approach to an important problem in medical imaging, namely the reconstruction of time sequences of images. We focus specifically on two emission tomography methods: positron emission tomography (PET) and single-photon emission computed tomography (SPECT) [2].

For purposes of computation, the imaging model for PET and SPECT sequences can be approximated by the set of matrix equations

$$E\{g_k\} = Hf_k, \quad k = 1, 2, \ldots, K. \tag{45}$$

This corresponds exactly to the previously discussed multichannel linear imaging model for the special case in which $\mathbf{H} = \mathrm{diag}\{H, H, \ldots, H\}$. In this application, H represents a tomographic projection operator that is not shift invariant. In an idealized model, the projection operator is the discrete Radon transform [8] but more-realistic models include blur caused by various physical factors in the imaging process.

In dynamic PET, one obtains a time sequence of data g_k, $k = 1, \ldots, K$, from which an image sequence f_k, $k = 1, \ldots, K$ is to be reconstructed. Usually the reconstruction is performed image

by image, but recent research [9, 12, 13, 16, 17] has shown that it is preferable to reconstruct all of the images collectively as a multichannel image $\mathbf{f}$ from all of the data in the multichannel observation vector $\mathbf{g}$. The following example from PET brain imaging illustrates this principle. The images shown depict slices of the brain of a monkey; the bright areas indicate tissues rich in dopamine receptors, which are part of the brain's chemical communication system.

In the implicit approach, one begins by applying a KL transformation along the time axis of the data (across the channels of $\mathbf{g}$). Figure 8 shows example frames from a time sequence of 44 frames of tomographic projection data; Fig. 9 shows the first six frames following the KL transformation. The KL transform eliminates redundancy in the observations and compresses the useful information into the first three frames. The remaining frames, which all look similar, are dominated by noise and can be discarded. The importance of the first three frames is depicted quantitatively by the eigenvalue spectrum shown in Fig. 10. Figure 11 shows the result of reconstructing images from the first three KL-domain observations. The inverse KL transform is applied to these three KL-domain images to obtain a sequence of images. Examples of these results are shown in Fig. 12, where they are compared with the results obtained by more-conventional approaches. Note that, not only

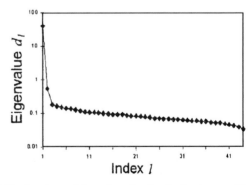

FIGURE 10   Spectrum of eigenvalues showing the dominance of the first two KL components. Usually the next component contains significant signal content as well.

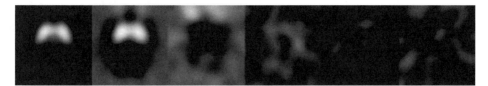

FIGURE 11 Images reconstructed from the first six KL components of the data (see Fig. 9). All but the first three are dominated by noise and can be omitted from the computations.

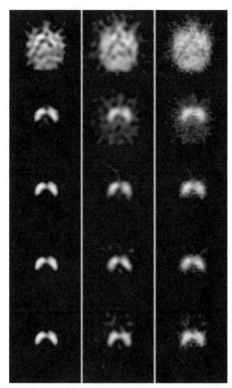

FIGURE 12 Example frames from the sequence of dynamic PET images reconstructed by the implicit approach (left column), by separate single-channel reconstruction by PWLS (center column), and filtered backprojection (right column). Because of the high noise level, single-channel PWLS fails to produce significantly better results than filtered back projection, but multichannel regularization, provided by the implicit approach, yields more-accurate images.

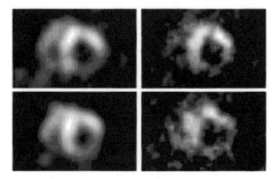

FIGURE 13 Example frames from a sequence of gated SPECT images of the heart. Images reconstructed by the implicit multichannel approach (left column) are less noisy than those obtained by single-channel reconstruction (right column). These images were obtained without accounting for blur in the system matrix **H**, so no deblurring effect is apparent. (Image results courtesy of V. Manoj Naryanan and Michael A. King.)

are the images obtained by the implicit multichannel approach superior to the others, but they were obtained in less time because only three KL frames required reconstruction instead of 44 time-domain image frames. Quantitative performance evaluations of the implicit approach for image reconstruction can be found in [9, 17].

Figure 13 shows another application of the implicit approach to cardiac SPECT imaging. Two example frames are shown, each reconstructed by both a single-channel approach and by the implicit approach. Image features that are normally obscured by noise are clearly visible when reconstructed by the implicit approach. Because of the separability assumption in Eq. (26), one might expect the implicit approach to perform poorly when there is motion; however, these images of the beating heart show that the KL decomposition can capture motion information in some cases.

Acknowledgments

N. P. Galatsanos aknowledges the financial support of the National Science Foundation under grant MIP-9309910 for work on the explicit multichannel approach described in this chapter. He also recognizes Roland Chin, his Ph.D. thesis adviser, for introducing him to the multichannel problem, and Mun Gi Choi and Yongyi Yang for their significant contributions to this research.

M. N. Wernick is grateful to the National Institutes of Health for supporting his research on reconstruction of dynamic nuclear medicine images under grant R29 NS35273, and the Whitaker Foundation for their support of his initial work in this area. He also recognizes the research efforts of his former students Chien-Min Kao, E. James Infusino, and Miloš Milošević, who made substantial contributions to the work. He also thanks his collaborators V. Manoj Narayanan and Michael A. King for contributing the research results shown in Fig. 13, and Jogeshwar Mukherjee for providing the PET data shown in Fig. 8 and for helpful discussions about neuroreceptor imaging.

References

[1] Mun Gi Choi, N. P. Galatsanos, and A. K. Katsaggelos, "Multichannel regularized iterative restoration of motion compensated image sequences," *J. Vis. Commun. Image Rep.* **7**, 244–258 (1996).

[2] Z. H. Cho, J. P. Jones, and M. Singh, *Foundations of Medical Imaging* (Wiley, New York, 1993).

[3] E. Z. P. Chong and S. H. Zak, *An Introduction to Optimization* (Wiley, New York, 1996).

[4] G. Demoment, "Image reconstruction and restoration: Overview of common estimation structures and problems," *IEEE Trans. Acoust. Speech Signal Process.* **37**, 2024–2036 (1989).

[5] N. P. Galatsanos and R. T. Chin, "Digital restoration of multichannel images," *IEEE Trans. Acoust. Speech Signal Process.* **37** (1989).

[6] N. P. Galatsanos, A. K. Katsaggelos, R. T. Chin, and A. D. Hillery, "Least squares restoration of multichannel images," *IEEE Trans. Signal Process.* **39**, 2222–2236 (1991).

[7] N. P. Galatsanos and A. K. Katsaggelos, "Methods for choosing the regularization parameter and estimating the noise variance in image restoration and their relation," *IEEE Trans. Image Process.* **1**, 322–336 (1992).

[8] A. K. Jain, *Fundamentals of Digital Image Processing* (Prentice-Hall, Englewood Cliffs, NJ, 1989).

[9] C.-M. Kao, J. T. Yap, J. Mukherjee, and M. N. Wernick, "Image reconstruction for dynamic PET based on low-order approximation and restoration of the sinogram," *IEEE Trans. Med. Imag.* **16**, 738–749 (1997).

[10] A. K. Katsaggelos, K. T. Lay, and N. P. Galatsanos, "A general framework for frequency domain multichannel signal processing," *IEEE Trans. Image Process.* **2**, 417–420 (1993).

[11] S. M. Kay, *Fundamentals of Statistical Signal Processing* (Prentice-Hall, Englewood Cliffs, NJ, 1993).

[12] D. S. Lalush and B. M. W. Tsui, "A priori motion models for four-dimensional reconstruction in gated cardiac SPECT," 1996 Conference Record of the Nuclear Science Symposium Medical Imaging Conference, 1996.

[13] V. M. Narayanan, M. A. King, E. Soares, C. Byrne, H. Pretorius, and M. N. Wernick, "Application of the Karhunen-Loeve transform to 4D reconstruction of gated cardiac SPECT images," 1997 Conference Record of the Nuclear Science Symposium Medical Imaging Conference, 1997.

[14] R. W. Schafer, R. M. Mersereau, and M. A. Richards, "Constrained iterative restoration algorithm," *Proc. IEEE.* **69**, 432–450 (1981).

[15] Y. Yang, M. C. Choi, and N. P. Galatsanos, "Image recovery from compressed video using multichannel regularization," in *Signal Recovery Techniques for Image and Video Compression and Transmission*, A. K. Katsaggelos and N. P. Galatsanos, eds. (Kluwer, Boston, 1998).

[16] M. N. Wernick, G. Wang, C.-M. Kao, J. T. Yap, J. Mukherjee, M. Cooper, and C.-T. Chen, "An image reconstruction method for dynamic PET," 1995 Conference Record of the Nuclear Science Symposium Medical Imaging Conference, 1718–1722, 1995.

[17] M. N. Wernick, E. J. Infusino, and M. Milošević, "Fast spatio-temporal image reconstruction for dynamic PET," *IEEE Trans. Med. Imag.* **18**, 185–195 (1999).

3.8

Multiframe Image Restoration

Timothy J. Schulz
Michigan Technological University

1 Introduction

Multiframe image restoration is concerned with the improvement of imagery acquired in the presence of varying degradations. The degradations can arise from a variety of factors—common examples include undersampling of the image data, uncontrolled platform or scene motion, system aberrations and instabilities, and wave propagation through atmospheric turbulence. In a typical application, a sequence of images (frames) is recorded and a restored image is extracted through analog or digital signal processing. In most situations digital data are acquired, and the restoration processing is carried out by a general- or special-purpose digital computer. The general idea is depicted in Fig. 1, and the following examples illustrate applications for which multiframe restoration is utilized.

Example 1.1 (Resolution Improvement in Undersampled Systems) A critical factor in the design of visible and infrared imaging systems is often the tradeoff between field of view and pixel size. The pixel size for a fixed detector array becomes larger as the field of view is increased, and the need for a large field of view can result in undersampled imagery. This phenomenon is illustrated in Fig. 2. One way to overcome the effects of larger pixels while preserving field of view is to utilize controlled or uncontrolled pointing jitter. In the presence of subpixel translations, a sequence of image frames can be processed to estimate the image values on a grid much smaller than the physical size of the detector pixels. Uncontrolled motion, however, presents

the additional challenge of motion identification or the determination of optical flow [1]. Often referred to as microscanning [2], the idea of processing a sequence of undersampled image frames to restore resolution has received attention in a variety of applications [3, 4].

Example 1.2 (Imaging Through Turbulence) Spatial and temporal variations in the temperature of the Earth's atmosphere cause the refractive index at optical wavelengths to vary in a random and unpredictable manner. Because of this, imagery acquired with ground-based telescopes can exhibit severe, time-varying distortions [5]. A sequence of short exposure image frames will exhibit blurs such as those shown in Fig. 3, and the goal of a multiframe image restoration procedure is to form a fine-resolution estimate of the object's reflectance from the noisy, blurred frames. Because the point-spread functions are not easily measured or predicted, this problem is often referred to as one of multiframe blind deconvolution [6].

Many methods have been proposed and studied for solving multiframe restoration problems — see, for example, Refs. [7–16] and those cited within. Well-established restoration methods exist for situations in which all sources of blur and degradation are known or easily predicted. Some of the more popular techniques include regularized least squares and Wiener methods [12, 13, 17], and multiframe extensions of the iterative Richardson–Lucy method [18–21]. When some of the system parameters are unknown, however, the problem becomes much more difficult. In this situation, the recovery of the object

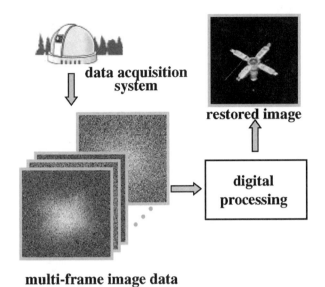

data acquisition system

restored image

digital processing

multi-frame image data

FIGURE 1   A general scenario in which multiframe data are recorded and a restored image is produced through digital image processing.

intensity can be called a multiframe *blind* restoration problem, because, in addition to the object intensity, the unknown system parameters must also be estimated [3, 4, 6, 22–27].

In the remainder of this chapter, we will develop mathematical models for the multiframe imaging process, pose the multiframe restoration problem as one of numerical optimization, provide an overview of restoration methods and illustrate the methods with some current examples.

2 Mathematical Models

The imaging problems discussed in this chapter all involve the detection and processing of electromagnetic fields after reflection or emission from a remote object or scene. Furthermore, the applications considered here are all examples of planar *incoherent imaging*, wherein the object or scene is characterized by its incoherent reflectance or emission function $f(\boldsymbol{x})$, $\boldsymbol{x} \in \mathcal{R}^2$. Throughout this chapter we will refer to f as the *image intensity* — a nonnegative function that represents an object's ability to

original scene

large field of view – coarse pixels

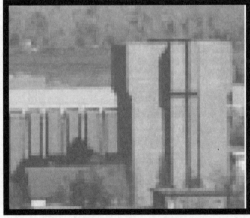

small field of view – finer pixels

FIGURE 2   An illustration of the tradeoff between field of view and pixel size. For a fixed number of pixels, the larger field of view results in coarse sampling — finer sampling leads to a smaller field of view.

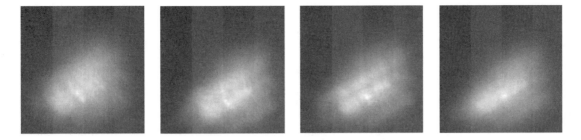

FIGURE 3   Imagery of the Hubble Space Telescope as acquired by a 1.6-m telescope at the Air Force Maui Optical Station.

reflect or emit light (or other electromagnetic radiation). The central task of a multiframe image restoration problem, then, is the estimation of this intensity function from a sequence of noisy, blurred images.

2.1 Image Blur and Sampling

As illustrated in Fig. 4, the need for image restoration is, in general, motivated by two factors: i) system and environmental blur; and ii) detector sampling. In the absence of noise, these two stages of image formation are described as follows.

System and Environmental Blur

In all imaging applications, the signal available for detection is not the image intensity f. Instead, f is blurred by the imaging system, and the observable signal is

$$g_c(\boldsymbol{y};\theta_t) = \int h(\boldsymbol{y},\boldsymbol{x};\theta_t)\,f(\boldsymbol{x})\,\mathrm{d}\boldsymbol{x}, \qquad (1)$$

where $h(\boldsymbol{y},\boldsymbol{x};\theta_t)$ denotes the (time-varying) system and environmental point-spread function, $g_c(\boldsymbol{y};\theta_t)$ denotes the (time-varying) continuous-domain intensity that results because of the blur, $\boldsymbol{x}$ and $\boldsymbol{y}$ are continuous-domain spatial coordinates, and θ_t denotes a set of time-varying parameters that determine the form of the point-spread function. The role of these parameters is discussed in more detail later in this chapter. Many applications involve *space-invariant* blurs for which the point-spread function depends only on the spatial difference $\boldsymbol{y}-\boldsymbol{x}$, and not on the absolute positions $\boldsymbol{y}$ and $\boldsymbol{x}$. When this occurs the

point spread is written as a function of only one spatial variable, and the continuous-domain intensity is formed through a convolution relationship with the image intensity:

$$g_c(\boldsymbol{y};\theta_t) = \int h(\boldsymbol{y}-\boldsymbol{x};\theta_t)\,f(\boldsymbol{x})\,\mathrm{d}\boldsymbol{x}. \qquad (2)$$

Diffraction is the most common form of image blur, and its effects are present in every application involving remotely sensed image data. For narrow-band, incoherent imaging systems such as telescopes, microscopes, and infrared or visible cameras, the point-spread function for diffraction is modeled by the space-invariant function:

$$h(\boldsymbol{y}) = \left| \int A(\boldsymbol{u})e^{-j\frac{2\pi}{\lambda f}\boldsymbol{u}\cdot\boldsymbol{y}}\mathrm{d}\boldsymbol{u} \right|^2, \qquad (3)$$

where $A(\boldsymbol{u})$ is the system's aperture function, $\boldsymbol{u}$ is a two-dimensional spatial variable in the aperture plane, λ is the nominal wavelength of the detected radiation, and f is the system focal length [28]. The notation $\boldsymbol{u}\cdot\boldsymbol{y}$ denotes the inner product operation, and it is defined for two-dimensional spatial variables as

$$u \cdot y = u_1 y_1 + u_2 y_2. \qquad (4)$$

The use of this model for diffraction implicitly requires that the image intensity be spatially magnified by the factor $-f/r$, where r is the distance from the object or scene to the sensor. For a circular aperture of diameter D, the diffraction-limited point-spread function is the isotropic Airy pattern whose one-

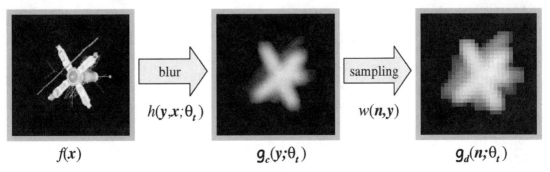

$$f(\boldsymbol{x}) \qquad\qquad g_c(\boldsymbol{y};\theta_t) \qquad\qquad g_d(\boldsymbol{n};\theta_t)$$

FIGURE 4   Pictorial representation of the image degradations caused by system/environmental blur and detector sampling.

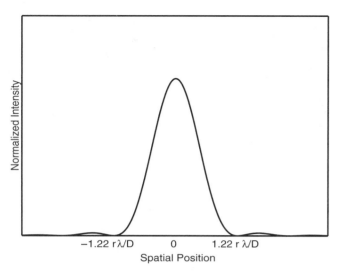

FIGURE 5   Cross section of the Airy diffraction pattern for a circular aperture.

dimensional cross section is shown in Fig. 5. As a result of the location of the first zero relative to the central peak, the resolution of a diffraction-limited system with a circular aperture is often cited as $1.22r\lambda/D$. This definition of resolution is, however, very arbitrary. Nevertheless, decreasing the wavelength, increasing the aperture diameter, or decreasing the distance to the scene will result in a narrowing of the point-spread function and an improvement in imaging resolution.

Imaging systems often suffer from various types of optical aberrations — imperfections in the figure of the system's focusing element (usually a mirror or lens). When this happens, the point-spread function takes the form

$$h(\boldsymbol{y};\theta) = \left| \int A(\boldsymbol{u}) e^{j\theta(\boldsymbol{u})} e^{-j\frac{2\pi}{\lambda f}\boldsymbol{u}\cdot\boldsymbol{y}} \mathrm{d}\boldsymbol{u} \right|^2, \qquad (5)$$

where $\theta(\boldsymbol{u})$ is the aberration function, often measured in units of *waves*.[1] Here, the notation $h(\boldsymbol{y};\theta)$ explicitly shows the dependence of the aberrated point-spread function on the aberration function θ. An out-of-focus blur induces a quadratic aberration

function:

$$\theta(\boldsymbol{u}) = \frac{\pi}{\lambda}\left(\frac{1}{r}+\frac{1}{d}-\frac{1}{f}\right)|\boldsymbol{u}|^2, \qquad (6)$$

where r is the distance to the scene, d is the focal setting, and f is the focal length. This blur is reduced to diffraction when the "imaging equation" is satisfied and the system is in focus: $\frac{1}{r}+\frac{1}{d}=\frac{1}{f}$. Spherical aberration, such as that present in the Hubble Space Telescope's infamous primary mirror [29], induces a fourth-order aberration function:

$$\theta(\boldsymbol{u}) = B|\boldsymbol{u}|^4, \qquad (7)$$

where the constant B determines the strength of the aberration. By setting the aberration to

$$\theta(\boldsymbol{u}) = \frac{2\pi}{\lambda f}\boldsymbol{\Delta}\cdot\boldsymbol{u}, \qquad (8)$$

one can also use this model to represent a tilt or pointing error $\boldsymbol{\Delta}$, so that

$$h(\boldsymbol{y};\theta) = h(\boldsymbol{y}-\boldsymbol{\Delta}). \qquad (9)$$

Wave propagation through an inhomogeneous medium such as the Earth's atmosphere can induce additional distortions. These distortions are due to temperature-induced variations in the atmosphere's refractive index, and they are frequently modeled in a manner similar to that used for system aberrations:

$$h(\boldsymbol{y};\theta_t) = \left| \int A(\boldsymbol{u}) e^{j\theta_t(\boldsymbol{u})} e^{-j\frac{2\pi}{\lambda f}\boldsymbol{u}\cdot\boldsymbol{y}} \mathrm{d}\boldsymbol{u} \right|^2, \qquad (10)$$

where the aberration function $\theta_t(\boldsymbol{u})$ can now vary with time [5]. A typical diffraction-limited point-spread function along with a sequence of turbulence degraded point-spread functions are shown in Fig. 6.

Another interesting perturbation to the diffraction-limited point-spread function can arise because of time-varying translations and rotations between the sensor and scene. In this case,

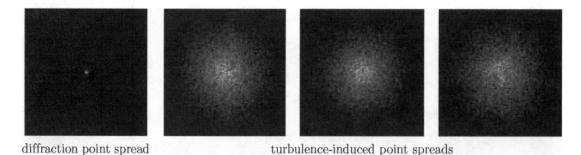

diffraction point spread   turbulence-induced point spreads

FIGURE 6   Diffraction-limited point-spread function and a typical sequence of turbulence-induced point-spread functions.

---

[1]One wave of aberration corresponds to $\theta(\boldsymbol{u}) = 2\pi$.

the continuous-domain intensity is modeled as

$$g_c(\boldsymbol{y}; \theta_t) = \int h(\boldsymbol{y} - \boldsymbol{x}) f[A(\phi_t)\boldsymbol{x} - \boldsymbol{\Delta}_t] d\boldsymbol{x}, \quad (11)$$

where $\boldsymbol{\Delta}_t$ represents a two-dimensional, time-varying translation, and

$$A(\phi_t) = \begin{bmatrix} \cos(\phi_t) & -\sin(\phi_t) \\ \sin(\phi_t) & \cos(\phi_t) \end{bmatrix} \quad (12)$$

is a time-varying rotation matrix (at angle ϕ_t). A simple change of variables leads to

$$g_c(\boldsymbol{y}; \theta_t) = \int h[\boldsymbol{y} - A^{-1}(\phi_t)(\boldsymbol{x} + \boldsymbol{\Delta}_t)] f(\boldsymbol{x}) d\boldsymbol{x}, \quad (13)$$

so that the shift variant point-spread function can be written as

$$h(\boldsymbol{y}, \boldsymbol{x}; \theta_t) = h[\boldsymbol{y} - A^{-1}(\phi_t)(\boldsymbol{x} + \boldsymbol{\Delta}_t)], \quad (14)$$

and the parameters characterizing the point-spread function are then $\theta_t = (\boldsymbol{\Delta}_t, \phi_t)$.

Without loss of generality, we will model the system and environmental point-spread function as the (possibly) space-variant function $h(\boldsymbol{y}, \boldsymbol{x}; \theta_t)$, and note that this model captures diffraction, system aberrations, time-varying translations and rotations, and environmental distortions such as atmospheric turbulence. The parameter θ_t may be a simple vector parameter, or a more complicated parameterization of a two-dimensional function. Many times θ_t will not be well known or predicted, and the identification of this parameter can be one of the most challenging aspects of a multiframe image restoration problem.

Sampling

The detection of imagery with discrete detector arrays results in the measurement of the (time-varying) sampled intensity:

$$g_d(\boldsymbol{n}; \theta_t) = \int w(\boldsymbol{n}, \boldsymbol{y}) g_c(\boldsymbol{y}; \theta_t) d\boldsymbol{y}, \quad (15)$$

where $w(\boldsymbol{n}, \boldsymbol{y})$ is the response function for the nth pixel in the image detector array, $\boldsymbol{n}$ is a discrete-domain spatial coordinate, and $g_d(\boldsymbol{n}; \theta_t)$ is the discrete-domain intensity that results due to sampling of the continuous-domain, blurred intensity. The response function for an incoherent detector element is often of the form

$$w(\boldsymbol{n}, \boldsymbol{y}) = \begin{cases} 1 & \boldsymbol{y} \in \mathcal{Y}_n \\ 0 & \boldsymbol{y} \notin \mathcal{Y}_n \end{cases}, \quad (16)$$

where $\mathcal{Y}_n$ denotes the spatial region of integration for the nth detector element. The regions of integration for most detectors are typically square or rectangular regions centered about the detector locations $\{\boldsymbol{y}_n\}$.

The combined effects of blur and sampling are modeled as

$$\begin{aligned} g_d(\boldsymbol{n}; \theta_t) &= \int w(\boldsymbol{n}, \boldsymbol{y}) \int h(\boldsymbol{y}, \boldsymbol{x}; \theta_t) f(\boldsymbol{x}) d\boldsymbol{x} \, d\boldsymbol{y} \\ &= \int \left[\int w(\boldsymbol{n}, \boldsymbol{y}) h(\boldsymbol{y}, \boldsymbol{x}; \theta_t) d\boldsymbol{y} \right] f(\boldsymbol{x}) d\boldsymbol{x} \\ &= \int h_{cd}(\boldsymbol{n}, \boldsymbol{x}; \theta_t) f(\boldsymbol{x}) d\boldsymbol{x}, \quad (17) \end{aligned}$$

where

$$h_{cd}(\boldsymbol{n}, \boldsymbol{x}; \theta_t) = \int w(\boldsymbol{n}, \boldsymbol{y}) h(\boldsymbol{y}, \boldsymbol{x}; \theta_t) d\boldsymbol{y} \quad (18)$$

denotes the mixed-domain (continuous/discrete) point-spread function. These equations establish the linear relationship between the unknown intensity function f and the multiframe, sampled image intensities $g_d(\boldsymbol{n}; \theta_t)$.

Throughout this chapter we will focus on applications for which the data collection interval for each frame is short compared with the fluctuation time for the parameter θ_t, so that a sequence of image frames

$$\{g_d(\boldsymbol{n}; k) = g_d(\boldsymbol{n}; \theta_{t_k}), \quad k = 1, 2, \dots, K\} \quad (19)$$

is available for detection. Each frame is recorded at the time $t = t_k$, and the blur parameter takes the value $\theta_k = \theta_{t_k}$ during the frame so that we write

$$h(\boldsymbol{y}, \boldsymbol{x}; k) = h(\boldsymbol{y}, \boldsymbol{x}; \theta_{t_k}), \quad (20)$$

and

$$h_{cd}(\boldsymbol{n}, \boldsymbol{x}; k) = h_{cd}(\boldsymbol{n}, \boldsymbol{x}; \theta_{t_k}). \quad (21)$$

2.2 Noise Models

Electromagnetic waves such as light interact with matter in a fundamentally random way, and quantum electrodynamics (QED) is the most sophisticated theory available for describing the detection of electromagnetic radiation. In most imaging applications, however, the semiclassical theory for the detection of radiation is sufficient for the development of practical and useful models. In accordance with this theory, electromagnetic energy is transported according to the classical theory of wave propagation, and the field energy is quantized only during the detection process [30].

When an optical field interacts with a photodetector, a quantum of energy is absorbed in the form of a *photon* and the absorption of this photon gives rise to the release of an excited electron. This interaction is referred to as a photoevent, and the number of photoevents occurring within a photodetector element during a collection interval is referred to as a photocount. Most detectors of light record these photocounts, and the number of photocounts recorded during an exposure interval is a fundamentally random quantity. The utilization of this theory leads to a statistical model for image detection in which the photocounts for each recorded frame are modeled as independent Poisson random variables, each with a conditional mean that is proportional to the sampled image intensity $g_d(\boldsymbol{n}; k)$ for the frame. Specifically, the expected photocount for the $\boldsymbol{n}$th detector during the kth frame is:

$$E[N_d(\boldsymbol{n}; k) \mid g_d(\boldsymbol{n}; k)] = \alpha_k g_d(\boldsymbol{n}; k), \qquad (22)$$

where the scale factor α_k is proportional to the frame exposure time. Because the variance of a Poisson variable is equal to its mean, the image contrast (mean-squared to variance ratio) for photon noise increases linearly with the exposure time.

The data recorded by charge coupled devices (CCD) and other detectors of optical radiation are usually subject to other forms of noise. The most common — *read-out* noise — is induced by the electronics used for the data acquisition. This noise is often modeled by additive, zero-mean Gaussian random variables so that the recorded data are modeled as

$$d(\boldsymbol{n}; k) = N_d(\boldsymbol{n}; k) + v(\boldsymbol{n}; k), \qquad (23)$$

where $v(\boldsymbol{n}; k)$ represents the read-out noise at the $\boldsymbol{n}$th detector for the kth frame. The read-out noise is usually statistically independent across detectors and frames, but the variance may be different for each detector element.

3 The Restoration Problem

Stated simply, the restoration problem is one of estimating the image intensity f from the blurred multiframe data $\{d(\boldsymbol{n}; k), \quad k = 1, 2, \ldots, K\}$. The intensity function f is, however, an infinite-dimensional parameter, and its estimation from finite data is a terribly ill-conditioned problem. As a way to overcome this problem, it is common to approximate the intensity function in terms of a finite-dimensional basis set:

$$f(\boldsymbol{x}) \simeq \sum_m f_d(\boldsymbol{m}) \psi_m(\boldsymbol{x}), \qquad (24)$$

where the basis functions $\{\psi_m(\boldsymbol{x})\}$ are selected in a manner that is appropriate for the application. Expression of the object function on a predetermined grid of pixels, for example, might require $\psi_m(\boldsymbol{x})$ to be an indicator function that denotes the location and spatial support of the $\boldsymbol{m}$th pixel. Alternatively, the basis functions might be selected as two-dimensional impulses colocated with the center of each pixel. Other basis sets are possible, and a clever choice here can have a great effect on estimator performance.

Using a basis as described in (24) results in the following approximation to the imaging equation:

$$g_d(\boldsymbol{n}; k) = \int h_{cd}(\boldsymbol{n}, \boldsymbol{x}; k) f(\boldsymbol{x}) \, d\boldsymbol{x}$$

$$\simeq \int h_{cd}(\boldsymbol{n}, \boldsymbol{x}; k) \sum_m f_d(\boldsymbol{m}) \psi_m(\boldsymbol{x}) \, d\boldsymbol{x}$$

$$= \sum_m \left[\int h_{cd}(\boldsymbol{n}, \boldsymbol{x}; k) \psi_m(\boldsymbol{x}) \, d\boldsymbol{x} \right] f_d(\boldsymbol{m})$$

$$= \sum_m h_d(\boldsymbol{n}, \boldsymbol{m}; k) f_d(\boldsymbol{m}), \quad k = 1, 2, \ldots, K, \quad (25)$$

where

$$h_d(\boldsymbol{n}, \boldsymbol{m}; k) = \int h_{cd}(\boldsymbol{n}, \boldsymbol{x}; k) \psi_m(\boldsymbol{x}) \, d\boldsymbol{x} \qquad (26)$$

is the discrete-domain impulse response for the kth frame. This impulse response (or point-spread function) defines a linear relationship between the discrete-domain images $\{g_d(\boldsymbol{n}; k)\}$ and the discrete-domain intensity $f_d(\boldsymbol{m})$. For shift-invariant applications, h_d is a function of only the difference $\boldsymbol{n} - \boldsymbol{m}$. With a little thought on notation, the discrete-domain imaging equations can be written in matrix-vector form as

$$\boldsymbol{g}_d(k) = \boldsymbol{H}_d(k) \boldsymbol{f}_d, \quad k = 1, 2, \ldots, K, \qquad (27)$$

and when the point-spread functions are shift-invariant, the measurement matrices $\{\boldsymbol{H}_d(k), k = 1, 2, \ldots, K\}$ are Toeplitz. One potential advantage of multiframe restoration methods arises when the eigensystems for the measurement matrices are sufficiently different. In this situation, each image frame records different information about the object, and the system of multiframe measurements can be used to estimate more detail about the object than can a single image frame.

3.1 Restoration as an Optimization Problem

In this section we focus on restoration problems for which the point-spread parameters $\{\theta_k\}$ are well known or easily determined. In the following section we will address the challenges that are presented when these parameters must be identified from the measured data.

Statistical inference problems such as those encountered in multiframe image restoration are frequently classified as ill-posed problems [31], and, because of this, *regularization* methods play an important role in the estimation process. An image-restoration problem is ill posed if it is not well posed, and a problem is well posed in the classical sense of Hadamard if it has a unique solution and the solution varies continuously with the data. Multiframe image restoration problems that are formulated on infinite-dimensional parameter spaces are almost always ill posed, and their ill-posed nature is usually due to the discontinuity of the solution. Problems that are formulated on finite-dimensional spaces (as ours is here) are frequently well posed in the classical sense — they have a unique solution and the solution is continuous in the data. However, these problems are usually ill conditioned or badly behaved and are frequently classified as ill posed even though they are technically well posed.

For problems that are ill posed or practically ill posed, the original problem's solution is often replaced by the solution to a well-posed (or well-behaved) problem. This process is referred to as regularization, and the basic idea is to change the problem in a manner such that the solution is still meaningful but no longer badly behaved [32]. The consequence for multiframe restoration problems is that we do not seek to match the measured data perfectly. Instead, we settle for a more stable — but inherently biased — image estimate.

Most approaches to regularized image restoration are induced through attempts to solve an optimization problem of the following form:

$$\hat{f}_d = \underset{f_d \in \mathcal{F}}{\arg\min}\{D(g_d, d) + \gamma\psi(f_d)\}, \qquad (28)$$

where $D(g_d, d)$ is a discrepancy measure between the estimated image intensities $\{g_d(\boldsymbol{n}; k), \ k = 1, 2, \ldots, K\}$ and the measured data $\{d(\boldsymbol{n}; k), \ k = 1, 2, \ldots, K\}$, $\psi(f_d)$ is a penalty (or prior) function that penalizes undesirable attributes of the object estimate $f_d(\boldsymbol{m})$ (or rewards desirable ones), γ is a scale factor that determines the degree to which the penalty influences the estimate, and $\mathcal{F}$ is a constraint set of allowable object estimates. Methods that are covered by this general framework include the following.

Maximum-Likelihood Estimation

For maximum-likelihood estimation the penalty is not used ($\gamma = 0$), the constraint set is typically the set of nonnegative functions $\mathcal{F} = \{f_d : f_d \geq 0\}$, and the discrepancy measure is induced by the statistical model that is used for the data collection process. Discrepancy measures that result from various noise models are illustrated in the following examples.

Example 3.1 (Maximum-Likelihood for Gaussian Noise) When the measured data are corrupted only by additive, independent Gaussian noise of variance σ^2, the data are modeled as

$$d(\boldsymbol{n}; k) = g_d(\boldsymbol{n}; k) + v(\boldsymbol{n}; k), \qquad (29)$$

and the log-likelihood function [33] is of the form

$$L(d; g_d) = -\frac{1}{2\sigma^2}\sum_k\sum_n [d(\boldsymbol{n}; k) - g_d(\boldsymbol{n}; k)]^2. \qquad (30)$$

The discrepancy measure can then be selected as

$$D(g_d, d) = \sum_k\sum_n [d(\boldsymbol{n}; k) - g_d(\boldsymbol{n}; k)]^2, \qquad (31)$$

where the scale factor $1/2\sigma^2$ is omitted without affecting the optimization.

Example 3.2 (Maximum-Likelihood for Poisson Noise) When the measured data are corrupted only by Poisson (photon)

noise, the log-likelihood function is of the form

$$L(d; g_d) = -\sum_k\sum_n \alpha_k g_d(\boldsymbol{n}; k) \\ + \sum_k\sum_n d(\boldsymbol{n}; k)\ln \alpha_k g_d(\boldsymbol{n}; k), \qquad (32)$$

and the discrepancy measure is selected as

$$D(g_d, d) = \sum_k\sum_n \alpha_k g_d(\boldsymbol{n}; k) \\ - \sum_k\sum_n d(\boldsymbol{n}; k)\ln \alpha_k g_d(\boldsymbol{n}; k). \qquad (33)$$

Example 3.3 (Maximum-Likelihood for Poisson and Gaussian Noise) When the measured data are corrupted by both Poisson (photon) noise and additive Gaussian (read-out) noise as in Eq. (23), then the likelihood has a complicated form involving an infinite summation [34]. When the variance for the Gaussian noise is the same for all detector elements and sufficiently large (greater than 50 or so), however, the modified data,

$$\tilde{d}(\boldsymbol{n}; k) = d(\boldsymbol{n}; k) + \sigma^2, \qquad (34)$$

have an approximate log-likelihood of the form [34]

$$L(d; g_d) = -\sum_k\sum_n \{\alpha_k g_d(\boldsymbol{n}; k) + \sigma^2\} \\ + \sum_k\sum_n \tilde{d}(\boldsymbol{n}; k)\ln\{\alpha_k g_d(\boldsymbol{n}; k) + \sigma^2\}. \qquad (35)$$

The discrepancy measure is then

$$D(g_d, d) = \sum_k\sum_n \{\alpha_k g_d(\boldsymbol{n}; k) + \sigma^2\} \\ - \sum_k\sum_n \tilde{d}(\boldsymbol{n}; k)\ln\{\alpha_k g_d(\boldsymbol{n}; k) + \sigma^2\}. \qquad (36)$$

Sieve-Constrained Maximum-Likelihood Estimation

For sieve-constrained maximum-likelihood estimation [35], the discrepancy measure is again induced by the statistical model that is used for the data collection process and the penalty is not used ($\gamma = 0$). However, the constraint set is selected to be a "smooth" subset of nonnegative functions. A Gaussian kernel sieve [36], for example, is defined as

$$\mathcal{F} = \left\{ f_d : f_d(\boldsymbol{m}) = \sum_p \alpha(\boldsymbol{p})\frac{1}{\sqrt{2\pi a}}e^{-\frac{(\boldsymbol{p}-\boldsymbol{m})^2}{2a}}, \quad \alpha(\boldsymbol{p}) \geq 0 \right\}, \qquad (37)$$

where the parameter a determines the width of the Gaussian kernel and the smoothness of the sieve. Selection of this parameter for a particular application is part of the *art* of performing sieve-constrained estimation.

Penalized Maximum-Likelihood Estimation

For penalized maximum-likelihood estimation [37], the discrepancy measure is induced by the statistical model that is used for the data collection, and the constraint set is typically the set of nonnegative functions. However, the function ψ is chosen to penalize undesirable properties of the object estimate. Commonly used penalties include the weighted quadratic roughness penalty,

$$\psi(f_d) = \sum_{m} \sum_{m' \in \mathcal{N}_m} w(m, m') |f_d(m) - f_d(m')|^2, \quad (38)$$

where $\mathcal{N}_m$ denotes a neighborhood about the *m*th pixel and $w(m, m')$ is a nonnegative weighting function, and the divergence penalty,

$$\psi(f_d) = \sum_{m} f_d(m) \ln f_d(m) - \sum_{m} f_d(m) \sum_{m' \in \mathcal{N}_m} \ln f_d(m'), \quad (39)$$

where $\mathcal{N}_m$ is again a neighborhood about the *m*th pixel. As shown in Ref. [38], this penalty can also be viewed as a discretization of the roughness measure proposed by Good and Gaskins [39].

Many other roughness penalties are possible [38], and the proper choice can depend largely on the application. In all cases, selection of the parameter γ for a particular application is part of the *art* of using penalized maximum-liklihood methods.

Maximum *a Posteriori* Estimation

For maximum *a posteriori* (MAP) estimation, the discrepancy measure is induced by the statistical model for the data collection, and the constraint set is typically the set of nonnegative functions. However, the penalty term $\psi(f_d)$ and scale factor γ are induced by a prior statistical model for the unknown object intensity. MAP methods are mathematically, but not always philosophically, equivalent to penalty methods. Markov random fields [40] are commonly used for image priors, and, within this framework, Bouman and Sauer [41] have proposed and investigated the use of a generalized Gauss–Markov random field model for images:

$$\psi(f_d) = \sum_{m} a(m) |f_d(m)|^p$$
$$+ \sum_{m} \sum_{m' \in \mathcal{N}_m} b(m, m') |f_d(m) - f_d(m')|^p, \quad (40)$$

where $p \in [1, 2]$, and $a(m)$ and $b(m, m')$ are nonnegative parameters. A detailed discussion of the effect of p, a, and b on estimator performance is provided in Ref. [41].

Regularized Least-Squares Estimation

For regularized least-squares estimation, the discrepancy measure is selected as:

$$D(g_d, d) = \sum_{k} \sum_{n} [g_d(n; k) - d(n; k)]^2, \quad (41)$$

the constraint set is typically the set of nonnegative functions, and the penalty is selected and used as discussed for penalized maximum-likelihood estimation. For additive, white Gaussian noise, the regularized least-squares and penalized maximum-likelihood methods are mathematically equivalent.

Minimum I-Divergence Estimation

For problems involving nonnegative image measurements, the I divergence has also received attention as a discrepancy measure:

$$D(g_d, d) = \sum_{k} \sum_{n} [g_d(n; k) - d(n; k)]$$
$$- \sum_{k} \sum_{n} d(n; k) \ln \frac{g_d(n; k)}{d(n; k)}. \quad (42)$$

For problems in which the noise is Poisson, the minimum I-divergence and maximum-likelihood methods are mathematically equivalent.

After selecting an appropriate estimation methodology, multiframe image restoration — as we have posed the problem here — is a problem of constrained optimization. For most situations this optimization must be performed numerically, but in some cases a direct-form linear solution can be obtained. In these situations, however, the physical constraint that the intensity function f must be nonnegative is usually ignored.

3.2 Linear Methods

Linear methods for solving multiframe restoration problems are usually derived as solutions to the regularized least-squares problem:

$$\hat{f}_d = \arg\min_{f_d} \left\{ \sum_{k} \sum_{n} [d(n; k) - g_d(n; k)]^2 \right.$$
$$\left. + \gamma \sum_{s} \left[\sum_{m} C(s, m) f_d(m) \right]^2 \right\}, \quad (43)$$

where C is called the regularizing operator. A common choice for this operator is the two-dimensional Laplacian:

$$C(s, m) = \begin{cases} 1 & s = m \\ -1/4 & s = m + (0, 1) \\ -1/4 & s = m + (0, -1) \\ -1/4 & s = m + (1, 0) \\ -1/4 & s = m + (-1, 0) \\ 0 & \text{otherwise} \end{cases}, \quad (44)$$

but other operators can be used. In matrix-vector notation, the regularized least-squares optimization problem can be reposed as

$$\hat{\boldsymbol{f}}_d = \arg\min_{\boldsymbol{f}_d}\left\{ \sum_k \|\boldsymbol{d}(k) - \boldsymbol{H}(k)\boldsymbol{f}_d\|^2 + \gamma\|\boldsymbol{C}\boldsymbol{f}_d\|^2 \right\}, \quad (45)$$

with the minimum-norm solution $\hat{\boldsymbol{f}}_d$ satisfying:

$$\sum_k \boldsymbol{H}^T(k)\boldsymbol{d}(k) = \left(\sum_k \boldsymbol{H}^T(k)\boldsymbol{H}(k) + \gamma\boldsymbol{C}^T\boldsymbol{C}\right)\hat{\boldsymbol{f}}_d, \quad (46)$$

or

$$\hat{\boldsymbol{f}}_d = \left(\sum_k \boldsymbol{H}^T(k)\boldsymbol{H}(k) + \gamma\boldsymbol{C}^T\boldsymbol{C}\right)^{-1}\sum_k \boldsymbol{H}^T(k)\boldsymbol{d}(k), \quad (47)$$

when the inverse exists for $(\sum_k \boldsymbol{H}^T(k)\boldsymbol{H}(k) + \gamma\boldsymbol{C}^T\boldsymbol{C})$. A key to the derivation of a linear solution to Eq. (45) is the absence of a nonnegativity constraint. If the solution is constrained to be nonnegative, then a linear solution will not exist in general. Solutions for problems in which the nonnegativity constraint must be satisfied generally require iterative, nonlinear processing. Furthermore, because the processing required to solve Eqs. (46) and (47) can be prohibitive, iterative methods can be required even when attempting to solve the unconstrained problem.

3.3 Nonlinear (Iterative) Methods

The presence of a nonnegativity constraint or the use of non-quadratic discrepancy measures and penalties will, in general, prohibit the derivation of a closed-form, linear solution to the multiframe restoration problem. Instead, iterative methods are needed. Basic descent methods, coordinate descent methods, conjugate direction methods, and quasi-Newton methods [42] can all be applied to solve the general optimization problem:

$$\hat{\boldsymbol{f}}_d = \arg\min_{\boldsymbol{f}_d \in \mathcal{F}}\{D(g_d, d) + \gamma\psi(f_d)\}, \quad (48)$$

and the most appropriate method will, in general, depend on the application. For problems in which the discrepancy measure $D(g_d, d)$ is induced by a statistical procedure such as maximum-likelihood or MAP estimation, another class of optimization methods can also be applied. These are derived by means of the expectation-maximization (EM) procedure [43], the generalized EM procedure [43], or the space alternating generalized EM (SAGE) procedure [44]. For problems in incoherent imaging and statistical tomography [45], the SAGE procedure leads to methods that are similar in form to coordinate descent optimization [46], and both methods lead to highly efficient optimization procedures.

4 Nuisance Parameters and Blind Restoration

In the previous section, we addressed the restoration problem for situations in which the parameters $\{\theta_k\}$ that characterize the point-spread functions are well known or easily predicted. This is not the case for applications such as uncontrolled microscanning and imaging through turbulence, and in these situations the unknown parameters must be eliminated or estimated. Estimation of these parameters often leads to one of the following optimization problems:

$$\hat{\boldsymbol{f}}_d = \arg\min_{f_d \in \mathcal{F}}\{D(g_d, d; \hat{\theta}(f_d)) + \gamma\psi(f_d)\},$$
$$\hat{\theta}(f_d) = \arg\min_{\theta \in \Theta} E(f_d, d; \theta), \quad (49)$$

or

$$(\hat{\boldsymbol{f}}_d, \hat{\theta}) = \arg\min_{f_d \in \mathcal{F}, \; \theta \in \Theta} \{D(g_d, d; \theta) + \gamma\psi(f_d) + \beta\Upsilon(\theta)\}, \quad (50)$$

where the discrepancy measure D now shows an explicit dependence on the unknown parameters $\theta = \{\theta_k\}$. For the first approach, Eq. (49), we fix the unknown intensity f_d and estimate θ by minimizing some cost functional E over the constraint set Θ. The cost functional may be induced by estimation-theoretic principles, or it may be motivated through heuristic arguments. The resulting estimate for θ, which may be a function of the unknown intensity f_d, is then used in the original discrepancy measure D, and an estimator for f_d is determined. For the second approach, Eq. (50), the unknown parameters θ and intensity f_d are jointly estimated by minimizing a cost functional containing a discrepancy measure D, an intensity penalty ϕ, and a parameter penalty Υ. Determination of the discrepancy measure and penalty functionals can be approached in a manner similar to that discussed in the previous section.

Many methods have been proposed for solving multiframe blind restoration problems, and each has been tailored for specific blur models. For microscanning applications with under-sampled detectors, the blur models are often as in Eq. (14) with frame-to-frame translations and rotations of the diffraction-limited point-spread [3]. For ground-based imaging of astronomical and space objects, the blur models are usually as in Eq. (10), with phase aberrations distributed across the system's aperture [6]. Another class of solutions has addressed problems for which the discrete-domain point spreads are modeled as two-dimensional finite impulse response filters, for which "perfect" or "exact" methods can be extended from methods developed for one-dimensional applications [47, 48].

The successful application of a few methods developed for real-world imaging problems is illustrated in the following section.

5 Applications

We conclude this chapter by presenting an overview of three applications of multiframe blind restoration.

5.1 Fine-Resolution Imaging from Undersampled Image Sequences

For problems in which an image is undersampled by the system's detector array, multiframe restoration methods can be used to obtain a fine-resolution object estimate provided that a sequence of translated (or microscanned) images is obtained. An example considered by Hardie *et al.* [3, 49] concerns image formation with a forward-looking infrared (FLIR) imaging system. This system's continuous-domain point-spread function caused by diffraction is modeled as

$$h(\boldsymbol{y}) = \left| \int A(\boldsymbol{u}) e^{-j \frac{2\pi}{\lambda f} \boldsymbol{u} \cdot \boldsymbol{y}} \, d\boldsymbol{u} \right|^2, \qquad (51)$$

where $A(\boldsymbol{u})$ denotes the system's pupil function as determined by the physical dimensions of the camera's lens, λ is the operational wavelength, and f is the system focal length. Accordingly, the continuous-domain intensity caused only by diffraction is modeled as

$$g_c(\boldsymbol{y}) = \int h(\boldsymbol{y} - \boldsymbol{x}) f(\boldsymbol{x}) \, d\boldsymbol{x}. \qquad (52)$$

For a circular lens of diameter D, the highest spacial frequency present in the continous-domain image is $D/(\lambda f)$, so that critical sampling of the image is obtained on a grid whose spacing is $\lambda f/(2D)$.

The sampling operator for FLIR cameras is typically of the form

$$w(\boldsymbol{n}, \boldsymbol{y}) = \begin{cases} 1 & \boldsymbol{y} \in \mathcal{Y}_n \\ 0 & \text{otherwise} \end{cases}, \qquad (53)$$

where $\mathcal{Y}_n$ is a rectangular neighborhood around the center of the nth detector element $\boldsymbol{y}_n$. For a circular aperture of radius D, if the spacing between detector elements is greater than $\lambda f/(2D)$, as is often the case for current FLIR systems, then the image data will be undersampled and the full resolving power of the system will not be utilized. Frame-to-frame motion or camera jitter in conjunction with multiframe image restoration methods can, however, be used to restore resolution to an undersampled system.

Frame-to-frame motion or camera jitter, in the form of translations and rotations, can be modeled by modifying the continuous-domain imaging equation according to

$$g_c(\boldsymbol{y}; \theta_k) = \int h(\boldsymbol{y} - \boldsymbol{x}) f[A(\phi_k)\boldsymbol{x} - \boldsymbol{\Delta}_k] \, d\boldsymbol{x}$$

$$= \int h[\boldsymbol{y} - A^{-1}(\phi_k)(\boldsymbol{x} + \boldsymbol{\Delta}_k)] f(\boldsymbol{x}) \, d\boldsymbol{x}$$

$$= \int h(\boldsymbol{y}, \boldsymbol{x}; \theta_k) f(\boldsymbol{x}) \, d\boldsymbol{x}, \qquad (54)$$

where $\boldsymbol{\Delta}_k$ denotes the two-dimensional translation, and

$$A(\phi_k) = \begin{bmatrix} \cos(\phi_k) & -\sin(\phi_k) \\ \sin(\phi_k) & \cos(\phi_k) \end{bmatrix} \qquad (55)$$

is the rotation matrix (at angle ϕ_k) associated with the kth frame. These parameters, $\{\theta_k = (\boldsymbol{\Delta}_k, \phi_k)\}$, are often unknown at the time of data collection, and the accurate estimation of their values is essential for fine-resolution enhancement of multiframe FLIR imagery.

Hardie *et al.* [3] have addressed this problem for an application using a FLIR camera with an Amber AE-4 infrared focal plane array. The nominal wavelength for this system is $\lambda = 4 \, \mu\mathrm{m}$, and the aperture diameter is $D = 100 \, \mathrm{mm}$. With a focal length of 300 mm, the required sample spacing for critical sampling is $\lambda f/(2D) = 6 \, \mu\mathrm{m}$; however, the detector spacing for the Amber focal plane array is 50 μm with integration neighborhoods that are 40 μm square. This results in undersampling by a factor of 8.33.

Using an object expansion of the form

$$f(\boldsymbol{x}) = \sum_m f_d(\boldsymbol{m}) \psi_m(\boldsymbol{x}), \qquad (56)$$

where the basis functions $\{\psi_m(\boldsymbol{x})\}$ represent square indicator functions with spatial support that is five times smaller than the detector elements (10 $\mu m \times$ 10 μm), Hardie *et al.* used the method of Irani and Peteg [50] to estimate the frame-to-frame rotations and shifts $\{\theta_k = (A(\phi_k), \boldsymbol{\Delta}_k)\}$ followed by a regularized least-squares method to restore a fine-resolution scene estimate from a multiframe sequence of noisy microscanned images. This is the two-step procedure as described by Eq. (49). The regularization operator was the discretized Laplacian from Eq. (44), and the smoothing parameter γ was tuned in a heuristic manner. A conjugate-gradient approach, based on the Fletcher–Reeves method, was used to solve the multiframe optimization problem. A typical image frame is displayed in Fig. 7(a), showing a FLIR image of buildings and roads in the Dayton, Ohio area.[2] A multiframe image restoration obtained from 20 such frames, each with unknown translations and rotations, is shown in Fig. 7(b). Clearly, resolution has been improved in the imagery.

5.2 Ground-Based Imaging through Atmospheric Turbulence

The distorting effects of atmospheric turbulence give rise to continuous-domain point-spread functions of the form

$$h(\boldsymbol{y}; k) = \left| \int A(\boldsymbol{u}) e^{j\theta_k(u)} e^{-j \frac{2\pi}{\lambda f} \boldsymbol{u} \cdot \boldsymbol{y}} \, d\boldsymbol{y} \right|^2, \qquad (57)$$

where θ_k represents the turbulence-induced aberrations for the kth frame. The discrete-domain point-spread function is then

---

[2]These data were collected courtesy of the Infrared Threat Warning Laboratory, Threat Detection Branch at Wright Laboratory (WL/AAJP).

(a)   (b)

FIGURE 7 Demonstration of multiframe image restoration for undersampled FLIR images: (a) an undersampled image frame from the FLIR imagery; (b) the restored image from 20 undersampled image frames.

of the form

$$h_d(\boldsymbol{n}, \boldsymbol{m}; k) = \int h_{cd}(\boldsymbol{n}, \boldsymbol{x}; k)\phi_m(\boldsymbol{x})\,d\boldsymbol{x}$$

$$= \iint w(\boldsymbol{n}, \boldsymbol{y})h(\boldsymbol{y} - \boldsymbol{x}; k)\psi_m(\boldsymbol{x})\,d\boldsymbol{y}\,d\boldsymbol{x}, \quad (58)$$

and, if the spatial support for the detector elements $w(\boldsymbol{n}, \boldsymbol{y})$ and basis functions $\psi_m(\boldsymbol{x})$ are sufficiently small, then the discrete-domain point spread can be reasonable approximated as

$$h_d(\boldsymbol{n}, \boldsymbol{m}; k) \simeq h(\boldsymbol{y}_n - \boldsymbol{x}_m; k), \quad (59)$$

where $\boldsymbol{y}_n$ is the spatial location of the nth detector element and $\boldsymbol{x}_m$ is the spatial location of the mth object pixel. If the detector elements and object pixels are furthermore on the same grid ($\Delta_x = \Delta_y$), then the discrete-domain point spread can be further approximated as

$$h_d(\boldsymbol{n}, \boldsymbol{m}; k) \simeq \left| \sum_l A(\Delta_u \boldsymbol{l})e^{j\theta_k(\Delta_u \boldsymbol{l})}e^{-j\frac{2\pi}{\lambda f}\Delta_u\Delta_x \boldsymbol{l}\cdot(\boldsymbol{n}-\boldsymbol{m})} \right|^2, \quad (60)$$

where Δ_u is the pupil-plane discretization grid spacing. If the aperture and image planes are discretized on a grid of size $N \times N$ and if $\Delta_u\Delta_x/(\lambda f) = 1/N$, then the discrete-domain point spread can be approximated by the space-invariant function:

$$h_d(\boldsymbol{m}; k) \simeq \left| \sum_l A(\boldsymbol{l})e^{j\theta_k(\boldsymbol{l})}e^{-j\frac{2\pi}{N}\boldsymbol{l}\cdot\boldsymbol{m}} \right|^2, \quad (61)$$

where $A(\boldsymbol{l}) = A(\Delta_u\boldsymbol{l})$ and $\theta_k(\boldsymbol{l}) = \theta_k(\Delta_u\boldsymbol{l})$. Using these approximations, one can compute the discrete-domain point spread easily and efficiently by means of the fast Fourier transform al-

gorithm. The discrete-domain imaging equations are then

$$g_d(\boldsymbol{n}; k) = \sum_m h_d(\boldsymbol{n} - \boldsymbol{m}; k)\,f_d(\boldsymbol{m}), \quad (62)$$

and the joint estimation of the unknown object and the turbulence parameters in the presence of Poisson (photon) noise can be accomplished by solving the following maximum-likelihood problem:

$$(\hat{f}_d, \hat{\theta}) = \underset{f_d \in \mathcal{F}, \theta \in \Theta}{\arg\min} \left\{ \sum_k \sum_n g_d(\boldsymbol{n}; k) \right.$$

$$\left. - \sum_k \sum_n d(\boldsymbol{n}; k)\ln g_d(\boldsymbol{n}; k) \right\}. \quad (63)$$

The generalized expectation-maximization method has been used to derive an iterative solution to this joint-estimation problem. The algorithm derivation, and extensions to problems involving Gaussian (read-out) noise and nonuniform detector gain and bias, are presented in Refs. [6, 51]. The use of this method on real data is illustrated in Fig. 8. The four image frames of the Hubble Space Telescope were acquired by a 1.6-m telescope at the Air Force Maui Optical Station. The nominal wavelength for these images was 750 nm and the exposure time for each frame was 8 ms. The object estimate was obtained by processing 16 of these frames.

5.3 Ground-Based Solar Imaging with Phase Diversity

Phase-diverse speckle is a *measurement* and processing method for the simultaneous estimation of an object and the atmospheric phase aberrations from multiframe imagery acquired

Multiframe imagery of the Hubble Space Telescope

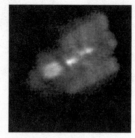

Restored image estimate

FIGURE 8 Multiframe imagery and restored image estimate of the Hubble Space Telescope as acquired by a 1.6-m telescope at the Air Force Maui Optical Station.

in the presence of turbulence-induced aberrations. By modifying a telescope to simultaneously record both an in-focus and out-of-focus image for each exposure frame, the phase-diverse speckle method records a sequence of discrete-domain images that are formed according to

$$g_{c1}(\boldsymbol{n}; k) = \sum_{\boldsymbol{m}} h_d(\boldsymbol{n} - \boldsymbol{m}; k) f_d(\boldsymbol{m}) \qquad (64)$$

and

$$g_{c2}(\boldsymbol{n}; k) = \sum_{\boldsymbol{m}} h_d(\boldsymbol{n} - \boldsymbol{m}; k, \theta_{df}) f_d(\boldsymbol{m}), \qquad (65)$$

where $h_d(\boldsymbol{m}; k)$ is the point-spread function for turbulence and diffraction, parameterized by the turbulence-induced aberration parameters θ_k for the kth frame as defined in Eq. (61), and $h_d(\boldsymbol{m}; k, \theta_{df})$ is the out-of-focus point-spread function for the same frame. The additional aberration that is due to the known defocus error θ_{df} is usually well modeled as a quadratic function

$$\theta_{df}(\boldsymbol{l}) = a\|\boldsymbol{l}\|^2, \qquad (66)$$

so that

$$h_d(\boldsymbol{m}; k, \theta_{df}) = \left| \sum_{\boldsymbol{l}} A(\boldsymbol{l}) e^{j[\theta_k(\boldsymbol{l}) + a\|\boldsymbol{l}\|^2]} e^{-j\frac{2\pi}{N}\boldsymbol{l} \cdot \boldsymbol{m}} \right|^2. \qquad (67)$$

For Poisson (photon) noise, the maximum-likelihood estimation of the object and aberrations is accomplished by solving the

following optimization problem:

$$(\hat{f}_d, \hat{\theta}) = \underset{f_d \in \mathcal{F}, \theta \in \Theta}{\operatorname{argmin}} \left\{ \sum_k \sum_{i=1}^{2} \sum_{\boldsymbol{n}} g_{di}(\boldsymbol{n}; k) \right.$$
$$\left. - \sum_k \sum_{i=1}^{2} \sum_{\boldsymbol{n}} d(\boldsymbol{n}; k, i) \ln g_{di}(\boldsymbol{n}; k) \right\}, \qquad (68)$$

where $d(\boldsymbol{n}; k, 1)$ and $d(\boldsymbol{n}; k, 2)$ are the the in-focus and out-of-focus images for the kth frame, respectively. Although the formation of two images for each frame generally leads to less light and an increased noise level in each recorded image, the addition of the defocused diversity channel can result in significant improvements in the ability to restore fine-resolution imagery from turbulence degraded imagery[52].

Paxman, Seldin, *et al.* [24, 53–56] have applied this method with great success to a problem in solar imaging by using a quasi-Newton method for the optimization of Eq. (68). Within their estimation procedure, they have modified the measurement model to account for nonuniform detector gain and bias, included a Gaussian-kernel sieve constraint for the object [as in Eq. (37)], and incorporated a polynomial expansion for the phase aberrations:

$$\Theta = \left\{ \theta_k(\boldsymbol{l}) = \sum_i a_{ki} z_i(\boldsymbol{l}/R), \quad k = 1, 2, \dots, K \right\}, \qquad (69)$$

where R is the radius of the telescope's aperture, and the polynomial functions $\{z_i(\boldsymbol{l})\}$ are the circle polynomials of Zernike, which are orthonormal over the interior of a unit circle [57]. These polynomials have found widespread use in optics because

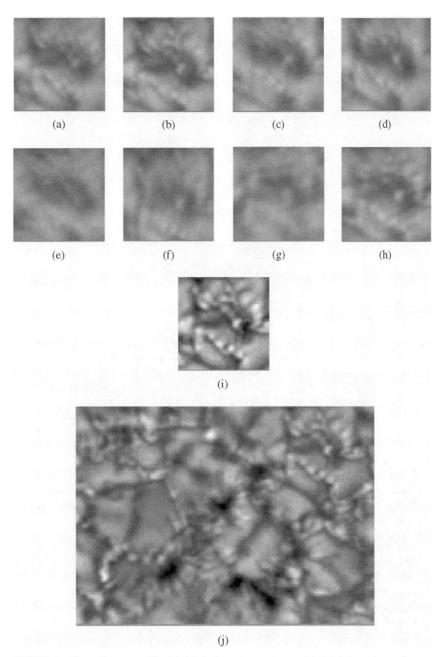

FIGURE 9 Phase diverse speckle: (a)–(d) in-focus image frames; (e)–(h) defocus image frames; (i) restoration from 10 in-focus and defocus image frames; (j) large field of view obtained from 35 small field-of-view restorations on a 5 × 7 grid.

they represent common aberration modes such as defocus, coma, and spherical aberration, and because they form a good approximation to the Karhunen–Loeve expansion for atmospheric aberrations that obey Kolmogorov statistics [5, 30].

The top row of Fig. 9 shows four in-focus image frames that were acquired by Dr. Christoph Keller, using a 76-cm vacuum tower telescope at the National Solar Observatory on Sacramento Peak, NM. Many processes in the solar atmosphere have typical spatial scales that are much smaller than the resolution of these

blurred images, and because of this, important solar features cannot be observed without some form of image restoration. The second row of Fig. 9 shows the corresponding out-of-focus image frames that were acquired for use with the phase-diverse speckle method. Using in-focus and defocused image pairs from 10 frames, Paxman and Seldin obtained the restored image shown in Fig. 9(i). The restored image for this field of view was blended with 34 others on a 5 × 7 grid across the solar surface to create the large field-of-view restoration shown in Fig. 9(j). By using the phase diversity method, the resolution of the large field-of-view restoration is now sufficient to perform meaningful inferences about solar processes.

References

[1] B. K. Horn and B. Schunk, "Determining optical flow," *Artificial Intell.* **17**, 185–203 (1981).

[2] J. C. Gillette, T. M. Stadtmiller, and R. C. Hardie, "Reduction of aliasing in staring infrared imagers utilizing subpixel techniques," *Opt. Eng.* **34**, 3130–3137 (1995).

[3] R. C. Hardie, K. J. Barnard, J. G. Bognar, E. E. Armstrong, and E. A. Watson, "High-resolution image reconstruction from a sequence of rotated and translated frames and its application to an infrared imaging system," *Opt. Eng.* **37**, 247–260 (1998).

[4] R. R. Schultz and R. L. Stevenson, "Extraction of high-resolution frames from video sequences," *IEEE Trans. Image Process.* **5**, 996–1011 (1996).

[5] M. C. Roggemann and B. Welsh, *Imaging Through Turbulence* (CRC Press, Boca Raton, FL, 1996).

[6] T. J. Schulz, "Multi-frame blind deconvolution of astronomical images," *J. Opt. Soc. Am. A* **10**, 1064–1073 (1993).

[7] D. Ghiglia, "Space-invariant deblurring given N independently blurred images of a common object," *J. Opt. Soc. Am. A* **1**, 398–402 (1982).

[8] B. R. Hunt and O. Kubler, "Karhunen-Loeve multispectral image restoration. Part I: Theory," *IEEE Trans. Acoust. Speech Signal Process.* **32**, 592–599 (1984).

[9] R. Tsai and T. Huang, "Multiframe image restoration and registration," in *Advance in Computer Vision and Image Processing* (JAI Press, 1984), Vol. 1.

[10] N. P. Galatsanos and R. T. Chin, "Digital restoration of multichannel images," *IEEE Trans. Acoust. Speech Signal Process.* **37**, 592–599 (1989).

[11] S. P. Kim, N. K. Bose, and H. M. Valenzuela, "Recursive reconstruction of high-resolution image from noisy undersampled multiframes," *IEEE Trans. Acoust. Speech Signal Process.* **38**, 1013–1027 (1990).

[12] N. P. Galatsanos, A. K. Katsaggelos, R. T. Chin, and A. D. Hillery, "Least squares restoration of multichannel images, "*IEEE Trans. Signal Process.* **39**, 2222–2236 (1991).

[13] M. K. Ozkan, A. T. Erdem, M. I. Sezan, and A. M. Tekalp, "Efficient multiframe Wiener restoration of blurred and noisy images," *IEEE Trans. Image Process.* **1**, 453–476 (1992).

[14] S. P. Kim and Wen-Yu Su, "Recursive high-resolution reconstruction from blurred multiframe images," *IEEE Trans. Image Process.* **2**, 534–539 (1993).

[15] L. P. Yaroslavsky and H. J. Caulfield, "Deconvolution of multiple images of the same object," *Appl. Opt.* **33**, 2157–2162 (1994).

[16] M. Elad and A. Feuer, "Restoration of a single superresolution image from several blurred, noisy, and undersampled measured images," *IEEE Trans. Image Process.* **6**, 1646–1658 (1997).

[17] M. K. Ozkan, M. I. Sezan, A. T. Erdem, and A. M. Tekalp, "Multiframe Wiener restoration of image sequences," in *Motion Analysis and Image Sequence Processing*, M. I. Sezan and R. L. Lagendijk, eds., (Kluwer, Boston, 1993), Chap. 13, pp. 375-410.

[18] W. H. Richardson, "Bayesian-based iterative method of image restoration," *J. Opt. Soc. Am.* **62**, 55–59 (1972).

[19] L. B. Lucy, "An iterative technique for the rectification of observed distributions," *Astronom. J.* **79**, 745–754 (1974).

[20] D. L. Snyder, T. J. Schulz, and J. A. O'Sullivan, "Deblurring subject to nonnegativity constraints," *IEEE Trans. Signal Process.* **40**, 1143–1150 (1992).

[21] Y. Vardi and D. Lee, "From image deblurring to optimal investments: Maximum likelihood solutions for positive linear inverse problems," *J. R. Stat. Soc. B* **55**, 569–612 (1993). (with discussion).

[22] R. A. Gonsalves, "Phase retrieval and diversity in adaptive optics," *Opt. Eng.* **21**, 829–832 (1982).

[23] R. L. Lagendijk, J. Biemond, and D. E. Boekee, "Identification and restoration of noisy blurred images using the expectation-maximization algorithm," *IEEE Trans. Acoust. Speech Signal Process.* **38**, 1180–1191 (1990).

[24] R. G. Paxman, T. J. Schulz, and J. R. Fienup, "Joint estimation of object and aberrations using phase diversity," *J. Opt. Soc. Am. A* **9**, 1072–1085 (1992).

[25] H. J. Trussell and S. Fogel, "Identification and restoration of spatially variant motion blurs in sequential images," *IEEE Trans. Image Process.* **1**, 375–391 (1992).

[26] N. Miura and N. Baba, "Extended-object reconstruction with sequential use of the iterative blind deconvolution method," *Opt. Commun.* **89**, 375–379 (1992).

[27] S. M. Jefferies and J. C. Christou, "Restoration of astronomical images by iterative blind deconvolution," *Astrophys. J.* **63**, 862–874 (1993).

[28] J. W. Goodman, *Introduction to Fourier Optics* (McGraw-Hill, New York, 1996), 2nd ed.

[29] J. R. Fienup, J. C. Marron, T. J. Schulz, and J. H. Seldin, "Hubble Space Telescope characterized by using phase-retrieval algorithms," *Appl. Opt.* **32**, 1747–1767 (1993).

[30] J. W. Goodman, *Statistical Optics* (Wiley, New York, 1985).

[31] A. Tikhonov and V. Arsenin, *Solutions of Ill-Posed Problems* (Winston, Washington, D.C., 1977).

[32] W. L. Root, "Ill-posedness and precision in object-field reconstruction problems," *J. Opt. Soc. Am. A* **4**, 171–179 (1987).

[33] B. Porat, *Digital Processing of Random Signals: Theory and Methods* (Prentice-Hall, Englewood Cliffs, NJ, 1993).

[34] D. L. Snyder, C. W. Helstrom, A. D. Lanterman, M. Faisal, and R. L. White, "Compensation for readout noise in CCD images," *J. Opt. Soc. Am. A* **12**, 272–283 (1995).

[35] U. Grenander, *Abstract Inference* (Wiley, New York, 1981).

[36] D. L. Snyder and M. I. Miller, "The use of sieves to stabilize images produced with the EM algorithm for emission tomography," *IEEE Trans. Nucl. Sci.* **NS-32**, 3864–3871 (1985).

[37] J. R. Thompson and R. A. Tapia, *Nonparametric Function Estimation, Modeling, and Simulation* (SIAM, Philadelphia, 1990).

[38] J. A. O'Sullivan, "Roughness penalties on finite domains," *IEEE Trans. Image Process.* **4**, 1258–1268 (1995).

[39] I. J. Good and R. A. Gaskins, "Nonparametric roughness penalties for probability densities," *Biometrika* **58**, 255–277 (1971).

[40] R. Chellappa and A. Jain, eds., *Markov Random Fields: Theory and Application* (Academic, New York, 1993).

[41] C. A. Bouman and K. Sauer, "A generalized Gaussian image model for edge-preserving MAP estimation," *IEEE Trans. Image Process.* **2**, 296–310, (1993).

[42] D. G. Luenberger, *Linear and Nonlinear Programming* (Addison-Wesley, Reading, MA, 1984).

[43] A. P. Dempster, N. M. Laird, and D. B. Rubin, "Maximum Likelihood from incomplete data via the EM algorithm," *J. R. Stat. Soc. B* **39**, 1–37 (1977). (with discussion)

[44] J. A. Fessler and A. O. Hero, "Space-alternating generalized expectation-maximimization algorithm," *IEEE Trans. Signal Process.* **42**, 2664–2677 (1994).

[45] J. A. Fessler and A. O. Hero, "Penalized maximum-likelihood image reconstruction using space-alternating generalized EM algorithms," *IEEE Trans. Image Process.* **4**, 1417–1429 (1995).

[46] C. A. Bouman and K. Sauer, "A unified approach to statistical tomography using coordinate descent optimization," *IEEE Trans. Image Process.* **5**, 480–492 (1996).

[47] H. T. Pai and A. C. Bovik, "Exact multichannel blind image restoration," *IEEE Signal Process. Lett.* **4**, 217–220 (1997).

[48] G. Harikumar and Y. Bresler, "Perfect blind restoration of images blurred by multiple filters: Theory and efficient algorithms," *IEEE Trans. Image Process.* **8**, 202–219 (1999).

[49] R. C. Hardie, K. J. Barnard, and E. E. Armstrong. "Joint MAP registration and high-resolution image estimation using a sequence of undersampled images," *IEEE Trans. Image Process.* **6**, 1621–1633 (1997).

[50] M. Irani and S. Peleg. "Improving resolution by image registration," *CVGIP: Graph. Mod. Image Process.* **53**, 231–239 (1991).

[51] T. J. Schulz, B. E. Stribling, and J. J. Miller, "Multiframe blind deconvolution with real data: Imagery of the Hubble Space Telescope," *Opt. Express* **1**, 355–362 (1997).

[52] D. W. Tyler, S. D. Ford, B. R. Hunt, R. G Paxman, M. C. Roggemann, J. C. Rountree, T. J. Schulz, K. J. Schulze, J. H. Seldin, D. G. Sheppard, B. E. Stribling, W. C. Van Kampen, and B. M. Welsh, "Comparison of image reconstruction algorithms using adaptive optics instrumentation," in *Adaptive Optical System Technologies*, Proc. SPIE **3353**, 160–171 (1998).

[53] R. G. Paxman, T. J. Schulz, and J. R. Fienup, "Phase-diverse speckle interferometry," in *Signal Recovery and Synthesis IV*, Technical Digest (Optical Society of America, Washington, D. C., 1992), pp. 5–7.

[54] J. H. Seldin and R. G. Paxman, "Phase-diverse speckle reconstructin of solar data, "in T. J. Schulz and D. L. Snyder, eds., *Image Reconstruction and Restoration*, Proc. SPIE **2302**, 268–280 (1994).

[55] J. H. Seldin, R. G. Paxman, and C. U. Keller, "Times series restoration from ground-based solar observations," in *Missions to the Sun*, Proc. SPIE **2804**, 166–174 (1996).

[56] R. G. Paxman, J. H. Seldin, M. G. Löfdahl, G. B. Scharmer, and C. U. Keller, "Evaluation of phase-diversity techniques for solar-image restoration," *Astrophys. J.* **467**, 1087–1099 (1996).

[57] M. Born and E. Wolf, *Principles of Optics* (Pergamon, Elmsford, New York, 1980), 6th ed.

Iterative Image Restoration

Aggelos K. Katsaggelos
and Chun-Jen Tsai
Northwestern University

1 Introduction

In this chapter we consider a class of iterative image restoration algorithms. Let $\mathbf{g}$ be the observed noisy and blurred image, $\mathbf{D}$ the operator describing the degradation system, $\mathbf{f}$ the input to the system, and $\mathbf{v}$ the noise added to the output image. The input–output relation of the degradation system is then described by [2]

$$\mathbf{g} = \mathbf{D}\mathbf{f} + \mathbf{v}. \qquad (1)$$

The image restoration problem therefore to be solved is the inverse problem of recovering $\mathbf{f}$ from knowledge of $\mathbf{g}$, $\mathbf{D}$, and $\mathbf{v}$.

There are numerous imaging applications which are described by (1) [2, 3, 15]. $\mathbf{D}$, for example, might represent a model of the turbulent atmosphere in astronomical observations with ground-based telescopes, or a model of the degradation introduced by an out-of-focus imaging device. $\mathbf{D}$ might also represent the quantization performed on a signal or a transformation of it, for reducing the number of bits required to represent the signal.

The success in solving any recovery problem depends on the amount of the available prior information. This information refers to properties of the original image, the degradation system (which is in general only partially known), and the noise process. Such prior information can, for example, be represented by the fact that the original image is a sample of a stochastic field, or that the image is "smooth," or that it takes only nonnegative values. Besides defining the amount of prior information, equally critical is the ease of incorporating it into the recovery algorithm.

After the degradation model is established, the next step is the formulation of a solution approach. This might involve the stochastic modeling of the input image (and the noise), the determination of the model parameters, and the formulation of a criterion to be optimized. Alternatively, it might involve the formulation of a functional that is to be optimized subject to constraints imposed by the prior information. In the simplest possible case, the degradation equation defines directly the solution approach. For example, if $\mathbf{D}$ is a square invertible matrix, and the noise is ignored in Eq. (1), $\mathbf{f} = \mathbf{D}^{-1}\mathbf{g}$ is the desired unique solution. In most cases, however, the solution of Eq. (1) represents an ill-posed problem [19]. Application of regularization theory transforms it to a well-posed problem, which provides meaningful solutions to the original problem.

There are a large number of approaches providing solutions to the image restoration problem. For recent reviews of such approaches refer, for example, to [3, 15]. This chapter concentrates on a specific type of iterative algorithms, the *successive approximations* algorithm, and its application to the image restoration problem.

2 Iterative Recovery Algorithms

Iterative algorithms form an important part of optimization theory and numerical analysis. They date back to Gauss's time, but they also represent a topic of active research. A large part of any textbook on optimization theory or numerical analysis deals with iterative optimization techniques or algorithms [17].

Out of all possible iterative recovery algorithms, we concentrate on the successive approximations algorithms, which have been successfully applied to the solution of a number of inverse problems ([18] represents a very comprehensive paper on the topic). The basic idea behind such an algorithm is that the solution to the problem of recovering a signal that satisfies certain constraints from its degraded observation can be found by the alternate implementation of the degradation and the constraint operator. Problems reported in [18] that can be solved with such an iterative algorithm are the phase-only recovery problem, the magnitude-only recovery problem, the bandlimited extrapolation problem, the image restoration problem, and the filter design problem [6]. Reviews of iterative restoration algorithms are also presented in [4, 12, 16]. There are a number of advantages associated with iterative restoration algorithms, among which [12, 18]: (i) there is no need to determine or implement the inverse of an operator; (ii) knowledge about the solution can be incorporated into the restoration process in a relatively straightforward manner; (iii) the solution process can be monitored as it progresses; and (iv) the partially restored signal can be utilized in determining unknown parameters pertaining to the solution.

In the following we first present the development and analysis of two simple iterative restoration algorithms. Such algorithms are based on a linear and spatially invariant degradation, when the noise is ignored. Their description is intended to provide a good understanding of the various issues involved in dealing with iterative algorithms. We adopt a "how-to" approach; it is expected that no difficulties will be encountered by anybody wishing to implement the algorithms. We then proceed with the matrix-vector representation of the degradation model and the iterative algorithms. The degradation systems described now are linear but not necessarily spatially invariant. The relation between the matrix-vector and scalar representation of the degradation equation and the iterative solution is also presented. Experimental results demonstrate the capabilities of the algorithms.

3 Spatially Invariant Degradation

3.1 Degradation Model

Let us consider the following degradation model,

$$g(n_1, n_2) = d(n_1, n_2) * f(n_1, n_2), \tag{2}$$

where $g(n_1, n_2)$ and $f(n_1, n_2)$ represent respectively the observed degraded and original image, $d(n_1, n_2)$ represents the impulse response of the degradation system, and $*$ denotes two-dimensional (2-D) convolution. It is mentioned here that the arrays $d(n_1, n_2)$ and $f(n_1, n_2)$ are appropriately padded with zeros, so that the result of 2-D circular convolution equals the result of 2-D linear convolution in Eq. (2) (see Chapter 2.3). Henceforth, in the following all the convolutions involved are circular convolutions and all the shifts are circular shifts.

We rewrite Eq. (2) as follows:

$$\Phi(f(n_1, n_2)) = g(n_1, n_2) - d(n_1, n_2) * f(n_1, n_2) = 0. \tag{3}$$

Therefore, the restoration problem of finding an estimate of $f(n_1, n_2)$ given $g(n_1, n_2)$ and $d(n_1, n_2)$, becomes the problem of finding a root of $\Phi(f(n_1, n_2)) = 0$.

3.2 Basic Iterative Restoration Algorithm

The following identity holds for any value of the parameter β:

$$f(n_1, n_2) = f(n_1, n_2) + \beta \Phi(f(n_1, n_2)). \tag{4}$$

Equation (4) forms the basis of the successive approximations iteration, by interpreting $f(n_1, n_2)$ on the left-hand side as the solution at the current iteration step, and $f(n_1, n_2)$ on the right-hand side as the solution at the previous iteration step. That is, with $f_0(n_1, n_2) = 0$,

$$\begin{aligned} f_{k+1}(n_1, n_2) &= f_k(n_1, n_2) + \beta \Phi(f_k(n_1, n_2)) \\ &= \beta g(n_1, n_2) + (\delta(n_1, n_2) \\ &\quad - \beta d(n_1, n_2)) * f_k(n_1, n_2), \end{aligned} \tag{5}$$

where $f_k(n_1, n_2)$ denotes the restored image at the kth iteration step, $\delta(n_1, n_2)$ denotes the discrete delta function, and β denotes the relaxation parameter that controls the convergence as well as the rate of convergence of the iteration. Iteration (5) is the basis of a large number of iterative recovery algorithms, and it is therefore analyzed in detail. Perhaps the earliest reference to iteration (5) with $\beta = 1$ was by Van Cittert [20] in the 1930's.

3.3 Convergence

Clearly if a root of $\Phi(f(n_1, n_2))$ exists, this root is a *fixed point* of iteration (5), that is, a point for which $f_{k+1}(n_1, n_2) = f_k(n_1, n_2)$. It is not guaranteed, however, that iteration (5) will converge, even if Eq. (3) has one or more solutions. Let us, therefore, examine under what condition (sufficient condition) iteration (5) converges. Let us first rewrite it in the discrete frequency domain, by taking the 2-D discrete Fourier transform (DFT) of both sides. It then becomes

$$F_{k+1}(u, v) = \beta G(u, v) + (1 - \beta D(u, v)) F_k(u, v), \tag{6}$$

where $F_k(u, v)$, $G(u, v)$, and $D(u, v)$ represent respectively the 2-D DFT of $f_k(n_1, n_2)$, $g(n_1, n_2)$, and $d(n_1, n_2)$. We express next

$F_k(u, v)$ in terms of $F_0(u, v)$. Clearly,

$$F_1(u, v) = \beta G(u, v)$$

$$F_2(u, v) = \beta G(u, v) + (1 - \beta D(u, v))\beta G(u, v)$$

$$= \sum_{\ell=0}^{1}(1 - \beta D(u, v))^{\ell}\beta G(u, v)$$

$$\vdots$$

$$F_k(u, v) = \sum_{\ell=0}^{k-1}(1 - \beta D(u, v))^{\ell}\beta G(u, v)$$

$$= H_k(u, v)G(u, v). \tag{7}$$

We therefore see that the restoration filter at the kth iteration step is given by

$$H_k(u, v) = \beta \sum_{\ell=0}^{k-1}(1 - \beta D(u, v))^{\ell}. \tag{8}$$

The obvious next question is, under what conditions does the series in Eq. (8) converge, and what is this convergence filter equal to? Clearly, if

$$|1 - \beta D(u, v)| < 1, \tag{9}$$

then

$$\lim_{k\to\infty} H_k(u, v) = \lim_{k\to\infty} \beta \frac{1 - (1 - \beta D(u, v))^k}{1 - (1 - \beta D(u, v))} = \frac{1}{D(u, v)}. \tag{10}$$

Notice that Eq. (9) is not satisfied at the frequencies for which $D(u, v) = 0$. At these frequencies

$$H_k(u, v) = k\beta, \tag{11}$$

and therefore, in the limit $H_k(u, v)$ is not defined. However, since the number of iterations run is always finite, $H_k(u, v)$ is a large but finite number.

Having a closer look at the sufficient condition for convergence, we see that Eq. (9) can be rewritten as

$$|1 - \beta \,\mathrm{Re}\{D(u, v)\} - j\beta \,\mathrm{Im}\{D(u, v)\}|^2 < 1$$
$$\Rightarrow (1 - \beta \,\mathrm{Re}\{D(u, v)\})^2 + (\beta \,\mathrm{Im}\{D(u, v)\})^2 < 1. \tag{12}$$

Inequality (12) defines the region inside a circle of radius $1/\beta$ centered at $c = (1/\beta, 0)$ in the $(\mathrm{Re}\{D(u, v)\}, \mathrm{Im}\{D(u, v)\})$ domain, as shown in Fig. 1. From this figure it is clear that the left half-plane is not included in the region of convergence. That is, even though by decreasing β the size of the region of convergence increases, if the real part of $D(u, v)$ is negative, the sufficient condition for convergence cannot be satisfied. Therefore, for the class of degradations for which this is

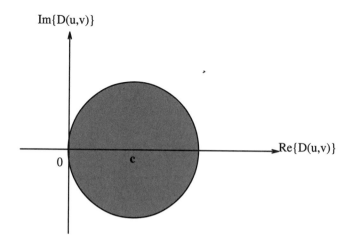

FIGURE 1   Geometric interpretation of the sufficient condition for convergence of the basic iteration, where $c = (1/\beta, 0)$.

the case, such as the degradation that is due to motion, iteration (5) is not guaranteed to converge.

The following form of Eq. (12) results when $\mathrm{Im}\{D(u, v)\} = 0$, which means that $d(n_1, n_2)$ is symmetric,

$$0 < \beta < \frac{2}{D_{\max}(u, v)}, \tag{13}$$

where $D_{\max}(u, v)$ denotes the maximum value of $D(u, v)$ over all frequencies (u, v). If we now also take into account that $d(n_1, n_2)$ is typically normalized, i.e., $\sum_{n_1, n_2} d(n_1, n_2) = 1$, and represents a low-pass degradation, then $D(0, 0) = D_{\max}(u, v) = 1$. In this case Eq. (12) becomes

$$0 < \beta < 2. \tag{14}$$

From the above analysis, when the sufficient condition for convergence is satisfied, the iteration converges to the original signal. This is also the inverse solution obtained directly from the degradation equation. That is, by rewriting Eq. (2) in the discrete frequency domain,

$$G(u, v) = D(u, v)F(u, v), \tag{15}$$

we obtain

$$F(u, v) = \begin{cases} \frac{G(u, v)}{D(u, v)} & \text{for } D(u, v) \neq 0 \\ 0 & \text{otherwise} \end{cases}, \tag{16}$$

which represents the *pseudo-inverse* or *generalized inverse* solution.

An important point to be made here is that, unlike the iterative solution, the inverse solution of Eq. (16) can be obtained without imposing any requirements on $D(u, v)$. That is, even if Eq. (2) or Eq. (15) has a unique solution, that is, $D(u, v) \neq 0$ for all (u, v), iteration (5) may not converge, if the sufficient condition for convergence is not satisfied. It is therefore not the appropriate iteration to solve the problem. Actually iteration (5) may not offer any advantages over the direct implementation of the inverse filter of Eq. (16), if no other features of the iterative algorithms are used, as will be explained later. One possible advantage of Eq. (5) over Eq. (16) is that the noise amplification in the restored image can be controlled by terminating the iteration before convergence, which represents another form of regularization, as will also be demonstrated experimentally. An iteration that will converge to the inverse solution of Eq. (2) for any $d(n_1, n_2)$ is described in the next section.

3.4 Reblurring

The degradation equation, Eq. (2), can be modified so that the successive approximations iteration converges for a larger class of degradations. That is, the observed data $g(n_1, n_2)$ are first filtered (reblurred) by a system with impulse response $d^*(-n_1, -n_2)$, where the asterisk denotes complex conjugation. Since circular convolutions have been adopted, the impulse response of the degradation system is equal to $d^*((N_1 - n_1)_{N_1}, (N_2 - n_2)_{N_2})$, where $(\cdot)_N$ denotes modulo N operation, assuming the images are of size $N_1 \times N_2$ pixels. Degradation Eq. (2) therefore becomes

$$
\begin{aligned}
\tilde{g}(n_1, n_2) &= g(n_1, n_2) * d^*(-n_1, -n_2) \\
&= d^*(-n_1, -n_2) * d(n_1, n_2) * f(n_1, n_2) \\
&= \tilde{d}(n_1, n_2) * f(n_1, n_2).
\end{aligned}
\tag{17}
$$

If we follow the same steps as in the previous section, substituting $g(n_1, n_2)$ by $\tilde{g}(n_1, n_2)$ and $d(n_1, n_2)$ by $\tilde{d}(n_1, n_2)$, the iteration providing a solution to Eq. (17) becomes

$$
\begin{aligned}
f_{k+1}(n_1, n_2) &= \beta d^*(-n_1, -n_2) * g(n_1, n_2) \\
&\quad + (\delta(n_1, n_2) - \beta \tilde{d}(n_1, n_2)) * f_k(n_1, n_2),
\end{aligned}
\tag{18}
$$

with $f_0(n_1, n_2) = 0$, and the restoration filter at the kth iteration step is now given by

$$
\begin{aligned}
H_k(u, v) &= \beta \sum_{\ell=0}^{k-1} (1 - \beta |D(u, v)|^2)^{\ell} D^*(u, v) \\
&= \beta \frac{1 - (1 - \beta |D(u, v)|^2)^k}{1 - (1 - \beta |D(u, v)|^2)} D^*(u, v).
\end{aligned}
\tag{19}
$$

Therefore, the sufficient condition for convergence, corresponding to condition (9), becomes

$$
|1 - \beta |D(u, v)|^2| < 1, \quad \text{or } 0 < \beta < \frac{2}{\max_{u,v} |D(u, v)|^2}.
\tag{20}
$$

In this case

$$
\lim_{k \to \infty} H_k(u, v) = \begin{cases} \frac{1}{D(u, v)} & D(u, v) \neq 0 \\ 0 & \text{otherwise} \end{cases}.
\tag{21}
$$

3.5 Experimental Results

In this section the performance of the iterative image restoration algorithms presented so far is demonstrated experimentally. We use a relatively simple degradation model in order to clearly analyze the behavior of the restoration filters. The degradation is due to one-dimensional (1-D) horizontal motion between the camera and the scene, caused by, for example, camera panning or fast object motion. The impulse response of the degradation system is given by

$$
d(n_1, n_2) = \begin{cases} \frac{1}{L} & -\frac{L-1}{2} \leq n_1 \leq \frac{L-1}{2}, & L \text{ odd, } n_2 = 0 \\ \frac{1}{L} & -\frac{L}{2} + 1 \leq n_1 \leq \frac{L}{2}, & L \text{ even, } n_2 = 0. \\ 0 & \text{otherwise} \end{cases}
\tag{22}
$$

The blurred signal-to-noise ratio (BSNR) is typically used in the restoration community to measure the degree of the degradation (blur plus additive noise). This figure is given by

$$
\text{BSNR} = 10 \log_{10} \frac{\sigma_{\text{Df}}^2}{\sigma_{\text{v}}^2},
\tag{23}
$$

where σ_{Df}^2 and σ_{v}^2 are respectively the variance of the blurred image and the additive noise.

For the purpose of objectively testing the performance of image restoration algorithms, the improvement in SNR (ISNR) is often used. This metric using the restored image at the kth iteration step is given by

$$
\text{ISNR} = 10 \log_{10} \frac{\sum_{n_1, n_2} (f(n_1, n_2) - g(n_1, n_2))^2}{\sum_{n_1, n_2} (f(n_1, n_2) - f_k(n_1, n_2))^2}.
\tag{24}
$$

Obviously, this metric can only be used for simulation cases when the original image is available. While mean squared error (MSE) metrics such as ISNR do not always reflect the visual quality of the restored image, they serve to provide an objective standard by which to compare different techniques. However, in all cases presented here, it is important to consider the behavior of the various algorithms from the viewpoint of ringing and noise amplification, which can be a key indicator of improvement in quality for subjective comparisons of restoration algorithms.

In Fig. 2(a) the image blurred by the 1-D motion blur of extent 8 pixels ($L = 8$ in Eq. (22)) is shown, along with $|D(u, 0)|$, a slice of the magnitude of the 256×256-point DFT of $d(n_1, n_2)$; notice that all slices of the DFT are the same, independently of the value of v. No noise has been added. The extent of the blur and the size of the DFT were chosen in such a way that exact zeros exist in $D(u, v)$. The next three images represent the restored images using Eq. (18) with $\beta = 1.0$, along with $|H_k(u, 0)|$ in Eq. (19), after 20, 50, and 221 iterations (at convergence). The criterion

$$
\frac{\sum_{n_1, n_2} (f_{k+1}(n_1, n_2) - f_k(n_1, n_2))^2}{\sum_{n_1, n_2} (f_k(n_1, n_2))^2} \leq 10^{-8}
\tag{25}
$$

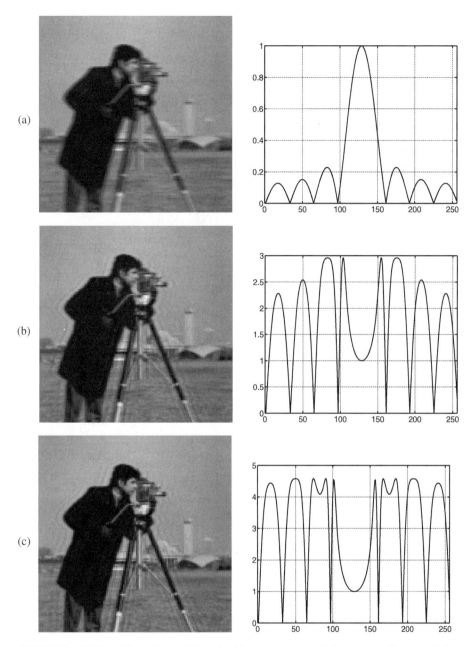

FIGURE 2 (a) Blurred image by an 1-D motion blur over 8 pixels and the corresponding magnitude of the frequency response of the degradation system; (b)–(d) images restored by iteration (18), after 20 iterations (ISNR = 4.03 dB), 50 iterations (ISNR = 6.22 dB), and at convergence after 221 iterations (ISNR = 9.92 dB), and the corresponding magnitude of $H_k(u, 0)$ in Eq. (19); (e) image restored by the direct implementation of the generalized inverse filter in Eq. (16) (ISNR = 15.50 dB), and the corresponding magnitude of the frequency response of the restoration filter. (*Continues.*)

is used for terminating the iteration. Notice that Eq. (5) is not guaranteed to converge for this particular degradation since $D(u, v)$ takes negative values. The restored image of Fig. 2(e) is the result of the direct implementation of the pseudo-inverse filter, which can be thought of as the result of the iterative restoration algorithm after infinitely many iterations, assuming infinite precision arithmetic. The corresponding ISNRs are as follows:

4.03 dB, Fig. 2(b); 6.22 dB, Fig. 2(c); 9.92 dB, Fig. 2(d); and 15.50 dB, Fig. 2(e). Finally, the normalized residual error shown in Eq. (25) versus the number of iterations is shown in Fig. 3. The iteration steps at which the restored images are shown in the previous figure are indicated by circles.

We repeat the same experiment when noise is added to the blurred image, resulting in a BSNR of 20 dB, as shown in

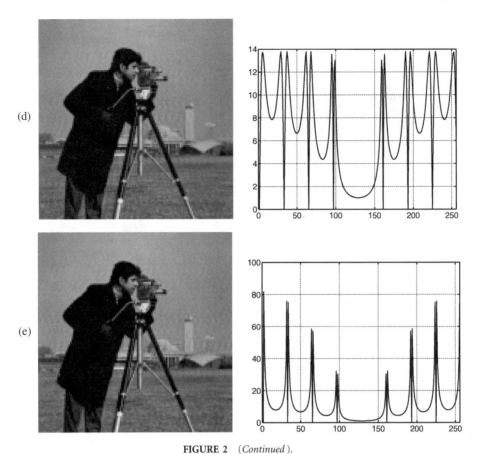

FIGURE 2 (*Continued*).

Fig. 4(a). The restored images after 20 iterations (ISNR = 1.83 dB), 50 iterations (ISNR = −0.40 dB), and at convergence after 1712 iterations (ISNR = −9.43 dB) are shown respectively in Figs. 4(b), 4(c), and 4(d). Finally, the restoration based on the direct implementation of the pseudo-inverse filter (ISNR = −12.09 dB) is shown in Fig. 4(e). The iterative algorithm converges slower in this case.

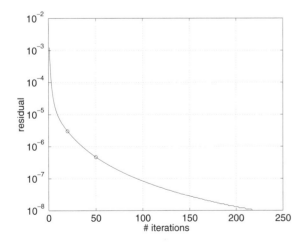

FIGURE 3 Normalized residual error as a function of the number of iterations.

What becomes evident from these experiments is the following.

1. As expected, for the noise-free case the visual quality as well as the objective quality, in terms of the ISNR, of the restored images increase as the number of iterations increases.

2. For the noise-free case the inverse filter outperforms the iterative restoration filter. Based on this experiment there is no reason to implement this particular filter iteratively, except possibly for computational reasons.

3. For the noisy-blurred image the noise is amplified and the ISNR decreases as the number of iterations increases. Noise completely dominates the image restored by the pseudo-inverse filter. In this case, the iterative implementation of the restoration filter offers the advantage that the number of iterations can be used to control the amplification of the noise, which represents a form of regularization. The restored image, for example, after 50 iterations (Fig. 4(c)) represents a reasonable restoration.

4. The iteratively restored image exhibits noticeable *ringing artifacts*, which will be further analyzed below. Such artifacts can be masked by noise, as demonstrated, for example, with the image in Fig. 4(d).

FIGURE 4 (a) Noisy-blurred image; 1-D motion blur over 8 pixels, BSNR = 20 dB; (b)–(d) images restored by iteration (18), after 20 iterations (ISNR = 1.83 dB), 50 iterations (ISNR = −0.30 dB), and at convergence after 1712 iterations (ISNR = −9.43 dB); (e) image restored by the direct implementation of the generalized inverse filter in Eq. (16) (ISNR = −12.09 dB).

Ringing Artifacts

Let us compare the magnitudes of the frequency response of the restoration filter after 221 iterations (Fig. 2(d)) and the inverse filter (Fig. 2(e)). First of all, it is clear that the existence of spectral zeros in $D(u, v)$ does not cause any difficulty in the determination of the restoration filter in both cases, since the restoration filter is also zero at these frequencies. The main difference is that the values of $|H(u, v)|$, the magnitude of the frequency response of the inverse filter, at frequencies close to the zeros of $D(u, v)$ are considerably larger than the corresponding values of $|H_k(u, v)|$. This is because the values of $H_k(u, v)$ are approximated by a series according to Eq. (19). The important term in this series is $(1 − \beta | D(u, v)|^2)$, since it determines whether the iteration converges or not (sufficient condition). Clearly, this term for values of $D(u, v)$ close to zero is close to one, and therefore, it approaches zero much slower when raised to the power of k, the number of iterations, than the terms for which $D(u, v)$ assumes larger values and therefore the term $(1 − \beta | D(u, v)|^2)$ is close to zero. This means that each frequency component is restored independently and with different convergence rates. Clearly, the larger the values of β, the faster the convergence, it is mentioned here that the quality of the restored

image at convergence depends on the value of β; in other words, two images restored with different β's but satisfying the same convergence criterion might differ considerably in terms of both visual quality and ISNR.

Let us denote by $h(n_1, n_2)$ the impulse response of the restoration filter and define

$$h_{\text{all}}(n_1, n_2) = d(n_1, n_2) * h(n_1, n_2). \tag{26}$$

Ideally, $h_{\text{all}}(n_1, n_2)$ should be equal to an impulse, or its DFT $H_{\text{all}}(u, v)$ should be a constant, that is, the restoration filter is precisely undoing what the degradation system did. Because of the spectral zeros, however, in $D(u, v)$, $H_{\text{all}}(u, v)$ deviates from a constant. For the particular example under consideration $|H_{\text{all}}(u, 0)|$ is shown in Figs. 5(a) and 5(c), for the inverse filter and the iteratively implemented inverse filter by Eq. (18), respectively. In Figs. 5(b) and 5(d) the corresponding impulse responses are shown. Because of the periodic zeros of $D(u, v)$ in this particular case, $h_{\text{all}}(n_1, n_2)$ consists of the sum of an impulse and an impulse train (of period 8 samples). The deviation from a constant or an impulse is greater with the iterative restoration filter than with the direct inverse filter.

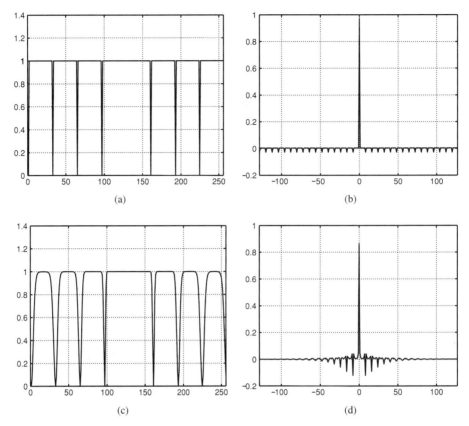

FIGURE 5   (a) $|H_{\text{all}}(u, 0)|$ for direct implementation of the inverse filter; (c) $|H_{\text{all}}(u, 0)|$ for the iterative implementation of the inverse filter; (b),(d) $h_{\text{all}}(n, 0)$ corresponding to (a) and (c).

Now, in the absense of noise the restored image $\hat{f}(n_1, n_2)$ is given by

$$\hat{f}(n_1, n_2) = h_{\text{all}}(n_1, n_2) * f(n_1, n_2). \tag{27}$$

Clearly, because of the shape of $h_{\text{all}}(n_1, n_2)$ shown in Figs. 5(b) and 5(d) (only $h_{\text{all}}(n_1, 0)$ is shown, since it is zero for the rest of the values of n_2), the existence of the periodic train of impulses gives rise to ringing. In the case of the inverse filter (Fig. 5(b)) the impulses of the train are small in magnitude and therefore ringing is not visible. In the case of the iterative filter, however, the few impulses close to zero have larger amplitude and therefore ringing is noticeable in this case.

4 Matrix-Vector Formulation

The presentation so far has followed a rather simple and intuitive path. We hope that it demonstrated some of the issues involved in developing and implementing an iterative algorithm. In this section we present the matrix-vector formulation of the degradation process and the restoration iteration. More general results are therefore obtained, since now the degradation can be spatially varying, while the restoration filter may be spatially varying as well, but even nonlinear. The degradation actually can be non-

linear as well (of course it is not represented by a matrix in this case), but we do not focus on this case, although most of the iterative algorithms discussed below would be applicable.

What became clear from the previous sections is that in applying the successive approximations iteration, the restoration problem to be solved is brought first into the form of finding the root of a function (see Eq. (3)). In other words, a solution to the restoration problem is sought that satisfies

$$\Phi(\mathbf{f}) = 0, \tag{28}$$

where $\mathbf{f} \in \mathcal{R}^N$ is the vector representation of the signal resulting from the stacking or ordering of the original signal, and $\Phi(\mathbf{f})$ represents a nonlinear in general function. The row by row, from left-to-right stacking of an image is typically referred to as *lexicographic ordering*. For a 256×256 image, for example, vector $\mathbf{f}$ is of dimension 64K × 1.

Then the successive approximations iteration that might provide us with a solution to Eq. (28) is given by

$$\mathbf{f}_{k+1} = \mathbf{f}_k + \beta \Phi(\mathbf{f}_k) = \Psi(\mathbf{f}_k), \tag{29}$$

with $\mathbf{f}_0 = 0$. Clearly if $\mathbf{f}'$ is a solution to $\Phi(\mathbf{f}) = 0$, i.e., $\Phi(\mathbf{f}') = 0$, then $\mathbf{f}'$ is also a fixed point of the above iteration, that is, $\mathbf{f}_{k+1} = \mathbf{f}_k = \mathbf{f}'$. However, as was discussed in the previous section, even if $\mathbf{f}'$ is the unique solution to Eq. (28), this does not imply that iteration (29) will converge. This again underlines the importance of convergence when dealing with iterative algorithms. The form iteration (29) takes for various forms of the function $\Phi(\mathbf{f})$ is examined next.

4.1 Basic Iteration

From Eq. (1) when the noise is ignored, the simplest possible form $\Phi(\mathbf{f})$ can take is

$$\Phi(\mathbf{f}) = \mathbf{g} - \mathbf{Df}. \tag{30}$$

Then Eq. (29) becomes

$$\mathbf{f}_{k+1} = \beta\mathbf{g} + (\mathbf{I} - \beta\mathbf{D})\mathbf{f}_k, \tag{31}$$

where $\mathbf{I}$ is the identity operator.

4.2 Least-Squares Iteration

According to the least-squares approach, a solution to Eq. (1) is sought by minimizing

$$M(\mathbf{x}) = \|\mathbf{g} - \mathbf{Df}\|^2. \tag{32}$$

A necessary condition for $M(\mathbf{f})$ to have a minimum is that its gradient with respect to $\mathbf{f}$ is equal to zero. That is, in this case

$$\Phi(\mathbf{f}) = \frac{1}{2}\nabla_\mathbf{f} M(\mathbf{f}) = \mathbf{D}^T(\mathbf{g} - \mathbf{Df}), \tag{33}$$

where ^{T} denotes the transpose of a matrix or vector. Application of iteration (29) then results in

$$\mathbf{f}_{k+1} = \beta\mathbf{D}^T\mathbf{g} + (\mathbf{I} - \beta\mathbf{D}^T\mathbf{D})\mathbf{f}_k. \tag{34}$$

It is mentioned here that the matrix-vector representation of an iteration does not necessarily determine the way the iteration is implemented. In other words, the pointwise version of the iteration may be more efficient, from the implementation point of view, than the matrix-vector form of the iteration. Now when Eq. (2) is used to form the matrix-vector equation $\mathbf{g} = \mathbf{Df}$, matrix $\mathbf{D}$ is a block-circulant matrix [2]. A square matrix is circulant when a circular shift of one row produces the next row, and the circular shift of the last row produces the first row. A square matrix is block circulant when it consists of circular submatrices, which when circularly shifted produce the next row of circulant matrices. The singular values of the block circulant matrix $\mathbf{D}$ are the DFT values of $d(n_1, n_2)$, and the eigenvectors are the complex exponential basis functions of the DFT. Iterations (31) and (34) can therefore be written in the discrete frequency domain, and they become identical to iteration (6) and the frequency domain version of iteration (18), respectively [12, 16].

4.3 Constrained Least-Squares Iteration

The image restoration problem is an ill-posed problem, which means that matrix $\mathbf{D}$ is ill conditioned. A regularization method replaces an ill-posed problem by a well-posed problem, whose solution is an acceptable approximation to the solution of the ill-posed problem [19]. Most regularization approaches transform the original inverse problem into a constrained optimization problem. That is, a functional has to be optimized with respect to the original image, and possibly other parameters. By using the necessary condition for optimality, the gradient of the functional with respect to the original image is set equal to zero, therefore determining the mathematical form of $\Phi(\mathbf{f})$. The successive approximations iteration becomes in this case a gradient method with a fixed step (determined by β).

As an example, a restored image is sought as the result of the minimization of [9]

$$\|\mathbf{Cf}\|^2, \tag{35}$$

subject to the constraint that

$$\|\mathbf{g} - \mathbf{Df}\|^2 \le \epsilon^2. \tag{36}$$

Operator $\mathbf{C}$ is a high-pass operator. The meaning then of the minimization of $\|\mathbf{Cf}\|^2$ is to constrain the high-frequency energy of the restored image, therefore requiring that the restored image is smooth. In contrast, by enforcing inequality (36) the fidelity to the data is preserved.

Following the Lagrangian approach that transforms the constrained optimization problem into an unconstrained one, the following functional is minimized:

$$M(\alpha, \mathbf{f}) = \|\mathbf{Df} - \mathbf{g}\|^2 + \alpha\|\mathbf{Cf}\|^2. \tag{37}$$

The necessary condition for a minimum is that the gradient of $M(\alpha, \mathbf{f})$ is equal to zero. That is, in this case

$$\Phi(\mathbf{f}) = \frac{1}{2}\nabla_\mathbf{f} M(\alpha, \mathbf{f}) = (\mathbf{D}^T\mathbf{D} + \alpha\mathbf{C}^T\mathbf{C})\mathbf{f} - \mathbf{D}^T\mathbf{g}, \tag{38}$$

is used in iteration (29). The determination of the value of the regularization parameter α is a critical issue in regularized restoration, since it controls the tradeoff between fidelity to the data and smoothness of the solution, and therefore the quality of the restored image. A number of approaches for determining its value are presented and compared in [8].

Since the restoration filter resulting from Eq. (38) is widely used, it is worth looking further into its properties. When the degradation matrices $\mathbf{D}$ and $\mathbf{C}$ are block circulant, Eq. (38), the resulting successive approximations iteration can be written in the discrete frequency domain. The iteration takes the form

$$F_{k+1}(u, v) = \beta D^*(u, v)G(u, v) + (1 - \beta(|D(u, v)|^2 + \alpha|C(u, v)|^2))F_k(u, v), \tag{39}$$

where $C(u, v)$ represents the 2-D DFT of the impulse response of a high-pass filter, such as the 2-D Laplacian. Following steps similar to the ones presented in Section 3.3, we find it straightforward to verify that in this case the restoration filter at the kth iteration step is given by

$$H_k(u, v) = \beta \sum_{\ell=0}^{k-1} (1 - \beta(|D(u, v)|^2 + \alpha|C(u, v)|^2))^\ell D^*(u, v). \tag{40}$$

Clearly if

$$|1 - \beta(|D(u, v)|^2 + \alpha|C(u, v)|^2)| < 1, \tag{41}$$

then

$$\lim_{k \to \infty} H_k(u, v)$$

$$= \lim_{k \to \infty} \beta \frac{1 - (1 - \beta(|D(u, v)|^2 + \alpha|C(u, v)|^2))^k}{1 - (1 - \beta(|D(u, v)|^2 + \alpha|C(u, v)|^2))} D^*(u, v)$$

$$= \frac{D^*(u, v)}{|D(u, v)|^2 + \alpha|C(u, v)|^2}. \tag{42}$$

Notice that condition (41) is not satisfied at the frequencies for which $H_d(u, v) = |D(u, v)|^2 + \alpha|C(u, v)|^2 = 0$. It is therefore now not the zeros of the degradation matrix that have to be considered, but the zeros of the regularized matrix, with DFT values $H_d(u, v)$. Clearly if $H_d(u, v)$ is zero at certain frequencies, this means that both $D(u, v)$ and $C(u, v)$ are zero at these frequencies. This demonstrates the purpose of regularization, which is to remove the zeros of $D(u, v)$ without altering the rest of its values, or in general to make the matrix $\mathbf{D}^T\mathbf{D} + \alpha\mathbf{C}^T\mathbf{C}$ better conditioned than the matrix $\mathbf{D}^T\mathbf{D}$.

For the frequencies at which $H_d(u, v) = 0$,

$$\lim_{k \to \infty} H_k(u, v) = \lim_{k \to \infty} k \beta D^*(u, v) = 0, \tag{43}$$

since $D^*(u, v) = 0$.

4.3.1 Experimental Results

The noisy and blurred image of Fig. 4(a) (1-D motion blur over 8 pixels, BSNR = 20 dB) is now restored using iteration (39), with $\alpha = 0.01$, $\beta = 1.0$, and C as the 2-D Laplacian operator. It is mentioned here that the regularization parameter is chosen to be equal to $\sigma_\mathbf{v}^2/\sigma_{\mathbf{Df}}^2$, as determined by a set theoretic restoration approach presented in [14]. The restored images after 20 iterations (ISNR = 2.12 dB), 50 iteration (ISNR = 0.98 dB), and at convergence after 330 iterations (ISNR = −1.01 dB) with the corresponding $|H_k(u, 0)|$ in Eq. (40), are shown respectively in Figs. 6(a), 6(b), and 6(c). In Fig. 6(d) the restored image (ISNR = −1.64 dB) by the direct implementation of the constrained least-squares filter in Eq. (42) is shown, along with the magnitude of the frequency response of the restoration filter. It is clear now by comparing the restoration filters of Figs. 2(d) and 6(c) and 2(e) and 6(d) that the high frequencies have been suppressed, because of regularization, that is, the addition in

the denominator of the filter of the term $\alpha|C(u, v)|^2$. Because of the iterative approximation of the constrained least-squares filter, however, the two filters shown in Figs. 6(c) and 6(d) differ primarily in the vicinity of the low-frequency zeros of $D(u, v)$. Ringing is still present, as it can be primarily seen in Figs. 6(a) and 6(b), although is not as visible in Figs. 6(c) and 6(d). Because of regularization the results in Figs. 6(c) and 6(d) are preferred over the corresponding results with no regularization ($\alpha = 0.0$), shown in Figs. 4(d) and 4(e).

It is emphasized here that, unlike the previous experiments, the magnitude of the frequency response of the restoration filter shown in Fig. 6 at zero vertical frequency, i.e., $|H(u, 0)|$ is not the same for all vertical frequencies v. This is because the Laplacian operator is two-dimensional, unlike the degradation operator, which is one-dimensional. To further illustrate this $|D(u, v)|^2$, $|C(u, v)|^2$, and $|H_\infty(u, v)|^2$ in Eq. (42), are shown respectively in Figs. 7(a), 7(b), and 7(c).

The value of the regularization parameter is very critical for the quality of the restored image. The restored images with three different values of the regularization parameter are shown in Figs. 8(a)–8(c), corresponding to $\alpha = 1.0$ (ISNR = 2.4 dB), $\alpha = 0.1$ (ISNR = 2.96 dB), and $\alpha = 0.01$ (ISNR = −1.80 dB). The corresponding magnitudes of the error images, i.e., |original-restored|, scaled linearly to the 32–255 range, are shown in Figs. 8(d)–8(f). What is observed is that for large values of α the restored image is "smooth" while the error image contains the high-frequency information of the original image (large bias of the estimate), while as α decreases the restored image becomes more noisy and the error image takes the appearance of noise (large variance of the estimate). It has been shown in [8] that the bias of the constrained least-squares estimate is a monotonically increasing function of the regularization parameter, while the variance of the estimate is a monotonically decreasing function of the estimate. This implies that the mean-squared error (MSE) of the estimate, the sum of the bias and the variance, has a unique minimum for a specific value of α.

4.4 Spatially Adaptive Iteration

Spatially adaptive image restoration is the next natural step in improving the quality of the restored images. There are various ways to argue the introduction of spatial adaptivity, the most commonly used ones being the nonhomogeneity or nonstationarity of the image field and the properties of the human visual system. In either case, the functional to be minimized takes the form [4, 12]

$$M(\alpha, \mathbf{f}) = \|\mathbf{Df} - \mathbf{g}\|_{\mathbf{W}_1}^2 + \alpha\|\mathbf{Cf}\|_{\mathbf{W}_2}^2, \tag{44}$$

in which case

$$\Phi(\mathbf{f}) = \frac{1}{2}\boldsymbol{\nabla}_\mathbf{f} M(\alpha, \mathbf{f})$$

$$= (\mathbf{D}^T\mathbf{W}_1^T\mathbf{W}_1\mathbf{D} + \alpha\mathbf{C}^T\mathbf{W}_2^T\mathbf{W}_2\mathbf{C})\mathbf{f} - \mathbf{D}^T\mathbf{W}_1^T\mathbf{W}_1\mathbf{g}. \tag{45}$$

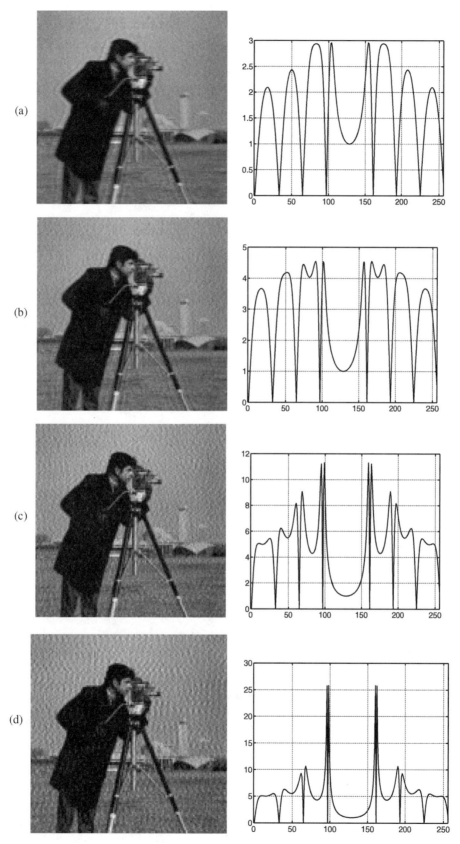

FIGURE 6 Restoration of the noisy-blurred image in Fig. 4(a) (motion over 8 pixels, BSNR = 20 dB), (a)–(c): images restored by iteration (39), after 20 iterations (ISNR = 2.12 dB), 50 iterations (ISNR = 0.98 dB) and at convergence after 330 iterations (ISNR = −1.01 dB), and the corresponding $|H_k(u, 0)|$ in Eq. (40); (d) image restored by the direct implementation of the constrained least-squares filter (ISNR = −1.64 dB), and the corresponding magnitude of the frequency response of the restoration filter (Eq. (42)).

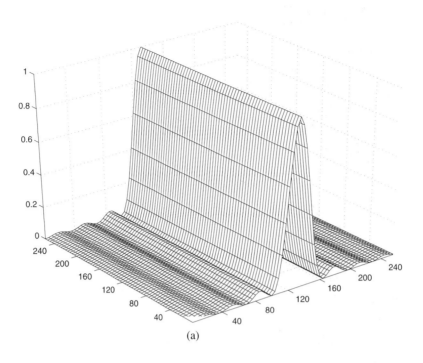

(a)

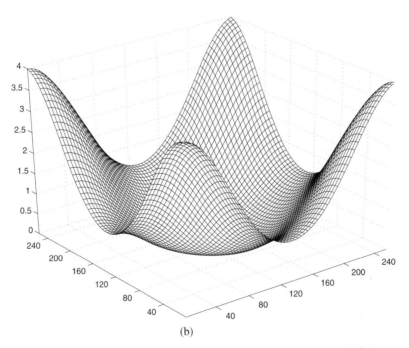

(b)

FIGURE 7 (a) $|D(u, v)|^2$ for horizontal motion blur over 8 pixels; (b) $|C(u, v)|^2$ for the 2-D Laplacian operator; (c) $|H_\infty(u, v)|^2$ in Eq. (42). (*Continues.*)

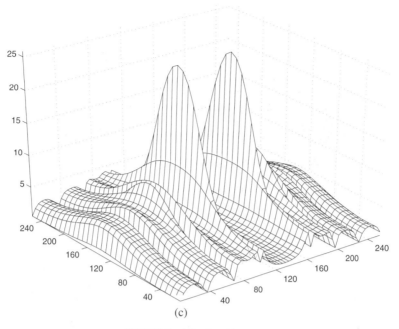

(c)

FIGURE 7 (*Continued*).

FIGURE 8 Direct constrained least-squares restorations of the noisy-blurred image in Fig. 4(a) (motion over 8 pixels, BSNR = 20 dB) with α equal to (a) 1, (b) 0.1, (c) 0.01. (d)–(f) Corresponding |original-restored| linearly mapped to the range [32, 255].

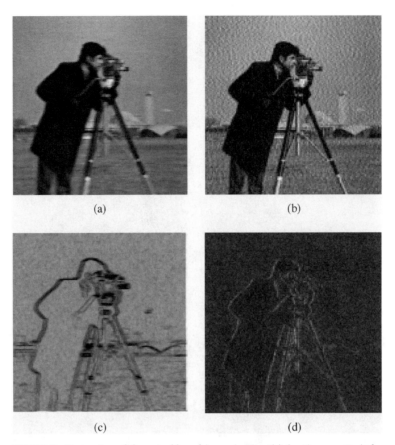

FIGURE 9 Restoration of the noisy-blurred image in Fig. 4(a) (motion over 8 pixels, BSNR = 20 dB), using (a) the adaptive algorithm of Eq. (45); (b) the nonadaptive algorithm of iteration Eq. (39); (c) values of the weight matrix in Eq. (46); (d) amplitude of the difference between (a) and (b) linearly mapped to the range [32, 255].

The choice of the diagonal weighting matrices W_1 and W_2 can be justified in various ways. In [12] both matrices are determined by the diagonal noise visibility matrix V [1]. That is, $W_2 = V$ and $W_1 = 1 - W_2$. The entries of V take values between 0 and 1. They are equal to 0 at the edges (noise is not visible and therefore smoothing is disabled), equal to 1 at the flat regions (noise is visible and therefore smoothing is fully enforced), and take values in between at the regions with moderate spatial activity. A study of the mapping between the level of spatial activity and the values of the visibility function appears in [7].

4.4.1 Experimental Results

The resulting successive approximations iteration from the use of $\Phi(f)$ in Eq. (45) has been tested with the noisy and blurred image we have been using so far in our experiments, which is shown in Fig. 4(a). It should be emphasized here that although matrices D and C are block circulant, the iteration cannot be implemented in the discrete frequency domain, since the weight matrices W_1 and W_2, are diagonal but not circulant. Therefore, the iterative algorithm is implemented exclusively in the spatial domain, or by switching between the frequency domain (where the

convolutions are implemented) and the spatial domain (where the weighting takes place). Clearly, from an implementation point of view the use of iterative algorithms offers a distinct advantage in this particular case.

The iteratively restored image with $W_1 = 1 - W_2$, $\alpha = 0.01$, and $\beta = 0.1$ is shown in Fig. 9(a), at convergence after 381 iterations and ISNR = 0.61 dB. The entries of the diagonal matrix W_2, denoted by $w_2(i)$, are computed according to

$$w_2(i) = \frac{1}{\theta\sigma^2(i) + 1}, \qquad (46)$$

where $\sigma^2(i)$ is the local variance at the ordered ith pixel location, and θ is a tuning parameter. The resulting values of $w_2(i)$ are linearly mapped into the [0, 1] range. These weights computed from the degraded image are shown in Fig. 9(c), linearly mapped to the [32, 255] range, using a 3×3 window to find the local variance and $\theta = 0.001$. The image restored by the nonadaptive algorithm, that is, $W_1 = W_2 = I$ and the rest of the parameters the same, is shown in Fig. 9(b) (ISNR = -0.20 dB). The absolute value of the difference between the images in Figs. 9(a) and 9(b), linearly mapped in the [32, 255] range, is shown in Fig. 9(d). It is clear that the two algorithms differ primarily at the vicinity of edges, where the smoothing is downweighted or disabled with the adaptive algorithm. Spatially adaptive algorithms in general can greatly improve the restoration results, since they can adopt to the local characteristics of each image.

5 Use of Constraints

Iterative signal restoration algorithms regained popularity in the 1970's because of the realization that improved solutions can be obtained by incorporating prior knowledge about the solution into the restoration process. For example, we may know in advance that **f** is bandlimited or space limited, or we may know on physical grounds that **f** can only have nonnegative values. A convenient way of expressing such prior knowledge is to define a constraint operator $\mathcal{C}$, such that

$$\mathbf{f} = \mathcal{C}\mathbf{f}, \qquad (47)$$

if and only if **f** satisfies the constraint. In general, $\mathcal{C}$ represents the concatenation of constraint operators. With the use of constraints, iteration (29) becomes [18]

$$
\begin{aligned}
\mathbf{f}_0 &= 0, \\
\tilde{\mathbf{f}}_k &= \mathcal{C}\mathbf{f}_k, \\
\mathbf{f}_{k+1} &= \Psi(\tilde{\mathbf{f}}_k).
\end{aligned}
\qquad (48)
$$

As already mentioned, a number of recovery problems, such as the bandlimited extrapolation problem, and the reconstruction from phase or magnitude problem, can be solved with the use of algorithms of the form of Eq. (48), by appropriately describing the distortion and constraint operators [18].

The *contraction mapping theorem* [17] usually serves as a basis for establishing convergence of iterative algorithms. Sufficient conditions for the convergence of the algorithms presented in Sec. 4 are presented in [12, 16]. Such conditions become identical to the ones derived in Sec. 3, when all matrices involved are block circulant. When constraints are used, the sufficient condition for convergence of the iteration is that at least one of the operators $\mathcal{C}$ and Ψ is contractive while the other is nonexpansive ($\mathcal{C}$ is nonexpansive, for example, when it represents a projection onto a convex set operator). Usually, it is harder to prove convergence and determine the convergence rate of the constrained iterative algorithm, taking also into account that some of the constraint operators are nonlinear, such as the positivity constraint operator.

5.1 Experimental Results

We demonstrate the effectiveness of the positivity constraint with the use of a simple example. A one-dimensional impulsive signal is shown in Fig. 10(a). Its degraded version by a motion blur over 8 samples is shown in Fig. 10(b). The blurred signal is restored by iteration (18) ($\beta = 1.0$) with the use of the

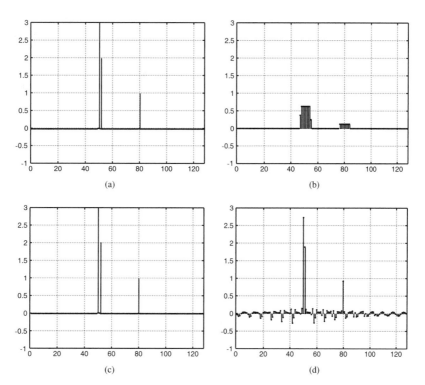

FIGURE 10 (a) original signal; (b) blurred signal by motion blur over 8 samples; signals restored by iteration (18); (c) with positivity constraint; (d) without positivity constraint.

positivity constraint (Fig. 10(c), 370 iterations, ISNR = 41.35 dB), and without the use of the positivity constraint (Fig. 10(d), 543 iterations, ISNR = 11.05 dB). The application of the positivity constraint, which represents a nonexpansive mapping, simply sets to zero all negative values of the signal. Clearly a considerably better restoration is represented by Fig. 10(c).

6 Discussion

In this chapter we briefly described the application of the successive approximations-based class of iterative algorithms to the problem of restoring a noisy and blurred image. We presented and analyzed in some detail the simpler forms of the algorithm. With this presentation we have simply touched the tip of the iceberg. We only covered a small amount of the material on the topic. More sophisticated forms of iterative image restoration algorithms were left out, since they were deemed to be beyond the scope and the level of this chapter. Examples of such algorithms are

- algorithms with higher rates of convergence [13];
- algorithms with a relaxation parameter β which depends on the iteration step (steepest descent and conjugate gradient algorithms are examples of this);
- algorithms that use a regularization parameter which depends on the partially restored image [11];
- algorithms that use a different regularization parameter for each discrete frequency component (which can also be iteration dependent) [10]; and
- algorithms that depend on more than one previous restoration step (multistep algorithms) [12].

It is our hope and expectation that the presented material will form a good introduction to the topic for the engineer or the student who would like to work in this area.

References

[1] G. L. Anderson and A. N. Netravali, "Image restoration based on a subjective criterion," *IEEE Trans. Sys. Man. Cybern.* **SMC-6**, 845–853 (1976).

[2] H. C. Andrews and B. R. Hunt, *Digital Image Restoration* (Prentice-Hall, NJ, 1977).

[3] M. R. Banham and A. K. Katsaggelos, "Digital image restoration," *IEEE Signal Process. Magazine*, **14**, 24–41 (1997).

[4] J. Biemond, R. L. Lagendijk, and R. M. Mersereau, "Iterative methods for image deblurring," *Proc. IEEE* **78**, 856–883 (1990).

[5] G. Demoment, "Image reconstruction and restoration: Overview of common estimation structures and problems," *IEEE Trans. Acoust. Speech Signal Process.* **37**, 2024–2036 (1989).

[6] D. E. Dudgeon and R. M. Mersereau, *Multidimensional Digital Signal Processing* (Prentice-Hall, NJ, 1984).

[7] S. N. Efstratiadis and A. K. Katsaggelos, "Adaptive iterative image restoration with reduced computational load," *Opt. Eng.* **29**, 1458–1468 (1990).

[8] N. P. Galatsanos and A. K. Katsaggelos, "Methods for choosing the regularization parameter and estimating the noise variance in image restoration and their relation," *IEEE Trans. Image Process.* **1**, 322–336 (1992).

[9] B. R. Hunt, "The application of constrained least squares estimation to image restoration by digital computers," *IEEE Trans. Comput.* **C-22**, 805–812 (1973).

[10] M. G. Kang and A. K. Katsaggelos, "Frequency domain adaptive iterative image restoration and evaluation of the regularization parameter," *Opt. Eng.* **33**, 3222–3232 (1994).

[11] M. G. Kang and A. K. Katsaggelos, "General choice of the regularization functional in regularized image restoration," *IEEE Trans. Image Process.* **4**, 594–602 (1995).

[12] A. K. Katsaggelos, "Iterative image restoration algorithms," *Opt. Eng.* **28**, 735–748 (1989).

[13] A. K. Katsaggelos and S. N. Efstratiadis, "A class of iterative signal restoration algorithms," *IEEE Trans. Acoust. Speech Signal Process.* **38**, 778–786 (1990).

[14] A. K. Katsaggelos, J. Biemond, R. M. Mersereau, and R. W. Schafer, "A regularized iterative image restoration algorithm," *IEEE Trans. Signal Process.* **39**, 914–929 (1991).

[15] A. K. Katsaggelos, Ed., *Digital Image Restoration*, Springer Series in Information Sciences, Vol. 23 (Springer-Verlag, Heidelberg, 1991).

[16] A. K. Katsaggelos, "Iterative image restoration algorithms," in V. K. Madisetti and D. B. Williams, eds., *The Digital Signal Processing Handbook* (CRC and IEEE, New York, 1998).

[17] J. M. Ortega and W. C. Rheinboldt, *Iterative Solution of Nonlinear Equations in Several Variables* (Academic, New York, 1970).

[18] R. W. Schafer, R. M. Mersereau, and M. A. Richards, "Constrained iterative restoration algorithms," *Proc. IEEE* **69**, 432–450 (1981).

[19] A. N. Tikhonov and V. Y. Arsenin, *Solution of Ill-Posed Problems* (Wiley, Winston, 1977).

[20] P. H. Van Citttert, "Zum einfluss der spaltbreite auf die intensitatswerteilung in Spektrallinien II," *Z. Phys.* **69**, 298–308 (1931).

3.10

Motion Detection and Estimation

Janusz Konrad

INRS Télécommunications

1 Introduction

A video sequence is a much richer source of visual information than a still image. This is primarily due to the capture of *motion*; while a single image provides a snapshot of a scene, a sequence of images registers the dynamics in it. The registered motion is a very strong cue for human vision; we can easily recognize objects as soon as they move even if they are inconspicuous when still. Motion is equally important for video processing and compression for two reasons. First, motion carries a lot of information about spatiotemporal relationships between image objects. This information can be used in such applications as traffic monitoring or security surveillance, for example to identify objects entering or leaving the scene or objects that just moved. Secondly, image properties, such as intensity or color, have a very high correlation in the direction of motion, i.e., they do not change significantly when tracked in the image (the color of a car does not change as the car moves across the image). This can be used for the removal of temporal video redundancy; in an ideal situation only the first image and the subsequent motion have to be transmitted. It can be also used for general temporal filtering of video. In this case, one-dimensional temporal filtering along a motion trajectory, e.g., for noise reduction or temporal interpolation, does not affect the spatial detail in the image.

The above applications require that image points be identified as moving or not (surveillance), or that it be measured how they move (compression, filtering). The first task is often referred to as *motion detection*, whereas the latter is referred to as *motion estimation*. The goal of this chapter is to present today's most promising approaches to solving both. Note that only two-dimensional (2-D) motion of intensity patterns in the image plane, often referred to as *apparent motion*, will be considered. Three-dimensional (3-D) motion of objects will not be treated here. *Motion segmentation*, i.e., the identification of groups of image points moving similarly, is treated in Chapter 4.9.

The discussion of motion in this chapter will be carried out from the point of view of video processing and compression. Necessarily, the scope of methods reported will not be complete. To present the methods in a consistent fashion, a classification will be made based on *models*, *estimation criteria*, and *search strategies* used. This classification will be introduced for two reasons. First, it is essential for the understanding of methods described here and elsewhere in the literature. Second, it should help the reader in the development of his or her own motion detection or estimation method.

Although motion detection and estimation still require specialized hardware for real-time execution, the present rapid growth of CPU power available in a personal computer will soon allow execution of motion-related tasks in software on a general CPU. This will certainly spawn new applications and an even greater need for robust, flexible, and fast motion detection and estimation algorithms. Hopefully, in designing a new algorithm or understanding an existing one, the reader will be able to exploit the variety of tools presented here.

The chapter is organized as follows. In the next section, the notation is established, followed by a brief review of some tools needed. Then, in Section 3, motion detection is discussed from the point of view of hypothesis testing and maximum *a posteriori* probability (MAP) detection. In Section 4, motion estimation is described in two parts. First, models, estimation criteria, and search strategies are discussed. Then, five motion estimation algorithms are described in more detail, of which three are based on models supported by the current video compression standards. Both motion detection and estimation are illustrated by numerous experimental results.

2 Notation and Preliminaries

In this chapter, both continuous and discrete representations of motion and images will be used, with bold characters denoting vectors. Let $\boldsymbol{x} = (x, y)^T$ be a spatial position of a pixel in continuous coordinates, i.e., $\boldsymbol{x} \in R^2$ within image limits, and let I_t denote image intensity at time t. Then, $I_t(\boldsymbol{x}) \in R$ is limited by the dynamic range of the sensing device (e.g., vidicon, CCD). Before images can be manipulated digitally, they have to be sampled and quantized. Let $\boldsymbol{n} = (n_1, n_2)^T$ be a discretized spatial position in the image that corresponds to $\boldsymbol{x}$. Similarly, let k be a discretized position in time, also denoted t_k. The triplet $(n_1, n_2, k)^T$ belongs to a 3-D sampling grid, for example a 3-D lattice (Chapter 7.2). It is assumed here that images are either continuous or discrete simultaneously in position and in amplitude. Consequently, the same symbol I will be used for continuous *and* quantized intensities; the nature of I can be inferred from its argument (continuous-valued for $\boldsymbol{x}$ and quantized for $\boldsymbol{n}$).

Motion in continuous images can be described by a *velocity* vector $\boldsymbol{v} = (v_1, v_2)^T$. Whereas $\boldsymbol{v}(\boldsymbol{x})$ is a velocity at spatial position $\boldsymbol{x}$, $\boldsymbol{v}_t$ will denote a *velocity field* or *motion field*, i.e., the set of all velocity vectors within the image, at time t. Often the computation of this *dense* representation is replaced by the computation of a small number of motion parameters $\boldsymbol{b}_t$ with the benefit of reduced computational complexity. Then, $\boldsymbol{v}_t$ is approximated by $\boldsymbol{b}_t$ by means of a known transformation. For discrete images, the notion of velocity is replaced by *displacement* $\boldsymbol{d}$.

2.1 Hypothesis Testing

Let y be an *observation* and let Y be the associated random variable. Suppose that there are two hypotheses H_0 and H_1 with

corresponding probability densities $p_Y(y \mid H_0)$ and $p_Y(y \mid H_1)$, respectively. The goal is to decide from which of the two densities a given y is more likely to have been drawn. Clearly, four possibilities exist (hypothesis/decision): H_0/H_0, H_0/H_1, H_1/H_0, H_1/H_1. Whereas H_0/H_0 and H_1/H_1 correspond to correct choices, H_0/H_1 and H_1/H_0 are erroneous. In order to make a decision, a *decision criterion* is needed that attaches some relative importance to the four possible scenarios.

Under the *Bayes criterion*, two a priori probabilities P_0 and $P_1 = 1 - P_0$ are assigned to the two hypotheses H_0 and H_1, respectively, and a cost is assigned to each of the four scenarios listed above. Naturally, one would like to design a decision rule so that on average the cost associated with making a decision based on y is minimal. By computing an average risk and by assuming that costs associated with erroneous decisions are higher than those associated with the corresponding correct decisions, one can show that an optimal decision can be made according to the following rule [24, Chapter 2]:

$$\frac{p_Y(y \mid H_1)}{p_Y(y \mid H_0)} \overset{H_1}{\underset{H_0}{\gtrless}} \theta \frac{P_0}{P_1}. \tag{1}$$

The quantity on the left is called the *likelihood ratio* and θ is a constant dependent on the costs of the four scenarios. Since these costs are determined in advance, θ is a fixed threshold. If P_0 and P_1 are predetermined as well, the above hypothesis test compares the likelihood ratio with a given threshold. Alternatively, the prior probabilities can be made variable; variable-threshold hypothesis testing results.

2.2 Markov Random Fields

A Markov random field (MRF) is a multidimensional random process that generalizes the notion of a one-dimensional (1-D) Markov process. Below, some essential properties of MRFs are described; for more details the reader is referred to Chapter 4.3 and to the literature (e.g., [9] and references therein).

Let Λ be a sampling grid in R^N and let $\eta(\boldsymbol{n})$ be a *neighborhood* of $\boldsymbol{n} \in \Lambda$, i.e., a set of such $\boldsymbol{n}$'s that $\boldsymbol{n} \notin \eta(\boldsymbol{n})$ and $\boldsymbol{n} \in \eta(\boldsymbol{l}) \Leftrightarrow \boldsymbol{l} \in \eta(\boldsymbol{n})$. The first-order neighborhood consists of the immediate top, bottom, left, and right neighbors of $\boldsymbol{n}$. Let $\mathcal{N}$ be a *neighborhood system*, i.e., a collection of neighborhoods of all $\boldsymbol{n} \in \Lambda$.

A random field Υ over Λ is a multidimensional random process such that each site $\boldsymbol{n} \in \Lambda$ is assigned a random variable. A random field Υ with the following properties,

$$P(\Upsilon = \upsilon) > 0, \quad \forall \upsilon \in \Gamma, \text{ and}$$

$$P(\Upsilon_{\boldsymbol{n}} = \upsilon_{\boldsymbol{n}} \mid \Upsilon_l = \upsilon_l, \forall l \neq \boldsymbol{n})$$

$$= P(\Upsilon_{\boldsymbol{n}} = \upsilon_{\boldsymbol{n}} \mid \Upsilon_l = \upsilon_l, \forall l \in \eta(\boldsymbol{n})), \quad \forall \boldsymbol{n} \in \Lambda, \forall \upsilon \in \Gamma,$$

where P is a probability measure, is called a Markov random field with state space Γ.

In order to define the Gibbs distribution, the concepts of clique and potential function are needed. A *clique c* defined over Λ with respect to $\mathcal{N}$ is a subset of Λ such that either c consists of a single site or every pair of sites in c are neighbors, i.e., belong to η. The set of all cliques is denoted by C. Examples of a two-element spatial clique $\{n, l\}$ are two immediate horizontal, vertical, or diagonal neighbors. A *Gibbs distribution* with respect to Λ and $\mathcal{N}$ is a probability measure π on Γ such that

$$\pi(v) = \frac{1}{Z} e^{-U(v)/T},$$

where the constants Z and T are called the *partition function* and *temperature*, respectively, and the *energy function U* is of the form

$$U(v) = \sum_{c \in C} V(v, c).$$

$V(v, c)$ is called a *potential function*, and depends only on the value of v at sites that belong to the clique c.

The equivalence between Markov random fields and Gibbs distributions is provided through the important *Hammersley–Clifford theorem*, which states that Υ is a MRF on Λ with respect to $\mathcal{N}$ if and only if its probability distribution is a Gibbs distribution with respect to Λ and $\mathcal{N}$. The equivalence between MRFs and Gibbs distributions results in a straightforward relationship between qualitative properties of a MRF and its parameters by means of the potential functions V. Extension of the Hammersley–Clifford theorem to vector MRFs is straightforward (a new definition of a state is needed).

2.3 MAP Estimation

Let Y be a random field of observations and let Υ be a random field that we want to estimate based on Y. Let y, v be their respective realizations. For example, y could be a difference between two images while v could be a field of motion detection labels. In order to compute v based on y, a powerful tool is the MAP estimation, expressed as follows:

$$\hat{v} = \arg\max_v P(\Upsilon = v \mid y)$$
$$= \arg\max_v P(Y = y \mid v) P(\Upsilon = v), \quad (2)$$

where $\max_v P(\Upsilon = v \mid y)$ denotes the maximum of the posterior probability $P(\Upsilon = v \mid y)$ with respect to v and arg denotes the argument of this maximum, i.e., such $\hat{v}$ that $P(\Upsilon = \hat{v} \mid y) \geq P(\Upsilon = v \mid y)$ for any v. Above, the Bayes rule was used and $P(Y = y)$ was omitted since it does not depend on v. If $P(\Upsilon = v)$ is the same for all realizations v, then only the likelihood $P(Y = y \mid v)$ is maximized, resulting in the *maximum likelihood* (ML) estimation.

3 Motion Detection

Motion detection is, arguably, the simplest of the three motion-related tasks, i.e., detection, estimation, and segmentation. Its goal is to identify which image points, or more generally which regions, of the image have moved between two time instants. As such, motion detection applies only to images acquired with a static camera. However, if camera motion can be counteracted, e.g., by global motion estimation and compensation, then the method equally applies to images acquired with a moving camera [14, Chapter 8].

It is essential to realize that the motion of image points is not perceived directly but rather through intensity changes. However, such intensity changes over time may be also induced by camera noise or illumination changes. Moreover, object motion itself may induce small intensity variations or even none at all. The latter will happen in the rare case of exactly constant luminance and color within the object. Clearly, motion detection from time-varying images is not as easy as may seem initially.

3.1 Hypothesis Testing with a Fixed Threshold

Fixed-threshold hypothesis testing belongs to the simplest motion detection algorithms, as it requires very few arithmetic operations. Several early motion detection methods belong to this class, although originally they were not developed as such.

Let $H_{\mathcal{M}}$ and $H_{\mathcal{S}}$ be two hypotheses declaring an image point at n as moving ($\mathcal{M}$) and stationary ($\mathcal{S}$), respectively. Let us assume that $I_k(n) = I_{k-1}(n) + q$ and that q is a noise term (Chapter 4.5), zero-mean Gaussian with variance σ^2 in stationary areas and zero-mean uniformly distributed in $[-L, L]$ in moving areas. The motivation is that in stationary areas only camera noise will distinguish same-position pixels at t_{k-1} and t_k, while in moving areas this difference is attributed to motion and therefore is unpredictable. Then, let

$$\rho_k(n) = I_k(n) - I_{k-1}(n)$$

be an observation [Eq. (1)] upon which we intend to select one of the two hypotheses. With the above assumptions and after taking the natural logarithm of both sides of Eq. (1), we can write the hypothesis test as follows:

$$\rho_k^2(n) \underset{\mathcal{S}}{\overset{\mathcal{M}}{\gtrless}} \theta, \quad (3)$$

where the threshold θ equals $2\sigma^2 \ln[(2L P_{\mathcal{S}})/(\sqrt{2\pi\sigma^2} P_{\mathcal{M}})]$. A similar test can be derived for a Laplacian-distributed noise term q; in Eq. (3) $\rho_k^2(n)$ is replaced by $|\rho_k(n)|$ and θ is computed accordingly. Such a test was used in the early motion detection algorithms. Note that both the Laplacian and Gaussian cases are equivalent under the appropriate selection of θ. Although θ

includes the prior probabilities, they are usually fixed in advance as is the noise variance, and thus the test is parameterized by one constant.

The above pixel-based hypothesis test is not robust to noise in the image; for small θ's "noisy" detection masks result (many isolated small regions or even pixels), whereas for large θ's only object boundaries and the most textured parts are detected. To attenuate the impact of noise, the method can be extended by measuring the temporal differences over a spatial window $\mathcal{W}$ with N points:

$$\frac{1}{N} \sum_{n \in \mathcal{W}} \rho_k^2(n) \underset{\mathcal{S}}{\overset{\mathcal{M}}{\gtrless}} \theta.$$

This approach exploits the fact that typical variations of camera characteristics, such as camera noise, can be closely approximated by an additive white noise model; by averaging over $\mathcal{W}$ the noise impact is reduced. Still, the method is not very robust and therefore is usually followed by some postprocessing (such as median filtering, suppression of small regions, etc.). Moreover, since the classification at position n is done based on all pixels within $\mathcal{W}$, the resolution of the method is reduced; a moving image point affects the decision of many of its neighbors. This method can be further improved by modeling the intensities I_{k-1} and I_k within $\mathcal{W}$ by a polynomial function, e.g., linear or quadratic [11].

The methods discussed thus far cannot deal with illumination changes between t_{k-1} and t_k; any intensity change caused by illumination variation is interpreted as motion. In the case of a global illumination change (across the whole image) a normalization of intensities can be used to match the second-order statistics in I_{k-1} and I_k. By allowing a linear transformation $\hat{I}_k(n) = a I_k(n) + b$, one can find coefficients a and b so that such statistics at t_k and t_{k-1} are equal, i.e., $\hat{\mu}_k = \mu_{k-1}$ and $\hat{\sigma}_k^2 = \sigma_{k-1}^2$ where $\hat{\mu}$ and $\hat{\sigma}$ are, respectively, mean and variance of the normalized image and μ and σ are those for the original image. Solving for a and b yields

$$\hat{I}_k(n) = \frac{\sigma_{k-1}}{\sigma_k} (I_k(n) - \mu_k) + \mu_{k-1}.$$

This transformation helps in the case of global or almost-global illumination change only. However, a typical illumination change varies with n and moreover is often localized. The above normalization can be made adaptive on a region-by-region basis, but for fixed regions (e.g., blocks) the improvements are marginal and it is unclear how to adapt the region shape.

A different approach to handling illumination change is to compare intensity gradients rather than intensities themselves, i.e., to construct the following test:

$$\frac{1}{N} \sum_{n \in \mathcal{W}} \| \nabla I_k(n) - \nabla I_{k-1}(n) \| \underset{\mathcal{S}}{\overset{\mathcal{M}}{\gtrless}} \theta,$$

where $\nabla = (\partial/\partial x, \partial/\partial y)^T$ is the spatial gradient and $\| \cdot \|$ is a suitable norm, e.g., Euclidean or city-block distance. This approach, applied to the polynomial-based intensity model [11], has been shown to increase robustness in the presence of illumination changes [18]. However, the method can handle the multiplicative nature of illumination only approximately (the intensity gradient above is not invariant under intensity scaling). To address this issue in more generality, shading models extensively used in computer graphics must be employed [18].

3.2 Hypothesis Testing with Adaptive Threshold

The motion detection methods presented thus far were based solely on image intensities and made no *a priori* assumptions about the nature of moving areas. However, moving 3-D objects usually create compact, closed boundaries in the image plane, i.e., if an image point is declared moving, it is likely that its neighbor is moving as well (unless the point is on a boundary) and the boundary is smooth rather than rough. To take advantage of this *a priori* information, hypothesis testing can be combined with Markov random field models.

Let E_k be a MRF of all labels assigned at time t_k, and let e_k be its realization. Let us assume for the time being that $e_k(n)$ is known for all n except l. Since the estimation process is iterative, this assumption is not unreasonable; previous estimates are known at $n \neq l$. Thus, the estimation process is reduced to a decision between $e_k(l) = \mathcal{M}$ and $e_k(l) = \mathcal{S}$. Let the label field resulting from $e_k(l) = \mathcal{M}$ be denoted by $e_k^{\mathcal{M}}$ and that produced by $e_k(l) = \mathcal{S}$ be $e_k^{\mathcal{S}}$. Then, based on Eq. (1) the decision rule for $e_k(l)$ can be written as follows:

$$\frac{p(\rho_k \mid e_k^{\mathcal{M}})}{p(\rho_k \mid e_k^{\mathcal{S}})} \underset{\mathcal{S}}{\overset{\mathcal{M}}{\gtrless}} \theta \frac{P(E_k = e_k^{\mathcal{S}})}{P(E_k = e_k^{\mathcal{M}})}, \qquad (4)$$

where P is a probability distribution governing the MRF E_k. By making the simplifying assumption that the temporal differences $\rho_k(n)$ are conditionally independent, i.e., $p(\rho_k \mid e_k) = \prod_n p(\rho_k(n) \mid e_k(n))$, we can further rewrite Eq. (4) as

$$\frac{p(\rho_k(l) \mid H_{\mathcal{M}})}{p(\rho_k(l) \mid H_{\mathcal{S}})} \underset{\mathcal{S}}{\overset{\mathcal{M}}{\gtrless}} \theta \frac{P(E_k = e_k^{\mathcal{S}})}{P(E_k = e_k^{\mathcal{M}})}. \qquad (5)$$

The hypothesis $H_{\mathcal{M}}$ means that $e_k(l) = \mathcal{M}$ and $H_{\mathcal{S}}$ means that $e_k(l) = \mathcal{S}$. All constituent probability densities from the left-hand side of Eq. (4), except at l, cancel out since $e_k^{\mathcal{M}}$ and $e_k^{\mathcal{S}}$ differ only at l. Although the conditional independence assumption is reasonable in stationary areas (temporal differences are mostly due to camera noise), it is less so in the moving areas. However, a convincing argument based on experimental results can be made in favor of such independence [1].

To increase the detection robustness to noise, the temporal differences should be pooled together, for example within a spatial window $\mathcal{W}_l$ centered at l. This leads to the evaluation of the likelihood for all ρ_k within $\mathcal{W}_l$ given the hypothesis $H_\mathcal{M}$ or $H_\mathcal{S}$ at l. Under the assumption of zero-mean Gaussian density p with variances $\sigma_\mathcal{M}^2$ and $\sigma_\mathcal{S}^2$ for $H_\mathcal{M}$ and $H_\mathcal{S}$, respectively, and assuming that $\sigma_\mathcal{M}^2 \gg \sigma_\mathcal{S}^2$, the final hypothesis becomes:

$$\sum_{n \in \mathcal{W}_l} \rho_k^2(\boldsymbol{n}) \underset{\mathcal{S}}{\overset{\mathcal{M}}{\gtrless}} 2\sigma_\mathcal{S}^2 \left(-\ln\theta + N\ln\frac{\sigma_\mathcal{M}}{\sigma_\mathcal{S}} + \ln\frac{P\left(E_k = e_k^\mathcal{S}\right)}{P\left(E_k = e_k^\mathcal{M}\right)} \right),$$

(6)

where N is the number of pixels in $\mathcal{W}_l$. In case the *a priori* probabilities are identical or fixed (independent of the realization e_k), the overall threshold depends only on model variances. Then, for increasing $\sigma_\mathcal{S}^2$, the overall threshold rises as well, thus discouraging $\mathcal{M}$ labels (as camera noise increases only large temporal differences should induce moving labels). Conversely, for decreasing $\sigma_\mathcal{S}^2$, the threshold falls, thus biasing the decision toward moving labels. In the limit, as $\sigma_\mathcal{S}^2 \to 0$, the threshold becomes 0; for a noiseless camera even the slightest temporal difference will induce a moving label.

By suitably defining the *a priori* probabilities one can adapt the threshold in response to the properties of e_k. Since the required properties are object compactness and smoothness of its boundaries, a simple MRF model supported on a first-order neighborhood [9] with two-element cliques $c = \{\boldsymbol{n}, \boldsymbol{l}\}$ and the following potential function,

$$V_{nl} = \begin{cases} 0 & \text{if } e_k(\boldsymbol{n}) = e_k(\boldsymbol{l}) \\ \beta & \text{if } e_k(\boldsymbol{n}) \neq e_k(\boldsymbol{l}) \end{cases},$$

(7)

is appropriate. Whenever a neighbor of $\boldsymbol{n}$ has a different label than $e_k(\boldsymbol{n})$, a penalty $\beta > 0$ is incurred; summed over the whole field it is proportional to the length of the moving mask boundary. Thus, the resulting prior (Gibbs) probability

$$P(E_k = e_k) = \frac{1}{Z} \exp\left(-\frac{1}{T} \sum_{\{n,l\}} V_{nl} \right)$$

will increase for configurations with smooth boundaries and will reduce for those with rough boundaries. More advanced models, for example, based on second-order neighborhood systems with diagonal cliques, can be used similarly [1].

Note that the MRF model facilitates the use of adaptive thresholds: if $P(E_k = e_k^\mathcal{S}) > P(E_k = e_k^\mathcal{M})$, the fixed part of the threshold in Eq. (6), i.e., the first two terms, will be augmented by a positive number, thus biasing the decision toward a static label. Conversely, for $P(E_k = e_k^\mathcal{S}) < P(E_k = e_k^\mathcal{M})$, the bias is in favor of a moving label.

3.3 MAP Detection

The MRF label model introduced in the previous section can be combined with another Bayesian criterion, namely the MAP

criterion (Section 2.3). To find the MAP estimate of the label field E_k, we must maximize the posterior probability $P(E_k = e_k \mid \rho_k)$, or its Bayes equivalent $p(\rho_k | e_k) P(E_k = e_k)$.

Let us consider the likelihood $p(\rho_k | e_k)$. One of the questionable assumptions made in the previous section was the conditional independence of the ρ_k given e_k [Eq. (5)]. To alleviate this problem, let $|I_k(\boldsymbol{n}) - I_{k-1}(\boldsymbol{n})|$ be an observation modeled as $\rho_k(\boldsymbol{n}) = \xi(e_k(\boldsymbol{n})) + q(\boldsymbol{n})$ where q is zero-mean uncorrelated Gaussian noise with variance σ^2 and

$$\xi(e_k(\boldsymbol{n})) = \begin{cases} 0 & \text{if } e_k(\boldsymbol{n}) = \mathcal{S} \\ \alpha & \text{if } e_k(\boldsymbol{n}) = \mathcal{M} \end{cases}$$

Above, α is considered to be an average of the observations in moving areas. For example, α could be computed as an average temporal difference for previous-iteration moving labels e_k or previous-time moving labels e_{k-1}. Clearly, ξ attempts to closely model the observations since for a static image point it is zero, while for a moving point it tracks average temporal intensity mismatch; the uncorrelated q should be a better approximation here than in the previous section.

Under the uncorrelated Gaussian assumption for the likelihood $p(\rho_k | e_k)$ and a Gibbs distribution for the *a priori* probability $P(E_k = e_k)$, the overall energy function can be written as follows:

$$\begin{aligned} U(\rho_k, e_{k-1}, e_k) = &\frac{1}{2\sigma^2} \sum_n (\rho_k(\boldsymbol{n}) - \xi(e_k(\boldsymbol{n})))^2 \\ &+ \sum_{\{n,l\}} V_s(e_k(\boldsymbol{n}), e_k(\boldsymbol{l})) \\ &+ \sum_{\{t_{k-1}, t_k\}} V_t(e_{k-1}(\boldsymbol{n}), e_k(\boldsymbol{n})). \end{aligned}$$

The first term measures how well the current label $e_k(\boldsymbol{n})$ explains the observation $\rho_k(\boldsymbol{n})$. The other terms measure how contiguous are the labels in the image plane (V_s) and in time (V_t). Both V_s and V_t are specified similarly to Eq. (7), thus favoring spatial and temporal similarity of the labels [6]. This basic model can be enhanced by selecting a more versatile model for ξ [6] or a more complete prior model for including spatiotemporal, as opposed to purely spatial and temporal, cliques [15]. The above cost function can be optimized by using various approaches, such as those discussed in Section 4.3: simulated annealing, iterated conditional modes, or highest confidence first. The latter method, based on an adaptive selection of visited labels according to their impact on the energy U (most influential visited first), gives the best compromise between performance (final energy value) and computing time.

3.4 Experimental Comparison of Motion Detection Methods

Figure 1 shows the original images as well as the binary label fields obtained by using the MAP detection mechanism discussed in

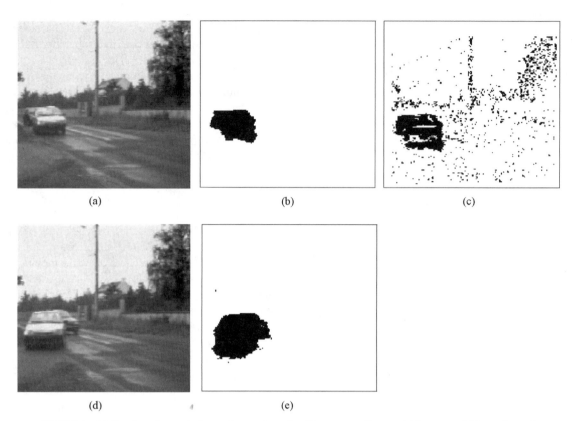

FIGURE 1   Motion detection results for two frames of a road traffic sequence of images: (a) frame 137; (b) MAP-detected motion for frame 137; (c) motion in frame 137 detected by the fixed-threshold hypothesis test; (d) frame 143; (e) MAP-detected motion for frame 143. (From Bouthemy and Lalande [6]. Reproduced with permission of the SPIE.)

Section 3.3, as well as those obtained by using the fixed-threshold hypothesis test Eq. (3). Note the compactness of the detection mask and the smoother boundary in case of the MAP estimate as opposed to the noisy detection result obtained by thresholding. The improvement is due primarily to the *a priori* knowledge incorporated into the algorithm.

4 Motion Estimation

As mentioned in the introduction, knowledge of motion is essential for both the compression and processing of image sequences. Although compression is often considered to be encompassed by processing, a clear distinction between these two terms will be made here. Methods explicitly reducing the number of bits needed to represent a video sequence will be classified as video compression techniques. For example, motion-compensated hybrid (predictive/DCT) coding is exploited today in all video compression standards, i.e., H.261, H.263, MPEG-1, MPEG-2, MPEG-4 (Chapters 6.4 and 6.5). In contrast, methods that do not attempt such a reduction but transform the video sequence, e.g., to improve quality, will be considered to belong to video processing methods. Examples of video processing are motion-compensated noise reduction (Chapter 3.11),

motion-compensated restoration (Chapter 3.11), and motion-based video segmentation (Chapter 4.9).

The above classification is important from the point of view of the goals of motion estimation, which in turn influence the choice of models and estimation criteria. In the case of video compression, the estimated motion parameters should lead to the highest compression ratio possible (for a given video quality). Therefore, the computed motion need not resemble the *true* motion of image points as long as some minimum bit rate is achieved. In video processing, however, it is the *true* motion of image points that is sought. For example, in motion-compensated temporal interpolation (Fig. 2) the task is to compute new images located between the existing images of a video sequence (e.g., video frame rate conversion between NTSC and PAL scanning standards). In order for the new images to be consistent with the existing ones, image points must be displaced according to their true motion, as otherwise a "jerky" motion of objects would result. This is a very important difference that influences the design of motion estimation algorithms and, most importantly, that usually precludes a good performance of a compression-optimized motion estimation algorithm in video processing and vice versa.

In order to develop a motion estimation algorithm, three important elements have to be considered: *models, estimation*

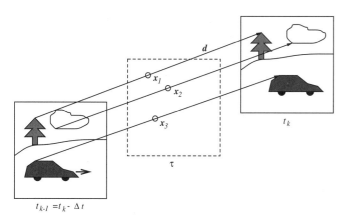

$t_{k-1} = t_k - \Delta t$

FIGURE 2 Motion-compensated interpolation between images at time $t_k - \Delta t$ and t_k. Motion compensation is essential for smooth rendition of moving objects. Shown are three motion vectors that map the corresponding image points at time $t_k - \Delta t$ and t_k onto image at time τ.

criteria, and *search strategies*. They will be discussed next, but no attempt will be made to include an exhaustive list pertaining to each of them. Clearly, this cannot be considered a universal classification scheme of motion estimation algorithms, but it is very useful in understanding the properties and merits of various approaches. Then, five practical motion estimation algorithms will be discussed in more detail.

4.1 Motion Models

There are two essential models in motion estimation: a *motion model*, i.e., how to represent motion in an image sequence, and a model relating motion parameters to image intensities, called an *observation model*. The latter model is needed since, as mentioned before, the computation of motion is carried out indirectly by the examination of intensity changes.

Spatial Motion Models

The goal is to estimate the motion of image points, i.e., the *2-D motion* or *apparent motion*. Such a motion is a combination of projections of the motion of objects in a 3-D scene and of 3-D camera motion. Whereas camera motion affects the movements of all or almost all image points, the motion of 3-D objects only affects a subset of image points corresponding to objects' projections. Since, in principle, the camera-induced motion can be compensated for by either estimating it (Section 5.1) or by physically measuring it at the camera, we need to model the object-induced motion only. Such a motion depends on

- the image formation model, e.g., perspective, orthographic projection [23],
- the motion model of the 3-D object, e.g., rigid-body with 3-D translation and rotation, 3-D affine motion, and
- the surface model of the 3-D object, e.g., planar, parabolic.

In general these relationships are fairly complex, but two cases are relatively simple and have been used extensively in practice.

For an orthographic projection and arbitrary 3-D surface undergoing 3-D translation, the resulting 2-D instantaneous velocity at position x in the image plane is described by a 2-D vector:

$$\boldsymbol{v}(x) = \begin{pmatrix} b_1 \\ b_2 \end{pmatrix}, \qquad (8)$$

where parameters $\boldsymbol{b} = (b_1, b_2)^T = (v_1, v_2)^T$ depend on camera geometry and 3-D translation parameters. This 2-D translational model has proven very powerful in practice, especially in video compression, since locally it provides a close approximation for most natural images.

The second powerful, yet simple, parametric model is that of orthographic projection combined with 3-D affine motion of a planar surface. It leads to the following six-parameter *affine* model [22, Chapter 6]:

$$\boldsymbol{v}(x) = \begin{pmatrix} b_1 \\ b_2 \end{pmatrix} + \begin{pmatrix} b_3 & b_4 \\ b_5 & b_6 \end{pmatrix} x, \qquad (9)$$

where, again, $\boldsymbol{b} = (b_1, \ldots, b_6)^T$ is a vector of parameters related to the camera as well as 3-D surface and motion parameters. Clearly, the translational model above is a special case of the affine model. More complex models have been proposed as well but, depending on the application, they do not always improve the precision of motion estimates. In general, the higher the number of motion parameters, the more precise the description of motion. However, an excessive number of parameters may be detrimental to the performance. This depends on the number of degrees of freedom, i.e., model complexity (dimensionality of $\boldsymbol{b}$ and the functional dependence of $\boldsymbol{v}$ on x, y) versus the size of the region of support (see below). A complex model applied to a small region of support may lead to an actual increase in the estimation error compared to a simpler model such as in Eq. (9).

Temporal Motion Models

The trajectories of individual image points drawn in the (x, y, t) space of an image sequence can be fairly arbitrary since they depend on object motion. In the simplest case, trajectories are linear, such as the ones shown in Fig. 2. Assuming that the velocity $\boldsymbol{v}_t(x)$ is constant between $t = t_{k-1}$ and τ ($\tau > t$), a linear trajectory can be expressed as follows:

$$x(\tau) = x(t) + \boldsymbol{v}_t(x)(\tau - t) = x(t) + \boldsymbol{d}_{t,\tau}(x), \qquad (10)$$

where $\boldsymbol{d}_{t,\tau}(x) = \boldsymbol{v}_t(x) \cdot (\tau - t)$ is a displacement vector[1] measured in the positive direction of time, i.e., from t to τ. Consequently, for linear motion the task is to find the two components of the velocity $\boldsymbol{v}$ or displacement $\boldsymbol{d}$ for each x. This simple motion model embedding the two-parameter spatial model of Eq. (8) has proven to be a powerful motion estimation tool in practice.

---

[1] In the sequel, the dependence of $\boldsymbol{d}$ on t and τ will be dropped whenever it is clear between what time instants $\boldsymbol{d}$ applies.

A natural extension of the linear model is a quadratic trajectory model, accounting for acceleration of image points, which can be described by

$$x(\tau) = x(t) + v_t(x)(\tau - t) + \frac{1}{2}a_t(x)(\tau - t)^2. \qquad (11)$$

The model is based on two velocity (linear) variables and two acceleration (quadratic) variables $a = (a_1, a_2)^T$, thus accounting for second-order effects. This relatively new model has recently been demonstrated to greatly benefit such motion-critical tasks as frame rate conversion [7] because of its improved handling of variable-speed motion present in typical videoconferencing images (e.g., hand gestures, facial expressions).

The above models require two [Eq. (10)] or four [Eq. (11)] parameters at each position x. To reduce the computational burden, parametric (spatial) motion models can be combined with the temporal models above. For example, the affine model [Eq. (9)] can be used to replace v_t [Eq. (10)] and then applied over a suitable region of support. This approach has been successfully used in various region-based motion estimation algorithms. Recently, a similar parametric extension of the quadratic trajectory model (v_t and a_t replaced by affine expressions) has been proposed [14, Chapter 4], but its practical importance remains to be verified.

Region of Support

The set of points x to which a spatial and temporal motion model applies is called a *region of support*, denoted $\mathcal{R}$. The selection of a motion model and a region of support is one of the major factors determining the precision of the resulting motion parameter estimates. Usually, for a given motion model, the smaller the region of support $\mathcal{R}$, the better the approximation of motion. This is because over a larger area, motion may be more complicated and thus require a more complex model. For example, the translational model of Eq. (8) can fairly well describe the motion of one car in a highway scene, while this very model would be quite poor for the complete image. Typically, the region of support for a motion model belongs to one of the four types listed below. Figure 3 shows schematically each type of region.

1. $\mathcal{R}$ = the whole image: A single motion model applies to all image points. This model is suitable for the estimation of camera-induced motion (Section 5.1) as very few parameters describe the motion of all image points. This is the most constrained model (a relatively small number of motion fields can be represented), but with the fewest parameters to estimate.

2. $\mathcal{R}$ = one pixel: This model applies to a single image point (position x). Typically, the translational spatial model [Eq. (8)] is used jointly with the linear [Eq. (10)] or the quadratic temporal model [Eq. (11)]. This pixel-based or *dense* motion representation is the least constrained one since at least two parameters describe the movement of each image point. Consequently, a very large number of motion fields can be represented by all possible combinations of parameter values, but computational complexity is, in general, high.

3. $\mathcal{R}$ = rectangular block of pixels: This motion model applies to a rectangular (or square) block of image points. In the simplest case the blocks are nonoverlapping and their union covers the whole image. A spatially translational [Eq. (8)] and temporally linear [Eq. (10)] motion of a rectangular block of pixels has proven to be a very powerful model and is used today in all digital video compression standards, i.e., H.261, H.263, MPEG-1 and MPEG-2 (Chapters 6.4 and 6.5). It can be also argued that a spatially translational but temporally quadratic [Eq. (11)] motion has been implicitly exploited in the MPEG standards since in the B frames two independent motion vectors are used (two noncollinear motion vectors starting at x can describe both velocity and acceleration). Although very successful in hardware implementations, because of its simplicity, the translational model lacks precision for images with rotation, zoom, deformation, and is often replaced by the affine model [Eq. (9)].

4. $\mathcal{R}$ = irregularly shaped region: This model applies to all pixels in a region $\mathcal{R}$ of arbitrary shape. The reasoning is that for objects with a sufficiently smooth 3-D surface and 3-D motion, the induced 2-D motion can be closely approximated by the affine model [Eq. (9)] applied linearly

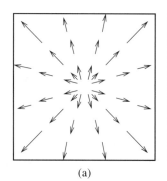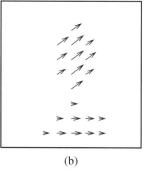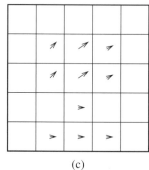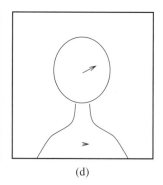

(a) (b) (c) (d)

FIGURE 3 Schematic representation of motion for the four regions of support $\mathcal{R}$: (a) whole image, (b) pixel, (c) block, and (d) arbitrarily shaped region. The implicit underlying scene is "head and shoulders" as captured by the region-based model in (d).

over time [Eq. (10)] to the image area arising from object projection. Thus, regions $\mathcal{R}$ are expected to correspond to object projections. This is the most advanced motion model that has found its way into standards; a square block divided into arbitrarily shaped parts, each with independent translational motion, is used in the MPEG-4 standard (Chapter 6.5).

Observation Models

Since motion is estimated (and observed by the human eye) based on the variations of intensity, color, or both, the assumed relationship between motion parameters and image intensity plays a very important role. The usual, and reasonable, assumption made is that image intensity remains constant along a motion trajectory, i.e., that objects do not change their brightness and color when they move. For temporally sampled images, this means that $I_k(\boldsymbol{x}(t_k)) = I_{k-1}(\boldsymbol{x}(t_{k-1}))$. Using the relationship [Eq. (10)] with $t = t_{k-1}$ and $\tau = t_k$, and assuming spatial sampling of the images, we can express this condition as follows:

$$I_k(\boldsymbol{n}) = I_{k-1}(\boldsymbol{n} - \boldsymbol{d}).\qquad(12)$$

Equation (12), however, cannot be used directly to solve for $\boldsymbol{d}$ since in practice it does not hold exactly because of noise q, aliasing, etc., present in the images, i.e., $I_k(\boldsymbol{n}) = I_{k-1}(\boldsymbol{n} - \boldsymbol{d}) + q(\boldsymbol{n})$. Therefore, $\boldsymbol{d}$ must be found by minimizing a function of the error between $I_k(\boldsymbol{n})$ and $I_{k-1}(\boldsymbol{n} - \boldsymbol{d})$. Moreover, Eq. (12) applied to a single image point $\boldsymbol{n}$ is insufficient since $\boldsymbol{d}$ contains at least two unknowns. Both issues will be treated in depth in the next section.

Let us consider now the continuous case. Let s be a variable along a motion trajectory. Then, the constant-intensity assumption translates into the following constraint equation:

$$\frac{dI}{ds} = 0,\qquad(13)$$

i.e., the directional derivative in the direction of motion is zero. By applying the chain rule, one can write the above equation as the well-known *motion constraint equation* [10],

$$\frac{\partial I}{\partial x}v_1 + \frac{\partial I}{\partial y}v_2 + \frac{\partial I}{\partial t} = (\nabla I)^T \boldsymbol{v} + \frac{\partial I}{\partial t} = 0,\qquad(14)$$

where $\nabla = (\partial/\partial x, \partial/\partial y)^T$ denotes the spatial gradient and $\boldsymbol{v} = (v_1, v_2)^T$ is the velocity to be estimated. The above constraint equation, whether in the above continuous form or as a discrete approximation, has served as the basis for many motion estimation algorithms. Note that, similarly to Eq. (12), Eq. (14) applied at single position (x, y) is underconstrained (one equation, two unknowns) and allows to determine the component of velocity $\boldsymbol{v}$ in the direction of the image gradient ∇I only [10]. Thus, additional constraints are needed in order to uniquely solve for $\boldsymbol{v}$ [10]. Also, Eq. (14) does not hold exactly for real images and usually a minimization of a function of $(\nabla I)^T \boldsymbol{v} + \partial I/\partial t$ is performed.

Since color is a very important attribute of images, a possible extension of the above models would be to include chromatic image components into the constraint equation. The assumption is that in the areas of uniform intensity but substantial color detail, the inclusion of a color-based constraint could prove beneficial. In such a case, Eqs. (13) and (14) would hold with a multicomponent (vector) function replacing I.

The assumption about intensity constancy is usually only approximately satisfied, but it is particularly violated when scene illumination changes. As an alternative, a constraint based on the spatial gradient's constancy in the direction of motion can be used [2]:

$$\frac{d\nabla I}{ds} = \vec{0}.\qquad(15)$$

This equation can be rewritten as follows:

$$\begin{bmatrix} \partial^2 I/\partial x^2 & \partial^2 I/\partial x \partial y \\ \partial^2 I/\partial x \partial y & \partial^2 I/\partial y^2 \end{bmatrix} \boldsymbol{v} + \frac{\partial(\nabla I)}{\partial t} = \vec{0}.\qquad(16)$$

It relaxes the constant-intensity assumption but requires that the amount of dilation or contraction, and rotation in the image be negligible,[2] a limitation often satisfied in practice. Although both Eqs. (15) and (16) are linear vector equations in two unknowns, in practice they do not lend themselves to the direct computation of motion, but need to be further constrained by a motion model. The primary reason for this is that Eq. (16) holds only approximately. Furthermore, it is based on second-order image derivatives that are difficult to compute reliably as a result of the high-pass nature of the operator; usually image smoothing must be performed first.

The constraints discussed above find different applications in practice. A discrete version of the constant-intensity constraint [Eq. (14)] is often applied in video compression since it yields small motion-compensated prediction error. Although motion can be computed also based on color using a vector equivalent of Eq. (14), experience shows that the small gains achieved do not justify the substantial increase in complexity. However, motion estimation from color data is useful in video processing tasks (e.g., motion-compensated filtering, resampling), in which motion errors may result in visible distortions. Moreover, the multicomponent (color) constraint is interesting for estimating motion from multiple data sources (e.g., range and intensity data).

4.2 Estimation Criteria

The models discussed have to be incorporated into an estimation criterion that will be subsequently optimized. There is no unique criterion for motion estimation since its choice depends on the task at hand. For example, in compression an average performance (prediction error) of a motion estimator is important,

---

[2]Even when the constant-intensity assumption is valid, the intensity gradient changes its amplitude under dilation or contraction and its direction under rotation.

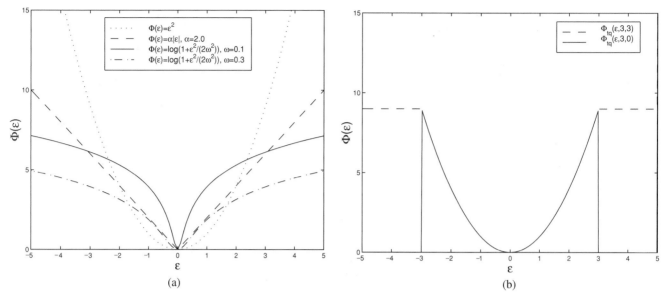

FIGURE 4   Comparison of estimation criteria: (a) quadratic, absolute value, Lorentzian functions (two different ω's); and (b) truncated-quadratic functions.

whereas in motion-compensated interpolation the worst case performance (maximum interpolation error) may be of concern. Moreover, the selection of a criterion may be guided by the processor capabilities on which the motion estimation will be implemented.

Pixel-Domain Criteria

Most of the criteria arising from the constant-intensity assumption [Eq. (12)] aim at the minimization of a function (e.g., absolute value) of the following error:

$$\varepsilon_k(\boldsymbol{n}) = I_k(\boldsymbol{n}) - \tilde{I}_k(\boldsymbol{n}), \quad \forall \boldsymbol{n} \in \Lambda \qquad (17)$$

where $\tilde{I}_k(\boldsymbol{n}) = I_{k-1}(\boldsymbol{n} - \boldsymbol{d}(\boldsymbol{n}))$ is called a motion-compensated prediction of $I_k(\boldsymbol{n})$. Since, in general, $\boldsymbol{d}$ is real valued, intensities at positions $\boldsymbol{n} - \boldsymbol{d}$ outside of the sampling grid Λ must be recovered by a suitable interpolation. For estimation methods based on matching, C^0 interpolators that ensure continuous interpolated intensity (e.g., bilinear) are sufficient, whereas for methods based on gradient descent, C^1 interpolators giving both continuous intensity *and* its derivative are preferable for stability reasons.

Motion fields calculated solely by the minimization of the prediction error are sensitive to noise if the number of pixels in $\mathcal{R}$ is not large compared to the number of motion parameters estimated or if the region is poorly textured. However, such a minimization may yield good estimates for parametric motion models with few parameters and reasonable region size.

A common choice for the estimation criterion is the following sum:

$$\mathcal{E}(\boldsymbol{d}) = \sum_{\boldsymbol{n} \in \mathcal{R}} \Phi(I_k(\boldsymbol{n}) - \tilde{I}_k(\boldsymbol{n})), \qquad (18)$$

where Φ is a nonnegative real-valued function. The often-

used quadratic function $\Phi(\varepsilon) = \varepsilon^2$ is not a good choice since a single large error ε (an outlier) overcontributes to $\mathcal{E}$ and biases the estimate of $\boldsymbol{d}$. A more robust function is the absolute value $\Phi(\varepsilon) = \alpha|\varepsilon|$, since the cost grows linearly with error [Fig. 4(a)]. Since it does not require multiplications, it is the criterion of choice in practical video encoders today. An even more robust criterion is based on the Lorentzian function $\Phi(\varepsilon) = \log(1 + \varepsilon^2/2\omega^2)$ that grows slower than $|x|$ for large errors. The growth of the cost for increasing errors ε is adjusted by the parameter ω [a Lorentzian function for two different values of ω is shown in Fig. 4(a)].

Since for matching algorithms the continuity of Φ is not important (no gradient computations), non-continuous functions based on the concept of the truncated quadratic,

$$\Phi_{tq}(\varepsilon, \theta, \beta) = \begin{cases} \varepsilon^2 & |\varepsilon| < \theta \\ \beta & \text{otherwise} \end{cases}, \qquad (19)$$

are often used [Fig. 4(b)]. If $\beta = \theta^2$, the usual truncated quadratic results, fixing the cost of outliers at θ^2. An alternative is to set $\beta = 0$ with the consequence that the outliers have zero cost and do not contribute to the overall criterion $\mathcal{E}$. In other words, the criterion is defined only for nonoutlier pixels, and therefore the estimate of $\boldsymbol{d}$ will be computed solely on the basis of reliable pixels.

The similarity between $I_k(\boldsymbol{n})$ and its prediction $\tilde{I}_k(\boldsymbol{n})$ can be also measured by a cross-correlation function:

$$\mathcal{C}(\boldsymbol{d}) = \sum_{\boldsymbol{n}} I_k(\boldsymbol{n}) I_{k-1}(\boldsymbol{n} - \boldsymbol{d}(\boldsymbol{n})). \qquad (20)$$

Although more complex computationally than the absolute value criterion because of the multiplications, this criterion is an interesting and practical alternative to the prediction error-based criteria (Section 5.3). Note that a cross-correlation criterion requires maximization, unlike the prediction-based criteria.

For a detailed discussion of robust estimation criteria in the context of motion estimation, the reader is referred to the literature (e.g., [5] and references therein).

Frequency-Domain Criteria

Although the frequency-domain criteria are less used in practice today than the space/time-domain methods, they form an important alternative. Let $\hat{I}_k(\boldsymbol{u}) = \mathcal{F}[I_k(\boldsymbol{n})]$ be a spatial (2-D) discrete Fourier transform (DFT) of the intensity signal $I_k(\boldsymbol{n})$, where $\boldsymbol{u} = (u, v)^T$ is a 2-D discrete frequency (see Chapter 2.3). Suppose that the image I_{k-1} has been uniformly shifted to create the image I_k, i.e., that $I_k(\boldsymbol{n}) = I_{k-1}(\boldsymbol{n}-z)$. This means that only translational global motion exists in the image and all boundary effects are neglected. Then, by the shift property of the Fourier transform,

$$\mathcal{F}[I_{k-1}(\boldsymbol{n}-z)] = \hat{I}_{k-1}(\boldsymbol{u})e^{-j2\pi\boldsymbol{u}^T z}, \tag{21}$$

where $\boldsymbol{u}^T$ denotes a transposed vector $\boldsymbol{u}$. Since the amplitudes of both Fourier transforms are independent of z while the argument difference

$$\arg\{\mathcal{F}[I_k(\boldsymbol{n})]\} - \arg\{\mathcal{F}[I_{k-1}(\boldsymbol{n})]\} = -2\pi\boldsymbol{u}^T z$$

depends linearly on z, the global motion can be recovered by evaluating the phase difference over a number of frequencies and solving the resulting overconstrained system of linear equations. In practice, this method will work only for single objects moving across a uniform background. Moreover, the positions of image points to which the estimated displacement z applies are not known; this assignment must be performed in some other way. Also, care must be taken of the non-uniqueness of the Fourier phase function, which is periodic.

A Fourier-domain representation is particularly interesting for the cross-correlation criterion [Eq. (20)]. Based on the Fourier transform properties (Chapter 2.3) and under the assumption that the intensity function I is real valued, it is easy to show that:

$$\mathcal{F}[\mathcal{C}(\boldsymbol{d})] = \mathcal{F}\left[\sum_{\boldsymbol{n}} I_k(\boldsymbol{n})\,I_{k-1}(\boldsymbol{n}-\boldsymbol{d})\right] = \hat{I}_k(\boldsymbol{u})\,\hat{I}_{k-1}^*(\boldsymbol{u}), \tag{22}$$

where the transform is applied in spatial coordinates only and $\hat{I}^*$ is the complex conjugate of $\hat{I}$. This equation expresses spatial cross-correlation in the Fourier domain, where it can be efficiently evaluated by using the DFT.

Regularization

The criteria described thus far deal with the underconstrained nature of Eq. (14) by applying the motion measurement to either a region, such as a block of pixels, or to the whole image (frequency-domain criteria). In consequence, resolution of the computed motion may suffer.

To maintain motion resolution at the level of original images, the pixelwise motion constraint equation, Eq. (14), can be used, but to address its underconstrained nature, we must combine it with another constraint. In typical real-world images the moving objects are close to being rigid. Upon projection onto the image plane this induces very similar motion of neighboring image points within the object's projection area. In other words, the motion field is locally smooth. Therefore, a motion field $\boldsymbol{v}_t$ must be sought that satisfies the motion constraint of Eq. (14) as closely as possible and simultaneously is as smooth as possible. Since gradient is a good measure of local smoothness, this may be achieved by minimizing the following criterion [10]:

$$\mathcal{E}(\boldsymbol{v}) = \int_{\mathcal{D}} \left(\nabla^T I(\boldsymbol{x})\boldsymbol{v}(\boldsymbol{x}) + \frac{\partial I(\boldsymbol{x})}{\partial t} \right)^2$$
$$+ \lambda(\|\nabla(v_1(\boldsymbol{x}))\|^2 + \|\nabla(v_2(\boldsymbol{x}))\|^2)\,d\boldsymbol{x}, \tag{23}$$

where $\mathcal{D}$ is the domain of the image. This formulation is often referred to as *regularization* [2] (see also the discussion of regularized image recovery in Chapter 3.6). Note that the smoothness constraint may be also viewed as an alternative spatial motion model to those described in Section 4.1.

Bayesian Criteria

Bayesian criteria form a very powerful probabilistic alternative to the deterministic criteria described thus far. If motion field $\boldsymbol{d}_k$ is a realization of a vector random field $\boldsymbol{D}_k$ with a given *a posteriori* probability distribution, and image I_k is a realization of a scalar random field $\mathcal{I}_k$, then the MAP estimate of $\boldsymbol{d}_k$ (Section 2.3) can be computed as follows [12]:

$$\hat{\boldsymbol{d}}_k = \arg\max_{\boldsymbol{d}} P(\boldsymbol{D}_k = \boldsymbol{d}_k \,|\, \mathcal{I}_k = I_k; I_{k-1})$$
$$= \arg\max_{\boldsymbol{d}} P(\mathcal{I}_k = I_k \,|\, \boldsymbol{D}_k = \boldsymbol{d}_k; I_{k-1})\, P(\boldsymbol{D}_k = \boldsymbol{d}_k; I_{k-1}). \tag{24}$$

In this notation, the semicolon indicates that subsequent variables are only deterministic parameters. The first (conditional) probability distribution denotes the likelihood of image I_k given displacement field $\boldsymbol{d}_k$ and the previous image I_{k-1}, and therefore it is closely related to the observation model. In other words, this term quantifies how well a motion field $\boldsymbol{d}_k$ explains the change between the two images. The second probability $P(\boldsymbol{D}_k = \boldsymbol{d}_k; I_{k-1})$ describes the prior knowledge about the random field $\boldsymbol{D}_k$, such as its spatial smoothness, and therefore can be thought of as a motion model. It becomes particularly interesting when $\boldsymbol{D}_k$ is a MRF. By maximizing the product of the likelihood and the prior probabilities, one attempts to strike a balance between motion fields that give a small prediction error and those that are smooth. It will be shown in Section 5.5 that maximization [Eq. (24)] is equivalent to an energy minimization.

4.3 Search Strategies

Once models have been identified and incorporated into an estimation criterion, the last step is to develop an efficient (complexity) and effective (solution quality) strategy for finding the estimates of motion parameters.

For a small number of motion parameters and a small state space for each of them, the most common search strategy when minimizing a prediction error, like Eq. (17), is *matching*. In this approach, motion-compensated predictions $\tilde{I}_k(n) = I_{k-1}(n - d(n))$ for various motion candidates d are compared (matched) with the original images $I_k(n)$ within the region of support of the motion model (pixel, block, etc.). The candidate yielding the best match for a given criterion becomes the optimal estimate. For small state spaces, as is the case in block-constant motion models used in today's video coding standards, the full state space of each motion vector can be examined (an exhaustive search), but a partial search often gives almost as good results (Section 5.2).

As opposed to matching, *gradient-based techniques* require an estimation criterion $\mathcal{E}$ that is differentiable. Since this criterion depends on motion parameters by means of the image function, as in $I_{k-1}(n - d(n))$, to avoid nonlinear optimization I is usually linearized by using a Taylor expansion with respect to $d(n)$. Because of the Taylor approximation, the model is applicable only in a small vicinity of the initial d. Since initial motion is usually assumed to be zero, it comes as no surprise that gradient-based estimation is reported to yield accurate estimates only in regions of small motion; the approach fails if motion is large. This deficiency is usually compensated for by a hierarchical or multiresolution implementation [17, Chapter 1], (Chapter 4.2). An example of hierarchical gradient-based method is reported in Section 5.1.

For motion fields using a spatial noncausal model, such as that based on a MRF, the simultaneous optimization of thousands of parameters may be computationally prohibitive.[3] Therefore, *relaxation techniques* are usually employed to construct a series of estimates such that consecutive estimates differ in one variable at most. In case of estimating a motion field d, a series of motion fields $d^{(0)}, d^{(1)}, \ldots$ is constructed so that any two consecutive estimates $d^{(k-1)}, d^{(k)}$ differ at most at a single site n. At each step of the relaxation procedure the motion vector at a single site is computed; vectors at other sites remain unchanged. Repeating this process results in a propagation of motion properties, such as smoothness, that are embedded in the estimation criterion. Relaxation techniques are most often used in dense motion field estimation, but they equally apply to block-based methods.

In *deterministic relaxation*, such as Jacobi or Gauss–Seidel, each motion vector is computed with probability 1, i.e., there is no uncertainty in the computation process. For example, a new local estimate is computed by minimizing the given

criterion; variables are updated one after another and the criterion is monotonically improved step by step. Deterministic relaxation techniques are capable of correcting spurious motion vectors in the initial state $d^{(0)}$, but they often get trapped in a local optimum near $d^{(0)}$. Therefore, the availability of a good initial state is crucial.

The *highest confidence first* (HCF) algorithm [8] is an interesting variant of deterministic relaxation that is insensitive to the initial state. The distinguishing characteristic of the method is its site visiting schedule, which is not fixed but driven by the input data. Without going into the details, the HCF algorithm initially selects motion vectors that have the largest potential for reducing the estimation criterion $\mathcal{E}$. Usually, these are vectors in highly textured parts of an image. Later, the algorithm includes more and more motion vectors from low-texture areas, thus building on the neighborhood information of sites already estimated. By the algorithm's construction, the final estimate is independent of the initial state. The HCF is capable of finding close to optimal MAP estimates at a fraction of the computational cost of the globally optimal methods.

A deterministic algorithm specifically developed to deal with MRF formulations is called *iterated conditional modes* (ICMs) [3]. Although it does not maximize the *a posteriori* probability, it finds reasonably close approximations. The method is based on the division of sites of a random field into N sets such that each random variable associated with a site is independent of other random variables in the same set. The number of sets and their geometry depend on the selected cliques of the MRF. For example, for the first-order neighborhood system (Section 2.2), $N = 2$ and the two sets look like a chess board. First, all the sites of one set are updated to find the optimal solution. Then, the sites of the other set are examined with the state of the first set already known. The procedure is repeated until a convergence criterion is met. The method converges quickly but does not lead to as good solutions as the HCF approach.

The dependence on a good initial state is eliminated in *stochastic relaxation*. In contrast to the deterministic relaxation, the motion vector v under consideration is selected randomly (both its location x and parameters b), thus allowing (with a small probability) a momentary deterioration of the criterion [12]. In the context of minimization, such as in *simulated annealing* [9], this allows the algorithm to "climb" out of local minima and eventually reach the global minimum. Stochastic relaxation methods, although easy to implement and capable of finding excellent solutions, are very slow in convergence.

5 Practical Motion Estimation Algorithms

5.1 Global Motion Estimation

As discussed in Section 4.1 camera motion induces motion of all image points and therefore is often an obstacle to solving

---

[3]There exist methods based on causal motion models that are computationally inexpensive, e.g., pel-recursive motion estimation, but their accuracy is usually lower than that of methods based on noncausal motion models.

various video processing problems. For example, for motion to be detected in images obtained by a mobile camera, camera motion must be compensated first [14, Chapter 8]. Global motion compensation (GMC) plays also an important role in video compression, since only a few motion parameters are sufficient to greatly reduce the prediction error when images to be encoded are acquired, for example, by a panning camera. GMC has been included in version 2 of the MPEG-4 video compression standard (Chapter 6.5).

Since camera motion is limited to translation and rotation, and affects all image points, a spatially parametric, e.g., affine [Eq. (9)] and temporally linear [Eq. (10)] motion model supported on the whole image is appropriate. Under the constant-intensity observation model [Eq. (12)], the pixel-based quadratic criterion [Eq. (18)] leads to the following minimization:

$$\min_{b} \mathcal{E}(\boldsymbol{v}), \quad \mathcal{E}(\boldsymbol{v}) = \sum_{n} \varepsilon^2(\boldsymbol{n}),$$

$$\varepsilon(\boldsymbol{n}) = I_k(\boldsymbol{n}) - I_{k-1}(\boldsymbol{n} - \boldsymbol{v}(\boldsymbol{n})(t_k - t_{k-1})), \qquad (25)$$

where the dependence of $\boldsymbol{v}$ on $\boldsymbol{b}$ is implicit [Eq. (9)] and $t_k - t_{k-1}$ is usually assumed to equal 1. As a way to perform the above minimization, gradient descent can be used. However, since this method gets easily trapped in local minima, an initial search for approximate translation components b_1 and b_2 [Eq. (9)], which can be quite large, has to be performed. This search can be executed, for example, by using the three-step block matching (Section 5.2).

Since the dependence of the cost function $\mathcal{E}$ on $\boldsymbol{b}$ is nonlinear, an iterative procedure is typically used:

$$\boldsymbol{b}^{n+1} = \boldsymbol{b}^n + \boldsymbol{H}^{-1}\boldsymbol{c},$$

where $\boldsymbol{b}^n$ is the parameter vector $\boldsymbol{b}$ at iteration n, $\boldsymbol{H}$ is a $K \times K$ matrix equal to 1/2 of the Hessian matrix of $\mathcal{E}$ (i.e., matrix with elements $\partial^2 \mathcal{E}/\partial b_k \partial b_l$), $\boldsymbol{c}$ is a K-dimensional vector equal to $-1/2$ of $\nabla\mathcal{E}$, and K is the number of parameters in the motion model (six for affine). The above equation can be equivalently written as $\sum_l H_{kl}\Delta b_l = c_k$, where $\Delta \boldsymbol{b} = \boldsymbol{b}^{n+1} - \boldsymbol{b}^n$ and

$$H_{kl} = \frac{1}{2}\sum_{n}\frac{\partial^2 \varepsilon^2(\boldsymbol{n})}{\partial b_k \partial b_l}$$

$$= \sum_{n}\left(\frac{\partial \varepsilon(\boldsymbol{n})}{\partial b_k}\frac{\partial \varepsilon(\boldsymbol{n})}{\partial b_l} + \varepsilon(\boldsymbol{n})\frac{\partial^2 \varepsilon(\boldsymbol{n})}{\partial b_k \partial b_l}\right)$$

$$\approx \sum_{n}\left(\frac{\partial \varepsilon(\boldsymbol{n})}{\partial b_k}\frac{\partial \varepsilon(\boldsymbol{n})}{\partial b_l}\right),$$

$$c_k = -\frac{1}{2}\sum_{n}\frac{\partial \varepsilon^2(\boldsymbol{n})}{\partial b_k} = -\sum_{n}\varepsilon(\boldsymbol{n})\frac{\partial \varepsilon(\boldsymbol{n})}{\partial b_k}.$$

The approximation above is due to dropping the second-order derivatives; see [16, page 683] for justification.

In order to handle large velocities and to speed up computations, the method has to be implemented hierarchically. Thus, an image pyramid is built with spatial prefiltering and subsampling applied between each two levels. The computation starts at the top level of the pyramid (lowest resolution) with b_1 and b_2 estimated in the initial step and the other parameters set to zero. Then, gradient descent is performed by solving for $\Delta \boldsymbol{b}$, e.g., using singular value decomposition, and updating $\boldsymbol{b}^{n+1} = \boldsymbol{b}^n + \Delta \boldsymbol{b}$ until a convergence criterion is met. The resulting motion parameters are projected onto a lower level of the pyramid[4] and the gradient descent is repeated. This cycle is repeated until the bottom of the pyramid is reached.

Since the global motion model applies to all image points, it cannot account for local motion. Thus, points moving independently of the global motion may generate large errors $\varepsilon(\boldsymbol{n})$ and thus bias an estimate of the global motion parameters. The corresponding pixels are called *outliers* and, ideally, should be eliminated from the minimization [Eq. (25)]. This can be achieved by using a robust criterion (Fig. 4) instead of the quadratic. For example, a Lorentzian function or a truncated quadratic can be used, but both provide a nonzero cost for outliers. This reduces the impact of outliers on the estimation but does not eliminate it completely. To exclude the impact of outliers altogether, a modified truncated quadratic should be used such as $\Phi_{tg}(\varepsilon, \theta, 0)$ defined in Eq. (19). This criterion effectively limits the summation in Eq. (25) to the nonoutlier pixels and is used only in the gradient descent part of the algorithm. The threshold θ can be fixed or it can be made adaptive, e.g., by limiting the false alarm rate.

Figure 5 shows outlier pixels for two images "Foreman" and "Coastguard" declared by using the above method based on the eight-parameter perspective motion model [22, Chapter 6]. Note the clear identification of outliers in the moving head, on the boats, and in the water. The outliers tend to appear at intensity transitions since it is there that any inaccuracy in global motion caused by a local (inconsistent) motion will induce large error ε; in uniform areas of local motion the error ε remains small. By the exclusion of the outliers from the estimation, the accuracy of computed motion parameters is improved. Since the true camera motion is not known for these two sequences, the improvement was measured in the context of the GMC mode of MPEG-4 compression.[5] In comparison with nonrobust global motion estimation ($\Phi_{tq}(\varepsilon, \infty, \cdot)$), the robust method presented resulted in a bit rate reduction of 8% and 15% for "Foreman" and "Coastguard," respectively [13].

---

[4]The projection is performed by scaling the translation parameters b_1 and b_2 by 2 and leaving the other four parameters unchanged.

[5]MPEG-4 encoder (version 2) can send parameters of global motion for each frame. Consequently, for each macroblock it can make a decision as to whether to perform the temporal prediction based on the global motion parameters or the local macroblock motion. The benefit of GMC is that only few motion parameters (e.g., eight) are sent for the whole frame. The GMC mode is beneficial for sequences with camera motion or zoom.

FIGURE 5   (a), (b) Original images from CIF sequences "Foreman" and "Coastguard," and (c), (d) pixels declared as outliers (black) according to a global motion estimate.

5.2 Block Matching

Block matching is the simplest algorithm for the estimation of local motion. It uses a spatially constant [Eq. (8)] and temporally linear [Eq. (10)] motion model over a rectangular region of support. Although, as explained in Section 4.1, the translational 2-D motion is only valid for the orthographic projection and 3-D object translation, this model applied locally to a small block of pixels is quite accurate for a large variety of 3-D motions. It has proven accurate enough to serve as a basis for most of the practical motion estimation algorithms used today. Because of its simplicity and regularity (the same operations are performed for each block of the image), block matching can be relatively easily implemented in VLSI. Today, block matching is the only motion estimation algorithm massively implemented in VLSI and used for encoding within *all* video compression standards (see Chapters 6.4 and 6.5).

In video compression, motion vectors d are used to eliminate temporal video redundancy by means of motion-compensated prediction [Eq. (17)]. Hence, the goal is to achieve as low prediction error $\varepsilon_k(n)$ as possible, which is equivalent to the constant-intensity observation model. By applying this model within a pixel-based criterion, we can describe the method by the following minimization:

$$\min_{d_m \in \mathcal{P}} \mathcal{E}(d_m),$$

$$\mathcal{E}(d_m) = \sum_{n \in \mathcal{B}_m} \Phi(I_k(n) - I_{k-1}(n - d_m)) \quad \forall m, \quad (26)$$

where $\mathcal{P}$ is the search area to which d_m belongs, defined as follows:

$$\mathcal{P} = \{n = (n_1, n_2) : -P \le n_1 \le P, -P \le n_2 \le P\},$$

and $\mathcal{B}_m$ is an $M \times N$ block of pixels with the top-left corner coordinate at $m = (m_1, m_2)$. The goal is to find the best, in the sense of the criterion Φ, displacement vector d_m for each block $\mathcal{B}_m$. This is illustrated graphically in Fig. 6(a); a block is sought within image I_{k-1} that best matches the current block in I_k.

Estimation Criterion

Although an average error is used in Eq. (26), other measures are possible, such as a maximum error (min-max estimation). To fully define the estimation criterion, the function Φ must be established. Originally, $\Phi(x) = x^2$ was often used in block

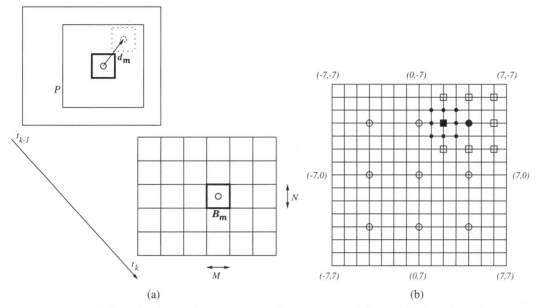

FIGURE 6   (a) Block matching between block $\mathcal{B}_m$ at time t_k (current image) and all possible blocks in the search area $\mathcal{P}$ at time t_{k-1}. (b) Three-step search method. Large circles denote level 1 ($\mathcal{P}_1$), squares denote level 2 ($\mathcal{P}_2$), and small circles denote level 3 ($\mathcal{P}_3$). The filled-in elements denote the best match found at each level. The final vector is (2,5).

matching, but it was soon replaced by an absolute error criterion $\Phi(x) = |x|$ for its simplicity (no multiplications) and robustness in the presence of outliers. Other improved criteria have been proposed, such as those based on the median of squared errors; however, their computational complexity is significantly higher.

Also, simplified criteria have been proposed to speed up the computations, for example, based on adaptive quantization to 2 bits or on pixel subsampling [4]. Usually, a simplification of the original criterion Φ leads to a suboptimal performance. However, with an adaptive adjustment of the criterion's parameters (e.g., quantization levels, decimation patterns) a close-to-optimal performance can be achieved at significantly reduced complexity.

Search Methods

An exhaustive search for $d_m \in \mathcal{P}$ that gives the lowest error $\mathcal{E}$ is computationally costly. An "intelligent" search, whereby only the more likely candidates from $\mathcal{P}$ are evaluated, usually results in substantial computational savings. One popular technique for reducing the number of candidates is the *logarithmic search*. Assuming that $P = 2^k - 1$ and denoting $P_l = (P+1)/2^l$, where k and l are integers, we establish the new reduced-size search area as follows:

$$\mathcal{P}_l = \{ n : n = (\pm P_l, \pm P_l) \text{ or }$$
$$n = (\pm P_l, 0) \text{ or } n = (0, \pm P_l) \text{ or } n = (0, 0) \},$$

i.e., $\mathcal{P}_l$ is reduced to the vertices, midway points between vertices, and the central point of the half-sized original rectangle $\mathcal{P}$. For example, for $P = 7$, $\mathcal{P}_1$ consists of the following candidates:

$(-4, -4), (-4, 4), (4, -4), (4, 4), (-4, 0), (4, 0), (0, -4), (0, 4),$ and $(0, 0)$. The search starts with the candidates from $\mathcal{P}_1$. Once the best match is found, the new search area $\mathcal{P}_2$ is centered around this match and the procedure is repeated for candidates from $\mathcal{P}_2$. Note that the error $\mathcal{E}$ does not have to be evaluated for the $(0, 0)$ candidate since it had been evaluated at the previous level. The procedure is continued with subsequently reduced search spaces. Since typically only three levels ($l = 1, 2, 3$) are used, such a method is often referred to as the *three-step search* [Fig. 6(b)].

In the search above, at each step a 2-D search is performed. An alternative approach is to perform 1-D searches only, usually in orthogonal directions. Examples of block matching algorithms based on 1-D search methods are as follows.

1. *One-at-a-time search* [19]: In this method, first a minimum of $\mathcal{E}$ is sought in one, for example, the horizontal, direction. Then, given the horizontal estimate, a vertical search is performed. Subsequently, a horizontal search is performed given the previous vertical estimate, and so on. In the original proposal, 1-D minima closest to the origin were examined only, but later a 1-D full search was used to avoid the problem of getting trapped close to the origin. Note that the searches are not independent since each relies on the result of the previous one.

2. *Parallel hierarchical one-dimensional search* [4]: This method also performs 1-D searches in orthogonal directions (usually horizontal and vertical) but independently of each other, i.e., the horizontal search is performed simultaneously with the vertical search since it does not depend on the outcome of the latter. In addition, the 1-D search is implemented hierarchically. First, every Kth location

from $\mathcal{P}$ is taken as a candidate for a 1-D search. Once the minima in both directions are identified, new 1-D searches begin with every $K/2$th location from $\mathcal{P}$, and so on. Typically, horizontal and vertical searches using every 8th, 4th, 2nd and finally every pixel (within the limits of $\mathcal{P}$) are performed.

A remark is in order at this point. All the fast search methods are based on the assumption that the error $\mathcal{E}$ has a single minimum for all $\boldsymbol{d}_m \in \mathcal{P}$, or in other words, that $\mathcal{E}$ increases monotonically when moving away from the best-match position. In practice this is rarely true since $\mathcal{E}$ depends on $\boldsymbol{d}_m$ by means of the intensity I, which can be arbitrary. Therefore, multiple local minima often exist within $\mathcal{P}$ and a fast search method can be easily trapped in anyone of them, whereas an exhaustive search will always find the "deepest" minimum. This is not a very serious problem in video coding since a suboptimal motion estimate translates into an increased prediction error [Eq. (17)] that will be entropy coded and at most will result in a rate increase. It is a serious problem, however, in video processing, in which *true* motion is sought and any motion errors may result in uncorrectable distortions. A good review of block matching algorithms can be found in [4].

5.3 Phase Correlation

As discussed above, block matching can precisely estimate local displacement but must examine many candidates. At the same time methods based on the frequency-domain criteria (Section 4.2) are capable of identifying global motion but cannot localize it in the space/time domain. By combining the two approaches, the *phase correlation* method [21] is able to exploit advantages of both approaches. First, likely candidates are computed by using a frequency-domain approach, and then they are assigned a spatial location by local block matching.

Recall the cross-correlation criterion $\mathcal{C}(\boldsymbol{d})$ expressed in the Fourier domain [Eq. (22)]. By normalizing $\mathcal{F}[\mathcal{C}(\boldsymbol{d})]$ and taking the inverse transform, one obtains

$$\Psi_{k-1,k}(\boldsymbol{n}) = \mathcal{F}^{-1} \left\{ \frac{\hat{I}_k(\boldsymbol{u}) \hat{I}_{k-1}^*(\boldsymbol{u})}{|\hat{I}_k(\boldsymbol{u}) \hat{I}_{k-1}^*(\boldsymbol{u})|} \right\}. \qquad (27)$$

Here $\Psi_{k-1,k}(\boldsymbol{n})$ is a normalized correlation computed between two images I_k and I_{k-1}. In the special case of a global translation ($I_k(\boldsymbol{n}) = I_{k-1}(\boldsymbol{n} - \boldsymbol{z})$), by using transformation [Eq. (21)] one can easily show that the correlation surface becomes a Kronecker delta function ($\delta(\boldsymbol{x})$ equals 0 for $\boldsymbol{x} \neq 0$ and 1 for $\boldsymbol{x} = 0$):

$$\Psi_{k-1,k}(\boldsymbol{n})|_{I_k(\boldsymbol{n}) = I_{k-1}(\boldsymbol{n}-\boldsymbol{z})} = \mathcal{F}\{e^{-j2\pi \boldsymbol{u}\boldsymbol{z}}\} = \delta(\boldsymbol{n} - \boldsymbol{z}).$$

In practice, when neither the global translation nor intensity constancy hold, $\Psi_{k-1,k}$ is a surface with numerous peaks. These peaks correspond to dominant displacements between I_{k-1} and I_k, and, if identified, are very good candidates for fine tuning by, for example, block matching. Note that no explicit motion model has been used thus far, while the observation model, as

usual, is that of constant intensity and the estimation criterion is the cross-correlation function. In practice, the method can be implemented as follows [21].

1. Divide I_{k-1} and I_k into large blocks, e.g., 64×64 (motion range of ± 32 pixels), and take a fast Fourier transform (FFT) of each block.
2. Compute $\Psi_{k-1,k}$, using same-position blocks in I_{k-1} and I_k.
3. Take the inverse FFT of $\Psi_{k-1,k}$ and identify most dominant peaks.
4. Use the coordinates of the peaks as the candidate vectors for block matching of 16×16 blocks.

The phase correlation can also initialize pixel-based based estimation. Moreover, the correlation over a large area (64×64) permits the recovery of subpixel displacements by interpolating the correlation surface. Note that since the discrete Fourier transform (implemented by means of FFT) assumes signal periodicity, intensity discontinuities between left and right and between top and bottom block boundaries may introduce spurious peaks.

The phase correlation method is basically an efficient maximization of a correlation-based error criterion. The shape of the maxima of the correlation surface is weakly dependent on the image content, and the measurement of their locations is relatively independent of illumination changes. This is due, predominantly, to the normalization in Eq. (27). However, rotations and zooms cannot be easily handled since the peaks in $\Psi_{k-1,k}$ are hard to distinguish as a result of the spatial smoothness of the corresponding motion fields.

5.4 Optical Flow by Means of Regularization

Recall the regularized estimation criterion Eq. (23). It uses a translational/linear motion model at each pixel under the constant-intensity observation model and a quadratic error criterion.

To find the functions v_1 and v_2, implicitly dependent on $\boldsymbol{x}$, the functional has to be minimized, which is a problem in the calculus of variations. The Euler–Lagrange equations yield [10]

$$\lambda \nabla^2 v_1 = \left(\frac{\partial I}{\partial x} v_1 + \frac{\partial I}{\partial y} v_2 + \frac{\partial I}{\partial t} \right) \frac{\partial I}{\partial x},$$

$$\lambda \nabla^2 v_2 = \left(\frac{\partial I}{\partial x} v_1 + \frac{\partial I}{\partial y} v_2 + \frac{\partial I}{\partial t} \right) \frac{\partial I}{\partial y},$$

where $\nabla^2 = \partial^2/\partial x^2 + \partial^2/\partial y^2$ is the Laplacian operator. This pair of elliptic partial differential equations can be solved iteratively by using finite-difference or finite-element discretization (see Chapter 3.6 for other examples of regularization).

An alternative is to formulate the problem directly in the discrete domain. Then, the integral is replaced by a summation while the derivatives are replaced by finite differences. In [10], for example, an average of first-order differences computed over a $2 \times 2 \times 2$ cube was used. By differentiation of this discrete cost function, a system of equations can be computed and subsequently solved by Jacobi or Gauss–Seidel relaxation. This discrete approach to regularization is a special case of the MAP estimation presented next.

5.5 MAP Estimation of Dense Motion

The MAP formulation (24) is very general and requires further assumptions. The likelihood relates one image to another by means of d_k. Since $I_k(n) = I_{k-1}(n - d(n)) + \varepsilon_k(n)$, the characteristics of this likelihood reside in ε_k. It is clear that for good displacement estimates ε_k should behave like a noise term. It can be shown that the statistics of ε_k are reasonably close to those of a zero-mean Gaussian distribution, although a generalized Gaussian is a better fit [20]. Therefore, assuming no correlation among $\varepsilon_k(n)$ for different n, $P(\mathcal{I}_k = I_k \mid D_k = d_k; I_{k-1})$ can be fairly accurately modeled by a product of zero-mean Gaussian distributions.

The prior probability is particularly flexible when D_k is assumed to be a MRF. Then, $P(D_k = d_k; I_{k-1})$ is a Gibbs distribution (Section 2.2) uniquely specified by cliques and a potential function. For example, for two-element cliques $\{n, l\}$ the smoothness of D_k can be expressed as follows:

$$V_s(d(n), d(l)) = \|d(n) - d(l)\|^2, \quad \forall \{n, l\} \in C.$$

Clearly, for similar $d(n)$ and $d(l)$ the potential V_s is small and thus the prior probability is high, whereas for dissimilar vectors this probability is small.

Since both likelihood and prior probability distributions are exponential in this case, the MAP estimation (24) can be rewritten as energy minimization:

$$\hat{d}_k = \arg \min_d \left(\frac{1}{2\sigma^2} \sum_n (I_k(n) - I_{k-1}(n - d(n)))^2 \right.$$
$$\left. + \sum_{l \in \eta(n)} \|d(n) - d(l)\|^2 \right). \tag{28}$$

The above energy can be minimized in various ways. To attain the global minimum, simulated annealing (Section 4.3) should be used. Given sufficiently many iterations, the method is theoretically capable of finding the global minimum, although at considerable computational cost. In contrast, the method is easy to implement [12]. A faster alternative is the deterministic ICM method that does not find a true MAP estimate, although usually finds a close enough solution in a fraction of time taken by simulated annealing. An even more effective method is the HCF method, although its implementation is a little bit more complex.

It is worth noting that formulation (28) comprises, as a special case, the discrete formulation of the optical flow computation described in Section 5.4. Consider constraint (14). Multiplying both sides by $\Delta t = t_k - t_{k-1}$, it becomes $I^x d_1 + I^y d_2 + I^t \Delta t = 0$,

where I^x, I^y, and I^t are discrete approximations to horizontal, vertical, and temporal derivatives, respectively. This constraint is not satisfied exactly, and as it turns out $I^x d_1 + I^y d_2 + I^t \Delta t$ is a noiselike term with similar characteristics to the prediction error ε_k. This is not surprising since both originate from the same constant-intensity hypothesis. By replacing the prediction error in Eq. (28) with this new term, one obtains a cost function equivalent to the discrete formulation of the optical flow problem (Section 5.4) [10].

Minimization (28) leads to smooth displacement fields d_k, also at object boundaries, which is undesirable. To relax the smoothness constraint at object boundaries, explicit models of motion discontinuities (line field) [12] or of motion segmentation labels (segmentation field) [20] can be easily incorporated into the MRF formulation, although their estimation is far from trivial.

5.6 Experimental Comparison of Motion Estimation Methods

To demonstrate the impact of various motion models, Fig. 7 shows results for the QCIF sequence "Carphone." Both the estimated displacements and the resulting motion-compensated prediction errors are shown for pixel-based (dense), block-based, and segmentation-based motion models. The latter motion estimate was obtained by minimizing the mean squared error ($\Phi(\varepsilon) = \varepsilon^2$ in Eq. (18)) within each region $\mathcal{R}$ from Fig. 7(c) for the affine motion model, Eq. (9).

Note the lack of detail caused by the low resolution (16×16 blocks) of the block-based approach, but the approximately correct motion of objects. The pixel-based model results in a smooth estimate with more spatial detail but at the cost of reduced precision. The segmentation-based motion estimate shows both better accuracy and detail. Although the associated segmentation [Fig. 7(c)] does not correspond exactly to the objects as perceived by humans, it nevertheless closely matches object boundaries. The motion of the head and of the upper body is well captured, but the motion of landscape in the car window is exaggerated because of the lack of image detail. As for the prediction error, note the blocking artifacts for the block-based motion model (31.8 dB[6]) but a very small error for the pixel-based model (35.9 dB). The region-based model results in a slightly higher prediction error (35.5 dB) than the pixel-based model, but one that is significantly lower than that of the block model.

6 Perspectives

In the past two decades, motion detection and estimation have moved from research laboratories to specialized products. This has been made possible by two factors. First, enormous advances in VLSI have facilitated practical implementation of CPU-

---

[6]The prediction error is defined as follows: $10 \log(255^2/\mathcal{E}(d))$ [dB] for $\mathcal{E}$ from Eq. (18), with quadratic Φ and $\mathcal{R}$ being the whole image.

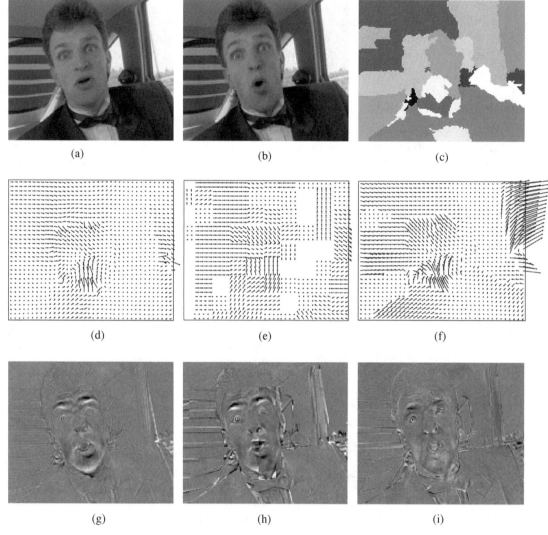

(a)   (b)   (c)

(d)   (e)   (f)

(g)   (h)   (i)

FIGURE 7   Original frames (a) 168 and (b) 171 from QCIF sequence "Carphone". (c) Motion-based segmentation of frame 171. Motion estimates (subsampled by four) and the resulting motion-compensated prediction error (magnified by two) at frame 171 for: (d), (g) dense-field MAP estimation; (e), (h) 16×16 block matching; (f), (i) region-based estimation for segments from (c). (From Konrad and Stiller [14, Chapter 4]. Reproduced with permission of Kluwer.)

hungry motion algorithms. Second, new models and estimation algorithms have lead to an improved reliability and accuracy of the estimated motion. With the continuing advances in VLSI, the complexity constraints plaguing motion algorithms will become less of an issue. This should allow practical implementation of more advanced motion models and estimation criteria, and, in turn, further improve the accuracy of the computed motion. One of the promising approaches studied today is the joint motion segmentation and estimation that effectively combines the detection and estimation discussed separately in this chapter.

References

[1] T. Aach and A. Kaup, "Bayesian algorithms for adaptive change detection in images sequences using Markov random fields," *Signal Process. Image Commun.* **7**, 147–160 (1995).

[2] M. Bertero, T. Poggio, and V. Torre, "Ill-posed problems in early vision," *Proc. IEEE* **76**, 869–889 (1988).

[3] J. Besag, "On the statistical analysis of dirty pictures," *J. Roy. Stat. Soc. B* **48**, 259–279 (1986).

[4] V. Bhaskaran and K. Konstantinides, *Image and Video Compression Standards: Algorithms and Architectures* (Kluwer, Boston, MA, 1997).

[5] M. Black, "Robust incremental optical flow," Ph.D. dissertation (Yale University, New Haven, CT, 1992).

[6] P. Bouthemy and P. Lalande, "Recovery of moving object masks in an image sequence using local spatiotemporal contextual information," *Opt. Eng.* **32**, 1205–1212 (1993).

[7] M. Chahine and J. Konrad, "Estimation and compensation of accelerated motion for temporal sequence interpolation," *Signal Process. Image Commun.* **7**, 503–527 (1995).

[8] P. Chou and C. Brown, "The theory and practice of Bayesian image labelling," *Intern. J. Comput. Vis.* **4**, 185–210 (1990).

[9] S. Geman and D. Geman, "Stochastic relaxation, Gibbs distributions, and the Bayesian restoration of images," *IEEE Trans. Pattern Anal. Machine Intell.* **6**, 721–741 (1984).

[10] B. Horn, *Robot Vision* (MIT Press, Cambridge, MA, 1986).

[11] Y. Hsu, H.-H. Nagel, and G. Rekers, "New likelihood test methods for change detection in image sequences," *Comput. Vis. Graph. Image Process.* **26**, 73–106 (1984).

[12] J. Konrad and E. Dubois, "Bayesian estimation of motion vector fields," *IEEE Trans. Pattern Anal. Machine Intell.* **14**, 910–927 (1992).

[13] J. Konrad and F. Dufaux, "Improved global motion estimation for N3," Tech. Rep. MPEG97/-M3096, ISO/IEC JTC1/SC29/WG11, Feb. 1998.

[14] H. Li, S. Sun, and H. Derin, eds., *Video Compression for Multimedia Computing — Statistically Based and Biologically Inspired Techniques* (Kluwer, Boston, MA, 1997).

[15] F. Luthon, A. Caplier, and M. Liévin, "Spatiotemporal MRF approach with application to motion detection and lip segmentation in video sequences," *Signal Process.* **76**, 61–80 (1999).

[16] W. Press, S. Teukolsky, W. Vetterling, and B. Flannery, *Numerical Recipes in C: The Art of Scientific Computing* (Cambridge U. Press, New York, 1992).

[17] I. Sezan and R. Lagendijk, eds., *Motion Analysis and Image Sequence Processing* (Kluwer, Boston, MA, 1993).

[18] K. Skifstad and R. Jain, "Illumination independent change detection for real world image sequences," *Comput. Vis. Graph. Image Process.* **46**, 387–399 (1989).

[19] R. Srinivasan and K. Rao, "Predictive coding based on efficient motion estimation," *IEEE Trans. Commun.* **33**, 888–896 (1985).

[20] C. Stiller, "Object-based estimation of dense motion fields," *IEEE Trans. Image Process.* **6**, 234–250 (1997).

[21] A. Tekalp, *Digital Video Processing* (Prentice-Hall, Englewood Cliffs, NJ, 1995).

[22] L. Torres and M. Kunt, eds., *Video Coding: Second Generation Approach* (Kluwer, Boston, MA, 1996).

[23] E. Trucco and A. Verri, *Introductory Techniques for 3-D Computer Vision* (Prentice-Hall, Englewood Cliffs, NJ, 1998).

[24] H. van Trees, *Detection, Estimation and Modulation Theory* (Wiley, New York, 1968).

3.11

Video Enhancement and Restoration

Reginald L. Lagendijk,
Peter M. B. van
Roosmalen,
and Jan Biemond
*Delft University of Technology
The Netherlands*

1 Introduction

Even with the advancing camera and digital recording technology, there are many situations in which recorded image sequences — or *video* for short — may suffer from severe degradations. The poor quality of recorded image sequences may be due to, for instance, the imperfect or uncontrollable recording conditions such as one encounters in astronomy, forensic sciences, and medical imaging. Video enhancement and restoration has always been important in these application areas, not only to improve the visual quality but also to increase the performance of subsequent tasks such as analysis and interpretation.

Another important application of video enhancement and restoration is that of preserving motion pictures and videotapes recorded over the last century. These unique records of historic, artistic, and cultural developments are deteriorating rapidly because of aging effects of the physical reels of film and magnetic tapes that carry the information. The preservation of these fragile archives is of interest not only to professional archivists, but also to broadcasters as a cheap alternative to fill the many television channels that have come available with digital broadcasting. Reusing old film and video material is, however, only feasible if the visual quality meets the standards of today. First, the archived film and video is transferred from the original film reels or magnetic tape to digital media. Then, all kinds of degradations are removed from the digitized image sequences, in this way increasing the visual quality and

commercial value. Because the objective of restoration is to remove irrelevant information such as noise and edges, it restores the original spatial and temporal correlation structure of digital image sequences. Consequently, restoration may also improve the efficiency of the subsequent MPEG compression of image sequences.

An important difference between the enhancement and restoration of two-dimensional (2-D) images and of video is the amount of data to be processed. Whereas for the quality improvement of important images elaborate processing is still feasible, this is no longer true for the absolutely huge amounts of pictorial information encountered in medical sequences and film/video archives. Consequently, enhancement and restoration methods for image sequences should be fit for — at least partial — implementation in hardware, should have a manageable complexity, and should be semiautomatic. The term semiautomatic indicates that in the end professional operators control the visual quality of the restored image sequences by selecting values for some of the critical restoration parameters.

The most common artifact encountered in the above-mentioned applications is noise. Over the past two decades an enormous amount of research has focused on the problem of enhancing and restoring 2-D images. Clearly, the resulting spatial methods are also applicable to image sequences, but such an approach implicitly assumes that the individual pictures of the image sequence, or *frames*, are temporally independent. By ignoring the temporal correlation that exists, one may obtain suboptimal results, and the spatial *intraframe* filters tend to introduce temporal artifacts in the restored image sequences. In this chapter we focus our attention specifically on exploiting temporal dependencies, yielding *interframe* methods. In this respect the material offered in this chapter is complementary to that on image enhancement in Chapters 3.1 to 3.4 of this *Handbook*. The resulting enhancement and restoration techniques operate in the temporal dimension by definition, but they often have a spatial filtering component as well. For this reason, video enhancement and restoration techniques are sometimes referred to as *spatiotemporal* filters or three-dimensional (3-D) filters. Section 2 of this chapter presents three important classes of noise filters for video frames, namely *linear temporal filters, order-statistic filters, and multiresolution filters.*

In forensic sciences and in film and video archives, a large variety of artifacts are encountered. Besides noise, we discuss the removal of two other important impairments that rely on temporal processing algorithms, namely blotches (Section 3) and intensity flicker (Section 4). Blotches are dark and bright spots that are often visible in damaged film image sequences. The removal of blotches is essentially a temporal detection and interpolation problem. Intensity flicker refers to variations in intensity in time, caused by aging of film, by copying and format conversion (for instance, from film to video), and — in the case of earlier film — by variations in shutter time. Whereas blotches are spatially highly localized artifacts in video frames, intensity flicker is usually a spatially global, but not stationary, artifact.

In practice, image sequences may be degraded by multiple artifacts. In principle, a single method for restoring all of the artifacts simultaneously is conceivable. More usual is, however, to follow a sequential procedure, in which artifacts are removed one by one. As an example, Fig. 1 illustrates the order in which the removal of flicker, blotches, and noise takes place. The reasons for this modular approach are the necessity to judge the success of the individual steps (for instance, by an operator), and the algorithmic and implementation complexity.

As already suggested in Fig. 1, most temporal filtering techniques require an estimate of the motion in the image sequence. Motion estimation has been discussed in detail in Chapters 3.7 and 6.1 of this *Handbook*. The estimation of motion from degraded image sequences is, however, problematic. We are faced with the problem that the impairments of the video disturb the motion estimator, but that at the same time correct motion estimates are assumed in developing enhancement and restoration algorithms. In this chapter we will not discuss the design of new motion estimators that are robust to the various artifacts, but we will assume that existing motion estimators can be modified appropriately such that sufficiently correct and smooth motion fields are obtained. The reason for this approach is that even under ideal conditions, motion estimates are never perfect. Incorrect or unreliable motion vectors are dealt with in two ways. In the first place, clearly incorrect or unreliable motion vectors can be repaired. In the second, the enhancement and restoration algorithms should be robust against the less obviously incorrect or unreliable motion vectors.

2 Spatiotemporal Noise Filtering

Any recorded signal is affected by noise, no matter how precise the recording equipment. The sources of noise that can corrupt an image sequence are numerous (see Chapter 4.4 of this *Handbook*). Examples of the more prevalent ones include camera noise, shot noise originating in electronic hardware and the storage on magnetic tape, thermal noise, and granular noise on film. Most recorded and digitized image sequences contain a mixture of noise contributions, and often the (combined) effects of the noise are nonlinear of nature. In practice, however, the aggregated effect of noise is modeled as an additive white (sometimes Gaussian) process with zero mean and variance σ_w^2 that is independent from the ideal uncorrupted image sequence $f(\mathbf{n}, k)$. The recorded image sequence $g(\mathbf{n}, k)$ corrupted by noise $w(\mathbf{n}, k)$ is then given by

$$g(\mathbf{n}, k) = f(\mathbf{n}, k) + w(\mathbf{n}, k), \tag{1}$$

where $\mathbf{n} = (n_1, n_2)$ refers to the spatial coordinates and k to the frame number in the image sequence. More accurate models are often much more complex but lead to little gain compared to the added complexity.

The objective of noise reduction is to make an estimate $\hat{f}(\mathbf{n}, k)$ of the original image sequence given only the observed noisy image sequence $g(\mathbf{n}, k)$. Many different approaches toward noise reduction are known, including optimal linear filtering, nonlinear filtering, scale-space processing, and Bayesian techniques. In this section we discuss successively the class of linear image sequence filters, order-statistic filters, and multiresolution filters. In all cases the emphasis is on the temporal filtering aspects. More rigorous reviews of noise filtering for image sequences are given in [2, 3, 15].

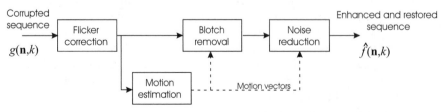

FIGURE 1   Some processing steps in the removal of noise, blotches, and intensity flicker from video.

2.1 Linear Filters

Temporally Averaging Filters

The simplest temporal filter carries out a weighted averaging of successive frames. That is, the restored image sequence is obtained by [6]

$$\hat{f}(\mathbf{n}, k) = \sum_{l=-K}^{K} h(l) g(\mathbf{n}, k - l). \qquad (2)$$

Here $h(l)$ are the temporal filter coefficients used to weight $2K+1$ consecutive frames. In case the frames are considered equally important we have $h(l) = 1/(2K + 1)$. Alternatively, the filter coefficients can be optimized in a minimum mean-squared error fashion,

$$h(l) \leftarrow \min_{h(l)} \mathrm{E}[(f(\mathbf{n}, k) - \hat{f}(\mathbf{n}, k))^2], \qquad (3)$$

yielding the well-known temporal Wiener filtering solution:

$$
\begin{pmatrix} h(-K) \\ \vdots \\ h(0) \\ h(1) \\ \vdots \\ h(K) \end{pmatrix}
$$

$$
= \begin{pmatrix} R_{gg}(0) & \cdots & R_{gg}(-K) & \cdots & \cdots & R_{gg}(-2K) \\ \vdots & \ddots & & & & \vdots \\ R_{gg}(K) & & R_{gg}(0) & & & \vdots \\ \vdots & & & R_{gg}(0) & & \vdots \\ \vdots & & & & \ddots & \vdots \\ R_{gg}(2K) & \cdots & \cdots & \cdots & \cdots & R_{gg}(0) \end{pmatrix}^{-1}
$$

$$
\times \begin{pmatrix} R_{fg}(-K) \\ \vdots \\ R_{fg}(0) \\ R_{fg}(1) \\ \vdots \\ R_{fg}(K) \end{pmatrix}, \qquad (4)
$$

where $R_{gg}(m)$ is the *temporal* autocorrelation function defined as $R_{gg}(m) = \mathrm{E}[g(\mathbf{n}, k) g(\mathbf{n}, k - m)]$, and $R_{fg}(m)$ is the temporal cross-correlation function defined as $R_{fg}(m) = \mathrm{E}[f(\mathbf{n}, k) g(\mathbf{n}, k - m)]$. The temporal window length, i.e., the parameter K, determines the maximum degree by which the noise power can be reduced. The larger the window the greater the reduction of the noise; at the same time, however, the more visually noticeable the artifacts resulting from motion between the video frames. A dominant artifact is blur of moving objects caused by the averaging of object and background information.

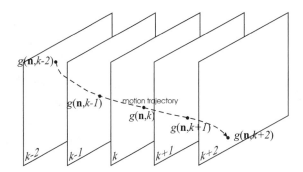

FIGURE 2 Noise filter operating along the motion trajectory of the picture element $(\mathbf{n}, k)$.

The motion artifacts can greatly be reduced by operating the filter, Eq. (2), along the picture elements (pixels) that lie on the same motion trajectory [5]. Equation (2) then becomes a *motion-compensated* temporal filter (see Fig. 2):

$$\hat{f}(\mathbf{n}, k)$$

$$= \sum_{l=-K}^{K} h(l) g(n_1 - d_x(n_1, n_2; k, l), n_2 - d_y(n_1, n_2; k, l), k - l).$$

$$(5)$$

Here $\mathbf{d}(\mathbf{n}; k, l) = (d_x(n_1, n_2; k, l), d_y(n_1, n_2; k, l))$ is the motion vector for spatial coordinate (n_1, n_2) estimated between the frames k and l. It is pointed out here that the problems of noise reduction and motion estimation are inversely related as far as the temporal window length K is concerned. That is, as the length of the filter is increased temporally, the noise reduction potential increases, but so are the artifacts caused by incorrectly estimated motion between frames that are temporally far apart.

In order to avoid the explicit estimation of motion, which might be problematic at high noise levels, two alternatives are available that turn Eq. (2) into a motion-adaptive filter. In the first place, in areas where motion is detected (but not explicitly estimated) the averaging of frames should be kept to a minimum. Different ways exist to realize this. For instance, temporal filter (2) can locally be switched off entirely, or it can locally be limited to using only future or past frames, depending on the temporal direction in which motion was detected. Basically the filter coefficients $h(l)$ are spatially adapted as a function of detected motion between frames. Second, filter (2) can be operated along M a priori selected motion directions at each spatial coordinate. The finally estimated value $\hat{f}(\mathbf{n}, k)$ is subsequently chosen from the M partial results according to some selection criterion, for instance as the median [6]:

$$\hat{f}_{i,j}(\mathbf{n}, k) = 1/3(g(n_1 - i, n_2 - j, k - 1)$$

$$+ g(n_1, n_2, k) + g(n_1 + i, n_2 + j, k + 1)) \qquad (6a)$$

$$\hat{f}(\mathbf{n}, k) = \mathrm{median}\,(\hat{f}_{-1,-1}(\mathbf{n}, k), \hat{f}_{1,-1}(\mathbf{n}, k),$$

$$\hat{f}_{-1,1}(\mathbf{n}, k), \hat{f}_{1,1}(\mathbf{n}, k), \hat{f}_{0,0}(\mathbf{n}, k), g(\mathbf{n}, k)). \qquad (6b)$$

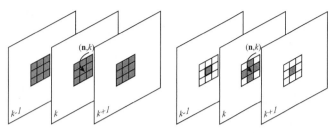

FIGURE 3 Examples of spatiotemporal windows to collect data for noise filtering of the picture element $(\mathbf{n}, k)$.

Clearly cascading Eqs. (6a) and (6b) turns the overall estimation procedure into a nonlinear one, but the partial estimation results are still obtained by the linear filter operation [Eq. (6a)].

It is easy to see that filter (2) can be extended with a spatial filtering part. There exist many variations to this concept, basically as many as there are spatial restoration techniques for noise reduction. The most straightforward extension of Eq. (2) is the following 3-D weighted averaging filter [15]:

$$\hat{f}(\mathbf{n}, k) = \sum_{(\mathbf{m}, l) \in S} h(\mathbf{m}, l) g(\mathbf{n} - \mathbf{m}, k - l) \qquad (7)$$

Here S is the spatiotemporal support or *window* of the 3-D filter (see Fig. 3). The filter coefficients $h(\mathbf{m}, l)$ can be chosen to be all equal, but a performance improvement is obtained if they are adapted to the image sequence being filtered, for instance by optimizing them in the mean-squared error sense of Eq. (3). In the latter case, Eq. (7) becomes the theoretically optimal 3-D Wiener filter.

There are, however, two disadvantages with the 3-D Wiener filter. The first is the requirement that the 3-D autocorrelation function for the original image sequence is known *a priori*. The second is the 3-D wide-sense stationarity assumptions, which are virtually never true because of moving objects and scene changes. These requirements are detrimental to the performance of the 3-D Wiener filter in practical situations of interest. For these reasons, simpler ways of choosing the 3-D filter coefficients are usually preferred, provided that they allow for adapting the filter coefficients. One such choice for adaptive filter coefficients is the following [10]:

$$h(\mathbf{m}, l; \mathbf{n}, k) = \frac{c}{1 + \max(\alpha, (g(\mathbf{n}, k) - g(\mathbf{n} - \mathbf{m}, k - l))^2)}. \qquad (8)$$

Here $h(\mathbf{m}, l; \mathbf{n}, k)$ weights the intensity at spatial location $\mathbf{n} - \mathbf{m}$ in frame $k - l$ for the estimation of the intensity $\hat{f}(\mathbf{n}, k)$. The adaptive nature of the resulting filter can immediately be seen from Eq. (8). If the difference between the pixel intensity $g(\mathbf{n}, k)$ being filtered and the intensity $g(\mathbf{n} - \mathbf{m}, k - l)$ for which the filter coefficient is calculated is less than α, this pixel is included in the filtering with weight $c/1 + \alpha$; otherwise it is weighted with a much smaller factor. In this way, pixel intensities that seem to deviate too much from $g(\mathbf{n}, k)$ — for instance, due to moving objects within the spatiotemporal window S — are excluded from Eq. (7). As with temporal filter (2), spatiotemporal filter (7) can be carried out in a motion-compensated way by arranging the window S along the estimated motion trajectory.

Temporally Recursive Filters

A disadvantage of temporal filter (2) and spatiotemporal filter (7) is that they have to buffer several frames of an image sequence. Alternatively, a recursive filter structure can be used that generally has to buffer fewer (usually only one) frames. Furthermore, these filters are easier to adapt since there are fewer parameters to control. The general form of a recursive temporal filter is as follows:

$$\hat{f}(\mathbf{n}, k) = \hat{f}_b(\mathbf{n}, k) + \alpha(\mathbf{n}, k)[g(\mathbf{n}, k) - \hat{f}_b(\mathbf{n}, k)]. \qquad (9)$$

Here $\hat{f}_b(\mathbf{n}, k)$ is the prediction of the original kth frame on the basis of previously filtered frames, and $\alpha(\mathbf{n}, k)$ is the filter gain for updating this prediction with the observed kth frame. Observe that for $\alpha(\mathbf{n}, k) = 1$ the filter is switched off, i.e., $\hat{f}(\mathbf{n}, k) = g(\mathbf{n}, k)$. Clearly, a number of different algorithms can be derived from Eq. (9) depending on the way the predicted frame $\hat{f}_b(\mathbf{n}, k)$ is obtained and the gain $\alpha(\mathbf{n}, k)$ is computed. A popular choice for the prediction $\hat{f}_b(\mathbf{n}, k)$ is the previously restored frame, either in direct form

$$\hat{f}_b(\mathbf{n}, k) = \hat{f}(\mathbf{n}, k - 1) \qquad (10a)$$

or in motion-compensated form:

$$\hat{f}_b(\mathbf{n}, k) = \hat{f}(\mathbf{n} - \mathbf{d}(\mathbf{n}; k, k - 1), k - 1). \qquad (10b)$$

More elaborate variations of Eq. (10) make use of a local estimate of the signals mean within a spatiotemporal neighborhood. Furthermore, Eq. (9) can also be cast into a formal 3-D motion-compensated Kalman estimator structure [16]. In this case the prediction $\hat{f}_b(\mathbf{n}, k)$ depends directly on the dynamic spatiotemporal state-space equations used for modeling the image sequence.

The simplest case for selecting $\alpha(\mathbf{n}, k)$ is by using a globally fixed value. As with the filter structures of Eqs. (2) and (7), it is generally necessary to adapt $\alpha(\mathbf{n}, k)$ to the presence or correctness of the motion in order to avoid filtering artifacts. Typical artifacts of recursive filters are "comet tails" that moving objects leave behind.

A *switching filter* is obtained if the gain takes on the values α and 1, depending on the difference between the prediction $\hat{f}_b(\mathbf{n}, k)$ and the actually observed signal value $g(\mathbf{n}, k)$:

$$\alpha(\mathbf{n}, k) = \begin{cases} 1 & \text{if } |g(\mathbf{n}, k) - \hat{f}_b(\mathbf{n}, k)| > \varepsilon \\ \alpha & \text{if } |g(\mathbf{n}, k) - \hat{f}_b(\mathbf{n}, k)| \le \varepsilon \end{cases}. \qquad (11)$$

For areas that have a lot of motion [if prediction (10a) is used] or for which the motion has been estimated incorrectly [if prediction (10b) is used], the difference between the predicted intensity value and the noisy intensity value is large, causing the filter to switch off. For areas that are stationary or for which the motion has been estimated correctly, the prediction differences are small, yielding the value α for the filter coefficient.

A finer adaptation is obtained if the prediction gain is optimized to minimize the mean-squared restoration error (3), yielding

$$\alpha(\mathbf{n}, k) = \max\left(1 - \frac{\sigma_w^2}{\sigma_g^2(\mathbf{n}, k)}, 0\right). \quad (12)$$

Here $\sigma_g^2(\mathbf{n}, k)$ is an estimate of the image sequence variance in a local spatiotemporal neighborhood of $(\mathbf{n}, k)$. If this variance is high, it indicates large motion or incorrectly estimated motion, causing the noise filter to switch off, i.e., $\alpha(\mathbf{n}, k) = 1$. If $\sigma_g^2(\mathbf{n}, k)$ is of the same order of magnitude as the noise variance σ_w^2, the observed noisy image sequence is obviously very unreliable so that the predicted intensities are used without updating it, i.e., $\alpha(\mathbf{n}, k) = 0$. The resulting estimator is known as the local linear minimum mean-squared error (LLMMSE) estimator. A drawback of Eq. (12), as with any noise filter that requires the calculation of $\sigma_g^2(\mathbf{n}, k)$, is that outliers in the windows used to calculate this variance may cause the filter to switch off. Order-statistic filters are more suitable for handling data in which outliers are likely to occur.

2.2 Order-Statistic Filters

Order-statistic (OS) filters are nonlinear variants of weighted-averaging filters. The distinction is that in OS filters the observed noisy data — usually taken from a small spatiotemporal window — are ordered before being used. Because of the ordering operation, correlation information is ignored in favor of magnitude information. Examples of simple OS filters are the minimum operator, maximum operator, and median operator. OS filters are often applied in *directional filtering*. In directional filtering, different filter directions are considered that correspond to different spatiotemporal edge orientations. Effectively this means that the filtering operation takes place along the spatiotemporal edges, avoiding the blurring of moving objects.

The directional filtering approach may be superior to adaptive or switching filters, since noise around spatiotemporal edges can effectively be eliminated by filtering along those edges, as opposed to turning off the filter in the vicinity of edges [15].

The general structure of an OS restoration filter is as follows:

$$\hat{f}(\mathbf{n}, k) = \sum_{r=1}^{|S|} = h_{(r)}(\mathbf{n}, k)g_{(r)}(\mathbf{n}, k). \quad (13)$$

Here $g_{(r)}(\mathbf{n}, k)$ are the ordered intensities, or *ranks*, of the corrupted image sequence, taken from a spatiotemporal window S with finite extent centered around $(\mathbf{n}, k)$; see Fig. 3. The number of intensities in this window is denoted by $|S|$. As with linear filters, the objective is to choose appropriate filter coefficients $h_{(r)}(\mathbf{n}, k)$ for the ranks.

The most simple order-statistic filter is a straightforward temporal median, for instance taken over three frames:

$$\hat{f}(\mathbf{n}, k) = \text{median}(g(\mathbf{n}, k - 1), g(\mathbf{n}, k), g(\mathbf{n}, k + 1)). \quad (14)$$

Filters of this type are very suitable for removing shot noise. In order to avoid artifacts at the edges of moving objects, Eq. (14) is normally applied in a motion-compensated way. A more elaborate OS filter is the multistage median filter (MMF) [1]. In the MMF the outputs of basic median filters with different spatiotemporal support are combined. An example of the spatiotemporal supports is shown in Fig. 4. The outputs of these intermediate median filter results are then combined as follows:

$$\hat{f}(\mathbf{n}, k) = \text{median}(g(\mathbf{n}, k), \max(\hat{f}_1(\mathbf{n}, k)), \dots, \hat{f}_9(\mathbf{n}, k)),$$
$$\min(\hat{f}_1(\mathbf{n}, k), \dots, \hat{f}_9(\mathbf{n}, k))). \quad (15)$$

The advantage of this class of filters is that although it does not incorporate motion estimation explicitly, artifacts on edges of moving objects are significantly reduced. Nevertheless, the intermediate medians can also be computed in a motion-compensated way by positioning the spatiotemporal windows in Fig. 4 along motion trajectories.

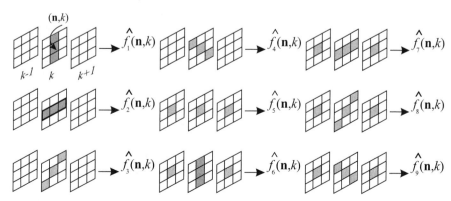

FIGURE 4   Spatiotemporal windows used in the multistage median filter.

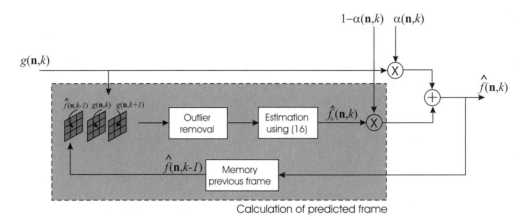

FIGURE 5 Overall filtering structure combining Eqs. (9) and (16) and an outlier-removing rank order test.

The filter coefficients $h_{(r)}(\mathbf{n}, k)$ in Eq. (13) can also be statistically designed, as described in Chapter 4.4 of this *Handbook*. If the coefficients are optimized in the mean-squared error sense, the following general solution for the restored image sequence is obtained [7]:

$$
\begin{pmatrix} \hat{f}(\mathbf{n}, k) \\ \hat{\sigma}_w^2(\mathbf{n}, k) \end{pmatrix}
$$

$$
= \begin{pmatrix} h_{(1)}(\mathbf{n}, k) & h_{(2)}(\mathbf{n}, k) & \cdots & h_{(|S|)}(\mathbf{n}, k) \\ n_{(1)}(\mathbf{n}, k) & n_{(2)}(\mathbf{n}, k) & \cdots & n_{(|S|)}(\mathbf{n}, k) \end{pmatrix} \begin{pmatrix} g_{(1)}(\mathbf{n}, k) \\ \vdots \\ g_{(|S|)}(\mathbf{n}, k) \end{pmatrix}
$$

$$
= \left(\mathbf{A}^t \mathbf{C}_{(w)}^{-1} \mathbf{A}\right)^{-1} \mathbf{A}^t \mathbf{C}_{(w)}^{-1} \begin{pmatrix} g_{(1)}(\mathbf{n}, k) \\ \vdots \\ g_{(|S|)}(\mathbf{n}, k) \end{pmatrix}.
$$

(16a)

This expression formulates the optimal filter coefficients $h_{(r)}(\mathbf{n}, k)$ in terms of a matrix product involving the $|S| \times |S|$ autocovariance matrix of the ranks of the noise, denoted by $\mathbf{C}_{(w)}$, and a matrix $\mathbf{A}$ defined as

$$
\mathbf{A} = \begin{pmatrix} 1 & E[w_{(1)}(\mathbf{n}, k)] \\ 1 & E[w_{(2)}(\mathbf{n}, k)] \\ \vdots & \vdots \\ 1 & E[w_{(|S|)}(\mathbf{n}, k)] \end{pmatrix}.
$$

(16b)

Here $E[w_{(r)}(\mathbf{n}, k)]$ denotes the expectation of the ranks of the noise. The result in Eq. (16a) gives an estimate not only of the filtered image sequence, but also for the local noise variance. This quantity is of use by itself in various noise filters to regulate the noise reduction strength. In order to calculate $E[w_{(r)}(\mathbf{n}, k)]$ and $\mathbf{C}_{(w)}$, the probability density function of the noise has to be assumed known. In case the noise $w(\mathbf{n}, k)$ is uniformly distributed, Eq. (16a) becomes the average of the minimum and maximum observed intensity. For Gaussian distributed noise, Eq. (16a) degenerates to Eq. (2) with equal weighting coefficients.

An additional advantage of ordering the noisy observation prior to filtering is that *outliers* can easily be detected. For instance, with a statistical test, such as the rank order test [7], the observed noisy values within the spatiotemporal window S that are significantly different from the intensity $g(\mathbf{n}, k)$ can be detected. These significantly different values originate usually from different objects or different motion patterns in the image sequence. By letting the statistical test reject these values, filters (13) and (16) use locally only data from the observed noisy image sequence that are close — in intensity — to $g(\mathbf{n}, k)$. This further reduces the sensitivity of noise filter (13) to outliers that are due to motion or incorrectly compensated motion.

Estimator (16) can also be used in a recursive structure such as the one in Eq. (9). Essentially (16) is then interpreted as an estimate for the local mean of the image sequence, and the filtered value resulting from Eq. (16) is used as the predicted value $\hat{f}_b(\mathbf{n}, k)$ in Eq. (9). Furthermore, instead of using only noisy observations in the estimator, previously filtered frames can be used by extending the spatiotemporal window S over the current noisy frame $g(\mathbf{n}, k)$ and the previously filtered frame $\hat{f}(\mathbf{n}, k-1)$. The overall filter structure thus obtained is shown in Fig. 5.

2.3 Multiresolution Filters

The multiresolution representation of 2-D images has become quite popular for analysis and compression purposes. This signal representation is also useful for image sequence restoration. The fundamental idea is that if an appropriate decomposition into bands of different spatial and temporal resolutions and orientations is carried out, the energy of the structured signal will locally be concentrated in selected bands whereas the noise is spread out over all bands. The noise can therefore effectively be removed by mapping all small (noise) components in all bands to zero, while leaving the remaining larger components relatively unaffected. Such an operation on signals is also known as *coring* [4]. Figure 6 shows two coring functions, namely soft and hard thresholding. Chapter 3.4 of this *Handbook* discusses 2-D wavelet-based thresholding methods for image enhancement.

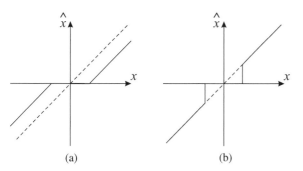

FIGURE 6 Coring functions: (a) soft thresholding, (b) hard thresholding. Here x is a signal amplitude taken from one of the spatiotemporal bands (which carry different resolution and orientation information), and $\hat{x}$ is the resulting signal amplitude after coring.

The discrete wavelet transform has been widely used for decomposing one-dimensional and multidimensional signals into bands. A problem with this transform for image sequence restoration is, however, that the decomposition is not shift invariant. Slightly shifting the input image sequence in spatial or temporal sense can cause significantly different decomposition results. For this reason, in [14] a shift-invariant, but overcomplete, decomposition was proposed, known as the Simoncelli pyramid. Figure 7(a) shows the 2-D Simoncelli pyramid decomposition scheme. The filters $L_i(\omega)$ and $H_i(\omega)$ are linear phase low- and high-pass filters, respectively. The filters $F_i(\omega)$ are fanfilters that decompose the signal into four directional bands. The resulting spectral decomposition is shown in Fig. 7(b). From this spectral tessellation, the different resolutions and orientations of the spatial bands obtained by Fig. 7(a) can be inferred. The radial bands have a bandwidth of 1 octave.

The Simoncelli pyramid gives a spatial decomposition of each frame into bands of different resolution and orientation. The extension to temporal dimension is obtained by temporally decomposing each of the spatial resolution and orientation bands using a regular wavelet transform. The low-pass and high-pass filters are operated along the motion trajectory in order to avoid blurring of moving objects. The resulting motion-compensated spatiotemporal wavelet coefficients are filtered by one of the coring functions, followed by the reconstruction of the video frame by an inverse wavelet transformation and Simoncelli pyramid reconstruction. Figure 8 shows the overall scheme.

Though multiresolution approaches have been shown to outperform the filtering techniques described in Sections 2.1 and 2.2 for some types of noise, they generally require much more processing power because of the spatial and temporal decomposition, and—depending on the temporal wavelet decomposition—they require a significant number of frame stores.

3 Blotch Detection and Removal

Blotches are artifacts that are typically related to film. Dirt particles covering film introduce bright or dark spots on the

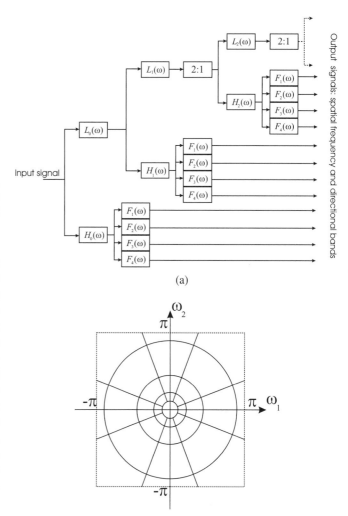

FIGURE 7 (a) Simoncelli pyramid decomposition scheme. (b) Resulting spectral decomposition, illustrating the spectral contents carried by the different resolution and directional bands.

frames, and the mishandling or aging of film causes loss of gelatin covering the film. Figure 11(a) on page 235 shows a film frame containing dark and bright spots: the blotches. A model for this artifact is the following [11]:

$$g(\mathbf{n}, k) = (1 - b(\mathbf{n}, k)) f(\mathbf{n}, k) + b(\mathbf{n}, k)c(\mathbf{n}, k) + w(\mathbf{n}, k).$$
(17)

Here $b(\mathbf{n}, k)$ is a binary mask that indicates for each spatial location in each frame whether or not it is part of a blotch. The (more or less constant) intensity values at the corrupted spatial locations are given by $c(\mathbf{n}, k)$. Though noise is not considered to be the dominant degrading factor in the section, it is still included in Eq. (17) as the term $w(\mathbf{n}, k)$. The removal of blotches is a two-step procedure. In the first, most complicated step, the blotches have to be detected; i.e., an estimate for the mask $b(\mathbf{n}, k)$ is made [8]. In the second step, the incorrect intensities $c(\mathbf{n}, k)$

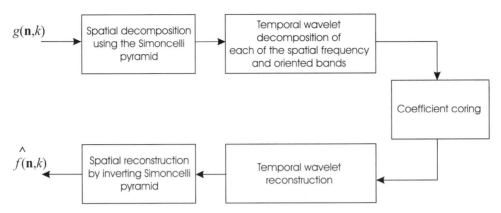

FIGURE 8 Overall spatiotemporal multiresolution filtering, using coring.

at the corrupted locations are spatiotemporally interpolated [9]. In case a motion-compensated interpolation is carried out, the second step also involves the local repair of motion vectors estimated from the blotched frames. The overall blotch detection and removal scheme is shown in Fig. 9.

3.1 Blotch Detection

Blotches have three characteristic properties that are exploited by blotch detection algorithms. In the first place, blotches are temporally independent and therefore hardly ever occur at the same spatial location in successive frames. In the second, the intensity of a blotch is significantly different from its neighboring uncorrupted intensities. Finally, blotches form coherent regions in a frame, as opposed to, for instance, spatiotemporal shot noise.

There are various blotch detectors that exploit these characteristics. The first is a pixel-based blotch detector known as the spike-detector index (SDI). This method detects temporal discontinuities by comparing pixel intensities in the current frame with motion-compensated reference intensities in the previous and following frame:

$$\text{SDI}(\mathbf{n}, k) = \min((g(\mathbf{n}, k) - g(\mathbf{n} - \mathbf{d}(\mathbf{n}; k, k-1), k-1))^2,$$
$$(g(\mathbf{n}, k) - g(\mathbf{n} - \mathbf{d}(\mathbf{n}; k, k+1), k+1))^2) \quad (18)$$

Since blotch detectors are pixel oriented, the motion field $\mathbf{d}(\mathbf{n}; k; l)$ should have a motion vector per pixel; i.e., the motion field is dense. Observe that any motion-compensation procedure must be robust against the presence of intensity spikes; this will

be discussed later in this Section. A blotch pixel is detected if $\text{SDI}(\mathbf{n}, k)$ exceeds a threshold:

$$b(\mathbf{n}, k) = \begin{cases} 1 & \text{if } \text{SDI}(\mathbf{n}, k) > T \\ 0 & \text{otherwise} \end{cases}. \quad (19)$$

Since blotch detectors are essentially searching for outliers, order-statistic-based detectors usually perform better. The rank-order difference (ROD) detector is one such method. It takes $|S|$ reference pixel intensities from a motion-compensated spatiotemporal window S (see, for instance, Fig. 10), and finds the deviation between the pixel intensity $g(\mathbf{n}, k)$ and the reference pixel r_i ranked by intensity value as follows:

$$\text{ROD}_i(\mathbf{n}, k) = \begin{cases} r_i - g(\mathbf{n}, k) & \text{if } g(\mathbf{n}, k) \leq \text{median}(r_i) \\ g(\mathbf{n}, k) - r_i & \text{if } g(\mathbf{n}, k) > \text{median}(r_i) \end{cases}$$
$$\text{for } i = 1, 2, \dots, \frac{|S|}{2}. \quad (20)$$

A blotch pixel is detected if any of the rank order differences exceeds a specific threshold T_i:

$$b(\mathbf{n}, k) = \begin{cases} 1 & \text{if } \text{ROD}_i(\mathbf{n}, k) > T_i \\ 0 & \text{otherwise} \end{cases} \quad (21)$$

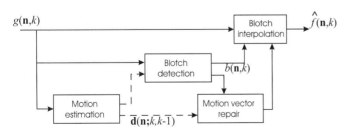

FIGURE 9 Blotch detection and removal system.

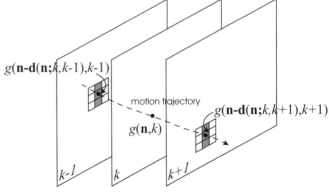

FIGURE 10 Example of motion-compensated spatiotemporal window for obtaining reference intensities in the ROD detector.

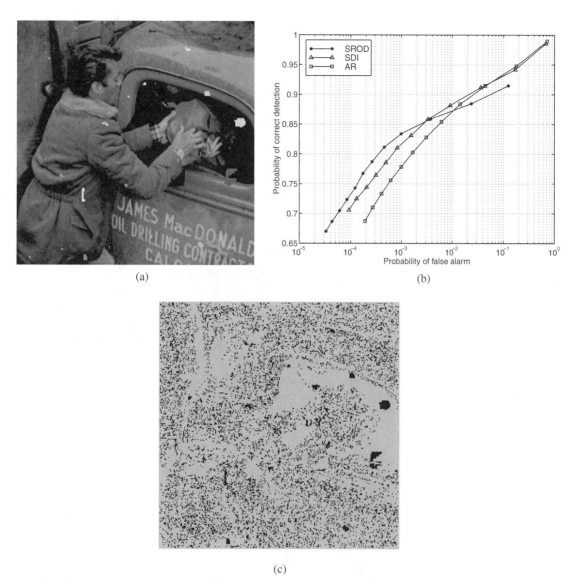

FIGURE 11 (a) Video frame with blotches. (b) Correct detection versus false detection for three different blotch detectors. (c) Blotch detection mask, using the sROD ($T = 0$).

More complicated blotch detectors explicitly incorporate a model for the uncorrupted frames, such as a two- or three-dimensional autoregressive model or a Markov random field to develop the maximum *a posteriori* detector for the blotch mask $b(\mathbf{n}, k)$. Figure 11(b) illustrates the detection probability versus the false detection probability of three different detectors, on a sequence of which a representative blotched frame is shown in Figure 11(a). These results indicate that for reasonable detection probabilities the false detection probability is fairly high. False detections are detrimental to the restoration process because the interpolation process itself is fallible and may introduce disturbing artifacts that were not present in the blotched image sequence.

The blotch detectors described so far are essentially pixel-based detectors. They do not incorporate the spatial coherency of the detected blotches. The effect is illustrated by Figure 11(c), which shows the detected blotch mask $b(\mathbf{n}, k)$, using a simplified version of the ROD detector (with $T_i \rightarrow \infty$, $i \geq 2$). The simplified ROD (sROD) is given by the following relations [11]:

$$\text{sROD}(\mathbf{n}, k) = \begin{cases} \min(r_1, \ldots, r_{|S|}) - g(\mathbf{n}, k) \\ \quad \text{if } \min(r_1, \ldots, r_{|S|}) - g(\mathbf{n}, k) > 0 \\ g(\mathbf{n}, k) - \max(r_1, \ldots, r_{|S|}) \\ \quad \text{if } g(\mathbf{n}, k) - \max(r_1, \ldots, r_{|S|}) > 0, \\ 0 \quad \text{elsewhere} \end{cases} \quad (22a)$$

$$b(\mathbf{n}, k) = \begin{cases} 1 & \text{if sROD}(\mathbf{n}, k) > T \\ 0 & \text{otherwise} \end{cases}. \quad (22b)$$

The sROD basically looks at the range of the reference pixel intensities obtained from the motion-compensated window, and it compares the range with the pixel intensity under investigation. A blotch pixel is detected if the intensity of the current pixel $g(\mathbf{n}, k)$ lies far enough outside that range.

The performance of even this simple pixel-based blotch detector can be improved significantly by exploiting the spatial coherence of blotches. This is done by postprocessing the blotch mask in Fig. 11(c) in two ways, namely by removing small blotches and by completing partially detected blotches. We first discuss the removal of small blotches. Detector output (22a) is not only sensitive to intensity changes caused by blotches corrupting the image sequence, but also to noise. If the probability density function of the noise — denoted by $f_W(w)$ — is known, the probability of false detection for a single pixel can be calculated. Namely, if the sROD uses $|S|$ reference intensities in evaluating Eq. (22a), the probability that $sROD(\mathbf{n}, k)$ for a single pixel is larger than T due to noise only is [11]:

$$P(sROD(\mathbf{n}, k) > T \mid no\ blotch)$$
$$= 2P(g(\mathbf{n}, k) - \max(r_i) > T \mid no\ blotch)$$
$$= 2 \int_{-\infty}^{\infty} \left[\int_{-\infty}^{u-T} f_W(w)dw \right]^S f_W(u)du. \qquad (23)$$

In the detection mask $b(\mathbf{n}, k)$ blotches may consist of single pixels or of multiple connected pixels. A set of connected pixels that are all detected as (being part of a) blotch is called a spatially coherent blotch. If a coherent blotch consists of N connected pixels, the probability that this blotch is due to noise

only is

$$P(sROD(\mathbf{n}, k) > T\ for\ N\ connected\ pixels \mid no\ blotch)$$
$$= (P(sROD(\mathbf{n}, k) > T \mid no\ blotch))^N. \qquad (24)$$

When this false detection probability is bounded to a certain maximum, the minimum number of pixels identified by the sROD detector as being part of a blotch can be computed. Consequently, coherent blotches consisting of fewer pixels than this minimum are removed from the blotch mask $b(\mathbf{n}, k)$.

A second postprocessing technique for improving the detector performance is hysteresis thresholding. First a blotch mask is computed by using a very low detection threshold T, for instance $T = 0$. From the detection mask the small blotches are removed as described above, yielding the mask $b_0(\mathbf{n}, k)$. Nevertheless, because of the low detection threshold this mask still contains many false detections. Then a second detection mask $b_1(\mathbf{n}, k)$ is obtained by using a much higher detection threshold. This mask contains fewer detected blotches and the false detection rate in this mask is small. The second detection mask is now used to validate the detected blotches in the first mask: only those spatially coherent blotches in $b_0(\mathbf{n}, k)$ that have a corresponding blotch in $b_1(\mathbf{n}, k)$ are preserved; all others are removed. The result of the above two postprocessing techniques on the frame shown in Fig. 11(a) is shown in Fig. 12(a). In Fig. 12(b) the detection and false detection probabilities are shown.

3.2 Motion Vector Repair and Interpolating Corrupted Intensities

Block-based motion estimators will generally find the correct motion vectors even in the presence of blotches, provided that

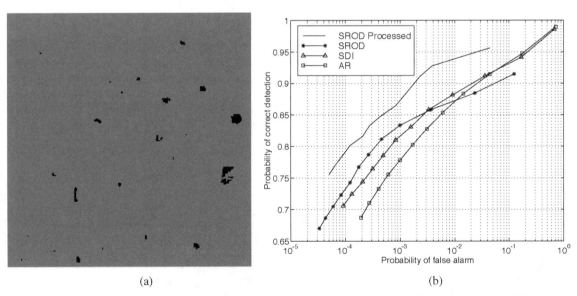

(a)   (b)

FIGURE 12   (a) Blotch detection mask after postprocessing. (b) Correct detection versus false detections obtained for sROD with postprocessing (top curve), compared to results from Fig. 11(b).

the blotches are small enough. The disturbing effect of blotches is usually confined to small areas of the frames. Hierarchical motion estimators will experience little influence of the blotches at the lower resolution levels. At higher resolution levels, blotches covering larger parts of (at those levels) small blocks will significantly influence the motion estimation result. If the blotch mask $b(\mathbf{n}, k)$ has been estimated, it is also known which estimated motion vectors are unreliable.

There are two strategies in recovering motion vectors that are known to be unreliable. The first approach is to take an average of surrounding motion vectors. This process — known as *motion vector interpolation or motion vector repair* — can be realized by using, for instance, the median or average of the motion vectors of uncorrupted regions adjacent to the corrupted blotch. Though simple, the disadvantages of averaging are that motion vectors may be created that are not present in the uncorrupted part of the image and that no validation of the selected motion vector on the actual frame intensities takes place.

The second, more elaborate, approach circumvents this disadvantage by validating the corrected motion vectors using intensity information directly neighboring the blotched area. As a validation criterion the motion-compensated mean squared intensity difference can be used [3]. Candidates for the corrected motion vector can be obtained either from motion vectors taken from adjacent regions or by motion reestimation using a spatial window containing only uncorrupted data such as the pixels directly bordering the blotch.

The estimation of the frame intensities labeled by the mask as being part of a blotch can be done either by a spatial or temporal interpolation, or a combination of both. We concentrate on spatiotemporal interpolation. Once the motion vector for a blotched area has been repaired, the correct temporally neighboring intensities can be obtained. In a multistage median interpolation filter, five interpolated results are computed by using the (motion-compensated) spatiotemporal neighborhoods shown in Fig. 13. Each of the five interpolated results is computed as the median over the corresponding neighborhood S_i:

$$\hat{f}_i(\mathbf{n}, k) = \text{median}\big(\{\mathbf{n} \in S_i^{k-1} \mid f(\mathbf{n}, k-1)\},$$
$$\{\mathbf{n} \in S_i^k \mid f(\mathbf{n}, k)\}, \{\mathbf{n} \in S_i^{k+1} \mid f(\mathbf{n}, k+1)\}\big). \quad (25)$$

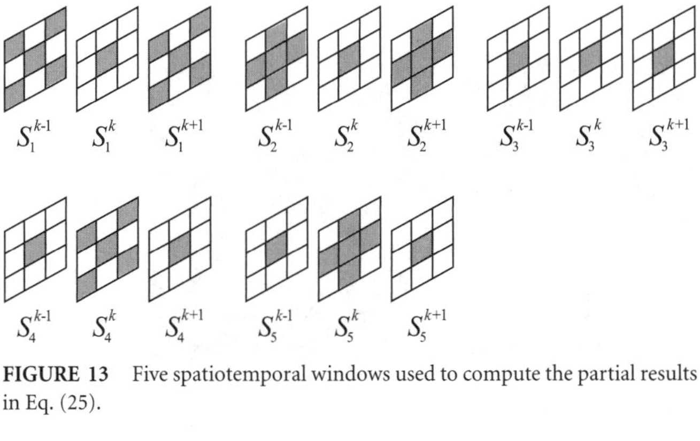

FIGURE 13 Five spatiotemporal windows used to compute the partial results in Eq. (25).

The final result is computed as the median over the five intermediate results:

$$\hat{f}(\mathbf{n}, k) = \text{median}(\hat{f}_1(\mathbf{n}, k), \hat{f}_2(\mathbf{n}, k),$$
$$\hat{f}_3(\mathbf{n}, k), \hat{f}_4(\mathbf{n}, k), \hat{f}_5(\mathbf{n}, k)). \quad (26)$$

The multistage median filter does not rely on any model for the image sequence. Though simple, this is at the same time a drawback of median filters. If a model for the original image sequence can be assumed, it is possible to find statistically optimal values for the missing intensities. For the sake of completeness we mention here that if one assumes the popular Markov random field, the following complicated expression has to be optimized:

$$P(\hat{f}(\mathbf{n}, k) \mid f(\mathbf{n} - \mathbf{d}(\mathbf{n}; k, k-1), k-1), f(\mathbf{n}, k),$$
$$f(\mathbf{n} - \mathbf{d}(\mathbf{n}; k, k+1), k+1))$$

$$\propto \exp\left(-\sum_{\mathbf{m}:d(\mathbf{m},k)=1}\sum_{\mathbf{s}\in S^k}(\hat{f}(\mathbf{m}, k) - \hat{f}(\mathbf{s}, k))^2+\right.$$
$$\lambda \sum_{\mathbf{s}\in S^{k-1}}(\hat{f}(\mathbf{m}, k) - f(\mathbf{n} - \mathbf{d}(\mathbf{n}; k, k-1), k-1))^2$$
$$\left.+ \sum_{\mathbf{s}\in S^{k+1}}(\hat{f}(\mathbf{m}, k) - f(\mathbf{n} - \mathbf{d}(\mathbf{n}; k, k+1), k+1))^2\right). \quad (27)$$

The first term on the right-hand side of Eq. (27) forces the interpolated intensities to be spatially smooth, while the second and third term enforce temporal smoothness. The sets S^{k-1}, S^k, and S^{k+1} denote appropriately chosen spatial windows in the frames $k-1$, k, and $k+1$. The temporal smoothness is calculated along the motion trajectory using the repaired motion vectors. The optimization of Eq. (27) requires an *iterative optimization technique*. If a simpler 3-D autoregressive model for the image sequence is assumed, the interpolated values can be calculated by solving a set of linear equations.

Instead of interpolating the corrupted intensities, it is also possible to directly copy and paste intensities from past or future frames. The simple copy-and-paste operation instead of a full spatiotemporal data regeneration is motivated by the observation that, at least on local and motion-compensated basis, image sequences are heavily correlated. Furthermore, straightforward interpolation is not desirable in situations in which part of the information in the past and future frames itself is unreliable, for instance if it was part of a blotch itself or if it is situated in an occluded area. The objective is now to determine — for each pixel being part of a detected blotch — if intensity information from the previous or next frame should be used. This decision procedure can again be cast into a statistical framework [11]. As an illustration, Fig. 14 shows the interpolated result of the blotched frame in Fig. 11(a).

FIGURE 14 Blotch-corrected frame resulting from Fig. 11(a).

4 Intensity Flicker Correction

Intensity flicker is defined as unnatural temporal fluctuations of frame intensities that do not originate from the original scene. Intensity flicker is a spatially localized effect that occurs in regions of substantial size. Figure 15 shows three successive frames from a sequence containing flicker. A model describing the intensity flicker is the following:

$$g(\mathbf{n}, k) = \alpha(\mathbf{n}, k) f(\mathbf{n}, k) + \beta(\mathbf{n}, k) + w(\mathbf{n}, k). \quad (28)$$

Here, $\alpha(\mathbf{n}, k)$ and $\beta(\mathbf{n}, k)$ are the multiplicative and additive *unknown* flicker parameters, which locally scale the intensities of the original frame. The model includes a noise term $w(\mathbf{n}, k)$ that is assumed to be flicker independent. In the absence of flicker we have $\alpha(\mathbf{n}, k) = 1$ and $\beta(\mathbf{n}, k) = 0$. The objective of flicker correction is the estimation of the flicker parameters, followed by the inversion of Eq. (28). Since flicker always affects fairly large areas of a frame in the same way, the flicker parameters $\alpha(\mathbf{n}, k)$ and $\beta(\mathbf{n}, k)$ are assumed to be spatially smooth functions. Temporally the flicker parameters in one frame may not be correlated at all with those in a subsequent frame.

The earliest attempts to remove flicker from image sequences applied intensity histogram equalization or mean equalization on frames. These methods do not form a general solution to the problem of intensity flicker correction because they ignore changes in scene contents, and they do not appreciate that intensity flicker is a localized effect. In Section 4.1 we show how the flicker parameters can be estimated on stationary image sequences. Section 4.2 addresses the more realistic case of parameter estimation on image sequences with motion [12].

4.1 Flicker Parameter Estimation

When removing intensity flicker from an image sequence, we essentially make an estimate of the original intensities, given the observed image sequence. Note that the undoing of intensity flicker is only relevant for image sequences, since flicker is a temporal effect by definition. From a single frame intensity flicker cannot be observed nor be corrected.

If the flicker parameters were known, then one could form an estimate of the original intensity from a corrupted intensity by using the following straightforward linear estimator:

$$\hat{f}(\mathbf{n}, k) = h_1(\mathbf{n}, k) g(\mathbf{n}, k) + h_0(\mathbf{n}, k). \quad (29)$$

In order to obtain estimates for the coefficients $h_i(\mathbf{n}, k)$, the mean-squared error between $f(\mathbf{n}, k)$ and $\hat{f}(\mathbf{n}, k)$ is minimized, yielding the following optimal solution:

$$h_0(\mathbf{n}, k) = -\frac{1}{\alpha(\mathbf{n}, k)} \left(\beta(\mathbf{n}, k) + \frac{\sigma_w^2(\mathbf{n}, k)}{\sigma_g^2(\mathbf{n}, k)} E[g(\mathbf{n}, k)] \right), \quad (30a)$$

$$h_1(\mathbf{n}, k) = \frac{1}{\alpha(\mathbf{n}, k)} \frac{\sigma_g^2(\mathbf{n}, k) - \sigma_w^2(\mathbf{n}, k)}{\sigma_g^2(\mathbf{n}, k)}. \quad (30b)$$

If the observed image sequence does not contain any noise, then Eq. (30) degenerates to the obvious solution:

$$h_0(\mathbf{n}, k) = -\frac{\beta(\mathbf{n}, k)}{\alpha(\mathbf{n}, k)}, \qquad h_1(\mathbf{n}, k) = \frac{1}{\alpha(\mathbf{n}, k)}. \quad (31)$$

FIGURE 15 Three successive frames that contain intensity flicker.

In the extreme situation that the variance of the corrupted image sequence is equal to the noise variance, the combination of Eqs. (29) and (30) shows that the estimated intensity is equal to the expected value of the original intensities $E[f(\mathbf{n}, k)]$.

In practice, the true values for the intensity-flicker parameters $\alpha(\mathbf{n}, k)$ and $\beta(\mathbf{n}, k)$ are unknown and have to be estimated from the corrupted image sequence itself. Since the flicker parameters are spatially smooth functions, we assume that they are locally constant:

$$\left.\begin{array}{l} \alpha(\mathbf{n}, k) = \alpha_{\mathbf{m}}(k) \\ \beta(\mathbf{n}, k) = \beta_{\mathbf{m}}(k) \end{array}\right\} \forall \mathbf{n} \in S_{\mathbf{m}}, \tag{32}$$

where $S_{\mathbf{m}}$ indicates a small frame region. This region can, in principle, be arbitrarily shaped, but in practice rectangular blocks are chosen. By computing the averages and variances of both sides of Eq. (28), one can obtain the following analytical expressions for the estimates of $\alpha_{\mathbf{m}}(k)$ and $\beta_{\mathbf{m}}(k)$:

$$\hat{\alpha}_{\mathbf{m}}(k) = \sqrt{\frac{\sigma_g^2(\mathbf{n}, k) - \sigma_w^2(\mathbf{n}, k)}{\sigma_f^2(\mathbf{n}, k)}}, \tag{33}$$

$$\hat{\beta}_{\mathbf{m}}(k) = E[g(\mathbf{n}, k)] - \hat{\alpha}_{\mathbf{m}}(k) E[f(\mathbf{n}, k)].$$

To solve Eq. (33) in a practical situation, the mean and variance of $g(\mathbf{n}, k)$ are estimated within the region $S_{\mathbf{m}}$. The only quantities that remain to be estimated are the mean and variance of the original image sequence $f(\mathbf{n}, k)$. If we assume that the flicker correction is done frame by frame, we can estimate these values from the previous corrected frame $k - 1$ in the temporally corresponding frame region $S_{\mathbf{m}}$:

$$E[f(\mathbf{n}, k)] \approx \frac{1}{|S_{\mathbf{m}}|} \sum_{\mathbf{m} \in S_{\mathbf{m}}} \hat{f}(\mathbf{m}, k - 1),$$

$$\sigma_f{}^2(\mathbf{n}, k) \approx \frac{1}{|S_{\mathbf{m}}|} \sum_{\mathbf{m} \in S_{\mathbf{m}}} (\hat{f}(\mathbf{m}, k - 1) - E[f(\mathbf{n}, k)])^2. \tag{34}$$

There are situations in which the above estimates are unreliable. The first case is that of uniform intensity areas. For any original image intensity in a uniform regions, there are an infinite number of combinations of $\alpha_{\mathbf{m}}(k)$ and $\beta_{\mathbf{m}}(k)$ that lead to the observed intensity. The estimated flicker parameters are also potentially unreliable because of ignoring the noise $w(\mathbf{n}, k)$ in Eqs. (33) and (34). The reliability of the estimated flicker parameters can be assessed by the following measure:

$$W_{\mathbf{m}}(k) = \begin{cases} 0 & \text{if } \sigma_g^2(\mathbf{m}, k) < T \\ \sqrt{\frac{\sigma_g^2(\mathbf{m}, k) - T}{T}} & \text{otherwise} \end{cases}. \tag{35}$$

The threshold T depends on the noise variance. Large values of $W_{\mathbf{m}}(k)$ indicate reliable estimates, whereas for the most unreliable estimates $W_{\mathbf{m}}(k) = 0$.

4.2 Estimation on Sequences with Motion

Results (33) and (34) assume that the image sequence intensities do not change significantly over time. Clearly, this is an incorrect assumption if motion occurs. The estimation of motion on image sequences that contain flicker is, however, problematic because virtually all motion estimators are based on the constant luminance constraint. Because of the intensity flicker this assumption is violated heavily. The only motion that can be estimated with sufficient reliability is global motion such as camera panning or zooming. In the following we assume that in the evaluation of Eqs. (34) and (35), possible global motion is compensated for. At that point we still need to detect areas with any remaining — and uncompensated — motion, and areas that were previously occluded. For both of these cases the approximation in Eq. (34) leads to incorrect estimates, which in turn lead to visible artifacts in the corrected frames.

There are various approaches for detecting local motion. One possibility is the detection of large differences between the current and previously (corrected) frame. If local motion occurs, the frame differences will be large. Another possibility to detect local motion is to compare the estimated intensity-flicker parameters to threshold values. If disagreeing temporal information has been used for computing Eq. (34), we will locally find flicker parameters that do not correspond with the spatial neighbors or with the *a priori* expectations of the range of the flicker parameters. An outlier detector can be used to localize these incorrectly estimated parameters.

For frame regions $S_{\mathbf{m}}$ where the flicker parameters could not be estimated reliably from the observed image sequence, the parameters are estimated on the basis of the results in spatially neighboring regions. At the same time, for the regions in which the flicker parameters could be estimated, a smoothing postprocessing step has to be applied to avoid sudden parameter changes that lead to visible artifacts in the corrected image sequence. Such an interpolation and smoothing postprocessing step may exploit the reliability of the estimated parameters, as for instance given by Eq. (35). Furthermore, in those frame regions where insufficient information was available for reliably estimating the flicker parameters, the flicker correction should switch off itself. Therefore, smoothed and interpolated parameters are biased toward $\alpha_{\mathbf{m}}(k) = 1$ and $\beta_{\mathbf{m}}(k) = 0$.

In Fig. 16 below, an example of smoothing and interpolating the estimated flicker parameter for $\alpha_{\mathbf{m}}(k)$ is shown as a 2-D matrix [12]. Each entry in this matrix corresponds to a 30×30 pixel region $\Omega_{\mathbf{m}}$ in the frame shown in Fig. 15. The interpolation technique used is successive overrelaxation (SOR). Successive overrelaxation is a well-known iterative interpolation technique based on repeated low-pass filtering. Starting off with an initial estimate $\alpha_{\mathbf{m}}^0(k)$ found by solving Eq. (33), at each iteration a new

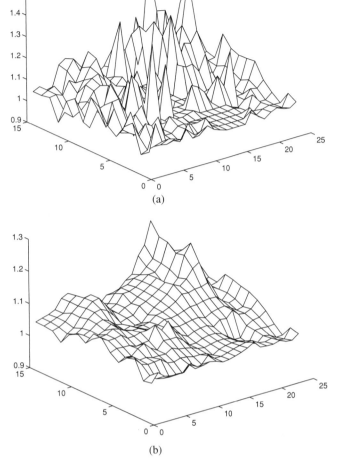

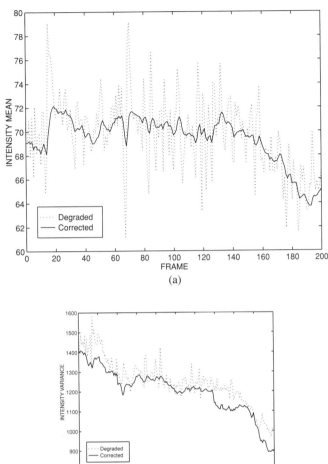

FIGURE 16 (a) Estimated intensity flicker parameter $\alpha_{\mathbf{m}}(k)$, using Eq. (33) and local motion detection. (b) Smoothed and interpolated $\alpha_{\mathbf{m}}(k)$, using SOR.

FIGURE 17 (a) Mean of the corrupted and corrected image sequence. (b) Variance of the corrupted and corrected image sequence.

estimate is formed as follows:

$$r_{\mathbf{m}}^{i+1}(k) = W_{\mathbf{m}}(k)\big(\alpha_{\mathbf{m}}^{i}(k) - \alpha_{\mathbf{m}}^{0}(k)\big) + \lambda C\big(\alpha_{\mathbf{m}}^{i}(k)\big),$$

$$\alpha_{\mathbf{m}}^{i+1}(k) = \alpha_{\mathbf{m}}^{i}(k) + \omega \frac{r_{\mathbf{m}}^{i+1}(k)}{W_{\mathbf{m}}(k) + \lambda}. \qquad (36)$$

Here $W_{\mathbf{m}}(k)$ is the reliability measure, computed by Eq. (35), and $C(\alpha_{\mathbf{m}}(k))$ is a function that measures the spatial smoothness of the solution $\alpha_{\mathbf{m}}(k)$. The convergence of iteration (36) is determined by the parameter ω, while the smoothness is determined by the parameter λ. For those estimates that have a high reliability, the initial estimates $\alpha_{\mathbf{m}}^{0}(k)$ are emphasized, whereas for the initial estimates that are deemed less reliable, i.e., $\lambda \gg W_{\mathbf{m}}(k)$, emphasis is on achieving a smooth solution. Other smoothing and interpolation techniques include dilation and 2-D polynomial interpolation. The smoothing and interpolation has to be applied not only to the multiplicative parameter $\alpha_{\mathbf{m}}(k)$, but also to the additive parameter $\beta_{\mathbf{m}}(k)$.

As an example, Fig. 17 shows the mean and variance as a function of the frame index k of the corrupted and corrected image sequence, "Tunnel." Clearly, the temporal fluctuations of the mean and variance have been greatly reduced, indicating the suppression of flicker artifacts. An assessment of the resulting visual quality, as with most results of video processing algorithms, has been done by actually viewing the corrected image sequences. Although the original sequence cannot be recovered, the flicker-corrected sequences have a much higher visual quality and they are virtually without any remaining visible flicker.

5 Concluding Remarks

This chapter has described methods for enhancing and restoring corrupted video and film sequences. The material that was offered in this chapter is complementary to the spatial enhancement and restoration techniques described in other chapters of the *Handbook*. For this reason, the algorithmic details concentrated on the temporal processing aspects of image sequences.

Although the focus has been on noise removal, blotch detection and correction, and flicker removal, the approaches and tools described in this chapter are of a more general nature, and they can be used for developing enhancement and restoration methods for other types of degradation.

References

[1] G. R. Arce, "Multistage order statistic filters for image sequence processing," *IEEE Trans. Signal Process.* **39**, 1147–1163 (1991).

[2] J. C. Brailean, R. P. Kleihorst, S. Efstratiadis, A. K. Katsaggelos, and R. L. Lagendijk, "Noise reduction filters for dynamic image sequences: A review," *Proc. IEEE* **83**, 1272–1291 (1995).

[3] M. J. Chen, L. G. Chen, and R. Weng, "Error Concealment of lost motion vectors with overlapped Motion compensation," *IEEE Trans. Circuits Sys. Video Technol.* **7**, 560–563 (1997).

[4] D. L. Donoho and I. M. Johnstone, "Ideal spatial adaptation via wavelet shrinkage," *Biometrika* **81**, 425–455 (1994).

[5] E. Dubois and S. Sabri, "Noise reduction using motion compensated temporal filtering," *IEEE Trans. Commun.* **32**, 826–831 (1984).

[6] T. S. Huang (ed.), *Image Sequence Analysis.* (Springer-Verlag, Berlin, 1991).

[7] R. P. Kleihorst., R. L. Lagendijk, and J. Biemond, "Noise reduction of image sequences using motion compensation and signal decomposition," *IEEE Trans. Image Process.* **4**, 274–284 (1995).

[8] A. C. Kokaram, R. D. Morris, W. J. Fitzgerald, and P. J. W. Rayner, "Detection of missing data in image sequences," *IEEE Trans. Image Process.* **4**, 1496–1508 (1995).

[9] A. C. Kokaram, R. D. Morris, W. J. Fitzgerald, and P. J. W. Rayner, "Interpolation of missing data in image sequences," *IEEE Trans. Image Process.* **4**, 1509–1519 (1995).

[10] M. K. Ozkan, M. I. Sezan, and A. M. Tekalp, "Adaptive motion-compensated filtering of noisy image sequences," *IEEE Trans. Circuits Sys. Video Technol.* **3**, 277–290 (1993).

[11] P. M. B. Roosmalen, "Restoration of archived film and video," Ph.D. dissertation (Delft University of Technology, The Netherlands, 1999).

[12] P. M. B. van Roosmalen, R. L. Lagendijk, and J. Biemond, "Correction of intensity flicker in old film sequences," *IEEE Trans. Circuits Sys. Video Technol.* **9**(7), 1013–1019 (1999).

[13] M. I. Sezan and R. L. Lagendijk, eds., *Motion Analysis and Image Sequence Processing* (Kluwer, Boston, MA, 1993).

[14] E. P. Simoncelli, W. T. Freeman, E. H. Adelson, and D. J. Heeger, "Shiftable multiscale transform," *IEEE Trans. Inf. Theory* **38**, 587–607 (1992).

[15] A. M. Tekalp, *Digital Video Processing* (Prentice-Hall, Upper Saddle River, NJ, 1995).

[16] J. W. Woods and J. Kim, "Motion-compensated spatiotemporal Kalman filter," in M. I. Sezan and R. L. Lagendijk, eds., *Motion Analysis and Image Sequence Processing* (Kluwer, Boston, MA, 1993).

3.12

3-D Shape Reconstruction from Multiple Views

Huaibin Zhao and
J. K. Aggarwal
*The University of Texas
at Austin*

Chhandomay Mandal
Sun Microsystems, Inc.

Baba C. Vemuri
University of Florida

1 Problem Definition and Applications

One of the recurring problems in computer vision is the inference of the three-dimensional (3-D) structure of an object or a scene from its two-dimensional (2-D) projections. Analysis of multiple images of the same scene taken from different viewpoints has emerged as an important method for extracting the 3-D structure of a scene. Generally speaking, extracting structure from multiple views of a scene involves determination of the 3-D shape of visible surfaces in the static scene from images acquired by two or more cameras (stereo sequences) or from one camera at multiple positions (monocular sequences). That is, we identify the 3-D description of a scene through images of the scene obtained from different viewpoints. With this 3-D description, we can create models of terrain and other natural environments for use in robot navigation, flight simulation, virtual reality, human-computer interactions, stereomicroscopy, and so on.

In the classical stereo problem [1, 2], after the initial camera calibration, correspondence is found among a set of points in the multiple images by using either a flow-based or a feature-based approach. Disparity computation for the matched points is then performed, followed by interpolation to produce piecewise smooth surfaces. Establishing correspondences between point locations in images acquired from multiple views (matching) is the key problem in reconstruction from multiple view images as well as in stereo image analysis. Two types of approaches are

used for the computation of correspondences from a sequence of frames — the optical flow-based approach and the feature-based approach. The flow-based approach uses the brightness constancy assumption to find a transformation between the images that maps corresponding points in these images into one another. The feature-based approach involves detecting the feature points and tracking their positions in multiple views of the scene. Dhond and Aggarwal [1] presented an excellent review of the problem in which they discussed the developments in establishing stereo correspondence for the extraction of the 3-D structure of a scene up to the end of the 1980's. A few well-known algorithms representing widely different approaches were presented. The focus of the review was stereo matching.

In this chapter, we will present a state of the art review of the major developments and techniques that have evolved in the past decade for recovering depth by using images from multiple views. We will not only include the stereo computation methods developed in this decade, but also describe a self-contained procedure to reconstruct a 3-D scene from multiple images of a scene acquired from different views taken by either calibrated or uncalibrated cameras. Our purpose is to guide those readers who are setting up their own 3-D environment from multiple view images, as well as to present a critical overview of the current stereo and multiview image analysis techniques.

The rest of the chapter is organized as follows. Section 2 briefly reviews the algebraic projective geometry of cameras, which is the foundation for the rest of the chapter. Knowing the projective geometry among various coordinate systems (world coordinate system, camera coordinate systems, and image coordinate systems), we can calculate the 3-D information from accurate correspondences between the images. In Section 3, we discuss optic flow-based as well as feature-based matching techniques. We present reconstruction techniques, including camera calibration, and an uncalibrated stereo analysis in Section 4. After demonstrating a simple example to show the performance of a stereo reconstruction process, we finally conclude in Section 6.

2 Preliminaries: The Projective Geometry of Cameras

In this section, we briefly review the basic geometric material that is essential for stereo and multiview image analysis. The reader is referred to the book by Faugeras [3] for a detailed introduction. We assume the *pinhole* camera model, which can ideally be modeled as a linear projection from 3-D space into each 2-D image. Consider four coordinate systems: the fixed reference coordinate system (the world coordinate system), the camera coordinate system, the ideal image coordinate system, and the real image coordinate system, shown in Fig. 1. The camera coordinate system (C, X', Y', Z') is centered at the optical center C, and the Z axis coincides with the optical axis of the camera. The ideal image coordinate system (c, x, y) is defined such that the origin c (called the *principal point*) is at the intersection of the optical axis with the image plane and that the X and Y axes are aligned with the axes of the camera-centered coordinate system. For a 3-D point $\mathbf{M}$, its coordinates $\mathbf{M}_c = [X', Y', Z']$ in the camera coordinate system and the coordinates $\mathbf{m} = [x, y]$ of its projection in the ideal image coordinate system can be related by

$$s \begin{bmatrix} x \\ y \\ 1 \end{bmatrix} = \begin{bmatrix} f & 0 & 0 & 0 \\ 0 & f & 0 & 0 \\ 0 & 0 & 1 & 0 \end{bmatrix} \begin{bmatrix} X' \\ Y' \\ Z' \\ 1 \end{bmatrix}, \tag{1}$$

$$s\tilde{\mathbf{m}} = \mathbf{P}\tilde{\mathbf{M}}_c, \tag{2}$$

where $\tilde{\mathbf{m}} = [x, y, 1]^T$ and $\tilde{\mathbf{M}}_c = [X, Y, Z, 1]^T$ are the augmented vector of $\mathbf{m}$ and $\mathbf{M}_c$ by adding 1 as the last element. The 3×4 matrix $\mathbf{P}$ is called the *camera perspective projection matrix*, which is determined by the camera focal length f. Here s represents the depth, i.e., $s = Z'$, and therefore cannot be determined from a single image.

In practice, we usually express a 3-D point in a fixed 3-D coordinate system referred to as the *world coordinate system* or the *reference coordinate system*. For a single point $\mathbf{M}$, the relation between its coordinates in the camera system $\mathbf{M}_c = [X', Y', Z']^T$

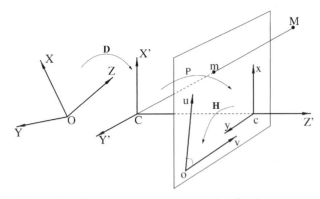

FIGURE 1 Coordinate systems and camera extrinsic and intrinsic parameters: (O, X, Y, Z), world coordinate system; (C, X', Y', Z'), camera coordinate system; (c, x, y), ideal image coordinate system; (o, u, v), real image coordinate system.

and in the world system $\mathbf{M}_w = [X, Y, Z]^T$ can be written as

$$\tilde{\mathbf{M}}_c = \mathbf{D}\tilde{\mathbf{M}}_w, \tag{3}$$

$$[X' \ Y' \ Z' \ 1]^T = \begin{bmatrix} \mathbf{R}_{3\times3} & \mathbf{t}_{3\times1} \\ 0 \ \ 0 \ \ 0 & 1 \end{bmatrix}^{-1} [X \ Y \ Z \ 1]^T. \tag{4}$$

The 4×4 matrix $\mathbf{D}$ is called the *extrinsic matrix*, which is specified by the 3×3 rotation matrix $\mathbf{R}$ and 3×1 translation vector $\mathbf{t}$. $\mathbf{R}$ gives axis orientations of the camera in the reference coordinate system; $\mathbf{t}$ gives the pose of the camera center in the reference coordinate system.

In practical applications, the image coordinate system (o, u, v) in which we address the pixels in an image is decided by the camera sensing array and is usually not the same as the ideal image coordinate system (c, x, y). The origin o of the actual image plane generally does not coincide with the optical principle point C, because of possible misalignment of the sensing array. Determined by the sampling rates of the image acquisition devices, the scale factors of the image coordinate axes are not necessarily equal. Additionally, the two axes of the real image may not form a right angle as a result of the lens distortion. The following transformation is used to handle these effects:

$$[u \ v \ 1]^T = \mathbf{H}[x \ y \ 1]^T,$$

$$\mathbf{H} = \begin{bmatrix} k_u & k_u \cos\theta & u_0 \\ 0 & k_v/\sin\theta & v_0 \\ 0 & 0 & 1 \end{bmatrix}. \tag{5}$$

$\mathbf{H}$ is composed of the parameters characterizing the inherent properties of the camera and optics: the coordinates of the point $\mathbf{c}(u_0, v_0)$ in the real image coordinate system (o, u, v), the scale factors k_u and k_v along the u and v axes with respect to the units used in (c, x, y), and the angle θ between the u and v axes caused by nonperpendicularity of axes. These parameters

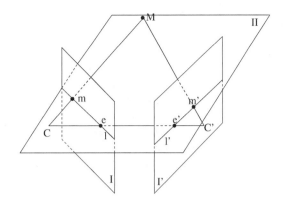

FIGURE 2   Epipolar geometry.

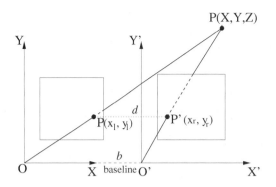

FIGURE 3   Parallel axes stereo imaging system.

do not depend on the position and orientation of the cameras, and are thus called the camera *intrinsic* or *internal parameters*. **H** is thus called the *intrinsic* or *internal matrix*. More generally, the perspective projection matrix **P** is integrated into **H** with $a_u = fk_u$, $a_v = fk_v$.

Combining Eqs. (2), (3), and (5) leads to the following expression that relates the pixel position $\mathbf{m} = [u, v]^T$ with the 3-D world coordinates $\mathbf{M} = [X, Y, Z]^T$:

$$s\tilde{\mathbf{m}} = \mathbf{Q}\tilde{\mathbf{M}}, \quad \mathbf{Q} = \mathbf{HPD}. \tag{6}$$

Clearly, Eq. (6) indicates that from a point in one image plane, only a 3-D location up to an arbitrary scale can be computed.

The geometric relationship between two projections of the same physical point can be expressed by means of the *epipolar geometry* [4], which is the only geometric constraint between a stereo pair of images of a single scene. Let us consider the case of two cameras as shown in Fig. 2. Let **C** and **C'** be the optical centers of the first and second camera, and let plane I and I' be the first and second image plane. According to the epipolar geometry, for a given image point **m** in the first image, its corresponding point **m'** in the second image is constrained to lie on a line l'. This line is called the *epipolar line* of **m**. The line l' is the intersection of the *epipolar plane* Π, defined by **m**, **C**, and **C'**, with the second image plane I'. This epipolar constraint can be formulated as

$$\tilde{\mathbf{m}}\mathbf{F}\tilde{\mathbf{m}}' = 0, \tag{7}$$

where **F** is called the *fundamental matrix*. It is a 3×3 matrix, determined by the intrinsic matrix of the two cameras and the relative position of the two cameras (explicit or extrinsic parameters), and it can be written as

$$\mathbf{F} = \mathbf{H}^{-T}[\mathbf{t}]_\times \mathbf{R}_{3\times3}\mathbf{H}'^{-1}, \tag{8}$$

where $\mathbf{t}_\times$ is a *skew symmetric matrix* defined by the translation

vector $\mathbf{t} = [t_1, t_2, t_3]^T$,

$$[\mathbf{t}]_\times = \begin{bmatrix} t_1 \\ t_2 \\ t_3 \end{bmatrix}_\times = \begin{bmatrix} 0 & -t_3 & t_2 \\ t_3 & 0 & -t_1 \\ -t_2 & t_1 & 0 \end{bmatrix}.$$

H and **H'** are the intrinsic matrices of camera 1 and camera 2, respectively. $\mathbf{R}_{3\times3}$ and $[\mathbf{t}]_\times$ are the rotation and translation transformations between the two camera coordinate systems.

Observe that all epipolar lines of the points in the first image pass through a common point **e'** on the second image. This point is called the *epipole* of the first image, which is the intersection of the line **CC'** with the second image plane I'. Similarly, **e** on the first image plane I is the epipole of the second image through which all epipolar lines of points in the second image pass through. For the epipole **e'**, the epipolar geometry suggests

$$\mathbf{Fe'} = \mathbf{0}. \tag{9}$$

As shown in Fig. 3, when the two image planes are parallel to each other, the epipoles are at infinity, and the geometric relationship between the two projections becomes very simple — the well-known parallel axes stereo system. The optical axes of the pair of cameras are mutually parallel and are separated by a horizontal distance known as the stereo baseline. The optical axes are perpendicular to the stereo baseline, and the image scanlines are parallel to the horizontal baseline. In the conventional parallel-axis geometry, all epipolar planes intersect the image planes along horizontal lines, i.e., $y_r = y_l$.

3 Matching

As mentioned earlier, computing the correspondence among a given set of images is one of the most important issues in 3-D shape reconstruction from multiple views. Establishing correspondence between two views of a scene involves either finding a match between the location of points in the two images or finding a transformation between the two images that maps corresponding points between the two images into one another. The former is known as feature- based matching technique, whereas the latter is known as the direct method of finding correspondences from raw images. In the following, we will first pose the problem of direct image matching as one of determining the optimal transformation between the two images, and then discuss the feature-based matching techniques.

3.1 Optical Flow-Based Matching Techniques

The problem[1] of finding the transformation between two images is equivalent to estimating the motion between them. There are numerous motion estimation algorithms in the computer vision literature [6–9] that are relevant in this context. We draw upon this large body of literature of motion estimation techniques for making the problem formulation, and we present a numerical algorithm for robustly and efficiently computing the motion parameters. Among motion estimation schemes, the most general is the optical flow formulation. We therefore treat the problem of finding the transformation between the two images as equivalent to computing the flow between the data sets. There are numerous techniques for computing the optical flow from a pair of images. Readers are referred to Chapter 3.10 ("Motion Detection and Estimation") in this handbook for more details.

The motion model proposed by Szeliski *et al.* [9] can be used to compute the interframe registration transformation to establish the correspondence. This model consists of globally parameterized motion flow models at one end of the "spectrum" and a local motion flow model at the other end. The global motion model is defined by associating a single global motion model with each patch of a recursively subdivided input image. The flow field corresponding to the displacement at each pixel/voxel is represented by a B-spline basis. Using this spline-based representation of the flow field (u_i, v_i) in the popular sum of squared differences (SSD) error term, i.e.,

$$E_{\text{SSD}}(u_i, v_i) = \sum_i [I_2(x_i + u_i, y_i + v_i) - I_1(x_i, y_i)]^2, \quad (10)$$

one may estimate the unknown flow field at each pixel/voxel by means of the numerical iterative minimization of E_{SSD}. Here I_2 and I_1 denote the target and initial reference images, respectively. In the local flow model, the flow field is not parameterized. It may be noted that the sum of squared differences error term gets simplified to $E_{\text{SSD}}(d_i) = \sum_i [I_L(x_i + d_i, y_i) - I_R(x_i, y_i)]^2$ when the two given images are in stereo geometry. We present a formulation of the global/local flow field computation model and the development of a robust numerical solution technique [5] in the following sections.

3.1.1 Local/Global Motion Model

Optical flow computation has been a very active area of computer vision research for over 15 years. This model of motion

---

[1] The material in this section was originally published in [5]. Copyright Oxford University Press, 1998. Used with permission.

computation is very general, especially when set in a hierarchical framework. In this framework, at one extreme, each pixel/voxel is assumed to undergo an independent displacement. This is considered as a local motion model. At the other extreme, we have global motion wherein the flow field model is expressed parametrically by a small set of parameters e.g., rigid motion, affine motion, and so on.

A general formulation of the image registration problem can be posed as follows: Given a pair of images (possibly from a sequence) I_1 and I_2, we assume that I_2 was formed by locally displacing the reference image I_1 as given by $I_2(x + u, y + v) = I_1(x, y)$. The problem is to recover the displacement field (u, v) for which the maximum likelihood solution is obtained by minimizing the error given by Eq. (10), which is popularly known as the *sum of squared differences* formula. In this motion model, the key underlying assumption is that intensity at corresponding pixels in I_1 and I_2 is unchanged and that I_1 and I_2 differ by local displacements. Other error criteria that take into account global variations in brightness and contrast between the two images and that are nonquadratic can be designed, as in Szeliski *et al.* [9],

$$E'_{\text{SSD}}\{(u_i, v_i)\} = \sum_i [I_2(x_i + u_i, y_i + v_i) - c I_1(x_i, y_i) + b]^2, \quad (11)$$

where b and c are the intensity and uniform contrast correction terms per frame, which have to be recovered concurrently with the flow field. The just-described objective function has been minimized in the past by several techniques, some of them using regularization on (u, v) [8, 10]. We subdivide a single image into several patches, each of which can be described by either a local motion or a global parametric motion model. The tiling process is made recursive. The decision to tile a region further is made based on the error in computed motion or registration.

2-D Local Flow. We represent the displacement fields $u(x, y)$ and $v(x, y)$ by B splines with a small number of control points $\hat{u}_j$ and $\hat{v}_j$ as in Szeliski *et al.* [9]. Then the displacement at a pixel location i is given by

$$u(x_i, y_i) = \sum_j \hat{u}_j w_{ij} = \sum_j \hat{u}_j B_j(x_i, y_i),$$
$$v(x_i, y_i) = \sum_j \hat{v}_j w_{ij} = \sum_j \hat{v}_j B_j(x_i, y_i), \quad (12)$$

where $w_{ij} = B_j(x_i, y_i)$ are the *basis functions* with finite support. In our implementation, we have used bilinear basis $B(x, y) = (1 - |x|)(1 - |y|)$ for (x, y) in $[-1, 1]^2$ as shown in Fig. 4, and we also assumed that the spline control grid is a subsampled version of the image pixel grid $(\hat{x}_j = mx_i, \hat{y}_j = my_i)$, as in Fig. 5. This spline-based representation of the motion field possesses several advantages. First, it imposes a built-in smoothness on the motion field and thus removes the need for further regularization. Second, it eliminates the need for correlation windows centered at

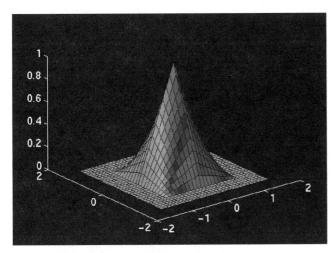

FIGURE 4 Bilinear basis function. (See color section, p. C–7.)

each pixel, which are computationally expensive. In this scheme, the flow field $(\hat{u}_j, \hat{v}_j)$ is estimated by a weighted sum from all the pixels beneath the support of its basis functions. In the correlation window-based scheme, each pixel contributes to m^2 overlapping windows, where m is the size of the window. However, in the spline-based scheme, each pixel contributes its error only to its four neighboring control vertices, which influence its displacement. Therefore, the latter achieves computation savings of $O(m^2)$ over the correlation window-based approaches.

We use a slightly different error measurement from the one described herein. Given two gray-level images $I_1(x, y)$ and $I_2(x, y)$, where I_1 is the model and I_2 is the target, to compute an estimate $\hat{\mathbf{T}} = (\hat{u}_1, \hat{v}_1, \ldots, \hat{u}_n, \hat{v}_n)^T$ of the true flow field $\mathbf{T} = (u_1, v_1, \ldots, u_n, v_n)^T$ at n control points, first an intermediate image I_m is introduced and the motion is modeled in terms of the B-spline control points as

$$I_m(\mathbf{X}_i, \hat{\mathbf{T}}) = I_1\left(x_i + \sum_j \hat{u}_j w_{ij}, \, y_i + \sum_j \hat{v}_j w_{ij}\right), \quad (13)$$

where $\mathbf{X}_i = (x_i, y_i)$ and w_{ij} are the basis functions as before. The expectation E of the squared difference, E_{SD}, is chosen to

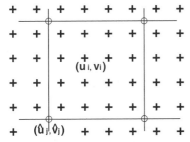

FIGURE 5 Spline-based flow field representation: the spline control points $\{(\hat{u}_j, \hat{v}_j)\}$ are indicated as circles and the pixel grid $\{(u_j, v_j)\}$ is shown as crosses.

be the error criterion $J\{E_{SD}(\hat{\mathbf{T}})\}$

$$J\{E_{SD}(\hat{\mathbf{T}})\} = E\{(I_m(\mathbf{X}_i, \hat{\mathbf{T}}) - I_2(\mathbf{X}_i))^2\} \quad (14)$$

2-D Global Flow. When a global motion model is used to model the motion between I_1 and I_2, it is possible to parameterize the flow by a small set of parameters describing the motion for rigid, affine, quadratic, and other types of transformations. The affine flow model is defined in the following manner:

$$\begin{bmatrix} u(x, y) \\ v(x, y) \end{bmatrix} = \begin{bmatrix} t_0 & t_1 \\ t_3 & t_4 \end{bmatrix} \begin{bmatrix} x \\ y \end{bmatrix} + \begin{bmatrix} t_2 \\ t_5 \end{bmatrix} - \begin{bmatrix} x \\ y \end{bmatrix}, \quad (15)$$

where the parameters $\mathbf{T} = (t_0, \ldots, t_5)^T$ are called *global motion parameters*. To compute an estimate of the global motion, we first define the spline control vertices $\hat{\mathbf{u}}_\mathbf{j} = (\hat{u}_j, \hat{v}_j)^T$ in terms of the global motion parameters:

$$\hat{\mathbf{u}}_\mathbf{j} = \begin{bmatrix} \hat{x}_j & \hat{y}_j & 1 & 0 & 0 & 0 \\ 0 & 0 & 0 & \hat{x}_j & \hat{y}_j & 1 \end{bmatrix} \mathbf{T} - \begin{bmatrix} \hat{x}_j \\ \hat{y}_j \end{bmatrix},$$

$$= \mathbf{S_j}\mathbf{T} - \hat{\mathbf{X}}_\mathbf{j}. \quad (16)$$

where $\mathbf{S_j}$ is the 2×6 matrix shown earlier. We then define the flow at each pixel by interpolation, using our spline representation. The error criterion $J\{E_{SD}(\hat{\mathbf{T}})\}$ becomes

$$J\{E_{SD}(\hat{\mathbf{T}})\}$$
$$= E\left\{\left\{I_1\left[\mathbf{X}_i + \sum_j w_{ij}(\mathbf{S_j}\mathbf{T} - \hat{\mathbf{X}}_\mathbf{j})\right] - I_2(\mathbf{X}_i)\right\}^2\right\}.$$

3.1.2 Numerical Solution

We now describe a novel adaptation of an elegant numerical method by Burkardt and Diehl [11] that is a modification of the standard Newton method for solving a system of nonlinear equations. The modification involves precomputation of the Hessian matrix at the optimum without starting the iterative minimization process. Our adaptation of this idea to the framework of optical flow computation with spline-based flow field representations leads to a very efficient and robust flow computation technique.

We present the modified Newton method based on the work of Burkardt and Diehl [11] to minimize the error term $J\{E_{SD}(\hat{\mathbf{T}})\}$. In the following, we will essentially adopt the notation from Burkardt and Diehl [11] to derive the modified Newton iteration and develop new notation as necessary. The primary structure of the algorithm is given in the following iteration formula:

$$\hat{\mathbf{T}}^{k+1} = \hat{\mathbf{T}}^k - \mathbf{H}^{-1}(\hat{\mathbf{T}} = \hat{\mathbf{T}}^*)g(\hat{\mathbf{T}}^k), \quad (17)$$

where $\mathbf{H}$ is the Hessian matrix and g is the gradient vector of the objective function $J\{E_{SD}(\hat{\mathbf{T}})\}$. Unlike in a typical Newton iteration, in Eq. (17) the *Hessian* matrix is always computed at the optimum $\hat{\mathbf{T}} = \hat{\mathbf{T}}^* = \mathbf{T}$ instead of the iteration point $\hat{\mathbf{T}}^k$. So, one of the key problems is how to calculate the Hessian at the optimum prior to beginning the iteration, i.e., without actually knowing the optimum.

Let the vector $\mathbf{X}$ denote the coordinates (x, y) in any image and $h : \mathbf{X} \to \mathbf{X}'$ denote a transformation from $\mathbf{X}$ to another set of coordinates $\mathbf{X}'$, characterized by a set of parameters collected into a vector $\mathbf{T}$, i.e., $\mathbf{X}' = h(\mathbf{X}, \mathbf{T})$. The parameter vector $\mathbf{T}$ can represent any of rigid, affine, shearing, and projective (etc.) transformations. Normally the Hessian at the optimum will explicitly depend on the optimum motion vector and hence cannot be computed directly. However, a clever technique was introduced in Burkhardt *et al.* [11], involving a moving coordinate system $\{\mathbf{X}^k\}$ and an intermediate motion vector $\tilde{\mathbf{T}}$ to develop the formulas for precomputing the Hessian. This intermediate motion vector gives the relationship between $\{\mathbf{X}^k\}$ of iteration step k and $\{\mathbf{X}^{k+1}\}$ of iteration step $k + 1$:

$$\mathbf{X}^k = h(\mathbf{X}, \hat{\mathbf{T}}^k) \tag{18}$$

$$\mathbf{X}^{k+1} = h(\mathbf{X}^k, \tilde{\mathbf{T}}^{k+1}) = h[h(\mathbf{X}, \hat{\mathbf{T}}^k), \tilde{\mathbf{T}}^{k+1}]$$

$$= h(\mathbf{X}, \hat{\mathbf{T}}^{k+1}) \tag{19}$$

After some tedious derivations, it can be shown [5] that the Hessian at the optimum is given by

$$\tilde{\mathbf{H}} = 2E \left\{ \left(\frac{\partial h(\mathbf{X}, \mathbf{T})}{\partial \mathbf{T}} \right)^T \frac{\partial I_1(\mathbf{X})}{\partial \mathbf{X}} \left(\frac{\partial I_1(\mathbf{X})}{\partial \mathbf{X}} \right)^T \frac{\partial h(\mathbf{X}, \mathbf{T})}{\partial \mathbf{T}} \right\} \Bigg|_{\mathbf{T}=0}, \tag{20}$$

whereas, the gradient vector with respect to $\tilde{\mathbf{T}}$ is given by:

$$\tilde{g}(\hat{\mathbf{T}}^k) = 2E \left\{ e \left[\left(\frac{\partial I_2(\mathbf{X})}{\partial \mathbf{X}} \right)^T \left(\frac{\partial \mathbf{X}^k}{\partial \mathbf{X}} \right)^{-1} \left(\frac{\partial \mathbf{X}^{k+1}}{\partial \mathbf{T}^{k+1}} \right) \Bigg|_{\tilde{\mathbf{T}}^{k+1}=0} \right]^T \right\}, \tag{21}$$

where $e = I_m(\mathbf{X}, \hat{\mathbf{T}}^k) - I_2(\mathbf{X})$.

Thus the modified Newton algorithm consists of the following iteration:

$$\tilde{\mathbf{T}}^{k+1} = -\tilde{\mathbf{H}}^{-1} \tilde{g}(\hat{\mathbf{T}}^k), \tag{22}$$

and the estimate at step $k + 1$ is given by

$$\hat{\mathbf{T}}^{k+1} = f(\hat{\mathbf{T}}^k, \tilde{\mathbf{T}}^{k+1}), \tag{23}$$

where f is a function that depends on the type of motion model used. One of the advantages of the modified Newton method is an increase in the size of the region of convergence. Note that normally the Newton method requires that the initial guess for starting the iteration be reasonably close to the optimum. However, in all our experiments — described in Vemuri *et al.* [5] — with the modified Newton scheme described here, we always used the zero vector as the initial guess for the motion vector to start the iterations. For more details on the convergence behavior of this method, we refer the reader to Burkhardt and Diehl [11].

In the following sections, we describe a reliable way of precomputing the Hessian matrix and the gradient vector at the optimum for the local and global motion models.

Hessian Matrix and Gradient Vector Computation for 2-D Local Flow. Let $\hat{\mathbf{X}}_j$, $(j = 1, 2, \ldots, n)$ be the control points. Then the flow vector $\mathbf{T}$ is $(\hat{u}_1, \hat{v}_1, \ldots, \hat{u}_n, \hat{v}_n)^T$. Actually, local flow is equivalent to pure translation at each pixel, and hence the Hessian at the optimum is only related to the derivatives with respect to the original coordinates and does not depend on the flow vector. Therefore, it can be calculated without introducing $\tilde{\mathbf{T}}$ as shown below.

Let

$$\partial = \left(w_1 \frac{\partial}{\partial x}, w_1 \frac{\partial}{\partial y}, \ldots, w_n \frac{\partial}{\partial x}, w_n \frac{\partial}{\partial y} \right). \tag{24}$$

The Hessian at the optimum is then given by

$$\mathbf{H}^*_{ij} = \tilde{\mathbf{H}}_{ij} = 2E\{\partial_i I_1(\mathbf{X}) \partial_j I_1(\mathbf{X})\}, \tag{25}$$

whereas the gradient vector is

$$\tilde{g}(\hat{\mathbf{T}}^k) = 2E \left\{ (I_m - I_2) N^T M^{-T} \frac{\partial I_2(X)}{\partial X} \right\}, \tag{26}$$

where the matrices $\mathbf{M}$ and $\mathbf{N}$ are given by

$$\mathbf{M} = \frac{\partial \mathbf{X}^k}{\partial \mathbf{X}} = \begin{bmatrix} 1 + \sum_j (w_j)'_x \hat{u}_j^k & \sum_j (w_j)'_y \hat{u}_j^k \\ \sum_j (w_j)'_x \hat{v}_j^k & 1 + \sum_j (w_j)'_y \hat{v}_j^k \end{bmatrix} \tag{27}$$

and

$$\mathbf{N} = \left(\frac{\partial \mathbf{X}^{k+1}}{\partial \tilde{\mathbf{T}}^{k+1}} \right) \Bigg|_{\tilde{\mathbf{T}}^{k+1}=0} = \begin{bmatrix} w_1 & 0 & \ldots & w_n & 0 \\ 0 & w_1 & \ldots & 0 & w_n \end{bmatrix},$$

respectively. We can now substitute Eqs. (25) and (26) into Eq. (22), yielding $\tilde{\mathbf{T}}^{k+1}$ which upon substitution into Eq. (23) results in $\tilde{\mathbf{T}}^{k+1}$. Hence, numerical iterative formula (23) used in computing the local motion becomes

$$\hat{\mathbf{T}}^{k+1} = \hat{\mathbf{T}}^k - \tilde{\mathbf{H}}^{-1} \tilde{g}(\hat{\mathbf{T}}^k). \tag{28}$$

The size of $\tilde{\mathbf{H}}$ is determined by how many control points are used in representing the flow field (u, v). For 3-D problems, $\tilde{\mathbf{H}}$ is $(3n \times 3n)$ where n is the number of control points. For such large problems, numerical iterative solvers are quite attractive and we use a preconditioned conjugate gradient (PCG) algorithm [8, 9] to solve the linear system $\tilde{\mathbf{H}}^{-1} \tilde{g}(\hat{\mathbf{T}}^k)$. The specific preconditioning we use in our implementation of the local flow is a simple diagonal Hessian preconditioning. More sophisticated preconditioners can be used in place of this simple preconditioner, and the reader is referred to Lai *et al.* [8] for the source.

Hessian Matrix and Gradient Vector Computation for 2-D Rigid Flow. We now derive the Hessian matrix and the gradient vector for the case in which the flow field is expressed by using a global parametric form, specifically, a rigid motion parameterization, i.e., $\mathbf{T} = (\phi, d_1, d_2)^T$, with ϕ being the rotation angle and d_1, d_2 being the components of the translation in the x and y direction, respectively.

Let $\hat{\mathbf{X}}_j$, $(j = 1, 2, \ldots, n)$ be the control points and let

$$\mathbf{A} = \left(\frac{\partial h(\mathbf{X}, \mathbf{T})}{\partial \mathbf{T}} \right)^T = \begin{bmatrix} -\sum_j w_j \hat{y}_j & \sum_j w_j \hat{x}_j \\ -\sum_j w_j & 0 \\ 0 & -\sum_j w_j \end{bmatrix}. \quad (29)$$

The Hessian at the optimum can then be written as

$$\tilde{\mathbf{H}} = 2E \left\{ \mathbf{A} \frac{\partial I_1(\mathbf{X})}{\partial \mathbf{X}} \left(\frac{\partial I_1(\mathbf{X})}{\partial \mathbf{X}} \right)^T \mathbf{A}^T \right\}. \quad (30)$$

The gradient vector $\tilde{\mathbf{g}}(\hat{\mathbf{T}}^k)$ at the optimum is given by

$$\tilde{\mathbf{g}}(\hat{\mathbf{T}}^k) = 2E \left\{ (I_m - I_2) \mathbf{N} \mathbf{M}^{-T} \frac{\partial I_2(\mathbf{X})}{\partial \mathbf{X}} \right\}, \quad (31)$$

where the matrices $\mathbf{M}$ and $\mathbf{N}$ are

$$\mathbf{M} = \frac{\partial \mathbf{X}^k}{\partial \mathbf{X}} = \begin{bmatrix} 1 + \sum_j (w_j)'_x \hat{u}_j^k & \sum_j (w_j)'_y \hat{u}_j^k \\ \sum_j (w_j)'_x \hat{v}_j^k & 1 + \sum_j (w_j)'_y \hat{v}_j^k \end{bmatrix} \quad (32)$$

and

$$\mathbf{N} = \left(\frac{\partial \mathbf{X}^{k+1}}{\partial \tilde{\mathbf{T}}^{k+1}} \right)^T \Bigg|_{\tilde{\mathbf{T}}^{k+1}=0} = \begin{bmatrix} -\sum_j w_j \hat{y}_j^k & \sum_j w_j \hat{x}_j^k \\ -\sum_j w_j & 0 \\ 0 & -\sum_j w_j \end{bmatrix}, \quad (33)$$

respectively. The basic steps in our algorithm for computing the global rigid flow are as follows.

1. Precompute the Hessian at the optimum $\tilde{\mathbf{H}}$ by using Eq. (30).
2. At iteration k, compute the gradient vector by using Eq. (31).
3. Compute the innovation $\tilde{\mathbf{T}}^{k+1}$ by using Eq. (22).
4. Update the motion parameter $\hat{\mathbf{T}}^{k+1}$ by using the following equation:

$$\begin{bmatrix} \hat{\phi} \\ \hat{d}_1 \\ \hat{d}_2 \end{bmatrix}_{k+1} = \begin{bmatrix} 1 & 0 & 0 \\ 0 & \cos\tilde{\phi} & -\sin\tilde{\phi} \\ 0 & \sin\tilde{\phi} & \cos\tilde{\phi} \end{bmatrix}_{k+1} \begin{bmatrix} \hat{\phi} \\ \hat{d}_1 \\ \hat{d}_2 \end{bmatrix}_k + \begin{bmatrix} \tilde{\phi} \\ \tilde{d}_1 \\ \tilde{d}_2 \end{bmatrix}_{k+1}. \quad (34)$$

Once the transformation between the images is known, one can set up the n point matches and reconstruct the 3-D geometry from this information. The details of this reconstruction process will be discussed in detail in a subsequent section.

3.2 Feature-Based Matching Techniques

Feature-based matching techniques establish correspondences among *homologous* features, i.e., features that are projections of the same physical entity in each view. It is a difficult task to decide which features can be used to effectively represent the projection of a point or an area, and how to find their correspondences in different images. Generally speaking, the feature-based stereo matching approach is divided into two steps. The first step is to extract a set of salient 2-D features from a sequence of frames. Commonly used features include points, lines, or curves corresponding to corners, boundaries, region marks, occluding boundaries of surfaces, or shadows of objects in 3-D space. The second step is to find correspondences between features, usually called the *correspondence problem*; i.e., find the points in the images that are the projections of the same physical point in the real world. This problem is recognized as being difficult and continues to be the bottleneck in most stereo applications. In the later parts of this section, we review the commonly used matching strategies for finding unique correspondences.

3.2.1 Feature Extraction

In this stage, image locations satisfying certain well-defined feature characteristics are identified in each image. The choice of features is very important because the subsequent matching strategy will be based on and make extensive use of these characteristic features.

Low-level tokens such as edge points have been used for matching in early work in stereo vision [12]. Feature points based on gray level, intensity gradient, disparity, and so on are extracted and later used as attributes for point-based matching. Marr and Poggio [13] used 12 filters in different orientations to extract the zero-crossing points and recorded their contrast sign and orientation as attributes. Lew *et al.* [14] used intensity, gradient in both x and y directions, gradient magnitude, gradient orientation, Laplacian of intensity, and curvature as the attributes for each edge point.

There are some intrinsic disadvantages in using solely point-based matching. It demands very large computational resources and usually results in a large number of ambiguous candidate matches that must be explored further. Because of these problems, edge segments are used as primitives more prevalently, especially in applications in structured indoor or urban scenes. Compared with edge points, edge segments are fewer and are able to provide rich attribute information, such as length, orientation, middle and end points, contrast, and so on. Hence, matching based on edge segments is expected to be much more stable than point-based matching in the presence of changes in contrast and ambient lighting. There is abundant literature on stereo matching based on edge segments, including but not limited to [15, 16].

3.2.2 Feature Matching

The task of feature matching is to identify correspondences by analyzing extracted primitives from two or more images captured from multiple viewpoints. The simplest method for identifying matches is merely to test the *similarity* of the attributes of the matched tokens, and accept the match if the similarity probability is greater than some threshold. However, since tokens rarely contain sufficiently unique sets of token attributes, a simple comparison strategy is unlikely to lead to a unique correspondence for every image token, particularly for less complex features, such as edge points. Some more complicated constraints and searching strategies have to be exploited to limit the matching candidates, thereby reducing the matching numbers and possibly the ambiguity. In the following paragraphs, we summarize the most common constraints and the integrating strategies used for finding the matches.

- **Similarity constraints:** Certain geometric similarity constraints, such as similarity of edge orientation or edge strength, are usually used to find the preliminary matches. For example, in Marr and Poggio's original paper [13] on a feature-point-based computational model of human stereopsis, similarity of zero crossings is defined based on them having the same contrast sign and approximately the same orientation. More generally, the similarity constraint (as well as the other constraints stated later) is formulated in a statistical framework; i.e., a similarity measure is exploited to quantify the confidence of a possible match. For each attribute, a probability density function $p(|a_k - a_k'|)$ can be empirically derived and parameterized by attribute disparities $|a_k - a_k'|$ [17]. A similarity measure is defined as a weighted combination (e.g., an average) of multiple attributes

$$(m, m') = \omega \mathbf{f}(\mathbf{a}, \mathbf{a}'), \qquad (35)$$

where (m, m') are candidate match token pairs (edge points, line segments, etc.) from two images, such as the left and right images in the binocular case. Attribute vector $\mathbf{a} = (a_1, a_2, \ldots)$ is composed of the attributes of the token features such as line length, grey level, curvature, etc. Here $\mathbf{f}$ is a similarity function vector, which normalizes each disparity component relative to some explicit or implicit disparity variance. Weight ω_i defines the relative influence of each attribute on the similarity score. Provided that the values taken by token attributes are independent, a match confidence based on similarity may be computed [18] as $S(m, m') = p_{sim}(m, m') = \prod_{k=1}^{M} p(a_k | a_k')$, where given M attributes, $p(a_k | a_k')$ is a conditional probability relating the kth attribute value of a token in the second image to the attribute value of the first. The values of these conditional probabilities are usually determined from a training data set. For instance, Boyer and Kak [19] measure the information content (entropy) of feature attributes from training data, and determine the relative influence of each attribute.

- **Epipolar constraint:** The only geometric constraint between two stereo images of a single scene is the *epipolar constraint*. Specifically, the epipolar constraint implies that *the epipolar line in the second image corresponding to a point* $\mathbf{m}$ *in the first image defines the search space within which the corresponding match point* $\mathbf{m}'$ *should lie in the second image, and vice versa*. When the projection (epipolar) geometry is known, epipolar constraints (formulated by Eq. (7)) can be used to limit the search space of correspondence.

  In the conventional parallel-axis geometry shown in Fig. 3, for each feature point $\mathbf{m}(x, y)$ in the left image, possible candidate matches $\mathbf{m}'(x', y')$ can be searched for along the horizontal scan line (epipolar line) in the right image such that

$$x + d - \sigma \leq x' \leq x + d + \sigma, \quad y' = y, \qquad (36)$$

  where d is the estimated disparity and $2\sigma + 1$ is the width of the search region. However, local distortions due to perspective effects, noise in early processing, and digitization effects can cause deterioration in matching performance at finer resolutions. To consider this distortion, the second equation in Eq. (36) can be modified to $y - \epsilon \leq y' \leq y + \epsilon$ to include the vertical disparity, where $(2\epsilon + 1)$ is the height of the search space in the vertical direction. The epipolar search for matching edge points is usually aided by certain geometric similarity constraints, e.g., the similarity of edge orientation or edge strength.

- **Disparity gradient limit constraint:** The image *disparity* of matched points/line segments is the difference in their respective features, such as the difference of their positions, the difference between their orientations, etc. For any pair of matches, the *disparity gradient* is defined as the ratio of the difference in disparity of the two matches to the average separation of the tokens in each image or local area. It is suggested that for most natural scene surfaces, including jagged ones, the *disparity gradient* between correct matches is usually <1, whereas it is very rare among incorrect matches obtained for the same set of images or areas.

The above-mentioned constraints (similarity, epipolar constraint and disparity constraints) are often called *local constraints* or *unary constraints* since they are specific to each *individual* match. They are usually applied in the first matching stage and used to identify the set of candidate matches. The *global consistency* of the local matches is then tested by figural continuity or other global constraints. In the following, we describe the different types of global constraints and cite examples from pertinent literature.

- **Uniqueness constraint:** This constraint requires each item in an image to be assigned to one and only one disparity value. It is a very simple but general constraint used in many matching strategies.

- **Continuity constraints:** The continuity constraints depend on the observation that points adjacent in 3-D space remain adjacent in each image projection. They would be used to determine the consistency of the disparities obtained as a result of the local matching or to guide the local searching to avoid inconsistent or false matches by supporting compatible or inhibiting incompatible matches. In practice, this observation of the nature of the surface in the 3-D scene can be formulated in a number of different ways. For example, Horaud and Skordas [20] impose a continuity constraint on edges (*edge connectivity constraint*), which states that connected edge points in one image must match to connected edge points in the other image. Prazdny [21] suggested a Gaussian similarity function $s(i, j) = 1/(c\|i - j\|\sqrt{2\pi})$ $\exp[(-\|d_i - d_j\|^2)/(2c^2\|i - j\|^2)]$, which quantifies the similarity between neighboring disparities. When counting on the various continuity constraints, a measure of support from neighboring matches is computed from the compatibility relations and used to modify the match probability.

- **Topological constraints:** The most popular topological constraint is the *relative position constraint*, which assumes that the relative positions of tokens remain similar between images. The *Left_of* and *Right_of* relations applied by Horaud and Skordas [20] is a simple but popular form of this kind of topology constraint. For horizontally mounted cameras, near-horizontal lines have a very narrow disparity distribution, whereas vertical lines have a greater disparity distribution.

Both *continuity* and *topological* constraints are usually called *compatibility constraints* [18], since they are used to decide the *mutual compatibility* among the matches and their neighbors. Until now, we introduced several of the most common constraints imposed on extracted tokens to limit the correspondence search space. A single constraint is usually not powerful enough to locate all the matches uniquely and correctly. Almost all current algorithms use a combination of two or more constraints to extract a final set of matches. These can be classified into two categories: relaxation labeling and hierarchical schemes. Relaxation labeling groups neighbor information iteratively to update the match probability, while the hierarchical methods usually follow a coarse-to-fine procedure. Typical methods from these categories will be described next. A feature-based matching example integrating several of the above mentioned constraints is presented in Section 5.

3.2.3 Relaxation Labeling Algorithms

Relaxation labeling, also called *graph matching*, is a fairly general model proposed for scene labeling. It has often been employed to solve correspondence problems, whether in stereo, image registration, or object recognition. The basic structure of the relaxation labeling algorithm is illustrated in Fig. 6.[2] In the

²From Chang and Aggarwal's paper [15].

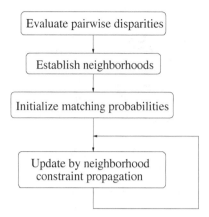

FIGURE 6   Basic structure of the relaxation labeling algorithm.

paradigm of matching a stereo pair of images by using relaxation labeling, a set of feature points (nodes) are identified in each image, and the problem involves assigning unique labels (or matches) to each node out of a discrete space (list of possible matches). For each candidate pair of matches, a matching probability is updated iteratively depending upon the matching probabilities of neighboring nodes so that stronger neighboring matches improve the chances of weaker matches in a globally consistent manner. At the end of iteration, the node assignments with the highest score are chosen as the matched tokens. This interaction between neighboring matches is motivated by the existence of cooperative processes in the biological vision systems.

A general matching score update formula presented in Ranade and Rosenfeld [22] is given by

$$S^{(r+1)}\left(N_l^i, N_r^j\right) =$$

$$\frac{1}{n} \sum_{N_l^m \in K} \sum_{m=1}^{N_l} \left[\max_{n=1}^{N_l} C\left(N_l^i, N_r^j; N_l^m, N_r^n\right) S^{(r)}\left(N_l^m, N_r^n\right) \right],$$

where $C(N_l^i, N_r^j; N_l^m, N_r^n) = 1$ if $j = n$ and 0 otherwise. In their relaxation technique, the initial score S^0 is updated by the maximum support from neighboring pair of nodes. For each node N_l, its node assignment score is updated; only the nodes that form a node assignment that are within a distance of K pixels from it can contribute to its node assignment score. The pair of nodes that have the same disparity will contribute significantly, whereas the nodes that have different disparities will contribute very little. As the iteration progresses, the node assignment score is decreased; however, the score decreases faster for the less likely matches than for the most likely ones.

Despite differences in terminology, most methods are highly similar but often offer novel developments in some aspect of the correspondence process. We refer the reader to Dhond and Aggarwal's review paper [1] and the references contained therein for the details.

3.2.4 Hierarchical Schemes

The objective of a hierarchical computational structure for stereo matching is to reduce the complexity of matching by using a coarse-to-fine strategy. Matching is performed at each level in the hierarchy consecutively from the top down. Coarse features are matched first, and the results are used to converge and guide the matching of finer features.

Basically, there are two ways used to extract the coarser and finer features. One popular way is to use several filters to extract features in different resolutions, like the Laplacian of Gaussian ($\nabla^2 G$) operators used by Marr *et al.* [13] and Nasrabadi [23], the spatial-frequency-tuned channels of Mayhew and Frisby [24], and so on. Another way is to choose 2-D structural-description tokens with different complexity, such as Lim and Binford's from-objects-to-surfaces-to-edges method [25]. The imposed requirement that if any two tokens matched, then the subcomponents of these tokens are also matched, is also called the *hierarchical constraint*. Dhond and Aggarwal [26] employed hierarchical matching techniques in the presence of narrow occluding objects to reduce false-positive matches. A comprehensive review of hierarchical matching is given in Jones[18].

4 3-D Reconstruction

In this section, we will discuss how to reconstruct the 3-D geometry once the matching (correspondence) problem is solved. The reconstruction strategies usually fall into two categories. One is based on traditional camera calibration, which usually assumes that the camera geometry is known or starts from the computation of the projection matrix for well-controlled camera systems. The projective geometry is obtained by estimating the perspective projection matrix for each camera, using an apparatus with a known shape and size, and then computing the epipolar geometry from the projection matrices [3, 27]. Another technique is called uncalibrated stereo analysis, which reconstructs the perspective structure of a 3-D scene without knowing the camera positions. After presenting the simple idea of triangulation, we will introduce the camera calibration techniques that are used by the traditional calibrated stereo reconstruction to establish the camera models. At the end of this section, we will discuss the uncalibrated stereo analysis for 3-D reconstruction.

4.1 3-D Reconstruction Geometry

The purpose of structure analysis in stereo is to find the accurate 3-D location of those image points. Assuming we have full knowledge of the perspective projection matrix $\mathbf{Q}$, for a point $\mathbf{m} = [u, v]^T$ in image 1, which corresponds to the point $\mathbf{M} = [X, Y, Z]^T$ in the real world, we can rewrite Eq. (6) as

$$s[u \quad v \quad 1]^T = [\mathbf{q}_1 \quad \mathbf{q}_2 \quad \mathbf{q}_3]^T [X \quad Y \quad Z \quad 1]^T, \quad (37)$$

where $\mathbf{q}_i$ is the ith row vector in the perspective projection matrix $\mathbf{Q}$. The scalar s can be obtained from Eq. (37) as

$$s = \mathbf{q}_3 \tilde{\mathbf{M}} = \mathbf{q}_3 [X \quad Y \quad Z \quad 1]^T.$$

Eliminating s from Eq. (37) gives

$$\begin{bmatrix} \mathbf{q}_1 - \mathbf{q}_3 u \\ \mathbf{q}_2 - \mathbf{q}_3 v \end{bmatrix} \tilde{\mathbf{M}} = 0.$$

We get a similar equation for the corresponding point $\mathbf{m}' = [u', v']$ in the second image, which is given by

$$\begin{bmatrix} \mathbf{q}'_1 - \mathbf{q}'_3 u' \\ \mathbf{q}'_2 - \mathbf{q}'_3 v' \end{bmatrix} \tilde{\mathbf{M}} = 0.$$

Combining these two equations, we get

$$\mathbf{A} \tilde{\mathbf{M}} = 0, \quad (38)$$

where, $\mathbf{A} = [\mathbf{q}_1 - \mathbf{q}_3 u \quad \mathbf{q}_2 - \mathbf{q}_3 v \quad \mathbf{q}'_1 - \mathbf{q}'_3 u' \quad \mathbf{q}'_2 - \mathbf{q}'_3 v']^T$ is a 4×4 matrix that depends only on the camera parameters and the coordinates of the image points. $\tilde{\mathbf{M}} = [X \quad Y \quad Z \quad 1]^T$ is the unknown 3-D location of the point $\mathbf{M}$, which has to be calculated. Here Eq. (38) indicates that the vector $\tilde{\mathbf{M}}$ should be in the null space of $\mathbf{A}$ or the symmetric matrix $\mathbf{A}^T \mathbf{A}$. Practically, this can be solved by singular value decomposition. In detail, $[X \quad Y \quad Z \quad 1]^T$ can be calculated as the eigenvector corresponding to the least eigenvalue of $\mathbf{A}^T \mathbf{A}$. Let $\mathbf{A}^T \mathbf{A} = \mathbf{U} \mathbf{S} \mathbf{V}^T$ be the SVD of matrix $\mathbf{A}^T \mathbf{A}$, where $\mathbf{S} = \text{diag}(\sigma_1, \sigma_2, \sigma_3)$ satisfying $\sigma_1 > \sigma_2 > \sigma_3 > \sigma_4$, $\mathbf{U}^T = \mathbf{V} = [\mathbf{v}_1 \quad \mathbf{v}_2 \quad \mathbf{v}_3 \quad \mathbf{v}_4]^T$ are orthogonal matrices, and $\mathbf{v}_i$ are eigenvectors corresponding to eigenvalues σ_i. The coordinate vector can be computed as $[X \quad Y \quad Z \quad 1]^T = \mathbf{v}_4 / v_{44}$, where v_{44} is the last element in vector $\mathbf{v}_4$.

Particularly, in the conventional baseline stereo system (Fig. 3), the reconstruction mathematics becomes more simple. Suppose the axes are parallel between the camera coordinate systems and the reference coordinate system. The origin of the reference coordinate system is right at the midway between the focal centers of the left and right cameras. Ignoring the intrinsic distortion of the cameras, one can find the object space coordinates from the following:

$$\begin{aligned} X &= b(x_l + x_r)/(2d), \\ Y &= b(y_l + y + r)/(2d), \quad (39) \\ Z &= bf/d, \end{aligned}$$

where $d = (x_l - x_r)$ is referred to as disparity and b is the baseline.

Since a point in 3-D space and its projections in two images always compose a triangle, the reconstruction problem, which is to estimate the position of a point in 3-D space, given its projections and camera geometry, is referred to as *triangulation*.

4.2 Camera Calibration

Camera calibration is the problem of determining the elements that govern the relationship between the 2-D image that a camera perceives and the 3-D information of the imaged object. In other words, the task of camera calibration is to estimate the intrinsic and extrinsic parameters of a camera. This problem has been a major issue in photogrammetry and computer vision for many years. The main reason for such an interest is that the knowledge of the imaging parameters allows one to relate the image measurements to the spatial structure of the observed scene. Although the intrinsic camera parameters are known from the manufacturer's specifications, it is advisable to estimate them from images of known 3-D points in a scene. This is primarily to account for a variety of aberrations sustained by the camera during use.

There are six extrinsic parameters, describing the orientation and position of a camera with respect to the world coordinate system, and five intrinsic parameters that depend on the projection of the camera optical axis on the real image and the sampling rates of imaging devices. One can reduce the number of parameters by a specially set up camera system, such as the conventional parallel binocular stereo system whose imaging geometry is shown in Fig. 3. But this mechanical setup is a tedious task, and no matter how carefully it is set up, there is no guarantee that it will be error free. To eliminate expensive system setup costs and simplify image feature extraction, most classical camera calibration processes proceed by analyzing an image of one or several reference objects whose geometry is *accurately known*. Figure 7 shows a popular calibration pattern[3] that is often used for calibration and testing by the stereo vision community. Many other common shapes could also be used as

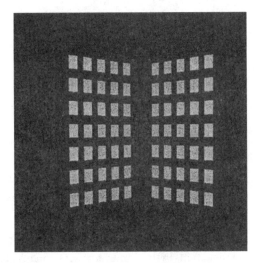

FIGURE 7 Example of calibration pattern: a flat plate with rectangle marks on it.

the calibration pattern, as long as the image coordinates of the projected reference points can be measured with great accuracy.

These pattern-based approaches proceed in two steps. First, some features, generally points or lines, are extracted from the image by means of standard image analysis techniques. Then, these features are used as input to an optimization process that searches for the projection parameters **P** that best project the 3-D model onto them. The solution to the optimization process can be achieved by means of a nonlinear iterative minimization process or in a closed form based on the camera model considered. A general criterion to be minimized is the distance (e.g., mean square discrepancy) between the observed image points and their inferred image projections computed with the estimated calibration parameters, i.e., $\min_p d(\mathbf{P}(\mathbf{A}^3), \mathbf{A}^2)$, where $\mathbf{A}^2$ is a set of calibration features extracted from the images, $\mathbf{A}^3$ is the set of known 3-D model features, and **P** is the estimated projection matrix. One typical and popular camera calibration method proposed by Tsai [28] is implemented by R. Willson and and can be downloaded from the web (http://www.cs.cmu.edu/afs/cs.cmu.edu/user/rgw/www/TsaiCode.html). Detailed reviews of the main existing approaches can be found in Tsai [27] and Weng *et al.* [29].

4.3 Uncalibrated Stereo Analysis

Although camera calibration is widely used in the fields of photogrammetry and computer vision, it is a very tedious task and is sometimes unfeasible. In many applications, on-line calibration is required, a calibration pattern may not be available, or both. For instance, in the reconstruction of a scene from a sequence of video images where the parameters of the video lens are subject to continuous change, camera calibration in the classical sense is not possible. Faugeras [30] pointed out that, from point correspondences in pairs of images when *no* initial assumption is made about either the intrinsic or extrinsic parameters of the camera, it is possible to compute a *projective representation* of the world. This representation is defined up to certain transformations of the environment, which we assume is 3-D and Euclidean. This concept of constructing the projective representation of a 3-D object instead of an Euclidean projection is called *uncalibrated stereo*. Since the work reported by Faugeras [30], several approaches for uncalibrated stereo have been proposed that permit projective reconstructions from multiple views. These approaches use *weak calibration*, which is represented by the epipolar geometry, and hence requires no knowledge of the intrinsic or extrinsic camera parameters. Faugeras *et al.* [31] recovered a realistic texture model of an urban scene from a sequence of video images, using uncalibrated stereo technique, without any prior knowledge of the camera parameters or camera motion. The structure of their vision system is shown in Fig. 8. In the following sections, we introduce the basic ideas of uncalibrated stereo and present a technique for deriving 3-D scene structure from video sequences or a number of snapshots. First we introduce the results of *weak calibration*, which refers to algorithms that find

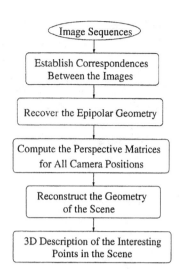

FIGURE 8 General paradigm of 3-D reconstruction from uncalibrated multiviews.

the projective structure of the scene with a given epipolar geometry (fundamental matrix) between cameras. Then, the theory for constructing the fundamental matrix from correspondences in multiple images will be presented.

4.3.1 Weak Calibration: Projective Reconstruction

Projective reconstruction algorithms can be classified into two distinct classes: explicit strategies and implicit strategies that are due to the way in which the 3-D projective coordinates are computed. Explicit algorithms are essentially similar to traditional stereo algorithms in the sense that the explicit estimation of camera projective matrices is always involved in the initial phase of the processing. Implicit algorithms are based on implicit image measurements, which are used to compute *projective invariants* from image correspondences. The invariants are functionally dependent on the 3-D coordinates, for example, the *projective depth*, the *Cayley algebra* or *Double algebra* invariants, *cross ratios*, etc.

Rothwell *et al.* [32] compared three explicit weak calibration methods (pseudo-inverse-based, singular value decomposition-based, and intersecting ray-based algorithms) and two implicit methods (Cayley algebra-based and cross ratio-based approaches). They found that the singular value decomposition-based approach provides the best results. Here, we will only present the principle of the explicit weak calibration method based on singular value decomposition.

Given the fundamental matrix $\mathbf{F}$ for the two cameras there are an infinite number of projective bases, which all satisfy the epipolar geometry. Luong and Vieville [33] derive a canonical solution set for the cameras projective matrix that is consistent with the epipolar geometry,

$$\mathbf{Q} = \left[-\frac{1}{\|\mathbf{e}\|^2} [\mathbf{e}]_\times \mathbf{F} \quad \mathbf{e} \right] \mathbf{G}_{4\times 4}, \qquad (40)$$

$$\mathbf{Q}' = [\mathbf{I}_{3\times 3} \quad \mathbf{O}_{3\times 1}] \, \mathbf{G}_{4\times 4}, \qquad (41)$$

where $\mathbf{e}_\times$ is a *skew symmetric matrix* defined by the epipole $\mathbf{e} = [e_1, e_2, e_3]^T$.

According to Eq. (9), the epipole $\mathbf{e}$ can be computed by the eigenvector of the matrix $\mathbf{FF}^T$ associated with the smallest eigenvalue. A detailed robust procedure is provided in Xu *et al.* [34] to compute the epipole from fundamental matrix.

The projective matrices $\mathbf{Q}$ and $\mathbf{Q}'$ allow us to triangulate 3-D structure from image correspondences from Eq. (38), *up to a projective transformation* $\mathbf{G}$. Here, $\mathbf{G}$ defines a transformation from a three-dimensional *virtual* coordinate system to the image planes.

4.3.2 Recovery of the Fundamental Matrix

The fundamental matrix was first described in Longuet-Higgins [4] for uncalibrated images. It determines the positions of the two epipoles and the epipolar transformation mapping an epipolar line from the first image to its counterpart in the second image. It is the key concept in the case of two uncalibrated cameras, because it contains all the geometrical information relating two images of a single object. The fundamental matrix can be computed from a certain number of point correspondences obtained from a pair of images without using the camera calibration process. Correspondences in stereo sequences can be established by using methods demonstrated in Section 3 as well as correspondences from motion techniques such as tracking methods. In this section, we will specify a method to estimate the fundamental matrix from point correspondences. For fundamental matrix recovery from line segments, we refer the reader to Xu *et al.* [34].

The basic theory of recovering the epipolar geometry is essentially given in Eq. (7). If we are given n point matches $(\mathbf{m}_i, \mathbf{m}'_i)$, $\mathbf{m}_i = (u_i, v_i)$, $\mathbf{m}'_i = (u'_i, v'_i)$, then using Eq. (7), we get the following linear system to solve:

$$\mathbf{U}_n \mathbf{f} = \mathbf{0}, \qquad (42)$$

where

$$\mathbf{U}_n = \begin{bmatrix} \mathbf{u}_1^T & \dots & \mathbf{u}_n^T \end{bmatrix}^T,$$

$$\mathbf{u}_i = [u_i u'_i, \, u_i v'_i, \, u_i, \, v_i u'_i, \, v_i v'_i, \, v_i, \, u'_i, \, v'_i, \, 1]^T,$$

$$\mathbf{f} = [F_{11}, F_{12}, F_{13}, F_{21}, F_{22}, F_{23}, F_{31}, F_{32}, F_{33},]^T.$$

Here, F_{ij} is the element of fundamental matrix $\mathbf{F}$ at row i and column j. These sets of linear homogeneous equations, together with the rank constraint of the matrix $\mathbf{F}$ (i.e., rank($\mathbf{F}$) = 2), lead to epipolar geometry estimation.

Given eight or more matches, from Eq. (7) we can write down a set of linear equations in the nine unknown elements of matrix $\mathbf{F}$. In general, we will be able to determine a unique solution for $\mathbf{F}$, defined *up to a scale factor*. For example, the singular value decomposition technique can be used for this purpose.

The simplest way to solve Eq. (42) under the rank-2 constraint is to use the *linear least-squares* technique. The entire problem

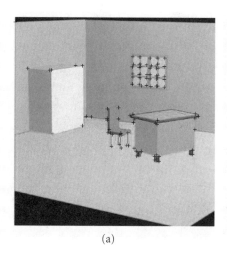 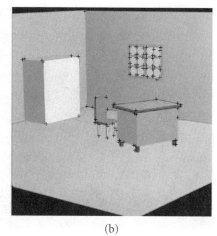

|(a)|(b)|

FIGURE 9 Stereo pair: (a) left and (b) right views.

can be transformed to a minimization problem:

$$\min_{\mathbf{f}} \psi(\mathbf{f}, \lambda), \qquad (43)$$

where

$$\psi(\mathbf{f}, \lambda) = \|\mathbf{U}_n \mathbf{f}\|^2 + \lambda(1 - \|\mathbf{f}\|^2). \qquad (44)$$

It can be shown that the solution is the unit eigenvector of matrix $\mathbf{U}_n^T \mathbf{U}_n$ associated with the *smallest* eigenvalue. Several natural nonlinear minimization criteria are discussed in Xu and Zhang's book [34].

5 Experiments

In this section, we demonstrate the performance of the feature detection and matching algorithms and 3-D recovery on two synthetic images. Figure 9 shows the original images acquired from left and right cameras.[4] All intrinsic and extrinsic parameters of both cameras are known. The center of the left camera is coincident with the origin of the reference coordinate system. The right camera has a slight rotation and translation in the x axis with respect to the left camera. Corner points with high curvature are located as feature primitives, marked with "+" in left and right views shown in Fig. 10. Geometry ordering, intensity similarity, and epipolar constraints (by thresholding the distance from the corresponding epipolar line) are employed consecutively to narrow the correspondence search space. The final decision on correspondence is made by applying a uniqueness constraint that selects the candidate match point with the highest window-based correlation [35]. As an exercise, readers can formulate these matching constraints and the confidence of correct matching to a statistic framework as described in Eq. (35). Figure 11 shows all the reported correspondence pairs in left and right views. Most of the significant token pairs (e.g., the corners of the furniture) are successfully identified. By using the

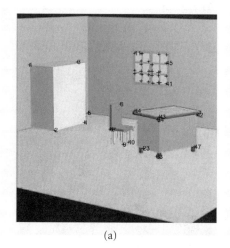 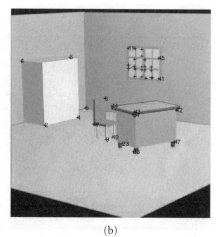

|(a)|(b)|

FIGURE 10 Stereo pair with detected corners: (a) left and (b) right views.

[4]Copyright, Institute National de Recherche en Informatique et Automatique, 1994, 1995, 1996.

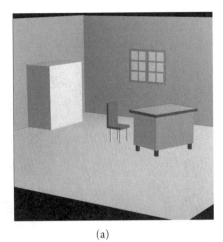 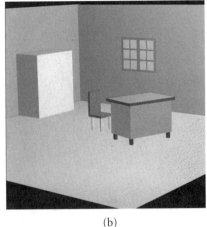

(a)   (b)

FIGURE 11   Stereo pair with established correspondences: (a) left and (b) right views.

triangulation method introduced in Section 4.1, we can recover the 3-D description of those feature points. Figure 12(a) shows the result to reproject those recovered 3-D points to the left camera, while Fig. 12(b) shows the mapping of those points to a camera slightly rotating from the Y axis of the left camera. The projection of each corner point is marked with "+," overlapping on the original image taken by the left camera. Note that there is no significant vertical translation while remarkable horizontal disparities are produced as a result of the rotation of the camera. We actually succeed in a good reconstruction for those important corner points. In this example, we considered only individual points. If we can combine the connectivity and coplanarity information from those feature points (obtained from segmentation or from the prior knowledge of the scene structure), a more accurate and reliable understanding of the 3-D geometry in the scene can be recovered. Reconstruction using line segments is expected to achieve a more robust scene reconstruction.

6 Conclusions

In this chapter, we have discussed in detail how to reconstruct 3-D shapes from multiple 2-D images taken from different viewpoints. The most important step in solving this problem is matching, i.e., finding the correspondence among multiple images. We have presented various optic flow-based and feature-based techniques used for the purpose of matching. Once the correspondence problem is solved, we can reconstruct the 3-D shape by using a calibrated or uncalibrated stereo analysis.

Acknowledgments

H. B. Zhao and J. K. Aggarwal were supported in part by the U.S. Army Research Office under contracts DAAH04-95-I-0494 and DAAG55-98-1-0230, and the Texas Higher Education Coordinating Board Advanced Research Project 97-ARP-275.

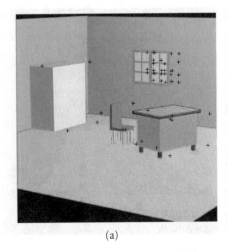 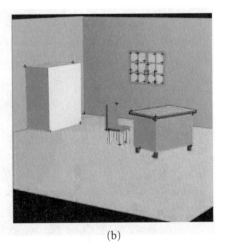

(a)   (b)

FIGURE 12   Reprojections from two different viewpoints; projected (a) to the left camera and (b) to a camera with slight rotation from the left camera.

C. Mandal and B. C. Vemuri were supported in part by the NSF 9811042 & NIH R01-RR13197 grants. Special thanks go to Ms. Debi Paxton and Mr. Umesh Dhond for their generous help and suggestions in editing and commenting on the paper.

References

[1] U. R. Dhond and J. K. Aggarwal, "Structure from stereo — a review," *IEEE Trans. Syst. Man Cybernet.* **19**, 1489–1510 (1989).

[2] S. D. Cochran and G. Medioni, "3-D surface description from binocular stereo," *IEEE Trans. Pattern Anal. Machine Intell.* **14**, 981–994 (1992).

[3] O. D. Faugeras, *Three-Dimensional Computer Vision: A Geometric Viewpoint* (MIT Press, Cambridge, MA, 1993).

[4] H. C. Longuet-Higgins, "A computer algorithm for reconstructing a scene from two projections," *Nature* **293**, 133–135 (1981).

[5] B. C. Vemuri, S. Huang, S. Sahni, C. M. Leonard, C. Mohr, R. Gilmore, and J. Fitzsimmons, "An efficient motion estimator with application to medical image registration," *Med. Image Anal.* **2**, 79–98 (1998).

[6] J. K. Aggarwal and N. Nandhakumar, "On the computation of motion from sequences of images — a review," *Proc. IEEE* **76**, 917–935 (1988).

[7] J. L. Barron, D. J. Fleet, and S. S. Beauchemin, "Performance of optical flow techniques," *Int. J. Comput. Vis.* **12**, 43–77 (1994).

[8] S. H. Lai and B. C. Vemuri, "Reliable and efficient computation of optical flow," *Int. J. Comput. Vis.* **29**, 87–105 (1998).

[9] R. Szeliski and J. Coughlan, "Hierarchical spline-based image registration," *IEEE Conf. Comput. Vis. Pattern Recog.* **1**, 194–201 (Los Alamitos, 1994).

[10] B. K. P. Horn and B. G. Schunk, "Determining optical flow," *Artificial Intell.* **17**, 185–203 (1981).

[11] H. Burkhardt and N. Diehl, "Simultaneous estimation of rotation and translation in image sequences," in *Proceedings of the European Signal Processing Conference* (Elsevier Science Publishers, The Hague, Netherlands 1986), pp. 821–824.

[12] Y. C. Kim and J. K. Aggarwal, "Finding range from stereo images," in *Proc. IEEE Comput. Soc. Conf. Comput. Vis. Pattern Recog.* **1**, 289–294, (1985).

[13] D. Marr and T. A. Poggio, "A computational theory of human stereo vision," *Proc. Royal Soc. London* **B204**, 301–328 (1979).

[14] M. S. Lew, T. S. Huang, and K. Wong, "Learning and feature selection in stereo matching," *IEEE Trans. Pattern Anal. Machine Intell.* **16**, 869–881 (1994).

[15] Y. L. Chang and J. K. Aggarwal, "Line correspondences from cooperating spatial and temporal grouping processes for a sequence of images," *Comput. Vis. Image Understanding* **67**, 186–201 (1997).

[16] W. J. Christmas, J. Kittler, and M. Petrou, "Structural matching in computer vision using probabilistic relaxation," *IEEE Trans. Pattern Anal. Machine Intell.* **17**, 749–764 (1995).

[17] P. Fornland, G. A. Jones, G. Matas, and J. Kittler, "Stereo correspondence from junction," in *Proceedings of the 8th Scandinavians Conference on Image Analysis* (NOBIM–Norwegian Soc. Image Process. Pattern Recognitions, **1**, 1993), 449–455.

[18] G. A. Jones, "Constraint, optimization, and hierarchy: Reviewing stereoscopic correspondence of complex features," *Comput. Vis. Image Understanding* **65**, 57–78 (1997).

[19] K. L. Boyer and A. C. Kak, "Structural stereopsis for 3-D vision," *IEEE Trans. Pattern Anal. Machine Intell.* **10**(2), 144–166 (1988).

[20] R. Horaud and T. Skordas, "Stereo correspondences through feature grouping and maximal cliques," *IEEE Trans. Pattern Anal. Machine Intell.* **11**, 1168–1180 (1989).

[21] K. Prazdny, "Detection of binocular disparities," *Biol. Cybernetics* **52**, 93–99 (1985).

[22] S. Ranade and A. Rosenfeld, "Point pattern matching by relaxation," *Pattern Recog.* **12**, 269–275 (1980).

[23] N. M. Nasrabadi, "A stereo vision technique using curve-segments and relaxation matching," *IEEE Trans. Pattern Anal. Machine Intell.* **14**, 566–572 (1992).

[24] J. E. W. Mayhew and J. P. Frisby, "Psychophysical and computational studies towards a theory of human stereopsis," *Artificial Intell.* **17**, 349–385 (1981).

[25] H. S. Lim and T. O. Binford, "Stereo correspondence: A hierarchical approach," in *Proceedings of the DARPA Image Understanding Workshop* (Los Altos. CA, 1987), pp. 234–241.

[26] U. R. Dhond and J. K. Aggarwal, "Stereo matching in the presence of narrow occluding objects using dynamic disparity search," *IEEE Trans. Pattern Anal. Machine Intell.* **17**, 719–724 (1995).

[27] R. Tsai, "Synopsis of recent progress on camera calibration for 3-D machine vision," in *The Robotics Review* (MIT Press, Cambridge, MA, 1989), pp. 147–159.

[28] R. Y. Tsai, "A versatile camera calibration technique for high-accuracy 3-D machine vision metrology using off-the-shelf TV cameras and lenses," *IEEE J. Robot. Automat.* **3**, 323–344 (1987).

[29] J. Weng, P. Cohen, and M. Herniou, "Camera calibration with distortion models and accuracy evaluation," *IEEE Trans. Pattern Anal. Machine Intell.* **14**, 965–980 (1992).

[30] O. D. Faugeras, "What can be seen in three dimensions with an uncalibrated stereo rig," in *Proceedings of ECCV92* (Santa Margherita Ligure, Italy, 1992), pp. 563–578.

[31] O. Faugeras, L. Robert, S. Laveau, G. Csurka, C. Zeller, C. Gauclin, and I. Zoghlami, "3-D reconstruction of urban scenes from image sequences," *Comput. Vis. Image Understanding* **69**, 292–309 (1998).

[32] C. Rothwell, O. Faugeras, and G. Csurka, "A comparison of projective reconstruction methods for pairs of views," *Comput. Vis. Image Understanding* **68**, 37–58 (1997).

[33] Q. T. Luong and T. Vieville, "Canonical representations for the geometries of multiple projective views," *Comput. Vis. Image Understanding* **64**, 193–229 (1996).

[34] G. Xu and Z. Y. Zhang, *Epipolar Geometry in Stereo, Motion and Object Recognition*, Vol. 6 (Kluwer, The Netherlands, 1996).

[35] C. M. Sun, "A fast stereo matching method," in *Digital Image Computing: Techniques and Applications* (Massey U. Press, Auckland, New Zealand, 1997), pp. 95–100.

3.13

Image Sequence Stabilization, Mosaicking, and Superresolution

S. Srinivasan
Sensar Corporation

R. Chellappa
University of Maryland

1 Introduction

A sequence of temporal images gathered from a single sensor adds a whole new dimension to two-dimensional image data. Availability of an image sequence permits the measurement of quantities such as subpixel intensities, camera motion and depth, and detection and tracking of moving objects that is not possible from any single image. In turn, the processing of image sequences necessitates the development of sophisticated techniques to extract this information. With the recent availability of powerful yet inexpensive computing equipment, data storage systems, and imagery acquisition devices, image sequence analysis has moved from an esoteric research domain to a practical area with significant commercial interest.

Motion problems in which the scene motion largely conforms to a smooth, low-order motion model are termed global motion problems. Electronically stabilizing video, creating mosaics from image sequences, and performing motion superresolution are examples of global motion problems. Applications of these processes are often encountered in surveillance, navigation, teleoperation of vehicles, automatic target recognition (ATR), and forensic science. Reliable motion estimation is critical to these tasks, which is particularly challenging when the sequences display random as well as highly structured systematic errors. The former is primarily a result of sensor noise, atmospheric turbulence, and lossy compression, whereas the latter is caused by

occlusion, shadows, and independently moving foreground objects. The goal in global motion problems is to maintain the integrity of the solution in the presence of both types of errors.

Temporal variation in the image luminance field is caused by several factors, including camera motion, rigid object motion, nonrigid deformation, illumination and reflectance change, and sensor noise. In several situations, it can be assumed that the imaged scene is rigid, and temporal variation in the image sequence is only due to camera and object motion. Classical motion estimation characterizes the local shifts in the image luminance patterns. The global motion that occurs across the entire image frame is typically a result of camera motion and can often be described in terms of a low-order model whose parameters are the unknowns. Global motion analysis is the estimation of these model parameters.

The computation of global motion has seldom attained the center stage of research as a result of the (often incorrect) assumption that it is a linear or otherwise well-conditioned problem. In practice, an image sequence displays phenomena that voids the assumption of Gaussian noise in the motion field data. The presence of moving foreground objects or occlusion locally invalidates the global motion model, giving rise to outliers. Robustness to such outliers is required of global motion estimators. Researchers [4–19] have formulated solutions to global motion problems, usually with an application perspective. These can be broadly classified as *feature-based* and *flow-based* techniques.

Feature-based methods extract and match discrete features between frames, and the trajectories of these features is fit to a global motion model. In flow-based algorithms, the optical flow of the image sequence is an intermediate quantity that is used in determining the global motion. Chapter 3.8 provides an extended discussion of optical flow.

The focus of this chapter is a flow-based solution to the global motion problem. First, the optical flow field of the image sequence is modeled in terms of a linear combination of basis functions. Next, the model weights that describe the flow field are computed. Finally, these weights are combined by using an iterative refinement mechanism to identify outliers, and they provide a robust global motion estimate. This algorithm is described in Section 3. The three primary applications of this procedure are two-dimensional (2-D) stabilization, mosaicking, and motion superresolution. These are described in Sections 4, 5, and 6. A related but theoretically distinct problem, three-dimensional (3-D) stabilization, is introduced in Section 7.

Prior to examining solutions to the global motion problem, it is advisable to verify whether the apparent motion on the image plane induced by camera motion can indeed be approximated by a global model. This study takes into consideration the 3-D structure of the scene being viewed, and its corresponding image. The moving camera has 6 degrees of freedom, determining its three translational and three rotational velocities. It remains to be seen whether the motion field generated by such a system can be parametrized in terms of a global model largely independent of scene depth. This is analyzed in Section 2.

2 Global Motion Models

The imaging geometry of a perspective camera is shown in Fig. 1. The origin of the 3-D coordinate system (X, Y, Z) lies at the optical center C of the camera. The *retinal plane* or *image plane* is normal to the optical axis Z and is offset from C by the focal length f. Images of unoccluded 3-D objects in front of

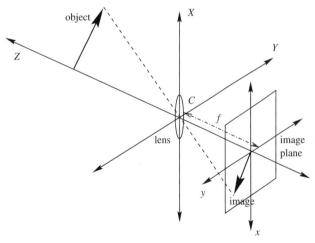

FIGURE 1   3-D imaging geometry.

the camera are formed on the image plane. The 2-D image plane coordinate system (x, y) is centered at the *principal point*, which is the intersection of the optical axis with the image plane. The orientation of (x, y) is flipped with respect to (X, Y) in Fig. 1, because of an inversion caused by simple transmissive optics. For this system, the image plane coordinate (x_i, y_i) of the image of the unoccluded 3-D point (X_i, Y_i, Z_i) is given by

$$x_i = f\frac{X_i}{Z_i}, \qquad y_i = f\frac{Y_i}{Z_i}. \tag{1}$$

Projective relation (1) assumes a rectilinear system, with an isotropic optical element. In practice, the plane containing the sensor elements may be misaligned from the image plane, and the camera lens may suffer from optical distortions, including nonisotropy. However, these effects can be compensated for by calibrating the camera or remapping the image. In the remainder of this chapter, it is assumed that linear dimensions are normalized to the focal length, i.e., $f = 1$.

When a 3-D scene is imaged by a moving camera, with translation $t = (t_x, t_y, t_z)$ and rotation $\omega = (\omega_x, \omega_y, \omega_z)$, the optical flow of the scene (Chapter 3.8) is given by

$$u(x, y) = (-t_x + xt_z)g(x, y) + xy\omega_x - (1 + x^2)\omega_y + y\omega_z,$$
$$v(x, y) = (-t_y + yt_z)g(x, y) + (1 + y^2)\omega_x - xy\omega_y - x\omega_z, \tag{2}$$

for small ω. Here, $g(x, y) = 1/Z(x, y)$ is the inverse scene depth. Clearly, the optical flow field can be arbitrarily complex and does not necessarily obey a low-order global motion model. However, several approximations to Eq. (2) exist that reduce the dimensionality of the flow field. One possible approximation is to assume that translations are small compared to the distance of the objects in the scene from the camera. In this situation, image motion is caused purely by camera rotation and is given by

$$u(x, y) = xy\omega_x - (1 + x^2)\omega_y + y\omega_z,$$
$$v(x, y) = (1 + y^2)\omega_x - xy\omega_y - x\omega_z. \tag{3}$$

Equation (3) represents a true global motion model, with 3 degrees of freedom $(\omega_x, \omega_y, \omega_z)$. When the field of view (FOV) of the camera is small, i.e., when $|x|, |y| \ll 1$, the second-order terms can be neglected, giving a further simplified three-parameter global motion model:

$$u(x, y) = \omega_y + y\omega_z,$$
$$v(x, y) = \omega_x - x\omega_z. \tag{4}$$

Alternatively, the 3-D world being imaged can be assumed to be approximately planar. It can be shown that the inverse scene depth for an arbitrarily oriented planar surface is a planar function of the image coordinates (x, y),

$$g(x, y) = ax + by + c. \tag{5}$$

Substituting Eq. (5) into Eq. (2) gives the eight-parameter global motion model:

$$u(x, y) = a_0 + a_1 x + a_2 y + a_6 x^2 + a_7 xy,$$
$$v(x, y) = a_3 + a_4 x + a_5 y + a_6 xy + a_7 y^2, \tag{6}$$

for appropriately computed $\{a_i, i = 0, \ldots, 7\}$. Equation (6) is called the *pseudo-perspective* model or transformation.

Equation (2) relating the optical flow with structure and motion assumes that the interframe rotation is small. If this is not the case, the effect of camera motion must be computed by using projective geometry [1, 2]. Assume that an arbitrary point in the 3-D scene lies at (X_0, Y_0, Z_0) in the reference frame of the first camera and moves to (X_1, Y_1, Z_1) in the second. The effect of camera motion relates the two coordinate systems according to

$$\begin{pmatrix} X_1 \\ Y_1 \\ Z_1 \end{pmatrix} = \begin{pmatrix} r_{xx} & r_{xy} & r_{xz} \\ r_{yx} & r_{yy} & r_{yz} \\ r_{zx} & r_{zy} & r_{zz} \end{pmatrix} \begin{pmatrix} X_1 \\ Y_1 \\ Z_1 \end{pmatrix} + \begin{pmatrix} t_x \\ t_y \\ t_z \end{pmatrix}, \tag{7}$$

where the rotation matrix $[r_{ij}]$ is a function of ω. Combining Eqs. (1) and (7) permits the expression of the projection of the point in the second image in terms of that in the first as

$$x_1 = \frac{r_{xx} x_0 + r_{xy} y_0 + r_{xz} + t_x / Z_0}{r_{zx} x_0 + r_{zy} y_0 + r_{zz} + t_z / Z_0},$$
$$y_1 = \frac{r_{yx} x_0 + r_{yy} y_0 + r_{yz} + t_y / Z_0}{r_{zx} x_0 + r_{zy} y_0 + r_{zz} + t_z / Z_0}. \tag{8}$$

Assuming either that (a) points are distant compared to the interframe translation, i.e., neglecting the effect of translation, or (b) a planar embedding of the real world of Eq. (5), the *perspective* transformation is obtained:

$$x_1 = \frac{p_{xx} x_0 + p_{xy} y_0 + p_{xz}}{p_{zx} x_0 + p_{zy} y_0 + p_{zz}},$$
$$y_1 = \frac{p_{yx} x_0 + p_{yy} y_0 + p_{yz}}{p_{zx} x_0 + p_{zy} y_0 + p_{zz}}. \tag{9}$$

The flow field (u, v) is the difference between image plane coordinates $(x_1 - x_0, y_1 - y_0)$ across the entire image. When the FOV is small, it can be assumed that $|p_{zx} x_0|, |p_{zy} y_0| \ll |p_{zz}|$. Under this assumption, the flow field, as a function of image coordinate, is given by

$$u(x, y) = \frac{(p_{xx} - p_{zz})x + p_{xy} y + p_{xz}}{p_{zx} x + p_{zy} y + p_{zz}},$$
$$v(x, y) = \frac{p_{yx} x + (p_{yy} - p_{zz})y + p_{yz}}{p_{zx} x + p_{zy} y + p_{zz}}, \tag{10}$$

which is also a perspective transformation, albeit with different parameters. Here $p_{zz} = 1$, without loss of generality, giving 8 degrees of freedom for the perspective model.

Other popular global deformations mapping the projection of a point between two frames are the similarity and affine transformations, which are given by

$$\begin{pmatrix} x_1 \\ y_1 \end{pmatrix} = s \begin{pmatrix} \cos\theta & \sin\theta \\ -\sin\theta & \cos\theta \end{pmatrix} \begin{pmatrix} x_0 \\ y_0 \end{pmatrix} + \begin{pmatrix} b_0 \\ b_1 \end{pmatrix}, \tag{11}$$

$$\begin{pmatrix} x_1 \\ y_1 \end{pmatrix} = \begin{pmatrix} a_0 & a_1 \\ a_2 & a_3 \end{pmatrix} \begin{pmatrix} x_0 \\ y_0 \end{pmatrix} + \begin{pmatrix} b_0 \\ b_1 \end{pmatrix}, \tag{12}$$

respectively. Free parameters for the similarity model are the scale factor s, image plane rotation θ, and translation (b_0, b_1). Taking the difference between interframe coordinates of the similarity transform gives the optical flow field model of Eq. (4) with one constraint on the free parameters. The affine transformation is a superset of the similarity operator, and it incorporates shear and skew as well. The optical flow field corresponding to the coordinate affine transform, Eq. (12), is also a 6 degrees of freedom affine model. The perspective operator is a superset of the affine, as can be readily verified by setting $p_{zx} = p_{zy} = 0$ in Eq. (9).

The similarity, affine, and perspective transformations are *group operators*, which means that each family of transformations constitutes an equivalence class. The following four properties define group operators.

1. Closure: if $A, B \in \mathcal{G}$ where $\mathcal{G}$ is a group, then the composition $AB \in \mathcal{G}$.
2. Associativity: for all $A, B, C \in \mathcal{G}$, $(AB)C = A(BC)$.
3. Identity: $\exists I \in \mathcal{G}$ such that $AI = IA = A$.
4. Inverse: for each operator $A \in \mathcal{G}$, there exists an inverse $A^{-1} \in \mathcal{G}$ such that $AA^{-1} = A^{-1}A = I$.

The utility of the closure property is that a sequence of images can be rewarped to an arbitrarily chosen "origin" frame by using any single class of operators, and flows computed only between adjacent frames. Since the inverse of each transformation exists, the origin need not necessarily be the first frame of the sequence. Note that pseudo-perspective transformation (6) is not a group operator. Therefore, in order to warp an image under a pseudo-perspective global deformation, one must register each new image directly to the origin. This can get tricky when the displacement between them is large, and worse yet when the overlap between them is small.

In the process of global motion estimation, each data point is the optical flow at a specified pixel, described by the data vector (u, v, x, y). For the affine and pseudo-perspective transformations, it is obvious that the unknowns form a set of linear equations with coefficients that are functions of the data vector components. The same is true for the perspective and similarity operators, although not obvious. For the perspective transform, the denominators of Eq. (10) are multiplied out, while for the similarity transform, the substitutions $s_0 = s\cos\theta$ and $s_1 = s\sin\theta$ give rise to linear equations. In particular, the coefficients of the unknowns in the linear equations for the similarity, affine, and pseudo-perspective models are functions of the coordinate (x, y) of the data point. With the assumption that errors in data are present only in u, v, this implies that errors in the linear system for the similarity, affine, and pseudo-perspective transforms are present only on the "right-hand side." In contrast, errors exist in all terms for the perspective model. When errors in u, v are Gaussian, the *least-squares* (LS) solution of a system of equations of the form of Eqs. (6), (11), or (12) yields the minimum mean-squared error estimate. For the perspective case, the presence of errors on the "left-hand side" calls for a *total least-squares* (TLS) [3] approach. In practice, errors in u, v are seldom Gaussian, and simple linear techniques are not sufficient.

3 Algorithm

The computation of optical flow by using image derivatives hinges on the preservation of the image luminance pattern $\psi(x, y, t)$ over time. This translates into the gradient constraint equation (Chapter 3.8 and [21]),

$$\frac{\partial \psi}{\partial t} + u \frac{\partial \psi}{\partial x} + v \frac{\partial \psi}{\partial y} = 0, \quad \forall x, y, t, \tag{13}$$

in the first-order approximation. The flow field (u, v) is a function of location (x, y). For smooth motion fields encountered in typical global motion problems, it is meaningful to model (u, v) as a weighted sum of basis functions:

$$u = \sum_{k=0}^{K-1} u_k \phi_k, \quad v = \sum_{k=0}^{K-1} v_k \phi_k. \tag{14}$$

The basis function $\phi_k(x, y)$ is typically a locally supported interpolator generated by shifts of a prototype function $\phi_0(x, y)$ along a square grid of spacing w. An example of linear basis function modeling in one dimension is shown in Example 1. Additional requirements are imposed on ϕ_0, to ensure computational ease and an intuitive appeal for modeling a flow field. These are:

1. Separability: $\phi_0(x, y) = \phi_0(x)\phi_0(y)$.
2. Differentiability: $[d\phi_0(x)/dx]$ exists $\forall x$.
3. Symmetry about the origin: $\phi_0(x) = \phi_0(-x)$.
4. Peak at the origin: $|\phi_0(x)| \le \phi_0(0) = 1$.
5. Compact support: $\phi_0(x) = 0 \; \forall |x| > w$.

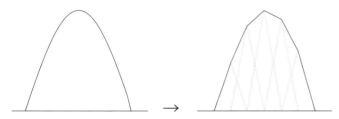

EXAMPLE 1   Function (left) and its modeled version (right). The model is the linear interpolator or triangle function; the contribution of each model basis function is denoted by the dotted curves.

The cosine window,

$$\phi_0(x) = \frac{1}{2}\left[1 + \cos\left(\frac{\pi x}{w}\right)\right], \quad x \in [-w, w], \tag{15}$$

is one such choice of basis that has been shown to accurately model typical optical flow fields associated with global motion problems. A useful range for w is between 8 and 32.

It can be shown that an unbiased estimate for the basis function model parameters $\{u_k, v_k\}$ is obtained by solving the following $2K$ equations [19]:

$$\sum_k u_k \int \frac{\partial \phi_k \phi_l (\partial \hat{\psi} / \partial x)}{\partial x} \psi + \sum_k v_k \int \frac{\partial \phi_k \phi_l (\partial \hat{\psi} / \partial x)}{\partial y} \psi$$

$$= \int \phi_l \frac{\partial \hat{\psi}}{\partial t} \frac{\partial \hat{\psi}}{\partial x},$$

$$\sum_k u_k \int \frac{\partial \phi_k \phi_l (\partial \hat{\psi} / \partial y)}{\partial x} \psi + \sum_k v_k \int \frac{\partial \phi_k \phi_l (\partial \hat{\psi} / \partial y)}{\partial y} \psi$$

$$= \int \phi_l \frac{\partial \hat{\psi}}{\partial t} \frac{\partial \hat{\psi}}{\partial y}, \quad \forall l = 0, 1, \ldots, K-1. \tag{16}$$

Each pair of equations of the type of Eq. (16) characterizes the solution around the image area covered by the basis function ϕ_l. The dominant unknowns, which are the corresponding model weights, are u_l, v_l. The finite support requirement on basis function ϕ_l ensures that only the center weights u_l, v_l and their immediate neighbors in the cardinal and diagonal directions enter each equation. In practice, sampled differentiations and integrations are performed on the sequence. Each equation pair is computed as follows.

1. First, the X, Y and temporal gradients are computed for the observed frame of the sequence. Smoothing is performed prior to gradient estimation, if the images are dominated by sharp edges.
2. Three templates, each of size $2w \times 2w$, are formed. The first template is the prototype function ϕ_0, with its support coincident with the template. The other two are its X and Y gradients. Knowledge of the analytical expression for ϕ_0 means that its gradients can be determined with no error.
3. Next, a square tile of size $2w \times 2w$ of the original and spatiotemporal gradient images, coincident with the support of ϕ_l, is extracted.
4. The 18 left-hand-side terms of each equation, and one right-hand-side term, are computed by overlaying the templates as necessary and computing the sum of products.
5. Steps 3 and 4 are repeated for all K basis functions.
6. Since the interactions are only between spatially adjacent basis function weights, the resulting matrix is sparse, block tridiagonal, with tridiagonal submatrices, each entry of which is a 2×2 matrix. This permits convenient storage of the left-hand-side matrix.

7. The resulting sparse system is solved rapidly by using the preconditioned biconjugate gradients algorithm [22, 23].

The procedure described above produces a set of model parameters $\{u_k, v_k\}$ that largely conforms to the appropriate global motion model, where one exists. In the second phase, these parameters are simultaneously fit to the global motion model while outliers are identified, using the iterated weighted LS technique outlined below.

1. Initialization:
   (a) All flow field parameters whose support regions show a sufficiently large high-frequency energy (quantified in terms of the determinant and condition number of the covariance matrix of the local spatial gradient) are flagged as valid data points.
   (b) A suitable global motion model is specified.
2. Model fitting:
   (a) If there are an insufficient number of valid data points, the algorithm signals an inability to compute the global motion. In this event, a more restrictive motion model must be specified.
   (b) If there are sufficient data points, model parameters are computed to be the LS solution of the linear system relating observed model parameters with the global motion model of choice.
   (c) When a certain number of iterations of this step are complete, the LS solution of valid data points is output as the global motion model solution.
3. Model consistency check:
   (a) The compliance of the global motion model to the overlapped basis flow vectors is computed at all grid points flagged as valid, using a suitable error metric.
   (b) The mean error $\bar{\epsilon}$ is computed. For a suitable multiplier f, all grid points with errors larger than $f\bar{\epsilon}$ are declared invalid.
   (c) Step 2 is repeated.

Typically, three to four iterations are sufficient. Since this system is open loop, small errors do tend to build up over time. It is also conceivable to use a similar approach to refine the global motion estimate by registering the current image with a suitably transformed origin frame.

4 Two-Dimensional Stabilization

Image stabilization is a differential process that compensates for the unwanted motion in an image sequence. In typical situations, the term "unwanted" refers to the motion in the sequence resulting from the kinematic motion of the camera with respect to an inertial frame of reference. For example, consider high-magnification handheld binoculars. The jitter introduced by an unsteady hand causes unwanted motion in the scene being viewed. Although this jitter can be eliminated by anchoring the binoculars on a tripod, this is not always feasible. Gyro-

(a)　　　　　　　　(b)

(c)　　　　　　　　(d)

EXAMPLE 2 The first sequence was gathered by a Texas Instruments infrared camera with a relatively narrow field of view. The scene being imaged is a road segment with a car, a cyclist, two pedestrians and foliage. The car, cyclist, and pedestrians move across the scene, and the foliage ruffles mildly. The camera is fixated on the cyclist, throwing the entire background into motion. It is difficult for a human observer to locate the cyclist without stabilizing for camera motion. The camera undergoes panning with no rotation about the optical axis, and no translation. The first and forty-second frames are shown in (a) and (b). The difference between these frames with no stabilization is shown in (c), with the zero difference offset to 50% gray intensity. Large difference magnitudes can be seen for several foreground and background objects in the scene. In contrast, the cyclist disappears in the difference image. The same difference, after stabilization, is shown in (d). Background areas disappear almost entirely, and all moving foreground objects including the evasive cyclist appear in the stabilized difference. The position of the cyclist in the first and forty-second frames is indicated by the white and black arrows, respectively.

scopic stabilizers are employed by professional videographers, but their bulk and cost are a deterrent to several users. Simpler inertial mechanisms are often found in cheaper "image stabilizing" optical equipment. These work by perturbing the optical path of the device to compensate for unsteady motion. The same effect can be realized in electronic imaging systems by rewarping the generated sequence in the digital domain, with no need for expensive transducers or moving parts.

The unwanted component of motion does not carry any information of relevance to the observer, and is often detrimental to the image understanding process. For general 3-D motion of a camera imaging a 3-D scene, the translational component of the velocity cannot be annulled because of motion parallax. Compensating for 3-D rotation of the camera or components thereof

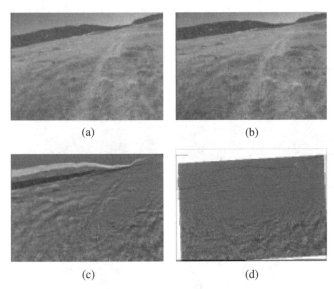

(a)   (b)

(c)   (d)

EXAMPLE 3 The second image sequence (b) portrays a navigation scenario where a forward-looking camera is mounted on a vehicle. The platform translates largely along the optical axis of the camera and undergoes pitch, roll, and yaw. The camera has a wide FOV, and the scene shows significant depth variation. The lower portion of the image is the foreground, which diverges rapidly as the camera advances. The horizon and distantly situated hills remain relatively static. The third and twentieth frames of this sequence are shown in (a) and (b). Clearly, forward translation of the camera is not insignificant and full stabilization is not possible. However, the affine model performs a satisfactory job of stabilizing for pitch, roll, and yaw. This is verified by looking at the unstabilized and stabilized frame differences, shown in (c) and (d). In (d), the absolute difference around the hill areas is visibly very small. The foreground does show change caused by forward translation parallax that cannot be compensated for. (Courtesy of Martin Marietta.)

is referred to as 3-D stabilization and is discussed in Section 7. More commonly, the optical flow field is assumed to obey a global model, and the rewarping process using computed global motion model parameters is known as em 2-D stabilization. Under certain conditions, for example when there is no camera translation, the 2-D and 3-D stabilization processes produce identical results.

The similarity, affine, and perspective models are commonly used in 2-D stabilization. Algorithms, such as the one described in Section 3, compute the model unknowns. The interframe transformation parameters are accumulated to estimate the warping with respect to the first or arbitrarily chosen origin frame. Alternatively, the registration parameters of the current frame with respect to the origin frame can be directly estimated. For smooth motion, the former approach allows the use of gradient-based flow techniques for motion computation. However, the latter approach usually has better performance since errors in the interframe transformation tend to accumulate in the former. Two sequences, reflecting disparate operating conditions, are presented here for demonstrating the effect of 2-D stabilization. It must be borne in mind that the output of a stabilizer is an image sequence whose full import cannot be conveyed by means of still images.

5 Mosaicking

Mosaicking is the process of compositing or piecing together successive frames of an image sequence so as to virtually increase the FOV of the camera [24]. This process is especially important for remote surveillance, tele-operation of unmanned vehicles, rapid browsing in large digital libraries, and in video compression. Mosaics are commonly defined only for scenes viewed by a pan/tilt camera. However, recent studies look into qualitative representations, nonplanar embeddings, [25] and layered models [26]. The newer techniques permit camera translation and gracefully handle the associated parallax. Mosaics represent the real world in two dimensions, on a plane or other manifold like the surface of a sphere or "pipe." Mosaics that are not true projections of the 3-D world, yet present extended information on a plane, are referred to as *qualitative* mosaics.

Several options are available while building a mosaic. A *simple* mosaic is obtained by compositing several views of a static 3-D scene from the same view point and different view angles. Two alternatives exist, when the imaged scene has moving objects, or when there is camera translation. The *static* mosaic is generated by aligning successive images with respect to the first frame of a batch, and performing a temporal filtering operation on the stack of aligned images. Typical filters are the pixelwise mean and median over the batch of images, which have the effect of blurring out moving foreground objects. Alternatively, the mosaic image can be populated with the first available information in the batch.

Unlike the static mosaic, the *dynamic* mosaic is not a batch operation. Successive images of a sequence are registered to either a fixed or a changing origin, referred to as the *backward* and *forward stabilized* mosaics, respectively. At any time instant, the mosaic contains all the new information visible in the most recent input frame. The fixed coordinate system generated by a backward stabilized dynamic mosaic literally provides a snapshot into the transitive behavior of objects in the scene. This finds use in representing video sequences using still frames. The forward stabilized dynamic mosaic evolves over time, providing a view port with the latest past information supplementing the current image. This procedure is useful for virtual field of view enlargement in the remote operation of unmanned vehicles.

In order to generate a mosaic, the global motion of the scene is first estimated. This information is then used to rewarp each incoming image to a chosen frame of reference. Rewarped frames are combined in a manner suitable to the end application. The algorithm presented in Section 3 is an efficient means of computing the global motion model parameters. Results using this algorithm are presented in the following examples.

6 Motion Superresolution

Besides being used to eliminate foreground objects, data redundancy in a video sequence can be exploited for enhancing the resolution of an image mosaic, especially when the overlap between

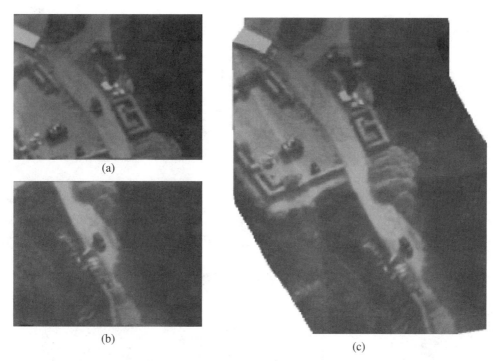

(a)

(b)   (c)

EXAMPLE 4 Images (a) and (b) show the first and 180th frames of the *Predator F* sequence. The vehicle near the center moves as the camera pans across the scene in the same general direction. Poor contrast is evident in the top right of (a) and in most of (b). The use of basis functions for computing optical flow pools together information across large areas of the sequence, thereby mitigating the effect of poor contrast. Likewise, the iterative process of obtaining model parameters successfully eliminates outliers caused by the moving vehicle. The mosaic constructed from this sequence is shown in (c).

frames is significant. This process is known as motion superresolution. Each frame of the image sequence is assumed to represent a warped subsampling of the underlying high-resolution original. In addition, blur and noise effects can be incorporated into the image degradation model. Let ψ_u represent the underlying image, and $K(x_u, y_u, x, y)$ be a multirate kernel that incorporates the effect of global deformation, subsampling, and blur. The observed low resolution image ψ is given by

$$\psi(x, y) = \sum_{x_u, y_u} \psi_u(x_u, y_u) K(x_u, y_u, x, y) + \eta(x, y), \quad (17)$$

where η is a noise process.

Example 1 To illustrate the operation of Eq. (17), consider a simple example. Let the observed image be a 4:1 downsampled representation of the original, with a global translation of $(2, -3)$ pixels and no noise. Also assume that the downsampling kernel is a perfect anti-aliasing filter. The observed image ψ formed by this process is given by

$$\psi_4(x_u, y_u) = \psi_u(x_u, y_u) * K_0(x_u - 2, y_u + 3),$$

$$\mathcal{F}(K_0)(\omega_x, \omega_y) = \begin{cases} 1 & |\omega_x|, |\omega_y| < \frac{\pi}{4} \\ 0 & \text{otherwise} \end{cases}, \quad (18)$$

$$\psi(x, y) = \psi_4(4x, 4y),$$

with K_0 being the anti-aliasing filter and $\mathcal{F}(K_0)$ its Fourier transform. The process defined in Eq. (18) represents, in some ways, the worst-case scenario. For this case, it can be shown that the original high-pass frequencies can never be estimated, since they are perfectly filtered out in the image degradation process. Thus,

EXAMPLE 5 The TI car sequence is reintroduced here to demonstrate the generation of static mosaics. After realignment with the first frame of the sequence, a median filter is applied to the stack of stabilized images, generating the static mosaic shown above. Moving objects, e.g., the car, cyclist, and pedestrians, are virtually eliminated, giving a pure background image.

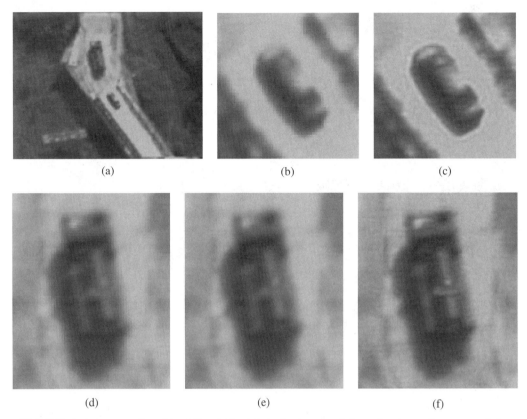

(a)   (b)   (c)

(d)   (e)   (f)

EXAMPLE 6   A demonstration of the ability of this relatively simple approach for performing motion superresolution is presented here. The *Predator B* sequence data are gathered from an aerial platform (the predator unmanned air vehicle) and compressed with loss. One frame of this sequence is shown in (a). Forty images of this sequence are coregistered by using an affine global motion model, upsampled by a factor of 4, and are combined and sharpened to generate the superresolved image. (b) and (d) show the car and truck present in the scene, at the original resolution; (e) shows the truck image upsampled by a factor of 4, using a bilinear interpolator. The superresolved images of the car and truck are shown in (c) and (f), respectively. The significant improvement in visual quality is evident. It must be mentioned here that for noisy input imagery, much of the data redundancy is expended in combating compression noise. More dramatic results can be expected when noise-free input data are available to the algorithm.

on one hand, multiple high-resolution images produce the same low-resolution images after Eq. (18). On the other hand, when the kernel K is a finite support filter, the high-frequency information is attenuated but not eliminated. In theory it is now possible to restore the original image content, at almost all frequencies, given sufficient low-resolution frames.

Motion superresolution algorithms usually comprise three distinct stages of processing — (a) registration, (b) blur estimation, and (c) refinement. Registration is the process of computing and compensating for image motion. More often than not, the blur is assumed to be known, although in theory the motion superresolution problem can be formulated to perform blind deconvolution. The kernel K is specified given the motion and blur. The process of reconstructing the original image from this information and the image sequence data is termed as refinement. Often, these stages are performed iteratively and the high-resolution image estimate evolves over time.

The global motion estimation algorithm outlined in Section 3 can be used to perform rapid superresolution. It can be shown

that superresolution can be approximated by first constructing an upsampled static mosaic, followed by some form of inverse filtering to compensate for blur. This approximation is valid when the filter K has a high attenuation over its stopband, and thereby minimizes aliasing. Moreover, such a procedure is highly efficient to implement and provides reasonably detailed superresolved frames. Looking into the techniques used in mosaicking, the median filter emerges as an excellent procedure for robustly combining a sequence of images prone to outliers. The superresolution process is defined in terms of the following steps.

1. Compute the global motion for the image sequence.
2. For an upsampling factor M, scale up the relevant global motion parameters.
3. Using a suitable interpolation kernel and scaled motion parameters, generate a stabilized, upsampled sequence.
4. Build a static mosaic by using a robust temporal operator such as the median filter.
5. Apply a suitable sharpening operator to the static mosaic.

7 Three-Dimensional Stabilization

Three-Dimensional stabilization is the process of compensating an image sequence for the true 3-D rotation of the camera. Extracting the rotation parameters for the image sequence under general conditions involves solving the *structure from motion* (SFM) problem, which is the simultaneous recovery of full 3-D camera motion and scene structure. A mathematical analysis of SFM shows the nonlinear interdependence of structure and motion given observations on the image plane. Solutions to SFM are based on elimination of the depth field by cross multiplication [1, 7, 29–32], differentiation of flow fields [33, 34], nonlinear optimization [4, 35], and other approaches. For a comprehensive discussion of SFM algorithms, the reader is encouraged to refer to [1, 2, 19, 36]. Alternatively, camera rotation can be measured by using transducers.

Upon computation of the three rotation angles, i.e., the pitch, roll, and yaw of the camera, the original sequence can be rewarped to compensate for these effects. Alternatively, one can perform *selective stabilization*, by compensating the sequence for only one or two of these components. Extending this concept, one can selectively stabilize for certain frequencies of motion so as to eliminate handheld jitter, while preserving deliberate camera pan.

8 Summary

Image stabilization, mosaicking, and motion superresolution are processes operating on a temporal sequence of images of a largely static scene viewed by a moving camera. The apparent motion observed in the image can be approximated to comply with a global motion model under a variety of circumstances. A simple and efficient algorithm for recovering the global motion parameters is presented here. The 2-D stabilization, mosaicking, and superresolution processes are described, and experimental results are demonstrated. The estimation of 2-D and 3-D motion has been studied for over two decades now, and the following references provide a useful set of starting material for the interested reader.

Acknowledgment

R. Chellappa is supported in part by the MURI ONR grant N00014-95-1-0521.

References

[1] A. Mitiche, *Computational Analysis of Visual Motion* (Plenum, New York, 1994).

[2] O. D. Faugeras, *Three-Dimensional Computer Vision* (MIT Press, Cambridge, MA, 1993).

[3] S. V. Huffel and J. Vandewalle, *The Total Least Squares Problem — Computational Aspects and Analysis* (SIAM, Philadelphia, PA, 1991).

[4] G. Adiv, "Determining 3-D motion and structure from optical flow generated by several moving objects," *IEEE Trans. Pattern Anal. Machine Intell.* **7**, 384–401 (1985).

[5] M. Hansen, P. Anandan, P. J. Burt, K. Dana, and G. van der Wal, "Real-time scene stabilization and mosaic construction," in *DARPA Image Understanding Workshop* (Morgan Kaufmann, San Francisco, CA, 1994), pp. I:457–465.

[6] S. Negahdaripour and B. K. P. Horn, "Direct passive navigation," *IEEE Trans. Pattern Anal. Machine Intell.* **9**, 168–176 (1987).

[7] N. C. Gupta and L. N. Kanal, "3-D motion estimation from motion field," *Artif. Intell.* **78**, 45–86 (1995).

[8] R. Szeliski and J. Coughlan, "Spline-based image registration," *Int. J. Comput. Vis.* **22**, 199–218 (1997).

[9] Y. S. Yao, "Electronic stabilization and feature tracking in long image sequences," Ph.D. dissertation (University of Maryland, 1996), available as Tech. Rep. CAR-TR-790.

[10] C. Morimoto and R. Chellappa, "Fast 3-D stabilization and mosaic construction," in *IEEE Conference on Computer Vision and Pattern Recognition* (IEEE, New York, 1997), pp. 660–665.

[11] H. Y. Shum and R. Szeliski, "Construction and refinement of panoramic mosaics with global and local alignment," in *International Conference on Computer Vision* (Narosa Publishing House, New Delhi, India, 1998), pp. 953–958.

[12] D. Capel and A. Zisserman, "Automated mosaicing with super-resolution zoom," in *IEEE Computer Vision and Pattern Recognition* (IEEE, New York, 1998), pp. 885–891.

[13] M. Irani and S. Peleg, "Improving resolution by image registration," *Graph. Models Image Process.* **53**, 231–239 (1991).

[14] M. S. Alam, *et al.*, "High-resolution infrared image reconstruction using multiple randomly shifted low-resolution aliased frames," in *Proc. SPIE* **3063**, (1997).

[15] R. C. Hardie, K. J. Barnard, and E. E. Armstrong, "Joint map registration and high resolution image estimation using a sequence of undersampled images," *IEEE Trans. Image Process.* **6**, 1621–1633 (1997).

[16] M. Irani, B. Rousso, and S. Peleg, "Recovery of ego-motion using region alignment," *IEEE Trans. Pattern Anal. Machine Intell.* **19**, 268–272 (1997).

[17] S. Peleg and J. Herman, "Panoramic mosaics with videobrush," in *DARPA Image Understanding Workshop* (Morgan Kaufmann, San Francisco, CA, 1997), pp. 261–264.

[18] M. Irani and P. Anandan, "Robust multi-sensor image alignment," in *International Conference on Computer Vision* (Narosa Publishing House, New Delhi, India, 1998), pp. 959–966.

[19] S. Srinivasan, "Image sequence analysis: estimation of optical flow and focus of expansion, with applications," Ph.D. dissertation (University of Maryland, 1998), available as Tech. Rep. CAR-TR-893, www.cfar.umd.edu/~shridhar/Research.

[20] A. K. Jain, *Fundamentals of Digital Image Processing* (Prentice-Hall, Englewood Cliffs, NJ, 1989).

[21] C. L. Fennema and W. B. Thompson, "Velocity determination in scenes containing several moving objects," *Comput. Graph. Image Process.* **9**, 301–315 (1979).

[22] O. Axelsson, *Iterated Solution Methods* (Cambridge University Press, Cambridge, UK, 1994).

[23] W. H. Press, S. A. Teukolsky, W. T. Vetterling, and B. P. Flannery, *Numerical Recipes in C*, 2nd ed. (Cambridge University Press, Cambridge, UK, 1992).

[24] M. Irani, P. Anandan, and S. Hsu, "Mosaic based representations of video sequences and their applications, in *International Conference on Computer Vision* (IEEE Computer Society Press, Washington, D.C., 1995), pp. 605–611.

[25] B. Rousso, S. Peleg, I. Finci, and A. Rav-Acha, "Universal mosaicing using pipe projection," in *International Conference on Computer Vision* (Narosa Publishing House, New Delhi, India, 1998), pp. 945–952.

[26] J. Y. A. Wang and E. H. Adelson, "Representing moving images with layers," *IEEE Trans. Image Process.*, **3**, 625–638 (1994).

[27] S. Kim, N. Bose, and H. Valenzuela, "Recursive reconstruction of high resolution image from noisy undersampled multiframes," *IEEE Trans. Acoust. Speech Signal Process.* **38**, 1013–1027 (1990).

[28] S. Kim and W. Y. Su, "Recursive high resolution reconstruction of blurred multiframe images," *IEEE Trans. Image Process.* **2**, 534–539 (1993).

[29] R. Y. Tsai and T. S. Huang, "Estimating 3-D motion parameters of a rigid planar patch I," *IEEE Trans. Acoust. Speech Signal Process.* **29**, 1147–1152 (1981).

[30] X. Zhuang, T. S. Huang, N. Ahuja, and R. M. Haralick, "A simplified linear optical flow-motion algorithm," *Comput. Vis. Graph. Image Process.* **42**, 334–344 (1988).

[31] X. Zhuang, T. S. Huang, N. Ahuja, and R. M. Haralick, " Rigid body motion and the optic flow image," in *First IEEE Conference on AI Applications* (IEEE, New York, 1984), pp. 366–375.

[32] A. M. Waxman, B. Kamgar-Parsi, and M. Subbarao, "Closed-form solutions to image flow equations for 3D structure and motion," *Int. J. Comput. Vis.* **1**, 239–258 (1987).

[33] H. C. Longuet-Higgins and K. Prazdny, "The interpretation of a moving retinal image," *Proc. Roy. Soc. London B.* **208**, 385–397 (1980).

[34] A. M. Waxman and S. Ullman, "Surface structure and three-dimensional motion from image flow kinematics," *Int. J. Robot. Res.* **4**, 72–94 (1985).

[35] A. R. Bruss and B. K. P. Horn, "Passive navigation," *Comput. Vis. Graph. Image Process.* **21**, 3–20 (1983).

[36] J. Weng, T. S. Hwang, and N. Ahuja, *Motion and Structure from Image Sequences* (Springer-Verlag, Berlin, 1991).

Image and Video Analysis

Edge and Boundary Detection in Images

Algorithms for Image Processing

4.1

Computational Models of Early Human Vision

Lawrence K. Cormack
*The University of Texas
at Austin*

"The nature of things, hidden in darkness, is revealed only by analogizing. This is achieved in such a way that by means of simpler machines, more easily accessible to the senses, we lay bare the more intricate." *Marcello Malpighi, 1675*

1 Introduction

1.1 Aim and Scope

The author of a short chapter on computational models of human vision is faced with an *embarras de richesse*. One wishes to make a choice between breadth and depth, but even this is virtually impossible within a reasonable space constraint. It is hoped that this chapter will serve as a brief overview for engineers interested in processing done by the early levels of the human visual system. We will focus on the representation of luminance information at three stages: the optics and initial sampling, the representation at the output of the eyeball itself, and the representation at primary visual cortex. With apologies, I have allowed us a very brief foray into the historical roots of the quantitative analysis of vision, which I hope may be of interest to some readers.

1.2 A Brief History

The first known quantitative treatment of image formation in the eyeball by Alhazan predated the Renaissance by four centuries.

In 1604, Kepler codified the fundamental laws of physiological optics, including the then-controversial inverted retinal image, which was then verified by direct observation of the image *in situ* by Pater Scheiner in 1619 and later (and more famously) by Rene Descarte. Over the next two centuries there was little advancement in the study of vision and visual perception *per se*, with the exception of Newton's formulation of laws of color mixture, However, Newton's seemingly innocuous suggestion that "the Rays to speak properly are not coloured" [1] anticipated the core feature of modern quantitative models of visual perception: the computation of higher *perceptual* constructs (e.g., color) based upon the activity of peripheral receptors differentially sensitive to a *physical* dimension (e.g., wavelength).[1]

In 1801, Thomas Young proposed that the eye contained but three classes of photoreceptor, each of which responded with a sensitivity that varied over a broad spectral range [2]. This theory, including its extensions by Helmholtz, was arguably the first modern computational theory of visual perception. The Young/Helmholtz theory *explicitly* proposed that the properties of objects in the world are not sampled directly, but that certain properties of light are encoded by the nervous system, and that

---

[1]Newton was pointing out that colors must arise in the brain, because a given color can arise from many wavelength distributions, and some colors can *only* arise from multiple wavelengths. The purples, for example, and even unique red (red that observers judge as tinged with neither orange nor violet), are colors that cannot be experienced by viewing a monochromatic light.

the resulting neural activity was transformed and combined by the nervous system to result in perception. Moreover, the neural activity was assumed to be quantifiable in nature, and thus the output of the visual system could be precisely predicted by a mathematical model. In the case of color, it could be firmly stated that sensation "may always be represented as simply a function of three variables" [3]. While not a complete theory of color perception, this has been borne out for a wide range of experimental conditions.

Coincident with the migration of trichromatic theory from England to Central Europe, some astronomical data made the same journey, and this resulted in the first *applied* model of visual processing. The data were observations of stellar transit times from the Greenwich Observatory taken in 1796. There was a half-second discrepancy between the observations by Maskelyne (the director) and Kinnebrook (his assistant), and for this Kinnebrook lost his job. The observations caught the notice of Bessel in Prussia at a time when the theory of variability was being given a great deal of attention because of the work of Laplace, Gauss, and others. Unable to believe that such a large, systematic error could be due to sloppy astronomy, Bessel developed a linear model of observers' reaction times to visual stimuli (i.e., stellar transits) relative to one another. These models, which Bessel called "personal equations," could then be used to correct the data for the individual making the observations.

It was no accident that the nineteenth century saw the genesis of models of visual behavior, for it was at that time that several necessary factors came together. First, it was realized that an understanding of the eyeball itself begged rather than yielded an explanation of vision.

Second, the brain had be to viewed as explainable, that is, viewed in a mechanistic fashion. While this was not entirely new to the nineteenth century, the measurement of the conduction velocity of a neural impulse by Helmholtz in 1850 probably did more than any other single experiment to demonstrate that the senses did not give rise to immediate, qualitative (and therefore incalculable) impressions, but rather transformed and conveyed information by means that were ultimately quantifiable.

Third, the stimulus had to be understood to some degree. To make tangible progress in modeling the *early* levels of the visual system, it was necessary to think not in terms of objects and meaningful structures in the environment, but in terms of light, of wavelength, of intensity, and its spatial and temporal derivatives. The enormous progress in optics in the nineteenth century created a climate in which vision could be thought of quantitatively; light was not understood, but its veils of magic were quickly falling away.

Finally, theories of vision would have to constrained and testable in a quantitative manner. Experiments would have to be done in which observers made well-defined responses to well-controlled stimuli in order to establish quantitative input–output relationships for the visual system, which could then in turn be modeled. This approach, called *psychophysics*, was born with the publication of *Elemente der Psychophysik* by Gustav Fechner in 1860.

With the historical backdrop painted, we can now proceed to a selective survey of quantitative treatments of early human visual processing.

1.3 A Short Overview

Figure 1 shows a schematic overview of the major structures of the early visual system and some of the functions they perform. We start with the visual world, which varies with space, time, and wavelength, and which has an amplitude spectrum roughly proportional to $1/f$, where f is the spatial frequency of luminance variation. The first major operations by the visual system are passive: low-pass filtering by the optics and sampling by the receptor mosaic, and both of these operations, and the relationship between them, vary with eccentricity.

The retina of the eyeball filters the image further. The photoreceptors themselves filter along the dimensions of time and wavelength, and the details of the filtering varies with receptor type. The output cells of the retina, the retinal ganglion cells, synapse onto the lateral geniculate nucleus of the thalamus (known as the LGN). We will consider the LGN primarily as a relay station to cortex, and the properties of retinal ganglion cells and LGN cells will be treated as largely interchangeable.

LGN cells come in two major types in primates, magnocellular ("M") and parvocellular ("P"); the terminology was adopted for morphological reasons, but important functional properties distinguish the cell types. To grossly simplify, M cells are tuned to low spatial frequencies and high temporal frequencies, and they are insensitive to wavelength variation. In contrast, P cells are tuned to high spatial frequencies and low temporal frequencies, and they encode wavelength information. These two cell types work independently and in parallel, emphasizing different aspects of the same visual stimuli. In the two-dimensional (2-D) Fourier plane, both are essentially circularly-symmetric bandpass filters.

In the primary visual cortex, several properties emerge. Cells become tuned to orientation; they now inhabit something like a Gaussian blob on the spatial Fourier plane. Cells also become tuned to direction of motion (displacement across time) and binocular disparity (displacement across eyeballs). A new dichotomy also emerges, that between so-called simple and complex cells. Simple cells behave much as wavelet-like linear filters, although they demonstrate some response nonlinearities critical to their function. The complex cells are more difficult to model, as their sensitivity shows no obvious spatial structure.

We will now explore the properties of each of these functional divisions, and their consequences, in turn.

2 The Front End

A scientist in biological vision is likely to refer to anything between the front of the cornea and the area on which he or she is

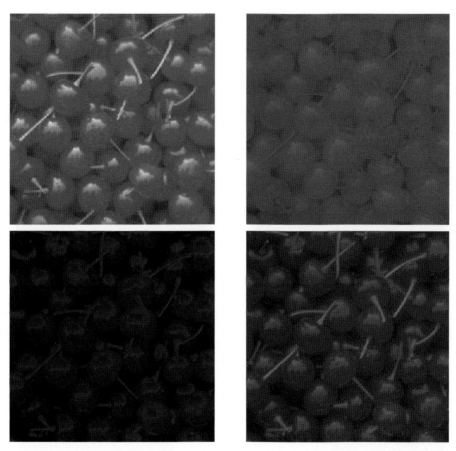

FIGURE 1.13 Color image of "cherries" (top left), and (clockwise) its red, green, and blue components.

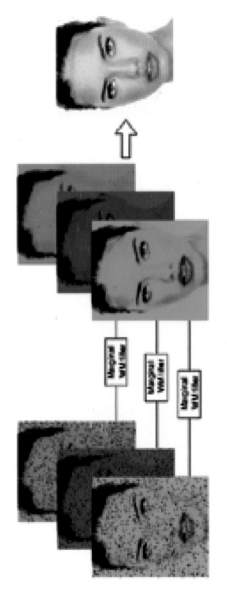

FIGURE 3.2.7 Center WM filter applied to each component independently.

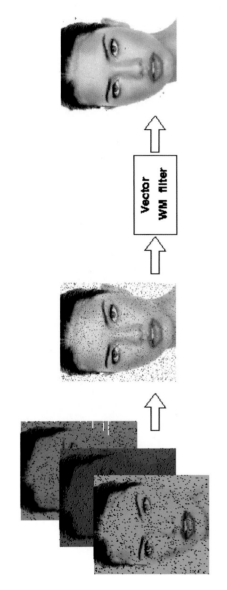

FIGURE 3.2.8 Center vector WM filter applied in the three-dimensional space.

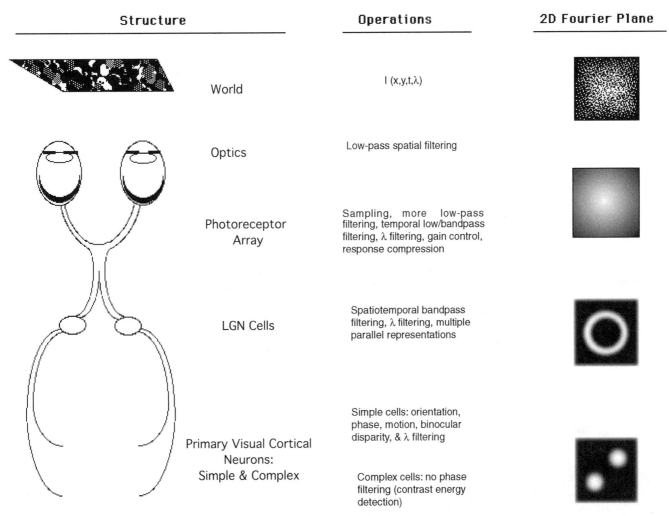

Structure	Operations	2D Fourier Plane

World

$I(x,y,t,\lambda)$

Optics

Low-pass spatial filtering

Photoreceptor Array

Sampling, more low-pass filtering, temporal low/bandpass filtering, λ filtering, gain control, response compression

LGN Cells

Spatiotemporal bandpass filtering, λ filtering, multiple parallel representations

Primary Visual Cortical Neurons: Simple & Complex

Simple cells: orientation, phase, motion, binocular disparity, & λ filtering

Complex cells: no phase filtering (contrast energy detection)

FIGURE 1 Schematic overview of the processing done by the early visual system. On the left, are some of the major structures to be discussed; in the middle, are some of the major operations done at the associated structure; in the right, are the 2-D Fourier representations of the world, retinal image, and sensitivities typical of a ganglion and cortical cell.

working as "the front end." Herein, we use the term to refer to the optics and sampling of the visual system and thus take advantage of the natural division between optical and neural events.

2.1 Optics

The optics of the eyeball are characterized by its 2-D spatial impulse response function, the point-spread function [4]:

$$h(r) = 0.952e^{-2.59|r|^{1.36}} + 0.048e^{-2.43|r|^{1.74}}, \qquad (1)$$

in which r is the radial distance in minutes of arc from the center of the image.

This function, plotted in Fig. 2 (or its Fourier transform, the modulation-transfer function), completely characterizes the optics of the eye within the central visual field. The optics deteriorate substantially in the far periphery, so a spatially variant point-spread function is actually required to fully characterize

image formation in the human eyeball. For most purposes, however, the point-spread function may be simply convolved with an input image,

$$i(x, y) = I(x, y) * h(x, y), \qquad (2)$$

to compute the central retinal image for an arbitrary stimulus, and thus derive the starting point of vision.

2.2 Sampling

While sampling by the retina is a complex spatiotemporal neural event, it is often useful to consider the spatial sampling to be a passive event governed only by the geometry of the receptor grid and the stationary probability of a single receptor absorbing a photon. In the human retina, there are two parallel sampling grids to consider, one comprising the rod photoreceptors and operating in dim light, and the other comprising the

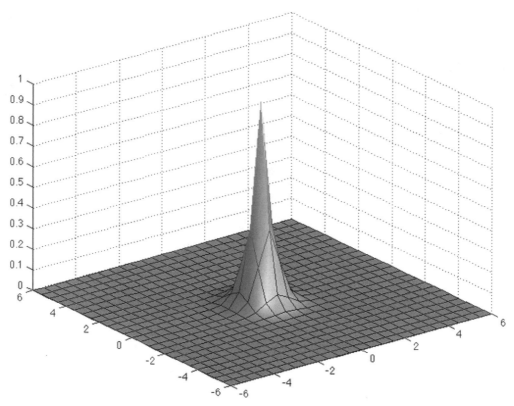

FIGURE 2   Point-spread function of the human eyeball. The x and y axes are in minutes of arc, and the z axis is in
arbitrary units. The spacing of the grid lines is equal to the spacing of the photoreceptors in the central visual field of the
human eyeball, which is approximately 30 arc sec.

cone photoreceptors (on which we concentrate) and operating
in bright light. Shown in Fig. 3(a) are images of the cone sampling grid $1°$ from the center of the fovea taken in two living, human eyes, using aberration-correcting adaptive optics (similar to
those used for correcting atmospheric distortions for terrestrial
telescopes) [5]. The short-, medium-, and long-wavelength sensitive cones have been pseudo-colored blue, green, and red, respectively. At the central fovea, the average interreceptor distance
is ∼2.5 μm, which is ∼30 arc sec in the human eyeball. Locally,
the lattice is roughly hexagonal, but it is irregular over large areas
and seems to become less regular as eccentricity increases. Theoretical performance has been compared in various visual tasks
using both actual foveal receptor lattices taken from anatomical
studies of the macaque[2] retina and idealized hexagonal lattices of
the same receptor diameter, and little difference was found [6].

While the use of a regular hexagonal lattice is convenient for
calculations in the space domain, it is often more efficient to
work in the frequency domain. In the central retina, one can
take the effective sampling frequency to be $\sqrt{3/2}$ times the average interreceptor distance (due to the hexagonal lattice), and
then treat the system as sampling with an equivalent 2-D comb

(sampling) function. In the peripheral retina, where the optics
of the eye pass frequencies above the theoretical sampling limits
of the retina, it is possible that the irregular nature of the array
helps prevent some of the effects of aliasing. However, visual
discriminations in the periphery can be made above the Nyquist
frequency by the detection of aliasing [7], so a 2-D comb function of appropriate sampling density can probably suffice for
representing the peripheral retina under some conditions.

The photoreceptor density as a function of eccentricity for
the rod and cone receptor types in the human eye is shown in
Fig. 3(b). The cone lattice is *foveated*, peaking in density at a
central location and dropping off rapidly away from this point.
Also shown is the variation in the density of retinal ganglion cells
that transmit the information out of the eyeball. The ganglion
cells effectively sample the photoreceptor array in *receptive fields*,
whose size also varies with eccentricity. This variation for the two
main types of ganglion cells (which will be discussed below) is
shown in Fig. 3(c). The ganglion cell density falls more rapidly
than cone density, indicating that ganglion cell receptive fields in
the periphery summate over a larger number of receptors, thus
sacrificing spatial resolution. This is reflected in measurements
of visual acuity as a function of eccentricity, which fall in accord
with the ganglion cell data.

The other main factor to consider is the probability of given
receptor absorbing a photon, which is governed by the area of

[2]The macaque is an old-world monkey, *macaca fascicularis*, commonly used in
vision research because of the great similarity between the macaque and human
visual systems.

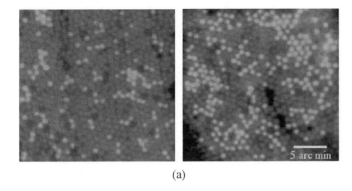

(a)

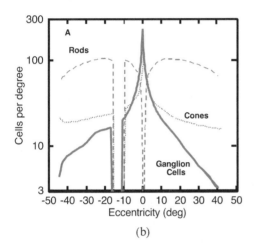

(b)

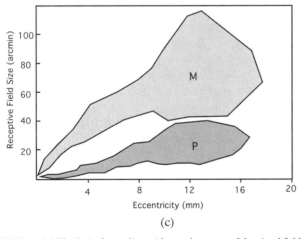

(c)

FIGURE 3 (a) The Retinal sampling grid near the center of the visual field of two living human eyeballs. The different cone types are color coded (from Roorda and Williams, 1999, reprinted with permission). (b) The density of various cell types in the human retina. The rods and cones are the photoreceptors that do the actual sampling in dim and bright light, respectively. The ganglion cells pool the photoreceptor responses and transmit information out of the eyeball (from Geisler and Banks, 1995) reprinted with permission. (c) The dendritic field size (assumed to be roughly equal to the receptive field size) of the two main types of ganglion cell in the human retina (redrawn from Dacy, 1993). The gray shaded region shows the parasol (or M) cells, and the green region shows the midget (or P) cells. The two cell types seem to independently and completely tile the visual world. The functional properties of the two cell types are summarized in Table 1. (See color section, p. C–7.)

the effective aperture of the photoreceptor and the probability that a photon entering the aperture will be absorbed. This latter probability is obtained from Beer's Law, which gives the ratio of radiant flux reaching the back of the receptor outer segment to that entering the front [8]:

$$v(\lambda) = 10^{-lc\varepsilon(\lambda)} \qquad (3)$$

in which l is the length of the receptor outer segment, c is the concentration of unbleached photopigment, and $\varepsilon(\lambda)$ is the absorption spectrum of the photopigment.

For many modeling tasks, it is most convenient to express the stimulus in terms of $n(\lambda)$, the number of quanta per second as a function of wavelength. This is given by [9]

$$n(\lambda) = 2.24 \times 10^3 A \frac{L(\lambda)}{V(\lambda)} t(\lambda)\lambda, \qquad (4)$$

in which A is area of the entrance pupil, $L(\lambda)$ is the spectral luminance distribution of the stimulus, $V(\lambda)$ is the standard spectral sensitivity of human observers, and $t(\lambda)$ is the transmittance of the ocular media. Values of these functions are tabulated in [8].

Thus, for any receptor, the number of absorptions per second, N, is given approximately by

$$N = \int a(1 - v(\lambda))n(\lambda) \, d\lambda \qquad (5)$$

in which a is the receptor aperture.

These equations are of fundamental import because they describe the data that the visual system collects about the world. Any comprehensive model of the visual system must ultimately use these data as input. In addition, since these equations specify the information available to the visual system, they allow us to specify how well a particular visual task could be done in principle. This specification is done with a special type of model called an *ideal observer*.

2.3 Ideal Observers

An ideal observer is a mathematical model that performs a given task as well as possible given the information in the stimulus. It is included in this section because it was traditionally used to assess the visual system in terms of quantum efficiency, f, which is the ratio of the number of quanta theoretically required to do a task to the number actually required [e.g., 10]. It is therefore more natural to introduce the topic in terms of optics. However, ideal observers have been used to assess the information loss at various neurophysiological sites in the visual system [6, 11]; the only requirement is that the information present at a given site can be quantitatively expressed.

An ideal observer performs a given task optimally (in the Bayesian sense), and it thus provides an *absolute* theoretical limit on performance in any given task (it thus gives to psychophysics and neuroscience what absolute zero gives to thermodynamics: a fundamental baseline). For example, the smallest offset between a pair of abutting lines (such as on a vernier scale on a pair of calipers) that a human observer can reliably discriminate (75% correct, say) from a stimulus with no offset is almost unbelievably low – a few *seconds* of arc. Recalling from above that foveal cone diameters and receptor spacing are of the order of a half a *minute* of arc, such performance seems rather amazing. But amazing relative to what? The ideal observer gives us the answer by defining what the best possible performance is. In our example, a human observer would be less than 1% efficient as measured at the level of the photoreceptors, meaning that the human observer would require of the order of 10^3 more quanta to achieve the same level of discrimination performance. In this light, human performance ceases to appear quite so amazing, and attention can be directed toward determining how and where the information loss is occurring.

An ideal observer consists of two main parts, a model of the visual system and a Bayesian classifier. The latter is usually expressed as a likelihood ratio:

$$l(s) = \frac{P(s \mid a)}{P(s \mid b)}, \qquad (6)$$

in which the numerator and denominator are the conditional probabilities of making observations given that the stimulus was actually *a* or *b*, respectively. If the likelihood ratio, or more commonly its logarithm, exceeds a certain amount, stimulus *a* is judged to have occurred. For a simple discrimination, *s* would be a vector containing observed quantum catches in a set of photoreceptors, and the probability of this observation given hypotheses *a* and *b* would be calculated with the Poisson distribution of light and the factors described above in Sections 2.1 and 2.2.

The beauty of the ideal observer is that it can be used to parse the visual system into layers, and to examine the information loss at each layer. Thus, it becomes a tool by which we can learn which patterns of behavior result from the physics of the stimulus and the structure of the early visual system, and which patterns of behavior result from nonoptimal strategies or algorithms employed by the human visual system. For example, there exists an asymmetry in visual search in which a patch of low-frequency texture in a background of high-frequency texture is much easier to find than when the figure and ground are reversed. It is intuitive to think that if only low-level factors were limiting performance, detecting A on a background of B should be equivalent to detecting B on a background of A (by almost any measure, the contrast of A on B would be equal to that of B on A). However, an ideal-observer analysis proves this intuition false, and an ideal-observer based model of visual search produces the aforementioned search asymmetry [12].

3 Early Filtering and Parallel Pathways

In this section, we discuss the nature of the information that serves as the input to visual cortex. This information is contained in the responses of the retinal ganglion cells (the output of the eyeball) and the LGN.[3] Arguably, this is the last stage that can be comfortably modeled as a strictly data-driven system in which neural responses are independent of activity from other cells in the same or subsequent layers.

3.1 Spatiotemporal Filtering

One difficulty with modeling neural responses in the visual system, particularly for someone new to reading the physiology literature, is that people have an affinity for dichotomies. This is especially evident from a survey of the work on retinogeniculate processing. Neurons have been dichotomized a number of dimensions. In most studies, only one or perhaps two of these dimensions are addressed, which leaves the relationships between the various dimensions somewhat unclear.

With that caveat in mind, the receptive field shown in Fig. 4 is fairly typical of that encountered in retinal ganglion cells or cells of the lateral geniculate nucleus. Figure 4(a) shows the hypothetical cell's sensitivity as a function of spatial position. The receptive field profile shown is a difference of Gaussians, which agrees well with physiological recordings of the majority of ganglion cell receptive field profiles [13, 14], and it is given by

$$DOG(x, y) = a_1 e^{[(x^2 - y^2)/(s_1^2)]} - a_2 e^{[(x^2 - y^2)/(s_2^2)]}, \qquad (7)$$

in which a_1 and a_2 normalize the areas, and s_1 and s_2 are space constants in a ratio of about 1:1.6. Their exact values will vary as a function of eccentricity as per Fig. 3(c).

This representation is fairly typical of that seen in the early work on ganglion cells [e.g., 15], in which the peak response of a neuron to a small stimulus at a given location in the receptive field was recorded, but the location in *time* of this peak response was somewhat indefinite. Thus, a receptive field profile as shown represents a slice in time of the neuron's response some tens of milliseconds after stimulation and, further, the slice of time represented in one spatial location isn't necessarily the same as that represented in another (although for the majority of ganglion cells, the discrepancy would not be too large).

Since the receptive field is spatially symmetric, we can get a more complete picture by looking at a plot of one spatial dimension against time. Such an x–t plot is shown in Fig. 4(b), in which the x dimension is in arcminutes and the t dimension is in milliseconds. The response is space–time separable; the value at any given point is simply the value of the spatial impulse response at that spatial location scaled by the value of the temporal impulse response at that point in time. Thus, the response is given by

$$r(x, t) = DOG(x)[h(t)] \qquad (8)$$

---

[3]Thus we regretfully omit a discussion of the response properties of the photoreceptors *per se* and of the circuitry of the retina. These are fascinating topics — the retina is a marvelous computational structure — and interested readers are referred to [40].

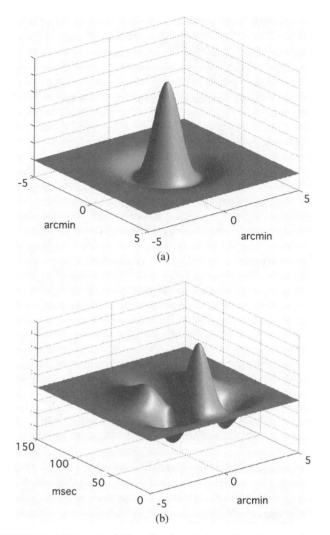

(a)

(b)

FIGURE 4 (a) Receptive field profile of a retinal ganglion cell modeled as a difference of Gaussians. The x and y axes are in minutes of arc, so this cell would be typical of an M cell near the center of the retina, or a P cell at an eccentricity of $10°$ to $15°$ (see Fig. 2). (b) Space–time plot of the same receptive field, illustrating its biphasic temporal impulse response. (The x-axis is in minutes of arc, and the y-axis is in milliseconds).

in which $h(t)$ is a biphasic temporal impulse response function. This response function, $h(t)$, was constructed by subtracting two cascaded low-pass filters of different order [cf. 16]. These low-pass filters are constructed by successive autocorrelation of an impulse response function of the form

$$h(t) = H(t)e^{-t/\tau}, \qquad (9)$$

in which $H(t)$ is the Heaviside unit step:

$$H(t) = \begin{cases} 1, & t \geq 0 \\ 0, & t < 0 \end{cases}. \qquad (10)$$

A succession of n autocorrelations gives

$$h_n(t) = \frac{H(t)(t/\tau)^n e^{-t/\tau}}{\tau n!}, \qquad (11)$$

which is a monophasic (low-pass) filter of order n. A difference of two filters of different orders produces the biphasic bandpass response function, and the characteristics of the filter can be adjusted by using component filters of various orders.

The most important implication of this receptive field structure, obvious from the figure, is that the cell is bandpass in both spatial and temporal frequency. As such, the cell discards information about absolute luminance and emphasizes change across space (likely to denote the edge of an object) or change across time (likely to denote the motion of an object). Also obvious from the receptive field structure is that the cell is not selective for orientation (the direction of the spatial change) or the direction of motion.

The cell depicted in the figure is representative in terms of its qualitative characteristics, but the projection from retina to cortex comprises of the order of 10^6 such cells that vary in their specific spatiotemporal tuning properties. Rather than being continuously distributed, however, the cells seem to form functional subgroups that operate on the input image in parallel.

3.2 Early Parallel Representations

The early visual system carries multiple representations of the visual scene. The earliest example of this is at the level of the photoreceptors, where the image can be sampled by rods, cones, or both (at intermediate light levels). An odd aspect of the rod pathway is that it ceases to exist as a separate entity at the output of the retina; there is no such thing as a "rod retinal ganglion cell." This is an interesting example of a need for a separate sensor system for certain conditions combined with a need for neural economy. The pattern analyzing mechanisms in primary visual cortex and beyond are used for both rod and cone signals with (probably) no information about which system is providing the input.

Physiologically, the most obvious example of separate, parallel projections from the retina to the cortex is the presence of the so-called ON and OFF pathways. All photoreceptors have the same sign of response. In the central primate retina, however, each photoreceptor makes direct contact with at least two *bipolar* cells — cells intermediate between the receptors and the ganglion cells — one of which preserves the sign of the photoreceptor response, and the other of which inverts it. Each of these bipolar cells in turn serves as the excitatory center of a ganglion cell receptive field, thus forming two parallel pathways: an ON pathway, which responds to increases in light in the receptive field center, and an OFF pathway, which responds to decreases in light in the receptive field center. Each system forms an independent tiling of the retina, resulting in two complete, parallel neural images being transmitted to the brain.

Another fundamental dichotomy is between midget (or "P" for reasons to become clear in a moment) and parasol (or "M") ganglion cells. Like the ON–OFF subsystems, the midget and parasol ganglion receptive fields perform a separate and parallel tiling of the retina. On average, the receptive fields of parasol

TABLE 1 Important properties of the two major cell types providing input to the visual cortex

Property	P cells	M cells	Comments
Percent of cells	80	10	The remainder project to subcortical streams.
Receptive field size	relatively small, single cone center in fovea, increases with eccentricity (see Fig. 3)	Relatively large, $\sim3\times$ larger than P cells at any given eccentricity	RF modeled well by a difference of Gaussians.
Contrast sensitvity	poor (factor of 8–10 lower than for M cells), driven by high contrast	good, saturation at high contrasts	
Contrast gain	low	high ($\sim6\times$ higher)	Possible gain control in M cells.
Spatial frequency response	peak and high-frequency cutoff at relatively low spatial frequency	Peak and high-frequency cutoff at relatively high spatial frequency	Unclear dichotomy: physiological differences tend to be less pronounced than predicted by anatomy.
temporal frequency response	low-pass, fall off at 20–30 Hz.	bandpass, peaking at or above 20 Hz	
Spatial linearity	almost all have linear summation	most have linear summation, some show marked nonlinearities	Estimated proportion of nonlinear neurons depends on how the distinction is made.
Wavelength opponency	yes	no	
Conduction velocity	slow (6 m/s)	fast (15 m/s)	

ganglion cells are about a factor of 3 larger than those of midget ganglion cells at any given eccentricity, as shown in Fig. 3(c), so the two systems can be thought of as operating in parallel at different spatial scales. This separation is strictly maintained in the projection to the LGN, which is layered like a wedding cake. The midget cells project exclusively to what are termed the *parvocellular* layers of the LGN (the dorsalmost four layers), and the parasol cells project exclusively to the *magnocellular* layers (the ventralmost two layers). Because of this separation and the important physiological distinctions that exist, visual scientist now generally speak in terms of the parvocellular (or "P") pathway, and the magnocellular (or "M") pathway.

There is a reliable difference in the temporal frequency response between the cells of the M and P pathways [17]. In general, the parvocellular cells peak at a lower temporal frequency than magnocellular cells (<10 Hz vs. 10–20 Hz), have a lower high-frequency cutoff (~20 Hz vs. ~60 Hz), and shallower low-frequency rolloff (with many P cells showing a DC response). The temporal frequency response envelopes of both cell types can be functionally modeled as a difference of exponentials in the frequency domain.

Another prevalent distinction is based upon linear versus non-linear summation within a cell's receptive field. Two major classes of retinal ganglion cell have been described in the cat, termed X and Y cells, based on the presence or absence of a null phase when stimulated with a sinusoidal grating [15]. The response of a cell such as shown in Fig. 4 will obviously depend strongly on the spatial phase of the stimulus. For such a cell, a spatial phase of a grating can be found such that the grating can be exchanged with a blank field of equal mean luminance with no effect on

the output of the cell. These X cells compose the majority. For other cells, termed Y cells, no such null phase can be found, indicating that something other than linear summation across space occurs.

In the primate, nonlinear spatial summation is much less prevalent at the level of the LGN; although nonlinear cells do exist, and are more prevalent in M cells than in P cells [17]. It may be that nonlinear processing, which is very important, has largely shifted to the cortex in primates, just as have other important functions such as motion processing, which occurs much earlier in the visual systems of more phylogenically-challanged species.

At this point, there is a great body of evidence suggesting that the M–P distinction is a fundamental one in primates, and that most of the above dichotomies are either an epiphenomenon of it, or at least best understood in terms of it. We can summarize the important parameters of M and P cells as follows. Table 1 (cf. [18]) provides a fairly comprehensive, albeit qualitative, overview of what we could term the magnocellular and parvocellular "geniculate transforms" that serve as the input to the cortex. If, in fact, work on the visual cortex continues to show effects such as malleability of receptive fields, it may be that models of geniculate function will actually increase in importance, because it may be the last stage at which we can confidently rely on a relatively linear transform-type model. Attempts in this direction have been made [19, 20] but most modeling efforts seem to have been concentrated on either cortical cells or psychophysical behavior (i.e., modeling the output of the human as a whole, e.g., contrast threshold in response to some stimulus manipulation).

4 The Primary Visual Cortex and Fundamental Properties of Vision

4.1 Neurons of the Primary Visual Cortex

The most striking feature of neurons in the visual cortex is the presence of several emergent properties. We begin to see, for example, orientation tuning, binocularity, and selectivity for the direction of motion. The distinction between the magnocellular and parvocellular pathways remains — they synapse at different input layers in the visual cortex — but interactions between them begin to occur.

Perhaps the most obvious and fundamental physiological distinction in the cortex is between so-called simple and complex cells [21, 22]. This terminology was adopted (prior to wide application of linear systems analysis in vision) because the simple cells made sense. Much as with ganglion cells, mapping the receptive field was straightforward and, once the receptive field was mapped, the response of the cell to a variety of patterns could be intuitively predicted. Complex cells, in contrast, were more complex. The simple/complex distinction seems to have no obvious relationship with the magnocellular/parvocellular distinction, but it seems to be a manifestation of a computational scheme used within both processing streams.

The spatial receptive field of a generic simple cell is shown in Fig. 5(a). The cell is modeled as a Gabor function, in which sensitivity is given by

$$s(x, y) = ae^{-(x^2/\sigma_x^2 + y^2/\sigma_y^2)} \sin(2\pi\omega x + \phi) \quad (12)$$

As the axes are in arcminutes, the cell is most sensitive to horizontal Fourier energy at $\sim$3 cycles/deg. In this case, the cell is odd symmetric. While it would be elegant if cells were always even or odd symmetric, it seems that phase is continuously represented [23, 24], although this certainly does not preclude the use of pairs of cells in quadrature phase in subsequent processing.

As in Fig. 4, Fig. 5(b) shows the spatiotemporal receptive field of the model cell: the cell's sensitivity at $y = 0$ plotted as a function of x and t. Notice that, in this case, the cell is spatiotemporally inseparable; it is oriented in space–time and is directionally selective [25, 26]. Thus, the optimal stimulus would be drifting sinusoidal grating, in this case a 3 cycle/deg grating drifting at approximately 5 deg/s. Many, but not all, cortical cells are directionally selective (see below).

Cells in the primary visual cortex can be thought of as a bank or banks of spatiotemporal filters that tile the visual world on several dimensions and, in so doing, determine the envelope of information to which we have access. We can get a feel for this envelope by looking at the distribution of cell tuning along various dimensions. This is done in Fig. 6 using data from cells in the Macaque primary visual cortex reported in Geisler and Albrecht [27]. In the upper row, the response of a typical cell is shown as a function of the spatial frequency of a counterphasing grating (left column), the temporal frequency of same stimulus

at optimal spatial frequency (middle column), or the orientation of a drifting grating of optimal spatiotemporal frequency (right column). The middle and lower rows show the normalized frequency distributions of the parameters of the tuning functions for the population of cells surveyed ($n = 71$).[4]

At this point, we can sketch a sort of standard model of the spatial response properties of simple and complex cortical cells [e.g., 27, 28]. The basic elements of such a model are illustrated in Fig. 7(a). The model comprises four basic components, the first of which is a contrast gain control, which causes a response saturation to occur (see below). Typically, it takes the form of

$$r(c) = \frac{c^n}{c^n + c_{50}^n}, \quad (13)$$

in which c is the image contrast, c_{50} is the contrast at which half the maximum response is obtained, and n is the response exponent, which averages $\sim$2.5 for Macaque cortical cells.

Next is the sampling of the image by a Gabor or Gabor-like receptive field, which is a linear spatial summation:

$$f(x, y) = \sum c(x, y)h(x, y), \quad (14)$$

in which $h(x, y)$ is the spatial receptive field profile, and $c(x, y)$ is the effective contrast of the pixel at (x, y), i.e., the departure of the pixel value from the average pixel value in the image.

The third stage is a half-wave rectification (unlike ganglion cells, cortical cells have a low maintained discharge and thus can signal in only one direction) and an expansive nonlinearity, which serves to enhance the response disparity between optimal and nonoptimal stimuli. Finally, Poisson noise is incorporated, which provides a good empirical description of the response variability of cortical cells. The variance of the response of a cortical cell is proportional to the mean response with an average constant of proportionality of $\sim$1.7.

A model complex cell is adequately constructed by summing (or averaging) the output of two quadrature pairs of simple cells with opposite sign, as shown in Fig. 7(b) [e.g., 28]. Whether complex cells are actually constructed out of simple cells this way in primary visual cortex is not known; they could be constructed directly from LGN input. For modeling purposes, using simple cells to construct them is simply convenient. The important aspect is that their response is phase independent, and thus they behave as detectors of local contrast energy.

The contrast response of cortical cells deserves a little additional discussion. At first glance, the saturating contrast response function described above seems to be a rather mundane response limit, perhaps imposed by metabolic constraints. However, a subtle but key feature is that the response of a given cortical

---

[4]While these distributions are based on real data, they are schematized using a Gaussian assumption, which is probably not strictly valid. They do, however, convey a fairly accurate portrayal of the variability of the various parameters.

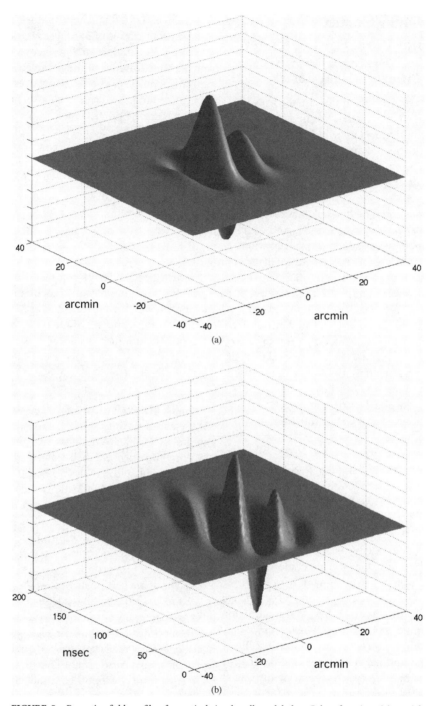

FIGURE 5 Receptive field profile of a cortical simple cell modeled as Gabor function: (a) spatial receptive field profile with the x and y axes in minutes of arc, and the z axis in arbitrary units of sensitivity; (b) space–time plot of the same receptive field with the x axis in minutes of arc and the y axis in milliseconds. The receptive field is space–time inseparable and the cell would be sensitive to rightward motion.

neuron saturates at the same *contrast*, regardless of overall response level (as opposed to saturating at some given *response* level). Why is this important? Neurons have a multidimensional sensitivity manifold, but a unidimensional output. Thus, if the output of a neuron increases from 10 to 20 spikes per second,

say, then any number of things could have occurred to cause this. The contrast may have increased, the spatial frequency may have shifted to a more optimal one, etc., or any combination of such factors may have occurred. There is no way to identify which may have occurred from the output of the neuron.

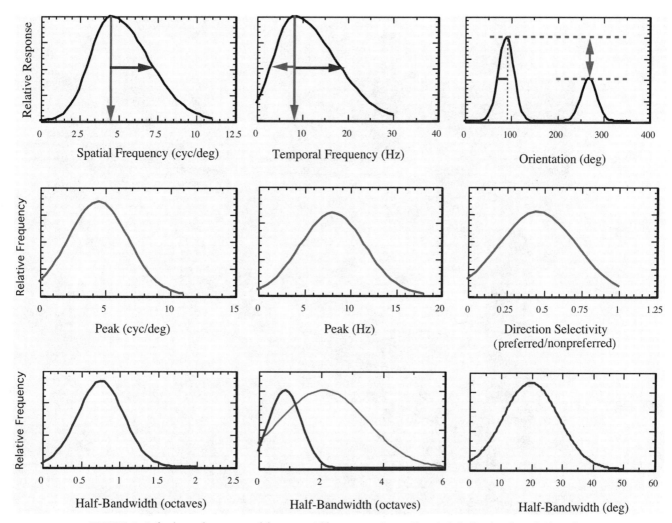

FIGURE 6 Left column: the upper panel shows a spatial frequency tuning profile typical of cell such as shown in Fig. 5; the middle and lower panels show distribution estimates of the two parameters of peak sensitivity (middle) and half-bandwidth in octaves (lower) for cells in macaque visual cortex. Middle column: same as the left column, but showing the temporal frequency response. As the response is asymmetric in octave bandwidth, the lower figure shows separate distributions for the upper and lower half-bandwidths (blue and green, respectively). Right column: the upper panel shows the response of a typical cortical cell to the orientation of a drifting sinusoidal grating. The ratio of responses between the optimal direction and its reciprocal is taken as an index of directional selectivity; the estimated distribution of this ratio is plotted in the middle panel (the index cannot exceed unity by definition). The estimate of half-bandwidth for Macaque cortical cells is shown in the lower panel. (See color section, p. C–8.)

But now consider the effect of the contrast saturation on the output of the neuron for both an optimal and a nonoptimal stimulus. Since the optimal stimulus is much more effective at driving the neuron, the saturation will occur at a higher *response* rate for the optimal stimulus. This effectively defeats the response ambiguity: *because of the contrast saturation, only an optimal stimulus is capable of driving the neuron to its maximum output.* Thus, if a neuron is firing at or near its maximum output, the stimulus is specified fairly precisely. Moreover, the expansive nonlinearity magnifies this by enhancing small differences in output. Thus, 95% confidence regions for cortical neurons on, for example, the contrast/spatial frequency plane are much narrower than the spatial frequency tuning curves themselves [29].

This suggests that it is important to rethink the manner in which subsequent levels of the visual system may use the information conveyed by neurons in primary visual cortex. Over the past $2^1/_2$ decades, linear system analysis has dominated the thinking in vision science. It has been assumed that the act of perception would involve a large-scale comparison of the outputs of many linear filters, outputs which would individually be very ambiguous. While such across-filter comparison is certainly necessary, it may be that the filters of primary visual cortex behave much more like "feature detectors" than we have been assuming.

I doubt that anyone reading a volume on image processing could look at receptive profiles in the cortex (such as shown in Fig. 5) and not be reminded of schemes such as a wavelet

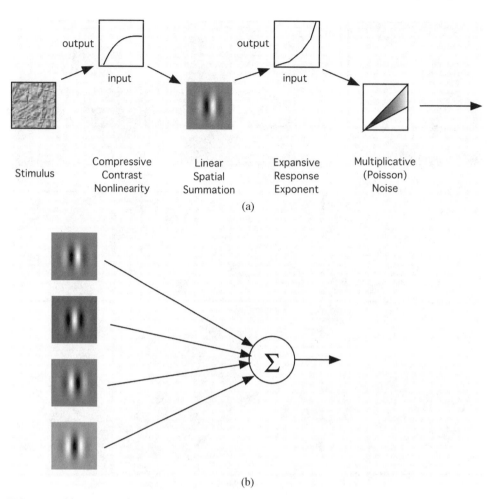

FIGURE 7 (a) Overview of a model neuron similar to that proposed by Heeger and colleagues (1991, 1996) and Geisler and Albrecht (1997). An early contrast saturation precedes linear spatial summation across the Gabor-like receptive field; the contrast saturation ensures that only optimal stimuli can maximally stimulate the cell (see text). An expansive nonlinearity such as half-squaring enhances small differences in output. Multiplicative noise is then added; the variance of cortical cell output is proportional to the mean response (with the constant of proportionality $\sim$1.7), so the signal-to-noise ratio grows as the square root of output. (b) Illustration of the construction of a phase-independent (i.e., energy detecting) complex cell from simple cell outputs.

transform or Laplacian pyramid. Not surprisingly, then, most models of the neural image in primary visual cortex share the property of encoding the image in parallel at multiple spatial scales, and several such models have been developed. One model that is computationally very efficient and easy to implement is the cortex transform [30]. The cortex transform is not, nor was it meant to be, a full model of the cortical representation. For example, response nonlinearities, the importance of which were discussed above, are omitted. It does, however, produce a simulated neural image that shares many of the properties of the simple cell representation in primary visual cortex. Models such as this have enormous value in that they give vision scientists a sort of testbed that can be used to investigate other aspects of visual function, e.g., possible interactions between the different frequency and orientation bands, in subsequent visual processes such as the computation of depth from stereopsis.

4.2 Motion and Cortical Cells

As mentioned previously, ganglion cell receptive fields are space–time separable. The resulting symmetry around a constant-space axis [Fig. 4(b)] makes them incapable of coding the direction of motion. Many cortical cells, in contrast, are directionally selective.

In the analysis of motion, a representation in space–time is often most convenient. Figure 8 (top) shows three frames of a moving spot. The continuous space–time representation is shown beneath, and it is simply an oriented bar in space–time. The next row of the figure shows the space–time representation of both a rightward and leftward moving bar. The third row of the figure shows a space–time receptive field of a typical cortical cell as was also shown in Fig. 5 (for clarity, it is shown enlarged relative to the stimulus). The orientation of the

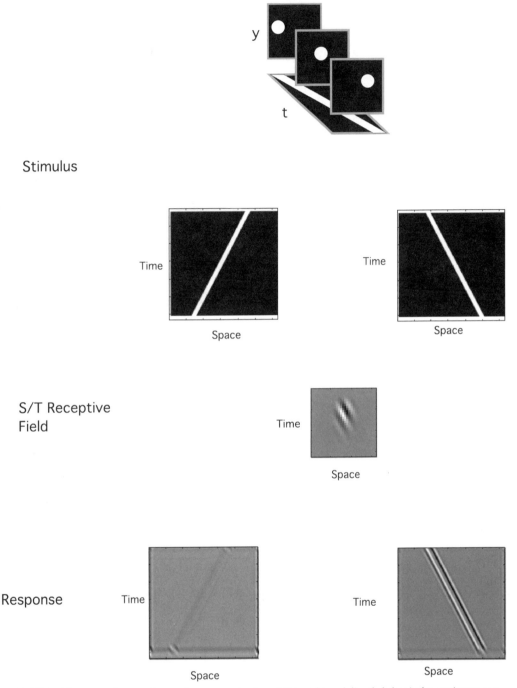

FIGURE 8 Three *x–y* slices are shown of a spot moving from left to right, and directly below is the continuous *x–t* representation: a diagonal bar. Below this are the space–time representations of a leftward and rightward moving bar, the receptive field of a directionally selective cortical cell (shown enlarged for clarity), and the response of the cell to the leftward and rightward stimuli.

receptive field in space–time gives it a fairly well defined velocity tuning; it effectively performs an autocorrelation along a space–time diagonal. Such space–time inseparable receptive fields are easily constructed from ganglion cell inputs by summing pairs of space–separable receptive fields (such as those shown in Fig. 4),

which are in quadrature in both the space and time domains [25, 26].

The bottom row of the figure shows the response of such cells to the stimuli shown in the second row obtained by convolution. In these panels, each column represents the output of a cell as a function of time (row), and each cell has a receptive field centered at the spatial location represented by its column. Clearly, each cell produces vigorous output modulation in response to motion in the preferred direction (with a relative time delay proportional to its spatial position, obviously), and almost no output in response to motion in the opposite direction.

For most purposes, it would be desirable to sense "motion energy." That is, one desires units that would respond to motion in one direction regardless of the sign of contrast or the phase of the stimulus. Indeed, such motion energy units may be thought of as the spatiotemporal equivalent of the complex cells described above. Similar to the construction of complex cells, such energy detectors are easily formed by, for example, summing the squared output of quadrature pairs of simple velocity sensitive units. Such a model captures many of the basic attributes of human motion perception, as well as a some common motion illusions [25].

Motion sensing is vital. If nothing else, a primitive organism asks its visual system to sense moving things, even if it is only the change in a shadow which triggers a sea scallop to close. It is perhaps not surprising, then, that there seems to be a specialized cortical pathway, an extension of the magnocellular pathway an earlier levels, for analyzing motion in the visual field. A review of the physiology and anatomy of this pathway is clearly beyond the scope of this chapter. One aspect of the pathway worth mentioning here, however, is the behavior of neurons in an area of the cortex known as MT, which receives input from primary visual cortex (it also receives input from other areas, but for our purposes, we can consider only its V1 inputs).

Consider a "plaid" stimulus, as illustrated in Fig. 9(a) composed of two drifting gratings differing in orientation by 90° — one drifting up and to the right and the other up and to the left. When viewing such a stimulus, a human observer sees an array of alternating dark and light areas — the intersections of the plaid — drifting upward. The response of cells such as pictured in Fig. 5, however, would be quite different. Such cells would respond in a straightforward way according to the Fourier energy in the pattern, and would thus signal a pair of motion vectors corresponding to the individual grating components of the stimulus. Obviously, then, the human visual system incorporates some mechanism that is capable of combining motion estimates from filters such as the cells in primary visual cortex to yield estimates of motion for more complex structures. These mechanisms, corresponding to cells in area MT, can be parsimoniously modeled by combining complex cell outputs in manner similar to that by which complex cells can be constructed from simple cell outputs [31, 32]. These cells effectively perform a local sum over the set of cells tuned to the appropriate orientation and spatiotemporal frequency combinations consistent with a real object moving in a given direction at a given rate. In effect, then, these cells are a neural implementation of the intersection-of-constraints solution to the aperture problem of edge (or grating) motion [33]. This problem is illustrated in Figs. 9(b) and 9(c). In Fig. 9(b), an object is shown moving to the right with some velocity. Various edges along the object will stimulate receptive fields with

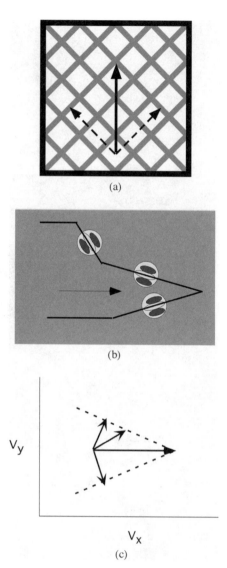

FIGURE 9 (a) Two gratings drifting obliquely (dashed arrows) generate a percept of a plaid pattern moving upward (solid arrow). (b) Illustration of the aperture problem and the ambiguity of motion sensitive cells in primary visual cortex. Each cell is unable to distinguish a contour moving rapidly to the right from a contour moving more slowly perpendicular to its orientation. (c) Intersection of constraints that allows cells that integrate over units such as in (b) to resolve the motion ambiguity.

the appropriate orientation. Clearly, these individual cells have no way of encoding the true motion of the *object*. All they can sense is the motion of the *edge*, be it almost orthogonal to the motion of the object at a relatively low speed, or in the direction of the object at a relatively high speed. The set of motion vectors generated by the edges, however, must satisfy the intersection of motion constraint as illustrated in Fig. 9(c). The endpoints of the motion vectors generated by the moving edges lie on a pair of lines that intersect at the true motion of the object. Thus, a cell summing (or averaging) the outputs of receptive fields of the appropriate orientation and spatiotemporal frequency (i.e., speed) combinations will effectively be tuned to a particular velocity and largely independent of the structure moving at that velocity.

4.3 Stereopsis and Cortical Cells

Stereopsis refers to the computation of depth from the image displacements that result from the horizontal separation of the eyeballs. Computationally, stereopsis is closely related to motion, the former involving displacements across viewpoint rather than across time. For this reason, the development of models in the two domains has much in common. Early models tended to focus on local correlations between the images, and excitatory or inhibitory interactions in order to filter out false matches (spurious correlations).

As with motion, however, neurophysiological and psychophysical findings [e.g., 34] have served to concentrate efforts on models based on receptive field structures similar to those found in Fig. 5. Of course, this is not incompatible with disparity domain interactions, but ambiguity is more commonly eliminated via interactions between spatial scales.

The primary visual cortex is the first place along the visual system in which information from the two eyes converges on single cells; as such, it represents the beginning of the binocular visual processing stream. Traditionally, it has been assumed that in order to encode horizontal disparities, these binocular cells received monocular inputs from cells with different receptive field *locations* in the two eyes, thus being maximally stimulated by an object off the plane of fixation. It is now clear, however, that binocular simple cells in the primary visual cortex often have receptive fields like that shown in Fig. 5, but with different *phases* between the two eyes [35].[5] The relative phase relation between the receptive fields in the two eyes is distributed uniformly (not in quadrature pairs) for cells tuned to vertical orientations, whereas there is little phase difference for cells tuned to horizontal orientations, indicating that these phase differences are almost certainly involved in stereopsis. Just as in motion, however, these simple cells have many undesirable properties, such as phase sensitivity and phase ambiguity (a phase disparity $k\pi$ being indistinguishable from a phase disparities of $2nk\pi$, where n is an integer).

To obviate the former difficulty, an obvious solution would be to build a binocular version of the complex cell by summing across simple cells with the same disparity tuning but various monocular phase tunings [e.g., 36]. Such construction is analogous to the construction of phase-independent, motion-sensitive complex cells discussed earlier, except that the displacement of interest is across eyeballs instead of time. This has been shown to occur in cortical cells and, in fact, these cells show more precise disparity tuning than 2-D position tuning [37].

Yet, because these cells are tuned to a certain phase disparity of a given *spatial frequency*, there remains an ambiguity concerning the absolute disparity of a stimulus. This can be seen in Fig. 10, which plots the output (as brightness) of a hypothetical collection of cells tuned to various values of phase disparity, orientation, and spatial frequency. The tuning of the cell is given by its position in the volume; in Fig. 10(a) orientation is ignored, and only a single spatial frequency/disparity surface is shown. In Fig.10(a), note that the output of cells tuned to a single spatial frequency contains multiple peaks along the dimension of disparity, indicating the phase ambiguity of the output. It has been suggested that this ambiguity could be resolved by units that sum the outputs of disparity units across spatial frequency and orientation [e.g., 36]. Such units would solve the phase ambiguity in a manner very analogous to the intersection-of-constraints solution to motion ambiguity described above. In the case of disparity, as a broadband stimulus is shifted along the disparity axis, it yields a sinusoidal variation in output at all spatial frequencies, but the frequency of modulation is proportional to the spatial frequency to which the cells are tuned. The resolution to the ambiguity lies in the fact that there is only *one* disparity at which peak output is obtained at *all* spatial frequencies, and that is the true disparity of the stimulus. This is shown in Fig. 10(a) by the white ridge running down the spatial frequency — disparity plane.

The pattern of outputs of cells tuned to a single spatial frequency but to a variety of orientations as a function of disparity is shown on the floor of Fig. 10(b). Summing across cells tuned to different orientations will also disambiguate disparity information because a Fourier component at an oblique orientation will behave as a vertical component with a *horizontal* frequency proportional to the cosine of the angle of its orientation from the vertical.

Figure 10(b) is best thought of as a volume of cells whose sensitivity is given by their position in the volume (for visualization convenience, the phase information is repeated for the higher spatial frequencies, so the phase tuning is giving by the position on the disparity axis modulo 2π). The combined spatial frequency and disparity information results in a surface of maximum activity at the true disparity of a broadband stimulus, so a cell that sums across surfaces in this space will encode for physical disparity independent of spatial frequency and orientation.

Very recent work indicates that cells in MT might perform just such a task [38]. Recall from above that cells in MT decouple velocity information from the spatial frequency and orientation sensitivity of motion selective cells. DeAngelis *et al.* [38] have discovered a patterned arrangement of disparity sensitive cells in the same area and have demonstrated their consequence in perceptual judgments. Given the conceptually identical nature of the ambiguities to be resolved the domains of motion and disparity, it would seem likely that the disparity-sensitive cells in MT perform role in stereopsis analogous to that which the velocity-sensitive cells play in motion perception.

---

[5]Many recent studies have not measured the absolute receptive field position in the two eyes, as it is very difficult to do. Thus, the notion that absolute monocular receptive field position plays a role in stereopsis cannot be rejected.

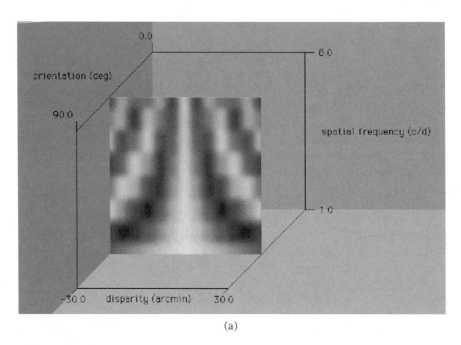

(a)

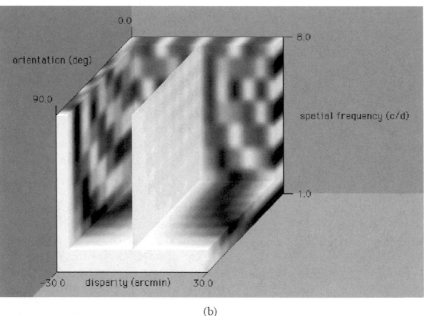

(b)

FIGURE 10   (a) Output of cortical cells on the spatial-frequency/disparity plane. The output of any one cell uniquely specifies only a phase disparity, but summation across spatial frequencies at the appropriate phase disparities uniquely recovers absolute disparity. (b) Orientation is added to this representation.

5 Concluding Remarks

Models are wonderful tools and have an indispensable role in vision science. Neuroscientists must reverse-engineer the brain, and for this the methods of engineering are required. But the tools themselves can lead to biases (when all you have is a hammer, everything looks like a nail). There is always a danger of carrying too much theory, often implicitly, into an analysis of the visual system. This is particularly true in the case of modeling, because a model must have a quantitative output and thus must be specified, whether intentionally or not, at what Marr called the level of computational theory [7]. Tools, like categories, make wonderful servants but horrible masters.

Yet without quantitative models, it would be almost impossible to compare psychophysics (human behavior) and physiology except in trivial ways.[6] This may seem like a strong statement, but there are subtle flaws in simple comparisons between the results of human experiments and single-cell response profiles. Consider an example taken from [39]. The experiment was designed to reveal the underlying mechanisms of disparity processing. A "mechanism" is assumed to comprise many neurons with similar tuning properties (peak location and bandwidth) on the dimension of interest working in parallel to encode that dimension. The tuning of the mechanism then reflects the tuning of the underlying neurons. This experiment used the typical psychophysical technique of *adaptation*. In this technique, one first measures the sensitivity of human observers along a dimension; in this case, we measured the sensitivity to the interocular correlation of binocular white noise signals as a function of binocular disparity. Following this, the subjects adapted to a signal at a given disparity. This adaptation fatigues the neurons sensitive to this disparity and therefore reduces the sensitivity of any mechanism comprising these neurons. Retesting sensitivity, we found that it was systematically elevated in the region of the adaptation, and a difference between the pre- and postadaptation sensitivity yielded a "tuning profile" of the adaptation, for which a peak location, bandwidth, etc. can be defined.

But what *is* this tuning profile? In these types of experiments, it is tempting to assume that it directly reflects the sensitivity profile of an underlying mechanism, but this would be a dangerous and generally wrong assumption. The tuning profile actually reflects the combined outputs of numerous mechanisms in response to the adaptation. The degree to which the tuning profile itself resembles any one of the individual underlying mechanisms depends on a number of factors involving the nature of the mechanisms themselves, their interaction, and how they are combined at subsequent levels to determine overall sensitivity.

If one cannot get a direct glimpse of the underlying mechanisms using psychophysics, how does one reveal them? This is where computational models assert their value. We constructed various models incorporating different numbers of mechanisms, different mechanism characteristics, and different methods of combining the outputs of mechanisms. We found that with a small number of disparity-sensitive mechanisms (e.g., three, as had been proposed by earlier theories of disparity processing) we were unable to simulate our psychophysical data. With a larger number of mechanisms, however, we able to reproduce our data rather precisely, and the model became much less sensitive to the manner in which the outputs of the mechanisms were combined.

So although we are unable to get a *direct* glimpse at underlying mechanisms using psychophysics, models can guide us

in determining what kinds of mechanisms can and cannot be used to produce sets of psychophysical data. As more physiological data become available, more precise models of the neurons themselves can be constructed, and these can be used, in turn, within models of psychophysical behavior. It is thus that models sew together psychophysics and physiology, and I would argue that without them the link could never be but tenuously established.

References

[1] I. Newton, *Opticks* (G. Bell & Sons, London, 1931).

[2] T. Young, "On the theory of light and color," *Phil. Trans. Roy. Soc.* **73**, 12–48 (1802).

[3] H. V. Helmholtz, *Treatise on Physiological Optics* (Dover, New York, 1962).

[4] G. Westheimer, "The eye as an optical instrument," in K. R. Boff, L. Kaufman, and J. P. Thomas, eds., *Handbook of Human Perception and Performance* (Wiley, New York, 1986).

[5] A. Roorda and D. R. Williams, "The arrangement of the three cone classes in the living human eye, *Nature* **397**, 520–522 (1999).

[6] W. S. Geisler, "Sequential ideal-observer analysis of visual discriminations," *Psychol. Rev.* **96**, 267–314 (1989).

[7] D. R. Williams and N. J. Coletta, "Cone spacing and the visual resolution limit," *J. Opt. Soc. Am.* A **4**, 1514–1523 (1987).

[8] G. Wyszecki and W. S. Stiles, *Color Vision* (Wiley, New York, 1982).

[9] W. S. Geisler and M. S. Banks, "Visual performance," in M. Bass, ed., *Handbook of Optics* (McGraw-Hill, New York, 1995).

[10] H. B. Barlow, "Measurements of the quantum efficiency of descrimination in human scotopic vision," *J. Physiol.* **150**, 169–188 (1962).

[11] D. G. Pelli, "The quantum efficiency of vision," in C. Blakemore, ed., *Vision: Coding and Efficiency* (Cambridge U. Press, Cambridge, 1990).

[12] W. S. Geisler and K. Chou, "Separation of low-level and high-level factors in complex tasks: visual search," *Psychol. Rev.* **102**, 356–378 (1995).

[13] D. Marr, *Vision* (Freeman, New York, 1982).

[14] R. W. Rodieck, "Quantitative analysis of cat retinal ganglion cell response to visual stimuli," *Vis. Res.* **5**, 583–601 (1965).

[15] C. Enroth-Cugel and J. G. Robson, "The contrast sensitivity of retinal ganglion cells in the cat," *J. Physiol.* **187**, 517–522 (1966).

[16] A. B. Watson, "Temporal sensitivity," in K. R. Boff, L. Kauffman, and J. P. Thomas, eds., *Handbook of Perception and Human Performance* (Wiley, New York, 1986).

[17] A. M. Derrington and P. Lennie, "Spatial and temporal contrast sensitivities of neurons in the lateral geniculate nucleus of Macaque," *J. Physiol.* **357**, 2219–240 (1984).

[18] P. Lennie, "Roles of M and P pathways," in R. Shapley and D. M. K. Lam, eds., *Contrast Sensitivity* (MIT, Cambridge, 1993).

[19] J. B. Troy, "Modeling the receptive fields of mammalian retinal ganglion cells," in R. Shapley and D. M. K. Lam, eds., *Contrast Sensitivity* (MIT Press, Cambridge, 1993).

[20] K. Donner and S. Hemila, "Modeling the spatio-temporal modulation response of ganglion cells with difference-of-Gaussians receptive fields: relation to photoreceptor response kinetics," *Visual Neurosci.* **13**, 173–186 (1996).

---

[6]Psychophysicists, such as myself, attempt to quantify the performance of human sensory and perceptual systems. Psychophysics encompasses a host of experimental techniques used to determine the ability of sensory systems (e.g., the visual system) to detect, discriminate, and/or identify well-defined and tightly-controlled input stimuli. These techniques share a general grounding in signal detection theory, which itself grew out of electronic communication theory and statistical decision theory.

[21] D. H. Hubel and T. N. Weisel, "Receptive fields, binocular interaction and functional archetecture in the cat's visual cortex," *J. Physiol.* **160**, 106–154 (1962).

[22] B. C. Skottun, R. L. DeValois, D. H. Grosof, J. A. Movshon, D. G. Albrecht, and A. B. Bonds, "Classifying simple and complex cells on the basis of response modulation," *Vis. Res.* **31**, 1079–1086 (1991).

[23] D. B. Hamilton, D. G. Albrecht, and W. S. Geisler, "Visual cortical receptive fields in monkey and cat: spatial and temporal phase transfer function," *Vis. Res.* **29**, 1285–1308 (1989).

[24] D. J. Field and D. J. Tolhurst, "The structure and symmetry of simple-cell receptive-field profiles in the cat's visual cortex," *Proc. Roy. Soc. London* **228**, 379–400 (1986).

[25] E. H. Adelson and J. R. Bergen, "Spatiotemporal energy models for the perception of motion," *J. Opt. Soc. Am.* A **2**, 284–299 (1985).

[26] A. B. Watson and A. J. Ahumada, "Spatiotemporal energy models for the perception of motion," *J. Opt. Soc. Am.* A **2**, 322–341 (1985).

[27] W. S. Geisler and D. G. Albrecht, "Visual cortex neurons in monkeys and cats: detection, discrimination, and stimulus certainty," *Visual Neurosci.* **14**, 897–919 (1997).

[28] D. J. Heeger, "Nonlinear model of neural responses in cat visual cortex," in M. S. Landy and J. A. Movshon, eds., *Computational Models of Visual Processing* (MIT Press, Cambridge, 1991).

[29] W. S. Geisler and D. G. Albrecht, "Bayesian analysis of identification performance in monkey visual cortex: nonlinear mechanisms and stimulus certainty," *Vis. Res.* **35**, 2723–2730 (1995).

[30] A. B. Watson, "The cortex transform: rapid computation of simulated neural images," *Comput. Vis. Graph. Image Process.* **39**, 311–327 (1987).

[31] E. P. Simoncelli and D. J. Heeger, "A model of neuronal responses in visual area MT," *Vis. Res.* **38**, 743–761 (1998).

[32] D. J. Heeger, E. P. Simoncelli, and J. A. Movshon, "Computational models of cortical visual processing," *Proc. Nat. Acad. Sci.* **93**, 623–627 (1996).

[33] E. H. Adelson and J. A. Movshon, "Phenomenal coherence of visual moving pattens," *Nature* **300**, 523–525 (1982).

[34] G. C. DeAngelis, I. Ohzawa, and R. D. Freeman, "Depth is encoded in the visual cortex by a specialized receptive field structure," *Nature* **352**, 156–159 (1991).

[35] I. Ohzawa, G. C. DeAngelis, and R. D. Freeman, "Encoding of binocular disparity by simple cells in the cat's visual cortex," *J. Neurophysiol.* **75**, 1779–1805 (1996).

[36] D. J. Fleet, H. Wagner, and D. J. Heeger, "Neural encoding of binocular disparity: energy models, position shifts, and phase shifts," *Vis. Res.* **36**, 1839–1858 (1996).

[37] I. Ohzawa, G. C. DeAngelis, and R. D. Freeman, "Encoding of binocular disparity by complex cells in the cat's visual cortex," *J. Neurophysiol.* **76**, 2879–2909 (1997).

[38] G. C. DeAngelis, B. G. Cumming, and W. T. Newsome, "Cortical area MT and the perception of stereoscopic depth," Nature **394**, 677–680 (1998).

[39] S. B. Stevenson, L. K. Cormack, C. M. Schor, and C. W. Tyler, "Disparity-tuned mechanisms of human stereopsis," *Vis. Res.* **32**, 1685–1689 (1992).

[40] R. W. Rodiek, *The First Steps in Seeing* (Sunderland, Sinauer Associates, 1998).

Multiscale Image Decompositions and Wavelets

Pierre Moulin
*University of Illinois
at Urbana-Champaign*

1 Overview

The concept of scale, or resolution of an image, is very intuitive. A person observing a scene perceives the objects in that scene at a certain level of resolution that depends on the distance to these objects. For instance, walking toward a distant building, she or he would first perceive a rough outline of the building. The main entrance becomes visible only in relative proximity to the building. Finally, the doorbell is visible only in the entrance area. As this example illustrates, the notions of resolution and scale loosely correspond to the size of the details that can be perceived by the observer. It is of course possible to formalize these intuitive concepts, and indeed signal processing theory gives them a more precise meaning.

These concepts are particularly useful in image and video processing and in computer vision. A variety of digital image processing algorithms decompose the image being analyzed into several components, each of which captures information present at a given scale. While the main purpose of this chapter is to introduce the reader to the basic concepts of multiresolution image decompositions and wavelets, applications will also be briefly discussed throughout the chapter. The reader is referred to other chapters of this book for more details.

Throughout, let us assume that the images to be analyzed are rectangular with $N \times M$ pixels. While there exist several types of multiscale image decompositions, we consider three main methods [1–6].

1. In a Gaussian pyramid representation of an image [Fig. 1(a)], the original image appears at the bottom of a pyramidal stack of images. This image is then low-pass filtered and subsampled by a factor of 2 in each coordinate. The resulting $N/2 \times M/2$ image appears at the second level of the pyramid. This procedure can be iterated several times. Here resolution can be measured by the size of the image at any given level of the pyramid. The pyramid in Fig. 1(a) has three resolution levels, or scales. In the original application of this method to computer vision, the low-pass filter used was often a Gaussian filter[1]; hence the terminology *Gaussian pyramid*. We shall use this terminology even when a low-pass filter is not a Gaussian filter. Another possible terminology in that case is simply *low-pass pyramid*. Note that the total number of pixels in a pyramid representation is $NM + NM/4 + NM/16 + \cdots \approx 4/3 NM$. This is said to be an *overcomplete* representation of the original image, caused by an increase in the number of pixels.

2. The Laplacian pyramid representation of the image is closely related to the Gaussian pyramid, but here the

---

[1]This design was motivated by analogies to the Human Visual System; see Section 3.6.

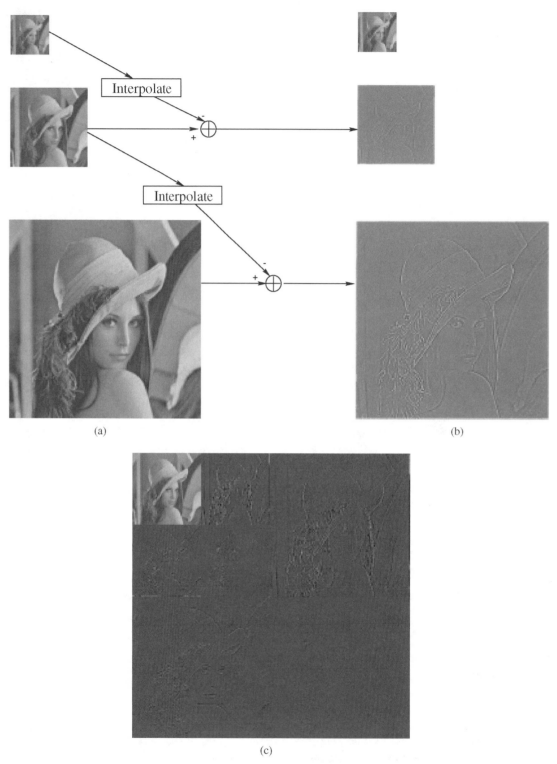

(a)

(b)

(c)

FIGURE 1   Three multiscale image representations applied to Lena: (a) Gaussian pyramid, (b) Laplacian pyramid, (c) wavelet representation.

difference between approximations at two successive scales is computed and displayed for different scales; see Fig. 1(b). The precise meaning of the *Interpolate* operation in the figure will be given in Section 2.1. The displayed images represent details of the image that are significant at each scale. An equivalent way to obtain the image at a given scale is to apply the difference between two Gaussian filters to the original image. This is analogous to filtering the image by using a Laplacian filter, a technique commonly employed for edge detection (see Chapter 4.20). Laplacian filters are bandpass; hence the name Laplacian pyramid, also termed *bandpass pyramid*.

3. In a wavelet decomposition, the image is decomposed into a set of subimages (or subbands) that also represent details at different scales [Fig. 1(c)]. Unlike pyramid representations, the subimages also represent details with different spatial orientations (such as edges with horizontal, vertical, and diagonal orientations). The number of pixels in a wavelet decomposition is only NM. As we shall soon see, the signal processing operations involved here are more sophisticated than those for pyramid image representations.

The pyramid and wavelet decompositions are presented in more detail in Sections 2 and 3, respectively. The basic concepts underlying these techniques are applicable to other multiscale decomposition methods, some of which are listed in Section 4.

Hierarchical image representations such as those in Fig. 1 are useful in many applications. In particular, they lend themselves to effective designs of reduced-complexity algorithms for texture analysis and segmentation, edge detection, image analysis, motion analysis, and image understanding in computer vision. Moreover, the Laplacian pyramid and wavelet image representations are *sparse* in the sense that most detail images contain few significant pixels (little significant detail). This sparsity property is very useful in image compression, as bits are allocated only to the few significant pixels; in image recognition, because the search for significant image features is facilitated; and in the restoration of images corrupted by noise, as images and noise possess rather distinct properties in the wavelet domain.

2 Pyramid Representations

In this section, we shall explain how the Gaussian and Laplacian pyramid representations in Fig. 1 can be obtained from a few basic signal processing operations. To this end, we first describe these operations in Section 2.1 for the case of one-dimensional (1-D) signals. The extension to two-dimensional (2-D) signals is presented in Sections 2.2 and 2.3 for Gaussian and Laplacian pyramids, respectively.

2.1 Decimation and Interpolation

Consider the problem of decimating a 1-D signal by a factor of 2, namely, reducing the sample rate by a factor of 2. This operation

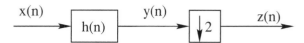

FIGURE 2 Decimation of a signal by a factor of 2, obtained by cascade of a low-pass filter $h(n)$ and a subsampler $\downarrow 2$.

generally entails some loss of information, so it is desired that the decimated signal retain as much fidelity as possible to the original. The basic operations involved in decimation are low-pass filtering (using a digital anti-aliasing filter) and subsampling, as shown in Fig. 2. The impulse response of the low-pass filter is denoted by $h(n)$, and its Discrete-time Fourier transform [7] by $H(e^{j\omega})$. The relationship between input $x(n)$ and output $y(n)$ of the filter is the convolution equation:

$$y(n) = x(n) * h(n) = \sum_k h(k)x(n-k).$$

The downsampler discards every other sample of its input $y(n)$. Its output is given by

$$z(n) = y(2n).$$

Combining these two operations, we obtain

$$z(n) = \sum_k h(k)x(2n-k). \qquad (1)$$

Downsampling usually implies a loss of information, as the original signal $x(n)$ cannot be exactly reconstructed from its decimated version $z(n)$. The traditional solution for reducing this information loss consists in using an "ideal" digital anti-aliasing filter $h(n)$ with cutoff frequency $\omega_c = \pi/2$ [7][2]. However such "ideal" filters have infinite length. In image processing, short finite impulse response (FIR) filters are preferred for obvious computational reasons. Furthermore, approximations to the ideal filters herein have an oscillating impulse response, which unfortunately results in visually annoying ringing artifacts in the vicinity of edges. The FIR filters typically used in image processing are symmetric, with a length between three and 20 taps. Two common examples are the three-tap FIR filter $h(n) = (1/4, 1/2, 1/4)$, and the length-$(2L + 1)$ truncated Gaussian, $h(n) = Ce^{-n^2/(2\sigma^2)}$, $|n| \le L$, where $C = 1/\sum_{|n| \le L} e^{-n^2/(2\sigma^2)}$. The coefficients of both filters add up to one: $\sum_n h(n) = 1$, which implies that the DC response of these filters is unity.

Another common image processing operation is *interpolation*, which increases the sample rate of a signal. Signal processing theory tells us that interpolation may be performed by cascading two basic signal processing operations: upsampling and low-pass filtering (see Fig. 3). The upsampler inserts a zero between every

---

[2]The paper [8] derives the filter that actually minimizes this information loss in the mean-square sense, under some assumptions on the input signal.

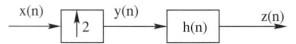

FIGURE 3 Interpolation of a signal by a factor of 2, obtained by cascade of an upsampler ↑2 and a low-pass filter $h(n)$.

other sample of the signal $x(n)$:

$$y(n) = \begin{cases} x(n/2) & n \text{ even} \\ 0 & n \text{ odd} \end{cases}.$$

The upsampled signal is then filtered by using a low-pass filter $h(n)$. The interpolated signal is given by $z(n) = h(n) * y(n)$ or, in terms of the original signal $x(n)$,

$$z(n) = \sum_k h(k)x(n - 2k). \qquad (2)$$

The so-called ideal interpolation filters have infinite length. Again, in practice, short FIR filters are used.

2.2 Gaussian Pyramid

The construction of a Gaussian pyramid involves 2-D low-pass filtering and subsampling operations. The 2-D filters used in image processing practice are *separable*, which means that they can be implemented as the cascade of 1-D filters operating along image rows and columns. This is a convenient choice in many respects, and the 2-D decimation scheme is then separable as well. Specifically, 2-D decimation is implemented by applying 1-D decimation to each row of the image [using Eq. (1)] followed by 1-D decimation to each column of the resulting image [using Eq. (1) again]. The same result would be obtained by first processing columns and then rows. Likewise, 2-D interpolation is obtained by first applying Eq. (2) to each row of the image, and then again to each column of the resulting image, or vice versa.

This technique was used at each stage of the Gaussian pyramid decomposition in Fig. 1(a). The low-pass filter used for both horizontal and vertical filtering was the three-tap filter $h(n) = (1/4, 1/2, 1/4)$.

Gaussian pyramids have found applications to certain types of image storage problems. Suppose for instance that remote users access a common image database (say an Internet site) but have different requirements with respect to image resolution. The representation of image data in the form of an image pyramid would allow each user to directly retrieve the image data at the desired resolution. While this storage technique entails a certain amount of redundancy, the desired image data are available directly and are in a form that does not require further processing. This technique has been used in the Kodak CD-I application, where image data are transferred from a CD-ROM and displayed on a television set at a user-specified resolution level [9]. Another application of Gaussian pyramids is in motion estimation for video [1, 2]: in a first step, coarse motion

estimates are computed based on low-resolution image data, and in subsequent steps, these initial estimates are refined based on higher-resolution image data. The advantages of this multiresolution, coarse-to-fine approach to motion estimation are a significant reduction in algorithmic complexity (as the crucial steps are performed on reduced-size images) and the generally good quality of motion estimates, as the initial estimates are presumed to be relatively close to the ideal solution. Another closely related application that benefits from a multiscale approach is pattern matching [1].

2.3 Laplacian Pyramid

We define a *detail image* as the difference between an image and its approximation at the next coarser scale. The Gaussian pyramid generates images at multiple scales, but these images have different sizes. In order to compute the difference between a $N \times M$ image and its approximation at resolution $N/2 \times M/2$, one should interpolate the smaller image to the $N \times M$ resolution level before performing the subtraction. This operation was used to generate the Laplacian pyramid in Fig. 1(b). The interpolation filter used was the three-tap filter $h(n) = (1/2, 1, 1/2)$.

As illustrated in Fig. 1(b), the Laplacian representation is *sparse* in the sense that most pixel values are zero or near zero. The significant pixels in the detail images correspond to edges and textured areas such as Lena's hair. Just like the Gaussian pyramid representation, the Laplacian representation is also *overcomplete*, as the number of pixels is greater (by a factor of $\approx 33\%$) than in the original image representation.

Laplacian pyramid representations have found numerous applications in image processing, and in particular in texture analysis and segmentation [1]. Indeed, different textures often present very different spectral characteristics which can be analyzed at appropriate levels of the Laplacian pyramid. For instance, a nearly uniform region such as the surface of a lake contributes mostly to the coarse-level image, whereas a textured region like grass often contributes significantly to other resolution levels. Some of the earlier applications of Laplacian representations include image compression [10, 11], but the emergence of wavelet compression techniques has made this approach somewhat less attractive. However, a Laplacian-type compression technique was adopted in the hierarchical mode of the lossy JPEG image compression standard [12]; also see Chapter 5.5.

3 Wavelet Representations

Although the sparsity of the Laplacian representation is useful in many applications, overcompleteness is a serious disadvantage in applications such as compression. The wavelet transform offers both the advantages of a sparse image representation and a complete representation. The development of this transform and its theory has had a profound impact on a variety of applications. In this section, we first describe

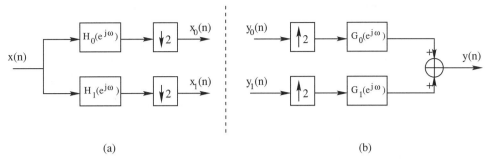

FIGURE 4 (a) Analysis filter bank, with low-pass filter $H_0(e^{j\omega})$ and high-pass filter $H_1(e^{j\omega})$. (b) Synthesis filter bank, with low-pass filter $G_0(e^{j\omega})$ and high-pass filter $G_1(e^{j\omega})$.

the basic tools needed to construct the wavelet representation of an image. We begin with filter banks, which are elementary building blocks in the construction of wavelets. We then show how filter banks can be cascaded to compute a wavelet decomposition. We then introduce wavelet bases, a concept that provides additional insight into the choice of filter banks. We conclude with a discussion of the relation of wavelet representations to the human visual system, and a brief overview of some applications.

3.1 Filter Banks

Figure 4(a) depicts an *analysis filter bank*, with one input $x(n)$ and two outputs $x_0(n)$ and $x_1(n)$. The input signal $x(n)$ is processed through two paths. In the upper path, $x(n)$ is passed through a low-pass filter $H_0(e^{j\omega})$ and decimated by a factor of 2. In the lower path, $x(n)$ is passed through a high-pass filter $H_1(e^{j\omega})$ and also decimated by a factor of 2. For convenience, we make the following assumptions. First, the number N of available samples of $x(n)$ is even. Second, the filters perform a circular convolution (see Chapter 2.3), which is equivalent to assuming that $x(n)$ is a periodic signal. Under these assumptions, the output of each path is periodic with period equal to $N/2$ samples. Hence the analysis filter bank can be thought of as a *transform* that maps the original set $\{x(n)\}$ of N samples into a new set $\{x_0(n), x_1(n)\}$ of N samples.

Figure 4(b) shows a *synthesis filter bank*. Here there are two inputs $y_0(n)$ and $y_1(n)$, and one single output $y(n)$. The input signal $y_0(n)$ (resp. $y_1(n)$) is upsampled by a factor of 2 and filtered by using a low-pass filter $G_0(e^{j\omega})$ (resp. high-pass filter $G_1(e^{j\omega})$). The output $y(n)$ is obtained by summing the two filtered signals. We assume that the input signals $y_0(n)$ and $y_1(n)$ are periodic with period $N/2$. This implies that the output $y(n)$ is periodic with period equal to N. Thus the synthesis filter bank can also be thought of as a transform that maps the original set of N samples $\{y_0(n), y_1(n)\}$ into a new set of N samples $\{y(n)\}$.

What happens when the output $x_0(n)$, $x_1(n)$ of an analysis filter bank is applied to the input of a synthesis filter bank? As it turns out, under some specific conditions on the four filters

$H_0(e^{j\omega})$, $H_1(e^{j\omega})$, $G_0(e^{j\omega})$, and $G_1(e^{j\omega})$, the output $y(n)$ of the resulting *analysis/synthesis* system is *identical* (possibly up to a constant delay) to its input $x(n)$. This condition is known as *perfect reconstruction*. It holds, for instance, for the following trivial set of one-tap filters: $h_0(n)$ and $g_1(n)$ are unit impulses, and $h_1(n)$ and $g_0(n)$ are unit delays. In this case, the reader can verify that $y(n) = x(n-1)$. In this simple example, all four filters are all pass. It is, however, not obvious to design more useful sets of FIR filters that also satisfy the perfect reconstruction condition. A general methodology for doing so was discovered in the mid-1980s. We refer the reader to [4, 5] for more details.

Under some additional conditions on the filters, the transforms associated with both the analysis and the synthesis filter banks are orthonormal. Orthonormality implies that the energy of the samples is preserved under the transformation. If these conditions are met, the filters possess the following remarkable properties: the synthesis filters are a time-reversed version of the analysis filters, and the high-pass filters are modulated versions of the low-pass filters, namely, $g_0(n) = (-1)^n h_1(n)$, $g_1(n) = (-1)^{n+1} h_0(n)$, and $h_1(n) = (-1)^{-n} h_0(K - n)$, where K is an integer delay. Such filters are often known as quadrature mirror filters (QMFs), or conjugate quadrature filters (CQFs), or power-complementary filters [5], because both low-pass (resp. high-pass) filters have the same frequency response, and the frequency responses of the low-pass and high-pass filters are related by the power-complementary property $|H_0(e^{j\omega})|^2 + |H_1(e^{j\omega})|^2 = 2$, valid at all frequencies. The filter $h_0(n)$ is viewed as a prototype filter, because it automatically determines the other three filters.

Finally, if the prototype low-pass filter $H_0(e^{j\omega})$ has a zero at frequency $\omega = \pi$, the filters are said to be *regular filters*, or *wavelet filters*. The meaning of this terminology will become apparent in Section 3.4. Figure 5 shows the frequency responses of the four filters generated from a famous four-tap filter designed by Daubechies [4, p. 195]:

$$h_0(n) = \frac{1}{4\sqrt{2}}(1 + \sqrt{3},\ 3 + \sqrt{3},\ 3 - \sqrt{3},\ 1 - \sqrt{3}).$$

This filter is the first member of a family of FIR wavelet filters that has been constructed by Daubechies and possess nice properties

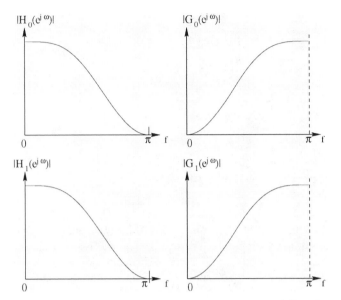

FIGURE 5 Magnitude frequency response of the four subband filters for a QMF filter bank generated from the prototype Daubechies' four-tap low-pass filter.

(such as shortest support size for a given number of vanishing moments; see Section 3.4).

There also exist *biorthogonal wavelet filters*, a design that sets aside degrees of freedom for choosing the synthesis low-pass filter $h_1(n)$ given the analysis low-pass filter $h_0(n)$. Such filters are subject to regularity conditions [4]. The transforms are no longer orthonormal, but the filters can have linear phase (unlike nontrivial QMF filters).

3.2 Wavelet Decomposition

An analysis filter bank decomposes 1-D signals into low-pass and high-pass components. One can perform a similar decomposition on images by first applying 1-D filtering along rows of the image and then along columns, or vice versa [13]. This operation is illustrated in Fig. 6(a). The same filters $H_0(e^{j\omega})$ and $H_1(e^{j\omega})$ are used for horizontal and vertical filtering. The output of the analysis system is a set of four $N/2 \times M/2$ subimages: the so-called LL (low low), LH (low high), HL (high low), and HH (high high) *subbands*, which correspond to different spatial frequency bands in the image. The decomposition of Lena into four such subbands is shown in Fig. 6(b). Observe that the LL subband is a coarse (low resolution) version of the original image, and that the HL, LH, and HH subbands respectively contain details with vertical, horizontal, and diagonal orientations. The total number of pixels in the four subbands is equal to the original number of pixels, NM.

In order to perform the wavelet decomposition of an image, one recursively applies the scheme of Fig. 6(a) to the LL subband. Each stage of this recursion produces a coarser version of the image as well as three new detail images at that particular scale. Figure 7 shows the cascaded filter banks that implement

this wavelet decomposition, and Fig. 1(c) shows a three-stage wavelet decomposition of Lena. There are seven subbands, each corresponding to a different set of scales and orientations (different spatial frequency bands).

Both the Laplacian decomposition in Fig. 1(b) and the wavelet decomposition in Fig. 1(c) provide a coarse version of the image as well as details at different scales, but the wavelet representation is complete and provides information about image components at different spatial orientations.

3.3 Discrete Wavelet Bases

So far we have described the mechanics of the wavelet decomposition in Fig. 7, but we have yet to explain what wavelets are, and how they relate to the decomposition in Fig. 7. In order to do so, we first introduce discrete wavelet bases. Consider the following representation of a signal $x(t)$ defined over some (discrete or continuous) domain $\mathcal{T}$:

$$x(t) = \sum_k a_k \varphi_k(t), \quad t \in \mathcal{T}. \tag{3}$$

Here $\varphi_k(t)$ are termed *basis functions*, and a_k are the coefficients of the signal $x(t)$ in the *basis* $\mathcal{B} = \{\varphi_k(t)\}$. A familiar example of such signal representations is the Fourier series expansion for periodic real-valued signals with period T, in which case the domain $\mathcal{T}$ is the interval $[0, T)$, $\varphi_k(t)$ are sines and cosines, and k represents frequency. It is known from Fourier series theory that a very broad class of signals $x(t)$ can be represented in this fashion.

For discrete $N \times M$ images, we let the variable t in Eq. (3) be the pair of integers (n_1, n_2), and the domain of x be $\mathcal{T} = \{0, 1, \ldots, N-1\} \times \{0, 1, \ldots, M-1\}$. The basis $\mathcal{B}$ is then said to be discrete. Note that the wavelet decomposition of an image, as described in Section 3.2, can be viewed as a linear transformation of the original NM pixel values $x(t)$ into a set of NM wavelet coefficients a_k. Likewise, the synthesis of the image $x(t)$ from its wavelet coefficients is also a linear transformation, and hence $x(t)$ is the sum of contributions of individual coefficients. The contribution of a particular coefficient a_k is obtained by setting all inputs to the synthesis filter bank to zero, except for one single sample with amplitude a_k, at a location determined by k. The output is a_k times the response of the synthesis filter bank to a unit impulse at location k. We now see that the signal $x(t)$ takes the form (3), where $\varphi_k(t)$ are the spatial impulse responses above.

The index k corresponds to a given location of the wavelet coefficient within a given subband. The discrete basis functions $\varphi_k(t)$ are translates of each other for all k within a given subband. However, the shape of $\varphi_k(t)$ depends on the scale and orientation of the subband. Figures 8(a)–8(d) show discrete basis functions in the four coarsest subbands. The basis function in the LL subband [Fig. 8(a)] is characterized by a strong central bump, while the basis functions in the other three subbands (detail images) have zero mean. Notice that the basis functions in the HL and LH subbands are related through a simple 90° rotation. The

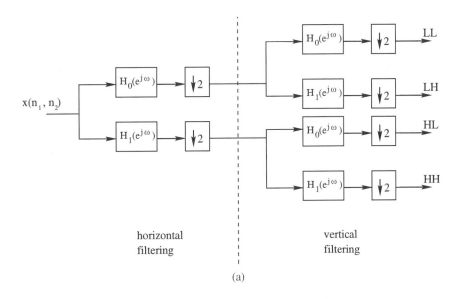

(a)

(b)

FIGURE 6 Decomposition of $N \times M$ image into four $N/2 \times M/2$ subbands: (a) basic scheme, (b) application to Lena, using Daubechies' four-tap wavelet filters.

orientation of these basis functions makes them suitable to represent patterns with the same orientation. For reasons that will become apparent in Section 3.4, the basis functions in the low subband are called *discrete scaling functions*, while those in the other subbands are called *discrete wavelets*. The size of the support set of the basis functions is determined by the length of the

wavelet filter, and essentially quadruples from one scale to the next.

3.4 Continuous Wavelet Bases

Basis functions corresponding to different subbands with the *same orientation* have a similar shape. This is illustrated in

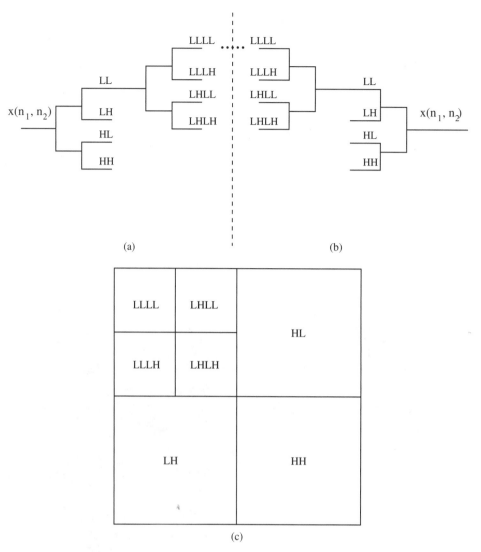

FIGURE 7   Implementation of wavelet image decomposition using cascaded filter banks: (a) wavelet decomposition of input image $x(n_1, n_2)$; (b) reconstruction of $x(n_1, n_2)$ from its wavelet coefficients; (c) nomenclature of subbands for a three-level decomposition.

Fig. 9, which shows basis functions corresponding to two subbands with vertical orientation [Figs. 9(a)–9(c)]. The shape of the basis functions converges to a limit [Fig. 9(d)] as the scale becomes coarser. This phenomenon is due to the regularity of the wavelet filters used (Section 3.1). One of the remarkable results of Daubechies' wavelet theory [4] is that under regularity conditions, the shape of the impulse responses corresponding to subbands with the same orientation does converge to a limit shape at coarse scales. Essentially the basis functions come in four shapes, which are displayed in Figs. 10(a)–10(d). The limit shapes corresponding to the vertical, horizontal, and diagonal orientations are called *wavelets*. The limit shape corresponding to the coarse scale is called *scaling function*. The three wavelets and the scaling function depend on the wavelet filter $h_0(n)$ used (in Fig. 10, Daubechies' four-tap filter). The four functions in Figs. 10(a)–10(d) are separable and are respectively of the form

$\phi(x)\phi(y)$, $\phi(x)\psi(y)$, $\psi(x)\phi(y)$, and $\psi(x)\psi(y)$. Here (x, y) are horizontal and vertical coordinates, and $\phi(x)$ and $\psi(x)$ are respectively the 1-D scaling function and the 1-D wavelet generated by the filter $h_0(n)$. These two functions are shown in Fig. 11, respectively. While the aspect of these functions is somewhat rough, Daubechies' theory shows that the smoothness of the wavelet increases with the number K of zeros of $H_0(e^{j\omega})$ at $\omega = \pi$. In this case, the first K moments of the wavelet $\psi(x)$ are

$$\int x^k \psi(x)dx = 0, \quad 0 \leq k < K.$$

The wavelet is then said to possess K vanishing moments.

3.5 More on Wavelet Image Representations

The connection between wavelet decompositions and bases for image representation shows that images are sparse linear

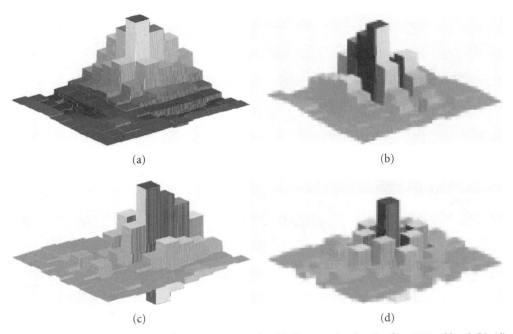

FIGURE 8 Discrete basis functions for image representation: (a) discrete scaling function from LLLL subband; (b)–(d) discrete wavelets from LHLL, LLLH, and LHLH subbands. These basis functions are generated from Daubechies' four-tap filter. (See color section, p. C–9.)

combinations of elementary images (discrete wavelets and scaling functions) and provide valuable insights for selecting the wavelet filter. Some wavelets are better able to compactly represent certain types of images than others. For instance, images with sharp edges would benefit from the use of short wavelet filters, because of the spatial localization of such edges. Conversely, images with mostly smooth areas would benefit from the use of longer wavelet filters with several vanishing moments, as such filters generate smooth wavelets. See [14] for a performance comparison of wavelet filters in image compression.

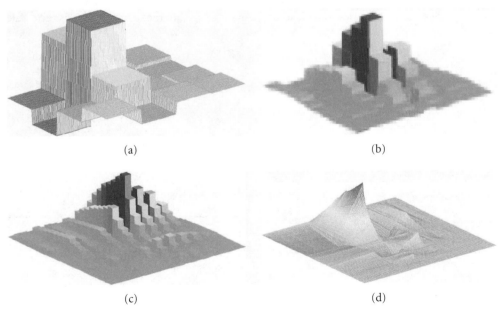

FIGURE 9 Discrete wavelets with vertical orientation at three consecutive scales: (a) in HL band; (b) in LHLL band; (c) in LLHLLL band. (d) Continuous wavelet is obtained as a limit of (normalized) discrete wavelets as the scale becomes coarser. (See color section, p. C–9.)

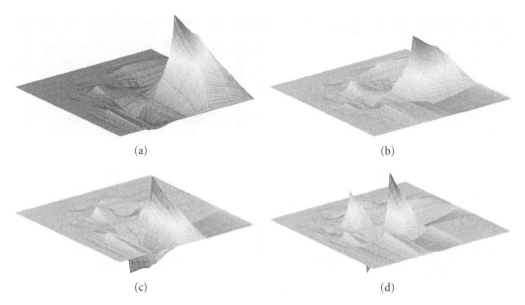

FIGURE 10 Basis functions for image representation: (a) scaling function; (b)–(d) wavelets with horizontal, vertical, and diagonal orientations. These four functions are tensor products of the 1-D scaling function and wavelet in Fig. 11. The horizontal wavelet has been rotated by 180° so that its negative part is visible on the display. (See color section, p. C–10.)

3.6 Relation to Human Visual System

Experimental studies of the human visual system (HVS) have shown that the eye's sensitivity to a visual stimulus strongly depends upon the spatial frequency contents of this stimulus. Similar observations have been made about other mammals. Simplified linear models have been developed in the psychophysics community to explain these experimental findings. For instance, the modulation transfer function describes the sensitivity of the HVS to spatial frequency; see Chapter 1.2. Additionally, several

experimental studies have shown that images sensed by the eye are decomposed into bandpass channels as they move toward and through the visual cortex of the brain [15]. The bandpass components correspond to different scales and spatial orientations. Figure 5 in [16] shows the spatial impulse response and spatial frequency response corresponding to a channel at a particular scale and orientation. While the Laplacian representation provides a decomposition based on scale (rather than orientation), the wavelet transform has a limited ability to distinguish between patterns at different orientations, as each scale comprises three

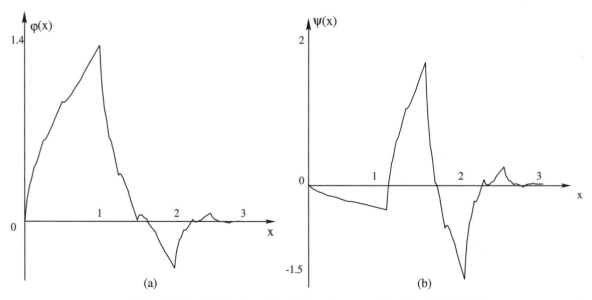

FIGURE 11 (a) 1-D scaling function and (b) 1-D wavelet generated from Daubechies' D4 filter.

channels that are respectively associated with the horizontal, vertical, and diagonal orientations. This may not be not sufficient to capture the complexity of early stages of visual information processing, but the approximation is useful. Note there exist linear multiscale representations that more closely approximate the response of the HVS. One of them is the *Gabor transform*, for which the basis functions are Gaussian functions modulated by sine waves [17]. Another one is the *cortical transform* developed by Watson [18]. However, as discussed by Mallat [19], the goal of multiscale image processing and computer vision is not to design a transform that mimics the HVS. Rather, the analogy to the HVS motivates the use of multiscale image decompositions as a front end to complex image processing algorithms, as nature already contains successful examples of such a design.

3.7 Applications

We have already mentioned several applications in which a wavelet decomposition is useful. This is particularly true of applications in which the completeness of the wavelet representation is desirable. One such application is image and video compression; see Chapters 5.4 and 6.2. Another one is image denoising, as several powerful methods rely on the formulation of statistical models in an orthonormal transform domain [20]; also see Chapter 3.4. There exist other applications in which wavelets present a plausible (but not necessarily superior) alternative to other multiscale decomposition techniques. Examples include texture analysis and segmentation [3, 21, 22] which is also discussed in Chapter 4.7, recognition of handwritten characters [23], inverse image halftoning [24], and biomedical image reconstruction [25].

4 Other Multiscale Decompositions

For completeness, we also mention two useful extensions of the methods covered in this chapter.

4.1 Undecimated Wavelet Transform

The wavelet transform is not invariant to shifts of the input image, in the sense that an image and its translate will in general produce different wavelet coefficients. This is a disadvantage in applications such as edge detection, pattern matching, and image recognition in general. The lack of translation invariance can be avoided if the outputs of the filter banks are not decimated. The undecimated wavelet transform then produces a set of bandpass images that have the same size as the original dataset ($N \times M$).

4.2 Wavelet Packets

Although the wavelet transform often provides a sparse representation of images, the spatial frequency characteristics of some images may not be best suited for a wavelet representation. Such is the case of fingerprint images, as ridge patterns constitute relatively narrow-band bandpass components of the image. An even sparser representation of such images can be obtained by recursively splitting the appropriate subbands (instead of systematically splitting the low-frequency band as in a wavelet decomposition). This scheme is simply termed *subband decomposition*. This approach was already developed in signal processing during the 1970s [5]. In the early 1990s, Coifman and Wickerhauser developed an ingenious algorithm for finding the subband decomposition that gives the sparsest representation of the input signal (or image) in a certain sense [26]. The idea has been extended to find the best subband decomposition for compression of a given image [27].

5 Conclusion

We have introduced basic concepts of multiscale image decompositions and wavelets. We have focused on three main techniques: Gaussian pyramids, Laplacian pyramids, and wavelets. The Gaussian pyramid provides a representation of the same image at multiple scales, using simple low-pass filtering and decimation techniques. The Laplacian pyramid provides a coarse representation of the image as well as a set of detail images (bandpass components) at different scales. Both the Gaussian and the Laplacian representation are overcomplete, in the sense that the total number of pixels is approximately 33% higher than in the original image.

Wavelet decompositions are a more recent addition to the arsenal of multiscale signal processing techniques. Unlike the Gaussian and Laplacian pyramids, they provide a complete image representation and perform a decomposition according to both scale and orientation. They are implemented using cascaded filter banks in which the low-pass and high-pass filters satisfy certain specific constraints. While classical signal processing concepts provide an operational understanding of such systems, there exist remarkable connections with work in applied mathematics (by Daubechies, Mallat, Meyer and others) and in psychophysics, which provide a deeper understanding of wavelet decompositions and their role in vision. From a mathematical standpoint, wavelet decompositions are equivalent to signal expansions in a wavelet basis. The regularity and vanishing-moment properties of the low-pass filter affect the shape of the basis functions and hence their ability to efficiently represent typical images. From a psychophysical perspective, early stages of human visual information processing apparently involve a decomposition of retinal images into a set of bandpass components corresponding to different scales and orientations. This suggests that multiscale/multiorientation decompositions are indeed natural and efficient for visual information processing.

Acknowledgments

I thank Juan Liu for generating all figures and plots in this chapter.

References

[1] A. Rosenfeld, ed., *Multiresolution Image Processing and Analysis* (Springer-Verlag, New York, 1984).

[2] P. Burt, "Multiresolution techniques for image representation, analysis, and 'Smart' transmission," in *Visual Communications and Image Processing* IV, Proc. SPIE **1199**, W. A. Pearlman, ed. (1989).

[3] S. G. Mallat, "A theory for multiresolution signal decomposition: The wavelet transform," *IEEE Trans. Pattern Anal. Machine Intell.* **11**, 674–693 (1989).

[4] I. Daubechies, *Ten Lectures on Wavelets*, CBMS-NSF Regional Conference Series in Applied Mathematics, Vol. 61 (SIAM, Philadelphia, 1992).

[5] M. Vetterli and J. Kovačević, *Wavelets and Subband Coding* (Prentice-Hall, Englewood Cliffs, NJ, 1995).

[6] S. G. Mallat, *A Wavelet Tour of Signal Processing* (Academic, San Diego, CA, 1998).

[7] J. Proakis and Manolakis, *Digital Signal Processing: Principles, Algorithms, and Applications*, 3rd ed. (Prentice-Hall, Englewood Cliffs, NJ, 1996).

[8] M. K. Tsatsanis and G. B. Giannakis, "Principal component filter banks for optimal multiresolution analysis," *IEEE Trans. Signal Process.* **43**, 1766–1777 (1995).

[9] N. D. Richards, "Showing photo CD pictures on CD–I," in *Digital Video: Concepts and Applications Across Industries*, T. Rzeszewski, ed. (IEEE Press, New York, 1995).

[10] P. Burt and A. H. Adelson, "The Laplacian pyramid as a compact image code," *IEEE Trans. Commun.* **31**, 532–540 (1983).

[11] M. Vetterli and K. M. Uz, "Multiresolution coding techniques for digital video: a review," *Multidimen. Syst. Signal Process.* Special Issue on Multidimensional Proc. of Video Signals, Vol. 3, pp. 161–187 (1992).

[12] W. B. Pennebaker and J. L. Mitchell, *JPEG: Still Image Data Compression Standard* (Van Nostrand Reinhold, New York, 1993).

[13] M. Vetterli, "Multi-dimensional sub-band coding: some theory and algorithms," *Signal Process.* **6**, 97–112 (1984).

[14] J. D. Villasenor, B. Belzer and J. Liao, "Wavelet filter comparison for image compression," *IEEE Trans. Image Process.* **4**, 1053–1060 (1995).

[15] F. W. Campbell and J. G. Robson, "Application of Fourier analysis to cortical cells," *J. Physiol.* **197**, 551–566 (1968).

[16] M. Webster and R. De Valois, "Relationship between spatial-frequency and orientation tuning of striate-cortex cells," *J. Opt. Soc. Am.* (1985).

[17] J. G. Daugmann, "Two-dimensional spectral analysis of cortical receptive field profile," *Vis. Res.* **20**, 847–856 (1980).

[18] A. B. Watson, "The cortex transform: rapid computation of simulated neural images," *Comput. Graph. Image Process.* **39**, 2091–2110 (1987).

[19] S. G. Mallat, "Multifrequency channel decompositions of images and wavelet models," *IEEE Trans. Acoust. Speech Signal Process.* **37**, (1989).

[20] P. Moulin and J. Liu, "Analysis of multiresolution image denoising schemes using generalized-gaussian and complexity priors," *IEEE Trans. Info. Theory*, Special Issue on Multiscale Analysis, Apr. (1999).

[21] M. Unser, "Texture classification and segmentation using wavelet frames," *IEEE Trans. Image Process.* **4**, 1549–1560 (1995)

[22] R. Porter and N. Canagarajah, "A robust automatic clustering scheme for image segmentation using wavelets," *IEEE Trans. Image Process.* **5**, 662–665 (1996).

[23] Y. Qi and B. R. Hunt, "A multiresolution approach to computer verification of handwritten signatures," *IEEE Trans. Image Process.* **4**, 870–874 (1995).

[24] J. Luo, R. de Queiroz, and Z. Fan, "A robust technique for image descreening based on the wavelet transform," *IEEE Trans. Signal Process.* **46**, 1179–1184 (1998).

[25] A. H. Delaney and Y. Bresler, "Multiresolution tomographic reconstruction using wavelets," *IEEE Trans. Image Process.* **4**, 799–813 (1995).

[26] R. R. Coifman and M. V. Wickerhauser, "Entropy–based algorithms for best basis selection," *IEEE Trans. Inf. Theory*, Special Issue on Wavelet Tranforms and Multiresolution Signal Analysis, Vol. 38, No.2, pp. 713–718 (1992).

[27] K. Ramchandran and M. Vetterli, "Best wavelet packet bases in a rate-distortion sense," *IEEE Trans. Image Process.* **2**, 160–175 (1993).

4.3

Random Field Models

J. Zhang
*University of
Wisconsin-Milwaukee*

P. Fieguth
University of Waterloo

D. Wang
Samsung Electronics

1 Introduction

Random fluctuations in intensity, color, texture, object boundary, or shape can be seen in most real-world images, as shown in Fig. 1. The causes for these fluctuations are diverse and complex, and they are often due to factors such as non-uniform lighting, random fluctuations in object surface orientation and texture, complex scene geometry, and noise.[1] Consequently, the processing of such images becomes a problem of statistical inference [1], which requires the definition of a statistical model corresponding to the image pixels.

Although simple image models can be obtained from image statistics such as the mean, variance, histogram, and correlation function (e.g., see [2, 3]), a more general approach is to use *random fields*. Indeed, as a two-dimensional extension of the one-dimensional random process, a random field model provides a complete statistical characterization for a given class of images — all statistical properties of the images can, in principle, be derived from this random field model. Combined with various frameworks for statistical inference, such as maximum-likelihood (ML) and Bayesian estimation, random field models have in recent years led to significant advances in many statistical image processing applications. These include image restoration, enhancement, classification, segmentation, compression, and synthesis.

Early studies of random fields can be traced to the 1970s, with many of the results summarized in [4]. Among the wide variety of proposed models, the most used is perhaps the AR (autoregressive) model and its various extensions (e.g., [3]). A landmark paper by Geman and Geman [5] in 1984 addressed Markov random field (MRF) models and has attracted great attention and invigorated research in image modeling; indeed the MRF, coupled with the Bayesian framework, has been the focus of many studies [6, 7]. Section 2 will introduce notation and provide an overview of random field models, emphasizing the autoregressive and Markov fields.

With the advent of multiresolution processing techniques, such as the pyramid [8] and wavelets [9], much of the current research in random field models focuses on multiscale models [10–22]. This interest has been motivated by the significant advantages they may have in *computational power* and *representational power* over the single-resolution/single-scale models. Specifically, multiresolution/multiscale processing can provide drastic computation reduction and represent a highly complicated model by a set of simpler models.

Multiresolution/multiscale models that aim at computation reduction include various multiresolution/multiscale MRFs [14–16] and multiscale tree models [11–13]. Through their connection to multigrid methods [26], these models often can improve convergence in iterative procedures. The multiscale tree model is described in more detail in Section 3. Multiresolution/multiscale models that aim at representing highly complicated random fields (e.g., those with high-order and nonlinear interactions) include various hierarchical/multiresolution/multiscale texture models [18–23]. Section 4 describes a wavelet-based nonlinear texture model.

---

[1] Chapter 4.5 of this book by Boncelet is devoted to noise and noise models.

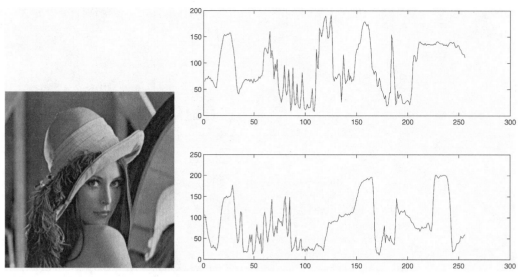

FIGURE 1   Typical image (left), two rows of which are plotted (right); the fine-scale details appear nearly "random" in nature.

2 Random Fields: Overview

A random field x is a collection of random variables arranged on a lattice L:

$$x = \{x_i, i \in L\}. \tag{1}$$

In principle the lattice can be any (possibly irregular) collection of discrete points; however it is most convenient and intuitive to visualize the lattice as a rectangular, regular array of sites:

$$L = \{(i, j) | 1 \leq i \leq N, 1 \leq j \leq M\}, \tag{2}$$

in which case a random field is just a set of random pixels

$$x = \{x_{i,j}, (i, j) \in L\}. \tag{3}$$

As with random variables or random vectors, any random field can, in principle, be completely characterized by its associated probability measure $p_x(x)$. The detailed form of $p(\cdot)$ will depend on whether the elements $x_{i,j}$ are discrete, in which case $p_x(x)$ denotes a probability distribution, or continuous, in which case $p_x(x)$ denotes a probability density function.

However, suppose we take an image of modest size, say $N = M = 256$; this implies that $p(\cdot)$ must explicitly characterize the joint statistics of 65,536 elements. Often the function $p(\cdot)$ is a cumbersome and computationally inefficient means of defining the statistics of the random field. Indeed, a great part of the research into random fields involves the discovery or definition of *implicit* statistical forms that lead to effective or faithful representations of the true statistics, while admitting computationally efficient algorithms.

Broadly speaking, there are five typical problems associated with random fields.

1. Representation: how is the random field represented and parameterized? In general the probability distribution $p(\cdot)$ is not computable for large fields, except in pathological cases, for example, in which all of the elements are independent:

$$p(x) = \prod_{(i,j) \in L} p(x_{i,j}). \tag{4}$$

2. Synthesis: generate "typical" realizations, known as sample paths, of the random field (e.g., used in stochastic image compression, random texture synthesis, lattice-physics simulations). That is, generate fields $y_1, y_2, \ldots$, statistically sampled from $p(\cdot)$, such that the probability of generating y is $P(y)$.[2]

3. Parameter estimation: given a parameterized statistical model (e.g., of the form $p(x | \theta)$) and sample image y, estimate the parameters θ. Typically we are interested in the ML estimates

$$\hat{\theta}_{ML} = \arg\max_{\theta} \left\{ p(y | \theta) \right\}. \tag{5}$$

This can be used to estimate any continuous parameter on which the field statistics depend; for example, correlation length, temperature, or ambient color.

---

[2]Assuming that y is discrete. If y is continuous, sample paths still exist and are intuitively the same as in the discrete case; however, a more careful formulation is required.

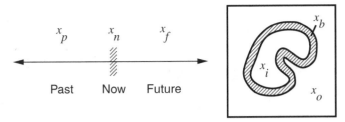

Past   Now   Future

FIGURE 2 The Markov property: given a boundary, the two separated portions of the field are conditionally independent.

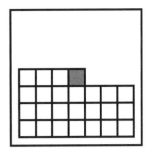 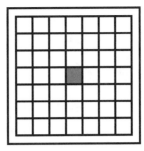

FIGURE 3 Regions of support for causal (left) and *acausal* (right) neighborhoods of the shaded element.

4. Least-squares estimation: given a statistical model $p(\boldsymbol{x})$ and observations of the random field

$$\boldsymbol{y} = f(\boldsymbol{x}) + \boldsymbol{v}, \qquad (6)$$

where $\boldsymbol{v}$ is a noise signal with known statistics, find the least-squares estimate

$$\hat{\boldsymbol{x}} = \arg\min_{\hat{\boldsymbol{x}}} E\left[|(\boldsymbol{x} - \hat{\boldsymbol{x}})|^2 | \boldsymbol{y}\right]. \qquad (7)$$

Least-squares estimates are of interest in reconstruction or inference problems; for example, denoising images or interpolation.

5. More general versions of the above; for example, Bayesian estimation of $\boldsymbol{x}$ subject to a criterion other than least squares, or the deduction of expectations $E[g(\boldsymbol{x})]$ by Monte Carlo sampling of synthesized fields.

2.1 Markov Random Fields

The fundamental notion associated with Markovianity is one of conditional independence: a one-dimensional process x_n is Markovian (Fig. 2) if the knowledge of the process at some point x_n decouples the "past" $\boldsymbol{x}_p$ and the "future" $\boldsymbol{x}_f$:

$$p(\boldsymbol{x}_f \mid x_n, \boldsymbol{x}_p) = p(\boldsymbol{x}_f \mid x_n), \quad p(\boldsymbol{x}_p \mid x_n, \boldsymbol{x}_f) = p(\boldsymbol{x}_p \mid x_n). \quad (8)$$

The decoupling extends perfectly into two dimensions, except that the natural concepts of "past" and "future" are lost, since there is no natural ordering of the elements in a grid; instead, a random field $\boldsymbol{x}$ is Markov (Fig. 2) if the knowledge of the process on a boundary set b decouples the inside and outside of the set:

$$p(\boldsymbol{x}_i \mid \boldsymbol{x}_b, \boldsymbol{x}_o) = p(\boldsymbol{x}_i \mid \boldsymbol{x}_b), \quad p(\boldsymbol{x}_o \mid \boldsymbol{x}_b, \boldsymbol{x}_i) = p(\boldsymbol{x}_o \mid \boldsymbol{x}_b). \quad (9)$$

This boundary concept, although elegantly intuitive, is lacking in details (e.g., how "thick" does the boundary have to be?). It is often simpler, and more explicit, to talk about separating a single element $x_{i,j}$ from the entire field $\boldsymbol{x}$ conditioned on a local neighborhood $\mathcal{N}_{i,j}$ [5,6]:

$$p(x_{i,j} \mid \{x_{k,l}, (k,l) \in L \backslash (i,j)\}) = p(x_{i,j} \mid \{x_{k,l}, (k,l) \in \mathcal{N}_{i,j}\}). \tag{10}$$

The shape and extent of $\mathcal{N}_{i,j}$ is one aspect that characterizes the nature of the random field.

- Causal/acausal: A neighborhood structure is causal (Fig. 3) if all elements of the neighborhood live in one half of the plane; e.g.,

$$(k,l) \in \mathcal{N}_{i,j} \Longrightarrow l < j, \text{ or } l = j, k < i. \quad (11)$$

That is, if the field can be reordered into a one-dimensional random vector which is Markov, satisfying Eq. (8). Otherwise, more typically, the neighborhood is acausal (Fig. 3).

- Order: The order of a neighborhood reflects its extent. A first-order neighborhood is shown in Fig. 4, which also illustrates the pattern followed by higher-order neighborhoods.

The above discussion is entirely formulated in terms of the joint probability density $p(\boldsymbol{x})$, which is often impractical for large random fields; the following sections summarize the two, broad alternatives that have been developed.

1. Gauss–Markov random fields: $\boldsymbol{x}$ is Gaussian, in which case the field can be characterized explicitly in terms of expectations rather than probability densities.

2. Gibbs random fields: an energy $E(\boldsymbol{x})$ is associated with each possible field $\boldsymbol{x}$; a probability density is then constructed implicitly from $E(\boldsymbol{x})$ in such a way that $p(\boldsymbol{x})$ satisfies Eqs. (9) and (10).

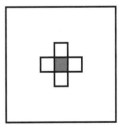 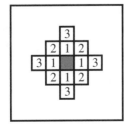

FIGURE 4 Left: The region of support of a first-order neighborhood. Right: Neighborhood size as a function of model order.

2.2 Gauss–Markov Random Fields

When the random field x is Gaussian [6], then conditional independence is equivalent to conditional uncorrelatedness, so instead of Eq. (2.1) we write

$$\hat{x}_{i,j} = E[x_{i,j}|\{x_{k,l}, (k, l) \in L\backslash(i, j)\}]$$
$$= E[x_{i,j}|\{x_{k,l}, (k, l) \in \mathcal{N}_{i,j}\}], \quad (12)$$

where $\hat{x}_{i,j}$ is the estimated value of $x_{i,j}$. However if x is Gaussian the expectation is known to be linear, so the right side of Eq. (12) can be rewritten as

$$\hat{x}_{i,j} = \sum_{(k,l) \in \mathcal{N}_{i,j}} a_{i,j,k,l} x_{k,l}. \quad (13)$$

Alternatively, we can describe the field elements directly, instead of their estimates:

$$x_{i,j} = \hat{x}_{i,j} + w_{i,j} = \sum_{(k,l) \in \mathcal{N}_{i,j}} a_{i,j,k,l} x_{k,l} + w_{i,j}, \quad (14)$$

where $w_{i,j}$ is the estimation error process. If the random field is stationary then the coefficients simplify as

$$a_{i,j,k,l} = a_{i-k,j-l}. \quad (15)$$

Causal GMRFs

If a random field x is causal, each neighborhood $\mathcal{N}_{i,j}$ must limit its support to one-half of the plane, as sketched in Fig. 3. These are known as nonsymmetric half-plane (NSHP) models, and they lead to very simple, $\mathcal{O}(NM)$ autoregressive equations for sample paths and estimation.

Specifically, there must exist an ordering of the field elements into a vector $\vec{x}$ such that each element depends only on the values of elements lying earlier in the ordering, in which case Eq. (14) is the autoregressive equation to generate sample paths and the Kalman filter can be used for estimation. These models have limited applicability, since most random fields are not well represented causally. The limitations of the causal model are most obvious when computing estimates from sparse observations, as shown in Fig. 5, since the arrangement of the estimates is obviously asymmetric with respect to the observations.

Toroidally Stationary GMRFs

A second special case is that of toroidally stationary fields; that is, rectangular fields in which the left and right edges are considered adjacent, as are the top and bottom.[3] In other words, the field is periodic. The correlation structure of a toroidally stationary

[3]That is, topologically, the wrapping of a rectangular sheet onto a torus or doughnut.

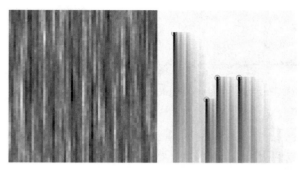

FIGURE 5 Causal model produces a reasonable sample (left), but shows obvious limitations when computing estimates (right) from sparse observations (circled).

$N \times M$ field therefore takes the form

$$E[x_{i,j} x_{i+\Delta i, j+\Delta j}] = E[x_{0,0} x_{\Delta i \bmod N, \Delta j \bmod M}]. \quad (16)$$

The significance of this stationarity is that *any* such covariance (matrix) is diagonalized by the two-dimensional FFT, leading to extremely fast algorithms. Specifically, let Λ be the correlation structure of the field,

$$\Lambda_{i,j} = E[x_{0,0} x_{i,j}] \; 1 \leq i \leq N, \; 1 \leq j \leq M. \quad (17)$$

Then sample paths may be computed as

$$s = \text{IFFT}_2\{\text{sqrt}(\text{FFT}_2(\Lambda)) \cdot \text{FFT}_2(q)\} \quad (18)$$

where sqrt( ) and $\cdot$ are element-by-element operations, and q is an array of unit variance, independent Gaussian samples. Similarly, given a set of observations with Gaussian error,

$$y = x + v, \quad \text{cov}(v) = \sigma^2 I, \quad (19)$$

the least-squares estimates may be computed as

$$\hat{x} = \text{IFFT}_2\{\text{FFT}_2(\Lambda) \cdot \text{FFT}_2(y)/(\text{FFT}_2(\Lambda) + \sigma^2)\} \quad (20)$$

where again $\cdot$ and / are performed element-by-element.

In general the circumstances of Eqs. (16) and (19) are restrictive: the field must be toroidally stationary, regularly sampled, and densely measured with constant error variance; however, the FFT approach is fast, $\mathcal{O}(NM \log(NM))$, when these circumstances are satisfied (for example, texture synthesis, image denoising). The FFT was used to generate the "wood" texture of Fig. 6.

Acausal GMRFs

In general, we can rewrite Eq. (14), stacking the random field into a vector $\vec{x}$ (by rows or by columns):

$$G\vec{x} = \vec{w}. \quad (21)$$

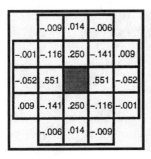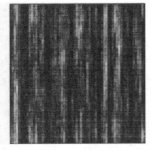

FIGURE 6 Fourth-order stationary MRF, synthesized by using the FFT method, based on the coefficients[4] $a_{i,j}$.

If the field is small (a few thousand elements) and A is invertible we can, in principle, solve for the covariance P of the entire field:

$$G\vec{x} = \vec{w} \Longrightarrow GPG^T = W \Longrightarrow P = G^{-1}WG^{-T}. \quad (22)$$

Sample paths can be computed using the Cholesky decomposition[5] of P, and least-squares estimates computed by inverting P; however, neither of these operations is practical for large fields.

Instead, methods of *domain decomposition* (e.g., nested-dissection or Marching methods) [11, 26–28] are often used. The goal is to somehow break a large field into smaller pieces. Suppose we have a first-order GMRF; although the individual field elements cannot be ordered causally, if we divide the field into columns

$$\boldsymbol{x} = [\vec{x}_1 \vec{x}_2 \cdots \vec{x}_N], \quad (23)$$

then the sequence of columns is a one-dimensional first-order Markov process; that is,

$$E[\vec{x}_{i+1} \mid \vec{x}_1, \ldots, \vec{x}_i] = E[\vec{x}_{i+1} \mid \vec{x}_i]. \quad (24)$$

Since the field is Gaussian, we can express $\vec{x}_i$ by means of a linear model,

$$\vec{x}_{i+1} = A_i \vec{x}_i + B_i w_i. \quad (25)$$

The estimates $\hat{\vec{x}}_i$ still have to be computed acausally; however, this can be accomplished efficiently, $\mathcal{O}(\min(NM)^3)$, using the RTS smoother [40]. A tremendous number of variations exist: different decomposition schemes, approximate or partial matrix inversion, reduced update, etc.

2.3 Gibbs Random Fields

Gibbs random fields (GRFs) are random fields characterized by neighboring-site interactions. These were originally used in

---

[4]The coefficients in the figure are rounded; the exact values may be found in [13].

[5]A common matrix operation, available in standard mathematics packages such as MATLAB.

statistical physics [24, 30] to study the thermodynamic properties of interacting particle systems, such as lattice gases, and their use in image processing was popularized by papers of Geman and Geman [5] and Besag [31, 32]. The neighboring interactions in GRFs lead to effective and intuitive image models — for example, to assert the piecewise-continuity of image intensity. Hence, the GRF is often used as a prior model in a Bayesian formulation to enforce image constraints.

Mathematically, a GRF $\boldsymbol{x}$ is described or defined by a Gibbs distribution:

$$p(\boldsymbol{x}) = \frac{1}{Z} e^{-\beta E(\boldsymbol{x})}. \quad (26)$$

Here $E(\boldsymbol{x})$, the *energy function*, is a sum of neighboring-interaction terms, called *clique potentials*, i.e.,

$$E(\boldsymbol{x}) = \sum_{c \in \mathbf{C}} V_c(\boldsymbol{x}), \quad (27)$$

where c is a *clique*, i.e., either a single site or a set of sites that are all neighbors of each other; $V_c(\cdot)$ is the clique potential, a function of the random variables associated with c; and $\mathbf{C}$ is the set of all possible cliques. Finally, $\beta > 0$ is a constant, also known as the temperature parameter, and

$$Z = \sum_{x'} e^{-\beta E(x')} \quad (28)$$

is a normalization constant, known as the *partition function*.

As an example of the GRF, consider a binary Ising model [30], which has the following energy function:

$$E(\boldsymbol{x}) = a \sum_{i,j} h_{i,j} x_{i,j} + b_1 \sum_{i,j} x_{i,j} x_{i,j-1} + b_2 \sum_{i,j} x_{i,j} x_{i-1,j}, \quad (29)$$

where a, b_1, b_2 are model parameters, $h_{i,j}$ are constants sometimes called the *external field*, and $x_{i,j} = +1$ or -1. In this example, a clique is either a single site $\{(i, j)\}$, or two neighboring sites $\{(i, j), (i, j-1)\}$, $\{(i, j), (i-1, j)\}$, with respective clique potentials $ah_{i,j} x_{i,j}$, $b_1 x_{i,j} x_{i,j-1}$, and $b_2 x_{i,j} x_{i-1,j}$. Typical realizations of this GRF are shown in Fig. 7.

Given this brief introduction to GRFs, it is natural to ask how they relate to the MRFs and how to address the basic random field problems (Section 2). First, according to the Hammersley–Clifford theorem [31], the GRF and MRF are equivalent. As a result, the Ising model described above is an MRF. Similarly, the Gauss–Markov models described in Section 2.2 have associated the energy functions and clique potentials; for example, the energy function for a first-order acausal GMRF model is [38]

$$E(\boldsymbol{x}) = \frac{1}{2\sigma^2} \sum_{i,j} \left[x_{i,j} - \sum_{(k,l) \in \mathcal{N}_{i,j}} a_{i-k,j-l} x_{k,l} \right]^2,$$

 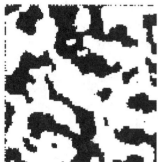

FIGURE 7 Typical sample paths of the Ising model. left, $a = 0$, $b_1 = b_2 = -0.4$; Right, $a = 0$, $b_1 = b_2 = -1.0$. [See Eq. (29)].

from which one can identify the clique potentials, which are of the form $x_{i,j}^2/2\sigma^2$ and $-a_{k,l}x_{i,j}x_{k,l}/\sigma^2$.

Second, of the basic random field problems, the most important in the GRF context are parameter estimation and sample generation. In terms of parameter estimation, the typical ML approach

$$\hat{\boldsymbol{\theta}}_{ML} = \arg\max_{\boldsymbol{\theta}} p(\boldsymbol{x} \,|\, \boldsymbol{\theta})$$

is impractical for GRFs, because evaluating $p(\boldsymbol{x} \,|\, \boldsymbol{\theta})$ requires the calculation of Z in Eq. (28), which sums over all possible realizations of the GRF. As an alternative, Besag proposed to maximize the *pseudo-likelihood* [31],

$$q(\boldsymbol{x} \,|\, \boldsymbol{\theta}) = \prod_{i,j} p(x_{i,j} \,|\, \{x_{k,l}, (k,l) \in \mathcal{N}_{i,j}\}, \boldsymbol{\theta}), \qquad (30)$$

which is made up by a set of conditional probabilities, with respect to $\boldsymbol{\theta}$. These conditional probabilities, also called the *local characteristics*, are easily calculated since the partition function no longer appears and each term in Eq. (30) can be evaluated locally:

$$p(x_{i,j} \,|\, \{x_{k,l}, (k,l) \in \mathcal{N}_{i,j}\}, \boldsymbol{\theta}) = \frac{e^{-\beta \sum_{c \ni (i,j)} V_c(\boldsymbol{x})}}{\sum_{x_{i,j}} e^{-\beta \sum_{c \ni (i,j)} V_c(\boldsymbol{x})}}. \qquad (31)$$

Finally, in the problem of producing samples of a GRF, we need to differentiate two cases: (1) to produce a sample according to the Gibbs distribution $p(\boldsymbol{x})$, a synthesis problem; and (2) to produce a sample that will maximize $p(\boldsymbol{x})$, an optimization problem.

Synthesis Problem

A number of techniques have been developed for the synthesis problem, such as the Metropolis algorithm [37] or the Gibbs sampler [5]. The Gibbs sampler is summarized here for the discrete binary-valued case (for the continuous case, see [38]).

Step 0: Start with a sample from an i.i.d binary random field with $p(x_{i,j}) = 1/2$.

Step 1: Scan the image from left to right, top to bottom. At each site (i, j), sample $x_{i,j}$ from $p(x_{i,j} \,|\, \{x_{k,l}, (k, l) \in \mathcal{N}_{i,j}\})$.[6]

Step 2: Repeat Step 1 many times; after many iterations $\boldsymbol{x}$ is a statistical sample of the random field [5].

Optimization Problem

The optimization problem is closely related to the synthesis problem. Rather than sampling the individual local elements $x_{i,j}$ at random, we want to bias our selection in the direction of maximizing $p(\boldsymbol{x})$. Define $T = 1/\beta$ to be the temperature; then the Gibbs sampler for optimization is as follows.

Step 0: Start with an i.i.d. sample image and an initial temperature $T = T_0$.

Step 1: Perform one iteration of the Gibbs Sampler for Synthesis (see Synthesis Problem).

Step 2: Lower the temperature T and repeat Step 1 till convergence.

Because of its close relation to the original simulated annealing algorithm, this algorithm is called the simulated annealing algorithm for GRFs/MRFs. Theoretically, it produces a global optimum when $k \longrightarrow \infty$ and $T \longrightarrow 0$ [5], where k is the number of iterations. In practice, however, to achieve good results, the temperature has to be lowered very slowly, e.g., according to $T(k) = C/\log(1 + k)$. This is usually computation intensive. Hence, suboptimal techniques are often used. Among them are Besag's iterative conditional mode (ICM) [32] and the mean field theory [35, 36].

Finally, we provide an example of how the GRF can be used in a Bayesian formulation. Specifically, consider the problem of segmenting an image into two types of image regions, labeled -1 and $+1$. Suppose the true labels are described by binary field $\boldsymbol{x}$, and we are given corrupted measurements $\boldsymbol{r} = m\boldsymbol{x} + \boldsymbol{v}$, where $\boldsymbol{v}$ is an additive zero-mean white Gaussian noise. The problem of segmentation, then, is to label each pixel of $\boldsymbol{r}$ to either -1 or $+1$; that is, we seek to estimate $\boldsymbol{x}$. In a Bayesian formulation, we find $\boldsymbol{x}$ as

$$\hat{\boldsymbol{x}} = \arg\max_{\boldsymbol{x}} \log p(\boldsymbol{x} \,|\, \boldsymbol{r}) \propto \arg\max_{\boldsymbol{x}} \{\log p(\boldsymbol{r} \,|\, \boldsymbol{x}) + \log p(\boldsymbol{x})\}.$$

$$(32)$$

If an Ising model is adopted for $p(\boldsymbol{x})$ to enforce the region continuity constraint (i.e., neighboring pixels are likely to be in the same region), it can be shown easily that $p(\boldsymbol{x} \,|\, \boldsymbol{r})$ is also an Ising model (with an external field) and the segmentation can be obtained by simulated annealing (Fig. 8) or ICM or the MFT. For more details and more realistic examples, see [36].

---

[6]This can be implemented as follows: generate a random number, r, from a uniform pdf over $[0, 1]$. If $r < p(x_{i,j} = -1 | \{x_{k,l}, (k, l) \in \mathcal{N}_{i,j}\})$, assign $x_{i,j} = -1$; otherwise, $x_{i,j} = +1$.

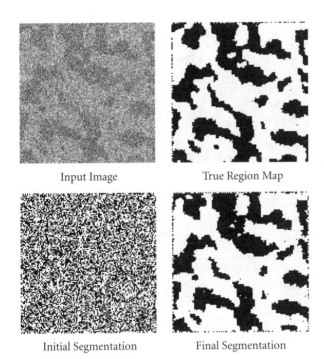

Input Image   True Region Map

Initial Segmentation   Final Segmentation

FIGURE 8   Segmentation by simulated annealing.

3 Multiscale Random Fields

This section will discuss the multiscale statistical modeling of random fields based on a particular multiscale tree structure that has been the focus of research for several years [10–13]. The motivations driving this research are broad, including $1/f$ processes, stochastic realization theory, and a variety of application areas (computer vision, synthetic aperture radar, groundwater hydrology, ocean altimetry, and hydrography). Although the method is applicable more broadly, we will focus on the GMRF case.

A statistical characterization of a random field in multiscale form (detailed below) possesses the following attributes.

- an efficient least-squares estimation algorithm [i.e., computing $\hat{x}$ in Eq. (7)]

- an efficient likelihood calculation algorithm (i.e., computing $p(y \mid \theta)$ in Eq. (5))
- the ability to accommodate nonlocal or distributed measurements
- computational complexity is unaffected by nonstationarities in the random field or in the measurements (compare with the FFT, Section 2.2).

3.1 GMRF Models on Trees

Multiscale models for GMRFs can be developed as a straight forward, recursive extension of the discussion in Section 2.2. Specifically, consider the generalization of the boundary in Fig. 2 to the boundary x_b shown in Fig. 9. From Eq. (9) we see that each of the quadrant fields x_i satisfies

$$p(x_i \mid x_b, x_j, \, j \neq i) = p(x_i \mid x_b). \tag{33}$$

That is, the effect of the rest of the field on x_i is captured by x_b. If we stack each field into a vector, the Gaussianity of the field allows us to write

$$E[\vec{x}_i \mid \vec{x}_b, \vec{x}_j, \, j \neq i] = E[\vec{x}_i \mid \vec{x}_b] = \mathcal{A}_i \vec{x}_b. \tag{34}$$

In other words,

$$\vec{x}_i = \mathcal{A}_i \vec{x}_b + \vec{\omega}_i, \tag{35}$$

where $\vec{\omega}_i$ is uncorrelated with $\vec{\omega}_j$ for $i \neq j$. There is, however, no reason to content ourselves with limiting the decomposition of the field into four quadrants. We can proceed further, creating a boundary x_{b_i} within quadrant i; from Eq. (35) it follows that we can write the boundary as

$$\vec{x}_{b_i} = A_i \vec{x}_b + \vec{w}_i. \tag{36}$$

We can continue the successive subdivision of the field into smaller pieces; Fig. 9 shows such a set of boundaries organized onto a tree structure. Now *every* vector on our tree is a boundary;

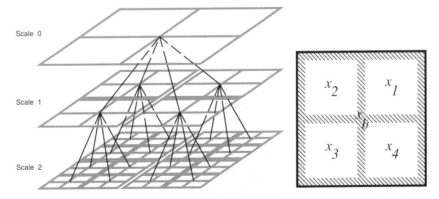

Scale 0

Scale 1

Scale 2

x_2 x_1

x_{b}

x_3 x_4

FIGURE 9   A quad tree (left) can be used to model an MRF, if the state at each node (right) is chosen to decorrelate the four quadrants represented by that node.

that is, our random field is described in terms of a set of hierarchical boundaries, terminating with an individual pixel at each tree element at the finest level of the tree.

Let $\vec{x}_{s,i}$ be the ith boundary at scale s on the tree; then

$$\vec{x}_{s,i} = A_{s,i}\vec{x}_{s-1,p(i)} + B_{s,i}\vec{w}_{s,i}, \qquad (37)$$

where $p(i)$ represents the "parent" of the ith boundary, and $\vec{w}$ is a white-noise process.

Having developed a multiscale model, Eq. (37), for a random field, we can also introduce measurements

$$\vec{y}_{s,i} = C_{s,i}\vec{x}_{s,i} + D_{s,i}\vec{v}_{s,i}, \qquad (38)$$

where $\vec{v}$ is a white-noise process. Local point measurements are normally associated with the individual pixels at the finest level of the tree; however, with the appropriate definition of $\vec{x}_{s,i}$ at coarser levels of the tree, nonlocal measurements can also be accommodated.

It is Eqs. (37) and (38) that form the basis for the multiscale environment; *any* random field (whether Markov or not) that can be written in this form leads to efficient algorithms for sample paths, estimation, and likelihood calculations, as mentioned earlier.

It should be noted that Eq. (37) is essentially a distributed marching algorithm, here marching over boundaries on a tree rather than across space (Section 2.2). Indeed, the marching principle applies to determining $A_{s,i}, B_{s,i}$: if we write each boundary as $\vec{x}_{s,i} = L_{s,i}\vec{x}$ where $\vec{x}$ is the original random field with covariance P, then

$$A_{s,i} = \left\{ L_{s,i}PL_{s-1,p(i)} \right\} \left\{ L_{s-1,p(i)}PL_{s-1,p(i)} \right\}^{-1} \qquad (39)$$

$$B_{s,i} = \{L_{s,i}PL_{s,i}\} - A_{s,i}\{L_{s-1,p(i)}PL_{s-1,p(i)}\}^{-1}A_{s,i}^T. \qquad (40)$$

With the determination of the parameters $A_{s,i}, B_{s,i}$, above, the definition of the multiscale random-field is complete. There are three issues to address before such a model can be put into practice.

- How does the structure of Eqs. (37) and (38) lead to an efficient estimator? The Kalman filter and Rauch–Tung–Striebel smoother [40] can be applied to Eq. (37) to compute estimates; however since every state $x_{s,i}$ is a boundary — only a small subset of the random field — the matrices operations at each tree node are modest. Algorithms have been published and are available on the Internet [11, 12].
- How are the boundaries $L_{s,i}$ determined? For Gauss–Markov random fields $\vec{x}$, Section 2.2 discusses the criteria for a boundary to conditionally decorrelate subsets of the field. For non-Markov fields the "optimal" choice of $L_{s,i}$ is much more complicated; further discussion may be found in [13].
- Can further computational efficiency be gained by use of approximations? For large random fields (10^6 pixels), an "exact" solution may require orders of magnitude more

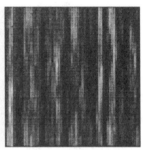

FIGURE 10   Multiscale estimation of random-field textures: left, reducedorder model with texture artifacts; right, overlapped multiscale model. (From [22]).

memory and computational effort than an approximate method yielding essentially the same estimates. This is typically accomplished by selecting $L_{s,i}$ to very nearly, although not perfectly, decorrelate subsets of the field. Such approximations, and methods of dealing with possible resulting artifacts, are discussed at length in [13].

3.2 Examples

We will briefly survey three examples; many more are available in the literature. Our first example [11, 13] continues with the MRF texture of Fig. 6. Figure 10 shows two estimates of this texture based on noisy measurements: using a reduced-order multiscale model, which illustrates the artifacts that may be present with poorly approximated models, and using an overlapped multiscale model, having the same computational complexity, but free of artifacts.

Figure 11 shows a related example, which illustrates the ability to estimate nonstationary random fields with the multiscale approach, not possible using the FFT.

Finally, Fig. 12 highlights two random fields of significant interest in oceanography [12] and remote sensing: the estimation of ocean height (altimetry) and temperature (hydrography) fields from sparse, nonstationary, noisy measurements.

4 Wavelet Multiresolution Models

Many real-world images, especially textures, contain long range and nonlinear spatial interactions (correlations) that can only

FIGURE 11   Overlapped multiscale estimation of nonstationary random fields: left, observations; right, estimates.

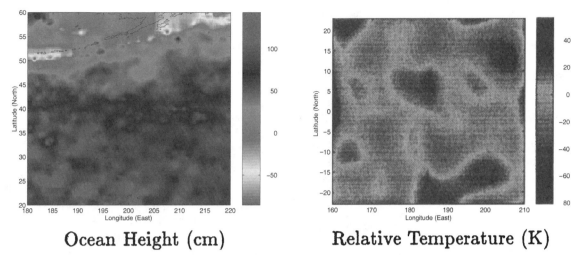

Ocean Height (cm)　　　　**Relative Temperature (K)**

FIGURE 12 Multiscale estimation of remotely sensed fields: left, North-Pacific altimetry based on Topex/Poseidon data; right, equatorial Pacific temperature estimates based on *in situ* ship data. (See color section, p. C–10.)

be adequately captured by high-order nonlinear models [17]. In such cases, the AR and MRF, like many others, may run into difficulties. Specifically, high-order models require large neighborhood structures and a large number of parameters, and this makes parameter estimation and applications computation intensive, if not impossible. Similarly, linear models, such as the AR and GMRF, tend to have trouble capturing nonlinear interactions (e.g., patterns containing sharp intensity changes). In this section, we describe wavelet-based multiresolution models that may overcome these problems. The basic idea here is that a high-order, and possibly nonlinear, model can be constructed through a set of low-order models living in the subbands of a wavelet decomposition. Related work can also be found in [17–23].

We assume that the reader is familiar with the theory of orthonormal wavelets (see, e.g., [39]). Let L be a square lattice and $\boldsymbol{x} = \{x_{i,j}, (i, j) \in L\}$ be a random field used to model a class of images. Suppose L represents the finest resolution and denote $\boldsymbol{x}$ by $\boldsymbol{x}^0$. Suppose that for some positive integer M, $\boldsymbol{x}^0$ has a wavelet expansion

$$\boldsymbol{x}^0 \sim \{\boldsymbol{w}^{-1}, \boldsymbol{w}^{-2}, \ldots, \boldsymbol{w}^{-M}, \boldsymbol{x}^{-M}\}. \tag{41}$$

As shown in Fig. 13, $\boldsymbol{w}^m$, $m = -1, -2, \ldots, -M$ are the wavelet coefficients at various levels, obtained from bandpass filtering

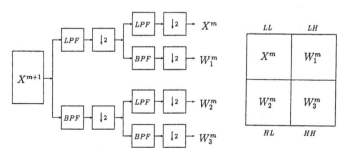

FIGURE 13 Wavelet decomposition.

and subsampling. Each $\boldsymbol{w}^m$ contains three subsets $\boldsymbol{w}_1^m$, $\boldsymbol{w}_2^m$, $\boldsymbol{w}_3^m$ corresponding to LH, HL, and HH components, and $\boldsymbol{x}^{-M}$ contains the scaling function coefficients at level M, obtained by lowpass filtering and subsampling. The $\sim$ sign denotes equivalence in the sense that the wavelet coefficients can be used to reconstruct $\boldsymbol{x}^0$.

4.1 The Model

Since $\boldsymbol{x}$ is completely determined from its wavelet coefficients, we can model $\boldsymbol{x}$ by modeling the wavelet coefficients. In other words, $\boldsymbol{x}$ can be modeled by specifying the joint probability density:

$$
\begin{aligned}
p(\boldsymbol{x}^0) &\sim p(\boldsymbol{w}^{-1}, \boldsymbol{w}^{-2}, \ldots, \boldsymbol{w}^{-M}, \boldsymbol{x}^{-M}), \\
&= p(\boldsymbol{w}^{-1} \mid \boldsymbol{w}^{-2}, \ldots, \boldsymbol{w}^{-M}, \boldsymbol{x}^{-M}) \\
&\quad \times p(\boldsymbol{w}^{-2} \mid \boldsymbol{w}^{-3}, \ldots, \boldsymbol{w}^{-M}, \boldsymbol{x}^{-M}) \cdots p(\boldsymbol{x}^{-M}), \\
&= p(\boldsymbol{w}^{-1} \mid \boldsymbol{x}^{-1}) \, p(\boldsymbol{w}^{-2} \mid \boldsymbol{x}^{-2}) \cdots p(\boldsymbol{x}^{-M}), \tag{42}
\end{aligned}
$$

where we have used the fact that $(\boldsymbol{w}^m, \ldots, \boldsymbol{w}^{-M}, \boldsymbol{x}^{-M}) \sim \boldsymbol{x}^{(m+1)}$ for $-M < m < -1$. Now, the problem of specifying $p(\boldsymbol{x}^0)$ becomes that of specifying densities $p(\boldsymbol{x}^{-M})$ and $p(\boldsymbol{w}^m \mid \boldsymbol{x}^m)$, for $m = -1, -2, \ldots, -M$. As described previously, the complexity (e.g., model order) for these latter models can be considerably lower than that for the fine resolution model.

Suppose parametric models are used. Then,

$$p(\boldsymbol{w}^m \mid \boldsymbol{x}^m) = p(\boldsymbol{w}^m \mid \boldsymbol{x}^m, \Phi_m), \quad m = -1, -2, \ldots, -M, \tag{43}$$

$$p(\boldsymbol{x}^{-M}) = p(\boldsymbol{x}^{-M} \mid \Theta_{-M}), \tag{44}$$

where Φ_m and Θ_{-M} are parameter vectors. Generally, the model for $\boldsymbol{x}^{-M}$ is similar to that of $\boldsymbol{x}^0$ but at a lower order. The models for $p(\boldsymbol{w}^m \mid \boldsymbol{x}^m, \Phi_m)$, on the other hand, are different from that of $\boldsymbol{x}^0$.

Furthermore, since it is well known that the wavelet transform reduces correlation within and between resolutions, two assumptions can be made to simplify the modeling of $p(\boldsymbol{w}^m \mid \boldsymbol{x}^m, \Phi_m)$:

- *Assumption 1* (Interband Conditional Independence): $\boldsymbol{w}_n^m$, $n = 1, 2, 3$ are independent given $\boldsymbol{x}^m$.
- *Assumption 2* (Interband Independence): $\boldsymbol{w}_n^m$, $n = 1, 2, 3$ are independent to each other as well as to $\boldsymbol{x}^m$.

Assumption 2 is stronger than assumption 1 and seems to work well for textures that are characterized by random microstructures. However, when the texture contains long range correlation and large structural elements, the weaker and more realistic assumption 1 is needed [22].

4.2 Wavelet AR and Wavelet RBF

In this section, we provide two examples. The first is a wavelet-AR model. Here, assumption 2 is used and $\boldsymbol{w}^m$'s and $\boldsymbol{x}^{-M}$ are AR models with the nonsymmetrical half-plane (NSHP) neighborhood system shown in (Fig. 14). For example, a first-order AR model can be used for $\boldsymbol{x}^{-M}$ with

$$x_{i,j}^{-M} = \theta_{0,1} x_{i,j-1}^{-M} + \theta_{1,1} x_{i-1,j-1}^{-M} + \theta_{1,0} x_{i-1,j}^{-M}$$
$$+ \theta_{1,-1} x_{i-1,j+1}^{-M} + \sigma_{-M}^2 n_{i,j}^{-M}, \qquad (45)$$

where $n_{i,j}^{-M}$ is a zero-mean white Gaussian noise. Similar first-order ARs can be used for the $\boldsymbol{w}^m$'s, completing perhaps the simplest nontrivial wavelet multiresolution model.

In the second example, the random field at each resolution is an MRF characterized by its conditional densities. For example, $\boldsymbol{x}^{-M}$ can be described by $p(x_{i,j}^{-M} \mid \{x_{k,l}^{-M}, (k, l) \in \mathcal{N}_{i,j}^q\})$, where $\mathcal{N}_{i,j}^q$ can either be a NSHP (Fig. 3), or a noncausal (Fig. 3) neighborhood of order q. In general, $p(x_{i,j}^{-M} \mid \{x_{k,l}^{-M}, (k, l) \in \mathcal{N}_{i,j}^q\})$ is a high-dimensional function in $x_{i,j}^{-M}$ and $x_{k,l}^{-M}, (k, l) \in \mathcal{N}_{i,j}^q$, and the learning of a high-dimensional function is a difficult problem. Among various proposed learning techniques, neural

network based techniques, in particular the radial basis function (RBF) network [42], have been shown to be competitive. Using this technique, we first specify the joint density as a *mixture density*

$$p\big(x_{i,j}^{-M}, x_{k,l}^{-M}, (k, l) \in \mathcal{N}_{i,j}^q\big)$$
$$= \sum_{\kappa=1}^K \pi_\kappa\, p_\kappa\big(x_{i,j}^{-M}, x_{k,l}^{-M}, (k, l) \in \mathcal{N}_{i,j}^q\big) \qquad (46)$$

where $p_\kappa(\cdot)$ are individual density functions, e.g., Gaussians, and π_κ are positive weights that sum to one [42]. The conditional density $p(x_{i,j}^{-M} \mid \{x_{k,l}^{-M}, (k, l) \in \mathcal{N}_{i,j}^q\})$ can be derived by applying Bayes's formula to Eq. (46).

For $\boldsymbol{w}^m$'s, the conditional densities can be obtained in a similar way. The derivation is slightly more involved because of the possible interband conditioning of Eq. (44). Specifically, suppose assumption 1 is adopted to increase representation power. Without loss of generality, consider $\boldsymbol{w}_1^m$, the LH band, and suppose a causal neighborhood system is used. Since $\boldsymbol{w}_1^m$ is obtained by horizontal lowpass and vertical bandpass filtering, it is reasonable to assume, as a first-order approximation, that $\boldsymbol{w}_1^m$ and $\boldsymbol{x}^m$ are similar along horizontal directions, i.e.,

$$p\big(w_{i,j}^m \mid \{w_{k,l}^m, (k, l) \in \mathcal{N}_{i,j}^q\}, \boldsymbol{x}^m\big)$$
$$= p\big(w_{i,j}^m \mid \{w_{k,l}^m, x_{k',l'}^m, ((k, l), (k', l')) \in N_{i,j}^{q,r}\}\big) \qquad (47)$$

where $\mathcal{N}_{i,j}^{q,r}$ is the combined neighborhood of Fig. 14. Notice that the horizontal coefficients of $\boldsymbol{x}^m$ are dropped since they are redundant given the horizontal coefficients of $\boldsymbol{w}_1^m$. The conditional density of Eq. (47) can be derived from the joint density $p(w_{i,j}^m, w_{k,l}^m, x_{k',l'}^m, ((k, l), (k', l')) \in \mathcal{N}_{i,j}^{q,r})$, which, like Eq. (46), is assumed to be a mixture distribution. Finally, the wavelet RBF model introduced above has an intuitive interpretation: a complicated random field can be represented by a set spatially correlated patterns, each characterized by the individual densities in the mixture densities [e.g., Eq. (46)].

4.3 Examples in Texture Synthesis

The efficacy of the wavelet AR and RBF models is illustrated through the texture synthesis in Figs. 15 and 16, where model parameters were estimated from texture images and then used to generate "copies" of the originals. In all cases, the images are of size 128×128 and the original textures are from Brodatz [41]. The techniques for parameter estimation and sample synthesis for the wavelet AR and RBF models are relatively straightforward and are described in detail in [21, 22].

Figure 15 shows the wood grain texture that contains long-range correlation in both horizontal and vertical directions. As a result, a single-scale AR model, even with a high model order, does not capture the correlation well. A wavelet AR model with low model orders in the subbands, on the other hand, provides a good correlation match.

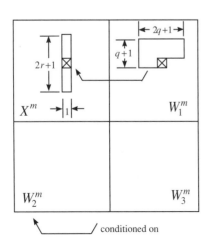

FIGURE 14 Neighborhood system for the LH band. (Adapted from [22].)

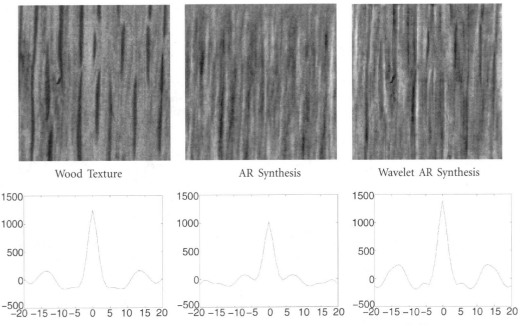

FIGURE 15   Texture synthesis using the wavelet AR model. (Adapted from [22].)

The wavelet AR model, however, does not work well when the texture contains nonlinear interaction (e.g., when pixels "switch" quickly from black to white). This is illustrated in Fig. 16. In this case, better results can be obtained by using the wavelet RBF model, which is nonlinear and has higher complexity. Finally, we would like to point out that, in addition to texture synthesis, the wavelet AR and RBF models can also be used for other random field problems described in Section 2.

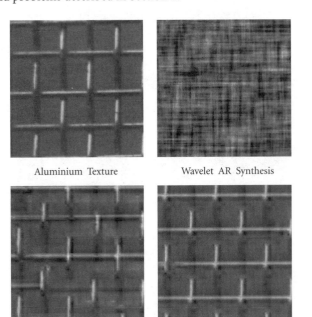

FIGURE 16   Texture synthesis using the wavelet RBF model. (Adapted from [22].)

References

[1] G. Casella and R. L. Berger, *Statistical Inference* (Wadsworth & Brooks, Pacific Grove, CA, 1990).

[2] A. Rosenfeld and A. C. Kak, *Digital Picture Processing*, 2nd ed. (Academic, New York, 1982).

[3] A. K. Jain, *Fundamentals of Digital Image Processing* (Prentice-Hall, Englewood-Cliffs, NJ, 1989).

[4] A. Rosenfeld, *Image Modeling* (Academic, New York, 1981).

[5] S. Geman and D. Geman, "Stochastic relaxation, Gibbs distribution, and the Bayesian restoration of images," *IEEE Trans. Pattern Anal. Machine Intell.* **6**, 721–741 (1984).

[6] H. Derin and P. A. Kelly, "Discrete-index Markov-type random processes," *Proc. IEEE* **77**, 1485–1511 (1989).

[7] R. Chellappa and A. Jain, eds., *Markov Random Fields — Theory and Applications* (Academic, New York, 1993).

[8] P. J. Burt and E. H. Adelson, "The Laplacian pyramid as a compact image code," *IEEE Trans. Commun.* **31**, 532–540 (1983).

[9] Mallat, S., "A theory for multiresolution signal decomposition: the wavelet representation," *IEEE Trans. Pattern Anal. Machine Intell.* **11**, 674–693 (1989).

[10] K. Chou, A. Willsky, and A. Benveniste, "Multiscale recursive estimation, data fusion, and regularization," *IEEE Trans. Automat. Control* **39**, 464–478 (1994).

[11] M Luettgen, W. Karl, A. Willsky, and R. Tenney, "Multiscale representations of Markov random fields," *IEEE Trans. Signal Process.* **41**, 3377–3396 (1993).

[12] P. Fieguth, W. Karl, A. Willsky, and C. Wunsch, "Multiresolution optimal interpolation and statistical analysis of TOPEX/ POSEIDON satellite altimetry," *IEEE Trans. Geosci. Remote Sens.* **33**, 280–292 (1995).

[13] W. Irving, P. Fieguth, and A. Willsky, "An overlapping tree approach to multiscale stochastic modeling and estimation," *IEEE Trans. Image Process.* **6**, 1517–1529 (1997).

[14] C. Bouman and B. Liu, "Multiple resolution segmentation of textured images," *IEEE Trans. Pattern Anal. Machine Intell.* **13**, 99–113 (1991).

[15] C. Bouman and M. Shapiro, "A multiscale random field model for Bayesian image segmentation," *IEEE Trans. Image Process.* **3**, 162–177 (1994).

[16] F. Heitz, P. Perez, and P. Bouthemy, "Parallel visual motion analysis using multiscale Markov random fields," presented at the IEEE Workshop on Visual Motion, Princeton, NJ, Oct., 1991.

[17] K. Popat and R. Picard, "Novel cluster-based probability model for texture synthesis," in *Visual Communications and Image Processing '93*, B. G. Haskell and H. M. Hang, eds., *Proc. SPIE* **2094**, 756–768 (1993).

[18] J. M. Francos, A. Z. Meiri, and B. Porat, "A unified texture model based on a 2-D Wold-like decomposition," *IEEE Trans. Signal Process.* 2665–2678 (1993).

[19] S. C. Zhu, Y. Wu, and D. Mumford, "Frame: filters, random fields and maximum entropy: towards a unified theory for texture modeling" *Intell. J. Comput. Vis.* **27**, 1–20 (1998).

[20] J. De Bonet and P. Viola, "A non-parametric multiscale statistical model for natural images," in *Advances in Neural Information Processing*, Vol. 9 (MIT Press, Cambridge, MA, 1997).

[21] J. Zhang, D. Wang, and Q. Tran, "Wavelet-based stochastic image modeling," in *Nonlinear Image Processing* VIII, E. R. Dougherty and J. T. Aspola, eds., *Proc. SPIE* **3026**, 293–304 (1997).

[22] J. Zhang, D. Wang, and Q. Tran, "Wavelet-based multiresolution stochastic models," *IEEE Trans. Image Process.* **7**, 1621–1626 (1998).

[23] E. P. Simoncelli and J. Portilla, "Texture characterization via joint statistics of wavelet coefficient magnitudes," presented at ICIP '98, Chicago, Illnois, Oct. 1998.

[24] K. Wilson, "Problems in physics with many scales of length," *Scientif. Am.* **241**, 158–179 (1979).

[25] J. Goodman and A. Sokal, "Multigrid Monte Carlo method. Conceptual foundations," *Phys. Rev. D* **40**, 2035–2071 (1989).

[26] W. Hackbusch, *Multi-Grid Methods and Applications* (Springer-Verlag, New York, 1985).

[27] A. George, *Computer Solution of Large Sparse Positive Definite Systems*, (Prentice-Hall, Englewood Cliffs, NJ, 1981).

[28] P. Roache, *Elliptic Marching Methods and Domain Decomposition*, (CRC Press, Boca Raton, FL, 1995).

[29] G. Winkler, *Image Analysis, Random Fields, and Dynamic Monte Carlo Methods: a Mathematical Introduction*, (Springer-Verlag, New York, 1995).

[30] D. Chandler, *Introduction to Modern Statistical Mechanics* (Oxford U. Press, New York, 1987).

[31] J. Besag, "Spatial interaction and the statistical analysis of lattice systems," *J. Roy. Stat. Soc. B* **36**, 192–226 (1974).

[32] J. Besag, "On the statistical analysis of dirty pictures," *J. Royal Stat. Soc. B* **48**, 259–302 (1986).

[33] J. Goutsias, "A theoretical analysis of Monte Carlo algorithms for the simulation of Gibbs random field images," *IEEE Inf. Theory* **37**, 1618–1628 (1991).

[34] R. Kindermann and J. L. Snell, *Markov Random Fields and Their Applications* (American Mathematical Society, Providence, RI, 1980).

[35] G. L. Bilbro and W. E. Snyder, "Applying of mean field annealing to image noise removal," *J. Neural Net. Comput.* **Fall**, 5–17 (1990).

[36] J. Zhang, "The mean field theory in EM procedures for Markov random fields," *IEEE Trans. Acoust. Speech Signal Process.* **40**, 2570–2583 (1992).

[37] N. Metropolis, A. W. Rosenbluth, M. N. Rosenbluth, A. H. Teller, and E. Teller, "Equations of state calculations by fast computing machines," *J. Chem. Phys.* **21**, 1087–1091 (1953).

[38] F.-C. Jeng and J. W. Woods, "Simulated annealing in compound Gauss-Markov random fields," *IEEE Trans. Inf. Theory* **36**, 94–107 (1990).

[39] I. Daubechies, *Ten Lectures on Wavelet* (SIAM, Philadelphia, 1992).

[40] H. Rauch, F. Tung, and C. Striebel, "Maximum likelihood estimates of linear dynamic systems," *AIAA J.* **3** (1965).

[41] P. Brodatz, *Textures; A Photographic Album for Artists and Designers* (Dover, New York, 1966).

[42] B. D. Ripley, *Pattern Recognition and Neural Networks* (Cambridge U. Press, New York, 1996).

[43] A. P. Dempster, N. M. Laird, and D. B. Rubin, "Maximum likelihood from incomplete data via the EM algorithm," *J. Roy. Soc. Stat. B* **1**, 1–38 (1977).

4.4

Image Modulation Models

J. P. Havlicek
University of Oklahoma

A. C. Bovik
The University of Texas at Austin

1 Introduction

In this chapter we describe image modulation models that may be used to represent a complicated image with spatially varying amplitude and frequency characteristics as a sum of joint amplitude-frequency modulated AM–FM components. Ideally, each AM–FM component has an instantaneous amplitude (AM function) and an instantaneous frequency (FM function) that are locally smooth but may contain substantial, wideband variations on a global scale. Intuitively, the AM function of a component may be interpreted as the instantaneous envelope, which carries image contrast information, while the FM function is the vector-valued derivative of the instantaneous phase and describes the local texture orientation and granularity.

For a given image, demodulation is concerned with computing estimates of the AM and FM functions for one or more components. In one-dimensional (1-D) cases, such computed modulations are used for time-frequency analysis and in the study of nonlinear air flow in human speech [1–4]. In two-dimensional (2-D) cases, the computed modulations provide a rich description of the local texture structure. They can be used for analysis [5,6], for texture segmentation and classification [7,8], for edge detection and image enhancement [9], for estimating three-dimensional (3-D) shape from texture [7,10], and for performing texture-based computational stereopsis [11]. Techniques for computing AM–FM image representations and for reconstructing an image from its computed representation have also emerged recently [12–14].

Although this article is primarily concerned with discrete 2-D techniques and algorithms, we temporarily focus on the 1-D case for simplicity in motivating the use of modulation models.

Consider the discrete-time 1-D chirp signal

$$f(k) = \cos\left(\frac{0.4\pi}{512}k^2\right), \tag{1}$$

defined for $0 \le k < 512$. The graph of this signal appears in Fig. 1(a). With the discrete Fourier transform (DFT), the signal can be represented as a weighted sum of 512 complex exponentials with frequencies uniformly spaced between −0.5 and +0.5 cycles per sample (cps). The magnitude of the DFT is shown in Fig. 1(b) and provides little intuition about the nature of the signal. In the Fourier representation, signal structure with time-varying frequency content is created by constructive and destructive interference between Fourier components that each have a constant frequency. Often, this interference can be both complicated and subtle. A modulation model for the signal of Eq. (1) computed with the algorithms described in [15] is shown in Figs. 1(c) and 1(d). In contrast to the DFT, the modulation model is both easy to interpret and intuitively appealing. It indicates that the signal is a single-component AM–FM function with constant amplitude and a frequency that increases linearly from DC to 0.4 cps. A similar modulation model could have been obtained using the discrete Teager–Kaiser Energy Operator and energy separation algorithm described in [1].

An interesting aspect of discrete modulation models is that they depend on the notion that the discrete signal being modeled was obtained, or at least in theory could have been obtained, by sampling a continuous-time signal. In cases in which the continuous signal does not actually exist, one may assume without loss of generality that the sampling was done with respect to a unity sampling interval. Thus, if we let $a(k)$ denote the AM

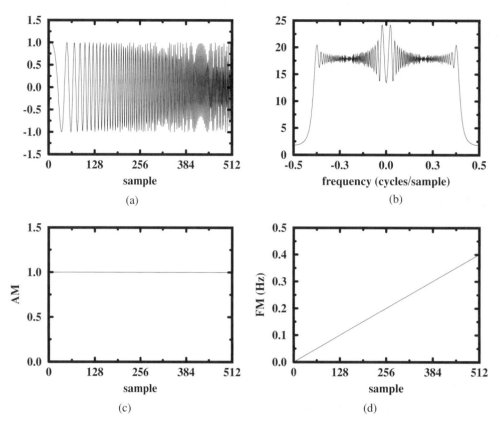

FIGURE 1   1-D chirp example: (a) signal, (b) DFT magnitude, (c) computed AM function, (d) computed FM function.

function in Fig. 1(c) and let $\dot{\varphi}(k)$ denote the FM function in Fig. 1(d), then we assume that these discrete modulating functions contain the samples of their continuous counterparts $a_c(t)$ and $\dot{\varphi}_c(t)$, where $\dot{\varphi}_c(t) = (\mathrm{d}/\mathrm{d}t)\varphi_c(t)$. We also assume that $\varphi(k)$ contains the samples of $\varphi_c(t)$. The relationship between the discrete signal of Eq. (1) and the computed modulation model of Figs. 1(c) and 1(d) is then given by

$$f(k) = a(k)\cos[\varphi(k)] = \mathrm{Re}\{a(k)\exp[\,j\varphi(k)]\}. \qquad (2)$$

In theory, any signal can be modeled as a single AM–FM function like the one appearing on the right-hand side of Eq. (2). There is a problem with doing this in practice, however. Single-component AM–FM models for the types of complicated signals that are often encountered in real-world applications generally require AM and FM functions that are not locally smooth. All AM–FM demodulation algorithms akin to those given in [1–3, 5, 7, 9, 12–16] are based on approximations of one form or another, and they can suffer from large approximation errors when the modulations are not locally smooth. Thus, while single-component models theoretically exist for complicated real-world signals, they generally cannot be computed. For this reason, it is preferable to model such signals as a sum of AM–FM components wherein each component has AM and FM functions that are locally smooth. Models of this type, which involve multiple

AM–FM components, are referred to as *multicomponent models*. Bandpass filtering, be it explicit or implicit as in [4], is generally used to isolate the multiple AM–FM components in the signal from one another on a pointwise basis prior to computation of the individual component modulating functions.

For a 2-D image with N rows and M columns, the DFT is a trivial multicomponent modulation model with NM AM–FM components that each have constant AM and FM functions (see Chapter 2.3 for a discussion of the 2-D DFT). The goal of general AM–FM modeling is to compute alternative modulation models involving fewer than NM components, where each component has spatially varying modulating functions that are locally smooth. In contrast to the DFT, such models provide a local characterization of the image texture structure. The dominant modulations can be extracted on a spatially local basis and used to solve a variety of classical machine vision problems, including texture segmentation, 3-D surface reconstruction, and stereopsis [7–11].

The organization of the article is as follows. In Section 2, we examine the discrete single-component demodulation problem in some detail. Demodulation algorithms based on the Teager–Kaiser energy operator and the complex-valued analytic image are presented in Sections 2.2 and 2.3, respectively. In Section 3, these algorithms are extended to the general multicomponent case. A technique called dominant component analysis for extracting the dominant modulations on a spatially local basis is described in Section 3.1, and the channelized components analysis paradigm for computing multicomponent AM–FM representations is presented in Section 3.2. Finally, conclusions appear in Section 4.

2 Single-Component Demodulation

In this section we describe demodulation algorithms applicable to an image that is modeled as a single AM–FM component. As we mentioned in Section 1, single-component modulation models are rarely appropriate for images encountered in real-world applications. Nevertheless, single-component demodulation techniques are important because they form the foundation upon which the multicomponent techniques to be presented in Section 3 are based. Suppose that $f(n_1, n_2)$ is an $N \times M$ image, where n_1 and n_2 are integer indices satisfying $0 \leq n_1 < N$ and $0 \leq n_2 < M$. Suppose further that $f(n_1, n_2)$ takes real floating point values. We model the image according to

$$f(n_1, n_2) = a(n_1, n_2) \cos[\varphi(n_1, n_2)], \qquad (3)$$

where we assume that $a(n_1, n_2) \geq 0$. With this assumption, the AM function $a(n_1, n_2)$ may be interpreted as the image contrast function. An example of the type of image that can be modeled well by Eq. (3) is the fingerprint image in Fig. 8 of Chapter 1.1.

We assume that $f(n_1, n_2)$ contains the samples of a continuous image

$$f_c(x, y) = a_c(x, y) \cos[\varphi_c(x, y)], \qquad (4)$$

where $a(n_1, n_2)$ and $\varphi(n_1, n_2)$ in Eq. (3) contain the samples of their continuous counterparts in Eq. (4). The instantaneous frequency of $f_c(x, y)$ is given by $\nabla \varphi_c(x, y)$. This quantity is a vector having components $\partial \varphi_c(x, y)/\partial x$ and $\partial \varphi_c(x, y)/\partial y$, which are referred to respectively as the horizontal and vertical (instantaneous) frequencies. By definition, $\nabla \varphi_c(x, y)$ is the FM function of $f_c(x, y)$ in Eq. (4). The FM function of $f(n_1, n_2)$ in (3) is $\nabla \varphi(n_1, n_2)$, which contains the samples of $\nabla \varphi_c(x, y)$. Given the image $f(n_1, n_2)$, the single-component demodulation problem is to compute estimated AM and FM functions $\hat{a}(n_1, n_2)$ and $\nabla \hat{\varphi}(n_1, n_2)$ such that $f(n_1, n_2) \approx \hat{a}(n_1, n_2) \cos[\hat{\varphi}(n_1, n_2)]$. As we shall see in Section 2.1, this problem does not have a unique solution. Henceforth, we will write $U_c(x, y)$ for the horizontal frequencies $\partial \varphi_c(x, y)/\partial x$ and $V_c(x, y)$ for the vertical frequencies $\partial \varphi_c(x, y)/\partial y$. The samples of these functions will be denoted $U(n_1, n_2)$ and $V(n_1, n_2)$.

2.1 Resolving Ambiguities in the Model

For any given image $f(n_1, n_2)$, there are infinitely many distinct pairs of functions $\{a(n_1, n_2), \varphi(n_1, n_2)\}$ that satisfy Eq. (3) exactly. As an extreme example, we could interpret the variations in $f(n_1, n_2)$ exclusively as frequency modulations

by setting $a(n_1, n_2) = \max |f(n_1, n_2)|$ and $\varphi(n_1, n_2) = \arccos[f(n_1, n_2)/a(n_1, n_2)]$. Equally extreme, we might take $a(n_1, n_2) = |f(n_1, n_2)|$ and $\varphi(n_1, n_2) = \arccos[\operatorname{sgn} f(n_1, n_2)]$, in which case we would interpret the variations in the image exclusively as amplitude modulations. In either case, $f(n_1, n_2) = a(n_1, n_2) \cos[\varphi(n_1, n_2)]$. Moreover, an infinite set of possible choices for $a(n_1, n_2)$ and $\varphi(n_1, n_2)$ exist between these two extremes. To disambiguate the demodulation problem, we consider two approaches. Both are based on the fact that demodulating a real-valued image is precisely equivalent to adding an imaginary part to create a complex-valued image. Indeed, for any complex-valued image $z(n_1, n_2)$, the modulating functions $a(n_1, n_2)$ and $\nabla \varphi(n_1, n_2)$ are unique.

The first approach is to demodulate $f(n_1, n_2)$ by specifying a well-defined algorithm that uses the real values of the image to calculate estimates of a particular pair of AM and FM functions. This approach is used by the demodulation algorithms described in Section 2.2. Any such technique that associates a particular $a(n_1, n_2)$ and $\nabla \varphi(n_1, n_2)$ with $f(n_1, n_2)$ implicitly specifies a complex image $z(n_1, n_2)$ with real part $f(n_1, n_2) = a(n_1, n_2) \cos[\varphi(n_1, n_2)]$ and imaginary part $g(n_1, n_2) = a(n_1, n_2) \sin[\varphi(n_1, n_2)]$. The second approach is to disambiguate the demodulation problem by specifying a well-defined algorithm that uses the real values of $f(n_1, n_2)$ to calculate a complex extension $z(n_1, n_2) = f(n_1, n_2) + jg(n_1, n_2)$. Estimates of $a(n_1, n_2)$ and $\nabla \varphi(n_1, n_2)$ can then be computed from $z(n_1, n_2)$. This latter approach is used by the demodulation algorithms described in Section 2.3.

2.2 Multidimensional Energy Separation

For a 1-D signal $f(k)$, the discrete Teager–Kaiser energy operator (TKEO) is defined by [17]

$$\Psi[f(k)] = f^2(k) - f(k+1) f(k-1). \qquad (5)$$

When applied to the pure cosine signal $f(k) = A \cos(w_0 k + \phi)$, the TKEO yields $\Psi[f(k)] = A^2 w_0^2$, a quantity that is proportional to the *energy* required to generate the displacement $f(k)$ in a mass-spring harmonic oscillator. For many locally smooth signals such as chirps, damped sinusoids, and human speech formants, the TKEO delivers

$$\Psi[f(k)] = \hat{a}^2(k) \dot{\hat{\varphi}}^2(k), \qquad (6)$$

where $\hat{a}(k)$ and $\dot{\hat{\varphi}}(k)$ are good estimates of an intuitively appealing and physically meaningful pair of modulations $\{a(k), \dot{\varphi}(k)\}$ satisfying $f(k) = a(k) \cos[\varphi(k)]$ [2,17]. The quantity $a^2(k) \dot{\varphi}^2(k)$ is known as the *Teager energy* of the signal $f(k)$. One-dimensional *energy separation algorithms* (ESAs) for obtaining estimates of the magnitudes of the individual AM and FM functions from the Teager energy were described in [1].

By applying Eq. (5) independently to the rows and columns of an image and summing the results, one obtains the 2-D discrete

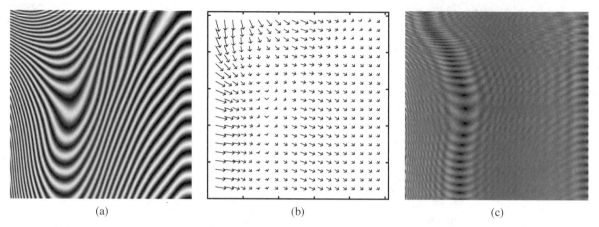

(a)   (b)   (c)

FIGURE 2   Energy separation example: (a) synthetic single-component AM–FM image; (b) estimated frequency vectors (FM function) obtained with the TKEO and ESA; (c) estimated AM function.

TKEO [5]:

$$\Phi\left[f(n_1, n_2)\right] = 2 f^2(n_1, n_2) - f(n_1 - 1, n_2) f(n_1 + 1, n_2)$$
$$- f(n_1, n_2 - 1) f(n_1, n_2 + 1). \tag{7}$$

For a particular pair of modulating functions $\{a(n_1, n_2), (n_1, n_2)\nabla\varphi\}$ satisfying $f(n_1, n_2) = a(n_1, n_2) \cos[\varphi(n_1, n_2)]$, the operator $\Phi\left[f(n_1, n_2)\right]$ approximates the multidimensional Teager energy $a^2(n_1, n_2)|\nabla\varphi(n_1, n_2)|^2$. For images that are reasonably locally smooth, the modulating functions selected by the 2-D TKEO are generally consistent with intuitive expectations [5]. With the TKEO, the magnitudes of the individual amplitude and frequency modulations can be estimated using the ESA [5]

$$|\hat{U}(n_1, n_2)| = \arcsin\left\{\frac{\Phi\left[f(n_1 + 1, n_2) - f(n_1 - 1, n_2)\right]}{4\Phi\left[f(n_1, n_2)\right]}\right\}^{1/2}, \tag{8}$$

$$|\hat{V}(n_1, n_2)| = \arcsin\left\{\frac{\Phi\left[f(n_1, n_2 + 1) - f(n_1, n_2 - 1)\right]}{4\Phi\left[f(n_1, n_2)\right]}\right\}^{1/2}, \tag{9}$$

$$\hat{a}(n_1, n_2) = |\hat{a}(n_1, n_2)|$$
$$= \left\{\frac{\Phi\left[f(n_1, n_2)\right]}{\sin^2(|\hat{U}(n_1, n_2)|) + \sin^2(|\hat{V}(n_1, n_2)|)}\right\}^{1/2}. \tag{10}$$

Algorithms (8)–(10) are straightforward to implement digitally, either in software or in hardware. Furthermore, these algorithms are well localized spatially, which makes them particularly suitable for implementation in sections or on a parallel computing engine. Two additional comments are in order. First, all three of these algorithms involve square root operations that are subject to failure if the corresponding TKEO outputs are negative at

some image pixels. Estimates of the modulating functions at such points can be obtained by simple spatial interpolation. Conditions for positivity of the energy operator were studied in [18]. Second, Eqs. (8) and (9) deliver estimates of the *magnitudes* of the horizontal and vertical frequencies. Thus some auxiliary technique must generally be used to determine the relative signs of $U(n_1, n_2)$ and $V(n_1, n_2)$, which embody local orientation in the image. One such technique will be examined in Section 3.

An example of applying the ESA of Eqs. (8)–(10) to a synthetic single-component AM–FM image is shown in Fig. 2. The original image, which is shown in Fig. 2(a), had 256×256 pixels, each taking an integer value in the range [0, 255]. Note that image model (3) is the product of a nonnegative AM function with a cosine that oscillates between -1 and $+1$. For accordance to be achieved with this model, it is generally necessary to rescale the pixel values so that the mean of the image is zero. Prior to application of the ESA, the image of Fig. 2(a) was converted to a zero-mean floating point image. The ESA was applied at every pixel, and edge effects were handled by replication.

The computed frequency estimates $\nabla\hat{\varphi}(n_1, n_2)$ are depicted in the needle diagram of Fig. 2(b), where one needle is displayed for each block of 12×12 pixels. Each needle points in the direction $\arctan[\hat{V}(n_1, n_2)/\hat{U}(n_1, n_2)]$, where the positive U axis points to the right and the positive V axis points down. With this convention, needles are normal to the corresponding wavefronts in the image. The lengths of the needles are proportional to the magnitudes of the instantaneous frequency vectors. The frequency vectors shown in Fig. 2(b) generally agree with our intuitive expectations, except for one notable exception. In the leftmost region of the image, many of the actual frequency vectors lie in the second or third quadrants of the U–V plane, where $U(n_1, n_2)$ and $V(n_1, n_2)$ have different signs. Because of the inability of the ESA to estimate signed frequency, the orientations of the estimated frequency vectors are incorrect in these instances.

The computed AM function $\hat{a}(n_1, n_2)$ is shown in Fig. 2(c), and it is nearly constant over much of the image. The large spikes visible in the outermost three rows and columns of Fig. 2(c) are edge effects that occur because the spatial support of the ESA is 5×5 pixels. Amplitude spikes appearing elsewhere in Fig. 2(c) are a consequence of approximation errors in the ESA. In fact, the raw floating point AM estimate delivered by Eq. (8) had a maximum exceeding 2×10^6, and was clipped for display. The main advantages to using the ESA are that it is computationally efficient and that it estimates the AM and FM functions from the values of the real image alone. The main disadvantage is the inability of the ESA to estimate signed frequency.

2.3 Demodulation by Complex Extension

In this section we describe a technique that estimates the AM and FM functions of a real-valued single-component image by first computing a complex-valued extension of the image, and then demodulating the complex image. The particular complex extension we use is called the *analytic image* [19]. The DFT of the discrete analytic image is equal to zero over half of the 2-D frequency plane. Over the other half of the frequency plane, it is equal to twice the DFT of the original real-valued image (except for four frequency samples that are identical to their counterparts in the DFT of the original image).

For an $N \times M$ real-valued image $f(n_1, n_2)$, the analytic image is given by $z(n_1, n_2) = f(n_1, n_2) + jg(n_1, n_2)$, where $g(n_1, n_2)$ is the discrete 2-D directional Hilbert transform of $f(n_1, n_2)$. If $\tilde{F}(u, v)$ is the DFT of $f(n_1, n_2)$, where u and v are integer indices, then the DFT of $g(n_1, n_2)$ is given by $\tilde{G}(u, v) = \mathcal{H}(u, v) \tilde{F}(u, v)$, where [19]

$$
\mathcal{H}(u, v) = \begin{cases} -j, & u = 1, 2, \ldots, \frac{N}{2} - 1 \\ j, & u = \frac{N}{2} + 1, \frac{N}{2} + 2, \ldots, N - 1 \\ -j, & u = 0, \ v = 1, 2, \ldots, \frac{M}{2} - 1 \\ -j, & u = \frac{N}{2}, \ v = 1, 2, \ldots, \frac{M}{2} - 1 \\ j, & u = 0, \ v = \frac{M}{2} + 1, \frac{M}{2} + 2, \ldots, M - 1 \\ j, & u = \frac{N}{2}, \ v = \frac{M}{2} + 1, \frac{M}{2} + 2, \ldots, M - 1 \\ 0, & \text{otherwise} \end{cases}
$$
(11)

Thus, $z(n_1, n_2)$ may be computed by the following straightforward procedure. First, the DFT is used to obtain $\tilde{F}(u, v)$ from $f(n_1, n_2)$. Second, $\tilde{G}(u, v)$ is computed by taking the pointwise product of $\tilde{F}(u, v)$ with $\mathcal{H}(u, v)$ as given in Eq. (11). Third, the DFT of $z(n_1, n_2)$ is computed according to $\tilde{Z}(u, v) = \tilde{F}(u, v) + j\tilde{G}(u, v)$. Finally, $z(n_1, n_2)$ is obtained by taking the inverse DFT of $\tilde{Z}(u, v)$. A more efficient algorithm for calculating $\tilde{Z}(u, v)$ may be derived by realizing that, for each u and each v, $\tilde{Z}(u, v)$ assumes one of only three possible values: zero, $2\tilde{F}(u, v)$, or $\tilde{F}(u, v)$. However, the details of this derivation are beyond the scope of the present chapter.

Once $z(n_1, n_2)$ has been calculated, the AM function $a(n_1, n_2)$ can easily be estimated by using the algorithm [16]

$$
\hat{a}(n_1, n_2) = |z(n_1, n_2)|. \tag{12}
$$

Estimates of the magnitudes and signs of the FM functions can then be obtained by using [16]

$$
|\hat{U}(n_1, n_2)| = \arccos\left[\frac{z(n_1 + 1, n_2) + z(n_1 - 1, n_2)}{2z(n_1, n_2)}\right],
$$
(13)

$$
\text{sgn}\,\hat{U}(n_1, n_2) = \text{sgn}\,\arcsin\left[\frac{z(n_1 + 1, n_2) - z(n_1 - 1, n_2)}{2jz(n_1, n_2)}\right],
$$
(14)

$$
|\hat{V}(n_1, n_2)| = \arccos\left[\frac{z(n_1, n_2 + 1) + z(n_1, n_2 - 1)}{2z(n_1, n_2)}\right],
$$
(15)

$$
\text{sgn}\,\hat{V}(n_1, n_2) = \text{sgn}\,\arcsin\left[\frac{z(n_1, n_2 + 1) - z(n_1, n_2 - 1)}{2jz(n_1, n_2)}\right].
$$
(16)

Like the ESA of Section 2.2, demodulation algorithms (12)–(16) are easily implemented in hardware or software and are well suited for implementation in sections or on a parallel processor. Two comments are in order concerning frequency algorithms (13)–(16). First, these algorithms cannot be applied at pixels where $z(n_1, n_2) = 0$. At such pixels, frequency estimates may be obtained by interpolating the frequency estimates from neighboring pixels. Second, the arguments of the transcendentals in Eqs. (13)–(16) are guaranteed to be real up to approximation errors; any nonzero imaginary component should be discarded prior to the evaluation of the arccos and arcsin functions.

Figure 3 shows an example in which the analytic image-based demodulation technique was applied to the synthetic single-component AM–FM image of Fig. 2(a), which appears again in Fig. 3(a). As before with the TKEO, the image was converted to floating point and normalized to have zero mean. Equation (11) was used to generate the complex-valued analytic image, and demodulation algorithms (12)–(16) were applied at every pixel, where edge effects were handled by replication.

The computed frequency estimates $\nabla\hat{\varphi}(n_1, n_2)$ are shown in the needle diagram of Fig. 3(b), where one needle is shown for each block of 12×12 pixels. These frequency estimates are generally in good agreement with those obtained with the ESA, as shown in Fig. 2(b). Note, however, that the estimated frequency vectors in the leftmost region of Fig. 3(b), which were obtained by using the signed frequency algorithms (13)–(16), agree with intuitive expectations and do not suffer from the orientation errors clearly visible in Fig. 2(b). The amplitude estimates $\hat{a}(n_1, n_2)$ delivered by Eq. (12) appear in Fig. 3(c), where no postprocessing was applied in this case. The main advantages of using the analytic image-based technique described in this section are that it is computationally efficient and that it estimates signed frequency. The main disadvantage is that the complex image $z(n_1, n_2)$ must be calculated explicitly.

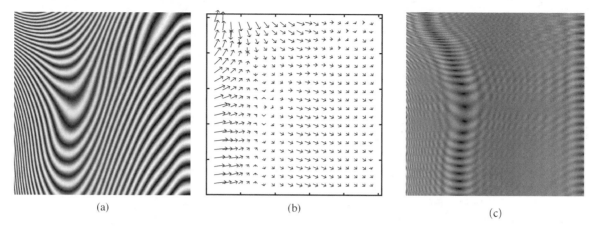

(a)   (b)   (c)

FIGURE 3   Demodulation example using an explicit complex extension: (a) synthetic single-component AM–FM image; (b) estimated frequency vectors (FM function); (c) estimated AM function.

3 Multicomponent Demodulation

In this section we will analyze real-valued images $f(n_1, n_2)$ against the multicomponent modulation model

$$f(n_1, n_2) = \sum_{q=1}^{Q} a_q(n_1, n_2) \cos[\varphi_q(n_1, n_2)] = \sum_{q=1}^{Q} f_q(n_1, n_2),$$
(17)

where $f_q(n_1, n_2) = a_q(n_1, n_2) \cos[\varphi_q(n_1, n_2)]$ is one of Q AM–FM image components. There are two main reasons for considering such multicomponent models. First, for many signals, intuitively satisfying and physically meaningful interpretations in terms of a single pair of amplitude and frequency modulation functions simply do not exist. Second, even in cases in which such a single-component interpretation does exist, there is no guarantee in general that the single-component modulating functions will be locally smooth. Thus, the single-component demodulation algorithms presented in Section 2 may suffer from large approximation errors that render the computed modulating function estimates meaningless. For many images of practical interest, however, it is possible to compute a multicomponent model wherein each individual component has modulating functions that are locally smooth almost everywhere.

For any given image $f(n_1, n_2)$, note that the componentwise decomposition indicated in Eq. (17) could be done in many different ways. Each different decomposition into components would yield different solutions for $\hat{a}_q(n_1, n_2)$ and $\nabla \hat{\varphi}_q(n_1, n_2)$, and would therefore lead to a different multicomponent interpretation of the image. One popular approach for estimating the modulating functions of the individual components is to pass the image $f(n_1, n_2)$ or its complex extension $z(n_1, n_2)$ through a bank of bandpass linear Gabor filters [1, 3, 5, 7, 10–12, 20]. This bank of filters, or *filterbank*, produces filter outputs that are similar to a wavelet decomposition using Gabor functions for the wavelet filters (see Chapter 4.1). Each filter in the filterbank is called a *channel*, and, for a given input image, each channel

produces a filtered output called the *channel response*. With this approach, the structure of the filterbank determines the multicomponent interpretation of the image. Provided that they are modified to account for the scaling effects incurred during filtering, the single-component demodulation algorithms presented in Section 2 can be applied directly to the channel responses to estimate the component modulating functions.

Suppose that $h_i(n_1, n_2)$ and $H_i(U, V)$ are, respectively, the unit pulse response and frequency response of a particular one of the filterbank channels. Under mild and realistic assumptions, one may show that, at pixels where the channel response $y_i(n_1, n_2)$ is dominated by a particular AM–FM component $f_q(n_1, n_2)$, the output of the TKEO is well approximated by [3]

$$\Phi[y_i(n_1, n_2)] \approx a_q^2(n_1, n_2)|\nabla \varphi_q(n_1, n_2)|^2 |H_i[\nabla \varphi_q(n_1, n_2)]|^2$$
$$= \Phi[f_q(n_1, n_2)]|H_i[\nabla \varphi_q(n_1, n_2)]|^2.$$
(18)

Note that the energy operator appears in both the numerators and denominators of Eqs. (8) and (9). If these frequency demodulation algorithms are applied to $y_i(n_1, n_2)$, then the scaling by $|H_i[\nabla \varphi_q(n_1, n_2)]|^2$ indicated in Eq. (18) is approximately canceled by division. Thus, Eqs. (8) and (9) may be applied directly to a channel response to estimate the FM function of the component that dominates that response at any given pixel.

Moreover, the multiband filtering provides a means of approximating the relative signs of the frequency estimates of Eqs. (8) and (9). Suppose that $|\hat{U}_q(n_1, n_2)|$ and $|\hat{V}_q(n_1, n_2)|$ are the magnitude frequency estimates obtained by demodulating $y_i(n_1, n_2)$. Since $h_i(n_1, n_2)$ will be real valued in this case, the frequency response $H_i(U, V)$ will be conjugate symmetric. Hence, the bandpass characteristic $|H_i(U, V)|$ will have two main lobes with center frequencies located either in quadrants one and three or in quadrants two and four of the 2-D frequency plane. If the signs of the horizontal and vertical components of these center frequencies agree, then we take sgn $\hat{U}_q(n_1, n_2)$ = sgn $\hat{V}_q(n_1, n_2)$ = $+1$. Otherwise, we set sgn $\hat{U}_q(n_1, n_2) = +1$ and sgn $\hat{V}_q(n_1, n_2) = -1$. This admittedly simplistic approach often works well enough to be effective in practical implementations.

For the demodulation algorithms described in Section 2.3, similar arguments can be used to establish the validity of applying frequency estimation algorithms (13)–(16) directly to the filterbank channel responses as well [14–16]. Note that the Hilbert transform described in Section 2.3 is a linear operator; therefore, the complex-valued analytic image $z(n_1, n_2)$ associated with $f(n_1, n_2)$ in Eq. (17) may be expressed as

$$z(n_1, n_2) = \sum_{q=1}^{Q} a_q(n_1, n_2) \exp[j\varphi_q(n_1, n_2)] = \sum_{q=1}^{Q} z_q(n_1, n_2).$$
(19)

Thus, when $z(n_1, n_2)$ is input to the filterbank, the channel responses are given by $y_i(n_1, n_2) = z(n_1, n_2) * h_i(n_1, n_2) \approx z_q(n_1, n_2) * h_i(n_1, n_2)$, where filterbank channel i is dominated by component $z_q(n_1, n_2)$. However, since the Hilbert transform is implemented by using spectral multiplier (11) and since 2-D multiband linear filtering is almost always implemented by using pointwise multiplication of DFTs, great computational savings can be realized by combining the linear filtering and analytic image generation into a single operation. If $\tilde{H}_i(u, v)$ is the DFT of $h_i(n_1, n_2)$, then the channel response $y_i(n_1, n_2)$ can be obtained by taking the inverse DFT of

$$\tilde{Y}_i(u, v) = \tilde{H}_i(u, v)\tilde{F}(u, v)[1 + j\mathcal{H}(u, v)].$$
(20)

Since half of the frequency samples in $\tilde{Z}(u, v)$ are identically zero, Eq. (20) actually saves half of the complex multiplies required to implement the convolution performed by each filterbank channel.

Unlike the frequency algorithms discussed above, the amplitude demodulation algorithms, Eqs. (10) and (12), require explicit modification before they can be applied directly to the filterbank channel responses. This is because the image components in models (17) and (19) are individually scaled as they pass through the filterbank. In particular, the amplitude estimates obtained by applying Eq. (10) or (12) to $y_i(n_1, n_2)$ must be divided by $|H_i[\nabla\hat{\varphi}_q(n_1, n_2)]|$, where $\nabla\hat{\varphi}_q(n_1, n_2)$ is the FM estimate obtained by performing frequency demodulation on $y_i(n_1, n_2)$. Thus, the modified amplitude estimation algorithms for the ESA-based approach and the analytic image-based approach are given by

Gabor filters are a common choice for the channel filters $H_i(U, V)$. These filters have Gaussian spectra that fall rapidly toward zero away from the center frequency. Consequently, moderate to severe approximation errors in the estimated frequencies $\hat{U}_q(n_1, n_2)$ and $\hat{V}_q(n_1, n_2)$ can cause the denominators of Eqs. (21) and (22) to approach zero. This often produces large-scale errors in the amplitude estimates and can lead to numerical instability in the amplitude estimation algorithms. Similar problems can also occur at pixels where the image $f(n_1, n_2)$ contains phase discontinuities. In the neighborhoods of such pixels, the FM functions $\nabla\varphi_q(n_1, n_2)$ may contain large-scale frequency excursions that lie far outside the filter passband. A popular approach for mitigating these effects is to postprocess the frequency estimates with a low-pass filter such as a Gaussian (see Chapter 4.4) or an order statistic filter such as a median filter (see Chapter 3.1) [1,13]. The smoothed frequency estimates can then be used in Eqs. (21) and (22). It is often beneficial to subsequently apply the same type of post processing to the amplitude estimates themselves.

3.1 Dominant Component Analysis

In this section, we describe a multicomponent computational technique called dominant component analysis, or DCA, which at every pixel delivers modulating function estimates $\hat{a}_D(n_1, n_2)$ and $\nabla\hat{\varphi}_D(n_1, n_2)$ corresponding to the AM–FM component that is locally dominant at that pixel [8, 14, 20]. The dominant frequency vectors $\nabla\hat{\varphi}_D(n_1, n_2)$ are often referred to as the *emergent frequencies* of the image. Generally, different components in sums (17) and (19) are expected to be dominant in different image regions. A block diagram of DCA is shown in Fig. 4. The real image $f(n_1, n_2)$ or the analytic image $z(n_1, n_2)$ is passed through a multiband linear filterbank. Demodulation algorithms (8), (9), and (21) or (13)–(16) and (22) are then applied to the response of every filterbank channel in the blocks labeled "DEMOD" in Fig. 4.

The dominant component at each pixel is defined as the one that dominates the response of the channel that maximizes a channel selection criterion $\Gamma_i(n_1, n_2)$. For the ESA-based and analytic image-based demodulation approaches, $\Gamma_i(n_1, n_2)$ is given respectively by

$$\Gamma_i(n_1, n_2) = \frac{\Phi[y_i(n_1, n_2)]}{|\nabla\hat{\varphi}_q(n_1, n_2)|^2 \max_{U,V} |H_i(U, V)|^2},$$
(23)

$$\hat{a}_q(n_1, n_2) = \frac{\sqrt{\Phi[y_i(n_1, n_2)]}}{|H_i[\hat{U}_q(n_1, n_2), \hat{V}_q(n_1, n_2)]|\sqrt{\sin^2(|\hat{U}_q(n_1, n_2)|) + \sin^2(|\hat{V}_q(n_1, n_2)|)}},$$
(21)

$$\hat{a}_q(n_1, n_2) = \frac{|y_i(n_1, n_2)|}{|H_i[\hat{U}_q(n_1, n_2), \hat{V}_q(n_1, n_2)]|},$$
(22)

respectively.

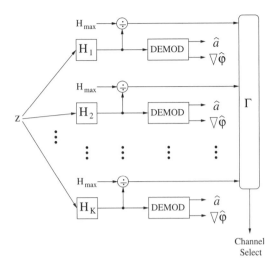

FIGURE 4   Block diagram of DCA.

$$\Gamma_i(n_1, n_2) = \frac{|y_i(n_1, n_2)|}{\max_{U, V} |H_i(U, V)|}, \qquad (24)$$

where $\nabla\hat{\varphi}_q(n_1, n_2)$ in Eq. (23) is the frequency estimate obtained by demodulating $y_i(n_1, n_2)$. Modulating function estimates

$\hat{a}_D(n_1, n_2)$ and $\nabla\hat{\varphi}_D(n_1, n_2)$ are extracted from the channel that maximizes $\Gamma_i(n_1, n_2)$ on a pointwise basis.

The dominant modulations provide a rich description of the local texture structure of the image, and, as we pointed out in Section 1, they can be used for a number of important applications including texture segmentation, 3-D surface reconstruction, and stereopsis [7–11]. An example of DCA is shown in Fig. 5. The ESA-based and analytic image-based demodulation algorithms were both applied to the 256×256 texture image *Tree* shown in Fig. 5(a). Prior to analysis, the image was converted to a zero-mean floating point image. A bank of 43 Gabor filters was used to isolate components from one another, as depicted in the 2-D frequency plane in Fig. 5(d). Detailed descriptions of this filterbank are given in [7] and [20]. Since most natural images are dominated by low frequencies that describe large-scale shading and contrast variations rather than local texture features, the response of the baseband filter appearing at the center of Fig. 5(d) was not considered in the dominant component analysis. Also, for the ESA-based approach, the imaginary components of the channel filter unit pulse responses were set to zero, producing frequency responses that were both real valued and even symmetric. For postprocessing, median filters of sizes 5×5 and 7×7 pixels were applied to the frequency and amplitude estimates of

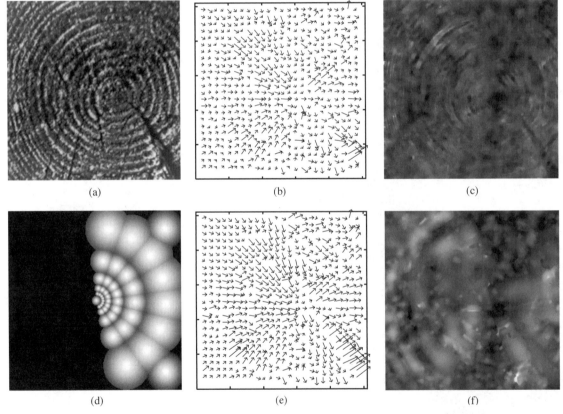

FIGURE 5   DCA example. (a) Texture image *Tree*. (b), (c) Dominant FM function $\nabla\hat{\varphi}_D(n_1, n_2)$ and AM function $\hat{a}_D(n_1, n_2)$ estimated by the TKEO and ESA. (d) Frequency response of multiband Gabor filterbank; for DCA, the baseband channel was not used. (e), (f) Dominant FM function and AM function estimated by the analytic image-based approach.

the ESA, respectively. The modulating functions estimated by the analytic image-based approach were smoothed with low-pass Gaussian filters having linear bandwidths identical to the corresponding channel filters.

The dominant amplitude and frequency modulations estimated by the ESA are shown in Figs. 5(b) and 5(c), while those obtained using Eqs. (13)–(16) and (22) are shown in Figs. 5(e) and 5(f). Although the interpretations delivered by the two approaches are different, they are in good qualitative agreement. The lengths of the needles in Figs. 5(b) and 5(e) are inversely proportional to the magnitudes of the dominant frequency vectors, so that longer needles correspond to larger features in the image, while shorter needles correspond to smaller features. Additional nonlinear scaling has been applied for display to accentuate the differences between the highest and lowest frequencies. Note that the relative signs of the frequency vectors delivered by the ESA in Fig. 5(b) have been corrected by setting them equal to the relative signs of the appropriate channel filter center frequencies.

Figure 6 shows three examples illustrating how the dominant modulations computed by DCA can be used to perform texture segmentation. The 256×256 image *Paper-Burlap* shown in Fig. 6(a) was created by removing the central region from one texture image and replacing it with the corresponding region from another image. DCA was applied to this image, and the computed dominant component AM function $\hat{a}_D(n_1, n_2)$ is shown in Fig. 6(b). A Laplacian-of-Gaussian (LoG) edge detection filter with space constant $\sigma = 46.5$ pixels was applied to the dominant AM image. The resulting edge map contained only one closed contour. This contour, which effectively segments the image, is shown overlaid on the original image in Fig. 6(c).

DCA was also applied to the *Mica-Burlap* image shown in Fig. 6(d). In this case, a LoG edge detector with gradient magnitude thresholding and a space constant of $\sigma = 15$ pixels was applied to the emergent frequency magnitudes $|\nabla \hat{\varphi}_D(n_1, n_2)|$, which are displayed as a gray-scale image in Fig. 6(e). The threshold value was adjusted until the resulting edge map contained only one closed contour, which is shown overlaid on the original image in Fig. 6(f). Finally, the *Wood-Wood* image of Fig. 6(g) was obtained by rotating the central portion of a texture image counterclockwise by 45°. The emergent frequency orientations delivered by DCA are shown as a gray-scale image in Fig. 6(h). The contour that is shown overlaid on the image in Fig. 6(i) was obtained by applying a LoG edge detector with space constant $\sigma = 14$ pixels to the emergent frequency orientations and adjusting the gradient magnitude threshold value until only a single closed contour remained.

3.2 Channelized Component Analysis

One of the most exciting emerging application areas of multidimensional AM–FM modeling lies in the development of modulation domain image representations, which are similar in many respects to 2-D time-frequency distributions. While the DFT is a trivial example of such a representation, the objective of more general AM–FM representations is to capture the essential structure of an image using a relatively small number of components by allowing each component to have spatially varying but locally smooth amplitude and frequency modulations.

Channelized components analysis, or CCA, is perhaps the simplest approach for computing general AM–FM image representations [13, 14, 20]. In CCA, the image is passed through a bank of bandpass filters such as the one depicted in Fig. 5(d). The componentwise decomposition of the image is carried out by assuming that the filterbank isolates components on a global scale, so that demodulating each channel response delivers modulating function estimates for one component in sums (17) and (19). Thus, CCA representations provide a dense description characterizing not only the dominant image structures, but subtle subemergent texture features as well.

Under the assumption that each filterbank channel is globally dominated by a single AM–FM component, a CCA image representation computed using the filterbank of Fig. 5(d) will necessarily comprise 43 components. Since the Gabor filters that are frequently used for the filterbank are not orthogonal, such representations are not invertible in general. In fact, adjacent filters in Fig. 5(d) intersect at frequencies where each is at precisely half-peak. Nevertheless, reasonably high-quality image reconstructions can often be obtained by substituting the modulating function estimates in a computed CCA representation back into models (17) and (19).

Three examples of CCA image representations are presented in Fig. 7. The original images *Peppers*, *Salesman*, and *Mandrill* appear in Figs. 7(a)–7(c), respectively. Each was converted to a floating point zero-mean complex-valued analytic image and passed through the 43-channel Gabor filterbank of Fig. 5(d). Demodulation algorithms (13)–(16) and (22) were applied to each channel response to compute modulating function estimates for a single AM–FM image component. Gaussian postfilters were used to smooth the estimated frequencies prior to application of Eq. (22), and the amplitude estimates were then postfiltered. The linear bandwidth of each postfilter was identical to that of the corresponding channel filter.

For each CCA component, phase reconstruction was performed by using the simple algorithm [12]

$$\hat{\varphi}_q(n_1, n_2) = \frac{1}{2}[\hat{\varphi}_q(n_1 - 1, n_2) + \hat{U}_q(n_1 - 1, n_2)$$
$$+ \hat{\varphi}_q(n_1, n_2 - 1) + \hat{V}_q(n_1, n_2 - 1)]. \quad (25)$$

In order to reduce the propagation of frequency estimation errors, phase reconstruction was carried out independently on blocks of 4×4 pixels. Within each block, Eq. (25) was initialized by saving the phase of the pixel located in the upper left-hand corner of the block. This approach is straightforward to implement, since Gabor filters have real-valued spectra. Thus, for any given channel, the phase of the channel response is equal to the phase of the input image component that dominates that channel. When this approach is used to reconstruct the phase of an image block

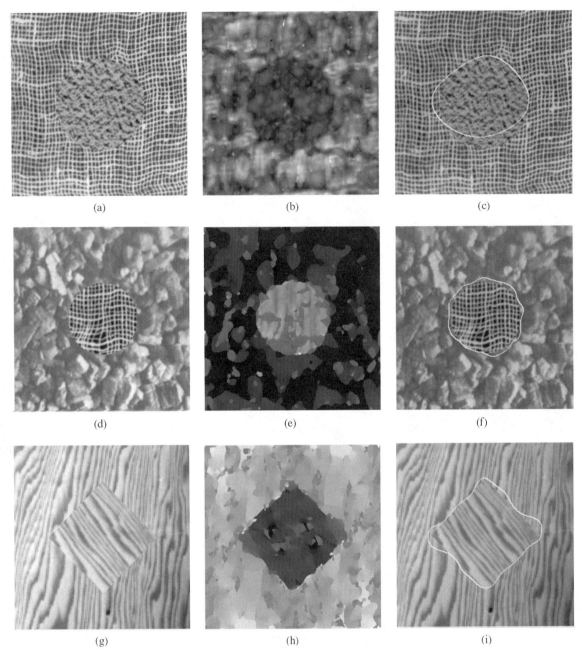

FIGURE 6   DCA texture segmentation examples: (a) *Paper-Burlap* image; (b) estimated dominant AM function; (c) segmentation obtained by applying a LoG edge detector to the image in (b); (d) *Mica-Burlap* image; (e) magnitude of estimated dominant FM function; (f) segmentation obtained by applying a LoG edge detector to the image in (e); (g) *Wood-Wood* image; (h) orientations of estimated dominant frequency vectors; (i) segmentation obtained by applying a LoG edge detector to the image in (h).

independently from the other blocks, note that Eq. (25) cannot be applied on the top row and leftmost column of the block. Instead, Eq. (25) should be replaced by $\hat{\varphi}_q(n_1, n_2) = \hat{\varphi}_q(n_1 - 1, n_2) + \hat{U}_q(n_1 - 1, n_2)$ along the top row of each block. Similarly, the equation $\hat{\varphi}_q(n_1, n_2) = \hat{\varphi}_q(n_1, n_2 - 1) + \hat{V}_q(n_1, n_2 - 1)$ should be used along the first column of each block.

Subsequent to phase reconstruction, the amplitude and phase estimates for the channelized components of the images in Figs. 7(a)–7(c) were substituted into Eq. (19) to obtain the image reconstructions shown in Figs. 7(d)–7(f). In each case, the reconstructions are of remarkably high quality for such a small number of AM–FM components. Reconstructions of one individual

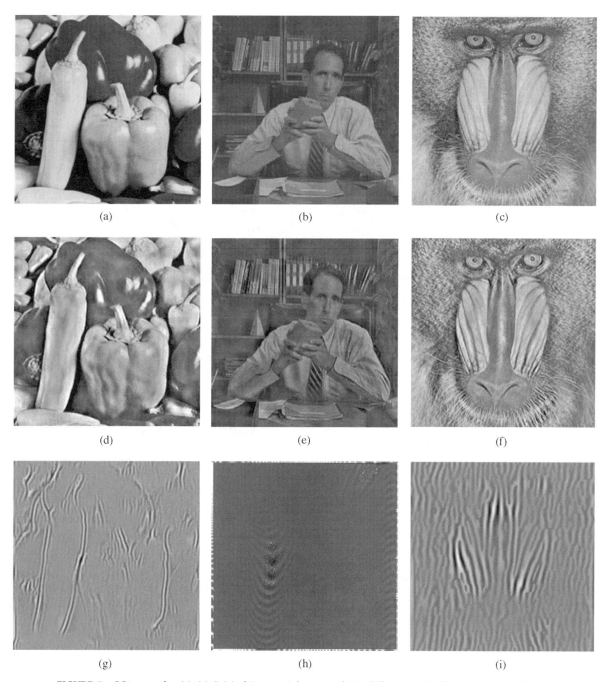

FIGURE 7 CCA examples: (a)–(c) Original *Peppers*, *Salesman*, and *Mandrill* images; (d)–(f) 43-component CCA reconstructions; (g)–(i) reconstructions of one channelized component from each image.

channelized AM–FM component from each image are shown in Figs. 7(g)–7(i).

4 Conclusion

In this article we have presented two recently developed approaches for demodulating images modeled as sums of AM–FM functions having spatially varying but locally smooth amplitude and frequency modulations. Computed estimates of the component modulating functions can be used with great success in a wide variety of applications, including analysis, image enhancement, edge detection, segmentation and classification, shape from texture, and stereopsis.

The Teager–Kaiser energy operator and its associated energy separation algorithm operate on the real values of the image alone to estimate a unique pair of modulating functions for each image component, while the approach based on the analytic image estimates the component modulating functions from an explicit complex extension of the image. Although the modulating functions delivered by the ESA also implicitly determine a unique complex-valued image, the imaginary components of these two complex images generally differ. However, for a given filterbank used to effect the separation into components, the interpretations delivered by the two approaches are often in close agreement for image components that are reasonably locally smooth. Perhaps the most notable difference between the ESA and the analytic image-based approach is that the latter estimates the magnitudes and relative signs of the horizontal and vertical frequencies, whereas the ESA estimates only the frequency magnitudes. Thus, the ESA must generally be supplemented with auxiliary techniques to estimate the relative signs of the horizontal and vertical frequencies, which characterize the local texture orientation in an image.

Multidimensional modulation modeling is a relatively new area, and a veritable wealth of open problems remain to be investigated. Particularly exciting among these are the design of new efficient, quasi-invertible multicomponent AM–FM image representations and the development of general theories for image processing in the modulation domain

References

[1] P. Maragos, J. F. Kaiser, and T. F. Quatieri, "Energy separation in signal modulations with applications to speech analysis," *IEEE Trans. Signal Proc.* **41**, 3024–3051 (1993).

[2] P. Maragos, J. F. Kaiser, and T. F. Quatieri, "On amplitude and frequency demodulation using energy operators," *IEEE Trans. Signal Proc.* **41**, 1532–1550 (1993).

[3] A. C. Bovik, P. Maragos, and T. F. Quatieri, "AM–FM energy detection and separation in noise using multiband energy operators," *IEEE. Trans. Signal Proc.* **41**, 3245–3265 (1993).

[4] S. Lu and P. C. Doerschuk, "Nonlinear modeling and processing of speech based on sums of AM–FM formant models," *IEEE Trans. Signal Proc.* **44**, 773–782 (1996).

[5] P. Maragos and A. C. Bovik, "Image demodulation using multidimensional energy separation," *J. Opt. Soc. Am. A* **12**, 1867–1876 (1995).

[6] B. Friedlander and J. M. Francos, "An estimation algorithm for 2-D polynomial phase signals," *IEEE Trans. Image Proc.* **5**, 1084–1087 (1996).

[7] A. C. Bovik, N. Gopal, T. Emmoth, and A. Restrepo, "Localized measurement of emergent image frequencies by Gabor wavelets," *IEEE. Trans. Inf. Theory* **38**, 691–712 (1992).

[8] J. P. Havlicek, "The evolution of modern texture processing," *Elektrik, Turk. J. Electric. Eng. Comput. Sci.* **5**, 1–28 (1997).

[9] S. K. Mitra, S. Thurnhofer, M. Lightstone, and N. Strobel, "Two-dimensional Teager operators and their image processing applications," in *Proceedings of the 1995 IEEE Workshop on Nonlinear Signal and Image Processing* (IEEE, New York, 1995), pp. 959–962.

[10] B. J. Super and A. C. Bovik, "Shape from texture using local spectral moments," *IEEE. Trans. Pattern Anal. Machine Intell.* **17**, 333–343 (1995).

[11] T. Y. Chen, A. C. Bovik, and L. K. Cormack, "Stereoscopic ranging by matching image modulations," *IEEE Trans. Image Proc.* **8**, 785–797 (1999).

[12] J. P. Havlicek, D. S. Harding, and A. C. Bovik, "The multicomponent AM–FM image representation," *IEEE Trans. Image Proc.* **5**, 1094–1100 (1996).

[13] J. P. Havlicek, D. S. Harding, and A. C. Bovik, "Extracting essential modulated image structure," in *Proceedings of the 30th IEEE Asilomar Conference on Signals, Systems, and Computers* (IEEE, New York, 1996), pp. 1014–1018.

[14] J. P. Havlicek, D. S. Harding, and A. C. Bovik, "Multidimensional quasi-eigenfunction approximations and multicomponent AM-FM models," *IEEE Trans. Image Proc.* **9**, (to appear Feb. 2000).

[15] A. C. Bovik, J. P. Havlicek, D. S. Harding, and M. D. Desai, "Limits on discrete modulated signals," *IEEE Trans. Signal Proc.* **45**, 867–879 (1997).

[16] J. P. Havlicek, D. S. Harding, and A. C. Bovik, "Discrete quasi-eigenfunction approximation for AM–FM image analysis," in *Proceedings of the IEEE International Conference on Image Processing*, (IEEE, New York, 1996), pp. 633–636.

[17] J. F. Kaiser, "On a simple algorithm to calculate the 'energy' of a signal," in *Proceedings of the IEEE International Conference on Acoustics, Speech, and Signal Processing*, (IEEE, New York, 1990), pp. 381–384.

[18] A. C. Bovik and P. Maragos, "Conditions for positivity of an energy operator," *IEEE Trans. Signal Process* **42**, 469–471 (1994).

[19] J. P. Havlicek, J. W. Havlicek, and A. C. Bovik, "The analytic image," in *Proceedings of the IEEE International Conference on Image Processing*, (IEEE, New York, 1997).

[20] J. P. Havlicek, A. C. Bovik, and D. Chen, "AM-FM image modeling and Gabor analysis," in *Visual Information Representation, Communication, and Image Processing*, C. W. Chen and Y. Zhang, eds. (Marcel Dekker, New York, 1999), pp. 343–385.

4.5

Image Noise Models

Charles Boncelet
University of Delaware

1 Introduction

This chapter reviews some of the more commonly used *image noise models*. Some of these are naturally occurring, e.g., Gaussian noise; some are sensor induced, e.g., photon counting noise and speckle; and some result from various processing, e.g., quantization and transmission.

2 Preliminaries

2.1 What Is Noise?

Just what is noise, anyway? Somewhat imprecisely, we will define *noise* as an unwanted component of the image. Noise occurs in images for many reasons. Gaussian noise is a part of almost any signal. For example, the familiar white noise on a weak television station is well modeled as Gaussian. Since image sensors must count photons — especially in low light situations — and the number of photons counted is a random quantity, images often have photon counting noise. The grain noise in photographic films is sometimes modeled as Gaussian and sometimes as Poisson. Many images are corrupted by salt and pepper noise, as if someone had sprinkled black and white dots on the image. Other noises include quantization noise and speckle in coherent light situations.

Let $f(\cdot)$ denote an image. We will decompose the image into a desired component, $g(\cdot)$, and a noise component, $q(\cdot)$. The most common decomposition is *additive*:

$$f(\cdot) = g(\cdot) + q(\cdot) \tag{1}$$

For instance, Gaussian noise is usually considered to be an additive component.

The second most common is *multiplicative:*

$$f(\cdot) = g(\cdot)q(\cdot). \tag{2}$$

An example of a noise often modeled as multiplicative is speckle.

Note, the multiplicative model can be transformed into the additive model by taking logarithms and the additive model into the multiplicative one by exponentiation. For instance, Eq. (1) becomes

$$e^{f} = e^{g+q} = e^{f}e^{q}. \tag{3}$$

Similarly, Eq. (2) becomes

$$\log f = \log(gq) = \log g + \log q. \tag{4}$$

If the two models can be transformed into one another, what is the point? Why do we bother? The answer is that we are looking for *simple* models that properly describe the behavior of the system. The additive model, Eq. (1), is most appropriate when the noise *in that model* is independent of f. There are many applications of the additive model. Thermal noise, photographic noise, and quantization noise, for instance, obey the additive model well.

The multiplicative model is most appropriate when the noise in that model is independent of f. One common situation in

FIGURE 1 Original picture of the San Francisco skyline.

which the multiplicative model is used is for speckle in coherent imagery.

Finally, there important situations in which neither the additive nor the multiplicative model fits the noise well. Poisson counting noise and salt and pepper noise fit neither model well.

The questions about noise models one might ask include the following: What are the properties of $\mathbf{q}(\cdot)$? Is $\mathbf{q}$ related to g or are they independent? Can $\mathbf{q}(\cdot)$ be eliminated or at least mitigated? As we will see in this chapter and in others, it is only occasionally true that $\mathbf{q}(\cdot)$ will be independent of $g(\cdot)$. Furthermore, it is usually impossible to remove all the effects of the noise.

Figure 1 is a picture of the San Francisco, California skyline. It will be used throughout this chapter to illustrate the effects of various noises. The image is 432×512, 8 bits per pixel, gray scale. The largest value (the whitest pixel) is 220 and the minimum value is 32. This image is relatively noise free with sharp edges and clear details.

2.2 Notions of Probability

The various noises considered in this chapter are random in nature. Their exact values are random variables whose values are best described by using probabilistic notions. In this section, we will review some of the basic ideas of probability. A fuller treatment can be found in many texts on probability and randomness, including Feller [6], Billingsley [2], and Woodroofe [16].

Let $\mathbf{a} \in R^n$ be an n-dimensional *random vector* and $a \in R^n$ be a point. Then the *distribution function* of $\mathbf{a}$ (also known as the cumulative distribution function) will be denoted as $P_\mathbf{a}(a) = \mathrm{Pr}(\mathbf{a} \leq a)$ and the corresponding *density function* as $p_\mathbf{a}(a) = dP_\mathbf{a}(a)/da$. Probabilities of events will be denoted as $\mathrm{Pr}(A)$.

The *expected value* of a function, $\psi(\mathbf{a})$, is

$$E\psi(\mathbf{a}) = \int_{-\infty}^{\infty} \psi(a)\, p_\mathbf{a}(a)\, da. \tag{5}$$

Note for discrete distributions the integral is replaced by the corresponding sum:

$$E\psi(\mathbf{a}) = \sum_k \psi(a_k)\, \mathrm{Pr}(\mathbf{a} = a_k). \tag{6}$$

The *mean* is $\mu_\mathbf{a} = E\mathbf{a}$ (i.e., $\psi(\mathbf{a}) = \mathbf{a}$), the *variance* of a single random variable is $\sigma_\mathbf{a}^2 = E(\mathbf{a} - \mu_\mathbf{a})^2$, and the *covariance matrix* of a random vector $\Sigma_\mathbf{a} = E(\mathbf{a} - \mu_\mathbf{a})(\mathbf{a} - \mu_\mathbf{a})^T$.

Related to the covariance matrix is the *correlation matrix*,

$$\mathbf{R}_\mathbf{a} = E\mathbf{a}\mathbf{a}^T. \tag{7}$$

The various moments are related by the well-known relation, $\Sigma = \mathbf{R} - \mu\mu^T$.

The *characteristic function*, $\Phi_\mathbf{a}(u) = E(\exp(ju\mathbf{a}))$, has two main uses in analyzing probabilistic systems: calculating moments and calculating the properties of sums of independent random variables. For calculating moments, consider the power series of $\exp(ju\mathbf{a})$:

$$e^{ju\mathbf{a}} = 1 + ju\mathbf{a} + \frac{(ju\mathbf{a})^2}{2!} + \frac{(ju\mathbf{a})^3}{3!} + \cdots . \tag{8}$$

After taking expected values, one finds

$$E e^{ju\mathbf{a}} = 1 + juE\mathbf{a} + \frac{(ju)^2 E\mathbf{a}^2}{2!} + \frac{(ju)^3 E\mathbf{a}^3}{3!} + \cdots . \tag{9}$$

One can isolate the kth moment by taking k derivatives with respect to u and then setting $u = 0$:

$$E\mathbf{a}^k = \frac{1}{j^k} \frac{dE e^{ju\mathbf{a}}}{du}\bigg|_{u=0}. \tag{10}$$

Consider two independent random variables, $\mathbf{a}$ and $\mathbf{b}$, and their sum $\mathbf{c}$. Then,

$$\Phi_\mathbf{c}(u) = E e^{ju(\mathbf{c})}, \tag{11}$$

$$= E e^{ju(\mathbf{a}+\mathbf{b})}, \tag{12}$$

$$= E e^{ju\mathbf{a}} e^{ju\mathbf{b}}, \tag{13}$$

$$= E e^{ju\mathbf{a}} E e^{ju\mathbf{b}}, \tag{14}$$

$$= \Phi_\mathbf{a}(u)\Phi_\mathbf{b}(u), \tag{15}$$

where Eq. (14) used the independence of $\mathbf{a}$ and $\mathbf{b}$. Since the characteristic function is the (complex conjugate of the) Fourier transform of the density, the density of $\mathbf{c}$ is easily calculated by taking an inverse Fourier transform of $\Phi_\mathbf{c}(u)$.

3 Elements of Estimation Theory

As we said in the Introduction, noise is generally an *unwanted* component in an image. In this section, we review some of the techniques to eliminate — or at least minimize — the noise.

The basic estimation problem is to find a good estimate of the noise-free image, g, given the noisy image, $\mathbf{f}$. Some authors refer to this as an *estimation problem*, whereas others say it is a *filtering* problem. Let the estimate be denoted $\hat{g} = \hat{g}(\mathbf{f})$. The most common performance criterion is the mean-squared error (MSE):

$$\mathrm{MSE}(g, \hat{\mathbf{g}}) = E(g - \hat{\mathbf{g}})^2 \tag{16}$$

The estimator that minimizes the MSE is called the *minimum mean-squared error estimator* (MMSE). Many authors prefer to measure the performance in a positive way using the *peak signal-to-noise ratio* (PSNR) measured in dB:

$$\mathrm{PSNR} = 10 \log_{10} \left(\frac{\max^2}{\mathrm{MSE}} \right) \tag{17}$$

where max is the maximum pixel value, e.g., 255 for 8-bit images.

Although the MSE is the most common error criterion, it is by no means the only one. Many researchers argue that MSE results are not well correlated with the human visual system. For instance, the mean absolute error (MAE) is often used in motion compensation in video compression. Nevertheless, MSE has the advantages of easy tractablility and intuitive appeal since MSE can be interpreted as "noise power."

Estimators can be classified in many different ways. The primary division we will consider here is into linear versus nonlinear estimators.

The *linear estimators* form estimates by taking linear combinations of the sample values. For example, consider a small region of an image modeled as a constant value plus additive noise:

$$\mathbf{f}(x, y) = \mu + \mathbf{q}(x, y). \tag{18}$$

A linear estimate of μ is

$$\hat{\mu} = \sum_{x, y} \alpha(x, y) \mathbf{f}(x, y), \tag{19}$$

$$= \mu \sum_{x, y} \alpha(x, y) + \sum_{x, y} \alpha(x, y) \mathbf{q}(x, y). \tag{20}$$

An estimator is called *unbiased* if $E(\mu - \hat{\mu}) = 0$. In this case, assuming $E\mathbf{q} = 0$, unbiasedness requires $\sum_{x,y} \alpha(x, y) = 1$. If the $\mathbf{q}(x, y)$ are independent and identically distributed (i.i.d.), meaning that the random variables are independent and each has the same distribution function, then the MMSE for this example is the sample mean:

$$\hat{\mu} = \frac{1}{M} \sum_{(x, y)} \mathbf{f}(x, y), \tag{21}$$

where M is the number of samples averaged over.

Linear estimators in image filtering get more complicated primarily for two reasons: First, the noise may not be i.i.d., and, second and more commonly, the noise-free image is not well modeled as a constant. If the noise-free image is Gaussian and the noise is Gaussian, then the optimal estimator is the well-known Weiner filter [10].

In many image filtering applications, linear filters do not perform well. Images are not well modeled as Gaussian, and linear filters are not optimal. In particular, images have small details and sharp edges. These are blurred by linear filters. It is often true that the filtered image is more objectionable than the original. The blurriness is worse than the noise.

Largely because of the blurring problems of linear filters, nonlinear filters have been widely studied in image filtering. While there are many classes of nonlinear filters, we will concentrate on the class based on order statistics. Many of these filters were invented to solve image processing problems.

Order statistics are the result of sorting the observations from smallest to largest. Consider an image *window* (a small piece of an image) centered on the image to be estimated. Some windows are square, some are "**X**" shaped, some are "+" shaped, and some more oddly shaped. The choice of a window size and shape is usually up to the practitioner. Let the samples in the window be denoted simply as $\mathbf{f}_i$ for $i = 1, \ldots, N$. The order statistics are denoted $\mathbf{f}_{(i)}$ for $i = 1, \ldots, N$ and obey the ordering $\mathbf{f}_{(1)} \leq \mathbf{f}_{(2)} \leq \cdots \leq \mathbf{f}_{(N)}$.

The simplest order statistic based estimator is the sample median, $\mathbf{f}_{((N+1)/2)}$. For example, if $N = 9$, the median is $\mathbf{f}_{(5)}$. The median has some interesting properties. Its value is one of the samples. The median tends to blur images much less than the mean. The median can pass an edge without any blurring at all.

Some other order statistic estimators are the following.

Linear combinations of order statistics, $\hat{\mu} = \sum_{i=1}^{N} \alpha_i \mathbf{f}_{(i)}$: The α_i determine the behavior of the filter. In some cases, the coefficients can be determined optimally; see Lloyd [14] and Bovik *et al.* [5].

Weighted medians and the LUM filter: Another way to weight the samples is to repeat certain samples more than once before the data are sorted. The most common situation is to repeat the center sample more than once. The center weighted median does "less filtering" than the ordinary median and is suitable when the noise is not too severe. (See Salt and Pepper noise below.) The LUM filter [9] is a rearrangement of the center weighted median. It has the advantages of being easy to understand and extensible to image sharpening applications.

Iterated and recursive forms: The various filtering operations can be combined or iterated upon. One might first filter horizontally, then vertically. One might compute the outputs of three or more filters and then use "majority rule" techniques to choose between them.

To analyze or optimally design order statistics filters, we need descriptions of the probability distributions of the order statistics. Initially, we will assume the $\mathbf{f}_i$ are i.i.d. Then the $\Pr(\mathbf{f}_{(i)} \le x)$ equals the probability that at least i of the $\mathbf{f}_i$ are less than or equal to x. Thus,

$$\Pr(\mathbf{f}_{(i)} \le x) = \sum_{k=i}^{N} \binom{N}{k} (P_{\mathbf{f}}(x))^k (1 - P_{\mathbf{f}}(x))^{N-k}. \quad (22)$$

We see immediately that the order statistic probabilities are related to the binomial distribution.

Unfortunately, Eq. (22) does not hold when the observations are not i.i.d. In the special case in which the observations are independent (or Markov), but not identically distributed, there are simple recursive formulas to calculate the probabilities [3,4]. For example, even if the additive noise in Eq. (1) is i.i.d, the image may not be constant throughout the window. One may be interested in how much blurring of an edge is done by a particular order statistics filter.

4 Types of Noise and Where They Might Occur

In this section, we present some of the more common image noise models and show sample images illustrating the various degradations.

4.1 Gaussian Noise

Probably the most frequently occurring noise is additive Gaussian noise. It is widely used to model thermal noise and, under some often reasonable conditions, is the limiting behavior of other noises, e.g., photon counting noise and film grain noise. Gaussian noise is used in many places in this book.

The density function of univariate Gaussian noise, $\mathbf{q}$, with mean μ and variance σ^2 is

$$p_{\mathbf{q}}(x) = (2\pi\sigma^2)^{-1/2} e^{-(x-\mu)^2/2\sigma^2} \quad (23)$$

for $-\infty < x < \infty$. Notice that the support, which is the range of values of x where the probability density is nonzero, is infinite in both the positive and negative directions. But, if we regard an image as an intensity map, then the values must be nonnegative. In other words, the noise cannot be strictly Gaussian. If it were, there would be some nonzero probability of having negative values. In practice, however, the range of values of the Gaussian noise is limited to approximately $\pm 3\sigma$ and the Gaussian density is a useful and accurate model for many processes. If necessary, the noise values can be truncated to keep $f > 0$.

In situations in which $\mathbf{a}$ is a random vector, the multivariate Gaussian density becomes

$$p_{\mathbf{a}}(a) = (2\pi)^{-n/2} |\Sigma|^{-1/2} e^{-(a-\mu)^T \Sigma^{-1} (a-\mu)/2}, \quad (24)$$

where $\mu = E\mathbf{a}$ is the mean vector and $\Sigma = E(\mathbf{a} - \mu)(\mathbf{a} - \mu)^T$

is the covariance matrix. We will use the notation $\mathbf{a} \sim N(\mu, \Sigma)$ to denote that $\mathbf{a}$ is Gaussian (also known as *normal*) with mean μ and covariance Σ.

The Gaussian characteristic function is also Gaussian in shape:

$$\Phi_{\mathbf{a}}(u) = e^{u^T \mu - u^T \Sigma u/2}. \quad (25)$$

The Gaussian distribution has many convenient mathematical properties — and some not so convenient ones. Certainly the least convenient property of the Gaussian distribution is that the cumulative distribution function cannot be expressed in closed form by using elementary functions. However, it is tabulated numerically. See almost any text on probability, e.g., [15].

Linear operations on Gaussian random variables yield Gaussian random variables. Let $\mathbf{a}$ be $N(\mu, \Sigma)$ and $\mathbf{b} = G\mathbf{a} + h$. Then a straightforward calculation of $\Phi_{\mathbf{b}}(u)$ yields

$$\Phi_{\mathbf{b}}(u) = e^{ju^T (G\mu+h) - u^T G\Sigma G^T u/2}, \quad (26)$$

which is the characteristic function of a Gaussian random variable with mean, $G\mu + h$, and covariance, $G\Sigma G^T$.

Perhaps the most significant property of the Gaussian distribution is called the Central Limit Theorem, which states that the distribution of a sum of a large number of independent, small random variables has a Gaussian distribution. Note the individual random variables do not have to have a Gaussian distribution themselves, nor do they even have to have the same distribution. For a detailed development, see, e.g., Feller [6] or Billingsley [2]. A few comments are in order.

1. There must be a large number of random variables that contribute to the sum. For instance, thermal noise is the result of the thermal vibrations of an astronomically large number of tiny electrons.
2. The individual random variables in the sum must be independent, or nearly so.
3. Each term in the sum must be small, negligible compared to the sum.

As one example, thermal noise results from the vibrations of a very large number of electrons, the vibration of any one electron is independent of that of another, and no one electron contributes significantly more than the others. Thus, all three conditions are satisfied and the noise is well modeled as Gaussian. Similarly, binomial probabilities approach the Gaussian. A binomial random variable is the sum of N independent Bernoulli (0 or 1) random variables. As N gets large, the distribution of the sum approaches a Gaussian distribution.

In Fig. 2 we see the effect of a small amount of Gaussian noise ($\sigma = 10$). Notice the "fuzziness" overall. It is often counterproductive to try to use signal processing techniques to remove this level of noise; the filtered image is usually visually less pleasing than the original noisy one.

In Fig. 3, the noise has been increased by a factor of 3 ($\sigma = 30$). The degradation is much more objectionable. Various filtering

FIGURE 2 San Francisco image corrupted by additive Gaussian noise, with the standard deviation equal to 10.

techniques can improve the quality, though usually at the expense of some loss of sharpness.

4.2 Heavy-Tailed Noises

In many situations, the conditions of the Central Limit Theorem are almost, but not quite, true. There may not be a large enough number of terms in the sum, or the terms may not be sufficiently independent, or a small number of the terms may contribute a disproportionate amount to the sum. In these cases, the noise may only be approximately Gaussian. One should be careful.

FIGURE 3 San Francisco image corrupted by additive Gaussian noise, with the standard deviation equal to 30.

TABLE 1 Comparison of tail probabilities for the Gaussian and Double Exponential distributions[a]

x_0	Gaussian	Double Exponential
1	0.32	0.37
2	0.046	0.14
3	0.0027	0.050

[a]Specifically, the values of $\Pr(|\mathbf{x}| > x_0)$ are listed for both distributions.

Even when the center of the density is approximately Gaussian, the tails may not be.

The *tails* of a distribution are the areas of the density corresponding to large x, i.e., as $|x| \to \infty$. A particularly interesting case is that in which the noise has *heavy tails*. "Heavy tails" means that for large values of x, the density, $p_{\mathbf{a}}(x)$, approaches 0 more slowly than the Gaussian. For example, for large values of x, the Gaussian density goes to 0 as $\exp(-x^2/2\sigma^2)$; the double exponential density (described below) goes to 0 as $\exp(-|x|/\sigma)$. The double exponential density is said to have heavy tails.

In Table 1, we present the tail probabilities, $\Pr(|\mathbf{x}| > x_0)$, for the Gaussian and double exponential distributions (both with mean 0 and variance 1). Note the probability of exceeding 1 is approximately the same for both distributions, while the probability of exceeding 3 is ~20 times greater for the double exponential than for the Gaussian.

An interesting example of heavy tailed noise that should be familiar is static on a weak, broadcast AM radio station during a lightning storm. Most of the time, the conditions of the Central Limit Theorem are well satisfied and the noise is Gaussian. Occasionally, however, there may be a lightning bolt. The lightning bolt overwhelms the tiny electrons and dominates the sum. During the time period of the lightning bolt, the noise is non-Gaussian and has much heavier tails than the Gaussian.

Some of the heavy-tailed models that arise in image processing include the following.

Double exponential:

$$p_{\mathbf{a}}(x) = \frac{1}{2\sigma} e^{-|x-\mu|/\sigma}. \tag{27}$$

The mean is μ and the variance σ^2. The double exponential is interesting in that the best estimate of μ is the median, not the mean, of the observations.

Negative exponential:

$$p_{\mathbf{a}}(x) = \frac{1}{\mu} e^{-x/\mu} \tag{28}$$

for $x > 0$ and $E\mathbf{a} = \mu > 0$ and variance, μ^2. The negative exponential is used to model speckle, for example, in SAR systems (Chapter 10.1).

Alpha stable: In this class, appropriately normalized sums of independent and identically distributed random variables have the same distribution as the individual random variables. We have already seen that sums of Gaussian random variables are Gaussian, so the Gaussian is in the class of alpha-stable distributions. In general, these distributions have characteristic functions that look like $\exp(-|u|^\alpha)$ for $0 < \alpha \leq 2$. Unfortunately, except for the Gaussian ($\alpha = 2$) and the Cauchy ($\alpha = 1$), it is not possible to write the density functions of these distributions in closed form.

As $\alpha \to 0$, these distributions have very heavy tails. Gaussian mixture models:

$$p_\mathbf{a}(x) = (1 - \alpha) p_0(x) + \alpha p_1(x), \tag{29}$$

where $p_0(x)$ and $p_1(x)$ are Gaussian densities with differing means, μ_0 and μ_1, or variances, σ_0^2 and σ_1^2. In modeling heavy-tailed distributions, it is often true that α is small, say $\alpha = 0.05$, $\mu_0 = \mu_1$, and $\sigma_1^2 \gg \sigma_0^2$.

In the "static in the AM radio" example above, at any given time, α would be the probability of a lightning strike, σ_0^2 the average variance of the thermal noise, and σ_1^2 the variance of the lightning induced signal. Sometimes this model is generalized further and $p_1(x)$ is allowed to be non-Gaussian (and sometimes completely arbitrary). See Huber [11].

One should be careful to use estimators that behave well in heavy-tailed noise. The sample mean, optimal for a constant signal in additive Gaussian noise, can perform quite poorly in heavy-tailed noise. Better choices are those estimators designed to be robust against the occasional outlier [11]. For instance, the median is only slightly worse than the mean in Gaussian noise, but can be much better in heavy-tailed noise.

4.3 Salt and Pepper Noise

Salt and pepper noise refers to a wide variety of processes that result in the same basic image degradation: only a few pixels are noisy, but they are *very* noisy. The effect is similar to sprinkling white and black dots — salt and pepper — on the image.

One example where salt and pepper noise arises is in transmitting images over noisy digital links. Let each pixel be quantized to B bits in the usual fashion. The value of the pixel can be written as $X = \sum_{i=0}^{B-1} b_i 2^i$. Assume the channel is a binary symmetric one with a crossover probability of ϵ. Then each bit is flipped with probability ϵ. Call the received value Y. Then

$$\Pr(|X - Y| = 2^i) = \epsilon \tag{30}$$

for $i = 0, 1, \ldots, B - 1$. The MSE due to the most significant bit is $\epsilon 4^{B-1}$ compared to $\epsilon(4^{B-1} - 1)/3$ for all the other bits combined. In other words, the contribution to the MSE from the most significant bit is approximately 3 times that of all the other bits. The pixels whose most significant bits are changed will likely appear as black or white dots.

Salt and pepper noise is an example of (very) heavy-tailed noise. A simple model is the following. Let $f(x, y)$ be the original

FIGURE 4 San Francisco image corrupted by salt and pepper noise, with a probability of occurrence of 0.05.

image and $\mathbf{q}(x, y)$ be the image after it has been altered by salt and pepper noise:

$$\Pr(\mathbf{q} = f) = 1 - \alpha, \tag{31}$$

$$\Pr(\mathbf{q} = \max) = \alpha/2, \tag{32}$$

$$\Pr(\mathbf{q} = \min) = \alpha/2, \tag{33}$$

where max and min are the maximum and minimum image values, respectively. For 8-bit images, min = 0 and max = 255. The idea is that with probability $1 - \alpha$ the pixels are unaltered; with probability α the pixels are changed to the largest or smallest values. The altered pixels look like black and white dots sprinkled over the image.

Figure 4 shows the effect of salt and pepper noise. Approximately 5% of the pixels have been set to black or white (95% are unchanged). Notice the sprinkling of the black and white dots. Salt and pepper noise is easily removed with various order statistic filters, especially the center weighted median and the LUM filter [1].

Salt and pepper noise appears in Chapter 3.2.

4.4 Quantization and Uniform Noise

Quantization noise results when a continuous random variable is converted to a discrete one or when a discrete random variable is converted to one with fewer levels. In images, quantization noise often occurs in the acquisition process. The image may be continuous initially, but to be processed it must be converted to a digital representation.

As we shall see, quantization noise is usually modeled as uniform. Various researchers use uniform noise to model other impairments, e.g., dither signals. Uniform noise is the opposite of the heavy-tailed noises just discussed. Its tails are very light (zero!).

Let $\mathbf{b} = Q(\mathbf{a}) = \mathbf{a} + \mathbf{q}$, where $-\Delta/2 \leq \mathbf{q} \leq \Delta/2$ is the quantization noise and $\mathbf{b}$ is a discrete random variable usually represented with β bits. In the case in which the number of quantization levels is large (so Δ is small), $\mathbf{q}$ is usually modeled as being uniform between $-\Delta/2$ and $\Delta/2$ and independent of $\mathbf{a}$. The mean and variance of $\mathbf{q}$ are

$$E\mathbf{q} = \frac{1}{\Delta} \int_{-\Delta/2}^{\Delta/2} s\,ds = 0, \tag{34}$$

$$E(\mathbf{q} - E\mathbf{q})^2 = \frac{1}{\Delta} \int_{-\Delta/2}^{\Delta/2} s^2\,ds = \Delta^2/12. \tag{35}$$

Since $\Delta \sim 2^{-\beta}$, $\sigma_\nu^2 \sim 2^{2\beta}$. The signal-to-noise ratio increases by 6 dB for each additional bit in the quantizer.

When the number of quantization levels is small, the quantization noise becomes signal dependent. In an image of the noise, signal features can be discerned. Also, the noise is correlated on a pixel-by-pixel basis and is not uniformly distributed.

The general appearance of an image with too few quantization levels may be described as "scalloped." Fine graduations in intensities are lost. There are large areas of constant color separated by clear boundaries. The effect is similar to transforming a smooth ramp into a set of discrete steps.

In Fig. 5, the San Francisco image has been quantized to only 4 bits. Note the clear "stair stepping" in the sky. The previously smooth gradations have been replaced by large constant regions separated by noticeable discontinuities.

FIGURE 5   San Francisco image quantized to 4 bits.

4.5 Photon Counting Noise

Fundamentally, most image acquisition devices are photon counters. Let $\mathbf{a}$ denote the number of photons counted at some location (a pixel) in an image. Then, the distribution of $\mathbf{a}$ is usually modeled as Poisson with parameter, λ. This noise is also called *Poisson noise* or Poisson counting noise. Poisson noise in the human visual system is discussed in Chapter 1.2.

$$P(\mathbf{a} = k) = e^{-\lambda}\lambda^k/k! \tag{36}$$

for $k = 0, 1, 2, \dots$.

The Poisson distribution is one for which calculating moments by using the characteristic function is much easier than by the usual sum.

$$\Phi(u) = \sum_{k=0}^{\infty} \frac{e^{juk}e^{-\lambda}\lambda^k}{k!}, \tag{37}$$

$$= e^{-\lambda} \sum_{k=0}^{\infty} \frac{(\lambda e^{ju})^k}{k!}, \tag{38}$$

$$= e^{-\lambda}e^{\lambda e^{ju}}, \tag{39}$$

$$= e^{\lambda(e^{ju}-1)}. \tag{40}$$

Although this characteristic function does not *look* simple, it does yield the moments:

$$E\mathbf{a} = \frac{1}{j}\frac{\mathrm{d}}{\mathrm{d}u}e^{\lambda(e^{ju}-1)}\bigg|_{u=0}, \tag{41}$$

$$= \frac{1}{j}\lambda j e^{ju}e^{\lambda(e^{ju}-1)}\bigg|_{u=0}, \tag{42}$$

$$= \lambda. \tag{43}$$

Similarly, $E\mathbf{a}^2 = \lambda + \lambda^2$ and $\sigma^2 = (\lambda + \lambda^2) - \lambda^2 = \lambda$. We see one of the most interesting properties of the Poisson distribution, that the variance is equal to the expected value.

Consider two different regions of an image, one brighter than the other. The brighter one has a higher λ and therefore a higher noise variance.

As another example of Poisson counting noise, consider the following.

Example: Effect of Shutter Speed on Image Quality   Consider two pictures of the same scene, one taken with a shutter speed of 1 unit time and the other with $\Delta > 1$ units of time. Assume that an area of an image emits photons at the rate λ per unit time. The first camera measures a random number of photons, whose expected value is λ and whose variance is also λ. The second, however, has an expected value and variance equal to $\lambda\Delta$. When time averaged (divided by Δ), the second now has an expected value of λ and a variance of $\lambda/\Delta < \lambda$. Thus, we are led to the

FIGURE 6   San Francisco image corrupted by Poisson noise.

intuitive conclusion: all other things being equal, slower shutter speeds yield better pictures.

For example, astrophotographers traditionally used long exposures to average over a long enough time to get good photographs of faint celestial objects. Today's astronomers use CCD arrays and average many short photographs, but the principal is the same.

Figure 6 shows the image with Poisson noise. It was constructed by taking each pixel value in the original image and generating a Poisson random variable with λ equal to that value. Careful examination reveals that the white areas are noisier than the dark areas. Also, compare this image with Fig. 2, which shows Gaussian noise of almost the same power.

4.6 Photographic Grain Noise

Photographic grain noise is a characteristic of photographic films. It limits the effective magnification one can obtain from a photograph. A simple model of the photography process is as follows:

A photographic film is made up from millions of tiny *grains*. When light strikes the film, some of the grains absorb the photons and some do not. The ones that do change their appearance by becoming metallic silver. In the developing process, the unchanged grains are washed away.

We will make two simplifying assumptions: (1) the grains are uniform in size and character and (2) the probability that a grain changes is proportional to the number of photons incident upon it. Both assumptions can be relaxed, but the basic answer is the same. In addition, we will assume the grains are independent of each other.

Slow film has a large number of small fine grains, whereas fast film has a smaller number of larger grains. The small grains give slow film a better, less grainy picture; the large grains in fast film cause a grainier picture.

In a given area, A, assume there are L grains, with the probability of each grain changing, equal to p. Then the number of grains that change, N, is binomial:

$$\Pr(\mathbf{N} = k) = \binom{L}{k} p^k (1 - p)^{L-k}. \tag{44}$$

Since L is large, when p small but $\lambda = Np = E\mathbf{N}$ moderate, this probability is well approximated by a Poisson distribution

$$\Pr(\mathbf{N} = k) = \frac{e^{-\lambda}\lambda^k}{k!}, \tag{45}$$

and by a Gaussian when p is larger:

$$\Pr(k \leq \mathbf{N} < k + \Delta_k)$$
$$= \Pr\left(\frac{k - Lp}{\sqrt{Lp(1 - p)}} \leq \frac{\mathbf{N} - Lp}{\sqrt{Lp(1 - p)}} \leq \frac{k + \Delta_k - Lp}{\sqrt{Lp(1 - p)}}\right), \tag{46}$$

$$\approx (2\pi Lp(1 - p))^{-1/2} e^{-0.5\left(\frac{k - Lp}{Lp(1 - p)}\right)^2} \Delta_k. \tag{47}$$

The probability interval on the right-hand side of Eq. (46) is exactly the same as that on the left except that it has been normalized by subtracting the mean and dividing by the standard deviation. Equation (47) results from Eq. (46) by an application of the Central Limit Theorem. In other words, the distribution of grains that change is approximately Gaussian with mean Lp and variance $Lp(1 - p)$. This variance is maximized when $p = 0.5$. Sometimes, however, it is sufficiently accurate to ignore this variation and model grain noise as additive Gaussian with a constant noise power.

4.7 Speckle in Coherent Light Imaging

Speckle is one of the more complex image noise models. It is signal dependent, non-Gaussian, and spatially dependent. Much of this discussion is taken from [8, 12]. We will first discuss the origins of speckle, then derive the first-order density of speckle, and conclude this section with a discussion of the second-order properties of speckle.

In coherent light imaging, an object is illuminated by a coherent source, usually a laser or a radar transmitter. For the remainder of this discussion, we will consider the illuminant to be a light source, e.g., a laser, but the principles apply to radar imaging as well.

When coherent light strikes a surface, it is reflected back. Because of the microscopic variations in the surface roughness within one pixel, the received signal is subjected to random variations in phase and amplitude. Some of these variations in phase add constructively, resulting in strong intensities, and others add deconstructively, resulting in low intensities. This variation is called *speckle*.

Of crucial importance in the understanding of speckle is the point-spread function of the optical system. There are three regimes.

- The point-spread function is so narrow that the individual variations in surface roughness can be resolved. The reflections off the surface are random (if, indeed, we can model the surface roughness as random in this regime), but we cannot appeal to the central limit theorem to argue that the reflected signal amplitudes are Gaussian. Since this case is uncommon in most applications, we will ignore it further.
- The point-spread function is broad compared to the feature size of the surface roughness, but small compared to the features of interest in the image. This is a common case and leads to the conclusion, presented below, that the noise is exponentially distributed and uncorrelated on the scale of the features in the image. Also, in this situation, the noise is often modeled as multiplicative.
- The point-spread function is broad compared to both the feature size of the object and the feature size of the surface roughness. Here, the speckle is correlated and its size distribution is interesting and is determined by the point-spread function.

The development will proceed in two parts. First we will derive the first-order probability density of speckle and, second we will discuss the correlation properties of speckle.

In any given macroscopic area, there are many microscopic variations in the surface roughness. Rather than trying to characterize the surface, we will content ourselves with finding a statistical description of the speckle.

We will make the (standard) assumptions that the surface is very rough on the scale of the optical wavelengths. This roughness means that each microscopic reflector in the surface is at a random height (distance from the observer) and a random orientation with respect to the incoming polarization field. These random reflectors introduce random changes in the reflected signal's amplitude, phase, and polarization. Further, we assume these variations at any given point are independent from each other and independent from the changes at any other point.

These assumptions amount to assuming that the system cannot resolve the variations in roughness. This is generally true in optical systems, but may not be so in some radar applications.

The above assumptions on the physics of the situation can be translated to statistical equivalents: the amplitude of the reflected signal at any point, (x, y), is multiplied by a random amplitude, denoted $\mathbf{a}(x, y)$, and the polarization, $\boldsymbol{\phi}(x, y)$, is uniformly distributed between 0 and 2π.

Let $\mathbf{u}(x, y)$ be the complex phasor of the incident wave at a point (x, y), $\mathbf{v}(x, y)$ be the reflected signal, and $\mathbf{w}(x, y)$ be the received phasor. From the above assumptions,

$$\mathbf{v}(x, y) = \mathbf{u}(x, y)\mathbf{a}(x, y)e^{j\boldsymbol{\phi}(x, y)}, \tag{48}$$

and, letting $h(\cdot, \cdot)$ denote the two-dimensional point-spread function of the optical system,

$$\mathbf{w}(x, y) = h(x, y) * \mathbf{v}(x, y). \tag{49}$$

One can convert the phasors to rectangular coordinates:

$$\mathbf{v}(x, y) = \mathbf{v}_R(x, y) + j\mathbf{v}_I(x, y), \tag{50}$$

$$\mathbf{w}(x, y) = \mathbf{w}_R(x, y) + j\mathbf{w}_I(x, y). \tag{51}$$

Since the change in polarization is uniform between 0 and 2π, $\mathbf{v}_R(x, y)$ and $\mathbf{v}_I(x, y)$ are statistically independent. Similarly, $\mathbf{w}_R(x, y)$ and $\mathbf{w}_I(x, y)$ are statistically independent. Thus,

$$\mathbf{w}_R(x, y) = \int_{-\infty}^{\infty} \int_{-\infty}^{\infty} h(\alpha, \beta) v_R(x - \alpha, y - \beta) \, d\alpha \, d\beta, \tag{52}$$

and similarly for $\mathbf{w}_I(x, y)$.

The integral in Eq. (52) is basically a sum over many tiny increments in x and y. By assumption, the increments are independent of one another. Thus, we can appeal to the Central Limit Theorem and conclude that the distributions of $\mathbf{w}_R(x, y)$ and $\mathbf{w}_I(x, y)$ are each Gaussian with mean 0 and variance σ^2. Note, this conclusion does not depend on the details of the roughness, as long as the surface is rough on the scale of the wavelength of the incident light and the optical system cannot resolve the individual components of the surface.

The measured intensity, $\mathbf{F}(x, y)$, is the squared magnitude of the received phasors:

$$\mathbf{F}(x, y) = \mathbf{w}_R(x, y)^2 + \mathbf{w}_I(x, y)^2. \tag{53}$$

The distribution of $\mathbf{F}$ can be found by integrating the joint density of $\mathbf{w}_R$ and $\mathbf{w}_I$ over a circle of radius $f^{0.5}$:

$$\Pr(\mathbf{F}(x, y) \le f) = \int_0^{2\pi} \int_0^{f^{0.5}} \frac{1}{2\pi\sigma^2} e^{-\rho/2\sigma^2} \rho \, d\rho \, d\phi, \tag{54}$$

$$= 1 - e^{-f/2\sigma^2}. \tag{55}$$

The corresponding density is $p_{\mathbf{f}}(f)$:

$$p_{\mathbf{f}}(f) = \begin{cases} \frac{1}{g} e^{-f/g} & f \ge 0 \\ 0 & f < 0 \end{cases}, \tag{56}$$

where we have taken the liberty to introduce the mean intensity, $g = g(x, y) = 2\sigma^2(x, y)$. A little rearrangement can put this into a multiplicative noise model:

$$\mathbf{f}(x, y) = g(x, y)\mathbf{q}, \tag{57}$$

where $\mathbf{q}$ has a exponential density

$$p_{\mathbf{q}}(x) = \begin{cases} e^{-x} & x \geq 0 \\ 0 & x < 0 \end{cases}. \tag{58}$$

The mean of $\mathbf{q}$ is 0 and the variance is 1.

The exponential density is much heavier tailed than the Gaussian density, meaning that much greater excursions from the mean occur. In particular, the standard deviation of $\mathbf{f}$ equals $E\mathbf{f}$, i.e., the typical deviation in the reflected intensity is equal to the typical intensity. It is this large variation that causes speckle to be so objectionable to human observers.

It is sometimes possible to obtain multiple images of the same scene with independent realizations of the speckle pattern, i.e., the speckle in any one image is independent of the speckle in the others. For instance, there may be multiple lasers illuminating the same object from different angles or with different optical frequencies. One means of speckle reduction is to average these images:

$$\hat{\mathbf{F}}(x, y) = \frac{1}{M} \sum_{i=1}^{M} \mathbf{F}_i(x, y), \tag{59}$$

$$= g(x, y) \frac{\sum_{i=1}^{M} \mathbf{q}_i(x, y)}{M}. \tag{60}$$

Now, the average of the negative exponentials has mean 1 (the same as each individual negative exponential) and variance $1/M$. Thus, the average of the speckle images has a mean equal to $g(x, y)$ and variance $g(x, y)/M$.

Figure 7 shows an uncorrelated speckle image of San Francisco. Notice how severely degraded this image is. Careful

FIGURE 8   San Francisco image with correlated speckle.

examination will show that the light areas are noisier than the dark areas. This image was created by generating an "image" of exponential variates and multiplying each by the corresponding pixel value. Intensity values beyond 255 were truncated to 255.

See also Fig. 4(b) of Chapter 1.1 for an example of a SAR image with speckle.

The correlation structure of speckle is largely determined by the width of the point-spread function. As above, the real and imaginary components (or, equivalently, the X and Y components) of the reflected wave are independent Gaussian. These components ($\mathbf{w_R}$ and $\mathbf{w_I}$ above) are individually filtered by the point-spread function of the imaging system. The intensity image is formed by taking the complex magnitude of the resulting filtered components.

Figure 8 shows a correlated speckle image of San Francisco. The image was created by filtering $\mathbf{w_R}$ and $\mathbf{w_I}$ with a 2-D square filter of size 5×5. This size filter is too big for the fine details in the original image, but it is convenient to illustrate the correlated speckle. As above, intensity values beyond 255 were truncated to 255. Notice the correlated structure to the "speckles." The image has a pebbly appearance.

We will conclude this discussion with a quote from Goodman [7]:

> The general conclusions to be drawn from these arguments are that, in any speckle pattern, large-scale-size fluctuations are the most populous, and no scale sizes are present beyond a certain small-size cutoff. The distribution of scale sizes in between these limits depends on the autocorrelation function of the object geometry, or on the autocorrelation function of the pupil function of the imaging system in the imaging geometry.

FIGURE 7   San Francisco image with uncorrelated speckle.

4.8 Atmospheric Speckle

The twinkling of stars is similar in cause to speckle in coherent light but has important differences. Averaging multiple frames of independent coherent imaging speckle results in an image estimate whose mean equals the underlying image and whose variance is reduced by the number of frames averaged over. However, averaging multiple images of twinkling stars results in a blurry image of the star.

From the Earth, stars (except the Sun!) are point sources. Their light is spatially coherent and planar when it reaches the atmosphere. Because of thermal and other variations, the diffusive properties of the atmosphere changes in an irregular way. This causes the index of refraction to change randomly. The star appears to twinkle. If one averages multiple images of the star, one obtains a blurry image.

Until recent years, the preferred way to eliminate atmospheric induced speckle (the "twinkling") was to move the observer to a location outside the atmosphere, i.e., in space. In recent years, new techniques to estimate and track the fluctuations in atmospheric conditions have allowed astronomers to take excellent pictures from the earth. One class is called "speckle interferometry" [13]. It uses multiple short duration (typically less than 1 s each) images and a nearby star to estimate the random speckle pattern. Once estimated, the speckle pattern can be removed, leaving the unblurred image.

5 Conclusions

In this chapter, we have tried to summarize the various image noise models and give some recommendations for minimizing the noise effects. Any such summary is, by necessity, limited. We do, of course, apologize to any authors whose work we may have omitted.

For further information, the interested reader is urged to consult the references for this and other chapters.

References

[1] J. Astola and P. Kuosmanen, *Fundamentals of Nonlinear Digital Filtering* (CRC Press, Boca Raton, FL, 1997).

[2] P. Billingsley, *Probability and Measure* (Wiley, New York, 1979).

[3] C. G. Boncelet, Jr., "Algorithms to compute order statistic distributions," *SIAM J. Sci. Stat. Comput.* **8**, 868–876 (1987).

[4] C. G. Boncelet, Jr., "Order statistic distributions with multiple windows," *IEEE Trans. Inf. Theory* **IT-37** (1991).

[5] A. C. Bovik, T. S. Huang, and D. C. Munson, Jr., "A generalization of median filtering using linear combinations of order statistics," *IEEE Trans. Acoust. Speech Signal Proc.* **ASSP-31**, 1342–1350 (1983).

[6] W. Feller, *An Introduction to Probability Theory and Its Applications* (Wiley, New York, 1968).

[7] J. Goodman, "Some fundamental properties of speckle," *J. Opt. Soc. Am.* **66**, 1145–1150 (1976).

[8] J. Goodman, *Statistical Optics* (Wiley-Interscience, New York, 1985).

[9] R. C. Hardie and C. G. Boncelet, Jr., "LUM filters: a class of order statistic based filters for smoothing and sharpening," *IEEE Trans. Signal Process.* **41**, 1061–1076 (1993).

[10] C. Helstrom, *Probability and Stochastic Processes for Engineers* (Macmillan, New York, 1991).

[11] P. J. Huber, *Robust Statistics* (Wiley, New York, 1981).

[12] D. Kuan, A. Sawchuk, T. Strand, and P. Chavel, "Adaptive restoration of images with speckle," *IEEE Trans. Acoust. Speech Signal Proc.* **ASSP-35**, 373–383 (1987).

[13] A Labeyrie, "Attainment of diffraction limited resolution in large telescopes by fourier analysis speckle patterns in star images," *Astron. Astrophys.* **VI**, 85–87 (1970).

[14] E. H. Lloyd, "Least-squares estimations of location and scale parameters using order statistics," *Biometrika* **39**, 88–95 (1952).

[15] P. Peebles, *Probability, Random Variables, and Random Signal Principles* (McGraw-Hill, New York, 1993).

[16] M. Woodroofe, *Probability with Applications* (McGraw-Hill, New York, 1975).

4.6

Color and Multispectral Image Representation and Display

H. J. Trussell
North Carolina State University

1 Introduction

One of the most fundamental aspects of image processing is the representation of the image. The basic concept that a digital image is a matrix of numbers is reinforced by virtually all forms of image display. It is another matter to interpret how that value is related to the physical scene or object that is represented by the recorded image and how closely displayed results represent the data obtained from digital processing. It is these relationships to which this chapter is addressed.

Images are the result of a spatial distribution of radiant energy. The most common images are two-dimensional (2-D) color images seen on television. Other everyday images include photographs, magazine and newspaper pictures, computer monitors, and motion pictures. Most of these images represent realistic or abstract versions of the real world. Medical and satellite images form classes of images for which there is no equivalent scene in the physical world. Because of the limited space in this chapter, we will concentrate on the pictorial images.

The representation of an image goes beyond the mere designation of independent and dependent variables. In that limited case, an image is described by a function

$$f(x, y, \lambda, t), \tag{1}$$

where x, y are spatial coordinates (angular coordinates can also be used), λ indicates the wavelength of the radiation, and t represents time. It is noted that images are inherently two-dimensional spatial distributions. Higher dimensional functions can be represented by a straightforward extension. Such applications include medical CT and MRI, as well as seismic surveys. For this chapter, we will concentrate on the spatial and wavelength variables associated with still images. The temporal coordinate will be left for another chapter.

In addition to the stored numerical values in a discrete coordinate system, the representation of multidimensional information includes the relationship between the samples and the real world. This relationship is important in the determination of appropriate sampling and subsequent display of the image.

Before the fundamentals of image presentation are presented, it necessary define our notation and to review the prerequisite knowledge that is required to understand the following material. A review of rules for the display of images and functions is presented in Section 2, followed by a review of mathematical preliminaries in Section 3. Section 4 will cover the physical basis for multidimensional imaging. The foundations of colorimetry are reviewed in Section 5. This material is required to lay a foundation for a discussion of color sampling. Section 6 describes multidimensional sampling with concentration on sampling color spectral signals. We will discuss the fundamental differences between sampling the wavelength and spatial dimensions of the multidimensional signal. Finally, Section 7 contains a mathematical description of the display of multidimensional data. This area is often neglected by many texts. The section will emphasize the requirements for displaying data in a fashion that is both accurate and effective. The final Section briefly considers future needs in this basic area.

2 Preliminary Notes on Display of Images

One difference between one-dimensional (1-D) and 2-D functions is the way they are displayed. One-dimensional functions are easily displayed in a graph where the scaling is obvious. The observer need only examine the numbers that label the axes to determine the scale of the graph and get a mental picture of the function. With two-dimensional scalar-valued functions the display becomes more complicated. The accurate display of vector-valued two-dimensional functions, e.g., color images,

will be discussed after the necessary material on sampling and colorimetery is covered.

Two-dimensional functions can be displayed in several different ways. The most common are supported by MATLAB [1]. The three most common are the isometric plot, the gray-scale plot, and the contour plot. The user should choose the right display for the information to be conveyed. Let us consider each of the three display modalities. As simple example, consider the two-dimensional Gaussian functional form

$$f(m, n) = \mathrm{sinc}\!\left(\frac{m^2}{a^2} + \frac{n^2}{b^2}\right)$$

where, for the following plots, $a = 1$ and $b = 2$.

The isometric or surface plots give the appearance of a three-dimensional (3-D) drawing. The surface can be represented as a wire mesh or as a shaded solid, as in Fig. 1. In both cases, portions of the function will be obscured by other portions; for example, one cannot see through the main lobe. This representation is reasonable for observing the behavior of mathematical functions, such as point-spread functions, or filters in the space or frequency domains. An advantage of the surface plot is that it gives a good indication of the values of the function since a scale is readily displayed on the axes. It is rarely effective for the display of images.

Contour plots are analogous to the contour or topographic maps used to describe geographical locations. The sinc function is shown using this method in Fig. 2. All points that have a specific value are connected to form a continuous line. For a continous function the lines must form closed loops. This type

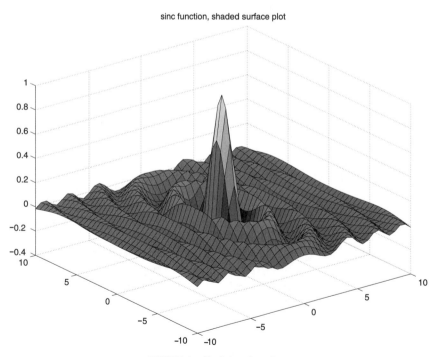

FIGURE 1   Shaded surface plot.

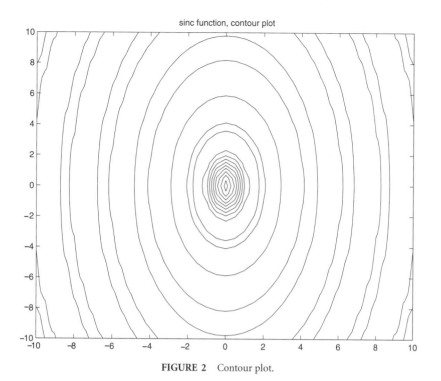

FIGURE 2 Contour plot.

of plot is useful in locating the position of maxima or minima in images or two-dimensional functions. It is used primarily in spectrum analysis and pattern recognition applications. It is difficult to read values from the contour plot and it takes some effort to determine whether the functional trend is up or down. The filled contour plot, available in MATLAB, helps in this last task.

Most monochrome images are displayed by using the gray-scale plot, in which the value of a pixel is represented by it relative lightness. Since in most cases, high values are displayed as light and low values are displayed as dark, it is easy to determine functional trends. It is almost impossible to determine exact values. For images, which are nonnegative functions, the display is natural; but for functions, which have negative values, it can be quite artificial.

In order to use this type of display with functions, the representation must be scaled to fit in the range of displayable gray levels. This is most often done using a min/max scaling, in which the function is linearly mapped such that the minimum value appears as black and the maximum value appears as white. This method was used for the sinc function shown in Fig. 3. For the display of functions, the min/max scaling can be effective to indicate trends in the behavior. Scaling for images is another matter.

Let us consider a monochrome image that has been digitized by some device, e.g., a scanner or camera. Without knowing the physical process of digitization, it is impossible to determine the best way to display the image. The proper display of images requires calibration of both the input and output devices. For now, it is reasonable to give some general rules about the display of monochrome images.

1. For the comparison of a sequences of images, it is *imperative* that all images be displayed with the same scaling.

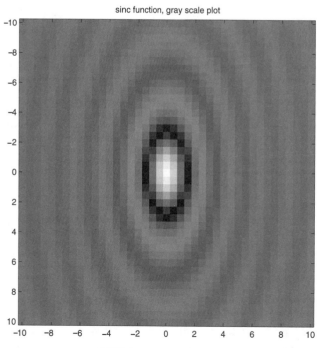

FIGURE 3 Gray-scale plot.

It is hard to emphasize this rule sufficiently and hard to count all the misleading results that have occurred when it has been ignored. The most common violation of this rule occurs when comparing an original and processed image. The user scales both images independently, using min/max scaling. In many cases the scaling can produce a significant enhancement of low contrast images, which can be mistaken for improvements produced by an algorithm under investigation. For example, consider an algorithm designed to reduce noise. The noisy image modelled by

$$\mathbf{g} = \mathbf{f} + \mathbf{n}.$$

Since the noise is both positive and negative, the noisy image, $\mathbf{g}$, has a larger range than the clean image, $\mathbf{f}$. Almost any noise reduction method will reduce the range of the processed image; thus, the output image undergoes additional contrast enhancement if min/max scaling is used. The result is greater apparent dynamic range and a better looking image.

There are several ways to implement this rule. The most appropriate way will depend on the application. The scaling may be done using the min/max of the collection of all images to be compared. In some cases, it is appropriate to truncate values at the limits of the display, rather than force the entire range into the range of the display. This is particularly true of images containing a few outliers. It may be advantageous to reduce the region of the image to a particular region of interest, which will usually reduce the range to be reproduced.

2. Display a step wedge, a strip of sequential gray levels from minimum to maximum values, with the image to show how the image gray levels are mapped to brightness or density. This allows some idea of the quantitative values associated with the pixels. This is routinely done on images that are used for analysis, such as the digital photographs from space probes.

3. Use a graytone mapping, which allows a wide range of gray levels to be visually distinguished. In software such as MATLAB, the user can control the mapping between the continuous values of the image and the values sent to the display device. For example, consider the CRT monitor as the output device. The visual tonal qualities of the output depend on many factors, including the brightness and contrast setting of the monitor, the specific phosphors used in the monitor, the linearity of the electron guns, and the ambient lighting. It is recommended that adjustments be made so that a user is able to distinguish all levels of a step wedge of $\sim$32 levels from min black to max white.

Most displays have problems with gray levels at the ends of the range being indistinguishable. This can be overcome by proper adjustment of the contrast and gain controls and an appropriate mapping from image values to display values. For hardcopy devices, the medium should be taken into account. For example, changes in paper type or manufacturer can results in significant tonal variations.

3 Notation and Prerequisite Knowledge

In most cases, the multidimensional process can be represented as a straightforward extension of one-dimensional processes. Thus, it is reasonable to mention the one-dimensional operations that are prerequisite to the chapter and will form the basis of the multidimensional processes.

3.1 Practical Sampling

Mathematically, ideal sampling is usually represented with the use of a *generalized function*, the Dirac delta function, $\delta(t)$ [2]. The entire sampled sequence can be represented using the comb function,

$$\text{comb}(t) = \sum_{n=-\infty}^{\infty} \delta(t-n), \qquad (2)$$

where the sampling interval is unity. The sampled signal is obtained by multiplication:

$$s_d(t) = s(t)\text{comb}(t) = s(t) \sum_{n=-\infty}^{\infty} \delta(t-n)$$

$$= \sum_{n=-\infty}^{\infty} s(t)\delta(t-n). \qquad (3)$$

It is common to use the notation of $\{s(n)\}$ or $s(n)$ to represent the collection of samples in discrete space. The arguments n and t will serve to distinguish the discrete or continuous space.

Practical imaging devices, such as video cameras, CCD arrays, and scanners, must use a finite aperture for sampling. The comb function cannot be realized by actual devices. The finite aperture is required to obtain a finite amount of energy from the scene. The engineering tradeoff is that large apertures receive more light and thus will have higher signal-to-noise ratios (SNRs) than smaller apertures, while smaller apertures have a higher spatial resolution than larger ones. This is true for apertures larger than the order of the wavelength of light. Below that point, diffraction limits the resolution.

The aperture may cause the light intensity to vary over the finite region of integration. For a single sample of a one-dimensional signal at time, nT, the sample value can be obtained by

$$s(n) = \int_{(n-1)T}^{nT} s(t)a(nT-t)\,\mathrm{d}t, \qquad (4)$$

where $a(t)$ represents the impulse response (or light variation) of the aperture. This is simple convolution. The sampling of the signal can be represented by

$$s(n) = [s(t) * a(t)]\text{comb}(t/T), \tag{5}$$

where the asterisk represents convolution. This model is reasonably accurate for the spatial sampling of most cameras and scanning systems.

The sampling model can be generalized to include the case in which each sample is obtained with a different aperture. For this case, the samples, which need not be equally spaced, are given by

$$s(n) = \int_l^u s(t)a_n(t)\,\mathrm{d}t, \tag{6}$$

where the limits of integration correspond to the region of support for the aperture. While there may be cases in which this form is used in spatial sampling, its main use is in sampling the wavelength dimension of the image signals. That topic will be covered later. The generalized signal reconstruction equation has the form

$$s(t) = \sum_{n=-\infty}^{\infty} s(n)g_n(t), \tag{7}$$

where the collection of functions, $\{g_n(t)\}$, provide the interpolation from discrete to continuous space. The exact form of $\{g_n(t)\}$ depends on the form of $\{a_n(t)\}$.

3.2 One-Dimensional Discrete System Representation

Linear operations on signals and images can be represented as simple matrix multiplications. The internal form of the matrix may be complicated, but the conceptual manipulation of images is very easy. Let us consider the representation of a one-dimensional convolution before going on to multidimensions. Consider the linear, time-invariant system

$$g(t) = \int_{-\infty}^{\infty} h(u)s(t-u)\,\mathrm{d}u.$$

The discrete approximation to continuous convolution is given by

$$g(n) = \sum_{k=0}^{L-1} h(k)s(n-k), \tag{8}$$

where the indices n and k represent sampling of the analog signals, e.g., $s(n) = s(nT)$. Since it is assumed that the signals under investigation have finite support, the summation is over a finite number of terms. If $s(n)$ has M nonzero samples and $h(n)$ has L nonzero samples, then $g(n)$ can have at most $N = M + L - 1$ nonzero samples. It is assumed the reader is familiar with what conditions are necessary so that we can represent the analog system by discrete approximation. Using the definition of the

signal as a vector, $\mathbf{s} = [s(0), s(1), \ldots, s(M-1)]$, we can write the summation of Eq. (8) as

$$\mathbf{g} = \mathbf{Hs}, \tag{9}$$

where the vectors $\mathbf{s}$ and $\mathbf{g}$ are of length M and N, respectively and the $N \times M$ matrix $\mathbf{H}$ accordingly [3]. It is often desirable to work with square matrices. In this case, the input vector can be padded with zeros to the same size as $\mathbf{g}$ and the matrix $\mathbf{H}$ modified to produce an $N \times N$ Toeplitz form. It is often useful, because of the efficiency of the FFT, to approximate the Toeplitz form by a circulant form by forcing appropriate elements into the upper-right region of the square Toeplitz matrix. This approximation works well with impulse responses of short duration and autocorrelation matrices with small correlation distances.

3.3 Multidimensional System Representation

The images of interest are described by two spatial coordinates and a wavelength coordinate, $f(x, y, \lambda)$. This continuous image will be sampled in each dimension. The result is a function defined on a discrete coordinate system, $f(m, n, l)$. This would usually require a three-dimensional matrix. However, to allow the use of standard matrix algebra, it is common to use stacked notation [3]. Each band, defined by wavelength λ_l or simply l, of the image is a $P \times P$ image. Without loss of generality, we will assume a square image for notational simplicity. This image can be represented as a $P^2 \times 1$ vector. The Q bands of the image can be stacked in a like manner forming a $QP^2 \times 1$ vector.

Optical blurring is modeled as convolution of the spatial image. Each wavelength of the image may be blurred by a slightly different point-spread function (PSF). This is represented by

$$\mathbf{g}_{(QP^2 \times 1)} = \mathbf{H}_{(QP^2 \times QP^2)}\mathbf{f}_{(QP^2 \times 1)}, \tag{10}$$

where the matrix $\mathbf{H}$ has a block form

$$\mathbf{H} = \begin{bmatrix} \mathbf{H}_{1,1} & \mathbf{H}_{1,2} & \cdots & \mathbf{H}_{1,Q} \\ \mathbf{H}_{2,1} & \mathbf{H}_{2,2} & \cdots & \mathbf{H}_{2,Q} \\ \vdots & \vdots & \cdots & \vdots \\ \mathbf{H}_{Q,1} & \mathbf{H}_{Q,2} & \cdots & \mathbf{H}_{Q,Q} \end{bmatrix}. \tag{11}$$

The submatrix $\mathbf{H}_{i,j}$ is of dimension $P^2 \times P^2$ and represents the contribution of the jth band of the input to the ith band of the output. Since an optical system does not modify the frequency of an optical signal, $\mathbf{H}$ will be block diagonal. There are cases, e.g., imaging using color filter arrays, in which the diagonal assumption does not hold.

Algebraic representation using stacked notation for 2-D signals is more difficult to manipulate and understand than for 1-D signals. An example of this is illustrated by considering the autocorrelation of multiband images that are used in multispectral restoration methods. This is easily written in terms of the matrix notation reviewed earlier:

$$\mathbf{R}_{ff} = E\{\mathbf{ff}^T\},$$

where $\mathbf{f}$ is a $QP^2 \times 1$ vector. In order to compute estimates we must be able to manipulate this matrix. While the $QP^2 \times QP^2$ matrix is easily manipulated symbolically, direct computation with the matrix is not practical for realistic values of P and Q, e.g. $Q = 3$, $P = 256$. For practical computation, the matrix form is simplified by using various assumptions, such as separability, circularity, and independence of bands. These assumptions result in block properties of the matrix that reduce the dimension of the computation. A good example is shown in the multidimensional restoration problem [4].

4 Analog Images as Physical Functions

The image that exists in the analog world is a spatiotemporal distribution of radiant energy. As mentioned earlier, this chapter will not discuss the temporal dimension but will concentrate on the spatial and wavelength aspects of the image. The function is represented by $f(x, y, \lambda)$. While it is often overlooked by students eager to process their first image, it is fundamental to define what the value of the function represents. Since we are dealing with radiant energy, the value of the function represents energy flux, exactly like electromagnetic theory. The units will be energy per unit area (or angle) per unit time per unit wavelength. From the imaging point of view, the function is described by the spatial energy distribution at the sensor. It does not matter whether the object in the image emits light or reflects light.

To obtain a sample of the analog image we must integrate over space, time, and wavelength to obtain a finite amount of energy. Since we have eliminated time from the description, we can have watts per unit area per unit wavelength. To obtain overall lightness, the wavelength dimension is integrated out using the luminous efficiency function discussed in the following section on colorimetry. The common units of light intensity are lux (lumens/m^2) or footcandles. See [5] for an exact definition of radiometric quantities. A table of typical light levels is given in Table 1. The most common instrument for measuring light intensity is the light meter used in professional and amateur photography.

In order to sample an image correctly, we must be able to characterize its energy distribution in each of the dimensions.

TABLE 1   Qualitative description of luminance levels

Description	Lux (cd/m^2)	Footcandles
Moonless night	$\sim 10^{-6}$	$\sim 10^{-7}$
Full moon night	$\sim 10^{-3}$	$\sim 10^{-4}$
Restaurant	~ 100	~ 9
Office	~ 350	~ 33
Overcast day	~ 5000	~ 465
Sunny day	$\sim 200,000$	~ 18600

There is little that can be said about the spatial distribution of energy. From experience, we know that images vary greatly in spatial content. Objects in an image usually may appear at any spatial location and at any orientation. This implies that there is no reason to vary the sample spacing over the spatial range of an image. In the cases of some very restricted ensembles of images, variable spatial sampling has been used to advantage. Since these examples are quite rare, they will not be discussed here.

Spatial sampling is done by using a regular grid. The grid is most often rectilinear, but hexagonal sampling has been thoroughly investigated [6]. Hexagonal sampling is used for efficiency when the images have a natural circular region of support or circular symmetry. All the mathematical operations, such as Fourier transforms and convolutions, exist for hexagonal grids. It is noted that the reasons for uniform sampling of the temporal dimension follow the same arguments.

The distribution of energy in the wavelength dimension is not as straightforward to characterize. In addition, we are often not as interested in reconstructing the radiant spectral distribution as we are the spatial distribution. We are interested in constructing an image that appears to the human to have the same colors as the original image. In this sense, we are actually using color aliasing to our advantage. Because of this aspect of color imaging, we need to characterize the color vision system of the eye in order to determine proper sampling of the wavelength dimension.

5 Colorimetry

To understand the fundamental difference in the wavelength domain, we must describe some of the fundamentals of color vision and color measurement. What is presented here is only a brief description that will allow us to proceed with the description of the sampling and mathematical representation of color images. A more complete description of the human color visual system can be found in [7, 8].

The retina contains two types of light sensors, rods and cones. The rods are used for monochrome vision at low light levels; the cones are used for color vision at higher light levels. There are three types of cones. Each type is maximally sensitive to a different part of the spectrum. They are often referred to as long, medium, and short wavelength regions. A common description refers to them as red, green, and blue cones, although their maximal sensitivity is in the yellow, green, and blue regions of the spectrum. Recall that the visible spectrum extends from ~ 400 nm (blue) to ~ 700 nm (red). Cones sensitivities are related to the absorption sensitivity of the pigments in the cones. The absorption sensitivity of the different cones has been measured by several researchers. An example of the curves is shown in Fig. 4. Long before the technology was available to measure the curves directly, they were estimated from a clever color matching experiment. A description of this experiment, which is still used today, can be found in [8, 5].

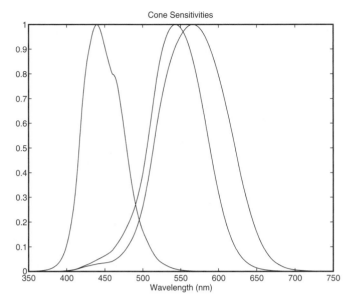

FIGURE 4 Cone sensitivities.

Grassmann formulated a set of laws for additive color mixture in 1853 [9, 10, 5]. Additive in this sense refers to the addition of two or more radiant sources of light. In addition, Grassmann conjectured that any additive color mixture could be matched by the proper amounts of three primary stimuli. Considering what was known about the physiology of the eye at that time, these laws represent considerable insight. It should be noted that these "laws" are not physically exact but represent a good approximation under a wide range of visibility conditions. There is current research in the vision and color science community on the refinements and reformulations of the laws.

Grassmann's laws are essentially unchanged as printed in recents texts on color science [5]. With our current understanding of the physiology of the eye and a basic background in linear algebra, Grassmann's laws can be stated more concisely. Furthermore, extensions of the laws and additional properties are easily derived by using the mathematics of matrix theory. There have been several papers that have taken a linear systems approach to describing color spaces as defined by a standard human observer [11, 12, 13, 14]. This section will briefly summarize these results and relate them to simple signal processing concepts. For the purposes of this work, it is sufficient to note that the spectral responses of the three types of sensors are sufficiently different so as to define a three-dimensional vector space.

5.1 Color Sampling

The mathematical model for the color sensor of a camera or the human eye can be represented by

$$v_k = \int_{-\infty}^{\infty} r_a(\lambda) m_k(\lambda) \, d\lambda, \quad k = 1, 2, 3, \quad (12)$$

where $r_a(\lambda)$ is the radiant distribution of light as a function of wavelength and $m_k(\lambda)$ is the sensitivity of the kth color sensor. The sensitivity functions of the eye were shown in Fig. 4.

Note that sampling of the radiant power signal associated with a color image can be viewed in at least two ways. If the goal of the sampling is to reproduce the spectral distribution, then the same criteria for sampling the usual electronic signals can be directly applied. However, the goal of color sampling is not often to reproduce the spectral distribution but to allow reproduction of the color sensation. This aspect of color sampling will be discussed in detail below. To keep this discussion as simple as possible, we will treat the color sampling problem as a subsampling of a high resolution discrete space; that is, the N samples are sufficient to reconstruct the original spectrum, using the uniform sampling of Section 3.

It has been assumed in most research and standards work that the visual frequency spectrum can be sampled finely enough to allow the accurate use of numerical approximation of integration. A common sample spacing is 10 nm over the range 400–700 nm, although ranges as wide as 360–780 nm have been used. This is used for many color tables and lower priced instrumentation. Precision color instrumentation produces data at 2-nm intervals. Finer sampling is required for some illuminants with line emitters. Reflective surfaces are usually smoothly varying and can be accurately sampled more coarsely. Sampling of color signals is discussed in Section 6 and in detail in [15].

Proper sampling follows the same bandwidth restrictions that govern all digital signal processing. Following the assumption that the spectrum can be adequately sampled, the space of all possible visible spectra lies in an N-dimensional vector space, where $N = 31$ is the range if 400–700 nm is used. The spectral response of each of the eye's sensors can be sampled as well, giving three linearly independent N vectors that define the visual subspace.

Under the assumption of proper sampling, the integral of Eq. (12) can be well approximated by a summation

$$v_k = \sum_{n=L}^{U} r_a(n\Delta\lambda) s_k(n\Delta\lambda), \quad k = 1, 2, 3, \quad (13)$$

where $\Delta\lambda$ represents the sampling interval and the summation limits are determined by the region of support of the sensitivity of the eye. This equation can be generalized to represent any color sensor by replacing $s_k(\cdot)$ with $m_k(\cdot)$. This discrete form is easily represented in matrix/vector notation. This will be done in the following sections.

5.2 Discrete Representation of Color Matching

The response of the eye can be represented by a matrix, $\mathbf{S} = [\mathbf{s}_1, \mathbf{s}_2, \mathbf{s}_3]$, where the N vectors, $\mathbf{s}_i$, represent the response of the ith type sensor (cone). Any visible spectrum can be represented by an N vector, $\mathbf{f}$. The response of the sensors to the input spectrum is a three vector, **t**, obtained by

$$\mathbf{t} = \mathbf{S}^T \mathbf{f}. \tag{14}$$

Two visible spectra are said to have the same color if they appear the same to the human observer. In our linear model, this means that if **f** and **g** are two N vectors representing different spectral distributions, they are equivalent colors if

$$\mathbf{S}^T \mathbf{f} = \mathbf{S}^T \mathbf{g}. \tag{15}$$

It is clear that there may be many different spectra that appear to be the same color to the observer. Two spectra that appear the same are called metamers. Metamerism (meh·tam·er·ism) is one of the greatest and most fascinating problems in color science. It is basically color "aliasing" and can be described by the generalized sampling described earlier.

It is difficult to find the matrix, **S**, that defines the response of the eye. However, there is a conceptually simple experiment that is used to define the human visual space defined by **S**. A detailed discussion of this experiment is given in [8, 5]. Consider the set of monochromatic spectra $\mathbf{e}_i$, for $i = 1, 2, \ldots, N$. The N vectors, $\mathbf{e}_i$, have a one in the ith position and zeros elsewhere. The goal of the experiment is to match each of the monochromatic spectra with a linear combination of primary spectra. Construct three lighting sources that are linearly independent in N space. Let the matrix, $\mathbf{P} = [\mathbf{p}_1, \mathbf{p}_2, \mathbf{p}_3]$, represent the spectral content of these primaries. The phosphors of a color television are a common example (Fig. 5).

An experiment is conducted in which a subject is shown one of the monochromactic spectra, $\mathbf{e}_i$, on one-half of a visual field. On the other half of the visual field appears a linear combination of the primary sources. The subject attempts to visually match an input monochromatic spectrum by adjusting the relative in-tensities of the primary sources. Physically, it may impossible to match the input spectrum by adjusting the intensities of the primaries. When this happens, the subject is allowed to change the field of one of the primaries so that it falls on the same field as the monochromatic spectrum. This is mathematically equivalent to subtracting that amount of primary from the primary field. De-noting the relative intensities of the primaries by the three vector $\mathbf{a}_i = [a_{i1}, a_{i2}, a_{i3}]^T$, we write the match mathematically as

$$\mathbf{S}^T \mathbf{e}_i = \mathbf{S}^T \mathbf{P} \mathbf{a}_i. \tag{16}$$

Combining the results of all N monochromatic spectra, we can write Eq. (5) as

$$\mathbf{S}^T \mathbf{I} = \mathbf{S}^T = \mathbf{S}^T \mathbf{P} \mathbf{A}^T, \tag{17}$$

where $\mathbf{I} = [\mathbf{e}_1, \mathbf{e}_2, \ldots, \mathbf{e}_N]$ is the $N \times N$ identity matrix.

Note that because the primaries, **P**, are not metameric, the product matrix is nonsingular, i.e., $(\mathbf{S}^T \mathbf{P})^{-1}$ exists. The human visual subspace (HVSS) in the N-dimensional vector space is defined by the column vectors of **S**; however, this space can be equally well defined by any nonsingular transformation of those basis vectors. The matrix,

$$\mathbf{A} = \mathbf{S}(\mathbf{P}^T \mathbf{S})^{-1}, \tag{18}$$

is one such transformation. The columns of the matrix **A** are called the color matching functions associated with the primaries **P**.

To avoid the problem of negative values that cannot be real-ized with transmission or reflective filters, the CIE developed a standard transformation of the color matching functions that yields no negative values. This set of color matching functions is known as the *standard observer*, or the CIE *XYZ* color matching functions. These functions are shown in Fig. 6. For the remainder of this chapter, the matrix, **A**, can be thought of as this standard set of functions.

5.3 Properties of Color Matching Functions

Having defined the human visual subspace, we find it worthwhile to examine some of the common properties of this space. Because of the relatively simple mathematical definition of color match-ing given in the last section, the standard properties enumerated by Grassmann are easily derived by simple matrix manipulations [13]. These properties play an important part in color sampling and display.

Property 1 (Dependence of Color on A)

Two visual spectra, **f** and **g**, appear the same if and only if $\mathbf{A}^T \mathbf{f} = \mathbf{A}^T \mathbf{g}$. Writing this mathematically, we have $\mathbf{S}^T \mathbf{f} = \mathbf{S}^T \mathbf{g}$ iff $\mathbf{A}^T \mathbf{f} = \mathbf{A}^T \mathbf{g}$. Metamerism is color aliasing. Two signals **f** and **g** are sampled by the cones or equivalently by the color matching func-tions and produce the same tristimulus values. The importance

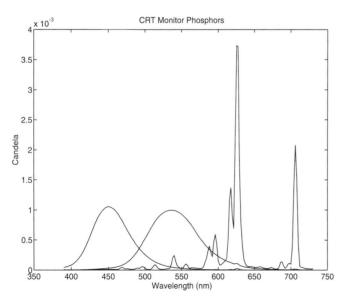

FIGURE 5 CRT monitor phosphors.

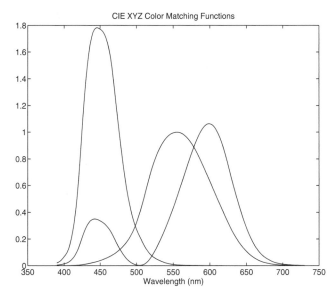

FIGURE 6 CIE *XYZ* color matching functions.

of this property is that any linear transformation of the sensitivities of the eye or the CIE color matching functions can be used to determine a color match. This gives more latitude in choosing color filters for cameras and scanners as well as for color measurement equipment. It is this property that is the basis for the design of optimal color scanning filters [16, 17].

A note on terminology is appropriate here. When the color matching matrix is the CIE standard [5], the elements of the three vector defined by $\mathbf{t} = \mathbf{A}^T \mathbf{f}$ are called tristimulus values and usually denoted by *X, Y, Z*; i.e., $\mathbf{t}^T = [X, Y, Z]$. The chromaticity of a spectrum is obtained by normalizing the tristimulus values,

$$x = X/(X + Y + Z),$$

$$y = Y/(X + Y + Z),$$

$$z = Z/(X + Y + Z).$$

Since the chromaticity coordinates have been normalized, any two of them are sufficient to characterize the chromaticity of a spectrum. The *x* and *y* terms are the standard for describing chromaticity. It is noted that the convention of using different variables for the elements of the tristimulus vector may make mental conversion between the vector space notation and notation in common color science texts more difficult.

The CIE has chosen the $\mathbf{a}_2$ sensitivity vector to correspond to the luminance efficiency function of the eye. This function, shown as the middle curve in Fig. 6, gives the relative sensitivity of the eye to the energy at each wavelength. The *Y* tristimulus value is called luminance and indicates the perceived brightness of a radiant spectrum. It is this value that is used to calculate the effective light output of light bulbs in lumens. The chromaticities *x* and *y* indicate the hue and saturation of the color. Often the color is described in terms of [*x*, *y*, *Y*] because of the ease of

interpretation. Other color coordinate systems will be discussed later.

Property 2 (Transformation of Primaries)

If a different set of primary sources, **Q**, are used in the color matching experiment, a different set of color matching functions, **B**, are obtained. The relation between the two color matching matrices is given by

$$\mathbf{B}^T = (\mathbf{A}^T \mathbf{Q})^{-1} \mathbf{A}^T. \tag{19}$$

The more common interpretation of the matrix $\mathbf{A}^T \mathbf{Q}$ is obtained by a direct examination. The *j*th column of **Q**, denoted $\mathbf{q}_j$, is the spectral distribution of the *j*th primary of the new set. The element $[\mathbf{A}^T \mathbf{Q}]_{i,j}$ is the amount of the primary $\mathbf{p}_i$ required to match primary $\mathbf{q}_j$. It is noted that the above form of the change of primaries is restricted to those that can be adequately represented under the assumed sampling discussed previously. In the case that one of the new primaries is a Dirac delta function located between sample frequencies, the transformation $\mathbf{A}^T \mathbf{Q}$ must be found by interpolation. The CIE RGB color matching functions are defined by the monochromatic lines at 700 nm, 546.1 nm, and 435.8 nm and shown in Fig. (7). The negative portions of these functions are particularly important, since they imply that all color matching functions associated with realizable primaries have negative portions.

One of the uses of this property is in determining the filters for color television cameras. The color matching functions associated with the primaries used in a television monitor are the ideal filters. The tristimulus values obtained by such filters would directly give the values to drive the color guns. The NTSC standard [*R, G, B*] are related to these color matching functions. For coding purposes and efficient use of bandwidth, the *RGB* values are transformed to *YIQ* values, where *Y* is the CIE *Y* (luminance)

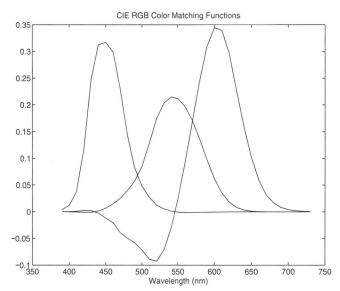

FIGURE 7 CIE *XYZ* color matching functions.

and *I* and *Q* carry the hue and saturation information. The transformation is a 3×3 matrix multiplication [3] (see property 3 below).

Unfortunately, since the TV primaries are realizable, the color matching functions which correspond to them are not. This means that the filters which are used in TV cameras are only an approximation to the ideal filters. These filters are usually obtained by simply clipping the part of the ideal filter that falls below zero. This introduces an error which cannot be corrected by any postprocessing.

Property 3 (Transformation of Color Vectors)

If **c** and **d** are the color vectors in three space associated with the visible spectrum, **f**, under the primaries **P** and **Q** respectively, then

$$\mathbf{d} = (\mathbf{A}^T \mathbf{Q})^{-1} \mathbf{c}, \tag{20}$$

where **A** is the color matching function matrix associated with primaries **P**. This states that a 3×3 transformation is all that is required to go from one color space to another.

Property 4 (Metamers and the Human Visual Subspace)

The *N*-dimensional spectral space can be decomposed into a 3-D subspace known as the HVSS and an *N*-3-D subspace known as the black space. All metamers of a particular visible spectrum, **f**, are given by

$$\mathbf{x} = \mathbf{P}_v \mathbf{f} + \mathbf{P}_b \mathbf{g}, \tag{21}$$

where $\mathbf{P}_v = \mathbf{A}(\mathbf{A}^T\mathbf{A})^{-1}\mathbf{A}^T$ is the orthogonal projection operator to the visual space, $\mathbf{P}_b = [\mathbf{I} - \mathbf{A}(\mathbf{A}^T\mathbf{A})^{-1}\mathbf{A}^T]$ is the orthogonal projection operator to the black space, and **g** is any vector in *N* space.

It should be noted that humans cannot see (or detect) all possible spectra in the visual space. Since it is a vector space, there exist elements with negative values. These elements are not realizable and thus cannot be seen. All vectors in the black space have negative elements. While the vectors in the black space are not realizable and cannot be seen, they can be combined with vectors in the visible space to produce a realizable spectrum.

Property 5 (Effect of Illumination)

The effect of an illumination spectrum, represented by the *N* vector **l**, is to transform the color matching matrix **A** by

$$\mathbf{A}_l = \mathbf{L}\mathbf{A}, \tag{22}$$

where **L** is a diagonal matrix defined by setting the diagonal elements of **L** to the elements of the vector **l**. The emitted spectrum for an object with reflectance vector, **r**, under illumination, **l**, is given by multiplying the reflectance by the illuminant at each wavelength, $\mathbf{g} = \mathbf{L}\mathbf{r}$. The tristimulus values associated with this emitted spectrum are obtained by

$$\mathbf{t} = \mathbf{A}^T\mathbf{g} = \mathbf{A}^T\mathbf{L}\mathbf{r} = \mathbf{A}_l^T\mathbf{r}. \tag{23}$$

The matrix $\mathbf{A}_l$ will be called the color matching functions under illuminant **l**.

Metamerism under different illuminants is one of the greatest problems in color science. A common imaging example occurs in making a digital copy of an original color image, e.g., a color copier. The user will compare the copy to the original under the light in the vicinity of the copier. The copier might be tuned to produce good matches under the fluorescent lights of a typical office but may produce copies that no longer match the original when viewed under the incandescent lights of another office or viewed near a window that allows a strong daylight component.

A typical mismatch can be expressed mathematically by relations

$$\mathbf{A}^T\mathbf{L}_f\mathbf{r}_1 = \mathbf{A}^T\mathbf{L}_f\mathbf{r}_2, \tag{24}$$

$$\mathbf{A}^T\mathbf{L}_d\mathbf{r}_1 \neq \mathbf{A}^T\mathbf{L}_d\mathbf{r}_2, \tag{25}$$

where $\mathbf{L}_f$ and $\mathbf{L}_d$ are diagonal matrices representing standard fluorescent and daylight spectra, respectively, and $\mathbf{r}_1$ and $\mathbf{r}_2$ represent the reflectance spectra of the original and copy, respectively. The ideal images would have $\mathbf{r}_2$ matching $\mathbf{r}_1$ under all illuminations, which would imply they are equal. This is virtually impossible since the two images are made with different colorants. The conditions for obtaining a match are discussed next.

5.4 Notes on Sampling for Color Aliasing

Sampling of the radiant power signal associated with a color image can be viewed in at least two ways. If the goal of the sampling is to reproduce the spectral distribution, then the same criteria for sampling the usual electronic signals can be directly applied. However, the goal of color sampling is not often to reproduce the spectral distribution but to allow reproduction of the color sensation. To illustrate this problem, let us consider the case of a television system. The goal is to sample the continuous color spectrum in such a way that the color sensation of the spectrum can be reproduced by the monitor.

A scene is captured with a television camera. We will consider only the color aspects of the signal, i.e., a single pixel. The camera uses three sensors with sensitivities **M** to sample the radiant spectrum. The measurements are given by

$$\mathbf{v} = \mathbf{M}^T\mathbf{r}, \tag{26}$$

where **r** is a high-resolution sampled representation of the radiant spectrum and $\mathbf{M} = [\mathbf{m}_1, \mathbf{m}_2, \mathbf{m}_3]$ represent the high-resolution sensitivities of the camera. The matrix **M** includes the effects of the filters, detectors, and optics.

These values are used to reproduce colors at the television receiver. Let us consider the reproduction of color at the receiver by a linear combination of the radiant spectra of the three phosphors on the screen, denoted $\mathbf{P} = [\mathbf{p}_1, \mathbf{p}_2, \mathbf{p}_3]$, where $\mathbf{p}_k$ represent the spectra of the red, green, and blue phosphors. We will also assume that the driving signals, or control values, for the phosphors to be linear combinations of the values measured by the camera, $\mathbf{c} = \mathbf{Bv}$. The reproduced spectrum is $\hat{\mathbf{r}} = \mathbf{Pc}$.

The appearance of the radiant spectra is determined by the response of the human eye,

$$\mathbf{t} = \mathbf{S}^T \mathbf{r}, \tag{27}$$

where $\mathbf{S}$ is defined by Eq. (14). The tristimulus values of the spectrum reproduced by the TV are obtained by

$$\hat{\mathbf{t}} = \mathbf{S}^T \hat{\mathbf{r}} = \mathbf{S}^T \mathbf{PBM}^T \mathbf{r}. \tag{28}$$

If the sampling is done correctly, the tristimulus values can be computed, that is, $\mathbf{B}$ can be chosen so that $\mathbf{t} = \hat{\mathbf{t}}$. Since the three primaries are not metameric and the eye's sensitivities are linearly independent, $(\mathbf{S}^T\mathbf{P})^{-1}$ exists and from the equality we have

$$(\mathbf{S}^T\mathbf{P})^{-1}\mathbf{S}^T = \mathbf{BM}^T, \tag{29}$$

Since equality of tristimulus values holds for all $\mathbf{r}$. This means that the color spectrum is sampled properly if the sensitivities of the camera are within a linear transformation of the sensitivities of the eye, or equivalently the color matching functions.

Considering the case in which the number of sensors, Q, in a camera or any color measuring device is larger than three, the condition is that the sensitivities of the eye must be linear combination of the sampling device sensitivities. In this case,

$$(\mathbf{S}^T\mathbf{P})^{-1}\mathbf{S}^T = \mathbf{B}_{3 \times Q}\mathbf{M}_{Q \times N}^T. \tag{30}$$

There are still only three types of cones that are described by $\mathbf{S}$. However, the increase in the number of basis functions used in the measuring device allows more freedom to the designer of the instrument. From the vector space viewpoint, the sampling is correct if the three-dimensional vector space defined by the cone sensitivity functions lies within the N-dimensional vector space defined by the device sensitivity functions.

Let us now consider the sampling of reflective spectra. Since color is measured for radiant spectra, a reflective object must be illuminated to be seen. The resulting radiant spectra is the product of the illuminant and the reflection of the object,

$$\mathbf{r} = \mathbf{Lr}_0, \tag{31}$$

where $\mathbf{L}$ is diagonal matrix containing the high resolution sampled radiant spectrum of the illuminant and the elements of the reflectance of the object are constrained, $0 \le \mathbf{r}_0(k) \le 1$.

To consider the restrictions required for sampling a reflective object, we must account for two illuminants: the illumination under which the object is to be viewed, and the illumination under which the measurements are made. The equations for computing the tristimulus values of reflective objects under the viewing illuminant $\mathbf{L}_v$ are given by

$$\mathbf{t} = \mathbf{A}^T \mathbf{L}_v \mathbf{r}_0, \tag{32}$$

where we have used the CIE color matching functions instead of the sensitivities of the eye (Property 1). The equation for estimating the tristimulus values from the sampled data is given by

$$\hat{\mathbf{t}} = \mathbf{BM}^T \mathbf{L}_d \mathbf{r}_0, \tag{33}$$

where $\mathbf{L}_d$ is a matrix containing the illuminant spectrum of the device. The sampling is proper if there exists a $\mathbf{B}$ such that

$$\mathbf{BM}^T\mathbf{L}_d = \mathbf{A}^T\mathbf{L}_v. \tag{34}$$

It is noted that in practical applications the device illuminant usually places severe limitations on the problem of approximating the color matching functions under the viewing illuminant. In most applications the scanner illumination is a high-intensity source, so as to minimize scanning time. The detector is usually a standard CCD array or photomultiplier tube. The design problem is to create a filter set $\mathbf{M}$ that brings the product of the filters, detectors, and optics to within a linear transformation of $\mathbf{A}_l$. Since creating a perfect match with real materials is a problem, it is of interest to measure the goodness of approximations to a set of scanning filters that can be used to design optimal realizable filter sets [16, 17].

5.5 A Note on the Nonlinearity of the Eye

It is noted here that most physical models of the eye include some type of nonlinearity in the sensing process. This nonlinearity is often modelled as a logarithm; in any case, it is always assumed to be monotonic within the intensity range of interest. The nonlinear function, $\mathbf{v} = V(\mathbf{c})$, transforms the three vector in an element independent manner; that is,

$$[v_1, v_2, v_3]^T = [V(c_1),\, V(c_2),\, V(c_3)]^T. \tag{35}$$

Since equality is required for a color match by Eq. (2), the function $V(\cdot)$ does not affect our definition of equivalent colors. Mathematically,

$$V(\mathbf{S}^T\mathbf{f}) = V(\mathbf{S}^T\mathbf{g}) \tag{36}$$

is true if, and only if, $\mathbf{S}^T\mathbf{f} = \mathbf{S}^T\mathbf{g}$. This nonlinearity does have a definite effect on the relative sensitivity in the color matching process and is one of the causes of much searching for the "uniform color space" discussed next.

5.6 Uniform Color Spaces

It has been mentioned that the psychovisual system is known to be nonlinear. The problem of color matching can be treated by linear systems theory since the receptors behave in a linear mode and exact equality is the goal. In practice, it is seldom that an engineer can produce an exact match to any specification. The nonlinearities of the visual system play a critical role in the determination of a color sensitivity function. Color vision is too complex to be modeled by a simple function. A *measure* of sensitivity that is consistent with the observations of arbitrary scenes is well beyond present capability. However, much work has been done to determine human color sensitivity in matching two color fields that subtend only a small portion of the visual field.

Some of the first controlled experiments in color sensitivity were done by MacAdam [18]. The observer viewed a disk made of two semicircles of different colors on a neutral background. One color was fixed; the other could be adjusted by the user. Since MacAdam's pioneering work, there have been many additional studies of color sensitivity. Most of these have measured the variability in three dimensions that yields sensitivity ellipsoids in tristimulus space. The work by Wyszecki and Felder [19] is of particular interest, as it shows the variation between observers and between a single observer at different times. The large variation of the sizes and orientation of the ellipsoids indicates that mean square error in tristimulus space is a very poor measure of color error. A common method of treating the nonuniform error problem is to transform the space into one where the Euclidean distance is more closely correlated with perceptual error. The CIE recommended two transformations in 1976 in an attempt to standardize measures in the industry.

Neither of the CIE standards exactly achieve the goal of a uniform color space. Given the variability of the data, it is unreasonable to expect that such a space could be found. The transformations do reduce the variations in the sensitivity ellipses by a large degree. They have another major feature in common: the measures are made relative to a reference white point. By using the reference point the transformations attempt to account for the adaptive characteristics of the visual system. The CIELAB (see-lab) space is defined by

$$L^* = 116 \left(\frac{Y}{Y_n} \right)^{1/3} - 16, \tag{37}$$

$$a^* = 500 \left[\left(\frac{X}{X_n} \right)^{1/3} - \left(\frac{Y}{Y_n} \right)^{1/3} \right], \tag{38}$$

$$b^* = 200 \left[\left(\frac{Y}{Y_n} \right)^{1/3} - \left(\frac{Z}{Z_n} \right)^{1/3} \right], \tag{39}$$

for $X/X_n, Y/Y_n, Z/Z_n > 0.01$. The values X_n, Y_n, Z_n are the tristimulus values of the reference white under the reference illumination, and X, Y, Z are the tristimulus values that are to be mapped to the Lab color space. The restriction that the nor-

malized values be greater than 0.01 is an attempt to account for the fact that at low illumination the cones become less sensitive and the rods (monochrome receptors) become active. A linear model is used at low light levels. The exact form of the linear portion of CIELAB and the definition of the CIELUV (see-luv) transformation can be found in [9, 17].

The color error between two colors $\mathbf{c}_1$ and $\mathbf{c}_2$ are measured in terms of

$$\Delta E_{ab} = [(L_1^* - L_2^*)^2 + (a_1^* - a_2^*)^2 + (b_1^* - b_2^*)^2]^{1/2}, \tag{40}$$

where $\mathbf{c}_i = [L_i^*, a_i^*, b_i^*]$. A useful rule of thumb is that two colors cannot be distinguished in a scene if their ΔE_{ab} value is less than 3. The ΔE_{ab} threshold is much lower in the experimental setting than in pictorial scenes. It is noted that the sensitivities discussed above are for flat fields. The sensitivity to modulated color is a much more difficult problem.

6 Sampling of Color Signals and Sensors

It has been assumed in most of this chapter that the color signals of interest can be sampled sufficiently well to permit accurate computation by using discrete arithmetic. It is appropriate to consider this assumption quantitatively. From the previous sections, it is seen that there are three basic types of color signals to consider: reflectances, illuminants, and sensors. Reflectances usually characterize everyday objects, but occasionally manmade items with special properties such as filters and gratings are of interest. Illuminants vary a great deal from natural daylight or moonlight to special lamps used in imaging equipment. The sensors most often used in color evaluation are those of the human eye. However, because of their use in scanners and cameras, CCDs and photomultiplier tubes are of great interest.

The most important sensor characteristics are the cone sensitivities of the eye or, equivalently, the color matching functions, e.g., Fig. 6. It is easily seen that functions in Figs. 4, 6, and 7 are very smooth functions and have limited bandwidths. A note on bandwidth is appropriate here. The functions represent continuous functions with finite support. Because of the finite support constraint, they cannot be bandlimited. However, they are clearly smooth and have very low power outside of a very small frequency band. With the use of 2-nm representations of the functions, the power spectra of these signals are shown in Fig. 8. The spectra represent the Welch estimate in which the data are first windowed, and then the magnitude of the DFT is computed [2]. It is seen that 10-nm sampling produces very small aliasing error.

In the context of cameras and scanners, the actual photoelectric sensor should be considered. Fortunately, most sensors have very smooth sensitivity curves that have bandwidths comparable to those of the color matching functions. See any handbook on CCD sensors or photomultiplier tubes. Reducing the variety of sensors to be studied can also be justified by the fact that filters

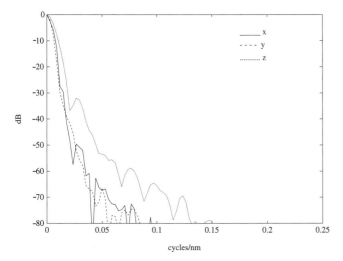

FIGURE 8   Power spectrum of CIE *XYZ* color matching functions.

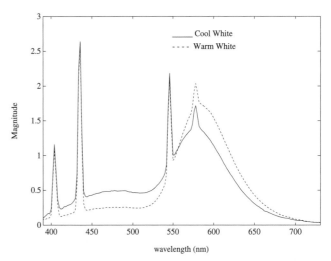

FIGURE 9   Cool white fluorescent and warm white fluorescent.

can be designed to compensate for the characteristics of the sensor and bring the combination within a linear combination of the CMFs.

The function $r(\lambda)$, which is sampled to give the vector **r** used in the Colorimetry section, can represent either reflectance or transmission. Desktop scanners usually work with reflective media. There are, however, several film scanners on the market that are used in this type of environment. The larger dynamic range of the photographic media implies a larger bandwidth. Fortunately, there is not a large difference over the range of everyday objects and images. Several ensembles were used for a study in an attempt to include the range of spectra encountered by image scanners and color measurement instrumentation [20]. The results showed again that 10-nm sampling was sufficient [15].

There are three major types of viewing illuminants of interest for imaging: daylight, incandescent, and fluorescent. There are many more types of illuminants used for scanners and measurement instruments. The properties of the three viewing illuminants can be used as a guideline for sampling and signal processing, which involves other types. It has been shown that the illuminant is the determining factor for the choice of sampling interval in the wavelength domain [15].

Incandescent lamps and natural daylight can be modeled as filtered blackbody radiators. The wavelength spectra are relatively smooth and have relatively small bandwidths. As with previous color signals, they are adequately sampled at 10 nm. Office lighting is dominated by fluorescent lamps. Typical wavelength spectra and their frequency power spectra are shown in Figs. 9 and 10.

It is with the fluorescent lamps that the 10-nm sampling becomes suspect. The peaks that are seen in the wavelength spectra are characteristic of mercury and are delta function signals at 404.7, 435.8, 546.1, and 578.4 nm. The fluorescent lamp can be modeled as the sum of a smoothly varying signal produced by

the phosphors and a delta function series:

$$l(\lambda) = l_d(\lambda) + \sum_{k=1}^{q} \alpha_k \delta(\lambda - \lambda_k), \qquad (41)$$

where α_k represents the strength of the spectral line at wavelength λ_k. The wavelength spectra of the phosphors is relatively smooth, as seen from Fig. 9.

From Fig. 10, it is clear that the fluorescent signals are not bandlimited in the sense used previously. The amount of power outside of the band is a function of the positions and strengths of the line spectra. Since the lines occur at known wavelengths, it remains only to estimate their power. This can be done by signal restoration methods, which can use the information about this specific signal. With the use of such methods, the frequency spectrum of the lamp may be estimated by combining the frequency

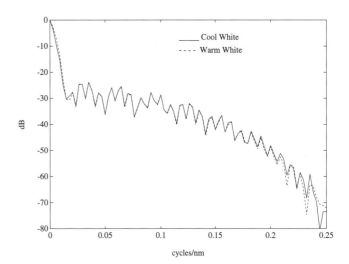

FIGURE 10   Power spectra of cool white fluorescent and warm white fluorescent.

spectra of its components

$$L(\omega) = L_d(\omega) + \sum_{k=1}^{q} \alpha_k e^{j\omega(\lambda_0 - \lambda_k)}, \qquad (42)$$

where λ_0 is an arbitrary origin in the wavelength domain. The bandlimited spectra $L_d(\omega)$ can be obtained from the sampled restoration and is easily represented by 10-nm sampling.

7 Color I/O Device Calibration

In Section 2, we briefly discussed control of gray-scale output. Here, a more formal approach to output calibration will be given. We can apply this approach to monochrome images by considering only a single band, corresponding to the CIE Y channel. In order to mathematically describe color output calibration, we need to consider the relationships between the color spaces defined by the output device control values and the colorimetric space defined by the CIE.

7.1 Calibration Definitions and Terminology

A *device-independent color space* is defined as any space that has a one-to-one mapping onto the CIE *XYZ* color space. Examples of CIE device-independent color spaces include *XYZ*, LAB, LUV, and *Yxy*. Current image format standards, such as JPEG, support the description of color in LAB. By definition, a *device-dependent color space* cannot have a one-to-one mapping onto the CIE *XYZ* color space. In the case of a recording device (e.g., scanners), the device-dependent values describe the response of that particular device to color. For a reproduction device (e.g., printers), the device dependent values describe only those colors the device can produce.

The use of device dependent descriptions of color presents a problem in the world of networked computers and printers. A single *RGB* or *CMYK* vector can result in different colors on different display devices. Transferring images colorimetrically between multiple monitors and printers with device dependent descriptions is difficult since the user must know the characteristics of the device for which the original image is defined, in addition to those of the display device.

It is more efficient to define images in terms of a CIE color space and then transform these data to device-dependent descriptors for the display device. The advantage of this approach is that the same image data are easily ported to a variety of devices. To do this, it is necessary to determine a mapping, $\mathcal{F}_{\text{device}}(\cdot)$, from device-dependent control values to a CIE color space.

Modern printers and display devices are limited in the colors they can produce. This limited set of colors is defined as the *gamut* of the device. If Ω_{CIE} is the range of values in the selected CIE color space and Ω_{print} is the range of the device control

values, then the set

$$G = \{t \in \Omega_{\text{CIE}} \mid \text{there exists } \mathbf{c} \in \Omega_{\text{print}} \text{ where } \mathcal{F}_{\text{device}}(\mathbf{c}) = \mathbf{t}\}$$

defines the gamut of the color output device. For colors in the gamut, there will exist a mapping between the device-dependent control values and the CIE *XYZ* color space. Colors that are in the complement, G^c, cannot be reproduced and must be *gamut-mapped* to a color that is within G. The gamut mapping algorithm $\mathcal{D}$ is a mapping from Ω_{CIE} to G, that is, $\mathcal{D}(\mathbf{t}) \in G \;\; \forall \mathbf{t} \in \Omega_{\text{CIE}}$. A more detailed discussion of gamut mapping is found in [21].

The mappings $\mathcal{F}_{\text{device}}$, $\mathcal{F}_{\text{device}}^{-1}$, and $\mathcal{D}$ make up what is defined as a *device profile*. These mappings describe how to transform between a CIE color space and the device control values. The International Color Commission (ICC) has suggested a standard format for describing a profile. This standard profile can be based on a physical model (common for monitors) or a look-up table (LUT) (common for printers and scanners) [22]. In the next sections, we will mathematically discuss the problem of creating a profile.

7.2 CRT Calibration

A monitor is often used to provide a preview for the printing process, as well as comparison of image processing methods. Monitor calibration is almost always based on a physical model of the device [23–25]. A typical model is

$$r' = (r - r_0)/(r_{\max} - r_0)^{\gamma_r},$$
$$g' = (g - g_0)/(g_{\max} - g_0)^{\gamma_g},$$
$$b' = (b - b_0)/(b_{\max} - b_0)^{\gamma_b},$$
$$\mathbf{t} = \mathbf{H}[r', g', b']^T,$$

where $\mathbf{t}$ is the CIE value produced by driving the monitor with control value $\mathbf{c} = [r, g, b]^T$. The value of the tristimulus vector is obtained by using a colorimeter or spectrophotometer.

Creating a profile for a monitor involves the determination of these parameters where $r_{\max}$, $g_{\max}$, $b_{\max}$ are the maximum values of the control values (e.g., 255). To determine the parameters, a series of color patches is displayed on the CRT and measured with a colorimeter, which will provide pairs of CIE values $\{t_k\}$ and control values $\{c_k\}$, $k = 1, \ldots, M$.

Values for γ_r, γ_g, γ_b, r_0, g_0, and b_0 are determined such that the elements of $[r', g', b']$ are linear with respect to the elements of *XYZ* and scaled between the range $[0, 1]$. The matrix $\mathbf{H}$ is then determined from the tristimulus values of the CRT phosphors at maximum luminance. Specifically, the mapping is given by

$$\begin{bmatrix} X \\ Y \\ Z \end{bmatrix} = \begin{bmatrix} X_{R\max} & X_{R\max} & X_{R\max} \\ Y_{G\max} & Y_{G\max} & Y_{G\max} \\ Z_{B\max} & Z_{B\max} & Z_{B\max} \end{bmatrix} \begin{bmatrix} r' \\ b' \\ g' \end{bmatrix},$$

where $[X_{R\max} Y_{R\max} Z_{R\max}]^T$ is the CIE *XYZ* tristimulus value of the red phosphor for control value $\mathbf{c} = [r_{\max}, 0, 0]^T$.

This standard model is often used to provide an approximation to the mapping $\mathcal{F}_{\text{monitor}}(\mathbf{c}) = \mathbf{t}$. Problems such as spatial variation of the screen or electron gun dependence are typically ignored. A LUT can also be used for the monitor profile in a manner similar to that described below for the scanner calibration.

7.3 Scanners and Cameras

Mathematically, the recording process of a scanner or camera can be expressed as

$$\mathbf{z}_i = \mathcal{H}(\mathbf{M}^T \mathbf{r}_i),$$

where the matrix $\mathbf{M}$ contains the spectral sensitivity (including the scanner illuminant) of the three (or more) bands of the device, $\mathbf{r}_i$ is the spectral reflectance at spatial point i, $\mathcal{H}$ models any nonlinearities in the scanner (invertible in the range of interest), and $\mathbf{z}_i$ is the vector of recorded values.

We define *colorimetric recording* as the process of recording an image such that the CIE values of the image can be recovered from the recorded data. This reflects the requirements of ideal sampling in Section 5.4. Given such a scanner, the calibration problem is to determine the continuous mapping $\mathcal{F}_{\text{scan}}$ that will transform the recorded values to a CIE color space:

$$\mathbf{t} = \mathbf{A}^T \mathbf{L} \mathbf{r} = \mathcal{F}_{\text{scan}}(\mathbf{z}) \quad \text{for all } \mathbf{r} \in \Omega_r.$$

Unfortunately, most scanners and especially desktop scanners are not colorimetric. This is caused by physical limitations on the scanner illuminants and filters that prevent them from being within a linear transformation of the CIE color matching functions. Work related to designing optimal approximations is found in [26, 27].

For the noncolorimetric scanner, there will exist spectral reflectances that look different to the standard human observer but when scanned produce the same recorded values. These colors are defined as being metameric to the scanner. This cannot be corrected by any transformation $\mathcal{F}_{\text{scan}}$.

Fortunately, there will always (except for degenerate cases) exist a set of reflectance spectra over which a transformation from scan values to CIE *XYZ* values will exist. Such a set can be expressed mathematically as

$$B_{\text{scan}} = \{\mathbf{r} \in \Omega_r \mid \mathcal{F}_{\text{scan}}(\mathcal{H}(\mathbf{M}\mathbf{r})) = \mathbf{A}^T L r\},$$

where $\mathcal{F}_{\text{scan}}$ is the transformation from scanned values to colorimetric descriptors for the set of reflectance spectra in B_{scan}. This is a restriction to a set of reflectance spectra over which the continuous mapping $\mathcal{F}_{\text{scan}}$ exists.

Look-up tables, neural nets, and nonlinear and linear models for $\mathcal{F}_{\text{scan}}$ have been used to calibrate color scanners [28–31, 32].

In all of these approaches, the first step is to select a collection of color patches that span the colors of interest. These colors should not be metameric to the scanner or to the standard observer under the viewing illuminant. This constraint ensures a one-to-one mapping between the scan values and the device-independent values across these samples. In practice, this constraint is easily obtained. The reflectance spectra of these M_q color patches will be denoted by $\{\mathbf{q}\}_k$ for $1 \leq k \leq M_q$.

These patches are measured by using a spectrophotometer or a colorimeter, which will provide the device-independent values

$$\{\mathbf{t}_k = \mathbf{A}^T \mathbf{q}_k\} \quad \text{for } 1 \leq k \leq M_q.$$

Without loss of generality, $\{\mathbf{t}_k\}$ could represent any colorimetric or device-independent values, e.g., CIE LAB, CIE LUV in which case $\{\mathbf{t}_k = \mathcal{L}(\mathbf{A}^T \mathbf{q}_k)\}$ where $\mathcal{L}(\cdot)$ is the transformation from CIE *XYZ* to the appropriate color space. The patches are also measured with the scanner to be calibrated providing $\{\mathbf{z}_k = \mathcal{H}(\mathbf{M}^T \mathbf{q}_k)\}$ for $1 \leq k \leq M_q$. Mathematically, the calibration problem is to find a transformation $\mathcal{F}_{\text{scan}}$ where

$$\mathcal{F}_{\text{scan}} = \arg\left(\min_{\mathcal{F}} \sum_{i=1}^{M_q} \|\mathcal{F}(\mathbf{z}_i) - \mathbf{t}_i\|^2\right)$$

and $\|\cdot\|^2$ is the error metric in the CIE color space. In practice, it may be necessary and desirable to incorporate constraints on $\mathcal{F}_{\text{scan}}$ [31].

7.4 Printers

Printer calibration is difficult because of the nonlinearity of the printing process, and the wide variety of methods used for color printing (e.g., lithography, inkjet, dye sublimation, etc.). Thus, printing devices are often calibrated with an LUT with the continuum of values found by interpolating between points in the LUT [28, 33].

For a profile of a printer to be produced, a subset of values spanning the space of allowable control values, $\mathbf{c}_k$ for $1 \leq k \leq M_p$, for the printer is first selected. These values produce a set of reflectance spectra that are denoted by $\mathbf{p}_k$ for $1 \leq k \leq M_p$.

The patches $\mathbf{p}_k$ are measured by using a colorimetric device that provides the values

$$\{\mathbf{t}_k = \mathbf{A}^T \mathbf{p}_k\} \quad \text{for } 1 \leq k \leq M_p.$$

The problem is then to determine a mapping $\mathcal{F}_{\text{print}}$, which is the solution to the optimization problem

$$\mathcal{F}_{\text{print}} = \arg\left(\min_{\mathcal{F}} \sum_{i=1}^{M_p} \|\mathcal{F}(\mathbf{c}_i) - \mathbf{t}_i\|^2\right),$$

FIGURE 11   Original Lena. (See color section, p. C–11.)

FIGURE 12   Calibrated Lena. (See color section, p. C–11.)

FIGURE 13   New scan Lena. (See color section, p. C–12.)

where as in the scanner calibration problem, there may be constraints which $\mathcal{F}_{\text{print}}$ must satisfy.

7.5 Calibration Example

Before an example of the need for calibrated scanners and displays is presented, it is necessary to state some problems with the display to be used, i.e., the color printed page. Currently, printers and publishers do not use the CIE values for printing but judge the quality of their prints by subjective methods. Thus, it is impossible to numerically specify the image values to the publisher of this book. We have to rely on the experience of the company to produce images that faithfully reproduce those given them. Every effort has been made to reproduce the images as accurately as possible. The tiff image format allows the specification of CIE values, and the images defined by those values can be found on the ftp site, ftp.ncsu.edu in directory pub/hjt/calibration. Even in the tiff format, problems arise because of quantization to 8 bits.

The original color "Lena" image is available in many places as an *RGB* image. The problem is that there is no standard to which the *RGB* channels refer. The image is usually printed to an *RGB* device (one that takes *RGB* values as input) with no transformation. An example of this is shown in Fig. 11. This image compares well with current printed versions of this image, e.g., those shown in papers in the special issue on color image processing of the *IEEE Transactions on Image Processing* [34]. However, the displayed image does not compare favorably with the original. An original copy of the image was obtained and scanned by using a calibrated scanner, and then printed by using a calibrated printer. The result, shown in Fig. 12, does compare well with the original. Even with the display problem mentioned above, it is clear that the images are sufficiently different to make the point that calibration is necessary for accurate comparisons of any processing method that uses color images. To complete the comparison, the *RGB* image that was used to create the corrected image shown in Fig. 12 was also printed directly on the *RGB* printer. The result, shown in Fig. 13, further demonstrates the need for calibration. A complete discussion of this calibration experiment is found in [21].

8 Summary and Future Outlook

The major portion of the chapter emphasized the problems and differences in treating the color dimension of image data. Understanding of the basics of uniform sampling is required to proceed to the problems of sampling the color component. The phenomenon of aliasing is generalized to color sampling by noting that the goal of most color sampling is to reproduce the sensation of color and not the actual color spectrum. The calibration of recording and display devices is required for accurate representation of images. The proper recording and display outlined in Section 7 cannot be overemphasized.

Although the fundamentals of image recording and display are well understood by experts in that area, they are not well appreciated by the general image processing community. It is hoped that future work will help widen the understanding of this aspect of image processing. At present, it is fairly difficult to calibrate color image I/O devices. The interface between the devices and the interpretation of the data is still problematic. Future work can make it easier for the average user to obtain, process, and display accurate color images.

Acknowledgments

The author acknowledges Michael Vrhel for his contribution to the section on color calibration. Most of the material in that section was the result of a joint paper with him [21].

References

[1] MATLAB, High Performance Numeric Computation and Visualization Software, The Mathworks Inc., Natick, MA.

[2] A. V. Oppenheim and R. W. Schafer, *Discrete-Time Signal Processing* (Prentice-Hall, Englewood Cliffs, NJ, 1989).

[3] A. K. Jain, *Fundamentals of Digital Image Processing* (Prentice-Hall, Englewood Cliffs, NJ, 1989).

[4] N. P. Galatsanos and R. T. Chin, "Digital restoration of multichannel images," *IEEE Trans. Acoust. Speech Signal Process.* **37**, 415–421 (1989).

[5] G. Wyszecki and W. S. Stiles, *Color Science: Concepts and Methods, Quantitative Data and Formulae*, 2nd ed. (Wiley, New York, 1982).

[6] D. E. Dudgeon and R. M. Mersereau, *Multidimensional Digital Signal Processing* (Prentice-Hall, Englewood Cliffs, NJ, 1984).

[7] H. B. Barlow and J. D. Mollon, *The Senses* (Cambridge U. Press, Cambridge, U.K., 1982).

[8] B. A Wandell, *Foundations of Vision* (Sinauer, Sunderland, MA, 1995).

[9] H. Grassmann, "Zur Therorie der Farbenmischung," Ann. Physik Chem. **89**, 69–84 (1853).

[10] H. Grassmann, "On the theory of compound colours," *Philos. Mag.* **7**, 254–264 (1854).

[11] J. B. Cohen and W. E. Kappauf, "Metameric color stimuli, fundamental metamers, and Wyszecki's metameric blacks," *Am. J. Psychol.* **95**, 537–564 (1982).

[12] B. K. P. Horn, "Exact reproduction of colored images," *Comput. Vis. Graph. Image Process.* **26**, 135–167 (1984).

[13] H. J. Trussell, "Application of set theoretic methods to color systems," *Color: Res. Appl.* **16**, 31–41 (1991).

[14] B. A. Wandell, "The synthesis and analysis of color images," *IEEE Trans. Pattern Anal. Machine Intell.* **9**, 2–13 (1987).

[15] H. J. Trussell and M. S. Kulkarni, "Sampling and processing of color signals," *IEEE Trans. Image Process.* **5**, 677–681 (1996).

[16] P. L. Vora and H. J. Trussell, "Measure of goodness of a set of colour scanning filters," *J. Opt. Soc. Am.* **10**, 1499–1508 (1993).

[17] M. J. Vrhel, and H. J. Trussell, "Optimal color filters in the presence of noise," *IEEE Trans. Image Process.* **4**, 814–823 (1995).

[18] D. L. MacAdam, "Visual sensitivies to color differences in daylight," *J. Opt. Soc. Am.* **32**, 247–274 (1942).

[19] G. Wyszecki and G. H. Felder, "New color matching ellipses," *J. Opt. Soc. Am.* **62**, 1501–1513 (1971).

[20] M. J. Vrhel, R. Gershon, and L. S. Iwan, "Measurement and analysis of object reflectance spectra," *Color Res. Appl.* **19**, 4–9 (1994).

[21] M. J. Vrhel, and H. J. Trussell, "Color device calibration: a mathematical formulation," *IEEE Trans. Image Process.* **8**, 1796–1806 (1999).

[22] International Color Consortium, International Color Consortium Profile Format Version 3.4, available at http://color.org/.

[23] W. B. Cowan, "An inexpensive scheme for calibration of a color monitor in terms of standard CIE coordinates," *Comput. Graph.* **17**, 315–321 (1983).

[24] R. S. Berns, R. J. Motta, and M. E. Grozynski, "CRT colorimetry. Part I: Theory and practice," *Color Res. Appl.* **18**, 5–39 (1988).

[25] R. S. Berns, R. J. Motta, and M. E. Grozynski, "CRT colorimetry. Part II: Metrology," *Color Res. Appl.* **18**, 315–325 (1988).

[26] P. L. Vora and H. J. Trussell, "Mathematical methods for the design of color scanning filters," *IEEE Trans. Image Process.* **6**, 312–320 (1997).

[27] G. Sharma, H. J. Trussell, and M. J. Vrhel, "Optimal nonnegative color scanning filters," *IEEE Trans. Image Process.* **7**, 129–133 (1998).

[28] P. C. Hung, "Colorimetric calibration in electronic imaging devices using a look-up table model and interpolations," *J. Electron. Imag.* **2**, 53–61 (1993).

[29] H. R Kang and P. G. Anderson, "Neural network applications to the color scanner and printer calibrations," *J. Electron. Imag.* **1**, 125–134 (1992).

[30] H. Haneishi, T. Hirao, A. Shimazu, and Y. Mikaye, "Colorimetric precision in scanner calibration using matrices," in *Procedings of the Third IS&T/SID Color Imaging Conference: Color Science, Systems and Applications* (Society for Imaging Science and Technology (IS&T), Springfield, VA; Scottsdale, AZ, 1995), pp. 106–108.

[31] H. R. Kang, "Color scanner calibration," *J. Imag. Sci. Technol.* **36**, 162–170 (1992).

[32] M. J. Vrhel, and H. J. Trussell, "Color scanner calibration via neural networks," presented at the Conference on Acoustics, Speech and Signal Processing, Phoenix, AZ, March 15–19, 1999.

[33] J. Z. Chang, J. P. Allebach, and C. A. Bouman, "Sequential linear interpolation of multidimensional functions," *IEEE Trans. Image Process.* **6**, 1231–1245 (1997).

[34] *IEEE Trans. Image Process.* **6** (1997).

4.7

Statistical Methods
for Image Segmentation

Sridhar Lakshmanan
*University of Michigan —
Dearborn*

1 Introduction

Segmentation is a fundamental low-level operation on images. If an image is already partitioned into segments, where each segment is a "homogeneous" region, then a number of subsequent image processing tasks become easier. A homogeneous region refers to a group of connected pixels in the image that share a common feature. This feature could be brightness, color, texture, motion, etc. (see Fig. 1). References [1–5] contain exploratory articles on image segmentation, and they provide an excellent place to start for any newcomer researching this topic.

Boundary detection is the dual goal of image segmentation. After all, if the boundaries between segments are specified then it is equivalent to identifying the individual segments themselves. However, there is one important difference. In the process of image segmentation, one obtains regionwise information regarding the individual segments. This information can then be subsequently used to classify the individual segments. Unfortunately, detection of the boundaries between segments does not automatically yield regionwise information about the individual segments. So, further image analysis is necessary before any segment-based classification can be attempted. Since segmentation, and not classification, is the focus of this chapter, from here on image segmentation is meant to include the dual problem of boundary detection as well. Note

that boundary detection is distinctly different from edge detection. Edges are typically detected by examining the local variation of image intensity or color. Edge detection is covered in two separate chapters, Chapters 4.11 and 4.12, of this handbook.

The importance of segmentation is clear by the central role it plays in a number of applications that involve image and video processing — remote sensing, medical imaging, intelligent vehicles, video compression, and so on. The success or failure of segmentation algorithms in any of these applications is heavily dependent on the type of feature(s) used,[1] the reliability with which these features are extracted, and the criteria used for merging pixels based on the similarity of their features.

As one can gather from [1–5], there are many ways to segment an image. So, the question is, Why statistical methods? Statistical methods are a popular choice for image segmentation because they involve image features that are simple to interpret by using a model, features that are easy to compute from a given image, and merging methods that are firmly rooted in statistical/mathematical inference. Although there is no explicit consensus in the image processing community that statistical methods are the way to go as far image segmentation is concerned, the volume and diversity of publications certainly seem

---

[1] Features refer to image attributes such brightness, color, texture, motion, etc.

FIGURE 1   Examples of images that could be segmented based on brightness (top left), color (top right), motion (middle row), and texture (bottom row). (See color section, p. C–13.)

to indicate that they are a very popular choice. An example serves to illustrate the point. Consider the images in Fig. 2. It is clear that each of these images consists of four homogenous regions delineated by spatially contiguous boundaries, and pixels belonging to each region share a common texture. So the question becomes, How does one represent, identify, and model the spatial continuity/relationship that exists between pixels that share a common texture? This is not a trivial question, because two blocks of pixels from two entirely different parts of the image

may exhibit the same type of distribution[2] that makes them indistinguishable. In contrast, if one examines statistics other than the block mean and variance, the textural differences between pixels belonging those two blocks become more apparent. The rest of this chapter is devoted to describing a handful of these "other" statistics and their role in image segmentation.

[2]For example, the blocks may have the same average pixel intensity values (means) and spread of intensity values about the mean (variances).

FIGURE 2   Collection of images; in each there are four clearly distinguishable segments. (See color section, p. C–14.)

2 Image Segmentation: The Mathematical Problem

Let $\Omega = \{(m, n): 1 \leq m \leq M$ and $1 \leq n \leq N\}$ denote an $M \times N$ lattice of points (m, n). An observed image f is a function defined on this domain Ω, and for any given point (m, n) the observation $f(m, n)$ at that point takes a value from a set Λ. Two common examples for the set Λ are $\Lambda = \{\lambda: 0 \leq \lambda \leq 255\}$ for black–white images, and $\Lambda = \{(\lambda_1, \lambda_2, \lambda_3): 0 \leq \lambda_1 \leq 255, 0 \leq \lambda_2 \leq 255,$ and $0 \leq \lambda_3 \leq 255\}$ for color (red, green, and blue channel) images. A segmented image g is also a function on the same domain Ω, but for any given point (m, n) the segmentation $g(m, n)$ at that point takes a value from a different set Γ. Two common examples for the set Γ are: $\Gamma = \{\gamma: \gamma = 0$ or $1\}$, denoting the two segments in a binary segmentation, and $\Gamma = \{\gamma: \gamma = 1, 2, 3, 4, \ldots, k\}$, denoting the k different segments in the case of a multiclass segmentation. Of course, g could also denote a boundary image. For any given point (m, n), $g(m, n) = 1$ could denote the presence of a boundary at that point and $g(m, n) = 0$ the absence.

Given a particular realization of the observed image $f = f_0$, the problem of image segmentation is one of estimating the corresponding segmented image using $g_0 = h(f_0)$. Statistical methods for image segmentation provide a coherent derivation of this estimator function $h(.)$.

3 Image Statistics for Segmentation

To understand the role of statistics in image segmentation, let us examine some preliminary functions that operate on images. Given an image f_0 that is observed over the lattice Ω, suppose that $\Omega_1 \subseteq \Omega$ and f_1 is a restriction of f_0 to only those pixels that belong to Ω_1. Then, one can define a variety of statistics that capture the spatial continuity of the pixels that comprise f_1. Here are some common examples.

3.1 Gaussian Statistics

$$T_{f_1}(p, q) = \sum_{(m, n) \in \Omega_1} [f_1(m, n) - f_1(m + p, n + q)]^2,$$

$$\text{for } (p, q) \in \{(0, 1), (1, 0), (1, 1), (1, -1), \ldots\} \quad (1)$$

measures the amount of variability in the pixels that comprise f_1 along the (p, q)th direction. For a certain f_1, if $T_{f_1}(0, 1)$ is very small, for example, then that implies that f_1 has a little or no variability along the $(0, 1)$th (i.e., horizontal) direction. Computation of this statistic is straightforward, as it is merely a quadratic operation on the difference between intensity values of adjacent (neighboring) pixels. $T_{f_1}(p, q)$ and minor variation thereof is referred to as the Gaussian statistic and is widely used in statistical methods for segmentation of gray-tone images; see [6, 7].

3.2 Fourier Statistics

$$
\begin{aligned}
F_{f_1}(\alpha, \beta) = \sum_{(m, n) \in \Omega_1} & \left[f_1(m, n) e^{-\sqrt{-1}(m\alpha + n\beta)} \right] \\
& \times \left[f_1(m, n) e^{\sqrt{-1}(m\alpha + n\beta)} \right], \\
& \text{for } (\alpha, \beta) \in [-\pi, \pi]^2 \quad (2)
\end{aligned}
$$

measures the amount of energy in frequency bin (α, β) that the pixels that comprise f_1 possess. For a certain f_1, if $F_{f_1}(0, 20\pi/N)$ has a large value, for example, then that implies that f_1 has a significant cyclical variation of the $(0, 20\pi/N)$ (i.e., horizontally every 10 pixels) frequency. Computation of this statistic is more complicated that the Gaussian one. The use of fast Fourier transform algorithms, however, can significantly reduce the associated burden. $F_{f_1}(\alpha, \beta)$, called the periodogram statistic, is also used in statistical methods for segmentation of textured images; see [8, 9].

3.3 Covariance Statistics

$$
\begin{aligned}
K_{f_1} &= \sum_{(m, n) \in \Omega_1} (f_1(m, n) - \mu_{f_1})^T (f_1(m, n) - \mu_{f_1}), \quad \text{where} \\
\mu_{f_1} &= \sum_{(m, n) \in \Omega_1} f_1(m, n) \quad (3)
\end{aligned}
$$

measures the correlation between the various components that comprise each pixel of f_1. If K_{f_1} is a 3×3 matrix and $K_{f_1}(1, 2)$ has large value, for example, then that means that components 1 and 2 (could be the red and green channels) of the pixels that make up f_1 are highly correlated. Computation of this statistic is very time consuming, even more so than the Fourier one, and there are no known methods to alleviate this burden. K_{f_1} is called the covariance matrix of f_1, and this too has played a substantial role in statistical methods for segmentation of color images; see [10, 11].

3.4 Label Statistics

$$
L_{g_1}(m, n) = \Psi[g_1(m, n), g_1(m + p, n + q)], \quad \text{where}
$$

$$
\Psi(a, b) = \begin{cases} 1 & \text{if } a = b \\ -1 & \text{if } a \neq b \end{cases}
$$

$$
\text{for } (p, q) \in \{(0, 1), (1, 0), (1, 1), (1, -1), \ldots\} \quad (4)
$$

measures the amount of homogeneity in the pixels that comprise g_1 along the (p, q)th direction. For a certain g_1, if $L_{g_1}(1, 1)$ is very large, for example, then that implies that g_1 has a little or no variability along the $(1, 1)$th (i.e., 135° diagonal) direction. Computation of this statistic is straightforward, as it is merely an indicator operation on the difference between label values of adjacent (neighboring) pixels. $L_{g_1}(p, q)$ and minor variation thereof is referred to as the label statistic and is widely used in statistical methods for restoration of gray-tone images; see [12, 13].

4 Statistical Image Segmentation

Computation of image statistics of the type defined in Section 3 tremendously facilitates the task of image segmentation. As a way to illustrate their utility, three image segmentation problems that arise in three distinctly different applications are presented in the paragraphs that follow. In each case, a description of the how a solution was arrived at using statistical methods is given.

4.1 Vehicle Segmentation

Today, there is a desire among consumers worldwide for automotive accessories that make their driving experience safer and more convenient. Studies have shown that consumers believe that safety and convenience accessories are important in their new-car purchasing decision. In response to this growing market, the automotive industry, in cooperation with government agencies, has embarked on programs to develop new safety and convenience technologies. These include, but are not limited to, collision warning (CW) systems, lane departure warning (LDW) systems, and intelligent cruise control (ICC) systems. These systems and others comprise an area of study referred to as intelligent transportation systems (ITS), or more broadly, intelligent vehicle highway systems (IVHS).

An important image segmentation problem within ITS is one of segmenting vehicles from their background; see [14]. Figure 3 contains a typical image in which a vehicle has to be segmented

FIGURE 3 Typical image in which a vehicle has to be segmented from the background. (See color section, p. C–14.)

FIGURE 4 Fisher color distance between pixels inside and outside of a square template placed on top of the image in Fig. 3. The template hypothesis on the right has higher merit than the one on the left. (See color section, p. C–15.)

from its background. In the following paragraphs, a statistical method for this segmentation is described. The vehicle of interest, it is assumed,[3] is merely a square that is described by three parameters (V_b, V_l, V_w) corresponding to the bottom edges, left edges, and width of the square. Different values of these three parameters yield vehicles of different sizes and positions within the image.

Vehicles seldom tend to be too big or small,[4] and so depending on the distance of the vehicle from the camera, it is possible to expect the width of the vehicle to be within a certain range. Suppose that W_{min} and W_{max} denote this range; then

$$P(V_w) = \begin{cases} 1/(W_{max} - W_{min}) & \text{if } W_{min} \le V_w \le W_{max} \\ 0 & \text{otherwise} \end{cases} \quad (5)$$

is a probability density function (pdf) that enforces the strict constraint that V_w be between W_{min} and W_{max}. Since it is a probability over (one of) the quantities being estimated, it is commonly referred to as a prior pdf, or simply a prior.

Let (v_b, v_l, v_w) denote a specific hypothesis of the unknown vehicle parameters (V_b, V_l, V_w). The merit of this hypothesis is decided by another probability, called the likelihood pdf, or simply the likelihood. In this application, it is appealing to decide the merit of a hypothesis by evaluating the difference in color between pixels that are inside the square (i.e., the pixels that are hypothesized to be the vehicle) and those that immediately surround the square (i.e., the pixels that are in the immediate background of the hypothesized vehicle). The specific color difference evaluator that is employed is called the Fisher distance:

$$\text{FishDist}\,(v_b, v_l, v_w) = (\mu_1 - \mu_2)^T (K_1 + K_2)^{-1} (\mu_1 - \mu_2), \quad (6)$$

where μ_1 and K_1 are the mean and covariance of the pixels

that are inside the hypothesized square — computed by using Eq. (3) — and μ_2 and K_2 are the mean and covariance of pixels that are immediately surrounding the hypothesized square. Hypotheses corresponding to a large color difference between pixels inside and immediately surrounding the square have more merit (and hence a higher probability of occurrence) than those with smaller color difference; see Fig. 4.

The problem of segmenting a vehicle from its background boils down to estimating the three parameters (V_b, V_l, V_w) from the given color image. An optimal[5] estimate of these parameters is the one that maximizes the product of the prior and likelihood probabilities in Eqs. (5) and (6), respectively — the so-called maximum *a posteriori* (MAP) estimate. Figure 5 shows a few examples of estimating the correct (V_b, V_l, V_w) by using this procedure. This same procedure can also be adapted to segment images in other applications. Figure 6 shows a few examples in which the procedure has been used to segment images that are entirely different from those in Figs. 3–5.

4.2 Aerial Image Segmentation

Accurate maps have widespread uses in modern day-to-day living. Maps of urban and rural areas are regularly used in an entire spectrum of civilian and military tasks, starting from simple ones like obtaining driving directions all the way to complicated ones like highway planning. Maps themselves are just a portion of the information, and they are typically used to index other important geophysical attributes such as weather, traffic, population, size, and so on. Large systems called geographical information systems (GISs) collate, maintain, and deliver maps, weather, population, and the like on demand.

Image segmentation is a tool that finds widespread use in the creation and maintenance of a GIS. One example pertains to the operator-assisted updating of old maps by using aerial images, in which segmentation is used to supplement or complement the human operator; see [15]. Shown in Fig. 7 is an aerial image

---

[3]This is a valid assumption when the rear view of the vehicle is obtained from a camera placed at ground level.

[4]Even accounting for the variations in the actual physical dimensions of the vehicle.

[5]Optimal in the sense that among all estimates of the parameters, this is the one that minimizes the probability of making an error.

FIGURE 5 Correct estimation of the vehicle ahead, using the MAP procedure. (See color section, p. C–15.)

FIGURE 6 Segmentation of other images, using the same Fisher color distance. Top: A segmentation that yields all segments that contains the color white. Bottom: A segmentation that yields all segments that do not contain the color green. (See color section, p. C–16.)

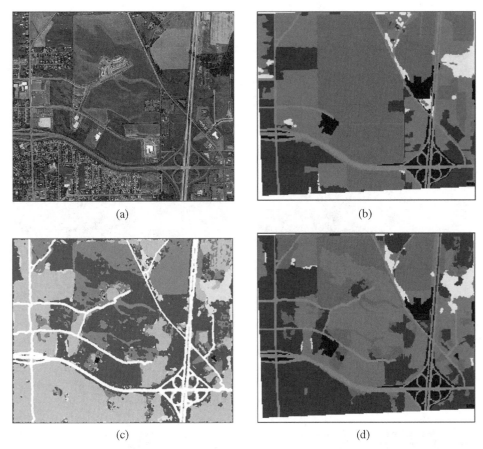

FIGURE 7 Updating old maps using image segmentation. (a) Aerial image of Eugene, Oregon in 1993. (b) Map of the same area in 1987. (c) Operator-assisted segmentation of the 1993 aerial image. (d) Updated map in 1993. (See color section, p. C–16 and 17.)

of the Eugene, Oregon area taken in 1993. Accompanying the aerial image is an old (1987) map of the same area that indicates what portion of that area contains brown crops (in red), grass (in green), development (in blue), forest (in yellow), major roads (in gray), and everything else (in black). The aerial image indicates a significant amount of change in the area's composition from the time the old map was constructed. Especially noticeable is the new development of a road network south of the highway, in an area that used to be a large brown field of crops. The idea is to use the new 1993 aerial image in order to update or correct the old 1987 map. The human operator examines the aerial image and chooses a collection of polygons corresponding to various homogeneous segments of the image. By use of the pixels with these polygons as a training sample, a statistical segmentation of the aerial image is effected; the segmentation result is also shown in Fig. 7. Regions in the old map are compared to segments of the new image, and where they are different, the old map is updated or corrected. The resulting new map is shown Fig. 7 as well.

The segmentation procedure used for this map updating application is based on Gaussian statistics; see Eq. (1). Specifically, for each homogeneous polygonal region selected in the aerial image by the human operator, the Gaussian statistics for that polygon are automatically computed. With these statistics, a

model of probable variation in the pixels' intensities within the polygon is subsequently created[6]:

$$P(f_l \mid \underline{\theta}_l) = \frac{1}{Z(\underline{\theta}_l)} \exp\left\{ -\sum_{p,q} T_{f_l}(p,q)\, \theta_l(p,q) \right\}, \qquad (7)$$

where f_l denotes the pixels within the lth polygon, $Z(\underline{\theta}_l)$ is a normalizing constant that makes $\sum_{f_l} P(f_l \mid \underline{\theta}_l) = 1$, and $\theta_l(p,q)$ are parameters chosen so that $P(f_l \mid \underline{\theta}_l) \geq P(f_l \mid \phi)$ for all $\underline{\phi} \neq \underline{\theta}_l$. Equation (7) forms the basis for segmenting the aerial image in Fig. 7. Suppose that there are k distinctly different polygonal segments, corresponding to k distinctly different $\underline{\theta}_l$ values; then each pixel (m, n) in the aerial image is classified according to a maximum likelihood rule. The probability of how likely $f(m, n)$ is if it were classified as belonging to the lth class is assessed according to Eq. (7), and the pixel is classified as belonging to class l if $P(f(m, n) \mid \underline{\theta}_l) \geq P(f(m, n) \mid \underline{\theta}_r)$ for all $r \neq l$. Shown in Fig. 8 is another example of segmenting an aerial image by using this same maximum likelihood statistical procedure.

---

[6]This model is referred to as the Gaussian Markov random field model; see [6–9].

FIGURE 8 Segmentation of another aerial image, this time of a rural crop field area, using the same texture-based maximum likelihood procedure employed in Fig. 7. (See color section, p. C–17.)

4.3 Segmentation for Image Compression

The enormous amount of image and video data that typifies many modern multimedia applications mandates the use of encoding techniques for their efficient storage and transmission.

The use of such encoding is standard in new personal computers, video games, digital video recorders, players, and disks, digital television, and so on. Image and video encoding schemes that are object based are most efficient (i.e., achieve the best compression rates), and they also facilitate many advanced multimedia

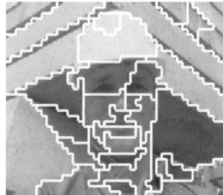

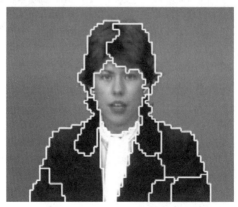

FIGURE 9 Block-based segmentation of images into large "homogeneous" objects, using a MAP estimation method that employs Fourier statistics.

functionalities. Object-based encoding of images and video, however, requires that the objects be delineated *a priori*. An obvious method for extracting objects in an image is by segmenting it.

Reference [16] describes a statistical image segmentation method that is particularly geared for object-based encoding of images and video. A given image is first divided into 8×8 blocks of pixels, and for each block, the Fourier statistics of the pixels in that block is computed. If the pixels f_1 within a single block have little or no variation, then $F_{f_1}(0, 0)$ will have a very large value; similarly, if the block contains a vertical edge, then $\sum_{\beta} F_{f_1}(0, \beta)$ will have a very large value, and so on. There are six such categories, corresponding to uniform/monotone, vertical edge, horizontal edge, 45° diagonal edge, 135° diagonal edge, and texture (randomly oriented edge). Let $t_{f_1}(1), t_{f_1}(2), \ldots, t_{f_1}(6)$, be the Fourier statistics-based quantities — one of their values will be large, corresponding to which of these six categories f_1 belongs.

If g denotes the collection of unknown block labels, then an estimate of g from f would correspond to an object-based segmentation of f. Reference [16] pursues a MAP estimate of g from f, where the prior pdf

$$P(g = g_0) = \frac{1}{Z} \exp \left\{ -\sum_{p,q} L_{g_0}(p, q) \right\}, \qquad (8)$$

and the likelihood pdf

$$P(f = f_0 \mid g = g_0) = \frac{1}{C(g_0)} \exp \left\{ -\sum_{m,n} t_{f_{m,n}}(g_0(m, n)) \right\}. \qquad (9)$$

Here Z and $C(g_0)$ are the normalizing constants for the prior and the likelihood pdfs, respectively, the index (m, n) denotes the 8×8 blocks, and $L_{g_1}(p, q)$ is the label statistic defined in Eq. (4). Figure 9 shows a few examples of image segmentation using this procedure.

5 Discussion

The previous four sections provide a mere sampling of the various statistical methods that are employed for image segmentation. References [17–20] contain some of the other methods. The main differences between those and the methods described in this chapter lie in the type of prior or likelihood pdfs employed.

In particular, [20] contains a method for image segmentation that is based on elastic deformation of templates. Rather than specify a prior pdf as probability over the space of all images, [20] specifies a prior pdf over the space of all deformations of a prototypical image. The space of deformation of the prototype image is a very rich one and even includes images that are quite distinctly different from the original. More importantly, the deformation space's dimension is significantly smaller than

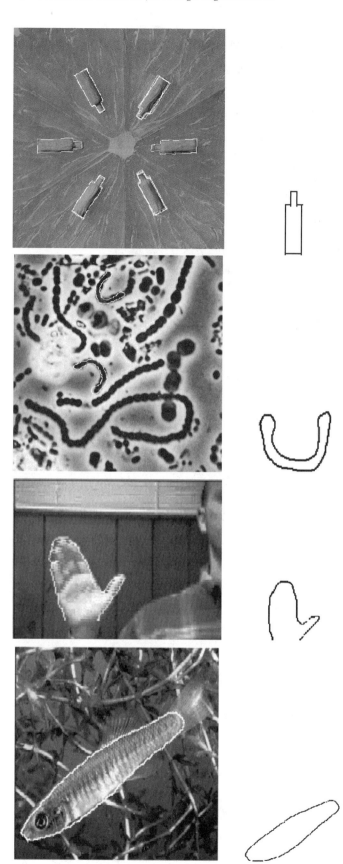

FIGURE 10 Segmenting objects out of images when they "resemble" the query.

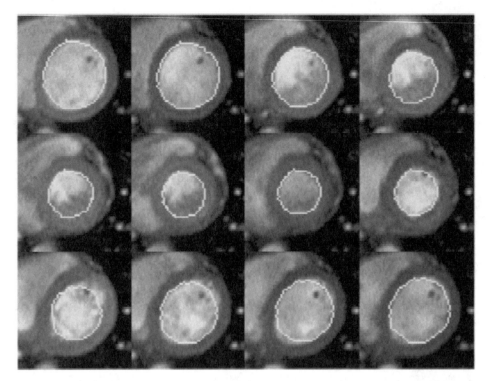

FIGURE 11   Tracking an object of interest, in this case a human heart, from frame to frame by using the elastic deformation method described in [20]. (See color section, p. C–18.)

the conventional space of all images that "resemble" the prototype. This smaller dimension pays tremendous dividends when it comes to image segmentation.

A query of image databases provides an important application where a prototype of an object to be segmented from a given image is readily available. A user may provide a typical object of interest — its approximate shape, color, and texture — and ask to retrieve all database images that contain objects similar to the one of interest. Figure 10 shows a few examples of the object(s) of interest being segmented out of a given image by using the elastic deformation method described in [20]. Figure 11 shows an example of tracking an object from frame to frame, using the same method.

As one can gather from this chapter, when statistical methods are employed for image segmentation, there is always an associated multivariate optimization problem. The number of variables involved in the problem varies according to the dimensionality of the prior pdfs domain space. For example, the MAP estimation procedure in the vehicle segmentation application has an associated three-parameter optimization problem, whereas the MAP estimation procedure in the segmentation for image compression application has an associated 64×64 parameter optimization problem. The functions that have to be maximized with respect to these variables are typically nonconcave and contain many local maxima. This implies that simple gradient-based optimization algorithms cannot be employed, as they are prone to converge to a local (as opposed to the global) maxima. Statistical methods for image segmentation abound

with a wide variety of algorithms to address such multivariate optimization problems. The reference list that follows this section contains several distinct examples: [12] contains the greedy iterated conditional maximum (ICM) algorithm; [9, 13, 18] contain a stochastic algorithm called Gibbs sampler (a simulated annealing procedure); [2] contains a randomized jump-diffusion algorithm; and finally, [20] contains a multiresolution algorithm. For a given application, there always appears to be a "most appropriate" algorithm, although any of the existing global optimization algorithms can conceptually be employed.

Acknowledgment

The figures used in this article are courtesy of C.-S. Won, M.-P. Dubuisson Jolly, and Y. Zhong.

References

[1] D. Geiger and A. Yuille, "A common framework for image segmentation," *Int. J. Comput. Vis.* **6**, 227–243 (1990).

[2] U. Grenander and M. I. Miller, "Representation of knowledge in complex systems," *J. Roy. Stat. Soc. B* **56**, 1–33 (1994).

[3] M. J. Swain and D. H. Ballard, "Color indexing," *Int. J. Comput. Vis.* **7**, 11–32 (1991).

[4] T. R. Reed and J. M. H. Du Buf, "A review of recent texture segmentation and feature extraction techniques," *Comput. Vis. Graph. Image Process. Image Understand.* **57**, 359–372 (1993).

[5] H.-H. Nagel, "Overview on image sequence analysis," in

T. S. Huang, ed. *Image Sequence Processing and Dynamic Scene Analysis* (Springer-Verlag, New York, 1983), pp. 2–39.

[6] J. Zhang, "Two-dimensional stochastic model-based image analysis," Ph.D. dissertation (Rensselaer Polytechnic Institute, Troy, NY, 1988).

[7] C.-S. Won and H. Derin, "Unsupervised segmentation of noisy and textured images using Markov random field models," *Comput. Vis. Graph. Image Process. Graph. Mod. Image Process.* **54**, 308–328 (1992).

[8] R. Chellappa, "Two-dimensional discrete Gaussian Markov random fields for image processing," in *Progress in Pattern Recognition*, L. N. Kanal and A. Rosenfeld, eds., Vol. 2 (North-Holland, Amsterdam, 1985).

[9] F. C. Jeng and J. W. Woods, "Compound Gauss–Markov random fields for image estimation," *IEEE Trans. Signal Process.* **39**, 683–691 (1991).

[10] D. Panjwani and G. Healy, "Markov random fields for unsupervised segmentation of textured color images," *IEEE Trans. Pattern Anal. Machine Intell.* **17**, 939–954 (1995).

[11] A. C. Hurlbert, "*The computation of color*," Ph.D. dissertation (Massachusetts Institute of Technology, Cambridge, MA, 1989).

[12] J. E. Besag, "On the statistical analysis of dirty pictures," *J. Roy. Stat. Soc. B* **48**, 259–302 (1986).

[13] S. Geman and D. Geman, "Stochastic relaxation, Gibbs distributions, and the Bayesian restoration of images," *IEEE Trans. Pattern Anal. Machine Intell.* **6**, 721–741 (1984).

[14] M. Beauvais and S. Lakshmanan, "CLARK: A heterogeneous sensor fusion method for finding lanes and obstacles," *Proc. IEEE Int. Conf. Intell. Veh.* **2**, 475–480 (1998).

[15] M-P. Dubuisson Jolly and A. Gupta, "Color and texture fusion: application to aerial image segmentation and GIS updating," *Proc. 3rd IEEE Workshop Appl. Comput. Vis.* 2–7 (1996).

[16] C. S. Won, "A block-based MAP segmentation for image compressions," *IEEE Trans. Circuit Syst. Video Technol.* **8**, 592–601 (1998).

[17] R. Chellappa and A. K. Jain, eds., *Markov Random Fields: Theory and Application* (Academic, New York, 1992).

[18] D. Geman and G. Reynolds, "Constrained restoration and the recovery of discontinuities," *IEEE Trans. Pattern Anal. Mach. Intell.* **14**, 367–383 (1992).

[19] J. Subrahmonia, D. Keren, and D. B. Cooper, "Recognizing mice, vegetables and hand printed characters based on implicit polynomials, invariants and Bayesian methods," *Proc. IEEE 4th Int. Conf. Comput. Vis.*, 320–324 (1993).

[20] Y. Zhong, "Object matching using deformable templates," Ph.D. dissertation (Michigan State University, East Lansing, MI, 1997).

4.8

Multiband Techniques for Texture Classification and Segmentation

B. S. Manjunath
University of California

G. M. Haley
Ameritech

W. Y. Ma
Hewlett-Packard Laboratories

1 Introduction

1.1 Image Texture

Texture as an image feature is very useful in many image processing and computer vision applications. There is extensive research on texture analysis in the image processing literature where the primary focus has been on classification, segmentation, and synthesis. Texture features have been used in diverse applications such as satellite and aerial image analysis, medical image analysis for the detection of abnormalities, and more recently, in image retrieval, using texture as a descriptor. In this chapter, we present an approach to characterizing texture by using a multiband decomposition of the image with application to classification, segmentation, and image retrieval.

In texture classification and segmentation, the objective is to partition the given image into a set of homogeneous textured regions. Aerial images are excellent examples of textured regions where different areas such as water, sand, vegetation, etc. have distinct texture signatures. In many other cases, such as in the classification of tissues in the magnetic resonance images of the brain, homogeneity is not that well defined. If an image consists of multiple textured regions, as is the case with most natural imagery, segmentation can be achieved through classification. This, however, is a chicken-and-egg problem, as classification requires an estimate of the region boundaries — note that texture is a region property and individual pixels are labeled based on information in a small neighborhood around the pixels. This may lead to problems near region boundaries, as the computed texture descriptors are corrupted from pixels that do not belong to the same region.

Early work on texture classification focused on spatial image statistics. These include image correlation [9], energy features [27], features from co-occurrence matrices [22], and run-length statistics [19]. During the past 15 years, much attention has been given to generative models, such as those using the Markov random fields (MRF) [7, 8, 11, 12, 15, 16, 23–26, 33]; also see Chapter 4.2 on MRF models. MRF-based methods have proven to be quite effective for texture synthesis, classification, and segmentation. Since general MRF models are inherently dependent on rotation, several methods have been introduced to obtain rotation invariance. Kashyap and Khotanzad [24] developed the "circular autoregressive" model with parameters that are invariant to image rotation. Choe and Kashyap [10] introduced an autoregressive fractional difference model that has rotation (as well as tilt and slant) invariant parameters. Cohen, Fan, and Patel [11] extended a likelihood function to incorporate rotation (and scale) parameters. To classify a sample, an estimate of its rotation (and scale) is required.

Much of the work in MRF models uses the image intensity as the primary feature. In contrast, spatial filtering methods derive the texture descriptors by using the filtered coefficient values. A compact representation of the filtered outputs is needed for classification or segmentation purposes. The first few moments of the filtered images are often used as feature vectors. For segmentation, one may consider abrupt transitions in the filtered image space or transformations of the filtered images. Malik and Perona [31], for example, argue that a nonlinear transformation of the filtered coefficients is necessary to model preattentive segmentation by humans.

Laws [27] is perhaps among the first to propose the use of energy features for texture classification. In recent years, multiscale decomposition of the images has been extensively used in deriving image texture descriptors and for segmentation [3, 4, 6, 17, 20, 21, 28, 34, 35, 38–40, 42]. Orthogonal wavelets (see Chapter 4.1) and Gabor wavelets have been widely used for computing such multiscale decompositions. Gabor functions are modulated Gaussians, and Section 2 describes the design of Gabor filters in detail.

For feature-based approaches, rotation invariance is achieved by using anisotropic features. Porat and Zeevi [40] use first- and second-order statistics based upon three spatially localized features, two of which (dominant spatial frequency and orientation of dominant spatial frequency) are derived from a Gabor-filtered image. Leung and Peterson [28] present two approaches, one that transforms a Gabor-filtered image into rotation-invariant features and the other of which rotates the image before filtering; however, neither utilizes the spatial resolving capabilities of the Gabor filter. You and Cohen [43] use filters that are tuned over a training set to provide high discrimination among its constituent textures. Greenspan *et al.* [20] use rotation-invariant structural features obtained by multiresolution Gabor filtering. Rotation invariance is achieved by using the magnitude of a DFT in the rotation dimension.

Many researchers have used the Brodatz album [5] for evaluating the performance of their texture classification and segmentation schemes. However, there is such a large variance in the actual subsets of images used and in the performance evaluation methodology that it is practically impossible to compare the evaluations presented in various papers. For example, Porter and Canagarajah [41] discuss several schemes for rotation-invariant classification using wavelets, Gabor filters, and GMRF models. They conclude, based on experiments on 16 images from the Brodatz set, that the wavelet features provide better classification performance compared with the other two texture features. A similar study by Manian and Vasquez [32] also conclude that orthogonal wavelet features provide better invariant descriptors. A different study, by Pichler *et al.* [37], from an image segmentation point of view, concludes that Gabor features provide better segmentation results compared with orthogonal wavelet features. Perhaps the most comprehensive study to date on evaluating different texture descriptors is provided by Manjunath and Ma [35], in the context of image retrieval. They use the entire Brodatz texture set and compare features derived from wavelet decomposition, tree-structured decomposition, Gabor wavelets, and multiresolution simultaneous autoregressive (MRSAR) models. They conclude that Gabor features and MRSAR model features outperform features from orthogonal or tree-structured wavelet decomposition. More recently, the study presented by Haley and Manjunath [21] indicates that the rotation-invariant features from Gabor filtering compare favorably with GMRF based schemes. They also provide results on the entire Brodatz dataset.

1.2 Gabor Features for Texture Classification and Image Segmentation

The following sections describe this rotation-invariant texture feature set, and for detailed experimental results, we refer the reader to [21]. The texture feature set is derived by filtering the image through a bank of modified Gabor kernels. The particular set of filters forms a multiresolution decomposition of the image. While there are several viable options, including orthogonal wavelet transforms, Gabor wavelets were chosen for their desirable properties.

1. Gabor functions achieve the theoretical minimum space-frequency bandwidth product [13, 14, 18]; that is, spatial resolution is maximized for a given bandwidth.
2. A narrow-band Gabor function closely approximates an analytic (frequency causal) function (see also Chapter 4.3 for a discussion on analytic signals). Signals convolved with an analytic function are also analytic, allowing separate analysis of the magnitude (envelope) and phase characteristics in the spatial domain.
3. The magnitude response of a Gabor function in the frequency domain is well behaved, having no sidelobes.
4. Gabor functions appear to share many properties with the human visual system [36].

While Gabor functions are a good choice, the standard forms can be further improved. Under certain conditions, very low frequency effects (e.g., caused by illumination and shading variations) can cause a significant response in a Gabor filter, leading to misclassification. An analytic form is introduced (see Section 2.2) to minimize these undesirable effects. When the center frequencies are evenly spaced on concentric circles, the polar form of the 2-D Gabor function allows for superior frequency domain coverage, improves rotation invariance, and simplifies analysis, compared with the standard 2-D form.

1.3 Chapter Organization

This chapter is organized as follows: In Section 2 we introduce an analytic Gabor function and a polar representation for the 2-D Gabor filters. A multiresolution representation of the image samples using Gabor functions is presented. In Section 3, the Gabor space samples are then transformed into a *microfeature* space, where a rotation-independent feature set is identified. Section 4 describes a texture model based on *macrofeatures* that are computed from the texture *microfeatures*. These macrofeatures provide a global description of the image sample and are useful for classification and segmentation. Section 5 gives experimental results on rotation-invariant texture classification. Section 6 outlines a new segmentation scheme, called *EdgeFlow* [29], that uses the texture energy features to partition the image. Finally, Section 7 gives an application of using texture descriptors to image retrieval [30, 35]. Some retrieval examples in the context of aerial imagery are shown.

2 Gabor Functions

2.1 One-Dimensional Gabor Function

A Gabor function is the product of a Gaussian function and a complex sinusoid. Its general one-dimensional form is

$$g_S(x, \omega_C, \sigma) = \frac{1}{\sqrt{2\pi}\sigma} \exp\left(\frac{-x^2}{2\sigma^2}\right) \exp(j\omega_C x), \quad (1)$$

$$G_S(\omega, \omega_C, \sigma) = \exp\left[\frac{-\sigma^2(\omega - \omega_C)^2}{2}\right]. \quad (2)$$

Thus, Gabor functions are bandpass filters. Gabor functions are used as complete, albeit nonorthogonal, basis sets. It has been shown that a function $i(x)$ is represented exactly [18] as

$$i(x) = \sum_{n=-\infty}^{\infty} \sum_{k=-\infty}^{\infty} \beta_{n,k} h_{n,k}(x), \quad (3)$$

where $h_{n,k}(x) = g_S(x - nX, k\Omega, \sigma)$, and σ, X, and Ω are all parameters and $X\Omega = 2\pi$.

2.2 Analytic Gabor Function

$G_S(\omega, \omega_C, \sigma)$ exhibits a potentially significant response at $\omega = 0$ and at very low frequencies. The response to a constant-valued input (i.e., $\omega = 0$) relative to the response to an input of equal magnitude at $\omega = \omega_C$ can be computed as a function of octave bandwidth [3]:

$$|G_S(0)|/|G_S(\omega_C)| = 2^{-\gamma}, \quad (4)$$

where $\gamma = (2^B + 1)/(2^B - 1)$ and $B = \log_2((\omega_C + \delta)/(\omega_C - \delta))$ and δ is the half-bandwidth. It is interesting to note that the response at $\omega = 0$ depends upon B but not ω_C. This behavior manifests itself as an undesirable response to interimage and intraimage variations in contrast and intensity as a result of factors unrelated to the texture itself, potentially causing misclassification. Cases include

- sample images of a texture with differences in average intensity
- images with texture regions having differences in contrast or intensity (Bovik [3] has demonstrated that region boundaries defined in segmentation using unmodified Gabor filters vary according to these differences between the regions)
- images with uneven illumination

There are two approaches to avoiding these problems: preprocessing the image or modifying the Gabor function. Normalizing each image to have a standard average intensity and contrast corrects for interimage, but not intraimage, variations. Alternative methods of image preprocessing are required to compensate for intraimage variations, such as point logarithmic processing [3] or local normalization.

An equally effective and more straightforward approach is to modify the Gabor function to be analytic[1] (see also Chapter 4.3 on analytic signals) by forcing the real and imaginary parts to become a Hilbert transform pair. This is accomplished by replacing the real part of $g_S(x)$, $g_{S,Re}(x)$, with the inverse Hilbert transform of the imaginary part, $-\hat{g}_{S,Im}(x)$:

$$g_A(x) = -\hat{g}_{S,Im}(x) + j g_{S,Im}(x). \quad (5)$$

The Fourier transforms of the real and imaginary parts of $g_S(x)$ are respectively conjugate symmetric and conjugate antisymmetric, resulting in cancellation for $\omega \leq 0$:

$$G_A(\omega) = \begin{cases} G_S(\omega) - G_S^*(-\omega), & \omega > 0 \\ 0, & \omega \leq 0 \end{cases}. \quad (6)$$

Because it is analytic, $G_A(\omega)$ possesses several advantages over for $G_S(\omega)$ for many applications including texture analysis:

- improved low frequency response since $|G_A(\omega)| < |G_S(\omega)|$ for small ω and $|G_A(0)| = 0$

---

[1]Since $G_s(\omega) \neq 0$ for $\omega \leq 0$, a Gabor function only approximates an analytic function.

- simplified frequency domain analysis since $G_A(\omega) = 0$ for $\omega \le 0$
- reduced frequency domain computations since $G_A(\omega) = 0$ for $\omega \le 0$

These advantages are achieved without requiring additional processing. Thus, it is an attractive alternative for most texture analysis applications.

2.3 Two-Dimensional Gabor Function: Cartesian Form

The Gabor function is extended into two dimensions as follows. In the spatial frequency domain, the Cartesian form is a 2-D Gaussian formed as the product of two 1-D Gaussians from Eq. (2):

$$G_C(\omega_x, \omega_y, \omega_{Cx'}, \omega_{Cy'}, \theta, \sigma_{x'}, \sigma_{y'})$$
$$= G(\omega_{x'}, \omega_{Cx'}, \sigma_{x'},)G(\omega_{y'}, \omega_{Cy'}, \sigma_{y'}), \tag{7}$$

where θ is the orientation angle of G_C, $x' = x\cos\theta + y\sin\theta$, and $y' = -x\sin\theta + y\cos\theta$. In the spatial domain, G_C is separable into two orthogonal 1-D Gabor functions from Eq. (1) that are respectively aligned to the x' and y' axes:

$$g_C(x, y, \omega_{Cx'}, \omega_{Cy'}, \theta, \sigma_{x'}, \sigma_{y'})$$
$$= g(x', \omega_{Cx'}, \sigma_{x'})g(y', \omega_{Cy'}, \sigma_{y'}). \tag{8}$$

As in Eq. (3), an image is represented exactly $[1, 2]^2$ as

$$i(x, y) = \sum_{n_x=-\infty}^{\infty} \sum_{n_y=-\infty}^{\infty} \sum_{k_x=-\infty}^{\infty}$$
$$\times \sum_{k_y=-\infty}^{\infty} \beta_{k_x,k_y,n_x,n_y} h_{k_x,k_y,n_x,n_y}(x, y), \tag{9}$$

where $h_{k_x,k_y,n_x,n_y}(x, y) = g(x - n_x X, k_x\Omega_x, \sigma_x)g(y - n_y Y, k_y\Omega_y, \sigma_y)$; $\sigma_x, \sigma_y, X, Y, \Omega_x$, and Ω_y are constants; and $X\Omega_x = Y\Omega_y = 2\pi$. Approximations to β_{k_x,k_y,n_x,n_y} are obtained by using [36]

$$\hat{\beta}_{k_x,k_y,n_x,n_y} = i(x, y) * h_{k_x,k_y,n_x,n_y}(x, y) \approx \beta_{k_x,k_y,n_x,n_y}, \tag{10}$$

provided that the parameters are chosen appropriately.

2.4 Two-Dimensional Gabor Function: Polar Form

An alternative approach to extending the Gabor function into two dimensions is to form, in the frequency domain, the product of a 1-D analytic Gabor function $G(\omega)$ (the subscript is omitted

^{2}The proofs in the references are based on the standard, not analytic, form of the Gabor function.

to indicate that the concepts are generally applicable to the standard form as well) of radial frequency ω and a Gaussian function of orientation θ:

$$G_P(\omega, \theta, \omega_C, \theta_C, \sigma_\rho, \sigma_\theta)$$
$$= G(\omega, \omega_C, \sigma_\rho) \exp\left[\frac{-\sigma_\theta^2(\theta - \theta_C)^2}{2}\right] \tag{11}$$

$$g_P(x, y, \omega_C, \theta_C, \sigma_\rho, \sigma_\theta)$$
$$= \iint G_P\left(\sqrt{\omega_x^2 + \omega_y^2}, \tan^{-1}(\omega_y/\omega_x), \omega_C, \theta_C, \sigma_\rho, \sigma_\theta\right)$$
$$\times \exp[j2\pi(\omega_x x + \omega_y y)]\,d\omega_x\,d\omega_y \tag{12}$$

where $\omega = \sqrt{\omega_x^2 + \omega_y^2}$ and $\tan(\theta) = \omega_y/\omega_x$. Thus, Eq. (11) is a 2-D Gaussian in the polar, rather than Cartesian, spatial frequency domain. The frequency domain regions of both polar and Cartesian forms of Gabor functions are compared in Fig. 1.

In the Cartesian spatial frequency domain, the -3 dB contour of the Cartesian form is an ellipse, while the polar form has a narrower response at low ω and a wider response at high ω. When arranged as "flower petals" (equally distributed along a circle centered at the origin), the polar form allows for more uniform coverage of the frequency domain, with less overlap at low frequencies and smaller gaps at high frequencies. The polar form is more suited for rotation-invariant analysis since the response always varies as a Gaussian with rotation. The Cartesian form varies with rotation in a more complex manner, introducing an obstacle to rotation invariance and complicating analysis.

2.5 Multiresolution Representation with Gabor Wavelets

The Gabor function is used as the basis for generating a wavelet family for multiresolution analysis (see Chapter 4.1 on wavelets). Wavelets have two salient properties: the octave bandwidth B and the octave spacing $\Delta = \log_2(\omega_{s+1}/\omega_s)$ are both constant, where ω_s is the center frequency. The filter spacing is achieved by defining

$$\omega_s = \omega_0 2^{-s\Delta}, \quad S \in \{0, 1, 2, \ldots\} \tag{13}$$

where ω_0 is the highest frequency in the wavelet family. Constant bandwidth requires that σ_ρ be inversely proportional to ω_S:

$$\sigma_{\rho_s} = \frac{1}{\kappa\omega_s}, \tag{14}$$

where

$$\kappa = \frac{2B - 1}{\sqrt{2\ln 2}(2B + 1)}$$

is a constant. The orientations of the wavelets are defined as

$$\theta_r = \theta_0 + \frac{2\pi r}{R}, \tag{15}$$

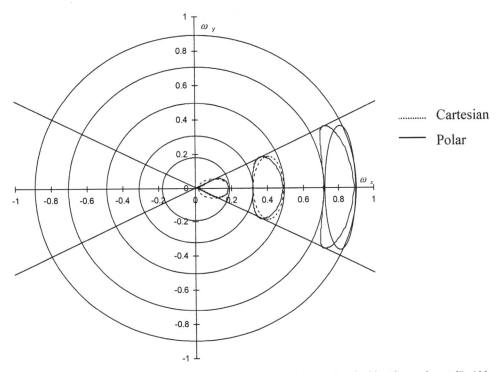

FIGURE 1 −3 dB contours of Cartesian and polar Gabor functions of varying bandwidths. The angular −3 dB width of the polar Gabor functions is 45°.

where θ_0 is the starting angle, the second term is the angular increment, and r and R are both integers such that $0 \leq r < R$. Using Eqs. (13), (14) and (15) in Eq. (11), we define the 2-D Gabor wavelet family as

$$
\begin{aligned}
H_{s,r}&(\omega_x, \omega_y) \\
&= G_P\left(\sqrt{\omega_x^2 + \omega_y^2}, \tan^{-1}(\omega_y/\omega_x), \omega_s, \theta_r, \frac{1}{\kappa\omega_s}, \sigma_\theta\right) \\
&= G\left(\sqrt{\omega_x^2 + \omega_y^2}, \omega_S, \frac{1}{\kappa\omega_s}\right) \\
&\quad \times \exp\left[\frac{-\sigma_\theta^2(\tan^{-1}(\omega_y/\omega_x) - \theta_r)^2}{2}\right]
\end{aligned}
\tag{16}
$$

$$
h_{s,r,n_x,n_y}(x, y) \\
= g_P\left(x - n_x X_s, y - n_y Y_s, \omega_s, \theta_r, \frac{1}{\kappa\omega_s}, \sigma_\theta\right),
\tag{17}
$$

where X_S and Y_S, the sampling intervals, are inversely proportional to the bandwidths corresponding to s. As in Eq. (9), an image is represented by using the polar wavelet form of the Gabor function from Eq. (17):

$$
i(x, y) = \sum_{n_x=-\infty}^{\infty} \sum_{n_y=-\infty}^{\infty} \sum_{s=0}^{\infty} \sum_{r=0}^{R-1} \beta_{s,r,n_x,n_y} h_{s,r,n_x,n_y}(x, y).
\tag{18}
$$

Approximations to β_{s,r,n_x,n_y} are obtained as in Eq. (10):

$$
\hat{\beta}_{s,r,n_x,n_y} = i(x, y) * h_{s,r,n_x,n_y}(x, y) \approx \beta_{s,r,n_x,n_y},
\tag{19}
$$

and parameters X_s, Y_s, ω_0, κ, and σ_θ are chosen appropriately. Instead of a rectangular lattice, a polar Gabor wavelet representation has the shape of a cone.

3 Microfeature Representation

3.1 Transformation into Gabor Space

As described in Section 2, a set of two-dimensional Gabor wavelets can represent an image. Assuming that the image is spatially limited to $0 \leq x < N_x X_s, 0 \leq y < N_y Y_s$, where N_x and N_y represent the number of samples in their respective dimensions, and is bandlimited to $0 < \omega \leq \omega_H$,[3] the number of Gabor wavelets needed to represent the image is finite. Substituting $\hat{\beta}_{s,r,n_x,n_y}$ from Eq. (19) for β_{s,r,n_x,n_y} in Eq. (18), we approximately represent a texture image by using the polar wavelet form of the Gabor function as

$$
i(x, y) \approx \sum_{n_x=0}^{N_{sx}-1} \sum_{n_y=0}^{N_{sy}-1} \sum_{s=0}^{S-1} \sum_{r=0}^{R-1} \hat{\beta}_{s,r,n_x,n_y} h_{s,r,n_x,n_y}(x, y),
\tag{20}
$$

---

[3] For sampled texture images, the upper frequency bound is enforced, although aliasing may be present since natural textures are generally not bandlimted. It is both reasonable and convenient to assume that, for textures of interest, a lower frequency bound $\omega_L > 0$ exists below which there is no useful discriminatory information.

where parameters S, R, X_s, Y_s, ω_0, κ and σ_θ are chosen appropriately. Note that the s subscript is added to N_x and N_y to indicate their dependencies on X_s and Y_s. Thus, a texture image is represented with relatively little information loss by the coefficients $\hat{\beta}_{s,r,n_x,n_y}$.

Following Bovik *et al.* [4], $\hat{\beta}_{s,r,n_x,n_y}$ is interpreted as a channel or band $b_{s,r}(n_x, n_y)$ of the image $i(x, y)$ tuned to the carrier frequency $\omega_s = \omega_0 2^{-s\Delta}$, Eq. (13), oriented at angle $\theta_r = \theta_0 + 2\pi r / R$, Eq. (15), and sampled in the spatial domain at intervals of X_s and Y_s. Since $b_{s,r}(n_x, n_y)$ is formed by convolution with a narrow band, analytic function, Eq. (19), $b_{s,r}(n_x, n_y)$ is also narrow band and analytic and is therefore decomposable into amplitude and phase components that can be independently analyzed:

$$b_{s,r}(n_x, n_y) = a_{s,r}(n_x, n_y) \exp(j\psi_{s,r}(n_x, n_y)), \quad (21)$$

where $a_{s,r}(n_x, n_y) = b_{s,r}(n_x, n_y)$ and $\psi_{s,r}(n_x, n_y) = \arg(b_{s,r}(n_x, n_y))$. Here $a_{s,r}(n_x, n_y)$ contains information about the amplitude and amplitude modulation (AM) characteristics of the texture's periodic features within the band, and $\psi_{s,r}(n_x, n_y)$ contains information about the phase, frequency, and frequency modulation (FM) characteristics (see Chapter 4.3 for a discussion on AM/FM signals). For textures with low AM in band (s, r), $a_{s,r}(n_x, n_y)$ is approximately constant over (n_x, n_y). For textures with low FM in band s, r, the slope of $\psi_{s,r}(n_x, n_y)$ with respect to (n_x, n_y) is nearly constant.

Both $a_{s,r}(n_x, n_y)$ and $\psi_{s,r}(n_x, n_y)$ are rotation dependent and periodic in r such that

$$a_{s,((r+R/2))_R}(n_x, n_y) = a_{s,r}(n_x, n_y), \quad (22)$$

$$\psi_{s,((r+R))_R}(n_x, n_y) = \psi_{s,r}(n_x, n_y), \quad (23)$$

$$\psi_{s,((r+R/2))_R}(n_x, n_y) = \psi_{s,r}(n_x, n_y). \quad (24)$$

Rotating $i(x, y)$ by $\theta°$ produces a circular shift in r of $-R\theta/180°$ for $a_{s,r}(n_x, n_y)$ and $-R\theta/360°$ for $\psi_{s,r}(n_x, n_y)$.

3.2 Local Frequency Estimation

While $\psi_{s,r}(n_x, n_y)$ contains essential information about a texture, it is not directly usable for classification. However, local frequency information can be extracted from $\psi_{s,r}(n_x, n_y)$ as follows:

$$\phi_{s,r}(n_x, n_y) = \begin{cases} \theta_\nabla, & |\theta_r - \theta_\nabla| \le 90° \\ \theta_\nabla + 180°, & |\theta_r - \theta_\nabla| > 90° \end{cases}, \quad (25)$$

$$u_{s,r}(n_x, n_y) = \sqrt{\nabla_x^2(\psi_{s,r}(n_x, n_y)) + \nabla_y^2(\psi_{s,r}(n_x, n_y))}$$
$$\times \cos(\theta_r - \phi_{s,r}(n_x, n_y))$$
$$= \sqrt{\nabla_x^2(\psi_{s,r}(n_x, n_y)) + \nabla_y^2(\psi_{s,r}(n_x, n_y))}$$
$$\times |\cos(\theta_r - \theta_\nabla)|, \quad (26)$$

where $\nabla_x()$ and $\nabla_y()$ are gradient estimation functions, θ_r is the orientation of the Gabor function, and $\theta_\nabla = \tan^{-1}(\nabla_y(\psi_{s,r}(n_x, n_y))/\nabla_x(\psi_{s,r}(n_x, n_y)))$ is the direction of the gradient vector. Here $u_{s,r}(n_x, n_y)$ is a spatially localized estimate of the frequency along the direction ϕ_r, and $\phi_{s,r}(n_x, n_y)$ is the direction of maximal phase change rate, i.e., highest local frequency.

3.3 Transformation into Microfeatures

To facilitate discrimination between textures, $b_{s,r}(n_x, n_y)$ is further decomposed into *microfeatures* that contain local amplitude, frequency, phase, direction, and directionality characteristics. In the following, for simplicity, R is assumed to be even. The microfeatures are defined to be as follows.

$$f_{As,p}(n_x, n_y) = \sum_{r=0}^{R/2-1} a_{s,r}(n_x, n_y) a_{s,((r+p))_R}(n_x, n_y),$$
$$0 \le p \le R/4; \quad (27)$$

$$f_{Fs,q}(n_x, n_y) = \sum_{r=0}^{R/2-1} u_{s,r}(n_x, n_y) \exp\left(-\frac{2\pi jrq}{R/2}\right),$$
$$0 \le q \le R/4; \quad (28)$$

$$f_{Ys,q}(n_x, n_y) = \sum_{r=0}^{R-1} \exp(j\phi_{s,r}(n_x, n_y)) \exp\left(-\frac{2\pi jrq}{R}\right),$$
$$q = 1, 3, \ldots, R-1; \quad (29)$$

$$f_{DAs,q}(n_x, n_y) = \arg\left[\sum_{r=0}^{R/2-1} a_{s,r}(n_x, n_y) \exp\left(-\frac{2\pi jrq}{R/2}\right)\right],$$
$$1 \le q \le R/4; \quad (30)$$

$$f_{DFs,q}(n_x, n_y) = \arg\left[\sum_{r=0}^{R/2-1} u_{s,r}(n_x, n_y) \exp\left(-\frac{2\pi jrq}{R/2}\right)\right],$$
$$1 \le q \le R/4; \quad (31)$$

$$f_{DYs,q}(n_x, n_y) = \arg\left[\sum_{r=0}^{R-1} \exp(j\phi_{s,r}(n_x, n_y)) \exp\left(-\frac{2\pi jrq}{R}\right)\right],$$
$$q = 1, 3, \ldots, R-1. \quad (32)$$

Here $f_{As,p}(n_x, n_y)$ contains the amplitude envelope information from $b_{s,r}(n_x, n_y)$. Because of the $R/2$ periodicity of $a_{s,r}$ (22), only $R/2$ components are needed in the sum in Eq. (27). Eliminating the redundant components from the circular autocorrelation allows complete representation by the $0 \le p \le R/4$ components of $f_{As,p}(n_x, n_y)$. It is rotation invariant because the autocorrelation operation eliminates the dependence on r, and thus on θ.

We see that $f_{Fs,q}(n_x, n_y)$ contains the frequency envelope information from $b_{s,r}(n_x, n_y)$. Similar to $a_{s,r}(n_x, n_y)$, $u_{s,r}(n_x, n_y)$

has $R/2$ periodicity. Since $u_{s,r}(n_x, n_y)$ is real, $f_{Fs,q}(n_x, n_y)$ is conjugate symmetric in q, and consequently, its $0 \le q \le R/4$ components are sufficient for complete representation. It is rotation invariant because the DFT operation maps rotationally induced shifts into the complex numbers' phase components, which are removed when the magnitude operation is performed. $f_{Ys,q}(n_x, n_y)$ contains the directionality information from $b_{s,r}(n_x, n_y)$. Since $\phi_{s,(r+R/2)_R}(n_x, n_y) = \phi_{s,r}(n_x, n_y) + 180°$, only the components with odd q are nonzero. For the same reason as $f_{Fs,q}(n_x, n_y)$, $f_{Ys,q}(n_x, n_y)$ is rotation invariant.

We see that $f_{DAs,q}(n_x, n_y)$, $f_{DFs,q}(n_x, n_y)$, and $f_{DYs,q}(n_x, n_y)$ contain the direction information from $b_{s,r}(n_x, n_y)$. Because $f_{DAs,q}(n_x, n_y)$ and $f_{DFs,q}(nx, ny)$ are conjugate symmetric in q, they are represented completely by their $0 \le q \le R/4$ components. However, the $q = 0$ component is always zero since the DFTs are on real sequences in both cases. Here $f_{DYs,q}(n_x, n_y)$ has the same nonzero indexes as $f_{Ys,q}(n_x, n_y)$. Furthermore, $f_{DAs,q}(n_x, n_y)$, $f_{DFs,q}(n_x, n_y)$, and $f_{DYs,q}(n_x, n_y)$ are inherently rotation variant since the phases of the DFT contain all of the direction information.

Since all transformations in this decomposition are invertible (assuming boundary conditions are available), it is possible to exactly reconstruct $b_{s,r}(n_x, n_y)$ from their microfeatures. Thus, $f_{As,p}(n_x, n_y)$, $f_{Fs,q}(n_x, n_y)$, $f_{Ys,q}(n_x, n_y)$, $f_{DAs,q}(n_x, n_y)$, $f_{DFs,q}(n_x, n_y)$, and $f_{DYs,q}(n_x, n_y)$ provide a nearly exact representation of $i(x, y)$.

4 The Texture Model

4.1 The Texture Micromodel

A texture may be modeled as a vector-valued random field $\mathbf{f} = [\mathbf{f}_A\ \mathbf{f}_F\ \mathbf{f}_Y\ \mathbf{f}_{DA}\ \mathbf{f}_{DF}\ \mathbf{f}_{DY}]^T$, where $\mathbf{f}_A$, $\mathbf{f}_F$, $\mathbf{f}_Y$, $\mathbf{f}_{DA}$, $\mathbf{f}_{DF}$ and $\mathbf{f}_{DY}$ are vectors containing the microfeature components for all s and p or q indexes. It is assumed that $\mathbf{f}$ is stationary. Accurate modeling of $\mathbf{f}$ is not practical from a computational point of view. Such modeling is also not needed if the objective is only texture classification (and not synthesis). Further, we assume a Gaussian distribution of $\mathbf{f}$ strictly for mathematical tractability and simplicity, although many sample distributions were observed to be very non-Gaussian.

Given these assumptions, the *micromodel* for texture t is stated as the multivariate Gaussian probability distribution function:

$$p(f \mid t) = \frac{1}{\sqrt{(2\pi)^{N_f} |C_{ft}|}} \exp\left(\frac{-(f - \mu_{ft}) C_{ft}^{-1} (f - \mu_{ft})^T}{2}\right), \tag{33}$$

where $\mu_{ft} = E\{\mathbf{f} \mid t\}$ and $C_{ft} = E\{\mathbf{f} \cdot \mathbf{f}^T \mid t\} - E\{\mathbf{f} \mid t\} \cdot E\{\mathbf{f}^T \mid t\}$ are the mean and covariance of $\mathbf{f}$, respectively, and N_f is the number of microfeatures.

Texture A   Texture B

FIGURE 2 Textures with similar microfeatures.

4.2 Macrofeatures

While microfeatures can be used to represent a texture sample, microfeatures are spatially localized and do not characterize global attributes of textures. For instance, consider the textures in Fig. 2. Most of the spatial samples in the upper-right and lower-left quadrants of texture A would be classified as texture B based on microfeatures alone. Furthermore, $\mathbf{f}_{DA}$, $\mathbf{f}_{DF}$, and $\mathbf{f}_{DY}$ are rotation dependent, making them unsuitable for rotation-invariant classification.

For classification, a better texture model is derived from the micromodel parameters, μ_{ft} and C_{ft}. For instance, for the two textures shown in Fig. 2, the standard deviations of $\mathbf{f}_{DA}$, $\mathbf{f}_{DF}$, and $\mathbf{f}_{DY}$ provide excellent discrimination information not available in the microfeatures themselves. A texture t's *macrofeatures* are defined to be $\mathbf{F} = [F_{CA}\ F_{CF}\ F_{CY}\ F_{AM}\ F_{FM}\ F_{YM}\ F_{DMA}\ F_{DMF}\ F_{DMY}]^T$, where

$$\begin{bmatrix} F_{CA} \\ F_{CF} \\ F_{CY} \\ F_{AM} \\ F_{FM} \\ F_{YM} \\ F_{DMA} \\ F_{DMF} \\ F_{DMY} \end{bmatrix} = \begin{bmatrix} E\{\mathbf{f}_A \mid t\} \\ E\{\mathbf{f}_F \mid t\} \\ E\{\mathbf{f}_Y \mid t\} \\ \sqrt{E\{\mathbf{f}_A^2\} - E\{\mathbf{f}_A\}^2} \\ \sqrt{E\{\mathbf{f}_F^2\} - E\{\mathbf{f}_F\}^2} \\ \sqrt{E\{\mathbf{f}_Y^2\} - E\{\mathbf{f}_Y\}^2} \\ \sqrt{E\{\mathbf{f}_{DA}^2\} - E\{\mathbf{f}_{DA}\} \times E\{\mathbf{f}_{DA}\}^*} \\ \sqrt{E\{\mathbf{f}_{DF}^2\} - E\{\mathbf{f}_{DF}\} \times E\{\mathbf{f}_{DF}\}^*} \\ \sqrt{E\{\mathbf{f}_{DY}^2\} - E\{\mathbf{f}_{DY}\} \times E\{\mathbf{f}_{DY}\}^*} \end{bmatrix}, \tag{34}$$

where $\mathbf{f}^2 = \langle \mathbf{f}, \mathbf{f} \rangle = [(f_{CA\,0,0} \cdot f_{CA\,0,0})\ (f_{CA\,0,1} \cdot f_{CA\,0,1}) \cdots (f_{DY\,S-1,R-1} \cdot f_{DY\,S-1,R-1})]^T$. For a texture t, F_{CA}, F_{CF}, and F_{CY} describe amplitude, frequency, and directionality characteristics, respectively, of the "carrier." F_{AM}, F_{FM}, and F_{YM} describe a texture's amplitude modulation, frequency modulation, and directionality modulation characteristics, respectively. F_{CA}, F_{CF}, F_{CY}, F_{AM}, F_{FM}, and F_{YM} are all rotation invariant because the microfeatures upon which they are based are rotation invariant. F_{DMA}, F_{DMF}, and F_{DMY} capture the directional modulation characteristics. While $\mathbf{f}_{DA}$, $\mathbf{f}_{DF}$, and $\mathbf{f}_{DY}$ are rotation dependent, their variances are not. Means of $\mathbf{f}_{DA}$, $\mathbf{f}_{DF}$, and $\mathbf{f}_{DY}$ are directional in nature and are not used as classification features.

For simplicity, off-diagonal covariances are not used, although they may contain useful information.

The expected values of **f** are estimated by using the mean and variance of a texture sample's microfeatures.

4.3 The Texture Macromodel

For purposes of classification, a texture t is modeled as a vector-valued Gaussian random vector **F** with the conditional probability density function

$$p(\hat{F} \mid t) = \frac{1}{\sqrt{(2\pi)^{N_F} \mid C_{Ft} \mid}} \exp\left(\frac{-(\hat{F} - \mu_{Ft}) C_{Ft}^{-1} (\hat{F} - \mu_{Ft})^T}{2} \right),$$

(35)

where $\mu_{Ft} = E\{\mathbf{F} \mid t\}$ and $C_{Ft} = E\{\mathbf{F} \cdot \mathbf{F}^T \mid t\} - E\{\mathbf{F} \mid t\} \cdot E\{\mathbf{F}^T \mid t\}$ are the mean and covariance of **F**, respectively, N_F is the number of macrofeatures, and $\hat{F}$ is an estimate of F based on a sample of texture t. This is the texture *macromodel*.

The parameters μ_{Ft} and C_{Ft} are estimated from statistics over M samples for each texture t:

$$\hat{\mu}_{Ft} = \frac{1}{M} \sum_{m=1}^{M} \hat{F}_m \quad \text{and}$$

$$\hat{C}_{Ft} = \frac{1}{M-1} \sum_{m=1}^{M} (\hat{F}_m - \hat{\mu}_{Ft}) \cdot (\hat{F}_m - \hat{\mu}_{Ft})^T,$$

(36)

where $\hat{F}_m$ is the estimate of F based on sample m of texture t.

5 Experimental Results

Experiments were performed on two groups of textures. The first group comprises 13 texture images [44] digitized from the Brodatz album [5] and other sources. Each texture was digitized at rotations of 0, 30, 60, 90, 120, and 150° as 512×512 pixels, each of which was then subdivided into sixteen 128×128 subimages. Figure 3 presents the 120° rotations of these images. The second group comprises 109 texture images from the Brodatz album digitized at 0° with 512×512 pixels at a 300 DPI resolution, each of which was then subdivided into sixteen 128×128 subimages. A polar, analytic Gabor transform was used with parameter values of $\omega_0 = 0.8\pi$, $\theta_0 = 0°$, $S = 4$, $R = 16$, $\kappa = 0.283$ ($B = 1$ octave), and $\sigma_\theta = 0.0523/°$ (-3 dB width of $90°$).

Classification performance was demonstrated with both groups of textures. Half of the subimages (separated in a checkerboard pattern) were used to estimate the model parameters (mean and covariance of the macrofeatures) for each type of texture, while the other half were used as test samples. Features were extracted from all of the subimages in an identical manner. To reduce filter sampling effects at high frequencies caused by rotation, the estimation of model parameters was based on

TABLE 1

Classification performance for the first group of textures

Sample type	%Classified correctly
Bark	87.5
Sand	97.9
Pigskin	95.8
Bubbles	100
Grass	95.8
Leather	93.8
Wool	91.7
Raffia	100
Weave	100
Water	97.9
Wood	97.9
Straw	100
Brick	100

the features from subimages at all rotations in the first group of images.

5.1 Classification

A model of each type of texture was established by using half of its samples to estimate mean and covariance, the parameters required by Eq. (34). For the other half of the samples, each was classified as the texture t that maximized $p(F \mid t)$. Because of rank deficiency problems in the covariance matrix that were due to high interfeature correlation, off-diagonal terms in the covariance matrix were set to zero.

The classification performance for the first group of textures is summarized in Table 1. Out of a total of 624 sample images, 604 were correctly classified (96.8%). The misclassification rate per competing texture type is $(100–96.8\%)/12 = 0.27\%$. Bark was misclassified as brick, bubbles, pigskin, sand, and straw; sand as bark; pigskin as bark and wool; grass as leather; leather as grass and straw; wool as bark and pigskin; water as straw; and wood as straw.

The classification performance for the second group of textures (the complete Brodatz album) was as follows. Out of a total of 872 sample images, 701 were classified correctly (80.4%). The misclassification rate per competing texture type is $(100\%–80.4\%)/108 = 0.18\%$. Perhaps some comments are in order regarding the classification rate. Many of the textures in the Brodatz album are not homogeneous. Although one can use a selected subset of textures, it will make comparisons between different algorithms more difficult. Finally, for comparison purposes, when using the same subset of the Brodatz album used by Chang and Kuo [6], 100% of the samples were correctly classified.

6 Image Segmentation Using Texture

Image segmentation can be achieved either by classification or by considering the gradient in the texture feature space. Here we outline a novel technique, called *EdgeFlow*, that uses the texture

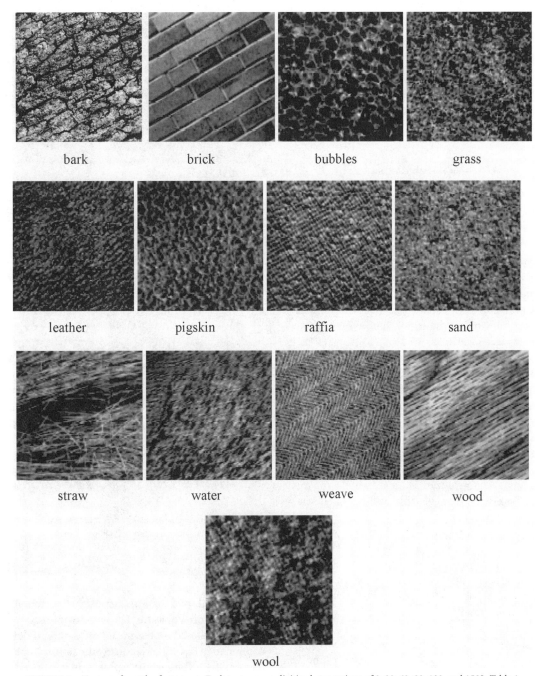

FIGURE 3 Textures from the first group. Each texture was digitized at rotations of 0, 30, 60, 90, 120, and 150°. Table 1 summarizes the results for rotation-invariant classification for these textures.

feature as input to partition the image. A detailed description of this technique can be found in [29].

The *EdgeFlow* method utilizes a predictive coding model to identify and integrate the direction of change in a given set of image attributes, such as color and texture, at each image pixel location. Toward this objective, the following values are computed: $E(\mathbf{x}, \theta)$, which measures the edge energy at pixel $\mathbf{x}$ along the orientation θ; $P(\mathbf{x}, \theta)$, which is the probability of finding an edge in the direction θ from $\mathbf{x}$; and $P(\mathbf{x}, \theta + \pi)$, which is the

probability of finding an edge along $\theta + \pi$ from $\mathbf{x}$. These edge energies and the associated probabilities are computed by using the features of interest.

Consider the Gabor filtered outputs represented by Eq. (21):

$$b_{s,r}(\mathbf{x}) = a_{s,r}(\mathbf{x}) \exp(J \psi_{s,r}(\mathbf{x})).$$

By taking the amplitude of the filtered output across different filters at the location represented by $\mathbf{x}$, a texture feature vector characterizing the local spectral energies in different spatial frequency bands is formed:

$$\mathbf{A}(\mathbf{x}) = [a_1(\mathbf{x}), a_2(\mathbf{x}), a_3(\mathbf{x}), \ldots, a_N(\mathbf{x})], \qquad (37)$$

where, for simplicity, the combination of s and r indices is numbered from 1 through N. The texture edge energy, which is used to measure the change in local texture, is computed as

$$E(\mathbf{x}, \theta) = \sum_{1 \leq i \leq N} |a_i(\mathbf{x}) * \mathrm{GD}_{\sigma, \theta}(\mathbf{x})| w_i, \qquad w_i = 1 \Big/ \sum_{\mathbf{x}} a_i(\mathbf{x}),$$

$$(38)$$

where GD is the first derivative of the Gaussian along the orientation θ. The weights w_i normalize the contribution of edge energy from the various frequency bands. The error in predicting the texture energies in the neighboring pixel locations is used to compute the probabilities $\{P(s, \theta)\}$. For example, a large prediction error in a certain direction implies a higher probability of finding the region boundary in that direction. Thus, at each location $\mathbf{x}$ we have $\{[E(\mathbf{x}, \theta), P(\mathbf{x}, \theta), P(\mathbf{x}, \theta + \pi)]|_{0 \leq \theta \leq \pi}\}$. From these measurements, an *edge flow* vector is constructed whose direction represents the *flow* direction along which a boundary is likely to be found, and whose magnitude is an estimate of the total edge energy along that direction.

The distribution of the edge flow vectors in the image forms a flow field that is allowed to propagate. At each pixel location the flow is in the estimated direction of the boundary pixel. A boundary location is characterized by flows in opposing directions toward it. On a discrete image grid, the flow typically takes a few iterations to converge.

Figure 4 shows two images, one with different textures and another with an illusory boundary. For the textured image, the edge flow vectors are constructed at each location as outlined above, and the final segmentation result is shown in the figure. It turns out that the phase information in the filtered outputs is quite useful in detecting illusory contours, as illustrated. The details of computing the phase discontinuities can be found in [29]. Figure 5 shows another example of texture based segmentation, illustrating the results at two different choices for the scale parameter that controls the *EdgeFlow* segmentation. A few other examples of using color, texture, and phase in detecting image boundaries are shown in Fig. 6.

7 Image Retrieval Using Texture

In recent years, texture descriptor has emerged as an important visual feature for content-based image retrieval. In [30] we present an image retrieval system for browsing a collection of large aerial imagery using texture. Texture turns out to be a surprisingly powerful descriptor for aerial imagery, and many of the geographically salient features, such as vegetation, water, urban

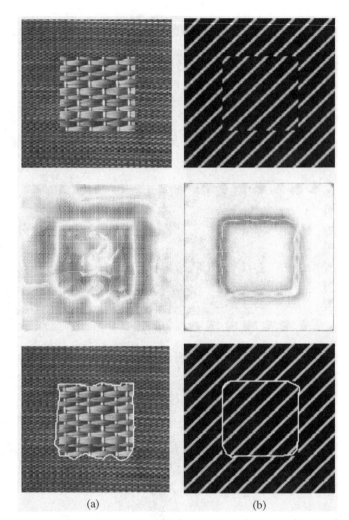

FIGURE 4 Segmentation using *EdgeFlow*. From top to bottom are the original image, edge flow vectors, and detected boundaries: (a) texture image example; (b) an illusory boundary detected by using the texture phase component from the Gabor filtered images.

development, parking lots, airports, etc., are well characterized by their texture signature. The particular texture descriptor used in [30] was based on the mean and standard deviation of $\mathbf{A}(\mathbf{x})$ computed in Eq. (37). Measuring the similarity between two patterns in the texture feature space is an important issue in image retrieval. A hybrid neural network algorithm was used to learn this similarity and thus construct a *texture thesaurus* that would facilitate fast search and retrieval. Figures 7 and 8 show two *query by example* results, wherein the input to the search engine was an image region, and the system was asked to retrieve similarly looking patterns in the image database.

8 Summary

We have presented schemes for texture classification and segmentation using features computed from Gabor-filtered images.

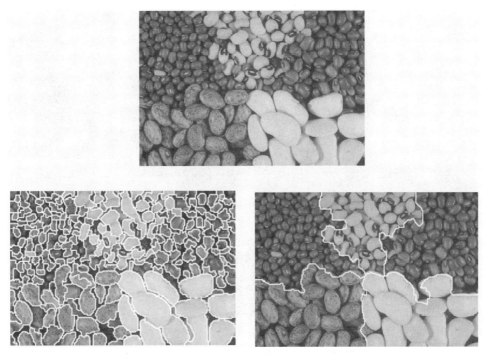

FIGURE 5 The choice of *scale* plays a critical role in the *EdgeFlow* segmentation. Two different segmentation results shown above are the result of two different choices for the scale parameter in the algorithm.

Image texture research has seen much progress during the past two decades, and both random field model-based approaches and multiband filtering methods will have applications to texture analysis. Model-based methods are particularly useful for synthesis and rendering. Filtering methods compare favorably to the random field methods for classification and segmentation, and they can be efficiently implemented on dedicated hardware. Finally, texture features appear quite promising for image database applications such as search and retrieval, and the current MPEG-7 documents list texture as among the normative components of the standard feature set that the MPEG plans to standardize.

Acknowledgment

This work was supported in part by grants 94-1130 and 97-04795 from the National Science Foundation and by the UC Micro program (1997–1998), with a matching grant from Spectron Microsystems.

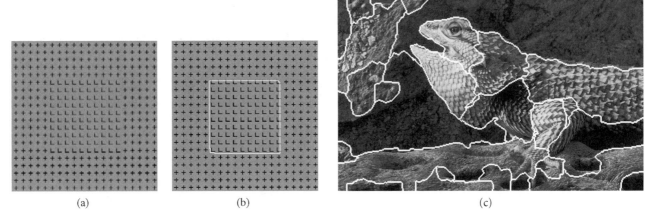

(a)   (b)   (c)

FIGURE 6 Two other examples of segementation: (a) an illusory boundary, (b) segementation using texture phase in the EdgeFlow algorithm, and (c) segmentation using color and texture energy. (See color section, p. C–18.)

FIGURE 7   Example of a texture-based search, using a dataset of aerial photographs of the Santa Barbara area taken over a 30-y period. Each photograph is approximately 5,000 × 5,000 pixels in size. (a) The downsampled version of the aerial photograph from which the query is derived. (b) A full-resolution detail of the region used for the query. The region contains a housing development. (c)–(e) The ordered three best results of the query. The black line indicates the boundaries of the regions that were retrieved. The results come from three different aerial photographs that were taken the same year as the photograph used for the query.

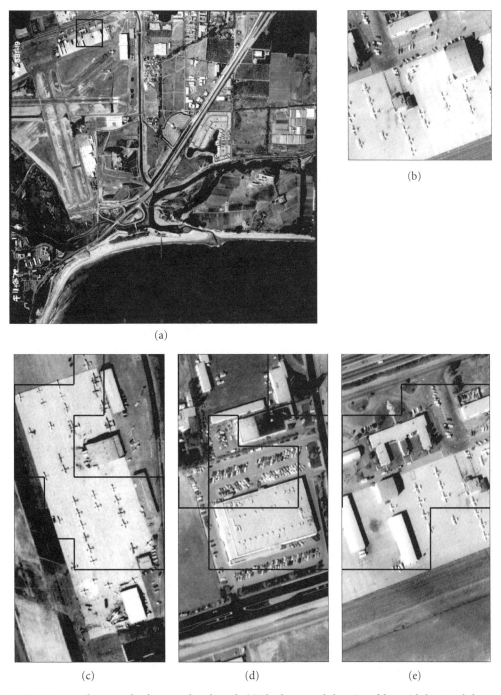

FIGURE 8 Another example of a texture based search. (a) The downsampled version of the aerial photograph from which the query is derived. (b) A full-resolution detail of the region used for the query. The region contains aircraft, cars, and buildings. (c)–(e) The ordered three top matching retrievals. Once again, the results come from three different aerial photographs. This time, the second and third results are from the same year (1972) as the query photograph, but the first match is from a different year (1966).

References

[1] M. J. Bastiaans, "A sampling theorem for the complex spectrogram, and Gabor's expansion of a signal in Gaussian elementary signals," *Opt. Eng.* **20**, 594–598 (1981).

[2] M. J. Bastiaans, "Gabor's signal expansion and degrees of freedom of a signal," *Opt. Acta.* **29**, 1223–1229 (1982).

[3] A. C. Bovik, "Analysis of multichannel narrow-band filters for image texture segmentation," *IEEE Trans. Signal Process.* **39**, 2025–2042 (1991).

[4] A. C. Bovik, M. Clark, and W. S. Geisler, "Multichannel texture analysis using localized spatial filters," *IEEE Trans. Pattern Anal. Machine Intell.* **12**, 55–73 (1990).

[5] P. Brodatz, *Textures: A Photographic Album for Artists and Designers* (Dover, New York, 1966).

[6] T. Chang and C. C. J. Kuo, "Texture analysis and classification with tree-structured wavelet transform," *IEEE Trans. Image Process.* **2**, 429–441 (1993).

[7] R. Chellappa and S. Chatterjee, "Classification of textures using Gaussian Markov random field models," *IEEE Trans. Acoust. Speech Signal Process.* **33**, 959–963 (1985).

[8] R. Chellappa, R. L. Kashyap, and B.S. Manjunath, "Model-based texture segmentation and classification," in *Handbook of Pattern Recognition and Computer Vision*, C. H. Chen, L. F. Pau, and P. F. P. Wang, eds. (World Scientific, Teaneck, NJ, 1992).

[9] P. C. Chen and T. Pavlidis, "Segmentation by texture using correlation," *IEEE Trans. Pattern Anal. Machine Intell.* **5**, 64–69 (1983).

[10] Y. Choe and R. L. Kashyap, "3-D shape from a shaded and textural surface image," *IEEE Trans. Pattern Anal. Machine Intell.* **13**, 907–918 (1991).

[11] F. S. Cohen, Z. Fan and M. A. Patel, "Classification of rotated and scaled textured image using Gaussian Markov random field models," *IEEE Trans. Pattern Anal. Machine Intell.* **13**, 192–202 (1991).

[12] G. R. Cross and A. K. Jain, "Markov random field texture models," *IEEE Trans. Pattern Anal. Machine Intell.* **5**, 25–39 (1983).

[13] J. G. Daugman, "Complete discrete 2-D Gabor transforms by neural networks for image analysis and compression," *IEEE Trans. Acoust. Speech Signal Process.* **36**, 1169–1179 (1988).

[14] J. G. Daugman, "Uncertainty relation for resolution in space, spatial frequency and orientation optimized by two-dimensional visual cortical filters," *J. Opt. Soc. Am.* **2**, 1160–1169 (1985).

[15] H. Derin and H. Elliott, "Modeling and segmentation of noisy and textured images using Gibbs random fields," *IEEE Trans. Pattern Anal. Machine Intell.* **9**, 39–55 (1987).

[16] H. Derin, H. Elliott, R. Cristi, and D. Geman, "Bayes smoothing algorithms for segmentation of binary images modeled by Markov random fields," *IEEE Trans. Pattern Anal. Machine Intell.* **6**, 707–720 (1984).

[17] D. Dunn, W. E. Higgins, and J. Wakeley, "Texture segmentation using 2-D Gabor elementary functions," *IEEE Trans. Pattern Anal. Machine Intell.* **16** (1994).

[18] D. Gabor, "Theory of communication," *J. Inst. Elect. Eng.* **93**, 429–457 (1946).

[19] M. M. Galloway, "Texture analysis using gray level run lengths," *Computer Graph. Image Process.* **4**, 172–179 (1975).

[20] H. Greenspan, S. Belongie, R. Goodman, and P. Perona, "Rotation invariant texture recognition using asteerable pyramid," presented at the IEEE International Conference on Image Processing, Jerusalem, Israel, October 1994.

[21] G. M. Haley and B. S. Manjunath, "Rotation invariant texture classification using a complete space-frequency model," *IEEE Trans. Image Process.* **8**, 256–269 (1999).

[22] R. Haralick and R. Bosley, "Texture features for image classification, in *Third ERTS Symposium*, NASA SP-351 (NASA; Washington, D.C., 1973), pp. 1219–1228.

[23] R. L. Kashyap, R. Chellappa, and A. Khotanzad, "Texture classification using features derived from random field models," *Pattern Recog. Lett.* **Oct**, 43–50 (1982).

[24] R. L. Kashyap and A. Khotanzad, "A model-based method for rotation invariant texture classification," *IEEE Trans. Pattern Anal. Machine Intell.* **8**, 472–481 (1986).

[25] S. Krishnamachari and R. Chellappa, "Multiresolution Gauss-Markov random fields for texture segmentation," *IEEE Trans. Image Process.* **6**, 251–267 (1997).

[26] S. Lakshmanan and H. Derin, "Simultaneous parameter estimation and segmentation of Gibbs random fields using simulated annealing," *IEEE Tran. Pattern Anal. Machine Intell.* **11**, 799–813 (1989).

[27] K. Laws, "Textured image segmentation," Ph.D. thesis (University of Southern California, 1978).

[28] M. M. Leung and A. M. Peterson, "Scale and rotation invariant texture classification," presented at the 26th Asilomar Conference on Signals, Systems and Computers, Pacific Grove, CA, October 1992.

[29] W. Y. Ma and B. S. Manjunath, "EdgeFlow: a framework of boundary detection and image segmentation," presented at the *IEEE International Conference on Computer Vision and Pattern Recognition*, San Juan, Puerto Rico, June 1997.

[30] W. Y. Ma and B. S. Manjunath, "A texture thesaurus for browsing large aerial photographs," *J. Am. Soc. Info. Sci.*, special issue on AI techniques to emerging information systems, **49**, 633–648 (1998).

[31] J. Malik and P. Perona, "Preattentive texture segmentation with early vision mechanisms," *J. Opt. Soc. Am.* **7**, 923–932 (1990).

[32] V. Maniyan and R. Vasquez, "Scaled and rotated texture classification using a class of basis function," *Pattern Recog.* **31**, 1937–1948 (1998).

[33] B. S. Manjunath, T. Simchony, and R. Chellappa, "Stochastic and deterministic networks for texture segmentation," *IEEE Trans. Acoust. Speech Signal Process.* **38**, 1039–1049 (1990).

[34] B. S. Manjunath and R. Chellappa, "A Unified approach to boundary perception: edges, textures and illusory contours," *IEEE Trans. Neural Net.* **4**, 96–108 (1993).

[35] B. S. Manjunath and W. Y. Ma, "Texture features for browsing and retrieval of image data," *IEEE Trans. Pattern Anal. Machine Intell.* **18**, 837–842 (1996).

[36] S. Marcelja, "Mathematical description of the responses of simple cortical cells," *J. Opt. Soc. Am.* **70**, 1297–1300 (1980).

[37] O. Pichler, A. Teuner, and B. J. Hosticka, "A comparison of texture feature extraction using adaptive Gabor filtering, pyramidal and tree structured wavelet transforms," *Pattern Recog.* **29**, 733–742 (1996).

[38] O. Pichler, A. Teuner, and B. J. Hosticka, "An unsupervised texture

segmentation algorithm for feature space reduction and knowledge feedback," *IEEE Trans. Image Process.* **7**, 53–61 (1998).

[39] M. Porat and Y. Y. Zeevi, "The generalized scheme of image representation in biological and machine vision," *IEEE Trans. Pattern Anal. Machine Intell.* **10**, 452–468 (1988).

[40] M. Porat and Y. Zeevi, "Localized texture processing in vision: analysis and synthesis in the Gaborian space," *IEEE Trans. Biomed. Eng.* **36**, 115–129 (1989).

[41] R. Porter and N. Canagarajah, "Robust rotation invariant texture classification: wavelets, Gabor filter, and GMRF based schemes," *IEE Proc. Vis. Image Signal Process.* **144**, 180–188 (1997).

[42] M. R. Turner, "Texture discrimination by Gabor functions," *Biol. Cybernet.* **55**, 71–82 (1986).

[43] J. You and H. A. Cohen, "Classification and segmentation of rotated and scaled textured images using 'tuned' masks," *Pattern Recog.* **26**, 245–258 (1993).

[44] Signal and Image Processing Institute, University of Southern California; http://sipi.usc.edu.

4.9

Video Segmentation

A. Murat Tekalp
University of Rochester

1 Introduction

Video segmentation refers to the identification of regions in a frame of video that are homogeneous in some sense. Different features and homogeneity criteria generally lead to different segmentations of the same data; for example, color segmentation, texture segmentation, and motion segmentation usually result in different segmentation maps. Furthermore, there is no guarantee that any of the resulting segmentations will be semantically meaningful, since a semantically meaningful region may have multiple colors, multiple textures, or multiple motion. In this chapter, we are primarily concerned with labeling independently moving image regions (motion segmentation) or semantically meaningful image regions (video object plane segmentation). Motion segmentation (also known as optical flow segmentation) methods label pixels (or optical flow vectors) at each frame that are associated with independently moving part of a scene. These regions may or may not be semantically meaningful. For example, a single object with articulated motion may be segmented into multiple regions. Although it is possible to achieve fully automatic motion segmentation with some limited accuracy, semantically meaningful video object segmentation generally requires user to define the object of interest in at least some key frames.

Motion segmentation is closely related to two other problems, motion (change) detection and motion estimation. Change detection is a special case of motion segmentation with only two regions, namely changed and unchanged regions (in the case of a static camera) or global and local motion regions (in the case of a moving camera) [1–3]. An important distinction between change detection and motion segmentation is that the former can be achieved without motion estimation if the scene is recorded with a static camera. Change detection in the case of a moving camera and general motion segmentation, in contrast, require some sort of global or local motion estimation, either explicitly or implicitly. Motion detection and segmentation are also plagued with the same two fundamental limitations associated with motion estimation: occlusion and aperture problems [25]. For example, pixels in a flat image region may appear stationary even if they are moving as a result of an aperture problem (hence the need for hierarchical methods); or erroneous labels may be assigned to pixels in covered or uncovered image regions as a result of an occlusion problem.

It should not come as a surprise that motion/object segmentation is an integral part of many video analysis problems, including (i) improved motion (optical flow) estimation, (ii) three-dimensional (3-D) motion and structure estimation in the presence of multiple moving objects, and (iii) description of the temporal variations or the content of video. In the former case, the segmentation labels help to identify optical flow boundaries (motion edges) and occlusion regions where the smoothness constraint should be turned off. Segmentation is required in the second case, because distinct 3-D motion and structure parameters are needed to model the flow vectors associated with each independently moving object. Finally, in the third case, segmentation information may be employed in an object-level description of frame-to-frame motion, as opposed to a pixel-level description provided by individual flow vectors.

As with any segmentation problem, proper feature selection facilitates effective motion segmentation. In general, the application of standard image segmentation methods directly to estimated optical flow vectors may not yield meaningful results, since an object moving in 3-D usually generates a spatially varying optical flow field. For example, in the case of a rotating object, there is no flow at the center of the rotation, and the magnitude of the flow vectors grows as we move away from the center of rotation. Therefore, a parametric model-based approach, where we assume that the motion field can be described by a set of K parametric models, is usually adopted. In parametric motion segmentation, the model parameters are the motion features. Then, motion segmentation algorithms aim to determine the number of motion models that can adequately describe a scene, type/complexity of these motion models, and the spatial support of each motion model. The most commonly used types of parametric models are affine, perspective, and quadratic mappings, which assume a 3-D planar surface in motion. In the case of a nonplanar object, the resulting optical flow can be modeled by a piecewise affine, perspective, or quadratic flow field if we approximate the object surface by a union of a small number of planar patches. Because each independently moving object or planar patch will best fit a different parametric model, the parametric approach may lead to an oversegmentation of motion in the case of nonplanar objects.

It is difficult to associate a generic figure of merit with a video segmentation result. If segmentation is employed to improve the compression efficiency or rate control, then oversegmentation may not be a cause of concern. The occlusion and aperture problems are mainly responsible for misalignment of motion and actual object boundaries. Furthermore, model misfit possibly as a result of a deviation of the surface structure from a plane generally leads to oversegmentation of the motion field. In contrast, if segmentation is needed for object-based editing and composition as in the upcoming MPEG-4 standard, then it is of utmost importance that the estimated boundaries align with actual object boundaries perfectly. Even a single pixel error may not be tolerable in this case. Although elimination of outlier motion estimates and imposing spatio temporal smoothness constraints on the segmentation map improve the chances of obtaining more meaningful segmentation results, semantic object segmentation in general requires specialized capture methods (chroma keying) or user interaction (semi-automatic methods).

We start our discussion of video segmentation methods with change detection in Section 2, where we study both two-frame methods and methods employing memory, spatial segmentation, or both. Motion segmentation methods can be classified as sequential and simultaneous methods. The dominant motion segmentation approach, which aims to label independently moving regions sequentially (one at a time), is investigated in Section 3, where we discuss the estimation of the parameters and detection of the support of the dominant motion. Section 4 presents methods for simultaneous multiple motion segmentation, including clustering in the motion parameter space, maximum likelihood segmentation, maximum *a posteriori* probability segmentation, and region labeling methods. Since the accuracy of segmentation results depends on the accuracy of the estimated motion field, optical flow estimation and segmentation should be addressed simultaneously for best results. This is addressed in Section 5. Finally, Section 6 deals with semantically meaningful object segmentation with emphasis on chroma keying and semi-automatic methods.

2 Change Detection

Change detection methods segment each frame into two regions, namely changed and unchanged regions in the case of a static camera or global and local motion regions in the case of a moving camera. This section deals only with the former case, in which unchanged regions correspond to the background (null hypothesis) and changed regions to the foreground object(s) or uncovered (occlusion) areas. The case of moving camera is identical to the former, once the global motion between successive frames that is due to camera motion is estimated and compensated. However, an accurate estimation of the camera motion requires scene segmentation; hence, there is a chicken–egg problem. Fortunately, the dominant motion segmentation approach, presented in the next section, offers a solution to the estimation of the camera motion without prior scene segmentation. Hence, the discussion of the case of moving camera is deferred until Section 3. In the following, we first discuss change detection using two frames. Temporal integration (using more than two frames) and the combination of spatial and temporal segmentation are also studied to obtain spatially and temporally coherent regions.

2.1 Detection Using Two Frames

The simplest method to detect changes between two registered frames would be to analyze the frame difference (FD) image, which is given by

$$\text{FD}_{k,r}(\mathbf{x}) = s(\mathbf{x}, k) - s(\mathbf{x}, r), \tag{1}$$

where $\mathbf{x} = (x_1, x_2)$ denotes pixel location and $s(\mathbf{x}, k)$ stands for the intensity value at pixel $\mathbf{x}$ in frame k. The FD image shows the pixel-by-pixel difference between the current image k and the reference image r. The reference image r may be taken as the previous image $k - 1$ (successive frame difference) or an image at a fixed time. For example, if we are interested in monitoring a hallway by using a fixed camera, an image of the hallway when it is empty may be used as a fixed reference image. Assuming that we have a static camera and the illumination remains more or less constant between the frames, the pixel locations where $FD_{k,r}(\mathbf{x})$ differs from zero indicate regions "changed" as a result of local motion. In order to distinguish the nonzero differences that are due to noise from those that are due to local motion, segmentation can be achieved by thresholding the FD as

$$z_{k,r}(\mathbf{x}) = \begin{cases} 1 & \text{if } |FD_{k,r}(\mathbf{x})| > T \\ 0 & \text{otherwise} \end{cases}, \qquad (2)$$

where T is an appropriate threshold. Here, $z_{k,r}(\mathbf{x})$ is called a segmentation label field, which is equal to "1" for changed regions and "0" otherwise. The value of the threshold T can be chosen by an optimal threshold determination algorithm (see Chapter 2.2). This pixelwise thresholding is generally followed by one or more postprocessing steps to eliminate isolated labels. Postprocessing operations include forming four- or eight-connected regions and discarding labels with less than a predetermined number of entries, and morphological filtering of the changed and unchanged region masks.

In practice, a simple FD image analysis is not satisfactory for two reasons: first, a uniform intensity region may be interpreted as stationary even if it is moving (aperture problem). It may be possible to avoid the aperture problem by using a multiresolution decision procedure, since uniform intensity regions are smaller at lower resolution levels. Second, the intensity difference caused by motion is affected by the magnitude of the spatial gradient in the direction of motion. This problem can be addressed by considering a locally normalized frame difference function [4], or locally adaptive thresholding [3]. An improved change detection algorithm that addresses both concerns can be summarized as follows.

1. Construct a Gaussian pyramid in which each frame is represented in multiple resolutions. Start processing at the lowest resolution level.
2. For each pixel at the present resolution level, compute the normalized frame difference given by [4]

$$FDN_{k,r}(\mathbf{x}) = \frac{\sum_{\mathbf{x} \in \mathcal{N}} |s(\mathbf{x}, k) - s(\mathbf{x}, r)||\nabla s(\mathbf{x}, r)|}{\sum_{\mathbf{x} \in \mathcal{N}} |\nabla s(\mathbf{x}, r)|^2 + c}, \qquad (3)$$

where $\mathcal{N}$ denotes a local neighborhood of the pixel $\mathbf{x}$, $\nabla s(\mathbf{x}, r)$ denotes the gradient of image intensity at pixel $\mathbf{x}$, and c is a constant to avoid numerical instability. If the normalized difference is high (indicating that the pixel is moving), replace the normalized difference from the previous resolution level at that pixel with the new value. Otherwise, retain the value from the previous resolution level.
3. Repeat step 2 for all resolution levels.

Finally, we threshold the normalized motion detection function at the highest resolution level.

2.2 Temporal Integration

An important consideration is to add memory to the motion detection process in order to ensure both spatial and temporal continuity of the changed regions at each frame. This can be achieved in a number of different ways, including temporal filtering (integration) of the intensity values across multiple frames before thresholding and postprocessing of labels after thresholding.

A variation of the successive frame difference and normalized frame difference is the frame difference with memory $FDM_k(\mathbf{x})$, which is defined as the difference between the present frame $s(\mathbf{x}, k)$ and a weighted average of past frames $\bar{s}(\mathbf{x}, k)$, given by

$$FDM_k(\mathbf{x}) = s(\mathbf{x}, k) - \bar{s}(\mathbf{x}, k), \qquad (4)$$

where

$$\bar{s}(\mathbf{x}, k) = (1 - \alpha)s(\mathbf{x}, k) + \alpha\bar{s}(\mathbf{x}, k - 1), \quad k = 1, \ldots, \quad (5)$$

and

$$\bar{s}(\mathbf{x}, 0) = s(\mathbf{x}, 0).$$

Here $0 < \alpha < 1$ is a constant. After processing a few frames, the unchanged regions in $\bar{s}(\mathbf{x}, k)$ maintain their sharpness with a reduced level of noise, while the changed regions are blurred. The function $FDM_k(\mathbf{x})$ is thresholded either by a global or a spatially adaptive threshold as in the case of two frame methods. The temporal integration increases the likelihood of eliminating spurious labels, thus resulting in spatially contiguous regions.

Accumulative differences can be employed when detecting changes between a sequence of images and a fixed reference image (as opposed to successive frame differences). Let $s(\mathbf{x}, k)$, $s(\mathbf{x}, k - 1), \ldots, s(\mathbf{x}, k - N)$ be a sequence of N frames, and let $s(\mathbf{x}, r)$ be a reference image. An accumulative difference image is formed by comparing every frame in the sequence with this reference image. For every pixel location, the accumulative image is incremented if the difference between the reference image and the current image in the sequence at that pixel location is bigger than a threshold. Thus, pixels with higher counter values are more likely to correspond to changed regions.

An alternative procedure that was proposed to MPEG-4 considers the postprocessing of labels [5]. First, an initial change detection mask is estimated between successive pairs of frames by global thresholding of the frame difference function. Next, the boundary of the changed regions is smoothed by a relaxation method using local adaptive thresholds [1]. Then, memory is incorporated by relabeling unchanged pixels that correspond to changed locations in one of the last L frames. This step ensures temporal continuity of changed regions from frame to frame. The depth of the memory L may be adapted to scene content to limit error propagation. Finally, postprocessing to obtain the final changed and unchanged masks eliminates small regions.

2.3 Combination with Spatial Segmentation

Another consideration is to enforce consistency of the boundaries of the changed regions with spatial edge locations at each frame. This may be accomplished by first segmenting each frame into uniform color or texture regions. Next, each region resulting from the spatial segmentation is labeled as changed or unchanged as a whole, as opposed to labeling each pixel independently. Region labeling decisions may be based on the number of changed and unchanged pixels within each region or thresholding the average value of the frame differences within each region [6].

3 Dominant Motion Segmentation

Segmentation by dominant motion analysis refers to extracting one object (with the dominant motion) from the scene at a time [4, 7–9]. Dominant motion segmentation can be considered as a hierarchically structured top-down approach, which starts by fitting a single parametric motion model to the entire frame, and then partitions the frame into two regions, those pixels that are well represented by this dominant motion model and those that are not. The process converges to the dominant motion model in a few iterations, each time fitting a new model to only those pixels that are well represented by the motion model in the previous iteration. The dominant motion may correspond to the camera (background) motion or a foreground object motion, whichever occupies a larger area in the frame. The dominant motion approach may also handle separation of individually moving objects. Once the first dominant object is segmented, it is excluded from the region of analysis, and the entire process is repeated to define the next dominant object. This is unlike the multiple motion segmentation approaches that are discussed in the next section, which start with an initial segmentation mask (usually with many small regions) and refine them according to some criterion function to form the final mask. It is worth noting that the dominant motion approach is a direct method that is based on spatiotemporal image intensity gradient information. This is in contrast to first estimating the optical flow field between two frames and then segmenting the image based on the estimated optical flow field.

3.1 Segmentation by Using Two Frames

Motion estimation in the presence of more than one moving objects with unknown supports is a difficult problem. It was Burt *et al.* [7] who first showed that the motion of a two-dimensional (2-D) translating object can be accurately estimated by using a multiresolution iterative approach, even in the presence of other independently moving objects without prior knowledge of their supports. This is, however, not always possible with more

sophisticated motion models (e.g., affine and perspective), which are more sensitive to presence of other moving objects in the region of analysis.

To this effect, Irani *et al.* [4] proposed multistage parametric modeling of dominant motion. In this approach, first a translational motion model is employed over the whole image to obtain a rough estimate of the support of the dominant motion. The complexity of the model is then gradually increased to affine and projective models with refinement of the support of the object in between. The parameters of each model are estimated only over the support of the object based on the previously used model. The procedure can be summarized as follows.

1. Compute the dominant 2-D translation vector (d_x, d_y) over the whole frame as the solution of

$$\begin{bmatrix} \sum I_x^2 & \sum I_x I_y \\ \sum I_x I_y & \sum I_y^2 \end{bmatrix} \begin{bmatrix} d_x \\ d_y \end{bmatrix} = \begin{bmatrix} -\sum I_x I_t \\ -\sum I_y I_t \end{bmatrix} \quad (6)$$

where I_x, I_y, and I_t denote partials of image intensity with respect to x, y, and t. In case the dominant motion is not a translation, the estimated translation becomes a first-order approximation of the dominant motion.

2. Label all pixels that correspond to the estimated dominant motion as follows.
   (a) Register the two images by using the estimated dominant motion model. The dominant object appears stationary between the registered images, whereas other parts of the image are not.
   (b) Then, the problem reduces to labeling stationary regions between the registered images, which can be solved by the multiresolution change detection algorithm given in Section 2.1.
   (c) Here, in addition to the normalized frame difference, Eq. (3), define a motion reliability measure as the reciprocal of the condition number of the coefficient matrix in Eq. (6), given by [4]

$$R(\mathbf{x}, k) = \frac{\lambda_{\min}}{\lambda_{\max}}, \quad (7)$$

   where $\lambda_{\min}$ and $\lambda_{\max}$ are the smallest and largest eigenvalues of the coefficient matrix, respectively. A pixel is classified as stationary at a resolution level if its normalized frame difference is low, and its motion reliability is high. This step defines the new region of analysis.

3. Estimate the parameters of a higher-order motion model (affine, perspective, or quadratic) over the new region of analysis as in [4]. Iterate over steps 2 and 3 until a satisfactory segmentation is attained.

3.2 Temporal Integration

Temporal continuity of the estimated dominant objects can be facilitated by extending the temporal integration scheme introduced in Section 2.2. To this effect, we define an internal representation image [4]

$$\bar{s}(\mathbf{x}, k) = (1 - \alpha)s(\mathbf{x}, k) + \alpha \operatorname{warp}(\bar{s}(\mathbf{x}, k - 1), s(\mathbf{x}, k)),$$
$$k = 1, \ldots, \quad (8)$$

where

$$\bar{s}(\mathbf{x}, 0) = s(\mathbf{x}, 0)$$

and warp(A, B) denotes warping image A toward image B according to the dominant motion parameters estimated between images A and B, and $0 < \alpha < 1$. As in the case of change detection, the unchanged regions in $\bar{s}(\mathbf{x}, k)$ maintain their sharpness with a reduced level of noise, while the changed regions are blurred after processing a few frames.

The algorithm to track the dominant object across multiple frames can be summarized as follows [4]. For each frame, do the following.

1. Compute the dominant motion parameters between the internal representation image $\bar{s}(\mathbf{x}, k)$ and the new frame $s(\mathbf{x}, k)$ within the support M_{k-1} of the dominant object at the previous frame.
2. Warp the internal representation image at frame $k - 1$ toward the new frame according to the computed motion parameters.
3. Detect the stationary regions between the registered images as described in Section 3.1, using M_{k-1} as an initial estimate to compute the new mask M_k.
4. Update the internal representation image by using Eq. (8).

Comparing each new frame with the internal representation image as opposed to the previous frame allows the method to track the same object. This is because the noise is significantly filtered in the internal representation image of the tracked object, and the image gradients outside the tracked object are lowered because of blurring. Note that there is no temporal motion constancy assumption in this tracking scheme.

3.3 Multiple Motions

Multiple object segmentation can be achieved by repeating the same procedure on the residual image after each object is extracted. Once the first dominant object is segmented and tracked, the procedure can be repeated recursively to segment and track the next dominant object after excluding all pixels belonging to the first object from the region of analysis. Hence, the method is capable of segmenting multiple moving objects in a top-down fashion if a dominant motion exists at each stage.

Some difficulties with the dominant motion approach were reported when there was no overwhelmingly dominant motion. Then, in the absence of competing motion models, the dominant motion approach could lead to arbitrary decisions (relying upon absolute threshold values) that are irrevocable, especially when the motion measure indicates unreliable motion vectors (in low spatial gradient regions). Sawhney *et al.* [10] proposed the use of robust estimators to partially alleviate this problem.

4 Multiple Motion Segmentation

Multiple motion segmentation methods let multiple motion models compete against each other at each decision site. They consist of three basic steps, which are strongly interrelated: estimation of the number K of independent motions, estimation of model parameters for each motion, and determination of support of each model (segmentation labels). If we assume that we know the number K of motions and the K sets of motion parameters, then we can determine the support of each model. The segmentation procedure then assigns the label of the parametric motion vector that is closest to the estimated flow vector at each site. Alternatively, if we assume that we know the value of K and a segmentation map consisting of K regions, the parameters for each model can be computed in the least-squares sense (either from estimated flow vectors or from spatiotemporal intensity values) over the support of the respective region. But because both the parameters and supports are unknown in reality, we have a chicken–egg problem; that is, we need to know the motion model parameters to find the segmentation labels, and the segmentation labels are needed to find the motion model parameters.

Various approaches exist in the literature for solving this problem by iterative procedures. They may be grouped as follows: segmentation by clustering in the motion parameter space [11–13], maximum likelihood (ML) segmentation [9, 14, 15], and maximum *a posteriori* probability (MAP) segmentation [16], which are covered in Sections 4.1–4.3, respectively. Pixel-based segmentation methods suffer from the drawback that the resulting segmentation maps may contain isolated labels. Spatial continuity constraints in the form of Gibbs random field (GRF) models have been introduced to overcome this problem within the MAP formulation [16]. However, the computational cost of these algorithms may be prohibitive. Furthermore, they do not guarantee that the estimated motion boundaries coincide with spatial color edges (object boundaries). Section 4.4 presents an alternative region labeling approach to address this problem.

4.1 Clustering in the Motion Parameter Space

A simple segmentation strategy is to first determine the number K of models (motion hypotheses) that are likely to be observed in a sequence, and then perform clustering in the model parameter space (e.g., a six-dimensional space for the case of affine models) to find K models representing the motion. In the following, we study two distinct approaches in this class: the K-means method and the Hough transform method.

4.1.1 *K*-Means Method

Wang and Adelson (W-A) [12] employed K-means clustering for segmentation in their layered video representation. The W-A method starts by partitioning the image into nonoverlapping blocks uniformly distributed over the image, and it fits an affine model to the estimated motion field (optical flow) within each block. In order to determine the reliability of the parameter estimates at each block, the sum of squared distances between the synthesized and estimated flow vectors is computed as

$$\bar{\eta}^2 = \sum_{\mathbf{x} \in \mathcal{B}} \| \mathbf{v}(\mathbf{x}) - \tilde{\mathbf{v}}(\mathbf{x}) \|^2, \tag{9}$$

where $\mathcal{B}$ refers to a block of pixels. Obviously, on one hand, if the flow within the block complies with a single affine model, the residual will be small. On the other hand, if the block falls on the boundary between two distinct motions, the residual will be large. The motion parameters for blocks with acceptably small residuals are selected as the seed models. Then, the seed model parameter vectors are clustered to find the K representative affine motion models. The clustering procedure can be described as follows. Given N seed affine parameter vectors $\mathbf{A}_1, \mathbf{A}_2, \ldots, \mathbf{A}_N$, where

$$\mathbf{A}_n = \begin{bmatrix} a_{n,1} \\ a_{n,2} \\ a_{n,3} \\ a_{n,4} \\ a_{n,5} \\ a_{n,6} \end{bmatrix}, \quad n = 1, \ldots, N, \tag{10}$$

find K cluster centers $\bar{\mathbf{A}}_1, \bar{\mathbf{A}}_2, \ldots, \bar{\mathbf{A}}_K$, where $K \ll N$, and the label k, $k = 1, \ldots, K$, assigned to each affine parameter vector $\mathbf{A}_n$ that minimizes

$$\sum_{n=1}^{N} \mathcal{D}(\mathbf{A}_n, \bar{\mathbf{A}}_k). \tag{11}$$

The distance measure $\mathcal{D}$ between two affine parameter vectors $\mathbf{A}_n$ and $\mathbf{A}_k$ is given by

$$\mathcal{D}(\mathbf{A}_n, \mathbf{A}_k) = \mathbf{A}_n^T \mathbf{M} \mathbf{A}_k, \tag{12}$$

where $\mathbf{M}$ is a 6×6 scaling matrix.

The solution to this problem is given by the well-known K-means algorithm, which consists of the following iteration.

1. Initialize $\bar{\mathbf{A}}_1, \bar{\mathbf{A}}_2, \ldots, \bar{\mathbf{A}}_K$ arbitrarily.
2. For each seed block n, $n = 1, \ldots, N$, find k given by

$$k = \arg \min_s \mathcal{D}(\mathbf{A}_n, \bar{\mathbf{A}}_s), \tag{13}$$

where s takes values from the set $\{1, 2, \ldots, K\}$. It should be noted that if the minimum distance exceeds a threshold, then the site is not labeled, and the corresponding flow vector is ignored in the parameter update that follows.

3. Define $\mathcal{S}_k$ as the set of seed blocks whose affine parameter vector is closest to $\bar{\mathbf{A}}_k$, $k = 1, \ldots, K$. Then, update the class means

$$\bar{\mathbf{A}}_k = \frac{\sum_{n \in \mathcal{S}_k} \mathbf{A}_n}{\sum_{n \in \mathcal{S}_k} 1}. \tag{14}$$

4. Repeat steps 2 and 3 until the class means $\bar{\mathbf{A}}_k$ do not change by more than a predefined amount between successive iterations.

Statistical tests can be applied to eliminate some parameter vectors that are deemed as outliers.

Once the K cluster centers are determined, a label assignment procedure is employed to assign a segmentation label $z(\mathbf{x})$ to each pixel $\mathbf{x}$ as

$$z(\mathbf{x}) = \arg \min_k \| \mathbf{v}(\mathbf{x}) - \mathcal{P}(\bar{\mathbf{A}}_k; (\mathbf{x})) \|^2, \tag{15}$$

where k is from the set $\{1, 2, \ldots, K\}$, the operator $\mathcal{P}$ is defined as

$$\mathcal{P}(\bar{\mathbf{A}}_k; (\mathbf{x})) = \begin{bmatrix} (\bar{a}_{k,1} - 1)x_1 + \bar{a}_{k,2}x_2 + \bar{a}_{k,3} \\ \bar{a}_{k,4}x_1 + (\bar{a}_{k,5} - 1)x_2 + \bar{a}_{k,6} \end{bmatrix}, \tag{16}$$

and $\mathbf{v}(\mathbf{x})$ is the dense motion vector at pixel $\mathbf{x}$ given by

$$\mathbf{v}(\mathbf{x}) = \begin{bmatrix} v_1(\mathbf{x}) \\ v_2(\mathbf{x}) \end{bmatrix}, \tag{17}$$

where v_1 and v_2 denote the horizontal and vertical components, respectively. All sites without labels are assigned one according to the motion compensation criterion, which assigns the label of the parameter vector that gives the best motion compensation at that site. This feature ensures more robust parameter estimation by eliminating the outlier vectors. Several postprocessing operations may be employed to improve the accuracy of the segmentation map. The procedure can be repeated by estimating new seed model parameters over the regions estimated in the previous iteration. Furthermore, the number of clusters can be varied by splitting or merging of clusters between iterations. The K-means method requires a good initial estimate of the number of classes K. The Hough transform methods do not require this information but are more expensive.

4.1.2 Hough Transform Methods

The Hough transform is a well-known clustering technique in which the data samples "vote" for the most representative feature values in a quantized feature space. In a straightforward application of the Hough transform method to optical flow segmentation, using the six-parameter affine flow model, Eq. (16), the six-dimensional feature space $a_1, \ldots, a_6$ would be quantized to certain parameter states after the minimal and maximal values for each parameter are determined. Then, each flow vector $\mathbf{v}(\mathbf{x}) = [v_1(\mathbf{x}) \ v_2(\mathbf{x})]^T$ votes for a set of quantized parameters that minimizes

$$\eta^2(\mathbf{x}) \doteq \eta_1^2(\mathbf{x}) + \eta_2^2(\mathbf{x}), \qquad (18)$$

where $\eta_1(\mathbf{x}) = v_1(\mathbf{x}) - a_1 - a_2 x_1 - a_3 x_2$ and $\eta_2(\mathbf{x}) = v_2(\mathbf{x}) - a_4 - a_5 x_1 - a_6 x_2$. The parameter sets that receive at least a predetermined amount of votes are likely to represent candidate motions. The number of classes K and the corresponding parameter sets to be used in labeling individual flow vectors are hence determined. The drawback of this scheme is the significant amount of computation and memory requirements involved.

In order to keep the computational burden at a reasonable level, several modified Hough methods have been presented. Proposed simplifications to ease the computational load include [11] (a) decomposition of the parameters space into two disjoint subsets $\{a_1, a_2, a_3\} \times \{a_4, a_5, a_6\}$ to perform two 3-D Hough transforms, (b) a multiresolution Hough transform, in that at each resolution level the parameter space is quantized around the estimates obtained at the previous level, and (c) a multipass Hough technique, in which the flow vectors that are most consistent with the candidate parameters are grouped first. In the second stage, those components formed in the first stage that are consistent with the same flow model in the least-squares sense are merged together to form segments. Several merging criteria have been proposed. In the third and final stage, ungrouped flow vectors are assimilated into one of their neighboring segments. Other simplifications that are proposed include the probabilistic Hough transform [17] and the randomized Hough transform [13].

Clustering in the parameter space has some drawbacks: (a) both methods rely on precomputed optical flow as an input representation, which is generally blurred at motion boundaries and may contain outliers, (b) clustering based on distances in the parameter space can lead to clustered parameters that are not physically meaningful and the results are sensitive to the choice of the weight matrix $\mathbf{M}$ and small errors in the estimation of affine parameters, and (c) parameter clustering and label assignment procedures are decoupled; hence, *ad hoc* postprocessing operations that depend on some threshold values are needed to clean up the final segmentation map. The following section proposes a maximum likelihood segmentation method, which addresses all of these shortcomings.

4.2 Maximum Likelihood Segmentation

Motion segmentation approaches in general are classified as optical flow segmentation methods, which operate on precomputed optical flow estimates as an input representation, and direct methods, which operate on spatiotemporal intensity values. We present here a unified formulation that covers both cases. The ML method finds the segmentation labels that maximize the likelihood function, which models the deviation of the observations (estimated dense motion vectors or observed intensity values) from a parametric description of them (parametric

motion vectors or motion compensated intensity values, respectively) for a given motion model.

We start by defining the log likelihood function as

$$L(\mathbf{o} \,|\, \mathbf{z}) = \log(p(\mathbf{o} \,|\, \mathbf{z})), \qquad (19)$$

where $\mathbf{z}$ denotes the lexicographical ordering of the segmentation labels $\mathbf{z}(\mathbf{x})$, which takes values from the set $1, 2, \ldots, K$ at each pixel $\mathbf{x}$. The vector $\mathbf{o}$ stands for the lexicographic ordering of the observations, which are either estimated dense motion (optical flow) vectors or image intensity values. The conditional probability $p(\mathbf{o} \,|\, \mathbf{z})$ quantifies how well piecewise parametric motion modeling fits the observations $\mathbf{o}$ given the segmentation labels $\mathbf{z}$. If we model the mismatch between the observations $\mathbf{o}(\mathbf{x})$ and their parametric representations computed by the operator $\mathcal{O}(\mathbf{A}_{z(\mathbf{x})}; \mathbf{x})$,

$$\eta = \mathbf{o}(\mathbf{x}) - \mathcal{O}\big(\mathbf{A}_{z(\mathbf{x})}; \mathbf{x}\big), \qquad (20)$$

where $\mathbf{A}_k$ denotes the kth parametric motion model, by white Gaussian noise with zero mean and variance σ^2, then the conditional pdf of the observations given the segmentation labels can be expressed as

$$p(\mathbf{o} \,|\, \mathbf{z}) = \frac{1}{(2\pi\sigma^2)^{M/2}} \exp\left\{ -\sum_{i=1}^{M} \eta^2(\mathbf{x}_i)/2\sigma^2 \right\}, \qquad (21)$$

where M is the number of observations available at the sites $\mathbf{x}_i$. Assuming that the parametric flow model is more or less accurate, this deviation is due to presence of observation noise (given correct segmentation labels). Then the problem is to find K motion models $\mathbf{A}_1, \mathbf{A}_2, \ldots, \mathbf{A}_K$, and a label field $z(\mathbf{x})$ to maximize the log likelihood function $L(\mathbf{o} \,|\, \mathbf{z})$.

We consider two cases:

I. Precomputed optical flow segmentation: The observation $\mathbf{o}(\mathbf{x})$ stands for the estimated dense motion vectors $\mathbf{v}(\mathbf{x})$, and the operator $\mathcal{O}$ stands for the parametric motion operator $\mathcal{P}$ given by Eq. (16) or a higher-order model given by

$$\begin{aligned}
\tilde{v}_1(\mathbf{x}) &= a_1 x_1 + a_2 x_2 - a_3 + a_7 x_1^2 + a_8 x_1 x_2, \\
\tilde{v}_2(\mathbf{x}) &= a_4 x_1 + a_5 x_2 - a_6 + a_7 x_1 x_2 + a_8 x_2^2.
\end{aligned} \qquad (22)$$

Then,

$$\eta^2(\mathbf{x}_i) = (v_1(\mathbf{x}_i) - \tilde{v}_1(\mathbf{x}_i))^2 + (v_2(\mathbf{x}_i) - \tilde{v}_2(\mathbf{x}_i))^2 \qquad (23)$$

is the norm-squared deviation of the actual flow vectors from what is predicted by the quadratic flow model. This case concerns motion segmentation by motion vector matching.

II. Direct segmentation: The observation $\mathbf{o}(\mathbf{x})$ stands for the scalar pixel intensities $I_t(\mathbf{x})$ at frame t, and the operator $\mathcal{O}$ is the motion compensation operator $\mathcal{Q}$, defined by

$$\mathcal{Q}(\mathbf{A}_{z(\mathbf{x})}; \mathbf{x}) = \mathbf{I}_{t-1}(\mathbf{x}'), \qquad (24)$$

where

$$\mathbf{x}' = [x_1' \ x_2']^T = [x_1 \ x_2]^T + \mathcal{P}\big(\mathbf{A}_{z(\mathbf{x})}; \mathbf{x}\big). \qquad (25)$$

Then,

$$\eta^2(\mathbf{x}_i) = (I_t(\mathbf{x}) - I_{t-1}(\mathbf{x}'))^2. \qquad (26)$$

This case corresponds to motion segmentation by motion-compensated intensity matching. The motion parameters $\mathbf{A}_k$ are estimated over the support of model k by using direct methods (see step (3) below).

In either case, assuming that the variances for all classes are the same, maximization of the log likelihood function is equivalent to minimization of the cost function

$$\sum_{x_1, x_2} \|\mathbf{o}(\mathbf{x}) - \mathcal{O}\big(\mathbf{A}_{z(\mathbf{x})}; \mathbf{x}\big)\|^2, \qquad (27)$$

or equivalently

$$\sum_{k=1}^{K} \sum_{\mathbf{x} \in \mathcal{Z}_k} \|\mathbf{o}(\mathbf{x}) - \mathcal{O}_k(\mathbf{x})\|^2 \qquad (28)$$

where $\mathcal{Z}_k$ is the set of pixels $\mathbf{x}$ with the motion label $z(\mathbf{x}) = k$, and $\mathcal{O}_k(\mathbf{x}) \doteq \mathcal{O}(\mathbf{A}_k; \mathbf{x})$.

A two-step iterative solution to this problem is given as follows.

1. Initialize $\mathbf{A}_1, \mathbf{A}_2, \ldots, \mathbf{A}_K$.
2. Assign a motion label $z(\mathbf{x})$ to each pixel $\mathbf{x}$ as

$$z(\mathbf{x}) = \arg \min_k \|\mathbf{o}(\mathbf{x}) - \mathcal{O}(\mathbf{A}_k; \mathbf{x})\|^2 \qquad (29)$$

where k takes values from the set $\{1, 2, \ldots, K\}$.
3. Update $\mathbf{A}_1, \mathbf{A}_2, \ldots, \mathbf{A}_K$ as

$$\mathbf{A}_k = \arg \min_{\mathbf{A}} \sum_{\mathbf{x} \in \mathcal{Z}_k} \|\mathbf{v}(\mathbf{x}) - \mathcal{P}(\mathbf{A}; \mathbf{x})\|^2 \qquad (30)$$

This minimization is equivalent to a least-squares estimation of the affine motion model fit to the motion vectors with the label $z(\mathbf{x}) = k$. A closed-form solution to this problem can be expressed in terms of a linear matrix equation

$$\begin{bmatrix} x_1 & x_2 & 1 & 0 & 0 & 0 \\ 0 & 0 & 0 & x_1 & x_2 & 1 \end{bmatrix} \begin{bmatrix} a_{k,1} \\ a_{k,2} \\ a_{k,3} \\ a_{k,4} \\ a_{k,5} \\ a_{k,6} \end{bmatrix} = \begin{bmatrix} x_1 + v_1(\mathbf{x}) \\ x_2 + v_2(\mathbf{x}) \end{bmatrix}$$

$$(31)$$

for all $\mathbf{x}$ such that $z(\mathbf{x}) = k$.
4. Repeat steps (2) and (3) until the class means $\mathbf{A}_k$ do not change by more than a predefined amount between successive iterations.

This method does not require gradient-based optimization or other numeric search procedures for optimization of a cost function. Thus, it is robust and computationally efficient. Extensions of this formulation using mixture modeling and robust estimators, which require a gradient-based optimization, have also been proposed [9].

Motion vector matching is a good motion segmentation criterion when the estimated motion field is accurate; that is, all outlier motion estimates are properly eliminated. Motion-compensated intensity matching is a more suitable criterion when spatial intensity (color) variations are sufficient and/or a multiresolution labeling procedure is employed. A possible limitation of the ML segmentation framework is that it lacks constraints to enforce spatial and temporal continuity of the segmentation labels. Thus, rather *ad hoc* steps are needed to eliminate small, isolated regions in the segmentation label field. The MAP segmentation strategy promises to impose continuity constraints in an optimization framework.

4.3 Maximum *a posteriori* Probability Segmentation

The MAP method is a Bayesian approach that searches for the maximum of the *a posteriori* pdf of the segmentation labels given the observations (either precomputed optical flow or spatio-temporal intensity data). This pdf is not only a measure of how well the segmentation labels explain the observed data, but also how well they conform with our prior expectations. The MAP formulation differs from the maximum likelihood approach in that it includes smoothness terms to enforce spatial continuity of the output motion segmentation map.

The *a posteriori* pdf $p(\mathbf{z} \,|\, \mathbf{o})$ of the segmentation label field $\mathbf{z}$ given the observed data $\mathbf{o}$ can be expressed, using the Bayes theorem, as

$$p(\mathbf{z} \,|\, \mathbf{o}) = \frac{p(\mathbf{o} \,|\, \mathbf{z})\, p(\mathbf{z})}{p(\mathbf{o})} \qquad (32)$$

where $p(\mathbf{o} \,|\, \mathbf{z})$ is the conditional pdf of the optical flow data given the segmentation $\mathbf{z}$, and $p(\mathbf{z})$ is the *a priori* pdf of the segmentation. Observe that, (a) $\mathbf{z}$ is a discrete-valued random vector with a finite sample space Ω, and (b) $p(\mathbf{o})$ is constant with respect to the segmentation labels and hence can be ignored for the purpose of computing $\mathbf{z}$. The MAP estimate, then, maximizes the numerator of Eq. (32) over all possible realizations of the segmentation field $\mathbf{z} = \omega$, $\omega \in \Omega$.

Modeling of the conditional pdf $p(\mathbf{o} \,|\, \mathbf{z})$ has been discussed in detail in Section 4.2 through Eqs. (21) and (23) or Eq. (26). The prior pdf is modeled by a Gibbs distribution, which effectively introduces local constraints on the segmentation. It is given by

$$p(\mathbf{z}) = \frac{1}{Q} \sum_{\omega \in \Omega} \exp\{-U(\mathbf{z})\}\delta(\mathbf{z} - \omega), \qquad (33)$$

where Ω denotes the sample space of the discrete-valued random vector $\mathbf{z}$, Q is the partition function (normalization constant) given by

$$Q = \sum_{\omega \in \Omega} \exp\{-U(\omega)\}, \qquad (34)$$

$U(\mathbf{z})$ is the potential function given by

$$U(\mathbf{z}) = \sum_{\mathbf{x}_i} \sum_{\mathbf{x}_j \in \mathcal{N}_{\mathbf{x}_i}} V_C(z(\mathbf{x}_i), z(\mathbf{x}_j)), \qquad (35)$$

which can be expressed as a sum of local clique potential functions, such as

$$V_C(z(\mathbf{x}_i), z(\mathbf{x}_j)) = \begin{cases} -\gamma & \text{if } z(\mathbf{x}_i) = z(\mathbf{x}_j) \\ +\gamma & \text{otherwise} \end{cases}, \qquad (36)$$

and $\mathcal{N}_{\mathbf{x}_i}$ denotes the neighborhood system for the label field. Prior constraints on the structure of the segmentation labels, such as spatial smoothness, can be specified in terms of the clique potential function. Temporal continuity of the labels can similarly be modeled [16]

Substituting Eqs. (21) and (33) into the criterion (32) and taking the logarithm of the resulting expression, maximization of the *a posteriori* probability distribution can be performed by minimizing the cost function

$$E = \frac{1}{2\sigma^2} \sum_{i=1}^{M} \eta^2(\mathbf{x}_i) + U(\omega). \qquad (37)$$

The first term describes how well the predicted data fit the actual measurements (estimated optical flow vectors or image intensity values), and the second term measures how well the segmentation conforms to our prior expectations.

Because the motion model parameters corresponding to each label are not known *a priori*, the MAP segmentation must alternate between estimation of the model parameters and assignment of the segmentation labels to optimize the cost function, Eq. (37). Murray and Buxton [16] were the first to propose a MAP segmentation method in which the optical flow was modeled by a piecewise quadratic flow field, Eq. (22), and the segmentation labels, were assigned based on a simulated annealing (SA) procedure. Given the estimated flow field $\mathbf{v}$ and the number of independent motion models K, the MAP segmentation using the Metropolis algorithm can be summarized as follows:

1. Start with an initial labeling $\mathbf{z}$ of the optical flow vectors. Calculate the model parameters $\mathbf{a} = [a_1 \cdots a_8]^T$ for each region, using least-squares fitting (similar to that in Section 4.2). Set the initial temperature for SA.
2. Update the segmentation labels at each site $\mathbf{x}_i$ as follows.
   (a) Perturb the label $z_i = z(\mathbf{x}_i)$ randomly.

(b) Decide whether to accept or reject this perturbation, based on the change ΔE in the cost function, Eq. (37),

$$\Delta E = \frac{1}{2\sigma^2} \Delta \eta^2(\mathbf{x}_i) + \sum_{\mathbf{x}_j \in \mathcal{N}_{\mathbf{x}_i}} \Delta V_C(z(\mathbf{x}_i), z(\mathbf{x}_j)).$$

$$(38)$$

where $\mathcal{N}_{\mathbf{x}_i}$ denotes a neighborhood of the site $\mathbf{x}_i$ and $V_C(z(\mathbf{x}_i), z(\mathbf{x}_j))$ is given by Eq. (36). The first term indicates whether or not the perturbed label is more consistent with the given flow field determined by the residual, Eq. (23), and the second term reflects whether or not it is in agreement with the prior segmentation field model.

Because the update at each site is dependent on the labels of the neighboring sites, the order in which the sites are visited affects the result of this step.

3. After all pixel sites are visited once, reestimate the mapping parameters for each region based on the new segmentation label configuration.
4. Exit if a stopping criterion is satisfied. Otherwise, lower the temperature according to a predefined temperature schedule, and go to step 2.

We can make the following observations. First, the MAP method carries a high computational cost. Second, the procedure proposed by Murray-Buxton suggests performing step 3 above, the model parameter update, after each and every perturbation. We did not notice a significant difference in performance if motion parameter updates are done after all sites are visited once. Third, the method can be applied with any parametric motion model, although the original formulation has been developed on the basis of the eight-parameter model.

4.4 Region-Based Label Assignment

In this section, we extend the ML approach (Section 4.2) to region-based motion segmentation, where the image is first divided into predefined homogeneous regions, and then, at every iteration, each region is assigned a single motion label. This region-based label assignment strategy facilitates obtaining spatially continuous segmentation maps that are closely related to actual object boundaries, without the heavy computational burden of statistical Markov random field (MRF) model-based approaches. The predefined regions should be such that each region has a single motion. It is generally true that motion boundaries coincide with color segment boundaries, but not vice versa; that is, color segments are almost always a subset of motion segments, as illustrated in Fig. 1. Therefore, one can first perform a color segmentation to obtain a set of candidate motion segments. Other approaches to region definition include mesh-based partitioning of the scene [18] and macropixels (N × N blocks) to improve the robustness of the ML motion segmentation. Here, we assume that each frame of video has been subject to a region

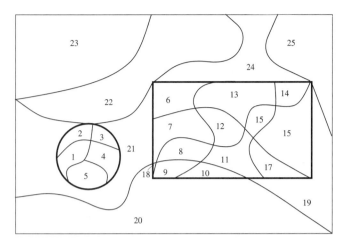

FIGURE 1 Illustration of the observation that color segments are generally subsets of motion segments. The bold lines indicate motion segment boundaries, and each motion segment is composed of many color regions.

formation procedure. We let $C(\mathbf{x})$ denote the region map of a frame consisting of M mutually exclusive and exhaustive regions, and we define $\mathcal{C}_m$ as the set of pixels $\mathbf{x}$ with the region label $C(\mathbf{x}) = m$, $m = 1, \ldots, M$.

We wish to find the motion segmentation map $\mathbf{z}$ [a vector formed by lexicographic ordering of $z(\mathbf{x})$] and the corresponding affine parameter vectors $\mathbf{A}_1, \mathbf{A}_2, \ldots, \mathbf{A}_K$ that best fit the dense motion-vector field, such that [15]

$$\sum_{m=1}^{M} \sum_{\mathbf{x} \in \mathcal{C}_m} \left\| \mathbf{v}(\mathbf{x}) - \mathcal{P}\left(\mathbf{A}_{z(m)}; \mathbf{x}\right) \right\|^2 \qquad (39)$$

is minimized. Here $z(m)$ refers to the motion label of all pixels within $\mathcal{C}_m$ and takes one of the values $1, 2, \ldots, K$; $\mathcal{P}$ is an operator defined by Eq. (16), and $\mathbf{v}(\mathbf{x})$ is the dense motion vector at pixel $\mathbf{x}$ as defined by Eq. (17). The procedure is given as follows.

1. Initialize the motion segmentation map $\mathbf{z}$ by assigning a single motion label k, $k = 1, \ldots, K$ to each $\mathcal{C}_m$.
2. Update the parameter vectors $\mathbf{A}_1, \mathbf{A}_2, \ldots, \mathbf{A}_K$ as

$$\mathbf{A}_k = \arg \min_{\mathbf{A}} \sum_{\mathbf{x} \in \mathcal{Z}_k} \left\| \mathbf{v}(\mathbf{x}) - \mathcal{P}(\mathbf{A}; \mathbf{x}) \right\|^2, \qquad (40)$$

where $\mathcal{Z}_k$ is the set of pixels $\mathbf{x}$ with the label $z(\mathbf{x}) = k$. This minimization can be achieved by solving the linear matrix equation

$$\begin{bmatrix} x_1 & x_2 & 1 & 0 & 0 & 0 \\ 0 & 0 & 0 & x_1 & x_2 & 1 \end{bmatrix} \begin{bmatrix} a_{k,1} \\ a_{k,2} \\ a_{k,3} \\ a_{k,4} \\ a_{k,5} \\ a_{k,6} \end{bmatrix} = \begin{bmatrix} v_1(\mathbf{x}) \\ v_2(\mathbf{x}) \end{bmatrix} \quad (41)$$

for all $\mathbf{x}$ in $\mathcal{Z}_k$.

3. Assign a motion label to each region $\mathcal{C}_m$, $m = 1, 2, \ldots, M$, such that

$$z(\mathcal{C}_m) = \arg \min_k \sum_{\mathbf{x} \in \mathcal{C}_m} \left\| \mathbf{o}(\mathbf{x}) - \mathcal{O}(\mathbf{A}_k; \mathbf{x}) \right\|^2, \qquad (42)$$

where $k = 1, 2, \ldots, K$ and $\mathbf{o}(\mathbf{x})$ and $\mathcal{O}(\mathbf{x})$ are as defined in Section 4.2. This allows region-based affine motion segmentation with pixel-based motion-vector or intensity matching.
4. Repeat steps 2 and 3 until the class means $\mathbf{A}_k$ do not change by more than a predefined amount between successive iterations.

We note that the pixel-based ML motion segmentation method presented in Section 4.2 is a special case of this region-based framework. If each region $\mathcal{C}_m$ contains a single pixel, then the iterations are carried over individual pixels, and the motion label assignment is performed at each pixel independently.

We conclude this section by observing that the methods discussed here that used precomputed optical flow as an input representation are limited by the accuracy of the available optical flow estimates. Next, we introduce a framework, in which optical flow estimation and segmentation interact in a mutually beneficial manner.

5 Simultaneous Estimation and Segmentation

Until now, we discussed methods to compute the segmentation labels from either precomputed optical flow or directly from intensity values, but we did not address how to compute an improved dense motion field along with the segmentation map. It is clear that the success of optical flow segmentation is closely related to the accuracy of the estimated optical flow field (in the case of using precomputed flow values), and vice versa. It follows that optical flow estimation and segmentation have to be addressed simultaneously for best results. Here we present a simultaneous Bayesian approach based on a representation of the motion field as the sum of a parametric field and a residual field. The interdependence of optical flow and segmentation fields is expressed in terms of a Gibbs distribution within the MAP framework. The resulting optimization problem, to find estimates of a dense set of motion vectors, a set of segmentation labels, and a set of mapping parameters, is solved by using the highest confidence first (HCF) and iterated conditional mode (ICM) algorithms.

5.1 Modeling

We model the optical flow field $\mathbf{v}(\mathbf{x})$ as the sum of a parametric flow field $\tilde{\mathbf{v}}(\mathbf{x})$ and a nonparametric residual field $\mathbf{v}_r(\mathbf{x})$, which accounts for local motion and other modeling errors; that is,

$$\mathbf{v}(\mathbf{x}) = \tilde{\mathbf{v}}(\mathbf{x}) + \mathbf{v}_r(\mathbf{x}) \tag{43}$$

The parametric component of the motion field clearly depends on the segmentation label $z(\mathbf{x})$, which takes on the values $1, \ldots, K$.

The simultaneous MAP framework aims at maximizing the *a posteriori* pdf

$$
\begin{aligned}
& p(\mathbf{v}_1, \mathbf{v}_2, \mathbf{z} \mid \mathbf{g}_k, \mathbf{g}_{k+1}) \\
& = \frac{p(\mathbf{g}_{k+1} \mid \mathbf{g}_k, \mathbf{v}_1, \mathbf{v}_2, \mathbf{z}) \, p(\mathbf{v}_1, \mathbf{v}_2 \mid \mathbf{z}, \mathbf{g}_k) \, p(\mathbf{z} \mid \mathbf{g}_k)}{p(\mathbf{g}_{k+1} \mid \mathbf{g}_k)}
\end{aligned} \tag{44}
$$

with respect to the optical flow $\mathbf{v}_1$, $\mathbf{v}_2$, and the segmentation labels $\mathbf{z}$, where $\mathbf{v}_1$ and $\mathbf{v}_2$ denote the lexicographic ordering of the first and second components of the flow vectors $\mathbf{v}(\mathbf{x}) = [v_1(\mathbf{x}) \ v_2(\mathbf{x})]^T$ at each pixel $\mathbf{x}$. Through careful modeling of these pdfs, we can express an interrelated set of constraints that help improve both optical flow and segmentation estimates.

The first conditional pdf $p(\mathbf{g}_{k+1} \mid \mathbf{g}_k, \mathbf{v}_1, \mathbf{v}_2, \mathbf{z})$ provides a measure of how well the present displacement and segmentation estimates conform with the observed frame $k + 1$ given frame k. It is modeled by a Gibbs distribution as

$$p(\mathbf{g}_{k+1} \mid \mathbf{g}_k, \mathbf{v}_1, \mathbf{v}_2, \mathbf{z}) = \frac{1}{Q_1} \exp\{-U_1(\mathbf{g}_{k+1} \mid \mathbf{g}_k, \mathbf{v}_1, \mathbf{v}_2, \mathbf{z})\} \tag{45}$$

where Q_1 is the partition function (normalizing constant), and

$$U_1(\mathbf{g}_{k+1} \mid \mathbf{g}_k, \mathbf{v}_1, \mathbf{v}_2, \mathbf{z}) = \sum_{\mathbf{x}} [g_k(\mathbf{x}) - g_{k+1}(\mathbf{x} + \mathbf{v}(\mathbf{x})\Delta t)]^2 \tag{46}$$

is called the Gibbs potential. Here, the Gibbs potential corresponds to the norm square of the displaced frame difference (DFD) between the frames $\mathbf{g}_k$ and $\mathbf{g}_{k+1}$. Thus, maximization of Eq. (45) imposes the constraint that $\mathbf{v}(\mathbf{x})$ minimizes the DFD.

The second term in the numerator in Eq. (44) is the conditional pdf of the displacement field given the motion segmentation and the search image. It is also modeled by a Gibbs distribution

$$p(\mathbf{v}_1, \mathbf{v}_2 \mid \mathbf{z}, \mathbf{g}_k) = p(\mathbf{v}_1, \mathbf{v}_2 \mid \mathbf{z}) = \frac{1}{Q_2} \exp\{-U_2(\mathbf{v}_1, \mathbf{v}_2 \mid \mathbf{z})\} \tag{47}$$

where Q_2 is a constant, and

$$
\begin{aligned}
U_2(\mathbf{v}_1, \mathbf{v}_2 \mid \mathbf{z}) = {} & \alpha \sum_{\mathbf{x}} \|\mathbf{v}(\mathbf{x}) - \tilde{\mathbf{v}}(\mathbf{x})\|^2 \\
& + \beta \sum_{\mathbf{x}_i} \sum_{\mathbf{x}_j \in \mathcal{N}_{\mathbf{x}_i}} \|\mathbf{v}(\mathbf{x}_i) - \mathbf{v}(\mathbf{x}_j)\|^2 \\
& \times \delta(z(\mathbf{x}_i) - z(\mathbf{x}_j))
\end{aligned} \tag{48}
$$

is the corresponding Gibbs potential, $\|\cdot\|$ denotes the Euclidian distance, and $\mathcal{N}_{\mathbf{x}}$ is the set of neighbors of site $\mathbf{x}$. The first term in Eq. (48) enforces a minimum norm estimate of the residual motion field $\mathbf{v}_r(\mathbf{x})$; that is, it aims to minimize the deviation of the motion field $\mathbf{v}(\mathbf{x})$ from the parametric motion field $\tilde{\mathbf{v}}(\mathbf{x})$ while minimizing the DFD. Note that the parametric motion field $\tilde{\mathbf{v}}(\mathbf{x})$ is calculated from the set of model parameters $\mathbf{a}_i$, $i = 1, \ldots, K$, which in turn is a function of $\mathbf{v}(\mathbf{x})$ and $z(\mathbf{x})$. The second term in Eq. (48) imposes a piecewise local smoothness constraint on the optical flow estimates without introducing any extra variables such as line fields. Observe that this term is active only for those pixels in the neighborhood $\mathcal{N}_{\mathbf{x}}$ that share the same segmentation label with the site $\mathbf{x}$. Thus, spatial smoothness is enforced only on the flow vectors generated by a single object. The parameters α and β allow for relative scaling of the two terms.

The third term in Eq. (44) models the *a priori* probability of the segmentation field in a manner similar to that in MAP segmentation. It is given by

$$p(\mathbf{z} \mid \mathbf{g}_k) = p(\mathbf{z}) = \frac{1}{Q_3} \sum_{\omega \in \Omega} \exp\{-U_3(\mathbf{z})\}\delta(\mathbf{z} - \omega), \tag{49}$$

where Ω denotes the sample space of the discrete-valued random vector $\mathbf{z}$, and Q_3 and $U_3(\mathbf{z})$ are as defined in Eqs. (34) and (35), respectively. The dependence of the labels on the image intensity is usually neglected, although region boundaries generally coincide with intensity edges.

5.2 An Algorithm

Maximizing the *a posteriori* pdf, Eq. (44), is equivalent to minimizing the cost function,

$$E = U_1(\mathbf{g}_{k+1} \mid \mathbf{g}_k, \mathbf{v}_1, \mathbf{v}_2, \mathbf{z}) + U_2(\mathbf{v}_1, \mathbf{v}_2 \mid \mathbf{z}) + U_3(\mathbf{z}) \tag{50}$$

which is composed of the potential functions in Eqs. (45), (47), and (49). Direct minimization of Eq. (50) with respect to all unknowns is an exceedingly difficult problem, because the motion and segmentation fields constitute a large set of unknowns. To this effect, we perform the minimization of Eq. (50) through the following two-step iteration [20]:

1. Given the best available estimates of the parameters $\mathbf{a}_i$, $i = 1, \ldots, K$, and $\mathbf{z}$, update the optical flow field $\mathbf{v}_1$, $\mathbf{v}_2$. This step involves the minimization of a modified cost function

$$
\begin{aligned}
E_1 = {} & \sum_{\mathbf{x}} [g_k(\mathbf{x}) - g_{k+1}(\mathbf{x} + \mathbf{v}(\mathbf{x})\Delta t)]^2 \\
& + \alpha \sum_{\mathbf{x}} \|\mathbf{v}(\mathbf{x}) - \tilde{\mathbf{v}}(\mathbf{x})\|^2 \\
& + \beta \sum_{\mathbf{x}_i} \sum_{\mathbf{x}_j \in \mathcal{N}_{\mathbf{x}_i}} \|\mathbf{v}(\mathbf{x}_i) - \mathbf{v}(\mathbf{x}_j)\|^2 \delta(z(\mathbf{x}_i) - z(\mathbf{x}_j)),
\end{aligned} \tag{51}
$$

which is composed of all terms in Eq. (50) that contain $\mathbf{v}(\mathbf{x})$. While the first term indicates how well $\mathbf{v}(\mathbf{x})$ explains our observations, the second and third terms impose prior constraints on the motion estimates that they should conform with the parametric flow model, and that they should vary smoothly within each region. To minimize this energy function, we employ the HCF method recently proposed by Chou and Brown [19]. HCF is a deterministic method designed to efficiently handle the optimization of multivariable problems with neighborhood interactions.

2. Update the segmentation field $\mathbf{z}$, assuming that the optical flow field $\mathbf{v}(\mathbf{x})$ is known. This step involves the minimization of all the terms in Eq. (50) that contain $\mathbf{z}$ as well as $\tilde{\mathbf{v}}(\mathbf{x})$, given by

$$E_2 = \alpha \sum_{\mathbf{x}} \|\mathbf{v}(\mathbf{x}) - \tilde{\mathbf{v}}(\mathbf{x})\|^2 + \sum_{\mathbf{x}_i} \sum_{\mathbf{x}_j \in \mathcal{N}_{\mathbf{x}_i}} V_C(z(\mathbf{x}_i), z(\mathbf{x}_j)) \tag{52}$$

The first term in Eq. (52) quantifies the consistency of $\tilde{\mathbf{v}}(\mathbf{x})$ and $\mathbf{v}(\mathbf{x})$. The second term is related to the *a priori* probability of the present configuration of the segmentation labels. We use an ICM procedure to optimize E_2 [20]. The mapping parameters $\mathbf{a}_i$ are updated by a least-squares a estimation within each region.

An initial estimate of the optical flow field can be found by using the Bayesian approach with a global smoothness constraint. Given this estimate, the segmentation labels can be initialized by a procedure similar to Wang and Adelson's [12]. The determination of the free parameters α, β, and γ is a design problem. One strategy is to choose them to provide a dynamic range correction so that each term in cost function (50) has equal emphasis. However, because the optimization is implemented in two steps, the ratio α/γ also becomes of consequence. We recommend the selection of $1 \le \alpha/\gamma \le 5$, depending on how well the motion field can be represented by a piecewise parametric model and whether we have a sufficient number of classes.

A hierarchical implementation of this algorithm is also possible by forming successive low-pass filtered versions of the images $\mathbf{g}_k$ and $\mathbf{g}_{k+1}$. Thus, the quantities $\mathbf{v}_1$, $\mathbf{v}_2$, and $\mathbf{z}$ can be estimated at different resolutions. The results of each hierarchy are used to initialize the next lower level. Note that the Gibbsian model for the segmentation labels has been extended to include neighbors in scale by Kato *et al.* [21].

Several other motion analysis approaches can be formulated as special cases of this framework. If we retain only the first and the third terms in Eq. (50), and assume that all sites possess the same segmentation label, then we have Bayesian motion estimation with a global smoothness constraint. The motion estimation algorithm proposed by Iu [22] utilizes the same two terms, but it replaces the $\delta(\cdot)$ function by a local outlier rejection function. The motion estimation and region labeling algorithm proposed by Stiller [23] involves all terms in Eq. (50), except the

first term in Eq. (48). Furthermore, the segmentation labels in Stiller's algorithm are used merely as tokens to allow for a piecewise smoothness constraint on the flow field, and they do not attempt to enforce consistency of the flow vectors with a parametric component. We also note that the motion estimation method of Konrad and Dubois [24], which uses line fields, is fundamentally different in that they model discontinuities in the motion field, rather than modeling regions that correspond to different parametric motions. In contrast, the motion segmentation algorithm of Murray and Buxton [16] (Section 4.3) employs only the second term in Eq. (48) and third term in Eq. (50) to model the conditional and prior pdf, respectively. Wang and Adelson [12] rely on the first term in Eq. (48) to compute the motion segmentation (Section 4.2). However, they also take the DFD of the parametric motion vectors into consideration when the closest match between the estimated and parametric motion vectors, represented by the second term, exceeds a threshold.

6 Semantic Video Object Segmentation

So far we discussed methods for automatic motion segmentation. However, it is difficult to achieve semantically meaningful object segmentation by using fully automatic methods based on low-level features such as motion, color, and texture. This is because a semantic object may contain multiple motions, colors, textures, and so on, and the definition of semantic objects may depend on the context, which may not be possible to capture by using low-level features. Thus, in this section, we present two approaches that can extract semantically meaningful objects by using capture-specific information or user interaction.

6.1 Chroma Keying

Chroma keying is an object-based video technology in which each video object is captured individually in a special studio against a key color. The key color is selected such that it does not appear on the object to be captured. Then, the problem of extracting the object from each frame of video becomes one of color segmentation. Chroma-keyed video capture requires special attention to avoid shadows and other nonuniformity in the key color within a frame; otherwise, the segmentation of key color may become a nontrivial problem.

6.2 Semi-Automatic Segmentation

Because chroma keying requires special studios or equipment to capture video objects, an alternative approach is interactive segmentation, using automated tools to aid a human operator. To this effect, we assume that the contour of the first occurrence of the semantic object of interest is marked interactively by a human operator. In many instances this is indeed the only way to define a semantically meaningful object unambiguously,

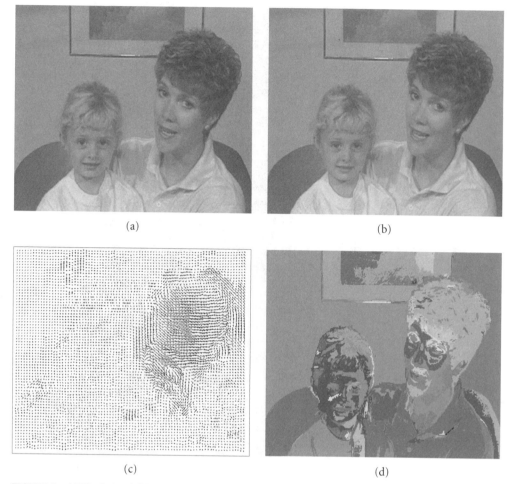

FIGURE 2   (a) The first and (b) second frames of the Mother and Daughter sequence; (c) 2-D dense motion field from the second frame to the first frame; (d) region map obtained segmentation. (See color section, p. C–19.)

because only the user can know what is semantically meaningful in the context of an application. For example, if we have the video clip of a person carrying a ball, whether the ball and the person are two separate objects or a single object may depend on the application. Once the boundary of the object of interest is interactively determined in one or more key frames, its boundary in all other frames can be automatically computed by 2-D motion tracking until the object exits the field of view. Two-dimensional tracking methods include boundary tracking and region tracking, which are discussed in [25]. This tracking step defines a polygonal or spline approximation of the boundary of the video object, which may be further refined interactively by using appropriate software tools.

7 Examples

Examples are shown for automatic motion segmentation using the pixel-based and region-based ML methods on two MPEG-4 test sequences: "Mother and Daughter"(frames 1–2) and

"Mobile and Calendar" (frames 136–137). The former is an example of a slowly moving object against a still background, where the mother's head is rotating while her body, the background, and the child are stationary. The latter is a challenging sequence, with several distinctly moving objects such as a rotating ball, a moving train, and a vertically translating calendar against a background that moves as a result of camera pan. Figures 2(a) and 2(b) show the first and second frames of the Mother and Daughter sequence, and Fig. 2(c) shows the estimated motion field between these frames. Figures 4(a), 4(b), 4(c) below show the corresponding pictures for frames 136 and 137 of the Mobile and Calendar sequence. Motion estimation was performed by using the hierarchical version of the Lucas-Kanade method [25] with three levels of hierarchy. In both cases, region definition by color segmentation is performed on the temporally second frame by using the fuzzy c-means technique [26]. Each spatially disconnected piece of the color segmentation map was defined as an individual region. The resulting region maps are shown in Figs. 2(d) and 4(d), respectively.

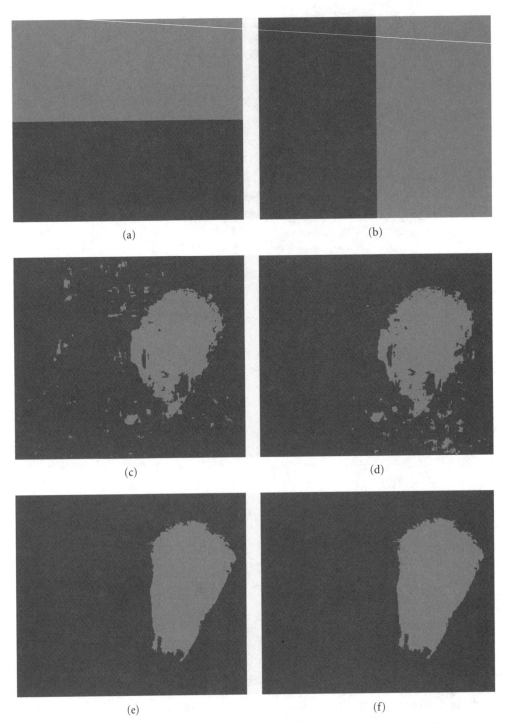

FIGURE 3   Results of the ML method with two different initializations: (a), (b) initial map; (c), (d) pixel-based motion-vector matching; (e), (f) region-based motion-vector matching.

Figure 3 demonstrates the performance of the ML method for foreground/background separation (i.e., $K = 2$) with two different initializations. Figures 3(a) and 3(b) show two possible initial segmentation maps, where the segmentation map is divided into two horizontal and vertical parts, respectively. Figures 3(b) and 3(c) show the segmentation maps using pixel-based labeling by motion-vector matching after 10 iterations starting from Figs. 3(a) and 3(b), respectively. Figures 3(d) and 3(e) show the results of region labeling by motion-vector matching starting with the affine parameter sets obtained from the maps of

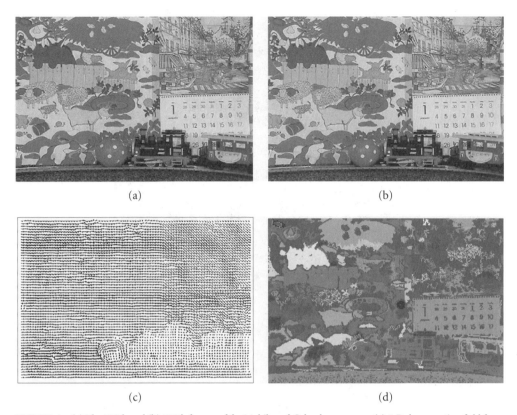

(a)

(b)

(c)

(d)

FIGURE 4 (a) The 136th and (b) 137th frames of the Mobile and Calendar sequence; (c) 2-D dense motion field from the 137th to 136th frame; (d) region map obtained by color segmentation. (See color section, p. C–20.)

Figs. 3(b) and 3(c), respectively. Observe that the segmentation maps obtained by pixel labeling contain many misclassified pixels, whereas the maps obtained by color-region labeling are more coherent with the moving object in the scene.

Figure 5 illustrates the performance of the ML method with different number of initial segments, K. Figures 5(a) and 5(b) show two initial segmentation maps with $K = 4$ and $K = 6$, respectively. The results of pixel-based labeling by motion vector-matching after 10 iterations for both initializations are depicted

in Figs. 5(c) and 5(d), respectively. Figure 5(e) shows the result of region-based labeling using the color regions depicted in Fig. 4(d) and the affine model parameters initialized by those computed from the map in Fig. 5(c) with $K = 4$. We observed that this procedure results in oversegmentation when repeated with $K = 6$. Therefore, we employ motion-compensated intensity matching and region merging to reduce the number of the motion classes if necessary. In this step, a region is merged with another if the latter set of affine parameters gives a comparable

(a)

(b)

FIGURE 5 Results of the ML method: initial map (a) $K = 4$, (b) $K = 6$; pixel-based labeling (c) $K = 4$, (d) $K = 6$; region-based labeling (e) $K = 4$, (f) $K = 6$. (See color section, p. C–21.) *(Continues.)*

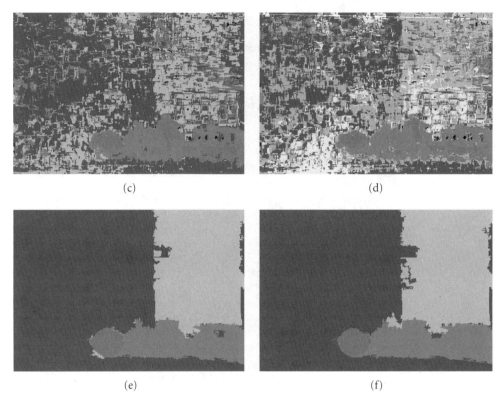

FIGURE 5   *(Continued).*

DFD as the former. The result of this final step is depicted in Fig. 5(f) for $K = 6$, where two of the six classes are eliminated by motion-compensated intensity matching. The ML segmentation method is computationally efficient, because it does not require gradient-based optimization or any numeric search. It converges within approximately 10 iterations, and each iteration involves solution of only two 3×3 matrix equations. The complete procedure takes less than a minute to implement on a SparcStation 20.

Acknowledgment

The author acknowledges Yucel Altunbasak and P. Erhan Eren for their contributions to the section on the maximum likelihood segmentation and Michael Chang for his contributions to the section on the maximum *a posteriori* probability segmentation. This work was supported by grants from National Science Foundation, New York State Science and Technology Foundation, and Eastman Kodak Company.

References

[1] T. Aach, A. Kaup, and R. Mester, "Statistical model-based change detection in moving video," *Signal Process*, **31**, 165–180 (1993).

[2] F. Dufaux, F. Moscheni, and A. Lippman, "Spatio-temporal segmentation based on motion and static segmentation, *Proc. IEEE Int. Conf. Image Proc.* **1**, 306–309 (1995).

[3] A. Neri, S. Colonnese, G. Russo, and P. Talone, "Automatic moving object and background separation," *Signal Process*. (special issue), **66**, 219–232 (1998).

[4] M. Irani, B. Rousso, and S. Peleg, "Computing occluding and transparent motions," *Int. J. Comput. Vision* **12**, 5–16 (1994).

[5] ISO 14496-2, MPEG-4 Draft International Standard, Nov. 98.

[6] C. Gu, T. Ebrahimi, and M. Kunt, "Morphological moving object segmentation and tracking for content-based video coding," presented at the International Symposium on Multimedia Communication and Video Coding, New York, NY, Oct. 1995.

[7] P. J. Burt, R. Hingorani, and R. Kolczynski, "Mechanisms for isolating component patterns in the sequential analysis of multiple motion," in *IEEE Workshop on Visual Motion* (IEEE, New York, 1971), pp. 187–193.

[8] J. R. Bergen, P. J. Burt, K. Hanna, R. Hingorani, P. Jeanne, and S. Peleg, "Dynamic multiple-motion computation," in *Artificial Intelligence and Computer Vision*, Y. A. Feldman and A. Bruckstein, eds. (Elsevier, Holland, 1991), pp. 147–156.

[9] S. Ayer and H. Sawhney, "Layered representation of motion video using robust maximum-likelihood estimation of mixture models and MDL coding," presented at the IEEE International Conference on Computer Vision, Cambridge, MA, June 1995.

[10] H. Sawhney, S. Ayer, and M. Gorkani, "Model-based 2-D and 3-D dominant motion estimation for mosaicing and video representation," presented at the IEEE International Conference on Computer Vision, Cambridge, MA, June 1995.

[11] G. Adiv, "Determining three-dimensional motion and structure from optical flow generated by several moving objects," *IEEE Trans. Pattern Anal. Machine Intell.* **7**, 384–401 (1985).

[12] J. Y. A. Wang and E. Adelson, "Representing moving images with layers," *IEEE Trans. Image Process.* **3**, 625–638 (1994).

[13] S.-M. Kruse, "Scene segmentation from dense displacement vector fields using randomized Hough transform," *Signal Process. Image Commun.* **9**, 29–41 (1996).

[14] Y. Weiss and E. H. Adelson, "A unified mixture framework for motion segmentation: incorporating spatial coherence and estimating the number of models," presented at the IEEE International Conference on Computer Vision and Pattern Recognition, June 1996.

[15] Y. Altunbasak, E. Eren, and A. M. Tekalp, "Region-based affine motion segmentation using color information," *Graphic. Models Image Process.* **60**, 13–23 (1998).

[16] D. W. Murray and B. F. Buxton, "Scene segmentation from visual motion using global optimization," *IEEE Trans. Pattern Anal. Machine Intell.* **9**, 220–228 (1987).

[17] N. Kiryati *et al.*, "A probabilistic Hough transform," *Pattern Recog.* **24**, 303–316 (1991).

[18] A. M. Tekalp, P. J. L. van Beek, C. Toklu, and B. Gunsel, "2-D mesh-based visual object representation for interactive synthetic/natural video," *Proc. IEEE* (special issue), **86**, 1029–1051 (1998).

[19] P. B. Chou and C. M. Brown, "The theory and practice of Bayesian image labeling," *Int. J. Comp. Vision* **4**, 185–210 (1990).

[20] M. M. Chang, A. M. Tekalp, and M. I. Sezan, "Simultaneous motion estimation and segmentation," *IEEE Trans. Image Process.* **6**, 1326–1333 (1997).

[21] Z. Kato, M. Berthod, and J. Zerubia, "Parallel image classification using multiscale Markov random fields," presented at the IEEE International Conference on ASSP, Minneapolis, MN, April 1993.

[22] S.-L. Iu, "Robust estimation of motion vector fields with discontinuity and occlusion using local outliers rejection," *Proc. SPIE* **2094**, 588–599 (1993).

[23] C. Stiller, "Object-oriented video coding employing dense motion fields," Conference on ASSP, Adelaide, Australia, April 1994.

[24] E. Dubois and J. Konrad, "Estimation of 2-D motion fields from image sequences with application to motion-compensated processing," in *Motion Analysis and Image Sequence Processing*, M. I. Sezan and R. L. Lagendijk, eds. (Kluwer, Norwell, MA, 1993).

[25] A. M. Tekalp, *Digital Video Processing* (Prentice Hall, NJ, 1995).

[26] Y. W. Lim and S. U. Lee, "On the color image segmentation algorithm based on the thresholding and the fuzzy c-means techniques," *Pattern Recog.* **23**, 935–952 (1990).

[27] J. R. Bergen, P. J. Burt, R. Hingorani, and S. Peleg, "A three-frame algorithm for estimating two-component image motion," *IEEE Trans. Pattern Anal. Machine Intell.* **14**, 886–896 (1992).

[28] N. Diehl, "Object-oriented motion estimation and segmentation in image sequences," *Signal Process. Image Commun.* **3**, 23–56 (1991).

[29] M. Hoetter and R. Thoma, "Image segmentation based on object oriented mapping parameter estimation," *Signal Process.* **15**, 315–334 (1988).

[30] S. Hsu, P. Anandan, and S. Peleg, "Accurate computation of optical flow by using layered motion representations," presented at the International Conference on Pattern Recognition, Jerusalem, Israel, Oct. 1994.

[31] R. Mech and M. Wollborn, "A noise robust method for 2-D shape estimation of moving objects in video sequences considering a moving camera," *Signal Process.* (special issue), **66**, 203–217 (1998).

[32] J.-M. Odobez and P. Bouthemy, "Direct model-based image motion segmentation for dynamic scene analysis," presented at the Second Asian Conference on Computer Vision (ACCV), Dec. 1995.

[33] P. Salembier, "Morphological multiscale segmentation for image coding," *Signal Process* **38**, 339–386 (1994).

[34] P. Schroeter and S. Ayer, "Multi-frame based segmentation of moving objects by combining luminance and motion," in *Signal Processing VII, Theories and Applications*.

[35] W. B. Thompson, "Combining motion and contrast for segmentation," *IEEE Trans. Pattern Anal. Mach. Intell.* **2**, 543–549 (1980).

[36] S. F. Wu and J. Kittler, "A gradient-based method for general motion estimation and segmentation," *J. Vis. Comm. Image Rep.* **4**, 25–38 (1993).

Adaptive and Neural Methods
for Image Segmentation

Joydeep Ghosh
The University of Texas, Austin

1 Introduction

The tremendous amount of research in image processing and analysis over the past three decades has been influenced not only by physiological or psychophysical discoveries and psychological observations about perception by living beings, but also by advances in signal processing, computational mathematics, pattern recognition, and artificial intelligence. Some researchers in the rejuvenated field of neural networks are also attempting to develop useful models of biological and machine vision. With the human visual system serving as a common source of inspiration, it is not surprising that neural network approaches to image processing and understanding often have commonalities with more traditional techniques. However, they also bring new elements of nonlinear processing with adaptation or learning, they bring some additional insights, and they promise breakthroughs through massively parallel and distributed implementations in VLSI [1, 2].

In this chapter, we highlight some *artificial* neural network techniques for image segmentation — the process of partitioning an image into regions that are contiguous and relatively homogeneous in image properties. Image segmentation has been studied in great depth ever since satellite images first became available. Segmentation is a key step in any image-based recognition system, and it fundamentally limits the success of all higher level subsystems [3]. The host of sophisticated techniques that have evolved over the past 30 years can be largely grouped into three categories.

1. *Edge-based* methods: these make use of local and global gradient information to provide boundaries to regions of interest, and thus indirectly segment the image.
2. *Region-based* methods: these group together local regions with relatively uniform image properties, using methods such as region growing, region splitting, and region splitting with merging. Segmentation methods based on *clustering* indirectly use region-based properties, but they are often considered as a separate category [4].
3. *Model-based* methods: these are guided by semantic attributes given to parts of the image, say, based on perceived match with (projections of) a set of object models of interest. In this view, segmentation is tightly coupled with image classification or object recognition. This is in contrast with methods in the first two categories, which are primarily based on image attributes rather than on what objects the images may be representing.

Techniques inspired by neural networks provide additional insights into, as well as performance improvements for, all three categories. In Section 2, we briefly describe common characteristics of neural network models and introduce some popular models. The next section highlights a prominent edge-based neural technique. Section 4 describes representative region-based methods, especially those that use textural cues. The study of region-based methods is continued in Section 5, where we examine optimization-based approaches to segmenting textured images. Section 6 describes adaptive clustering techniques for segmentation, and the next section summarizes biological-based methods in which segmentation is indicated by groups of neurons oscillating in synchrony. Finally, model-based methods are studied in Section 8, where integrated segmentation and recognition techniques are described.

2 Artificial Neural Networks

For our purposes, an artificial neural network (ANN) is a collection of computing cells (artificial neurons) interconnected through weighted links (synapses with varying strengths). The cells perform simple computations by using information available locally or from topologically adjacent cells through the weighted links. The knowledge of the system is embodied in the pattern of interconnects and their strengths, which vary as the system learns or adapts itself. In this setting we shall see that several ANN models are closely related to established image processing methods such as relaxation labeling, nonlinear filtering, and various feature extraction routines.

There is a large variety of ANNs that differ in their topology, cell behavior, weight update mechanisms, amount of supervision or feedback required, etc. Networks with three layers of cells (input, "hidden" and output), and with unidirectional or feedforward connections going from one layer to the next, as indicated in Fig. 1, are among the most popular topologies. This

class includes the multilayered perceptron (MLP) with a single hidden layer.

In an MLP, given a d-dimensional input $\mathbf{x} = [x_1, x_2, \ldots, x_d]^T$, the ith network output, y_i, is given by

$$y_i(\mathbf{x}) = f\left(\sum_{j=1}^{M} w_{ij} z_j(\mathbf{x}) + \theta_i \right), \qquad (1)$$

where z_j is the output of the jth hidden unit:

$$z_j = g\left(\sum_{k=1}^{d} v_{jk} x_k + \theta_j \right). \qquad (2)$$

In Eq. (2), v_{jk} denotes the weight of the connection between the jth input and kth hidden unit, and w_{ij} denotes the weight between the jth hidden unit and the ith output. The "activation function" or transfer function $g(\cdot)$ is S shaped or *sigmoidal*: nonlinear, monotonically increasing, and bounded. The typical choice is either the logistic map (sometimes called the sigmoid), $g(a) = 1/(1 + e^{-a})$, which is bounded between 0 and 1; or the hyperbolic tangent, $\tanh(a) = (e^a - e^{-a})/(e^a + e^{-a})$, bounded between -1 and 1. The output transfer function $f(\cdot)$ can be linear or sigmoidal as needed.

The MLP realizes a static map between inputs $\mathbf{x}$ and corresponding outputs $\mathbf{y}(\mathbf{x})$. This map depends on the u, w, and θ parameters (weights). These weights are trained or adjusted based on training samples $\{\mathbf{x}(n), \mathbf{t}(n)\}$, where $\mathbf{t}(n)$ is the desired target vector for the nth input vector, $\mathbf{x}(n)$. This adjustment is based on the error $\mathbf{t}(n) - \mathbf{y}(\mathbf{x}(n))$, typically using a stochastic gradient descent on the mean-squared value of this error, or by second-order methods.

Another popular feedforward network for realizing static maps is the radial basis function network (RBFN). A radial basis function (RBF), ϕ, is one whose output is symmetric around an associated *center*, $\boldsymbol{\mu}_c$. That is, $\phi_c(\mathbf{x}) = \phi(\|\mathbf{x} - \boldsymbol{\mu}_c\|)$, where $\|\cdot\|$ is a distance norm. For example, selecting the Euclidean norm and letting $\phi(r) = e^{-r^2/\sigma^2}$, one sees that the Gaussian function is an RBF. Note that Gaussian functions are also characterized by a *width* or scale parameter, σ, and this is true for many other popular RBF classes as well.

A set of RBFs can serve as a basis for representing a wide class of functions that are expressible as linear combinations of the chosen RBFs:

$$y_i(\mathbf{x}) = \sum_{j=1}^{M} w_{ij} \phi(\|\mathbf{x} - \boldsymbol{\mu}_j\|). \qquad (3)$$

A RBFN is nothing but an embodiment of Eq. (3) as a feedforward network with three layers: the inputs, the hidden layer, and the output node(s), i.e., they also have the topology in Fig. 1. Each hidden unit represents a single radial basis function, with associated center position and width. Such hidden units are sometimes referred to as centroids or kernels. Each output unit performs a weighted summation of the hidden

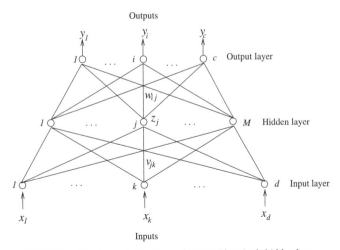

FIGURE 1   Topology of a feedforward ANN with a single hidden layer.

units, using the w_js as weights. Powerful *universal* approximation properties of Eq. (3) have been demonstrated for various settings.

From Eq. (3), one can see that designing an RBFN involves selecting the type of basis functions ϕ, with associated widths σ, the number of functions M, the center locations μ_j, and the weights w_j. Typically Gaussians or other *bell-shaped* functions with compact support are used. The choice of M is related to the complexity of the desired map. Given that the number of basis functions and their type have been selected, training an RBFN involves determining the values of three sets of parameters: the centers, the widths, and the weights, in order to minimize a suitable cost function. In general, this is a nonconvex optimization problem.

One can perform stochastic gradient descent on a mean-squared error cost function to iteratively update all three sets of parameters, once per training sample presentation. This may be suitable for nonstationary environments or on-line settings. But for static maps, RBFNs with localized basis functions offer a very attractive alternative, namely that in practice, the estimation of parameters can be decoupled into a *two-stage procedure*: (i) determine the μ_js and σ_js, and (ii) for the centers and widths obtained in step (i), determine the weights to the output units. Both subproblems allow for very efficient batch mode solutions. In the first stage only the input values $\{\mathbf{x}(n)\}$ are used for determining the centers μ_j and the widths σ_j of the basis functions. Thus learning is unsupervised and can even use unlabelled data. Once the basis function parameters are fixed, supervised training (i.e., training using target information) can be employed for determining the second layer weights.

We now briefly describe a third network, one that has a flat topology instead of the layered one of Fig. 1. Each cell is connected to all the cells, including itself, and is also capable of receiving direct input signals. One of the simplest such "fully recurrent" networks is the binary Hopfield model.[1] In a network of n cells, the kth cell receives a constant input I_k and updates its state by using

$$x_k(t+1) = \text{sgn}\left(\sum_{j=1}^{n} w_{kj}x_j(t) + I_k\right), \quad (4)$$

where sgn denotes the signum function, equal to 1 if x is nonnegative, and -1 otherwise. The weight matrix should be symmetric and is fixed, i.e., there is no weight adaptation. Starting from an initial state, given by the values of all cells at $t = 0$, cells update their state asynchronously till a final state is reached in which the left- and right-hand sides of Eq. (4) are equal for each of the cells. One can show that such a final state is guaranteed by

---

[1]Though work on this and related models has been carried out previously by many researchers, this name is most popular because of a paper by Hopfield [5] that sparked widespread interest in these networks.

constructing an energy or cost function:

$$E = -\frac{1}{2}\sum_l \sum_j w_{lj}x_l x_j - \sum_l I_l x_l. \quad (5)$$

By showing that E is bounded from below, and moreover that E is reduced by a finite amount every time a cell changes its value on an update, one concludes that updates have to terminate in finite time.

Clearly this model can serve as an associative memory since it maps the initial state and input to a final state. Also, if one desires to solve an optimization problem in which the cost function is quadratic in binary variables, it can be solved on the basis of Eq. (5), using one cell per variable, as shown by Hopfield and Tank [6] among others. We shall see such a use in Section 5, where texture segmentation is posed as a suitable optimization problem. Note that there is a generalization of Eq. (4) to continuous variables and continuous time. The signum function is now replaced by $\tanh(\cdot)$ or the logistic map, so that the cells can have graded responses, and the cell update is now given by a first-order differential equation, which can be readily implemented in VLSI using R-C circuits. Moreover, an analogous energy function exists for this generalization too. Indeed, even for binary optimization problems, it is preferable to use the continuous form, and then consider limiting values of the cell states to obtain the binary solution. Unfortunately, even then this approach to optimization is often practicable only for small problems. This is because the cost reduction only leads us to a local minima, and the probability that this minima is a poor or even invalid solution increases rapidly with increase in problem size (number of variables). Fortunately, related but more sophisticated and powerful schemes for optimization have emerged recently, and these can be readily applied for texture segmentation [7,8].

3 Perceptual Grouping and Edge-Based Segmentation

We perceive an image not as an array of pixels but as agglomerations or *groupings* of more abstract entities. A sharp spatial gradient in gray scale at an image location may not lead to the perception of an edge at that location if similar gradients do not occur in nearby locations. However, if there are sharp gradients with similar orientations in contiguous regions, then one typically perceives a line or contour formed by series of smoothly connected small edges. Such a contour is a common example of a perceptual grouping.

Indeed, there is a wide variety of grouping mechanisms in human perception. Interestingly, some of these groupings are illusory in that they appear to be very real, though there may not be evidence at the pixel level. As an example, Fig. 2(a) shows a Moiré pattern that distinctly gives the appearance of a circular flow pattern, even though there are no individual dotted

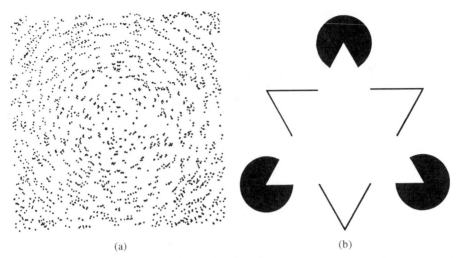

(a)   (b)

FIGURE 2 Examples of perceptual grouping: (a) random-dot Moiré pattern created by taking a pattern of 1000 dots and superimposing it with a copy rotated ~2°; (b) The Kanizsa subjective triangle.

contours running through. The grouping is a Gestalt phenomenon, occurring at a high level of abstraction.

Our visual mechanisms tend to perceptually connect edges that seem part of a longer edge, even across small regions with little gradient changes. This kind of grouping is dramatically indicated by Fig. 2(b), which shows a Kanizsa subjective triangle, demonstrating the formation of illusory contours (in this case an upright triangle).

Detecting edges and then eliminating irrelevant ones and connecting (grouping) the others are key to successful edge-based segmentation. To this end, cooperative processes such as relaxation labeling have been explored by the vision community for over a decade, without explicitly casting them in a neural network framework. The idea behind relaxation labeling is that local intensity edges typically form a part of a global line or boundary rather than occurring in isolation. Thus the presence or absence of a nearby edge of similar angular orientation would tend to reinforce the hypothesis of the existence of an edge at a given point in the intensity field. Detected line segments are assigned line-orientation labels are iteratively updated by a relaxation process, such that they become more compatible with neighboring labels. Thus adjacent "no-line-detected" labels support one another, and so do lines with similar orientation, while two adjacent labels corresponding to orthogonal orientations antagonize each other.

A neurallike scheme called the boundary contour system (BCS) has been proposed by Grossberg and Mingolla [9] to explain how edges are filled in when part of a boundary is missing, and how illusory contours, such as as in Fig. 2(b), can emerge from appropriately positioned line terminations. This real-time visual processing model explains a variety of perceptual grouping and segmentation phenomena, including the grouping of textured images. The BCS consists of a hierarchy of locally tuned interactions that controls the emergence of image segmentation and also detects, enhances, and completes boundaries. The interaction of BCS with a feature contour system and an object recog-

nition system attempts to attain a unifying precept for form, color, and brightness. The BCS is largely preattentive in that it is primarily driven by image properties. However, the model does allow feedback from the object recognition system to guide the segmentation process.

The BCS consists of several stages arranged in an approximately hierarchical organization. The image to be processed forms the input to the earliest stage. Here, elongated and oriented receptive fields or masks are employed for local contrast detection at each image position and each orientation. Thus there is a family of masks centered at each location, and that respond to a prescribed region around that location. These elliptical masks respond to the amount of luminance contrast over their elongated axis of symmetry, regardless of whether image contrasts are due to differences in textural distribution, a step change in luminance, or a smoother intensity gradient. The elongated receptive field makes the masks less sensitive to differences in average contrast in a direction orthogonal to the major axis. However, the penalty for making them sensitive to contrasts in the preferred orientation is the increased uncertainty in the exact locations of contrast. This positional uncertainty becomes acute during the processing of image line ends and corners. The authors assert that all line ends are illusory in the sense that they are not directly extracted from the retinal image, but are created by some process that generates line terminations. One such mechanism that is hypothesized by them is based on two short-range competitive stages followed by long-range cooperation, as described next.

First, each pair of masks at the same location that are sensitive to the same orientation but opposing direction of contrasts are input to a common cell. The output of such a cell at position (i, j) and orientation k is J_{ijk}, which is related to the two directional mask outputs, U_{ijk} and V_{ijk}, by

$$J_{ijk} = \frac{[U_{ijk} - \alpha V_{ijk}]^+ [V_{ijk} - \alpha U_{ijk}]^+}{1 + \beta(U_{ijk} + V_{ijk})}, \qquad (6)$$

where the notation $[p]+$ stands for $\max(p, 0)$. These oriented cells are sensitive to the amount of contrast, but not the direction. They in turn feed two short-range competitive stages. In the first stage, for cells of the same orientation, there is mutual support (by means of positive or excitatory connections) among very nearby cells, and competition (by means of negative or inhibitory connections) among cells that are at an intermediate distance. If the strength of a connection is plotted against the distance between the two cells being connected in two dimensions, a Mexican-hat or "on center, off surround" pattern is observed. Subsequently, in the second competitive stage, there is competition among orthogonally oriented masks at each position.

Let u_{ijk} represent the output signal for the cell corresponding to position (i, j) and orientation k, and u_{ijK} be the output for the cell at the same location but with orientation orthogonal to k, at the end of the first stage. The u_{ijk}s are obtained from

$$\frac{d}{dt} u_{ijk} = -u_{ijk}\left(1 + B\sum_{(p,q)\in R} BJ_{pqk}w_{pqij}\right) + I + BJ_{ijk}. \tag{7}$$

In Eq. (7), I is the external input (pixel value), B is a constant, R a neighborhood of (i, j), and w_{pqij} is the strength of the negative (inhibitory) connection between positions (p, q) and (i, j). The activity potentials y_{ijk} of cell outputs in the second stage are governed by

$$\frac{d}{dt} y_{ijk} = -Ay_{ijk} + (E - y_{ijk})O_{ijk} - y_{ijk}\sum_{m\neq k} O_{ijm}, \tag{8}$$

where $O_{ijk} = C[u_{ijk} - u_{ijK}]^+$, and A, C, and E are constants.

The behavior of the orientation field is shown in Fig. 3, in which adjacent lattice points are one unit apart. Each mask has a total exterior dimension of 16×8 units. Figure 3(b) shows the y_{ijk} responses at the end of the second competitive stage for the same input stimulus. The two competitive stages together have generated end cuts, as can be seen clearly on comparison with Fig. 3(a). Note that the second competitive stage has the property that inhibition of a vertical orientation excites the horizontal orientation at the same position, and vice versa.

The outputs of the second stage are also used for the boundary completion process that involves long-range cooperation between similarly oriented pairs of input groupings. This mechanism is able to complete boundaries across regions that receive no bottom-up inputs from the oriented receptive fields, and thus it accounts for illusory line phenomena such as the completion of the edges in a reverse-contrast Kanizsa triangle of Fig. 2(b). The process of boundary completion occurs discontinuously across space, using the gating properties of the cooperative cells to successively interpolate boundaries within progressively finer intervals. Unlike a low spatial frequency filter, this process does not sacrifice spatial resolution to achieve a broad spatial range. The cooperative cells used in this stage also provide positive feedback to the cells of the second competitive stage so as to

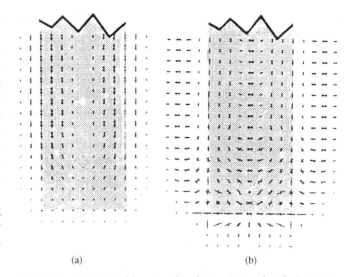

(a)   (b)

FIGURE 3 (a) Output of the oriented masks superimposed on the input pattern (shaded area). Lengths and orientations of lines encode the relative sizes of the activations and orientations of the masks at the corresponding positions. (b) Output of the second competitive stage for the same input as in (a) (Grossberg and Mingolla, 1985).

increase the activity of cells of favored orientation and position, thereby providing them with a competitive edge over other orientations and positions. This feedback helps in reducing the fuzziness of boundaries. The detailed architecture, equations, and simulation results can be found in [9].

The BCS approach can also form the basis for a hierarchical neural network for texture segmentation and labeling, as shown by Dupaguntla and Vemuri. The underlying premise is that textural segmentation can be achieved by recognizing local differences in texels. The architecture consists of a feature extraction network whose outputs are used by a texture discrimination network. The feature extraction network is a multilayer hierarchical network governed by the BCS theory. The image intensities input is first preprocessed by an array of cells whose receptive fields correspond to a difference of Gaussian filters, and that follow the feedforward shunting equations of Grossberg. The output of this array of cells form the input to a BCS system and are processed by oriented masks according to Eq. (6). These masks then feed into the two competitive stages of the BCS theory, governed by Eqs. (7) and (8). However, the long-range cooperative processes described above are not used. Instead, the outputs of the second competitive stage activate region encoding (RE) cells at the next level. Each RE cell gathers its activity from a region of orientation masks of the previous layer, as well as from a neighborhood of adjacent RE nodes of the same orientation. The activity potential of an RE node is given by the following equation, where the y_{lmk}'s are obtained from the previous layer according to Eq. (8), (l, m) is in the neighborhood of (p, q), and the activation function, f, is sigmoidal.

$$\frac{d}{dt} z_{ijk} = -\mu z_{ijk} + \mu \sum_{(p,q)} (f(z_{pqk}) - f(z_{ijk})) + \sum_{(l,m)} y_{lmk}. \tag{9}$$

The RE cells appear to be functionally analogous to the complex cells in the visual cortex, with the intralayer connections helping to propagate orientation information across this layer of cells. The outputs (z_{ijk}'s) of the feature extraction network are used by a texture discrimination network that is essentially Kohonen's single-layered self-organizing feature map [10]. At each position, there are T outputs, one for each possible texture type, which is assumed to be known a priori. Model (known) textures are passed through the feature extraction network. For a randomly selected position (i, j), the output cell of the texture discrimination network that responds maximally is given the known texture-type label. The weights in the texture discrimination network for that position are adapted according to the feature-map equations. Since these weights are the same for all positions, one can simply replicate the updated weights for all positions. The hierarchical scheme described above has been applied to natural images with good results. However, it is very computationally intensive, since there are cells corresponding to each orientation and position at every hierarchical level.

4 Adaptive Multichannel Modeling for Texture-Based Segmentation

Image texture provides useful information for segmentation of scenes, classification of surface materials, and computation of shape, and it is exploited by sophisticated biological vision systems for image analysis [11]. In 1980, Marcelja observed that highly oriented simple cell receptive fields in the cortex can be accurately modeled by one-dimensional (1-D) Gabor functions, which are Gaussian modulated sine wave functions. The Gabor functions play an important role in functional analysis and in physics, since they are the unique functions that satisfy the uncertainty principle, which is a measure of the function's simultaneous localization in space and in frequency. Daugman [12] successfully extended Marcelja's neuronal model to the two-dimensional (2-D) one, also extending Gabor's result by showing that the 2-D Gabor functions are the unique minimum-uncertainty 2-D functions. The implication of this for texture analysis purposes, and perhaps for neuronal processing of textured images, is that highly accurate measurements of textured image spectra can be made on a highly localized spatial basis. This simultaneous localization is important, since then it is possible to accurately identify sudden spatial transitions between texture types, which is important for segmenting images based on texture, and for detecting gradual variations within a textured region.

Based on these observations, a multiple-channel Gabor filter bank has been used to segment textured images [11]. Each filter's response is localized in the frequency (u–v) plane. A large set of these channel filters is used to sample the frequency plane densely to ensure that a filter exists that will respond strongly to any dominant texture frequency component. Segmentation

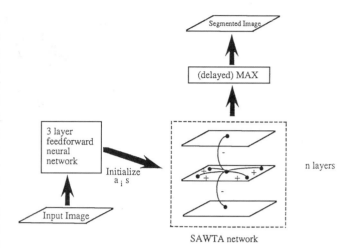

FIGURE 4   SAWTA network for the segmentation of textured images.

can be performed by assigning each pixel the label of the maximally responsive filter centered at that pixel. The success of this technique is quite impressive, given that no use is made of any sophisticated pattern classification superimposed on the basic segmentation structure. More details on multichannel image segmentation can be found in Chapter 4.7 of this handbook.

Some smoothing of the filter outputs before doing the max operation provides better results on texture segmentation. Further improvements can be achieved by using a cooperative-competitive feedback network called the smoothing, adaptive winner-take-all network (SAWTA) [11]. This network consists of n layers of cells, with each layer corresponding to one Gabor filter, as shown in Fig. 4. On the presentation of an image, a feedforward network using local receptive fields enables each cell plane to reach an activation level corresponding to the amplitude envelope of the Gabor filter that it represents, as outlined in the preceding paragraphs. Let $m_i(x, y), 1 \leq i \leq n$ be the activation of the cell in the ith layer with retinotopic coordinates (x, y). Initially, the n cell activations at each point (x, y) are set proportional to the amplitude responses of n Gabor filters.

To implement the SAWTA mechanism, each cell receives constant inhibition from all other cells in the same column, along with excitatory inputs from neighboring cells in the same row or plane. The synaptic strengths of the excitatory connections exhibit a 2-D Gaussian profile centered at (x, y). The network is mathematically characterized by shunting cooperative-competitive dynamics [9] that model on-center off-surround interactions among cells that obey membrane equations. Thus, at each point (x, y), the evolution of the cell in the ith layer is governed by

$$\tau \frac{\mathrm{d}}{\mathrm{d}t}(m_i) = -m_i + (A - m_i)J^+ - (B + Cm_i)J^-, \quad (10)$$

where J^+, J^- are the net excitatory and inhibitory inputs,

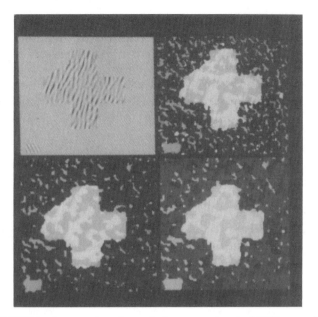

FIGURE 5 Segmentation of a synthetic texture using the SAWTA network (clockwise from top left): (a) original Image; (b) result of the original multichannel segmentation model [13]; (c) the results of this model with output smoothing; (d) segmentation after 10 iterations of the SAWTA network.

respectively, and are given by

$$J^+ = \alpha \sum_{(x_n, y_n) \in R} m_i(x_n, y_n) e^{\frac{-(x-x_n)^2 + (y-y_n)^2}{2\sigma^2}};$$

$$J^- = \sum_{j \neq i} f(m_j(x, y)). \tag{11}$$

Here, R is the neighboring region of support and f is a sigmoidal transfer function. A sigmoidal transfer function is needed to keep the response bounded between 0 and 1 while still maintaining a monotonically increasing response with the argument.

The convergence of a system described by Eq. (10) has been shown for the case in which the region of support R consists of the single point (x, y). The network is allowed to run for 10 iterations before region assignment is performed by selecting the most responsive filter.

Figures 5 and 6 shows comparative experimental result using the SAWTA network for segmentation. The 256×256 gray-level images are prefiltered by using Laplacian-of-Gaussian filters[2] to remove high dc components, low-frequency illumination effects, and to suppress aliasing. Then, only sixteen circularly symmetric Gabor filters are used to detect narrow-band components as follows. Sets of three filters with center frequencies increasing in geometric progression (ratio = 2:1) are arranged in a daisy-petal configuration along five orientations, while the sixteenth filter is centered at the origin. Figure 5 shows the segmentation achieved for a synthetic texture using three different techniques.

[2] For a description of such filters, see Chapter 4.10.

Figure 5(a) is the original image; Fig. 5(b) is the result of the original multichannel segmentation model [13]; Fig. 5(c) the results of this model with output smoothing; and Fig. 5(d) is the segmentation after 10 iterations of the SAWTA network. The constants A, B, and C in Eq. (10) were taken to be 1, 0, and 10 respectively. The activation function used is $f(x) = \tanh(2x)$. The results are seen to be superior to that obtained by the original multichannel based segmentation scheme.

Figure 6 shows the effect of varying the number of iteration steps, and the inhibition factor C, on the segmentation obtained. We observe that the SAWTA network achieves a more smooth segmentation in regions where the texture shows small localized variations, while preserving the boundaries between drastically different textures. Usually, 10 iterations suffice to demarcate the segment boundaries, and any changes after that are confined to arbitration among neighboring filters.

The SAWTA network does not require a feature extraction stage as in [14] or computationally expensive masking fields. The incremental and adaptive nature of the SAWTA network enables it to avoid making early decisions about texture boundaries. The dynamics of each cell is affected by the image characteristics in its neighborhood as well by the formation of more global hypotheses. It has been observed that usually four spatial frequencies are dominant at any given time in the human visual system. This suggests the use of a mechanism for postinhibitory response that suppress cells with activation below a threshold and speeds up the convergence of a SAWTA network. The SAWTA network can be easily extended to allow for multiple "winners." Then, it can cater to multicomponent textures, since a region that contains two predominant frequencies of comparable amplitude will not be segmented but rather viewed as a whole.

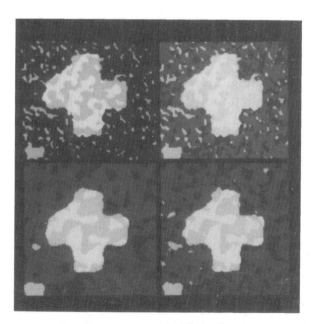

FIGURE 6 Effect of iteration steps and inhibition factor on segmentation: (a) same as Fig. 5(d); (b) segmentation with $C = 3$; (c) after 100 iterations; (d) after 100 iterations.

Learning and adaptation is also useful in the multichannel image model for determining the channels (filters) themselves. Indeed, the results of Gabor filtering can be obtained in an iterative fashion, by performing stochastic gradient descent on a suitable cost function. While these filters are useful for a large variety of images, one may wonder whether more customized filters may yield better results for a specific class of images, such as images of barcodes, or MRI scans for the brain. This leads to the concept of "texture discrimination masks," which may be learned in order to improve performance in the subsequent classification task [15].

First, note that the multichannel framework does not restrict one to using Gabor filters. Other filters reported include Laplacians of Gaussians, wavelets, and general IIR and FIR filters. Each filter can be considered as a localized feature detector, and after performing spatial smoothing if needed, the filter outputs for each pixel can serve as inputs to a multilayered feedforward network such as the MLP that performs the desired classification task. Thus we effectively have an MLP classifier with an additional hidden layer, i.e., the layer whose inputs are a the pixel values in a small image window, and whose afferent weights represent the mask coefficients. While training this network, the filter weights get modified to better perform texture classification. Moreover, by applying node pruning techniques, less important filters can be eliminated. Thus, instead of the usual large set of generic filters, a smaller set of task-specific filters is evolved. Details of this method, along with superior results obtained on page layout segmentation and bar-code localization, can be found in [15]. It is speculated that the efficacy of the learned masks stems from their ability to combine different frequency and directionality responses in the same masks, so that high discrimination information can be captured by a smaller number of filters. In contrast, if the problem domain changes substantially, a new set of filters has to be learned for the new set of images.

5 An Optimization Framework

The use of Markov random field (MRF) models for modeling texture has been investigated by several researchers (see Chapter 4.2). They can be used to model the texture intensity process as well as to describe the texture labeling process. In this framework, segmentation of textured images is posed as an optimization problem. Two optimality criteria considered in [16] are (i) to maximize the posterior distribution of the texture label field given the intensity field, and (ii) to minimize the expected percentage of misclassification per pixel by maximizing the posterior marginal distribution. Corresponding to each criteria, an energy (cost) function can be derived that is a function of $M \times M \times K$ binary labels, one for each of the K possible texture labels that a pixel in an $M \times M$ image can take.

A neural network solution to minimizing this cost function is provided by means of the discrete Hopfield-Tank formulation described in Section 2 [6]. A 3-D lattice of binary (ON/OFF) neurons is used, with one neuron for each of the $M \times M \times K$ labels. The cost function chosen imposes a severe penalty unless exactly one neuron is ON at each of the $M \times M$ positions. The location of this neuron in the third dimension provides the label to be given to the corresponding pixel. The other terms in the cost function encourage solutions in which the same label is given to neighboring pixels, and at the same time this class has a high probability of occurring given the initial gray-level values of the pixels in the neighborhood [16]. The cost function is quadratic in the neuron output values, and indeed has the form of Eq. (5). In Section 2 we saw that for such cost functions, a network with simple computing cells and local connections can be specified such that the cost is steadily reduced as the cells update their state, until a local minimum of the cost function is realized. This usually happens in 20–30 iterations, but the quality of the texture labeling thus obtained is quite sensitive to initial conditions, as it has a penchant for settling into local optima. Alternatively, a stochastic algorithm such as simulated annealing can be used to minimize the energy function. Indeed, any problem formulated in terms of minimizing an energy function can be given a probabilistic interpretation by use of the Gibbs distribution. The two approaches are related in that a mean field approximation of the stochastic algorithm yields the update equations of the network described above, with the free parameter being proportional to the inverse of the annealing temperature [6].

For the segmentation problem, a constraint on a valid solution is that each image position should have only one of the K labels "on." This constraint is usually incorporated in a soft fashion by adding bias terms to the energy function. Peterson and Soderberg have incorporated the 1-of-K constraint in a Potts glass, and they derived a mean field solution for that formulation. The alternative of putting global constraints on the set of allowable states in the corresponding stochastic formulation leads to significantly better solutions. An iterated hill climbing algorithm that combines fast convergence of the deterministic relaxation with the sustained exploration of the stochastic approach has also been proposed in [16] for the segmentation problem. Here, two-stage cycles are used, with the equilibrium state of the relaxation process providing the initial state for a stochastic learning automaton within each cycle.

The relation between neural network techniques and MRFs was explored in detail in [16, 17]. Since the optimization techniques applied were largely rooted in the Hopfield-Tank formulation, they were plagued by large training times, and a high possibility of being caught in local minima, leading to poor solutions. Fortunately, more sophisticated and powerful schemes for optimization have emerged recently and can be readily applied for texture segmentation [7, 8].

The MRF framework can also directly leverage the powerful mapping capabilities of feedforward networks. For example, Hwang and Chen have used an MLP to directly obtain the class distributions conditional on the neighborhood image statistics (needed for the MRF), based on training image samples. This obviates restrictive parametric representation and tedious parameter estimation for the MRF. Of note is the use of the Karhunen-Loeve cost criterion instead of the popular mean squared error, since the former tends to give more accurate estimates of low probability values.

6 Image Segmentation by Means of Adaptive Clustering

The use of clustering for image segmentation dates back to the late 1960s, and many of the techniques developed then are still in popular use today. In this approach each pixel is represented by a vector of features based on information derived from image characteristics in its neighborhood, as well as positional information. Similar feature vectors are then grouped together in clusters, and each cluster is given a texture label. Thus pixels that are nearby and have similar local image properties will tend to get grouped together and get the same label.

Clustering can be fruitfully applied to a variety of image domains, including multispectral images, range images, textured images, and intensity images from dot patterns to gray-level or colored images. Key issues in the design of any clustering-based segmenter are the choice of the number and type of features used, the distance metric chosen to measure similarity, data reduction techniques used, and the pre- and postprocessing routines applied. If the design choices are made suitably, the feature vectors will form more compact and well separated regions in the multidimensional feature space, and one can thus reliably segment the images based on both these regions and on image connectivity. A nice overview of this area can be found in [4].

Neural network research has spawned a variety of adaptive clustering techniques, from competitive learning — an iterative version of K-means clustering, to learned vector quantization (LVQ) [10] — a supervised clustering and classification technique related to classical vector quantization. In learned vector quantization, a set of labeled cluster centers (the codebook vectors) are first chosen by random subsampling or by K-means clustering of the data. Then, for every training sample the position of the nearest codebook vector is moved toward or away from that sample, depending on whether the two labels match or not. Several variations exist.

Instead of placing each sample in a unique cluster, one can "softly" associate a sample with multiple clusters. The resulting clusters are sometimes called fuzzy clusters, as their boundaries are not sharply delineated. Let the association of sample x_i with cluster j be denoted by $a_{i,j}$. We want this association value to decrease as the distance between x_i and the center of the jth cluster increases. Also, it is desired that the associations be nonnegative, and that $\sum_j a_{i,j} = 1, \forall i$. The jth cluster center is simply $\sum_i a_{i,j} x_i$. Depending on how the associations are formed and updated, a variety of powerful fuzzy clustering approaches have been obtained [18].

A critical issue in clustering is the choice of an appropriate scale, which determines the number of clusters obtained and hence the amount of segmentation obtained. For a given image, there are some natural scales for which the clusters are relatively well defined and stable in the sense that the optimum center locations change little with small variations in scale. In fact, just as scale-space theory views salient edges to be those that survive over multiple scales, one can view salient segments in the same way. Statistical mechanics based formulations for clustering provide a nice approach to the issue of scale, which is naturally related to the temperature parameter. At high temperature there are many clusters, and as the temperature is lowered, some of the clusters coalesce. Stable clusters are those that survive over a wide range of temperatures [19]. It turns out that if we adjust only the kernel locations by using gradient descent in Gaussian radial basis function network, maintaining fixed and equal output layer weights, fixed widths, and a constant target function, then these locations converge to "optimum" cluster locations for the chosen scale, now indicated by the widths (σs) of the Gaussian units [20]. Moreover, different scales may be indicated as being appropriate for different parts of an image for segmentation purposes. Thus such networks are promising for segmentation with locally adaptive resolution.

7 Oscillation-Based Segmentation

It would be remiss not to mention that there are several biologically oriented approaches to segmentation. In such approaches, it becomes clear that segmentation is closely tied with several other mechanisms. For example, when we view an apple, we regard it not as a smear of red amidst a riot of undifferentiated color, but recognize it as distinct desired object. How exactly we do this constitutes the *sensory segmentation* problem. When we thus discern an apple, we naturally separate it from the uninteresting background — this is the *figure–ground separation* problem. When we take a further step, like biting into the apple, then our experience is not merely a jumble of tactile, olfactory, and visual sensations; these different modes of sensations correspond to a single object — an apple; this is the *binding* problem. The aforementioned problems are intimately interrelated.

A single physical stimulus usually activates several groups of neurons corresponding to different sensory modalities. How does the brain then realize that these groups correspond to the same object? A popular hypothesis for answering this query is that neurons responding to aspects of a single object fire in synchrony. Applied to visual perception, the hypothesis states that cortical neurons corresponding to a distinct homogeneous area oscillate in phase, and those corresponding to different areas are out of phase.

Supportive of this hypothesis, stimulus-dependent oscillations that are correlated temporally and spatially have been found in the visual cortex of cats and monkeys. These experiments have motivated several oscillating neuron models of sensory segmentation, which largely fall into two broad categories: (i) those in which synchronization is achieved by cooperation/competition among neurons via fixed excitatory and inhibitory couplings, and (ii) those in which synchronization evolves by locally Hebbian-like synaptic modification, i.e., synaptic strength increases if both presynaptic and postsynaptic activity are simultaneous high, and decreases otherwise. In the former, the visual input is merely transformed into segregated oscillations, whereas in the latter, the input is encoded in modifiable synapses. For example, in [21], oscillating units, each consisting of excitatory and inhibitory cells, are connected by weights, modulated in a "pre-postsynaptic" fashion. Some of these essential ingredients can be seen in several of the subsequent models, e.g., the approach of Konig, wherein synchronization among oscillating units depends on similarity of local features, and segmentation can be achieved by local learning rules.

In [22], a model is proposed in which synchronization of neural oscillations is produced *both* by (i) cooperation and (ii) synaptic modification. It is demonstrated that either of these mechanisms is sufficient to generate coherent oscillations, but the two can be viewed as components of a more integral mechanism for neural synchronization.

In this model, the state of an oscillating unit or a neuron is described by a complex number, z, and each unit is connected to every other unit. The dynamics of the model is a generalization of the Hopfield equations in the complex plane and is described by

$$\frac{dz_j}{dt} = \sum_k T_{jk} V_k - (\nu + i(1 - \nu))z_j + I_j, \qquad (12)$$

$$V_j = \tanh\big(\lambda(\nu + i(1 - \nu))z_j^*\big), \qquad (13)$$

where the quantities, z_j, T_{jk}, and V_k are complex numbers. The real part of the neuron state, Re[z], is analogous to the transmembrane potential of the real neuron; the real part of V is the output firing rate; I_j is the sum of the external currents entering the jth neuron, and T_{jk} is the *weight* connecting jth and kth neurons. The mode parameter ν governs the qualitative nature of the above model. For ν close to 0, model of Eqs. (12) and (13) exhibits oscillatory behavior, and for ν near 1, it has fixed-point dynamics.

One can show that oscillations are produced in the model above if the cells are arranged in a 2-D grid, and the weights T_{jk} are real and have a "Mexican hat" profile. Suppose the 32×32 image of a plus (+) symbol with noise added (Fig. 7) forms the input to a 32×32 grid of cells. The image is presented for an interval of time (210 iterations in this example) and then removed, and the subsequent evolution of the network output is followed. It is seen that cells over the plus region start oscillating more or less in phase, and are $\sim 180°$ out of phase with the rest of the cells. Figure 8 depicts the oscillation of a typical neuron. Subsequent to input removal, network response reveals excellent noise removal with precise figure boundaries, as indicated in

Original "Plus (+)" with noise

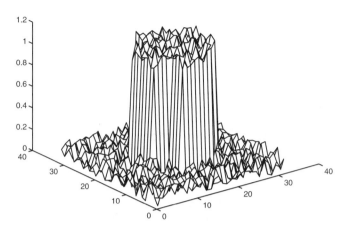

FIGURE 7   Original plus (+) image corrupted with uniformly distributed noise.

Fig. 9, which was taken after 450 iterations. It is also observed that the amount of noise in the interpreted image decreases right from the first iteration, and Fig. 9 shows almost no noise.

If the weights are random rather than "Mexican hat," the network exhibits coherent oscillations only while the input is present, and coherency is destroyed subsequent to input removal.

Alternatively, synchronization can be produced, without any special predetermined neighborhood, if the weights are not fixed but modified by Hebbian learning. In the Hebbian form of learning, the connection between a pair of simultaneously active neurons is strengthened, and it is expressed in this model as

$$\dot{T}_{jk} = -\gamma\, T_{jk} + g\big((\nu + i(1 - \nu))z_j^*\big) g((\nu + i(1 - \nu))z_k), \qquad (14)$$

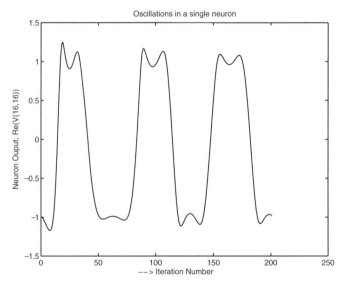

FIGURE 8   Output of neuron (16, 16) for a sequence of 2.5 iterations.

Co-operation with "Mexican hat" coupling weights (input removed)

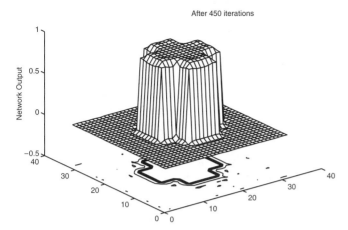

FIGURE 9 Network output (after 450 iterations) when the input image (noisy) is removed. Fixed "Mexican hat" neighborhood connections are used.

where $g(\cdot)$ is the sigmoid nonlinearity introduced in Eq. (13). Equation (14) is always simulated together with Eqs. (12) and (13). The external input I in Eq. (12) produces changes in T_{jk} indirectly via neuron outputs, V_k. As a result, the input pattern is encoded in the weights, and input-dependent synchronization takes place as an emergent effect. Figure 10 shows the result (for the same input image of Fig. 7) after 530 iterations, using random initial weights that are adapted with the Hebbian learning equations.

The two mechanisms described above for synchronized neural oscillations seem unconnected. However, Linsker and others have shown that Mexican-hat-like neighborhoods can develop automatically in a multilayered network with weights adapted by Hebbian learning. Such neighborhoods seem to be canonical for producing stimulus-specific oscillations, obtained as an

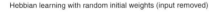

Hebbian learning with random initial weights (input removed)

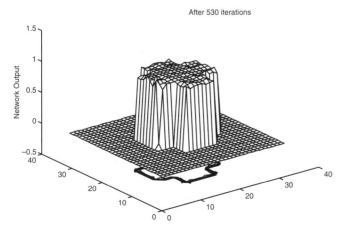

FIGURE 10 Network output (after 530 iterations) when the input image (noise corrupted) is removed and the weight adaptation is continued.

average effect of learning a large number of patterns. This is supported by experimental results [22].

8 Integrated Segmentation and Recognition

Often segmentation is an intermediate step toward object recognition or classification from 2-D images. For example, segmentation may be used for figure-ground separation or for isolating image regions that indicate objects of interest as differentiated from background or clutter. Even small images have lots of pixels — there are over 64,000 pixels in a 256×256 image. So, it is impractical to consider the raw image as an input to an object recognition or classification system. Instead, a small number of descriptive features are extracted from the image, and then the image is classified or further analyzed based on the values of these features. ANNs provide powerful methods for both the feature extraction and classification steps and have been used with much success in integrated segmentation and recognition applications.

Feature Extraction

The quality of feature selection/extraction limits the performance of the overall pattern recognition system. One desires the number of features to be small but highly representative of the underlying image classes, and highly indicative of the differences among these classes. Once the features are chosen, different methods typically give comparable classification rates when used properly. Thus feature extraction is the most crucial step. In fact, the Bayes error is defined for a given choice of features, and a poor choice can lead to a high Bayes rate.

Perhaps the most popular linear technique for feature extraction is principal component analysis (PCA) (sometimes referred to as the Karhunen–Loeve transform), wherein data are projected in the directions of the principal eigenvectors of the input covariance matrix. There are several iterative "neural" techniques in which weight vectors associated with linear cells converge to the principal eigenvectors under certain conditions. The earliest and most well known of these is Oja's rule, in which the weights w_i of a linear cell with single output $y = \sum_i x_i w_i$ are adapted according to

$$w_i(n) = w_i(n-1) + \eta(n)y(n)[x_i(n) - y(n)w_i(n-1)]. \quad (15)$$

The learning rate is $\eta(n)$ and should satisfy the Robbins–Munro conditions for convergence, and $x_i(n)$ is the ith component of the input presented at the nth instant. The inputs are presented at random. Then it can be shown that, if $\mathbf{x}$ is a zero-mean random variable, the weight vector converges to unit magnitude with its direction the same as that of the principal eigenvector of the input covariance matrix. In other words, the output y is nothing but the principal component after convergence! Moreover, when the "residual," $x_i(n) - y(n)w_i(n-1)$, is fed into another similar cell, the second principal component is iteratively obtained, and so on. Moreover, to make the iterative procedure robust against outliers, one can vary the learning rate so that it has a lower value if the current input is less probable.

The nonlinear discriminant analysis network proposed by Webb and Lowe [23] is a good example of a nonlinear feature extraction method. They use a multilayer perceptron with sigmoid hidden units and linear output units. The nonlinear transformation implemented by the subnetwork from the input layer to the final hidden layer of such networks tries to maximize the so-called network discriminant function, $\mathrm{Tr}\{S_B S_T^+\}$, where S_T^+ is the pseudo-inverse of the total scatter matrix of the patterns at the output of the final hidden layer, and S_B is the weighted between-class scatter matrix of the output of the final hidden layer. The role of the hidden layers is to implement a nonlinear transformation that projects input patterns from the original space to a space in which patterns are more easily separated by the output layer.

A nice overview of neural based feature techniques is given by Mao and Jain [24], who have also compared the performance of five feature extraction techniques using eight different data sets. They note that while several such techniques are nothing but online versions of some classical methods, they are more suitable for mildly nonstationary environments, and often provide better generalization. Some techniques, such as Kohonen's feature map, also provide a nice way of visualizing higher dimensional data in 2-D or 3-D space.

Classification

Several feedforward neural networks have properties that make them promising for image classification based on the extracted features. ANN approaches have led to the development of various "neural" classifiers using feedforward networks. These include the MLP as well as kernel-based classifiers such as those employing radial basis functions, both of which are described in Section 2. Such networks serve as adaptive classifiers that learn through examples. Thus, they do not require a good *a priori* mathematical model for the underlying physical characteristics. A good review of probabilistic, hyperplane, kernel and exemplar-based classifiers that discusses the relative merit of various schemes within each category is available in [25]. It is observed that, if trained and used properly, several neural networks show comparable performance over a wide variety of classification problems, while providing a range of tradeoffs in training time, coding complexity and memory requirements. Some of these networks, including the multilayered perceptron when augmented with regularization, and the elliptical basis function network, are quite insensitive to noise and to irrelevant inputs. Moreover, a firmer theoretical understanding of the pattern recognition properties of feedforward neural networks has emerged that can relate their properties to Bayesian decision making and to information theoretic results [26].

Neural networks are not magical. They do require that the set of examples used for training should come from the same (possibly unknown) distribution as the set used for testing the networks, in order to provide valid generalization and good performance on classifying unknown signals [27]. Also, the number of training examples should be adequate and comparable to the number of *effective* parameters in the neural network, for valid results. Interestingly, the complexity of the network model, as measured by the number of effective parameters, is not fixed, but increases with the amount of training. This provides an important knob: one can start with an adequately powerful network and keep on training until its complexity is of appropriate size. In practice, the latter may be readily arrived at by monitoring the network's performance on a validation set.

Training sufficiently powerful multilayer feedforward networks (e.g. MLP, RBF) by minimizing the expected mean square error (MSE) at the outputs and using a 0/1 teaching function yields network outputs that approximate posterior class probabilities [26]. In particular, the MSE is shown to be equivalent to

$$\mathrm{MSE} = K_1 + \sum_{i=1}^{c} \int_{\mathbf{x}} D_i(\mathbf{x})(P(C_i \mid \mathbf{x}) - f_i(\mathbf{x}))^2 \, \mathrm{d}\mathbf{x}, \qquad (16)$$

where K_1 and $D_i(\mathbf{x})$ depend on the class distributions only, $f_i(\mathbf{x})$ is the output of the node representing class C_i given an input $\mathbf{x}$, $P(C_i \mid \mathbf{x})$ denotes the posterior probability, and the summation is over all classes. Thus, minimizing the (expected) MSE corresponds to a weighted least squares fit of the network outputs to the posterior probabilities. Somewhat similar results are obtained by using other cost functions such as cross entropy.

The above result is exciting because it promises a direct way of obtaining posterior class probabilities and hence attaining the Bayes optimum decision. In practice, of course, the exact posterior probabilities may not be obtained, but only an approximation thereof. (If they had, the Bayes error rate could have been attained.) This is because in order to minimize Eq. (16), one needs to (i) use an adequately powerful network so that $P(C_i \mid \mathbf{x})$ can be realized, (ii) have enough number of training samples, and (iii) find the global minima in weight space. If any of the above conditions are violated, different classification techniques will have different inductive biases, and a single method cannot give the best results for all problems. Rather, more accurate and robust classification can obtained by combining the outputs (evidences) of multiple classifiers based on neural network and/or statistical pattern recognition techniques.

Pattern Recognition Techniques Specific to Segmentation

Although this discussion applies to generic pattern recognition systems, image segmentation has specific characteristics that may call for custom approaches. First, some invariance to (small) changes in rotation, scale, or translation is often desired. For example, in OCR, tolerances to such minor distortions is a must. Invariance can be achieved by (i) extracting invariant features such as Zernicke moments, (ii) by providing additional examples with different types of distortion, e.g., for each character in OCR also present various rotated, scaled, or shifted versions, or (iii) making the mapping robust to invariances by means of weight replications or symmetries. The last alternative is the most popular and has led to specialized feedforward networks with *two* or more hidden layers. Typically, the early hidden layers have cells with local receptive fields, and their weights are shared among all cells with similar purpose (i.e., extracting the same features) but acting on different portions of the image. A good example is the *convolutional net* [28], in which the first hidden layer may be viewed as a 3-D block of cells. Each column of cells extract different features from the corresponding *localized* portion of the image. These feature extractors essentially perform convolution by using nonlinear FIR filters. They are replicated at other localized portions by having identical weights among all cells in the same layer of the 3-D block. Multiple layers with subsampling is proposed to form an image processing pyramid. Higher layers are fully connected to extract more global information. For on-line handwriting recognition, a hidden Markov model postprocessor can be used. Remarkable results on document recognition are given in [28].

In an integrated segmentation and recognition scheme, it may even be possible to avoid the segmentation step altogether. For example, it is well known that presegmented characters are relatively easy to classify, but isolating such individual characters from handwriting is difficult. One can, however, develop a network that avoids this segmentation by making decisions only if the current window is centered on a valid character, and otherwise giving a "noncentered" verdict. Such networks can also be trained with handwriting that is not presegmented, thus saving substantial labor.

9 Concluding Remarks

Neural network based methods can be fruitfully applied in several approaches to image segmentation. While many of these methods are closely related to classical techniques involving distributed iterative computation, new elements of learning and adaptation are added. On one hand, such elements are particularly useful when the relevant properties of images are nonstationary, so continuous adaptation can yield better results and robustness than a fixed solution. On the other hand, most of the methods have not been fully developed as products with friendly GUI that a nonexpert end user can obtain off the shelf and readily use. Moreover, a detailed comparative analysis is desired for several of the techniques described in this chapter, to further understand when they are most applicable. Thus, further analysis, benchmarking, product development, and system integration is necessary if these methods are to gain widespread acceptance.

An exciting aspect of neural network based image processing is the prospect of parallel hardware realization in analog VLSI chips such as the silicon retina [1]. Such analog chips use networks of resistive grids and operational amplifiers to perform edge detection, smoothing, segmentation, compute optic flow, etc., and they can be readily embedded in a variety of smart platforms, from toy autonomous vehicles that can track edges, movements, etc., to security systems to retinal replacements [2]. Further progress toward the development of low power, realtime vision hardware requires an integrated approach encompassing image modeling, parallel algorithms, and the underlying implementation technology.

Acknowledgments

I thank Al Bovik for his friendship and inspiration over the years, and for numerous perceptive and humorous comments on this chapter that greatly helped in improving its quality. This work was funded in part by ARO contracts DAAH04-95-10494 and DAAG55-98-1-0230.

References

[1] C. A. Mead, *Analog VLSI and Neural Systems* (Addison-Wesley, Reading, MA, 1989).

[2] C. Koch and B. Mathur, "Neuromorphic vision chips," *IEEE Spectrum*, 38–46 (1996).

[3] N. R. Pal and S. K. Pal, "A review on image segmentation techniques," *Pattern Recog.* **26**, 1277–1294 (1993).

[4] A. K. Jain and P. J. Flynn, "Image segmentation using clustering," in *Advances In Image Understanding: A Festschrift for Azriel Rosenfeld*, K. Boyer and N. Ahuja, eds. (IEEE, New York, 1996), pp. 65–83.

[5] J. J. Hopfield, "Neural networks and physical systems with emergent collective computational abilities," *Proc. Natl. Acad. Sci.*, **79** 2554–2558 (1982).

[6] J. J. Hopfield and David W. Tank, "Neural computation of decisions in optimization problems," *Biol. Cybernet.* **52**, 141–152 (1985).

[7] A. Rangarajan, S. Gold, and E. Mjolsness, "A novel optimizing network architecture with applications," *Neural Comput.* **8**, 1041–1060 (1996).

[8] K. Smith, M. Paliniswami, and M. Krishnamoorthy, "Neural techniques for combinatorial optimization with applications," *IEEE Trans. on Neural Networks* **9**, 1301–1309 (1998).

[9] S. Grossberg and E. Mingolla, "Neural dynamics of perceptual grouping: textures, boundaries and emergent segmentations," *Perception Psychophys.* **38**, 141–171 (1985).

[10] T. Kohonen, *Self-Organizing Maps*, 2nd ed. (Springer-Verlag, Berlin, 1997).

[11] J. Ghosh and A. C. Bovik, "Processing of textured images using neural networks," in *Artificial Neural Networks and Statistical Pattern Recognition*, I. K. Sethi and A. Jain, eds. (Elsevier Science, Amsterdam, 1991), pp. 133–154.

[12] J. G. Daugman, "Complete discrete 2-D Gabor transforms by neural networks for image analysis and compression," *IEEE Trans. Acoust. Speech Signal Process.* **36**, 1169–1179 (1988).

[13] A. C. Bovik, M. Clark, and W. S. Geisler, "Multichannel texture analysis using localized spatial filters. I: Segmentation by channel demodulation," *IEEE Trans. Pattern Anal. Machine Intell.* **12**, 55–73 (1990).

[14] N. R. Dupaguntla and V. Vemuri, "A neural network architecture for texture segmentation and labeling," in *International Joint Conference on Neural Networks* (IEEE Press, Piscataway, NJ, 1989), pp. 127–144.

[15] A. K. Jain and K. Karu, "Learning texture discrimination masks," *IEEE Trans. Pattern Anal. Machine Intell.* **18**, 195–205 (1996).

[16] B. S. Manjunath, T. Simchony, and R. Chellappa, "Stochastic and deterministic networks for texture segmentation," *IEEE Trans. Acoust. Speech Signal Process.* **38**, 1039–1049 (1990).

[17] A. Rangarajan, R. Chellappa, and B. S. Manjunath, "Markov random fields and neural networks with applications to early vision problems," in *Artificial Neural Networks and Statistical Pattern Recognition*, I. K. Sethi and A. Jain, eds. (Elsevier Science, Amsterdam, 1991), pp. 155–174.

[18] J. C. Bezdek and S. K. Pal, *Fuzzy Models for Pattern Recognition* (IEEE Press, Piscataway, NJ, 1992).

[19] K. Rose, "Deterministic annealing for clustering, compression, classification, regression, and related optimization problems," *Proc. IEEE* **86**, 2210–2239 (1998).

[20] S. V. Chakaravathy and J. Ghosh, "Scale based clustering using a radial basis function network," *IEEE Trans. Neural Net.* **2**, 1250–1261 (1996).

[21] C. von der Malsburg and W. Schneider, "A neural cocktail-party processor," *Biol. Cybernet.* **54**, 29–40 (1986).

[22] S. V. Chakravarthy, V. Ramamurti, and J. Ghosh, "A network of oscillating neurons for image segmentation," in *Intelligent Engineering Systems Through Artificial Neural Networks*, P. Chen, B. Fernandez, C. Dagli, M. Akay, and J. Ghosh, eds. (ASME Press, 1995), Vol. 5.

[23] D. Lowe and A. R. Webb, "Optimized feature extraction and the bayes decision in feed-forward classifier networks," *IEEE Trans. Pattern Anal. Machine Intell.* **13**, 355–364 (1991).

[24] J. Mao and A. K. Jain, "Artificial neural networks for feature extraction and multivariate data projection," *IEEE Trans. Neural Net.* **6**, 296–317 (1995).

[25] K. Ng and R. P. Lippmann, "Practical characteristics of neural network and conventional pattern classifiers," in *Advances in Neural Information Processing Systems-3*, J. E. Moody, R. P. Lippmann, and D. S. Touretzky, eds. (Morgan Kaufmann, San Mateo, CA, 1991), pp. 970–976.

[26] C. M. Bishop, *Neural Networks for Pattern Recognition* (Oxford U. Press, New York, 1995).

[27] J. Ghosh and K. Tumer, "Structural adaptation and generalization in supervised feed-forward networks," *J. Art. Neural Net.* **1**, 431–458 (1994).

[28] Y. LeCun, L. Bottou, Y. Bengio, and P. Haffner, "Gradient-based learning applied to document recognition," *Proc. IEEE* **86**, 2278–2324 (1998).

4.11

Gradient and Laplacian-Type Edge Detection

Phillip A. Mlsna
Northern Arizona University

Jeffrey J. Rodríguez
The University of Arizona

1 Introduction

One of the most fundamental image analysis operations is edge detection. Edges are often vital clues toward the analysis and interpretation of image information, both in biological vision and computer image analysis. Some sort of edge detection capability is present in the visual systems of a wide variety of creatures, so it is obviously useful in their abilities to perceive their surroundings.

For this discussion, it is important to define what is and is not meant by the term "edge." The everyday notion of an edge is usually a physical one, caused by either the shapes of physical objects in three dimensions or by their inherent material properties. Described in geometric terms, there are two types of physical edges: (1) the set of points along which there is an abrupt change in local orientation of a physical surface, and (2) the set of points describing the boundary between two or more materially distinct regions of a physical surface. Most of our perceptual senses, including vision, operate at a distance and gather information by using receptors that work in, at most, two dimensions. Only the sense of touch, which requires direct contact to stimulate the skin's pressure sensors, is capable of direct perception of objects in three-dimensional (3-D) space. However, some physical edges of the second type may not be perceptible by touch because material differences — for instance different colors of paint — do not always produce distinct tactile sensations. Everyone first develops a working understanding of physical edges

in early childhood by touching and handling every object within reach.

The imaging process inherently performs a projection from a 3-D scene to a two-dimensional (2-D) representation of that scene, according to the viewpoint of the imaging device. Because of this projection process, edges in images have a somewhat different meaning than physical edges. Although the precise definition depends on the application context, an edge can generally be defined as a boundary or contour that separates adjacent image regions having relatively distinct characteristics according to some feature of interest. Most often this feature is gray level or luminance, but others, such as reflectance, color, or texture, are sometimes used. In the most common situation where luminance is of primary interest, edge pixels are those at the locations of abrupt gray-level change. To eliminate single-point impulses from consideration as edge pixels, one usually requires that edges be sustained along a contour; i.e., an edge point must be part of an edge structure having some minimum extent appropriate for the scale of interest. Edge detection is the process of determining which pixels are the edge pixels. The result of the edge detection process is typically an edge map, a new image that describes each original pixel's edge classification and perhaps additional edge attributes, such as magnitude and orientation.

There is usually a strong correspondence between the physical edges of a set of objects and the edges in images containing views of those objects. Infants and young children learn this as they develop hand–eye coordination, gradually associating visual patterns with touch sensations as they feel and handle items in their vicinity. There are many situations, however, in which edges in an image do not correspond to physical edges. Illumination differences are usually responsible for this effect — for example, the boundary of a shadow cast across an otherwise uniform surface.

Conversely, physical edges do not always give rise to edges in images. This can also be caused by certain cases of lighting and surface properties. Consider what happens when one wishes to photograph a scene rich with physical edges — for example, a craggy mountain face consisting of a single type of rock. When this scene is imaged while the Sun is directly behind the camera, no shadows are visible in the scene and hence shadow-dependent edges are nonexistent in the photo. The only edges in such a photo are produced by the differences in material reflectance, texture, or color. Since our rocky subject material has little variation of these types, the result is a rather dull photograph, because of the lack of apparent depth caused by the missing edges. Thus, images can exhibit edges having no physical counterpart, and they can also miss capturing edges that do. Although edge information can be very useful in the initial stages of such image processing and analysis tasks as segmentation, registration, and object recognition, edges are not completely reliable for these purposes.

If one defines an edge as an abrupt gray-level change, then the derivative, or gradient, is a natural basis for an edge detector. Figure 1 illustrates the idea with a continuous, one-dimensional (1-D) example of a bright central region against a dark background. The left-hand portion of the gray-level function $f_c(x)$ shows a smooth transition from dark to bright as x increases. There must be a point x_0 that marks the transition from the low-amplitude region on the left to the adjacent high-amplitude region in the center. The gradient approach to detecting this edge is to locate x_0 where $|f_c'(x)|$ reaches a local maximum or,

equivalently, $f_c'(x)$ reaches a local extremum, as shown in the second plot of Fig. 1. The second derivative, or Laplacian approach, locates x_0 where a zero crossing of $f_c''(x)$ occurs, as in the third plot of Fig. 1. The right-hand side of Fig. 1 illustrates the case for a falling edge located at x_1.

To use the gradient or the Laplacian approaches as the basis for practical image edge detectors, one must extend the process to two dimensions, adapt to the discrete case, and somehow deal with the difficulties presented by real images. Relative to the 1-D edges shown in Fig. 1, edges in 2-D images have the additional quality of direction. One usually wishes to find edges regardless of direction, but a directionally sensitive edge detector can be useful at times. Also, the discrete nature of digital images requires the use of an approximation to the derivative. Finally, there are a number of problems that can confound the edge detection process in real images. These include noise, crosstalk or interference between nearby edges, and inaccuracies resulting from the use of a discrete grid. False edges, missing edges, and errors in edge location and orientation are often the result.

Because the derivative operator acts as a high-pass filter, edge detectors based on it are sensitive to noise. It is easy for noise inherent in an image to corrupt the real edges by shifting their apparent locations and by adding many false edge pixels. Unless care is taken, seemingly moderate amounts of noise are capable of overwhelming the edge detection process, rendering the results virtually useless. The wide variety of edge detection algorithms developed over the past three decades exists, in large part, because of the many ways proposed for dealing with noise and its effects. Most algorithms employ noise-suppression filtering of some kind before applying the edge detector itself. Some decompose the image into a set of low-pass or bandpass versions, apply the edge detector to each, and merge the results. Still others use adaptive methods, modifying the edge detector's parameters and behavior according to the noise characteristics of the image data.

An important tradeoff exists between correct detection of the actual edges and precise location of their positions. Edge detection errors can occur in two forms: false positives, in which nonedge pixels are misclassified as edge pixels, and false negatives, which are the reverse. Detection errors of both types tend to increase with noise, making good noise suppression very important in achieving a high detection accuracy. In general, the potential for noise suppression improves with the spatial extent of the edge detection filter. Hence, the goal of maximum detection accuracy calls for a large-sized filter. Errors in edge localization also increase with noise. To achieve good localization, however, the filter should generally be of small spatial extent. The goals of detection accuracy and location accuracy are thus put into direct conflict, creating a kind of uncertainty principle for edge detection [20].

In this chapter, we cover the basics of gradient and Laplacian edge detection methods in some detail. Following each, we also describe several of the more important and useful edge detection

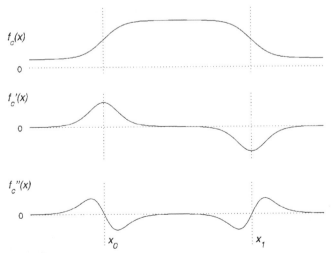

FIGURE 1   Edge detection in the 1-D continuous case; changes in $f_c(x)$ indicate edges, and x_0 and x_1 are the edge locations found by local extrema of $f_c'(x)$ or by zero crossings of $f_c''(x)$.

algorithms based on that approach. While the primary focus is on gray-level edge detectors, some discussion of edge detection in color and multispectral images is included.

2 Gradient-Based Methods

2.1 Continuous Gradient

The core of gradient edge detection is, of course, the gradient operator, ∇. In continuous form, applied to a continuous-space image, $f_c(x, y)$, the gradient is defined as

$$\nabla f_c(x, y) = \frac{\partial f_c(x, y)}{\partial x} i_x + \frac{\partial f_c(x, y)}{\partial y} i_y, \quad (1)$$

where i_x and i_y are the unit vectors in the x and y directions. Notice that the gradient is a vector, having both magnitude and direction. Its magnitude, $|\nabla f_c(x_0, y_0)|$, measures the maximum rate of change in the intensity at the location (x_0, y_0). Its direction is that of the greatest increase in intensity; i.e., it points "uphill."

To produce an edge detector, one may simply extend the 1-D case described earlier. Consider the effect of finding the local extrema of $\nabla f_c(x, y)$ or the local maxima of

$$|\nabla f_c(x, y)| = \sqrt{\left(\frac{\partial f_c(x, y)}{\partial x}\right)^2 + \left(\frac{\partial f_c(x, y)}{\partial y}\right)^2}. \quad (2)$$

The precise meaning of "local" is very important here. If the maxima of Eq. (2) are found over a 2-D neighborhood, the result is a set of isolated points rather than the desired edge contours. The problem stems from the fact that the gradient magnitude is seldom constant along a given edge, so finding the 2-D local maxima yields only the locally strongest of the edge contour points. To fully construct edge contours, it is better to apply Eq. (2) to a 1-D local neighborhood, namely a line segment, whose direction is chosen to cross the edge. The situation is then similar to that of Fig. 1, where the point of locally maximum gradient magnitude is the edge point. Now the issue becomes how to select the best direction for the line segment used for the search.

The most commonly used method of producing edge segments or contours from Eq. (2) consists of two stages: thresholding and thinning. In the thresholding stage, the gradient magnitude at every point is compared to a predefined threshold value, T. All points satisfying the following criterion are classified as candidate edge points:

$$|\nabla f_c(x, y)| \geq T. \quad (3)$$

The set of candidate edge points tends to form strips, which have positive width. Since the desire is usually for zero-width boundary segments or contours to describe the edges, a subsequent processing stage is needed to thin the strips to the final edge contours. Edge contours derived from continuous-space images should have zero width because any local maxima of

$|\nabla f_c(x, y)|$, along a line segment that crosses the edge, cannot be adjacent points. For the case of discrete-space images, the nonzero pixel size imposes a minimum practical edge width.

Edge thinning can be accomplished in a number of ways, depending on the application, but thinning by nonmaximum suppression is usually the best choice. Generally speaking, we wish to suppress any point that is not, in a 1-D sense, a local maximum in gradient magnitude. Since a 1-D local neighborhood search typically produces a single maximum, those points that are local maxima will form edge segments only one point wide. One approach classifies an edge-strip point as an edge point if its gradient magnitude is a local maximum in at least one direction. However, this thinning method sometimes has the side effect of creating false edges near strong edge lines [12]. It is also somewhat inefficient because of the computation required to check along a number of different directions. A better, more efficient thinning approach checks only a single direction, the gradient direction, to test whether a given point is a local maximum in gradient magnitude. The points that pass this scrutiny are classified as edge points. Looking in the gradient direction essentially searches perpendicular to the edge itself, producing a scenario similar to the 1-D case shown in Fig. 1. The method is efficient because it is not necessary to search in multiple directions. It also tends to produces edge segments having good localization accuracy. These characteristics make the gradient direction, local extremum method quite popular. The following steps summarize its implementation.

1. Using one of the techniques described in the next section, compute ∇f for all pixels.
2. Determine candidate edge pixels by thresholding all pixels' gradient magnitudes by T.
3. Thin by checking whether each candidate edge pixel's gradient magnitude is a local maximum along its gradient direction. If so, classify it as an edge pixel.

Consider the effect of performing the thresholding and thinning operations in isolation. If thresholding alone were done, the computational cost of thinning would be saved and the edges would show as strips or patches instead of thin segments. If thinning were done without thresholding, that is, if edge points were simply those having locally maximum gradient magnitude, many false edge points would likely result because of noise. Noise tends to create false edge points because some points in edge-free areas happen to have locally maximum gradient magnitudes. The thresholding step of Eq. (3) is often useful to reduce noise prior to thinning. A variety of adaptive methods have been developed that adjust the threshold according to certain image characteristics, such as an estimate of local signal-to-noise ratio. Adaptive thresholding can often do a better job of noise suppression while reducing the amount of edge fragmentation.

The edge maps in Fig. 3, computed from the original image in Fig. 2, illustrate the effect of the thresholding and subsequent thinning steps.

FIGURE 2   Original cameraman image, 512×512 pixels.

The selection of the threshold value T is a tradeoff between the wish to fully capture the actual edges in the image and the desire to reject noise. Increasing T decreases sensitivity to noise at the cost of rejecting the weakest edges, forcing the edge segments to become more broken and fragmented. By decreasing T, one can obtain more connected and richer edge contours, but the greater noise sensitivity is likely to produce more false edges. If only thresholding is used, as in Eq. (3) and Fig. 3(a), the edge strips tend to narrow as T increases and widen as it decreases. Figure 4 compares edge maps obtained from several different threshold values.

Sometimes a directional edge detector is useful. One can be obtained by decomposing the gradient into horizontal and vertical components and applying them separately. Expressed in the continuous domain, the operators become:

$$\left| \frac{\partial f_c(x, y)}{\partial x} \right| \geq T \quad \text{for edges in the } y \text{ direction,}$$

$$\left| \frac{\partial f_c(x, y)}{\partial y} \right| \geq T \quad \text{for edges in the } x \text{ direction.}$$

An example of directional edge detection is illustrated in Fig. 5.

A directional edge detector can be constructed for any desired direction by using the directional derivative along a unit vector $\boldsymbol{n}$,

$$\frac{\partial f_c}{\partial \boldsymbol{n}} = \boldsymbol{\nabla} f_c(x, y) \cdot \boldsymbol{n},$$

$$= \frac{\partial f_c(x, y)}{\partial x} \cos \theta + \frac{\partial f_c(x, y)}{\partial y} \sin \theta, \quad (4)$$

where θ is the angle of $\boldsymbol{n}$ relative to the positive x axis. The directional derivative is most sensitive to edges perpendicular to $\boldsymbol{n}$.

The continuous-space gradient magnitude produces an isotropic or rotationally symmetric edge detector, equally sensitive to edges in any direction [12]. It is easy to show why $|\boldsymbol{\nabla} f|$ is isotropic. In addition to the original X–Y coordinate system, let us introduce a new system, X'–Y', which is rotated by an angle of ϕ relative to X–Y. Let $\boldsymbol{n}_{x'}$ and $\boldsymbol{n}_{y'}$ be the unit vectors in the x' and y' directions, respectively. For the gradient magnitude to be isotropic, the same result must be produced in both coordinate systems, regardless of ϕ. Using Eq. (4) along with abbreviated notation, we find the partial derivatives with respect to the new coordinate axes are

$$f_{x'} = \boldsymbol{\nabla} f \cdot \boldsymbol{n}_{x'} = f_x \cos \phi + f_y \sin \phi,$$
$$f_{y'} = \boldsymbol{\nabla} f \cdot \boldsymbol{n}_{y'} = -f_x \sin \phi + f_y \cos \phi.$$

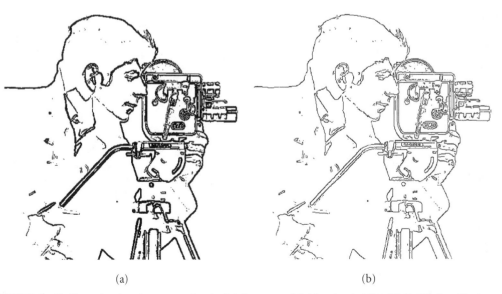

(a)   (b)

FIGURE 3   Gradient edge detection steps, using the Sobel operator: (a) After thresholding $|\boldsymbol{\nabla} f|$; (b) after thinning (a) by finding the local maximum of $|\boldsymbol{\nabla} f|$ along the gradient direction.

FIGURE 4   Roberts edge maps obtained by using various threshold values: (a) $T = 5$, (b) $T = 10$, (c) $T = 20$, (d) $T = 40$. As T increases, more noise-induced edges are rejected along with the weaker real edges.

Now let us examine the gradient magnitude in the new co-ordinate system:

So the gradient magnitude in the new coordinate system matches that in the original system, regardless of the rotation angle, ϕ.

$$|\nabla f| = \sqrt{f_{x'}^2 + f_{y'}^2},$$

$$= \sqrt{(f_x \cos \phi + f_y \sin \phi)^2 + (-f_x \sin \phi + f_y \cos \phi)^2},$$

$$= \sqrt{f_x^2 \cos^2 \phi + 2 f_x f_y \cos \phi \sin \phi + f_y^2 \sin^2 \phi + f_x^2 \sin^2 \phi - 2 f_x f_y \sin \phi \cos \phi + f_y^2 \cos^2 \phi},$$

$$= \sqrt{f_x^2 (\cos^2 \phi + \sin^2 \phi) + f_y^2 (\cos^2 \phi + \sin^2 \phi)},$$

$$= \sqrt{f_x^2 + f_y^2}.$$

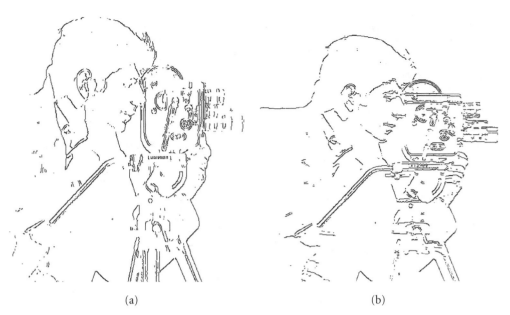

(a)   (b)

FIGURE 5   Directional edge detection comparison, using the Sobel operator: (a) results of horizontal difference operator; (b) results of vertical difference operator.

Occasionally, one may wish to reduce the computation load of Eq. (2) by approximating the square root with a computationally simpler function. Three possibilities are

$$|\nabla f_c(x, y)| \approx \max\left\{\left|\frac{\partial f_c(x, y)}{\partial x}\right|, \left|\frac{\partial f_c(x, y)}{\partial y}\right|\right\}, \qquad (5)$$

$$\approx \left|\frac{\partial f_c(x, y)}{\partial x}\right| + \left|\frac{\partial f_c(x, y)}{\partial y}\right|, \qquad (6)$$

$$\approx \max\left\{\left|\frac{\partial f_c(x, y)}{\partial x}\right|, \left|\frac{\partial f_c(x, y)}{\partial y}\right|\right\}$$
$$+ \frac{1}{4}\min\left\{\left|\frac{\partial f_c(x, y)}{\partial x}\right|, \left|\frac{\partial f_c(x, y)}{\partial y}\right|\right\}. \qquad (7)$$

One should be aware that approximations of this type may alter the properties of the gradient somewhat. For instance, the approximated gradient magnitudes of Eqs. (5), (6), and (7) are not isotropic and produce their greatest errors for purely diagonally oriented edges. All three estimates are correct only for the pure horizontal and vertical cases. Otherwise, Eq. (5) consistently underestimates the true gradient magnitude while Eq. (6) overestimates it. This makes Eq. (5) biased against diagonal edges and Eq. (6) biased toward them. The estimate of Eq. (7) is by far the most accurate of the three.

2.2 Discrete Gradient Operators

In the continuous-space image, $f_c(x, y)$, let x and y represent the horizontal and vertical axes, respectively. Let the discrete-space representation of $f_c(x, y)$ be $f(n_1, n_2)$, with n_1 describing the horizontal position and n_2 describing the vertical. For use

on discrete-space images, the continuous gradient's derivative operators must be approximated in discrete form. The approximation takes the form of a pair of orthogonally oriented filters, $h_1(n_1, n_2)$ and $h_2(n_1, n_2)$, which must be separately convolved with the image. Based on Eq. (1), the gradient estimate is

$$\hat{\nabla} f(n_1, n_2) = f_1(n_1, n_2)\mathbf{i}_{n_1} + f_2(n_1, n_2)\mathbf{i}_{n_2},$$

where

$$f_1(n_1, n_2) = f(n_1, n_2) * h_1(n_1, n_2),$$
$$f_2(n_1, n_2) = f(n_1, n_2) * h_2(n_1, n_2).$$

Two filters are necessary because the gradient requires the computation of an orthogonal pair of directional derivatives. The gradient magnitude and direction estimates can then be computed as follows:

$$|\hat{\nabla} f(n_1, n_2)| = \sqrt{f_1^2(n_1, n_2) + f_2^2(n_1, n_2)},$$

$$\angle\hat{\nabla} f(n_1, n_2) = \tan^{-1}\left(\frac{f_2(n_1, n_2)}{f_1(n_1, n_2)}\right). \qquad (8)$$

Each of the filters implements a derivative and should not respond to a constant, so the sum of its coefficients must always be zero. A more general statement of this property is described later in this chapter by Eq. (10).

There are many possible derivative-approximation filters for use in gradient estimation. Let us start with the simplest case. Two simple approximation schemes for the horizontal derivative are, for the first and central differences, respectively,

$$f_1(n_1, n_2) = f(n_1, n_2) - f(n_1 - 1, n_2),$$
$$f_1(n_1, n_2) = \tfrac{1}{2}[f(n_1 + 1, n_2) - f(n_1 - 1, n_2)].$$

The scaling factor of $1/2$ for the central difference is caused by the two-pixel distance between the nonzero samples. The origin positions for both filters are usually set at (n_1, n_2). The gradient magnitude threshold value can be easily adjusted to compensate for any scaling, so we omit the scale factor from here on. Both of these differences respond most strongly to vertically oriented edges and do not respond to purely horizontal edges. The case for the vertical direction is similar, producing a derivative approximation that responds most strongly to horizontal edges. These derivative approximations can be expressed as filter kernels, whose impulse responses, $h_1(n_1, n_2)$ and $h_2(n_1, n_2)$, are as follows for the first and central differences, respectively:

$$h_1(n_1, n_2) = \begin{bmatrix} 0 & 0 \\ \mathbf{-1} & 1 \end{bmatrix}, \quad h_2(n_1, n_2) = \begin{bmatrix} 1 & 0 \\ \mathbf{-1} & 0 \end{bmatrix},$$

$$h_1(n_1, n_2) = \begin{bmatrix} 0 & 0 & 0 \\ -1 & \mathbf{0} & 1 \\ 0 & 0 & 0 \end{bmatrix}, \quad h_2(n_1, n_2) = \begin{bmatrix} 0 & 1 & 0 \\ 0 & \mathbf{0} & 0 \\ 0 & -1 & 0 \end{bmatrix}.$$

Boldface elements indicate the origin position.

If used to detect edges, the pair of first difference filters above presents the problem that the zero crossings of its two $[-1 \ 1]$ derivative kernels lie at different positions. This prevents the two filters from measuring horizontal and vertical edge characteristics at the same location, causing error in the estimated gradient. The central difference, caused by the common center of its horizontal and vertical differencing kernels, avoids this position mismatch problem. This benefit comes at the costs of larger filter size and the fact that the measured gradient at a pixel (n_1, n_2) does not actually consider the value of that pixel.

Rotating the first difference kernels by an angle of $\pi/4$ and stretching the grid a bit produces the $h_1(n_1, n_2)$ and $h_2(n_1, n_2)$ kernels for the Roberts operator:

$$\begin{bmatrix} 0 & 1 \\ \mathbf{-1} & 0 \end{bmatrix} \quad \begin{bmatrix} 1 & 0 \\ \mathbf{0} & -1 \end{bmatrix}.$$

The Roberts operator's component filters are tuned for diagonal edges rather than vertical and horizontal ones. For use in an edge detector based on the gradient magnitude, it is important only that the two filters be orthogonal. They need not be aligned with the n_1 and n_2 axes. The pair of Roberts filters have a common zero-crossing point for their differencing kernels. This common center eliminates the position mismatch error exhibited by the horizontal–vertical first difference pair, as described earlier. If the origins of the Roberts kernels are positioned on the $+1$ samples, as is sometimes found in the literature, then no common center point exists for their first differences.

The Roberts operator, like any simple first-difference gradient operator, has two undesirable characteristics. First, the zero crossing of its $[-1 \ 1]$ diagonal kernel lies off grid, but the edge location must be assigned to an actual pixel location, namely the one at the filter's origin. This can create edge location bias that may lead to location errors approaching the interpixel distance. If we could use the central difference instead of the first difference, this problem would be reduced because the central difference operator inherently constrains its zero crossing to an exact pixel location.

The other difficulty caused by the first difference is its noise sensitivity. In fact, both the first- and central-difference derivative estimators are quite sensitive to noise. The noise problem can be reduced somewhat by incorporating smoothing into each filter in the direction normal to that of the difference. Consider an example based on the central difference in one direction for which we wish to smooth along the orthogonal direction with a simple three-sample average. To that end, let us define the impulse responses of two filters:

$$h_a(n_1) = [1 \ \mathbf{1} \ 1], \quad h_b(n_2) = [-1 \ \mathbf{0} \ 1].$$

Since h_a is a function only of n_1 and h_b depends only on n_2, one can simply multiply them as an outer product to form a separable derivative filter that incorporates smoothing:

$$h_a(n_1) h_b(n_2) = h_1(n_1, n_2),$$

$$\begin{bmatrix} 1 \\ 1 \\ 1 \end{bmatrix} \begin{bmatrix} -1 & \mathbf{0} & 1 \end{bmatrix} = \begin{bmatrix} -1 & 0 & 1 \\ -1 & 0 & 1 \\ -1 & 0 & 1 \end{bmatrix}.$$

Repeating this process for the orthogonal case produces the Prewitt operator:

$$\begin{bmatrix} -1 & 0 & 1 \\ -1 & \mathbf{0} & 1 \\ -1 & 0 & 1 \end{bmatrix} \begin{bmatrix} 1 & 1 & 1 \\ 0 & \mathbf{0} & 0 \\ -1 & -1 & -1 \end{bmatrix}.$$

The Prewitt edge gradient operator simultaneously accomplishes differentiation in one coordinate direction, using the central difference, and noise reduction in the orthogonal direction, by means of local averaging. Because it uses the central difference instead of the first difference, there is less edge-location bias.

In general, the smoothing characteristics can be adjusted by choosing an appropriate low-pass filter kernel in place of the Prewitt's three-sample average. One such variation is the Sobel operator, one of the most widely used gradient edge detectors:

$$\begin{bmatrix} -1 & 0 & 1 \\ -2 & \mathbf{0} & 2 \\ -1 & 0 & 1 \end{bmatrix} \begin{bmatrix} 1 & 2 & 1 \\ 0 & \mathbf{0} & 0 \\ -1 & -2 & -1 \end{bmatrix}.$$

Sobel's operator is often a better choice than Prewitt's because the low-pass filter produced by the $[1 \ 2 \ 1]$ kernel results in a smoother frequency response compared to that of $[1 \ 1 \ 1]$.

The Prewitt and Sobel operators respond differently to diagonal edges than to horizontal or vertical ones. This behavior is a consequence of the fact that their filter coefficients do not compensate for the different grid spacings in the diagonal and the horizontal directions. The Prewitt operator is less sensitive to diagonal edges than to vertical or horizontal ones. The opposite is true for the Sobel operator [16]. A variation designed for equal gradient magnitude response to diagonal, horizontal, and vertical edges is the Frei–Chen operator:

$$\begin{bmatrix} -1 & 0 & 1 \\ -\sqrt{2} & 0 & \sqrt{2} \\ -1 & 0 & 1 \end{bmatrix} \begin{bmatrix} 1 & \sqrt{2} & 1 \\ 0 & 0 & 0 \\ -1 & -\sqrt{2} & -1 \end{bmatrix}.$$

However, even the Frei–Chen operator retains some directional sensitivity in gradient magnitude, so it is not truly isotropic.

The residual anisotropy is caused by the fact that the difference operators used to approximate Eq. (1) are not rotationally symmetric. Merron and Brady [15] describe a simple method for greatly reducing the residual directional bias by using a set of four difference operators instead of two. Their operators are oriented in increments of $\pi/4$ radians, adding a pair of diagonal ones to the original horizontal and vertical pair. Averaging the gradients produced by the diagonal operators with those of the nondiagonal ones allows their complementary directional biases to reduce the overall anisotropy. However, Ziou and Wang [23] have described how an isotropic gradient applied to a discrete grid tends to introduce some anisotropy. They have also analyzed the errors of gradient magnitude and direction as a function of edge translation and orientation for several detectors. Figure 6 shows the results of performing edge detection on an example

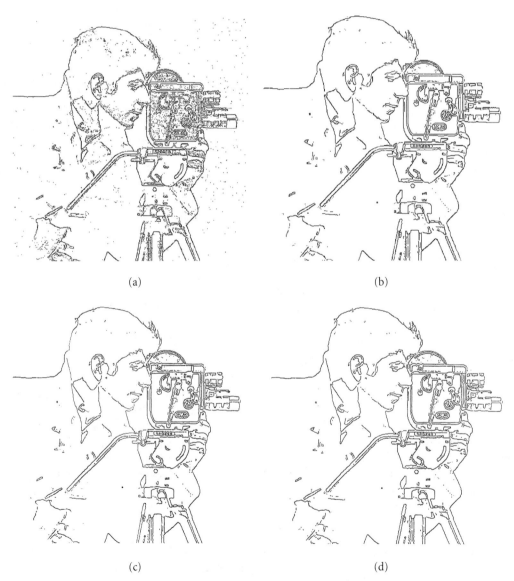

(a)   (b)

(c)   (d)

FIGURE 6   Comparison of edge detection using various gradient operators: (a) Roberts, (b) 3 × 3 Prewitt, (c) 3 × 3 Sobel, (d) 3 × 3 Frei–Chen. In each case, the threshold has been set to allow a fair comparison.

image by applying the discrete gradient operators discussed so far.

Haralick's facet model [8] provides another way of calculating the gradient in order to perform edge detection. In the sloped facet model, a small neighborhood is parameterized by $\alpha n_2 + \beta n_1 + \gamma$, describing the plane that best fits the gray levels in that neighborhood. The plane parameters α and β can be used to compute the gradient magnitude:

$$|\nabla f_c(n_1, n_2)| = \sqrt{\alpha^2 + \beta^2}.$$

The facet model also provides means for computing directional derivatives, zero crossings, and a variety of other useful operations.

Improved noise suppression is possible with increased kernel size. The additional coefficients can be used to better approximate the desired continuous-space noise-suppression filter. Greater filter extent can also be used to reduce directional sensitivity by more accurately modeling an ideal isotropic filter. However, increasing the kernel size will exacerbate edge localization problems and create interference between nearby edges. Noise suppression can be improved by other methods as well. Papers by Bovik [3] and Hardie and Boncelet [9] are just two that describe the use of edge-enhancing prefilters, which simultaneously suppress noise and steepen edges prior to gradient edge detection.

3 Laplacian-Based Methods

3.1 Continuous Laplacian

The Laplacian is defined as

$$\nabla^2 f_c(x, y) = \nabla \cdot \nabla f_c(x, y) = \frac{\partial^2 f_c(x, y)}{\partial x^2} + \frac{\partial^2 f_c(x, y)}{\partial y^2}.$$

$$(9)$$

The zero crossings of $\nabla^2 f_c(x, y)$ occur at the edge points of $f_c(x, y)$ because of the second derivative action (see Fig. 1). Laplacian-based edge detection has the nice property that it produces edges of zero thickness, making edge-thinning steps unnecessary. This is because the zero crossings themselves define the edge locations.

The continuous Laplacian is isotropic, favoring no particular edge orientation. Consequently, its second partial terms in Eq. (9) can be oriented in any direction as long as they remain perpendicular to each other. Consider an ideal, straight, and noise-free edge oriented in an arbitrary direction. Let us realign the first term of Eq. (9) parallel to that edge and the second term perpendicular to it. The first term then generates no response at all because it acts only along the edge. The second term produces a zero crossing at the edge position along its edge-crossing profile.

An edge detector based solely on the zero crossings of the continuous Laplacian produces closed edge contours if the image,

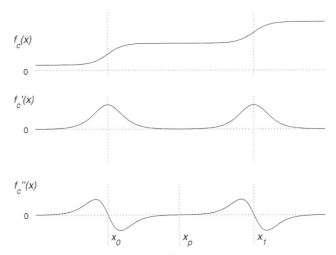

FIGURE 7 The zero crossing of $f_c''(x)$ at x_p creates a phantom edge.

$f(x, y)$, meets certain smoothness constraints [20]. The contours are closed because edge strength is not considered, so even the slightest, most gradual intensity transition produces a zero crossing. In effect, the zero-crossing contours define the boundaries that separate regions of nearly constant intensity in the original image. The second derivative zero crossings occur at the local extrema of the first derivative (see Fig. 1), but many zero crossings are not local maxima of the gradient magnitude. Some local minima of the gradient magnitude give rise to phantom edges, which can be largely eliminated by appropriately thresholding the edge strength. Figure 7 illustrates a 1-D example of a phantom edge.

Noise presents a problem for the Laplacian edge detector in several ways. First, the second-derivative action of Eq. (9) makes the Laplacian even more sensitive to noise than the first-derivative-based gradient. Second, noise produces many false edge contours because it introduces variation to the constant-intensity regions in the noise-free image. Third, noise alters the locations of the zero-crossing points, producing location errors along the edge contours. The problem of noise-induced false edges can be addressed by applying an additional test to the zero-crossing points. Only the zero crossings that satisfy this new criterion are considered edge points. One commonly-used technique classifies a zero crossing as an edge point if the local gray-level variance exceeds a threshold amount. Another method is to select the strong edges by thresholding the gradient magnitude or the slope of the Laplacian output at the zero crossing. Both criteria serve to reject zero crossing points which are more likely caused by noise than by a real edge in the original scene. Of course, thresholding the zero crossings in this manner tends to break up the closed contours.

Like any derivative filter, the continuous-space Laplacian filter, $h_c(x, y)$, has this important property:

$$\int_{-\infty}^{+\infty} \int_{-\infty}^{+\infty} h_c(x, y) \, dx \, dy = 0. \quad (10)$$

In other words, $h_c(x, y)$ is a surface bounding equal volumes above and below zero. Consequently, $\nabla^2 f_c(x, y)$ will also have equal volumes above and below zero. This property eliminates any response that is due to the constant or DC bias contained in $f_c(x, y)$. Without DC bias rejection, the filter's edge detection performance would be compromised.

3.2 Discrete Laplacian Operators

It is useful to construct a filter to serve as the Laplacian operator when applied to a discrete-space image. Recall that the gradient, which is a vector, required a pair of orthogonal filters. The Laplacian is a scalar. Therefore, a single filter, $h(n_1, n_2)$, is sufficient for realizing a Laplacian operator. The Laplacian estimate for an image, $f(n_1, n_2)$, is then

$$\hat{\nabla}^2 f(n_1, n_2) = f(n_1, n_2) * h(n_1, n_2)$$

One of the simplest Laplacian operators can be derived as follows. First needed is an approximation to the derivative in x, so let us use a simple first difference.

$$\frac{\partial f_c(x, y)}{\partial x} \rightarrow f_x(n_1, n_2) = f(n_1 + 1, n_2) - f(n_1, n_2). \quad (11)$$

The second derivative in x can be built by applying the first difference to Eq. (11). However, we discussed earlier how the first difference produces location errors because its zero crossing lies off grid. This second application of a first difference can be shifted to counteract the error introduced by the previous one:

$$\frac{\partial^2 f_c(x, y)}{\partial x^2} \rightarrow f_{xx}(n_1, n_2) = f_x(n_1, n_2) - f_x(n_1 - 1, n_2).$$

$$(12)$$

Combining the two derivative-approximation stages from Eqs. (11) and (12) produces

$$\frac{\partial^2 f_c(x, y)}{\partial x^2} \rightarrow f_{xx}(n_1, n_2)$$
$$= f(n_1 + 1, n_2) - 2f(n_1, n_2) + f(n_1 - 1, n_2)$$
$$= [1 \quad -2 \quad 1]. \quad (13)$$

Proceeding in an identical manner for y yields

$$\frac{\partial^2 f_c(x, y)}{\partial y^2} \rightarrow f_{yy}(n_1, n_2)$$
$$= f(n_1, n_2 + 1) - 2f(n_1, n_2) + f(n_1, n_2 - 1)$$
$$= \begin{bmatrix} 1 \\ -2 \\ 1 \end{bmatrix}. \quad (14)$$

Combining the x and y second partials of Eqs. (13) and (14)

produces a filter, $h(n_1, n_2)$, which estimates the Laplacian:

$$\nabla^2 f_c(x, y) \rightarrow \hat{\nabla}^2 f(n_1, n_2)$$
$$= f_{xx}(n_1, n_2) + f_{yy}(n_1, n_2)$$
$$= f(n_1 + 1, n_2) + f(n_1 - 1, n_2) + f(n_1, n_2 + 1)$$
$$+ f(n_1, n_2 - 1) - 4f(n_1, n_2)$$
$$= [1 \quad -2 \quad 1] + \begin{bmatrix} 1 \\ -2 \\ 1 \end{bmatrix} = \begin{bmatrix} 0 & 1 & 0 \\ 1 & -4 & 1 \\ 0 & 1 & 0 \end{bmatrix}.$$

Other Laplacian estimation filters can be constructed by using this method of designing a pair of appropriate 1-D second derivative filters and combining them into a single 2-D filter. The results depend on the choice of derivative approximator, the size of the desired filter kernel, and the characteristics of any noise-reduction filtering applied. Two other 3×3 examples are

$$\begin{bmatrix} 1 & 1 & 1 \\ 1 & -8 & 1 \\ 1 & 1 & 1 \end{bmatrix}, \quad \begin{bmatrix} -1 & 2 & -1 \\ 2 & -4 & 2 \\ -1 & 2 & -1 \end{bmatrix}.$$

In general, a discrete-space smoothed Laplacian filter can be easily constructed by sampling an appropriate continuous-space function, such as the Laplacian of Gaussian. When constructing a Laplacian filter, make sure that the kernel's coefficients sum to zero in order to satisfy the discrete form of Eq. (10). Truncation effects may upset this property and create bias. If so, the filter coefficients should be adjusted in a way that restores proper balance.

Locating zero crossings in the discrete-space image, $\nabla^2 f(n_1, n_2)$, is fairly straightforward. Each pixel should be compared to its eight immediate neighbors; a four-way neighborhood comparison, while faster, may yield broken contours. If a pixel, $\mathbf{p}$, differs in sign with its neighbor, $\mathbf{q}$, an edge lies between them. The pixel, $\mathbf{p}$, is classified as a zero crossing if

$$|\nabla^2 f(\mathbf{p})| \leq |\nabla^2 f(\mathbf{q})|. \quad (15)$$

3.3 The Laplacian of Gaussian (Marr–Hildreth Operator)

It is common for a single image to contain edges having widely different sharpnesses and scales, from blurry and gradual to crisp and abrupt. Edge scale information is often useful as an aid toward image understanding. For instance, edges at low resolution tend to indicate gross shapes, whereas texture tends to become important at higher resolutions. An edge detected over a wide range of scale is more likely to be physically significant in the scene than an edge found only within a narrow range of scale. Furthermore, the effects of noise are usually most deleterious at the finer scales.

Marr and Hildreth [14] advocated the need for an operator that can be tuned to detect edges at a particular scale. Their method is based on filtering the image with a Gaussian kernel selected for a particular edge scale. The Gaussian smoothing operation serves to band-limit the image to a small range of frequencies, reducing the noise sensitivity problem when detecting zero crossings. The image is filtered over a variety of scales and the Laplacian zero crossings are computed at each. This produces a set of edge maps as a function of edge scale. Each edge point can be considered to reside in a region of scale space, for which edge point location is a function of x, y, and σ. Scale space has been successfully used to refine and analyze edge maps [22].

The Gaussian has some very desirable properties that facilitate this edge detection procedure. First, the Gaussian function is smooth and localized in both the spatial and frequency domains, providing a good compromise between the need for avoiding false edges and for minimizing errors in edge position. In fact, Torre and Poggio [20] describe the Gaussian as the only real-valued function that minimizes the product of spatial- and frequency-domain spreads. The Laplacian of Gaussian essentially acts as a bandpass filter because of its differential and smoothing behavior. Second, the Gaussian is separable, which helps make computation very efficient.

Omitting the scaling factor, the Gaussian filter can be written as

$$g_c(x, y) = \exp\left(-\frac{x^2 + y^2}{2\sigma^2}\right). \tag{16}$$

Its frequency response, $G(\Omega_x, \Omega_y)$, is also Gaussian:

$$G(\Omega_x, \Omega_y) = 2\pi\sigma^2 \exp\left(-\frac{\sigma^2}{2}\left(\Omega_x^2 + \Omega_y^2\right)\right).$$

The σ parameter is inversely related to the cutoff frequency.

Because the convolution and Laplacian operations are both linear and shift invariant, their computation order can be interchanged:

$$\nabla^2[f_c(x, y) * g_c(x, y)] = [\nabla^2 g_c(x, y)] * f_c(x, y) \tag{17}$$

Here we take advantage of the fact that the derivative is a linear operator. Therefore, Gaussian filtering followed by differentiation is the same as filtering with the derivative of a Gaussian. The right-hand side of Eq. (17) usually provides for more efficient computation since $\nabla^2 g_c(x, y)$ can be prepared in advance as a result of its image independence. The Laplacian of Gaussian (LoG) filter, $h_c(x, y)$, therefore has the following impulse response:

$$\begin{aligned} h_c(x, y) &= \nabla^2 g_c(x, y) \\ &= \frac{x^2 + y^2 - 2\sigma^2}{\sigma^4} \exp\left(-\frac{x^2 + y^2}{2\sigma^2}\right). \end{aligned} \tag{18}$$

To implement the LoG in discrete form, one may construct a filter, $h(n_1, n_2)$, by sampling Eq. (18) after choosing a value

of σ, then convolving with the image. If the filter extent is not small, it is usually more efficient to work in the frequency domain by multiplying the discrete Fourier transforms of the filter and the image, then inverse transforming the result. The fast Fourier transform, or FFT, is the method of choice for computing these transforms.

Although the discrete form of Eq. (18) is a 2-D filter, Chen *et al.* [6] have shown that it is actually the sum of two separable filters because the Gaussian itself is a separable function. By constructing and applying the appropriate 1-D filters successively to the rows and columns of the image, the computational expense of 2-D convolution becomes unnecessary. Separable convolution to implement the LoG is roughly 1–2 orders of magnitude more efficient than 2-D convolution. If an image is $M \times M$ in size, the number of operations at each pixel is M^2 for 2-D convolution and only $2M$ if done in a separable, 1-D manner.

Figure 8 shows an example of applying the LoG using various σ values. Figure 8(d) includes a gradient magnitude threshold, which suppresses noise and breaks contours. Lim [12] describes an adaptive thresholding scheme that produces better results.

Equation (18) has the shape of a *sombrero* or "Mexican hat." Figure 9 shows a perspective plot of $\nabla^2 g_c(x, y)$ and its frequency response, $F\{\nabla^2 g_c(x, y)\}$. This profile closely mimics the response of the spatial receptive field found in biological vision. Biological receptive fields have been shown to have a circularly symmetric impulse response, with a central excitory region surrounded by an inhibitory band.

When sampling the LoG to produce a discrete version, it is important to size the filter large enough to avoid significant truncation effects. A good rule of thumb is to make the filter at least three times the width of the LoG's central excitory lobe [16]. Siohan [19] describes two approaches for the practical design of LoG filters. The errors in edge location produced by the LoG have been analyzed in some detail by Berzins [2].

3.4 Difference of Gaussian

The Laplacian of Gaussian of Eq. (18) can be closely approximated by the difference of two Gaussians having properly chosen scales. The difference of Gaussian (DoG) filter is

$$h_c(x, y) = g_{c1}(x, y) - g_{c2}(x, y),$$

where

$$\frac{\sigma_2}{\sigma_1} \approx 1.6$$

and g_{c1}, g_{c2} are evaluated by using Eq. (16). However, the LoG is usually preferred because it is theoretically optimal and its separability allows for efficient computation [14]. For the same accuracy of results, the DoG requires a slightly larger filter size [10].

The technique of unsharp masking, used in photography, is basically a difference of Gaussians operation done with light and negatives. Unsharp masking involves making a somewhat

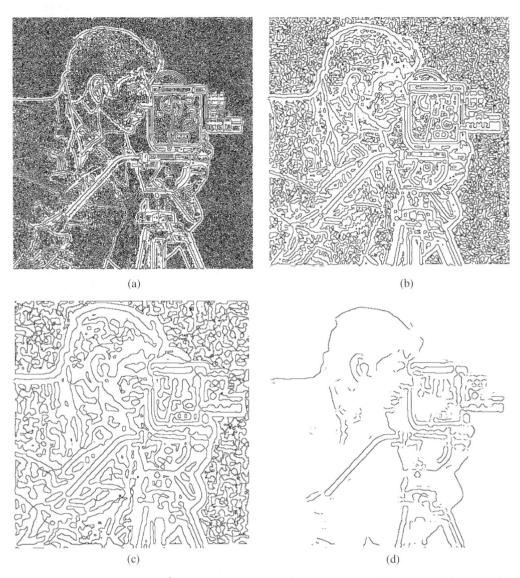

FIGURE 8   Zero crossings of $f * \nabla^2 g$ for several values of σ, with (d) also thresholded: (a) $\sigma = 1.0$, (b) $\sigma = 1.5$, (c) $\sigma = 2.0$, (d) $\sigma = 2.0$ and $T = 20$.

blurry exposure of an original negative onto a new piece of film. When the film is developed, it contains a blurred and inverted-brightness version of the original negative. Finally, a print is made from these two negatives sandwiched together, producing a sharpened image with the edges showing an increased contrast.

Nature uses the difference of Gaussians as a basis for the architecture of the retina's visual receptive field. The spatial-domain impulse response of a photoreceptor cell in the mammalian retina has a roughly Gaussian shape. The photoreceptor output feeds into horizontal cells in the adjacent layer of neurons. Each horizontal cell averages the responses of the receptors in its immediate neighborhood, producing a Gaussian-shaped impulse response with a higher σ than that of a single photoreceptor. Both layers send their outputs to the third layer, where bipolar

neurons subtract the high-σ neighborhood averages from the central photoreceptors' low-σ responses. This produces a biological realization of the difference-of-Gaussian filter, approximating the behavior of the Laplacian of Gaussian. The retina actually implements DoG bandpass filters at several spatial frequencies [13].

4 Canny's Method

Canny's method [4] uses the concepts of both the first and second derivatives in a very effective manner. His is a classic application of the gradient approach to edge detection in the presence of additive white Gaussian noise, but it also incorporates elements of the Laplacian approach. The method has three

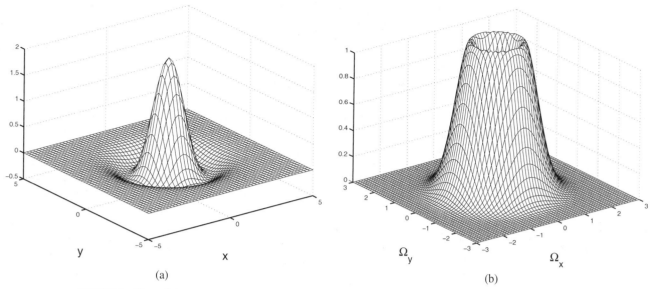

(a)   (b)

FIGURE 9   Plots of the LoG and its frequency response for $\sigma = 1$: (a) $-\nabla^2 g_c(x, y)$, the negative of Eq. (18);
(b) $F\{\nabla^2 g_c(x, y)\}$, the bandpass-shaped frequency response of Eq. (18).

simultaneous goals: low rate of detection errors, good edge localization, and only a single detection response per edge. Canny
assumed that false-positive and false-negative detection errors
are equally undesirable and so gave them equal weight. He further assumed that each edge has nearly constant cross section
and orientation, but his general method includes a way to effectively deal with the cases of curved edges and corners. With
these constraints, Canny determined the optimal 1-D edge detector for the step edge and showed that its impulse response can
be approximated fairly well by the derivative of a Gaussian.

An important action of Canny's edge detector is to prevent
multiple responses per true edge. Without this criterion, the
optimal step-edge detector would have an impulse response in
the form of a truncated signum function. (The signum function
produces $+1$ for any positive argument and -1 for any negative
argument.) But this type of filter has high bandwidth, allowing
noise or texture to produce several local maxima in the vicinity
of the actual edge. The effect of the derivative of Gaussian is to
prevent multiple responses by smoothing the truncated signum
in order to permit only one response peak in the edge neighborhood. The choice of variance for the Gaussian kernel controls
the filter width and the amount of smoothing. This defines the
width of the neighborhood in which only a single peak is to
be allowed. The variance selected should be proportional to the
amount of noise present. If the variance is chosen too low, the filter can produce multiple detections for a single edge; if too high,
edge localization suffers needlessly. Because the edges in a given
image are likely to differ in signal-to-noise ratio, a single-filter
implementation is usually not best for detecting them. Hence, a
thorough edge detection procedure should operate at different
scales.

Canny's approach begins by smoothing the image with a
Gaussian filter:

$$g_c(x, y) = \frac{1}{\sigma\sqrt{2\pi}} \exp\left(-\frac{x^2 + y^2}{2\sigma^2}\right). \qquad (19)$$

One may sample and truncate Eq. (19) to produce a finite-extent
filter, $g(n_1, n_2)$. At each pixel, Eq. (8) is used to estimate the
gradient direction. From a set of prepared edge detection filter
masks having various orientations, the one oriented nearest to
the gradient direction for the targeted pixel is then chosen. When
applied to the Gaussian-smoothed image, this filter produces an
estimate of gradient magnitude at that pixel. Next, the goal is
to suppress non-maxima of the gradient magnitude by testing
a 3×3 neighborhood, comparing the magnitude at the center
pixel with those at interpolated positions to either side along the
gradient direction.

The pixels that survive to this point are candidates for the
edge map. To produce an edge map from these candidate pixels, Canny applies thresholding by gradient magnitude in an
adaptive manner with hysteresis. An estimate of the noise in the
image determines the values of a pair of thresholds, with the
upper threshold typically two or three times that of the lower.
A candidate edge segment is included in the output edge map
if at least one of its pixels has a gradient magnitude exceeding
the upper threshold, but pixels not meeting the lower threshold
are excluded. This hysteresis action helps reduce the problem
of broken edge contours while improving the ability to reject
noise.

A set of edge maps over a range of scales can be produced
by varying the σ values used to Gaussian-filter the image. Since

smoothing at different scales produces different errors in edge location, an edge segment that appears in multiple edge maps at different scales may exhibit some position shift. Canny proposed unifying the set of edge maps into a single result by using a technique he called "feature synthesis," which proceeds in a fine-to-coarse manner while tracking the edge segments within their possible displacements.

The preoriented edge detection filters, mentioned previously, have some interesting properties. Each mask includes a derivative of Gaussian function to perform the nearly optimal directional derivative across the intended edge. A smooth, averaging profile appears in the mask along the intended edge direction in order to reduce noise without compromising the sharpness of the edge profile. In the smoothing direction, the filter extent is usually several times that in the derivative direction when the filter is intended for straight edges. Canny's method includes a "goodness of fit" test to determine if the selected filter is appropriate before it is applied. The test examines the gray-level variance of the strip of pixels along the smoothing direction of the filter. If the variance is small, then the edge must be close to linear, and the filter is a good choice. A large variance indicates the presence of curvature or a corner, in which case a better choice of filter would have smaller extent in the smoothing direction. There were six oriented filters used in Canny's work. The greatest directional mismatch between the actual gradient and the nearest filter is $15°$, which yields a gradient magnitude that is $\sim85\%$ of the actual value.

As discussed previously, edges can be detected from either the maxima of the gradient magnitude or the zero crossings of the second derivative. Another way to realize the essence of Canny's method is to look for zero crossings of the second directional derivative taken along the gradient direction. Let us examine the mathematical basis for this. If n is a unit vector in the gradient direction, and f is the Gaussian-smoothed image, then we wish to find

$$\frac{\partial^2 f}{\partial n^2} = \nabla\left(\frac{\partial f}{\partial n}\right) \cdot n$$
$$= \nabla(\nabla f \cdot n) \cdot n,$$

which can be expanded to the following form:

$$\frac{\partial^2 f}{\partial n^2} = \frac{f_x^2 f_{xx} + 2 f_x f_y f_{xy} + f_y^2 f_{yy}}{\sqrt{f_x^2 + f_y^2}}. \tag{20}$$

In Eq. (20), a concise notation has been used for the partial derivatives.

Like the Laplacian approach, Canny's method looks for zero crossings of the second derivative. The Laplacian's second derivative is nondirectional; it includes a component taken parallel to the edge and another taken across it. Canny's is evaluated only in the gradient direction, directly across the local edge. A derivative taken along an edge is counterproductive because it introduces noise without improving edge detection capability. By

being selective about the direction in which its derivatives are evaluated, Canny's approach avoids this source of noise and tends to produce better results.

Figures 10 and 11 illustrate the results of applying the Canny edge detector of Eq. (20) after Gaussian smoothing, then looking for zero crossings. Figure 10 demonstrates the effect of using the same upper and lower thresholds, T_U and T_L, over a range of σ values. The behavior of hysteresis thresholding is shown in Fig. 11. The partial derivatives were approximated using central differences. Thresholding was performed with hysteresis, but using fixed threshold values for each image instead of Canny's noise-adaptive threshold values. Zero-crossing detection was implemented in an eight-way manner, as described by Eq. (15) in the earlier discussion of discrete Laplacian operators. Also, Canny's preoriented edge detection filters were not used in preparing these examples, so it was not possible to adapt the edge detection filters according to the "goodness of fit" of the local edge profile as Canny did.

5 Approaches for Color and Multispectral Images

Edge detection for color images presents additional difficulties because of the three color components used. The most straightforward technique is to perform edge detection on the luminance component image while ignoring the chrominance information. The only computational cost beyond that for gray-scale images is incurred in obtaining the luminance component image, if necessary. In many color spaces, such as *YIQ*, *HSL*, *CIELUV*, and *CIELAB*, the luminance image is simply one of the components in that representation. For others, such as *RGB*, computing the luminance image is usually easy and efficient. The main drawback to luminance-only processing is that important edges are often not confined to the luminance component. Hence, a gray-level difference in the luminance component is often not the most appropriate criterion for edge detection in color images.

Another rather obvious approach is to apply a desired edge detection method separately to each color component and construct a cumulative edge map. One possibility for overall gradient magnitude, shown here for the *RGB* color space, combines the component gradient magnitudes [17]:

$$|\nabla f_c(x, y)| = |\nabla f_R(x, y)| + |\nabla f_G(x, y)| + |\nabla f_B(x, y)|.$$

The results, however, are biased according to the properties of the particular color space used. It is often important to employ a color space that is appropriate for the target application. For example, edge detection that is intended to approximate the human visual system's behavior should utilize a color space having a perceptual basis, such as *CIELUV* or perhaps *HSL*. Another complication is the fact that the components' gradient vectors may not always be similarly oriented, making the search for local

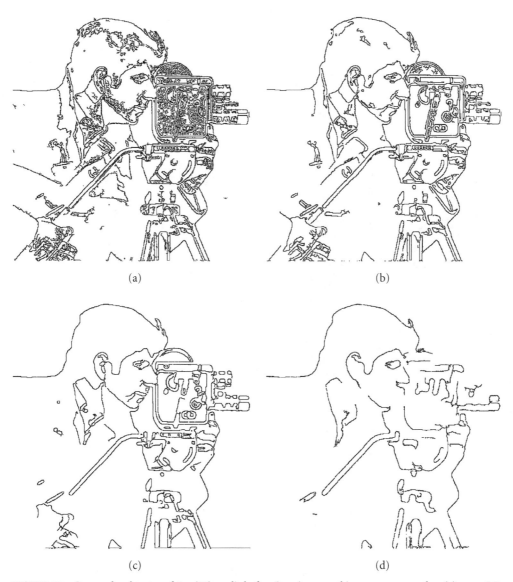

FIGURE 10 Canny edge detector of Eq. (20) applied after Gaussian smoothing over a range of σ: (a) σ = 0.5, (b) σ = 1, (c) σ = 2, (d) σ = 4. The thresholds are fixed in each case at $T_U = 10$ and $T_L = 4$.

maxima of $|\nabla f_c|$ along the gradient direction more difficult. If a total gradient image were to be computed by summing the color component gradient vectors, not just their magnitudes, then inconsistent orientations of the component gradients could destructively interfere and nullify some edges.

Vector approaches to color edge detection, while generally less computationally efficient, tend to have better theoretical justification. Euclidean distance in color space between the color vectors of a given pixel and its neighbors can be a good basis for an edge detector [17]. For the *RGB* case, the magnitude of the vector gradient is

$$|\nabla f_c(x, y)|$$
$$= \sqrt{|\nabla f_R(x, y)|^2 + |\nabla f_G(x, y)|^2 + |\nabla f_B(x, y)|^2}.$$

Trahanias and Venetsanopoulos [21] described the use of vector order statistics as the basis for color edge detection. A later paper by Scharcanski and Venetsanopoulos [18] furthered the concept. Although not strictly founded on the gradient or Laplacian, their techniques are effective and worth mention here because of their vector bases. The basic idea is to look for changes in local vector statistics, particularly vector dispersion, to indicate the presence of edges.

Multispectral images can have many components, complicating the edge detection problem even further. Cebrián [5] describes several methods that are useful for multispectral images having any number of components. His description uses the second directional derivative in the gradient direction as the basis for the edge detector, but other types of detectors can be used instead. The components-average method forms a gray-scale

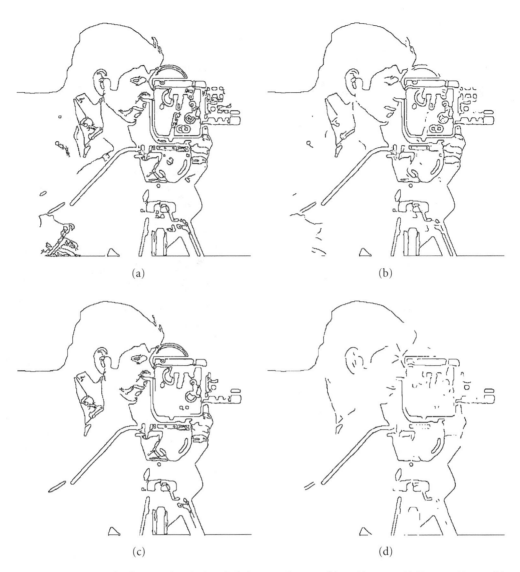

FIGURE 11   Canny edge detector of Eq. (20) applied after Gaussian smoothing with $\sigma = 2$: (a) $T_U = 10$, $T_L = 1$; (b) $T_U = T_L = 10$; (c) $T_U = 20$, $T_L = 1$; (d) $T_U = T_L = 20$. As T_L is changed, notice the effect on the results of hysteresis thresholding.

image by averaging all components, which have first been Gaussian smoothed, and then finds the edges in that image. The method generally works well because multispectral images tend to have high correlation between components. However, it is possible for edge information to diminish or vanish if the components destructively interfere.

Cumani [7] explored operators for computing the vector gradient and created an edge detection approach based on combining the component gradients. A multispectral contrast function is defined, and the image is searched for pixels having maximal directional contrast. Cumani's method does not always detect edges present in the component bands, but it better avoids the problem of destructive interference between bands.

The maximal gradient method constructs a single gradient image from the component images [5]. The overall gradient image's magnitude and direction values at a given pixel are those of the component having the greatest gradient magnitude at that pixel. Some edges can be missed by the maximal gradient technique because they may be swamped by differently oriented, stronger edges present in another band.

The method of combining component edge maps is the least efficient because an edge map must first be computed for every band. On the positive side, this method is capable of detecting any edge that is detectable in at least one component image. Combination of component edge maps into a single result is made more difficult by the edge location errors induced by Gaussian smoothing done in advance. The superimposed edges can become smeared in width because of the accumulated uncertainty in edge localization. A thinning step applied during the combination procedure can greatly reduce this edge blurring problem.

6 Summary

Gray-level edge detection is most commonly performed by convolving an image, f, with a filter that is somehow based on the idea of the derivative. Conceptually, edges can be revealed by locating either the local extrema of the first derivative of f or zero crossings of its second derivative. The gradient and the Laplacian are the primary derivative-based functions used to construct such edge-detection filters. The gradient, ∇, is a 2-D extension of the first derivative while the Laplacian, ∇^2, acts as a 2-D second derivative. A variety of edge detection algorithms and techniques have been developed that are based on the gradient or Laplacian in some way. Like any type of derivative-based filter, ones based on these two functions tend to be very sensitive to noise. Edge location errors, false edges, and broken or missing edge segments are often problems with edge detection applied to noisy images. For gradient techniques, thresholding is a common way to suppress noise and can be done adaptively for better results. Gaussian smoothing is also very helpful for noise suppression, especially when second-derivative methods such as the Laplacian are used. The Laplacian of Gaussian approach can also provide edge information over a range of scales, helping to further improve detection accuracy and noise suppression as well as providing clues that may be useful during subsequent processing.

References

[1] D. H. Ballard and C. M. Brown, *Computer Vision* (Prentice-Hall, Englewood Cliffs, NJ, 1982).

[2] V. Berzins, "Accuracy of Laplacian edge detectors," *Comput. Vis. Graph. Image Process.* **27**, 195–210 (1984).

[3] A. C. Bovik, T. S. Huang, and D. C. Munson Jr., "The effect of median filtering on edge estimation and detection," *IEEE Trans. Pattern Anal. Machine Intell.* **PAMI-9**, 181–194 (1987).

[4] J. Canny, "A computational approach to edge detection," *IEEE Trans. Pattern Anal. Machine Intell.* **PAMI-8**, 679–698 (1986).

[5] M. Cebrián, M. Perez-Luque, and G. Cisneros, "Edge detection alternatives for multispectral remote sensing images," in *Proceedings of the 8th Scandinavian Conference on Image Analysis* (NOBIM-Norwegian Soc. Image Process & Pattern Recognition, Tromso, Norway, 1993) **2**, pp. 1047–1054.

[6] J. S. Chen, A. Huertas, and G. Medioni, "Very fast convolution with Laplacian-of-Gaussian masks," in *Proceedings of the IEEE Computer Society Conference on Computer Vision and Pattern Recognition* (IEEE, New York, 1986), pp. 293–298.

[7] A. Cumani, "Edge detection in multispectral images," *Comput. Vis. Graph. Image Process. Graph. Models Image Process.* **53**, 40–51 (1991).

[8] R. M. Haralick and L. G. Shapiro, *Computer and Robot Vision* (Addison-Wesley, Reading, MA, 1992), Vol. 1.

[9] R. C. Hardie and C. G. Boncelet, "Gradient-based edge detection using nonlinear edge enhancing prefilters," *IEEE Trans. Image Process.* **4**, 1572–1577 (1995).

[10] A. Huertas and G. Medioni, "Detection of intensity changes with subpixel accuracy using Laplacian-Gaussian Masks," *IEEE Trans. Pattern Anal. Machine Intell.* **PAMI-8**, 651–664 (1986).

[11] A. K. Jain, *Fundamentals of Digital Image Processing* (Prentice-Hall, Englewood Cliffs, NJ, 1989).

[12] J. S. Lim, *Two-Dimensional Signal and Image Processing* (Prentice-Hall, Englewood Cliffs, NJ, 1990).

[13] D. Marr, *Vision* (W.H. Freeman, New York, 1982).

[14] D. Marr and E. Hildreth, "Theory of edge detection," *Proc. Roy. Soc. London B* **270**, 187–217 (1980).

[15] J. Merron and M. Brady, "Isotropic gradient estimation," in *Proceedings of the IEEE Computer Society Conference on Computer Vision and Pattern Recognition* (IEEE, New York, 1996), pp. 652–659.

[16] W. K. Pratt, *Digital Image Processing*, 2nd ed. (Wiley, New York, 1991).

[17] S. J. Sangwine and R. E. N. Horne, eds., *The Colour Image Processing Handbook* (Chapman and Hall, London, 1998).

[18] J. Scharcanski and A. N. Venetsanopoulos, "Edge detection of color images using directional operators," *IEEE Trans. Circuits Syst. Video Technol.* **7**, 397–401 (1997).

[19] P. Siohan, D. Pele, and V. Ouvrard, "Two design techniques for 2-D FIR LoG filters," in *Visual Communications and Image Processing*, M. Kunt, ed., *Proc. SPIE* **1360**, 970–981 (1990).

[20] V. Torre and T. A. Poggio, "On edge detection," *IEEE Trans. Pattern Anal. Machine Intell.* **PAMI-8**, 147–163 (1986).

[21] P. E. Trahanias and A. N. Venetsanopoulos, "Color edge detection using vector order statistics," *IEEE Trans. Image Process.* **2**, 259–264 (1993).

[22] A. P. Witkin, "Scale-space filtering," in *Proceedings of the International Joint Conference on Artificial Intelligence* (William Kaufmann Inc., Karlsruhe, Germany, 1983), pp. 1019–1022.

[23] D. Ziou and S. Wang, "Isotropic processing for gradient estimation," in *Proceedings of the IEEE Computer Society Conference on Computer Vision and Pattern Recognition* (IEEE, New York, 1996), pp. 660–665.

4.12

Diffusion-Based Edge Detectors

Scott T. Acton
Oklahoma State University

1 Introduction and Motivation

Sudden, sustained changes in image intensity are called *edges*. We know that the human visual system makes extensive use of edges to perform visual tasks such as object recognition [22]. Humans can recognize complex three-dimensional (3-D) objects by using only line drawings or image edge information. Similarly, the extraction of edges from digital imagery allows a valuable abstraction of information and a reduction in processing and storage costs. Most definitions of image edges involve some concept of feature scale. Edges are said to exist at certain scales — edges from detail existing at fine scales and edges from the boundaries of large objects existing at coarse scales. Furthermore, coarse scale edges exist at fine scales, leading to a notion of edge causality.

In order to locate edges of various scales within an image, it is desirable to have an image operator that computes a scaled version of a particular image or frame in a video sequence. This operator should preserve the position of such edges and facilitate the extraction of the edge map through the *scale space*. The tool of isotropic diffusion, a linear low-pass filtering process, is not able to preserve the position of important edges through the scale space. Anisotropic diffusion, however, meets this criterion and has been used in conjunction with edge detection during the last decade.

The main benefit of anisotropic diffusion is edge preservation through the image smoothing process. Anisotropic diffusion yields intraregion smoothing, not interregion smoothing, by impeding diffusion at the image edges. The anisotropic diffusion process can be used to retain image features of a specified scale. Furthermore, the localized computation of anisotropic diffusion allows efficient implementation on a locally interconnected computer architecture. Caselles *et al.* furnish additional motivation for using diffusion in image and video processing [14]. The diffusion methods use localized models in which discrete filters become partial differential equations (PDEs) as the sample spacing goes to zero. The PDE framework allows various properties to be proved or disproved including stability, locality, causality, and the existence and uniqueness of solutions. Through the established tools of numerical analysis, high degrees of accuracy and stability are possible.

In this chapter, we introduce diffusion for image and video processing. We specifically concentrate on the implementation of anisotropic diffusion, providing several alternatives for the diffusion coefficient and the diffusion PDE. Energy-based variational diffusion techniques are also reviewed. Recent advances in anisotropic diffusion processes, including multiresolution and multispectral techniques, are discussed. Finally, the extraction of image edges after anisotropic diffusion is addressed.

2 Background on Diffusion

2.1 Scale Space and Isotropic Diffusion

In order to introduce the diffusion-based processing methods and the associated processes of edge detection, let us define some notation. Let $\mathbf{I}$ represent an image with real-valued intensity $I(\mathbf{x})$

image at position $\mathbf{x}$ in the domain Ω. When defining the PDEs for diffusion, let $\mathbf{I}_t$ be the image at time t with intensities $I_t(\mathbf{x})$. Corresponding with image $\mathbf{I}$ is the edge map $\mathbf{e}$ — the image of "edge pixels" $e(\mathbf{x})$ with Boolean range (0 = no edge, 1 = edge), or real-valued range $e(\mathbf{x}) \in [0,1]$. The set of edge positions in an image is denoted by Ψ.

The concept of *scale space* is at the heart of diffusion-based image and video processing. A scale space is a collection of images that begins with the original, fine scale image and progresses toward more coarse scale representations. With the use of a scale space, important image processing tasks such as hierarchical searches, image coding, and image segmentation may be efficiently realized. Implicit in the creation of a scale space is the *scale generating filter*. Traditionally, linear filters have been used to scale an image. In fact, the scale space of Witkin [41] can be derived using a Gaussian filter:

$$\mathbf{I}_t = \mathbf{G}_\sigma * \mathbf{I}_0, \tag{1}$$

where $\mathbf{G}_\sigma$ is a Gaussian kernel with standard deviation (scale) of σ, and $\mathbf{I}_0 = \mathbf{I}$ is the initial image. If

$$\sigma = \sqrt{t}, \tag{2}$$

then the Gaussian filter result may be achieved through an isotropic diffusion process governed by

$$\frac{\partial \mathbf{I}_t}{\partial t} = \Delta \mathbf{I}_t, \tag{3}$$

where $\Delta \mathbf{I}_t$ is the Laplacian of $\mathbf{I}_t$ [21, 41]. To evolve one pixel of $\mathbf{I}$, we have the following PDE:

$$\frac{\partial I_t(\mathbf{x})}{\partial t} = \Delta I_t(\mathbf{x}). \tag{4}$$

The Marr–Hildreth paradigm uses a Gaussian scale space to define multiscale edge detection. Using the Gaussian-convolved (or diffused) images, one may detect edges by applying the Laplacian operator and then finding zero crossings [23]. This popular method of edge detection, called the Laplacian-of-Gaussian, or LoG, is strongly motivated by the biological vision system. However, the edges detected from isotropic diffusion (Gaussian scale space) suffer from artifacts such as corner rounding and from edge localization errors (deviations in detected edge position from the "true" edge position). The localization errors increase with increased scale, precluding straightforward multiscale image/video analysis. As a result, many researchers have pursued anisotropic diffusion as a practicable alternative for generating images suitable for edge detection. This chapter focuses on such methods.

2.2 Anisotropic Diffusion

The main idea behind anisotropic diffusion is the introduction of a function that inhibits smoothing at the image edges. This

function, called the diffusion coefficient $c(\mathbf{x})$, encourages intraregion smoothing over interregion smoothing. For example, if $c(\mathbf{x})$ is constant at all locations, then smoothing progresses in an isotropic manner. If $c(\mathbf{x})$ is allowed to vary according to the local image gradient, we have anisotropic diffusion. A basic anisotropic diffusion PDE is

$$\frac{\partial I_t(\mathbf{x})}{\partial t} = \text{div}\{c(\mathbf{x})\nabla I_t(\mathbf{x})\} \tag{5}$$

with $\mathbf{I}_0 = \mathbf{I}$ [30].

The discrete formulation proposed in [30] will be used as a general framework for implementation of anisotropic diffusion in this chapter. Here, the image intensities are updated according to

$$[I(\mathbf{x})]_{t+1} = \left[I(\mathbf{x}) + (\Delta T) \sum_{d=1}^{\Gamma} c_d(\mathbf{x}) \nabla I_d(\mathbf{x}) \right]_t, \tag{6}$$

where Γ is the number of directions in which diffusion is computed, $\nabla I_d(\mathbf{x})$ is the directional derivative (simple difference) in direction d at location $\mathbf{x}$, and time (in iterations) is given by t. ΔT is the time step — for stability, $\Delta T \leq {}^1/_2$ in the one-dimensional (1-D) case, and $\Delta T \leq {}^1/_4$ in the two-dimensional (2-D) case using four diffusion directions. For 1-D discrete-domain signals, the simple differences $\nabla I_d(\mathbf{x})$ with respect to the "western" and "eastern" neighbors, respectively (neighbors to the left and right), are defined by

$$\nabla I_1(x) = I(x - h_1) - I(x), \tag{7}$$

$$\nabla I_2(x) = I(x + h_2) - I(x). \tag{8}$$

The parameters h_1 and h_2 define the sample spacing used to estimate the directional derivatives. For the 2-D case, the diffusion directions include the "northern" and "southern" directions (up and down), as well as the western and eastern directions (left and right). Given the motivation and basic definition of diffusion-based processing, we will now define several implementations of anisotropic diffusion that can be applied for edge extraction.

3 Implementation of Diffusion

3.1 Diffusion Coefficient

The link between edge detection and anisotropic diffusion is found in the edge-preserving nature of anisotropic diffusion. The function that impedes smoothing at the edges is the diffusion coefficient. Therefore, the selection of the diffusion coefficient is the most critical step in performing diffusion-based edge detection. We will review several possible variants of the diffusion coefficient and discuss the associated positive and negative attributes.

To simplify the notation, we will denote the diffusion coefficient at location $\mathbf{x}$ by $c(\mathbf{x})$ in the continuous case. For the discrete-domain case, $c_d(\mathbf{x})$ represents the diffusion coefficient for direction d at location $\mathbf{x}$. Although the diffusion coefficients here are defined using $c(\mathbf{x})$ for the continuous case, the functions are equivalent in the discrete-domain case of $c_d(\mathbf{x})$. Typically $c(\mathbf{x})$ is a nonincreasing function of $|\nabla I(\mathbf{x})|$, the gradient magnitude at position $\mathbf{x}$. As such, we often refer to the diffusion coefficient as $c(|\nabla I(\mathbf{x})|)$. For small values of $|\nabla I(\mathbf{x})|$, $c(\mathbf{x})$ tends to unity. As $|\nabla I(\mathbf{x})|$ increases, $c(\mathbf{x})$ decreases to zero. Teboul *et al.* [38] establish three conditions for edge-preserving diffusion coefficients. These conditions are (1) $\lim_{|\nabla I(\mathbf{x})| \to 0} c(\mathbf{x}) = M$, where $0 < M < \infty$; (2) $\lim_{|\nabla I(\mathbf{x})| \to \infty} c(\mathbf{x}) = 0$; and (3) $c(\mathbf{x})$ is a strictly decreasing function of $|\nabla I(\mathbf{x})|$. Property 1 ensures isotropic smoothing in regions of similar intensity, while property 2 preserves edges. The third property is given in order to avoid numerical instability. Although most of the coefficients discussed here obey the first two properties, not all formulations obey the third property.

In [30], Perona and Malik propose

$$c(\mathbf{x}) = \exp\left\{ -\left[\frac{\nabla I(\mathbf{x})}{k} \right]^2 \right\}, \tag{9}$$

$$c(\mathbf{x}) = \frac{1}{1 + \left[\frac{\nabla I(\mathbf{x})}{k} \right]^2} \tag{10}$$

as diffusion coefficients. Diffusion operations using Eqs. (9) and (10) have the ability to sharpen edges (backward diffusion) and are inexpensive to compute. However, these diffusion coefficients are unable to remove heavy-tailed noise and create "staircase" artifacts [39, 44]. See the example of smoothing using Eq. (9) on the noisy image in Fig. 1(a), producing the result in Fig. 1(b). In this case, the anisotropic diffusion operation leaves several outliers in the resultant image. A similar problem is observed in Fig. 2(b), using the corrupted image in Fig. 2(a) as input. You *et al.* have also shown that diffusion algorithms using Eqs. (9) and (10) are ill posed — a small perturbation in the data may cause a significant change in the final result [43].

The inability of anisotropic diffusion to denoise an image has been addressed by Catte *et al.* [15] and Alvarez *et al.* [8]. Their regularized diffusion operation uses a modification of the gradient image used to compute the diffusion coefficients. In this case, a Gaussian-convolved version of the image is employed in computing diffusion coefficients. Using the same basic form as Eq. (9), we have

$$c(\mathbf{x}) = \exp\left\{ -\left[\frac{\nabla S(\mathbf{x})}{k} \right]^2 \right\}, \tag{11}$$

where $\mathbf{S}$ is the convolution of $\mathbf{I}$ and a Gaussian filter with

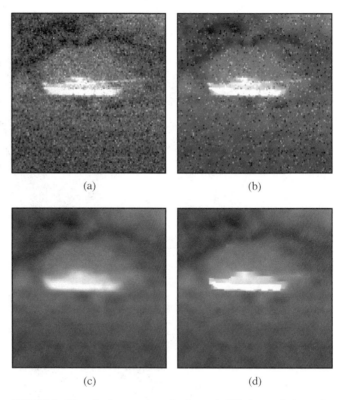

FIGURE 1 Three implementations of anisotropic diffusion applied to an infrared image of a tank: (a) original noisy image. (b) Results obtained using anisotropic diffusion with Eq. (9). (c) Results obtained using traditional modified gradient anisotropic diffusion with Eqs. (11) and (12). (d) Results obtained using morphological anisotropic diffusion with Eqs. (11) and (13).

standard deviation σ:

$$\mathbf{S} = \mathbf{I} * \mathbf{G}_\sigma. \tag{12}$$

This method can be used to rapidly eliminate noise in the image as shown in Fig. 1(c). The diffusion is also well posed and converges to a unique result, under certain conditions [15]. Drawbacks of this diffusion coefficient implementation include the additional computational burden of filtering at each step and the introduction of a linear filter into the edge-preserving anisotropic diffusion approach. The loss of sharpness due to the linear filter is evident in Fig. 2(c). Although the noise is eradicated, the edges are softened and blotching artifacts appear in the background of this example result.

Another modified gradient implementation, called *morphological anisotropic diffusion*, can be formed by substituting

$$\mathbf{S} = (\mathbf{I} \circ \mathbf{B}) \cdot \mathbf{B} \tag{13}$$

into Eq. (11), where $\mathbf{B}$ is a structuring element of size $m \times m$, $\mathbf{I} \circ \mathbf{B}$ is the morphological opening of $\mathbf{I}$ by $\mathbf{B}$, and $\mathbf{I} \cdot \mathbf{B}$ is the morphological closing of $\mathbf{I}$ by $\mathbf{B}$. In [36], the open–close and close–open filters were used in an alternating manner between iterations, thus reducing gray-scale bias of the open–close and close–open filters. As the result in Fig. 1(d) demonstrates, the

(a)

(b)   (c)

(d)   (e)

FIGURE 2   (a) Corrupted "cameraman" image (Laplacian noise, SNR = 13 dB) used as input for results in (b)–(e); (b) after eight iterations of anisotropic diffusion with Eq. (9), $k = 25$; (c) after eight iterations of anisotropic diffusion with Eqs. (11) and (12), $k = 25$; (d) after 75 iterations of anisotropic diffusion with Eq. (14), $T = 6$, $e = 1$, $p = 0.5$; (e) after 15 iterations of multigrid anisotropic diffusion with Eqs. (11) and (12), $k = 6$ [1].

morphological anisotropic diffusion method can be used to eliminate noise and insignificant features while preserving edges. Morphological anisotropic diffusion has the advantage of selecting feature scale (by specifying the structuring element **B**) and selecting the gradient magnitude threshold, whereas previous anisotropic diffusions, such as with Eqs. (9) and (10), only allowed selection of the gradient magnitude threshold. For this reason, we call anisotropic diffusion with Eqs. (11) and either (12) or (13) "scale aware" diffusion. Obviously, the need to specify feature scale is important in an edge-based application.

You *et al.* introduce the following diffusion coefficient in [43].

$$c(\mathbf{x}) = \begin{cases} 1/T + p(T+\varepsilon)^{p-1}/T, & \nabla I(\mathbf{x}) < T \\ 1/|\nabla I(\mathbf{x}) + p(|\nabla I(\mathbf{x})| \\ \quad +\varepsilon)^{p-1}/|\nabla I(\mathbf{x})|, & |\nabla I(\mathbf{x})| \geq T, \end{cases} \quad (14)$$

where the parameters are constrained by $\varepsilon > 0$ and $0 < p < 1$. Here T is a threshold on the gradient magnitude, similar to k in Eq. (9). This approach has the benefits of avoiding staircase artifacts and removing impulse noise. The main drawback is computational expense. As seen in Fig. 2(d), anisotropic diffusion with this diffusion coefficient succeeds in removing noise and retaining important features from Fig. 2(a) but requires a significant number of updates.

The diffusion coefficient

$$c(\mathbf{x}) = \frac{1}{|\nabla I(\mathbf{x})|} \quad (15)$$

is used in mean curvature motion formulations of diffusion [33], shock filters [28], and in locally monotonic diffusion [2]. One may notice that this diffusion coefficient is parameter free.

Designing a diffusion coefficient with robust statistics, Black *et al.* [9] model anisotropic diffusion as a robust estimation procedure that finds a piecewise smooth representation of an input image. A diffusion coefficient that utilizes the Tukey's biweight norm is given by

$$c(\mathbf{x}) = \frac{1}{2}\left\{ 1 - \left[\frac{\nabla I(\mathbf{x})}{\sigma}\right]^2 \right\}^2 \quad (16)$$

for $|\nabla I(\mathbf{x})| \leq \sigma$ and is 0 otherwise. Here, the parameter σ represents scale. Where the standard anisotropic diffusion coefficient as in Eq. (9) continues to smooth over edges while iterating, the robust formulation of Eq. (16) preserves edges of a prescribed scale σ and effectively stops diffusion.

Here, seven important versions of the diffusion coefficient are given that involve tradeoffs between solution quality, solution expense, and convergence behavior. Other research in the diffusion area focuses on the diffusion PDE itself. The next section reveals significant modifications to the anisotropic diffusion PDE that

affect fidelity to the input image, edge quality, and convergence properties.

3.2 Diffusion PDE

In addition to the basic anisotropic diffusion PDE given in Section 2.2, other diffusion mechanisms may be employed to adaptively filter an image for edge detection. Nordstrom [27] used an additional term to maintain fidelity to the input image, to avoid the selection of a stopping time, and to avoid termination of the diffusion at a trivial solution, such as a constant image. This PDE is given by

$$\frac{\partial I_t(\mathbf{x})}{\partial t} - \text{div}\{c(\mathbf{x})\nabla I_t(\mathbf{x})\} = I_0(\mathbf{x}) - I_t(\mathbf{x}). \quad (17)$$

Obviously, the right-hand side $I_0(\mathbf{x}) - I_t(\mathbf{x})$ enforces an additional constraint that penalizes deviation from the input image.

Just as Canny [13] modified the Laplacian-of-Gaussian edge detection technique by detecting zero crossings of the Laplacian only in the direction of the gradient, a similar edge-sensitive approach can be taken with anisotropic diffusion. Here, the boundary-preserving diffusion is executed only in the direction orthogonal to the gradient direction, whereas the standard anisotropic diffusion schemes *impede* diffusion across the edge. If the rate of change of intensity is set proportional to the second partial derivative in the direction orthogonal to the gradient (called τ), we have

$$\frac{\partial I_t(\mathbf{x})}{\partial t} = \frac{\partial^2 I_t(\mathbf{x})}{\partial \tau^2} = |\nabla I_t(\mathbf{x})|\text{div}\left\{ \frac{\nabla I_t(\mathbf{x})}{|\nabla I_t(\mathbf{x})|} \right\}. \quad (18)$$

This anisotropic diffusion model is called *mean curvature motion*, because it induces a diffusion in which the connected components of the image level sets of the solution image move in proportion to the boundary mean curvature. Several effective edge-preserving diffusion methods have arisen from this framework including [17] and [29]. Alvarez *et al.* [8] have used the mean curvature method in tandem with the regularized diffusion coefficient of Eqs. (11) and (12). The result is a processing method that preserves the causality of edges through scale space. For edge-based hierarchical searches and multiscale analyses, the edge causality property is extremely important.

The mean curvature method has also been given a graph theoretic interpretation [37, 42]. Yezzi [42] treats the image as a graph in $\Re^n$ — a typical 2-D gray-scale image would be a surface in $\Re^3$ where the image intensity is the third parameter, and each pixel is a graph node. Hence, a color image could be considered a surface in $\Re^5$. The curvature motion of the graphs can be used as a model for smoothing and edge detection. For example, let a 3-D graph **s** be defined by $\mathbf{s}(\mathbf{x}) = \mathbf{s}(x, y) = [x, y, I(x, y)]$ for the 2-D image **I** with $\mathbf{x} = (x, y)$. As a way to implement mean curvature motion on this graph, the PDE is given by

$$\frac{\partial \mathbf{s}(\mathbf{x})}{\partial t} = h(\mathbf{x})\mathbf{n}(\mathbf{x}), \quad (19)$$

where $h(\mathbf{x})$ is the mean curvature,

$$h(x, y)$$
$$= \frac{\partial^2 I(x, y)/\partial x^2[1 + (\partial I(x, y)/(\partial y)^2] - 2(\partial I(x, y)/\partial x)(\partial I(x, y)/\partial y)(\partial^2 I(x, y)/\partial x \partial y) + \partial^2 I(x, y)/\partial y^2[1 + (\partial I(x, y)/\partial x)^2]}{2[1 + (\partial I(x, y)/\partial y)^2 + (\partial I(x, y)/\partial y)^2]^{3/2}},$$

$$(20)$$

and $\mathbf{n}(\mathbf{x})$ is the unit normal of the surface:

$$\mathbf{n}(x, y) = \frac{[-\partial I(x, y)/\partial x, -\partial I(x, y)/\partial x, 1]}{\sqrt{1 + (\partial I(x, y)/\partial y)^2 + (\partial I(x, y)/\partial y)^2}}. \quad (21)$$

For a discrete (programmable) implementation, the partial derivatives of $I(x, y)$ may be approximated by using simple differences. One-sided differences or central differences may be employed. For example, a one-sided difference approximation for $\partial I(x, y)/\partial x$ is $I(x + 1, y) - I(x, y)$. A central difference approximation for the same partial derivative is given by $^1/_2[I(x + 1, y) - I(x - 1, y)]$.

The standard mean curvature PDE of Eq. (19) has the drawback of edge movement that sacrifices edge sharpness. To remedy this movement, Yezzi used projected mean curvature vectors to perform diffusion. Let $\mathbf{z}$ denote the unit vector in the vertical (intensity) direction on the graph $\mathbf{s}$. The projected mean curvature diffusion PDE can be formed by

$$\frac{\partial \mathbf{s}(\mathbf{x})}{\partial t} = \{[h(\mathbf{x})\mathbf{n}(\mathbf{x})] \cdot \mathbf{z}\}\mathbf{z}. \quad (22)$$

The PDE for updating image intensity is then

$$\frac{\partial I(\mathbf{x})}{\partial t}$$
$$= \frac{\Delta I(x, y) + k^2[(\partial I(x, y)/\partial x)^2(\partial^2 I(x, y)/\partial y^2) - 2(\partial I(x, y)/\partial x)(\partial I(x, y)/\partial y)(\partial^2 I(x, y)/\partial x \partial y) + (\partial I(x, y)/\partial y)^2(\partial^2 I(x, y)/\partial x^2)]}{\{1 + k^2[\nabla I(x, y)]^2\}^2}$$

$$(23)$$

where k scales the intensity variable. When k is zero, we have isotropic diffusion, and when k becomes larger, we have a damped geometric heat equation that preserves edges but diffuses more slowly. The projected mean curvature PDE gives edge preservation through scale space.

Another anisotropic diffusion technique leads to locally monotonic signals [2, 3]. Unlike previous diffusion techniques that diverge or converge to trivial signals, locally monotonic (LOMO) diffusion converges rapidly to well-defined LOMO signals of the desired degree — a signal is locally monotonic of degree d (LOMO-d) if each interval of length d is nonincreasing or nondecreasing. The property of local monotonicity allows both slow and rapid signal transitions (ramp and step edges) while excluding outliers due to noise. The degree of local monotonicity defines the signal scale. In contrast to other diffusion methods,

LOMO diffusion does not require an additional regularization step to process a noisy signal and uses no thresholds or *ad hoc* parameters.

On a 1-D signal, the basic LOMO diffusion operation is defined by Eq. (6) with $\Gamma = 2$ and using the diffusion coefficient Eq. (15), yielding

$$[I(x)]_{t+1} \leftarrow (I(x) + (1/2)\{\text{sgn}[\nabla I_1(x)] + \text{sgn}[\nabla I_2(x)]\})_t, \quad (24)$$

where a time step of $\Delta T = 1/2$ is used. Equation (24) is modified for the case in which the simple difference $\nabla I_1(x)$ or $\nabla I_2(x)$ is zero. Let $\nabla I_1(x) \leftarrow -\nabla I_2(x)$ in the case of $\nabla I_1(x) = 0$; $\nabla I_2(x) \leftarrow -\nabla I_1(x)$ when $\nabla I_2(x) = 0$. Let the fixed point of Eq. (24) be defined as $\text{ld}(\mathbf{I}, h_1, h_2)$, where h_1 and h_2 are the sample spacings used to compute the simple differences $\nabla I_1(x)$ and $\nabla I_2(x)$, respectively. Let $\text{ld}_d(\mathbf{I})$ denote the LOMO diffusion sequence that gives a LOMO-d (or greater) signal from input $\mathbf{I}$. For odd values of $d = 2m + 1$,

$$\text{ld}_d(\mathbf{I}) = \text{ld}(\dots \text{ld}(\text{ld}(\text{ld}(\mathbf{I}, m, m), m - 1, m),$$
$$m - 1, m - 1)\dots, 1, 1). \quad (25)$$

In Eq. (25), the process commences with $\text{ld}(\mathbf{I}, m, m)$ and continues with spacings of decreasing widths until $\text{ld}(\mathbf{I}, 1, 1)$ is implemented. For even values of $d = 2m$, the sequence of operations is similar:

$$\text{ld}_d(\mathbf{I}) = \text{ld}(\dots \text{ld}(\text{ld}(\text{ld}(\mathbf{I}, m - 1, m), m - 1, m - 1),$$
$$m - 2, m - 1)\dots, 1, 1). \quad (26)$$

For this method to be extended to two dimensions, the same procedure may be followed using Eq. (6) with $\Gamma = 4$[2]. Another possibility is diffusing orthogonal to the gradient direction at each point in the image, using the 1-D LOMO diffusion. Examples of 2-D LOMO diffusion and the associated edge detection results are given in Section 4.

3.3 Variational Formulation

The diffusion PDEs discussed thus far may be considered numerical methods that attempt to minimize a cost or energy functional. Energy-based approaches to diffusion have been effective for edge detection and image segmentation. Morel and Solimini [25] give an excellent overview of the variational methods. Isotropic diffusion by means of the heat diffusion equation leads to a minimization of the following energy:

$$E(\mathbf{I}) = \int_\Omega |\nabla I(\mathbf{x})|^2 \, d\mathbf{x}. \qquad (27)$$

If a diffusion process is applied to an image as in Eq. (4), the intermediate solutions may be considered a descent on

$$E(\mathbf{I}) = \lambda^2 \int_\Omega |\nabla I(\mathbf{x})|^2 \, d\mathbf{x} + \int_\Omega [I(\mathbf{x}) - I_0(\mathbf{x})]^2 \, d\mathbf{x}, \qquad (28)$$

where the regularization parameter λ denotes scale [25].

Likewise, anisotropic diffusion has a variational formulation. The energy associated with the Perona and Malik diffusion is

$$E(\mathbf{I}) = \lambda^2 \int_\Omega C[|\nabla I(\mathbf{x})|^2] \, d\mathbf{x} + \int_\Omega [I(\mathbf{x}) - I_0(\mathbf{x})]^2 \, d\mathbf{x}, \qquad (29)$$

where C is the integral of $c'(\mathbf{x})$ with respect to the independent variable $|\nabla I(\mathbf{x})|^2$. Here, $c'(\mathbf{x})$, as a function of $|\nabla I(\mathbf{x})|^2$, is equivalent to the diffusion coefficient $c(\mathbf{x})$ as a function of $|\nabla I(\mathbf{x})|$, so $c'(|\nabla I(\mathbf{x})|^2) = c(|\nabla I(\mathbf{x})|)$. The Nordstrom [27] diffusion PDE, Eq. (17), is a steepest descent on this energy functional.

Recently, Teboul *et al.* have introduced a variational method that preserves edges and is useful for edge detection. In their approach, image enhancement and edge preservation are treated as two separate processes. The energy functional is given by

$$E(\mathbf{I}, \mathbf{e}) = \lambda^2 \int_\Omega [e(\mathbf{x})^2 |\nabla I(\mathbf{x})|^2 + k(e(\mathbf{x}) - 1)^2] \, d\mathbf{x}$$
$$+ \frac{\alpha^2}{k} \int_\Omega \varphi(|\nabla e(\mathbf{x})|) \, d\mathbf{x} + \int_\Omega [I(\mathbf{x}) - I_0(\mathbf{x})]^2 \, d\mathbf{x}, \qquad (30)$$

where the real-valued variable $e(\mathbf{x})$ is the edge strength at $\mathbf{x}$, and $e(\mathbf{x}) \in [0, 1]$. In Eq. (30), the diffusion coefficient is defined by $c(|\nabla I(\mathbf{x})|) = \varphi'(|\nabla I(\mathbf{x})|)/2(|\nabla I(\mathbf{x})|)$. An additional regularization parameter α is needed, and k is essentially an edge threshold parameter.

The energy functional in Eq. (30) leads to a system of two coupled PDEs:

$$I_0(\mathbf{x}) - I_t(\mathbf{x}) - \lambda^2 \text{div}\{e(\mathbf{x})[\nabla I_t(\mathbf{x})]\nabla I_t(x)\} = 0, \qquad (31)$$

$$e(\mathbf{x})\left[\frac{|\nabla I(\mathbf{x})|^2}{k} + 1\right] - 1 + \frac{\alpha^2}{k^2}\text{div}[c(|\nabla e(\mathbf{x})|)\nabla e(\mathbf{x})] = 0. \qquad (32)$$

The coupled PDEs have the advantage of edge preservation within the adaptive smoothing process. An edge map can be directly extracted from the final state of $\mathbf{e}$.

This edge-preserving variational method is related to the segmentation approach of Mumford and Shah [26]. The energy functional to be minimized is

$$E(\mathbf{I}) = \lambda^2 \int_{\Omega \backslash \Psi} |\nabla I(\mathbf{x})|^2 \, d\mathbf{x} + \int_{\Omega \backslash \Psi} [I(\mathbf{x}) - I_0(\mathbf{x})]^2 \, d\mathbf{x}$$
$$+ \mu\lambda^2 \int_\Psi d\psi, \qquad (33)$$

where $\int_\Psi d\psi$ is the integrated length of the edges (Hausdorff measure), $\Omega \backslash \Psi$ is the set of image locations that exclude the edge positions, and μ is additional weight parameter. The additional edge-length term reflects the goal of computing a minimal edge map for a given scale λ. The Mumford–Shah functional has spurred several variational image segmentation schemes, including PDE-based solutions [25].

In edge detection, thin, contiguous edges are typically desired. With diffusion-based edge detectors, the edges may be "thick" or "broken" when a gradient magnitude threshold is applied after diffusion. The variational formulation allows the addition of additional constraints that promote edge thinning and connectivity. Black *et al.* used two additional terms, a hysteresis term for improved connectivity and a nonmaximum suppression term for thinning [9]. A similar approach was taken in [6]. The additional terms allow the effective extraction of spatially coherent outliers. This idea is also found in the design of line processes for regularization [18].

3.4 Multiresolution Diffusion

One drawback of diffusion-based edge detection is the computational expense. Typically, a large number (anywhere from 20 to 200) of iterative steps are needed to provide a high-quality edge map. One solution to this dilemma is the use of multiresolution schemes. Two such approaches have been investigated for edge detection: the anisotropic diffusion pyramid and multigrid anisotropic diffusion.

In the case of isotropic diffusion, the Gaussian pyramid has been used for edge detection and image segmentation [11, 12]. The basic idea is that the scale generating operator (a Gaussian filter, for example) can be used as an anti-aliasing filter before sampling. Then, a set of image representations of increasing scale and decreasing resolution (in terms of the number of pixels) can be generated. This image pyramid can be used for hierarchical searches and coarse-to-fine edge detection.

The anisotropic diffusion pyramids are born from the same fundamental motivation as their isotropic, linear counterparts. However, with a nonlinear scale-generating operator, the pre-sampling operation is constrained morphologically, not by the traditional sampling theorem. In the nonlinear case, the scale-generating operator should remove image features not supported in the subsampled domain. Therefore, morphological methods [24] for creating image pyramids have also been used in conjunction with the morphological sampling theorem [20].

To create level $L + 1$ of an anisotropic diffusion pyramid, one may do the following.

1. Perform ν diffusion steps on level L, starting with level 0, the original image.
2. Retain 1 of S samples from each row and column.

The filtering and subsampling operations are halted when the number of pixels in a row or column is smaller than S, or when a desired *root level* is attained. The root level represents the coarsest pyramid level that contains the features of interest. With the use of the morphological diffusion coefficient in Eq. (11) with Eq. (13), the number of diffusion steps ν performed on each pyramid level may be prescribed in order to remove all level set objects that are smaller (in terms of minimum diameter or width) than the sampling factor interval S [35].

For edge detection, one may implement coarse-to-fine edge detection by first detecting edges on the root level. On each descending pyramid level, causality is exploited where "children" can become edges only if the "parent" is an edge, in which case the child–parent relationship is defined through sampling. A superior method of edge detection and segmentation is achieved by means of pyramid node linking [11]. In this paradigm, each pixel on the original (or retinal) level is linked to a potential parent on the next ascending level by intensity similarity. This linking continues for each level until the root level is reached. Then, the root level values are propagated back to the original image level, and a segmentation is achieved. Edges are defined as the boundaries between the associated root level values on the original image. In this framework, step edges are sharpened and processing costs are decreased [4, 5]. Figure 3 provides an example of multiresolution anisotropic diffusion for edge detection. With the use of Fig. 3(a) as input, fixed-resolution anisotropic diffusion with Eq. (9) [see Fig. 3(b)] and pyramidal anisotropic diffusion with Eq. (9) [see Fig. 3(d)] are applied to form a segmented image. The fixed-resolution diffusion leads to the noisy edge map in Fig. 3(c) that requires thinning. Thin, contiguous contours that reflect the boundaries of the large scale objects are given in the edge map generated from the multiresolution approach [see Fig. 3(e)]. Another example of the edge-enhancing ability of the anisotropic diffusion pyramid is given in Fig. 4. The infrared image of the space shuttle produces thick, poorly localized edges when fixed-resolution anisotropic diffusion is applied [see Fig. 4(b)]. The multiresolution result yields a thin, contiguous contour suitable for edge-based object recognition and tracking [see Fig. 4(c)].

The anisotropic diffusion pyramids are, in a way, *ad hoc* multigrid schemes. A multigrid scheme can be useful for diffusion-based edge detectors in two ways. First, like the anisotropic diffusion pyramids, the number of diffusion updates may be decreased. Second, the multigrid approach can be used to eliminate low-frequency errors. The anisotropic diffusion PDEs are stiff; they rapidly reduce high-frequency errors (noise, small details), but they slowly reduce background variations and often create artifacts such as blotches (false regions) or staircases (false step edges). See Fig. 5 for an example of a staircasing artifact.

To implement a multigrid anisotropic diffusion operation [1], define $\mathbf{J}$ as an estimate of the image $\mathbf{I}$. A system of equations is defined by $A(\mathbf{I}) = 0$ where

$$[A(\mathbf{I})](\mathbf{x}) = (\Delta T) \sum_{d=1}^{\Gamma} c_d(\mathbf{x}) \nabla I_d(\mathbf{x}), \tag{34}$$

which is relaxed by the discrete anisotropic diffusion PDE, in Eq. (6). For this system of equations, the (unknown) algebraic error is $\mathbf{E} = \mathbf{I} - \mathbf{J}$, and the residual is $\mathbf{R} = -A(\mathbf{J})$ for image estimate $\mathbf{J}$. The residual equation $A(\mathbf{E}) = \mathbf{R}$ can be relaxed (diffused) in the same manner as Eq. (34), using Eq. (6) to form an estimate of the error.

The first step is performing ν diffusion steps on the original input image (level $L = 0$). Then, the residual equation at the coarser grid $L + 1$ is

$$A(\mathbf{E}_{L+1}) = -A[(\mathbf{J}_L)_{\downarrow S}], \tag{35}$$

where $\downarrow S$ represents downsampling by a factor of S. Now, residual Eq. (35) can be relaxed using the discrete diffusion PDE, Eq. (6), with an initial error estimate of $\mathbf{E}_{L+1} = 0$. The new error estimate $\mathbf{E}_{L+1}$ after relaxation can then be transferred to the finer grid to correct the initial image estimate $\mathbf{J}$ in a simple two-grid scheme. Or, the process of transferring the residual to coarser grids can be continued until a grid is reached in which a closed form solution is possible. Then, the error estimates are propagated back to the original grid.

Additional steps may be taken to account for the nonlinearity of the anisotropic diffusion PDE, such as implementing a full approximation scheme multigrid system, or by using a global linearization step in combination with a Newton method to solve for the error iteratively [10, 19].

The results of applying multigrid anisotropic diffusion are shown in Fig. 2(e). In just 15 updates, the multigrid anisotropic diffusion method was able to remove the noise from Fig. 2(b) while preserving the significant objects and avoiding the introduction of blotching artifacts.

3.5 Multispectral Anisotropic Diffusion

Color edge detection and boundary detection for multispectral imagery are important tasks in general image/video processing, remote sensing, and biomedical image processing. Applying

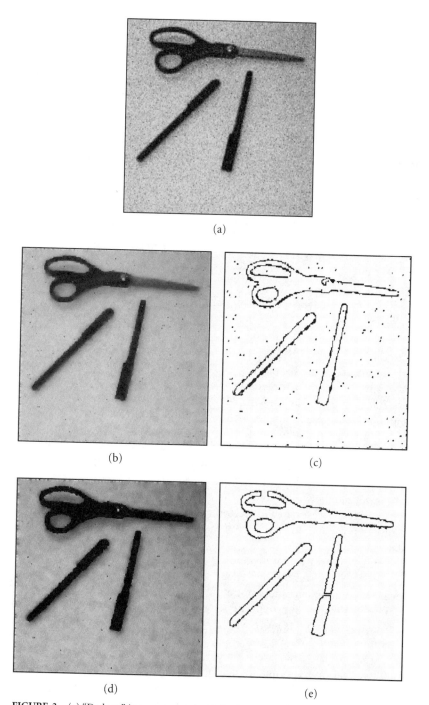

FIGURE 3 (a) "Desktop" image corrupted with Laplacian-distributed additive noise, SNR = 7.3 dB; (b) diffusion results using Eq. (9); (c) edges from result in (b); (d) multiresolution anisotropic diffusion pyramid segmentation; (e) edges from anisotropic diffusion pyramid segmentation in (d).

anisotropic diffusion to each channel or spectral band separately is one possible way of processing multichannel or multispectral image data. However, this single-band approach forfeits the richness of the multispectral data and provides individual edge maps that do not possess corresponding edges. Two solutions have

emerged for diffusing multispectral imagery. The first, called vector distance dissimilarity, utilizes a function of the gradients from each band to compute an overall diffusion coefficient. For example, to compute the diffusion coefficient in the "western" direction on an RGB color image, the following function could

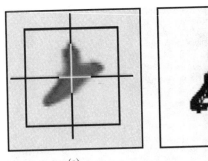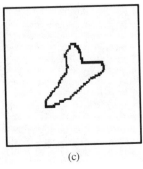

(a)   (b)   (c)

FIGURE 4 (a) IR image of shuttle with crosshairs marking the position as located by the anisotropic diffusion pyramid. (b) Edges found in (a) by thresholding the gradient magnitude. (c) Edges found after enhancement by the anisotropic diffusion pyramid.

be applied:

edge map in Fig. 6(e) shows improved resilience to impulse noise

$$\nabla I_1(\mathbf{x}) = \sqrt{[R(x - h_1, y) - R(x, y)]^2 + [G(x - h_1, y) - G(x, y)]^2 + [B(x - h_1, y) - B(x, y)]^2}, \tag{36}$$

where $R(\mathbf{x})$ is the red band intensity at $\mathbf{x}$, $G(\mathbf{x})$ is the green band, and $B(\mathbf{x})$ is the blue band. With the use of the vector distance dissimilarity method, the standard diffusion coefficients such as Eq. (9) can be employed. This technique was used in [40] for shape-based processing and in [7] for processing remotely sensed imagery. An example of multispectral anisotropic diffusion is shown in Fig. 6. Using the noisy multispectral image in Fig. 6(a) as input, the vector distance dissimilarity method produces the smoothed result shown in Fig. 6(b), which has an associated image of gradient magnitude shown in Fig. 6(c). As can be witnessed in Fig. 6(c), an edge detector based on vector distance dissimilarity is sensitive to noise and does not identify the important image boundaries.

The second method uses mean curvature motion and a multispectral gradient formula to achieve anisotropic, edge-preserving diffusion. The idea behind mean curvature motion, as discussed earlier, is to diffuse in the direction opposite to the gradient such that the image level set objects move with a rate in proportion to their mean curvature. With a gray-scale image, the gradient is always perpendicular to the level set objects of the image. In the multispectral case, this quality does not hold. A well-motivated diffusion is defined by Sapiro and Ringach [34], using DiZenzo's multispectral gradient formula [16]. In Fig. 6(d), results for multispectral anisotropic diffusion are shown for the mean curvature approach of [34] used in combination with the modified gradient approach of [15]. The

over the vector distance dissimilarity approach.

The implementation issues connected with anisotropic diffusion include specification of the diffusion coefficient and diffusion PDE, as discussed earlier. The anisotropic diffusion method can be expedited through multiresolution implementations. Furthermore, anisotropic diffusion can be extended to color and multispectral imagery. In the following section, we discuss the specific application of anisotropic diffusion to edge detection.

4 Application of Anisotropic Diffusion to Edge Detection

4.1 Edge Detection by Thresholding

Once anisotropic diffusion has been applied to an image $\mathbf{I}$, a procedure has to be defined to extract the image edges $\mathbf{e}$. The most typical procedure is to simply define a gradient magnitude threshold, T, that defines the location of an edge:

$$e(\mathbf{x}) = 1 \text{ if } |\nabla I(x)| > T, \tag{37}$$

and $e(\mathbf{x}) = 0$ otherwise. Of course, the question becomes one of a selecting a proper value for T. With typical diffusion coefficients such as those of Eqs. (9) and (10), $T = k$ is often asserted.

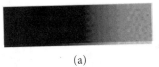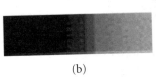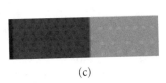

(a)   (b)   (c)

FIGURE 5 (a) Sigmoidal ramp edge; (b) after anisotropic diffusion with Eq. (9) ($k = 10$); (c) after multigrid anisotropic diffusion with Eq. (9) ($k = 10$) [1].

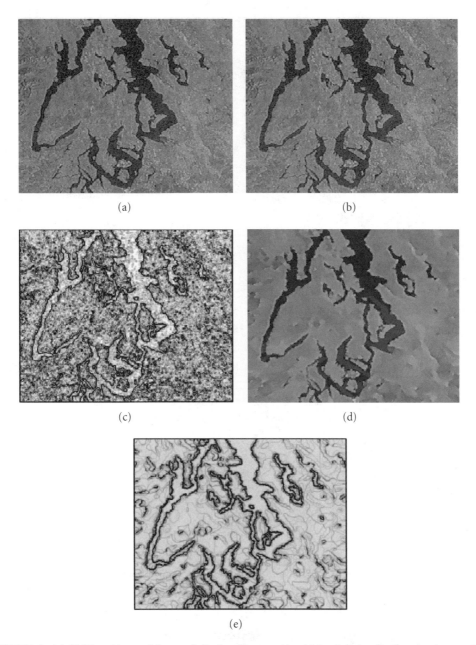

FIGURE 6 (a) SPOT multispectral image of the Seattle area, with additive Gaussian-distributed noise, $\sigma = 10$; (b) vector distance dissimilarity diffusion result, using the diffusion coefficient in Eq. (9); (c) edges (gradient magnitude) from the result in (b); (d) mean curvature motion [Eq. (18)] result using diffusion coefficients from Eqs. (11) and (12); (e) edges (gradient magnitude) from the result in (d). (See color section, p. C–22.)

In [9], edges are detected by finding gradient magnitudes that exceed the robust scale $T = \sigma_e$ of the image, as defined in [32]:

$$\sigma_e = 1.4826 \, \text{med}\{|\nabla I(\mathbf{x}) - \text{med}(|\nabla I(\mathbf{x})|)|\} \quad (38)$$

where the med operator, in Eq. (38), is the median performed over the entire image domain Ω. The constant used in Eq. (38) is derived from the mean absolute deviation of the normal distribution with unit variance [9].

4.2 Edge Detection From Image Features

Aside from thresholding the gradient magnitude of a diffusion result, a feature detection approach may be used. As with Marr's classical LoG detector, the inflection points of a diffused image may be located by finding the zero crossing in a Laplacian-convolved result. However, if the anisotropic diffusion operation produces piecewise constant images as in [9] and [43], the gradient magnitude is sufficient to define thin, contiguous edges.

With locally monotonic diffusion, other features that appear in the diffused image may be used for edge detection. An advantage of locally monotonic diffusion is that no threshold is required for edge detection. Locally monotonic diffusion segments each row and column of the image into ramp segments and constant segments. Within this framework, we can define *concave-down*, *concave-up*, and *ramp center* edge detection processes. Consider an image row or column. With a concave-down edge detection, the ascending (increasing intensity) segments mark the beginning of an object and the descending (decreasing intensity) segments terminate the object. With a concave-up edge detection, negative-going objects (in intensity) are detected. The ramp center edge detection sets the boundary points at the centers of the ramp edges, as the name implies. When no bias toward bright or dark objects is inferred, a ramp center edge detection can be utilized.

Figure 7 provides two examples of feature-based edge detection using locally monotonic diffusion. The images in Fig. 7(b) and Fig. 7(e) are the results of applying 2-D locally monotonic diffusion to Fig. 7(a) and Fig. 7(d), respectively. The concave-up edge detection given in Fig. 7(c) reveals the boundaries of the blood cells. In Fig. 7(f), a ramp center edge detection is used to find the boundaries between the aluminum grains of Fig. 7(d).

4.3 Quantitative Evaluation of Edge Detection by Anisotropic Diffusion

When choosing a suitable anisotropic diffusion process for edge detection, one may evaluate the results qualitatively or use an objective measure. Three such quantitative assessment tools include the percentage of edges correctly identified as edges, the percentage of false edges, and Pratt's edge quality metric. Given ground truth edge information, usually with synthetic data, one may measure the correlation between the ideal edge map and the computed edge map. This correlation leads to a classification of "correct" edges (in which the computed edge map and ideal version match) and "false" edges. Another method utilizes Pratt's edge quality measurement [31]:

$$F = \frac{\sum_{i=1}^{I_A} \frac{1}{1+\alpha(d(i)^2)}}{\max\{I_A, I_I\}}, \qquad (39)$$

where I_A is the number of edge pixels detected in the diffused image result, I_I is the number of edge pixels existing in the original, noise-free imagery, $d(i)$ is the Euclidean distance between an edge location in the original image and the nearest detected edge, and α is a scaling constant, with a suggested value of 1/9 [31]. A "perfect" edge detection result has value $F = 1$ in Eq. (39).

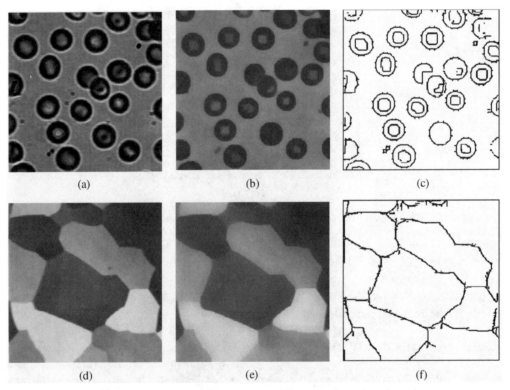

(a)   (b)   (c)

(d)   (e)   (f)

FIGURE 7 (a) Original "blood cells" image; (b) 2-D LOMO-3 diffusion result; (c) boundaries from concave-up segmentation of image in (b); (d) original "aluminum grains" image; (e) 2-D LOMO-3 diffusion result; (f) boundaries from ramp center segmentation of image in (e).

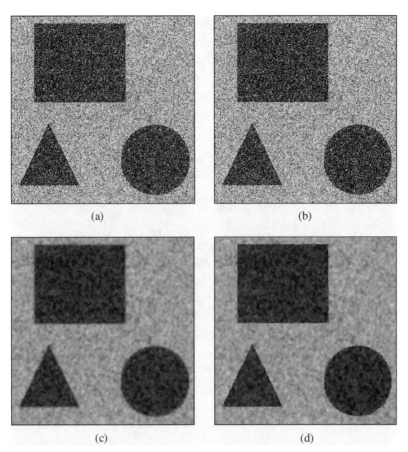

FIGURE 8 Three implementations of anisotropic diffusion applied to synthetic imagery: (a) original image corrupted with 40% salt-and-pepper noise; (b) results obtained using original anisotropic diffusion with Eq. (9); (c) results obtained using modified gradient anisotropic diffusion with Eqs. (11) and (12); (d) results obtained using morphological anisotropic diffusion with Eqs. (11) and (13) [36].

An example is given here in which a synthetic image is corrupted by 40% salt-and-pepper noise (Fig. 8). Three versions of anisotropic diffusion are implemented on the noisy imagery using the diffusion coefficients from Eqs. (9), from (11) and (12), and from (11) and (13). The threshold of the edge detector was defined to be equal to the gradient threshold of the diffusion coefficient, $T = k$. The results of the numerical experiment are presented in Fig. 9 for several solution times. It may be seen that the modified gradient coefficient [Eqs. (11) and (12)] initially outperforms the other diffusion methods in the edge quality measurement, but it produces the poorest identification percentage (because of the edge localization errors associated with the Gaussian filter). The morphological anisotropic diffusion method [Eqs. (11) and (13)] provides significant performance improvement, providing a 70% identification of true edges and a Pratt quality measurement of 0.95.

In summary, edges may be extracted from a diffused image by applying a heuristically selected threshold, by using a statistically motivated threshold, or by identifying features in the processed imagery. The success of the edge detection method can be evaluated qualitatively by visual inspection or quantitatively with edge quality metrics.

5 Conclusions and Future Research

Anisotropic diffusion is an effective precursor to edge detection. The main benefit of anisotropic diffusion over isotropic diffusion and linear filtering is edge preservation. By the proper specification of the diffusion PDE and the diffusion coefficient, an image can be scaled, denoised, and simplified for boundary detection. For edge detection, the most critical design step is specification of the diffusion coefficient. The variants of the diffusion coefficient involve tradeoffs between sensitivity to noise, the ability to specify scale, convergence issues, and computational cost. Different implementations of the anisotropic diffusion PDE result in improved fidelity to the original image, mean curvature motion, and convergence to locally monotonic signals. As the diffusion

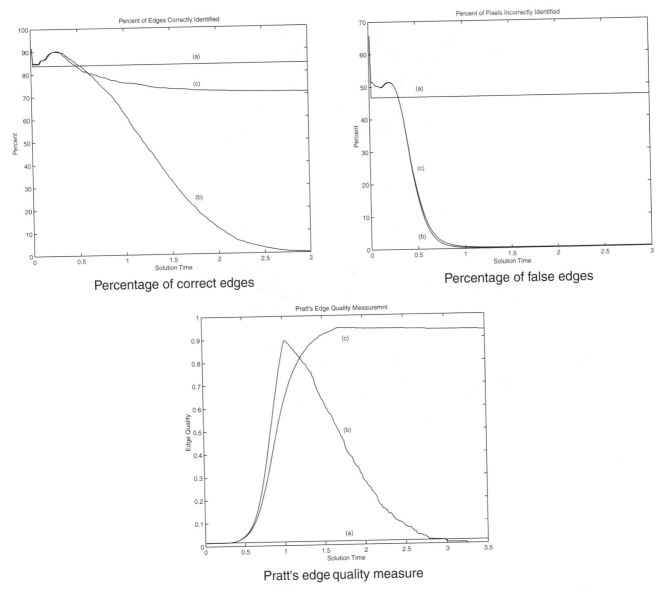

Percentage of correct edges

Percentage of false edges

Pratt's edge quality measure

FIGURE 9 Edge detector performance vs. diffusion time for the results shown in Fig. 8. In the graphs, curves (a) correspond to anisotropic diffusion with Eq. (9), curves (b) correspond to Eqs. (11) and (12), and curves (c) correspond to Eqs. (11) and (13) [36].

PDE may be considered a descent on an energy surface, the diffusion operation can be viewed in a variational framework. Recent variational solutions produce optimized edge maps and image segmentations in which certain edge-based features, such as edge length, curvature, thickness, and connectivity, can be optimized.

The computational cost of anisotropic diffusion may be reduced by using multiresolution solutions, including the anisotropic diffusion pyramid and multigrid anisotropic diffusion. The application of edge detection to color or other multispectral imagery is possible through techniques presented in the literature. In general, the edge detection step after anisotropic diffusion of the image is straightforward. Edges may be detected by using a simple gradient magnitude threshold, robust statistics, or a feature extraction technique.

Research in the area of PDEs and diffusion techniques for image and video processing continues. Important issues include the extension of discrete diffusion methods to multiple dimensions, differential morphology, and specialized hardware for PDE-based processing. With the edge preserving and scale-generating attributes, anisotropic diffusion methods have a promising future in application to image and video analysis tasks such as content-based retrieval, video tracking, and object recognition.

References

[1] S. T. Acton, "Multigrid anisotropic diffusion," *IEEE Trans. Image Process.* **7**, 280–291 (1998).

[2] S. T. Acton, "A PDE technique for generating locally monotonic

images," presented at the IEEE International Conference on Image Processing, Chicago, October 4–7, 1998.

[3] S. T. Acton, "Anisotropic diffusion and local monotonicity," presented at the IEEE International Conference on Acoustics, Speech and Signal Processing, Seattle, May 12–15, 1998.

[4] S. T. Acton, "A pyramidal edge detector based on anisotropic diffusion," presented at the IEEE International Conference on Acoustics, Speech and Signal Processing, Atlanta, May 7–10, 1996.

[5] S. T. Acton, A. C. Bovik, and M. M. Crawford, "Anisotropic diffusion pyramids for image segmentation," presented at the IEEE International Conference on Image Processing, Austin, Texas, Nov. 1994.

[6] S. T. Acton and A. C. Bovik, "Anisotropic edge detection using mean field annealing," presented at the IEEE International Conference on Acoustics, Speech and Signal Processing, San Francisco, March 23–26, 1992.

[7] S. T. Acton and J. Landis, "Multispectral anisotropic diffusion," *Int. J. Remote Sensing* **18**, 2877–2886 (1997).

[8] L. Alvarez, P.-L. Lions, and J.-M. Morel, "Image selective smoothing and edge detection by nonlinear diffusion II," *SIAM J. Numer. Anal.* **29**, 845–866 (1992).

[9] M. J. Black, G. Sapiro, D. H. Marimont, and D. Heeger, "Robust anisotropic diffusion" *IEEE Trans. Image Process.* **7**, 421–432 (1998).

[10] J. H. Bramble, *Multigrid Methods* (Wiley, New York, 1993).

[11] P. J. Burt, T. Hong, and A. Rosenfeld, "Segmentation and estimation of region properties through cooperative hierarchical computation," *IEEE Trans. Systems, Man, Cybernet.* **11**, 802–809 (1981).

[12] P. J. Burt, "Smart sensing within a pyramid vision machine," *Proc. IEEE* **76**, 1006–1015 (1988).

[13] J. Canny, "A computational approach to edge detection," *IEEE Trans. Pattern Anal. Mach. Intell.* **PAMI-8**, 679–698 (1986).

[14] V. Caselles, J.-M. Morel, G. Sapiro, and A. Tannenbaum, "Introduction to the special issue on partial differential equations and geometry-driven diffusion in image processing and analysis," *IEEE Trans. Image Process.* **7**, 269–273 (1998).

[15] F. Catte, P.-L. Lions, J.-M. Morel, and T. Coll, "Image selective smoothing and edge detection by nonlinear diffusion," *SIAM J. Numer. Anal.* **29**, 182–193 (1992).

[16] S. DiZenzo, "A note on the gradient of a multi-image," *Comput. Vis. Graph. Image Process.* **33**, 116–125 (1986).

[17] A. El-Fallah and G. Ford, "The evolution of mean curvature in image filtering," presented at the IEEE International Conference on Image Processing, Austin, TX, November 13–16, 1994.

[18] D. Geman and G. Reynolds, "Constrained restoration and the recovery of discontinuities," *IEEE Trans. Pattern Anal. Machine Intell.* **14**, 376–383 (1992).

[19] W. Hackbush and U. Trottenberg, eds. *Multigrid Methods* (Springer-Verlag, New York, 1982).

[20] R. M. Haralick, X. Zhuang, C. Lin, and J. S. J. Lee, "The digital morphological sampling theorem," *IEEE Trans. Acous. Speech Signal Process.* **37**, 2067–2090 (1989).

[21] J. J. Koenderink, "The structure of images," *Biol. Cybern.* **50**, 363–370 (1984).

[22] D. G. Lowe, *Perceptual Organization and Visual Recognition* (Kluwer, New York, 1985).

[23] D. Marr and E. Hildreth, "Theory of edge detection," *Proc. Roy. Soc. London* **B207**, 187–217 (1980).

[24] A. Morales, R. Acharya, and S. Ko, "Morphological pyramids with alternating sequential filters," *IEEE Trans. Image Process.* **4**, 965–977 (1996).

[25] J.-L. Morel and S. Solimini, *Variational Methods in Image Segmentation* (Birkhauser, Boston, 1995).

[26] D. Mumford and J. Shah, "Boundary detection by minimizing functionals," presented at the IEEE International Conference on Computer Vision and Pattern Recognition, San Francisco, June 11–13, 1985.

[27] K. N. Nordstrom, "Biased anisotropic diffusion — a unified approach to edge detection," *Image and Vision Computing*, **8**, 318–327 (1990).

[28] S. Osher and L.-I. Rudin, "Feature-oriented image enhancement using shock filters," *SIAM J. Numer. Anal.* **27**, 919–940 (1990).

[29] S. Osher and J. Sethian, "Fronts propagating with curvature dependent speed: algorithms based on the Hamilton-Jacobi formulation," *J. Comp. Phys.* **79**, 12–49 (1988).

[30] P. Perona and J. Malik, "Scale-space and edge detection using anisotropic diffusion," *IEEE Trans. Pattern Anal. Mach. Intell.* **PAMI-12**, 629–639 (1990).

[31] W. K. Pratt, *Digital Image Processing* (Wiley, New York, 1978), pp. 495–501.

[32] P. J. Rousseeuw and A. M. Leroy, *Robust Regression and Outlier Detection* (Wiley, New York, 1987).

[33] L.-I. Rudin, S. Osher, and E. Fatemi, "Nonlinear total variation noise removal algorithm," *Phys. D*, **60**, 1217–1229 (1992).

[34] G. Sapiro and D. L. Ringach, "Anisotropic diffusion of multivalued images with applications to color filtering, *IEEE Trans. Image Process.* **5**, 1582–1586 (1996).

[35] C. A. Segall, S. T. Acton, and A. K. Katsaggelos, "Sampling conditions for anisotropic diffusion," presented at the SPIE Symposium on Visual Communications and Image Processing, San Jose, January 23–29, 1999.

[36] C. A. Segall and S. T. Acton, "Morphological anisotropic diffusion," presented at the IEEE International Conference on Image Processing, Santa Barbara, CA, October 26–29, 1997.

[37] N. Sochen, R. Kimmel, and R. Malladi, "A general framework for low level vision," *IEEE Trans. Image Process.* **7**, 310–318 (1998).

[38] S. Teboul, L. Blanc-Feraud, G. Aubert, and M. Barlaud, "Variational approach for edge-preserving regularization using coupled PDEs," *IEEE Trans. Image Process.* **7**, 387–397 (1998).

[39] R. T. Whitaker and S. M. Pizer, "A multi-scale approach to nonuniform diffusion," *Comput. Vis. Graph. Image Process. Image Understand.* **57**, 99–110 (1993).

[40] R. T. Whitaker and G. Gerig, "Vector-valued diffusion," in Bart ter Haar Romeny, ed., *Geometry-Driven Diffusion in Computer Vision* (Kluwer, New York, 1994), pp. 93–134.

[41] A. P. Witkin, "Scale-space filtering," in *Proceedings of the International Joint Conference on Artificial Intelligence* (IJCAI, Inc. Karlsruhe, Germany, 1983), pp. 1019–1021.

[42] A. Yezzi, Jr., "Modified curvature motion for image smoothing and enhancement," *IEEE Trans. Image Process.* **7**, 345–352 (1998).

[43] Y.-L. You, W. Xu, A. Tannenbaum, and M. Kaveh, "Behavioral analysis of anisotropic diffusion in image processing," *IEEE Trans. Image Process.* **5**, 1539–1553 (1996).

[44] Y.-L. You, M. Kaveh, W. Xu, and A. Tannenbaum, "Analysis and design of anisotropic diffusion for image processing," presented at the IEEE International Conference on Image Processing, Austin, TX, November 13–16, 1994.

4.13

Software for Image and Video Processing

K. Clint Slatton and
Brian L. Evans
*The University of Texas
at Austin*

1 Introduction

Digital systems that process image data generally involve a mixture of software and hardware. For example, digital video disc (DVD) players employ audio and video processors to decode the compressed audio and visual data, respectively, in real time. These processors are themselves a mixture of embedded software and hardware. The DVD format is based on the MPEG-2 video compression and AC-3 audio compression standards, which took several years to finalize (refer to Chapter 6.4). Before these standards were established, several years of research went into developing the necessary algorithms for audio and video compression. This chapter describes some of the software that is available for developing image and video processing algorithms.

Once an algorithm has been developed and is ready for operational use, it is often implemented in one of the standard compiled languages such as C, C++, or Fortran for greater efficiency. Coding in these languages, however, can be time consuming because the programmer must iteratively debug compile-time and run-time errors. This approach also requires extensive knowledge of the programming language and the operating system of the computer platform on which the program is to be compiled and run. As a result, development time can be lengthy. To guarantee portability, the source code must be compiled and validated under different operating systems and compilers, which further

delays development time. In addition, output from programs written in these standard compiled languages must often be exported to a third-party product for visualization.

Many available software packages can help designers shorten the time required to produce an operational image and video processing prototype. Algorithm development environments (Section 2) can reduce development time by eliminating the compilation step, providing many high-level routines, and guaranteeing portability. Compiled libraries (Section 3) offer high-level routines to reduce the development time of compiled programs. Source codes (Section 4) are available for entire imaging applications. Visualization environments (Section 5) are especially useful when manipulating and interpreting large data sets. A wide variety of other software packages (Section 6) can also assist in the development of imaging applications.

2 Algorithm Development Environments

Algorithm development environments strive to provide the user with an interface that is much closer to mathematical notation and vernacular than are general-purpose programming languages. The idea is that a user should be able to write out the desired computational instructions in a native language that requires relatively little time to master. Also, graphical visualization

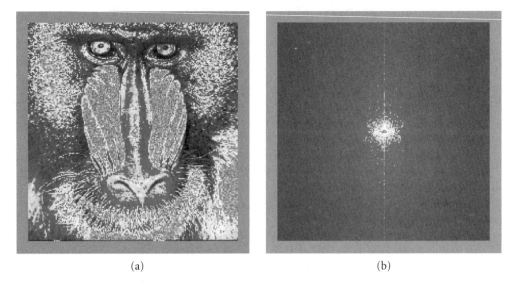

(a)   (b)

```
file_id = fopen('mandrill', 'r');
fsize = [512, 512];
[I1,count] = fread(file_id, fsize, 'unsigned char');
I1=I1';
figure, image(I1); axis off, axis square, colormap(gray(256))
map = 0:1/255:1;
map = [map', map', map'];
imwrite(I1, map, 'mandrilltiff', 'tiff')
I2=fft2(I1);
I2=abs(I2);
I2=log10(I2+1);
range=max(max(I2))-min(min(I2));
I2=(255/range)*(I2-min(min(I2)));
I2=fftshift(I2);
figure, image(I2); axis off, axis square, colormap(gray(256))
imwrite(I2, map, 'mandrillFFTtiff', 'tiff')
```

(c)

FIGURE 1   MATLAB example: (a) image, (b) FFT, and (c) code to display images, compute the FFT, and write out the TIFF images.

of the computations should be fully integrated so that the user does not have to leave the environment to observe the output. This section examines four widely used commercial packages: MATLAB, IDL, LabVIEW, and Khoros. For a comparison of the styles of specifying algorithms in these environments, Figs. 1–4 show examples of computing the same image processing operation by using MATLAB, IDL, LabVIEW, and Khoros, respectively.

2.1 MATLAB

MATLAB software is produced by The MathWorks, Inc. and has its origins in the command-line interface of the LINPACK and EISPACK matrix libraries developed by Cleve Moler in the late 1970s [1]. MATLAB interprets commands, which shortens programming time by eliminating compilation. The MATLAB programming language is a vectorized language, meaning that it can perform many operations on numbers grouped as vectors or matrices without explicit loop statements. Vectorized code is more compact, more efficient, and parallelizable.

Versions 1–4 of MATLAB assumed that every variable was a matrix. The matrix could be a real, complex, or string data type. Real and complex numbers were stored in a double-precision floating-point format. A scalar would have been represented as a 1×1 matrix of the appropriate data type. A vector is a matrix with either one row or one column. MATLAB 5 is also vectorized, but it adds signed and unsigned byte data types, which dramatically reduces storage in representing images. Version 5 also introduces other data types, such as signed and unsigned 16-bit, 32-bit, and 64-bit integers and 32-bit single-precision floating-point numbers. MATLAB 5 provides the ability to define data structures other than matrices and supports arrays of arbitrary dimension.

The MATLAB algorithm development environment interprets programs written in the MATLAB programming language, but a compiler for the MATLAB language is available as an add-on to the basic package. When developing algorithms, it is generally much faster to interpret code than to compile code because the developer can immediately test changes. In this sense, MATLAB can be used to rapidly prototype an algorithm. Once the algorithm is stable, it can be compiled for faster execution, which is

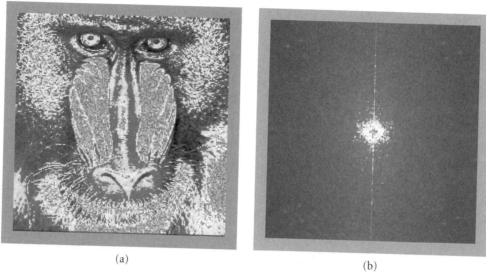

(a)

(b)

```
file1 = 'mandrill'
lun = 1;
openr, lun, file1
pic = bytarr(512,512)
readu, lun, pic
close, lun
picr = rotate(pic,7)
tiff_write,'mandrilltif', pic
window,0,xsize=512,ysize=512, title='mandrill 512x512'
tvscl,picr,0,0
picrf = fft(picr,-1)
picrfd = abs(picrf)
picrfd = alog10(picrfd+1.0)
picrfd = sqrt(sqrt(picrfd))
range = max(picrfd)-min(picrfd)
picrfd = ((255/range)*(picrfd-min(picrfd)))
picrfd = shift(picrfd,256,256)
tiff_write,'mandrillFFTtif', picrfd
window,1,xsize=512,ysize=512, title='fft 512x512'
tv, picrfd, 0, 0
return
end
```

(c)

FIGURE 2   IDL example: (a) image, (b) FFT, and (c) code to display images, compute the FFT, and write out the TIFF images.

especially important for large data sets. The MATLAB compiler MATCOM converts native MATLAB code into C++ code, compiles the C++ code, and links it with MATLAB libraries. The compiled code is up to ten times faster than the interpreted code when run on the same machine [2]. The more vectorized the MATLAB program is, the smaller the speedup in the compiled version. Highly optimized vectorized code may not experience any speedup at all.

The MATLAB algorithm development environment provides a command-line interface, an interpreter for the MATLAB programming language, an extensive set of common numerical and string manipulation functions, 2-D and 3-D plotting functions, and the ability to build custom graphical user interfaces (GUIs). A user-defined MATLAB function can be added by creating a file with a ".m" extension containing the interpreter commands.

Alternatively, a ".m" file can serve as a stand-alone program. For faster computation, users may dynamically link C routines as MATLAB functions through the MEX utility. As an alternative to the command-line interface, the MATLAB environment offers a "notebook" interface that integrates text and graphics into a single document.

MATLAB toolboxes are available as add-ons to the basic package and greatly extend its capabilities by providing application-specific function libraries [1, 3]. The *Signal Processing Toolbox* provides signal generation; finite impulse response (FIR) and infinite impulse response (IIR) filter design; linear system modeling; 1-D fast Fourier transforms (FFTs), discrete cosine transforms (DCTs), and Hilbert transforms; 2-D discrete Fourier transforms; statistical signal processing; and windows, spectral analysis, and time-frequency analysis. The *Image*

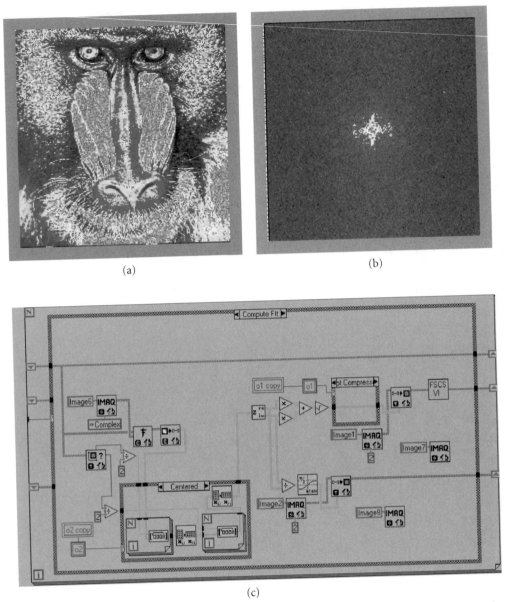

(a)

(b)

(c)

FIGURE 3 LabVIEW example: (a) image, (b) FFT, and (c) code to display images, compute the FFT, and write out the TIFF images.

Processing Toolbox represents an image as a matrix. It provides image file input/output for TIFF, JPEG, and other standard formats. Morphological operations, DCTs, FIR filter design in two dimensions, and color space manipulation and conversion are also provided. The Canny edge detector (refer to Chapter 4.10) is also available through the *Image Processing Toolbox*. The *Wavelet Toolbox* implements several wavelet families, 1-D and 2-D wavelet transforms, and coding of wavelet coefficients. Additional toolboxes with uses in imaging systems are available in control system design, neural networks, optimization, splines, and symbolic mathematics.

MATLAB's strength in developing signal and image processing algorithms lies in its ease of use, powerful functionality, and data visualization capabilities. Its programming syntax has similarities to C and Fortran. Because the MATLAB programming language is imperative, specifying algorithms in the MATLAB language biases the implementation toward software on a sequential machine. Using the *SIMULINK* add-on, designers can visually specify a system as a block diagram to expose the parallelism in the system. Each block is implemented as MATLAB or C code. *SIMULINK* is well suited for simulating and designing continuous, discrete, and hybrid discrete/continuous control systems [4]. *SIMULINK* has advanced ordinary differential equation solvers and supports discrete-event modeling. Because of the run-time scheduling overhead, simulations of digital signal, image, and video processing systems in *SIMULINK* are

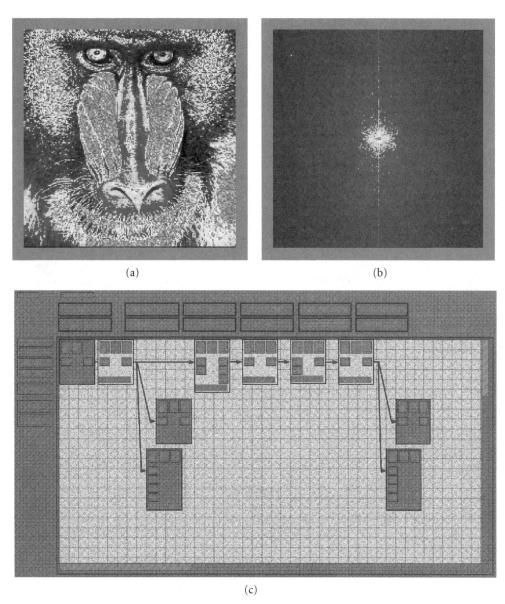

(a)

(b)

(c)

FIGURE 4 Khoros example: (a) image, (b) FFT, and (c) code to display images, compute the FFT, and write out the TIFF images.

extremely slow when compared to a simulation that uses the MATLAB programming language.

MATLAB runs on Windows, Macintosh, and Unix operating systems, including DEC Ultrix, HP-UX, IBM AIX, Linux, SGI, and Sun Solaris. The MathWorks Web site (http:// www. mathworks.com) contains freely distributable MATLAB add-ons. The MATLAB newsgroup is comp.soft-sys.matlab.

2.2 IDL

The Interactive Data Language (IDL), by Research Systems, Inc., is based on the APL computer language [5]. IDL provides a computer language with built-in data visualization routines and predefined mathematical functions. Of the algorithm development environments discussed in this chapter, IDL most closely resembles a low-level language such as C. Even in its interactive mode, IDL programs must be recompiled and executed each time a change is made to the code. Thus the advantage of IDL is not ease of algorithm development so much as the provision of a tremendously powerful integrated data visualization package. IDL is probably the best environment for flexible visualization of very large data sets.

Arrays are treated as a particular data type so that they may be operated on as single entities, reducing the need to loop through the array elements [5]. The basic IDL package consists of a command-line interface, low-level numerical and string manipulation operators (similar to C), high-level implicit functions such as frequency-domain operations, and many data display functions. IDL *Insight* provides a graphical user interface. IDL instructions and functions are put in ".pro" files. Although IDL syntax may not be as familiar to C and Fortran programmers as MATLAB's syntax, it offers streamlined file access, and scalar variables do not have to be explicitly declared.

IDL supports aggregate data structures in addition to scalars and arrays. Data types supported include 8-bit, 16-bit, 32-bit, and 64-bit integers, 32-bit and 64-bit floating-point numbers, and string data types [5]. Image formats supported include JPEG, TIFF, and GIF. Formatted I/O allows access to any user-defined ASCII or binary format. IDL supports arrays having up to eight dimensions. Many important image processing functions are provided, such as 2-D FFTs, wavelets, and median filters. IDL I/O supports MPEG video coding, but it does not provide an explicit DCT function.

IDL provides dynamic linking to external C and Fortran functions, which is analogous to the MEX utililty in MATLAB. IDL, however, does not have an automated method for converting code into another language. Most often, the users that work with those languages and wish to access IDLs capabilities must write output files from their programs and then read those files into IDL for analysis or visualization.

Research Systems, Inc. offers several complementary software packages written in IDL. Some of these can stand alone, while others are add-ons. Except for the Envi package, which is discussed in Section 5, these packages do not typically extend IDLs signal and image processing functionality. Instead, they provide capabilities such as database management and data sharing over the Internet.

IDL runs on Windows, Macintosh, and Unix operating systems including DEC Ultrix, HP-UX, SGI, and Sun Solaris. The Research Systems, Inc. Web site (http://www.rsinc.com) contains IDL libraries written by third parties, some of which are freely distributable. The IDL newsgroup is comp.lang.idl.

2.3 LabVIEW

LabVIEW, produced by National Instruments, is based on visual programming that uses block diagrams [6], unlike the default text-based interfaces for MATLAB and IDL. LabVIEW was originally developed for simulating electronic test equipment, so many of its icon and name conventions reflect that legacy. For example, it has many specialized I/O and data-handling routines for serial-port standards and hardware simulations.

LabVIEW block diagrams represent its own native language, called G. G may be either interpreted or compiled. G is a dataflow language, which is a natural representation for data-intensive computation for digital signal, image, and video processing systems [7].

LabVIEW is primarily an interactive environment. The basic interface is called a virtual instrument (VI). A VI is analogous to a function in a conventional programming language. Rather than being defined by lines of text as a MATLAB program, a VI consists of a graphical user interface with a dataflow diagram representing the source code and icon connections that allow the VI to be referenced by other VIs. This programming structure naturally lends itself to hierarchical and modular computing. Basic data structures available for use in the VIs are nodes, terminals, wires, and arrays [6]. LabVIEW supports 8-bit, 16-bit, 32-bit, and 64-bit integers and 32-bit and 64-bit floating-point numbers [6].

LabVIEW has limited data visualization capabilities. The add-on package *HiQ* is required for 2-D and 3-D graphics. The *LabWindows/CVI toolkit* allows the user to generate C code from VIs, which could be linked to LabVIEW libraries. A user can add a code interface node function to describe the operation of a node in the C language. The *Analysis VI toolkit* contains VIs for signal processing, linear algebra, and statistics [6]. The toolkit supports signal generation, frequency-domain filtering (based on the FFT), windowing, and statistical signal processing, in one dimension. Other signal processing and image processing toolkits are available. The *Signal Processing Suite* enables time-frequency and wavelet analysis. The *IMAQ* image processing toolkit contains formats for analog video standards like PAL and NTSC. *IMAQ* also provides 2-D frequency-domain and morphological operations, but it does not provide other important functions such as the DCT.

LabVIEW's image handling capabilities are limited compared to those of MATLAB and IDL. LabVIEW does not access raw binary image files. Images must be converted into standard formats such as TIFF before LabVIEW can access them. Many operations, such as logarithms, are not compatible with image data directly. Image data must first be converted into an array structure, then the operation is performed, and finally the array is converted back to an image. Also, operations on complex-valued data are limited. To take the absolute value of a complex-valued image, the user must explicitly multiply the data with its complex conjugate and then take the square root of the product.

Of the four algorithm development environments discussed in this section, LabVIEW would be the best suited for integration with hardware, especially for 1-D data acquisition. LabVIEW VIs can be compiled and downloaded into embedded real-time data acquisition systems. Neither of these capabilities, however, are available for imaging systems.

LabVIEW is available for Windows, Macintosh, and UNIX operating systems (e.g., HP-UX and Sun Solaris). However, many of the add-on packages such as *IMAQ* and *HiQ* are not available on UNIX platforms. The National Instruments Web site (http://www.natinst.com) contains freely distributable add-ons for LabVIEW and other National Instruments software packages. The LabVIEW newsgroup is comp.lang.labview.

2.4 Khoros

Khoros, by Khoral Research, Inc., is another visual programming environment for modeling and simulation [8]. The block diagrams use a mixture of data flow and control flow. Khoros supports 8-bit, 16-bit, and 32-bit integer and 32-bit and 64-bit float data types. Khoros is written in C, and supports calls to external C code. Khoros can also access external C++ code. A variety of toolboxes are available for Khoros that provide capabilities in I/O, data processing, data visualization, image processing, matrix operations, and software rendering.

Khoros libraries are effectively linked to the graphical coding workspace through a flow control tool called *Cantata*. When *Cantata* is run, a workspace window appears with several action buttons and pull-down menus along its periphery. The action buttons allow the user to run and reset the program. The pull-down menus access mathematical and I/O functions, called "subforms." Once the user selects a subform and specifies the input parameters, it is converted into an icon, referred to as a "glyph," and appears in the workspace. A particular glyph will perform a self-contained task such as generating an image or opening an existing file containing image data. Another glyph may perform a function such as a 2-D FFT. Glyphs can be written in C by using Khoros templates. Arrow buttons on the glyphs represent input and output connections. To perform an operation such as an FFT on an image, the user connects the output port of the image-accessing glyph to the input port of the FFT operator glyph. This is the primary manner in which larger algorithms are constructed.

The *Datamanip Toolbox* provides data I/O, data generation, trigonometric and nonlinear functions, bitwise and complex math, linear transforms (including FFTs), histogram and morphological operators, noise generation and processing, data clustering (data classification), and convolution. *Datamanip* requires that the *Bootstrap* and *Dataserv* toolboxes also be loaded. The *Image Toolbox* provides median filtering, 2-D frequency domain filtering, edge detectors, and geometric warping, but no DCT. Many of the matrix operations that are useful in image processing are only available in the *Matrix Toolbox*, and the *Geometry Toolbox* is required to provide 2-D and 3-D plotting capabilities. The Khoros Pro Software Development Kit comes bundled with most of the toolboxes relevant to signal and image processing.

The *Xprism Pro* package runs independently of Khoros, but it is meant to complement the Khoros product. *XPrism Pro* uses dynamic rendering so that large data sets can be viewed at variable resolution. Most other environments require large data sets to be explicitly downsampled to enable rapid plotting. Other add-on toolboxes offer wavelets and formats for accessing Geographic Information System (GIS) data. The strength of Khoros is that the user can develop complete algorithms very rapidly in the visual programming environment, which is significantly simpler than that of LabVIEW. The weakness is that this environment biases designs toward execution in software.

Khoros allows extensive integration with MATLAB through its *Mfile Toolbox*, making MATLAB functions and programs available to Khoros. This toolbox is available on most, but not all, of the platforms on which Khoros is supported. The MAT-LAB programs can be treated as source code inside Khoros objects. This toolbox includes the MATCOM compiler for converting MATLAB code to C++ code. It is based on Matrix, a C++ library consisting of over 500 mathematical functions. This in turn significantly increases the execution speed of interpreted MATLAB code. It also supports type single- and double-precision float calculations, but not all MATLAB functionality is supported.

Khoros runs on Windows and Unix (DEC Ultrix, Linux, SGI, and Sun Solaris) operating systems. The Khoral Research Web site (http://www.khoral.com) contains freely distributable add-ons for Khoros and other Khoral Research software packages. The Khoros newsgroup is comp.soft-sys.khoros.

3 Compiled Libraries

Whether users are working in an algorithm development environment or writing their own code, it is sometimes important to access mathematical functions that are written in low-level code for efficiency. Many libraries containing mathematical functions are available for this purpose. Often, a particular library will not be available in all languages or run on all operating systems. In general, the source code is not provided. Object files are supplied, which must be linked with the users' programs during compilation. When the source code is not available, the burden is on the documentation to inform the users about the speed and accuracy of each function.

3.1 Intel

Intel offers several free libraries (http://developer.intel.com/vtune/perflibst/) for signal processing, image processing, pattern recognition, general mathematics, and JPEG image coding functions. These functions have been compiled and optimized for a variety of Intel processors. The libraries require specific operating systems (Microsoft Windows 95, 98, or NT) and C/C++ compilers (Intel, Microsoft, or Borland). Signal processing functions include windows, FIR filters, IIR filters, FFTs, correlation, wavelets, and convolution. Image processing functions include morphological, thresholding, and filtering operations as well as 2-D FFTs and DCTs.

When the Intel library routines run on a Pentium processor with MMX, many of the integer and fixed-point routines will use MMX instructions [9]. MMX instructions compute integer and fixed-point arithmetic by applying the same operation on eight 8-bit words or four 16-bit words in parallel. In MMX, eight 8-bit additions, four 16-bit additions, or four 16-bit multiplications may be performed in parallel. Switching back and forth to the MMX instruction set incurs a 30-cycle penalty. The use of MMX generally reduces the accuracy of answers, primarily because Pentium processors do not have extended precision accumulation. Furthermore, many of the library functions make hidden function calls which reduces efficiency. When using the

Intel libraries on Pentium processors with MMX, the speedup for signal and image processing applications varies between 1.5 and 2.0, whereas the speedup for graphics applications varies from 4 to 6 [10].

3.2 IMSL

Other math libraries that are not specialized for signal and image processing applications may contain useful functions such as FFTs and median filters. The most prevalent is the family of IMSL libraries provided by Visual Numerics, Inc. (http://www.vni.com/). These libraries support Fortran, C, and Java languages. These libraries are very general. As a result, over 65 computer platforms are supported.

4 Source Code

Source codes for many mathematical functions and image processing applications are available. This section describes two sets of available source code besides those that come with the algorithm development environments listed in Section 2.

4.1 Numerical Recipes

Numerical Recipes by the Numerical Recipes Software Company (http://www.nr.com) provides source code in Fortran and C languages for a wide variety of mathematical functions. As long as users have a Fortran or C compiler on their machine, these programs can be run on any computer platform. It is the users' responsibility to write the proper I/O commands so that their programs can access the desired data. The tradeoff for this generality is the lack of optimization for any particular machine and the resulting lack of efficiency. The algorithms are not tailored for signal and image processing applications, but some common functions supported are 1-D and 2-D FFTs, wavelets, DCTs, Huffman encoding, and numerical linear algebra routines.

4.2 Encoding Standards

Information regarding the International Standards Organization (ISO) image coding standard developed by the Joint Photographic Experts Group (JPEG) is available at the Web site http://www.jpeg.org/. Links to the C source code for the JPEG encoding and decoding algorithms can be found at that Web site. Information regarding the ISO encoding standards for audio/video developed by the Moving Picture Experts Group (MPEG) is available at the Web site http://www.mpeg.org/. Links are available to the source code for the encoding and decoding algorithms. These programs can be used in conjunction with the algorithm development packages mentioned previously, or with low-level languages.

5 Specialized Processing and Visualization Environments

In addition to the general purpose algorithm development environments discussed earlier, many packages exist that are highly specialized for processing and visualizing large data sets. Some of these support user-written programs in limited native languages, but most of their functionality consists of predefined operations. The user can specify some parameters for these functions but typically cannot access the source code. Generally, these packages are specialized for certain applications, such as remote sensing, seismic analysis, and medical imaging. We examine packages that are specialized for remote sensing applications as examples.

Remote sensing data typically comprise electromagnetic (sometimes acoustic) energy that has been modulated through interaction with objects. The data are often collected by a sensor mounted on a moving platform, such as an airplane or satellite. The motivation for collecting remotely sensed data is to acquire information over large areas not accessible by means of *in situ* methods. This method of acquiring data results in very large data sets. When imagery is collected at more than one wavelength, there may be several channels of data per imaged scene. Remote sensing software packages must handle data sets of 1 Gb and larger. Although a multichannel image constitutes a multidimensional data set, these packages usually only display the data as images. These packages generally have very limited 2-D and 3-D graphics capabilities. They do, however, contain many specialized display and I/O routines for common remote sensing formats that other types of software do not have. They have many of the common image processing functions such as 2-D FFTs, median filtering, and edge detection. They are not very useful for generalized data analysis or algorithm development, but they can be ideal for processing data for interpretation without requiring the user to learn any programming languages or mathematical algorithms.

In addition to some of the common image processing functions, these packages offer functions particularly useful for remote sensing. In remote sensing, images of a given area are often acquired at different times, from different locations, and by different sensors. To facilitate an integrated analysis of the scene, the data sets must be coregistered so that a particular sample (pixel) will correspond to the same physical location in all of the channels. This is accomplished when control points are chosen in the different images that correspond to the same physical locations. Then 2-D polynomial warping functions or spline functions are created to resample the child images to the parent image. These packages contain the functions for coregistering so that the user does not need to be familiar with the underlying algorithms.

Another major class of functions that these packages contain is classification or pattern recognition. These algorithms can be either statistically based, neural-network based, or fuzzy-logic based. Classifying remote sensing imagery into homogeneous groups is very important for quantitatively assessing land cover types and monitoring environmental changes, such as deforestation and erosion.

When users have large remotely sensed data sets in sensor-specific formats and need to perform advanced operations on them, working with these packages will be quicker and easier than working with the algorithm development packages. Several remote sensing packages are available. We will discuss two of the most widely used and powerful packages: PCI and Envi. Other popular packages include ERDAS, ERMapper, PV-Wave, Raytheon, and ESRI.

5.1 PCI

PCI software, by the PCI company, is a geomatics software package. It supports imagery, GIS data, and many orthoprojection tools [11]. PCI supports many geographic and topographic formats such as Universal Transverse Mercator, and it can project the image data onto these non-uniform grids so that they match true physical locations.

Both command line and graphical interfaces are used depending on the operations to be performed. Low-level I/O routines make it easy to import and export data between PCI and other software packages in either ASCII or binary format. PCI provides a limited native language so that some user-defined operations can be performed without having to leave the PCI environment. PCI functions can be accessed by programs in other languages, such as C, by linking commands.

Most common image formats, such as JPEG and TIFF, are supported, as well as formats for particular sensors. Image files can have up to 1024 channels. Data represented by the image pixels are referred to as raster data. Raster data types supported include 8-bit and 16-bit (signed and unsigned) integers and 32-bit floating-point numbers. In addition to raster data, PCI also supports vector data. PCI vectors are collections of ordered pairs (vertices) corresponding to locations on the image. The vectors define piecewise linear paths that can be used to delineate exact boundaries among regions in the image. These lines are independent of the pixel size because they are defined by the mathematical lines between vertices. Vectors can be used to draw precise elevation contours and road networks on top of the imagery.

PCI can display images in a specialized 3-D perspective view, in which the gray levels of a particular channel correspond to heights. This format is useful for displaying topographic data. PCI also supports "fly throughs" in this perspective, allowing the user to scan over the data from different vantage points. PCI has a complete set of coregistering and mosaicking functions, and standard image filtering and FFT routines. PCI also includes its own drivers for accessing magnetic tape drives for reading data. Some applications for which PCI is well suited include watershed hydrological analysis, flight simulation, and land cover classification.

PCI is available on Windows, Macintosh, and Unix operating systems including DEC Utlrix, HP-UX, SGI, and Sun Solaris.

The Web site (http://www.pci.on.ca/) contains demonstration and image-handling freeware, as well as a subscriber discussion list, discussrequest@pci.on.ca.

5.2 Envi

The Environment for Visualizing Images (Envi), by Research Systems, Inc., is written in IDL. It is not necessary to acquire IDL separately to run Envi, because a basic IDL engine comes bundled with Envi. Envi has a menu-driven graphical user interface. Although batch operations are possible, it is best suited for interactive work.

Envi supports many of the same features and capabilities as PCI. PCI has more classification capability and more options for orthoprojection and hydrological analysis of the data. Envi has more user-friendly access to its functions and more up-to-date formatting for some sensors. Envi can also be easily integrated with external IDL code. Envi is accessible through the same Web site as IDL.

6 Other Software

Many other software tools are used in image and video processing. For image display, editing, and conversion, X windows tools xv and ImageMagick are often used. The xv program by John Bradley at the University of Pennsylvania (ftp://www.cis.upenn.edu/pub/xv/) is shareware. ImageMagick by John Cristy at E.I. du Pont de Nemours and Company, Inc. (http://www.wizards.dupont.com/cristy/) is freely distributable. ImageMagick can also compose images, and create and animate video sequences. Both tools run on Windows NT and Unix operating systems under X windows.

Symbolic mathematics environments are useful for deriving algebraic relationships and computing transformations algebraically, such as Fourier, Laplace, and z transforms. These environments include Mathematica [12] from Wolfram Research, Inc. (http://www.wolfram.com) and Maple [13] from Waterloo Maple Software (http://www.maplesoft.com). Mathematica has the following application packs related to signal and image processing: *Signals and Systems*, *Wavelet*, *Global Optimization*, *Time Series*, and *Dynamic Visualizer*. Commercial application packs are not available for Maple. A variety of notebooks on engineering and scientific applications are available on the Web site, but none of the Maple notebooks relates to signal or image processing. Maple is accessible in MATLAB through its symbolic toolbox. Mathematica and Maple run on Windows, Macintosh, and Unix operating systems. The newsgroup for symbolic mathematics environments is sci.math.symbolic. The Mathematica newsgroup is comp.soft-sys.math.mathematica.

System-level design tools, such as SPW by Cadence (http://www.cadence.com), COSSAP by Synopsys (http://www.synopsys.com), DFL by Mentor graphics (http://www.mentor.com), ADS by HP EEsof (http://www.tmo.hp.com/tmo/

hpeesof/), and Ptolemy by the University of California at Berkeley (http://ptolemy.eecs.berkeley.edu), are excellent at simulating and synthesizing 1-D signal processing systems. Using these tools, designers can specify a system using a mixture of graphical block diagrams and textual representations. The specification may be efficiently simulated or synthesized into software, hardware, or a mixture of hardware and software. These system-level design tools provide many basic image and video processing blocks for simulation. For example, Ptolemy provides image file I/O, median filtering, 2-D FIR filtering, 2-D FFTs, 2-D DCTs, motion vector computation, and matrix operations. These system-level design tools also provide an interface to MATLAB in which a block in a block diagram can represent a MATLAB function or script. These system-level design tools, however, currently have limited but evolving support for synthesizing image and video processing systems into hardware and/or software.

7 Conclusion

For image and video processing, we have examined algorithm development environments, function libraries, source code repositories, and specialized data processing packages. Algorithm development environments are useful when a user needs flexible and powerful coding capabilities for rapid prototyping of algorithms. Each of the four algorithm environments discussed provides much of the functionality needed for image processing and some of the functionality for video processing. When a user wants to code an algorithm in a compiled language for speed, then function libraries become extremely useful. A wide variety of source code upon which to draw exists as part of algorithm development environments and source code repositories. If there is no need to understand the underlying algorithms, but there is a need to perform specialized analysis of data, then the data interpretation and visualization packages should be used. We also surveyed a variety of other tools for small tasks. Electronic design automation tools for image and video processing systems are evolving.

References

[1] *The MATLAB 5 User's Guide* (MathWorks Inc., Natick, MA, 1997).
[2] "MATLAB compiler speeds up development," *MATLAB News Notes,* Winter 1996.
[3] R. Pratap, *Getting Started with MATLAB 5: A Quick Introduction for Scientists and Engineers* (Oxford U. Press, New York, 1999), ISBN 0-19-512947-4.
[4] *The SIMULINK User's Manual* (MathWorks Inc., Natick, MA, 1997).
[5] *The IDL User's Manual* (Research Systems, Inc., Boulder, CO, 1995).
[6] *The LabVIEW User's Manual* (National Instruments, Austin, TX, 1998).
[7] H. Andrade and S. Kovner, "Software synthesis from dataflow models for embedded software design in the G programming language and the LabVIEW development environment," presented at the IEEE Asilomar Conference on Signals, Systems, and Computers, Pacific Grove, CA, November 1998.
[8] *The Khoros User's Manual* (Khoral Research, Inc., Albuquerque, NM, 1997).
[9] *Intel Architecture Software Developer's Manual, Volume 1: Basic Architecture* (Intel Corp., http://developer.intel.com/design/PentiumII/manuals/243190.htm).
[10] R. Bhargava, R. Radhakrishnan, B. L. Evans, and L. K. John, "Evaluating MMX Technology Using DSP and Multimedia Applications," presented at the IEEE International Symposium on Microarchitecture Dallas, TX, Nov. 30–Dec. 2, 1998.
[11] *The PCI User's Manual* (PCI, Inc., Ontario, Canada, 1994).
[12] S. Wolfram, *The Mathematica Book,* 3rd ed. (Wolfram Media Inc., Champaign, IL, 1996).
[13] K. M. Heal, M. Hansen, and K. Rickard, *Maple V Learning Guide for Release 5* (Springer Verlag, 1997).

Image Compression

V

5.1

Lossless Coding

Lina J. Karam
Arizona State University

1 Introduction

The goal of lossless image compression is to represent an image signal with the smallest possible number of bits without loss of *any* information, thereby speeding up transmission and minimizing storage requirements. The number of bits representing the signal is typically expressed as an average bit rate (average number of bits per sample for still images, and average number of bits per second for video). The goal of lossy compression is to achieve the best possible fidelity given an available communication or storage bit-rate capacity, or to minimize the number of bits representing the image signal subject to some allowable loss of information. In this way, a much greater reduction in bit rate can be attained as compared to lossless compression, which is necessary for enabling many real-time applications involving the handling and transmission of audiovisual information. The function of *compression* is often referred to as *coding*, for short.

Coding techniques are crucial for the effective transmission or storage of data-intensive visual information. In fact, a single uncompressed color image or video frame with a medium resolution of 500×500 pixels would require 100 s for transmission over an ISDN (Integrated Services Digital Network) link having a capacity of 64,000 bits/s (64 Kbps). The resulting delay is intolerably large, considering that a delay as small as 1–2 s is needed to conduct an interactive "slide show," and a much smaller delay (of the order of 0.1 s) is required for video transmission or playback. Although a CD-ROM device has a storage capacity of few gigabits, its net throughput is only $\sim$1.5 Mbps. As a result, compression is essential for the storage and real-time transmission of digital audiovisual information, where large amounts of data must be handled by devices having a limited bandwidth and storage capacity.

Lossless compression is possible because, in general, there is significant redundancy present in image signals. This redundancy is proportional to the amount of correlation among the image data samples. For example, in a natural still image, there is usually a high degree of spatial correlation among neighboring image samples. Also, for video, there is additional temporal correlation among samples in successive video frames. In color images and multispectral imagery (Chapter 4.6), there is correlation, known as spectral correlation, between the image samples in the different spectral components.

In lossless coding, the decoded image data should be identical both quantitatively (numerically) and qualitatively (visually) to the original encoded image. Although this requirement preserves exactly the accuracy of representation, it often severely limits the amount of compression that can be achieved to a compression factor of 2 or 3. In order to achieve higher compression factors, perceptually lossless coding methods attempt to remove redundant as well as perceptually irrelevant information; these methods require that the encoded and decoded images be only visually, and not necessarily numerically, identical. So, in this case, some loss of information is allowed as long as the recovered image is perceived to be identical to the original one.

Although a higher reduction in bit rate can be achieved with lossy compression, there exist several applications that

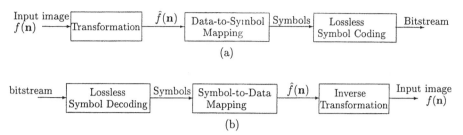

FIGURE 1   General lossless coding system: (a) encoder, (b) decoder.

require lossless coding, such as the compression of digital medical imagery and facsimile transmissions of bitonal images. These applications triggered the development of several standards for lossless compression, including the lossless JPEG standard (Chapter 5.6), facsimile compression standards and the JBIG compression standard. Furthermore, lossy coding schemes make use of lossless coding components to minimize the redundancy in the signal being compressed.

This chapter introduces the basics of lossless image coding and presents classical as well as some more recently developed lossless compression methods. This chapter is organized as follows. Section 2 introduces basic concepts in lossless image coding. Section 3 reviews concepts from information theory and presents classical lossless compression schemes including Huffman, arithmetic, and Lempel–Ziv–Welch codes. Standards for lossless compression are presented in Section 4. Section 5 introduces more recently developed lossless compression schemes and presents basics of perceptually lossless image coding.

2 Basics of Lossless Image Coding

The block diagram of a lossless coding system is shown in Fig. 1. The encoder, Fig. 1(a), takes as input an image and generates as output a compressed bit stream. The decoder, Fig. 1(b), takes as input the compressed bit stream and recovers the original uncompressed image. In general, the encoder and decoder can be each viewed as consisting of three main stages. In this section, only the main elements of the encoder will be discussed since the decoder performs the inverse operations of the encoder. As shown in Fig. 1(a), the operations of a lossless image encoder can be grouped into three stages: transformation, data-to-symbol mapping, and lossless symbol coding.

2.1 Transformation

This stage applies a reversible (one-to-one) transformation to the input image data. The purpose of this stage is to convert the input image data $f(\mathbf{n})$ into a form $\hat{f}(\mathbf{n})$ that can be compressed more efficiently. For this purpose, the selected transformation can aid in reducing the data correlation (interdependency, redundancy), alter the data statistical distribution, and/or pack a large amount of information into few data samples or subband regions.

Typical transformations include differential or predictive mapping (Chapter 5.6), unitary transforms such as the discrete cosine transform (DCT) (Chapter 5.5), subband decompositions such as wavelet transforms (Chapters 4.2 and 5.4), and color space conversions such as conversion from the highly correlated RGB representation to the less correlated luminance–chrominance representation. A combination of these transforms can be used at this stage. For example, an RGB color image can be transformed to its luminance–chrominance representation followed by DCT or subband decomposition followed by predictive–differential mapping. In some applications (e.g., low power), it might be desirable to operate directly on the original data without incurring the additional cost of applying a transformation; in this case, the transformation could be set to the identity mapping.

2.2 Data-to-Symbol Mapping

This stage converts the image data $\hat{f}(\mathbf{n})$ into entities called *symbols* that can be efficiently coded by the final stage. The conversion into symbols can be done through partitioning or run-length coding (RLC), for example.

The image data can be partitioned into blocks by grouping neighboring data samples together; in this case, each data block is a symbol. Grouping several data units together allows the exploitation of any correlation that might be present between the image data, and may result in higher compression ratios at the expense of increasing the coding complexity. In contrast, each separate data unit can be taken to be a symbol without any further grouping or partitioning.

The basic idea behind RLC is to map a sequence of numbers into a sequence of symbol pairs *(run, value)*, where *value* is the value of a data sample in the input data sequence and *run* or *run length* is the number of times that data sample is contiguously repeated. In this case, each pair *(run, value)* is a symbol. An example illustrating RLC for a binary sequence is shown in Fig. 2. Different implementations might use a slightly different format. For example, if the input data sequence has long runs of zeros, some coders such as the JPEG standard (Chapters 5.5 and 5.6),

FIGURE 2   Illustration of RLC for a binary input sequence.

use *value* to code only the value of the nonzero data samples and *run* to code the number of zeros preceding each nonzero data sample.

Appropriate mapping of the input data into symbols is very important for optimizing the coding. For example, grouping data points into small localized sets, where each set is coded separately as a symbol, allows the coding scheme to adapt to the changing local characteristics of the (transformed) image data. The appropriate data-to-symbol mapping depends on the considered application and the limitations in hardware/software complexity.

2.3 Lossless Symbol Coding

This stage generates a binary bitstream by assigning binary codewords to the input symbols. Lossless symbol coding, is commonly referred to as noiseless coding or just lossless coding, since this stage is where the actual lossless coding into the final compressed bitstream is performed. The first two stages can be regarded as preprocessing stages for mapping the data into a form that can be more efficiently coded by this lossless coding stage.

Lossless compression is usually achieved by using variable-length codewords, where the shorter codewords are assigned to the symbols that occur more frequently. This variable-length codeword assignment is known as *variable-length coding (VLC)* and also as *entropy coding*. Entropy coders, such as Huffman and arithmetic coders, attempt to minimize the average bit rate (average number of bits per symbol) needed to represent a sequence of symbols, based on the probability of symbol occurrence. Entropy coding will be discussed in more detail in Section 3. An alternative way to achieve compression is to code *variable-length strings* of symbols using fixed-length binary codewords. This is the basic strategy behind dictionary (Lempel–Ziv) codes, which are also described in Section 3.

The generated lossless code (bitstream) should be uniquely decodable; i.e., the bitstream can be decoded without ambiguity resulting in only one unique sequence of symbols (the original one). For VLC, unique decodability is achieved by imposing a prefix condition that states that no codeword can be a prefix of another codeword. Codes that satisfy the prefix condition are called *prefix codes* or *instantaneously decodable codes*, and they include Huffman and arithmetic codes. Binary prefix codes can be represented as a binary tree, and are also called *tree-structured codes*. For dictionary codes, unique decodability can be easily achieved since the generated codewords are of fixed length.

Selecting which lossless coding method to use depends on the application and usually involves a tradeoff between several factors, including the implementation hardware or software, the allowable coding delay, and the required compression level. Some of the factors that have to be considered when choosing or devising a lossless compression scheme are listed as follows.

1. Compression efficiency: Compression efficiency is usually given in the form of a compression ratio, C_R:

$$C_R = \frac{\text{Total size in bits of original input image}}{\text{Total size in bits of compressed bitstream}}$$
$$= \frac{\text{Total size in bits of encoder input}}{\text{Total size in bits of encoder output}}, \quad (1)$$

which compares the size of the original input image data with the size of the generated compressed bitstream. Compression efficiency is also commonly expressed as an average bit rate, B, in bits per pixel, or bpp for short:

$$B = \frac{\text{Total size in bits of compressed bitstream}}{\text{Total number of pixels in original input image}}$$
$$= \frac{\text{Total size in bits of encoder output}}{\text{Total size in pixels of encoder input}}. \quad (2)$$

As discussed in Section 3, for lossless coding, the achievable compression efficiency is bounded by the *entropy* of the finite set of symbols generated as the output of Stage 2, assuming these symbols are each coded separately, on a one-by-one basis, by Stage 3.

2. Coding delay: The coding delay can be defined as the minimum time required to both encode and decode an input data sample. The coding delay increases with the total number of required arithmetic operations. It also usually increases with an increase in memory requirements since memory usage usually leads to communication delays. Minimizing the coding delay is especially important for real-time applications.

3. Implementation complexity: Implementation complexity is measured in terms of the total number of required arithmetic operations and in terms of the memory requirements. Alternatively, implementation complexity can be measured in terms of the required number of arithmetic operations per second and the memory requirements for achieving a given coding delay or real-time performance. For applications that put a limit on power consumption, the implementation complexity would also include a measure of the level of power consumption. Higher compression efficiency can usually be achieved by increasing the implementation complexity, which would in turn lead to an increase in the coding delay. In practice, it is desirable to optimize the compression efficiency while keeping the implementation requirements as simple as possible. For some applications such as database browsing and retrieval, only a low decoding complexity is needed since the encoding is not performed as frequently as the decoding.

4. Robustness: For applications that require transmission of the compressed bitstream in error-prone environments, robustness of the coding method to transmission errors becomes an important consideration.

5. Scalability: Scalable encoders generates a layered bitstream embedding a hierarchical representation of the input image data. In this way, the input data can be recovered at different resolutions in a hierarchical manner (scalability in resolution), and the bit rate can be varied depending on the available resources using the same encoded bitstream (scalability in bit rate; the encoding does not have to be repeated to generate the different bit rates).

3 Lossless Symbol Coding

As mentioned in Section 2, lossless symbol coding is commonly referred to as lossless coding or lossless compression. The popular lossless symbol coding schemes fall into one of the following main categories: statistical schemes and dictionary-based schemes.

Statistical schemes (Huffman, Arithmetic) require knowledge of the source symbol probability distribution; shorter codewords are assigned to the symbols with higher probability of occurrence (VLC); a statistical source model (also called probability model) gives the symbol probabilities; the statistical source model can be fixed, in which case the symbol probabilities are fixed, or adaptive, in which case the symbol probabilities are calculated adaptively; sophisticated source models can provide more accurate modeling of the source statistics and, thus, achieve higher compression at the expense of an increase in complexity.

Dictionary-based schemes (Lempel-Ziv) do not require *a priori* knowledge of the source symbol probability distribution; they dynamically construct encoding and decoding tables (called dictionary) of variable-length symbol strings as they occur in the input data; as the encoding table is constructed, fixed-length binary codewords are generated by indexing into the encoding table.

Both the statistical and dictionary-based codes attempt to minimize the average bit rate without incurring any loss in fidelity. The field of information theory gives lower bounds on the achievable bit rates. This section presents the popular classical lossless symbol coding schemes, including Huffman, arithmetic, and Lempel–Ziv coding. In order to gain an insight into how the bit rate minimization is done by these different lossless coding schemes, some important basic concepts from information theory are reviewed first.

3.1 Basic Concepts from Information Theory

Information theory makes heavy use of probability theory since information is related to the degree of unpredictability and randomness in the generated messages. In here, the generated messages are the symbols output by Stage 2 (Section 2).

An information source is characterized by the set of symbols S it is capable of generating and the probability of occurrence of these symbols. For the considered lossless image coding application, the information source is a discrete-time discrete-amplitude source with a finite set of unique symbols; i.e., S consists of a finite number of symbols and is commonly called the *source alphabet*.

Let S consist of N symbols:

$$S = \{s_0, s_1, \ldots, s_{N-1}\}. \tag{3}$$

Then the information source outputs a sequence of symbols $\{x_1, x_2, x_3, \ldots, x_i, \ldots\}$ drawn from the set of symbols S, where x_1 is the first output source sample, x_2 is the second output sample, and x_i is the ith output sample from S. At any given time (given by the output sequence index), the probability that the source outputs symbol s_k is $p_k = P(s_k), 0 \leq k \leq N-1$. Note that $\sum_{k=0}^{N-1} p_k = 1$ since it is certain that the source outputs only symbols from its alphabet S. The source is said to be *stationary* if its statistics (set of probabilities) do not change with time.

The information associated with a symbol s_k ($0 \leq k \leq N-1$), also called *self-information*, is defined as:

$$I_{s_k} = \log_2 (1/p_k) \text{ (bits)} = -\log_2 (p_k) \text{ (bits)}. \tag{4}$$

From Eq. (4), it can be seen that $I_k = 0$ if $p_k = 1$ (certain event) and $I_k \to \infty$ if $p_k = 0$ (impossible event). Also, I_k is large when p_k is small (unlikely symbols), as expected.

The information content of the source can be measured by using the source *entropy* $H(S)$, which is a measure of the average amount of information per symbol. The source entropy $H(S)$, also known as *first-order entropy* or *marginal entropy*, is defined as the expected value of the self-information and is given by

$$H(S) = E\{I_k\} = -\sum_{k=0}^{N-1} p_k \log_2 (p_k) \text{ (bits per symbol)}. \tag{5}$$

Note that $H(S)$ is maximal if the symbols in S are equiprobable (flat probability distribution), in which case $H(S) = \log_2 (N)$ bits per symbol. A skewed probability distribution results in a smaller source entropy.

In case of memoryless coding, each source symbol is coded separately. For a given lossless code C, let l_k denote the length (number of bits) of the codeword assigned to code symbol s_k ($0 \leq k \leq N-1$). Then, the resulting average bit rate B_C corresponding to code C is

$$B_C = \sum_{k=0}^{N-1} p_k l_k \text{ (bits per symbol)}. \tag{6}$$

For any uniquely decodable lossless code C, the entropy $H(S)$ is a lower bound on the average bit rate B_C [1]:

$$B_C \geq H(S). \tag{7}$$

So, $H(S)$ puts a limit on the achievable average bit rate given that each symbol is coded separately in a memoryless fashion.

In addition, a uniquely decodable prefix code C can always be constructed (e.g., Huffman coding, Section 3.2) such that

$$H(S) \leq B_C \leq H(S) + 1. \tag{8}$$

An important result that can be used in constructing prefix codes is the Kraft inequality,

$$\sum_{k=0}^{N-1} 2^{-l_k} \leq 1. \tag{9}$$

Every uniquely decodable code has codewords with lengths satisfying Kraft inequality (9), and prefix codes can be constructed with any set of lengths satisfying inequality (9) [2].

Higher compression can be achieved by coding a block (subsequence, vector) of M successive symbols jointly. The coding can be done as in the case of memoryless coding by regarding each block of M symbols as one compound symbol $s^{(M)}$, drawn from the alphabet

$$S^{(M)} = \underbrace{S \times S \times \cdots \times S}_{M \text{ times}}, \tag{10}$$

where $\times$ in Eq. (10) denotes a Cartesian product, and the superscript (M) denotes the size of each compound block of symbols. Therefore, $S^{(M)}$ is the set of all possible compound symbols of the form $[x_1, x_2, \ldots, x_M]$ where $x_i \in S$, $1 \leq i \leq M$. Since S consists of N symbols, $S^{(M)}$ will contain $L = N^M$ compound symbols:

$$S^{(M)} = \{s_0^{(M)}, s_1^{(M)}, \ldots, s_{L-1}^{(M)}\}; \quad L = N^M. \tag{11}$$

The previous results and definitions directly generalize by replacing S with $S^{(M)}$ and replacing the symbol probabilities $p_k = P(s_k)$, $(0 \leq k \leq N-1)$ with the joint probabilities (compound symbol probabilities) $p_k^{(M)} = P(s_k^{(M)})$, $(0 \leq k \leq L-1)$. So, the entropy of the set $S^{(M)}$, which is the set of all compound symbols $s_k^{(M)}$, $(0 \leq k \leq L-1)$, is given by

$$H(S^{(M)}) = -\sum_{k=0}^{L-1} p_k^{(M)} \log_2 (p_k^{(M)}). \tag{12}$$

$H(S^{(M)})$ of Eq. (12) is also called the *Mth-order entropy* of S. If S corresponds to a stationary source (i.e., symbol probabilities do not change over time), $H(S^{(M)})$ is related to the source entropy $H(S)$ as follows [1]:

$$\frac{H(S^{(M)})}{M} \leq H(S), \tag{13}$$

with equality if and only if the symbols in S are statistically independent (memoryless source). The quantity

$$\lim_{M \to \infty} \frac{H(S^{(M)})}{M} \tag{14}$$

is called the *entropy rate* of the source S and gives the average information per output symbol drawn from S. For a stationary source, the limit in quantity (14) always exists. Also, from relation (13), the entropy rate is equal to the source entropy for the case of a memoryless source.

As before, each output (compound) symbol can be coded separately. For a given lossless code C, if $l_k^{(M)}$ is the length of the codeword assigned to code symbol $s_k^{(M)}$ $(0 \leq k \leq L-1)$, the resulting average bit rate $B_C^{(M)}$ in code bits per compound symbol is

$$B_C^{(M)} = \sum_{k=0}^{L-1} p_k^{(M)} l_k^{(M)} \text{ (bits per compound symbol).} \tag{15}$$

Also, as before, a prefix code C can be constructed such that

$$H(S^{(M)}) \leq B_C^{(M)} \leq H(S^{(M)}) + 1, \tag{16}$$

where $B_C^{(M)}$ is the resulting average bit rate per *compound* symbol. The desired average bit rate B_C in bits per *source* symbol is equal to $B_C^{(M)}/M$. So, dividing the terms in relation (16) by M, we obtain

$$\frac{H(S^{(M)})}{M} \leq B_C \leq \frac{H(S^{(M)})}{M} + \frac{1}{M}. \tag{17}$$

From relation (17), it follows that, by jointly coding very large blocks of source symbols (M very large), we can find a source code C with an average bit rate B_C approaching monotonically the entropy rate of the source as M goes to infinity. For a memoryless source, relation (17) becomes

$$H(S) \leq B_C \leq H(S) + \frac{1}{M}, \tag{18}$$

where $B_C = B_C^{(M)}/M$.

From this discussion, we see that the statistics of the considered source (given by the symbol probabilities) have to be known in order to compute the lower bounds on the achievable bit rate. In practice, the source statistics can be estimated from the histogram of a set of sample source symbols. For a nonstationary source, the symbol probabilities have to be estimated adaptively since the source statistics change over time.

3.2 Huffman Coding

In [3], D. Huffman presented a simple technique for constructing prefix codes that results in an average bit rate satisfying relation (8) when the source symbols are coded separately, or relation (17) in the case of joint M-symbol vector coding. A tighter upper bound on the resulting average bit rate is derived in [2].

The Huffman coding algorithm is based on the following optimality conditions for a prefix code [3]: (1) if $P(s_k) > P(s_j)$ (symbol s_k more probable than symbol s_j, $k \neq j$), then $l_k \leq l_j$, where l_k and l_j are the lengths of the codewords assigned to code symbols s_k and s_j, respectively; (2) if the symbols are listed in the order of decreasing probabilities, the last two symbols in the ordered list are assigned codewords that have the same length and are alike except for their final bit.

Given a source with alphabet S consisting of N symbols s_k with probabilities $p_k = P(s_k)$ $(0 \leq k \leq (N-1))$, we can construct a Huffman code corresponding to source S by iteratively constructing a binary tree as follows.

1. Arrange the symbols of S such that the probabilities p_k are in decreasing order; i.e.,

$$p_0 \geq p_1 \geq \cdots \geq p_{(N-1)}, \qquad (19)$$

and consider the ordered symbols $s_k, 0 \leq k \leq (N-1)$, as the leaf nodes of a tree. Let T be the set of the leaf nodes corresponding to the ordered symbols of S.
2. Take the two nodes in T with the smallest probabilities and merge them into a new node whose probability is the sum of the probabilities of these two nodes. For the tree construction, make the new resulting node the "parent" of the two least probable nodes of T by connecting the new node to each of the two least probable nodes. Each connection between two nodes form a "branch" of the tree; so, two new branches are generated. Assign a value of 1 to one branch and 0 to the other branch.
3. Update T by replacing the two least probable nodes in T with their parent node, and reorder the nodes (with their subtrees) if needed. If T contains more than one node, repeat from Step 2; otherwise, the last node in T is the "root" node of the tree.
4. The codeword of a symbol $s_k \in S$ $(0 \leq k \leq (N-1))$ can be obtained by traversing the linked path of the tree from the root node to the leaf node corresponding to s_k $(0 \leq k \leq (N-1))$ while reading sequentially the bit values assigned to the tree branches of the traversed path.

The Huffman code construction procedure is illustrated by the example shown in Fig. 3 for the source alphabet $S = \{s_0, s_1, s_2, s_3\}$ with symbol probabilities as given in Table 1. The resulting symbol codewords are listed in the 3rd column of Table 1. For this example, the source entropy is $H(S) = 1.84644$ and the resulting average bit rate is $B_H = \sum_{k=0}^{3} p_k l_k = 1.9$ (bits per symbol), where l_k is the length of the codeword assigned to symbol s_k of S. The symbol codewords are usually stored in a symbol-to-codeword mapping table that is made available to both the encoder and decoder.

If the symbol probabilities can be accurately computed, the above Huffman coding procedure is optimal in the sense that it results in the minimal average bit rate among all uniquely decodable codes assuming memoryless coding. Note that, for a given source S, more than one Huffman code is possible but they are all optimal in the above sense. In fact, another optimal Huffman code can be obtained by simply taking the complement of the resulting binary codewords.

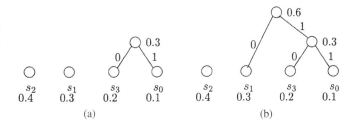

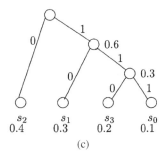

FIGURE 3 Example of Huffman code construction for the source alphabet of Table 1: (a) first, (b) second, (c) third and last iterations.

As a result of memoryless coding, the resulting average bit rate is within one bit of the source entropy since integer-length codewords are assigned to each symbol separately. The described Huffman coding procedure can be directly applied to code a group of M symbols jointly by replacing S with $S^{(M)}$ of Eq. (10). In this case, higher compression can be achieved (Section 3.1), but at the expense of an increase in memory and complexity since the alphabet becomes much larger and joint probabilities have to be computed.

While encoding can be simply done by using the symbol-to-codeword mapping table, the realization of the decoding operation is more involved. One way of decoding the bitstream generated by a Huffman code is to first reconstruct the binary tree from the symbol-to-codeword mapping table. Then, as the bitstream is read one bit at a time, the tree is traversed starting at the root until a leaf node is reached. The symbol corresponding to the attained leaf node is then output by the decoder. Restarting at the root of the tree, the above tree traversal step is repeated until all the bitstream is decoded. This decoding method produces a variable symbol rate at the decoder output since the codewords vary in length.

Another way to perform the decoding is to construct a lookup table from the symbol-to-codeword mapping table. The

TABLE 1 Example of Huffman code assignment

Source Symbol (s_k)	Probability (p_k)	Assigned Codeword
s_0	0.1	111
s_1	0.3	10
s_2	0.4	0
s_3	0.2	110

constructed lookup table has $2^{l_{max}}$ entries, where l_{max} is the length of the longest codeword. The binary codewords are used to index into the lookup table. The lookup table can be constructed as follows. Let l_k be the length of the codeword corresponding to symbol s_k. For each symbol s_k in the symbol-to-codeword mapping table, place the pair of values (s_k, l_k) in all the table entries, for which the l_k leftmost address bits are equal to the codeword assigned to s_k. Thus, there will be $2^{(l_{max}-l_k)}$ entries corresponding to symbol s_k. For decoding, l_{max} bits are read from the bitstream. The read l_{max} bits are used to index into the lookup table to obtain the decoded symbol s_k, which is then output by the decoder, and the corresponding codeword length l_k. Then the next table index is formed by discarding the first l_k bits of the current index and appending to the right the next l_k bits that are read from the bitstream. This process is repeated until all the bitstream is decoded. This approach results in a relatively fast decoding and in a fixed output symbol rate. However, the memory size and complexity grows exponentially with l_{max}, which can be very large.

In order to limit the complexity, procedures to construct *constrained-length Huffman codes* have been developed [4]. *Constrained-length Huffman codes* are Huffman codes designed while limiting the maximum allowable codeword length to a specified value l_{max}. The shortened Huffman codes result in a higher average bit rate compared to the unconstrained-length Huffman code.

Since the symbols with the lowest probabilities result in the longest codewords, one way of constructing shortened Huffman codes is to group the low-probability symbols into a compound symbol. The low-probability symbols are taken to be the symbols in S with a probability $\leq 2^{-l_{max}}$. The probability of the compound symbol is the sum of the probabilities of the individual low-probability symbols. Then the original Huffman coding procedure is applied to an input set of symbols formed by taking the original set of symbols and replacing the low-probability symbols with one compound symbol s_c. When one of the low-probability symbols is generated by the source, it is encoded with the codeword corresponding to s_c followed by a second fixed-length binary codeword corresponding to that particular symbol. The other "high probability" symbols are encoded as usual by using the Huffman symbol-to-codeword mapping table.

In order to avoid having to send an additional codeword for the low-probability symbols, an alternative approach is to use the original unconstrained Huffman code design procedure on the original set of symbols S with the probabilities of the low-probability symbols changed to be equal to $2^{-l_{max}}$. Other methods [4] involve solving a constrained optimization problem to find the optimal codeword lengths l_k $(0 \leq k \leq N-1)$ that minimize the average bit rate subject to the constraints $1 \leq l_k \leq l_{max}$ $(0 \leq k \leq N-1)$. Once the optimal codeword lengths have been found, a prefix code can be constructed by using Kraft inequality (9). In this case, the codeword of length l_k corresponding to s_k is given by the l_k bits to the right of the binary point in the binary representation of the fraction $\sum_{0 \leq i \leq k} 2^{-l_i}$.

This discussion assumes that the source statistics are described by a *fixed* (nonvarying) set of source symbol probabilities. As a result, only one fixed set of codewords has to be computed and supplied once to the encoder–decoder. This fixed model fails if the source statistics vary since the performance of Huffman coding depends on how accurately the source statistics are modeled. For example, images can contain different data types, such as text and picture data, with different statistical characteristics. Adaptive Huffman coding change the codeword set to match the locally estimated source statistics. As the source statistics change, the code changes remaining optimal for the current estimate of source symbol probabilities. One simple way for adaptively estimating the symbol probabilities is to maintain a count of the number of occurrences of each symbol [2]. The Huffman code can be dynamically changed by precomputing off-line different codes corresponding to different source statistics. The precomputed codes are then stored in symbol-to-codeword mapping tables that are made available to the encoder and decoder. The code is changed by dynamically choosing a symbol-to-codeword mapping table from the available tables based on the frequencies of the symbols that occurred so far. However, in addition to storage and the run-time overhead incurred for selecting a coding table, this approach requires *a priori* knowledge of the possible source statistics in order to predesign the codes. Another approach is to dynamically redesign the Huffman code while encoding based on the local probability estimates computed by the provided source model. This model is also available at the decoder, allowing it to dynamically alter its decoding tree or decoding table in synchrony with the encoder. Implementation details of adaptive Huffman coding algorithms can be found in [2, 5].

3.3 Arithmetic Coding

As indicated in Section 3.2, the main drawback of Huffman coding is that it assigns an integer-length codeword to each symbol separately. As a result, the bit rate cannot be less than 1 bit per symbol unless the symbols are coded jointly. However, joint symbol coding, which codes a block of symbols jointly as one compound symbol, results in delay and in an increased complexity in terms of source modeling, computation, and memory. Another drawback of Huffman coding is that the realization and the structure of the encoding and decoding algorithms depend on the source statistical model. It follows that any change in the source statistics would necessitate redesigning the Huffman codes and changing the encoding and decoding trees, which can render adaptive coding more difficult.

Arithmetic coding is a lossless coding method that does not suffer from the aforementioned drawbacks and that tends to achieve a higher compression ratio than Huffman coding. However, Huffman coding can generally be realized with simpler software and hardware.

In arithmetic coding, each symbol does not have to be mapped into an integral number of bits. Thus, an average fractional bit rate (in bits per symbol) can be achieved without the need for blocking the symbols into compound symbols. In addition, arithmetic coding allows the source statistical model to be separate from the structure of the encoding and decoding procedures; i.e., the source statistics can be changed without having to alter the computational steps in the encoding and decoding modules. This separation makes arithmetic coding more attractive than Huffman for adaptive coding.

The arithmetic coding technique is a practical extended version of Elias code and was initially developed by Pasco and Rissanen [6]. It was further developed by Rubin [7] to allow for incremental encoding and decoding with fixed-point computation. An overview of arithmetic coding is presented in [6] with C source code.

The basic idea behind arithmetic coding is to map the input sequence of symbols into one single codeword. Symbol blocking is not needed since the codeword can be determined and updated incrementally as each new symbol is input (symbol-by-symbol coding). At any time, the determined codeword uniquely represents all the past occurring symbols. Although the final codeword is represented by using an integral number of bits, the resulting average number of bits per symbol is obtained by dividing the length of the codeword by the number of encoded symbols. For a sequence of M symbols, the resulting average bit rate satisfies relation (17) and, therefore, approaches the optimum quantity (14) as the length M of the encoded sequence becomes very large.

In the actual arithmetic coding steps, the codeword is represented by a half-open subinterval $[L_c, H_c) \subset [0, 1)$. The half-open subinterval gives the set of all codewords that can be used to encode the input symbol sequence, which consists of all past input symbols. So, any real number within the subinterval $[L_c, H_c)$ can be assigned as the codeword representing all the past occurring symbols. The selected real codeword is then transmitted in binary form (fractional binary representation, where .1 represents 1/2, .01 represents 1/4, .11 represents 3/4, and so on). When a new symbol occurs, the current subinterval $[L_c, H_c)$ is updated by finding a new subinterval $[L'_c, H'_c) \subset [L_c, H_c)$ to represent the new change in the encoded sequence. The codeword subinterval is chosen and updated such that its length is equal to the probability of occurrence of the corresponding encoded input sequence. It follows that less probable events (given by the input symbol sequences) are represented with shorter intervals and, therefore, require longer codewords since more precision bits are required to represent the narrower subintervals. So, the arithmetic encoding procedure constructs, in a hierarchical manner, a code subinterval that uniquely represents a sequence of successive symbols.

In analogy with Huffman, in which the root node of the tree represents all possible occurring symbols, the interval $[0, 1)$ here represents all possible occurring sequences of symbols (all possible messages including single symbols). Also, considering the set of all possible M-symbol sequences having the same length M, the total interval $[0,1)$ can be subdivided into nonoverlapping

TABLE 2 Example of code subinterval assignment in arithmetic coding

Source Symbol (s_k)	Probability (p_k)	Symbol Subinterval $[L_{s_k}, H_{s_k})$
s_0	0.1	$[0, 0.1)$
s_1	0.3	$[0.1, 0.4)$
s_2	0.4	$[0.4, 0.8)$
s_3	0.2	$[0.8, 1)$

subintervals such that each M-symbol sequence is represented uniquely by one and only one subinterval whose length is equal to its probability of occurrence.

Let S be the source alphabet consisting of N symbols $s_0, \ldots, s_{(N-1)}$. Let $p_k = P(s_k)$ be the probability of symbol s_k, $0 \leq k \leq (N-1)$. Since, initially, the input sequence will consist of the first occurring symbol ($M = 1$), arithmetic coding begins by subdividing the interval $[0,1)$ into N nonoverlapping intervals, where each interval is assigned to a distinct symbol $s_k \in S$ and has a length equal to the symbol probability p_k. Let $[L_{s_k}, H_{s_k})$ denote the interval assigned to symbol s_k, where $p_k = H_{s_k} - L_{s_k}$. This assignment is illustrated in Table 2; the same source alphabet and source probabilities as in the example of Fig. 3 are used for comparison with Huffman. In practice, the subinterval limits L_{s_k} and H_{s_k} for symbol s_k can be directly computed from the available symbol probabilities and are equal to cumulative probabilities P_k as given here:

$$L_{s_k} = \sum_{i=0}^{k-1} p_k = P_{k-1}; \quad 0 \leq k \leq (N-1), \qquad (20)$$

$$H_{s_k} = \sum_{i=0}^{k} p_k = P_k; \quad 0 \leq k \leq (N-1). \qquad (21)$$

Let $[L_c, H_c)$ denote the code interval corresponding to the input sequence that consists of the symbols that occurred so far. Initially, $L_c = 0$ and $H_c = 1$; so, the initial code interval is set to $[0, 1)$. Given an input sequence of symbols, the calculation of $[L_c, H_c)$ is performed based on the following encoding algorithm:

1. $L_c = 0$; $H_c = 1$.
2. Calculate code subinterval length:

$$\text{length} = H_c - L_c. \qquad (22)$$

3. Get next input symbol s_k.
4. Update the code subinterval:

$$L_c = L_c + \text{length} \times L_{s_k},$$
$$H_c = L_c + \text{length} \times H_{s_k}. \qquad (23)$$

5. Repeat from Step 2 until all the input sequence has been encoded.

TABLE 3 Example of code subinterval construction in arithmetic coding

Iteration # No. (I)	Encoded Symbol (s_k)	Code Subinterval [L_c, H_c)
1	s_1	[0.1, 0.4)
2	s_0	[0.1, 0.13)
3	s_2	[0.112, 0.124)
4	s_3	[0.1216, 0.124)
5	s_3	[0.12352, 0.124)

As indicated before, any real number within the final interval [L_c, H_c) can be used as a valid codeword for uniquely encoding the considered input sequence. The binary representation of the selected codeword is then transmitted. The above arithmetic encoding procedure is illustrated in Table 3 for encoding the sequence of symbols $s_1 s_0 s_2 s_3 s_3$. Another representation of the encoding process within the context of the considered example is shown in Fig. 4. Note that arithmetic coding can be viewed as remapping, at each iteration, the symbol subintervals [L_{s_k}, H_{s_k}) ($0 \le k \le (N-1)$) to the current code subinterval [L_c, H_c). The mapping is done by rescaling the symbol subintervals to fit within [L_c, H_c), while keeping them in the same relative positions. So, when the next input symbol occurs, its symbol subinterval becomes the new code subinterval, and the process repeats until all input symbols are encoded.

In the arithmetic encoding procedure, the length of a code subinterval, "length" in Eq. (22), is always equal to the product of the probabilities of the individual symbols encoded so far, and it monotonically decreases at each iteration. As a result, the code interval shrinks at every iteration. So, longer sequences result in narrower code subintervals, which would require the use of high-precision arithmetic. Also, a direct implementation of the presented arithmetic coding procedure produces an output only after all the input symbols have been encoded. Implementations that overcome these problems are presented in [6,7]. The basic idea is to begin outputting the leading bit of the result as soon as it

can be determined (incremental encoding), and then to shift out this bit (which amounts to scaling the current code subinterval by 2). In order to illustrate how incremental encoding would be possible, consider the example in Table 3. At the second iteration, the leading part "0.1" can be output since it is not going to be changed by the future encoding steps. A simple test to check whether a leading part can be output is to compare the leading parts of L_c and H_c; the leading digits that are the same can then be output and they remain unchanged since the next code subinterval will become smaller. For fixed-point computations, overflow and underflow errors can be avoided by restricting the source alphabet size [4].

Given the value of the codeword, arithmetic decoding can be performed as follows:

1. $L_c = 0$; $H_c = 1$.
2. Calculate the code subinterval length:

$$length = H_c - L_c.$$

3. Find the symbol subinterval [L_{s_k}, H_{s_k}) ($0 \le k \le N-1$) such that

$$L_{s_k} \le \frac{codeword - L_c}{length} < H_{s_k}.$$

4. Output symbol s_k.
5. Update the code subinterval

$$L_c = L_c + length \times L_{s_k},$$
$$H_c = L_c + length \times H_{s_k}.$$

6. Repeat from Step 2 until the last symbol is decoded.

In order to determine when to stop the decoding (i.e., which symbol is the last symbol), a special end-of-sequence symbol

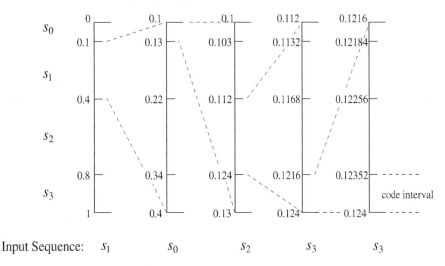

FIGURE 4 Arithmetic coding example.

is usually added to the source alphabet S and is handled like the other symbols. In the case in which fixed-length blocks of symbols are encoded, the decoder can simply keep a count of the number of decoded symbols and no end-of-sequence symbol is needed. As discussed before, incremental decoding can be achieved before all the codeword bits are output [6, 7].

3.4 Lempel–Ziv Coding

Huffman coding (Section 3.2) and arithmetic coding (Section 3.3) require *a priori* knowledge of the source symbol probabilities or of the source statistical model. In some cases, a sufficiently accurate source model is difficult to obtain, especially when several types of data (such as text, graphics, and natural pictures) are intermixed.

Universal coding schemes do not require *a priori* knowledge or explicit modeling of the source statistics. A popular lossless universal coding scheme is a dictionary-based coding method developed by Ziv and Lempel [8] and known as Lempel–Ziv (LZ) coding. Dictionary-based coders dynamically build a coding table (called dictionary) of variable-length symbol strings as they occur in the input data. As the coding table is constructed, fixed-length binary codewords are assigned to the variable-length input symbol strings by indexing into the coding table. In LZ coding, the decoder can also dynamically reconstruct the coding table and the input sequence as the code bits are received without any significant decoding delays. Although LZ codes do not explicitly make use of the source probability distribution, they asymptotically approach the source entropy rate for very long sequences [1]. Because of their adaptive nature, dictionary-based codes are ineffective for short input sequences since these codes initially result in a lot of bits being output. So, short input sequences can result in data expansion instead of compression.

There are several variations of LZ coding. They mainly differ in how the dictionary is implemented, initialized, updated, and searched. One popular LZ coding algorithm is known as the Lempel–Ziv–Welch (LZW) algorithm, a version of LZ coding developed by Welch [9]. This is the algorithm used for implementing the *compress* command in the UNIX operating system.

Let S be the source alphabet consisting of N symbols s_k ($1 \leq k \leq N$). The basic steps of the LZW algorithm can be stated as follows:

1. Initialize the first N entries of the dictionary with the individual source symbols of S, as shown:

Address	Entry
1	s_1
2	s_2
3	s_3
⋮	⋮
N	s_N

TABLE 4 Dictionary constructed while encoding the sequence $s_1 s_2 s_1 s_2 s_3 s_2 s_1 s_2^*$

Address	Entry
1	s_1
2	s_2
3	s_3
4	s_4
5	$s_1 s_2$
6	$s_2 s_1$
7	$s_1 s_2 s_3$
8	$s_3 s_2$
9	$s_2 s_1 s_2$

*This is emitted by a source with alphabet $S = \{s_1, s_2, s_3, s_4\}$.

2. Parse the input sequence and find the longest input string of successive symbols w (including the first still unencoded symbol s in the sequence) that has a matching entry in the dictionary.
3. Encode w by outputting the index (address) of the matching entry as the codeword for w.
4. Add to the dictionary the string ws formed by concatenating w and the next input symbol s (following w).
5. Repeat from Step 2 for the remaining input symbols starting with the symbol s, until the entire input sequence is encoded.

Consider the source alphabet $S = \{s_1, s_2, s_3, s_4\}$. The encoding procedure is illustrated for the input sequence $s_1 s_2 s_1 s_2 s_3 s_2 s_1 s_2$. The constructed dictionary is shown in Table 4. The resulting code is given by the fixed-length binary representation of the following sequence of dictionary addresses: 1 2 5 3 6 2. The length of the generated binary codewords depends on the maximum allowed dictionary size. If the maximum dictionary size is M entries, the length of the codewords would be $\log_2 (M)$ rounded to the next smallest integer.

The decoder constructs the same dictionary (Table 4) as the codewords are received. The basic decoding steps can be described as follows.

1. Start with the same initial dictionary as the encoder. Also, initialize w to be the empty string.
2. Get the next "codeword", and decode it by outputting the symbol string sm stored at address "codeword" in dictionary.
3. Add to the dictionary the string ws formed by concatenating the previous decoded string w (if any) and the first symbol s of the current decoded string.
4. Set $w = sm$ and repeat from Step 2 until all the codewords are decoded.

Note that the constructed dictionary has a prefix property; i.e., every string w in the dictionary has its prefix string (formed by removing the last symbol of *w*) also in the dictionary. Since the strings added to the dictionary can become very long, the actual LZW implementation exploits the prefix property to render the dictionary construction more tractable. To add a string *ws* to the dictionary, the LZW implementation only stores the pair of values (c, s), where *c* is the address where the prefix string *w* is stored and *s* is the last symbol of the considered string *ws*. So, the dictionary is represented as a linked list [1,9]

4 Lossless Coding Standards

The need for interoperability between various systems has led to the formulation of several international standards for lossless compression algorithms targeting different applications. Examples include the standards formulated by the International Standards Organization (ISO), the International Electrotechnical Commission (IEC), and the International Telecommunication Union (ITU), which was formerly known as the International Consultative Committee for Telephone and Telegraph (CCITT). A comparison of the lossless still image compression standards is presented in [10].

4.1 JBIG Standard

The JBIG (Joint Binary Image Experts Group) standard was developed jointly by the ITU and the ISO/IEC with the objective to provide improved lossless compression performance, for both business-type documents and binary halftone images, as compared to the existing standards. Another objective was to support progressive transmission. Gray-scale images are also supported by encoding separately each bit plane.

The JBIG standard consists of a context-based arithmetic encoder that takes as input the original binary image. The arithmetic encoder makes use of a context-based modeler that estimates conditional probabilities based on causal templates. A causal template consists of a set of already encoded neighboring pixels and is used as a context for the model to compute the symbol probabilities. Causality is needed to allow the decoder to recompute the same probabilities without the need to transmit side information.

Progressive transmission is supported by using a layered coding scheme. In this scheme, a low-resolution initial version of the image (initial layer) is first encoded. Higher-resolution layers can then be encoded and transmitted in the order of increasing resolution. In this case, the causal templates used by the modeler can include pixels from the previously encoded layers in addition to already encoded pixels belonging to the current layer.

Compared to the ITU Group 3 and Group 4 facsimile compression standards [4,10], the JBIG standard results in 20–50% more compression for business-type documents. For halftone images, JBIG results in compression ratios that are two to five times greater than those obtained from the ITU Group 3 and Group 4 facsimile standards [4,10].

4.2 Lossless JPEG Standard

The JPEG (Joint Photographic Experts Group) standard was developed jointly by the ITU and ISO/IEC for the lossy and lossless compression of continuous-tone, color or gray-scale, still images [11]. This section discusses very briefly the main components of the lossless mode of the JPEG standard (known as lossless JPEG).

The lossless JPEG coding standard can be represented in terms of the general coding structure of Fig. 1 as follows.

- Stage 1: linear prediction–differential (DPCM) coding is used to form prediction residuals. The prediction residuals have usually a lower entropy than the original input image. Thus, higher compression ratios can be achieved.
- Stage 2: the prediction residual is mapped into a pair of symbols (*category, magnitude*), where the symbol *category* gives the number of bits needed to encode *magnitude*.
- Stage 3: for each pair of symbols (*category, magnitude*), Huffman coding is used to code the symbol *category*. The symbol *magnitude* is then coded using a binary codeword whose length is given by the value *category*. Arithmetic coding can also be used in place of Huffman coding.

Complete details about the lossless JPEG standard and related recent developments, including JPEG-LS [12], are presented in Chapter 5.6.

5 Other Developments in Lossless Coding

Several recent lossless image coding systems have been proposed [13–15]. Most of these systems can be described in terms of the general structure of Fig. 1, and they make use of the lossless symbol coding techniques discussed in Section 3 or variations on those. Among the recently developed coding systems, LOCO-I [14] was adopted as part of the new JPEG-LS standard (Chapter 5.6) since it exhibits the best compression/complexity tradeoff. CALIC [13] achieves the best compression performance at a slightly higher complexity than LOCO-I. Perceptual-based coding schemes can achieve higher compression ratios at a much reduced complexity by removing perceptually irrelevant information in addition to the redundant information. In this case, the decoded image is required to only be visually, and not necessarily numerically, identical to the original image. In what follows, CALIC and perceptual-based image coding are introduced.

5.1 CALIC

CALIC (Context-based, adaptive, lossless image codec) represents one of the best performing practical and general purpose lossless image coding techniques.

CALIC encodes and decodes an image in raster scan order with a single pass through the image. For the purposes of context

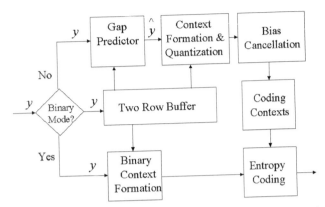

FIGURE 5 Schematic description of CALIC. (Courtesy of Nasir Memon.)

TABLE 5 Lossless bit rates with Intraband and Interband CALIC (courtesy of Nasir Memon)

Image	JPEG-LS	Intraband CALIC	Interband CALIC
Band	3.36	3.20	2.72
Aerial	4.01	3.78	3.47
Cats	2.59	2.49	1.81
Water	1.79	1.74	1.51
Cmpnd1	1.30	1.21	1.02
Cmpnd2	1.35	1.22	0.92
Chart	2.74	2.62	2.58
Ridgely	3.03	2.91	2.72

modeling and prediction, the coding process uses a neighborhood of pixel values taken only from the previous two rows of the image. Consequently, the encoding and decoding algorithms require a buffer that holds only two rows of pixels that immediately precede the current pixel. Figure 5 presents a schematic description of the encoding process in CALIC. Decoding is achieved by the reverse process. As shown in Fig. 5, CALIC operates in two modes: binary mode and continuous-tone mode. This allows the CALIC system to distinguish between binary and continuous-tone images on a local, rather than a global, basis. This distinction between the two modes is important because of the vastly different compression methodologies employed within each mode. The former uses predictive coding, whereas the latter codes pixel values directly. CALIC selects one of the two modes depending on whether or not the local neighborhood of the current pixel has more than two distinct pixel values. The two-mode design contributes to the universality and robustness of CALIC over a wide range of images.

In the binary mode, a context-based adaptive ternary arithmetic coder is used to code three symbols, including an escape symbol. In the continuous-tone mode, the system has four major integrated components: prediction, context selection and quantization, context-based bias cancellation of prediction errors, and conditional entropy coding of prediction errors. In the prediction step, a gradient-adjusted prediction (GAP) $\hat{y}$ of the current pixel y is made. The predicted value $\hat{y}$ is further adjusted by means of a bias cancellation procedure that involves an error feedback loop of one-step delay. The feedback value is the sample mean of prediction errors $\bar{e}$ conditioned on the current context. This results in an adaptive, context-based, nonlinear predictor $\check{y} = \hat{y} + \bar{e}$. In Fig. 5, these operations correspond to the blocks of context quantization, error modeling, and the error feedback loop.

The bias corrected prediction error $\check{y}$ is finally entropy coded based on a few estimated conditional probabilities in different conditioning states or coding contexts. A small number of coding contexts are generated by context quantization. The context quantizer partitions prediction error terms into few classes by the expected error magnitude. The described procedures in relation

to the system are identified by the blocks of context quantization and conditional probabilities estimation in Fig. 5. The details of this context quantization scheme in association with entropy coding are given in [13].

CALIC has also been extended to exploit interband correlations found in multiband images such as color images, multispectral images, and 3-D medical images. Interband CALIC can give 10–30% improvement over intraband CALIC, depending on the type of image. Table 5 shows bit rates achieved with intraband and interband CALIC on a set of multiband images. For the sake of comparison, results obtained with JPEG-LS, the new standard on lossless image coding, are also included.

5.2 Perceptually Lossless Image Coding

The lossless coding methods presented so far require the decoded image data to be identical both quantitatively (numerically) and qualitatively (visually) to the original encoded image. This requirement usually limits the amount of compression that can be achieved to a compression factor of 2 or 3, even when sophisticated adaptive models are used as discussed in Section 5.1. In order to achieve higher compression factors, perceptually lossless coding methods attempt to remove redundant as well as perceptually irrelevant information.

Perceptual-based algorithms attempt to discriminate between signal components that are and are not detected by the human receiver. They exploit the spatiotemporal masking properties of the human visual system and establish thresholds of *just-noticeable distortion* based on psychophysical contrast masking phenomena. The interest is in bandlimited signals because of the fact that visual perception is mediated by a collection of individual mechanisms in the visual cortex, denoted *channels* or *filters*, that are selective in terms of frequency and orientation [16].

Neurons respond to stimuli above a certain contrast. The necessary contrast to provoke a response from the neurons is defined as the *detection threshold*. The inverse of the detection threshold is the *contrast sensitivity*. Contrast sensitivity varies with frequency (including spatial frequency, temporal frequency, and orientation) and can be measured using *detection experiments* [17].

In *detection experiments*, the tested subject is presented with test images and needs only to specify whether the target stimulus is visible or not visible. They are used to derive just-noticeable-difference (JND) or detection thresholds in the absence or presence of a masking stimulus superimposed over the target. For the image coding application, the input image is the masker and the target (to be masked) is the quantization noise (distortion). JND contrast sensitivity profiles, obtained as the inverse of the measured detection thresholds, are derived by varying the target or the masker contrast, frequency and orientation. The common signals used in vision science for such experiments are sinusoidal gratings. For image coding, bandlimited subband components are used [17].

The detection experiments can be further subdivided into three types: *contrast sensitivity, luminance masking* (also known as light adaptation), and *contrast masking* experiments.

- The *contrast sensitivity* experiments measure the sensitivity of the eye in function of frequency (spatial and/or temporal) and orientation. In this case, a target sinusoidal stimulus at a selected frequency and orientation (u, θ) is presented over a flat background of constant luminance (corresponding to neutral gray) with no other masking stimulus present. The contrast of the target stimulus is varied until it becomes just visible. These experiments thus measure, for each frequency (u, θ), the smallest contrast $t_{(u,\theta)}$ that yields a visible signal. $t_{(u,\theta)}$ is often referred to as the *base detection threshold*. The inverse of the measured $t_{(u,\theta)}$ defines the sensitivity of the eye in function of frequency and orientation; this function is essentially known as the *Contrast Sensitivity Function* (CSF) which is a global characteristic independent of the input image. In perceptual image coding, a base detection threshold is measured for each subband at the center frequency.

- *Luminance masking* refers to the fact that the detection threshold values vary with the background intensity levels. In the *contrast sensitivity* experiments, the CSF function threshold values are measured based on a fixed background illumination. The variation of the threshold values in function of the background luminance can be determined by *luminance masking* experiments which vary the illumination level of the background over which a target stimulus is presented. For the image coding application, the detection thresholds will depend on the mean luminance of the local image region and, therefore, *luminance masking* experiments are used to determine the variation of $t_{(u,\theta)}$ in function of the mean luminance [17]. A brightness correction factor can be derived and applied to the contrast sensitivity profiles to account for this variation.

- Finally, *contrast masking* refers to the change in the visibility of one image component (the target) by the presence of another one (the masker). So, the *contrast masking* experiments measure the variation of the detection threshold of a target signal as a function of the contrast of the masker. In image coding, the masker signal is represented by the bandlimited subband components of the original visual data while the target signal is represented by the bandlimited components of the error or noise.

Several perceptual image coding schemes have been proposed [17–21]. These schemes differ in the way the perceptual thresholds are computed and used in coding the visual data. For example, not all the schemes account for contrast masking in computing the thresholds. One method, called DCTune [19], fits within the framework of JPEG. Based on a model of human perception that considers frequency sensitivity and contrast masking, it designs a DCT quantization matrix (three

(a)   (b)

FIGURE 6   Perceptually lossless image compression [17]: (a) Original Lena image, 8 bpp; (b) decoded Lena image, 0.361 bpp. The perceptual thresholds are computed for a viewing distance equal to 6 times the image height.

quantization matrices in the case of color images) for each image. The quantization matrix is selected to minimize an overall perceptual distortion which is computed in terms of the perceptual thresholds.

The perceptual image coder (PIC) proposed by Safranek and Johnston [18] works in a subband decomposition setting. Each subband is quantized using a uniform quantizer with a fixed step size. The step size is determined by the JND threshold for uniform noise at the most sensitive coefficient in the subband. The used model does not include contrast masking. A scalar multiplier in the range of 2–2.5 is applied to uniformly scale all step sizes in order to compensate for the conservative step size selection and to achieve good compression ratio.

Higher compression can be achieved by exploiting the varying perceptual characteristics of the input image in a locally adaptive fashion. Locally adaptive perceptual image coding requires computing and making use of image-dependent, locally varying, masking thresholds to adapt the quantization to the varying characteristics of the visual data. In [17, 21], locally adaptive perceptual image coders are presented without the need for side information for the locally varying perceptual thresholds. This is accomplished by using a low-order linear predictor, at both the encoder and decoder, for estimating the locally available amount of masking. Figure 6 presents coding results obtained by using the locally adaptive perceptual image coder of [17] for the Lena image. The original image is represented by 8 bits per pixel (bpp) and is shown in Fig. 6(a). The decoded perceptually lossless image is shown in Fig 6(b) and requires only 0.361 bpp (compression ratio $C_R = 22$).

References

[1] R. B. Wells, *Applied Coding and Information Theory for Engineers* (Prentice-Hall, Englewood Cliffs, NJ, 1999).

[2] R. G. Gallager, "Variations on a theme by huffman," *IEEE Trans. Inf. Theory* **IT-24**, 668–674 (1978).

[3] D. A. Huffman, "A method for the construction of minimum-redundancy codes," *Proc. IRE* **40**, 1098–1101 (1952).

[4] V. Bhaskaran and K. Konstantinides, *Image and Video Compression Standards: Algorithms and Architectures* (Kluwer, Norwell, MA, 1995).

[5] W. W. Lu and M. P. Gough, "A fast adaptive huffman coding algorithm," *IEEE Trans. Commun.* **41**, 535–538 (1993).

[6] I. H. Witten, R. M. Neal, and J. G. Cleary, "Arithmetic coding for data compression," *Commun. ACM* **30**, 520–540 (1987).

[7] F. Rubin, "Arithmetic stream coding using fixed precision registers," *IEEE Trans. Inf. Theory* **IT-25**, 672–675 (1979).

[8] J. Ziv and A. Lempel, "A universal algorithm for sequential data compression," *IEEE Trans. Inf. Theory* **IT-23**, 337–343 (1977).

[9] T. A. Welch, "A technique for high-performance data compression," *Computer* **17**, 8–19 (1987).

[10] R. B. Arps and T. K. Truong, "Comparison of international standards for lossless still image compression," *Proc. IEEE* **82**, 889–899 (1994).

[11] W. Pennebaker and J. Mitchell, *JPEG Still Image Data Compression Standard*. (Van Nostrand Rheinhold, New York, 1993).

[12] ISO/IEC JTC1/SC29 WG1 (JPEG/JBIG); ITU Rec. T. 87, "Information technology — lossless and near-lossless compression of continuous-tone still images — final draft international standard FDIS14495-1 (JPEG-LS)," Tech. Rep., ISO, 1998.

[13] X. Wu and N. Memon, "Context-based, adaptive, lossless image coding," *IEEE Trans. Commun.* **45**, 437–444 (1997).

[14] M. J. Weinberger, G. Seroussi, and G. Sapiro, "LOCO-I: A low complexity, context-based, lossless image compression algorithm," in *Data Compression Conference*, (IEEE Computer Society, Los Alamitos, CA, 1996), pp. 140–149.

[15] A. Said and W. A. Pearlman, "An image multiresolution representation for lossless and lossy compression," *IEEE Trans. Image Process.* **5**, 1303–1310 (1996).

[16] L. Karam, "An analysis/synthesis model for the human visual based on subspace decomposition and multirate filter bank theory," in *IEEE International Symposium on Time-Frequency and Time-Scale Analysis* (IEEE, New York, 1992), pp. 559–562.

[17] I. Hontsch and L. Karam, "APIC: Adaptive perceptual image coding based on subband decomposition with locally adaptive perceptual weighting," in *IEEE International Conference on Image Processing* (IEEE, New York, 1997), pp. 37–40.

[18] R. J. Safranek and J. D. Johnston, "A perceptually tuned subband image coder with image dependent quantization and postquantization," in *IEEE International Conference on Acoustics, Speech, and Signal Processing* (IEEE, New York, 1989), pp. 1945–1948.

[19] A. B. Watson, "Perceptual optimization of DCT color quantization matrices," *IEEE International Conference on Image Processing* (IEEE, New York, 1994), vol. 1, pp. 100–104.

[20] R. Rosenholtz and A. B. Watson, "Perceptual adaptive JPEG coding," in *IEEE International Conference on Image Processing* (IEEE, New York, 1996), pp. 901–904.

[21] I. Hontsch and L. Karam, "Locally-adaptive image coding based on a perceptual target distortion," in *IEEE International Conference on Acoustics, Speech, and Signal Processing* (IEEE, New York, 1998), pp. 2569–2572.

5.2

Block Truncation Coding

Edward J. Delp,
Martha Saenz,
and Paul Salama
Purdue University

1 Introduction and Historical Overview

The problem of how one stores and transmits a digital image has been a topic of research for more than 40 years and was initially driven by military applications and NASA. The problem, simply stated, is, How does one efficiently represent an image in binary form? This is the image compression problem. It is a special case of the source coding problem addressed by Shannon in his landmark paper [1] on communication systems. What is different about image compression is that techniques have been developed that exploit the unique nature of the image and the observer. These include the spatial nature of the data and of the human visual system. The "efficiency" of the representation depends on two properties of every image compression technique: data rate (in bits/pixel) and distortion in the decompressed image. The date rate is a measure of how much bandwidth one would require to transmit the image or how much space it would take to store the image.[1] Ideally one would like this to be as small as possible. If the decompressed image is exactly the same as the original image, the technique is said to be lossless. Otherwise the technique is lossy and the decompressed image has distortion or coding artifacts in it. Depending on the application, one can often trade distortion for data rate; hence, if a user is willing to accept images with more distortion, the data rate can often be lower.

Statistical and structural methods have been developed for image compression [2], the former being based on the principles of source coding with emphasis on the algebraic structure of the pixels in an image, whereas the latter methods exploit the geometric structure of the image. In recent years there has been a great deal of activity in formulating standards for image and video compression. The results are the JPEG and MPEG standards discussed in Chapters 5.5 and 6.4. Most statistical image compression methods are implemented by segmenting the image into nonoverlapping blocks, because dividing the images into blocks allows the image compression algorithm to adapt to local image statistics. The disadvantage, however, is that the borders of the blocks are often visible in the decoded image.[2]

In this chapter we describe a lossy image compression technique known as *Block Truncation Coding* (BTC). In the simplest possible terms, BTC is a block-adaptive binary encoder scheme based on moment preserving quantization. The basic concepts of BTC were born on March 17, 1977 in the office of O. Robert Mitchell at Purdue University during a conversation between Mitchell and his Ph.D. student, Edward J. Delp. Delp and Mitchell discussed many ideas relative to how one could exploit statistical moments in the context of image compression. Delp began working on this concept as part of his Ph.D. thesis.[3] The first papers on BTC appeared at the IEEE International Conference on Communications in 1978 [3] and 1979 [4]. The first journal articles also appeared in 1979 [5, 6] along with Delp's thesis [7]. Since 1977 a great deal of work has been done on BTC. There has been more than 200 journal papers, 400 conference papers, 40 Ph.D. theses, and one book [8] published on BTC. BTC was a final candidate for the JPEG compression standard in 1987.[4]

---

[1]One can also use the "compression ratio" when describing data rate efficiency. We find this term to be imprecise and prefer to use data rate in bits/pixel.

[2]The reader might be familiar with this problem when selecting a low "quality factor" when using JPEG.

[3]The term "block truncation coding" was coined by Delp in early 1978.

[4]See page 302 of [9].

In the next section we will describe the basic BTC algorithm followed by a description of moment preserving quantization. We then describe various extensions to BTC and applications.

2 Basics of BTC

The basic BTC algorithm is a lossy fixed length compression method that uses a Q-level quantizer to quantize a local region of the image. The quantizer levels are chosen such that a number of the moments of a local region in the image are preserved in the quantized output. In its simplest form, the objective of BTC is to preserve the sample mean and sample standard deviation of a gray-scale image. Additional constraints can be added to preserve higher-order moments. For this reason BTC is a block adaptive moment preserving quantizer.

The first step of the algorithm is to divide the image into nonoverlapping rectangular regions. For the sake of simplicity we let the blocks be square regions of size $n \times n$, where n is typically 4. For a two-level (1 bit) quantizer, the idea is to select two luminance values to represent each pixel in the block. These values are chosen such that the sample mean and standard deviation

of the reconstructed block are identical to those of the original block. An $n \times n$ bit map is then used to determine weather a pixel luminance value is above or below a certain threshold. In order to illustrate how BTC works, we will let the sample mean of the block be the threshold; a "1" would then indicate if an original pixel value is above this threshold, and "0" if it is below. Since BTC produces a bit map to represent a block, it is classified as a binary pattern image coding method [10]. The thresholding process makes it possible to reproduce a sharp edge with high fidelity, taking advantage of the human visual system's capability to perform local spatial integration and mask errors. Figure 1 illustrates the BTC encoding process for a block. Observe how the comparison of the block pixel values with a selected threshold produces the bit map.

By knowing the bit map for each block, the decompression/ reconstruction algorithm knows whether a pixel is brighter or darker than the average. Thus, for each block two gray-scale values, a and b, are needed to represent the two regions. These are obtained from the sample mean and sample standard deviation of the block, and they are stored together with the bit map. Figure 2 illustrates the decompression process. An explanation of how a and b are determined will be given below.

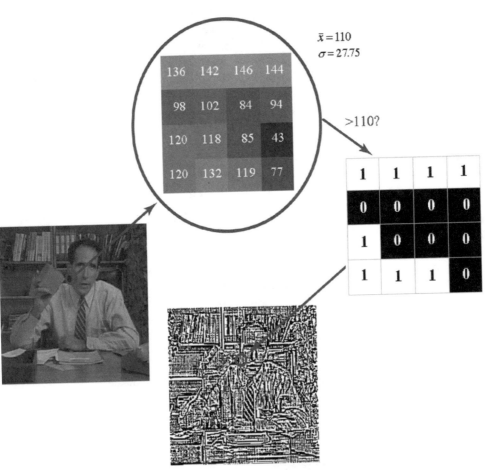

FIGURE 1 Illustration of the BTC compression process.

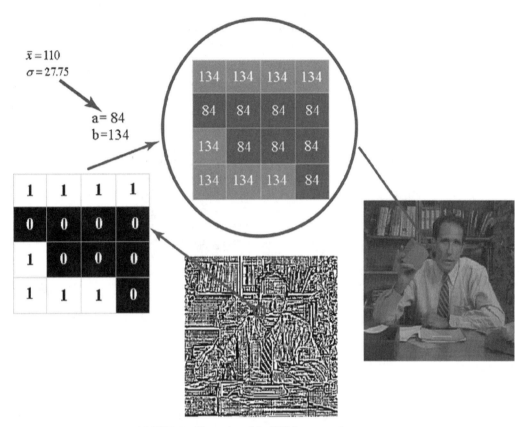

FIGURE 2　Illustration of the BTC decompression process.

For the example illustrated in Figs. 1 and 2, the image was compressed from 8 bits per pixel to 2 bits per pixel (bpp). This is done because BTC requires 16 bits for the bit map, 8 bits for the sample mean, and 8 bits for the sample standard deviation. Thus, the entire 4×4 block requires 32 bits, and hence the data rate is 2 bpp. From this example it is easy to understand how a smaller data rate can be achieved by selecting a bigger block size, or by allocating fewer bits for the sample mean or the sample standard deviation [5, 7]. We will discuss later how the data rate can be further reduced.

To understand how a and b are obtained, let k be the number of pixels of an $n \times n$ block ($k = n^2$) and $x_1, x_2, \ldots, x_k$ be the intensity values of the pixels in a block of the original image. The first two sample moments m_1 and m_2 are given by

$$m_1 = \bar{x} = \frac{1}{k} \sum_{i=1}^{k} x_i,$$

$$m_2 = \frac{1}{k} \sum_{i=1}^{k} x_i^2, \tag{1}$$

and the sample standard deviation σ is given by

$$\sigma^2 = m_2 - m_1^2. \tag{2}$$

The 1-bit quantizer for a block and threshold, x_{th}, as shown in Fig. 3, is defined by

$$\text{output} = \begin{cases} b & \text{if } x_i \geq x_{th} \\ a & \text{if } x_i < x_{th} \end{cases} \quad \text{for } i = 1, 2, \ldots, k. \tag{3}$$

As the example illustrated, the mean can be selected as the quantizer threshold. Other thresholds could also be used, such as the sample median. Another way to determine the threshold is to perform an exhaustive search over all possible intensity values to find a threshold that minimizes a distortion measure relative to the reconstructed image [7].

Once a threshold, x_{th}, is selected, the output levels of the quantizer (a and b) are found such that the first and second moments are preserved in the output. If we let q be the number

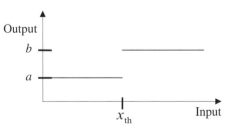

FIGURE 3　Binary quantizer.

of pixels in a block that are greater than or equal to x_{th} in value, we have

$$km_1 = (k - q)a + qb$$
$$km_2 = (k - q)a^2 + qb^2. \tag{4}$$

Solving for a and b:

$$a = m_1 - \sigma\sqrt{\frac{q}{k - q}},$$
$$b = m_1 + \sigma\sqrt{\frac{k - q}{q}}. \tag{5}$$

Rather than selecting the threshold to be the mean, we can add an additional constraint to Eq. (4) in order to determine the threshold of the quantizer. This is done by preserving the third sample moment (m_3):

$$km_3 = (k - q)a^3 + qb^3, \tag{6}$$

where m_3 is given by

$$m_3 = \frac{1}{k}\sum_{i=1}^{k} x_i^3. \tag{7}$$

Since q is defined as the number of x_i's greater than or equal to x_{th}, the threshold is then implicitly determined by q:

$$q = \frac{k}{2}\left[1 + A\sqrt{\frac{1}{A^2 + 4}}\right], \quad \text{where}$$
$$A = \frac{3 \times m_1 m_2 - m_3 - 2 \times m_1^3}{\sigma^3}(\sigma \neq 0). \tag{8}$$

It is evident how each block can be described by the sample mean (m_1), the sample standard deviation (σ), and a bit map where the ones and zeros indicate whether the pixel values are above or below the threshold. The data rate is then determined by the block size k and the number of bits f that are allocated to the sample mean and sample standard deviation of a block. The data rate is then given by $(k + f)/k = 1 + (f/k)$ bits, as shown in Fig. 4. For instance, for $k = 16$ and with the use of 10 bits to jointly quantize m_1 and σ, the image would be compressed to $1 + (10/16) = 1.625$ bpp.

The issue of how many bits to assign to the sample mean and sample standard deviation was discussed in detail in [7, 11]. The most important concept to note is that when the sample mean is small or large, the sample standard deviation must be small given the dynamic range of the pixel values. One can exploit this and assign fewer bits to the sample standard deviation. In [11] it was shown that one could also use spatial masking models to

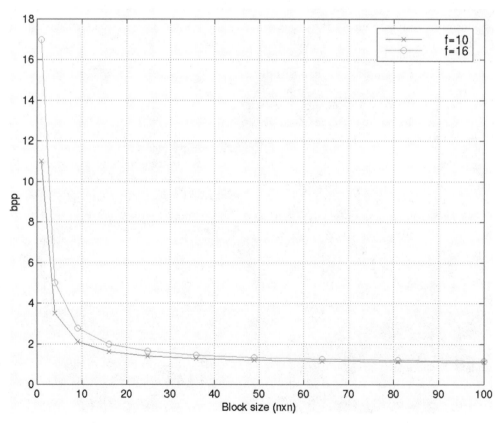

FIGURE 4   Data rate vs. block size. (See color section, p. C–23.)

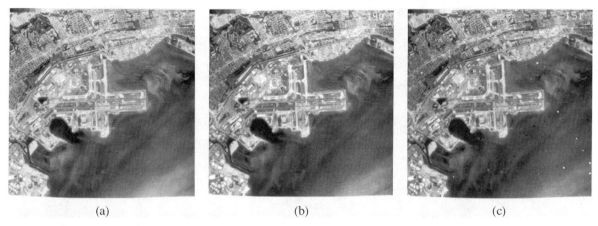

FIGURE 5   BTC with errors: (a) original image; (b) image compressed to 1.625 bpp; (c) performance of BTC in the presence of channel errors.

reduce the number of bits assigned to the mean and standard deviation, with 10 bits typically being enough to jointly quantize both values. The performance of BTC when the first three moments are preserved is illustrated in Fig. 5. The image shown in Fig. 5(b) is compressed to data rate of 1.625 bpp.

Another advantage to BTC is that channel errors do not propagate in the decompressed image because BTC produces a fixed length binary representation of each block. Figure 5(c) shows the performance of BTC in the presence of channel errors when the channel has a bit error probability of 10^{-3}.

Other techniques can be used to design a 1-bit quantizer; for instance, one can use a fidelity criterion such as mean square error (MSE) or mean absolute error (MAE). If we let $y_1, y_2, \ldots, y_k$ be the x_i's sorted in ascending order, that is, the order statistics of x_i's (see Chapter 4.4), the MSE is then given by

$$\text{MSE} = \sum_{i=1}^{k-q-1} (y_i - a)^2 + \sum_{i=k-q}^{m} (y_i - b)^2. \qquad (9)$$

By minimizing the MSE, a and b are

$$a = \frac{1}{k-q} \sum_{i=1}^{k-q-1} y_i,$$

$$b = \frac{1}{q} \sum_{i=k-q}^{m_1} y_i. \qquad (10)$$

When minimizing the MAE, Eq. (11), we find the values of a and b are given in Eq. (12):

$$\text{MAE} = \sum_{i=1}^{k-q-1} |y_i - a| + \sum_{i=k-q}^{m} |y_i - b|, \qquad (11)$$

$$a = \text{median of } (y_1, y_2, \ldots, y_{k-q-1}),$$

$$b = \text{median of } (y_{k-q}, \ldots, y_{k1}). \qquad (12)$$

A comparison between the use of MSE, MAE, and BTC is given in [5].

The main feature of BTC is the simplicity of its implementation, particularly because of low decompression complexity. Because of the block nature of the algorithm, the boundaries of adjacent blocks can sometimes be visible. The artifacts produced by BTC are usually seen around edges and in low contrast areas containing a sloping gray scale. In some images, edges may appear to be ragged despite being sharp, and some sloping gray levels may exhibit false contours [5].

3 Moment Preserving Quantization

In this section we will develop the moment preserving (MP) quantizer. We will show that quantizers that preserve moments can be derived in closed form when the input probability density function is symmetric and the number of levels is relatively small. We will discuss how a MP quantizer can be formulated as the classical Gauss–Jacobi mechanical quadrature problem.

Since the advent of the use of pulse code modulation systems, there has been great interest in the design of quantizers. It was observed that non-uniform quantizers possessed properties that could be used to achieve results such as a lower mean square error or enhanced subjective performance in the areas of speech and image compression. These types of quantizers are designed for a particular input probability distribution function relative to a particular performance index or fidelity criterion. The most popular fidelity criterion used is that of the MSE between the input and output, with the quantizer designed to minimize the mean square error. Other pointwise measures have also been proposed, such as the MAE criterion. Studies have shown that pointwise fidelity criteria cannot be used reliably in image coding [12].

Preserving the moments of the input and output of a quantizer has been proven to be a very successful approach for image coding [5, 11]. Block truncation coding, as described in the previous section, uses a small number of levels and a nonparametric form of a moment preserving quantizer. By nonparametric we mean that the quantizer was designed to fit the actual data; no a priori probability distribution function is assumed. We will approach the problem by first examining a two level MP quantizer and then generalize the result to Q-levels.

Let the random variable X denote the input to the quantizer, whose probability distribution function is $F(x)$, $x \in [c, d]$. The interval $[c, d]$ can be finite, infinite, or semi-infinite. Let Y denote the random variable at the output of the quantizer. For a two-level quantizer, the random variable Y is discrete and takes on the values $\{y_1, y_2\}$ with probabilities $P_1 = \text{prob}(Y = y_1)$ and $P_2 = \text{prob}(Y = y_2)$. The output Y takes on the value y_1 whenever the input x is below some threshold x_{th}; otherwise the output is y_2. Therefore, in general, to design any two-level quantizer one must choose the two output levels y_1 and y_2 (designated by a and b in the previous section), and the input threshold x_{th}, as illustrated in Fig. 3. It is necessary that the quantizer preserve the first three moments of the input; otherwise one of the three parameters would have to be known (or guessed) initially [7]. To specify the quantizer one must solve the following equations for y_1, y_2, and x_{th}:

$$E[Y] = E[X] = y_1 P_1 + y_2 P_2,$$
$$E[Y^2] = E[X^2] = y_1^2 P_1 + y_2^2 P_2,$$
$$E[Y^3] = E[X^3] = y_1^3 P_1 + y_2^3 P_2,$$
$$P_1 + P_2 = 1, \tag{13}$$

where the expectation operator is defined by

$$E[X^i] = \int_c^d x^i dF(x), \qquad y_1 \le x_{\text{th}} \le y_2.$$

We shall assume throughout this presentation that the moments exist and are finite. Equation (13) can be rewritten as

$$m_1 = y_1 F(x_{\text{th}}) + y_2[1 - F(x_{\text{th}})],$$
$$m_2 = y_1^2 F(x_{\text{th}}) + y_2^2[1 - F(x_{\text{th}})],$$
$$m_3 = y_1^3 F(x_{\text{th}}) + y_2^3[1 - F(x_{\text{th}})], \tag{14}$$

where $m_i = E[X^i]$,

$$P_1 = \text{prob}(X \le x_{\text{th}}) = F(x_{\text{th}}),$$
$$P_2 = \text{prob}(X > x_{\text{th}}) = 1 - F(x_{\text{th}}).$$

When Eq. (14) is solved for y_1, y_2, and x_{th}, the quantizer obtained is such that the first three moments of X and Y are identical. To find x_{th} we shall assume that $F^{-1}(\cdot)$ exists.

Without loss of generality we shall further assume that $m_1 = 0$ and $m_2 = 1$, i.e., X is zero mean and unit variance. Equation (14)

then becomes

$$0 = y_1 F(x_{\text{th}}) + y_2[1 - F(x_{\text{th}})],$$
$$1 = y_1^2 F(x_{\text{th}}) + y_2^2[1 - F(x_{\text{th}})],$$
$$m_3 = y_1^3 F(x_{\text{th}}) + y_2^3[1 - F(x_{\text{th}})]. \tag{15}$$

By solving the first two equations for y_1 and y_2 in terms of $F(x_{\text{th}})$ and using these solutions in the last equation, we arrive at the desired results:

$$y_1 = -\sqrt{\frac{1 - F(x_{\text{th}})}{F(x_{\text{th}})}} = \sqrt{\frac{P_2}{P_1}},$$

$$y_2 = \sqrt{\frac{F(x_{\text{th}})}{1 - F(x_{\text{th}})}} = \sqrt{\frac{P_1}{P_2}},$$

$$F(x_{\text{th}}) = \frac{1}{2} + \frac{m_3}{2}\sqrt{\frac{1}{4 + m_3^2}}. \tag{16}$$

This result is interesting in that the quantizer can be written in closed form. The above result in Eq. (16) also indicates that the threshold x_{th} is nominally the *median* of X and not the mean, as one would expect. The third moment m_3 is in general a signed number and can be thought of as a measure of skewness in the probability distribution function. This result indicates that the threshold is biased above or below the median according to the sign and magnitude of this skewness. These results are similar to those of BTC in the previous section, the difference being that BTC uses sample moments [5]. It should be noted that at this point we have no guarantee that $y_1 \le x_{\text{th}} \le y_2$. This problem will be addressed below.

The MP quantizer can be generalized to Q levels. One needs to recognize that for a Q-level quantizer there are Q output levels and $Q - 1$ thresholds. So if we desire an Q-level MP quantizer we need to know the first $2Q - 1$ moments, i.e., the Q-level MP quantizer preserves $2Q - 1$ moments. This, as shown in [13], guarantees the uniqueness of the quantizer. For large Q this does lead to the problem of knowing a large set of moments for a given distribution.

To arrive at the desired quantizer we need to know Q output levels $\{y_1, y_2, \ldots, y_Q\}$ and $Q - 1$ thresholds $\{x_1, x_2, \ldots, x_{Q-1}\}$ with $y_1 \le x_1 \le y_2 \cdots y_{Q-1} < x_{Q-1} \le y_Q$. We again assume $m_1 = 0$ and $m_2 = 1$, and solve

$$m_n = \int_c^d x^n dF(x) = \sum_{i=1}^Q y_i^n P_i \quad \text{for } n = 0, 1, 2, \ldots, 2Q-1, \tag{17}$$

where

$$x_0 = c,$$
$$x_Q = d,$$
$$m_n = E[X^n],$$
$$P_i = F(x_i) - F(x_{i-1}) = \text{prob}(Y = y_i).$$

For a large class of practical problems where $F(x)$ admits a probability density function $f(x)$ and if $f(x)$ is even, i.e., $f(x) = f(-x)$, then the complexity of Eq. (17) is simplified since $m_n \equiv 0$ for n odd and the quantizer itself is symmetric. For a symmetric probability density function, a closed form solution has been obtained for $Q = 2, 3, 4$ [13].

Equation (17) can be recognized as a form of the Gauss–Jacobi mechanical quadrature [14]. The output levels, y_i, of a Q-level MP quantizer are the zeros of the Qth degree orthogonal polynomial associated with $F(x)$. The P_i are the Christoffel numbers and the x_i and y_i alternate by the separation theorem of Chebyshev–Markov–Stieltjes [14]. A review of orthogonal polynomials, the Gauss–Jacobi mechanical quadrature, and the separation theorem are presented in [13].

Table 1 shows the MP quantizer thresholds and output levels for an input that has a zero mean, unit variance Gaussian probability density function (PDF). MP quantizer thresholds and output levels for uniform and Laplacian probability distribution functions and other distributions are given in [7, 13].

For comparison purposes the mean square error of the quantizer and the entropy of the output are shown.

The results for PDFs on an infinite interval exhibit one of the disadvantages of the MP quantizer. The outputs at y_1 and y_Q have a tendency to spread much further from the origin than a minimum MSE quantizer. What this says is that the quantizer assigns output levels that have a small probability of occurrence. These assignments of small probability output levels are reflected by the low values of the entropy for MP quantizers [13]. This indicates that it would be very hard to evaluate the MP quantizer for large values of Q (say larger than 30) because the output levels would be assigned such small probability of occurrence that one could have problems with computationally accuracy. Also it is no easy task to compute the zeros of a polynomial of high degree. These types of problems do not manifest themselves in the MSE quantizer because of the types of algorithms used to determine the output levels and input thresholds. Convergence properties of the MP quantizer for large Q are derived in [13]. It is also shown that the quantization error of the MP quantizer is negatively correlated with the input.

4 Variations and Applications of BTC

We will not attempt to list all the variations and extensions made to BTC over the years; rather we provide a general idea of the ways in which BTC has been used in image and video compression. Overviews of the many different variants of BTC are presented in [15, 16].

The first comparison study of the performance of BTC was done in 1980 [17]. In this study BTC was compared with the DCT and hybrid coding techniques in the context of high-resolution aerial reconnaissance imagery. This study showed that at data rates from 1–3 bits/pixel (monochrome images), BTC performed very favorably compared to the other techniques.

After the initial work on BTC and moment preserving quantizers [13], the group at Purdue worked on several enhancements and extensions to the basic algorithm. These include coding graphics images [11], predictive coding [18], coding color images [19], the use of absolute moments [19], video compression [20, 21], and hardware implementations [22]. Figure 6 illustrates one of the recent applications of BTC in coding color images [23]. Here BTC is used in a multiresolution decomposition of the image to achieve a data rate of 1.89 bpp.

A great deal of work has been done on the use of absolute moments [24]. The use of absolute moments is interesting in that the mean square performance is better than the standard BTC

(a)   (b)

FIGURE 6 Illustration of the use of BTC in color image compression: left, original image; right, image encoded at 1.89 bpp. (See color section, p. C–23.)

TABLE 1 Positive thresholds and output levels for a MP quantizer for a zero mean, unit variance Gaussian PDF

Quantizer	Output Levels	Thresholds	Quantizer	Output Levels	Thresholds
				0.4849	0.0000
$Q = 2$			$Q = 10$	1.4650	1.0137
Entropy 1.00	1.0	0.0	Entropy 2.0748	2.4843	2.0568
MSE 0.4042			MSE 0.0820	3.5818	3.1702
				4.8595	4.4491
				0.0000	
$Q = 3$	0.0000		$Q = 11$	0.9288	0.4805
Entropy 1.2516	1.7312	0.9673	Entropy 2.1419	1.8760	1.4537
MSE 0.2689			MSE 0.0745	2.8651	2.4620
				3.9361	3.5449
				5.1880	4.7951
				0.4444	0.0000
$Q = 4$			$Q = 12$	1.3404	0.9216
Entropy 1.4423	0.7419	0.0000	Entropy 2.2032	2.2595	1.8615
MSE 0.2032	2.3344	1.6866	MSE 0.06841	3.2237	2.8409
				4.2718	3.8979
				5.5009	5.1232
				0.0000	
				0.8567	0.4409
$Q = 5$	0.0000		$Q = 13$	1.7254	1.3309
Entropy 1.5936	1.3557	0.7277	Entropy 2.2598	2.6207	2.2429
MSE 0.1626	2.8570	2.2820	MSE 0.0631	3.5634	3.1978
				4.5914	4.2324
				5.8002	5.4358
				0.4126	0.0000
				1.2427	0.8509
$Q = 6$	6.6167	0.0000	$Q = 14$	2.0883	1.7142
Entropy 1.7188	1.8892	1.3338	Entropy 2.3123	2.9630	2.6026
MSE 0.1362	3.3242	2.8003	MSE 0.0587	3.8869	3.5363
				4.8969	4.5512
				6.0874	5.7349
				0.0000	
				0.7991	0.4096
$Q = 7$	0.0000	0.6081	$Q = 15$	1.6067	1.2352
Entropy 1.8255	1.1544	1.8624	Entropy 2.3611	2.4324	2.0755
MSE 0.1166	2.3667	3.2648	MSE 0.0547	3.2891	2.4435
	3.7504			4.1962	3.8586
				5.1901	4.8560
				6.3639	6.0221
				0.3868	0.0000
				1.1638	0.7943
$Q = 8$	0.5391	0.0000	$Q = 16$	1.9519	1.5977
Entropy 1.9185	1.6365	1.1408	Entropy 2.4060	2.7602	2.4182
MSE 0.1024	2.8025	2.3364	MSE 0.0519	3.6009	3.2683
	4.1445	3.6890		4.4929	4.1670
				5.4722	5.1485
				6.6308	6.2986
	0.0000	0.5332			
$Q = 9$	1.0233	1.6193			
Entropy 2.0008	2.0768	2.7694			
MSE 0.0909	3.2054	4.0818			

approach. A very interesting recent paper by Ma [25] examines the earlier work done at Purdue by Lema and Mitchell [19] and argues that this work is often improperly cited. BTC has also been used with vector quantization, nonlinear filters, and multilevel quantizers. Many video compression schemes have proposed using BTC, including HDTV [26].

Because of its low complexity, BTC is attractive for hardware or software implementation. The first paper describing an integrated circuit approach was prepared in 1978 [27], with more recent interest being in video [28]. Many software implementations have been proposed, including Sun's CellB video format [29], which is used in their XIL library and as part of the multicast transport used on the Internet. The XMovie [30] architecture that has been suggested for multimedia systems is an extension of the DECs Software Motion Pictures [31] system based on BTC. Perhaps one of the most interesting recent extensions of BTC is in the area of binary pattern image coding [10], whereby the BTC bit plane is extended so that only certain patterns in each block are encoded. An excellent example of this approach is visual pattern coding [32], which can preserve local gradients in each image block. These techniques have been shown to work quite well for video in multimedia applications at data rates below 100 kb/s.

5 Conclusions

Block truncation coding has come a long way since March 1977. Despite the recent work in image video compression standards, BTC is still attractive in many applications that require low complexity and moderate data rates. These include Internet video with software-only codecs, digital cameras, and printers. On the research side, work continues on combining BTC with other techniques and approaches to improve performance. As in all research, one never knows where this work will lead. We have no doubt that BTC will be of interest to the research community and applications engineers well into the next century.

Acknowledgments

This work was partially supported by grants from the AT&T Foundation, the Rockwell Foundation, Lucent Technologies,and Texas Instruments. Direct all correspondence relative to this chapter to E. J. Delp, at ace@ecn.purdue.edu, or http://www.ece.purdue.edu/~ace, or +1 765 494 1740.

References

[1] C. L. Shannon, "A mathematical theory of communication," *Bell Syst. Tech. J.* **27**, 379–423, 623–656 (1948).

[2] M. M. Reid, R. J. Millar, and N. D. Black, "Second-generation image coding: an overview," *ACM Comput. Surv.* **29**, 3–29 (1997).

[3] O. R. Mitchell, E. J. Delp, and S. G. Carlton, "Block truncation coding: a new approach to image compression," *Proc. IEEE Int. Conf. Commun.* **1**, 12B.1.1–12B.1.4 (1978).

[4] E. J. Delp and O. R. Mitchell, "Some aspects of moment preserving," *Proc. IEEE Int. Conf. on Commun.* **1**, 7.2.1–7.2.5 (1979).

[5] E. J. Delp and O. R. Mitchell, "Image compression using block truncation coding," *IEEE Trans. Commun.* **27**, 1335–1341 (1979).

[6] E. J. Delp, R. L. Kashyap, and O. R. Mitchell, "Image compression using autoregressive time series models," *Pattern Recog.* **11**, 313–323 (1979).

[7] E. J. Delp, "Moment preserving quantization and its application in block truncation coding," Ph.D. dissertation (Purdue University, Lafayette, IN, 1979).

[8] B. V. Dasarathy, *Image Data Compression: Block Truncation Coding* (IEEE Computer Society Press, Los Alamitos, CA, 1995).

[9] W. B. Pennebaker and J. L. Mitchell, *JPEG Still Image Compression Standard* (Van Nostrand Reinhold, New York, 1993).

[10] A. A. Rodriguez, C. E. Fogg, and E. J. Delp, "Video compression for multimedia applications," in *Image Technology: Advances in Image Processing, Multimedia, and Machine Vision,* Jorge L. C. Sanz, ed. (Springer, New York, 1996).

[11] O. R. Mitchell and E. J. Delp, "Multilevel graphics representation using block truncation coding," *Proc. IEEE* **68**, 868–873 (1980).

[12] D. J. Sakrison, "On the role of the observer and a distortion measure in image transmission," *IEEE Trans. Commun.* **COM-25**, 1251–1267 (1977).

[13] E. J. Delp and O. R. Mitchell, "Moment preserving quantization," *IEEE Trans. Commun.* **39**, 1549–1558 (1991).

[14] G. Szego, *Orthogonal Polynomials* (American Mathematical Society, Providence, RI, 1975), Vol. 23.

[15] H. B. Mitchell, N. Zilverberg, and M. Avraham, "A comparison of different block truncation coding algorithms for image compression," *Signal Process. Image Commun.* **6**, 77–82 (1994).

[16] P. Franti, O. Nevalainen, and T. Kaukoranta, "Compression of digital images by block truncation coding: a survey," *Comput.* **37**, 308–332 (1994).

[17] O. R. Mitchell, S. C. Bass, E. J. Delp, T. W. Goeddel, and T. S. Huang, "Image coding for photo analysis," *Proc. Soc. Inf. Display* **21**, 279–292 (1980).

[18] E. J. Delp and O. R. Mitchell, "The use of block truncation coding in DPCM image coding," *IEEE Trans. Signal Process.* **39**, 967–971 (1991).

[19] M. D. Lema and O. R. Mitchell, "Absolute moment block truncation coding and applications to color images," *IEEE Trans. Commun.* **32**, 1148–1157 (1984).

[20] D. J. Healy and O. R. Mitchell, "Digital video bandwidth compression using BTC," *IEEE Trans. Commun.* **29**, 1809–1817 (1981).

[21] M. D. Lema and O. R. Mitchell, "Compression of video sequences using AMBTC with motion compensated prediction," presented at the IEEE International Conference on Communications, Chicago, June 1985.

[22] T. N. Mudge, E. J. Delp, L. J. Siegel, and H. J. Siegel, "Image coding using the multimicroprocessor system PASM," *in Proceedings of the IEEE Computer Society Conference on Pattern Recognition and Image Processing* (IEEE, New York, 1982), pp. 200–205.

[23] L. A. Overturf, M. L. Comer, and E. J. Delp, "Color image-coding using morphological pyramid decomposition," *IEEE Trans. Image Process.* **4**, 177–185 (1995).

[24] K. K. Ma and S. A. Rajala, "New properties of AMBTC (absolute moment block truncation coding)," *IEEE Signal Process. Lett.* **2**, 34–36 (1995).

[25] K. K. Ma, "Put absolute moment block truncation coding in perspective," *IEEE Trans. Commun.* **45**, 284–286 (1997).

[26] N. M. Nasrabadi, C. Y. Choo, T. Harries, and J. Smallcomb, "Hierarchical block truncation coding of digital HDTV images," *IEEE Trans. Consumer Electron.* **36**, 254–261 (1990).

[27] W. L. Eversole, D. J. Mayer, F. B. Frazee, and T. F. Cheek, "Investigation of VLSI technologies for image processing," presented at the DARPA Image Understanding Workshop, Pittsburgh, PA, November 1978.

[28] L.-G. Chen, Y.-C. Liu, T.-D. Chiueh, and Y.-P. Lee, "A real-time video signal-processing chip," *IEEE Trans. Consumer Electron.* **39**, 82–92 (1993).

[29] W. K. Pratt, *Developing Visual Applications XIL: An Imaging Foundation Library* (Sun, 1998).

[30] R. Keller, W. Effelsberg and B. Lamparter, "XMovie: architecture and implementation of a distributed movie system," *ACM Trans. Inf. Syst.* **13**, 471–499 (1995).

[31] B. K. Neidecker-Lutz and R. Ulichney, "Software motion pictures," *Digital Tech. J.* **5**, 1–9 (1983).

[32] B. Barnett and A. C. Bovik, "Motion-compensated visual pattern image sequence coding for full motion multisession videoconferencing on multimedia workstations," *J. Electron. Imag.* **5**, 129–143 (1996).

5.3

Fundamentals of Vector Quantization

Mohammad A. Khan
and Mark J. T. Smith
Georgia Institute of Technology

1 Introduction

In this age of information, we see an increasing trend toward the use of digital representations for audio, speech, images, and video. Much of this trend is being fueled by the exploding use of computers and multimedia computer applications. The high volume of data associated with digital signals, particularly digital images and video, has stimulated interest in algorithms for data compression. Many such algorithms are discussed elsewhere in this book. At the heart of all these algorithms is quantization, a field of study that has matured over the past few decades. In simplest terms, quantization is a mapping of a large set of values to a smaller set of values. The concept is illustrated in Fig. 1(a), which shows on the left a sequence of unquantized samples with amplitudes assumed to be of infinite precision, and on the right that same sequence quantized to integer values. Obviously, quantization is an irreversible process, since it involves discarding information. If it is done wisely, the error introduced by the process can be held to a minimum.

The generalization of this notion is called *vector quantization*, commonly denoted VQ. It too is a mapping from a large set to a smaller set, but it involves quantizing blocks of samples together. The conceptual notion of VQ is illustrated in Fig. 1(b). Blocks of samples, which we view as vectors, are represented by codevectors stored in a codebook — a process called encoding. The codebook is typically a table stored in a digital memory, where each table entry represents a different codevector. A block diagram of the encoder is shown in Fig. 2. The output of the encoder is a binary index that represents the compressed form of the input vector. The reconstruction process, which is called decoding, involves looking up the corresponding codevector in a duplicate copy of the codebook, assumed to be available at the decoder.

The general concept of VQ can be applied to any type of digital data. For a one-dimensional signal as illustrated in Fig. 1(b), vectors can be formed by extracting contiguous blocks from the sequence. For two-dimensional signals (i.e., digital images) vectors can be formed by taking 2-D blocks, such as rectangular blocks, and unwrapping them to form vectors. Similarly, the same idea can be applied to 3-D data (i.e., video), color and multispectral data, transform coefficients, and so on.

2 Theory of Vector Quantization

Although conceptually simple, there are a number of issues associated with VQ that are technically complex and relevant for an in-depth understanding of the process. To address these issues, such as design and optimality, it is useful to treat VQ in a mathematical framework.

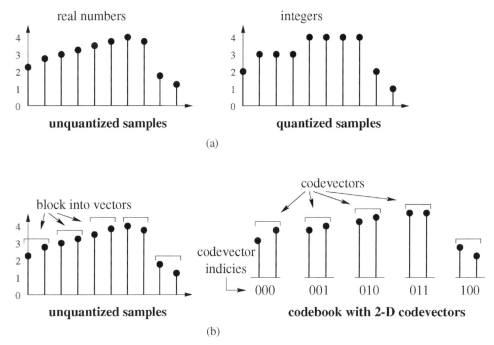

(a)

(b)

FIGURE 1   Illustration of (a) scalar and (b) vector quantization.

Toward this end, we can view VQ as two distinct operations — encoding and decoding — shown explicitly in Fig. 2. The encoder $\mathcal{E}$ performs a mapping from k-dimensional space $\mathcal{R}^k$ to the index set $\mathcal{I}$, and the decoder $\mathcal{D}$ maps the index set $\mathcal{I}$ into the finite subset $\mathcal{C}$, which is the codebook. The codebook has a positive integer number of codevectors that defines the codebook size. In this chapter, we will use N to denote the codebook size and $\mathbf{y}_i$ to denote the codevectors, which are the elements of $\mathcal{C}$. The bit rate R associated with the VQ depends on N (the number of codevectors in the codebook) and the vector dimension k. Since the bit rate is the number of bits per sample,

$$R = (\log_2 N)/k. \tag{1}$$

It is interesting to note that for VQ it is natural to have fractional bit rates such as $1/2$, $3/4$, $16/3$, etc. in contravention to basic

(fixed-rate) scalar quantization, in which noninteger rates do not arise naturally.

The operation associated with the decoder is extremely simple, involving no arithmetic at all. Conversely, the encoding procedure is complex, because a best matching vector decision must be made from among many candidate codevectors. To select a best matching codevector, we employ a numerically computable distortion measure $d(\mathbf{x}, \mathbf{y}_i)$, where low values of $d(\cdot, \cdot)$ imply a good match. There are many distortion measures that can be considered for quantifying the "quality of match" between two vectors $\mathbf{x}$ and $\mathbf{y}$, the most common of which is the *squared error* given by

$$d(\mathbf{x}, \mathbf{y}) = (\mathbf{x} - \mathbf{y})^t(\mathbf{x} - \mathbf{y}) = \sum_{\ell=1}^{k} (x[\ell] - y[\ell])^2,$$

where $x[\ell]$ and $y[\ell]$ are the elements of the vectors $\mathbf{x}$ and $\mathbf{y}$, respectively. For a vector $\mathbf{x}$ to be encoded, distortions are computed between it and each codevector $\mathbf{y}_i$ in the codebook. The codevector producing the smallest distortion is selected as the best match and the index associated with that codevector is used for the representation.

This process of encoding has an interesting and useful interpretation in the k-dimensional space. The set of codevectors defines a *partition* of $\mathcal{R}^k$ into N cells V_i, where $i = 1, 2, \ldots, N$. If we let $\mathcal{Q}(\cdot)$ represent the encoding operator, then the ith cell is defined by

$$V_i = \{\mathbf{x} \in \mathcal{R}^k : \mathcal{Q}(\mathbf{x}) = \mathbf{y}_i\}. \tag{2}$$

Partitions of this type that are formed uniquely from the codebook and a nearest neighbor distortion metric such as the

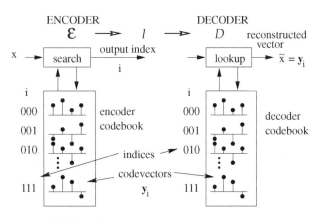

FIGURE 2   Block diagram of a VQ encoder and decoder.

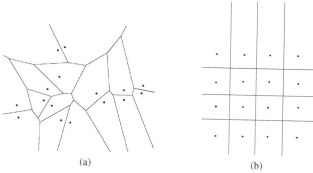

FIGURE 3 Illustration of the partition cells associated with VQ and scalar quantization: (a) partition cells for a 2-D VQ; (b) partition cells corresponding to scalar quantization.

squared error distortion are called Voronoi partitions. The notion of partitioning can be visualized easily in two dimensions, an illustration of which is shown in Fig. 3(a). Here, each vector has two elements (x_1, x_2) and consequently is a point in the two-dimensional space. That is, both the input vectors and codevectors are points in this space. The encoding procedure defines a unique partitioning of the space as shown in the figure, where the black dots denote codevectors.

The quality of performance of a VQ is typically measured by its average distortion for a given input source. In practice, sources are typically signal samples, image pixels, or some other data output associated with a signal that is being compressed. Whatever the source, average distortion measures are typically used to quantify the performance of a vector quantizer; the smaller the average distortion, the better the performance.

Vector quantizers are of interest because their performance is better than that of scalar quantization. The inherent advantage of VQ over scalar quantization can be understood through the concept of partitioning. Consider a fictitious input source for which we have designed an optimal 2-D codebook. Further, assume that the codevectors are those shown in Fig. 3(a). Observe that there are 16 2-D codevectors, implying a bit rate of 4 bits/vector, which is equivalent to 2 bits/sample. An optimal VQ design allows these codevectors to be positioned according to the statistical distribution of the input source vectors. Said another way, the codevectors are positioned to minimize the average distortion $\mathcal{D} = E\{(\mathbf{x}, \mathbf{y})\}$, where $\mathbf{x}$ and $\mathbf{y}$ are viewed as random vectors and E denotes the *expected value*. Assume the codevectors shown in Fig. 3(a) as black dots represent an optimal VQ for the input source in question. The associated partitions shown in the figure illustrate the diversity of cell shapes and sizes that VQ can realize.

Now consider quantizing our fictional input source with scalar quantization at an equivalent bit rate of 2 bits/sample. Two bits gives us four levels we can use to quantize the x_1 and x_2 axes. The cells implied by using a scalar quantizer for the input source are shown in Fig. 3(b). Notice that we have exactly 16 cells but that each cell is constrained to be rectangular. Moreover, scalar quantization imposes a structure that forces some cells to be placed in regions in the space where the input source may not be signif-

icantly populated. These observations lead to two immediately recognizable advantages of VQ over scalar quantization for the general k-dimensional case. First, VQ provides greater freedom to control the shapes of the cells to achieve more efficient tilings of the k-dimensional space. This property is often called *cell shape gain*. Second, VQ allows a greater number of cells to be concentrated in the k-dimensional regions where the source has the greatest density, which reduces the average distortion. Structural constraints associated with the scalar quantizer prevent it from capturing this property of the input. In general terms, because VQ operates on blocks of samples, it is able to exploit inherent statistical dependencies (both linear and nonlinear) within the blocks. The resulting gains in efficiency improve with higher vector dimension.

3 Design of Vector Quantizers

The key element in designing a VQ is determining the codebook for a given input source. In practice, the input source is represented by a large set of representative vectors called a training set. Over the years, there have been many algorithms proposed for VQ design. The most widely cited is the classical iterative method attributed to Linde, Buzo, and Gray, known as the LBG algorithm [2]. The LBG algorithm is fashioned around certain necessary conditions associated with the distinct encoder and decoder operations implicit in VQ. The first of these conditions states that for a fixed decoder codebook, an optimal encoder partition of $\mathcal{R}^k$ is the one that satisfies the *nearest neighbor rule*, which says that we map each input vector to the cell V_i producing the smallest distortion. By that measure, we are selecting the codevector that is nearest to the input vector.

The second optimality condition is the *centroid condition*. It states that for a given encoder partition cell, the optimal decoder codeword is the centroid of that cell, where the centroid of cell V_i is the vector $\mathbf{y}^*$ that minimizes $E\{d(\mathbf{x}, \mathbf{y})|\mathbf{x} \in V_i\}$, the average distortion in that cell. The centroid is a function of the distortion measure and is different for different distortion measures. For the popular squared error distortion, the centroid is simply the arithmetic average of the vectors in cell V_i, i.e.,

$$\mathbf{y}^* = \frac{1}{\|V_i\|} \sum_{\ell \in V_i} \mathbf{x}_\ell,$$

where $\|V_i\|$ denotes the number of vectors in cell V_i.

It can be shown that local optimality can be guaranteed by upholding these conditions, subject to some mild restrictions [1].

3.1 The LBG Design Algorithm

The necessary conditions for optimality provide the basis for the classical LBG VQ design algorithm. The LBG algorithm is a generalization of the scalar quantization design algorithm introduced by Lloyd, and hence it is also often called the generalized Lloyd algorithm, or GLA. Interestingly, this algorithm was known earlier in the pattern recognition community as the *k*-means algorithm.

The steps of the LBG algorithm for the design of an *N* vector codebook are straightforward and intuitive. Starting with a large training set (much larger than *N*), one first selects *N* initial codevectors. Initial codevectors can be selected randomly from the training set. There are two basic steps in the algorithm: encoding of other training vectors, and computation of the centroids. To begin, we first encode all the training vectors using the initial codebook. This process assigns a subset of the training vectors to each cell defined by the initial codevectors. Next, the centroid is computed for each cell. The centroids are then used to form an updated codebook. The process then repeats iteratively with a recoding of the training vectors and a new computation of the centroids to update the codebook. Ideally, at each iteration, the average distortion is reduced until convergence.

In practice, convergence is often slow near the point of convergence. Hence, in the interest of time, one often terminates the iterative algorithm when the codebook is very close to the local optimum. There are many stopping criteria that can be considered for this purpose. One approach in particular is to compute the average distortion $\mathcal{D}^{(\ell)}$ between the training vectors and the codevectors periodically during the design process, where the superscript ℓ denotes the ℓth iteration. If the normalized difference in distortion from one iteration to the next falls below a prespecified threshold, the design process can be terminated. For example, one could evaluate $\mathcal{D}$ at each iteration and compute the normalized difference,

$$\frac{\left(\mathcal{D}^{(\ell)} - \mathcal{D}^{(\ell-1)}\right)}{\mathcal{D}^{(\ell)}},$$

where forced termination is imposed when this normalized difference becomes less than the stopping threshold.

Often convergence proceeds smoothly. On occasion, the encoding stage of a given iteration may result in one or more cells' not being populated by any of the training vectors. This situation, known as the "empty cell" problem, effectively reduces the codebook size against our wishes. This condition when detected can be addressed in any one of a number of ways, one in particular consisting of splitting the cell with the greatest population in two to replace the lost empty cell.

3.2 Other Design Methods

Many methods of VQ design have appeared in the literature in recent years. Some focus on finding a good initial set of codevectors, which are then passed on to a classical LBG algorithm. By starting with a good initial codebook, one not only converges to a good solution, but generally converges in fewer iterations. Randomly selecting the initial codevectors from the training set is the easiest approach. This approach often works well, but sometimes it does not provide sufficient diversity to achieve a good locally optimal codebook. A simple variation that can be effective for certain sources is to select the *N* vectors (as initial codevectors)

from the training set that are farthest apart in terms of the distortion measure. This tends to assure that the initial codevectors are widely distributed in the *k*-dimensional space.

Alternatively, one can apply the splitting algorithm, which is a data dependent approach that systematically grows the initial codebook. The method, introduced in the original paper by Linde *et al.* [2], starts with a codebook consisting of the entire training set. First the centroid of the training set is computed. This centroid is then split into two codewords by perturbing the elements of the centroid. For instance, this could be done by adding some small value epsilon to each element. The original centroid and the perturbed centroid are used to encode the training set, after which centroids are computed to form a new initial codebook. These new centroids can then be perturbed and used to encode the training set. After centroids are computed, we have four codevectors in the codebook. The process can be repeated until *N* codevectors are obtained. At this point, the LBG algorithm can be applied as described earlier.

These approaches are intended primarily as a way to obtain initial codebooks for the LBG algorithms. Other methods have been proposed that attempt to find good codebooks directly, which may be optimized further by the LBG algorithm if so desired. One such algorithm in particular is the pairwise nearest neighbor, or PNN, algorithm [3]. In the PNN algorithm, we start with the training set and systematically merge vectors together until we arrive at a codebook of size *N*. The idea is to identify pairs of vectors that are closest together in terms of the distortion measure, and replace these two vectors with their mean, which reduces the codebook size at each stage. The PNN algorithm effectively merges those partitions that would result in the smallest increase in distortion.

The task of finding partitions to be merged is computationally demanding. In order to avoid this, a fast PNN method was developed that does not attempt to find the absolute smallest cost at each step. The interested reader is referred to the original paper by Equitz [3] for details. Codebooks designed by the PNN algorithm can be used directly for VQ or as initial codebooks for the LBG algorithm. It has been observed that using the PNN algorithm as a front end to the LBG algorithm (i.e., in lieu of the random selection or splitting methods) can lead to better locally optimal solutions. It is impossible to discuss all the design algorithms that have been proposed. However, it is appropriate to mention a few others in closing this section. There are a number of modifications to the LBG algorithm that can lead to an order of magnitude speedup in design time. One approach involves transforming all the training vectors into the discrete cosine transform domain and performing the VQ design in that domain. Because many of the transform coefficients are close to zero and hence can be neglected, codebook design can be performed effectively with lower dimensional vectors. Although there is overhead associated with performing the transform, it is offset by the efficiency concomitant with the design in a reduced dimensional space.

Neural nets have also been considered for VQ design. A number of researchers have successfully used neural nets to generate

VQ codebooks [5, 6]. Neural net algorithms can have advantages over the classical LBG algorithm, such as less sensitivity to the initialization of the codebook, better rate-distortion performance, and faster convergence.

The ultimate design algorithm is one that finds the global optimal. Several attempts at this have been reported, such as design by simulated annealing, by stochastic relaxation, and by genetic algorithms [7–10]. Algorithms of this type are perhaps the best in terms of performance, but they tend to have a very high computational complexity. Interestingly, amid all of these choices, the LBG algorithm still remains one of the most popular.

4 VQ Implementations

VQ is attractive because it has a performance advantage over scalar quantization. However, like all things in life, quality comes with a price. For VQ, that price comes in the form of increased encoder complexity and codebook memory. The number of codevectors that must be stored in a codebook grows exponentially with increasing bit rate. For example, a 16-dimensional VQ at a rate of 0.25 bits/sample requires a codebook of size 16, while the same VQ at a rate of 1 bit/sample requires 65,536 codevectors. Codebook memory also grows exponentially with vector dimension. For example, an eight-dimensional VQ at a rate of 1 bit/sample (with 1 byte codevector elements) would occupy 2048 bytes. Increasing the dimension to 32 causes the memory storage requirement to jump to over 34 Gbytes. Similarly, the same kind of exponential dependence exists for encoder complexity. Unlike scalar quantization, careful attention should be given to dimension and rate, because memory and complexity requirements can easily become prohibitively large. As a general rule, VQs that are employed in practice have a dimension of 16 or less, because complexity, memory, and performance tradeoffs are generally most attractive in this range.

A host of fast search methods have been reported for VQ that can be grouped into two general types. The first can be called *fast optimal search* methods, which are optimal in the sense that they guarantee that the encoder will find the best matching codevector for each input vector [1, 11, 12].

One of the simplest methods of this type is known as the *partial distortion* method. Consider the VQ encoder in which in the conventional paradigm the input vector **x** is compared to each of the codevectors by explicit computation of $d(\mathbf{x}, \mathbf{y}_i)$ for $i = 1, 2, \ldots, N$. The partial distortion method involves keeping track of the lowest distortion calculation to date as the codebook is being searched. To understand how complexity is reduced, assume that we have searched $N/4$ of the codevectors in the codebook and that the minimum distortion found thus far is $D[\min]$. For the next distortion calculation, we compute

$$D[i] = \sum_{\ell=1}^{k} (x[\ell] - y_i[\ell])^2,$$

where $x[\ell]$ and $y_i[\ell]$ are the elements of **x** and $\mathbf{y}_i$, respectively. If

during the process of evaluating the summation above, the value of $D[i]$ exceeds $D[\min]$, then we can terminate the calculation since we know that this vector is no longer a candidate. The net result of applying this procedure for encoding is that many of the vectors will be eliminated from further consideration prior to the full evaluation of the distortion calculation.

In addition, the triangle inequality can be used to reduce complexity, the idea being to use some reference points from which the distance to each code vector is precomputed and stored. The encoder then computes only the distance between the input vector and each reference point. Using these less complex comparisons in conjunction with precomputed data, one can achieve a reduction in complexity. The speed improvement realized by techniques of this type are clearly dependent on the codebook and input source; however, in general, one can expect a modest speedup.

Although every little bit helps, the complexity gains realized by optimal fast search algorithms fall short of addressing the exponential complexity growth associated with VQ. In this regard, efficient structured VQ encoding algorithms are attractive.

5 Structured VQ

A class of time-efficient methods has been studied extensively that sacrifice performance for substantial improvement in speed. The approach taken is to impose efficient structural constraints on the VQ codebook. These constraints are often formulated to make encoding complexity and/or memory *linearly* or *quadratically* dependent on the rate and dimension rather than *exponentially* dependent. The price paid, however, is usually inferior performance for the same rate and dimension. Nonetheless, the substantial reduction in complexity usually more than offsets the degradation in performance. To begin, we consider the most popular structured VQ of this class, tree-structured vector quantization (TSVQ).

5.1 Tree-Structured VQ

TSVQ consists of a hierarchical arrangement of codevectors, which allows the codebook to be searched efficiently. It has the property that search time grows linearly with rate instead of exponentially. Binary trees are often used for TSVQ because they are among the most efficient in terms of complexity. The concept of TSVQ can be illustrated by examining the binary tree shown in Fig. 4. As shown, the TSVQ has a root node at the top of the tree with many paths leading from it to the bottom. The codevectors of the tree,

$$\mathbf{y}_{000}, \mathbf{y}_{001}, \ldots, \mathbf{y}_{111},$$

are represented by the nodes at the bottom. The search path to reach any node (i.e., to find a codevector) is shown explicitly in the tree. In our particular example there are $N = 8$ codebook vectors and $N = 8$ paths in the tree, each leading to a different

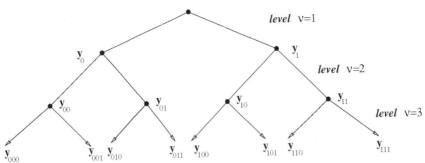

FIGURE 4 TSVQ diagram showing a three-level balance binary tree.

codevector. To encode an input vector **x**, we start at the top and move to the bottom of the tree. During that process, we encounter $\nu = 3$ (or $\log_2 N$) decision points (one at each level). The first decision (at level $\nu = 1$) is to determine whether **x** is closer to vector $\mathbf{y}_0$ or $\mathbf{y}_1$ by performing a distortion calculation. After a decision is made at the first level, the same procedure is repeated for the next levels until we have identified the codeword at the bottom of the tree. For a binary tree, it is apparent that $N = 2^\nu$, which means that for a codebook of size N, only $\log_2 N$ decisions have to be made. As presented, this implies the computation of two vector distortion calculations, $d(\cdot, \cdot)$, for each level, which results in only $2 \log_2 N$ distortion calculations per input vector.

Alternatively, one can perform the decision calculation explicitly in terms of hyperplane partitioning between the intermediate codevectors. The form of this calculation is the inner product between the hyperplane vector and input vector, where the sign of the output ($+$ or $-$) determines selection of either the right or left branch in the tree at that node. Implemented this way, only $\log_2 N$ distortion calculations are needed.

For the eight-vector TSVQ example above, this results in three instead of eight vector distortion calculations. For a larger (more realistic) codebook of size $N = 256$, the disparity is eight versus 256, which is quite significant.

TSVQ is a popular example of a constrained quantizer that allows implementation speed to be traded for increased memory and a small loss in performance. In many coding applications, such tradeoffs are often attractive.

5.2 Mean-Removed VQ

Mean-removed VQ is another popular example of a structured quantizer that leads to memory-complexity-performance tradeoffs that are often attractive in practice. It is a method for effectively reducing the codebook size by extracting the variation among vectors due specifically to the variation in the mean and coding that extracted component separately as a scalar. The motivation for this approach can be seen by recognizing that a codebook may have many similar vectors differing only in their mean values.

A functional block diagram of mean-removed VQ is shown in Fig. 5. First the mean of the input vector is computed and quantized with conventional scalar quantization. Then the mean-removed input vector is vector quantized in the conventional way by using a VQ that was designed with mean-removed training vectors. The outputs of the overall system are the VQ codewords and the mean values.

At the decoder, the mean-removed vectors are obtained by table loopup. These vectors are then added to a unit amplitude vector scaled by the mean, which in turn restores the mean to the mean-removed vector. This approach is really a hybrid of scalar quantization and VQ. The mean values, which are scalar quantized, effectively reduce the size of the VQ, making the overall system less memory and computation intensive.

One can represent the system as being a conventional VQ with codebook vectors consisting of all possible codewords obtainable by inserting the means in the mean-removed vectors. This representation is generally called a super codebook. The size of such a super codebook is potentially very large, but clearly it is also very constrained. Thus, better performance can always be achieved, in general, by using a conventional unconstrained codebook of the same size instead. However, since memory and complexity demands are often costly, mean-removed VQ is attractive.

5.3 Gain-Shape Vector Quantization

Gain-shape VQ is very similar to mean-removed VQ, but it involves extracting a gain term as the scalar component instead of a mean term. Specifically, the input vectors are decomposed

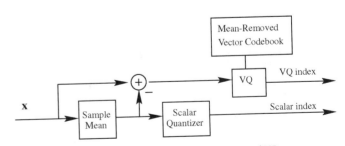

FIGURE 5 Block diagram of mean-removed VQ.

into a scalar *gain* term and a gain normalized vector term, which is commonly called the *shape*. The gain value is the Euclidean norm given by

$$g = \|\mathbf{x}\| = \sqrt{\sum_{i=1}^{k} x^2[i]}, \qquad (3)$$

and the shape vector **S** is given by

$$\mathbf{S} = \frac{\mathbf{x}}{g}. \qquad (4)$$

The gain term is quantized with a scalar quantizer, whereas the shape vectors are represented by a shape codebook designed specifically for the gain normalized shape vectors.

Perhaps not surprisingly, gain-shape VQ and mean-removed VQ can be combined effectively together to capture the complexity and memory reduction gains of both. Similarly, the implicit VQ could be designed as a TSVQ to achieve further complexity reduction if so desired.

To illustrate the performance of VQ in a printed medium such as a book, we find it convenient to use image coding as our application. Comparative examples are shown in Fig. 6. The image in Fig. 6(a) is an original eight bit/pixel 256×256 monochrome image. The image next to it is the same image coded with convention unstructured 4×4 VQ at a rate of 0.25 bits/pixel. The images on the bottom are results of the same coding using mean-removed and gain-shape VQ. From the example, one can observe distortion in all cases at this bit rate. The quality, however, for the unconstrained VQ case is better than that of the structured

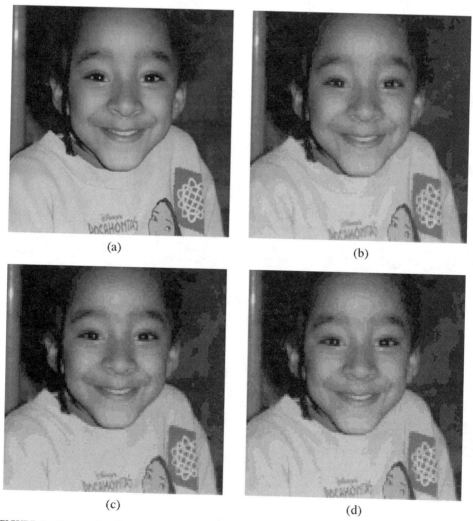

FIGURE 6 Comparative illustration of images coded using conventional VQ, mean-extraction VQ, and gain-shape VQ: (a) original image 256×256, Jennifer; (b) coded with VQ at 0.25 bpp (PSNR, 31.4 dB); (c) coded with mean-extraction VQ at 0.25 dB (PSNR = 30.85 dB); (d) coded with gain-shape VQ at 0.25 dB (PSNR = 30.56 dB). All coded images were coded at 0.25 bits/pixel using 4×4 vector blocks.

VQs in Fig. 6(c) and Fig. 6(d), both subjectively and in terms of the signal-to-noise ratio (SNR). For quantitative assessment of the quality, we can consider the peak SNR (PSNR) defined as

$$\text{PSNR} = 10 \log_{10}\left(\frac{255^2}{1/(N_1 N_2) \sum_{n_1=1}^{N_1} \sum_{n_2=1}^{N_2} (x[n_1, n_2] - \hat{x}[n_1, n_2])^2}\right). \tag{5}$$

Although the PSNR can be faulted easily as a good objective measure of quality, it can be useful if used with care. PSNRs are quoted in the examples shown, and they confirm the quality advantage of unconstrained VQ over the structured methods. However, the structured VQs have significantly reduced complexity.

5.4 Multistage Vector Quantization

A technique that has proven to be valuable for storage and complexity reduction is multistage VQ. This technique is also referred to as *residual* VQ, or RVQ. Multistage VQ divides the encoding task into a sequence of cascaded stages. The first stage performs a first-level approximation of the input vector. The approximation is refined by the second-level approximation that occurs in the second stage, and then is refined again in the third stage, and so on. The series of approximations or successive refinements is achieved by taking stage vector input and subtracting the coded vector from it, producing a residual vector. Thus, multistage VQ is simply a cascade of stage VQs that operate on stage residual vectors. At each stage, additional bits are needed to specify the new stage vector. At the same time, the quality of the representation is improved. A block diagram of a residual VQ is shown in Fig. 7.

Codebook design for multistage VQ can be performed in stages. First the original training set can be used to design the first-stage codebook. Residual vectors can then be computed for the training set using that codebook. The next stage codebook can then be designed using the first-stage residual training vectors, and so on until all stage codebooks are designed. This design approach is simple conceptually, but suboptimal. Improvement in performance can be achieved by designing the residual codebooks jointly as described in [13, 14].

The most dramatic advantage of residual VQ comes from its savings in memory and complexity, which for large VQs can be orders of magnitude less than that of the unconstrained VQ counterpart. In addition, residual VQ has the property that it al-

lows the bit rate to be controlled simply by specifying the number of VQ stage indices to be transmitted.

6 Variable-Rate Vector Quantization

The basic form of VQ alluded to thus far is more precisely called fixed rate VQ. That is, codevectors are represented by binary indices all with the same length. For practical data compression applications, we often desire variable rate coding, which allows statistical properties of the input to be exploited to further enhance the compression efficiency. Variable rate coding schemes of this type (entropy coders, an example of which is a Huffman coder) are based on the notion that codevectors that are selected infrequently on average are assigned longer indices, while codevectors that are used frequently are assigned short-length indices. Making the index assignments in this way (which is called entropy coding) results in a lower average bit rate in general and thus makes coding more efficient. Entropy coding the codebook indices can be done in a straightforward way. One only needs estimates of the codevector probabilities $P(i)$. With these estimates, methods such as Huffman coding will assign to the ith index a codeword whose length L_i is approximately $-\log_2 P(i)$ bits.

We can improve upon this approach by designing the VQ and the entropy coder together. This approach is called entropy-constrained VQ, or ECVQ. ECVQs can be designed by a modified LBG algorithm. Instead of finding the minimum distortion $d(\mathbf{x}, \mathbf{y_i})$ in the LBG iteration, one finds the minimum modified distortion

$$J_i = d(\mathbf{x}, \mathbf{y_i}) + \lambda L_i,$$

where $L_i = -\log_2 P(i)$. Employing this modified distortion J_i, which is a Lagrangian cost function, effectively enacts a Lagrangian minimization that seeks the minimum weighted cost of quantization error and bits.

To achieve rate control flexibility, one can design a set of codebooks corresponding to a discrete set of λs, which gives a set of VQs with a multiplicity of bit rates. The concept of ECVQ is powerful and can lead to performance gains in data compression systems. It can also be applied in conjunction with other structured VQs such as mean-removed VQ, gain-shape VQ, and residual VQ; the last of these is particularly interesting. Entropy constrained residual VQ, or EC-RVQ, and variations of it have proven to be among the most effective VQ methods for direct application to image compression. Schemes of this type involve the use of conditional probabilities in the entropy coding block

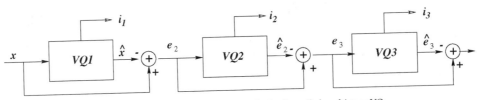

FIGURE 7   Block diagram of a residual VQ, also called multistage VQ.

where conditioning is performed on the previous stages and/or on adjacent stage vector blocks. Like ECVQ, the design is based on a Lagrangian cost function, but it is integrated into the RVQ design procedure. In the design algorithm reported in [14, 15], both the VQ stage codebooks and entropy coders are jointly optimized iteratively.

7 Closing Remarks

In the context of data compression, the concepts of optimality, partitioning, and distortion that we discussed are insightful and continue to inspire new contributions in the technical literature, particularly with respect to achieving useful tradeoffs among memory, complexity, and performance. Equally important are design methodologies and the use of variable length encoding for efficient compression. Although we have attempted to touch on the basics, the reader should be aware that the VQ topic area embodies much more than can be covered in a concise tutorial chapter. Thus, in closing, it is appropriate to at least mention several other classes of VQ that have received attention in recent years. First is the class of lattice VQs. Lattice VQs can be viewed as vector extensions of uniform quantizers, in the sense that the cells of a k- dimensional lattice VQ form a uniform tiling of the k-dimensional space. Searching such a codebook is highly efficient. The advantage achieved over scalar quantization is the ability of the lattice VQ to capture cell shape gain. The disadvantage, of course, is that cells are constrained to be uniform. Nonetheless, lattice VQ can be attractive in many practical systems.

Second is the general class of predictive VQs, which may include finite-state VQ (FSVQ), predictive VQ, vector predictive VQ, and several others. Some of these predictive approaches involve using neighboring vectors to define a state unambiguously at the encoder and decoder and then employing a specially designed codebook for that state. VQs of this type can exploit statistical dependencies (both linear and nonlinear) among adjacent vectors, but they have the disadvantage of being memory intensive.

Finally, it should be evident that VQ can be applied to virtually any lossy compression scheme. Prominent examples of this are transform VQ, in which the output of a linear block transform such as the DCT is quantized with VQ; and subband VQ, in which VQ is applied to the output of an analysis filter bank. The latter of these cases has proven to yield some of the best data compression algorithms currently known.

Interestingly, application of the principles of VQ extend far beyond our discussion. In addition to enabling construction of a wide variety of VQ data compression algorithms, VQ provides a useful framework in which one can explore fractal compression, motion estimation and motion compensated prediction, and automated classification. The inquisitive reader is challenged to explore this rich area of information theory in the literature [16].

References

[1] A. Gersho and R. Gray, *Vector Quantization and Signal Compression* (Kluwer, Boston, 1992).

[2] Y. Linde, A. Buzo, and R. Gray, "An algorithm for vector quantizer design," *IEEE Trans. Commun.* **C-28**, 84–95 (1980).

[3] W. Equitz, "A new vector quantization clustering algorithm," *IEEE Trans. Acoust. Speech Signal Process.* **37**, 1568–1575 (1989).

[4] R. King and N. Nasrabadi, "Image coding using vector quantization in the transform domain," *Pattern Recog. Lett.* **1**, 323–329 (1983).

[5] N. Nasrabadi and Y. Feng, "Vector quantization of images based upon the kohonen self-organizing feature map," in *IEEE International Conference on Neural Networks* (San Diego, CA, 1988), Vol. 1, pp. 101–105.

[6] J. McAuliffe, L. Atlas, and C. Rivera, "A comparison of the LBG algorithm and kohonen neural network paradigm for image vector quantization," in *IEEE International Conference on Acoustics, Speech, and Signal Processing* (Albuquerque, NM, 1990), pp. 2293–2296.

[7] J. Vaisey and A. Gersho, "Simulated annealing and codebook design," in *IEEE International Conference on Acoustics, Speech and Signal Processing* (New York, 1988), pp. 1176–1179.

[8] K. Zeger and A. Gersho, "A stochastic relaxation algorithm for improved vector quantizer design," *Electron. Lett.* **25**, 896–898 (1989).

[9] K. Zeger, J. Vaisey, and A. Gersho, "Globally optimal vector quantizer design by stochastic relaxation," *IEEE Trans. Signal Process.* **40**, 310–322 (1992).

[10] K. Krishna, K. Ramakrishna, and M. Thathachar, "Vector quantization using genetic k-means algorithm for image compression," in *Proceedings of the 1997 International Conference on Information, Communication, and Signal Processing, Part 3* (1997), Vol. 3, pp. 1585–1587.

[11] M. Soleymani and S. Morgera, "An efficient nearest neighbor search method," *IEEE Trans. Commun.* **COM-35**, 677–679 (1987).

[12] D. Cheng, A. Gersho, B. Ramamurthi, and Y. Shoham, "Fast search algorithms for vector quantization and pattern matching," in *Proceedings of the International Conference on Acoustics, Speech, and Signal Processing* (San Diego, CA, 1984), pp. 911.1–911.4

[13] C. Barnes, "Residual quantizers," Ph.D. thesis (Brigham Young University, Provo, UT, 1989).

[14] F. Kossentini, M. Smith, and C. Barnes, "Necessary conditions for the optimality of variable rate residual vector quantizers," *IEEE Trans. Inf. Theory*, November, **41**, 1903–1915 (1995).

[15] F. Kossentini, W. Chung, and M. Smith, "Conditional entropy-constrained residual VQ with application to image coding," *Trans. Image Process. Spec. Issue VQ*, February, **5**, 311–321 (1996).

[16] M. Barlaud, P. A. Chou, N. M. Nasrabadi, D. Neuhoff, M. Smith, and J. Woods, eds., *IEEE Transactions on Image Processing, Special Issue on Vector Quantization*, February 1996.

5.4

Wavelet Image Compression

Zixiang Xiong
Texas A&M University

Kannan Ramchandran
University of California, Berkeley

1 What Are Wavelets: Why Are They Good for Image Coding?

During the past decade, wavelets have made quite a splash in the field of image compression. In fact, the FBI has already adopted a wavelet-based standard for fingerprint image compression. The evolving next-generation image compression standard, dubbed JPEG-2000, which will dislodge the currently popular JPEG standard (see Chapter 5.5), will also be based on wavelets. Given these exciting developments, it is natural to ask why wavelets have made such an impact in image compression. This chapter will answer this question, providing both high-level intuition as well as illustrative details based on state-of-the-art wavelet-based coding algorithms. Visually appealing time-frequency based analysis tools are sprinkled generously to aid in our task.

Wavelets are tools for decomposing signals, such as images, into a hierarchy of increasing resolutions: as we consider more and more resolution layers, we get a more and more detailed look at the image. Figure 1 shows a three-level hierarchy wavelet decomposition of the popular test image *Lena* from coarse to fine resolutions (for a detailed treatment on wavelets and multiresolution decompositions, also see Chapter 4.1). Wavelets can be regarded as "mathematical microscopes" that permit one to "zoom in" and "zoom out" of images at multiple resolutions. The remarkable thing about the wavelet decomposition is that it enables this zooming feature at absolutely no cost in terms of excess redundancy: for an $M \times N$ image, there are exactly $M N$

wavelet coefficients — exactly the same as the number of original image pixels (see Fig. 2).

As a basic tool for decomposing signals, wavelets can be considered as duals to the more traditional Fourier-based analysis methods that we encounter in traditional undergraduate engineering curricula. Fourier analysis associates the very intuitive engineering concept of "spectrum" or "frequency content" of a signal. Wavelet analysis, in contrast, associates the equally intuitive concept of "resolution" or "scale" of the signal. At a functional level, Fourier analysis is to wavelet analysis as spectrum analyzers are to microscopes.

As wavelets and multiresolution decompositions have been described in greater depth in Chapter 4.1, our focus here will be more on the image compression application. Our goal is to provide a self-contained treatment of wavelets within the scope of their role in image compression. More importantly, our goal is to provide a high-level explanation for why they have made such an impact in image compression. Indeed, wavelets are ready to dislodge the more traditional Fourier-based method in the form of the discrete cosine transform (DCT) that is currently deployed in the popular JPEG image compression standard (see Chapter 5.5). Standardization activities are in full swing currently to deploy the next-generation JPEG-2000 standard in optimistic anticipation of when the supplanting is likely to occur. The JPEG-2000 standard will be a significant improvement over the current JPEG standard. While details of how it will evolve, and what features will be supported, are being worked out during the writing of this chapter, there is only one thing that is not

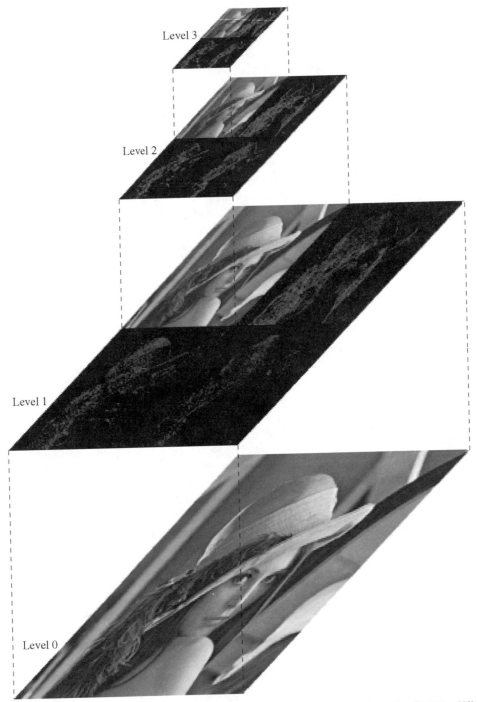

FIGURE 1 A three-level hierarchy wavelet decomposition of the 512×512 color *Lena* image. Level 1 (512×512) is the one-level wavelet representation of the original *Lena* at Level 0; Level 2 (256×256) shows the one-level wavelet representation of the low-pass image at Level 1; and Level 3 (128×128) gives the one-level wavelet representation of the low-pass image at Level 2. (See color section, p. C–24.)

in doubt: JPEG-2000 will be wavelet based. We will also cover powerful generalizations of wavelets, known as wavelet packets, that have already made an impact in the standardization world: the FBI fingerprint compression standard is based on wavelet packets.

Although this chapter is about image coding,[1] which involves two-dimensional (2-D) signals or images, it is much easier to

[1] We use the terms *image compression* and *image coding* interchangeably in this chapter.

FIGURE 2 A three-level wavelet representation of the *Lena* image generated from the top view of the three-level hierarchy wavelet decomposition in Fig. 1. It has exactly the same number of samples as in the image domain. (See color section, p. C–25.)

understand the role of wavelets in image coding using a one-dimensional (1-D) framework, as the conceptual extension to 2-D is straightforward. In the interests of clarity, we will therefore consider a 1-D treatment here. The story begins with what is known as the time-frequency analysis of the 1-D signal. As mentioned, wavelets are a tool for changing the coordinate system in which we represent the signal: we transform the signal into another domain that is much better suited for processing, e.g., compression. What makes for a good transform or analysis tool? At the basic level, the goal is to be able to represent all the useful signal features and important phenomena in as compact a manner as possible. It is important to be able to compact the bulk of the signal energy into the fewest number of transform coefficients: this way, we can discard the bulk of the transform domain data without losing too much information. For example, if the signal is a time impulse, then the best thing is to do no transforms at all! Keep the signal information in its original time-domain version, as that will maximize the temporal energy concentration or time resolution. However, what if the signal has a critical frequency component (e.g., a low-frequency background sinusoid) that lasts for a long time duration? In this case, the energy is spread out in the time domain, but it would be succinctly captured in a single frequency coefficient if one did a Fourier analysis of the signal. If we know that the signals of interest are pure sinusoids, then Fourier analysis is the way to go. But, what if we want to capture both the time impulse and the frequency impulse with good resolution? Can we get arbitrarily fine resolution in both time and frequency?

The answer is no. There exists an uncertainty theorem (much like what we learn in quantum physics), which disallows the existence of arbitrary resolution in time and frequency [1]. A good way of conceptualizing these ideas and the role of wavelet basis functions is through what is known as time-frequency "tiling"

plots, as shown in Fig. 3, which shows where the basis functions live on the time-frequency plane; i.e., where is the bulk of the energy of the elementary basis elements localized? Consider the Fourier case first. As impulses in time are completely spread out in the frequency domain, all localization is lost with Fourier analysis. To alleviate this problem, one typically decomposes the signal into finite-length chunks using windows or so-called short-time Fourier transform (STFT). Then, the time-frequency tradeoffs will be determined by the window size. An STFT expansion consists of basis functions that are shifted versions of one another in both time and frequency: some elements capture low-frequency events localized in time, and others capture high-frequency events localized in time, but the resolution or window size is constant in both time and frequency [see Fig. 3(a)]. Note that the uncertainty theorem says that the area of these tiles has to be nonzero.

Shown in Fig. 3(b) is the corresponding tiling diagram associated with the wavelet expansion. The key difference between this and the Fourier case, which is the critical point, is that the tiles are not all of the same size in time (or frequency). Some basis elements have short-time windows; others have short-frequency windows. Of course, the uncertainty theorem ensures that the area of each tile is constant and nonzero. It can be shown that

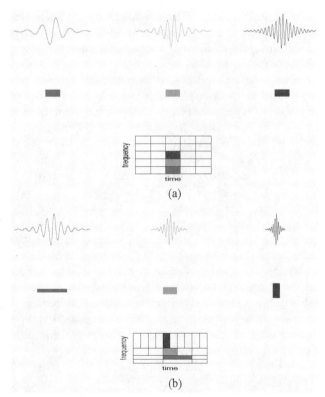

FIGURE 3 Tiling diagrams' associated STFT bases and wavelet bases. (a) STFT bases and the tiling diagram associated with a STFT expansion. STFT bases of different frequencies have the same resolution (or length) in time. (b) Wavelet bases and tiling diagram associated with a wavelet expansion. The time resolution is inversely proportional to frequency for wavelet basis. (See color section, p. C–25.)

the basis functions are related to one another by shifts and scales as is the key to wavelet analysis.

Why are wavelets well suited for image compression? The answer lies in the time-frequency (or more correctly, space-frequency) characteristics of typical natural images, which turn out to be well captured by the wavelet basis functions shown in Fig. 3(b). Note that the STFT tiling diagram of Fig. 3(a) is conceptually similar to what current commercial DCT-based image transform coding methods like JPEG use. Why are wavelets inherently a better choice? Looking at Fig. 3(b), one can note that the wavelet basis offers elements having good frequency resolution at lower frequency (the short and fat basis elements) while simultaneously offering elements that have good time resolution at higher frequencies (the tall and skinny basis elements).

This tradeoff works well for natural images and scenes that are typically composed of a mixture of important long-term low-frequency trends that have larger spatial duration (such as slowly varying backgrounds like the blue sky, and the surface of lakes, etc.) as well as important transient short-duration high-frequency phenomena such as sharp edges. The wavelet representation turns out to be particularly well suited to capturing both the transient high-frequency phenomena such as image edges (using the tall and skinny tiles) as well as long-spatial-duration low-frequency phenomena such as image backgrounds (the short and fat tiles). As natural images are dominated by a mixture of these kinds of events,[2] wavelets promise to be very efficient in capturing the bulk of the image energy in a small fraction of the coefficients.

To summarize, the task of separating transient behavior from long-term trends is a very difficult task in image analysis and compression. In the case of images, the difficulty stems from the fact that statistical analysis methods often require the introduction of at least some local stationarity assumption, i.e., that the image statistics do not change abruptly over time. In practice, this assumption usually translates into *ad hoc* methods to block data samples for analysis, methods that can potentially obscure important signal features: e.g., if a block is chosen too big, a transient component might be totally neglected when computing averages. The blocking artifact in JPEG decoded images at low rates is a result of the block-based DCT approach. A fundamental contribution of wavelet theory [2] is that it provides a unified framework in which transients and trends can be simultaneously analyzed without the need to resort to blocking methods.

As a way of highlighting the benefits of having a sparse representation, such as that provided by the wavelet decomposition, consider the lowest frequency band in the top level (Level 3) of the three-level wavelet hierarchy of *Lena* in Fig. 1. This band is just a downsampled (by a factor of $8^2 = 64$) and smoothed version of the original image. A very simple way of achieving compression is to simply retain this low-pass version and throw away the rest of the wavelet data, instantly achieving a compression ratio of 64:1. Note that if we want a full-size approximation to the original, we would have to interpolate the low-pass band by a factor of 64 — this can be done efficiently by using a three-stage synthesis filter bank (see Chapter 4.1). We may also desire better image fidelity, as we may be compromising high-frequency image detail, especially perceptually important high-frequency edge information. This is where wavelets are particularly attractive, as they are capable of capturing most image information in the highly subsampled low-frequency band, and additional localized edge information in spatial clusters of coefficients in the high-frequency bands (see Fig. 1). The bulk of the wavelet data is insignificant and can be discarded or quantized very coarsely.

Another attractive aspect of the coarse-to-fine nature of the wavelet representation naturally facilitates a transmission scheme that progressively refines the received image quality. That is, it would be highly beneficial to have an encoded bitstream that can be chopped off at any desired point to provide a commensurate reconstruction image quality. This is known as a progressive transmission feature or as an embedded bitstream (see Fig. 4). Many modern wavelet image coders have this feature, as will be covered in more detail in Section 5. This is ideally suited, for example, to Internet image applications. As is well known, the Internet is a heterogeneous mess in terms of the number of users and their computational capabilities and effective bandwidths. Wavelets provide a natural way to satisfy users having disparate bandwidth and computational capabilities: the low-end users can be provided a coarse quality approximation, whereas higher-end users can use their increased bandwidth to get better fidelity. This is also very useful for Web browsing applications, where having a coarse quality image with a short waiting time may be preferable to having a detailed quality with an unacceptable delay.

These are some of the high-level reasons why wavelets represent a superior alternative to traditional Fourier-based methods for compressing natural images: this is why the evolving JPEG-2000 standard will use wavelets instead of the Fourier-based DCTs.

In the sequel, we will review the salient aspects of the general compression problem and the transform coding paradigm in particular, and highlight the key differences between the class of early subband coders and the recent more advanced class of modern-day wavelet image coders. We pick the celebrated embedded zero-tree wavelet coder as a representative of this latter class, and we describe its operation by using a simple illustrative example. We conclude with more powerful generalizations of the basic wavelet image coding framework to wavelet packets, which are particularly well suited to handle special classes of images such as fingerprints.

2 The Compression Problem

Image compression falls under the general umbrella of data compression, which has been studied theoretically in the field of

---

[2] Typical images also contain textures; however, conceptually, textures can be assumed to be a dense concentration of edges, and so it is fairly accurate to model typical images as smooth regions delimited by edges.

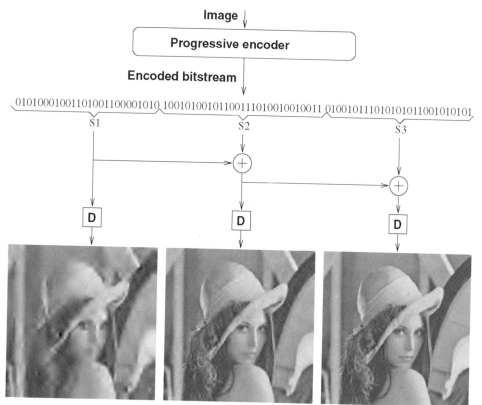

FIGURE 4 Multiresolution wavelet image representation naturally facilitates progressive transmission — a desirable feature for the transmission of compressed images over heterogeneous packet networks and wireless channels.

information theory [3], pioneered by Claude Shannon [4] in 1948. Information theory sets the fundamental bounds on compression performance theoretically attainable for certain classes of sources. This is very useful because it provides a theoretical benchmark against which one can compare the performance of more practical but suboptimal coding algorithms.

Historically, the lossless compression problem came first. Here the goal is to compress the source with no loss of information. Shannon showed that given any discrete source with a well-defined statistical characterization (i.e., a probability mass function), there is a fundamental theoretical limit to how well you can compress the source before you start to lose information. This limit is called the *entropy* of the source. In lay terms, entropy refers to the uncertainty of the source. For example, a source that takes on any of N discrete values $a_1, a_2, \ldots, a_N$ with equal probability has an entropy given by $\log_2 N$ bits per source symbol. If the symbols are not equally likely, however, then one can do better because more predictable symbols should be assigned fewer bits. The fundamental limit is the Shannon entropy of the source.

Lossless compression of images has been covered in Chapter 5.1 and Chapter 5.6. For image coding, typical lossless compression ratios are of the order of 2:1 or at most 3:1. For a 512×512 8-bit gray-scale image, the uncompressed representation is 256 Kbytes. Lossless compression would reduce this to at best ~80 Kbytes, which may still be excessive for many practical low-bandwidth transmission applications. Furthermore, lossless

image compression is for the most part overkill, as our human visual system is highly tolerant to losses in visual information. For compression ratios in the range of 10:1 to 40:1 or more, lossless compression cannot do the job, and one needs to resort to lossy compression methods.

The formulation of the lossy data compression framework was also pioneered by Shannon in his work on rate-distortion (R-D) theory [5], in which he formalized the theory of compressing certain limited classes of sources having well-defined statistical properties, e.g., independent, identically distributed (i.i.d.) sources having a Gaussian distribution subject to a fidelity criterion, i.e., subject to a tolerance on the maximum allowable loss or distortion that can be endured. Typical distortion measures used are mean square error (MSE) or peak signal-to-noise ratio (PSNR)[3] between the original and compressed versions. These fundamental compression performance bounds are called the theoretical R-D bounds for the source: they dictate the minimum rate R needed to compress the source if the tolerable distortion level is D (or alternatively, what is the minimum distortion D subject to a bit rate of R). These bounds are unfortunately not constructive; i.e., Shannon did not give an actual algorithm for attaining these bounds, and furthermore they are based on arguments that assume infinite complexity and delay, obviously impractical in real life. However, these bounds are useful in as

---

[3]The PSNR is defined as $10 \log_{10} \frac{255^2}{\text{MSE}}$ and measured in decibels (dB).

much as they provide valuable benchmarks for assessing the performance of more practical coding algorithms. The major obstacle of course, as in the lossless case, is that these theoretical bounds are available only for a narrow class of sources, and it is difficult to make the connection to real world image sources which are difficult to model accurately with simplistic statistical models.

Shannon's theoretical R-D framework has inspired the design of more practical *operational* R-D frameworks, in which the goal is similar but the framework is constrained to be more practical. Within the operational constraints of the chosen coding framework, the goal of operational R-D theory is to minimize the rate R subject to a distortion constraint D, or vice versa. The message of Shannon's R-D theory is that one can come close to the theoretical compression limit of the source, if one considers *vectors* of source symbols that get infinitely large in dimension in the limit; i.e., it is a good idea not to code the source symbols one at a time, but to consider chunks of them at a time, and the bigger the chunks the better. This thinking has spawned an important field known as *vector quantization*, or VQ [6], which, as the name indicates, is concerned with the theory and practice of quantizing sources using high-dimensional vector quantization (image coding using VQ is covered in Chapter 5.3). There are practical difficulties arising from making these vectors too high dimensional because of complexity constraints, so practical frameworks involve relatively small dimensional vectors that are therefore further from the theoretical bound.

Because of this reason, there has been a much more popular image compression framework that has taken off in practice: this is the transform coding framework [7] that forms the basis of current commercial image and video compression standards like JPEG and MPEG (see Chapters 6.4 and 6.5). The transform coding paradigm can be construed as a practical special case of VQ that can attain the promised gains of processing source symbols in vectors through the use of efficiently implemented high-dimensional source transforms.

3 The Transform Coding Paradigm

In a typical transform image coding system, the encoder consists of a linear transform operation, followed by quantization of transform coefficients, and lossless compression of the quantized coefficients using an entropy coder. After the encoded bitstream of an input image is transmitted over the channel (assumed to be perfect), the decoder undoes all the functionalities applied in the encoder and tries to reconstruct a decoded image that looks as close as possible to the original input image, based on the transmitted information. A block diagram of this transform image paradigm is shown in Fig. 5.

For the sake of simplicity, let us look at a 1-D example of how transform coding is done (for 2-D images, we treat the rows and columns separately as 1-D signals). Suppose we have a two-point signal: $x_0 = 216$, $x_1 = 217$. It takes 16 bits (8 bits for each sample) to store this signal in a computer. In transform coding, we first put x_0 and x_1 in a column vector $X = \begin{bmatrix} x_0 \\ x_1 \end{bmatrix}$ and apply an orthogonal transformation T to X to get $Y = \begin{bmatrix} y_0 \\ y_1 \end{bmatrix} = TX = \begin{bmatrix} 1/\sqrt{2} & 1/\sqrt{2} \\ 1/\sqrt{2} & -1/\sqrt{2} \end{bmatrix} \begin{bmatrix} x_0 \\ x_1 \end{bmatrix} = \begin{bmatrix} 1/\sqrt{2}[x_0+x_1] \\ 1/\sqrt{2}[x_0-x_1] \end{bmatrix} = \begin{bmatrix} 306.177 \\ -.707 \end{bmatrix}$. The transform T can be conceptualized as a counterclockwise rotation of the signal vector X by $45°$ with respect to the original (x_0, x_1) coordinate system. Alternatively and more conveniently, one can think of the signal vector as being fixed and instead rotate the (x_0, x_1) coordinate system by $45°$ clockwise to the new (y_1, y_0) coordinate system (see Fig. 6). Note that the abscissa for the new coordinate system is now y_1.

Orthogonality of the transform simply means that the length of Y is the same as the length of X (which is even more obvious when one freezes the signal vector and rotates the coordinate system as discussed above). This concept still carries over to the case of high-dimensional transforms. If we decide to use the simplest form of quantization known as uniform scalar quantization, where we round off a real number to the nearest integer multiple of a step size q (say $q = 20$), then the quantizer index vector $\hat{I}$, which captures *what* integer multiples of q are nearest to the

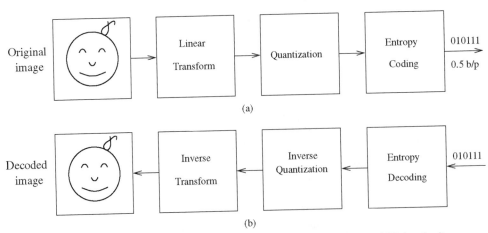

(a)

(b)

FIGURE 5 Block diagrams of a typical transform image coding system: (a) encoder and (b) decoder diagrams.

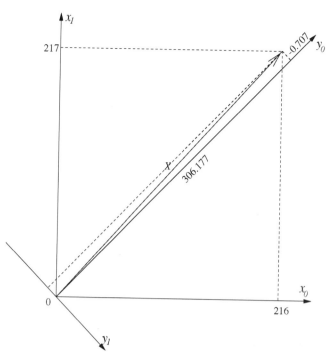

FIGURE 6 The transform T can be conceptualized as a counterclockwise rotation of the signal vector X by $45°$ with respect to the original (x_0, x_1) coordinate system.

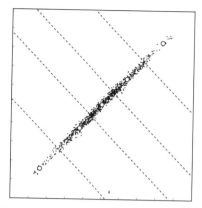

FIGURE 7 Linear transformation amounts to a rotation of the coordinate system, making correlated samples in the time domain less correlated in the transform domain.

entries of Y, is given by $\hat{I} = \begin{bmatrix} \text{round}(y_0/q) \\ \text{round}(y_1/q) \end{bmatrix} = \begin{bmatrix} 15 \\ 0 \end{bmatrix}$. We store (or transmit) $\hat{I}$ as the compressed version of X using 4 bits, achieving a compression ratio of 4:1. To decode X from $\hat{I}$, we first multiply $\hat{I}$ by $q = 20$ to dequantize, i.e., to form the quantized approximation $\hat{Y}$ of Y with $\hat{Y} = q \cdot \hat{I} = \begin{bmatrix} 300 \\ 0 \end{bmatrix}$, and then apply the inverse transform T^{-1} to $\hat{Y}$ [which corresponds in our example to a counterclockwise rotation of the (y_1, y_0) coordinate system by $45°$, just the reverse operation of the T operation on the original (x_0, x_1) coordinate system — see Fig. 6] to get $\hat{X} = T^{-1} \begin{bmatrix} q y_0 \\ q y_1 \end{bmatrix} = \begin{bmatrix} 1/\sqrt{2} & 1/\sqrt{2} \\ 1/\sqrt{2} & -1/\sqrt{2} \end{bmatrix} \begin{bmatrix} 300 \\ 0 \end{bmatrix} = \begin{bmatrix} 212.132 \\ 212.132 \end{bmatrix}$.

We see from the above example that, although we "zero out" or throw away the transform coefficient y_1 in quantization, the decoded version $\hat{X}$ is still very close to X. This is because the transform effectively compacts most of the energy in X into the first coefficient y_0, and renders the second coefficient y_1 considerably insignificant to keep. The transform T in our example actually computes a weighted sum and difference of the two samples x_0 and x_1 in a manner that preserves the original energy. It is in fact the *simplest* wavelet transform!

The energy compaction aspect of wavelet transforms was highlighted in Section 1. Another goal of linear transformation is decorrelation. This can be seen from the fact that, although the values of x_0 and x_1 are very close (highly correlated) before the transform, y_0 (sum) and y_1 (difference) are very different (less correlated) after the transform. Decorrelation has a nice geometric interpretation. A cloud of input samples of length 2 is shown along the $45°$ line in Fig. 7. The coordinates (x_0, x_1) at each point

of the cloud are nearly the same, reflecting the high degree of correlation among neighboring image pixels. The linear transformation T essentially amounts to a rotation of the coordinate system. The axes of the new coordinate system are parallel and perpendicular to the orientation of the cloud. The coordinates (y_0, y_1) are less correlated, as their magnitudes can be quite different and the sign of y_1 is random. If we assume x_0 and x_1 are samples of a stationary random sequence $X(n)$, then the correlation between y_0 and y_1 is $E\{y_0 y_1\} = E\{1/2(x_0^2 - x_1^2)\} = 0$. This decorrelation property has significance in terms of how much gain one can get from transform coding than from doing signal processing (quantization and coding) directly in the original signal domain, called pulse code modulation (PCM) coding.

Transform coding has been extensively developed for coding of images and video, where the DCT is commonly used because of its computational simplicity and its good performance. But as shown in Section 1, the DCT is giving way to the wavelet transform because of the latter's superior energy compaction capability when applied to natural images. Before discussing state-of-the-art wavelet coders and their advanced features, we address the functional units that comprise a transform coding system, namely the transform, quantizer, and entropy coder (see Fig. 5).

3.1 Transform Structure

The basic idea behind using a linear transformation is to make the task of compressing an image in the transform domain after quantization easier than direct coding in the spatial domain. A good transform, as has been mentioned, should be able to *decorrelate* the image pixels, and provide good *energy compaction* in the transform domain so that very few quantized nonzero coefficients have to be encoded. It is also desirable for the transform to be orthogonal so that the energy is conserved from the spatial domain to the transform domain, and the distortion in the spatial domain introduced by quantization of transform coefficients can be directly examined in the transform domain. What makes the wavelet transform special in all possible choices is that it offers an efficient space-frequency characterization for a broad class of natural images, as shown in Section 1.

3.2 Quantization

As the only source of information loss occurs in the quantization unit, efficient quantizer design is a key component in wavelet image coding. Quantizers come in many different shapes and forms, from very simple uniform scalar quantizers, such as the one in the example earlier, to very complicated vector quantizers. Fixed length uniform scalar quantizers are the simplest kind of quantizers: these simply round off real numbers to the nearest integer multiples of a chosen step size. The quantizers are fixed length in the sense that all quantization levels are assigned the same number of bits (e.g., an eight-level quantizer would be assigned all binary three-tuples between 000 and 111). Fixed length *non-uniform* scalar quantizers, in which the quantizer step sizes are not all the same, are more powerful: one can optimize the design of these non-uniform step sizes to get what is known as Lloyd–Max quantizers [8].

It is more efficient to do a joint design of the quantizer and the entropy coding functional unit (this will be described in the next subsection) that follows the quantizer in a lossy compression system. This joint design results in a so-called entropy-constrained quantizer that is more efficient but more complex, and results in variable length quantizers, in which the different quantization choices are assigned variable codelengths. Variable length quantizers can come in either scalar, known as entropy-constrained scalar quantization, or ECSQ [9], or vector, known as entropy-constrained vector quantization, or ECVQ [6], varieties. An efficient way of implementing vector quantizers is by the use of so-called trellis coded quantization, or TCQ [10]. The performance of the quantizer (in conjunction with the entropy coder) characterizes the operational R-D function of the source. The theoretical R-D function characterizes the fundamental lossy compression limit theoretically attainable [11], and it is rarely known in analytical form except for a few special cases, such as the i.i.d. Gaussian source [3]:

$$D(R) = \sigma^2 2^{-2R},$$

where the Gaussian source is assumed to have zero mean and variance σ^2 and the rate R is measured in bits per sample. Note from the formula that every extra bit reduces the expected distortion by a factor of 4 (or increases the signal to noise ratio by 6 dB). This formula agrees with our intuition that the distortion should decrease exponentially as the rate increases. In fact, this is true when quantizing sources with other probability distributions as well under high-resolution (or bit rate) conditions: the optimal R-D performance of encoding a zero mean stationary source with variance σ^2 takes the form of [6]

$$D(R) = h\sigma^2 2^{-2R},$$

where the factor h depends on the probability distribution of the source. For a Gaussian source, $h = \sqrt{3}\pi/2$ with optimal scalar quantization. Under high-resolution conditions, it can be shown that the optimal entropy-constrained scalar quantizer is a uniform one, whose average distortion is only approximately 1.53 dB worse than the theoretical bound attainable that is known as the Shannon bound [6, 9]. For low bit rate coding, most current subband coders employ a uniform quantizer with a "dead zone" in the central quantization bin. This simply means that the all-important central bin is wider than the other bins: this turns out to be more efficient than having all bins be of the same size. The performance of dead-zone quantizers is nearly optimal for memoryless sources even at low rates [12]. An additional advantage of using dead-zone quantization is that, when the dead zone is twice as much as the uniform step size, an embedded bitstream can be generated by successive quantization. We will elaborate more on embedded wavelet image coding in Section 5.

3.3 Entropy Coding

Once the quantization process is completed, the last encoding step is to use entropy coding to achieve the entropy rate of the quantizer. Entropy coding works like the Morse code in electric telegraph: more frequently occurring symbols are represented by short codewords, whereas symbols occurring less frequently are represented by longer codewords. On average, entropy coding does better than assigning the same codelength to all symbols. For example, a source that can take on any of the four symbols $\{A, B, C, D\}$ with equal likelihood has two bits of information or uncertainty, and its entropy is 2 bits per symbol (e.g., one can assign a binary code of 00 to A, 01 to B, 10 to C, and 11 to D). However if the symbols are not equally likely, e.g., if the probabilities of A, B, C, D are 0.5, 0.25, 0.125, 0.125, respectively, then one can do much better on average by not assigning the same number of bits to each symbol, but rather by assigning fewer bits to the more popular or predictable ones. This results in a variable length code. In fact, one can show that the optimal code would be one in which A gets 1 bit, B gets 2 bits, and C and D get 3 bits each (e.g., $A = 0$, $B = 10$, $C = 110$, $D = 111$). This is called an entropy code. With this code, one can compress the source with an average of only 1.75 bits per symbol, a 12.5% improvement in compression over the original 2 bits per symbol associated with having fixed length codes for the symbols. The two popular entropy coding methods are Huffman coding [13] and arithmetic coding [14]. A comprehensive coverage of entropy coding is given in Chapter 5.1. The Shannon entropy [3] provides a lower bound in terms of the amount of compression entropy coding can best achieve. The optimal entropy code constructed in the example actually achieves the theoretical Shannon entropy of the source.

4 Subband Coding: The Early Days

Subband coding normally uses bases of roughly equal bandwidth. Wavelet image coding can be viewed as a special case of subband coding with logarithmically varying bandwidth bases that satisfy certain properties.[4] Early work on wavelet image coding was thus hidden under the name of subband coding [7, 15], which builds upon the traditional transform coding paradigm of energy compaction and decorrelation. The main idea of subband coding is to treat different bands differently as each band can be modeled as a statistically distinct process in quantization and coding.

To illustrate the design philosophy of early subband coders, let us again assume for example that we are coding a vector source $\{x_0, x_1\}$, where both x_0 and x_1 are samples of a stationary random sequence $X(n)$ with zero mean and variance σ_x^2. If we code x_0 and x_1 directly by using PCM coding, from our earlier discussion on quantization, the R-D performance can be approximated as

$$D_{\text{PCM}}(R) = h\sigma_x^2 2^{-2R}.$$

In subband coding, two quantizers are designed: one for each of the two transform coefficients y_0 and y_1. The goal is to choose rates R_0 and R_1 needed for coding y_0 and y_1 so that the average distortion

$$D_{\text{SBC}}(R) = 1/2(D(R_0) + D(R_1))$$

is minimized with the constraint on the average bit rate

$$1/2(R_0 + R_1) = R.$$

Using the high rate approximation, we write $D(R_0) = h\sigma_{y_0}^2 2^{-2R_0}$ and $D(R_1) = h\sigma_{y_1}^2 2^{-2R_1}$; then the solutions to this *bit allocation* problem are [7]

$$R_0 = R + \frac{1}{2}\log_2\frac{\sigma_{y_0}}{\sigma_{y_1}}, \qquad R_1 = R - \frac{1}{2}\log_2\frac{\sigma_{y_0}}{\sigma_{y_1}},$$

with the minimum average distortion being

$$D_{\text{SBC}}(R) = h\sigma_{y_0}\sigma_{y_1} 2^{-2R}.$$

Note that, at the optimal point, $D(R_0) = D(R_1) = D_{\text{SBC}}(R)$. That is, the quantizers for y_0 and y_1 give the same distortion with optimal bit allocation. Since the transform T is orthogonal, we have $\sigma_x^2 = 1/2(\sigma_{y_0}^2 + \sigma_{y_1}^2)$. The *coding gain* of using subband coding over PCM is

$$\frac{D_{\text{PCM}}(R)}{D_{\text{SBC}}(R)} = \frac{\sigma_x^2}{\sigma_{y_0}\sigma_{y_1}} = \frac{1/2\left(\sigma_{y_0}^2 + \sigma_{y_1}^2\right)}{\left(\sigma_{y_0}^2\sigma_{y_1}^2\right)^{1/2}},$$

[4]Both wavelet image coding and subband coding are special cases of transform coding.

the ratio of arithmetic mean to geometric mean of coefficient variances $\sigma_{y_0}^2$ and $\sigma_{y_1}^2$. What this important result states is that subband coding performs no worse than PCM coding, and that the larger the disparity between coefficient variances, the bigger the subband coding gain, because $1/2(\sigma_{y_0}^2 + \sigma_{y_1}^2) \geq (\sigma_{y_0}^2\sigma_{y_1}^2)^{1/2}$, with equality if $\sigma_{y_0}^2 = \sigma_{y_1}^2$. This result can be easily extended to the case when $M > 2$ uniform subbands (of equal size) are used instead. The coding gain in this general case is

$$\frac{D_{\text{PCM}}(R)}{D_{\text{SBC}}(R)} = \frac{1/M \sum_{k=0}^{M-1}\sigma_k^2}{\left(\prod_{k=0}^{M-1}\sigma_k^2\right)^{1/M}},$$

where σ_k^2 is the sample variance of the kth band ($0 \leq k \leq M-1$). The above assumes that all M bands are of the same size. In the case of the subband or wavelet transform, the sizes of the subbands are not the same (see Fig. 8 below), but the above formula can be generalized pretty easily to account for this. As another extension of the results given in the above example, it can be shown that the necessary condition for optimal bit allocation is that all subbands should incur the same distortion at optimality — else it is possible to steal some bits from the lower-distortion bands to the higher-distortion bands in a way that makes the overall performance better.

Figure 8 shows typical bit allocation results for different subbands under a total bit rate budget of 1 bit per pixel for wavelet image coding. Since low-frequency bands in the upper-left corner have far more energy than high-frequency bands in the lower-right corner (see Fig. 1), more bits have to be allocated to low-pass bands than to high-pass bands. The last two frequency bands in the bottom half are not coded (set to zero) because of limited bit rate. Since subband coding treats wavelet coefficients according

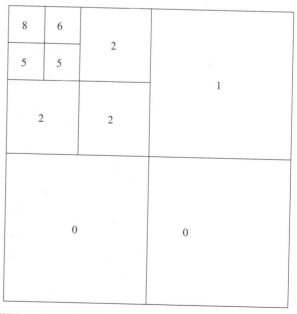

FIGURE 8 Typical bit allocation results for different subbands. The unit of the numbers is bits per pixel. These are designed to satisfy a total bit rate budget of 1 b/p, i.e., $1/4(1/4(1/4(8 + 6 + 5 + 5) + 2 + 2 + 2) + 1 + 0 + 0) = 1$.

to their frequency bands, it is effectively a *frequency* domain transform technique.

Initial wavelet-based coding algorithms, e.g., [16], followed exactly this subband coding methodology. These algorithms were designed to exploit the energy compaction properties of the wavelet transform only in the frequency domain by applying quantizers optimized for the statistics of each frequency band. Such algorithms have demonstrated small improvements in coding efficiency over standard transform-based algorithms.

5 New and More Efficient Class of Wavelet Coders

Because wavelet decompositions offer *space-frequency* representations of images, i.e., low-frequency coefficients have large spatial support (good for representing large image background regions), whereas high-frequency coefficients have small spatial support (good for representing spatially local phenomena such as edges), the wavelet representation calls for new quantization strategies that go beyond traditional subband coding techniques to exploit this underlying space-frequency image characterization.

Shapiro made a breakthrough in 1993 with his embedded zero-tree wavelet (EZW) coding algorithm [17]. Since then a new class of algorithms have been developed that achieve significantly improved performance over the EZW coder. In particular, Said and Pearlman's work on set partitioning in hierarchical trees (SPIHT) [18], which improves the EZW coder, has established zero-tree techniques as the current state-of-the-art of wavelet image coding since the SPIHT algorithm proves to be very successful for both lossy and lossless compression.

5.1 Zero-Tree-Based Framework and EZW Coding

A wavelet image representation can be thought of as a tree-structured spatial set of coefficients. *A wavelet coefficient tree* is defined as the set of coefficients from different bands that represent the same spatial region in the image. Figure 9 shows a three-level wavelet decomposition of the *Lena* image, together with a wavelet coefficient tree structure representing the eye region of *Lena*. Arrows in Fig. 9(b) identify the parent–children dependencies in a tree. The lowest frequency band of the decomposition is represented by the root nodes (top) of the tree, the highest frequency bands by the leaf nodes (bottom) of the tree, and each parent node represents a lower frequency component than its children. Except for a root node, which has only three children nodes, each parent node has four children nodes, the 2×2 region of the same spatial location in the immediately higher frequency band.

Both the EZW and SPIHT algorithms [17, 18] are based on the idea of using multipass zero-tree coding to transmit the largest wavelet coefficients (in magnitude) at first. We hereby use "zero coding" as a generic term for both schemes, but we focus on the popular SPIHT coder because of its superior performance. A set of tree coefficients is significant if the largest coefficient magnitude in the set is greater than or equal to a certain threshold (e.g., a power of 2); otherwise, it is insignificant. Similarly, a coefficient is significant if its magnitude is greater than or equal to the threshold; otherwise, it is insignificant. In each pass the significance of a larger set in the tree is tested at first: if the set is insignificant, a binary "zero-tree" bit is used to set all coefficients in the set to zero; otherwise, the set is partitioned into subsets (or child sets) for further significance tests. After all coefficients are tested in one pass, the threshold is halved before the next pass.

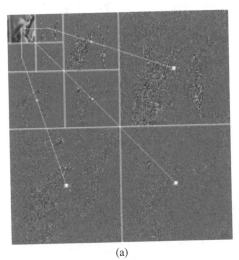

(a)

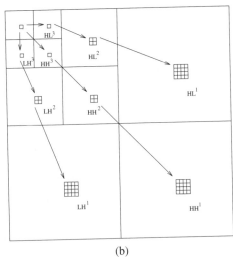

(b)

FIGURE 9 Wavelet decomposition offers a tree-structured image representation. (a) Three-level wavelet decomposition of the *Lena* image. (b) Spatial wavelet coefficient tree consisting of coefficients from different bands that correspond to the same spatial region of the original image (e.g., the eye of *Lena*). Arrows identify the parent–children dependencies.

63	-34	49	10	7	13	-12	7
-31	23	14	-13	3	4	6	-1
15	14	3	-12	5	-7	3	9
-9	-7	-14	8	4	-2	3	2
-5	9	-1	47	4	6	-2	2
3	0	-3	2	3	-2	0	4
2	-3	6	-4	3	6	3	6
5	11	5	6	0	3	-4	4

FIGURE 10 Example of a three-level wavelet representation of an 8×8 image.

The underlying assumption of the zero-tree coding framework is that most images can be modeled as having decaying power spectral densities. That is, if a parent node in the wavelet coefficient tree is insignificant, it is very likely that its descendents are also insignificant. The zero-tree symbol is used very efficiently in this case to signify a spatial subtree of zeros.

We give a SPIHT coding example to highlight the order of operations in zero-tree coding. Start with a simple three-level wavelet representation of an 8×8 image,[5] as shown in Fig. 10. The largest coefficient magnitude is 63. We can choose the threshold in the first pass between 31.5 and 63. Let $T_1 = 32$. Table 1 shows the first pass of the SPIHT coding process, with the following comments:

1. The coefficient value 63 is greater than the threshold 32 and positive, so a significance bit "1" is generated, followed by a positive sign bit "0". After decoding these symbols, the decoder knows the coefficient is between 32 and 64 and uses the midpoint 48 as an estimate.[6]
2. The descendant set of coefficient −34 is significant; a significance bit "1" is generated, followed by significance test of each of its four children {49, 10, 14, −13}.
3. The descendant set of coefficient −31 is significant; a significance bit "1" is generated, followed by significance test of each of its four children {15, 14, −9, −7}.
4. The descendant set of coefficient 23 is insignificant; an insignificance bit "0" is generated. This zero-tree bit is the only symbol generated in the current pass for the whole descendant set of coefficient 23.

TABLE 1 First pass of the SPIHT coding process at threshold $T_1 = 32$

Coefficient Coordinates	Coefficient Value	Binary Symbol	Reconstruction Value	Comments
(0,0)	63	1		(1)
		0	48	
(1,0)	−34	1		
		1	−48	
(0,1)	−31	0	0	
(1,1)	23	0	0	
(1,0)	−34	1		(2)
(2,0)	49	1		
		0	48	
(3,0)	10	0	0	
(2,1)	14	0	0	
(3,1)	−13	0	0	
(0,1)	−31	1		(3)
(0,2)	15	0	0	
(1,2)	14	0	0	
(0,3)	−9	0	0	
(1,3)	−7	0	0	
(1,1)	23	0		(4)
(1,0)	−34	0		(5)
(0,1)	−31	1		(6)
(0,2)	15	0		(7)
(1,2)	14	1		(8)
(2,4)	−1	0	0	
(3,4)	47	1		
		0	48	
(2,5)	−3	0	0	
(3,5)	2	0	0	
(0,3)	−9	0		(9)
(1,3)	−7	0		

5. The grandchild set of coefficient −34 is insignificant; a binary bit "0" is generated.[7]
6. The grandchild set of coefficient −31 is significant; a binary bit "1" is generated.
7. The descendant set of coefficient 15 is insignificant; an insignificance bit "0" is generated. This zero-tree bit is the only symbol generated in the current pass for the whole descendant set of coefficient 15.
8. The descendant set of coefficient 14 is significant; a significance bit "1" is generated, followed by significance test of each of its four children { −1, 47, −3, 2}.
9. Coefficient −31 has four children {15, 14, −9, −7}. Descendant sets of child 15 and child 14 were tested for significance before. Now descendant sets of the remaining two children −9 and −7 are tested.

In this example, the encoder generates 29 bits in the first pass. Along the process, it identifies four significant coefficients

---

[5]This set of wavelet coefficients is the same as the one used by Shapiro in an example to showcase EZW coding [17]. Curious readers can compare these two examples to see the difference between EZW and SPIHT coding.

[6]The reconstruction value can be anywhere in the uncertainty interval [32, 64). Choosing the midpoint is the result of a simple form of minimax estimation.

[7]In this example, we use the following convention: when a coefficient/set is significant, a binary bit "1" is generated; otherwise, a binary bit "0" is generated. In the actual SPIHT implementation [18], this convention was not always followed — when a grandchild set is significant, a binary bit "0" is generated; otherwise, a binary bit "1" is generated.

48	-48	48	0	0	0	0	0
0	0	0	0	0	0	0	0
0	0	0	0	0	0	0	0
0	0	0	0	0	0	0	0
0	0	0	48	0	0	0	0
0	0	0	0	0	0	0	0
0	0	0	0	0	0	0	0
0	0	0	0	0	0	0	0

(a)

56	-40	56	0	0	0	0	0
-24	24	0	0	0	0	0	0
0	0	0	0	0	0	0	0
0	0	0	0	0	0	0	0
0	0	0	40	0	0	0	0
0	0	0	0	0	0	0	0
0	0	0	0	0	0	0	0
0	0	0	0	0	0	0	0

(b)

FIGURE 11   Reconstructions after the (a) first and (b) second passes in SPIHT coding.

$\{63, -34, 49, 47\}$. The decoder reconstructs each coefficient based on these bits. When a set is insignificant, the decoder knows each coefficient in the set is between -32 and 32 and uses the midpoint 0 as an estimate. The reconstruction result at the end of the first pass is shown in Fig. 11(a).

The threshold is halved ($T_2 = T_1/2 = 16$) before the second pass, where insignificant coefficients/sets in the first pass are tested for significance again against T_2, and significant coefficients found in the first pass are refined. The second pass thus consists of the following.

1. Significance tests of the 12 insignificant coefficients found in the first pass — those having reconstruction value 0 in Table 1. Coefficients -31 at $(0, 1)$ and 23 at $(1, 1)$ are found to be significant in this pass; a sign bit is generated for each. The decoder knows the coefficient magnitude is between 16 and 32 and decode them as -24 and 24.
2. The descendant set of coefficient 23 at $(1, 1)$ is insignificant; so are the grandchild set of coefficient 49 at $(2, 0)$ and descendant sets of coefficients 15 at $(0, 2)$, -9 at $(0, 3)$, and -7 at $(1, 3)$. A zero-tree bit is generated in the current pass for each insignificant descendant set.
3. Refinement of the four significant coefficients $\{63, -34, 49, 47\}$ found in the first pass. The coefficient magnitudes are identified as being either between 32 and 48, which will be encoded with "0" and decoded as the midpoint 40, or between 48 and 64, which will be encoded with "1" and decoded as 56.

The encoder generates 23 bits (14 from step 1, five from step 2, and four from step 3) in the second pass. Along the process it identifies two more significant coefficients. Together with the four found in the first pass, the set of significant coefficients now

becomes $\{63, -34, 49, 47, -31, 23\}$. The reconstruction result at the end of the second pass is shown in Fig. 11(b).

The above encoding process continues from one pass to another and can stop at any point. For better coding performance, arithmetic coding [14] can be used to further compress the binary bitstream out of the SPIHT encoder.

From this example, we note that when the thresholds are powers of 2, zero-tree coding can be thought of as a bit-plane coding scheme. It encodes one bit-plane at a time, starting from the most significant bit. The effective quantizer in each pass is a dead-zone quantizer with the dead zone being twice the uniform step size. With the sign bits and refinement bits (for coefficients that become significant in previous passes) being coded on the fly, zero-tree coding generates an *embedded* bitstream, which is highly desirable for progressive transmission (see Fig. 4). A simple example of embedded representation is the approximation of an irrational number (say $\pi = 3.1415926535\cdots$) by a rational number. If we were only allowed two digits after the decimal point, then $\pi \approx 3.14$; if three digits after the decimal point were allowed, then $\pi \approx 3.141$; and so on. Each additional bit of the embedded bitstream is used to improve upon the previously decoded image for successive approximation, so rate control in zero-tree coding is exact and no loss is incurred if decoding stops at any point of the bitstream. The remarkable thing about zero-tree coding is that it outperforms almost all other schemes (such as JPEG coding) while being embedded. This good performance can be partially attributed to the fact that zero-tree coding captures across-scale interdependencies of wavelet coefficients. The zero-tree symbol effectively zeros out a set of coefficients in a subtree, achieving the coding gain of vector quantization [6] over scalar quantization.

Figure 12 shows the original *Lena* and *Barbara* images and their decoded versions at 0.25 bit per pixel (32:1 compression

FIGURE 12 Coding of the 512 × 512 *Lena* and *Barbara* images at 0.25 b/p (compression ratio of 32:1). Top: the original *Lena* and *Barbara* images. Middle: baseline JPEG decoded images, PSNR = 31.6 dB for *Lena*, and PSNR = 25.2 dB for *Barbara*. Bottom: SPIHT decoded images, PSNR = 34.1 dB for *Lena*, and PSNR = 27.6 dB for *Barbara*.

ratio) by baseline JPEG and SPIHT [18]. These images are coded at a relatively low bit rate to emphasize coding artifacts. The *Barbara* image is known to be hard to compress because of its insignificant high-frequency content (see the periodic stripe texture on *Barbara*'s trousers and scarf, and the checkerboard texture pattern on the tablecloth). The subjective difference in reconstruction quality between the two decoded versions of the same image is quite perceptible on a high-resolution monitor. The JPEG decoded images show highly visible blocking artifacts while the wavelet-based SPIHT decoded images have much sharper edges and preserve most of the striped texture.

5.2 Advanced Wavelet Coders: High-Level Characterization

We saw that the main difference between the early class of subband image coding algorithms and the zero-tree-based compression framework is that the former exploits only the frequency characterization of the wavelet image representation, whereas the latter exploits both the *spatial* and *frequency* characterization. To be more precise, the early class of coders were adept at exploiting the wavelet transform's ability to concentrate the image energy disparately in the different frequency bands, with the lower frequency bands having a much higher energy density. What these coders fail to exploit was the very definite spatial characterization of the wavelet representation. In fact, this is even apparent to the naked eye if one views the wavelet decomposition of the *Lena* image in Fig. 1, where the spatial structure of the image is clearly exposed in the high-frequency wavelet bands, e.g., the edge structure of the hat and face and the feather texture, etc. Failure to exploit this spatial structure limited the performance potential of the early subband coders.

In explicit terms, not only is it true that the energy density of the different wavelet subbands is highly disparate, resulting in gains by separating the data set into statistically dissimilar groupings of data, but it is also true that the data in the high-frequency subbands are highly spatially structured and clustered around the spatial edges of the original image. The early class of coders exploited the conventional coding gain associated with dissimilarity in the statistics of the frequency bands, but not the potential coding gain from separating individual frequency band energy into spatially localized clusters.

It is insightful to note that unlike the coding gain based on the frequency characterization, which is statistically predictable for typical images (the low-frequency subbands have much higher energy density than the high-frequency ones), there is a difficulty in going after the coding gain associated with the spatial characterization that is not statistically predictable; after all, there is no reason to expect the upper left corner of the image to have more edges than the lower right. This calls for a drastically different way of exploiting this structure — a way of pointing to the *spatial location* of significant edge regions within each subband. At a high level, a zero tree is no more than an efficient "pointing" data structure that incorporates the spatial characterization of wavelet coefficients by identifying tree-structured collections of insignificant spatial subregions across hierarchical subbands.

Equipped with this high-level insight, it becomes clear that the zero-tree approach is but only one way to skin the cat. Researchers in the wavelet image compression community have found other ways to exploit this phenomenon by using an array of creative ideas. The array of successful data structures in the research literature include (a) R-D optimized zero-tree-based structures, (b) morphology- or region-growing-based structures, (c) spatial context modeling-based structures, (d) statistical mixture modeling-based structure, (e) classification-based structures, and so on. As the details of these advanced methods are beyond the intended scope of this article, we refer the reader to "Wavelet image coding: PSNR results" (www.icsl.ucla.edu/~ipl/psnr_results.html) on the World Wide Web for the latest results [19] on wavelet image coding.

6 Adaptive Wavelet Transforms: Wavelet Packets

In noting how transform coding has become the *de facto* standard for image and video compression, it is important to realize that the traditional approach of using a transform with fixed frequency resolution (be it the logarithmic wavelet transform or the DCT) is good only in an ensemble sense for a typical *statistical* class of images. This class is well suited to the characteristics of the chosen fixed transform. This raises the natural question, Is possible to do better by being *adaptive* in the transformation so as to best match the features of the transform to the specific attributes of arbitrary individual images that may not belong to the typical ensemble?

To be specific, the wavelet transform is a good fit for typical natural images that have an exponentially decaying spectral density, with a mixture of strong stationary low-frequency components (such as the image background) and perceptually important short-duration high-frequency components (such as sharp image edges). The fit is good because of the wavelet transform's logarithmic decomposition structure, which results in its well-advertised attributes of good frequency resolution at low frequencies, and good time resolution at high frequencies (see Fig. 3(b)).

There are, however, important classes of images (or significant subimages) whose attributes go against those offered by the wavelet decomposition, e.g., images having strong high-pass components. A good example is the periodic texture pattern in the *Barbara* image of Fig. 12 — see the trousers and scarf textures, as well as the tablecloth texture. Another special class of images for which the wavelet is not a good idea is the class of fingerprint images (see Fig. 13 for a typical example), which has periodic high-frequency ridge patterns. These images are better matched with decomposition elements that have good frequency localization at high frequencies (corresponding to the texture patterns), which the wavelet decomposition does not offer in its menu.

This motivates the search for alternative transform descriptions that are more adaptive in their representation, and that are more robust to a large class of images of unknown or mismatched space-frequency characteristics. Although the task of finding an optimal decomposition for every individual image in the world is an ill-posed problem, the situation gets more interesting if we consider a large but finite library of desirable transforms, and match the best transform in the library

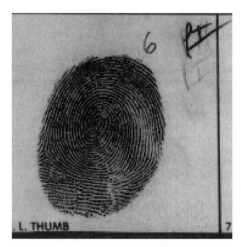

FIGURE 13 *Fingerprint* image: image coding using logarithmic wavelet transform does not perform well for fingerprint images such as this one with strong high-pass ridge patterns.

adaptively to the individual image. In order to make this feasible, there are two requirements. First, the library must contain a good representative set of entries (e.g., it would be good to include the conventional wavelet decomposition). Second, it is essential that there exists a fast way of searching through the library to find the best transform in an image-adaptive manner.

Both these requirements are met with an elegant generalization of the wavelet transform, called the *wavelet packet* decomposition, also known sometimes as the *best basis* framework. Wavelet packets were introduced to the signal processing community by Coifman and Wickerhauser in [20]. They represent a huge library of orthogonal transforms having a rich time-frequency diversity that also come with a easy-to-search capability, thanks to the existence of fast algorithms that exploit the tree-structured nature of these basis expansions. The tree structure comes from the cascading of multirate filter bank operations; see Chapter 4.1 and [2]. Wavelet packet bases essentially look like wavelet bases shown in Fig. 3(b), but they have more oscillations.

The wavelet decomposition, which corresponds to a logarithmic tree structure, is the most famous member of the wavelet packet family. Whereas wavelets are best matched to signals having a decaying energy spectrum, wavelet packets can be matched to signals having almost arbitrary spectral profiles, such as signals having strong high-frequency or midfrequency stationary components, making them attractive for decomposing images having significant texture patterns, as discussed earlier. There are an astronomical number of basis choices available in the typical wavelet packet library: for example, it can be shown that the library has over 10^{78} transforms for typical five-level 2-D wavelet packet image decompositions. The library is thus well equipped to deal efficiently with arbitrary classes of images requiring diverse spatial-frequency resolution tradeoffs.

Using the concept of time-frequency tilings introduced in Section 1, it is easy to see what wavelet packet tilings look like, and how they are a generalization of wavelets. We again start with 1-D signals. Tiling representations of several expansions are plotted in Fig. 14. Figure 14(a) shows a uniform STFT-like expansion, where the tiles are all of the same shape and size; Fig. 14(b) is the familiar wavelet expansion or the logarithmic subband decomposition; Fig. 14(c) shows a wavelet packet expansion where the bandwidths of the bases are neither uniformly nor logarithmically varying; and Fig. 14(d) highlights a wavelet packet expansion where the time-frequency attributes are exactly the reverse of the wavelet case: the expansion has good frequency resolution at higher frequencies, and good time localization at lower frequencies; we might call this the "antiwavelet" packet. There are a plethora of other options for the time-frequency resolution tradeoff, and these all correspond to admissible wavelet packet choices.

The extra adaptivity of the wavelet packet framework is obtained at the price of added computation in searching for the best wavelet packet basis, so an efficient fast search algorithm is the key in applications involving wavelet packets. The problem of searching for the best basis from the wavelet packet library for the compression problem using an R-D optimization framework and a fast tree-pruning algorithm was described in [21].

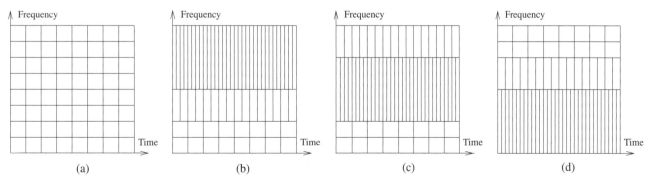

FIGURE 14 Tiling representations of several expansions for 1-D signals. (a) STFT-like decomposition, (b) wavelet decomposition, (c) wavelet packet decomposition, (d) antiwavelet packet.

(a)   (b)

FIGURE 15   (a) A wavelet packet decomposition for the *Barbara* image. White lines represent frequency boundaries. High-pass bands are processed for display. (b) Wavelet packet decoded *Barbara* at 0.1825 b/p. PSNR = 27.6 dB.

The 1-D wavelet packet bases can be easily extended to 2-D by writing a 2-D basis function as the product of two 1-D basis functions. In another words, we can treat the rows and columns of an image separately as 1-D signals. The performance gains associated with wavelet packets are obviously image dependent. For difficult images such as *Barbara* in Fig. 12, a wavelet packet decomposition shown in Fig. 15(a) gives much better coding performance than the wavelet decomposition. The wavelet packet decoded *Barbara* image at 0.1825 b/p is shown in Fig. 15(b), whose visual quality (or PSNR) is the same as the wavelet SPIHT decoded *Barbara* image at 0.25 b/p in Fig. 12. The bit rate saving achieved by using a wavelet packet basis instead of the wavelet basis in this case is 27% at the same visual quality.

An important practical application of wavelet packet expansions is the FBI wavelet scalar quantization (WSQ) standard for fingerprint image compression [22]. Because of the complexity associated with adaptive wavelet packet transforms, the FBI WSQ standard uses a fixed wavelet packet decomposition in the transform stage. The transform structure specified by the FBI WSQ standard is shown in Fig. 16. It was designed for 500 dots per inch fingerprint images by spectral analysis and trial and error. A total of 64 subbands are generated with a five-level wavelet packet decomposition. Trials by the FBI have shown that the WSQ standard benefited from having fine frequency partitions in the middle frequency region containing the fingerprint ridge patterns.

As an extension of adaptive wavelet packet transforms, one can introduce time-variation by segmenting the signal in time and allowing the wavelet packet bases to evolve with the signal. The result is a time-varying transform coding scheme that can adapt to signal nonstationarities. Computationally fast algorithms are again very important for finding the optimal signal expansions in such a time-varying system. For 2-D images, the simplest of these algorithms performs adaptive frequency segmentations over regions of the image selected through a quadtree decomposition. More complicated algorithms provide combinations of frequency decomposition and spatial segmentation. These jointly adaptive algorithms work particularly well for highly nonstationary images. Figure 17 shows the space-frequency tree segmentation and tiling for the *Building* image [23]. The image to the left shows the spatial segmentation result that separates the sky in the background from the building and the pond in the foreground. The image to the right gives the best wavelet packet decomposition for each spatial segment.

Finally we point out that, although this chapter is about wavelet coding of 2-D images, the wavelet coding framework and its extension to wavelet packets apply to 3-D video as well.

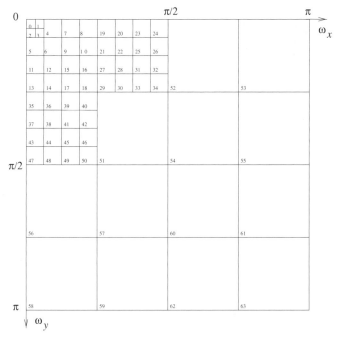

FIGURE 16   The wavelet packet transform structure given in the FBI WSQ specification. The number sequence shows the labeling of the different subbands.

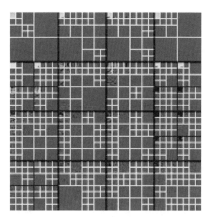

FIGURE 17 Space-frequency segmentation and tiling for the *Building* image. The image to the left shows that spatial segmentation separates the sky in background from the building and the pond in the foreground. The image to the right gives the best wavelet packet decomposition of each spatial segment. Dark lines represent spatial segments; white lines represent subband boundaries of wavelet packet decompositions. Note that the upper-left corners are the low-pass bands of wavelet packet decompositions.

We refer the reader to Chapter 6.2 for a detailed exposition of 3-D subband/wavelet video coding.

7 Conclusion

Since the introduction of wavelets as a signal processing tool in the late 1980s, a variety of wavelet-based coding algorithms have advanced the limits of compression performance well beyond that of the current commercial JPEG image coding standard. In this chapter, we have provided very simple high-level insights, based on the intuitive concept of time-frequency representations, into why wavelets are good for image coding. After introducing the salient aspects of the compression problem in general and the transform coding problem in particular, we have highlighted the key important differences between the early class of subband coders and the more advanced class of modern-day wavelet image coders. Selecting the embedded zero-tree wavelet coding structure embodied in the celebrated SPIHT algorithm as a representative of this latter class, we have detailed its operation by using a simple illustrative example. We have also described the role of wavelet packets as a simple but powerful generalization of the wavelet decomposition, in order to offer a more robust and adaptive transform image coding framework.

In response to the rapid progress in wavelet image coding research, the JPEG-2000 standardization committee will adopt the wavelet transform as its workhorse in the evolving next-generation image coding standard. A block-based embedded coding scheme is expected that will support a variety of coding functionalities such as spatial scalability, region of interest coding, error resilience, and spatial tiling [24]. The triumph of wavelet transform in the evolution of the JPEG-2000 standard underlines the importance of the fundamental insights provided in this chapter into why wavelets are so attractive for image compression.

References

[1] G. Strang and T. Nguyen, *Wavelets and Filter Banks* (Wellesley-Cambridge Press, New York, 1996).

[2] M. Vetterli and J. Kovačević, *Wavelets and Subband Coding* (Prentice-Hall, Englewood Cliffs, NJ, 1995).

[3] T. M. Cover and J. A. Thomas, *Elements of Information Theory* (Wiley, New York, 1991).

[4] C. E. Shannon, "A mathematical theory of communication," *Bell Syst. Tech. J.* **27**, 379–423, 623–656 (1948).

[5] C. E. Shannon, "Coding theorems for a discrete source with a fidelity criterion," *IRE Nat. Conv. Rec.* **4**, 142–163 (1959).

[6] A. Gersho and R. M. Gray, *Vector Quantization and Signal Compression* (Kluwer, Boston, MA, 1992).

[7] N. S. Jayant and P. Noll, *Digital Coding of Waveforms* (Prentice-Hall, Englewood Cliffs, NJ, 1984).

[8] S. P. Lloyd, "Least squares quantization in PCM," *IEEE Trans. Inf. Theory* **IT-28**, 127–135 (1982).

[9] H. Gish and J. N. Pierce, "Asymptotically efficient quantizing," *IEEE Trans. Inf. Theory* **IT-14**, 676–683 (1968).

[10] M. W. Marcellin and T. R. Fischer, "Trellis coded quantization of memoryless and Gauss-Markov sources," *IEEE Trans. Commun.* **38**, 82–93 (1990).

[11] T. Berger, *Rate Distortion Theory* (Prentice-Hall, Englewood Cliffs, NJ, 1971).

[12] N. Farvardin and J. W. Modestino, "Optimum quantizer performance for a class of non-Gaussian memoryless sources," *IEEE Trans. Inf. Theory* **30**, 485–497 (1984).

[13] D. A. Huffman, "A method for the construction of minimum redundancy codes," *Proc. IRE* **40**, 1098–1101 (1952).

[14] T. C. Bell, J. G. Cleary, and I. H. Witten, *Text Compression* (Prentice-Hall, Englewood Cliffs, NJ, 1990).

[15] J. W. Woods, *Subband Image Coding* (Kluwer, Boston, MA, 1991).

[16] M. Antonini, M. Barlaud, P. Mathieu, and I. Daubechies, "Image coding using wavelet transform," *IEEE Trans. Image Process.* **1**, 205–220 (1992).

[17] J. Shapiro, "Embedded image coding using zero-trees of wavelet coefficients," *IEEE Trans. Signal Process.* **41**, 3445–3462 (1993).

[18] A. Said and W. A. Pearlman, "A new, fast, and efficient image codec based on set partitioning in hierarchical trees," *IEEE Trans. Circuits Syst. Video Technol.* **6**, 243–250 (1996).

[19] University of California at Los Angeles (UCLA) Image Communications Laboratory, "Wavelet image coding: PSNR results," at http://www.icsl.ucla.edu/~ipl/psnr_results.html.

[20] R. R. Coifman and M. V. Wickerhauser, "Entropy based algorithms for best basis selection," *IEEE Trans. Inf. Theory* **32**, 712–718 (1992).

[21] K. Ramchandran and M. Vetterli, "Best wavelet packet bases in a rate-distortion sense," *IEEE Trans. Image Process.* **2**, 160–175 (1992).

[22] Criminal Justice Information Services, *WSQ Gray-Scale Fingerprint Image Compression Specification* (*ver. 2.0*), Federal Bureau of Investigation, Feb. 1993.

[23] K. Ramchandran, Z. Xiong, K. Asai, and M. Vetterli, "Adaptive transforms for image coding using spatially-varying wavelet packets," *IEEE Trans. Image Process.* **5**, 1197–1204 (1996).

[24] D. Taubman, C. Chrysafis, and A. Drukarev, "Embedded block coding with optimized truncation," ISO/IEC JTC1/SC29/WG1, JPEG-2000 Document WG1N1129, Nov. 1998.

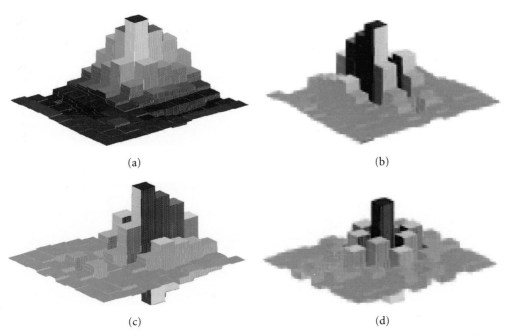

FIGURE 4.2.8 Discrete basis functions for image representation: (a) discrete scaling function from LLLL subband; (b)–(d) discrete wavelets from LHLL, LLLH, and LHLH subbands. These basis functions are generated from Daubechies' four-tap filter.

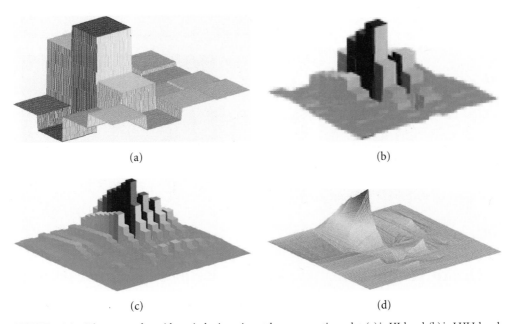

FIGURE 4.2.9 Discrete wavelets with vertical orientation at three consecutive scales: (a) in HL band; (b) in LHLL band; (c) in LLHLLL band. (d) Continuous wavelet is obtained as a limit of (normalized) discrete wavelets as the scale becomes coarser.

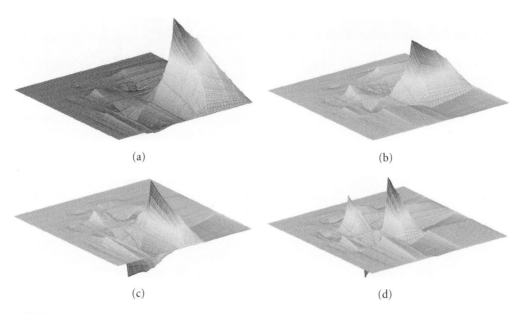

(a)   (b)

(c)   (d)

FIGURE 4.2.10   Basis functions for image representation: (a) scaling function; (b)–(d) wavelets with horizontal, vertical, and diagonal orientations. These four functions are tensor products of the 1-D scaling function and wavelet in Fig. 11. The horizontal wavelet has been rotated by 180° so that its negative part is visible on the display.

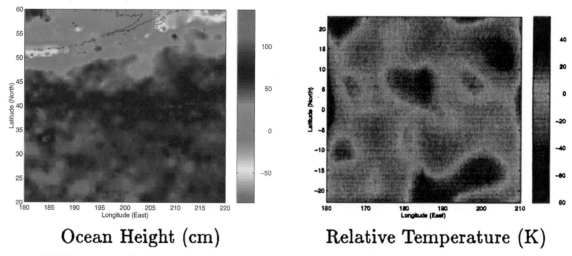

Ocean Height (cm)   Relative Temperature (K)

FIGURE 4.3.12   Multiscale estimation of remotely sensed fields: left, North-Pacific altimetry based on Topex/Poseidon data; right, equatorial Pacific temperature estimates based on *in situ* ship data.

FIGURE 4.6.11 Original Lena.

FIGURE 4.6.12 Calibrated Lena.

FIGURE 4.6.13 New scan Lena.

FIGURE 4.7.1 Examples of images that could be segmented based on brightness (top left), color (top right), motion (middle row), and texture (bottom row).

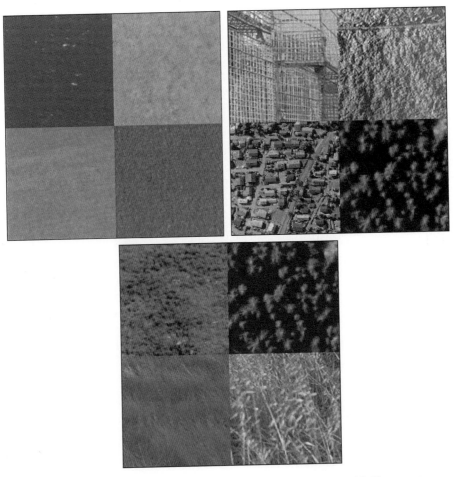

FIGURE 4.7.2 Collection of images; in each there are four clearly distinguishable segments.

FIGURE 4.7.3 Typical image in which a vehicle has to be segmented from the background.

FIGURE 4.7.4 Fisher color distance between pixels inside and outside of a square template placed on top of the image in Fig. 3. The template hypothesis on the right has higher merit than the one on the left.

FIGURE 4.7.5 Correct estimation of the vehicle ahead, using the MAP procedure.

FIGURE 4.7.6 Segmentation of other images, using the same Fisher color distance. Top: A segmentation that yields all segments that contains the color white. Bottom: A segmentation that yields all segments that do not contain the color green.

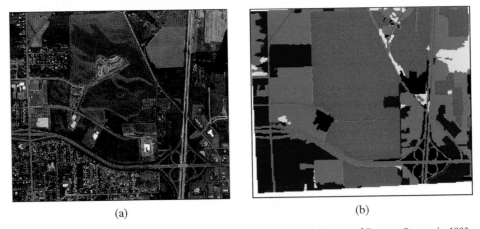

(a) (b)

FIGURE 4.7.7 Updating old maps using image segmentation. (a) Aerial image of Eugene, Oregon in 1993. (b) Map of the same area in 1987. (c) Operator-assisted segmentation of the 1993 aerial image. (d) Updated map in 1993. (*Continued*)

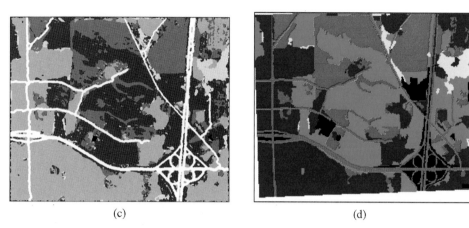

(c)   (d)

FIGURE 4.7.7   (*Continued*)

FIGURE 4.7.8   Segmentation of another aerial image, this time of a rural crop field area, using the same texture-based maximum likelihood procedure employed in Fig. 7.

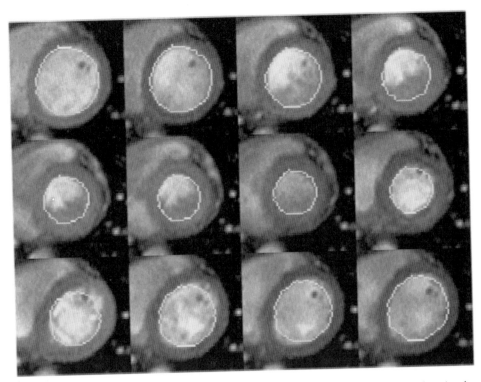

FIGURE 4.7.11 Tracking an object of interest, in this case a human heart, from frame to frame by using the elastic deformation method described in [20].

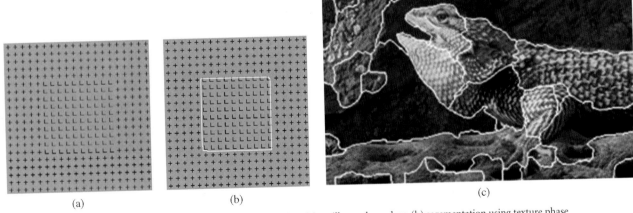

(a)   (b)   (c)

FIGURE 4.8.6 Two other examples of segementation: (a) an illusory boundary, (b) segementation using texture phase in the EdgeFlow algorithm, and (c) segmentation using color and texture energy.

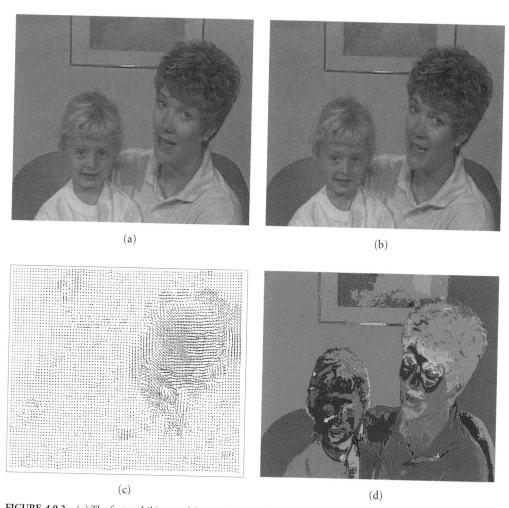

(a)

(b)

(c)

(d)

FIGURE 4.9.2 (a) The first and (b) second frames of the Mother and Daughter sequence; (c) 2-D dense motion field from the second frame to the first frame; (d) region map obtained segmentation.

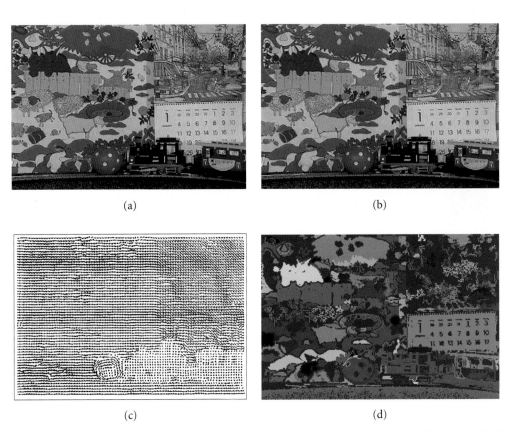

FIGURE 4.9.4 (a) The 136th and (b) 137th frames of the Mobile and Calendar sequence; (c) 2-D dense motion field from the 137th to 136th frame; (d) region map obtained by color segmentation.

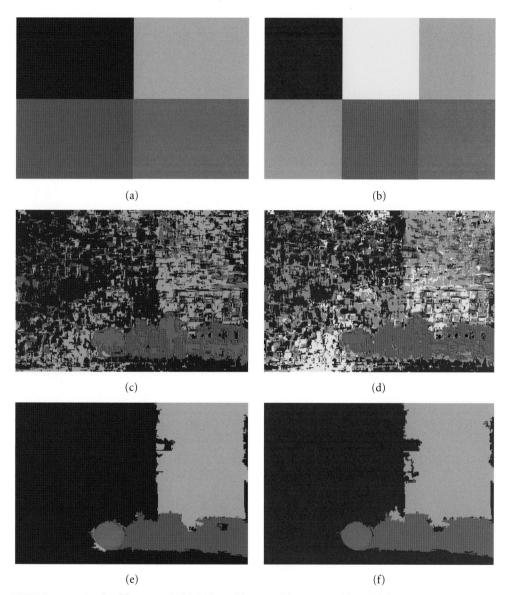

(a)

(b)

(c)

(d)

(e)

(f)

FIGURE 4.9.5 Results of the ML method: initial map (a) $K = 4$, (b) $K = 6$; pixel-based labeling (c) $K = 4$, (d) $K = 6$; region-based labeling (e) $K = 4$, (f) $K = 6$.

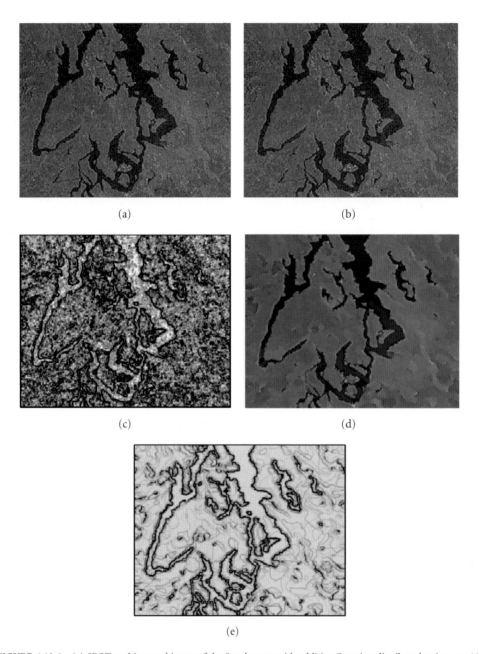

FIGURE 4.12.6 (a) SPOT multispectral image of the Seattle area, with additive Gaussian-distributed noise, $\sigma = 10$; (b) vector distance dissimilarity diffusion result, using the diffusion coefficient in Eq. (9); (c) edges (gradient magnitude) from the result in (b); (d) mean curvature motion [Eq. (18)] result using diffusion coefficients from Eqs. (11) and (12); (e) edges (gradient magnitude) from the result in (d).

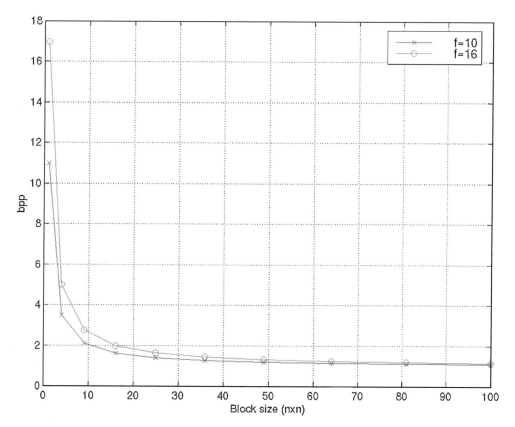

FIGURE 5.2.4 Data rate vs. block size.

(a)   (b)

FIGURE 5.2.6 Illustration of the use of BTC in color image compression: left, original image; right, image encoded at 1.89 bpp.

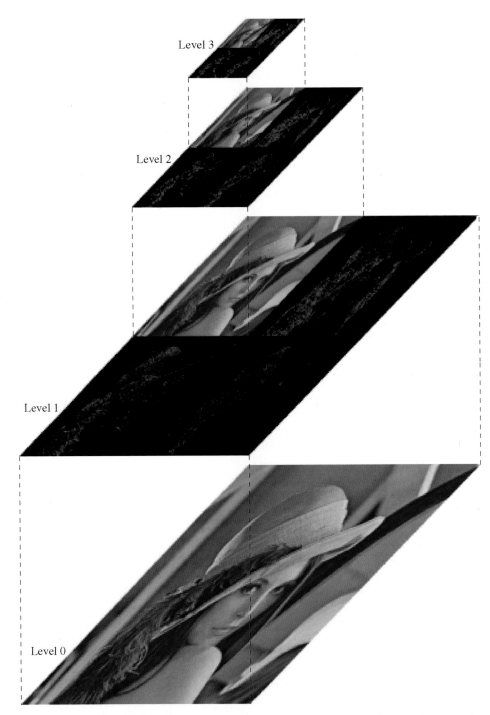

FIGURE 5.4.1 A three-level hierarchy wavelet decomposition of the 512×512 color *Lena* image. Level 1 (512×512) is the one-level wavelet representation of the original *Lena* at Level 0; Level 2 (256×256) shows the one-level wavelet representation of the low-pass image at Level 1; and Level 3 (128×128) gives the one-level wavelet representation of the low-pass image at Level 2.

5.5

The JPEG Lossy Image Compression Standard

Rashid Ansari
University of Illinois at Chicago

Nasir Memon
Polytechnic University

1 Introduction

JPEG is currently a worldwide standard for compression of digital images. The standard is named after the committee that created it and that continues to guide its evolution. This group, the Joint Photographic Experts Group (JPEG), consists of experts nominated by national standards bodies and by leading companies engaged in image-related work. The JPEG committee has an official title of ISO/IEC JTC1 SC29 Working Group 1, with a Web site at http://www.jpeg.org. The committee is charged with the responsibility of pooling efforts to pursue promising approaches to compression in order to produce an effective set of standards for still image compression. The lossy JPEG image compression procedure described in this chapter is part of the multipart set of ISO standards IS 10918-1,2,3 (ITU-T Recommendations T.81, T.83, T.84).

The JPEG standardization activity commenced in 1986, which generated 12 proposals for consideration by the committee in March 1987. The initial effort produced consensus that the compression should be based on the discrete cosine transform (DCT). Subsequent refinement and enhancement led to the Committee Draft in 1990. Deliberations on the JPEG Draft

International Standard (DIS) submitted in 1991 culminated in the approval of the International Standard (IS) in 1992.

Although the JPEG Standard defines both lossy and lossless compression algorithms, the focus in this chapter is on the lossy compression component of the JPEG standard. The JPEG lossless standards are described in detail in an accompanying chapter of this volume [7]. JPEG lossy compression entails an irreversible mapping of the image to a compressed bit stream, but the standard provides mechanisms for a controlled loss of information. Lossy compression produces a bit stream that is usually much smaller in size than that produced with lossless compression.

The key features of the lossy JPEG standard are as follows.

- Both sequential and progressive modes of encoding are permitted. These modes refer to the manner in which quantized DCT coefficients are encoded. In sequential coding, the coefficients are encoded on a block-by-block basis in a single scan that proceeds from left to right and top to bottom. In contrast, in progressive encoding only partial information about the coefficients is encoded in the first scan followed by encoding the residual information in successive scans.
- Low complexity implementations in both hardware and software are feasible.

- All types of images, regardless of source, content, resolution, color formats, etc., are permitted.
- A graceful tradeoff in bit rate and quality is offered, except at very low bit rates.
- A hierarchical mode with multiple levels of resolution is allowed.
- Bit resolution of 8–12 bits is permitted.
- A recommended file format, JPEG File Interchange Format (JFIF), enables the exchange of JPEG bit streams among a variety of platforms.

A JPEG compliant decoder has to support a minimum set of requirements, the implementation of which is collectively referred to as *baseline implementation*. Additional features are supported in the extended implementation of the standard. The features supported in the baseline implementation include the ability to provide

- a sequential buildup,
- custom or default Huffman tables,
- 8-bit precision per pixel for each component,
- image scans with 1–4 components, and
- both interleaved and noninterleaved scans.

A JPEG extended system includes all features in a baseline implementation and supports many additional features. It allows sequential buildup as well as an optional progressive buildup. Either Huffman coding or arithmetic coding can be used in the entropy coding unit. Precision of up to 12 bits per pixel is allowed. The extended system includes an option for lossless coding.

The rest of this chapter is organized as follows: in Section 2 we describe the structure of the JPEG codec and the units that it is made up of. In Section 3 the role and computation of the discrete cosine transform is examined. Procedures for quantizing the DCT coefficients are presented in Section 4. In Section 5, the mapping of the quantized DCT coefficients into symbols suitable for entropy coding is described. The use of Huffman coding and arithmetic coding for representing the symbols is discussed in Section 6. Syntactical issues and organization of data units are discussed in Section 7. Section 8 describes alternative modes of operation such as the progressive and hierarchical modes. In Section 9 some recent extensions made to the standard, collectively known as JPEG Part 3, are described. Finally, Section 10 lists further sources of information on the standard.

2 Lossy JPEG Codec Structure

It should be noted that in addition to defining an encoder and decoder, the JPEG standard also defines a syntax for representing the compressed data along with the associated tables and parameters. In this chapter, however, we largely ignore these syntactical issues and focus instead on the encoding and decoding procedures. We begin by examining the structure of the JPEG encoding and decoding systems. The discussion centers on the encoder

structure and the building blocks that an encoder is made up of. The decoder essentially consists of the inverse operations of the encoding process carried out in reverse.

2.1 Encoder Structure

The JPEG encoder and decoder are conveniently decomposed into units that are shown in Fig. 1. Note that the encoder shown in Fig. 1 is applicable in open-loop/unbuffered environments in which the system is not operating under a constraint of a prescribed bit rate/budget. The units constituting the encoder are described next.

2.1.1 Signal Transformation Unit: DCT

In JPEG image compression, each component array in the input image is first partitioned into 8×8 rectangular blocks of data. A signal transformation unit computes the DCT of each 8×8 block in order to map the signal reversibly into a representation that is better suited for compression. The object of the transformation is to reconfigure the information in the signal to capture the redundancies, and to present the information in a "machine-friendly" form that is convenient for disregarding the perceptually least relevant content. The DCT captures the spatial redundancy and packs the signal energy into a few DCT coefficients. The coefficient in the $[0, 0]$-th position in the 8×8 DCT array is referred to as the DC coefficient. The remaining 63 coefficients are called the AC coefficients.

2.1.2 Quantizer

If we wish to recover the original image exactly from the DCT coefficient array, then it is necessary to represent the DCT coefficients with high precision. Such a representation requires a large number of bits. In lossy compression the DCT coefficients are mapped into a relatively small set of possible values that are represented compactly by defining and coding suitable symbols. The quantization unit performs this task of a many-to-one mapping of the DCT coefficients, so that the possible outputs are limited in number. A key feature of the quantized DCT coefficients is that many of them are zero, making them suitable for efficient coding.

2.1.3 Coefficient-to-Symbol Mapping Unit

The quantized DCT coefficients are mapped to new symbols to facilitate a compact representation in the symbol coding unit that follows. The symbol definition unit can also be viewed as part of the symbol coding unit. However, it is shown here as a separate unit to emphasize the fact that the definition of symbols to be coded is an important task. An effective definition of symbols for representing AC coefficients in JPEG is the "runs" of zero coefficients followed by a nonzero terminating coefficient. For representing DC coefficients, symbols are defined by computing the difference between the DC coefficient in the current block and that in the previous block.

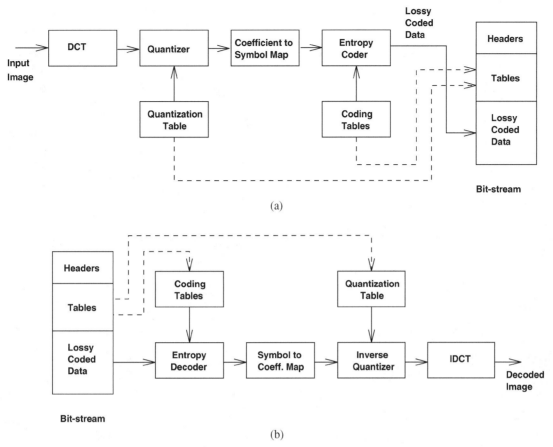

(a)

(b)

FIGURE 1   Constituent units of (a) JPEG encoder, (b) JPEG decoder.

2.1.4 Entropy Coding Unit

This unit assigns a codeword to the symbols that appear at its input, and it generates the bit stream that is to be transmitted or stored. Huffman coding is usually employed for variable-length coding of the symbols, with arithmetic coding allowed as an option.

2.2 Decoder Structure

In a decoder the inverse operations are performed and in an order that is the reverse of that in the encoder. The coded bit stream contains coding and quantization tables, which are first extracted. The coded data are then applied to the entropy decoder, which determines the symbols coded. The symbols are then mapped to an array of quantized DCT coefficients, which are then "dequantized" by multiplying each coefficient with the corresponding entry in the quantization table. The decoded image is then obtained by applying the inverse 2-D DCT to the array of the recovered DCT coefficients.

In the next three sections we consider each of the above encoder operations, DCT, quantization, and symbol mapping and coding, in more detail.

3 Discrete Cosine Transform

Lossy JPEG compression is based on transform coding that uses the DCT [2]. In DCT coding, each component of the image is subdivided into blocks of 8×8 pixels. A two-dimensional DCT is applied to each block of data to obtain an 8×8 array of coefficients. If $x[m, n]$ represents the image pixel values in a block, then the DCT is computed for each block of the image data as follows:

$$X[u, v] = \frac{C[u]C[v]}{4} \sum_{m=0}^{7} \sum_{n=0}^{7} x[m, n] \cos \frac{(2m+1)u\pi}{16}$$
$$\times \cos \frac{(2n+1)v\pi}{16}, \quad 0 \le u, v \le 7,$$

where

$$C[u] = \begin{cases} \frac{1}{\sqrt{2}} & u = 0, \\ 1 & 1 \le u \le 7. \end{cases}$$

The original image samples can be recovered from the DCT coefficients by applying the inverse discrete cosine transform

(IDCT) as follows:

$$x[m, n] = \sum_{u=0}^{7} \sum_{v=0}^{7} \frac{C[u]C[v]}{4} X[u, v] \cos \frac{(2m+1)u\pi}{16}$$
$$\times \cos \frac{(2n+1)v\pi}{16}, \quad 0 \le m, n \le 7.$$

The DCT, which belongs to the family of sinusoidal transforms, has received special attention because of its success in the compression of real-world images. It is seen from the definition of the DCT that an 8×8 image block being transformed is being represented as a linear combination of real-valued basis vectors that consist of samples of a product of one-dimensional cosinusoidal functions. The 2-D transform can be expressed as a product of 1-D DCT transforms applied separably along the rows and columns of the image block. The coefficients $X(u, v)$ of the linear combination are referred to as the DCT coefficients. For real-world digital images in which the interpixel correlation is reasonably high and that can be characterized with first-order autoregressive models, the performance of the DCT is very close to that of the Karhunen–Loeve transform [2]. The discrete Fourier transform (DFT) is not as efficient as DCT in representing an 8×8 image block. This is because when the DFT is applied to each row of the image, a periodic extension of the data, along with concomitant edge discontinuities, produces high-frequency DFT coefficients that are larger than the DCT coefficients of corresponding order. In contrast, there is a mirror-periodicity implied by the DCT that avoids the discontinuities at the edges when image blocks are repeated. As a result, the "high-frequency" or "high-order AC" coefficients are on the average smaller than the corresponding DFT coefficients.

We consider an example of the computation of the 2-D DCT of an 8×8 block in the 512×512 gray-scale image *Lena*. The specific block chosen is shown in the image in Fig. 2(a), where the block is indicated with a black boundary with one corner at [208, 296]. A closeup of the block enclosing part of the hat is shown in Fig. 2(b).

The 8-bit pixel values of the block chosen are shown in Fig. 3. After the DCT is applied to this block, the 8×8 DCT coefficient array obtained is shown in Fig. 4.

The magnitude of the DCT coefficients exhibits a pattern in their occurrences in the coefficient array. Also, their contribution to the perception of the information is not uniform across the array. The DCT coefficients corresponding to the lowest frequency basis functions are usually large in magnitude, and they are also deemed to be perceptually most significant. These features of the DCT coefficients are exploited in developing methods of quantization and symbol coding. The bulk of the compression achieved in transform coding occurs in the quantization step. The compression level is controlled by changing the total number of bits available to encode the blocks. The coefficients are quantized more coarsely when a large compression factor is required.

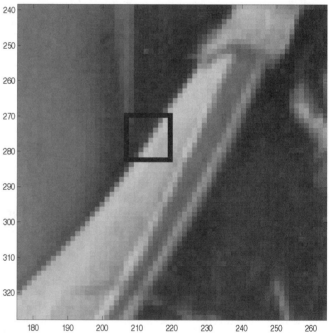

FIGURE 2 The original 512×512 *Lena* image (top) with an 8×8 block (bottom) identified with a black boundary and with one corner at [208, 296].

187	188	189	202	209	175	66	41
191	186	193	209	193	98	40	39
188	187	202	202	144	53	35	37
189	195	206	172	58	47	43	45
197	204	194	106	50	48	42	45
208	204	151	50	41	41	41	53
209	179	68	42	35	36	40	47
200	117	53	41	34	38	39	63

FIGURE 3 The 8×8 block identified in Fig. 2.

915.6	451.3	25.6	-12.6	16.1	-12.3	7.9	-7.3
216.8	19.8	-228.2	-25.7	23.0	-0.1	6.4	2.0
-2.0	-77.4	-23.8	102.9	45.2	-23.7	-4.4	-5.1
30.1	2.4	19.5	28.6	-51.1	-32.5	12.3	4.5
5.1	-22.1	-2.2	-1.9	-17.4	20.8	23.2	-14.5
-0.4	-0.8	7.5	6.2	-9.6	5.7	-9.5	-19.9
5.3	-5.3	-2.4	-2.4	-3.5	-2.1	10.0	11.0
0.9	0.7	-7.7	9.3	2.7	-5.4	-6.7	2.5

FIGURE 4   DCT of the 8 × 8 block in Fig. 3.

57	41	2	0	0	0	0	0
18	1	-16	-1	0	0	0	0
0	-5	-1	4	1	0	0	0
2	0	0	0	-1	0	0	0
0	-1	0	0	0	0	0	0
0	0	0	0	0	0	0	0
0	0	0	0	0	0	0	0
0	0	0	0	0	0	0	0

FIGURE 6   8 × 8 DCT block in Fig. 4 after quantization with the luminance quantization table shown in Fig. 5.

4 Quantization

Each DCT coefficient $X[m, n]$, $0 \leq m, n \leq 7$, is mapped into one of a finite number of levels determined by the compression factor desired. This is done by dividing each element of the DCT coefficient array by a corresponding element in an 8 × 8 *quantization matrix*, and rounding the result. Thus if the entry $q[m, n]$, $0 \leq m, n \leq 7$, in the mth row and nth column of the quantization matrix is large, then the corresponding DCT coefficient is coarsely quantized. The values of $q[m, n]$ are restricted to be integers with $1 \leq q[m, n] \leq 255$, and they determine the quantization step for the corresponding coefficient. The quantized coefficient is given by

$$q X[m, n] = \left[\frac{X[m, n]}{q[m, n]} \right]_{\text{round}} .$$

A quantization table (or matrix) is required for each image component. However, a quantization table can be shared by multiple components. For example in a luminance-plus-chrominance $Y–Cr–Cb$ representation, the two chrominance components usually share a common quantization matrix. JPEG quantization tables given in Annex K of the standard for luminance and components are shown in Fig. 5. These tables were obtained from a series of psychovisual experiments to determine the visibility thresholds for the DCT basis functions for a 760 × 576 image with chrominance components downsampled by 2 in the horizontal direction and at a viewing distance equal to six times the screen width. On examining the tables you will see that the quantization table for the chrominance components has larger values in general implying a coarser quantization of the chrominance planes as compared with the luminance plane. This is done to exploit the human visual system's relative insensitivity to chrominance components as compared to luminance components. The tables shown have been known to offer reasonable

performance, on the average, over a wide variety of applications and viewing conditions. Hence they have been widely accepted and over the years have become known as the "default" quantization tables.

Quantization tables can also be constructed by casting the problem as one of optimum allocation of a given budget of bits based on the coefficient statistics. The general principle is to estimate the variances of the DCT coefficients and assign more bits to coefficients with larger variances.

We now examine the quantization of the DCT coefficients given in Fig. 4, using the luminance quantization table in Fig. 5(a). Each DCT coefficient is divided by the corresponding entry in the quantization table, and the result is rounded to yield the array of quantized DCT coefficients in Fig. 6. We observe that a large number of quantized DCT coefficients are zero, making the array suitable for run-length coding which is described in Section 6. The block recovered after decoding is shown in Fig. 7.

4.1 Quantization Table Design

With lossy compression, the amount of distortion introduced in the image is inversely related to the number of bits (bit rate) used to encode the image. The higher the rate, the lower the distortion. Naturally, for a given rate we would like to incur the minimum possible distortion. Similarly, for a given distortion level, we would like to encode an image with the minimum rate possible. Hence lossy compression techniques are often studied in terms of their rate-distortion performance, i.e., the distortion they introduce at different bit rates. The rate-distortion performance of JPEG is determined mainly by the quantization tables. As mentioned before, the standard does not recommend any particular table or set of tables and leaves their design completely to the user. While the image quality obtained from the

16	11	10	16	24	40	51	61
12	12	14	19	26	58	60	55
14	13	16	24	40	57	69	56
14	17	22	29	51	87	80	62
18	22	37	56	68	109	103	77
24	35	55	64	81	104	113	92
49	64	78	87	103	121	120	101
72	92	95	98	112	100	103	99

17	18	24	47	99	99	99	99
18	21	26	66	99	99	99	99
24	26	56	99	99	99	99	99
47	66	99	99	99	99	99	99
99	99	99	99	99	99	99	99
99	99	99	99	99	99	99	99
99	99	99	99	99	99	99	99
99	99	99	99	99	99	99	99

FIGURE 5   Example quantization tables for luminance (left) and chrominance (right) components provided in the informative sections of the standard.

181	185	196	208	203	159	86	27
191	189	197	203	178	118	58	25
192	193	197	185	136	72	36	33
184	199	195	151	90	48	38	43
185	207	185	110	52	43	49	44
201	198	151	74	32	40	48	38
213	161	92	47	32	35	41	45
216	122	43	32	39	32	36	58

FIGURE 7   The block selected from the *Lena* image recovered after decoding.

use of the "default" quantization tables described earlier is very good, there is a need to provide flexibility to adjust the image quality by changing the overall bit rate. In practice, scaled versions of the "default" quantization tables are very commonly used to vary the quality and compression performance of JPEG. For example, the popular IJPEG implementation, freely available in the public domain, allows this adjustment through the use of *quality factor Q* for scaling all elements of the quantization table. The scaling factor is then computed as

$$\text{scale factor} = \begin{cases} \frac{5000}{Q} & \text{for } 1 \leq Q < 50 \\ 200 - 2 * Q & \text{for } 50 \leq Q \leq 99. \\ 1 & \text{for } Q = 100 \end{cases} \quad (1)$$

Although varying the rate by scaling a base quantization table according to some fixed scheme is convenient, it is clearly not optimal. Given an image and a bit rate, there exists a quantization table that provides the "optimal" distortion at the given rate. Clearly, the "optimal" table would vary with different images and different bit rates and even different definitions of distortion (for example MSE, perceptual distortion, etc.). In order to get the best performance from JPEG in a given application, custom quantization tables may have to be designed. Indeed, there has been a lot of work reported in the literature addressing the issue of quantization table design for JPEG. Broadly speaking, this work can be classified into three categories. The first deals with explicitly optimizing the rate-distortion performance of JPEG based on statistical models for DCT coefficient distributions. The second attempts to optimize the visual quality of the reconstructed image at a given bit rate, given a set of display conditions and a perception model.

An example of the first approach is provided by the work of Ratnakar and Livny [10], who propose RD-OPT, an efficient algorithm for constructing quantization tables with optimal rate-distortion performance for a given image. The RD-OPT algorithm uses DCT coefficient distribution statistics from any given image in a novel way to optimize quantization tables simultaneously for the entire possible range of compression-quality tradeoffs. The algorithm is restricted to the MSE related distortion measures as it exploits the property that the DCT is a unitary transform, that is, the mean-square error in the pixel domain is the same as the mean-square error in the DCT domain. The RD-OPT essentially consists of the following three stages.

1. Gather DCT statistics for given image or set of images. Essentially this step involves counting how many times the nth coefficient gets quantized to the value v when the quantization step size is q and computing what the MSE is for the nth coefficient at this step size.

2. Use statistics collected above to calculate $R_n(q)$, the rate for the nth coefficient when the quantization step size is q and $D_n(q)$, the corresponding distortion for each possible q. The rate $R_n(q)$ is estimated from the corresponding first-order entropy of the coefficient at the given quantization step size.

3. Compute $R(Q)$ and $D(Q)$ the rate and distortions for a quantization table Q as

$$R(Q) = \sum_{n=0}^{63} R_n(Q[n]), \qquad D(Q) = \sum_{n=0}^{63} D_n(Q[n]),$$

respectively. Use dynamic programming to optimize $R(Q)$ against $D(Q)$.

Optimizing quantization tables with respect to MSE may not be the best strategy when the end image is to be viewed by a human. A better approach is to match the quantization table to the Human Visual System (HVS) model. As mentioned before, the "default" quantization tables were arrived at in an *image independent* manner, based on the visibility of the DCT basis functions. Clearly, better performance could be achieved by an *image-dependent* approach that exploits HVS properties like frequency, contrast, and texture masking and sensitivity. A number of HVS model-based techniques for quantization table design have been proposed in the literature [3, 5, 15]. Such techniques perform an analysis of the given image and arrive at a set of thresholds, one for each coefficient, called the just noticeable distortion (JND) thresholds. The idea is that if the distortion introduced is at or just below these thresholds, the reconstructed image will be perceptually distortion free.

Optimizing quantization tables with respect to MSE may also not be appropriate when there are constraints on the type of distortion that can be tolerated. For example, on examining Fig. 5, we find that the "high-frequency" AC quantization factors, i.e., $q[m, n]$ for larger values of m and n, are significantly greater than the DC coefficient $q[0, 0]$ and the "low-frequency" AC quantization factors. There are applications in which the information of interest in an image may reside in the high-frequency AC coefficients. For example, in compression of radiographic images [12], the critical diagnostic information is often in the high-frequency components. The size of microcalcification in mammograms is often so small that a coarse quantization of the higher AC coefficients will be unacceptable. In such cases, JPEG allows custom tables to be provided in the bit streams.

Finally, quantization tables can also be optimized for hard copy devices such as printers. JPEG was designed for compressing images that are to be displayed on CRT-like display devices that can represent a large range of pixel intensities. Hence, when an image is rendered through a halftone device like a printer, the image quality could be far from optimal (For information on halftoning, please see Chapter 8.1). Vander Kam and Wong [13] give a closed-loop procedure to design a quantization table that is optimum for a given halftoning and scaling method chosen. The basic idea behind their algorithm is to code more coarsely frequency components that are corrupted by halftoning and to code more finely components that are left untouched by halftoning. Similarly, to take into account the effects of scaling, their design procedure assigns a higher bit rate to the frequency components that correspond to a large gain in the scaling filter response and a lower bit rate to components that are attenuated by the scaling filter.

5 Coefficient-to-Symbol Mapping and Coding

The quantizer makes the coding lossy, but it provides the major contribution in compression. However, the nature of the quantized DCT coefficients and the preponderance of zeros in the array leads to further compression with the use of lossless coding. This requires that the quantized coefficients be mapped to symbols in such a way that the symbols lend themselves to effective coding. For this purpose, JPEG treats the DC coefficient and the set of AC coefficients in a different manner. Once the symbols are defined, they are then represented with Huffman coding or arithmetic coding.

In defining symbols for coding, the DCT coefficients are scanned by traversing the quantized coefficient array in a zigzag fashion, shown in Fig. 8. The zigzag scan processes the DCT coefficients in increasing order of spatial frequency. Recall that the quantized high-frequency coefficients are zero with high probability; hence scanning in this order leads to a sequence that contains a large number of trailing zero values and these can be efficiently coded as described later.

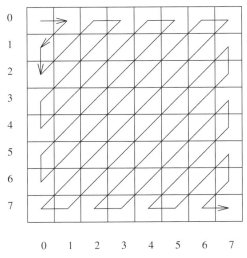

FIGURE 8 Zigzag scan procedure.

The [0,0]-th element or the quantized DC coefficient is separated from the remaining string of 63 AC coefficients, and symbols are defined next as shown in Fig. 1.

5.1 DC Coefficient Symbols

The DC coefficients in adjacent blocks are highly correlated. This fact is exploited to differentially code them. Let $qX_i[0,0]$ and $qX_{i-1}[0,0]$ denote the quantized DC coefficient in blocks i and $i-1$. The difference $\delta_i = qX_i[0,0] - qX_{i-1}[0,0]$ is computed. Assuming a precision of 8 bits/pixel for each component, it follows that the largest DC coefficient value, with $q[0,0] = 1$, is less than 2048, so that values of δ_i are in the range $[-2047, 2047]$. If Huffman coding is used, then these possible values would require a very large coding table. In order to limit the size of the coding table, the values in this range are grouped into 12 size categories, which are assigned labels 0 through 11. Category k contains 2^k elements $\{\pm 2^{k-1}, \ldots, \pm 2^k - 1\}$. The difference δ_i is mapped to a symbol described by a pair (category, amplitude). The 12 categories are Huffman coded. In order to distinguish values within the same category, extra k bits are used to represent one of the possible 2^k "amplitudes" of symbols within category k. On one hand, the amplitude of δ_i, $\{2^{k-1} \leq \delta_i \leq 2^k - 1\}$ is simply given by its their binary representation. On the other hand, the amplitude of δ_i, $\{-2^k - 1 \leq \delta_i \leq -2^{k-1}\}$ is given by the one's complement of the absolute value $|\delta_i|$, or simply by the binary representation of $\delta_i + 2^k - 1$.

5.2 Mapping AC Coefficient to Symbols

As observed before, most of the quantized AC coefficients are zero. The zigzag scanned string of 63 coefficients contains many consecutive occurrences or "runs of zeros", making the quantized AC coefficients suitable for run-length coding. The symbols in this case can be defined as [runs, nonzero terminating value], which can then be entropy coded. However the number of possible values of AC coefficients is large as is evident from the definition of DCT. For 8-bit pixels, the allowed range of AC coefficient values is $[-1023, 1023]$. In view of the large coding tables this entails, a procedure similar to that discussed above for DC coefficients is used. Categories are defined for suitable grouped values that terminate a run. Thus a run/category pair together with the amplitude within a category is used to define a symbol. The category definitions and amplitude bits generation use the same procedure as in DC difference coding. Thus a 4-bit category value is concatenated with a 4-bit run length to get an 8-bit [run/category] symbol. This symbol is then encoded by using either Huffman or arithmetic coding. There are two special cases that arise when coding the [run/category] symbol. First, since the run value is restricted to 15, the symbol (15/0) is used denote 15 zeros followed by a zero. A number of them can be cascaded to form specify larger runs. Second, if after a nonzero AC coefficient, all the remaining coefficients are zero, then a special symbol (0/0) denoting end of block is encoded. Figure 1 continues our example and shows the sequence of symbols generated for coding the quantized DCT block of Fig. 6.

5.3 Entropy Coding

The symbols defined for DC and AC coefficients can be entropy coded by using mostly Huffman coding, or optionally and infrequently, arithmetic coding based on the probability estimates of the symbols. Huffman coding is a method of variable-length coding (VLC) in which shorter codewords are assigned to the more frequently occurring symbols in order to achieve an average symbol codeword length that is as close to the symbol source entropy as possible. Huffman coding is optimal (meets the entropy bound) only when the symbol probabilities are integral powers of 1/2. The technique of *arithmetic coding* [16] provides a solution to attaining the theoretical bound of the source entropy. The baseline implementation of the JPEG standard uses Huffman coding only.

If Huffman coding is used, then Huffman tables, up to a maximum of eight in number, are specified in the bit stream. The tables constructed should not contain codewords that (a) are more that 16 bits long or (b) consist of all ones. Recommended tables are listed in annex K of the standard. If these tables are applied to the output of the quantizer shown in the first two columns of Fig. 1, then the algorithm produces output bits shown in the following columns of the figure. The procedures for specification and generation of the Huffman tables are identical to the ones used in the lossless standard explained in the accompanying Chapter [7].

6 Image Data Format and Components

The JPEG standard is intended for the compression of both greyscale and color images. In a gray-scale image there is a single "luminance" component. However, a color image is represented with multiple components and the JPEG standard sets stipulations on the allowed number of components and data formats. The standard permits a maximum of 255 color components, which are rectangular arrays of pixel values represented with 8- to 12-bit precision. For each color component, the largest dimension supported in either the horizontal or the vertical direction is $2^{16} = 65,536$.

All color component arrays do not necessarily have the same dimensions. Assume that an image contains K color components denoted by C_n, $n = 1, 2, \ldots, K$. Let the horizontal and vertical dimensions of the nth component be equal to X_n and Y_n, respectively. Define dimensions $X_{\max}$, $Y_{\max}$ and $X_{\min}$, $Y_{\min}$ as

$$X_{\max} = \max_{n=1}^{K}\{X_n\}, \qquad Y_{\max} = \max_{n=1}^{K}\{Y_n\},$$

$$X_{\min} = \min_{n=1}^{K}\{X_n\}, \qquad Y_{\min} = \min_{n=1}^{K}\{Y_n\}.$$

With each color component C_n, $n = 1, 2, \ldots, K$, one associates relative horizontal and vertical sampling factors, denoted by H_n and V_n, respectively, where

$$H_n = \frac{X_n}{X_{\min}}, \qquad V_n = \frac{Y_n}{Y_{\min}}.$$

The standard restricts the possible values of H_n and V_n to the set of four integers, 1, 2, 3, 4. The largest values of relative sampling factors are given by $H_{\max} = \max\{H_n\}$ and $V_{\max} = \max\{V_n\}$.

According to the JPEG file interchange format, the color information is specified by $[X_{\max}, Y_{\max}, H_n$ and $V_n, n = 1, 2, \ldots, K, H_{\max}, V_{\max}]$. The horizontal dimensions of the components are computed by the decoder as

$$X_n = \left\lceil X_{\max} \times \frac{H_n}{H_{\max}} \right\rceil.$$

Example 1: Consider a raw image in a luminance-plus-chrominance representation consisting of $K = 3$ components, $C_1 = Y$, $C_2 = Cr$, and $C_3 = Cb$. Let the dimensions of the luminance matrix (Y) be $X_1 = 720$ and $Y_1 = 480$, and the dimensions of the two chrominance matrices (Cr and Cb) be $X_2 = X_3 = 360$ and $Y_2 = Y_3 = 240$. In this case $X_{\max} = 720$ and $Y_{\max} = 480$. The relative sampling factors are $H_1 = V_1 = 2$, and $H_2 = V_2 = H_3 = V_3 = 1$.

When images have multiple components, the standard describes formats for organizing the data for the purpose of storage.

TABLE 1A DC coding

Difference δ_i	[Category, Amplitude]	Code
−2	[2, −2]	01101

Note: the code for a DC coefficient with a value of 57, assuming that the previous block has a DC coefficient of value 59.

TABLE 1B AC coding

Terminating Value	Run/Cat	Code Length	Code	Total Bits	Amplitude Bits
41	0/6	7	1111000	13	010110
18	0/5	5	11010	10	10010
1	1/1	4	1100	5	1
2	0/2	2	01	4	10
−16	1/5	11	11111110110	16	01111
−5	0/3	3	100	6	010
2	0/2	2	01	4	10
−1	2/1	5	11100	6	0
−1	0/1	2	00	3	0
4	3/3	12	111111110101	15	100
−1	1/1	4	1100	5	1
1	5/1	7	1111010	8	1
−1	5/1	7	1111010	8	0
EOB	EOB	4	1010	4	—
Total bits for block				112	

Rate = 112/64 = 1.75 bits per pixel

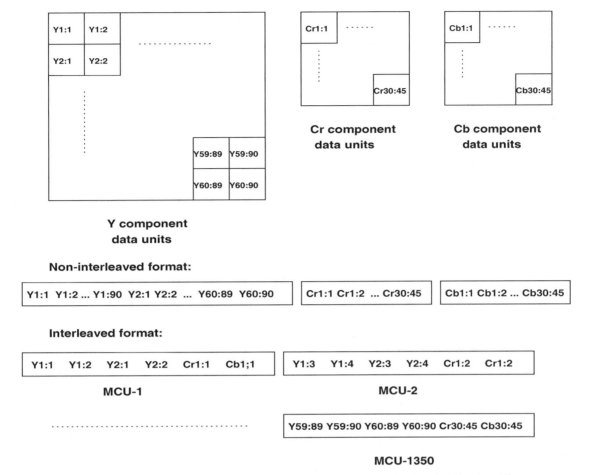

FIGURE 9 Organizations of the data units in the *Y, Cr, Cb* components into noninterleaved and interleaved formats.

In storing components, the standard provides the option of using either interleaved or noninterleaved formats. Processing and storage efficiency is aided, however, by interleaving the components where the data are read in a single scan. Interleaving is performed by defining a *data unit* for lossy coding as a single block of 8×8 pixels in each color component. This definition can be used to partition the *n*th color component C_n, $n = 1, 2, \ldots, K$ into rectangular blocks, each of which contains $H_n \times V_n$ data units. A *minimum coded unit* (MCU) is then defined as the smallest interleaved collection of data units obtained by successively picking $H_n \times V_n$ data units from *n*th color component. Certain restrictions are imposed on the data in order to be stored in the interleaved format.

- The number of interleaved components should not exceed four, and
- an MCU should contain no more than ten data units, i.e.,

$$\sum_{n=1}^{K} H_n V_n \leq 10.$$

If these restrictions are not met, then the data are stored in a noninterleaved format, where each component is processed in successive scans.

Example 2: We consider the case of storage of the *Y, Cr, Cb* components in Example 1. The luminance component contains 90×60 data units, and each of the two chrominance components contains 45×30 data units. Figure 9 shows both noninterleaved and an interleaved arrangement of the data for $K = 3$ components, $C_1 = Y$, $C_2 = Cr$, and $C_3 = Cb$, with $H_1 = V_1 = 2$, and $H_2 = V_2 = H_3 = V_3 = 1$. The MCU in this case contains six data units, consisting of $H_1 \times V_1 = 4$ data units of the *Y* component and $H_2 \times V_2 = H_3 \times V_3 = 1$ each of the *Cr* and *Cb* components.

7 Alternative Modes of Operation

What has been described thus far in this chapter represents the JPEG *sequential DCT mode*. The sequential DCT mode is the most commonly used mode of operation of JPEG and is required to be supported by any baseline implementation of the standard. However, in addition to the sequential DCT mode, JPEG also defines a *progressive DCT mode, sequential lossless mode*, and a

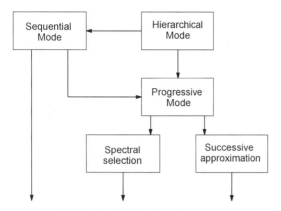

FIGURE 10   JPEG modes of operation.

hierarchical mode. In Fig. 10 we show how the different modes can be used. For example, the hierarchical mode could be used in conjunction with any of the other modes as shown in the figure. In the lossless mode, JPEG uses an entirely different algorithm based on predictive coding as described in detail in the next chapter. In this section we restrict ourselves to lossy compression and describe in more detail the DCT based progressive and hierarchical modes of operation.

7.1 Progressive Mode

In some applications it may be advantageous to transmit an image in multiple passes, such that after each pass an increasingly accurate approximation to the final image can be constructed at the receiver. In the first pass very few bits are transmitted and the reconstructed image is equivalent to one obtained with a very low quality setting. Each of the subsequent passes provides additional bits, which are used to refine the quality of the reconstructed image. The total number of bits transmitted is roughly the same as would be needed to transmit the final image in a sequential DCT mode. One example of an application that would benefit from progressive transmission is provided by the World Wide Web, where a user might want to start examining the contents of the entire page without waiting for each and every image contained in the page to be fully and sequentially downloaded. Other examples include remote browsing of image databases, telemedicine, and network-centric computing in general. JPEG contains a progressive mode of coding that is well suited to such applications. The disadvantage of progressive transmission of course is that the image has to be decoded a multiple number of times and only makes sense if the decoder is faster than the communication link.

In the progressive mode, the DCT coefficients are encoded in a series of scans. JPEG defines two ways for doing this: *spectral selection* and *successive approximation*. In the spectral selection mode, DCT coefficients are assigned to different groups according to their position in the DCT block and during each pass, the DCT coefficients belonging to a single group are transmitted. For example, consider the following grouping of the 64 DCT

coefficients numbered from 0 to 63 in the zigzag scan order

$$\{0\}, \{1, 2, 3\}, \{4, 5, 6, 7\}, \{8, \ldots, 63\}.$$

Here, only the DC coefficient is encoded in the first scan. This is a requirement imposed by the standard. In the progressive DCT mode, DC coefficients are always sent in a separate scan. The second scan of the example codes the first three AC coefficients in zigzag order, the third scan encodes the next four AC coefficients, and the fourth and the last scan encodes the remaining coefficients. JPEG provides the syntax for specifying the starting coefficient number and the final coefficient number being encoded in a particular scan. This limits a group of coefficients being encoded in any given scan to be successive in the zigzag order. The first few DCT coefficients are often sufficient to give a reasonable rendition of the image. In fact, just the DC coefficient can serve to essentially identify the contents of an image, although the reconstructed image contains severe blocking artifacts. It should be noted that after all the scans are decoded, the final image quality is the same as that obtained by a sequential mode of operation. The bit rate, however, can be different, as the entropy coding procedures for the progressive mode are different as described later in this section.

In successive approximation coding, the DCT coefficients are sent in successive scans with an increasing level of precision. The DC coefficient, however, is sent in the first scan with full precision, just as in spectral selection coding. The AC coefficients are sent bit plane by bit plane, starting from the most significant bit plane to the least significant bit plane.

The entropy coding techniques used in the progressive mode are slightly different than those used in the sequential mode. Since the DC coefficient is always sent as a separate scan, the Huffman and arithmetic coding procedures used remain the same as those in the sequential mode. However, coding of the AC coefficients is done a bit differently. In spectral selection coding (without selective refinement) and in the first stage of successive approximation coding, a new set of symbols are defined to indicate runs of end-of-block (EOB) codes. Recall, in the sequential mode, the EOB code indicates that the rest of the block contains zero coefficients. With spectral selection, each scan contains only a few AC coefficients and the probability of encountering EOB is significantly higher. Similarly, in successive approximation coding each block consists of reduced precision coefficients, leading again to the encoding of a large number of EOB symbols. Hence, to exploit this fact and acheive further reduction in bit rate, JPEG defines an additional set of 15 symbols as *EOBn*, each representing a run of 2^n EOB codes. After each *EOBi* run-length code, extra i bits are appended to specify the exact run length.

It should be noted that the two progressive modes, spectral selection and successive refinement, can be combined to give successive approximation in each spectral band being encoded. This results in quite a complex codec, which to our knowledge is rarely used.

It is possible to transcode between progressive JPEG and sequential JPEG without any loss in quality and approximately maintaining the same bit rate. Spectral selection results in bit rates slightly higher than the sequential mode, whereas successive approximation often results in lower bit rates. The differences, however, are small.

Despite the advantages of progressive transmission, there have not been many implementations of progressive JPEG codecs. There has been some interest in them recently because of the proliferation of images on the World Wide Web. It is expected that many more public domain progressive JPEG codecs will be available in the future.

7.2 Hierarchical Mode

The hierarchical mode defines another form of progressive transmission in which the image is decomposed into a pyramidal structure of increasing resolution. The topmost layer in the pyramid represents the image at the lowest resolution, and the base of the pyramid represents the image at full resolution. There is a doubling of resolutions both in the horizontal and vertical dimensions, between successive levels in the pyramid. Hierarchical coding is useful in applications where an image has to be displayed at different resolutions in units such as hand-held devices, computer monitors of varying resolutions, and high-resolution printers. In such a scenario, a multiresolution representation allows the transmission of the appropriate layer to each requesting device, thereby making full use of available bandwidth.

In the JPEG hierarchical mode, each image component is encoded as a sequence of frames. The lowest resolution frame (level 1) is encoded by using one of the sequential or progressive modes. The remaining levels are encoded differentially. That is, an estimate I_i' of the image, I_i, at the i-th level ($i \geq 2$) is first formed by upsampling the low-resolution image I_{i-1} from the layer immediately above. The difference between I_i' and I_i is then encoded by using modifications of the DCT based modes or the lossless mode. If lossless mode is used to code each refinement, then the final reconstruction at the base layer is lossless. The upsampling filter used is a bilinear interpolating filter that is specified by the standard and cannot be specified by the user. Starting from the high-resolution image, successive low-resolution images are created essentially by downsampling by 2 in each direction. The exact downsampling filter to be used is not specified, but the JPEG standard cautions that the downsampling filter used be consistent with the fixed upsampling filter. Note that the decoder does not need to know what downsampling filter was used in order to decode a bit stream. Figure 11 depicts the sequence of operations performed at each level of the hierarchy.

Since the differential frames are already signed values, they are not level-shifted prior to FDCT. Also, the DC coefficient is coded directly rather than differentially. Other than these two facts, the Huffman coding model in the progressive mode is the same as that used in the sequential mode. Arithmetic coding is, however, done a bit differently, with conditioning states based on differences with the pixel to the left as well as the one above being utilized. For details the user is referred to [9].

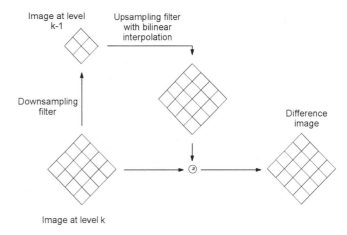

FIGURE 11 JPEG hierarchical mode.

8 JPEG Part 3

JPEG has made some recent extensions to the original standard described in [1]. These extensions are collectively known as JPEG Part 3. The most important elements of JPEG Part 3 are variable quantization and tiling, as described in more detail below.

8.1 Variable Quantization

One of the main limitations of the original JPEG standard was the fact that visible artifacts can often appear in the decompressed image at moderate to high compression ratios. This is especially true for parts of the image containing graphics, text, or some other such synthesized component. Artifacts are also common in smooth regions and in image blocks containing a single dominant edge. We consider compression of a 24 bits/pixel color version of the *Lena* image. In Fig. 12 we show the reconstructed *Lena* image with different compression ratios. At 24 to 1 compression we see little artifacts. But as the compression ratio is increased to 96 to 1, noticeable artifacts begin at appear. Especially annoying is the "blocking artifact" in smooth regions of the image.

One approach to deal with this problem is to change the "coarseness" of quantization as a function of image characteristics in the block being compressed. The latest extension of the JPEG standard, called JPEG Part 3, allows rescaling of quantization matrix Q on a block-by-block basis, thereby potentially changing the manner in which quantization is performed for each block. The scaling operation is not done on the DC coefficient $Y[0, 0]$, which is quantized in the same manner as baseline JPEG. The remaining 63 AC coefficients $Y[u, v]$ are quantized as follows:

$$\hat{Y}[u, v] = \left[\frac{Y[u, v] \times 16}{Q[u, v] \times \text{QScale}} \right].$$

Here QScale is a parameter that can take on values from 1 to

FIGURE 12   Lena image at 24 to 1 (top) and 96 to 1 (bottom) compression ratios. (See color section, p. C–26.)

It should be noted that the standard only specifies the syntax by means of which the encoding process can signal changes made to the QScale value. It does not specify how the encoder may determine if a change in QScale is desired and what the new value of QScale should be. Typical methods for variable quantization proposed in the literature utilize the fact that the human visual system is less sensitive to quantization errors in highly active regions of the image. Quantization errors are frequently more perceptible in blocks that are smooth or contain a single dominant edge. Hence, prior to quantization, they compute a few simple features for each block. These features are used to classify the block as either smooth, edge or texture, etc. Based on this classification, and a simple activity measure computed for the block, a QScale value is computed.

For example, Konstantinides and Tretter [6] give an algorithm for computing QScale factors for improving text quality on compound documents. They compute an activity measure M_i for each image block as a function of the DCT coefficients as follows:

$$M_i = \frac{1}{64} \left[\log_2 |Y_i[0,0] - Y_{i-1}[0,0]| + \sum_{j,k} \log_2 |Y_i[j,k]| \right]. \tag{2}$$

The QScale value for the block is then computed as

$$\text{QScale}_i = \begin{cases} a \times M_i + b & \text{if } 2 > a \times M_i + b \geq 0.4 \\ 0.4 & a \times M_i + b \geq 0.4. \\ 2 & a \times M_i + b > 2 \end{cases} \tag{3}$$

The technique is only designed to detect text regions and will quantize high-activity textured regions in the image part at the same scale as text regions. Clearly, this is not optimal, as high-activity textured regions can be quantized very coarsely, leading to an improved compression. In addition, the technique does not discriminate smooth blocks, where artifacts are often the first to appear.

Algorithms for variable quantization that perform a more extensive classification have been proposed for video coding but nevertheless are also applicable to still image coding. One such technique has been proposed by Chun *et al.* [4], who classify blocks as being either smooth, edge, or texture, based on several parameters defined in the DCT domain as shown below.

$$\begin{aligned} &E_h\text{: horizontal energy} &\quad E_a\text{: avg}(E_h, E_v, E_d) \\ &E_{m/M}\text{: ratio of } E_m \text{ and } E_M &\quad E_v\text{: vertical energy} \\ &E_m\text{: min}(E_h, E_v, E_d) &\quad E_d\text{: diagonal energy} \\ &\quad\quad E_M\text{: max}(E_h, E_v, E_d) \end{aligned}$$

Here E_a represents the average high-frequency energy of the block and is used to distinguish between low-activity blocks and high-activity blocks. Low-activity (smooth) blocks satisfy the relationship, $E_a \leq T_1$, where T_1 is a small constant. High-activity blocks are further classified into texture blocks and edge blocks.

112 (default 16). In order for the decoder to correctly recover the quantized AC coefficients, it has to know the value of QScale used by the encoding process. The standard specifies the exact syntax by which the encoder can specify change in QScale values. If no such change is signaled, then the decoder continues using the QScale value that is in current use. The overhead incurred in signaling a change in the scale factor is approximately 15 bits, depending on the Huffman table being employed.

Texture blocks are detected under the assumption that they have relatively uniform energy distribution in comparison with edge blocks. Specifically, a block is deemed to be a texture block if it satisfies the conditions $E_a > T_1$, $E_{min} > T_2$, and $E_{m/M} > T_3$, where T_1; T_2, and T_3 are experimentally determined constants. All blocks that fail to satisfy the smoothness and texture tests are classified as edge blocks.

8.2 Tiling

JPEG Part 3 defines a tiling capability whereby an image is subdivided into blocks or tiles, each coded independently. Tiling facilitates the following features:

- display of an image region on a given screen size,
- fast access to image subregions,
- region of interest refinement, and
- protection of large images from copying by giving access to only a part of it.

As shown in Fig. 13, the different types of tiling allowed by JPEG are as follows.

- Simple tiling: this form of tiling is essentially used for dividing a large image into multiple subimages, which are of the same size (except for edges) and are nonoverlapping. In this mode, all tiles are required to have the same sampling factors and components. Other parameters like quantization tables and Huffman tables are allowed to change from tile to tile.
- Pyramidal tiling: this is used for storing multiple resolutions of an image. Simple tiling as described above is used in each resolution. Tiles are stored in raster order, left to right, top to bottom, and low resolution to high resolution.

- Composite tiling: this allows multiple resolutions on a single image display plane. Tiles can overlap within a plane.

Another Part 3 extension is selective refinement. This feature permits a scan in a progressive mode, or a specific level of a hierarchical sequence, to cover only part of the total image area. Selective refinement could be useful, for example, in telemedicine applications in which a radiologist could request refinements to specific areas of interest in the image.

9 Additional Information

An excellent source of information on the JPEG compression standard is the book by Pennebaker and Mitchell [9]. This book also contains the entire text of the official committee draft international standard ISO DIS 10918-1 and ISO DIS 10918-2. The book has not been revised since its first publication in 1993, and hence later extensions to the standard, incorporated in JPEG Part 3, are not covered. The official standards document [1] is the only source for JPEG Part 3.

The JPEG committee maintains an official Web Site at www. jpeg.org, which contains general information about the committee and its activities, announcements, and other useful links related to the different JPEG standards. The JPEG FAQ is located at http://www.faqs.org/faqs/jpeg-faq/part1/preamble.html.

Free, portable C code for JPEG compression is available from the Independent JPEG Group (IJG). Source code, documentation, and test files are included. Version 6b is available from ftp.uu.net:/graphics/jpeg/jpegsrc.v6b.tar.gz and in ZIP archive format at ftp.simtel.net:/pub/simtelnet/msdos/graphics/jpegsr6b.zip.

The IJG code includes a reusable JPEG compression/decompression library, plus sample applications for compression, decompression, transcoding, and file format conversion. The package is highly portable and has been used successfully on many machines ranging from personal computers to supercomputers. The IJG code is free for both noncommercial and commercial use; only an acknowledgement in your documentation is required to use it in a product. A different free JPEG implementation, written by the PVRG group at Stanford, is available from have-fun.stanford.edu:/pub/jpeg/JPEGv1.2.1.tar.Z. The PVRG code is designed for research and experimentation rather than production use; it is slower, harder to use, and less portable than the IJG code, but the PVRG code is easier to understand.

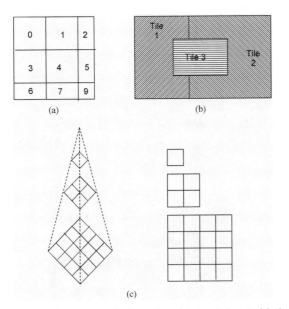

(a)
(b)
(c)

FIGURE 13 Different types of tilings allowed in JPEG Part 3: (a) simple, (b) composite, and (c) pyramidal.

References

[1] ISO/IEC JTC 1/SC 29/WG 1 N 993, "Information technology — digital compression and coding of continuous-tone still images," Recommendation T.84 ISO/IEC CD 10918-3, November, 1994.

[2] N. Ahmed, T. Natrajan, and K. R. Rao, "Discrete cosine transform," *IEEE Trans. Comput.* **C-23**, 90–93 (1974).

[3] A. J. Ahumada and H. A. Peterson, "Luminance model based DCT quantization for color image compression," in *Human Vision,*

Visual Processing, and Digital Display III, B. E. Rogowitz, ed., *Proc. SPIE* **1666**, 365–374 (1992).

[4] K. W. Chun, K. W. Lim, H. D. Cho and J. B. Ra, "An adaptive perceptual quantization algorithm for video coding," *IEEE Trans. Consumer Electron.* **39**, 555–558 (1993).

[5] N. Jayant, R. Safranek, and J. Johnson, "Signal compression based on models of human perception," *Proc. IEEE* **83**, 1385–1422 (1993).

[6] K. Konstantinides and D. Tretter, "A method for variable quantization in JPEG for improved text quality in compound documents," *Proceedings of the IEEE International Conference on Image Processing*, Chicago, IL, October 1998.

[7] N. Memon and R. Ansari, "The JPEG lossless compression standards," Chapter 5.6 in this Handbook. *Proc. ICIP* (1998).

[8] W. B. Pennebaker and J. L. Mitchell, "An overview of the basic principles of the Q-Coder adaptive binary arithmetic coder," *IBM J. Res. Devel.* **32**, 717–726 (1988); Tech. Rep. JPEG-18, ISO/IEC/JTC1/SC2/WG8, International Standards Organization, 1988. Working group document.

[9] W. B. Pennebaker and J. L. Mitchell, *JPEG Still Image Data Compression Standard*, Van Nostrand Reinhold, New York, 1993.

[10] V. Ratnakar and M. Livny, "RD-OPT: An efficient algorithm for optimizing DCT quantization tables," *IEEE Proceedings of the Data Compression Conference (DCC)*, Snowbird, UT, 332–341, March 1995.

[11] K. R. Rao and P. Yip, *Discrete Cosine Transform — Algorithms, Advantages, Applications* (Academic, San Diego, CA, 1990).

[12] B. J. Sullivan, R. Ansari, M. L. Giger, and H. MacMohan, "Relative effects of resolution and quantization on the quality of compressed medical images," *Proc. IEEE International Conference on Image Processing*, Austin, TX, 987–991, November 1994.

[13] R. VanderKam and P. Wong, "Customized JPEG compression for grayscale printing," *Proceedings of the Data Compression Conference (DCC)*, Snowbird, UT, 156–165, March 1994.

[14] G. K. Wallace, "The JPEG still picture compression standard," *Commun. ACM* **34**, 31–44 (1991).

[15] A. B. Watson, "Visually optimal DCT quantization matrices for indvidual images," *Proceedings of the IEEE Data Compression Conference (DCC)*, Snowbird, UT, 178–187, March 1993.

[16] I. H. Witten, R. M. Neal, and J. G. Cleary, "Arithmetic coding for data compression," *Commun. ACM* **30**, 520–540 (1987).

The JPEG Lossless Image Compression Standards

Nasir Memon
Polytechnic University

Rashid Ansari
University of Illinois at Chicago

1 Introduction

Although the Joint Photographic Expert Group (JPEG) commit-tee of the International Standards Organization is best known for the development of the ubiquitous *lossy* compression stan-dard, which is commonly known as JPEG compression today, it has also developed some important *lossless* image compression standards. The first lossless algorithm adopted by the JPEG com-mittee, known as JPEG lossless, was developed and standardized along with the well-known lossy standard. However, it had little in common with the lossy standard based on the discrete cosine transform (DCT). The original goals set by the JPEG committee in 1988 stated that the lossy standard should also have a lossless mode that gives about 2 to 1 compression on images similar to the original test set. Perhaps it was also envisioned that both lossy and lossless compression be achieved by a single algorithm working with different parameters. In fact, some of the proposals submitted did have this very same capability. However, given the superior performance of DCT-based algorithms for lossy com-pression, and given the fact that errors caused by implementing DCT with finite precision arithmetic preclude the possibility of lossless compression, an entirely different algorithm was adopted for lossless compression. The algorithm chosen was a very simple technique that uses differential pulse code modulation (DPCM) in conjunction with either Huffman or arithmetic coding for encoding prediction errors.

Although the JPEG lossless algorithm that uses Huffman cod-ing has seen some adoption and a few public domain imple-mentations of it are freely available, the JPEG lossless algorithm based on arithmetic coding has seen little use as of today, despite the fact that it provides about 10% to 15% better compression. Perhaps this is due to the intellectual property issues surround-ing arithmetic coding and to the perceived computational and conceptual complexity issues associated with it. To address this problem, the JPEG committee revisited the issue in 1994 and initiated the development of a new lossless image compression standard. A new work item proposal was approved in early 1994, titled *Next Generation Lossless Compression of Continuous-Tone Still Pictures*. A call was issued in March 1994 soliciting proposals specifying algorithms for lossless and near-lossless compression of continuous-tone (2 to 16 bits) still pictures. It was announced that the algorithms should:

- provide lossless and near-lossless compression,
- target 2- to 16-bit still images,
- be applicable over a wide variety of content,
- not impose any size or depth restrictions,
- be applicable to fields such as medical, satellite, archival, etc.,
- be amenable to implementation with reasonably low com-plexity,
- significantly improve upon the performance of current loss-less standards, and
- work with a single pass through data.

A series of additional requirements were imposed on submis-sions. The reader is referred to [1] for details. For instance, exploitation of interband correlations (in color and satellite images for example) was prohibited. This was done in order to facilitate fair comparison of competing schemes. Later extensions of the standard do incorporate interband coding.

In July of 1995, a total of nine proposals were submitted in response to this call. The nine submitted proposals were evaluated by ISO on a very large set of test images by using an objective performance measure that had been announced prior to the competition. Seven out of the nine proposals submitted employed a traditional DPCM-based approach very much like the original lossless JPEG standard, although they contained more sophisticated context modeling techniques for encoding prediction errors. The other two proposals were based on reversible integer wavelet transform coding. However, right from the first round of evaluations, it was clear that proposals based on transform coding did not provide compression ratios as good as those of the proposed algorithms based on the prediction-based DPCM techniques [15]. The best algorithm in the first round was CALIC, a context-based predictive technique [19].

After a few rounds of convergence the final baseline algorithm adopted for standardization was based largely on the revised Hewlett-Packard proposal LOCO-I_{1p}, and a DIS (Draft International Standard) was approved by the committee in 1997 [2]. The new draft standard was named JPEG-LS in order to distinguish it from the earlier lossy and lossless standards. JPEG-LS baseline is a modern and sophisticated lossless image compression algorithm that, despite its conceptual and computational simplicity, yields a performance that is surprisingly close to that of the best known techniques like CALIC. JPEG-LS baseline contains the core of the algorithm and many extensions to it are currently under standardization.

In the rest of this chapter the different lossless image compression standards developed by the JPEG committee are described in greater detail. In Section 2, both the Huffman and arithmetic coding versions of the original JPEG lossless standard are presented. In Section 3, JPEG-LS is described. In the same section we also briefly discuss different extensions that have been proposed to the baseline JPEG-LS algorithm and are currently under the process of standardization. Finally, in Section 4 we discuss the integration of lossless and lossy compression being proposed in JPEG 2000, another new standard currently under development by the JPEG committee.

2 The Original JPEG Lossless Standards

As mentioned before, the original JPEG lossless standards based on either Huffman or arithmetic coding both employ a predictive approach. That is, the algorithm scans an input image, row by row, left to right, predicting each pixel as a linear combination of previously processed pixels and encodes the prediction error. Since the decoder also processes the image in the same order, it can make the same prediction and recover the actual pixel value based on the prediction error. The standard allows the user to

TABLE 1 JPEG predictors for lossless coding

Mode	Prediction for $P[i, j]$
0	0 (No Prediction)
1	N
2	W
3	NW
4	$N + W - NW$
5	$W + (N - NW)/2$
6	$N + (W - NW/2)$
7	$(N + W)/2$

choose between eight different predictors, which are listed in Table 1. The notation used for specifying neighboring pixels used in arriving at a prediction is shown in Fig. 1 in the form of a template of two-dimensional neighborhood pixels. A subset of this neighborhood is generally used for prediction or context determination by most lossless image compression techniques. In the rest of the paper we shall consistently use this notation to denote specific neighbors of the pixel $P[i, j]$ in the ith row and jth column.

Prediction essentially attempts to capture the intuitive notion that the intensity function of typical images is usually quite "smooth" in a given local region and hence the value at any given pixel is quite similar to its neighbors. In any case, if the prediction made is reasonably accurate then the prediction error has significantly lower magnitude and variance when compared with the original signal, and it can be encoded efficiently with a suitable variable-length coding technique. In JPEG lossless, prediction errors can be encoded with either Huffman or arithmetic coding, codecs for both being provided by the standard. In the rest of this section we elaborate on the different procedures required or recommended by the standard for Huffman and arithmetic coding.

2.1 Huffman Coding Procedures

In the Huffman coding version, essentially no error model is used. Prediction errors are assumed to be independent and identically distributed (i.i.d.), and they are encoded with the Huffman

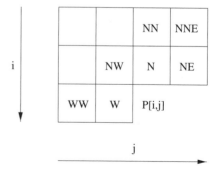

FIGURE 1 Notation used for specifying neighborhood pixels of current pixel $P[i, j]$.

TABLE 2 Mapping of prediction errors to magnitude category and extra bits

Category	Symbols	Extra Bits
0	0	—
1	$-1, 1$	$0, 1$
2	$-3, -2, 2, 3$	$00,01,10,11$
3	$-7, \ldots, -4, 4, \ldots, 7$	$000, \ldots, 011, 100, \ldots, 111$
4	$-15, \ldots, -8, 8, \ldots, 15$	$0000, \ldots, 0111, 1000, \ldots, 1111$
$\vdots$	$\vdots$	$\vdots$
15	$-32767, \ldots, -16384, 16384, \ldots, 32767$	$0 \ldots 00, \ldots, 01 \ldots 1, 10 \ldots 0, \ldots, 11 \ldots 1$
16	32768	

table provided in the bit stream using the specified syntax. The Huffman coding procedure specified by the standard for encoding prediction errors is identical to the one used for encoding DC coefficient differences in the lossy codec.

Since the alphabet size for the prediction errors is twice the original alphabet size, a Huffman code for the entire alphabet would require an unduly large code table. An excessively large Huffman table can lead to multiple problems. First of all, a larger code would require more bits to represent. In JPEG, this is not a problem, as a special length-limited Huffman code is used that can be specified by a small and fixed number of bits. More importantly, however, large Huffman tables can lead to serious difficulties in a hardware implementation of the codec. In order to reduce the size of the Huffman table, each prediction error (or DC difference in the lossy codec) is classified into a "magnitude category" and the label of this category is Huffman coded. Since each category consists of a multiple number of symbols, uncoded "extra bits" are also transmitted that identify the exact symbol (prediction error in the lossless codec, and DC difference in the lossy codec) within the category. Table 2 shows the 17 different categories that are defined. As can be seen, except for the 17th (last) category, each category k contains 2^k members $\{\pm 2^{k-1}, \ldots, \pm 2^k - 1\}$ and hence k extra bits would be required to identify a specific symbol within the category. The extra bits for specifying the prediction error e in the category k are given by the k-bit number n by the mapping

$$n = \begin{cases} e & \text{if } e \geq 0 \\ 2^k - 1 + e & \text{if } e < 0 \end{cases}.$$

For example, the prediction error -155 would be encoded by the Huffman code for category 8 and the eight extra bits (01100100) (integer 100) would be transmitted to identify -155 within the 256 different elements that fall within this category. If the prediction error was 155, then (10011011), integer 155, would be transmitted as extra bits. In practice, the above mapping can also be implemented by using the k-bit unsigned representation of e if e is positive and its one's complement if negative.

The Huffman code used for encoding the category label has to meet the following conditions.

1. The Huffman code is a length limited code. The maximum code length for a symbol is 16 bits.

2. The Huffman code is a canonical code. The k codewords of given length n are represented by the n-bit numbers $x+1, x+2, \ldots, x+k$, where x is obtained by left shifting the largest numerical value represented by an $(n-1)$-bit codeword.

The above two conditions greatly facilitate the specification of a Huffman table and a fast implementation of the encoding and decoding procedures. In a JPEG bit stream, a Huffman table is specified by two lists, BITS and HUFFVAL. BITS is a 16-byte array contained in the codeword stream, where byte n simply gives the number of codewords of length n that are present in the Huffman table. HUFFVAL is a list of symbol values in order of increasing codeword length. If two symbols have the same code length, then the symbol corresponding to the smaller numerical value is listed first. Given these two tables, the Huffman code table can be reconstructed in a relatively simple manner. The standard provides an example procedure for doing this in its informative sections but does not mandate its usage except in the functional sense. That is, given the lists BITS and HUFFVAL, different decoders need to arrive at the same reconstruction, irrespective of the procedure used. In practice, many hardware implementations of the lossy codec do not implement the reconstruction procedure and directly input the Huffman table.[1]

Furthermore, the standard only specifies the syntax used for representing a Huffman code. It does not specify how to arrive at the specific length-limited code to be used. One simple way to arrive at a length-limited code is to force probabilities of occurrence for any particular symbol not to be less than 2^{-l} and then run the regular Huffman algorithm. This will ensure that any given codeword does not contain more than l bits. It should be noted that although this procedure is simple, it does not necessary generate an optimal length-limited code. Algorithms for constructing an optimal length-limited code have been proposed in the literature. In practice, however, the above simple procedure works satisfactorily given the small alphabet size for the Huffman table used in JPEG. In addition, the standard also requires that the bit sequence of all 1s not be a codeword for any symbol.

---

[1] Actually, what is loaded into the ASIC implementing the lossy codec is not the Huffman code itself but a table that facilitates fast encoding and decoding of the Huffman code.

When an image consists of multiple components, like color images, separate Huffman tables can be specified for each component. The informative sections of the standard provide example Huffman tables for luminance and chrominance components. They work quite well for lossy compression over a wide range of images and are often used in practice. Most software implementations of the lossy standard permit the use of these "default" tables, allowing an image to be encoded in a single pass. However, since these tables were mainly designed for encoding DC coefficient differences for the lossy codec, they may not work well with lossless compression. For lossless compression, a custom Huffman code table can be specified in the bit stream along with the compressed image. Although this approach requires two passes through the data, it does give significantly better compression. Finally it should be noted that the procedures for Huffman coding are common to both the lossy and the lossless standards.

2.2 Arithmetic Coding Procedures

Unlike the version based on Huffman coding, which assumes the prediction error samples to be i.i.d., the arithmetic coding version uses quantized prediction errors at neighboring pixels as contexts for conditional coding of the prediction error. This is a simplified form of error modeling that attempts to capture the remaining structure in the prediction residual. Encoding within each context is done with a binary arithmetic coder by decomposing the prediction error into a sequence of binary decisions. The first binary decision determines if the prediction error is zero. If not zero, then the second step determines the sign of the error. The subsequent steps assist in classifying the magnitude of the prediction error into one of a set of ranges and the final bits that determine the exact prediction error magnitude within the range are sent uncoded.

The QM coder is used for encoding each binary decision. A detailed description of the coder and the standard can be found in [11]. Since the arithmetic coded version of the standard is rarely used, we do not dwell on the details of the procedures used for arithmetic coding and only provide a brief summary. The interested reader can find details in [11].

The QM coder is a modification of an adaptive binary arithmetic coder called the Q coder [10], which in turn is an extension of another binary adaptive arithmetic coder called the *skew* coder [13]. Instead of dealing directly with the 0s and 1s put out by the source, the QM coder maps them into a more probable symbol (MPS) and less probable symbol (LPS). If 1 represents black pixels, and 0 represents white pixels, then in a mostly black image, 1 will be the MPS, while in an image with mostly white regions 0 will be the MPS. In order to make the implementation simple, the committee recommended several deviations from the standard arithmetic coding algorithm. The update equations in arithmetic coding that keep track of the subinterval to be used for representing the current string of symbols involve multipli-

cations that are expensive in both hardware and software. In the QM coder, expensive multiplications are avoided and rescalings of the interval take the form of repeated doubling, which corresponds to a left shift in the binary representation. The probability q_c of the LPS for context C is updated each time a rescaling takes place and the context C is active. An ordered list of values for q_c is kept in a table. Every time a rescaling occurs, the value of q_c is changed to the next lower or next higher value in the table, depending on whether the rescaling was caused by the occurrence of an LPS or MPS. In a nonstationary situation, it may happen that the symbol assigned to LPS actually occurs more often than the symbol assigned to MPS. In this situation, the assignments are reversed; i.e., the symbol assigned the LPS label is assigned the MPS label and vice versa. The test is conducted every time a rescaling takes place. The decoder for the QM coder operates in much the same way as the encoder, by mimicking the encoder operation.

3 JPEG-LS — The New Lossless Standard

As mentioned earlier, the JPEG-LS algorithm, like its predecessor, is a predictive technique. However, there are significant differences, as described below.

1. Instead of using a simple linear predictor, JPEG-LS uses a nonlinear predictor that attempts to detect the presence of edges passing through the current pixel and accordingly adjusts prediction. This results in a significant improvement in performance in the prediction step.
2. Like JPEG lossless arithmetic, JPEG-LS uses some simple but very effective context modeling of the prediction errors prior to encoding.
3. Baseline JPEG-LS uses Golomb–Rice codes for encoding prediction errors. Golomb–Rice codes are Huffman codes for certain geometric distributions that serve well in characterizing the distribution of prediction errors. Although Golomb–Rice codes have been known for a long time, JPEG-LS uses some novel and highly effective techniques for adaptively estimating the parameter for the Golomb–Rice code to be used in a given context.
4. In order to effectively code low entropy images or regions, JPEG-LS uses a simple alphabet extension mechanism, by switching to a run-length mode when a uniform region is encountered. The run-length coding used is again an extension of Golomb codes and provides significant improvement in performance for highly compressible images.
5. For applications that require higher compression ratios, JPEG-LS provides a near-lossless mode that guarantees each reconstructed pixel to be within a distance k from its original value. Near-lossless compression is achieved by a simple uniform quantization of the prediction error.

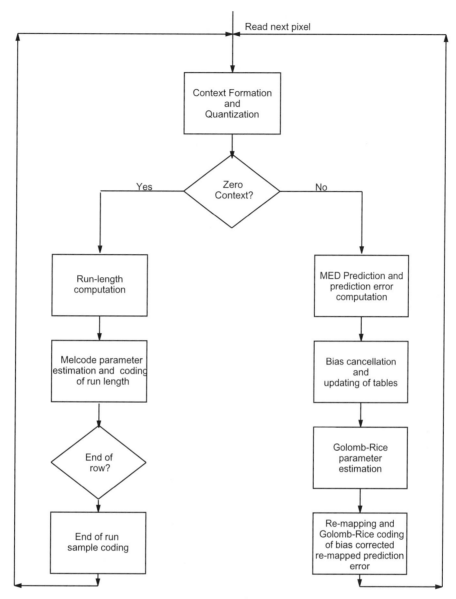

FIGURE 2   Overview of baseline JPEG-LS.

An overview of the JPEG-LS baseline algorithm is shown in Fig. 2. In the rest of this section we describe in more detail each of the steps involved in the algorithm and some of the extensions that are currently under the process of standardization. For a detailed description the reader is referred to the working draft [2].

3.1 The Prediction Step

JPEG-LS uses a very simple and effective predictor. The median edge detection (MED) predictor, which adapts in presence of local edges. MED detects horizontal or vertical edges by examining the North N, West W, and Northwest NW neighbors of the current pixel $P[i, j]$. The North (West) pixel is used as a predic-

tion in the case of a vertical (horizontal) edge. In case of neither, planar interpolation is used to compute the prediction value. Specifically, prediction is performed according to the following equations:

$$\hat{P}[i, j] = \begin{cases} \min(N, W) & \text{if } NW < \max(N, W) \\ \max(N, W) & \text{if } NW < \min(N, W) \\ N + W - NW & \text{otherwise} \end{cases}.$$

The MED predictor is essentially a special case of the median adaptive predictor (MAP), first proposed by Martucci in 1990 [6]. Martucci proposed the MAP predictor as a nonlinear adaptive predictor that selects the median of a set of three predictions in order to predict the current pixel. One way of interpreting such a predictor is that it always chooses either the best or the second-best predictor among the three candidate predictors. Martucci reported the best results with the following three predictors, in which case it is easy to see that MAP turns out to be the MED predictor.

1. $\hat{P}[i, j] = N$.
2. $\hat{P}[i, j] = W$.
3. $\hat{P}[i, j] = N + W - NW$.

In an extensive evaluation, Memon and Wu observed that the MED predictor gives a performance that is superior to or almost as good as that of several standard prediction techniques, many of which are significantly more complex [7, 8].

3.2 Context Formation

Gradients alone cannot adequately characterize some of the more complex relationships between the predicted pixel $P[i, j]$ and its surrounding area. Context modeling of the prediction error $e = \hat{P}[i, j] - P[i, j]$ can exploit higher-order structures such as texture patterns and local activity in the image for further compression gains. Contexts in JPEG-LS are formed by first computing the following differences:

$$D1 = NE - N,$$
$$D2 = N - NW,$$
$$D3 = NW - W, \tag{1}$$

where the notation for specifying neighbors is as shown in Fig. 1. The differences D1, D2, and D3 are then quantized into nine regions (labeled -4 to $+4$) symmetric about the origin with one of the quantization regions (region 0) containing only the difference value 0. Further, contexts of the type (q_1, q_2, q_3) and $(-q_1, -q_2, -q_3)$ are merged based on the assumption that

$$P(e|q_1, q_2, q_3) = P(-e| - q_1, -q_2, -q_3).$$

The total number of contexts turn out to be $9^3 - 1/2 = 364$. These contexts are then mapped to the set of integers $[0, 363]$ in a one-to-one fashion. The standard does not specify how contexts are mapped to indices and vice versa, leaving it completely to the implementation. In fact, two different implementations could use different mapping functions and even a different set of indices but nevertheless be able to decode files encoded by the other. The standard only requires that the mapping be one to one.

3.3 Bias Cancellation

As described earlier, in JPEG arithmetic, contexts are used as conditioning states for encoding prediction errors. Within each state the pdf of the associated set of events is adaptively estimated from events by keeping occurrence counts for each context. Clearly, to better capture the structure preset in the prediction residuals,

one would like to use a large number of contexts or conditioning states. However, the larger the number of contexts, the more the number of parameters (conditional probabilities in this case) that have to be estimated based on the same data set. This can lead to the "sparse context" or "high model cost" problem. In JPEG lossless arithmetic this problem is addressed by keeping the number of contexts small and decomposing the prediction error into a sequence of binary decisions, each requiring estimation of a single probability value. Although this results in alleviating the sparse context problem, there are two problems caused by such an approach. First, keeping the number of conditioning states to a small number fails to capture effectively the structure present in the prediction errors and results in poor performance. Second, binarization of the prediction error necessitates the use of an arithmetic coder, which adds to the complexity of the coder.

The JPEG-LS baseline algorithm employs a different solution for this problem. First of all it uses a relatively large number of contexts to capture the structure present in the prediction errors. However, instead of estimating the pdf of prediction errors, $p(e \mid C)$, within each context C, only the conditional expectation $E\{e \mid C\}$ is estimated, using the corresponding sample means $\bar{e}(C)$ within each context. These estimates are then used to further refine the prediction prior to entropy coding, by an error feedback mechanism that cancels prediction biases in different contexts. This process is called *bias cancellation*. Furthermore, for encoding the bias-cancelled prediction errors, instead of estimating the probabilities of each possible prediction error, baseline JPEG-LS essentially estimates a parameter that serves to characterize the specific pdf to be employed from a fixed set of pdfs. This is explained in greater detail in the next subsection.

A straightforward implementation of bias cancellation would require accumulating prediction errors within each context and keeping frequency counts of the number of occurrences for each context. The accumulated prediction error within a context divided by its frequency count would then be used as an estimate of prediction bias within the context. However, this division operation can be avoided by a simple and clever operation that updates variables in a suitable manner producing average prediction residuals in the interval $[-0.5, 0.5]$. For details the reader is referred to the proposal that presented this technique and also to the DIS [2, 18]. Also, since JPEG-LS uses Golomb–Rice codes, which assign shorter codes to negative residual values than to positive ones, the bias is adjusted such that it produces average prediction residuals in the interval $[-1, 0]$, instead of $[-0.5, 0.5]$. For a detailed justification of this procedure, and other details pertaining to bias estimation and cancellation mechanisms, see [18].

3.4 Rice–Golomb Coding

Before the development of JPEG-LS, the most popular compression standards, such as JPEG, MPEG, H263, and CCITT Group 4, have essentially used static Huffman codes in the entropy coding stage. This is because adaptive Huffman coding does not provide enough compression improvement in order to justify the additional complexity. Adaptive arithmetic coding, in contrast, despite being used in standards such as JBIG and JPEG arithmetic, has also seen little use because of concerns about intellectual property restrictions and also perhaps because of the additional computational resources that are needed. JPEG-LS is the first international compression standard that uses an adaptive entropy coding technique that requires only a single pass through the data and requires computational resources that are arguably lesser than what is needed by static Huffman codes.

In JPEG-LS, prediction errors are encoded using a special case of Golomb codes [5], which is also known as Rice coding [12]. Golomb codes of parameter m encode a positive integer n by encoding n mod m in binary followed by an encoding of n div m in unary. The unary coding of n is a string of n 0 bits, followed by a terminating 1 bit. When $m = 2^k$, Golomb codes have a very simple realization and have been referred to as Rice coding in the literature. In this case n mod m is given by the k least significant bits and n div m by the remaining bits. So, for example, the Rice code of parameter 3 for the 8-bit number 42 (00101010 in binary) is given by 010000001, where the first 3 bits 010 are the three least significant bits of 42 (which is the same as 42 mod 8) and the remaining 6 bits represent the unary coding of 42 div 8 = 5, which is represented by the remaining 5 bits 00101 in the binary representation of 42. Note that, depending on the convention being employed, the binary code can appear before the unary code and the unary code could have leading 1s terminated by a zero instead.

Clearly, the number of bits needed to encode a number n depend on the parameter k employed. For the example above, if $k = 2$ was used we would get a code length of 13 bits and the parameter 4 would result in 7 bits being used. It turns out that given an integer n the Golomb–Rice parameter that results in the minimum code length of n is $\lceil \log_2 n \rceil$.

From these facts, it is clear that the key factor behind the effective use of Rice codes is estimating the parameter k to be used for a given sample or block of samples. Rice's algorithm [12] tries codes with each parameter on a block of symbols and selects the one that results in the shortest code as suggested. This parameter is sent to the decoder as side information. However, in JPEG-LS the coding parameter k is estimated on the fly for each prediction error by using techniques proposed by Weinberger *et al.* [17]. Specifically, the Golomb parameter is estimated by maintaining in each context the count N of the prediction errors seen so far and the accumulated sum of magnitude of prediction errors A seen so far. The coding context k is then computed as

$$k = \min\{k' \mid 2^{k'} N \geq A\}.$$

The strategy employed is an approximation to optimal parameter selection for this entropy coder. Despite the simplicity of the coding and estimation procedures, the compression performance achieved is surprisingly close to that obtained by arithmetic coding. For details the reader is referred to [17].

Also, since Golomb–Rice codes are defined for positive integers, prediction errors have to be accordingly mapped. In JPEG-LS, prediction errors are first reduced to the range $[-128, 128]$ by the operation $e = y - x \bmod 256$ and then mapped to positive values by

$$m = \begin{cases} 2e & \text{if } e \geq 0 \\ -2e - 1 & \text{if } e < 0 \end{cases}.$$

3.5 Alphabet Extension

The use of Golomb–Rice codes is very inefficient when coding low-entropy distributions because the best coding rate achievable is 1 bit per symbol. Obviously for entropy values of less than 1 bit per symbol, such as would be found in smooth regions of an image, this can be very wasteful and lead to significant deterioration in performance. This problem can be alleviated by using *alphabet extension*, wherein blocks of symbols rather than individual symbols are coded, thus spreading the excess coding length over many symbols. The process of blocking several symbols together prior to coding produces less skewed distributions, which is desirable.

To implement alphabet extension, JPEG-LS first detects smooth areas in the image. Such areas in the image are characterized by the gradients $D1$, $D2$, and $D3$, as defined in 1, all being zero. In other words, this is the context $(0, 0, 0)$, which we call the *zero context*. When a zero context is detected, the encoder enters a run mode where a run of the west symbol B is assumed and the total run of the length is encoded. The end of run state is indicated by a new symbol $x \neq B$, and the new symbol is encoded by using its own context and special techniques described in the standard [2]. A run may also be terminated by the end of line, in which case only the total length of run is encoded.

The specific run-length coding scheme used is the MELCODE described in [9]. MELCODE is a binary coding scheme in which target sequences contain an MPS and a LPS. In JPEG-LS, if the current symbol is the same as the previous one, an MPS is encoded; otherwise, an LPS is encoded. Runs of the MPS of length n are encoded using only one bit. If the run is of length less than n (including 0), it is encoded by a zero bit followed by the binary value of the run length encoded using $\log n$ bits. The parameter n is constrained to be of the form 2^k and is adaptively updated while encoding a run. For details of the adaptation procedure and other details pertaining to the run mode, the reader is referred to the draft standard [2]. Again the critical factor behind effective usage of the MELCODE is the estimation of the parameter value n to be used.

3.6 Near-Lossless Compression

Although lossless compression is required in many applications, compression ratios obtained with lossless techniques are significantly lower than those possible with lossy compression.

Typically, on one hand, depending on the image, lossless compression ratios range from about 1.5 to 1 to 3 to 1. On the other hand, state-of-the-art lossy compression techniques give compression rations in excess of 20 to 1, with virtually no loss in visual fidelity. However, in many applications, the end use of the image is not human perception. In such applications, the image is subjected to postprocessing in order to extract parameters of interest like ground temperature or vegetation indices. The uncertainty about reconstruction errors introduced by a lossy compression technique is undesirable.

This leads to the notion of a *near-lossless* compression technique that gives quantitative guarantees about the type and amount of distortion introduced. Based on these guarantees, a scientist can be assured that the extracted parameters of interest will either not be affected or be affected only within a bounded range of error. Near-lossless compression could potentially lead to significant increase in compression, thereby giving more efficient utilization of precious bandwidth while preserving the integrity of the images with respect to the postprocessing operations that are carried out.

JPEG-LS has a near-lossless mode that guarantees a $\pm k$ reconstruction error for each pixel. Extension of the lossless baseline algorithm to the case of near-lossless compression is achieved by prediction error quantization according to the specified pixel value tolerance. In order for the predictor at the receiver to track the predictor at the encoder, the reconstructed values of the image are used to generate the prediction at both the encoder and the receiver. This is the classical DPCM structure. Specifically, the prediction error is quantized according to the following rule:

$$Q[e] = \left\lfloor \frac{e + k}{2k + 1} \right\rfloor (2k + 1), \qquad (2)$$

where e is the prediction error, k is the maximum reconstruction error allowed in any given pixel, and $\lfloor . \rfloor$ denotes the integer part of the argument. At the encoder, a label l is generated according to

$$l = \left\lfloor \frac{e + k}{2k + 1} \right\rfloor. \qquad (3)$$

This label is encoded, and at the decoder the prediction error is reconstructed according to

$$\hat{e} = l(2k + 1). \qquad (4)$$

This form of quantization, where all values in the interval $[nk - \lfloor \frac{k}{2} \rfloor, nk + \lfloor \frac{k}{2} \rfloor]$ are mapped to nk, is a special case of *uniform quantization*. It is well known that uniform quantization leads to a minimum entropy of the output, provided the step size is small enough for the constant pdf assumption to hold. For small values of k, as one would expect to be used in near-lossless compression, this assumption is reasonable.

It has been experimentally observed that for bit rates exceeding 1.5 bpp, JPEG-LS near-lossless actually gives better performance than baseline lossy JPEG. However, it should be noted that the uniform quantization performed in JPEG-LS near-lossless often gives rise to annoying "contouring" artifacts. Such artifacts are most visually obvious in smooth regions of the image. In Chapter 1.1 of this volume, such "false contouring" is examined in more detail and shown to possibly occur even from simple image quantization.

In many cases these artifacts can be reduced by some simple postprocessing operations. As explained in the next section, JPEG-LS Part 2 allows variation of the quantization step size spatially in a limited manner, thereby enabling some possible reduction in artifacts.

Finally, it may appear that the quantization technique employed in JPEG-LS is overly simplistic. In actuality, there is a complex dependency between the quantization error that is introduced and subsequent prediction errors. Clearly, quantization affects the prediction errors obtained. Although one can vary the quantization in an optimal manner by using a trellis-based technique and the Viterbi algorithm, it has been observed that such computationally expensive and elaborate optimizing strategies offer little advantage, in practice, over the simple uniform quantization used in JPEG-LS [3].

3.7 JPEG-LS Part 2

Even as the baseline algorithm was being standardized, the JPEG committee initiated development of JPEG-LS Part 2. Initially the motivation for Part 2 was to standardize an arithmetic coding version of the algorithm. As it evolves, however, Part 2 also includes many other features that improve compression but were considered to be too application specific to include in the baseline algorithm. Eventually, it appears that JPEG-LS Part 2 will be an algorithm that is substantially different from Part 1, although the basic approach, in terms of prediction followed by context-based modeling and coding of prediction errors, remains the same as the baseline. As of December 1998, Part 2 has been mostly, but not completely, finalized. In the subsections that follow we briefly describe some of the key features that are expected to be part of the standard.

3.7.1 Prediction

JPEG-LS baseline is not suitable for images with sparse histograms (prequantized images or images with less than 8 or 16 bits represented by 1 or 2 bytes, respectively). This is because predicted values of pixels do not actually occur in the image, which causes code space to be wasted during the entropy coding of prediction errors. In order to deal with such images, Part 2 defines an optional *prediction value control* mode, wherein it is ensured that a predicted value is always a symbol that has actually occurred in the past. This is done by forming the same prediction as JPEG baseline using the MED predictor, but by adjusting the predictor to a value that has been seen before.

A flag array is used to keep track of pixel values that have occurred thus far.

3.7.2 Context Formation

In order to better model prediction errors, Part 2 uses an additional gradient

$$D4 = WW - W, \qquad (5)$$

where W and WW are the neighboring pixels as shown in Fig. 1. D4 is quantized along with D1, D2, and D3, defined earlier in Eq. (1), just as in the baseline. D4 is, however, quantized only to three regions. Context merging is done only on the basis of D_1, D_2, and D_3, and not D_4. That is, contexts of type (q_1, q_2, q_3, q_4) and $(-q_1, -q_2, -q_3, q_4)$ are merged to arrive a total of $364 \times 3 = 1092$ contexts.

3.7.3 Bias Cancellation

In the bias-cancellation step of baseline JPEG-LS, prediction errors within each context are centered around -0.5 instead of 0. As explained, this was done because the prediction error mapping technique and Rice–Golomb coding used in the baseline algorithm assign shorter code words to negative errors as opposed to a positive error of the same magnitude. However, if arithmetic coding is employed, then there is no such imbalance and bias cancellation is used to center the prediction error distribution in each context around zero. This is exactly the bias-cancellation mechanism proposed in CALIC.

3.7.4 Alphabet Extension

If arithmetic coding is used, then alphabet extension is clearly not required. Hence, in the arithmetic coding mode, the coder does not switch to run mode on encountering the all-zeros context. However, in addition to this change, Part 2 also specifies some small changes to the run-length coding mode of the original baseline algorithm. For example, when the underlying alphabet is binary, Part 2 does away with the redundant encoding of sample values that terminated a run, as required by the baseline.

3.7.5 Arithmetic Coding

Even though the baseline algorithm has an alphabet extension mechanism for low-entropy images, performance can be significantly improved by the use of arithmetic coding. Hence the biggest difference between JPEG-LS and JPEG-LS Part 2 is in the entropy coding stage. Part 2 uses a binary arithmetic coder for encoding prediction errors that are binarized by a Golomb code. The Golomb code tree is produced based on an activity class of the context computed from its current average prediction error magnitude. Twelve activity levels are defined. In the arithmetic coding procedure, numerical data are treated in radix 255 representation, with each sample expressed as 8-bit data. Probabilities of the MPS and the LPS are estimated by keeping occurrence counts. Multiplication and division are avoided by approximate values stored in a look-up table.

3.7.6 Near-Lossless Mode

The near-lossless mode is another area where Part 2 differs significantly from the baseline. Essentially, Part 2 provides mechanisms for a more versatile application of the near-lossless mode. The two main features enabled by Part 2 in the near-lossless mode are as follows.

1. Visual quantization: as mentioned before, near-lossless compression can often lead to annoying artifacts at larger values of k. Furthermore, the baseline does not provide any graceful degradation mechanism between step size of k and $k + 1$. Hence JPEG-LS Part 2 defines a new "visual quantization" mode. In this mode, the quantization step size is allowed to be either k or $k + 1$, depending on the context. Contexts with a larger gradients use a step size of $k + 1$, and contexts with smaller gradients use a step size of k. The user specifies a threshold, based on which this decision is made. The standard does not specify how the threshold should be arrived at. It only provides a syntax for its specification.

2. Rate control: by allowing the user to change the quantization step size while encoding an image, Part 2 essentially provides a rate-control mechanism whereby the coder can keep track of the coded bytes, based on which appropriate changes to the quantization step size can be made. The encoder, for example, can compress the image to less than a bounded size with a single sequential pass over the image. Other uses of this feature are possible, including region-of-interest lossless coding, etc.

3.7.7 Fixed Length Coding

There is a possibility that a Golomb code will cause data expansion and result in a compressed image larger than the source image. To avoid such a case, an extension to the baseline is defined whereby the encoder can switch to a fixed length coding technique by inserting an appropriate marker in the bit stream. Another marker is used to signal the end of fixed length coding. The procedure for determining if data expansion is occurring and for selecting the size of the fixed length representation is left entirely up to the implementation. The standard does not make any recommendation.

3.7.8 Interband Correlations

Currently there is no mechanism for exploiting interband correlations in JPEG-LS baseline a well as JPEG-LS Part 2. The application is expected to decorrelate individual bands prior to encoding by JPEG-LS. The lack of informative or normative measures for exploiting interband correlations, in our opinion, is the most serious shortcoming of the JPEG-LS standard. It is very likely that the committee will incorporate some such mechanism into JPEG-LS Part 2 before its adoption.

4 The Future: JPEG 2000 and the Integration of Lossless and Lossy Compression

In prediction-based lossless image compression techniques, image pixels are processed in some fixed and predetermined order. The intensity of each pixel is modeled as being dependent on the intensities values in a fixed and predetermined neighborhood set of previously visited pixels. As a result, such techniques do not adapt well to the nonstationary nature of image data. Furthermore, such techniques form predictions and model the prediction error based solely on local information. Hence they usually do not capture "global patterns" that influence the intensity value of the current pixel being processed. As a consequence, recent years have seen techniques based on a predictive approach rapidly reach a point of diminishing returns. JPEG-LS, the new lossless standard provides testimony to this fact. Despite being extremely simple, it provides compression performance that is within a few percent of more sophisticated techniques such as CALIC [19] and UCM [16]. Experimentation suggests that an improvement of more than 10% is unlikely to be obtained by pushing the envelope on the current state-of-the-art predictive techniques like CALIC [8]. Furthermore, the complexity costs incurred for obtaining these improvements are enormous and usually not worth the marginal improvement in compression that is obtained.

An alternative approach to lossless image compression that has emerged recently is based on subband (or wavelet) decomposition. Subband decomposition provides a way to cope with the nonstationarity of image data by separating the information into several scales and exploiting correlations within each scale as well as across scales. A subband approach also provides a better framework for capturing global patterns in the image data. Finally, the wavelet transforms employed in the decomposition can be viewed as a prediction scheme, as in [4, 14], that is not restricted to a casual template but makes a prediction of the current pixel based on "past" and "future" pixels with respect to a spatial raster scan.

In addition to these advantages, there are other advantages offered by a subband approach for lossless image compression. The most important of these is perhaps the natural integration of lossy and lossless compression that the subband approach makes possible. By transmitting entropy-coded subband coefficients in an appropriate manner, one can produce an embedded bit stream that permits the decoder to extract a lossy reconstruction at the desired bit rate. This enables progressive decoding of the image that can ultimately lead to lossless reconstruction [14, 20]. The image can also be recovered at different spatial resolutions. These features are of great value for specific applications

like teleradiology and the World Wide Web, and for applications in "network-centric" computing in general. More details on these are given in Chapter 4.1 (wavelets), Chapter 5.4 (wavelet image coding), and Chapter 6.2 (wavelet video coding) of this handbook.

The above facts have caused an increasing popularity of the subband approach for lossless image compression. Some excellent work has already been done toward applying subband image coding techniques for lossless image compression, such as S + P [14] and CREW [20]. However, they do not perform as well for lossless compression as compared with predictive techniques, which are arguably simpler, both conceptually and computationally. Nevertheless, it should be noted that a subband decomposition approach for lossless image compression is still in its infancy, and hence it should be no surprise if it does not yet provide compression performance that matches state-of-the-art predictive techniques like CALIC, which took its current form after years of development and refinement.

The JPEG committee is currently going through the process of standardizing a state-of-the-art compression technique with many "modern" features like embedded quantization, region-of-interest decoding, etc. The new standard will be a wavelet-based technique, which among other features, will make lossy and lossless compression possible within a single framework. Although such a standard may receive wide adoption, it is not clear whether JPEG 2000 will eventually replace the other lossless standards that currently exist and were described in this chapter. Surely, certain application will require the computational simplicity of JPEG lossless Huffman and JPEG-LS and its extensions. For example, the memory requirements of a wavelet-based approach are typically very high. Such techniques are not suitable for printers and other applications in which additional memory adds to the fixed cost of the product.

5 Additional Information

A free implementation of the Huffman-based original JPEG lossless algorithm, written by the PVRG group at Stanford, is available from havefun.stanford.edu:/pub/jpeg/JPEGv1.2.1.tar.Z. The PVRG code is designed for research and experimentation rather than production use, but it is easy to understand. There's also a lossless-JPEG-only implementation available from Cornell, ftp.cs.cornell.edu:/pub/multimed/ljpg.tar.Z. Neither the PVRG nor Cornell codecs are being actively maintained. They are both written in the C language and can be ported to a variety of operating systems, including variants of UNIX and the different Microsoft platforms.

The JPEG committee maintains a Web site at www.jpeg.org. Currently this site contains a committee draft of the JPEG-LS baseline standard. This draft will be available to the general public until the standard is officially approved by ISO. Another site maintained by Hewlett-Packard at www.hpl.hp.com/loco/ contains an example decoder that is a public domain executable of their JPEG-LS implementation for Win95/NT, HP-UX, and SunOS. This site also contains literature on LOCO-I, the algorithm on which JPEG-LS baseline is largely based.

References

[1] ISO/IEC JTC 1/SC 29/WG 1, "Call for contributions — lossless compression of continous-tone still pictures," ISO Working Document ISO/IEC JTC1/SC29/WG1 N41, March, 1994.

[2] ISO/IEC JTC 1/SC 29/WG 1, "JPEG LS image coding system," ISO Working Document ISO/IEC JTC1/SC29/WG1 N399-WD14495, July, 1996.

[3] R. Ansari, N. Memon, and E. Ceran, "Near-lossless image compression techniques," *J. Electron. Imag.* **7**, 486–494 (1998).

[4] A. R. Calderbank, I. Daubechies, W. Sweldens, and B.-L. Yeo, "Wavelet transforms that map integers to integers," *Appl. Comput. Harmon. Anal.*, **5**, pp. 332–369 (1998).

[5] S. W. Golomb, "Run-length codings," *IEEE Trans. Inf. Theory* **12**, 399–401 (1966).

[6] S. A. Martucci, "Reversible compression of HDTV images using median adaptive prediction and arithmetic coding," in *IEEE International Symposium on Circuits and Systems* (IEEE, New York, 1990), pp. 1310–1313.

[7] N. D. Memon and K. Sayood, "Lossless image compression — a comparative study," in *Still Image Compression*, M. Rabbani, E. Delp, and S. Rajala, eds., *Proc. SPIE* **2418**, 8–20 (1995).

[8] N. D. Memon and X. Wu, "Recent developments in lossless image compression," *Comput. J.* **40**, 31–40 (1997).

[9] S. Ono, S. Kino, M. Yoshida, and T. Kimura, "Bi-level image coding with MELCODE — comparison of block type code and arithmetic type code," in *Proceedings Globecomm '89*, (1989)

[10] W. B. Pennebaker and. J. L. Mitchell, "An overview of the basic principles of the Q-Coder adaptive binary arithmetic coder," *IBM J. Res. Devel.* **32**, 717–726 (1988); Tech. Rep. JPEG-18, ISO/IEC/JTC1/SC2/WG8, International Standards Organization, 1988. Working group document.

[11] W. B. Pennebaker and J. L. Mitchell, *JPEG Still Image Data Compression Standard* (Van Nostrand Reinhold, New York, 1993).

[12] R. F. Rice, "Some practical universal noiseless coding techniques," Tech. Rep. 79-22, Jet Propulsion Laboratory, California Institute of Technology, Pasadena, 1979.

[13] J. J. Rissanen and G. G. Langdon, "Universal modeling and coding," *IEEE Trans. Inf. Theory* **27**, 12–22 (1981).

[14] A. Said and W. A. Pearlman, "An image multiresolution representation for lossless and lossy compression," *IEEE Trans. Image Process.* **5**, 1303–1310 (1996).

[15] S. Urban, "Compression results — lossless, lossy ±1, lossy ±3," ISO Working Document ISO/IEC JTC1/SC29/WG1 N281, 1995.

[16] M. J. Weinberger, J. Rissanen, and R. B. Arps, "Applications of universal context modeling to lossless compression of gray-scale images," *IEEE Trans. Image Process.* **5**, 575–586 (1996).

[17] M. J. Weinberger, G. Seroussi, and G. Sapiro, "LOCO-I: a low complexity context-based lossless image compression algorithm," in *Proceedings of the IEEE Data Compression Conference* (IEEE, New York, 1996), pp. 140–149.

[18] M. J. Weinberger, G. Seroussi, and G. Sapiro, "LOCO-I: a low complexity lossless image compression algorithm," ISO Working Document ISO/IEC JTC1/SC29/WG1 N203, July, 1995.

[19] X. Wu and N. D. Memon, "Context-based adaptive lossless image coding," *IEEE Trans. Commun.* **45**, 437–444 (1997).

[20] A. Zandi, J. D. Allen, E. L. Schwartz, and M. Boliek, "CREW: compression by reversible embedded wavelets," in *Proceedings of the Data Compression Conference* (IEEE, New York, 1995), pp. 212–221.

5.7

Multispectral Image Coding

Daniel Tretter
Hewlett Packard Laboratories

Nasir Memon
Polytechnic University

Charles A. Bouman
Purdue University

1 Introduction

Multispectral images are a particular class of images that require specialized coding algorithms. In multispectral images, the same spatial region is captured multiple times by using different imaging modalities. These modalities often consist of measurements at different optical wavelengths (hence the name multispectral), but the same term is sometimes used when the separate image planes are captured from completely different imaging systems. Medical multispectral images, for example, may combine MRI, CT, and X-ray images into a single multilayer data set [10]. Multispectral images are three-dimensional data sets in which the third (spectral) dimension is qualitatively different from the other two. Because of this, a straightforward extension of two-dimensional image compression algorithms is generally not appropriate. Also, unlike most two-dimensional images, multispectral data sets are often not meant to be viewed by humans. Remotely sensed multispectral images, for example, often undergo electronic computer analysis. As a result, the quality of decompressed images may be judged by a different criterion than for two-dimensional images.

The most common example of multispectral images are conventional RGB color images, which contain three spectral image planes. The image planes represent the red, green, and blue color channels, which all lie in the visible range of the optical band. These three spectral images can be combined to produce a full color image for viewing on a display. However, most printing systems use four colors, typically cyan, magenta, yellow, and black (CMYK), to produce a continuous range of colors. More

recently, many high-fidelity printing systems have begun to use more than four colors to increase the printer gamut, or range of printable colors. This is particularly common in photographic printing systems.

In fact, three colors are not sufficient to specify the appearance of an object under varying illuminants and viewing conditions. To accurately predict the perceived color of a physical surface, we must know the reflectance of the surface as a function of wavelength. Typically, spectral reflectance is measured at 31 wavelengths ranging from 400 to 700 nm; however, experiments indicate that the spectral reflectances of most physical materials can be accurately represented with eight or fewer spectral basis functions [13]. Therefore, some high-fidelity image capture systems collect and store more than three spectral measurements at each spatial location or pixel in the image [13]. For example, the VASARI imaging system developed at the National Gallery in London employs a seven-channel multispectral camera to capture paintings [20]. At this time, color image representations with more than four bands are only used in very high-quality and high-cost systems. However, such multispectral representations may become more common as the cost of hardware decreases and image quality requirements increase.

Another common class of multispectral data is remotely sensed imagery. Remote sensing consists of capturing image data from a remote location. The sensing platform is usually an aircraft or satellite, and the scene being imaged is usually the Earth's surface. Because the sensor and the target are so far apart, each pixel in the image can correspond to tens or even hundreds of square meters on the ground. Data gathered from remote sensing platforms are normally not meant primarily for human viewing. Instead, the images are analyzed electronically to determine factors such as land use patterns, local geography, and ground cover classifications. Surface features in remotely sensed imagery can be difficult to distinguish with only a few bands of data. In particular, a larger number of spectral bands are necessary if a single data set is to be used for multiple purposes. For instance, a geographical survey may require different spectral bands than a crop study. Older systems, like the French SPOT or American thematic mapper, use only a handful of spectral bands. More modern systems, however, can incorporate hundreds of spectral bands into a single image [17]. Compression is important for this class of images both to minimize transmission bandwidth from the sensing platform to a ground station and to archive the captured images.

Some medical images include multiple image planes. Although the image planes may not actually correspond to separate frequency bands, they are often still referred to as multispectral images. For example, magnetic resonance imaging (MRI) can simultaneously measure multiple characteristics of the medium being imaged [11]. Alternatively, multispectral medical images can be formed from different medical imaging modalities such as MRI, CT, and X-ray [10]. These multimodal images are useful for identifying and diagnosing medical disorders.

Most multispectral compression algorithms assume that the multispectral data can be represented as a two-dimensional image with vector-valued pixels. Each pixel then consists of one sample from each image plane (spectral band). This representation requires all spectral bands to be sampled at the same resolution and over the same spatial extent. Most multispectral compression schemes also assume the spectral bands are perfectly registered, so each pixel component corresponds to the same exact location in the scene. For instance, in a perfectly registered multispectral image, a scene feature that covers only a single image pixel will cover exactly the same pixel in all spectral bands. In actual physical systems, registration can be a difficult task, and misregistration can severely degrade the resulting compression ratio or decompressed image quality. Also, although the image planes may be resampled to have pixel values at the same spatial locations, the underlying images may not be of the same resolution.

As with monochrome image compression, multispectral image compression algorithms fall into two general categories: lossless and lossy. In lossless compression schemes, the decoded image is identical to the original. This gives perfect fidelity but limits the achievable compression ratio. For many applications, the required compression ratio is larger than can be achieved with lossless compression, so lossy algorithms are used. Lossy algorithms typically obtain much higher compression ratios but introduce distortions in the decompressed image. Popular approaches for lossy image coding are covered in Chapters 5.2–5.5 of this volume, whereas lossless image coding is discussed in Chapters 5.1 and 5.6. Lossy compression algorithms attempt to

introduce errors in such a way as to minimize the degradation in output image quality for a given compression ratio. In fact, the rate distortion curve gives the minimum bit rate (and hence maximum compression) required to achieve a given distortion. If the allowed distortion is taken to be zero, the resulting maximum compression is the limit for lossless coding. The limit obtained from the theoretical rate distortion curve can be useful for evaluating the effectiveness of a given algorithm. Although the bound is usually computed with respect to mean squared error (MSE) distortion, MSE is not a good measure of quality in all applications.

Most two-dimensional (2-D) image coding algorithms attempt to transform the image data so that the transformed data samples are largely uncorrelated. The samples can then be quantized independently and entropy coded. At the decoder, the quantized samples are recovered and inverse transformed to produce the reconstructed image. The optimal linear transformation for decorrelating the data is the well-known Karhunen–Loeve (KL) transform. Because the KL transformation is data dependent, it requires considerable computation and must be encoded along with the data so it is available at the decoder. As a result, frequency transforms such as the discrete cosine transform (used in JPEG) or the wavelet transform are used to approximate the KL transform along the spatial dimensions. In fact, it can be shown that frequency transforms approximate the KL transform when the image is a stationary 2-D random process. This is generally a reasonable assumption since, over a large ensemble of images, statistical image properties should not vary significantly with spatial position. A large number of frequency transforms can be shown to approach the optimal KL transform as the image size approaches infinity, but in practice, the discrete cosine and wavelet transforms approach this optimal point much more quickly than many other transforms, so they are preferred in actual compression systems.

Multispectral images complicate this scenario. The third (spectral) dimension is qualitatively different from the spatial dimensions, and it generally cannot be modeled as stationary. The correlation between adjacent spectral bands, for example, can vary widely depending on which spectral bands are being considered. In a remotely sensed image, for instance, two adjacent infrared spectral bands might have consistently higher correlation than adjacent bands in the visible range. The correlation is thus dependent on absolute position in the spectral dimension, which violates stationarity. This means that simple frequency transforms along the spectral dimension are generally not effective. Moreover, we will see that most multispectral compression methods work by treating each spectral band differently. This can be done by computing a KL transform across the spectral bands, using prediction filters which vary for each spectral band, or applying vector quantization methods which are trained for the statistical variation among bands.

Multispectral image compression algorithms can be roughly categorized by how they exploit the redundancies along the spatial and spectral dimensions. The simplest method for

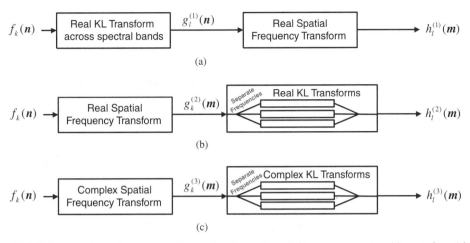

FIGURE 1 Three basic classes of transform coders for multispectral image compression: (a) spectral-spatial method, requiring that all image planes be registered and of the same resolution; (b) spatial-spectral method, requiring that all image planes be registered but not necessarily of the same resolution; (c) complex spatial-spectral method, not requiring registration or planes of the same resolution.

compressing multispectral data is to decompose the multispectral image into a set of monochrome images, and then to separately compress each image using conventional image compression methods. Other multispectral compression techniques concentrate solely on the spectral redundancy. However, the best compression methods exploit redundancies in both the spatial and spectral dimensions.

In [25], Tretter and Bouman categorized transform-based multispectral coders into three classes. These classes are important because they describe both the general structure of the coder and the assumptions behind its design. Figure 1 illustrates the structure of these three basic coding methods.

1.1 Spectral-Spatial Transform

In this method, a KL-transform is first applied across the spectral components to decorrelate them. Then each decorrelated component is compressed separately using a transform based image coding method. The image coding method can be based on either block DCTs (see Chapter 5.5 in this volume) or a wavelet transform (Chapter 5.4). This methods is asymptotically optimal if all image planes are properly registered and have the same spatial resolution.

1.2 Spatial-Spectral Transform

In this method, a spatial transform (i.e., block DCT or wavelet transform) is first applied. Then the spectral components of each spatial-frequency band are decorrelated using a **different** KL transform. So for example, a different KL transform is used for each coefficient of the DCT transform or each band of the wavelet transform. This method is useful when the different spectral components have different spatial-frequency content. For instance, an infrared band may have lower spatial resolution

than a visible band of the same multispectral image. In this case, the separate KL transforms result in better compression.

1.3 Complex Spatial-Spectral Transform

If, in addition, the image planes are not registered, the frequency transforms must be complex to retain phase information between planes. A spatial shift in one image plane relative to another (i.e. misregistration) corresponds to a phase shift in the frequency domain. In order to retain this information, frequency components must be stored as complex numbers. This method differs from the spatial-spectral method in that the transforms must be complex valued. This complex spatial-spectral transform has the advantage that it can remove the effect of misregistration between the spectral bands. However, because it requires the use of a DFT (instead of DCT), or a complex wavelet transform, it is more complicated to implement. If a real spatial-spectral transform is used to compress an image that has misregistered planes, much of the redundancy between image planes will be missed. The transform is unable to follow a scene feature as it shifts in location from one image plane to another, so the feature is essentially treated as a separate feature in each plane and is coded multiple times. As a result, the image will not compress well.

In the following sections, we will discuss a variety of methods for both lossy and lossless multispectral image coding. The most appropriate coding method will depend on the application and system constraints.

2 Lossy Compression

Many researchers have worked on the problem of compressing multispectral images. In the area of lossy compression, most of the work has concentrated on remotely sensed data and RGB color images rather than medical imagery or photographic images with more than three spectral bands. For diagnostic and legal reasons, medical images are often compressed losslessly, and the high-fidelity systems that use multispectral photographic data are still relatively rare.

Suppose we represent a multispectral image by $f_k(\mathbf{n})$, where

$$\mathbf{n} = (n_1, n_2), \quad n_1 \in \{0 \cdots M-1\}, \quad n_2 \in \{0 \cdots N-1\},$$
$$k \in \{0 \cdots K-1\}.$$

In this notation, $\mathbf{n}$ represents the two spatial dimensions and k is the spectral band number. In the development that follows, we will denote spectral band k by f_k, and $\mathbf{f}(\mathbf{n})$ will represent a single vector-valued pixel at spatial location $\mathbf{n} = (n_1, n_2)$.

Lossy compression algorithms attempt to introduce errors in such a way as to minimize the degradation in output image quality for a given compression ratio. To do this, algorithm designers must first decide on an appropriate measure of output image quality. Quality is often measured by defining an error metric relating the decompressed image to the original. The most popular error metric is the simple mean squared error (MSE) between the original image and the decompressed image. Although this metric does not necessarily correlate well with image quality, it is easy to compute and mathematically tractable to minimize when designing a coding algorithm. If a decompressed two-dimensional $M \times N$ image $\hat{f}(\mathbf{n})$ is compared with the original image $f(\mathbf{n})$, the mean squared error is defined as

$$\text{MSE} = \sum_{n_1=0}^{M-1} \sum_{n_2=0}^{N-1} [f(n_1, n_2) - \hat{f}(n_1, n_2)]^2.$$

For photographic images, quality is usually equated with visual quality as perceived by a human observer. The error metrics used thus often incorporate a human visual model. One popular choice is to use a visually weighted MSE between the original image and the decompressed image. This is normally computed in the frequency domain, since visual weighting of frequency coefficients is more natural than weightings in the spatial domain.

Some images are used for purposes other than viewing. Medical images may be used for diagnosis, and satellite photos are sometimes analyzed to classify surface regions or identify objects. For these images, other error metrics may be more appropriate as the measure of image quality is quite different. We will discuss this topic further with respect to multispectral images later in this chapter.

2.1 RGB Color Images

Lossy compression of RGB color images deserves special mention. These images are by far the most common type of multispectral image, and a considerable body of research has been devoted to development of appropriate coding techniques. Color images in uncompressed form typically consist of red, green,

and blue color planes, where the data in each plane have undergone a nonlinear gamma correction to make it appropriate for viewing on a CRT monitor [22]. Typical CRT monitors have a nonlinear response, so doubling the value of a pixel (the frame buffer value), for instance, will increase the luminance of the displayed pixel, but the luminance will not double. The nonlinear response approximates a power function, so digital color images are usually prewarped by using the inverse power function to make the image display properly. Different color imaging systems can have different definitions of red, green, and blue, different gamma curves, and different assumed viewing conditions. In recent years, however, many commercial systems are moving to the sRGB standard to provide better color consistency across devices and applications [2]. Before compression, color images are usually transformed from RGB to a luminance–chrominance representation. Each pixel vector $\mathbf{f}(\mathbf{n})$ is transformed by means of a reversible transformation to an equivalent luminance–chrominance vector $\mathbf{g}(\mathbf{n})$.

$$\mathbf{f}(\mathbf{n}) = \begin{bmatrix} R(n) \\ G(n) \\ B(n) \end{bmatrix} \xrightarrow{\;\; \substack{\text{RGB to} \\ \text{luminance-} \\ \text{chrominance}} \;\;} \mathbf{g}(\mathbf{n}) = \begin{bmatrix} \text{Lum}(n) \\ \text{Chr}\,1(n) \\ \text{Chr}\,2(n) \end{bmatrix}.$$

Two common luminance–chrominance color spaces used are YCrCb, a digital form of the YUV format used in NTSC color television, and CIELab [22]. YCrCb is obtained from sRGB by means of a simple linear transformation, whereas CIELab requires nonlinear computations and is normally computed with lookup tables.

The purpose of the transformation is to decorrelate the spectral bands visually so they can be treated separately. After transformation, the three new image planes are normally compressed independently by using a two-dimensional coding algorithm such as the ones described earlier in this chapter. The luminance channel Y (or L) is visually more important than the two chrominance channels, so the chrominance images are often subsampled by a factor of 2 in each dimension before compression [22]. Perhaps the most common color image compression algorithm uses the YCrCb (sometimes still referred to as YUV) color space in conjunction with chrominance subsampling and standard JPEG compression on each image plane. Many color devices refer to this entire scheme as JPEG, even though the standard does not specify color space or subsampling. Most JPEG images viewed across the World Wide Web by browsers have been compressed in this way. Figure 2 illustrates the artifacts introduced by JPEG compression. Figure 2(a) shows a detail from an original uncompressed image, Fig. 2(b) illustrates the decompressed image region after 30 : 1 JPEG compression with chrominance subsampling, and Fig. 2(c) illustrates the decompressed image after 30 : 1 JPEG compression with no chrominance subsampling. Figures 2(b) and 2(c) both show typical JPEG compression artifacts; the reconstructed images have blocking artifacts in the smooth

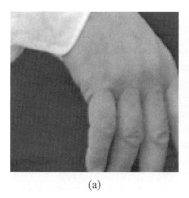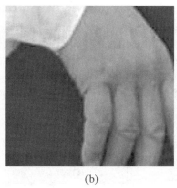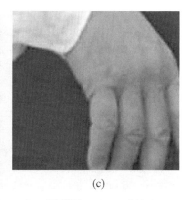

(a)   (b)   (c)

FIGURE 2 Detail illustrating JPEG compression artifacts (75 dpi): (a) original image data; (b) JPEG compressed $30:1$, using chrominance subsampling; (c) JPEG compressed $30:1$, using no chrominance subsampling. (See color section, p. C–27.)

regions such as the back of the hand, and both images show ringing artifacts along the edges. However, the artifacts are much more visible in Fig. 2(c), which was compressed without chrominance subsampling. Because the chrominance components are retained at full resolution, a larger percentage of the compressed data stream is required to represent the chrominance information, so fewer bits are available for luminance. The additional artifacts introduced by using fewer bits for luminance are more visible than the artifacts caused by chrominance subsampling, so Fig. 2(c) has more visible artifacts than Fig. 2(b).

One interesting approach to color image storage and compression is to use color palettization. In this approach, a limited palette of representative colors (usually no more than 256 colors) is stored as a lookup table, and each pixel in the image is replaced by an index into the table that indicates the best palette color to use to approximate that pixel. This is essentially a simple vector quantization scheme (vector quantization is covered in detail in Chapter 5.3). Palettization was first designed not for compression, but to match the capabilities of display monitors. Some display devices can only display a limited number of colors at a time, as a result of either a limited internal memory size or of characteristics of the display itself. As a result, images had to be palettized before display.

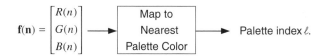

Palettization collapses the multispectral image into a single image plane, which can be further compressed if desired. Both lossy and lossless compression schemes for palettized images have been developed. The well-known GIF format, which is often used for images transmitted over the World Wide Web, is one example of this sort of image. As a compression technique, palettization is most useful for nonphotographic images, such as synthetically generated images, which often only use a limited number of colors.

2.2 Remotely Sensed Multispectral Images

Remotely sensed multispectral images have been in use for a long time. The Landsat 1 system, for example, was first launched in 1972. Aircraft-based systems have been in use even longer. Satellite and aircraft platforms can gather an extremely large amount of data in a short period of time, and remotely sensed data are often archived so changes in the Earth's surface can be tracked over long periods of time. As a result, compression has been of considerable interest since the earliest days of remote sensing, when the main purpose of compression was to reduce storage requirements and processing time [9,16]. Although processing and data storage facilities are becoming increasingly more powerful and affordable, recent remote sensing systems continue to stress state of the art technology. Compression is particularly important for spaceborne systems where transmission bandwidth reduction is a necessity [9,26]. Reviews of compression approaches for remotely sensed images can be found in [21,28].

The simplest type of lossy compression for multispectral images, known as spectral editing, consists of not transmitting all spectral bands. Some sort of algorithm is used to determine which bands are of lesser importance, and those bands are not sent. Because this amounts to simply throwing away some of the data, such a technique is obviously undesirable. For one thing, the choice of bands to eliminate is strongly dependent on the information desired from the image. Since a variety of researchers may want to extract entirely different information from the same image, all of the bands may be needed at one time or another. As a result, a number of researchers have proposed more sophisticated ways to combine the spectral bands and reduce the spectral dimensionality while retaining as much of the information as possible.

As in two-dimensional image compression, many algorithms attempt to first transform the image data such that the transformed data samples are largely uncorrelated. For multispectral images, the spectral bands are often modeled as a series of correlated random fields. If each spectral band f_k is a two-dimensional stationary random field, frequency transforms are appropriate across the spatial dimensions. However, redundancies between spectral bands are usually removed differently. A variety of schemes use a KL or similar transformation across spectral bands followed by a two-dimensional frequency transform like DCT or wavelets across the two spatial dimensions [1, 5–7, 24, 25]. These *spectral-spatial transform methods* are of the general form shown in Fig. 1(a) and have been shown to be asymptotically optimal for a MSE distortion metric as the size of the data set goes to infinity when the following three assumptions hold [25].

1. The spectral components can be modeled as a stationary Gaussian random field.
2. The spectral components are perfectly registered with one another.
3. The spectral components have similar frequency distributions (for instance, they are of the same resolution as one another).

If assumption 3 does not hold, a separate KL spectral transform must be used at every spatial frequency. Algorithms of this sort have been proposed by several researchers [1,25,27]. If assumption 2 does not apply either, a complex frequency transform must be used to preserve phase information if the algorithm is to remain asymptotically optimal [25]. However, the computational complexity involved makes this approach difficult, so it is generally preferable to add more preprocessing to better register the spectral bands. Some recent algorithms also get improved performance by adapting the KL transform spatially based on local data characteristics [6,7,24].

Figure 3 shows the result of applying two different coding algorithms to a thematic mapper multispectral data set. The data set consists of bands 1, 4, and 7 from a thematic mapper image. Figure 3(a) shows a region from the original uncompressed data. The image is shown in pseudo-color, with band 1 being mapped to red, band 4 to green, and band 7 to blue. Figure 3(b) shows the reconstructed data after 30 : 1 compression, using an algorithm from [25] that uses a single KL transform followed by a two-dimensional frequency subband transform across the two spatial dimensions (RSS algorithm), and Fig. 3(c) shows the reconstructed data after 30 : 1 compression, using a similar algorithm that first applies the frequency transform and then computes a separate KL transform for each frequency subband (RSM algorithm). For this imaging device, band 7 is of lower resolution than the other bands, so assumption 3 does not hold for this data set. As a result, we expect the RSM algorithm to outperform the RSS algorithm on this data set. Comparing Fig. 3(b) and 3(c), we can see that Fig. 3(c) has slightly fewer visual artifacts than Fig. 3(b). The mean squared error produced by the RSM algorithm was 27.44 for this image, compared with a mean squared error of 28.65 for the RSS algorithm. As expected, the RSM algorithm outperforms the RSS algorithm on this data set both in visual quality and mean squared error.

Rather than decorrelating the data samples by using a reversible transformation, some approaches use linear prediction to remove redundancy. The predictive algorithms are often used in conjunction with data transformations in one or more dimensions [9,14]. For instance, spectral redundancy may be removed using prediction, while spatial redundancies are removed via a decorrelating transformation.

Correlation in the data can also be accounted for by using clustering or vector quantization (VQ) approaches, often coupled with prediction. A number of predictive VQ and clustering techniques have been proposed [4,8,9,26]. As with predictive algorithms, VQ methods can be combined with decorrelating data transformations [1,27].

Finally, some compression algorithms have been devised specifically for multispectral images, where the authors assumed the images would be subjected to machine classification. These approaches, which are not strongly tied to two-dimensional image compression algorithms, use parametric modeling to approximate the relationships between spectral bands [15]. Classification accuracy is used to measure the effectiveness of these algorithms.

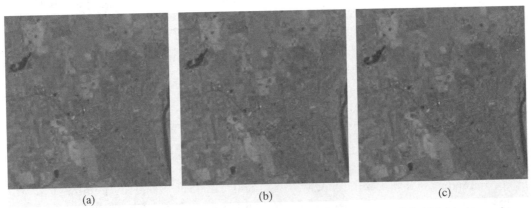

(a)	(b)	(c)

FIGURE 3 Detail illustrating transform-based compression on thematic mapper data (100 dpi): (a) Original image data in pseudo-color; (b) compressed 30 : 1, using the RSS algorithm (single KL transform); (c) compressed 30 : 1, using the RSM algorithm (multiple KL transforms). The RSM algorithm gives better compression for this result, with a MSE of 27.44 vs. the RSS algorithm at 28.65. (See color section, p. C–27.)

Two popular approaches for lossy compression of remotely sensed multispectral images have emerged in recent years. One approach is based on predictive VQ, and the other consists of a decorrelating KL transform across spectral bands in conjunction with frequency transforms in the spatial dimensions. We discuss a representative algorithm of each type in more detail below to help expand upon and illustrate the main ideas involved in each approach.

Gupta and Gersho propose a feature predictive vector quantization approach to the compression of multispectral images [8]. Vector quantization is a powerful compression technique, known to be capable of achieving theoretically optimal coding performance. However, straightforward VQ suffers from high encoding complexity, particularly as the vector dimension increases. Thus, Gupta and Gersho couple VQ with prediction to keep the vector dimension manageable while still accounting for all of the redundancies in the data. In particular, VQ is used to take advantage of spatial correlations, while spectral correlations are removed by using prediction. The authors propose several algorithm variants, but we only discuss one of them here.

Gupta and Gersho begin by partitioning each spectral band k into $P \times P$ nonoverlapping blocks, which will each be coded separately. Suppose $b_k(\mathbf{m})$ is one such set of blocks, with

$$\mathbf{m} = (m_1, m_2), \quad m_1 \in \{0 \cdots P - 1\}, \quad m_2 \in \{0 \cdots P - 1\},$$
$$k \in \{0 \cdots K - 1\}.$$

Figure 4 illustrates the operation of the algorithm for the two types of spectral blocks b_k and b_j in this set of blocks. A small subset L of the K spectral bands is chosen as feature bands. The number of feature bands will be chosen based on the total number of spectral bands and the correlations among them. Vector quantization is used to code each feature band separately, as illustrated for block b_k in the figure. Each feature band is coded with a separate VQ codebook. Each of the $K - L$ nonfeature bands is then predicted from one of the coded feature bands. The prediction is subtracted from the actual data values to get an error block e_j for each nonfeature band. If the energy (squared norm) of the error block exceeds a predefined threshold T, the error block is coded with yet another VQ codebook. This procedure is illustrated for block b_j in Fig. 4. A binary indicator flag I_j is set for each nonfeature band to indicate whether or not the error block was coded:

$$I_j = \begin{cases} 1 & \|e_j\|^2 > T \\ 0 & \text{else} \end{cases}.$$

To decode, simply add any decoded error blocks to the predicted block for nonfeature bands. Feature bands are decoded directly by using the appropriate VQ codebook. Gupta and Gersho also derive optimal predictors P_j for their algorithm and discuss how to design the various codebooks from training images. See [8] for details.

One example of a transform-based compression system is proposed by Saghri *et al.* in [19]. Their algorithm uses the KL transform to decorrelate the data spectrally, followed by JPEG compression on each of the transformed bands. Like Gupta and Gersho, they begin by partitioning each spectral band into nonoverlapping blocks, which will be coded separately. A separate KL transform is computed for each spatial block, so different regions of the image will undergo different transformations. This approach allows the scheme to adapt to varying terrain in the scene, producing better compression results.

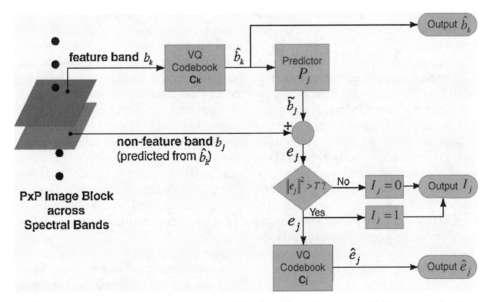

FIGURE 4 Gupta and Gersho's feature predictive VQ scheme encodes each spectral block separately, where feature blocks are used to predict nonfeature blocks, thus removing both spatial and spectral redundancy.

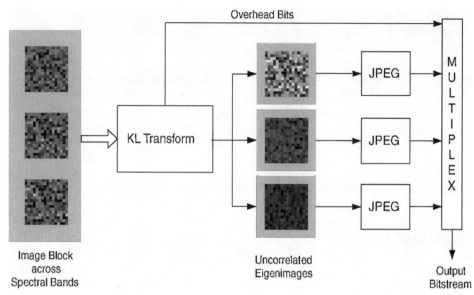

FIGURE 5   Algorithm by Saghri *et al.* uses the KL transform to decorrelate image blocks across spectral bands, followed by JPEG to remove spatial redundancies. The KL transform concentrates much of the energy into a single band, improving coding efficiency. Note that the KL transform is data dependent, so the transformation matrix must be sent to the decoder as overhead bits.

Figure 5 illustrates the algorithm for a single image block. In this example, the image consists of three highly correlated spectral bands. The KL transform concentrates much of the energy into a single band, improving overall coding efficiency. Saghri *et al.* have designed their algorithm for use on board an imaging platform, so they must consider a variety of practical details in their paper. The KL transform is generally a real-valued transform, so they use quantization to reduce the required overhead bits. Also, JPEG expects data of only 8 bits per pixel, so in order to use standard JPEG blocks, they scale and quantize the transformed bands (eigenimages) to 8 bits per pixel. This mapping is also sent to the decoder as overhead. The spatial block size is chosen to give good compression performance while keeping the number of overhead bits small.

The authors also discuss practical ways to select the JPEG parameters to get the best results. They use custom quantization tables and devise several possible schemes for selecting the appropriate quality factor for each transformed band.

Saghri *et al.* have found that their system can give approximately 40:1 compression ratios with visually lossless quality for test images containing eleven spectral bands. They give both measured distortion and classification accuracy results for the decompressed data to support this conclusion. They also examine the sensitivity of the algorithm to various physical system characteristics. They found that the coding results were sensitive to band misregistration, but were robust to changes in the dynamic range of the data, dead/saturated pixels, and calibration and preprocessing of the data. More details can be found in the paper [19].

Most coding schemes for remotely sensed multispectral images are designed for a MSE distortion metric, but many authors also consider the effect on scene classification accuracy

[9, 19, 25]. For fairly modest compression ratios, MSE is often a good indicator of classification accuracy. System issues in the design of compression for remotely sensed images are discussed in some detail in [28].

2.3 Multispectral Medical and Photographic Images

Relatively little work has been done on lossy compression for multispectral medical images and photographic images with more than three spectral bands. Medical images are usually compressed losslessly for legal and diagnostic reasons, although preliminary findings indicate that moderate degrees of lossy compression may not affect diagnostic accuracy. Because most diagnosis is performed visually, the visual quality of decompressed images correlates well with diagnostic accuracy. Hu *et al.* propose linear prediction algorithms for the lossy compression of multispectral MR images [11]. They compare MSE and visual quality of their algorithm versus several other common compression schemes and report preliminary results indicating that diagnostic accuracy is not affected by compression ratios up to 25:1. They do note that their results rely on the spectral bands' being well registered, so a preprocessing step may be necessary in some cases to register the planes before coding to get good results.

Multispectral photographic images with more than three spectral bands are still relatively rare. Most work with these images concentrates on determining the appropriate number and placement of the spectral bands and deriving the mathematical techniques to use the additional bands for improved color reproduction. Many researchers must design and build their own image capture systems as well, so considerable attention and effort is being spent on image capture issues. As a result, little work has been published on lossy coding of such data, although we expect this to be an interesting area for future research. One interesting approach that could be the first stage in a lossy compression algorithm is the multispectral system proposed by Imai and Berns, which combines a low spatial resolution multispectral image with a high-resolution monochrome image [12]. The monochrome image contains lightness information, while color information is obtained from the lower resolution multispectral image. Because human viewers are most sensitive to high frequencies in lightness, these images can be combined to give a high-resolution color image with little or no visible degradation resulting from the lower resolution of the multispectral data. In this way, the approach of Imai and Berns is analogous to the chrominance subsampling often done during RGB color image coding.

3 Lossless Compression

Because of the difference in goals, the best way of exploiting spatial and spectral redundancies for lossy and lossless compression is usually quite different. The decorrelating transforms used for lossy compression usually cannot be used for lossless compression, as they often require floating point computations that result in loss of data when implemented with finite precision arithmetic. This is especially true for "optimal" transforms such as the KL transform and the DCT transform. Also, techniques based on vector quantization are clearly of little utility for lossless compression. Furthermore, irrespective of the transform used, there is often a significant amount of redundancy that remains in the data after decorrelation, the modeling and capturing of which constitutes a crucial step in lossless compression.

There are essentially two main approaches used for lossless image compression. The first is the traditional DPCM approach based on prediction followed by context modeling of prediction errors. The second and more recent approach is based on reversible integer wavelet transforms followed by context modeling and coding of transform coefficients. For a detailed description of these techniques and specific algorithms that employ these approaches, the reader is referred to accompanying chapters in this volume on lossless image compression (Chapter 5.1) and wavelet-based coding (Chapter 5.4). In the rest of this section, we focus on how techniques based on each of these two approaches can be extended to provide lossless compression of multispectral data.

3.1 Predictive Techniques

When a predictive technique is extended to exploit interband correlations, the following new issues arise.

1. *Band ordering*: in what order does one encode the different spectral bands? This is related to the problem of determining which band(s) are the best to use as reference band(s) for predicting and modeling intensity values in a given band.
2. *Interband prediction*: how is it best to incorporate additional information available from pixels located in previously encoded spectral bands to improve prediction?
3. *Interband error modeling*: how does one exploit information available from prediction errors incurred at pixel locations in previously encoded spectral bands to better model and encode the current prediction error?

We examine typical approaches that have been taken to address these questions in the rest of this subsection.

3.1.1 Band Ordering

In [29], Wang *et al.* analyzed correlations between the seven bands of LANDSAT TM images, and proposed an order, based on heuristics, to code the bands that result in the best compression. According to their studies, bands 2, 4, and 6 should first be encoded by traditional intraband linear predictors optimized within individual bands. Then pixels in band 5 are predicted by using neighboring pixels in band 5 as well as those in bands 2, 4, and 6. Finally, bands 1, 3, and 7 are coded using pixels in the local neighborhood as well as selected pixels from bands 2, 4, 5, and 6.

If we restrict the number of reference bands that can be used to predict pixels in any given band, then Tate [23] showed that the problem of computing an optimal ordering can be formulated in graph theoretic terms, admitting an $O(N^2)$ solution for an N-band image. He also observed that using a single reference band is sufficient in practice as compression performance does not improve significantly with additional bands. Although significant improvements in compression performance were demonstrated, one major limitation of this approach is the fact that it is two pass. An optimal ordering and corresponding prediction coefficients are first computed by making an entire pass through the data set. This problem can be alleviated to some degree by computing an optimal ordering for different types of images. Another limitation of the approach is that it reorders entire bands. That is, it makes the assumption that spectral relationships do not vary spatially. The optimal spectral ordering and prediction coefficients will change spatially depending on the characteristics of the objects being imaged.

In the remainder of this subsection, for clarity of exposition, we assume that the image in question has been appropriately reordered, if necessary, and simply use the previous band as the reference band for encoding the current band. However, before we proceed, there is one further potential complication that should be addressed. The different bands in a multispectral image may be represented one pixel at a time (pixel interleaved), one row at a time (line interleaved), or an entire band at a time (band sequential). Because the coder needs to utilize at least one band (the reference band) in order to make compression gains on other bands, buffering strategy and requirements would vary with the different representations and should be taken into account before adopting a specific compression technique. We assume this to be the case in the remainder of this subsection and discuss prediction and error modeling techniques for lossless compression of multispectral images, irrespective of the band ordering and pixel interleaving employed.

3.1.2 Interband Prediction

Let Y denote the current band and X the reference band. In order to exploit interband correlations, it is easy to generalize a DPCM-like predictor from two-dimensional to three-dimensional sources. Namely, we predict the current pixel $Y[i, j]$ to be

$$\hat{Y}[i, j] = \sum_{a,b \in N_1} \theta_{a,b} Y[i-a, j-b] + \sum_{a',b' \in N_2} \theta'_{a',b'} X[i-a', j-b'], \quad (1)$$

$$\hat{Y}[i, j] = \frac{Y[i-1, j] + (X[i, j] - X[i-1, j]) + Y[i, j-1] + (X[i, j] - X[i, j-1])}{2} \quad (2)$$

where N_1 and N_2 are appropriately chosen neighborhoods that are causal with respect to the scan and the band interleaving being employed. The coefficients $\theta_{a,b}$ and $\theta'_{a',b'}$ can be optimized by standard techniques to minimize $\| Y - \hat{Y} \|$ over a given multispectral image. In [18] Roger and Cavenor performed a detailed study on AVIRIS[1] images with different neighborhood sets and found that a third-order spatial-spectral predictor based on the immediate two neighbors $Y[i, j-1]$, $Y[i-1, j]$ and the corresponding pixel $X[i, j]$ in the reference band is sufficient and larger neighborhoods provide very marginal improvements in prediction efficiency.

Since the characteristics of multispectral images often vary spatially, optimizing prediction coefficients over the entire image can be ineffective. Hence, Roger and Cavenor [18] compute optimal predictors for each row of the image and transmit them as side information. The motivation for adapting predictor coefficients a row at a time has to do with the fact that an AVIRIS image is acquired in a line-interleaved manner, and a real-time compression technique would have to operate under such constraints. However, for off-line compression, say for archival purposes, this may not be the best strategy, as one would expect spectral relationships to change significantly across the width of an image. A better approach to adapt prediction coefficients would be to partition the image into blocks, and compute optimal predictors on a block-by-block basis.

Computing an optimal least-squares multispectral predictor for different image segments does not always improve coding efficiency despite the high computational costs involved. This is because frequently changing prediction coefficients incur too

much side information (high model cost), especially for color images which have only three or four bands. In view of this, Wu and Memon [30] propose an adaptive interband predictor that exploits relationships between local gradients among adjacent spectral bands. Local gradients are an important piece of information that can help resolve uncertainty in high-activity regions of an image, and hence improve prediction efficiency. The gradient at the pixel being currently coded is known in the reference band but missing in the current band. Hence, the local waveform shape in the reference band can be projected to the current band to obtain a reasonably accurate prediction, particularly in the presence of strong edges. Although there are several ways in which one can interpolate the current pixel on the basis of local gradients in the reference band, in practice the following *difference based interband interpolation* works well:

Wu and Memon also observed that performing interband prediction in an unconditional manner does not always give significant improvements over intraband prediction and sometimes leads to a degradation in compression performance. This is because the correlation between bands varies significantly in different regions of the image, depending on the objects present in that specific region. Thus it is difficult to find an interband predictor that works well across the entire image. Hence, they propose a switched interband/intraband predictor that performs interband prediction only if the correlation in the current window is strong enough; otherwise intraband prediction is used. More specifically, they examine the correlation $\text{Cor}(X_w, Y_w)$ between the current and reference band in a local window w. If $\text{Cor}(X_w, Y_w)$ is high, then interband prediction is performed; otherwise intraband prediction is used. Since computing $\text{Cor}(X_w, Y_w)$ for each pixel can be computationally expensive, they give simple heuristics to approximate this correlation. They report that switched interband/intraband prediction gives significant improvement over optimal predictors using interband or intraband prediction alone.

3.1.3 Error Modeling and Coding

If the *residual image* consisting of prediction errors is treated as an source with independent identically distributed (i.i.d.) output, then it can be efficiently coded by using any of the standard variable length entropy coding techniques, such as Huffman coding or arithmetic coding. Unfortunately, even after applying the most sophisticated prediction techniques, the residual image generally has ample structure which violates the i.i.d. assumption. Hence, in order to encode prediction errors efficiently, we need a model that captures the structure that remains after prediction. This step is often referred to as

---

[1] Airborne Visible InfraRed Imaging Spectrometer. AVIRIS is a world-class instrument in the realm of Earth remote sensing. It delivers calibrated images in 224 contiguous spectral bands with wavelengths from 400 to 2500 nm.

error modeling. The error modeling techniques employed by most lossless compression schemes proposed in the literature can be captured within a *context modeling* framework. In this approach, the prediction error at each pixel is encoded with respect to a conditioning state or *context*, which is arrived at from the values of previously encoded neighboring pixels. Viewed in this framework, the role of the error model is essentially to provide estimates of the conditional probability of the prediction error, given the context in which it occurs. This can be done by estimating the probability density function by maintaining counts of symbol occurrences within each context or by estimating the parameters of an assumed probability density function. The accompanying chapter on lossless image compression (Chapter 5.1) gives more details on error modeling techniques. Here we look at examples of how each of these two approaches have been used for compression of multispectral images.

An example of the first approach used for multispectral image compression is provided in [18], where Roger and Cavenor investigate two different variations. First they assume prediction errors in a row belong to a single geometric probability mass function (pmf) and determine the optimal Rice–Golomb code by an exhaustive search over the parameter set. In the second technique they compute the variance of prediction errors for each row and based on this utilize one of eight predesigned Huffman codes. An example of the second approach is provided by Tate [23], who quantizes the prediction error in the corresponding location in the reference band and uses this as a conditioning state for arithmetic coding. Because this involves estimating the pmf in each conditioning state, only a small number of states (4–8) are used. An example of a hybrid approach is given by Wu and Memon [30], who propose an elaborate context formation scheme that includes gradients, prediction errors, and quantized pixel intensities from the current and reference band. They estimate the variance of prediction error within each context, and based on this estimate they select between one of eight different conditioning states for arithmetic coding. In each state they estimate the pmf of prediction errors by keeping occurrence counts of prediction errors.

Another simple technique for exploiting relationships between prediction errors in adjacent bands, which can be used in conjunction with any of the above error modeling techniques, follows from the observation that prediction errors in neighboring bands are correlated and just taking a simple difference between the prediction error in the current and reference band can lead to a significant reduction in the variance of the prediction error signal. This in turn leads to a reduction in bit rate produced by a variable length code such as a Huffman or an arithmetic code. The approach can be further improved by conditioning the differencing operation based on statistics gathered from contexts. However, it should be noted that the prediction errors would still contain enough structure to benefit from one of the error modeling and coding techniques described above.

3.2 Reversible Transform-Based Techniques

An alternative approach to lossless image compression that has emerged recently is based on subband decomposition. There are several advantages offered by a subband approach for lossless image compression, the most important of which is perhaps the natural integration of lossy and lossless compression that becomes possible. By transmitting entropy coded subband coefficients in an appropriate manner, one can produce an embedded bit stream that permits the decoder to extract a lossy reconstruction at a desired bit rate. This enables progressive decoding of the image that can ultimately lead to lossless reconstruction. The image can also be recovered at different spatial resolutions. These features are of great value for specific applications in remote sensing and "network-centric" computing in general. Although quite a few subband based lossless image compression have been proposed in recent literature, there has been very little work on extending them to multispectral images. Bilgin *et al.* [3] extend the well known zero-tree algorithm for compression of multispectral data. They perform a 3-D dyadic subband decomposition of the image and encode transform coefficients by using a zero-tree structure extended to three dimensions. They report an improvement of 15–20% over the best 2-D lossless image compression technique.

3.3 Near-Lossless Compression

Recent studies on AVIRIS data have indicated that the presence of sensor noise limits the amount of compression that can be obtained by any lossless compression scheme. This is supported by the fact that the best results reported in the literature on compression of AVIRIS data seem to be in the range of 5–6 bits per pixel. Increased compression can be obtained with lossy compression techniques that have been shown to provide very high compression ratios with little or no loss in *visual fidelity*. Lossy compression, however, may not be desirable in many circumstances because of the uncertainty of the effects caused by lossy compression on a subsequent scientific analysis that is performed with the image data. One compromise then is to use a *bounded distortion* (or *nearly lossless*) technique, which guarantees that each pixel in the reconstructed image is within $\pm k$ of the original.

Extension of a lossless predictive coding technique to a nearly lossless one can be done in a straightforward manner by prediction error quantization according to the specified pixel value tolerance. In order for the predictor at the receiver to track the predictor at the encoder, the reconstructed values of the image have to then be used to generate the prediction at both the encoder and the receiver. More specifically, the following uniform quantization procedure leads us to a nearly lossless compression technique:

$$Q[x] = \left\lfloor \frac{x+k}{2k+1} \right\rfloor (2k+1), \qquad (3)$$

where x is the prediction error, k is the maximum reconstruction error allowed in any given pixel, and $\lfloor . \rfloor$ denotes the integer part of the argument. At the encoder, a label l is generated according to

$$l = \left\lfloor \frac{x+k}{2k+1} \right\rfloor. \tag{4}$$

This label is encoded, and at the decoder the prediction error is reconstructed according to

$$\hat{x} = l \times (2k+1). \tag{5}$$

Nearly lossless compression techniques can yield significantly higher compression ratios as compared to lossless compression. For example, ± 1 near-lossless compression can usually lead to reduction in bit rates by approximately 1–1.3 bits per pixel.

4 Conclusion

In applications such as remote sensing, multispectral images were first used to store the multiple images corresponding to each band in a optical spectrum. More recently, multispectral images have come to refer to any image formed by multiple spatially registered scalar images, independent of the specific manner in which the individual images were obtained. This broader definition encompasses many emerging technologies such as multimodal medical images and high-fidelity color images. As these new sources of multispectral data become more common, the need for high-performance multispectral compression methods will increase.

In this chapter, we have described some of the current methods for both lossless and lossy coding of multispectral images. Effective methods for multispectral compression exploit the redundancy across spectral bands while also incorporating more conventional image coding methods based on spatial dependencies of the image data. Importantly, spatial and spectral redundancy differ fundamentally in that spectral redundancies generally depend on the specific choices and ordering of bands and are not subject to the normal assumptions of stationarity used in the spatial dimension.

We described some typical examples of lossy image coding methods. These methods use either a Karhunen–Loeve (KL) transform or prediction to decorrelate data along the spectral dimension. The resulting decorrelated images can then be coded by using more conventional image compression methods. Lossless multispectral image coding necessitates the use of prediction methods because general transformations result in undesired quantization error. For both the lossy and lossless compression, adaptation to the spectral dependencies is essential to achieve the best coding performance.

References

[1] F. Amato, C. Galdi, and G. Poggi, "Embedded zerotree wavelet coding of multispectral images," in *Proceedings of the IEEE Inter-*national Conference on Image Processing I* (IEEE, New York, 1997), pp. 612–615.

[2] M. Anderson, R. Motta, S. Chandrasekar, and M. Stokes, "Proposal for a standard default color space for the Internet — sRGB," [Conference Paper] Final Program and Proceedings of IS&T/SID Fourth Color Imaging Conference: Color Science, Systems and Applications. Soc. Imaging Sci. & Technol. 1996, pp. 238–246. Springfield, VA, USA.

[3] F. H. Imai and R. S. Berns, "High-resolution multi-spectral image archives: A hybrid approach" [Conference Paper] Final Program and Proceedings of IS&T/SID Sixth Color Imaging Conference: Color Science, Systems and Applications. Soc. Imaging Sci. & Technol. 1996, pp. 224–227. Springfield, VA, USA.

[4] G. R. Canta and G. Poggi, "Kronecker-product gain-shape vector quantization for multispectral and hyperspectral image coding," *IEEE Trans. Image Process.* **7**, 668–678 (1998).

[5] B. R. Epstein, R. Hingorani, J. M. Shapiro, and M. Czigler, "Multispectral KLT-wavelet data compression for Landsat thematic mapper images," in *Proceedings of the Data Compression Conference* (IEEE, New York, 1992), pp. 200–208.

[6] G. Fernandez and C. M. Wittenbrink, "Coding of spectrally homogeneous regions in multispectral image compression," in *Proceedings of the IEEE International Conference on Image Processing II* (IEEE, New York, 1996), pp. 923–926.

[7] M. Finelli, G. Gelli, and G. Poggi, "Multispectral-image coding by spectral classification," in *Proceedings of the IEEE International Conference on Image Processing II* (IEEE, New York, 1996), pp. 605–608.

[8] S. Gupta and A. Gersho, "Feature predictive vector quantization of multispectral images," *IEEE Trans. Geosci. Remote Sensing* **30**, 491–501 (1992).

[9] A. Habibi and A. S. Samulon, "Bandwidth compression of multispectral data," in *Efficient Transmission of Pictorial Information*, A. G. Tescher, ed., *Proc. SPIE* **66**, 23–35 (1975).

[10] S. K. Holland, A. Zavaljevsky, A. Dhawan, R. S. Dunn, and W. S. Ball, "Multispectral magnetic resonance image synthesis, 1998," CHMC Imaging Research Center website, http://www.chmcc.org/departments/misc/ircmsmris.htm.

[11] J.-H. Hu, Y. Wang, and P. T. Cahill, "Multispectral code excited linear prediction coding and its application in magnetic resonance images," *IEEE Trans. Image Process.* **6**, 1555–1566 (1997).

[12] F. H. Imai and R. S. Berns, "High-resolution multi-spectral image archives: a hybrid approach," presented at the Sixth Color Imaging Conference: Color Science, Systems, and Applications, Scottsdale, Arizona, November, 1998.

[13] T. Keusen, "Multispectral color system with an encoding format compatible with the conventional tristimulus model," *J. Imaging Sci. Technol.* **40**, 510–515 (1996).

[14] J. Lee, "Quadtree-based least-squares prediction for multispectral image coding," *Opt. Eng.* **37**, 1547–1552 (1998).

[15] C. Mailhes, P. Vermande, and F. Castanie, "Spectral image compression," *J. Opt. (Paris)* **21**, 121–132 (1990).

[16] N. Pendock, "Reducing the spectral dimension of remotely sensed data and the effect on information content," *Proc. Int. Symp. Remote Sensing Environ.* **17**, 1213–1222 (1983).

[17] J. A. Richards, *Remote Sensing Digital Image Analysis* (Springer-Verlag, Berlin, 1986).

[18] R. E. Roger and M. C. Cavenor, "Lossless compression of AVIRIS images," *IEEE Trans. Image Process.* **5**, 713–719 (1996).

[19] J. A. Saghri, A. G. Tescher, and J. T. Reagan, "Practical transform

coding of multispectral imagery," *IEEE Signal Process. Mag.* **12**, 32–43 (1995).

[20] D. Saunders and J. Cupitt, "Image processing at the National Gallery: the Vasari project," Tech. Rep. 14 : 72, National Gallery, 1993.

[21] K. Sayood, "Data compression in remote sensing applications," *IEEE Geosci. Remote Sensing Soc. Newslett.*, 7–15 (1992).

[22] G. Sharma and H. J. Trussell, "Digital color imaging," *IEEE Trans. Image Process.* **6**, 901–932 (1997).

[23] S. R. Tate, "Band ordering in lossless compression of multispectral images," In *Proceedings of the Data Compression Conference* (IEEE, New York, 1994), pp. 311–320.

[24] A. G. Tescher, J. T. Reagan, and J. A. Saghri, "Near lossless transform coding of multispectral images," in *Proc. of IGARSS'96 Symposium* (IEEE, New York, 1996), Vol. 2, pp. 1020–1022.

[25] D. Tretter and C. A. Bouman, "Optimal transforms for multispectral and multilayer image coding," *IEEE Trans. Image Process.* **4**, 296–308 (1995).

[26] Y. T. Tse, S. Z. Kiang, C. Y. Chiu, and R. L. Baker, "Lossy compression techniques of hyperspectral imagery," in *Proc. of IGARSS'90 Symposium* (IEEE, New York, 1990), pp. 361–364.

[27] J. Vaisey, M. Barlaud, and M. Antonini, "Multispectral image coding using lattice VQ and the wavelet transform," in *Proceedings of the IEEE International Conference on Images Processing II* (IEEE, New York, 1998).

[28] V. D. Vaughn and T. S. Wilkinson, "System considerations for multispectral image compression designs," *IEEE Signal Process. Mag.* **12**, 19–31 (1995).

[29] J. Wang, K. Zhang, and S. Tang, "Spectral and spatial decorrelation of Landsat-TM data for lossless compression," *IEEE Trans. Geosci. Remote Sensing* **33**, 1277–1285 (1995).

[30] X. Wu, W. K. Choi, and N. D. Memon, "Context-based lossless inter-band compression," in *Proceedings of the IEEE Data Compression Conference* (IEEE, New York, 1998), pp. 378–387.

VI

Video Compression

6.1

Basic Concepts and Techniques of Video Coding and the H.261 Standard

Barry Barnett

*The University of Texas
at Austin*

1 Introduction

The subject of video coding is of fundamental importance to many areas in engineering and the sciences. Video engineering is quickly becoming a largely digital discipline. The digital transmission of television signals via satellites is commonplace, and widespread HDTV terrestrial transmission is slated to begin in 1999. Video compression is an absolute requirement for the growth and success of the low-bandwidth transmission of digital video signals. Video encoding is being used wherever digital video communications, storage, processing, acquisition, and reproduction occur. The transmission of high-quality multimedia information over high-speed computer networks is a central problem in the design of *Quality of Services* (QOS) for digital transmission providers. The *Motion Pictures Expert Group* (MPEG) has already finalized two video coding standards, MPEG-1 and MPEG-2, that define methods for the transmission of digital video information for multimedia and television formats. MPEG-4 is currently addressing the transmission of very low bitrate video. MPEG-7 is addressing the standardization of video storage and retrieval services (Chapters 9.1 and 9.2 discuss video storage and retrieval). A central

aspect to each of the MPEG standards are the video encoding and decoding algorithms that make digital video applications practical. The MPEG Standards are discussed in Chapters 6.4 and 6.5.

Video compression not only reduces the storage requirements or transmission bandwidth of digital video applications, but it also affects many system performance tradeoffs. The design and selection of a video encoder therefore is not only based on its ability to compress information. Issues such as bitrate versus distortion criteria, algorithm complexity, transmission channel characteristics, algorithm symmetry versus asymmetry, video source statistics, fixed versus variable rate coding, and standards compatibility should be considered in order to make good encoder design decisions.

The growth of digital video applications and technology in the past few years has been explosive, and video compression is playing a central role in this success. Yet, the video coding discipline is relatively young and certainly will evolve and change significantly over the next few years. Research in video coding has great vitality and the body of work is significant. It is apparent that this relevant and important topic will have an immense affect on the future of digital video technologies.

2 Introduction to Video Compression

Video or visual communications require significant amounts of information transmission. Video compression, as considered here, involves the bitrate reduction of a digital video signal carrying visual information. Traditional video-based compression, like other information compression techniques, focuses on eliminating the redundant elements of the signal. The degree to which the encoder reduces the bitrate is called its *coding efficiency*; equivalently, its inverse is termed the *compression ratio*:

$$\text{coding efficiency} = (\text{compression ratio})^{-1}$$
$$= \text{encoded bitrate/decoded bitrate.} \quad (1)$$

Compression can be a lossless or lossy operation. Because of the immense volume of video information, lossy operations are mainly used for video compression. The loss of information or distortion measure is usually evaluated with the *mean square error* (MSE), *mean absolute error* (MAE) criteria, or *peak signal-to-noise ratio* (PSNR):

$$\text{MSE} = \frac{1}{MN}\sum_{i=1}^{M}\sum_{j=1}^{N}[I(i, j) - \hat{I}(i, j)]^2$$

$$\text{MAE} = \frac{1}{MN}\sum_{i=1}^{M}\sum_{j=1}^{N}|I(i, j) - \hat{I}(i, j)|$$

$$\text{PSNR} = 20\log_{10}\left(\frac{2^n}{\text{MSE}^{1/2}}\right), \quad (2)$$

for an image I and its reconstructed image $\hat{I}$, with pixel indices $1 \leq i \leq M$ and $1 \leq j \leq N$, image size $N \times M$ pixels, and n bits per pixel. The MSE, MAE, and PSNR as described here

are global measures and do not necessarily give a good indication of the reconstructed image quality. In the final analysis, the human observer determines the quality of the reconstructed image and video quality. The concept of distortion versus coding efficiency is one of the most fundamental tradeoffs in the technical evaluation of video encoders. The topic of perceptual quality assessment of compressed images and video is discussed in Section 8.2.

Video signals contain information in three dimensions. These dimensions are modeled as *spatial* and *temporal* domains for video encoding. Digital video compression methods seek to minimize information redundancy independently in each domain. The major international video compression standards (MPEG-1, MPEG-2, H.261) use this approach. Figure 1 schematically depicts a generalized video compression system that implements the spatial and temporal encoding of a digital image sequence. Each image in the sequence I_k is defined as in Eq. (1). The spatial encoder operates on image *blocks*, typically of the order of 8×8 pixels each. The temporal encoder generally operates on 16×16 pixel image blocks. The system is designed for two modes of operation, the *intraframe* mode and the *interframe* mode.

The single-layer feedback structure of this generalized model is representative of the encoders that are recommended by the *International Standards Organization* (ISO) and *International Telecommunications Union* (ITU) video coding standards, MPEG-1, MPEG-2/H.262, and H.261 [1–3]. The feedback loop is used in the interframe mode of operation and generates a *prediction error* between the blocks of the current frame and the current prediction frame. The prediction is generated by the *motion compensator*. The *motion estimation* unit creates *motion vectors* for each 16×16 block. The motion vectors and previously reconstructed frame are fed to the motion compensator to create the prediction.

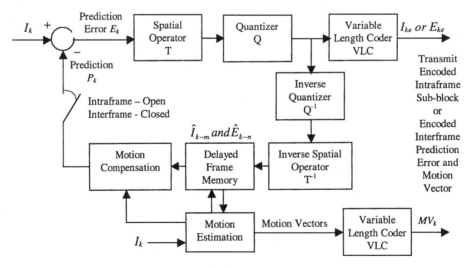

FIGURE 1 Generalized video compression system.

The intraframe mode spatially encodes an entire current frame on a periodic basis, e.g., every 15 frames, to ensure that systematic errors do not continuously propagate. The intraframe mode will also be used to spatially encode a block whenever the interframe encoding mode cannot meet its performance threshold. The intraframe versus interframe mode selection algorithm is not included in this diagram. It is responsible for controlling the selection of the encoding functions, data flows, and output data streams for each mode.

The Intraframe encoding mode does not receive any input from the feedback loop. I_k is spatially encoded, and losslessly encoded by the variable length coder (VLC) forming I_{ke}, which is transmitted to the decoder. The receiver decodes I_{ke}, producing the reconstructed image subblock $\hat{I}_k$. During the interframe coding mode, the current frame prediction P_k is subtracted from the current frame input I_k to form the current prediction error E_k. The prediction error is then spatially and VLC encoded to form E_{ke}, and it is transmitted along with the VLC encoded motion vectors MV_k. The decoder can reconstruct the current frame $\hat{I}_k$ by using the previously reconstructed frame $\hat{I}_{k-1}$ (stored in the decoder), the current frame motion vectors, and the prediction error. The motions vectors MV_k operate on $\hat{I}_{k-1}$ to generate the current prediction frame P_k. The encoded prediction error $\hat{E}_{ke}$ is decoded to produce the reconstructed prediction error $\hat{E}_k$. The prediction error is added to the prediction to form the current frame $\hat{I}_k$. The functional elements of the generalized model are described here in detail.

1. Spatial operator: this element is generally a unitary two-dimensional linear transform, but in principle it can be any unitary operator that can distribute most of the signal energy into a small number of coefficients, i.e., decorrelate the signal data. Spatial transformations are successively applied to small image blocks in order to take advantage of the high degree of data correlation in adjacent image pixels. The most widely used spatial operator for image and video coding is the *discrete cosine transform* (DCT). It is applied to 8×8 pixel image blocks and is well suited for image transformations because it uses real computations with fast implementations, provides excellent decorrelation of signal components, and avoids generation of spurious components between the edges of adjacent image blocks.

2. Quantizer: the spatial or transform operator is applied to the input in order to arrange the signal into a more suitable format for subsequent lossy and lossless coding operations. The quantizer operates on the transform generated coefficients. This is a lossy operation that can result in a significant reduction in the bitrate. The quantization method used in this kind of video encoder is usually scalar and non-uniform. The scalar quantizer simplifies the complexity of the operation as compared to *vector quantization* (VQ). The non-uniform quantization interval is sized according to the distribution of the transform coefficients in order to minimize the bitrate and the distortion created by the quantization process. Alternatively, the quantization interval size can be adjusted based on the performance of the *human Visual System* (HVS). The *Joint Pictures Expert Group* (JPEG) standard includes two (luminance and color difference) HVS sensitivity weighted quantization matrices in its "Examples and Guidelines" annex. JPEG coding is discussed in Sections 5.5 and 5.6.

3. Variable length coding: The lossless VLC is used to exploit the "symbolic" redundancy contained in each block of transform coefficients. This step is termed "entropy coding" to designate that the encoder is designed to minimize the source entropy. The VLC is applied to a serial bit stream that is generated by scanning the transform coefficient block. The scanning pattern should be chosen with the objective of maximizing the performance of the VLC. The MPEG encoder for instance, describes a zigzag scanning pattern that is intended to maximize transform zero coefficient run lengths. The H.261 VLC is designed to encode these run lengths by using a variable length *Huffman* code.

The feedback loop sequentially reconstructs the encoded spatial and prediction error frames and stores the results in order to create a current prediction. The elements required to do this are the inverse quantizer, inverse spatial operator, delayed frame memory, motion estimator, and motion compensator.

1. Inverse operators: The inverse operators Q^{-1} and T^{-1} are applied to the encoded current frame I_{ke} or the current prediction error E_{ke} in order to reconstruct and store the frame for the motion estimator and motion compensator to generate the next prediction frame.

2. Delayed frame memory: Both current and previous frames must be available to the motion estimator and motion compensator to generate a prediction frame. The number of previous frames stored in memory can vary based upon the requirements of the encoding algorithm. MPEG-1 defines a B frame that is a bidirectional encoding that requires that motion prediction be performed in both the forward and backward directions. This necessitates storage of multiple frames in memory.

3. Motion estimation: The temporal encoding aspect of this system relies on the assumption that rigid body motion is responsible for the differences between two or more successive frames. The objective of the motion estimator is to estimate the rigid body motion between two frames. The motion estimator operates on all current frame 16×16 image blocks and generates the pixel displacement or *motion vector* for each block. The technique used to generate motion vectors is called *block-matching motion estimation* and is discussed further in Section 5.4. The method uses the current frame I_k and the previous reconstructed frame

$\hat{I}_{k-1}$ as input. Each block in the previous frame is assumed to have a displacement that can be found by searching for it in the current frame. The search is usually constrained to be within a reasonable neighborhood so as to minimize the complexity of the operation. Search matching is usually based on a minimum MSE or MAE criterion. When a match is found, the pixel displacement is used to encode the particular block. If a search does not meet a minimum MSE or MAE threshold criterion, the motion compensator will indicate that the current block is to be spatially encoded by using the intraframe mode.

4. Motion compensation: The motion compensator makes use of the current frame motion estimates MV_k and the previously reconstructed frame $\hat{I}_{k-1}$ to generate the current frame prediction P_k. The current frame prediction is constructed by placing the previous frame blocks into the current frame according to the motion estimate pixel displacement. The motion compensator then decides which blocks will be encoded as prediction error blocks using motion vectors and which blocks will only be spatially encoded.

The generalized model does not address some video compression system details such as the bit-stream syntax (which supports different application requirements), or the specifics of the encoding algorithms. These issues are dependent upon the video compression system design.

Alternative video encoding models have also been researched. Three-dimensional (3-D) video information can be compressed directly using VQ or 3-D *wavelet* encoding models. VQ encodes a 3-D block of pixels as a codebook index that denotes its "closest or nearest neighbor" in the minimum squared or absolute error sense. However, the VQ codebook size grows on the order as the number of possible inputs. Searching the codebook space for the nearest neighbor is generally very computationally complex, but structured search techniques can provide good bitrates, quality, and computational performance. *Tree-structured* VQ (TSVQ) [13] reduces the search complexity from codebook size N to log N, with a corresponding loss in average distortion. The simplicity of the VQ decoder (it only requires a table lookup for the transmitted codebook index) and its bitrate-distortion performance make it an attractive alternative for specialized applications. The complexity of the codebook search generally limits the use of VQ in real-time applications. Vector quantizers have also been proposed for interframe, variable bitrate, and subband video compression methods [4].

Three-dimensional wavelet encoding is a topic of recent interest. This video encoding method is based on the *discrete wavelet transform* methods discussed in Section 5.4. The wavelet transform is a relatively new transform that decomposes a signal into a *multiresolution* representation. The multiresolution decomposition makes the wavelet transform an excellent signal analysis tool because signal characteristics can be viewed in a variety of time-frequency scales. The wavelet transform is implemented in practice by the use of multiresolution or *subband* filterbanks [5]. The wavelet filterbank is well suited for video encoding because of its ability to adapt to the multiresolution characteristics of video signals. Wavelet transform encodings are naturally hierarchical in their time-frequency representation and easily adaptable for *progressive transmission* [6]. They have also been shown to possess excellent bitrate-distortion characteristics.

Direct three-dimensional video compression systems suffer from a major drawback for real-time encoding and transmission. In order to encode a sequence of images in one operation, the sequence must be buffered. This introduces a buffering and computational delay that can be very noticeable in the case of interactive video communications.

Video compression techniques treating visual information in accordance with HVS models have recently been introduced. These methods are termed "second-generation or object-based" methods, and attempt to achieve very large compression ratios by imitating the operations of the HVS. The HVS model can also be incorporated into more traditional video compression techniques by reflecting visual perception into various aspects of the coding algorithm. HVS weightings have been designed for the DCT AC coefficients quantizer used in the MPEG encoder. A discussion of these techniques can be found in Chapter 6.3.

Digital video compression is currently enjoying tremendous growth, partially because of the great advances in VLSI, ASIC, and microcomputer technology in the past decade. The real-time nature of video communications necessitates the use of general purpose and specialized high–performance hardware devices. In the near future, advances in design and manufacturing technologies will create hardware devices that will allow greater adaptability, interactivity, and interoperability of video applications. These advances will challenge future video compression technology to support format-free implementations.

3 Video Compression Application Requirements

A wide variety of digital video applications currently exist. They range from simple low-resolution and low-bandwidth applications (multimedia, PicturePhone) to very high-resolution and high-bandwidth (HDTV) demands. This section will present requirements of current and future digital video applications and the demands they place on the video compression system.

As a way to demonstrate the importance of video compression, the transmission of digital video television signals is presented. The bandwidth required by a digital television signal is approximately one-half the number of picture elements (pixels) displayed per second. The analog pixel size in the vertical dimension is the distance between scanning lines, and the horizontal dimension is the distance the scanning spot moves during

$\frac{1}{2}$ cycle of the highest video signal transmission frequency. The bandwidth is given by Eq (3):

$$
\begin{aligned}
B_w &= (\text{cycles/frame})(F_R) \\
&= (\text{cycles/line})(N_L)(F_R) \\
&= \frac{(0.5)(\text{aspect ratio})(F_R)(N_L)(R_H)}{0.84} \\
&= (0.8)(F_R)(M_L)(R_H),
\end{aligned} \tag{3}
$$

where

B_W = system bandwidth,
F_R = number of frames transmitted per second (fps),
N_L = number of scanning lines per frame,
R_H = horizontal resolution (lines), proportional
       to pixel resolution.

The National Television Systems Committee (NTSC) aspect ratio is 4/3, the constant 0.5 is the ratio of the number of cycles to the number of lines, and the factor 0.84 is the fraction of the horizontal scanning interval that is devoted to signal transmission.

The NTSC transmission standard used for television broadcasts in the United States has the following parameter values: $F_R = 29.97$ fps, $N_L = 525$ lines, and $R_H = 340$ lines. This yields a video system bandwidth B_W of 4.2 MHz for the NTSC broadcast system. In order to transmit a color digital video signal, the digital pixel format must be defined. The digital color pixel is made of three components: one luminance (Y) component occupying 8 bits, and two color difference components (U and V) each requiring 8 bits. The NTSC picture frame has $720 \times 480 \times 2$ total luminance and color pixels. In order to transmit this information for an NTSC broadcast system at 29.97 frames/s, the following bandwidth is required:

$$
\begin{aligned}
\text{Digital } B_W \simeq \tfrac{1}{2}\text{bitrate} &= \tfrac{1}{2}(29.97 \text{ fps}) \times (24 \text{ bits/pixel}) \\
&\quad \times (720 \times 480 \times 2 \text{ pixels/frame}) \\
&= 249 \text{ MHz}.
\end{aligned}
$$

This represents an increase of ~59 times the available system bandwidth, and ~41 times the full transmission channel bandwidth (6 MHz) for current NTSC signals. HDTV picture resolution requires up to three times more raw bandwidth than this example! (Two transmission channels totaling 12 MHz are allocated for terrestrial HDTV transmissions.) It is clear from this example that terrestrial television broadcast systems will have to use digital transmission and digital video compression to achieve the overall bitrate reduction and image quality required for HDTV signals.

The example not only points out the significant bandwidth requirements for digital video information, but also indirectly brings up the issue of digital video quality requirements. The tradeoff between bitrate and quality or distortion is a funda-

mental issue facing the design of video compression systems. To this end, it is important to fully characterize an application's video communications requirements before designing or selecting an appropriate video compression system. Factors that should be considered in the design and selection of a video compression system include the following items.

1. Video characteristics: video parameters such as the dynamic range, source statistics, pixel resolution, and noise content can affect the performance of the compression system.
2. Transmission requirements: transmission bitrate requirements determine the power of the compression system. Very high transmission bandwidth, storage capacity, or quality requirements may necessitate lossless compression. Conversely, extremely low bitrate requirements may dictate compression systems that trade off image quality for a large compression ratio. *Progressive transmission* is a key issue for selection of the compression system. It is generally used when the transmission bandwidth exceeds the compressed video bandwidth. Progressive coding refers to a multiresolution, hierarchical, or subband encoding of the video information. It allows for transmission and reconstruction of each resolution independently from low to high resolution. In addition, channel errors affect system performance and the quality of the reconstructed video. Channel errors can affect the bit stream randomly or in burst fashion. The channel error characteristics can have different effects on different encoders, and they can range from local to global anomalies. In general, transmission error correction codes (ECC) are used to mitigate the effect of channel errors, but awareness and knowledge of this issue is important.
3. Compression system characteristics and performance: the nature of video applications makes many demands on the video compression system. Interactive video applications such as videoconferencing demand that the video compression systems have symmetric capabilities. That is, each participant in the interactive video session must have the same video encoding and decoding capabilities, and the system performance requirements must be met by both the encoder and decoder. In contrast, television broadcast video has significantly greater performance requirements at the transmitter because it has the responsibility of providing real-time high quality compressed video that meets the transmission channel capacity. Digital video system implementation requirements can vary significantly. Desktop televideo conferencing can be implemented by using software encoding and decoding, or it may require specialized hardware and transmission capabilities to provide a high-quality performance. The characteristics of the application will dictate the suitability of the video compression algorithm for particular system implementations.

The importance of the encoder and system implementation decision cannot be overstated; system architectures and performance capabilities are changing at a rapid pace and the choice of the best solution requires careful analysis of the all possible system and encoder alternatives.

4. Rate-distortion requirements: the rate-distortion requirement is a basic consideration in the selection of the video encoder. The video encoder must be able to provide the bitrate(s) and video fidelity (or range of video fidelity) required by the application. Otherwise, any aspect of the system may not meet specifications. For example, if the bitrate specification is exceeded in order to support a lower MSE, a larger than expected transmission error rate may cause a catastrophic system failure.

5. Standards requirements: video encoder compatibility with existing and future standards is an important consideration if the digital video system is required to interoperate with existing or future systems. A good example is that of a desktop videoconferencing application supporting a number of legacy video compression standards. This results in requiring support of the older video encoding standards on new equipment designed for a newer incompatible standard. Videoconferencing equipment not supporting the old standards would not be capable or as capable to work in environments supporting older standards.

These factors are displayed in Table 1 to demonstrate video compression system requirements for some common video communications applications. The video compression system designer at a minimum should consider these factors in making a determination about the choice of video encoding algorithms and technology to implement.

4 Digital Video Signals and Formats

Video compression techniques make use of signal models in order to be able to utilize the body of digital signal analysis/processing theory and techniques that have been developed over the past fifty or so years. The design of a video compression system, as represented by the generalized model introduced in Section 2, requires a knowledge of the signal characteristics, and the digital processes that are used to create the digital video signal. It is also highly desirable to understand video display systems, and the behavior of the HVS.

4.1 Sampling of Analog Video Signals

Digital video information is generated by sampling the *intensity* of the original continuous analog video signal $I(x, y, t)$ in three dimensions. The spatial component of the video signal is sampled in the horizontal and vertical dimensions (x, y), and the temporal component is sampled in the time dimension (t). This generates a series of digital images or image sequence $I(i, j, k)$. Video signals that contain colorized information are usually decomposed into three parameters (YC_rC_b, YUV, RGB) whose intensities are likewise sampled in three dimensions. The sampling process inherently quantizes the video signal due to the digital word precision used to represent the intensity values. Therefore the original analog signal can never be reproduced exactly, but for all intents and purposes, a high-quality digital video representation can be reproduced with arbitrary closeness to the original analog video signal. The topic of video sampling and interpolation is discussed in Chapter 7.2.

An important result of sampling theory is the *Nyquist sampling theorem*. This theorem defines the conditions under which

TABLE 1 Digital video application requirements

Application	Bitrate Req.	Distortion Req.	Transmission Req.	Computational Req.	Standards Req.
Network video on demand	1.5 Mbps 10 Mbps	High medium	Internet 100 Mbps LAN	MPEG-1 MPEG-2	MPEG-1 MPEG-2 MPEG-7
Video phone	64 Kbps	High distortion	ISDN p × 64	H.261 encoder H.261 decoder	H.261
Desktop multimedia video CDROM	1.5 Mbps	High distortion to medium	PC channel	MPEG-1 decoder	MPEG-1 MPEG-2 MPEG-7
Desktop LAN videoconference	10 Mbps	Medium distortion	Fast ethernet 100 Mbps	Hardware decoders	MPEG-2, H.261
Desktop WAN videoconference	1.5 Mbps	High distortion	Ethernet	Hardware decoders	MPEG-1, MPEG-4, H.263
Desktop dial-up videoconference	64 Kbps	Very high distortion	POTS and internet	Software decoder	MPEG-4, H.263
Digital satellite television	10 Mbps	Low distortion	Fixed service satellites	MPEG-2 decoder	MPEG-2
HDTV	20 Mbps	Low distortion	12-MHz terrestrial link	MPEG-2 decoder	MPEG-2
DVD	20 Mbps	Low distortion	PC channel	MPEG-2 decoder	MPEG-2

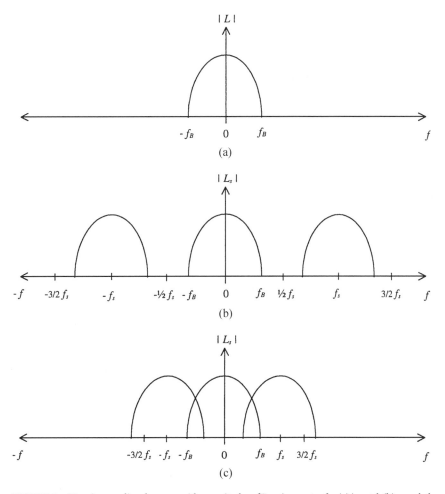

FIGURE 2 Nyquist sampling theorem, with magnitudes of Fourier spectra for (a) input l; (b) sampled input l_s, with $f_s > 2 f_B$; (c) sampled input l_s, with $f_s < 2 f_B$.

sampled analog signals can be "perfectly" reconstructed. If these conditions are not met, the resulting digital signal will contain *aliased* components which introduce artifacts into the reconstruction. The Nyquist conditions are depicted graphically for the one dimensional case in Fig. 2.

The one dimensional signal l is sampled at rate f_s. It is bandlimited (as are all real-world signals) in the frequency domain with an upper frequency bound of f_B. According to the Nyquist sampling theorem, if a bandlimited signal is sampled, the resulting *Fourier* spectrum is made up of the original signal spectrum $|L|$ plus replicates of the original spectrum spaced at integer multiples of the sampling frequency f_s. Diagram (a) in Fig. 2 depicts the magnitude $|L|$ of the Fourier spectrum for l. The magnitude of the *Fourier* spectrum $|L_s|$ for the sampled signal l_s is shown for two cases. Diagram (b) presents the case where the original signal l can be reconstructed by recovering the central spectral island. Diagram (c) displays the case where the Nyquist sampling criteria has not been met and spectral overlap occurs. The spectral overlap is termed *aliasing* and occurs when $f_s < 2 f_B$. When $f_s > 2 f_B$, the original signal can be reconstructed by using a low-pass digital filter whose passband

is designed to recover $|L|$. These relationships provide a basic framework for the analysis and design of digital signal processing systems.

Two-dimensional or spatial sampling is a simple extension of the one-dimensional case. The Nyquist criteria has to be obeyed in both dimensions; i.e., the sampling rate in the horizontal direction must be two times greater than the upper frequency bound in the horizontal direction, and the sampling rate in the vertical direction must be two times greater than the upper frequency bound in the vertical direction. In practice, spatial sampling grids are square so that an equal number of samples per unit length in each direction are collected. *Charge coupled devices* (CCDs) are typically used to spatially sample analog imagery and video. The sampling grid spacing of these devices is more than sufficient to meet the Nyquist criteria for most resolution and application requirements. The electrical characteristics of CCDs have a greater effect on the image or video quality than its sampling grid size.

Temporal sampling of video signals is accomplished by capturing a spatial or image frame in the time dimension. The temporal samples are captured at a uniform rate of ~60 fields/s for NTSC television and 24 fps for a motion film recording. These sampling rates are significantly smaller than the spatial sampling rate. The maximum temporal frequency that can be reconstructed according to the Nyquist frequency criteria is 30 Hz in the case of television broadcast. Therefore any rapid intensity change (caused, for instance, by a moving edge) between two successive frames will cause aliasing because the harmonic frequency content of such a steplike function exceeds the Nyquist frequency. Temporal aliasing of this kind can be greatly mitigated in CCDs by the use of low-pass temporal filtering to remove the high-frequency content. *Photoconductor storage tubes* are used for recording broadcast television signals. They are analog scanning devices whose electrical characteristics filter the high-frequency temporal content and minimize temporal aliasing. Indeed, motion picture film also introduces low-pass filtering when capturing image frames. The exposure speed and the response speed of the photo chemical film combine to mitigate high-frequency content and temporal aliasing. These factors cannot completely stop temporal aliasing, so intelligent use of video recording devices is still warranted. That is, the main reason movie camera panning is done very slowly is to minimize temporal aliasing.

In many cases in which fast motions or moving edges are not well resolved because of temporal aliasing, the HVS will interpolate such motion and provide its own perceived reconstruction. The HVS is very tolerant of temporal aliasing because it uses its own knowledge of natural motion to provide motion estimation and compensation to the image sequences generated by temporal sampling. The combination of temporal filtering in sampling systems and the mechanisms of human visual perception reduces the effects of temporal aliasing such that temporal undersampling (sub-Nyquist sampling) is acceptable in the generation of typical image sequences intended for general purpose use.

4.2 Digital Video Formats

Sampling is the process used to create the image sequences used for video and digital video applications. Spatial sampling and quantization of a natural video signal digitizes the image plane into a two-dimensional set of digital pixels that define a digital image. Temporal sampling of a natural video signal creates a sequence of image frames typically used for motion pictures and television. The combination of spatial and temporal sampling creates a sequence of digital images termed digital video. As described earlier, the digital video signal intensity is defined as $I(i, j, k)$, where $0 \leq i \leq M, 0 \leq j \leq N$ are the horizontal and vertical spatial coordinates, and $0 \leq k$ is the temporal coordinate.

The standard digital video formats introduced here are used in the broadcast for both analog and digital television, as well as computer video applications. Composite television signal digital broadcasting formats are introduced here because of their use in video compression standards, digital broadcasting, and standards format conversion applications. Knowledge of these digital

TABLE 2 Digital composite television parameters

Description	NTSC	PAL
Analog video bandwidth (MHz)	4.2	5.0
Aspect ratio; hor. size/vert. size	4/3	4/3
Frames/s	29.97	25
Lines/s	525	625
Interlace ratio; fields:frames	2:1	2:1
Subcarrier frequency (MHz)	3.58	4.43
Sampling frequency (MHz)	14.4	17.7
Samples/active line	757	939
Bitrate (Mbps)	114.5	141.9

video formats provides background for understanding the international video compression standards developed by the ITU and the ISO. These standards contain specific recommendations for use of the digital video formats described here.

Composite television digital video formats are used for the digital broadcasting, SMPTE digital recording, and conversion of television broadcasting formats. Table 2 contains both analog and digital system parameters for the NTSC and *Phase Alternating Lines* (PAL) composite broadcast formats.

Component television signal digital video formats have been defined by the *International Consultative Committee for Radio* (CCIR) Recommendation 601. It is based on component video with one luminance (Y) and two color difference signals (C_r and C_b). The raw bitrate for the CCIR 601 format is 162 Mbps. Table 3 contains important systems parameters of the CCIR 601 digital video studio component recommendation for both NTSC and PAL/SECAM (*Sequentiel Couleur avec Memoire*).

The ITU Specialist Group (SGXV) has recommended three formats that are used in the ITU H.261, H.263, and ISO MPEG video compression standards. They are the *Standard Input Format* (SIF), *Common Interchange Format* (CIF), and the low bitrate version of CIF, called *Quarter CIF* (QCIF). Together, these formats describe a comprehensive set of digital video formats that are widely used in current digital video applications. CIF and QCIF support the NTSC and PAL video formats using the

TABLE 3 Digital video component television parameters for CCIR 601

Description	NTSC	PAL/SECAM
Luminance channel		
Analog video bandwidth (MHz)	5.5	5.5
Sampling frequency (MHz)	13.5	13.5
Samples/active line	710	716
Bitrate (Mbps)	108	108
Color difference channels		
Analog video bandwidth (MHz)	2.2	2.2
Sampling frequency (MHz)	6.75	6.75
Samples/active line	355	358
Bitrate (Mbps)	54	54

TABLE 4 SIF, CIF, and QCIF digital video formats

Description	SIF NTSC/PAL	CIF	QCIF
Horizontal resolution (Y) pixels	352	360(352)	180(176)
Vertical resolution (Y) pixels	240/288	288	144
Horizontal resolution (C_r, C_b) pixels	176	180(176)	90(88)
Vertical resolution (C_r, C_b) pixels	120/144	144	72
Bits/pixel (bpp)	8	8	8
Interlace fields:frames	1:1	1:1	1:1
Frame rate (fps)	30	30, 15, 10, 7.5	30, 15, 10, 7.5
Aspect ratio; hor. size/ vert. size	4:3	4:3	4:3
Bitrate (Y) Mbps @ 30 fps	20.3	24.9	6.2
Bitrate (U, V) Mbps @ 30 fps	10.1	12.4	3.1

same parameters. The SIF format defines different vertical resolution values for NTSC and PAL. The CIF and QCIF formats also support the H.261 modified parameters. The modified parameters are integer multiples of eight in order to support the 8×8 pixel two-dimensional DCT operation. Table 4 lists this set of digital video standard formats. The modified H.261 parameters are listed in parentheses.

5 Video Compression Techniques

Video compression systems generally comprise two modes that reduce information redundancy in the spatial and the temporal domains. Spatial compression and quantization operates on a single image block, making use of the local image characteristics to reduce the bitrate. The spatial encoder also includes a VLC inserted after the quantization stage. The VLC stage generates a lossless encoding of the quantized image block. Lossless coding is discussed in Chapter 5.1. Temporal domain compression makes use of optical flow models (generally in the form of block-matching motion estimation methods) to identify and mitigate temporal redundancy.

This section presents an overview of some widely accepted encoding techniques used in video compression systems. *Entropy Encoders* are lossless encoders that are used in the VLC stage of a video compression system. They are best used for information sources that are *memoryless* (sources in which each value is independently generated), and they try to minimize the bitrate by assigning variable length codes for the input values according to the input *probability density function* (pdf). *Predictive coders* are suited to information sources that have memory, i.e., a source in which each value has a statistical dependency on some number of previous and/or adjacent values. Predictive coders can produce a new source pdf with significantly less statistical variation and entropy than the original. The transformed source can then

be fed to a VLC to reduce the bitrate. Entropy and predictive coding are good examples for presenting the basic concepts of statistical coding theory.

Block transformations are the major technique for representing spatial information in a format that is highly conducive to quantization and VLC encoding. Block transforms can provide a coding gain by packing most of the block energy into a fewer number of coefficients. The *quantization* stage of the video encoder is the central factor in determining the rate-distortion characteristics of a video compression system. It quantizes the block transform coefficients according to the bitrate and distortion specifications. Motion compensation takes advantage of the significant information redundancy in the temporal domain by creating current frame predictions based upon block matching motion estimates between the current and previous image frames. Motion compensation generally achieves a significant increase in the video coding efficiency over pure spatial encoding.

5.1 Entropy and Predictive Coding

Entropy coding is an excellent starting point in the discussion of coding techniques because it makes use of many of the basic concepts introduced in the discipline of *Information Theory* or *Statistical Communications Theory* [7]. The discussion of VLC and predictive coders requires the use of *information source* models to lay the statistical foundation for the development of this class of encoder. An information source can be viewed as a process that generates a sequence of symbols from a finite alphabet. Video sources are generated from a sequence of image blocks that are generated from a "pixel" alphabet. The number of possible pixels that can be generated is 2^n, when n is the number of bits per pixel. The order in which the image symbols are generated depends on how the image block is arranged or scanned into a sequence of symbols. Spatial encoders transform the statistical nature of the original image so that the resulting coefficient matrix can be scanned in a manner such that the resulting source or sequence of symbols contains significantly less information content.

Two useful information sources are used in modeling video encoders: the *discrete memoryless source* (DMS), and *Markov* sources. VLC coding is based on the DMS model, and the predictive coders are based on the Markov source models. The DMS is simply a source in which each symbol is generated independently. The symbols are *statistically independent* and the source is completely defined by its symbols/events and the set of probabilities for the occurrence for each symbol; i.e., $E = \{e_1, e_2, \ldots, e_n\}$ and the set $\{p(e_1), p(e_2), \ldots, p(e_n)\}$, where n is the number of symbols in the alphabet. It is useful to introduce the concept of entropy at this point. Entropy is defined as the average information content of the information source. The information content of a single event or symbol is defined as

$$I(e_i) = \log \frac{1}{p(e_i)}. \qquad (4)$$

The base of the logarithm is determined by the number of states used to represent the information source. Digital information sources use base 2 in order to define the information content using the number of bits per symbol or bitrate. The entropy of a digital source is further defined as the average information content of the source, i.e.,

$$H(E) = \sum_{i=1}^{n} p(e_i) \log_2 \frac{1}{p(e_i)}$$

$$= -\sum_{i=1}^{n} p(e_i) \log_2 p(e_i) \text{ bits/symbol.} \quad (5)$$

This relationship suggests that the average number of bits per symbol required to represent the information content of the source is the entropy. The *noiseless source coding theorem* states that a source can be encoded with an average number of bits per source symbol that is arbitrarily close to the source entropy. So called entropy encoders seek to find codes that perform close to the entropy of the source. *Huffman* and *arithmetic* encoders are examples of entropy encoders.

Modified Huffman coding [8] is commonly used in the image and video compression standards. It produces well performing variable length codes without significant computational complexity. The traditional Huffman algorithm is a two-step process that first creates a table of source symbol probabilities and then constructs codewords whose lengths grow according to the decreasing probability of a symbol's occurrence. Modified versions of the traditional algorithm are used in the current generation of image and video encoders. The H.261 encoder uses two sets of static Huffman codewords (one each for AC and DC DCT coefficients). A set of 32 codewords is used for encoding the AC coefficients. The zigzag scanned coefficients are classified according to the zero coefficient run length and first nonzero coefficient value. A simple table lookup is all that is then required to assign the codeword for each classified pair.

Markov and *random field* source models (discussed in Chapter 4.2) are well suited to describing the source characteristics of natural images. A Markov source has memory of some number of preceding or adjacent events. In a natural image block, the value of the current pixel is dependent on the values of some the surrounding pixels because they are part of the same object, texture, contour, etc. This can be modeled as an mth order Markov source, in which the probability of source symbol e_i depends on the last m source symbols. This dependence is expressed as the probability of occurrence of event e_i conditioned on the occurrence of the last m events, i.e., $p(e_i \mid e_{i-1}, e_{i-2}, \ldots, e_{i-m})$. The Markov source is made up of all possible n^m states, where n is the number of symbols in the alphabet. Each state contains a set of up to n conditional probabilities for the possible transitions between the current symbol and the next symbol. The *differential pulse code modulation* (DPCM) predictive coder makes use of the Markov source model. DPCM is used in the MPEG-1 and H.261 standards to encode the set of quantized DC coefficients generated by the discrete cosine transforms.

The DPCM predictive encoder modifies the use of the Markov source model considerably in order to reduce its complexity. It does not rely on the actual Markov source statistics at all, and it simply creates a linear weighting of the last m symbols (mth order) to predict the next state. This significantly reduces the complexity of using Markov source prediction at the expense of an increase in the bitrate. DPCM encodes the differential signal d between the actual value and the predicted value, i.e., $d = e - \hat{e}$, where the prediction $\hat{e}$ is a linear weighting of m previous values. The resulting differential signal d generally has reduced entropy as compared to the original source. DPCM is used in conjunction with a VLC encoder to reduce the bitrate. The simplicity and entropy reduction capability of DPCM makes it a good choice for use in real-time compression systems. Third order predictors ($m = 3$) have been shown to provide good performance on natural images [9].

5.2 Block Transform Coding: The Discrete Cosine Transform

Block transform coding is widely used in image and video compression systems. The transforms used in video encoders are *unitary*, which means that the transform operation has a inverse operation that uniquely reconstructs the original input. The DCT successively operates on 8×8 image blocks, and it is used in the H.261, H.263, and MPEG standards. Block transforms make use of the high degree of correlation between adjacent image pixels to provide *energy compaction* or coding gain in the transformed domain. The *block transform coding gain*, G_{TC}, is defined as the ratio of the arithmetic and geometric means of the transformed block variances, i.e.,

$$G_{TC} = 10 \log_{10} \left[\frac{\frac{1}{N} \sum_{i=0}^{N-1} \sigma_i^2}{\left(\prod_{i=0}^{N-1} \sigma_i^2 \right)^{1/N}} \right], \quad (6)$$

where the transformed image block contains N subbands, and σ_i^2 is the variance of each block subband i, for $0 \le i \le N-1$. G_{TC} also measures the gain of block transform coding over PCM coding. The coding gain generated by a block transform is realized by packing most the original signal energy content into a small number of transform coefficients. This results in a lossless representation of the original signal that is more suitable for quantization. That is, there may be many transform coefficients containing little or no energy that can be completely eliminated. Spatial transforms should also be orthonormal, i.e., generate uncorrelated coefficients, so that simple scalar quantization can be used to quantize the coefficients independently.

The *Karhunen–Loève transform* (KLT) creates uncorrelated coefficients, and it is optimal in the energy packing sense. But the KLT is not widely used in practice. It requires the calculation of the image block covariance matrix so that its unitary orthonormal eigenvector matrix can be used to generate the KLT coefficients. This calculation (for which no fast algorithms exist), and the transmission of the eigenvector matrix, is required for every transformed image block.

The DCT is the most widely used block transform for digital image and video encoding. It is an orthonormal transform, and it has been found to perform close to the KLT [10] for first-order Markov sources. The DCT is defined on an 8×8 array of pixels,

$$F(u, v) = \frac{1}{4} C_u C_v \sum_{i=0}^{7} \sum_{j=0}^{7} f(i, j) \cos\left(\frac{(2i + 1)u\pi}{16}\right)$$
$$\times \cos\left(\frac{(2j + 1)v\pi}{16}\right), \tag{7}$$

and the inverse DCT (IDCT) is defined as

$$f(i, j) = C_u C_v \sum_{u=v}^{7} \sum_{v=0}^{7} F(u, v) \cos\left(\frac{(2i + 1)u\pi}{16}\right)$$
$$\times \cos\left(\frac{(2j + 1)v\pi}{16}\right), \tag{8}$$

with

$$C_u = \frac{1}{\sqrt{2}} \quad \text{for } u = 0, \quad C_u = 1 \text{ otherwise,}$$

$$C_v = \frac{1}{2} \quad \text{for } v = 0, \quad C_v = 1 \text{ otherwise}$$

where i and j are the horizontal and vertical indices of the 8×8 spatial array, and u and v are the horizontal and vertical indices of the 8×8 coefficient array. The DCT is the chosen method for image transforms for a couple of important reasons. The DCT has fast $O(n \log n)$ implementations using real calculations. It is even simpler to compute than the DFT because it does not require the use of complex numbers.

The second reason for its success is that the reconstructed input of the IDCT tends not to produce any significant discontinuities at the block edges. Finite discrete transforms create a reconstructed signal that is periodic. Periodicity in the reconstructed signal can produce discontinuities at the periodic edges of the signal or pixel block. The DCT is not as susceptible to this behavior as the DFT. Since the cosine function is real and even, i.e., $\cos(x) = \cos(-x)$, and the input $F(u, v)$ is real, the IDCT generates a function that is even and periodic in $2n$, where n is the length of the original sequence. In contrast, the IDFT produces a reconstruction that is periodic in n, but and necessarily even. This phenomenon is illustrated in Fig .3 for the one-dimensional signal $f(i)$.

The original finite sequence $f(i)$ depicted in part Fig. 3(a) is transformed and reconstructed in Fig. 3(b) by using the DFT–

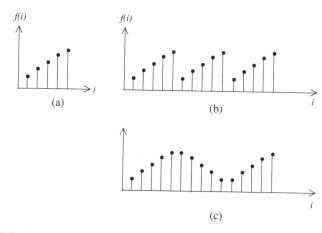

FIGURE 3 Reconstruction periodicity of DFT vs. DCT: (a) original sequence; (b) DFT reconstruction; (c) DCT reconstruction.

IDFT transform pairs, and in Fig. 3(c) by using the DCT–IDCT transform pairs. The periodicity of the IDFT in Fig. 3(b) is five samples long and illustrates the discontinuity introduced by the discrete transform. The periodicity of the IDCT in Fig. 3(c) is 10 samples long, as a result of the evenness of the DCT operation. Discontinuities introduced by the DCT are generally less severe than those of the DFT. The importance of this property of the DCT is that reconstruction errors and blocking artifacts are less severe in comparison to those of the DFT. Blocking artifacts are visually striking and occur because of the loss of high-frequency components that are either quantized or eliminated from the DCT coefficient array. The DCT minimizes blocking artifacts as compared to the DFT because it does not introduce the same level of reconstruction discontinuities at the block edges. Figure 4 depicts blocking artifacts introduced by gross quantization of the DCT coefficients.

This section ends with an example of the energy packing capability of the DCT. Figure 5 depicts the DCT transform operation. The original 8×8 image subblock from the Lena image is displayed in Fig. 5(a), and the DCT transformed coefficient array is displayed in Fig. 5(b).

The original image subblock in Fig. 5(a) contains large values in every position, and is not very suitable for spatial compression in this format. The coefficient matrix (b) concentrates most of the signal energy in the top left quadrant. The signal frequency coordinates $(u, v) = (0, 0)$ start at the upper left position. The DC component equals 1255 and contains the vast majority of the signal energy by itself. This dynamic range and concentration of energy should yield a significant reduction in nonzero values and bitrate after the coefficients are quantized.

5.3 Quantization

The quantization stage of the video encoder creates a lossy representation of the input. The input, as discussed earlier, should be

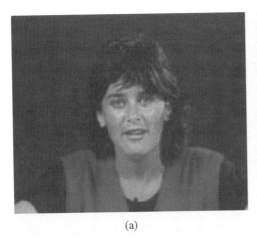 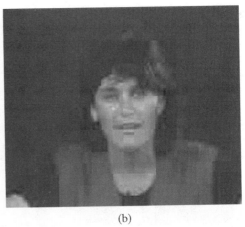

(a)   (b)

FIGURE 4   Severe blocking artifacts introduced by gross quantization of DCT coefficients: (a) original, (b) reconstructed. (See color section, p. C–27.)

conditioned with a particular method of quantization in mind. Vice versa, the quantizer should be well matched to the characteristics of the input in order to meet or exceed the rate-distortion performance requirements. As is always the case, the quantizer has an effect on system performance that must be taken under consideration. Simple scalar versus vector quantization implementations can have significant system performance implications.

Scalar and vector are the two major types of quantizers. These can be further classified as memoryless or containing memory, and symmetric or nonsymmetric. Scalar quantizers control the values taken by a single variable. The quantizer defined by the MPEG-1 encoder scales the DCT transform coefficients. Vector quantizers operate on multiple variables, i.e., a vector of

variables, and become very complex as the number of variables increases. This discussion will introduce the reader to the basic scalar and vector quantizer concepts that are relevant to image and video encoding.

The uniform scalar quantizer is the most fundamental scalar quantizer. It possesses a nonlinear staircase input–output characteristic that divides the input range into output levels of equal size. In order for the quantizer to effectively reduce the bitrate, the number of output values should be much smaller than the number of input values. The reconstruction values are chosen to be at the midpoint of the output levels. This choice is expected to minimize the reconstruction MSE when the quantization errors are uniformly distributed. The quantizers specified in the H.261, H.263, MPEG-1, and MPEG-2 video coders are *nearly* uniform. They have constant step sizes except for the larger *dead-zone* area (the input range for which the output is zero).

Non-uniform quantization is typically used for non-uniform input distributions, such as natural image sources. The scalar quantizer that produces the minimum MSE for a non-uniform input distribution will have non-uniform steps. Compared with the uniform quantizer, the non-uniform quantizer has increasingly better MSE performance as the number of quantization steps increases. The Lloyd–Max [11] is a scalar quantizer design that utilizes the input distribution to minimize the MSE for a given number of output levels. The Lloyd–Max places the reconstruction levels at the centroids of the adjacent input quantization steps. This minimizes the total absolute error within each quantization step based upon the input distribution.

$$f(i,j) = \begin{bmatrix} 136 & 141 & 143 & 153 & 152 & 154 & 154 & 156 \\ 143 & 150 & 153 & 156 & 160 & 156 & 155 & 155 \\ 149 & 155 & 161 & 163 & 158 & 155 & 156 & 155 \\ 158 & 161 & 162 & 161 & 160 & 158 & 160 & 157 \\ 157 & 161 & 160 & 162 & 161 & 157 & 154 & 155 \\ 160 & 160 & 161 & 160 & 160 & 156 & 156 & 156 \\ 160 & 161 & 160 & 161 & 161 & 157 & 157 & 156 \\ 162 & 162 & 161 & 161 & 162 & 157 & 157 & 157 \end{bmatrix}$$

(a)

$$F(u,v) = \begin{bmatrix} 1255 & -8 & -9 & -6 & 1 & -1 & -3 & 1 \\ -26 & -20 & -5 & 4 & -1 & 1 & 0 & 1 \\ -9 & -5 & 1 & -1 & 0 & 0 & -1 & 0 \\ -6 & -2 & 0 & 1 & -1 & 0 & 0 & 0 \\ 1 & 0 & 1 & 2 & 0 & -1 & -1 & 0 \\ -2 & 1 & 2 & 0 & 1 & 1 & 0 & -1 \\ -1 & 0 & 0 & -2 & 0 & 0 & 1 & -1 \\ 1 & 0 & -1 & -2 & 0 & 1 & -1 & 0 \end{bmatrix}$$

(b)

FIGURE 5   8 × 8 DCT: (a) original lena 8 × 8 image subblock; (b) DCT coefficients.

Vector quantizers (discussed in Chapter 5.3) decompose the input into a length n vector. An image for instance, can be divided into $M \times N$ blocks of n pixels each, or the image block can be transformed into a block of transform coefficients. The resulting vector is created by scanning the two-dimensional block elements into a vector of length n. A vector $\mathbf{X}$ is quantized by choosing a codebook vector representation $\hat{\mathbf{X}}$ that is its "closest match." The closest match selection can be made by minimizing an error measure, i.e., choose $\hat{\mathbf{X}} = \hat{\mathbf{X}}_i$ such that the MSE over all codebook vectors is minimized:

$$\hat{\mathbf{X}} = \hat{\mathbf{X}}_i: \quad \min_i \mathrm{MSE}(\mathbf{X}, \hat{\mathbf{X}}_i) = \min_i \frac{1}{n} \sum_{j=1}^{n} (x_j - \hat{x}_j)^2. \quad (9)$$

The index i of the vector $\hat{\mathbf{X}}_i$ denotes the codebook entry that is used by the receiver to decode the vector. Obviously the complexity of the decoder is much simpler than the encoder. The size of the codebook dictates both the coding efficiency and reconstruction quality. The raw bitrate of a vector quantizer is

$$\mathrm{bitrate}_{\mathrm{VQ}} = \frac{\log_2 m}{n} \text{ bits/pixel}, \quad (10)$$

where $\log_2 m$ is the number of bits required to transmit the index i of the codebook vector $\hat{\mathbf{X}}_i$. The codebook construction includes two important issues that are pertinent to the performance of the video coder. The set of vectors that are included in the codebook determine the bitrate and distortion characteristics of the reconstructed image sequence. The codebook size and structure determines the search complexity to find the minimum error solution for Eq. (9). Two important VQ codebook designs are the *Linde–Buzo–Gray* (LBG) [12] and TSVQ [13]. The LBG design is based on the Lloyd–Max scalar quantizer algorithm. It is widely used because the system parameters can be generated by the use of an input "training set" instead of the true source statistics. The TSVQ design reduces VQ codebook search time by using m-ary tree structures and searching techniques.

5.4 Motion Compensation and Estimation

Motion compensation [14] is a technique created in the 1960s that used to increase the efficiency of video encoders. Motion compensated video encoders are implemented in three stages. The first stage estimates objective motion (motion estimation) between the previously reconstructed frame and the current frame. The second stage creates the current frame prediction (motion compensation), using the motion estimates and the previously reconstructed frame. The final stage differentially encodes the prediction and the actual current frame as the prediction error. Therefore, the receiver reconstructs the current image only by using the VLC encoded motion estimates and the spatially and VLC encoded prediction error.

Motion estimation and compensation are common techniques used to encode the temporal aspect of a video signal. As discussed earlier, block-based motion compensation and motion estimation techniques used in video compression systems are capable of the largest reduction in the raw signal bitrate. Typical implementations generally outperform pure spatial encodings by a factor of 3 or more. The interframe redundancy contained in the temporal dimension of a digital image sequence accounts for the impressive signal compression capability that can be achieved by video encoders. Interframe redundancy can be simply modeled as static backgrounds and moving foregrounds to illustrate

the potential temporal compression that can be realized. Over a short period of time, image sequences can be described as a static background with moving objects in the foreground. If the background does not change between two frames, their difference is zero, and the two background frames can essentially be encoded as one. Therefore the compression ratio increase is proportional to two times the spatial compression achieved in the first frame. In general, unchanging or static backgrounds can realize additive coding gains, i.e.,

Static Background Coding Gain $\propto$

$N \bullet$ (Spatial Compression Ratio of Background Frame),

$$(11)$$

where N is the number of static background frames being encoded. Static backgrounds occupy a great deal of the image area and are typical of both natural and animated image sequences. Some variation in the background always occurs as a result of random and systematic fluctuations. This tends to reduce the achievable background coding gain.

Moving foregrounds are modeled as nonrotational rigid objects that move independently of the background. Moving objects can be detected by matching the foreground object between two frames. A perfect match results in zero difference between the two frames. In theory, moving foregrounds can also achieve additive coding gain. In practice, moving objects are subject to occlusion, rotational and nonrigid motion, and illumination variations that reduce the achievable coding gain. Motion compensation systems that make use of motion estimation methods leverage both background and foreground coding gain. They provide pure interframe differential encoding when two backgrounds are static; i.e., the computed motion vector is (0,0). The motion estimate computed in the case of moving foregrounds generates the minimum distortion prediction.

Motion estimation is an interframe prediction process falling in two general categories: pel-recursive algorithms [15] and block-matching algorithms (BMAs) [16]. The pel-recursive methods are very complex and inaccurate and restrict their use in video encoders. Natural digital image sequences generally display ambiguous object motion that adversely affects the convergence properties of pel-recursive algorithms. This has led to the introduction of *block-matching motion estimation*, which is tailored for encoding image sequences. Block-matching motion estimation assumes that the objective motion being predicted is rigid and nonrotational. The block size of the BMA for the MPEG, H.261, and H.263 encoders is defined as 16×16 luminance pixels. MPEG-2 also supports 16×8 pixel blocks.

BMAs predict the motion of a block of pixels between two frames in an image sequence. The prediction generates a pixel displacement or motion vector whose size is constrained by the search neighborhood. The search neighborhood determines the complexity of the algorithm. The search for the best prediction ends when the best block match is determined within the search neighborhood. The best match can be chosen as the minimum

Search Neighborhood

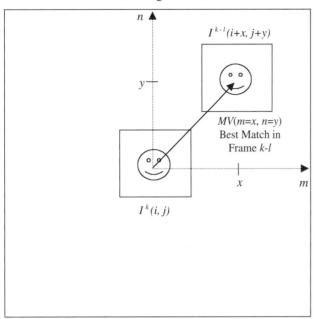

FIGURE 6   Best match motion estimate.

MSE, which for a full search is computed for each block in the search neighborhood:

BestMatch$_{\text{MSE}}$

$$= \min_{m,n} \frac{1}{N^2} \sum_{i=1}^{N} \sum_{j=1}^{N} [I^k(i,\ j) - I^{k-l}(i+m,\ j+n)]^2, \quad (12)$$

where k is the frame index, l is the temporal displacement in frames, N is the number of pixels in the horizontal and vertical directions of the image block, i and j are the pixel indices within the image block, and m and n are the indices of the search neighborhood in the horizontal and vertical directions. Therefore the best match motion vector estimate $MV(m = x, n = y)$ is the pixel displacement between the block $I^k(i, j)$ in frame k, and the best matched block $I^{k-l}(i+x, j+y)$ in the displaced frame $k - l$. The best match is depicted in Fig. 6.

In cases in which the block motion is not uniform or if the scene changes, the motion estimate may in fact increase the bitrate over a spatial encoding of the block. In the case in which the motion estimate is not effective, the video encoder does not use the motion estimate and encodes the block by using the spatial encoder.

The search space size determines the complexity of the motion estimation algorithm. Full search methods are costly and are not generally implemented in real-time video encoders. Fast searching techniques can considerably reduce computational complexity while maintaining good accuracy. These algorithms reduce the search process to a few sequential steps in which each sub-

sequent search direction is based upon the results of the current step. The procedures are designed to find local optimal solutions and cannot guarantee selection of the global optimal solution within the search neighborhood. The *logarithmic* search [17] algorithm proceeds in the direction of minimum distortion until the final optimal value is found. Logarithmic searching has been implemented in some MPEG encoders. The *three-step* search [18] is a very simple technique that proceeds along a best match path in three steps in which the search neighborhood is reduced for each successive step. Figure 7 depicts the three-step search algorithm.

A 14 × 14 pixel search neighborhood is depicted. The search area sizes for each step are chosen so that the total search neighborhood can be covered in finding the local minimum. The search areas are square. The length of sides of the search area for step 1 are chosen to be larger than or equal to $\frac{1}{2}$ the length of the range of the search neighborhood (in this example the search area is 8 × 8). The length of the sides are reduced by $\frac{1}{2}$ after each of the first two steps are completed. Nine points for each step are compared by using the matching criteria. These consist of the central point and eight equally spaced points along the perimeter of the search area. The search area for step 1 is centered on the search neighborhood. The search proceeds by centering the search area for the next step over the best match from the previous step. The overall best match is the pixel displacement chosen to minimize the matching criteria in step 3. The total number of required comparisons for the three-step algorithm is 25. That represents an 87% reduction in complexity versus the full search method for a 14 × 14 pixel search neighborhood.

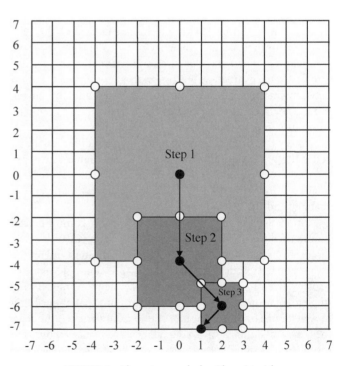

FIGURE 7   Three-step search algorithm pictorial.

6 Video Encoding Standards and H.261

The major internationally recognized video compression standards have been developed by the ISO, the International Electrotechnical Commission (IEC), and the ITU standards organizations.

The MPEG is a working group operating within ISO and IEC. Since starting its activity in 1988, MPEG has produced ISO/IEC 11172 (MPEG-1) and ISO/IEC 13818 (MPEG-2). The MPEG-1 specification was motivated by T1 transmission speeds, the CD-ROM, and the multimedia capabilities of the desktop computer. It is intended for video coding up to the rate of 1.5 Mbps, and it is composed of five sections: system configurations, video coding, audio coding, compliance testing, and software for MPEG-1 coding. The standard does not specify the actual video coding process, but only the syntax and semantics of the bit stream, and the video generation at the receiver. It does not accommodate interlaced video, and it only supports CIF quality format at 25 or 30 fps.

Activity for MPEG-2 was started in 1991. It was targeted for higher bitrates, broadcast video, and a variety of consumer and telecommunications video and audio applications. The syntax and technical contents of the standard were frozen in 1993. It is composed of four parts: systems, video, audio, and conformance testing. MPEG-2 was also recommended by the ITU as H.262.

MPEG is considering more advanced forms of video application interactivity that technology will make possible in the next few years. The MPEG-4 project is targeted to give users the possibility to achieve various forms of interactivity with the audiovisual content of a scene, and to mix synthetic and natural audio and video information in a seamless way. MPEG-4 technology comprises two major parts: a set of coding tools for audiovisual objects, and a syntactic language to describe both the coding tools and the coded objects. From a technical viewpoint, the most notable departure from traditional coding standards will be the possibility for a receiver to download the description of the syntax used to represent the audiovisual information. The visual information will not be restricted to the format of conventional video, i.e., it will not necessarily be frame based. This is expected to produce significant improvements in both encoder efficiency and functionality.

The ITU Recommendation H.261 was adopted in 1990 and specifies a video encoding standard for videoconferencing and videophone services for transmission over the Integrated Services Digital Network (ISDN) at $p \times 64$ Kbps, $p = 1, \ldots, 30$. H.261 describes the video compression methods that were later adopted by the MPEG standards and is presented in the following section. The ITU Experts Group for Very Low Bit-Rate Video Telephony (LBC) has produced the H.263 recommendation for *Public Switched Telephone Networks* (PSTN), which was finalized in December 1995 [18]. It is an extended version of H.261 supporting bidirectional motion compensation and sub-QCIF formats. The encoder is based on hybrid DPCM/DCT cod-

ing and improvements targeted to generate bitrates of less than 64 Kbps.

6.1 The H.261 Video Encoder

The H.261 recommendation [3] is targeted at the videophone and videoconferencing application market running on connection-based ISDN at $p \times 64$ kbps, $p = 1, \ldots, 30$. It explicitly defines the encoded bit stream syntax and *decoder*, while leaving the encoder design to be compatible with the decoder specification. The video encoder is required to carry a delay of less than 150 ms so that it can operate in real-time bidirectional videoconferencing applications. H.261 is part of a group of related ITU recommendations that define visual telephony systems. This group includes the following.

1. H.221: defines the frame structure for an audiovisual channel supporting 64–1920 Kbps.
2. H.230: defines frame control signals for audiovisual systems.
3. H.242: defines audiovisual communications protocol for channels supporting up to 2 Mbps.
4. H.261: defines the video encoder/decoder for audiovisual services at $p \times 64$ Kbps.
5. H.320: defines narrow-band audiovisual terminal equipment for $p \times 64$ Kbps transmission.

The H.261 encoder block diagrams are depicted in Fig. 8(a) and 8(b). An H.261 source coder implementation is depicted in Fig. 8(c). The source coder implements the video encoding algorithm that includes the spatial encoder, the quantizer, the temporal prediction encoder, and the VLC. The spatial encoder is defined to use the two-dimensional 8×8 pixel block DCT and a nearly uniform scalar quantizer, using up to a possible 31 step sizes to scale the AC and interframe DC coefficients. The resulting quantized coefficient matrix is zigzag scanned into a vector that is variable length coded using a hybrid modified run length and Huffman coder. Motion compensation is optional. Motion estimation is only defined in the forward direction because H.261 is limited to real-time videophone and videoconferencing. The recommendation does not specify the motion estimation algorithm or the conditions for the use of intraframe versus interframe encoding.

The video multiplex coder creates a H.261 bitstream that is based on the data hierarchy described below. The transmission buffer is chosen not to exceed the maximum coding delay of 150 ms, and it is used to regulate the transmission bitrate by means of the coding controller. The transmission coder embeds an ECC into the video bit stream that provides error resilience, error concealment, and video synchronization.

H.261 supports most of the internationally accepted digital video formats. These include CCIR 601, SIF, CIF, and QCIF. These formats are defined for both NTSC and PAL broadcast signals. The CIF and QCIF formats were adopted in 1984 by H.261

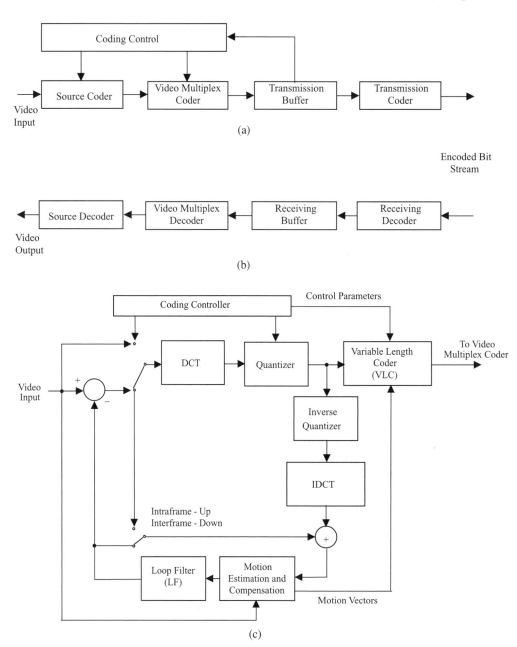

FIGURE 8   ITU-T H.261 block diagrams: (a) video encoder; (b) video decoder; (c) source encoder implementation.

in order to support 525-line NTSC and 625-line PAL/SECAM video formats. The CIF and QCIF operating parameters can be found in Table 4. The raw data rate for 30 fps CIF is 37.3 Mbps and 9.35 Mbps for QCIF. CIF is defined for use in channels in which $p \geq 6$ so that the required compression ratio for 30 fps is smaller than 98:1. CIF and QCIF formats support frame rates of 30, 15, 10, and 7.5 fps, which allows the H.261 encoder to achieve greater coding efficiency by skipping the encoding and transmission of whole frames. H.261 allows zero, one, two, three or more frames to be skipped between transmitted frames.

H.261 specifies a set of encoder protocols and decoder operations that every compatible system must follow. The H.261

video multiplex defines the data structure hierarchy that the decoder can interpret unambiguously. The video data hierarchy defined in H.261 is depicted in Fig. 9. They are the picture layer, group of block (GOB) layer, macroblock (MB) layer and the basic (8×8) block layer. Each layer is built from the previous or lower layer and contains its associated data payload, and header that describes the parameters used to generate the bit stream. The basic 8×8 block is used in intraframe DCT encoding. The MB is the smallest unit for selecting intraframe or interframe encoding modes. It is made up of four adjacent 8×8 luminance blocks and two subsampled 8×8 color difference blocks (C_B and C_R as defined in Table 4) corresponding to the luminance

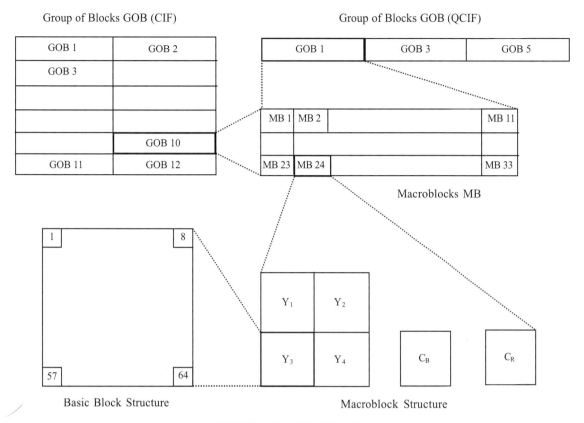

Group of Blocks GOB (CIF)   Group of Blocks GOB (QCIF)

FIGURE 9   H.261 block hierarchy.

blocks. The GOB is made up of 176×48 pixels (33 MBs) and is used to construct the 352×288 pixel CIF or 176×144 pixel QCIF picture layer.

The headers for the GOB and picture layers contain start codes so that the decoder can resynchronize when errors occur. They also contain other relevant information required to reconstruct the image sequence. The following parameters used in the headers of the data hierarchy complete the H.261 video multiplex.

Picture layer:

- Picture start code (PSC), 20-bit synchronization pattern (0000 0000 0000 0001 0000).
- Temporal reference (TR), 5-bit input frame number.
- Type information (PTYPE), indicates source format, CIF = 1 QCIF = 0, and other controls.
- User-inserted bits.

GOB layer:

- Group of blocks start code (GBSC), 16-bit synchronization code (0000 0000 0000 0001).
- Group number (GN), 4-bit address representing the 12 GOBs per CIF frame.
- Quantizer information (GQUANT), indicates one of 31 quantizer step sizes to be used in a GOB unless overridden by Macroblock MQUANT parameter.
- User-inserted bits.

Macroblock layer:

- Macroblock address (MBA) is the position of a MB within a GOB.
- Type information (MTYPE) for one of 10 encoding modes used for the MB. This includes permutations of intraframe, interframe, motion compensation (MC), and loop filtering (LF). A prespecified VLC is used to encode these modes.
- Quantizer (MQUANT), 5-bit normalized quantizer step size from 1–31.
- Motion vector data (MVD), up to 11-bit VLC describing the differential displacement.
- Coded block pattern (CBP), up to 9-bit VLC indicating the location of the encoded blocks in the MB.

Block layer:

- Transform coefficients (TCOEFF) are zigzag scanned and can be 8-bit fixed or up to 13-bit VLC.
- End of block (EOB), symbol.

The H.261 bit stream also specifies transmission synchronization and error code correction by using a BCH code [19] that is capable of correcting 2-bit errors in every 511-bit block. It inserts 18 parity bits for every 493 data bits. A synchronization bit is added to every codeword to be able to detect the BCH codeword boundaries. The transmission synchronization and encoding also operates on the audio and control information specified by the ITU H.320 Recommendation.

The H.261 video compression algorithm depicted in Fig. 8(c) is specified to operate in intraframe and interframe encoding modes. The intraframe mode provides spatial encoding of the 8×8 block and uses the two-dimensional DCT. Interframe mode encodes the prediction error, with motion compensation being optional. The prediction error is optionally DCT encoded. Both modes provide options that effect the performance and video quality of the system. The motion estimate method, mode selection criteria, and block transmission criteria are not specified, although the ITU has published reference models [20, 21] that make particular implementation recommendations. The coding algorithm used in the ITU-T *Reference Model* 8 (RM8) [21] is summarized in three steps, and is followed by an explanation of its important encoding elements.

1. The motion estimator creates a displacement vector for each MB. The motion estimator generally operates on the 16×16 pixel luminance MB. The displacement vector is an integer value between ± 15, which is the maximum size of the search neighborhood. The motion estimate is scaled by a factor of 2 and applied to the C_R and C_B component macroblocks.

2. The compression mode for each macroblock is selected by using a minimum error criteria that is based upon the *displaced macroblock difference* (DMD),

$$\text{DMD} (i, j, k) = b(i, j, k) - b(i - d_1, j - d_2, k - 1),$$
$$(13)$$

where b is a 16×16 MB, i and j are its spatial pixel indices, k is the frame index, and d_1 and d_2 are the pixel displacements of the MB in the previous frame. The displacements range from $-15 \leq d_1, d_2 \leq +15$. When d_1 and d_2 are set to zero, the DMD becomes the *macroblock difference* (MD). The compression mode determines the operational encoder elements that are used for the current frame. The H.261 compression modes are depicted in Table 5.

TABLE 5   H.261 MB video compression modes

Mode	MQUANT	MVD	CBP	TCOEFF
Intra				$\checkmark$
Intra	$\checkmark$			$\checkmark$
Inter			$\checkmark$	$\checkmark$
Inter	$\checkmark$		$\checkmark$	$\checkmark$
Inter + MC		$\checkmark$		
Inter + MC		$\checkmark$	$\checkmark$	$\checkmark$
Inter + MC	$\checkmark$	$\checkmark$	$\checkmark$	$\checkmark$
Inter + MC + LF		$\checkmark$		
Inter + MC + LF		$\checkmark$	$\checkmark$	$\checkmark$
Inter + MC + LF	$\checkmark$	$\checkmark$	$\checkmark$	$\checkmark$

3. The *video multiplex coder* processes each macroblock to generate the H.261 video bit stream whose elements are discussed above.

There are five basic MTYPE encoding mode decisions that are carried out in step 2. These are as follows.

- Use intraframe or interframe mode?
- Use motion compensation?
- Use a coded block pattern (CBP)?
- Use loop filtering?
- Change quantization step size MQUANT?

To select the macroblock compression mode, the *variances* (VAR) of the input macroblock, the MD, and the DMD (as determined by the best motion estimate) are compared as follows.

1. If VAR(DBD) < VAR(MD) then interframe + motion compensation (Inter + MC) coding is selected. In this case, the motion vector data (MVD) is transmitted. Table 5 indicates that there are three Inter + MC modes that allow for the transmission of the prediction error (DMD) with or without DCT encoding of some or all of the four 8×8 basic blocks.

2. "VAR input" is defined as the variance of the input macroblock. If VAR input < VAR(DMD) and VAR input < VAR(MD), then the intraframe mode (Intra) is selected. Intraframe mode uses DCT encoding of all four 8×8 basic blocks.

3. If VAR(MD) < VAR(DMD), then interframe mode (Inter) is selected. This mode indicates that the motion vector is zero, and that some or all of the 8×8 prediction error (MD) blocks can be DCT encoded.

The transform coefficient CBP parameter is used to indicate whether a basic block is reconstructed using the corresponding basic block from the previous frame, or if it is encoded and transmitted. In other words, no basic block encoding is used when the block content does not change significantly. The CPB parameter encodes 63 combinations of the four luminance blocks and two color difference blocks using a variable length code. The conditions for using CBP are not specified in the H.261 recommendation.

Motion compensated blocks can be chosen to be low-pass filtered before the prediction error is generated by the feedback loop. This mode is denoted as Inter + MC + LF in Table 5. The low-pass filter is intended to reduce the quantization noise in the feedback loop, as well as the high-frequency noise and artifacts introduced by the motion compensator. H.261 defines loop filtering as optional and recommends a separable two-dimensional spatial filter design, which is implemented by cascading two identical one-dimensional finite impulse response (FIR) filters. The coefficients of the 1-D filter are [1, 2, 1] for pixels inside the block, and [0, 1, 0] (no filtering) for pixels on the block boundary.

The MQUANT parameter is controlled by the state of the transmission buffer in order to prevent overflow or underflow conditions. The dynamic range of the DCT macroblock coefficients extends between $[-2047, \ldots, 2047]$. They are quantized to the range $[-127, \ldots, 127]$ using one of the 31 quantizer step sizes as determined by the GQUANT parameter. The step size is an even integer in the range of $[2, \ldots, 62]$. GQUANT can be overridden at the macroblock layer by MQUANT to clip or expand the range prescribed by GQUANT so that the transmission buffer is better utilized. The ITU-T RM8 *liquid level control model* specifies the inspection of 64-Kbit transmission buffers after encoding 11 macroblocks. The step size of the quantizer should be increased (decreasing the bitrate) if the buffer is full; vice versa, the step size should be decreased (increasing the bitrate) if the buffer is empty. The actual design of the rate control algorithm is not specified.

The DCT macroblock coefficients are subjected to variable thresholding before quantization. The threshold is designed to increase the number of zero valued coefficients, which in turn increases the number of the zero run lengths and VLC coding efficiency. The ITU-T RM8 provides an example thresholding algorithm for the H.261 encoder. Nearly uniform scalar quantization using a dead zone is applied after the thresholding process. All the coefficients in the luminance and chrominance macroblocks are subjected to the same quantizer, except for the intraframe DC coefficient. The intraframe DC coefficient is quantized by using a uniform scalar quantizer whose step size is 8. The quantizer decision levels are not specified, but the reconstruction levels are defined in H.261 as follows.

For case QUANT odd,

$$\text{REC\_LEVEL} = \text{QUANT} \times (2 \times \text{COEFF\_VALUE} + 1),$$
$$\text{for COEFF\_LEVEL} > 0,$$

$$\text{REC\_LEVEL} = \text{QUANT} \times (2 \times \text{COEFF\_VALUE} - 1),$$
$$\text{for COEFF\_LEVEL} < 0.$$

For case QUANT even,

$$\text{REC\_LEVEL} = \text{QUANT} \times (2 \times \text{COEFF\_VALUE} + 1) - 1,$$
$$\text{for COEFF\_LEVEL} > 0,$$

$$\text{REC\_LEVEL} = \text{QUANT} \times (2 \times \text{COEFF\_VALUE} - 1) + 1,$$
$$\text{for COEFF\_LEVEL} < 0.$$

If COEFF_VALUE $= 0$, then REC_LEVEL $= 0$, where REC_LEVEL is the reconstruction value, QUANT is $\frac{1}{2}$ the macroblock quantization step size ranging 1–31, and COEFF_VALUE is the quantized DCT coefficient.

To increase the coding efficiency, lossless variable length coding is applied to the quantized DCT coefficients. The coefficient matrix is scanned in a zigzag manner in order to maximize the number of zero coefficient run lengths. The VLC encodes events defined as the combination of a run length of zero coefficients preceding a nonzero coefficient, and the value of the nonzero coefficient, i.e., EVENT = (RUN, VALUE). The VLC EVENT tables are defined in [3].

7 Closing Remarks

Digital video compression, although only recently becoming a standardized technology, is strongly based upon the information coding technologies developed over the past 40 years. The large variety of bandwidth and video quality requirements for the transmission and storage of digital video information has demanded that a variety of video compression techniques and standards be developed. The major international standards recommended by ISO and the ITU make use of common video coding methods. The generalized digital video encoder introduced in Section 2 illustrates the spatial and temporal video compression elements that are central to the current MPEG-1, MPEG-2/H.262, H.261, and H.263 standards that have been developed over the past decade. They address a vast landscape of application requirements, from low- to high-bitrate environments, as well as stored video and multimedia to real-time videoconferencing and high-quality broadcast television.

The near future will drive video compression systems to incorporate support for more interactive functions, with the ability to define and download new functions to the encoder. New encoding methods currently being explored by the MPEG-4 and MPEG-7 standards look toward object-based encoding in which the encoder is not required to follow the international video transmission signal formats. These object-based techniques are expected to produce significant improvements in both encoder efficiency and functionality for the end user.

References

[1] ISO/IEC 11172 Information Technology: Coding of moving pictures and associated audio for digital storage media at up to ~1.5 Mbit/s, 1993.

[2] ISO/IEC JTC1/SC29/WG11, CD 13818: Generic coding of moving pictures and associated audio, 1993.

[3] CCITT Recommendation H.261: "Video Codec for Audio Visual Services at $p \times 64$ kbits/s," COM XV-R 37-E, 1990.

[4] H. Hseuh-Ming, and J. W. Woods, *Handbook of Visual Communications* (Academic, San Diego, CA, 1995), Chap. 6.

[5] J. W. Woods, *Subband Image Coding* (Kluwer, Norwell, MA, 1991).

[6] L. Wang and M. Goldberg, "Progressive image transmission using vector quantization on images in pyramid form," *IEEE Trans. Commun.*, **42(6)**, 1339–1349 (1989).

[7] C. E. Shannon, "A mathematical theory of communication," *Bell Syst. Tech. J.* **27**, 379–423, 623–656 (1948).

[8] D. Huffman, "A method for the construction of minimum redundancy codes," *Proc. IRE* **40**, 1098–1101 (1952).

[9] P. W. Jones and M. Rabbani, *Digital Image Compression Techniques*, (SPIE, Bellingham, WA, 1991), p. 60.

[10] N. Ahmed, T. R. Natarajan, and K. R. Rao, "On image processing and a discrete cosine transform," *IEEE Trans. Comput.* **IT-23**, 90–93 (1974).

[11] J. J. Hwang and K. R. Rao, *Techniques and Standards For Image, Video, and Audio Coding* (Prentice-Hall, Upper Saddle River, NJ, 1996), p. 22.

[12] R. M. Gray, "Vector quantization," *IEEE ASSP Mag.* **IT-1**, 4–29 (1984).

[13] W. H. Equitz, "A new vector quantization clustering algorithm," *IEEE Trans. Acoust. Speech Signal Process.* **37**, 1568–1575 (1989).

[14] B. G. Haskell and J. O. Limb, "Predictive video encoding using measured subjective velocity," U.S. Patent No. 3,632,865, January, 1972.

[15] A. N. Netravali and J. D. Robbins, "Motion-compensated television coding: Part I," *Bell Syst. Tech. J.* **58**, 631–670 (1979).

[16] J. R. Jain and A. K. Jain, "Displacement measurement and its application in interframe image coding," *IEEE Trans. Commun.* **COM-29**, 1799–1808 (1981).

[17] T. Koga, "Motion compensated interframe coding for video conferencing," presented at the National Telecommunications Conference, New Orleans, LA, November, 1981.

[18] ITU-T SG 15 WP 15/1, Draft Recommendation H.263 (Video coding for low bitrate communications), Document LBC-95-251, Oct., 1995.

[19] M. Roser, "Extrapolation of a MPEG-1 video-coding scheme for low-it-rate applications," in *Video Communications and PACS for Medical Applications Proc. SPIE* **1977**, 180–187 (1993).

[20] CCITT SG XV WP/1/Q4 Specialist Group on Coding for Visual Telephony, Description of Ref. Model 6 (RM6), Document 396, Oct. 1988.

[21] CCITT SG XV WP/1/Q4 Specialist Group on Coding for Visual Telephony, Description of Ref. Model 8 (RM8), Document 525, June, 1989.

6.2

Spatiotemporal Subband/ Wavelet Video Compression

John W. Woods,
Soo-Chul Han,
Shih-Ta Hsiang,
and T. Naveen
Rensselaer Polytechnic Institute

1 Introduction

This chapter is devoted to subband and wavelet video compression. We start out by showing the unity between these two approaches. They will be revealed to be essentially the same for digital video; hence our chapter title of subband/wavelet compression. Thus our chapter can be viewed as a companion to earlier chapters on wavelets (Chapter 4.1) and wavelet image compression (Chapter 5.4). We review image and video compression basics from the standpoint of subbands and wavelets. We treat subband/wavelet video compression itself in the next section, including the hybrid or recursive as well as nonrecursive methods that use a subband/wavelet transformation in the temporal direction also. There is the possibility of improving compression efficiency by performing the temporal filtering along the motion trajectory, if motion estimation is employed. In both cases efficiency can be improved by coding across the scales or subband levels by introducing a special zero symbol and forming a zero-tree structure.

For various reasons, initially related to compression efficiency, an object-based approach has been pursued. This means that the video is treated as being made up from separate objects

moving and deforming in time. The main advantages of object-based coding have turned out to be in the areas of providing additional functionalities, such as object-based scalability and compression capability for composited images and videos. We present an object-based version of spatiotemporal subband/ wavelet coding. Then we briefly present the topic of invertible motion compensated spatiotemporal coding. Here, even in the presence of half-pixel motion compensation, the synthesis operation can reconstruct the exact source video in the absence of quantization errors. These two techniques could be combined to achieve invertible subband/wavelet coding of spatiotemporal objects.

Currently with this writing, the JPEG 2000 standards body is adopting a subband/wavelet method for *image* coding. However, existent and emerging *video* compression standards are based on block processing using the discrete cosine transform (DCT) as the decorrelating transformation, followed by quantization and variable length coding. In this chapter we will review various methods for replacing the DCT by more general subband/wavelet transformations, both in hybrid coding employing spatial subbands, and in 3-D (spatiotemporal) subbands.

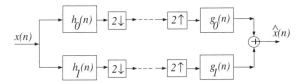

FIGURE 1   1-D subband/wavelet analysis and synthesis filter bank.

1.1 Subbands and Wavelets Reviewed

Subband methods started from work in digital signal processing in the area of speech compression. A major step was the invention of quadrature mirror filters (QMFs) by Esteban and Galland [1] in 1977. A very often used set of QMF filters appeared in Johnston's 1980 paper [2]. These filters, when applied in a 2-D separable manner, were found to be good for image coding too [3]. We summarize some results below. More details on subband/wavelet filters can be found in [4].

1.2 Subband/Wavelet Filter Sets

In Fig. 1, neglecting the coding errors and transmission losses, we can write

$$\hat{X}(\omega) = \tfrac{1}{2}\left[G_0(\omega)H_0(\omega) + G_1(\omega)H_1(\omega)\right]X(\omega)$$
$$+ \tfrac{1}{2}\left[G_0(\omega)H_0(\omega + \pi)\right.$$
$$\left. + G_1(\omega)H_1(\omega + \pi)\right]X(\omega + \pi). \tag{1}$$

or equivalently in the Z transform domain:

$$\hat{X}(z) = \tfrac{1}{2}\begin{bmatrix} G_0(z) \\ G_1(z) \end{bmatrix}^T \begin{bmatrix} H_0(z) & H_0(-z) \\ H_1(z) & H_1(-z) \end{bmatrix}\begin{bmatrix} X(z) \\ X(-z) \end{bmatrix}.$$

A common goal is to design this analysis/synthesis filter bank to have the perfect reconstruction (PR) property, i.e., $\hat{X}(\omega) = X(\omega)$.

The second term in Eq. (1) is due to aliasing, which can be made to disappear (necessary and sufficient) by setting

$$G_0(\omega)H_0(\omega + \pi) + G_1(\omega)H_1(\omega + \pi) = 0. \tag{2}$$

The necessary and sufficient solution to Eq. (2) in the Z-transform domain is

$$\begin{bmatrix} G_0(z) \\ G_1(z) \end{bmatrix} = C(z)\begin{bmatrix} H_1(-z) \\ -H_0(-z) \end{bmatrix} \tag{3}$$

for some $C(z)$, which is usually taken to be a constant c. In the Fourier domain, this is then equivalent to

$$G_0(\omega) = cH_1(\omega + \pi),$$
$$G_1(\omega) = -cH_0(\omega + \pi), \tag{4}$$

and in the spatial (time) domain,

$$g_0(n) = c(-1)^n h_1(n),$$
$$g_1(n) = -c(-1)^n h_0(n). \tag{5}$$

Upon cancellation of the aliased component in the output, the overall transfer function is given by

$$T(\omega) = \frac{\hat{X}(\omega)}{X(\omega)} = \frac{c}{2}\left[H_0(\omega)H_1(\omega + \pi) - H_0(\omega + \pi)H_1(\omega)\right],$$

$$T(z) = \frac{\hat{X}(z)}{X(z)} = \frac{c}{2}\left[H_0(z)H_1(-z) - H_0(-z)H_1(z)\right]. \tag{6}$$

Ideally the filter bank output should be a delayed replica of the input:

$$T(\omega) = e^{-j\omega D}. \tag{7}$$

The necessary and sufficient condition for this is [5]

$$\tfrac{c}{2}\left[H_0(z)H_1(-z) - H_0(-z)H_1(z)\right] = \text{const } z^{-2l-1}, \quad l \in \mathcal{Z}, \tag{8}$$

where $\mathcal{Z}$ denotes the set of integers.

Here, we summarize some design considerations for subband/wavelet filters, some of which are conflicting. For image and video coding, these criteria need not be satisfied exactly and approximations are sufficient.

1. Easy to implement (computationally efficient). This could be achieved through one or more: symmetric filter coefficients, short-length filters, multiplierless implementation of the convolution, and fast transform equivalent of the convolution.
2. The wavelet basis functions generated by h_0 are orthogonal. That is, the impulse response of filter h_0 and its shifted versions (by even shifts) form an orthogonal set. This ensures that there is no redundancy in the transform coefficients.
3. For the same reason as above, the wavelet basis functions generated by h_0 and h_1 should be orthogonal.
4. PR holds in the absence of coding and channel errors.
5. The aliased components in the subbands should be small. This is because the upper subband may have to be truncated because of bit-rate constraints. It is achieved by making the frequency response of h_0 as close as possible to that of an ideal half-band filter.
6. The filters should have linear phase, which is important in image compression.
7. The overall transfer function $T(\omega)$ should be maximally flat at zero frequency. This is important because the energy in images is concentrated near zero frequency, and it is undesirable to introduce much distortion there.
8. The coding gain should be maximized. This would involve signal adaptive design of the filters.
9. The filters should be such that the energy of the signal is concentrated in a single subband (as much as possible). This criterion is necessary for coding efficiency.
10. Step response of the filters should have small overshoots. Otherwise, ringing artifacts occur in the encoded image.
11. Regularity: Iterated synthesis applied to a sequence consisting of only one nonzero entry should look reasonably

nice, even after several iterations. This feature is needed when one subband is made zero while encoding at low bit rate.

12. For optimality, the subband signals should be uncorrelated. That is, for a zero-mean wide-sense stationary (WSS) input $x(n)$,

$$E\{x_i(k)x_j(l)\} = \sigma_i^2 \delta_{ij}\delta_{kl} \, \forall k, l, \quad i, j \in \{0, 1\}. \quad (9)$$

This would let us encode various subbands and their components independently, in the Gaussian case.

We say a subband filter set is orthogonal if criteria 2 and 3 are satisfied. We note that considerations 5 and 10 conflict. Though one cannot achieve zero overshoot and also an ideal frequency response, one can use a cost function for the optimization, which is a combination of both the step and frequency responses of the filters. Numerous approaches have been reported in the literature to design filters h_i and g_i, $i = 0, 1$, that satisfy exactly or approximately Eqs. (5) and (7), in addition to some of the other considerations given above. The QMF filters have the property that the high-pass and low-pass filters are mirror symmetric about $\omega = \pi/2$, but with only approximate perfect reconstruction. The biorthogonal case is a generalization of the PR orthogonal design wherein separate orthogonal basis filters can be used for analysis and synthesis. In such a case the h_i and g_i are less constrained than the orthogonal case in Eq. (6). It has been claimed that this extra freedom can result in significant improvement in coding efficiency. Biorthogonal filter design is considered in [6, 7], and the now widely used wavelet 9/7 biorthogonal filter set was first used for image coding in [6]. Both orthogonal and biorthogonal PR filter banks having the regularity property are related to wavelet theory (cf. Chapter 4.1). A wavelet transform splits the signal space in two, and then recursively splits the lower frequency half space in two, and so on. This is done for images by separable filtering, as mentioned above; i.e., filter the rows first and then the columns (cf. Chapter 5.4). For the video extension, one can continue this separable approach by addition of temporal domain filtering to accomplish an overall 3-D or spatiotemporal subband/wavelet transformation.

1.3 Optimal Subband/Wavelet Filters

Kronander [8] designed a linear phase biorthogonal filter, using a combination of step-response and frequency-response errors as the objective function.

References [9, 10] design a paraunitary filter bank that optimizes the coding gain for a given input signal. Assuming a constant quantizer performance factor [11], the coding gain over pulse code modulation (PCM) of a two-band subband scheme with an orthogonal filter bank is given by

$$G_{SBC} = \frac{\frac{1}{2}\left(\sigma_{x_0}^2 + \sigma_{x_1}^2\right)}{\left(\sigma_{x_0}^2 \sigma_{x_1}^2\right)^{1/2}}, \quad (10)$$

where $\sigma_{x_0}^2$ and $\sigma_{x_1}^2$ are the variances of subband signals $x_0(n)$ and

$x_1(n)$, respectively. Maximizing this gain involves finding the filter set that minimizes the variance of one of the two subbands. When the spectrum of the input signal $x(n)$ is nonincreasing and has components beyond $\omega = \pi/2$, which is true for many natural images, the design goal would be to approximate ideal half-band filters. A definitive treatment of this approach to optimal orthonormal coders is given by Vaidyanathan [12], where it is argued optimal biorthogonal coders cannot beat the performance of orthonormal coders if the power spectrum of the signal is flat over the subbands. Signal adapted finite-order filter design has been presented by Moulin et al. [13]. Again, by means of separable or row-column processing, this method can be extended to images and then onto image sequences or video.

1.4 Comparison of Two Subband/Wavelet Filter Sets

The power of filter sets for compression depends, of course, on the nature of the frequency decomposition as well as the nature of the filter. For this reason, it is of interest to compare the peak signal-to-noise ratio (PSNR) performance of some of these filters on a standard 10 subband wavelet decomposition versus a 16 subband full decomposition. The latter non-wavelet decomposition is sometimes called the 'wavelet packet' case. Some authors have found better PSNR performance of the full band case [14] over the wavelet, sometimes called dyadic or octave band, decomposition.

Knowing of the variety of filters that are available, there arises the question of how these various filters work in a coding context. Here we report on our test for the *Lena* image only and for two popular filters, the biorthogonal Daubechies 9/7 set [6] and from the oldest QMF design, we select Johnston's 16B [15]. The 16B is one of the first QMF filters and has been used for both audio and image coding from the early times. The more recent 9/7 filter has come from wavelet theory and is generally regarded as the best nonadapted filter to use currently for image compression. Both filters are used in a dyadic or octave band decomposition as well as a full or complete tree decomposition. The octave band decomposition is for three levels resulting in a traditional wavelet decomposition with 10 subbands. The full decomposition is for two levels and results in 16 subbands. A more thorough study of the effects of using different filters has been done by Villasenor [16]. Many notable individual coding results are posted at the website www.icsl.ucla.edu/~ipl/psnr_results.html.

We look at the *Lena* image and just two filters. The coder used is a one-class version of subband finite-state scalar quantization (SB-FSSQ) described in [17], and so does not implement prediction across the subbands or scales. The PSNR results are contained in Table 1, but can be summarized as follows. First the 16-band or nonwavelet decomposition is best at or above 0.5 bits/pixel, i.e., at higher qualities. At lower rates the 10-band octave or traditional wavelet decomposition works better. Having said this, we note that the PSNR difference between the

TABLE 1 PSNR comparison of Daubechies 9/7 vs. Johnston 16B filter sets on the *Lena* image

Daub 9/7				Johnston 16B			
10-band		16-band		10-band		16-band	
bpp	PSNR	bpp	PSNR	bpp	PSNR	bpp	PSNR
0.96	39.2	1.00	39.5	1.00	38.6	1.01	39.4
0.74	38.0	0.76	38.2	0.73	37.2	0.74	38.0
0.49	36.2	0.50	36.3	0.50	35.8	0.50	36.2
0.25	33.2	0.24	32.9	0.25	33.0	0.24	32.9
0.13	30.4	0.12	29.5	0.12	30.2	0.13	29.7

four cases at any given bit rate is never more than 1 dB and often much less. For example at 0.5 bpp, a 16-subband Johnston filter decomposition results in a PSNR of 36.2 dB, while the Daubechies 9/7 results in 36.3 dB, only a 0.1-dB difference. Both filter results are better for the full subband decomposition than for the wavelet decomposition. If we use a full 16-subband tree, the maximum observed difference is 0.2 dB. Visual differences are not judged as significant.

2 Video Compression Basics

Here we review some video compression basics relevant to spatiotemporal coding. We look at motion estimation and compensation first. This is followed by the transformation and quantization. Then we introduce the issue of scalability, which has been an interesting research topic, as well as being of concern to international standards bodies. The scalability properties of the spatiotemporal or 3-D filtering approaches has been considered one of their foremost advantages.

2.1 Motion Compensation

The motion estimation problem (cf. Chapter 6.1) for spatiotemporal subband/wavelet coding is somewhat different than that of hybrid coding. This is because in the scalable case, which is the main focus of the spatiotemporal coding, the lower frame-rate sequences are *created* by the motion compensated spatiotemporal filtering. Thus this is the ideal low frame-rate image sequence being communicated to the receiver. Any artifacts created by motion field errors will be seen directly in the lower frame rate output.

2.2 Transformation and Quantization

The role of the transformation is generally to reduce the dependence between the video samples. For example, linear transformations such as DCT and subband/wavelet filter trees and banks are known to reduce the correlation between transformed samples. In the Gaussian case, correlation and dependence are synonymous. More generally correlation and dependence typically reduce together, though this is not always the case. It is important to note that the transformation does not reduce entropy. A scalar or vector quantizer is then called on to provide the desired data compression. While the optimal quantizer (from a mean-square error viewpoint) will have the best performance, most modern coders use uniform step-size scalar quantizers, with a central dead zone to reject noise in the signal subband.

2.3 Scalabilities

Many applications of video coding require some sort of *scalability*, that is, the ability to usefully decode from only portions of the full compressed file. That is to say, we want one scalable coded file, consisting of a telescoping set of embedded files, that offers increasingly greater spatial resolution, higher frame rates, or a better signal-to-noise ratio (SNR). One motivation for scalability is for multicast on a heterogeneous computer network. If a certain part of the net contains only low-resolution terminals, then only that part of the scalable bit stream has to propagate there. Some receiving computers of varying clock speeds will only be able to keep up with lower resolution or lower frame-rate parts of the transmitted signal. Then an SNR scalable coder and appropriate decoder software will permit them all to get a usable image and keep up with the transmission. Alternatively, for a resolution or frame-rate scalable coder, we avoid the bandwidth inefficiency of the *ad hoc* solution of dropping frames at the receiver.

3 Subband/Wavelet Compression

There are basically two types of subband/wavelet video compression. One makes use of a frame-difference coder for the temporal direction amounting to a temporal differential PCM (DPCM) loop. Such a coder is called a *hybrid coder* when coupled with either a block transform or a subband/wavelet based coder in the spatial dimension. The other option is to use subband/wavelet coding for the temporal dimension too. Before presenting this 3-D or spatiotemporal subband/wavelet coder, we pause to look at the hybrid coder briefly.

3.1 Hybrid Subband/Wavelet Coder

This is motion compensated predictive coding with subband/wavelet filters used for the transformation instead of the DCT used in a standards-based coder like MPEG. Most subband/wavelet video coders are hybrid coders too. The class of hybrid coders is characterized by a very efficient one-frame recursive structure. While very efficient for implementation, and limiting the need for motion compensation to a frame-by-frame basis, this recursive structure can be a problem with regard to error propagation, scalability, picture in fastforward, and optimization of the coder. The latter arises because of the dependent frame nature of the hybrid coder's recursive structure.

3.2 Spatiotemporal Subband/Wavelet Coder

This type of coder is a spatiotemporal or 3-D subband/wavelet transformation, in contrast to the hybrid coder, which uses subband/wavelet filters only for the spatial transformation. There are two versions of these spatiotemporal subband/wavelet coders currently of interest; they differ in their use of motion compensation.

3.2.1 Without Motion Compensation

This is the simplest type of spatiotemporal subband/wavelet coder. The advantages of no motion compensation are

- computational simplicity,
- freedom from motion artifacts,
- easier transmission error concealment, and
- limited error propagation.

A number of such 3-D subband coders have been advanced in the literature [18–20]. Some have claimed to offer pictures equivalent to those of MPEG2 at similar rates [18, 21]. This is remarkable since no motion compensation is used. Of course the performance will vary with the motion content in the scene and whether the motion estimator is able to track it or not. For trackable moderate to high motion, we believe that motion compensation is still the best approach. Without motion compensation, the lower temporal video subbands will be blurred (or worse display multiple images or ghosts) when there is much motion. This is a serious disadvantage for scalable frame-rate coding.

3.2.2 With Motion Compensation

If we can afford to use motion compensation in our video coder, then we gain the added efficiencies of this method. There are two variants, the simpler of which uses just one global motion vector, which is suitable for camera-pan compensation [22]. Studies have indicated that camera pan constitutes a large portion of the motion seen in entertainment television. The next step is to get a fixed-size block-based motion estimate and compensation. More computation even can yield a variable block size and more accurate motion field [23–25], or still denser near-continuous motion fields. These latter two more accurate motion fields are important for spatiotemporal scalable methods, in which the motion compensated filter is used to generate the low frame-rate videos, which are the ideal videos that will be subject to the subsequent coding. Any motion compensation artifacts in these lower temporal subbands will inevitably show up in the received and decoded lower frame-rate videos.

3.3 Zero Coding and Embedding

Often, especially when high compression ratios are needed, the quantizer step size for the high-frequency subbands is large. Because of the central dead band of the normally used scalar quantizer, this makes its zero output value quite probable. Zero coding is a way to take advantage of this fact by attempting to code clusters or runs of these zero values together. This is done in MPEG by coding the run lengths in a so-called zigzag scan of the DCT coefficients. In subband/wavelet image coders, not only zero runs but zero clusters have been efficiently coded, most notably in the zero-tree image coder of Shapiro [26], who codes the zeros across spatial scales by introducing a special symbol called the *zero-tree root* for the often occurring situation in which a quantizer zero at one scale is associated with zero values at all finer spatial scales. This coder has been improved by Said and Pearlman in their set partitioning in hierarchical trees (SPIHT) [27], which processes lists of symbols related to significant and insignificant sets of wavelet coefficients. This image coder has been extended to video as 3-D SPIHT in [21]. These coders are made *embedded* by coding bit planes of coefficients in a most-significant-bit-first strategy, which results in a coded bit stream in which one can stop decoding after each bit plane and get the image or video represented to that level of significance. Thus this embedded property facilitates the SNR type of scalability that is desirable when compression is done once for many possible decodings at various quality levels, such as the computationally limited PC mentioned above. Resolution and frame-rate scalability were not addressed in these papers. Interestingly, the four class SB-FSSQ coder [17] has better PSNR than the embedded zero-tree wavelet (EZW) coder [26] on the *Lena* image.

4 Object-Based Subband/Wavelet Compression

At least two problems have prevented object-based video coding systems from outperforming standard block-based techniques. Object segmentation is a very difficult problem because of its sensitivity and complexity. Also, we have the additional need to transmit the contour or shape of the object, leading to additional bit rate. So, the gain in coding the objects must outweigh the need to transmit the additional contour information.

An object-based coder addressing these issues was presented in [28]. The extraction of the moving objects is performed by a joint motion estimation and segmentation algorithm based on Markov random field (MRF) models (cf. Chapter 4.2). In this approach, the object motion and shape are guided by the spatial color intensity information, thus utilizing the observation that in an image sequence, motion and intensity boundaries usually coincide. This not only improves the motion estimation/segmentation process itself in extracting meaningful objects true to the scene, but it also aids the process of coding the object intensities because a given object has a certain spatial cohesiveness. The MRF formulation also allows temporal linking of the objects, thus creating *spatiotemporal objects*. This helps stabilize the object segmentation process in time, and more importantly for coding, allows the object boundaries to be predicted temporally by using the motion information. An efficient temporal updating scheme to encode the object boundaries results in a significant reduction in bit rate while preserving the accuracy of the boundaries. With the linked objects, uncovered regions can be deduced in a systematic fashion. New objects are detected by utilizing both the motion and intensity information. The interiors of the objects are encoded adaptively, meaning that objects well described by the motion parameters are encoded in an "inter" mode, while those that cannot be predicted in time are encoded in an "intra" mode. This is analogous to P blocks and I blocks in the MPEG coding structure (cf. Chapters 6.4 and 6.5), where we now have P **objects** and I **objects**. I-object coding is feasible because the object segments are based on intensity information. The subband/wavelet approach [29] is adopted in spatial coding the objects. Both hybrid [28] and spatiotemporal [30] versions of this object-based subband/wavelet coder were developed.

4.1 Joint Motion Estimation and Segmentation

The main objective is to segment the video scene into objects that are undergoing distinct motion and to find the parameters that describe the motion. We have adopted a Bayesian formulation based on an MRF model to solve this challenging problem. The MRF approach was initially used in motion segmentation and motion estimation in separate works. Because of the interdependency of the two problems, algorithms to perform the motion estimation and segmentation jointly have been proposed [31, 32].

4.1.1 Problem Formulation

Let $\mathbf{I}^t$ represent the frame at time t of the discretized image sequence. The motion field $\mathbf{d}^t$ represents the displacement between $\mathbf{I}^t$ and $\mathbf{I}^{t-1}$ for each pixel. The segmentation field $\mathbf{z}^t$ consists of numerical labels at every pixel, with each label representing one moving object, i.e., $\mathbf{z}^t(\mathbf{x}) = n$ ($n = 1, 2, \ldots, N$), for each pixel location $\mathbf{x}$ on the lattice Λ. Here, N refers to the total number of moving objects. With the use of this notation, the goal of motion estimation/segmentation is to find $\{\mathbf{d}^t, \mathbf{z}^t\}$ given $\mathbf{I}^t$ and $\mathbf{I}^{t-1}$. We further assume that $\mathbf{d}^{t-1}$ and $\mathbf{z}^{t-1}$ are available, making it possible to impose temporal constraints and to link the object labels.

We adopt the maximum *a posteriori* (MAP) formulation to provide estimates $\hat{\mathbf{d}}^t, \hat{\mathbf{z}}^t$ by maximizing the joint conditional density $p(\mathbf{d}^t, \mathbf{z}^t \mid \mathbf{I}^t, \mathbf{I}^{t-1})$. With the use of Bayes' rule, this is simplified to the equivalent maximization of the product of mixed conditional densities and probabilities:

$$p(\mathbf{I}^{t-1} \mid \mathbf{d}^t, \mathbf{z}^t, \mathbf{I}^t)\, p(\mathbf{d}^t \mid \mathbf{z}^t, \mathbf{I}^t)\, P(\mathbf{z}^t \mid \mathbf{I}^t), \tag{11}$$

each of which will be explained in the paragraphs that follow, where we incorporate various assumptions and models about our motion and segmentation field in formulating these terms.

4.1.2 Probability Models

The first term of Eq. (11) is the likelihood functional that describes how well the observed images match the motion field

data. It reflects the relationship between the gray-level changes between frame $t-1$ and t that are corrupted by additive noise. Thus, the actual observed image $\mathbf{I}^t$ is regarded as a noisy version of the original image $\mathbf{G}^t$, or $\mathbf{I}^t(\mathbf{x}) = \mathbf{G}^t(\mathbf{x}) + \eta(\mathbf{x})$. Ignoring such factors as illumination changes, we assume the change of gray level between the two frames to be only due to object motion, and we have $\mathbf{G}^t(\mathbf{x}) = \mathbf{G}^{t-1}(\mathbf{x} - \mathbf{d}(\mathbf{x}))$. If the noise is assumed to be white Gaussian with zero mean and variance σ^2, $p(\mathbf{I}^{t-1} \mid \mathbf{d}^t, \mathbf{z}^t, \mathbf{I}^t)$ is also Gaussian with $p(\mathbf{I}^{t-1} \mid \mathbf{d}^t, \mathbf{z}^t, \mathbf{I}^t) = Q_l^{-1} \exp\{-U_l(\mathbf{I}^{t-1} \mid \mathbf{d}^t, \mathbf{I}^t)\}$ where the energy function $U_l(\mathbf{I}^{t-1} \mid \mathbf{d}^t, \mathbf{I}^t) = \sum_{\mathbf{x} \in \Lambda}(\mathbf{I}^t(\mathbf{x}) - \mathbf{I}^{t-1}(\mathbf{x} - \mathbf{d}^t(\mathbf{x})))^2/2^2$ and Q_l is a normalization constant.

The second term of Eq. (11) is the *a priori* density of motion $p(\mathbf{d}^t \mid \mathbf{z}^t, \mathbf{I}^t)$ and thus enforces prior constraints on the motion field. We adopted a coupled MRF model in [28] to govern the interaction between the motion field and segmentation field both spatially and temporally. The probability density and corresponding energy function is given as $p(\mathbf{d}^t \mid \mathbf{z}^t, \mathbf{I}^t) = Q_d^{-1} \exp\{-U_d(\mathbf{d}^t \mid \mathbf{z}^t)\}$ and

$$\begin{aligned} U_d(\mathbf{d}^t \mid \mathbf{z}^t) = {} & \lambda_1 \sum_{\mathbf{x}} \sum_{\mathbf{y} \in N_{\mathbf{x}}} \|\mathbf{d}^t(\mathbf{x}) - \mathbf{d}^t(\mathbf{y})\|^2 \delta(z^t(\mathbf{x}) - z^t(\mathbf{y})) \\ & + \lambda_2 \sum_{\mathbf{x}} \|\mathbf{d}^t(\mathbf{x}) - \mathbf{d}^{t-1}(\mathbf{x} - \mathbf{d}^t(\mathbf{x}))\|^2 \\ & - \lambda_3 \sum_{\mathbf{x}} \delta(z^t(\mathbf{x}) - z^{t-1}(\mathbf{x} - \mathbf{d}^t(\mathbf{x}))). \end{aligned} \tag{12}$$

where refers to the usual Kronecker delta function[1], $\|\cdot\|$ is Euclidean norm in $\mathbf{R}^2$, and $\mathcal{N}_{\mathbf{x}}$ indicates a small neighborhood of $\mathbf{x}$. The first term encourages motion vectors be locally smooth, but only *within* each object. The second term links the motion vectors along the motion trajectory. The last term encourages the object labels to be consistent along the motion trajectories.

Now as to the third term on the right-hand side of Eq. (11), $P(\mathbf{z}^t \mid \mathbf{I}^t)$, it models our *a priori* expectations for the object label field itself. In the temporal direction, we have already modeled the object labels to be consistent along the motion trajectories. Our model incorporates the spatial intensity information ($\mathbf{I}^t$) based on the reasonable assumption that object discontinuities coincide with spatial intensity boundaries. The segmentation field is a discrete-valued MRF, $P(\mathbf{z}^t \mid \mathbf{I}^t) = Q_z^{-1} \exp\{-U_z(\mathbf{z}^t \mid \mathbf{I}^t)\}$ with the energy function given as $U_z(\mathbf{z}^t \mid \mathbf{I}^t) = \sum_{\mathbf{x}} \sum_{\mathbf{y} \in \mathcal{N}_{\mathbf{x}}} V_c(z(\mathbf{x}), z(\mathbf{y}) \mid \mathbf{I}^t)$, where

$$V_c(z(\mathbf{x}), z(\mathbf{y}) \mid \mathbf{I}^t) = \begin{cases} -\gamma & \text{if } z(\mathbf{x}) = z(\mathbf{y}),\ s(\mathbf{x}) = s(\mathbf{y}) \\ 0 & \text{if } z(\mathbf{x}) = z(\mathbf{y}),\ s(\mathbf{x}) \neq s(\mathbf{y}) \\ +\gamma & \text{if } z(\mathbf{x}) \neq z(\mathbf{y}),\ s(\mathbf{x}) = s(\mathbf{y}) \\ 0 & \text{if } z(\mathbf{x}) \neq z(\mathbf{y}),\ s(\mathbf{x}) \neq s(\mathbf{y}). \end{cases} \tag{13}$$

Here, $\mathbf{s}$ refers to the spatial segmentation field that is predetermined from $\mathbf{I}$. As a simplification, we treat s as a *deterministic*

---

[1] The Kronecker delta function $\delta(\cdot)$ assigns the value $\delta = 1$ when its argument is 0 and $\delta = 0$ otherwise.

field that can be calculated uniquely from **I** alone. According to Eq. (13), if the spatial neighbors **x** and **y** belong to the same *intensity-based* object ($s(\mathbf{x}) = s(\mathbf{y})$), then the two pixels are encouraged to belong to the same *motion-based* object. This is achieved by the $\pm\gamma$ terms. In contrast, if **x** and **y** belong to *different* intensity-based objects ($s(\mathbf{x}) \neq s(\mathbf{y})$), we do not enforce z to be either way, and hence the 0 terms in Eq. (13). This slightly more complex model ensures that the moving object segments we extract have some sort of spatial cohesiveness as well. This is a very important property for our adaptive coding strategy, presented in the paragraphs that follow.

4.1.3 Maximation Approach

As a result of the equivalence of MRFs and *Gibbs densities*, i.e., those densities that can be written as the exponential of the negative of an energy function (cf. Chapter 4.2), the MAP solution amounts to a minimization of the sum of these energies. To ease the computation, a two-step iterative hierarchical procedure is implemented, in which the motion and segmentation fields are found in an alternating fashion, assuming the other is given. Mean field annealing is used for the motion field estimation, while the object label field is found by a deterministic iterated conditional modes (ICM) algorithm [33].

4.1.4 Video Object Segmentation Results

In Figs. 2 and 3, the segmentation results for *Miss America* are displayed in a horizontal versus time plot, corresponding to a fixed vertical position. We see that the segments generally follow the object in the scene and are coherent over time. We can see that the MRF model produced smooth vectors within the objects with definitive discontinuities at the intensity boundaries. Also, it can be observed that the object boundaries relate well to the "real" objects in the scene.

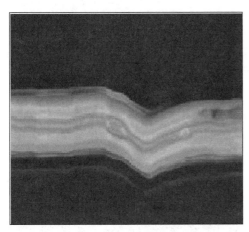

FIGURE 2 *Miss America* horizontal vs. time display. (Reprinted by permission from *Image and Video Compression*, P. Topiwala, ed., Kluwer Academic Publishers 1998.)

FIGURE 3 Segmented horizontal vs. time display. (Reprinted by permission from *Image and Video Compression*, P. Topiwala, ed., Kluwer Academic Publishers 1998.)

4.2 Coding of Video Objects

The coding of the object interior is performed by adaptive coding. Objects that can be described well by the motion were encoded by motion compensated predictive (MCP) coding in hybrid object-based (OB)-MCP [28], and those that cannot were encoded in the "intra" mode. The coding was done independently on each object, using spatial subband/wavelet coding. Since the objects are arbitrarily shaped, the efficient signal extension method proposed by Barnard [29] was applied.

Although the motion compensation was relatively good for most objects at most frames, the flexibility to switch to the intramode (I mode) in certain cases is inevitable. For instance, when a new object appears from outside the scene, it cannot be properly predicted from the previous frame. Thus, these new objects must be coded in the I mode. This includes the initial frame of the image sequence, where all the objects are considered new. Even for "continuing" objects, the motion might be too complex at certain frames for our model to describe properly, resulting in poor prediction. This is another case when objects should be encoded in the I mode. Such classification of objects into I objects and P objects is analogous to P blocks and I blocks in current MPEG video standards (cf. Chapters 6.4 and 6.5). Each of these linked spatiotemporal objects can also be coded by a 3-D spatiotemporal coder as in [30], offering scalability and robustness advantages over the hybrid OB-MCP method, and with, it turns out, almost the same performance.

4.2.1 Object Motion Field

The motion analysis provides us with the boundaries of the moving objects and a dense motion field within each object. An affine parametric representation can provide a smooth and efficient fit to each object's motion. Potential new objects can be found for regions where the fit fails. By modeling the motion of the temporally linked objects with affine parameters, one

TABLE 2 PSNR results for OB-3DSBC

Sequence	Bit Rate (kbps)	Channel	OB-3DSBC (PSNR)	H.263 (PSNR)
Miss America (15 fps)	20	Y	37.5	37.9
		U	38.9	38.5
		V	37.6	37.4
Carphone (15 fps)	40	Y	33.1	33.4
		U	38.3	38.6
		V	38.9	38.1

Source: From *Image and Video Compression*, P. Topiwala, ed., Kluwer Academic Publishers 1998.

reduces the bit rate to encode the object boundaries significantly [28,30]. Furthermore, one can extract uncovered regions simply by comparing the object location and motion parameters between two frames.

Because the objects are linked in time, covered/uncovered region extraction merely involves projecting the motion vectors in time and comparing labels. More specifically, for the uncovered regions in frame t to be found, each pixel is projected back to frame $t-1$ according to its synthesized motion vector. The uncovered pixels are simply those whose object labels don't match along the trajectory.

4.2.2 Coding the Object Boundaries

We have already seen that temporally linked objects in an object-based coding environment offer various advantages. However, the biggest advantage comes in reducing the contour information rate. Using the object boundaries from the previous frame and the affine transformation parameters, one can predict the boundaries with a good deal of accuracy. Some small error occurs near boundaries, and one can simply encode these by using 1-bit flags.

4.3 Object Motion/Segmentation Coding

The object-based 3-D subband/wavelet coding (OB-3DSBC) coder was tested on the QCIF resolution *Miss America* and *Carphone* sequences. Simulations were performed at the frame rate of 15 frames/s. The object segmentation and motion analysis from [28] was used. The target bit rate was 20 kbps at the full frame rate for *Miss America* and 40 kbps for *Carphone*, with the bits being divided equally among the group of pictures (GOPs) except for the first one. The first GOP was assigned twice as many bits as the other GOPs to account for the I-tLL band. For comparison, we obtained results at the same frame and bit rate with an H.263 standard coder (cf. Chapter 6.1). The average PSNRs are summarized in Table 2.

Figure 4 displays full-rate reconstruction results from the various methods for *Carphone*, with corresponding H.263 results shown in Fig. 5. In terms of the PSNR, we can see that the

FIGURE 4 OB-3DSBC coder result. (Reprinted by permission from *Image and Video Compression*, P. Topiwala, ed., Kluwer Academic Publishers 1998.)

OB-3DSBC is somewhat worse (by 0.2–0.4 dB) than the H.263 coder. The OB-MCP coder results are slightly better in PSNR and are shown in [28]; however, the difference in visual quality with OB-3DSBC is minimal. On the plus side, the OB-3DSBC gives us a natural scalability option in frame rate, i.e., the flexibility of decoding the given bit stream at multiple frame-rates [30].

5 Invertible Subband/Wavelet Compression

The spatiotemporal coding presented in Section 3 has the problem of requiring interpolation to *create* the lower frame-rate videos. Even in the absence of any quantization error, the interpolation step will cause some distortion in the lower frame-rate videos. The result is that the above presented technique does not work that well for high quality (read high bit rates). To extend the technique to high quality and also high resolution, we need to address this problem. The interpolation is only needed when motion compensation is used at subpixel accuracy, but this is necessary for high-efficiency coding. Also, the motion compensation itself is a big cause of artifacts at the lower frame rates, where it is more inaccurate.

FIGURE 5 H.263 coder result. (Reprinted by permission from *Image and Video Compression*. P. Topiwala, ed., Kluwer Academic Publishers 1998.)

Because of its high-energy compaction and nonrecursive coding structure, spatiotemporal (3-D) subband/wavelet coding with motion compensation (MC-3DSBC) has been demonstrated to outperform conventional hybrid coders in compression efficiency [23, 25, 34] and in robustness for video transmission.

It is widely acknowledged that motion compensation with half-pixel accuracy is necessary in order to effectively reduce the energy of the displaced frame difference (DFD). Since the high-frequency output of the temporal two-tap analysis filter bank utilized in [34, 35] is the scaled difference of the previous and current frames, they adopted half-pixel accuracy for MC temporal filtering in order to reduce the energy of the high-frequency band. The images therein had to be interpolated at both analysis and synthesis stages, and the resulting systems were thus not invertible. Therefore, reconstruction error was introduced even without any coding distortion. This excluded the technique from high-quality video coding applications and also limited the number of analysis/synthesis stages allowed. In [25, 34], two stages of temporal decomposition were applied in order to avoid buildup of reconstruction error from the analysis/synthesis system. For the HDTV application, only one stage could be used in [36]. To further enhance coding efficiency, the images of the lowest temporal band from the same GOP were encoded by temporal DPCM. Therefore, the overall system still could not fully avoid recursive coding structures and their disadvantages.

In [37], we presented an invertible 3-D or spatiotemporal subband/wavelet system with half-pixel-accurate motion compensation for video coding. We term it *invertible motion-compensated 3DSBC*, or IMC-3DSBC. There we looked at temporal decomposition of the progressively scanned image sequence as a kind of downconversion of the sampling lattice from the interlaced format to the progressive format, following the suggestion in [38]. We thus extended the method of [38], intended for interlaced/progressive scan conversion, to our analysis/synthesis system IMC-3DSBC. An important feature of the new system is that it guarantees perfect reconstruction while high-energy compaction is retained.

It is known that optimal bit allocation for conventional hybrid coders is very complex because of the frame-to-frame dependent quantization structure resulting from the DPCM coding loop [39]. In contrast, in a subband-based coder, coefficients of individual subbands are quantized and coded independently. Optimal bit allocation is therefore possible. However, since MC-DPCM was still used to encode frames of the lowest temporal band in the earlier MC-3DSBC [25, 34], bit allocation could not be fully optimized for the GOPs. In the new system, the input video is decomposed into four temporal stages without build-up of reconstruction error. The GOP consisting of 16 frames does not contain any dependent coding structure at all. Therefore, if the effects of side information are neglected, each GOP can be optimally encoded in an operational rate-distortion sense.

Figure 6 shows PSNR coding results versus bit rate of OB-3DSBC and MPEG-2 (TM5) for the color SIF version of the

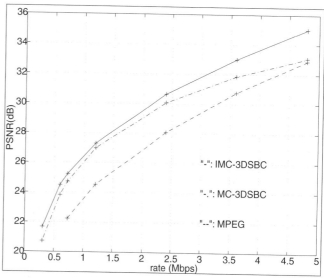

FIGURE 6 *Mobile Calendar* PSNR vs. bit rate for hybrid and spatiotemporal subband/wavelet object coders + MPEG2 (TM5).

Mobile Calendar test clip. Note that the improvement of MC-3DSBC drops off, and will actually saturate, at the higher bit rates, while IMC-3DSBC does not. Notice the 2–3 dB improvement over MPEG-2, which is largely due to optimization, but which in turn is easier for nonrecursive coders.

6 Summary and Look Forward

This chapter has presented 3-D or spatiotemporal coding using subband/wavelet methods. We have first reviewed available filters and compared results. We related the spatiotemporal methods to hybrid methods such as MPEG and hybrid subband/wavelet. We have presented spatiotemporal coding for a low bit-rate, object-based coder, and we addressed the needs for higher rates and resultant quality by showing a method for invertible motion compensated spatiotemporal coding. We believe that future work should extend the invertible coder to code objects and at higher qualities and bit rates.

References

[1] D. Esteban and C. Galand, "Application of quadrature mirror filters to split band voice coding schemes," in *Proceedings of the IEEE International Conference on Acoustics, Speech, and Signal Processing* (IEEE, New York, 1997), pp. 191–195.

[2] J. D. Johnston, "A filter family designed for use in quadrature mirror filter banks," in *Proceedings of the IEEE International Conference on Acoustics, Speech, and Signal Processing* (IEEE, New York, 1980), pp. 291–294.

[3] J. W. Woods and S. D. O'Neil, "Sub-band coding of images," *IEEE Trans. Acoust. Speech Signal Process.* **ASSP-34**, 1278–1288 (1986).

[4] T. Naveen and J. Woods, "Subband and wavelet filters for high-definition video compression," in *Handbook of Visual Communications*, H.-M. Hang and J. Woods, eds. (Academic, New York, 1995).

[5] M. Vetterli and C. Herley, "Wavelets and filter banks: theory and design," *IEEE Trans. Signal Process.* **40**, 2207–2232 (1992).

[6] M. Antonini, M. Barlaud, P. Mathieu, and I. Daubechies, "Image coding using wavelet transform," *IEEE Trans. Image Process.* **1**, 205–220 (1992).

[7] A. Cohen, I. Daubechies, and J.-C. Feauveau, "Biorthogonal bases of compactly supported wavelets," *Commun. Pure Appl. Math.* **45**, 485–560 (1992).

[8] T. Kronander, *Some Aspects of Perception Based Image Coding* (Linkoping U. Dissertations, Linkoping, Sweden, 1989).

[9] P. Desarte, B. Macq, and D. T. M. Slock, "Signal-adapted multiresolution transform for image coding," *IEEE Trans. Inf. Theory* **38**, 897–904 (1992).

[10] D. Taubman and A. Zakhor, "A multi-start algorithm for signal adaptive subband systems," in *Proceedings of the IEEE International Conference on Acoustics, Speech, and Signal Processing* (IEEE, New York, 1992), Vol. III, pp. 213–216.

[11] N. S. Jayant and P. Noll, *Digital Coding of Waveforms* (Prentice-Hall, Englewood Cliffs, NJ, 1984).

[12] P. P. Vaidyanathan, "Theory of optimal orthonormal subband coders," *IEEE Trans. Signal Process.* **46**, 1528–1543 (1998).

[13] P. Moulin, M. Anitescu, K. O. Kortnek, and F. A. Potra, "The role of linear semi-infinite programming in signal-adapted qmf band design," *IEEE Trans. Signal Process.* **45**, 2160–2174 (1997).

[14] M. Vetterli and J. Kovacevic, *Wavelets and Subband Coding* (Prentice-Hall, Englewood Cliffs, NJ, 1995).

[15] J. D. Johnston, "A filter family designed for use in quadrature mirror filter banks," in *Proceedings of the IEEE International Conference on Acoustics, Speech, and Signal Processing* (IEEE, New York, 1980), pp. 291–294.

[16] J. D. Villasenor, B. Belzer, and J. Liao, "Wavelet filter evaluation for image compression," *IEEE Trans. Image Process.* **2**, 1053–1060 (1995).

[17] T. Naveen and J. W. Woods, "Subband finite state scalar quantization," *IEEE Trans. Image Process.* **5**, 150–155 (1996).

[18] W. E. Glenn, J. Marcinka, and R. Dhein, "Simple scalable video compression using 3-D subband coding," *SMPTE J.* **Mar.**, **106**, 140–143 (1996).

[19] C. Podilchuk, N. Jayant, and N. Farvardin, "Three-dimensional subband coding of video," *IEEE Trans. Image Process.* **4**, 125–139 (1995).

[20] T. Meng, B. M. Gordon, E. K. Tsern, and A. C. Hung, "Portable video-on-demand in wireless communications," *Proc. IEEE* **83**, 659–680 (1995).

[21] W. A. Pearlman, B.-J. Kim, and Z. Xiong, "Embedded video coding with 3D SPIHT," in *Wavelet Image and Video Compression*, P. N. Topiwala, ed. (Kluwer, Boston, MA, 1998).

[22] D. Taubman and A. Zakhor, "Multirate 3-D subband coding of video," *IEEE Trans. Image Process.* **3**, 572–588 (1994).

[23] J. R. Ohm, "Three-dimensional subband coding with motion compensation," *IEEE Trans. Image Process.* **3**, 559–571 (1994).

[24] G. Lilienfield and J. W. Woods, "Scalable high-definition video coding," in *Proc. ICIP-95* (IEEE, Washington, DC) pp. 567–570.

[25] S.-J. Choi and J. W. Woods, "Motion-compensated 3-D subband coding of video," *IEEE Trans. Image Process.* **8**, 155–167 (1999).

[26] J. M. Shapiro, "Embedded image coding using zerotrees of wavelet coefficients," *IEEE Trans. Signal Process.* **41**, 3445–62 (1993).

[27] A. Said and W. A. Pearlman, "A new fast/efficient image codec based on set partitioning in hierarchical trees," *IEEE Trans. Video Technol.* **6**, 243–250 (1996).

[28] S.-C. Han and J. W. Woods, "Adaptive coding of moving objects for very low bit-rates," *IEEE Trans. Select. Areas in Commun. Spec. Issue* **16**, 56–70 (1998).

[29] H. J. Barnard, "Image and video coding using a wavelet decomposition," Ph.D. dissertation (Delft U. of Technology, The Netherlands, 1994).

[30] S.-C. Han and J. W. Woods, "Scalable object-based 3-D subband/wavelet coding of video," submitted, 1998.

[31] C. Stiller, "Object-oriented video coding employing dense motion fields," in *Proceedings of the IEEE International Conference on Acoustics, Speech, and Signal Processing* (IEEE, New York, 1994), Vol. V, pp. 273–276.

[32] M. Chang, I. Sezan, and A. Tekalp, "An algorithm for simultaneous motion estimation and scene segmentation," in *Proceedings of the IEEE International Conference on Acoustics, Speech, and Signal Processing* (IEEE, New York, 1994), Vol. V, pp. 221–224.

[33] A. M. Tekalp, *Digital Video Processing* (Prentice-Hall, Upper Saddle River, NJ, 1995).

[34] S.-J. Choi and J. W. Woods, "Three-dimensional subband/wavelet coding of video with motion compensation," in *VCIP-97, Proc. SPIE* **3024-17**, 96–104 (1997).

[35] J. Ohm, "Three-dimensional subband coding with motion compensation," *IEEE Trans. Image Process.* **3**, 559–571 (1994).

[36] G. Lilienfield and J. W. Woods, "Scalable high definition video coding," in *VCIP-98, Proc. SPIE* **3309**, 158–169 (1998).

[37] S.-T. Hsiang and J. W. Woods, "Invertible three-dimensional analysis/synthesis system for video coding with half-pixel accurate motion compensation," in *VCIP-99, Proc. SPIE* **3653**, 537–546 (1999).

[38] J.-R. Ohm, "Variable-raster multiresolution video processing with motion compenstion techniques," in *ICIP-97, Proc. IEEE* (Santa Barbara, CA, 1997), **1**, pp. 759–.

[39] K. Ramchandran, A. Ortega, and M. Vetterli, "Bit allocation for dependent quantization with applications to multiresolution and mpeg video coders," *IEEE Trans. Image Process.* **3**, 533–545 (1994).

6.3

Object-Based Video Coding

Touradj Ebrahimi
and Murat Kunt

*Swiss Federal Institute of
Technology — EPFL*

1 Introduction

Conventional digital image and image sequence coding has historically relied on a number of simple yet powerful concepts. An original image is converted into a digital format by sampling in space and time, and by quantizing in brightness or color. Messages defined by using this basic data format are referred to as being in "canonical form." Codewords have been assigned to messages in a variety of ways, motivated by the information theory framework. Examples of messages include pairs of adjacent pixels, groups of pixels within a geometrically simple data independent structure (e.g., a square image block), or a linear reversible transform of these pixels (such as the discrete cosine transform, or DCT). Statistical distributions of the messages have been used to determine optimal codeword assignments. The compression performance of these types of schemes saturated quickly. Natural images and image sequences are anything but stationary, meaning that the statistical properties of image data are variable over space and time. Although interesting, adaptive sampling is impractical. Furthermore, the entropy of a natural scene is hardly known and depends heavily, if not uniquely, on the model used to estimate image statistics and statistical dependencies. Last but not least, data independent structures such as Cartesian sampling grids (or square data blocks, as used in MPEG, for example) cannot describe nonstationarities and hence cannot serve as efficient data structures for images and image sequences.

Improvements have come by representing visual data in terms of regions, defined by their contour and texture, possibly corresponding to objects or to parts of objects. This approach closes the gap between technical systems and the human visual system (HVS), the latter usually being the last element of an image processing system. It also makes it possible to emphasize visually sensitive data while neglecting visually insignificant information. Of course, the raw data resulting from sampling and quantization must be transformed into this representation. Once the regions are obtained, there is still a challenging step to connect regions belonging to the same visual object. As a byproduct to compression and representation efficiency, this approach has paved the way to a number of new functionalities, such as interaction with regions and objects. This so-called second-generation concept is now widely accepted and has become the basic philosophy of the new MPEG-4 standard (Chapter 6.5).

Unfortunately, there is no single compression method or algorithm that can efficiently compress all possible image regions or objects, just as there is no single tool to repair a car. The ultimate representation is then to assign the tool to the information. Each type of visual information, region or object, can be compressed by the most efficient algorithm. The label of the algorithm is appended to the data and algorithms are accumulated in a tool box. This approach is called *dynamic coding*.

The chapter is organized as follows. In the next section, second-generation coding is presented as the basis for object-based coding. Section 3 describes an efficient and relatively simple way of encoding video information using objects. It has three main components based on the handling, respectively, of shape, motion, and texture. The components of the scheme are designed in such a way that each one allows progressive transmission or retrieval. Integrating these components into an overall progressive coding scheme is described in Section 4. The dynamic coding concept, together with a few illustrations, is developed in Section 5, before the conclusions in the last section.

2 Second-Generation Coding

The most widely used approach to represent still and moving pictures in the digital domain is based on pixels. This is mainly because pixel-based acquisition and reproduction of digital visual information are mature and relatively cheap technologies, as they produce uniformly sampled data. In parallel, we can view the lowest level of the HVS (rods and cones in the retina) as a non-uniformly sampled acquisition system [1, 2]. Compared to its technical counterpart, this system has, however, incredible complexity and sophistication in its higher levels. In a pixel-based representation, an image or a video is modeled as a set of pixels (with associated properties such as a given color or motion) the same way the physical world is made of atoms. Until recently, pixel-based image processing was the only digital representation available for the processing of visual information, and therefore the majority of techniques known today rely on this representation. It was in the mid-1980s that, for the first time, motivated by studies of the mechanisms of the human vision system, researchers developed other representation techniques [3]. The main idea behind this effort was that, since the HVS is in the majority of cases the final stage in the image processing chain, then a representation that matches the HVS will be more efficient in the design of image processing and coding systems. Non-pixel-based representation techniques for coding (also called second-generation coding) have been found to be superior in coding efficiency at very high compression ratios, when they are when compared with pixel-based representation methods [3].

Figure 1 depicts a representation pyramid illustrating various methods used to represent visual information and their relationships. Linear transform and (motion-compensated) predictive coding, which can be considered special cases of pixel-based representation techniques, have also shown outstanding results in compression efficiency for the coding of still images and video. One reason is that digital images and video are captured and therefore mainly available in a pixel-based form, as this is the only way we can acquire them today. In order to apply a non-pixel-based approach, either the input data should be captured in a non-pixel-based form, or the available pixel-based data have to be converted to a non-pixel-based representation, which brings additional complexity but also other inefficiencies. Examples of such conversions are depicted in Fig. 1 and can vary from simple visual primitive extraction methods to more sophisticated object segmentation and tracking techniques. An important class of non-pixel-based representation schemes is that of content-based representation. In this approach, an image is seen as a set of visual primitives (edges, contour, texture, etc.) containing the most salient visual information in the scene.

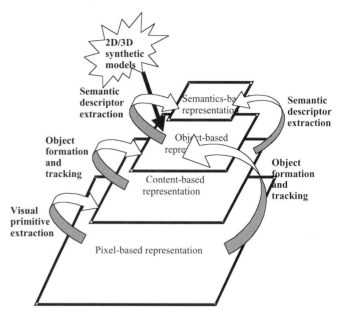

FIGURE 1 Visual information representation pyramid and its internal structure.

Among content-based representations, region-based and ultimately object-based visual data representations are very important classes. Here, regions are defined as segments in an image that share a common property, while objects are defined as sets of regions that represent a semantically meaningful entity in an image [4]. In object-based representations, objects replace pixels. An image or a video is seen as a set of objects that cannot be broken into smaller elements. In addition to texture (color) and motion properties, shape information is also needed in order to completely define any object. The shape in this case can be seen as a force field keeping together the elements of an image or video object like atoms in a molecule or a physical object. Once you grab a corner of an object, the rest comes with it because the force field has glued all atoms of the object together. The same is true in an object-based representation, where the role of the force field is played by shape. Thanks to this property, object-based representation brings a very important feature at no cost, called *interactivity*. Interactivity is defined by some as the element that defines multimedia [5]. This is one of the main reasons for which an object-based representation was adopted in the MPEG-4 standard; see Chapter 6.5 and [6]. As pointed out earlier, because of the fact that the majority of digital visual information is still in pixel-based representations, converters are needed in order to go from one representation to another. The passage from a pixel-based representation to an object-based representation can be performed by using manual, semiautomatic, or automatic segmentation techniques. This subject will not be covered here, as it is addressed in Chapter 4.8. The inverse operation is achieved by rendering, blending, or composition, which are typically used in computer graphics applications. Object-based representations are also very suitable to be cast in the same framework as natural and synthetic data coding, since synthetic objects (2-D or

3-D) can be treated in the same way as any natural object and added into the scene (see Fig. 1). A large number of object-based coding schemes have been proposed in the literature. The main differences among these techniques reside in one of the following points:

- the specific method used for the coding of shapes of objects
- the method used for the coding of texture and color information in objects
- the method used to estimate and to code the motion of objects
- the way in which the complete system is integrated, using the above components

In this chapter, we will not cover all possible variants and approaches to object-based video coding. Interested readers can refer to tutorial articles and books for this purpose [18, 20]. Rather, the remainder of this chapter will concentrate on object-based video coding algorithms that provide major functionalities expected from such an approach while providing other useful features.

In the data representation pyramid, one could think of yet another representation in which visual information is represented by describing its content. An example would be when you describe to someone a person he or she has never seen: She is tall, thin, has long black hair, blue eyes, etc. As this kind of representation would require some degree of semantic understanding, one could call it a "semantics-based representation." One way of building a semantics-based representation is to start from an object-based or content-based representation, as again, it seems that humans do it this way [1, 2]. An example of an implementation of a semantics-based representation would be a descriptor language that describes objects and their properties (position, dominant color, texture, shape, etc.), as well as their relations

to each other (close, far, connected, above, etc.). The semantic description can be based on other simpler semantic descriptors in a hierarchical manner. For instance, a house could be by itself a semantic descriptor, which can be also divided into other semantic descriptors such as doors, windows, roof, walls, etc., which could each be divided into simpler semantic descriptors (geometric objects with various shapes, colors and textures, etc.). The difficulty in a semantics-based representation is to make the description as application independent as possible. The coding scheme described in this chapter provides a mechanism that allows efficient access to the salient visual information in an image sequence that is useful for semantics-based representation, while still providing other features desired in a content-based and object-based representation, such as interactivity with objects and compression efficiency.

3 Object-Based Video Coding

This section describes a complete object-based video coding scheme that addresses many requirements desired in applications that would necessitate a content-, object-, or even semantics-based representation. It starts by giving a general overview of the algorithm used for the coding of arbitrarily shaped video objects. The general block diagram of this technique is depicted in Fig. 2. As in other object-based coding schemes, one would expect to distinguish three key components, namely, shape, motion, and texture coding blocks. In this scheme, shape coding is replaced by geometric coding, which refers to information about the outline of objects (shape) as well as its internal visual primitives (edges, corners, etc.).

In addition to the above, as in many video coding schemes, the algorithm operates either in intramode (I), when an object

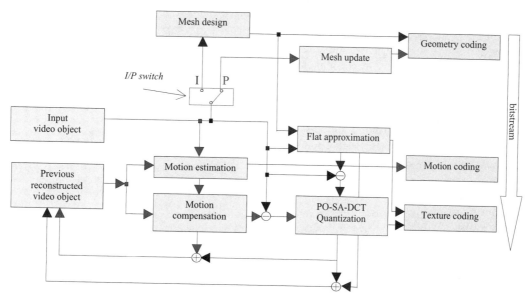

FIGURE 2   Overview of the object-based video object coding structure.

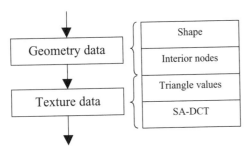

FIGURE 3   Overview of the intracoding mode syntax.

is coded independently, or in a predicted (P) intermode, when a video object is coded taking into account information available on its past. Intramode provides random access points in the bit stream, as well as some robustness to propagation of transmission errors as it does not refer to any prior information. Ideally, intracoded video objects should occur at each scene change, when new objects appear. In practice, they occur at a predefined, fixed rate, e.g., every 0.5 s. Figure 3 gives an overview of the intracoding mode syntax. The geometric information is coded first. The object shape is encoded by using a progressive polygonal approximation, which amounts to a simple vertex coding method when a lossy representation or quasi-lossless shape coding is good enough. Given the mesh outer boundary, interior nodes are selected at high-gradient points by using a minimum distance constraint. The object's outer boundary and inner vertices form a triangular mesh, which is described by coding each vertex position. The entropy associated with a vertex position is in general a function of the size of the video object. By taking into account the forbidden positions, one can reduce this entropy and consequently the amount of bits needed to code the geometry.

Once the geometry (mesh) is coded, the mean color value of each triangle is directly transmitted. The pointwise difference between the original image and the mean value constitutes the texture error image. A shape-adaptive DCT is applied to encode the resulting zero-mean triangular error patches. The transform is followed by uniform quantization of the AC coefficients, zigzag scan, run-length representation, and adaptive arithmetic coding, as in MPEG. At the decoder side, the inverse operations are applied.

Figure 4 gives an overview of the intermode coding syntax. First, the geometric update is encoded: a list of deleted boundary vertices, a list of inserted boundary vertices and their positions, shape motion vectors for predicted vertices, and texture motion vectors for every node (boundary as well as interior). The sign and the absolute value of each vector components are encoded separately with an adaptive arithmetic code (Chapter 5.1), and a special value is defined to indicate that the node is deleted. Then, texture updates are encoded. For each triangle, a one-bit flag indicates to the decoder whether it is updated or not. The choice of whether to update a triangle or not depends on its error measure and a threshold value that is a function of targeted

bitrate or desired quality. To perform an update, the same shape-adaptive DCT is applied to each error triangle, combined with uniform quantization of the AC coefficients, zigzag scan, run-length representation, and adaptive arithmetic coding, as in the intramode. At the decoder side, motion compensation and the inverse DCT are applied.

It is important to mention at this point that in addition to a mechanism to generate video objects (by manual, supervised, semiautomatic, or fully automatic segmentation), the encoder should also design a content-based mesh on the video objects by selecting nodes on high spatial gradient points such as those described in [7, 9, 13]. In this case, an adaptive triangular mesh partition is constructed from the resulting set of nodes by means of Delaunay triangulation [17]. Only the node positions (mesh geometry) need to be transmitted for the decoder to reconstruct this content-based partition. If an arbitrarily shaped video object is considered, its outer boundary is approximated by a polygon (vertices), transmitted to the decoder, and constrained Delaunay triangulation is applied. Consecutive occurrences of video objects are predicted by means of forward node tracking. Motion compensation is based on an affine triangular warping model where the motion of any pixel is linearly interpolated from that of surrounding triangle vertices. Only the node motion vectors need to be determined and transmitted to the decoder to track the mesh deformation along the video sequence.

The bitstream syntax is organized in a separable fashion, so as to allow efficient and independent access to geometry (and shape), motion, and texture information in a quality-scalable way, so that the salient information comes first and can be decoded without the need to reconstruct all of the data. Salient information includes a coarse shape description by polygon vertices; mesh node positions (which are selected based on specific image features, such as edges and corners); coarse texture data in intraframes (for instance, one DC component per mesh triangle—flat image approximation); and coarse mesh motion (defined by the tracking trajectories of a limited set of significant vertices). This codec provides many functionalities needed for video compression, video object coding, and manipulation, as well as content-based retrieval in video databases [13].

In the following paragraphs the major components of this coding algorithm are described in further detail, and more insights are provided.

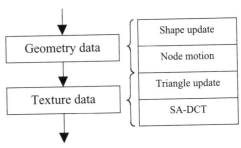

FIGURE 4   Overview of the intercoding mode syntax.

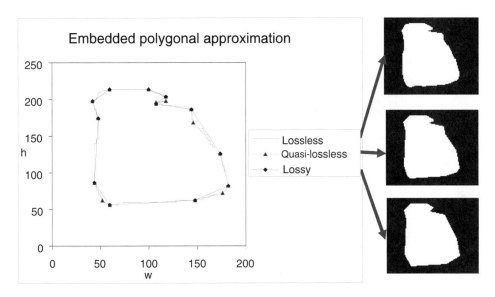

FIGURE 5 Example of video object decoding, using PPE from coarse to fine to lossless. (See color section, p. C–28.)

3.1 Object Shape and Geometry Coding

In object-based coding techniques, information about the shape of video objects has to be coded and made available in the bitstream. This component is the major difference between object-based and more conventional pixel-based techniques, as the shape information is not needed in the latter and is therefore not coded. A progressive contour coding based on a polygonal approximation of the shape boundary is used to code the outline of every video object [12]. The corresponding progressive polygonal encoding (PPE) method exploits the previously transmitted coarser polygons to achieve efficient compression of subsequent contour refinements, defined either geometrically or by a chain code (when lossless shape coding is desired). This representation offers several interesting features. First, being quality scalable, geometrical, and semantic, it is particularly suitable for sketch-based retrieval that is based on video object shapes, as well as for video manipulation. Indeed, the decoder can easily decode the first bits in the shape bit stream that correspond to the most salient vertices, typically high-curvature points along the shape contour. Figure 5 gives an example of a video object that has been decoded in a progressive manner from coarse to

fine, and up to a lossless level. Shape matching methods such as vertex-based modal matching or comparisons based on the Hausdorff distance can then directly exploit this coarse vertex representation [10, 13]. Second, a geometrical shape boundary description can be integrated into an object-based mesh coding scheme. Coarse vertices simply define the outer mesh boundary, and constrained Delaunay triangulation can be applied to define a corresponding arbitrarily shaped triangular mesh partition [1, 9, 17, 19].

In order to support lossless shape representation, as required by high-quality applications for appropriate object texture rendering, a solution has to be designed to efficiently and losslessly compress video object shapes while maintaining a reasonable complexity. Such a solution is based on *altered* boundary triangles, which is enabled by a specific property of the PPE representation: the lossless contour refinement is constrained into a geometrical stripe one or two pixels wide on both sides of the coarser polygonal approximation, if the latter was defined under an accuracy of one or two pixels respectively. The boundary triangles in the mesh can then be easily adapted to fit the lossless shape boundary, by checking only pixels in a thin stripe along the mesh boundary, as illustrated by Fig. 6. A detailed example of

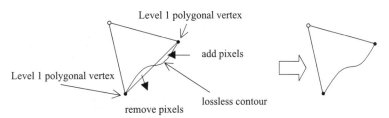

FIGURE 6 Lossless shape refinement with altered triangles. Left: mesh triangle and its corresponding original contour, which is no farther than one pixel away from the boundary edge; right: it is possible to obtain the original shape by adding and removing pixels where necessary.

(a)   (b)   (c)   (d)

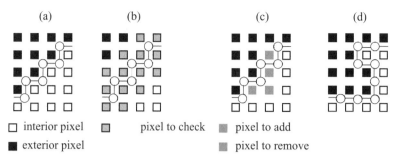

☐ interior pixel   ▣ pixel to check   ▪ pixel to add

■ exterior pixel   ▪ pixel to remove

FIGURE 7 Example of refinement along a boundary edge, resulting in an altered triangle. Squares indicate 2-D pixels; circles and lines represent the interpixel discrete segment. (a) Coarse shape boundary; (b) local stripe to refine; (c) refined pixels; (d) final lossless shape boundary.

this stripe-based boundary refinement is given in Fig. 7. Figure 8 shows the coarse mesh, the refined mesh, and the corresponding updated pixels for a mesh-based partition of a typical video object.

In the intermode, a temporally predicted shape can be used in order to reduce the shape coding overhead and to take advantage of temporal correlation regarding the contour information. To this end, the progressive polygonal approximation method is applied to each occurrence of video objects, and the resulting coarse vertices are matched to the previous corresponding video objects (polygon matching). Information about deleted, inserted, and tracked vertices is sent to the decoder in the form of binary lists, followed by the prediction motion vector or the inserted vertex position depending on the transmitted vertex status. Refinement vertices are still intra-encoded by means of the PPE algorithm, as they are likely to correspond to details that are expected to be temporally unstable. Experimental results show a gain of about 40% by using such predicted shape coding schemes when compared with intrashape coding for rigid and even slightly nonrigid video objects [13].

3.2 Object Motion Estimation, Compensation, and Coding

In triangular mesh-based video codecs, the motion at each node defining the mesh is determined and transmitted to the decoder, which applies affine warping as the motion compensation method to interpolate the motion in each mesh triangle [8, 9, 15, 16, 19]. Various node motion estimation methods have been

investigated so far in the literature. The simplest technique consists of performing block matching by defining a square block centered on the node to track. Forward or backward block matching may be used [15], the former being more suited to node trajectory tracking along the video sequence [19]. A variant of this method, called pixel matching, consists of weighting the error computation in the block matching process so that higher importance is given to the error at and immediately around the node itself, as the aim is a node motion estimation rather than a block motion estimation [15, 16]. Experimental results show that block matching methods outperform pixel matching algorithms in terms of motion compensation quality, especially in the presence of mild to complex motion [13]. The major drawback of block matching as well as pixel matching lies in the fact that they do not take into account the affine warping process in the motion optimization. Consequently, the compensation error is not exactly the computed error in the motion estimation procedure, and a suboptimal solution may be obtained.

To overcome this limitation, two major methods have been reported in the literature: closed-form connectivity-preserving solutions [8] and hexagonal matching refinement [14]. The first method operates on a dense optical flow field, possibly derived from a prior video segmentation and tracking stage. The dense motion field requirement together with its relative complexity explain its infrequent use in practice. The hexagonal matching refinement method aims at taking into account the warping-based motion compensation in the motion estimation process. It was initially applied to regular (hexagonal) triangular partitions [14], but several authors have adapted it to content-based

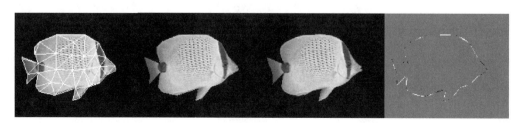

FIGURE 8 Left: triangular mesh partition; center left: coarse boundary; center right: exact boundary with altered triangles; right: pixels processed in a shape-refinement process (black: removed pixels; white: added pixels).

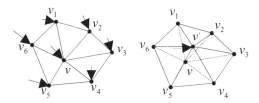

FIGURE 9 Hexagonal matching refinement. Left: triangulation is applied based on the positions of predicted refinement nodes and previously determined coarser nodes; right: the refinement node position is actually optimized to minimize the warping error in the corresponding surrounding polygons.

triangular meshes, fitting arbitrarily shaped video objects [9, 16]. This approach relies on an initial guess for mesh node motion, possibly provided by a block matching technique. Based on this initial solution, the motion vector at each node is optimized by minimizing the affine compensation error in connected triangles, assuming the motion of connected nodes is fixed (see Fig. 9). The optimization is repeated for each node and iterated over the whole set of nodes until stability is reached. Each pass of the algorithm guarantees that the error decreases when compared with the initial guess error. However, in addition to its complexity, the algorithm may also suffer from other limitations, such as its suboptimality and high sensitivity to initial guess values, as outlined in [22]. In practice, while being much more complex in terms of both implementation and computation, the hexagonal matching refinement method does not necessarily generate better results than the direct block matching method. The latter is therefore preferred as the motion estimation algorithm, and it is used here (and for the base layer in the progressive coding scheme described further).

3.3 Texture Representation

In a complete video compression scheme, texture approximation and encoding are needed in the intracoding mode, but also in intercoding modes when prediction residual errors should be coded. Until very recently, mesh-based video representations were often embedded in standard video coding schemes with little attention devoted to their efficient integration [21]. The present object-based coding algorithm is designed to provide a complete and consistent video compression scheme, where the texture representation method is suitable to the triangular partition associated with the warping motion model and applicable to both intracoding and intercoding modes.

In classical intraframe mesh-based image approximation methods, intensity values are transmitted only at the mesh nodes, and other pixel values are interpolated from them; for instance, by means of an affine model applied to the mesh triangles. The major drawback of this approach is the underlying assumption of a continuous image surface, which clearly does not support edge and contour discontinuities. As pointed out earlier, a flat approximation is used to represent the intensity of each triangle which results in coding one value per mesh triangle. Figure 10 shows

an example of such a coarse representation, based on nodes selected on features such as edges and corners. This coarse field corresponds to salient features easily accessed in the bit stream and suitable for fast discrimination between images, for instance in a content-based retrieval scheme [13]. The corresponding reconstructed image is then further refined by means of transform coding of the residual texture error.

In order to efficiently approximate and encode the texture data in intracoding as well as intercoding modes, a transform method is used. Such methods are very popular in image and video compression. Their major drawback lies in the fact that they were originally designed for pixel-based compression of rectangular images, as opposed to content-based approaches. However, with the emergence of the MPEG-4 standard, different solutions have been recently proposed that partly overcome this problem, such as padding and shape-adaptive transforms [6, 11].

In the framework of mesh-based compression, both Wang [21] and Altunbasak [7, 8] have reported the use of quadrilateral warping combined with conventional block-based DCT. However, the major drawback of this method lies in the additional low-pass filtering effect introduced in the compression scheme by the direct and inverse digital warping procedures [22]. Therefore, rather than transforming the triangles to fit a quadrilateral region over which conventional transforms may be applied, another approach consists of directly applying a transform to the triangular domain, such as the pseudo-orthonormal shape-adaptive discrete cosine transform (PO-SADCT) [11]. With such a transform, there are as many coefficients to code as there were pixels in the shape. In addition, these coefficients are gathered in the top-left part of the shape's bounding box, which makes further quantization and run-length coding similar to the conventional DCT coding scheme. The decoder only needs the shape information to apply the inverse operations and reconstruct the approximated segment. The efficiency of the SADCT has been assessed in the case of 8 × 8 boundary blocks (conventional blocks partly overlapping the border of a video object), for both intra-texture and displaced frame difference (DFD) coding. Variants of the SADCT method have been described in [11]. Among them,

FIGURE 10 Example of application of texture coding. Left: flat approximation (mean or DC intensity values of mesh triangles); Right: PO-SADCT coding of remaining AC coefficients per triangle (compression ratio, 32:1; PSNR, 30.6 dB).

the PO-SADCT method has been shown to be suitable for coding zero-mean texture data, such as DFD and error coding in general. Here, this transform is applied to partition triangles corresponding to intracoding mode as well as intercoding mode prediction residual errors. It is then followed by conventional uniform quantization, zigzag scan, run-length representation, and adaptive arithmetic coding, as in MPEG.

Figure 10 shows an example of the application of this texture coding scheme and its coding efficiency for compression of a still picture.

4 Progressive Object-Based Video Coding

The object-based video compression scheme discussed in the previous section may also be adapted to achieve progressive coding desired in a number of applications. As contour and texture coding techniques used in this coding algorithm are both inherently progressive, the key to achieve an overall progressive coding scheme would be to adopt a progressive geometry (mesh) and motion coding.

Let us consider a progressive mesh from its coarsest to finest levels. At the coarsest level, the mesh is designed as usual, using a minimum distance constraint. By progressively reducing this constraint, one may define new nodes along image edges. This technique enables a progressive mesh design. The shape accuracy may be improved at the same time by using the PPE method. Examples of a few levels of a progressive mesh built on top of a typical video object according to the above process are given in Fig. 11. When encoding the location of a node at a given resolution level, some positions within an 8-neighborhood around a previously transmitted node (contour or interior node, from coarser levels or from the current layer) are invalid. The entropy associated with this node location is therefore reduced accordingly, as well as the number of bits needed for its coding.

As the mesh node density progressively increases, the quality of the resulting motion approximation will improve, as long as the mesh has not reached the optimal size [9, 13]. The motion coding cost also increases with the number of motion vectors to transmit. It may therefore be interesting to first transmit the major node trajectories, then progressively refine them as more bits become available. In order to improve the rate-distortion performance, it is possible to exploit the previously transmitted motion information when encoding the refinement motion vectors. In particular, under the hypothesis of a reasonably smooth

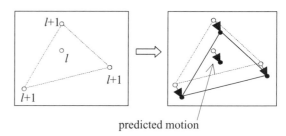

predicted motion

FIGURE 12 Example of progressive motion prediction. Once the motion vectors for level $l + 1$ nodes are known, it is possible to predict motion vectors for level l nodes.

motion field, coarse motion vectors can be used as predictors of the refinement trajectories, as illustrated by Fig. 12. In terms of compression, such a prediction will be efficient as long as the hypothesis of a smooth motion field remains true. Indeed, in this case, if motion estimation and resulting motion vector coding are performed relative to the predicted position, rather than the reference position, null and small displacements become more likely.

By enabling a suitable motion prediction for refinement data at both the encoder and decoder sides, progressive motion transmission also facilitates local optimization of the refinement motion vectors. Indeed, under the hypothesis that the initial prediction derived from coarser motion vectors provides a satisfactory initial guess, triangulation can be applied at this stage. Refinement motion vectors can then be further optimized by means of hexagonal matching refinement, as illustrated in Fig. 9. As opposed to direct block matching, this method takes into account the warping-based rendering process in the optimal displacement computation. As explained earlier, its major drawbacks lie in its inherent complexity and in the fact that it imposes *a priori* triangulation, which is suboptimal when the initial guess is far from the local optimum. By predicting the refinement motion from displacement vectors at surrounding coarser nodes, however, the initial guess is expected to be close enough to the final solution. In addition, many nodes corresponding to coarser motion and contour approximation are fixed, which reduces the search space and accordingly the necessary computation. In particular, if all the nodes connected to a refinement node are fixed (e.g., nodes v_1 to v_6 in Fig. 9), there is no need to iterate the optimization on the corresponding polygon, which accelerates convergence. For the reasons mentioned above, hexagonal matching is performed for estimation of motion vectors of a finer resolution mesh from a coarser one. Experiments show that this approach produces superior results in terms of rate distortion when compared with other motion estimation methods [13].

A complete progressive scheme can then be designed by combining progressive shape, geometric, and motion coding with texture refinement, applied at each layer to the error image generated at the previous coarser layer and cropped by the refined shape mask [13]. Such a progressive transmission scheme may be required when both high-quality decoders and low-cost decoders receive the encoded bit stream, so that the former can take advantage of all of the data while the latter only use the

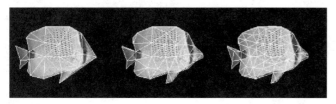

FIGURE 11 Example of a progressive geometry (mesh) construction by successive refinements from left to right.

FIGURE 13 Example of progressive decoding of a typical video object. From left to right are decoded results from lowest quality (70 kbps) to medium quality (170 kbps), to highest quality (320 kbps). The original video object is shown to the right side.

coarser part of it. Independent of the decoder performance, scalable transmission also provides decoders with the possibility to quickly browse and preview a coarse version of the video, at a fraction of the resources required by complete decoding.

Figure 13 depicts results from a progressive decoding of a typical video object using the described coding algorithm. When the progressive representation is achieved by means of a quality-scalable scheme carefully designed to this end, the availability of high-level, preorganized, and easy-to-access information at the coarse data layer enables specific processing at the decoder side. In particular, it facilitates sequence matching, indexing, retrieval, classification, and automatic event control. In such a scenario, part of the analysis stage performed at the encoder for the sake of compression can be exploited (and saved) at the decoder side, such as contour extraction or motion analysis. In this coding algorithm, examples of information potentially exploitable at this level include the following: the shape contour information, either accurate or approximated by a set of vertices; the mesh geometry, where nodes are defined along edges and in highly textured areas; the motion vectors; the node trajectories; and the coarse color representation.

5 Dynamic Coding

It is well known that visual information has a highly nonstationary nature. In multimedia applications, all sort of visual data could be transmitted between terminals. Among all the techniques already investigated in the literature, some perform better in particular regions of an image than others. Typically, subband/wavelet schemes are known to perform well in areas with texture, whereas techniques based on object representation, or morphological operators perform well in areas with sharp edges and contours. Similarly, methods using linear transforms produce poor results in areas with text or graphics. Dynamic coding is a solution to solve the drawbacks existing in a given scheme while still maintaining its strong performance where appropriate. The basic idea behind dynamic coding is simple yet powerful [23]. The visual information (a frame of video, or a video object) is subdivided into several regions with similar suitability for a given compression method. Each region is encoded by using a multitude of compression techniques. Among all these techniques, the one which is the most efficient is cho-

sen, and the compressed bit stream of the region using the best coding technique is sent to the decoder along with information specifying which technique was chosen for its coding. As an example, in areas with texture, a subband/wavelet technique would be used, while areas with strong edges and contours will be coded with morphological-based or other more appropriate techniques. Similarly, text areas will use an encoding technique more appropriate for an efficient compression of such data.

The concept of dynamic coding implicitly defines a general coding syntax. Video objects are further segmented into regions, each represented by their respective representation model. The syntax therefore relies on two degrees of freedom, namely, the video object partition into its constituting regions and their associated representation models.

As depicted in Fig. 14, the resulting syntax is both open and flexible. Indeed, different classes of partitioning can be considered, ranging from the whole image as a single object to arbitrarily shaped video objects segmented into regions of predefined or arbitrary shapes. Additionally, each region resulting from a particular segmentation can be coded with respect to a model chosen from a multitude of representation methods. Figure 15 gives an example of dynamic coding of a rectangular still image by putting in competition a linear and a nonlinear subband decomposition scheme. As it can be seen from this figure, the highly texture regions are best represented when a linear filter bank is used for subband decomposition, while sharp edges and contours are better maintained by using a nonlinear filter bank. A dynamic approach applied to this image allows the use of the best configuration in the region where it is appropriate and produces the best results.

6 Conclusions

In this chapter an object-based video coding scheme was presented that supports arbitrarily shaped video objects, possibly with a lossless shape accuracy. To this end, a progressive polygonal contour approximation is integrated in a complete, consistent, coding scheme. In this context, various node motion estimation methods are used, and the application of the shape-adaptive DCT transform to residual error representation in a content-based, triangular mesh partition is described. The adaptation of this scheme to achieve progressive compression was

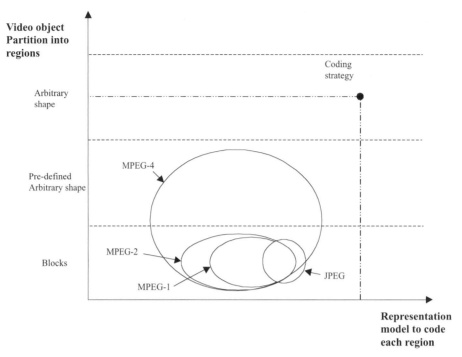

FIGURE 14 Dynamic coding principle.

also discussed and solutions were presented for geometric and motion representations. As opposed to standard compression schemes, including MPEG-4, the proposed scheme supports a separate, content based, quality-scalable syntax where the most salient information is transmitted first and the shape, motion, and texture information fields can be accessed separately in the bit stream. This hierarchical, semantic organization of the encoded data is of particular interest for content-based indexing and retrieval applications, while still allowing an efficient implementation of this scheme for compression and object-based interactivity.

The second part of the chapter briefly introduced the notion of dynamic coding of visual information. Dynamic coding offers the opportunity of combining several compression techniques on different objects or regions where most appropriate. It was shown in a simple example how a dynamic coding approach can provide superior results by a clever combination of algorithms in regions where specific compression techniques produce better results.

Acknowledgment

The object-based video coding scheme presented in this chapter is the result of the Ph.D. thesis of Corinne Le Buhan. Many materials of this chapter benefited from her work. The dynamic

(a) (b) (c)

FIGURE 15 Dynamic coding of a still picture at a compression factor of 50:1. (a) Compressed by a linear subband decomposition; (b) compressed by a nonlinear subband decomposition; (c) compressed by dynamic coding that pits the two methods against each other.

coding results were obtained in the framework of the Ph.D. theses of Olivier Egger and Emmanuel Reusens. Both authors acknowledge the valuable inputs from these sources.

References

[1] D. Marr, *Vision — A Computational Investigation into the Human Representation and Processing of Visual Information* (Freeman, San Francisco, 1982, ISBN 0-7167-1567-8).

[2] L. Spillmann and J. S. Werner, *Visual Perception — The Neurophysiological Foundations* (Academic, New York, 1990, ISBN 0-12-657676-9).

[3] M. Kunt, A. Ikonomopoulos, and M. Kocher, "Second generation image coding techniques," *Proc. IEEE* **73**, 549–675 (1985).

[4] R. Castagno, "Video segmentation based on multiple features for interactive and automatic multimedia applications," Ph.D. thesis 1894 (EPFL, Lausanne, Switzerland, 1998).

[5] N. Negroponte, *Being Digital* (Hodder & Stoughton, London, 1995).

[6] T. Ebrahimi, "MPEG-4 Video Verification Model: a video encoding/decoding algorithm based on content representation," *Signal Process. Image Commun.*, Special issue on MPEG-4, **9**, 367–384 (1997).

[7] Y. Altunbasak and A. M. Tekalp, "Scalable mesh-based interpolative coding of synthetic and natural image objects," in *Visual Communications and Image Processing, Proc. SPIE* **3024**, 1004–1011 (1997).

[8] Y. Altunbasak and A. M. Tekalp, "Closed-form connectivity preserving solutions for motion compensation using 2-D meshes," *IEEE Trans. Image Process.* **6**, 1255–1269 (1997).

[9] M. Dudon, O. Avaro, and C. Roux, "Triangular active mesh for motion estimation," *Signal Process. Image Commun.* **10**, 21–41 (1997).

[10] B. Gunsel, A. M. Tekalp, and P. J. L. van Beek, "Moving visual representations of video objects for content-based search and browsing," *IEEE Int. Conf. Image Proc.* **2**, 502–505 (1997).

[11] P. Kauff and K. Schuur, "Shape-adaptive DCT with block-based DC separation and ΔDC correction," *IEEE Trans. Circuits Syst. Video Technol.* **8**, 237–242 (1998).

[12] C. Le Buhan Jordan, T. Ebrahimi, and M. Kunt, "Progressive content-based shape compression for retrieval of binary images," *Computer Vis. Image Understand.*, Special issue on computer vision applications for network-centric computing, **71**, 198–212 (1998).

[13] C. Le Buhan, "Progressive geometrical compression of arbitrary shaped video objects," Ph.D. thesis 1891 (EPFL, Lausanne, Switzerland, 1998; http://ltswww.epfl.ch).

[14] Y. Nakaya and H. Hirashima, "Motion compensation based on spatial transformations," *IEEE Trans. Circuits Syst. Video Technol.* **4**, 339–356 (1994).

[15] J. Nieweglowski, T. G. Campbell, and P. Haavisto, "A novel video coding scheme based on temporal prediction using digital image warping," *IEEE Trans. Consumer Electron.* **39**, 141–150 (1993).

[16] K. Schröder, "Vertex tracking for grid-based motion compensation," in *Visual Communications and Image Processing, Proc. SPIE* **2952**, 264–275 (1996).

[17] J. R. Shewchuk, "Triangle: engineering a 2-D quality mesh generator and Delaunay triangulator," Tech. Rep., (Carnegie Mellon University, Pittsburgh, PA, 1996; http://www.cs.cmu.edu/~quake/triangle.html).

[18] L. Torres and M. Kunt, *Video Coding: The Second Generation Approach* (Kluwer, Boston, 1996).

[19] P. J. L. van Beek and A. M. Tekalp, "Object-based video coding using forward tracking 2-D mesh layers," in *SPIE Visual Communications and Image Processing, Proc. SPIE* **3024**, 699–710 (1997).

[20] T. Ebrahimi and M. Kunt, "Visual data compression for multimedia applications," *Proc. IEEE* **86**, 1109–1125 (1998).

[21] Y. Wang, O. Lee, and A. Vetro, "Use of two-dimensional deformable mesh structures for video coding, Part II – the analysis problem and a region-based coder employing an active mesh representation," *IEEE Trans. Circuits Syst. Video Technol.* **6**, 647–659 (1996).

[22] C. Le Buhan, T. Ebrahimi, and M. Kunt, "Progressive mesh-based coding of arbitrary-shaped video objects," in *Visual Communications and Image Processing, Proc. SPIE* **3653**, 1190–1201 (1999).

[23] E. Reusens, T. Ebrahimi, and M. Kunt, "Dynamic coding of visual information," *IEEE Trans. Circuits Syst. Video Technol.* **7**, 489–500 (1997).

6.4

MPEG-1 and MPEG-2 Video Standards

Supavadee Aramvith
and Ming-Ting Sun
University of Washington

1 MPEG-1 Video Coding Standard

1.1 Introduction

1.1.1 Background and Structure of MPEG-1 Standards Activities

The development of digital video technology in the 1980s has made it possible to use digital video compression in various kinds of applications. The effort to develop standards for coded representation of moving pictures, audio, and their combination is carried out in the Moving Picture Experts Group (MPEG). MPEG is a group formed under the auspices of the International Organization for Standardization (ISO) and the International Electrotechnical Commission (IEC). It operates in the framework of the Joint ISO/IEC Technical Committee 1 (JTC 1) on Information Technology, which was formerly Working Group 11 (WG11) of Sub-Committee 29 (SC29). The premise is to set the standard for coding moving pictures and the associated audio for digital storage media at ~1.5 Mbit/s so that a movie can be compressed and stored in a CD-ROM (Compact Disc-Read Only Memory). The resultant standard is the international standard for moving picture compression, ISO/IEC 11172 or MPEG-1 (Moving Picture Experts Group-Phase 1). MPEG-1 standards consist of five parts, including systems (11172-1), video (11172-2), audio (11172-3), conformance testing (11172-4), and software simulation (11172-5). In this chapter, we will focus only on the video part.

The activity of the MPEG committee started in 1988 based on the work of ISO JPEG (Joint Photographic Experts Group) [1] and CCITT Recommendation H.261: "Video Codec for Audiovisual Services at $p \times 64$ kbits/s" [2]. Thus, the MPEG-1 standard has much in common with the JPEG and H.261 standards. The MPEG development methodology was similar to that of H.261 and was divided into three phases: requirements, competition, and convergence [3]. The purpose of the requirements phase is to precisely set the focus of the effort and determine the rule for the competition phase. The document of this phase is a "proposal package description" [4] and a test methodology [5]. The next step is the competition phase, in which the goal is to obtain state of the art technology from the best of academic and industrial research. The criteria are based on the technical merits and the tradeoff between video quality and the cost of implementation of the ideas and the subjective test [5]. After the competition phase, various ideas and techniques are integrated into one solution in the convergence phase. The solution results in a document called the simulation model. The simulation model implements, in some sort of programming language, the operation of a reference encoder and a decoder. The simulation model is used to carry out simulations to optimize the performance of the coding scheme [6]. A series of fully documented experiments called core experiments are then carried out. The MPEG committee reached the Committee Draft (CD) status in September 1990 and the Committee Draft (CD 11172) was approved in December 1991. International Standard (IS) 11172 for the first three

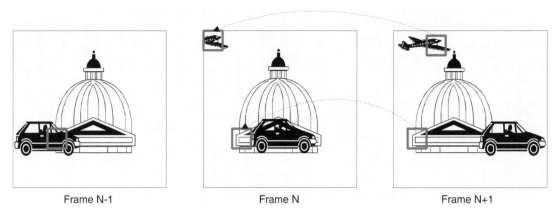

Frame N-1   Frame N   Frame N+1

FIGURE 1   A video sequence, showing the benefits of bidirectional prediction.

parts was established in November 1992. The IS for the last two parts was finalized in November 1994.

1.1.2 MPEG-1 Target Applications and Requirements

The MPEG standard is a generic standard, which means that it is not limited to a particular application. A variety of digital storage media applications of MPEG-1 have been proposed based on the assumptions that the acceptable video and audio quality can be obtained for a total bandwidth of ~1.5 Mbits/s. Typical storage media for these applications include CD-ROM, DAT (digital audio tape), Winchester-type computer disks, and writable optical disks. The target applications are asymmetric applications in which the compression process is performed once and the decompression process is required often. Examples of the asymmetric applications include video CD, video on demand, and video games. In these asymmetric applications, the encoding delay is not a concern. The encoders are needed only in small quantities, whereas the decoders are needed in large volumes. Thus, the encoder complexity is not a concern, whereas the decoder complexity has to be low in order to result in low-cost decoders.

The requirements for compressed video in digital storage media mandate several important features of the MPEG-1 compression algorithm. The important features include normal playback, frame-based random access and editing of video, reverse playback, fast forward/reverse play, encoding high-resolution still frames, robustness to uncorrectable errors, etc. The applications also require MPEG-1 to support flexible picture sizes and frame rates. Another requirement is that the encoding process can be performed in reasonable speed by using existing hardware technologies and that the decoder can be implemented by using a small number of chips at low cost.

Because MPEG-1 video coding algorithm is based heavily on H.261, in the following sections we will focus only on those that are different from H.261.

1.2 MPEG-1 Video Coding Versus H.261

1.2.1 Bidirectional Motion Compensated Prediction

In H.261, only the previous video frame is used as the reference frame for the motion compensated prediction (forward predic-

tion). MPEG-1 allows the future frame to be used as the reference frame for the motion compensated prediction (backward prediction), which can provide better prediction. For example, as shown in Fig. 1, if there are moving objects, and if only the forward prediction is used, there will be uncovered areas (such as the block behind the car in Frame N) for which we may not be able to find a good matching block from the previous reference picture (Frame $N-1$). In contrast, the backward prediction can properly predict these uncovered areas since they are available in the future reference picture, i.e., frame $N+1$ in this example. As also shown in Fig. 1, if there are objects moving into the picture (the airplane in the figure), then these new objects cannot be predicted from the previous picture but can be predicted from the future picture.

1.2.2 Motion Compensated Prediction with Half-Pixel Accuracy

The motion estimation in H.261 is restricted to only integer-pixel accuracy. However, a moving object often moves to a position that is not on the pixel grid but between the pixels. MPEG-1 allows half-pixel-accuracy motion vectors. By estimating the displacement at a finer resolution, we can expect improved prediction and, thus, better performance than motion estimation with integer-pixel accuracy. As shown in Fig. 2, since there is no pixel value at the half-pixel locations, interpolation is required to produce the pixel values at the half-pixel positions. Bilinear interpolation is used in MPEG-1 for its simplicity. As in H.261, the motion estimation is performed only on luminance blocks.

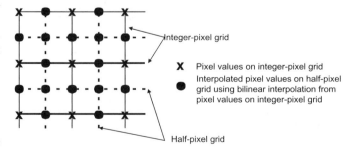

FIGURE 2   Half-pixel motion estimation.

The resulting motion vector is scaled by 2 and applied to the chrominance blocks. This reduces the computation but may not necessarily be optimal. Motion vectors are differentially encoded with respect to the motion vector in the preceding adjacent macroblock. The reason is that the motion vectors of adjacent regions are highly correlated, as it is quite common to have relatively uniform motion over areas of picture.

1.3 MPEG-1 Video Structure

1.3.1 Source Input Format

The typical MPEG-1 input format is the source input format (SIF). SIF was derived from CCIR601, a worldwide standard for digital TV studio. CCIR601 specifies the Y Cb Cr color coordinate, where Y is the luminance component (black and white information), and Cb and Cr are two color difference signals (chrominance components). A luminance sampling frequency of 13.5 MHz was adopted. There are several Y Cb Cr sampling formats, such as 4:4:4, 4:2:2, 4:1:1, and 4:2:0. In 4:4:4, the sampling rates for Y, Cb, and Cr are the same. In 4:2:2, the sampling rates of Cb and Cr are half of that of Y. In 4:1:1 and 4:2:0, the sampling rates of Cb and Cr are one quarter of that of Y. The positions of Y Cb Cr samples for 4:4:4, 4:2:2, 4:1:1, and 4:2:0 are shown in Fig. 3.

Converting analog TV signals to digital video with the 13.5-MHz sampling rate of CCIR601 results in 720 active pixels per line (576 active lines for PAL (Phase Alternating Line) and 480 active lines for NTSC (National Television System Committee)). This results in a 720×480 resolution for NTSC and a 720×576 resolution for PAL. With 4:2:2, the uncompressed bit rate for transmitting CCIR601 at 30 frames/s is then ~166 Mbits/s. Since it is difficult to compress a CCIR601 video to 1.5 Mb/s with good video quality, in MPEG-1, typically the source video resolution is decimated to a quarter of the CCIR601 resolution by filtering and subsampling. The resultant format is called source input format, which has a 360×240 resolution for NTSC and a 360×288 resolution for PAL. Since in the video coding algorithm the block size of 16×16 is used for motion compensated prediction, the number of pixels in both the horizontal and the vertical dimensions should be multiples of 16. Thus, the four leftmost and rightmost pixels are discarded to give a 352×240 resolution for NTSC systems (30 frames/s) and a 352×288 resolution for PAL systems (25 frames/s). The chrominance signals have half of the above resolutions in both the horizontal and vertical dimensions

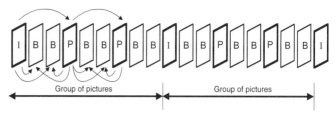

FIGURE 4   MPEG group of pictures.

(4:2:0, 176×120 for NTSC and 176×144 for PAL). The uncompressed bit rate for SIF (NTSC) at 30 frames/s is ~30.4 Mbits/s.

1.3.2 Group Of Pictures and I-B-P Pictures

In MPEG, each video sequence is divided into one or more groups of pictures (GOPs). There are four types of pictures defined in MPEG-1: I, P, B, and D pictures, of which the first three are shown in Fig. 4. Each GOP is composed of one or more pictures; one of these pictures must be an I picture. Usually, the spacing between two anchor frames (I or P pictures) is referred to as M, and the spacing between two successive I pictures is referred to as N. In Fig. 4, $M = 3$ and $N = 9$.

I pictures (intracoded pictures) are coded independently with no reference to other pictures. I pictures provide random access points in the compressed video data, since the I pictures can be decoded independently without referencing to other pictures. With I pictures, an MPEG bit stream is more editable. Also, error propagation due to transmission errors in previous pictures will be terminated by an I picture, since the I picture does not have a reference to the previous pictures. Since I pictures use only transform coding without motion compensated predictive coding, it provides only moderate compression.

P pictures (predictive-coded pictures) are coded by using the forward motion-compensated prediction similar to that in H.261 from the preceding I or P picture. P pictures provide more compression than the I pictures by virtue of motion-compensated prediction. They also serve as references for B pictures and future P pictures. Transmission errors in the I pictures and P pictures can propagate to the succeeding pictures, because the I pictures and P pictures are used to predict the succeeding pictures.

B pictures (bidirectional-coded pictures) allow macroblocks to be coded by using bidirectional motion-compensated prediction from both the past and future reference I or P pictures. In the B pictures, each bidirectional motion-compensated macroblock can have two motion vectors: a forward motion vector, which references to a best matching block in the previous I or P pictures, and a backward motion vector, which references to a best matching block in the next I or P pictures as shown in Fig. 5. The motion compensated prediction can be formed by the average of the two referenced motion compensated blocks. By averaging between the past and the future reference blocks, the effect of noise can be decreased. B pictures provide the best compression

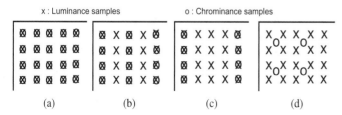

(a)	(b)	(c)	(d)

x : Luminance samples   o : Chrominance samples

FIGURE 3   Luminance and chrominance samples in formats for (a) 4:4:4, (b) 4:2:2, (c) 4:1:1, (d) 4:2:0.

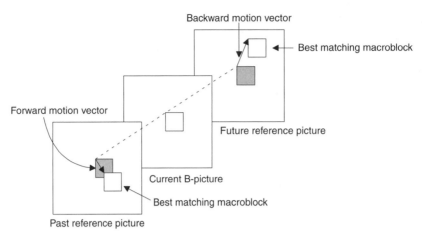

FIGURE 5   Bidirectional motion estimation.

compared to I and P pictures. I and P pictures are used as reference pictures for predicting B pictures. To keep the structure simple and since there is no apparent advantage to use B pictures for predicting other B pictures, the B pictures are not used as reference pictures. Hence, B pictures do not propagate errors.

D pictures (DC pictures) are low-resolution pictures obtained by decoding only the DC coefficient of the discrete cosine transform coefficients of each macroblock. They are not used in combination with I, P, or B pictures. D pictures are rarely used, but they are defined to allow fast searches on sequential digital storage media.

The tradeoff of having frequent B pictures is that it decreases the correlation between the previous I or P picture and the next reference P or I picture. It also causes coding delay and increases the encoder complexity. With the example shown in Fig. 4 and Fig. 6, at the encoder, if the order of the incoming pictures is 1, 2, 3, 4, 5, 6, 7, . . . , the order of coding the pictures at the encoder will be 1, 4, 2, 3, 7, 5, 6, ... At the decoder, the order of the decoded pictures will be 1, 4, 2, 3, 7, 5, 6, ... However, the display order after the decoder should be 1, 2, 3, 4, 5, 6, 7. Thus, frame memories have to be used to put the pictures in the correct order. This picture reordering causes delay. The computation of bidirectional motion vectors and the picture-reordering frame memories increase the encoder complexity.

In Fig. 6, two types of GOPs are shown. GOP1 can be decoded without referencing other GOPs. It is called a closed GOP. In GOP2, to decode the eighth B and ninth B pictures, the seventh

P picture in GOP1 is needed. GOP2 is called an open GOP, which means the decoding of this GOP has to reference other GOPs.

1.3.3 Slice, Macroblock, and Block Structures

An MPEG picture consists of slices. A slice consists of a contiguous sequence of macroblocks in a raster scan order (from left to right and from top to bottom). In an MPEG coded bit stream, each slice starts with a slice header, which is a clear codeword (a clear codeword is a unique bit pattern that can be identified without decoding the variable-length codes in the bit stream). As a result of the clear-codeword slice header, slices are the lowest level of units that can be accessed in an MPEG coded bit stream without decoding the variable-length codes. Slices are important in the handling of channel errors. If a bit stream contains a bit error, the error may cause error propagation because of the variable-length coding. The decoder can regain synchronization at the start of the next slice. Having more slices in a bit stream allows better error termination, but the overhead will increase.

A macroblock consists of a 16 × 16 block of luminance samples and two 8 × 8 block of corresponding chrominance samples as shown in Fig. 7. A macroblock thus consists of four 8 × 8 Y blocks, one 8 × 8 Cb block, and one 8 × 8 Cr block. Each coded macroblock contains motion-compensated prediction information (coded motion vectors and the prediction errors). There are four types of macroblocks: intra, forward predicted, backward predicted, and averaged macroblocks. The motion information consists of one motion vector for forward- and backward-predicted macroblocks and two motion vectors for bidirectionally predicted (or averaged) macroblocks. P pictures can have intra- and forward-predicted macroblocks. B pictures can have all four types of macroblocks. The first and last macroblocks in a slice must always be coded. A macroblock is designated as a skipped macroblock when its motion vector is zero and all the quantized DCT coefficients are zero. Skipped macroblocks are not allowed in I pictures. Nonintracoded macroblocks in P and B pictures can be skipped. For a skipped macroblock, the decoder just copies the macroblock from the previous picture.

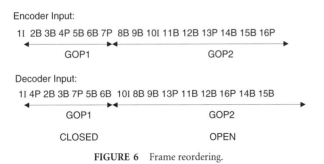

FIGURE 6   Frame reordering.

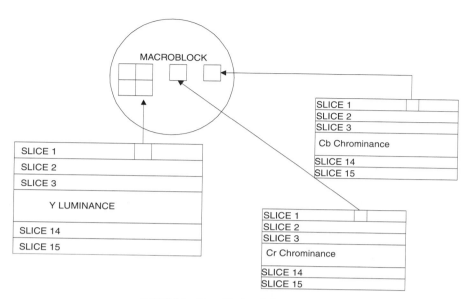

FIGURE 7 Macroblock and slice structures.

1.4 Summary of the Major Differences Between MPEG-1 Video and H.261

Compared with H.261, MPEG-1 video differs in the following aspects.

1. MPEG-1 uses bidirectional motion-compensated predictive coding with half-pixel accuracy, whereas H.261 has no bidirectional prediction (B pictures) and the motion vectors are always in integer-pixel accuracy.
2. MPEG-1 supports the maximum motion vector range of -512 to $+511.5$ pixels for half-pixel motion vectors and -1024 to $+1023$ for integer-pixel motion vectors, whereas H.261 has a maximum range of only ± 15 pixels.
3. MPEG-1 uses visually weighted quantization based on the fact that the human eye is more sensitive to quantization errors related to low spatial frequencies than to high spatial frequencies. MPEG-1 defines a default 64-element quantization matrix, but it also allows custom matrices appropriate for different applications. H.261 has only one quantizer for the intra-DC coefficient and 31 quantizers for all other coefficients.
4. H.261 only specifies two source formats: CIF (common intermediate format; 352×288 pixels) and QCIF (quarter CIF; 176×144 pixels). In MPEG-1, the typical source format is SIF (352×240 for NTSC, and 352×288 for PAL). However, the users can specify other formats. The picture size can be as large as $4k \times 4k$ pixels. There are certain parameters in the bit streams that are left flexible, such as the number of lines per picture (less than 4096), the number of pels per line (less than 4096), picture rate (24, 25, and 30 frames/s), and 14 choices of pel aspect ratios.
5. In MPEG-1, I, P, and B pictures are organized as a flexible group of pictures.

6. MPEG-1 uses a flexible slice structure instead of group of blocks (GOB) as defined in H.261.
7. MPEG-1 has D pictures to allow the fast-search option.
8. In order to allow cost-effective implementation of user terminals, MPEG-1 defines a constrained parameter set, which lays down specific constraints, as listed in Table 1.

1.5 Simulation Model

Similar to H.261, MPEG-1 specifies only the syntax and the decoder. Many detailed coding options such as the rate-control strategy, the quantization decision levels, the motion estimation schemes, and coding modes for each macroblock are not specified. This allows future technology improvement and product differentiation. In order to have a reference MPEG-1 video quality, simulation models were developed in MPEG-1. A simulation model contains a specific reference implementation of the MPEG-1 encoder and decoder, including all the details that are not specified in the standard. The final version of the MPEG-1 simulation model is "simulation model 3" (SM3) [7]. In SM3, the motion estimation technique uses one forward or one backward motion vector per macroblock with half-pixel accuracy. A two-step search scheme, which consists of a full-search in the range of ± 7 pixels with the integer-pixel precision, followed

TABLE 1 MPEG-1 constrained parameter set

Parameter	Constraint
Horiz. size	≤ 720 pels
Vert. size	≤ 576 pels
Total No. of Macroblocks/picture	≤ 396
Total No. of Macroblocks/second	$\leq 396 \times 25 = 330 \times 30$
Picture rate	≤ 30 frames/s
Bit rate	≤ 1.86 Mbits/s
Decoder Buffer	≤ 376832 bits

by a search in eight neighboring half-pixel positions, is used. The decision of the coding mode for each macroblock (whether or not it will use motion-compensated prediction and intra or inter coding), the quantizer decision levels, and the rate-control algorithm are all specified.

1.6 MPEG-1 Video Bit-Stream Structures

As shown in Fig. 8, there are six layers in the MPEG-1 video bit stream: the video sequence, group of pictures, picture, slice, macroblock, and block layers.

A video sequence layer consists of a sequence header, one or more groups of pictures, and an end-of-sequence code. It contains the setting of the following parameters: the picture size (horizontal and vertical sizes), pel aspect ratio, picture rate, bit rate, the minimum decoder buffer size (video buffer verifier size), constraint parameters flag (this flag is set only when the picture size, picture rate, decoder buffer size, bit rate, and motion parameters satisfy the constraints bound in Table 1), the control for the loading of sixty-four 8-bit values for intra and nonintra quantization tables, and the user data.

The GOP layer consists of a set of pictures that are in a continuous display order. It contains the setting of the following parameters: the time code, which gives the hours-minutes-seconds time interval from the start of the sequence; the closed GOP flag, which indicates whether the decoding operation requires pictures from the previous GOP for motion compensation; the broken link flag, which indicates whether the previous GOP can be used to decode the current GOP; and the user data.

The picture layer acts as a primary coding unit. It contains the setting of the following parameters: the temporal reference, which is the picture number in the sequence and is used to determine the display order; the picture types (I/P/B/D); the decoder buffer initial occupancy, which gives the number of bits that must be in the compressed video buffer before the idealized decoder model defined by MPEG decodes the picture (it is used to prevent the decoder buffer overflow and underflow); the forward motion vector resolution and range for P and B pictures; the backward motion vector resolution and range for B pictures; and the user data.

The slice layer acts as a resynchronization unit. It contains the slice vertical position where the slice starts, and the quantizer scale that is used in the coding of the current slice.

The macroblock layer acts as a motion compensation unit. It contains the setting of the following parameters: the optional stuffing bits, the macroblock address increment, the macroblock type, quantizer scale, motion vector, and the coded block pattern, which defines the coding patterns of the six blocks in the macroblock.

The block layer is the lowest layer of the video sequence and consists of coded 8×8 DCT coefficients. When a macroblock is encoded in the intra mode, the DC coefficient is encoded similar to that in JPEG (the DC coefficient of the current macroblock is predicted from the DC coefficient of the previous macroblock). At the beginning of each slice, predictions for DC coefficients for luminance and chrominance blocks are reset to 1024. The differential DC values are categorized according to their absolute values and the category information is encoded using VLC (variable-length code). The category information indicates the number of additional bits following the VLC to represent the prediction residual. The AC coefficients are encoded similar to that in H.261, using a VLC to represent the zero run length and the value of the nonzero coefficient. When a macroblock is encoded in nonintra modes, both the DC and AC coefficients are encoded similar to that in H.261.

Above the video sequence layer, there is a system layer in which the video sequence is packetized. The video and audio bit streams are then multiplexed into an integrated data stream. These are defined in the systems part.

1.7 Summary

MPEG-1 is mainly for storage media applications. Because of the use of B picture, it may result in a long end-to-end delay.

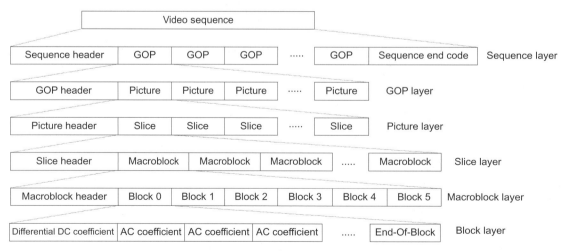

FIGURE 8 MPEG-1 bit-stream syntax layers.

The MPEG-1 encoder is much more expensive than the decoder because of the large search range, the half-pixel accuracy in motion estimation, and the use of the bidirectional motion estimation. The MPEG-1 syntax can support a variety of frame rates and formats for various storage media applications. Similar to other video coding standards, MPEG-1 does not specify every coding option (motion estimation, rate control, coding modes, quantization, preprocessing, postprocessing, etc.). This allows for continuing technology improvement and product differentiation.

2 MPEG-2 Video Coding Standard

2.1 Introduction

2.1.1 Background and Structure of MPEG-2 Standards Activities

The MPEG-2 standard represents the continuing efforts of the MPEG committee to develop generic video and audio coding standards after their development of MPEG-1. The idea of this second phase of MPEG work came from the fact that MPEG-1 is optimized for applications at ∼1.5 Mb/s with input source in SIF, which is a relatively low-resolution progressive format. Many higher quality, higher bit-rate applications require a higher resolution digital video source such as CCIR601, which is an interlaced format. New techniques can be developed to code the interlaced video better.

The MPEG-2 committee started working in late 1990 after the completion of the technical work of MPEG-1. The competitive tests of video algorithms were held in November 1991, followed by the collaborative phase. The Committee Draft (CD) for the video part was achieved in November 1993. The MPEG-2 standard (ISO/IEC 13818) [8] currently consists of nine parts. The first five parts are organized in the same fashion as MPEG-1: systems, video, audio, conformance testing, and simulation software technical report. The first three parts of MPEG-2 reached International Standard (IS) status in November 1994. Parts 4 and 5 were approved in March 1996. Part 6 of the MPEG-2 standard specifies a full set of digital storage media control commands (DSM-CC). Part 7 is the specification of a nonbackward compatible audio. Part 8 was originally planned to be the coding of 10-bit video but was discontinued. Part 9 is the specification of real-time interface (RTI) to transport stream decoders, which may be utilized for adaptation to all appropriate networks carrying MPEG-2 transport streams. Parts 6 and 9 have already been approved as International Standards in July 1996. Like the MPEG-1 standard, MPEG-2 video coding standard specifies only bit-stream syntax and the semantics of the decoding process. Many encoding options were left unspecified to encourage continuing technology improvement and product differentiation.

MPEG-3, which was originally intended for HDTV (high-definition digital television) at higher bit rates, was merged with MPEG-2. Hence there is no MPEG-3. The MPEG-2 video coding standard (ISO/IEC 13818-2) was also adopted by ITU-T, as ITU-T Recommendation H.262 [9].

2.1.2 Target Applications and Requirements

MPEG-2 is primarily targeted at coding high-quality video at 4–15 Mb/s for video on demand (VOD), digital broadcast television, and digital storage media such as DVD (digital versatile disc). It is also used for coding HDTV, cable/satellite digital TV, video services over various networks, two-way communications, and other high-quality digital video applications.

The requirements from MPEG-2 applications mandate several important features of the compression algorithm. Regarding picture quality, MPEG-2 has to be able to provide good NTSC quality video at a bit rate of approximately 4–6 Mbits/s and transparent NTSC quality video at a bit rate of approximately 8–10 Mbits/s. It also has to provide the capability of random access and quick channel switching by means of inserting I pictures periodically. The MPEG-2 syntax also has to support trick modes, e.g., fast forward and fast reverse play, as in MPEG-1. Low-delay mode is specified for delay-sensitive visual communications applications. MPEG-2 has scalable coding modes in order to support multiple grades of video quality, video formats, and frame rate for various applications. Error resilience options include intra motion vector, data partitioning, and scalable coding. Compatibility between the existing and the new standard coders is another prominent feature provided by MPEG-2. For example, MPEG-2 decoders should be able to decode MPEG-1 bit streams. If scalable coding is used, the base layer of MPEG-2 signals can be decoded by a MPEG-1 decoder. Finally, it should allow reasonable complexity encoders and low-cost decoders be built with mature technology. Since MPEG-2 video is based heavily on MPEG-1. In the following sections, we will focus only on those features which are different from MPEG-1 video.

2.2 MPEG-2 Profiles and Levels

MPEG-2 standard is designed to cover a wide range of applications. However, features needed for some applications may not be needed for other applications. If we put all the features into one single standard, it may result in an overly expensive system for many applications. It is desirable for an application to implement only the necessary features to lower the cost of the system. To meet this need, MPEG-2 classified the groups of features for important applications into profiles. A profile is defined as a specific subset of the MPEG-2 bit-stream syntax and functionality to support a class of applications (e.g., low-delay video conferencing applications, or storage media applications). Within each profile, levels are defined to support applications that have different quality requirements (e.g., different resolutions). Levels are specified as a set of restrictions on some of the parameters (or their combination) such as sampling rates, frame dimensions, and bit rates in a profile. Applications are implemented in the allowed range of values of a particular profile at a particular level.

TABLE 2 Profiles and levels

Level	Profile				
	Simple 4:2:0	Main 4:2:0	SNR scalable 4:2:0	Spatially scalable 4:2:0	High 4:2:0 or 4:2:2
High 1920 × 1152 (60 frames/s)		62.7 Ms/s 80 Mbit/s			100 Mbit/s for 3 layers
High-1440 1440 × 1152 (60 frames/s)		47 Ms/s 60 Mbit/s		47 Ms/s 60 Mbit/s for 3 layers	80 Mbit/s for 3 layers
Main 720 × 576 (30 frames/s)	10.4 Ms/s 15 Mbit/s	10.4 Ms/s 15 Mbit/s	10.4 Ms/s 15 Mbit/s for 2 layers		20 Mbit/s for 3 layers
Low 352 × 288 (30 frames/s)		3.04 Ms/s 4 Mbit/s	3.04 Ms/s 4 Mbit/s for 2 layers		

Table 2 shows the combination of profiles and levels that are defined in MPEG-2. MPEG-2 defines seven distinct profiles: simple, main, SNR scalable, spatially scalable, high, 4:2:2, and multiview. The last two profiles were developed after the final approval of MPEG-2 video in November 1994. Simple profile is defined for low-delay video conferencing applications. Main profile is the most important and widely used profile for general high-quality digital video applications such as VOD, DVD, Digital TV, and HDTV. SNR (signal-to-noise ratio) scalable profile supports multiple grades of video quality. Spatially scalable profile supports multiple grades of resolutions. High profile supports multiple grades of quality, resolution, and chroma format. Four levels are defined within the profiles: low (for SIF resolution pictures), main (for CCIR601 resolution pictures), high-1440 (for European HDTV resolution pictures), and high (for North American HDTV resolution pictures). The 11 combinations of profiles and levels in Table 2 define the MPEG-2 conformance points that cover most practical MPEG-2 target applications. The numbers in each conformance point indicate the maximum bound of the parameters. The number in the first line indicates the luminance rate in samples/s. The number in the second line indicates bit rate in bits/s. Each conformance point is a subset of the conformance point at the right or above. For example, a main-profile main-level decoder should also decode simple-profile main-level and main-profile low-level bit streams. Among the defined profiles and levels, main-profile at main-level (MP@ML) is used for digital television broadcast in CCIR601 resolution and DVD-video. The main-profile at high-level (MP@HL) is used for HDTV. The 4:2:2 profile is defined to support the pictures with a color resolution of 4:2:2 for higher bit-rate studio applications. Although the high profile supports 4:2:2 also, a high-profile codec has to support SNR scalable profile and spatially scalable profile. This makes the high-profile codec expensive. The 4:2:2 profile does not have to support the scalabilities and thus will be much cheaper to implement. Multiview profile is defined to support the efficient encoding of the application involving two video sequences from two cameras shooting the same scene with a small angle between them.

2.3 MPEG-2 Video Input Resolutions and Formats

Although the main concern of the MPEG-2 committee is to support the CCIR601 resolution, which is the digital TV resolution, MPEG-2 allows a maximum picture size of 16k × 16k pixels. It also supports the frame rates of 23.976, 24, 25, 29.97, 30, 50, 59.94 and 60 Hz, as in MPEG-1. MPEG-2 is suitable for coding progressive video format as well as interlaced video format. As for the color subsampling formats, MPEG-2 supports 4:2:0, 4:2:2, and 4:4:4. MPEG-2 uses the 4:2:0 format as in MPEG-1, except that there is a difference in the positions of the chrominance samples as shown in Figs. 9(a) and 9(b). On one hand, in MPEG-1, a slice can cross macroblock row boundaries. Therefore, a single slice in MPEG-1 can be defined to cover the entire picture. On the other hand, slices in MPEG-2 begin and end in the same horizontal row of macroblocks. There are two types of slice structure in MPEG-2: the general and the restricted slice structures. In the general slice structure, MPEG-2 slices need not cover the entire picture. Thus, only the regions enclosed in the slices are encoded. In the restricted slice structure, every macroblock in the picture shall be enclosed in a slice.

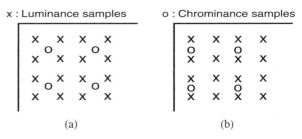

FIGURE 9 Position of luminance and chrominance samples for 4:2:0 format in (a) MPEG-1, (b) MPEG-2.

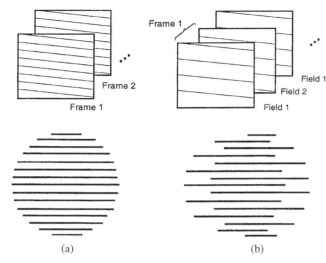

FIGURE 10 (a) Progressive scan, (b) interlaced scan.

2.4 MPEG-2 Video Coding Standard Compared with MPEG-1

2.4.1 Interlaced Versus Progressive Video

Figure 10 shows the progressive and interlaced video scan. In the interlaced video, each displayed frame consists of two interlaced fields. For example, frame 1 consists of field 1 and field 2, with the scanning lines in field 1 located between the lines of field 2. In contrast, the progressive video has all the lines of a picture displayed in one frame. There are no fields or half-pictures as with the interlaced scan. Thus, progressive video requires a higher picture rate than the frame rate of an interlaced video, to avoid a flickery display. The main disadvantage of the interlaced format is that when there are object movements, the moving object may appear distorted when we merge two fields into a frame. For example, Fig. 10 shows a moving ball. In the interlaced format, because the moving ball will be at different locations in the two fields, when we put the two fields into a frame, the ball will look distorted. Using MPEG-1 to encode the distorted objects in the frames of the interlaced video will not produce the optimal results. Interlaced video also tends to cause horizontal picture details to dither and thus introduces more high-frequency noises.

2.4.2 Interlaced Video Coding

Figure 11 shows the interlaced video format. As explained earlier, an interlaced frame is composed of two fields. From the figure, the top field (field 1) occurs earlier in time than the bottom field (field 2). Both fields together form a frame. In MPEG-2, pictures are coded as I, P, and B pictures, like in MPEG-1. To optimally encode the interlaced video, MPEG-2 can encode a picture either as a field picture or a frame picture. In the field-picture mode, the two fields in the frame are encoded separately. If the first field in a picture is an I picture, the second field in the picture can be either I or P pictures, as the second field can use the first field as a reference picture. However, if the first field in a picture is a P- or B-field picture, the second field has to be the same type of

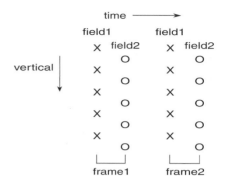

FIGURE 11 Interlaced video format.

picture. In a frame picture, two fields are interleaved into a picture and coded together as one picture, similar to the conventional coding of progressive video pictures. In MPEG-2, a video sequence is a collection of frame pictures and field pictures.

2.4.2.1 Frame-Based and Field-Based Motion-Compensated Prediction.
In MPEG-2, an interlaced picture can be encoded as a frame picture or as field pictures. MPEG-2 defines two different motion-compensated prediction types: frame-based and field-based motion-compensated prediction. Frame-based prediction forms a prediction based on the reference frames. Field-based prediction is made based on reference fields. For the simple profile in which the bidirectional prediction cannot be used, MPEG-2 introduced a dual-prime motion-compensated prediction to efficiently explore the temporal redundancies between fields. Figure 12 shows three types of motion-compensated prediction. Note that all motion vectors in MPEG-2 are specified with a half-pixel resolution.

Frame predictions in frame pictures: in the frame-based prediction for frame pictures, as shown in Fig. 12(a), the whole interlaced frame is considered as a single picture. It uses the same motion-compensated predictive coding method used in MPEG-1. Each 16 × 16 macroblock can have only one motion vector for each forward or backward prediction. Two motion vectors are allowed in the case of the bidirectional prediction.

Field prediction in a frame pictures: the field-based prediction in frame pictures considers each frame picture as two separate

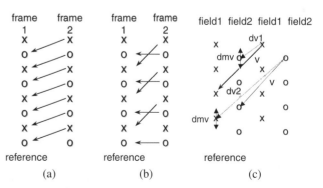

FIGURE 12 Three types of motion-compensated prediction: (a) frame, (b) field, (c) dual prime.

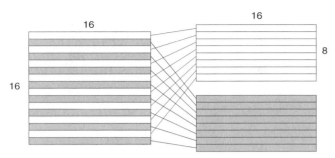

FIGURE 13   Blocks for frame-based or field-based prediction.

field pictures. Separate predictions are formed for each 16×8 block of the macroblock as shown in Fig. 13. Thus, field-based prediction in a frame picture requires two sets of motion vectors. A total of four motion vectors is allowed in case of bidirectional prediction. Each field prediction may select either the field 1 or the field 2 of the reference frame.

Field prediction in field pictures: in field-based prediction for field pictures, the prediction is formed from the two most recently decoded fields. The predictions are made from reference fields, independently for each field, with each field considered as an independent picture. The block size of prediction is 16×16; however, it should be noted that the 16×16 block in the field picture corresponds to a 16×32 pixel area in the frame picture. A field-based prediction in field pictures requires only one motion vector for each forward or backward prediction. Two motion vectors are allowed in the case of the bidirectional prediction.

16×8 prediction in field pictures: two motion vectors are used for each macroblock. The first motion vector is applied to the 16×8 block in field 1 and the second motion vector is applied to the 16×8 block in field 2. A total of four motion vectors is allowed in the case of bidirectional prediction.

Dual-prime motion-compensated prediction can be used only in P pictures. Once the motion vector "**v**" for a macroblock in a field of given parity (field 1 or field 2) is known relative to a reference field of the same parity, it is extrapolated or interpolated to obtain a prediction of the motion vector for the opposite parity reference field. In addition, a small correction is also made to the vertical component of the motion vectors to reflect the vertical shift between lines of the field 1 and field 2. These derived motion vectors are denoted by **dv1** and **dv2** (represented by dash line) in Fig. 12(c). Next, a small refinement differential motion vector, called "**dmv**", is added. The choice of **dmv** values $(-1, 0, +1)$ is determined by the encoder. The motion vector **v** and its corresponding **dmv** value are included in the bit stream so that the decoder can also derive **dv1** and **dv2**. In calculating the pixel values of the prediction, the motion-compensated predictions from the two reference fields are averaged, which tends to reduce the noise in the data. Dual-prime prediction is mainly for low-delay coding applications such as videophone and video conferencing. For low-delay coding using simple profile, B pictures should not be used. Without using bidirectional prediction, dual-prime prediction is developed for P pictures to provide a better prediction than the forward prediction.

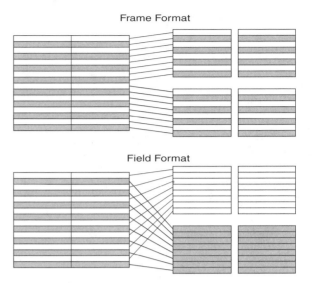

FIGURE 14   Frame/Field format block for DCT.

2.4.2.2 Frame/Field DCT.   MPEG-2 has two DCT modes: frame-based and field-based DCT, as shown in Fig. 14. In the frame-based DCT mode, a 16×16-pixel macroblock is divided into four 8×8 DCT blocks. This mode is suitable for the blocks in the background or in a still image that have little motion because these blocks have high correlation between pixel values from adjacent scan lines. In the field-based DCT mode, a macroblock is divided into four DCT blocks where the pixels from the same field are grouped together into one block. This mode is suitable for the blocks that have motion because, as explained, motion causes distortion and may introduce high-frequency noises into the interlaced frame.

2.4.2.3 Alternate Scan.   MPEG-2 defines two different zigzag scanning orders zigzag and alternate scans as shown in Fig. 15. The zigzag scan used in MPEG-1 is suitable for progressive images where the frequency components have equal importance in each horizontal and vertical direction. In MPEG-2, an alternate scan is introduced based on the fact that interlaced images tend to have higher frequency components in the vertical direction. Thus, the scanning order weighs more on the higher vertical frequencies than the same horizontal frequencies. In MPEG-2, the selection between these two zigzag scan orders can be made on a picture basis.

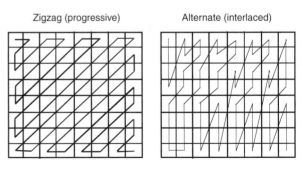

FIGURE 15   Progressive/interlaced scan.

2.5 Scalable Coding

Scalable coding is also called layered coding. In scalable coding, the video is coded in a base layer and several enhancement layers. If only the base layer is decoded, basic video quality can be obtained. If the enhancement layers are also decoded, enhanced video quality (e.g., higher SNR, higher resolution, higher frame rate) can be achieved. Scalable coding is useful for transmission over noisy channel since the more important layers (e.g., the base layer) can be better protected and sent over a channel with better error performance. Scalable coding is also used in video transport over variable-bit-rate channels. When the channel bandwidth is reduced, the less important enhancement layers will not be transmitted. It is also useful for progressive transmission, which means the users can get rough representations of the video fast with the base layer and then the video quality will be refined as more enhancement data arrive. Progress transmission is useful for database browsing and image transmission over the Internet.

MPEG-2 supports three types of scalability modes: SNR, spatial, and temporal scalability. Each of them is targeted at several applications with particular requirements. Different scalable modes can be combined into hybrid coding schemes such as hybrid spatial-temporal and hybrid spatial-SNR scalability. In a basic MPEG-2 scalability mode, there can be two layers of video: lower and enhancement layers. The hybrid scalability allows up to three layers.

2.5.1 SNR Scalability

MPEG-2 SNR scalability provides two different video qualities from a single video source while maintaining the same spatial and temporal resolutions. A block diagram of the two-layer SNR scalable encoder and decoder is shown in Figs. 16(a) and 16(b), respectively. In the base layer, the DCT coefficients are coarsely quantized and the coded bit stream is transmitted with moderate quality at a lower bit rate. In the enhancement layer, the difference between the nonquantized DCT coefficients and the coarsely quantized DCT coefficients from the lower layer is encoded with finer quantization step sizes. By doing this, the moderate video quality can be achieved by decoding only the lower-layer bit streams while the higher video quality can be achieved by decoding both layers.

2.5.2 Spatial Scalability

With Spatial scalability, the applications can support users with different resolution terminals. For example, the compatibility between SDTV (Standard Definition TV) and HDTV can be achieved with the SDTV being coded as the base layer. With the enhancement layer, the overall bit stream can provide the HDTV resolution. The input to the base layer usually is created by downsampling the original video to create a low-resolution video for providing the basic spatial resolution. The choice of video formats such as frame sizes, frame rate, or chrominance formats is flexible in each layer.

A block diagram of the two-layer spatial scalable encoder and decoder is shown in Figs. 17(a) and 17(b), respectively. In the base layer, the input video signal is downsampled by spatial decimation. To generate a prediction for the enhancement layer video signal input, the decoded lower layer video signal is upsampled by spatial interpolation and is weighted and combined with the motion-compensated prediction from the enhancement layer. The selection of weights is done on a macroblock basis and the selection information is sent as a part of the enhancement-layer bit stream.

The base- and enhancement-layer coded bit streams are then transmitted over the channel. At the decoder, the lower-layer bit streams are decoded to obtain the lower-resolution video. The lower-resolution video is interpolated and then weighted and added to the motion-compensated prediction from the enhancement layer. In the MPEG-2 video standard, the spatial interpolator is defined as a linear interpolation or a simple averaging for missing samples.

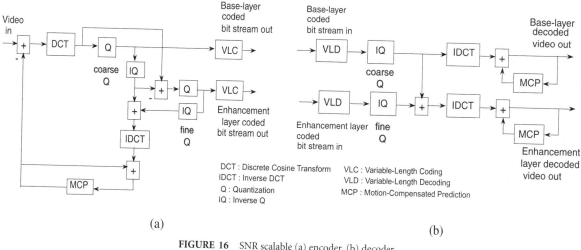

DCT : Discrete Cosine Transform
IDCT : Inverse DCT
Q : Quantization
IQ : Inverse Q
VLC : Variable-Length Coding
VLD : Variable-Length Decoding
MCP : Motion-Compensated Prediction

(a)   (b)

FIGURE 16 SNR scalable (a) encoder, (b) decoder.

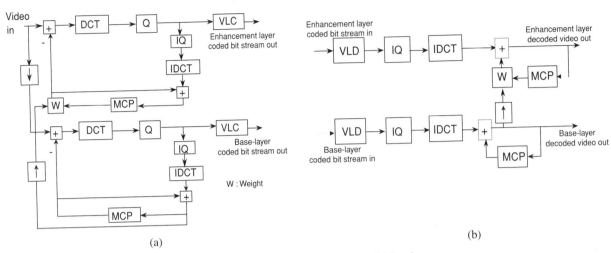

FIGURE 17   Spatial scalable (a) encoder, (b) decoder.

2.5.3 Temporal Scalability

The temporal scalability is designed for video services that require different temporal resolutions or frame-rates. The target applications include video over wireless channel where the video frame rate may need to be dropped when the channel condition is poor. It is also intended for stereoscopic video and coding of future HDTV format in which the baseline is to make the migration from the lower temporal resolution systems to the higher temporal resolution systems possible. In temporal scalable coding, the base layer is coded at a lower frame rate. The decoded base-layer pictures provide motion-compensated predictions for encoding the enhancement layer.

2.5.4 Hybrid Scalability

Two different scalable modes from the three scalability types, SNR, spatial, and temporal, can be combined into hybrid scalable coding schemes. Thus, this results in three combinations: hybrid of SNR and spatial, hybrid of spatial and temporal, and hybrid of SNR and temporal. Hybrid scalability supports up to three layers: the base layer, enhancement layer 1, and enhancement layer 2. The first combination, hybrid of SNR and spatial scalabilities, is targeted at applications such as HDTV/SDTV or SDTV/videophone at two different quality levels. The second combination, hybrid spatial and temporal scalability, can be used for applications such as high temporal resolution progressive HDTV with basic interlaced HDTV and SDTV. The last combination, hybrid SNR and temporal scalable mode, can be used for applications such as enhanced progressive HDTV with basic progressive HDTV at two different quality levels.

2.6 Data Partitioning

Data partitioning is designed to provide more robust transmission in an error-prone environment. Data partitioning splits the block of 64 quantized transform coefficients into partitions. The lower partitions contain more critical information, such as low-

frequency DCT coefficients. To provide more robust transmission, the lower partitions should be better protected or transmitted with a high priority channel with a low probability of error, while the upper partitions can be transmitted with a lower priority. This scheme has not been formally standardized in MPEG-2 but was specified in the information annex of the MPEG-2 DIS document [7]. One thing to note is that the partitioned data are not backward compatible with other MPEG-2 bit streams. Therefore, it requires a decoder that supports the decoding of data partitioning. Using the scalable coding and data partitioning may result in mismatch of reconstructed pictures in the encoder and the decoder and thus cause drift in video quality. In MPEG-2, since there are I pictures that can terminate error propagation, depending on the application requirements, it may not be a severe problem.

2.7 Other Tools for Error Resilience

The effect of bit errors in MPEG-2 coded sequences varies depending on the location of the errors in the bit stream. Errors occurring in the sequence header, picture header, and slice header can make it impossible for the decoder to decode the sequence, the picture, or the slice. Errors in the slice data that contains important information such as macroblock header, DCT coefficients, and motion vectors can cause the decoder to lose synchronization or cause spatial and temporal error propagation. There are several techniques to reduce the effects of errors besides the scalable coding. These include concealment motion vectors, the slice structure, and temporal localization by the use of intra pictures/slices/macroblocks.

The basic idea of concealment motion vector is to transmit motion vectors with the intra macroblocks. Since the intra macroblocks are used for future prediction, they may cause severe video quality degradations if they are lost or corrupted by transmission errors. With a concealment motion vector, a decoder can use the best matching block indicated by the concealment motion vector to replace the corrupted intra macroblock. This improves the concealment performance of the decoder.

In MPEG, each slice starts with a slice header, which is a unique pattern that can be found without decoding the variable-length codes. These slice headers represent possible resynchronization markers after a transmission error. A small slice size, i.e., a smaller number of macroblocks in a slice, can be chosen to increase the frequency of synchronization points, thus reducing the effects of the spatial propagation of each error in a picture. However, this can lead to a reduction in coding efficiency as the slice-header overhead information is increased.

The temporal localization is used to minimize the extent of error propagation from picture to picture in a video sequence, e.g., by using intra coding modes. For the temporal error propagation in an MPEG video sequence, the error from an I or P picture will stop propagating when the next error-free I picture occurs. Therefore, increasing the number of I pictures/slices/macroblocks in the coded sequence can reduce the distortion caused by the temporal error propagation. However, more I pictures/slices/macroblocks will result in a reduction of coding efficiency, and it is more likely that errors will occur in the I pictures, which will cause error propagation.

2.8 Test Model

Similar to other video coding standards such as H.261 and MPEG-1, MPEG-2 only specifies the syntax and the decoder. Many detailed coding options are not specified. In order to have a reference MPEG-2 video quality, test models were developed in MPEG-2. The final test model of MPEG-2 is called "test model 5" (TM5) [10]. TM5 was defined only for main profile experiments. The motion-compensated prediction techniques involve frame, field, and dual-prime prediction, and have forward and backward motion vectors as in MPEG-1. The dual-prime was kept in main profile but restricted to P pictures with no intervening B pictures. A two-step search, which consists of an integer-pixel full search followed by a half-pixel search, is used for motion estimation. The mode decision (intra/inter coding) is also specified. Main profile was restricted to only two quantization matrices, the default table specified in MPEG-1 and the nonlinear quantizer tables. The traditional zigzag scan is used for inter coding while the alternate scan is used for intra coding. The rate-control algorithm in TMN5 consists of three layers operating at the GOP, the picture, and the macroblock levels. A bit allocation per picture is determined at the GOP layer and updated based on the buffer fullness and the complexity of the pictures.

2.9 MPEG-2 Video and System Bit-Stream Structures

A high-level structure of the MPEG-2 video bit stream is shown in Fig. 18. Every MPEG-2 sequence starts with a sequence header and ends with an end of sequence. MPEG-2 syntax is a superset of the MPEG-1 syntax. The MPEG-2 bit stream is based on

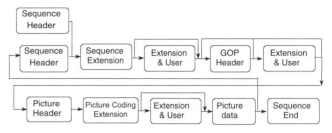

FIGURE 18   MPEG-2 data structure and syntax.

the basic structure of MPEG-1 (refer to Fig. 8). There are two types of bit stream syntax allowed: ISO/IEC 11172-2 video sequence syntax or ISO/IEC 13818-2 (MPEG-2) video sequence syntax.

If the sequence header is not followed by the sequence extension, the MPEG-1 bit-stream syntax is used. Otherwise, the MPEG-2 syntax is used, which accommodates more features but at the expense of higher complexity. The sequence extension includes a profile/level indication, a progressive/interlaced indicator, a display extension including choices of chroma formats and horizontal/vertical display sizes, and choices of scalable modes. The GOP header is located next in the bit-stream syntax with at least one picture following each GOP header. The picture header is always followed by the picture coding extension, the optional extension and user data fields, and picture data. The picture coding extension includes several important parameters such as the indication of intra-DC precision, picture structures (choices of the first/second fields or frame pictures), intra-VLC format, alternate scan, choices of updated quantization matrix, picture display size, display size of the base layer in the case of the spatial scalability extension, and indicator of forward/backward reference picture in the base layer in the case of the temporal scalability extension. The picture data consist of slices, macroblocks, and data for the coded DCT blocks. MPEG-2 defines six layers as MPEG-1. However, the specification of some data elements is different. The details of MPEG-2 syntax specification are documented in [8].

2.10 Summary

MPEG-2 is mainly targeted at general higher quality video applications at bit-rate greater than 2 Mbit/s. It is suitable for coding both progressive and interlaced video. MPEG-2 uses frame/field adaptive motion-compensated predictive coding and DCT. Dual-prime motion compensation for P pictures is used for low-delay applications with no intervening B pictures. In addition to the default quantization table, MPEG-2 defines a nonlinear quantization table with increased accuracy for small values. Alternate scan and new VLC tables are defined for DCT coefficient coding. MPEG-2 also supports compatibility and scalability with the MPEG-1 standard. MPEG-2 syntax is a superset of MPEG-1 syntax and can support a variety of rates and formats for various applications. Similar to other video coding standards, MPEG-2 defines only syntax and semantics. It does not specify every encoding options (preprocessing, motion estimation, quantizer, rate-quality control, and other coding options) and decoding options (postprocessing and error concealment) to allow continuing technology improvement and product differentiation. It is important to keep in mind that different implementations may lead to the different quality, bit rate, delay, and complexity tradeoffs with the different cost factors. An MPEG-2 encoder is much more expensive than an MPEG-2 decoder, because it has to perform many more operations (e.g., motion estimation, coding-mode decisions, and rate-control). An MPEG-2 encoder is also much more expensive than an H.261 or an MPEG-1 encoder as a result of the higher resolution and more complicated motion estimations (e.g., larger search range, frame/field bidirectional motion estimation). References [11–25] provide further information on the related MPEG-1 and MPEG-2 topics.

References

[1] ISO/IEC JTC1 CD 10918, "Digital compression and coding of continuous-tone still images," International Organization for Standardization (ISO), 1993.

[2] ITU-T Recommendation H.261, "Line transmission of nontelephone signals. Video codec for audio visual services at px64 kbits/s," March, 1993.

[3] S. Okubo, "Reference model methodology — a tool for the collaborative creation of video coding standards," *Proc. IEEE* **83**, 139–150 (1995).

[4] MPEG proposal package description, document ISO/WG8/MPEG/89-128, July, 1989.

[5] T. Hidaka and K. Ozawa, "Subjective assessment of redundancy-reduced moving images for interactive applications: test methodology and report," *Signal Process. Image Commun.* **2**, 201–219 (1990).

[6] ISO/IEC JTC1 CD 11172, "Coding of moving pictures and associated audio for digital storage media up to 1.5 Mbits/s," International Organization for Standardization (ISO), 1992.

[7] ISO/IEC JTC1/SC2/WG11, "MPEG video simulation model three (SM3)," MPEG 90/041, July, 1990.

[8] ISO/IEC JTC1 CD 13818, "Information technology — generic coding of moving pictures and associated audio information," International Organization for Standardization (ISO), 1994.

[9] ISO/IEC 13818-2-ITU-T Rec. H.262, "Generic coding of moving pictures and associated audio information: video," 1995.

[10] ISO/IEC JTC1/SC29/WG11, "Test model 5," MPEG 93/457, document AVC-491, April, 1993.

[11] M. L. Liou, "Visual telephony as an ISDN application," *IEEE Commun. Mag.* **28**, 30–38 (1990).

[12] A. Tabatabai, M. Mills, and M. L. Liou, "A review of CCITT px64 kbps video coding and related standards," in *Proceedings of International Electronic Imaging Exposition and Conference* (Boston, MA, 1990), pp. 58–61.

[13] D. J. Le Gall, "MPEG: A video compression standard for multimedia applications," *Commun. ACM*, **34**, 47–58 (1991).

[14] D. J. Le Gall, "The MPEG video compression algorithm," *Signal Process. Image Commun.* **4**, 129–140 (1992).

[15] L. Chiariglione, "Standardization of moving picture coding for interactive applications," in *Proceedings of IEEE GLOBECOM'89*, (Dallas, TX, 1989), pp. 559–563.

[16] A. Puri, *Video Coding Using the MPEG-1 Compression Standard* (Society for Information Display International Symposium, Boston, MA, 1992), pp. 123–126.

[17] A. Puri, "Video coding using the MPEG-2 compression standard," in *Proceedings SPIE Intl. Conf. Visual Communications and Image Processing (VCIP'93)* (Cambridge, MA, 1993), vol. SPIE-2094, pp. 1701–1713.

[18] S. Okubo, K. McCann, and A. Lippman, "MPEG-2 requirements, profiles, and performance verification," presented at the International Workshop on HDTV'93, Ottawa, Canada, October 25–28, 1993.

[19] A. Puri, R. Aravind, and B. Haskell, "Adaptive frame/field motion compensated video coding," *Signal Process. Image Commun.* **5**, 39–58 (1993).

[20] T. Naveen, C. Horne, A. Tabatabai, D. Messing, and R. Eifrig, "MPEG-2 4:2:2 profile at main level: an emerging high-quality video compression standard," in *Standards and Common Interfaces for Video Information Systems*, SPIE Proceedings vol. CR60, (Philadelphia, PA, 1995), pp. 288–308.

[21] A. Puri, R. Kollarits, and B. Haskell, "Compression of stereoscopic video using MPEG-2," in *Standards and Common Interfaces for Video Information Systems*, SPIE Proceedings vol. CR60, (Philadelphia, PA, 1995), pp. 309–334.

[22] R. J. Clarke, *Digital Compression of Still Images and Video* (Academic, New York, 1995).

[23] V. Bhaskaran and K. Konstantinides, *Image and Video Compression Standards: Algorithms and Architectures* (Kluwer, Boston, MA, 1995).

[24] J. L. Mitchell, W. B. Pennebaker, and D. J. Le Gall, *The MPEG Digital Video Compression Standard* (Van Nostrand Reinhold, New York, 1996).

[25] K. R. Rao and J. J. Hwang, *Techniques and Standards for Image, Video, and Audio Coding* (Prentice-Hall, Englewood Cliffs, NJ, 1996).

Emerging MPEG Standards:
MPEG-4 and MPEG-7*

Berna Erol,
Adriana Dumitraş,
and Faouzi Kossentini
University of British Columbia

1 Introduction

During the past two decades, we have witnessed an increasing number of multimedia applications and services in many areas, including entertainment, education, and medicine. Multimedia technologies improve interpersonal communication, promote faster understanding of complex ideas, provide increased access capabilities to information, and allow higher interactivity levels with the media.

The vast amount of digital data that are associated with multimedia applications, and the complex interactions between the different types of data, such as text, speech, music, images, graphics, and video, make the representation, exchange, storage, access, and manipulation of these data a challenging task. In order to provide interoperability between different multimedia applications and promote further use of multimedia data, there is a need to standardize the representation of, and access to, these data. There has already been significant work in the fields of efficient representation by means of compression, storage, and transmission [1–4]. However, there has been little emphasis on the content accessibility and manipulation. The new generation of highly interactive multimedia applications require that the users be able to access and manipulate multimedia data in both uncompressed and compressed forms. This has fueled several recent international standardization activities, such as those of

the Moving Picture Experts Group (MPEG), officially known as Working Group 11 of the ISO/IEC JTC1/SC29 technical committee. MPEG is currently developing two emerging standards: MPEG-4, which is standardizing an object-based coded representation of multimedia data, and MPEG-7, which is standardizing a multimedia content description interface.

MPEG-4, like the MPEG-1/2 [1, 2] and ITU-T H.263/H.263+ [3, 4] standards, which are discussed in Chapters 6.4 and 6.1, respectively, offers high compression performance levels, making much more efficient the storage and transmission of audiovisual data. However, the other key objectives of MPEG-4 are to enable content-based access and provide functionalities such as error resilience, scalability, and hybrid coding of synthetic and natural data [5,6]. On the other hand, MPEG-7 is expected to enable effective and efficient content-based access and manipulation of multimedia data, and to provide functionalities that are complementary to those of the MPEG-4 standard. With the use of an MPEG-4/MPEG-7 compliant system, it will be possible to randomly access, manipulate, and process individual objects within a scene. For example, consider the video scene given in Fig. 1. Using an MPEG-4/MPEG-7 compliant decoder, the user will be able to search for podiums that are similar to the one in the video scene, or search for fish that are similar to the one shown on the screen. The user can also search for curtains that have a texture similar to that of the background. Next, besides providing a comprehensive description of the emerging MPEG-4 and MPEG-7 visual standards, we show, through examples, how MPEG-4 and MPEG-7 will together enable

*This work was supported by the Natural Sciences and Engineering Research Council (NSERC) and the National Research Council (NRC) of Canada.

FIGURE 1   An audiovisual scene. (See color section, p. C–28.)

many desired functionalities and provide a complete multimedia solution.

2 The MPEG-4 Standard

The MPEG-4 standard addresses system issues such as the multiplexing and composition of audiovisual data in the systems part [7], the decoding of the visual data in the visual part [8], and the decoding of the audio data in the audio part [9]. The initial goal of MPEG-4 was to provide tools and algorithms for very low bit rate coding of audiovisual data. However, the scope has changed considerably in order to address the requirements of the new generation multimedia applications, which include multimedia communications (broadcast and interpersonal), Internet, interactive video games, video surveillance, and multimedia databases [10, 11]. Besides the need to achieve high compression performance levels, these applications require interactivity with individual objects, hybrid coding of natural and synthetic objects, and a high degree of scalability and error resilience [6, 12–14]. MPEG-4 addresses all of these requirements by providing the following functionalities: (1) improved coding efficiency by providing compression tools that are optimized for objects with a wide range of source material and bit rates, (2) object-based interactivity by enabling a high degree of user interaction with the individual audiovisual objects, (3) generic coding by providing tools for the efficient representation of both natural and synthetic objects, (4) object-based and temporal random access, (5) temporal, spatial, quality and object-based scalability, and (6) robust operation in error-prone environments.

2.1 Audiovisual Object Representation

An object-based representation is necessary to enable the above functionalities. MPEG-4 achieves object-based representation

by defining audiovisual objects and coding them into separate bit stream segments [6, 7, 15]. An audiovisual (AV) object (AVO) consists of a visual object component, an audio object component, or a combination of these components. The characteristics of the audio and visual components of individual AVOs can vary, such that the audio component can be (1) synthetic or natural, and (2) mono, stereo, or multichannel (e.g., surround sound), and the visual component can be natural or synthetic. Some examples of AVOs include a sound recorded with a microphone, a speech synthesized from a text, a person recorded by a video camera, and a 3-D image with text overlay.

MPEG-4 supports the composition of a set of audiovisual objects into a scene, also referred to as an audiovisual scene. In order to allow interactivity with individual AVOs within a scene, it is essential to transmit the information that describes each AVOs spatial and temporal coordinates. This information is referred to as the scene description information and is transmitted as a separate stream and multiplexed with AVO elementary bit streams so that the scene can be composed at the user's end. This functionality makes it possible to change the composition of AVOs without having to change the content of AVOs.

An example of an audiovisual scene, which is composed of natural and synthetic audio and visual objects, is presented in Fig. 1. AV objects can be organized in a hierarchical fashion. Elementary AVOs, such as the blue head and the associated voice, can be combined together to form a compound AVO, i.e., a talking head. It is possible to change the position of the AVOs, delete them or make them visible, or manipulate them in a number of ways depending on the nature of their characteristics. For example, if it is a visual object, the user can zoom and rotate it. If it is an audio object, the user can change its pitch, as well as his or her listening point. Also, the quality and spatial and temporal resolutions of the individual AVOs can be modified. For example, in a mobile video telephony application, the user can request a higher frame rate and spatial resolution for the talking person than those of the background objects.

Audiovisual scenes are reconstructed and presented by audiovisual terminals at the receiver's end. As seen from Fig. 2, an audiovisual terminal receives the bit stream from a network or a storage device, demultiplexes the bit stream to retrieve elementary streams, decompresses the primitive AV objects, and finally performs composition and rendering of the reconstructed AV objects by using the corresponding scene description information. An AV terminal also manages upstream data transfer for user commands that require server-side interaction.

2.2 The MPEG-4 Visual Standard: Technical Description

The emerging MPEG-4 visual standard, officially known as ISO/IEC 14496–2 [8], aims at providing standardized core processing elements that allow efficient storage, transmission, and manipulation of visual data [16]. While the MPEG-4 visual standard, like its predecessors, defines only the bit stream syntax and

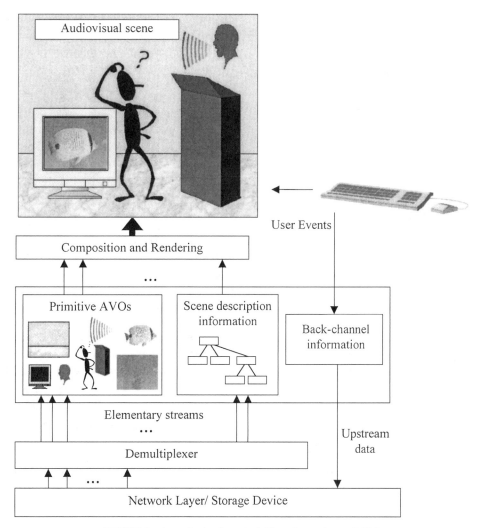

FIGURE 2   An audiovisual terminal. (See color section, p. C–29.)

the decoding process, the precise definitions of some compliant encoding algorithms are presented in two verification models: one for synthetic and natural hybrid coding (SNHC) [17], and the other one for natural video coding [18]. Although the MPEG-4 standard does not define the encoding process, both the encoding and decoding processes are discussed in this chapter.

Different representations and compression algorithms may offer optimum solutions for different applications, bit rates, and formats. Therefore, MPEG-4 provides four different types of coding tools: *Video object coding* for the coding of a naturally or synthetically originated, rectangular, or arbitrarily shaped video object; *mesh object coding* for the coding of a visual object represented with a mesh structure; *model-based coding* for the coding of a synthetic representation and animation of the human face and body; and *still texture coding* for the wavelet coding of still textures.

In the following sections, we first describe each of the MPEG-4 visual object coding tools. Next we discuss the scalability and the error resilience tools, followed by a presentation of the appli-

cations and profiles of the MPEG-4 visual standard. Finally, we provide an example that illustrates how MPEG-4 can be used for the coding of rectangular and arbitrarily shaped video objects.

2.2.1 Video Object Coding

A video object (VO) is an arbitrarily shaped video segment that has a semantic meaning. A 2-D snapshot of a VO at a particular time instant is called a video object plane (VOP). A VOP is defined by its texture (luminance and chrominance values) and its shape. MPEG-4 allows content-based access to not only the video objects, but also temporal instances of the video objects, i.e., VOPs. In general, MPEG-4 coding of a VOP involves coding of motion, texture, and shape information. However, when the VOP is a rectangularly shaped video frame, MPEG-4 video coding becomes quite similar to that specified in MPEG-1/ MPEG-2 [1, 2] and H.263 [3]. In fact, an MPEG-4 visual terminal must be able to decode all the bit streams of H.263 baseline encoders.

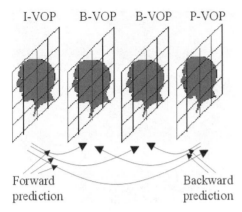

FIGURE 3 VOP prediction types.

To enable access to an arbitrarily shaped object, such an object has to be separated from the background and the other objects. This process is called segmentation, and it can be performed in real time during encoding (on line), or in nonreal time prior to encoding (off line). The segmentation process is not standardized in MPEG-4. However, there are a number of automatic and semiautomatic tools available for segmentation [19]. Also, it is possible to generate image sequences that are segmented initially by using techniques such as chroma keying [20], in which a unique color is used to separate the background from a video object.

MPEG-4 video object coding consists of shape coding (for arbitrarily shaped VOs), motion compensated prediction to reduce temporal redundancies, and DCT-based texture coding of the motion compensated prediction error data to reduce spatial redundancies. The video coding is performed at the macroblock level. VOPs are divided into macroblocks, such that they are rep-

resented with the minimum number of macroblocks within a bounding rectangle. Similar to MPEG-1 and MPEG-2, MPEG-4 supports intracoded (I), temporally predicted (P), and bidirectionally predicted (B) VOPs, all of which are illustrated in Fig. 3.

Figure 4 shows the basic VOP encoder structure. The encoder consists mainly of two parts: a hybrid of a motion compensated predictor and a DCT-based coder, and a shape coder. In the first part, motion estimation and compensation are performed (except for I-VOPs) on texture data, followed by DCT and quantization. Then, the difference between the predicted data and the original texture data is coded by variable length coding (VLC). Motion information is also encoded by using VLC. Then, the VOP is reconstructed as in the decoder, that is, by applying inverse quantization, applying inverse DCT (IDCT), and adding the resulting data to the motion compensated prediction data. The resulting VOP is then used for the prediction of future VOPs. The shape coder encodes the binary shape and the transparency information of the object. Since the shape of a VOP may not change significantly between consecutive VOPs, predictive coding is employed to reduce temporal redundancies. Thus, motion estimation and compensation are also performed for the shape of the object. Finally, motion, texture, and shape information is multiplexed with the headers to form the coded VOP bit stream. At the decoder end, the VOP is reconstructed by combining motion, texture, and shape data decoded from the bit stream.

2.2.1.1 Motion Vector Coding. In the bit stream, the motion data are transmitted in the form of motion vectors (MVs). MVs are predicted by using a spatial neighborhood of three MVs, and the prediction error is variable length coded. Motion vectors are transmitted only for P-VOPs and B-VOPs. MPEG-4 employs some advanced motion compensation techniques, such as

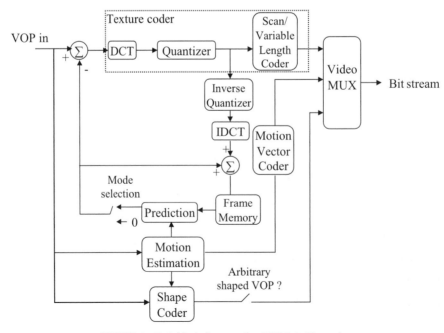

FIGURE 4 Basic block diagram of an MPEG-4 video coder.

the use of unrestricted MVs, where MVs are allowed to point outside the coded area of a reference VOP, overlapped motion compensation, and the use of four MVs per macroblock.

Since the VOPs are, in general, arbitrarily shaped, there may not be a corresponding pixel available for the prediction of the current VOP. In order to guarantee that every pixel of the current VOP can be predicted, some or all of the boundary and outside blocks of the reference VOP have to be padded by extrapolation. The boundary blocks are padded by first repeating the boundary pixels in the horizontal direction, and then repeating the boundary pixels in the vertical direction while averaging pixels whose values were obtained by horizontal padding. When a reference pixel belongs to a block that is completely outside the VOP, then the block is filled by extended padding, where pixels are assigned average values that are determined by the neighboring blocks.

2.2.1.2 Texture Coding.

Intrablocks, as well as motion compensation prediction error blocks, are texture coded. Similar to MPEG-1/MPEG-2 and H.263 (described in Chapters 6.4 and 6.1, respectively), DCT-based coding is employed to reduce spatial redundancies. That is, each VOP is divided into macroblocks as illustrated in Fig. 5, and DCT coding is applied to the four 8×8 luminance and two 8×8 chrominance blocks of the macroblocks. If a macroblock lies on the boundary of an arbitrarily shaped VOP, then the pixels that are outside the VOP are padded before DCT coding. For intra-VOP boundary macroblocks, padding is performed as described in the previous section, whereas for residual blocks, the region that is outside the VOP is padded with zeros. Alternatively, a shape-adaptive DCT (SA-DCT) coder can be used to encode only those pixels that belong to the VOP. This generally results in higher compression performance, but at the expense of an increased implementation complexity. Macroblocks that are completely inside the VOP are DCT transformed as in MPEG-1/MPEG-2 and H.263. The blocks that do not belong to the VOP are not coded. DCT transformation of the blocks is followed by quantization, zigzag scanning, and variable length coding. Note that adaptive DC/AC prediction methods and alternate scan techniques can be employed for efficient coding of the DCT coefficients of intra blocks.

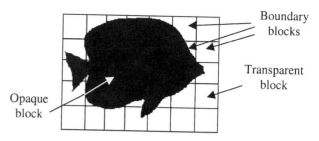

FIGURE 6 Binary alpha plane.

2.2.1.3 Shape Coding.

MPEG-4 supports coding of shape information to enable content-based access to individual video objects in a scene [6, 8, 18, 20]. MPEG-4 is the only video coding standard that supports shape coding, besides H.263+ [4], which provides some limited shape coding support by means of its chroma-keying coding technique. Because of its limitations on shape rate control and its unstable performance for complex shapes, the chroma-keying coding technique was not considered for shape coding in MPEG-4 [20]. Polygon-based and bitmap-based shape coding techniques were found to be better candidates. Because of its high compression performance and low complexity, a bitmap-based shape coder was adopted.

In bitmap-based shape coding, the shape and transparency of a VOP are defined by their binary and gray-scale (respectively) alpha planes. A binary alpha plane indicates whether or not a pixel belongs to a VOP. A gray-scale alpha plane indicates the transparency of each pixel within a VOP. MPEG-4 provides tools for both lossless and lossy coding of binary and gray-scale alpha planes. Furthermore, both intra shape and inter shape coding are supported.

Binary alpha planes are divided into 16×16 blocks, as illustrated in Fig. 6. The blocks that are inside the VOP are signaled as opaque blocks and the blocks that are outside the VOP are signaled as transparent blocks. The pixels in boundary blocks (i.e., blocks that contain pixels both inside and outside the VOP) are scanned in a raster scan order and coded by using context-based arithmetic coding. In intracoding, a context is computed for each pixel using 10 neighboring pixels, which are shown in Fig. 7(a), by using the equation $C = \sum_k c_k 2^k$, where k is the pixel index, and c_k is "0" for transparent pixels and "1" for opaque pixels. Pixels from neighboring blocks are used to build the context if

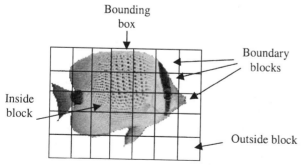

FIGURE 5 VOP enclosed in a rectangular bounding box and divided into macroblocks.

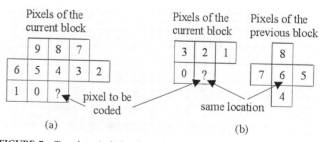

FIGURE 7 Template pixels that form the context of arithmetic coder for (a) intracoded and (b) intercoded shape blocks.

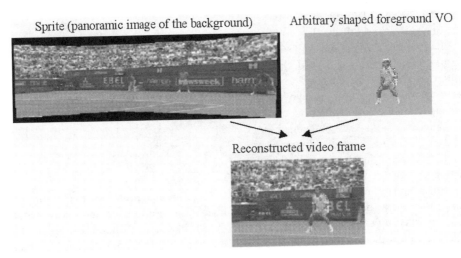

Sprite (panoramic image of the background)   Arbitrary shaped foreground VO

Reconstructed video frame

FIGURE 8   Sprite coding of a video sequence. (Courtesy of Dr. Thomas Sikora.) (See color section, p. C–30.)

the context pixels fall outside the current block. The computed context is used to access the table of probabilities. The selected probability is used to determine the appropriate code space for arithmetic coding. For each boundary block, the arithmetic encoding process is also applied to the transposed version of the block. The representation that results in less coding bits is conveyed in the bit stream.

In inter shape coding, the shape of the current block is first predicted from the shape of the temporally previous or future VOP (depending on the VOP coding type) by performing motion estimation and compensation in integer pixel accuracy. The shape motion vector is then coded predictively. Then, the difference between the current and the predicted shape block is arithmetically coded. The context for an intercoded shape block is computed by using a template of nine pixels from both the current and temporally previous VOP shape blocks, as shown in Fig. 7(b).

Lossy coding of the binary shape is achieved by either not transmitting the difference between the current and the predicted shape block (in inter shape coding), or by subsampling the binary alpha plane by a factor of 2 or 4 prior to arithmetic encoding (in both intra- and intercoding). In order to reduce the blocky appearance of the decoded shape caused by lossy coding, an upsampling filter is employed during the reconstruction.

Transparency of pixels can take values in the range of 0 (transparent) to 255 (opaque). If all of the pixels in a VOP block are opaque or transparent, then no transparency information is transmitted for that block. Otherwise, gray-scale alpha planes, which represent transparency information, are divided into 16×16 blocks and coded the same way as the texture in the luminance blocks.

2.2.1.4 *Sprite Coding.*
In MPEG-4, sprite coding is used for representation of video objects that are static throughout a video scene, or their changes can be approximated by warping the original object planes [8, 21]. Sprites are generally used for transmitting background in video sequences. They are coded in the

same way as intra VOPs and are saved in a buffer at the decoder to reconstruct the video sequences. An example of a sprite is shown in Fig. 8. As seen here, a sprite may consist of a panoramic image of the background, including the pixels that are occluded by the other video objects. Such a representation can increase coding efficiency, since the background image is coded only once at the beginning of the video segment, and the camera motion, such as panning and zooming, can be represented by a few transformation coefficients in the rest of the frames.

2.2.2 Mesh Object Coding

A mesh is a tessellation (partitioning) of an image into polygonal patches. Mesh representations have been successfully used in computer graphics for efficient modeling and rendering of 3-D objects. In order to benefit from functionalities provided by such representations, MPEG-4 supports 2-D mesh representations of natural and synthetic visual objects, and still texture objects, with triangular patches [8, 22]. The vertices of the triangular mesh elements are called node points, and they can be used to track the motion of a video object, as depicted in Fig. 9. Motion compensation is performed by spatially piecewise warping of the texture maps that correspond to the triangular patches. This representation provides a good model for spatially continuous motion fields.

An initial 2-D triangular mesh can be either a uniform mesh or a Delaunay mesh. An example of a uniform mesh is shown

A uniform, initial mesh object   Mesh object after motion trajectory

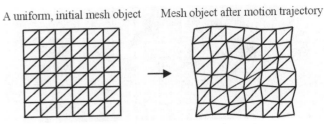

FIGURE 9   Mesh object with triangular patches.

FIGURE 10 Mesh representation of a video object with triangular patches. (Courtesy of Dr. Murat Tekalp.) (See color section, p. C–30.)

in Fig. 9. A uniform mesh can be represented by a small set of parameters: the width and height of the mesh rectangle, and the type of the mesh structure. On the other hand, Delaunay meshes provide more flexibility by allowing initial node points to be at any location. The locations of the node points are coded differentially with respect to the previously coded node point coordinate. The ordering of the node points is such that the boundary node points are coded first, followed by coding of the interior node points. As seen in Fig. 10, a Delaunay mesh can be adapted to the image content for a more accurate representation of the video object. The selection process of the node points for a Delaunay mesh and the tracking of mesh node points are not specified in the MPEG-4 standard.

Similar to VOPs, instances of mesh objects are called mesh object planes (MOPs). The structure (in the case of intracoding) and motion (in the case of intercoding) of MOPs are variable length coded into a nonscalable bit stream. The texture of the corresponding visual object has to be coded separately.

2.2.2.1 Functionalities.
A mesh-based representation of an object enables many functionalities. It improves content-based manipulation by enabling the merging of synthetic objects with natural objects. It also allows us to transmit only selected key frames, which can be animated to construct intermediate frames at the decoder. Moreover, mesh modeling can efficiently represent continuous motion, resulting in less blocking artifacts at low bit rates as compared with the block-based modeling. It also enables content-based retrieval of video objects by providing accurate object trajectory information and syntax for vertex-based object shape representation, which is more efficient than the bitmap representation.

2.2.3 Model-Based Coding
Model-based representation enables very low bit rate video coding applications by providing the syntax for the transmission of the parameters that describe the behavior of a human being,

rather than transmission of the video frames. MPEG-4 supports the coding of two types of models [8, 23, 24]: a *face object* model, which is a synthetic representation of the human face with 3-D polygon meshes that can be animated to have visual manifestations of speech and facial expressions, and a *body object* model, which is a virtual human body model represented with 3-D polygon meshes that can be rendered to simulate body movements.

2.2.3.1 Face Animation.
It is required that every MPEG-4 decoder that supports face object decoding has a default face model that can be replaced by downloading a new face model. Either model can be customized to have a different visual appearance by transmitting facial definition parameters (FDPs). FDPs can determine the shape (i.e., head geometry) and texture of the face model.

A face object consists of a collection of nodes, also called feature points, which are used to animate synthetic faces. The animation is controlled by face animation parameters (FAPs) that manipulate the displacements of feature points and angles of face features and expressions. MPEG-4 defines a set of 68 low-level animations, such as head and eye rotations, as well as motion of a total of 82 feature points for the jaw, lips, eye, eyebrow, cheek, tongue, hair, teeth, nose, and ear. These feature points are shown in Fig. 11. MPEG-4 also defines high-level expressions, such as joy, sadness, fear, and surprise, and visemes for determining the mouth movements for speech animation. High-level expressions consist of a set of low-level expressions. For example, the joy expression is defined by relaxed eyebrows and open mouth, with the mouth corners pulled back toward the ears. Figure 12 illustrates several video scenes that are constructed by using face animation parameters.

The FAPs are coded by quantization followed by arithmetic coding. The quantization is performed by taking into consideration the limited movements of the facial features. Alternatively, DCT coding can be applied to a vector of 16 temporal instances of the FAP, improving compression efficiency, but also increasing delay.

2.2.3.2 Body Animation.
Similar to the case of a face object, two sets of parameters are defined for a body object: body definition parameters (BDPs), which define the body through its

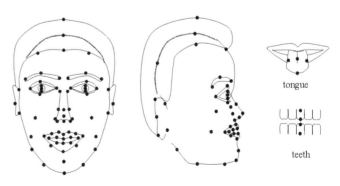

FIGURE 11 Feature points used for animation.

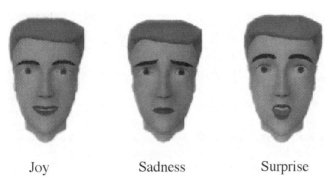

Joy   Sadness   Surprise

FIGURE 12 Examples of face expressions coded with FAPs. (Courtesy of Joern Ostermann.)

dimensions, surface and texture, and body animation parameters (BAPs), which define the posture and animation of a given body model. Body animation is being standardized in the Version 2 of the MPEG-4 standard.

2.2.4 Still Texture Coding

The block diagram of an MPEG-4 still texture coder is shown in Fig. 13. As depicted in the figure, the still texture is first decomposed using a 2-D separable wavelet transform, employing a Daubechies biorthogonal filter bank [8]. The discrete wavelet transform is performed using either integer or floating point operations. Also, a shape adaptive wavelet transform can be employed for coding arbitrarily shaped texture.

The DPCM coding method is applied to the coefficient values of the lowest frequency subband. A multiscale zero-tree coding method [26] is applied to the coefficients of the remaining subbands. Zero-tree modeling is used for encoding the location of nonzero wavelet coefficients by taking advantage of the fact that if a wavelet coefficient is quantized to zero, then all wavelet coefficients with the same orientation and the same spatial location at finer wavelet scales are also likely to be quantized to zero. Two different zero-tree scanning methods are employed to achieve spatial and SNR scalability. After DPCM coding of the coefficients of the lowest frequency subband, and zero-tree scanning of the remaining subbands, the resulting data are coded by using an adaptive arithmetic coder.

2.2.5 Scalability

Scalability means that a bit stream consists of a separately decodable base layer, and associated enhancement layers. This struc-

ture is especially desirable for heterogeneous environments to counter limitations such as constraints on bit rate, display resolution, network throughput, and decoder complexity. Moreover, scalability provides improved error resilience by allowing the syntax for prioritized transmission. MPEG-4 supports traditional frame-based temporal, spatial, and quality scalabilities, as well as object-based scalability. Object-based scalability allows one to add or remove video objects, as well as prioritize the objects within a scene. MPEG-4 supports both spatial and temporal object-based scalability. With the use of this functionality, it is possible to represent the objects of interest with a higher spatial or temporal resolution, while allocating less bandwidth and computational power to the objects that are not as important.

2.2.6 Error resilience

MPEG-4 offers error resilience tools to address the problem of robust operation over error-prone channels. These tools can be divided into three groups: resynchronization, data partitioning, and data recovery [8, 27].

If an error occurs during the transmission of the bit stream, then resynchronization is required to recover data and conceal the effects of errors. MPEG-4 allows resynchronization by employing a method that is similar to the group of macroblocks approach of H.263 [3]. The difference is that, in order to provide periodic resynchronization markers, the number of macroblocks in an MPEG-4 packet may be variable, depending on the number of bits required to represent each macroblock. Each video packet contains information such as macroblock number and quantizer, necessary to restart the decoding operation in case an error is encountered.

Data partitioning allows the separation between the motion and texture data, along with additional resynchronization markers in the bit stream to improve the ability to localize the errors. This technique provides enhanced concealment capabilities. For example, if texture information is lost, motion information can be used to conceal the errors. Error concealment, however, is not standardized in MPEG-4.

Reversible variable length codes (RVLCs) can be employed for the coding of macroblock texture information for improved error resilience. RVLCs can be decoded in both the forward and backward directions. Thus, if part of a bit stream cannot be decoded in the forward direction because of errors, data can be recovered partially by decoding in the backward direction.

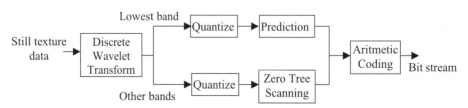

FIGURE 13 Block diagram of the still texture coder.

TABLE 1   MPEG-4 visual profiles

Profile Group	Profile Name	Supported Functionalities
Natural video	Simple	Error resilient coding of rectangular video objects
	Simple scalable	Simple profile + frame-based temporal and spatial scalability
	Core	Simple profile + coding of arbitrarily shaped objects
	Main	Core profile + interlaced video + transparency coding + sprite coding
	N-bit	Core profile + coding video objects with pixel depths between 4 and 12 bits
Synthetic video	Simple face animation	Basic coding of simple face animation
	Scalable texture	Spatially scalable coding of still texture objects
Hybrid (natural/synthetic) video	Simple basic animated 2-D texture	Simple face animation + spatial and quality scalability + mesh-based representation of still texture objects
	Hybrid	Coding of arbitrarily shaped objects + temporal scalability + face object coding + mesh coding of animated still texture objects

2.3 Applications and Profiles

MPEG-4 is designed to address a wide range of multimedia applications, which cover interactive video communications (e.g., video telephony and conferencing), noninteractive video communications (e.g., video e-mailing and multimedia broadcasting), digital storage media (e.g., optical disks), content-based image and video databases, video surveillance, and interactive video games. Since the MPEG-4 syntax is designed to be very generic and includes many tools to enable a wide variety of applications, the implementation of a decoder that supports the full syntax will most often be impractical. Therefore, the MPEG-4 visual standard defines a number of subsets of the syntax, referred to as "profiles" (Table 1), each targeting a specific group of applications. For example, the simple profile targets low-complexity and low-delay applications, such as mobile video communications, the main profile targets interactive broadcast and DVD applications, the N-bit profile targets surveillance applications, and the scalable texture profile targets applications that require multiple texture scalability levels, such as mapping texture onto objects in video games.

2.4 MPEG-4 Video Coding Example

In this section, we present an example to illustrate the capabilities and compression performance levels of an MPEG-4 compliant video encoder. We performed our simulations by using Microsoft's MPEG-4 encoder software [28] to encode the video sequence called *Bream*, which shows a fish that changes directions while swimming. The segmented sequence is coded following two different modes of operation that represent two distinct MPEG-4 profiles: the simple profile and the core profile. The simulation results are given in Fig. 14. The figure shows the reconstructed frames corresponding to the two used profiles, as well as the number of bits used to represent motion, texture, and shape information. In the example, the 100 frames of the *Bream* sequence are encoded at 10 frames per second (fps) and with a constant quantizer of 10. The first frame is intracoded and the rest of the frames are intercoded. In the

object-based coding case (i.e., core profile), lossless shape coding is employed. Figure 14(a) shows the original input frame (the first frame of *Bream*), and Fig. 14(b) shows the reconstructed frame after using an encoder that is compliant with the simple profile (no shape coding). In this example, the simple profile coder achieves a 56:1 compression ratio with relatively high reconstruction quality (34.4 dB). If the quantizer step size were larger, it would be possible to achieve up to a 200:1 compression ratio for this sequence, while still keeping the reconstruction quality above 30 dB.

The *Bream* video sequence consists mainly of two objects: a fish (foreground object), and a water background (background object). Using the core profile encoder, we encode these two objects into two separate bit streams. Figure 14(c) shows the shape of the foreground object. We encode only the pixels that are inside the shape, which are indicated by a darker color. Texture padding of the boundary blocks is shown in Fig. 14(d). Figure 14(e) shows the foreground object as it is decoded and displayed. In this example, 10%, 14%, and 73% of the total bits are spent to represent the shape, motion vectors, and texture information, respectively. The rest of the bits are used for headers and bit stuffing. These ratios would change depending on the sequence. For example, if the shape of the sequence is changing rapidly, then more bits will be spent for shape coding.

Figure 14(f) shows the background of the sequence. The combination of background and foreground objects is shown in Fig. 14(g). A compression ratio of 80:1 is obtained. Since the background object does not vary significantly with time, the number of bits spent for its representation is very small. Here, it is also possible to employ sprite coding by selecting background as a sprite.

The PSNR versus rate performance of the frame-based and object-based coders for the 100 frames of the *Bream* video sequence is presented in Fig. 15. As seen here, for this sequence, the PSNR bit-rate tradeoffs of object-based coding are better than those of frame-based coding. This is mainly due to the slowly varying foreground and background objects. However, for scenes with complex and quickly varying shapes, since a considerable amount of bits would be spent for shape coding, frame-based

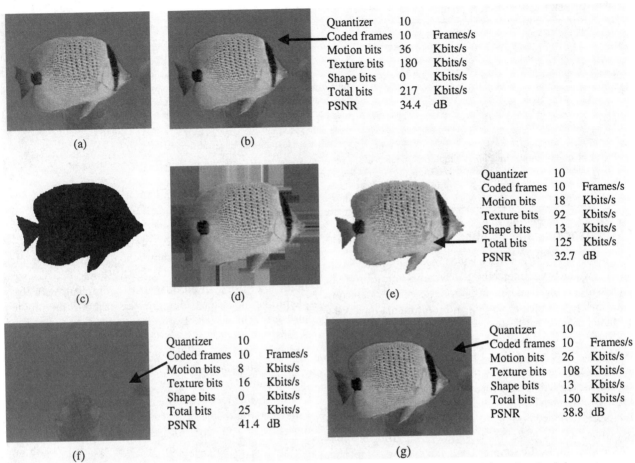

FIGURE 14 Illustration of MPEG-4 coding, simple profile vs. core profile: (a) original frame; (b) frame-based coded frame; (c) shape mask for the foreground object; (d) coded foreground object (boundary macroblocks are padded); (e) foreground object as it is decoded and displayed; (f) background object as it is decoded and displayed; (g) foreground + background objects (e + f). (Bream video sequence is courtesy of Matsushita Electric Industrial Co., Ltd.) (See color section, p. C–31.)

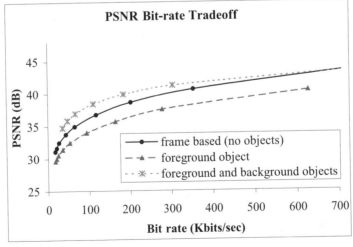

FIGURE 15 PSNR performance for the 100 frames of the *Bream* video sequence, using different profiles of the MPEG-4 video coder.

coding would achieve better compression levels, but at the cost of a limited content-based access capability.

3 The MPEG-7 Visual Standard

Many of the current multimedia applications require that the visual data be effectively and efficiently accessed and manipulated. Many text-based methods have been applied to the access and manipulation of visual content, where keywords are associated with each visual component. In order to overcome the limitations of the text-based methods, which typically require human assistance in describing visual content, feature-based methods have been introduced. Low-level features, such as texture, shape, and color, and high-level features, such as composition information, have been employed in many of the existing content-based access and manipulation (CBAM) systems. As they arise from different applications, these systems make use of various feature representations. For instance, the same "shape" feature may be represented by Fourier descriptors, geometric descriptors, etc. Therefore, data accessibility and interoperability between these systems are quite limited.

A unified framework for content representation can overcome the above problems. Moreover, such a framework would be very useful in the evaluation of current systems by various research and industry organizations, as well as for future research and development. Hence, it is not surprising that current international standardization committees, such as the MPEG committee, have focused on the standardization of a "multimedia content description interface" (MPEG-7). The major challenges facing the MPEG-7 standardization activity is that visual data can have different formats (e.g., uncompressed, compressed), different types (e.g., still pictures, audio, video), can be described by using heterogeneous feature representations, and can reside in different geographical locations.

MPEG-7 requirements for the systems, visual, and audio parts have already been developed [29, 30]. Here, we focus on the visual part of the MPEG-7 standard. We first describe the current work in the access and manipulation of visual data. Next, we present the objectives of the MPEG-7 visual standard and its normative components. Finally, we illustrate, through an example, how MPEG-7 will impact the CBAM of visual data.

3.1 Text-Based CBAM of Visual Data

Many of the text-based search methods that have been proposed are currently employed in the search engines of the World Wide Web. There are several types of text-based search engines, such as the full-text (e.g., Alta Vista, Lycos), catalogue-based (e.g., Excite, Yahoo!), meta- (e.g., Netsearch), and specialist (e.g., Bigfoot White Pages) search engines. *Full-text* search engines analyze the content of files in order to find the desired text. *Catalogue-based* search (also known as index-search) engines use classification systems in order to help the users identify the files that

TABLE 2 Examples of text-based search engines for visual content

Type	Name	Reference	Data Format
Still images	Icon Browser	33	GIF
	Image Surfer	34	JPEG
	Lycos Media	35	JPEG
	Virtual Image Archive	36	GIF, JPEG
	Yahoo Image Surfer	37	JPEG
Video	Whoopie	38	MPEG, AVI, and others
	Lycos Media	35	MOV

have been marked by human agents as being potentially useful to a particular topic. *Meta-search* (also known as multisearch) engines allow the users to search for keywords, using several search engines sequentially or simultaneously. *Specialist search* engines provide responses that are relevant to specific application areas.

All of the above text-based search methods can be applied to the access and manipulation of the visual content by assigning keywords to each visual component [31, 32]. Examples of such text search engines used for CBAM of visual content are shown in Table 2. Some of the existing standards, such as HTML, provide methods to associate a text descriptor with a still image. However, HTML does not provide a mechanism for attaching other sets of descriptors to images. The SGML standard overcomes this problem. Unfortunately, the vocabulary is restricted, and similarity-based retrieval cannot be performed. Moreover, human assistance in describing the content and entering the description in the database is required.

3.2 Feature-Based CBAM of Visual Data

Feature-based methods have been proposed in order to overcome the limitations of the text-based search methods for accessing visual content. The features that are employed by the CBAM methods can be divided into two classes: low-level and high-level features [39]. The low-level features can often be extracted automatically. However, the extraction of the high-level features usually requires human assistance.

Most of the current research in content-based access and manipulation of visual data has focused on using low-level features such as texture, shape, and color [40, 41]. Texture-based CBAM of visual data has been applied in [41, 42]. These systems use texture analysis methods that are based on structural, statistical, spectral, stochastic model-based, morphology-based, or multiresolution techniques [43–46]. Shape-based CBAM methods of visual data have been proposed [47, 48] that employ various boundary-based (e.g., chain codes, geometric, and Fourier descriptors) or region-based (e.g., area, roundness) shape models. Color features have been extensively used for the CBAM of image databases [49, 50], because of their invariance with respect to image scaling and rotation. The color features have been frequently represented by computing the average color, the dominant color, and the global/local histograms [49].

In many cases, using only one low-level feature may not be sufficient to discriminate between several objects. Therefore, combinations of two or several low-level features, as employed in [39, 51–55], can improve significantly the outcome of the CBAM of visual data.

3.3 Objectives of the MPEG-7 Visual Standard

MPEG-7 is the most recent standardization activity of the MPEG group. The goal of MPEG-7 is to provide a standardized description that allows effective and efficient access and manipulation of the multimedia content [29, 30, 56]. MPEG-7 will standardize a set of descriptors (Ds), a set of description schemes (DSs), a description definition language (DDL), and schemes for the coding of the descriptions [29, 56]. MPEG-7 will not standardize the tools that are used to generate the description (e.g., segmentation tools, feature extraction tools) and the tools that use the description (e.g., content recognition tools). The MPEG-7 requirements posed indirectly on the visual description tools would likely yield effective and efficient tools for segmentation, feature extraction, and visual recognition.

3.4 Visual Description

In this section, we describe the normative components, i.e., the Ds, the DSs, the DDL, and the coding schemes, of the visual part of MPEG-7.

3.4.1 Descriptors

For a given visual content (e.g., images, video), a set of features can be extracted. A feature is defined as a distinctive characteristic of the content. In order to compare several features, a meaningful representation of each feature (descriptor) and its instantiation for a given data set (descriptor value) are needed. Figure 16 illustrates the relationship between data, features, and descriptors. Using feature extraction, one projects the input visual content space onto the feature space. The result of the projection is a set of features $[f_1, f_2, \ldots, f_i, \ldots, f_N]$ associated with any item of the visual content, where N is the total number of features that are extracted. Then, each feature f_i of the feature vector can be represented by several descriptors. Examples of descriptors associated with the input features are pre-

TABLE 3 Examples of features and their descriptors

Feature	Descriptor
Texture	contrast, coarseness, directionality, Markov model, Co-occurrence matrix, DCT coefficients, wavelet coefficients, Wold coefficients
Shape	geometrical descriptors (area, perimeter, etc.), Fourier descriptors, chain code
Color	color histogram, color moments
Appearance	text, Fourier coefficients

sented in Table 3. For instance, the shape feature may be represented by geometric descriptors or Fourier descriptors. Some of these descriptors are standardized in MPEG-7 (i.e., they belong to the standardized descriptor space). The projection from the visual data space to the feature space, which is not standardized in MPEG-7, is not unique since different applications may require different features for describing the same visual content. The projection from the feature space to the descriptor space, which is being standardized in MPEG-7, is also not unique since several descriptors may be assigned to the same feature. An MPEG-7 descriptor should be relevant and effective. This guarantees that the descriptor expresses precisely and completely the associated feature. Moreover, it should have expression and processing efficiency. This guarantees the existence of an efficient method for computing the descriptor value. Descriptor scalability with the application and with the data are also required. Finally, the descriptor should provide a multilevel representation of the associated feature. Other requirements are included in [30].

3.4.2 Description Scheme

A description scheme (DS) is the pair $\{S, R\}$, where S is the structure consisting of several components, and R is the set of relationships between the components of S. These components are descriptors, descriptors and other description schemes, or description schemes. Similar to an MPEG-7 descriptor, an MPEG-7 description scheme must be relevant and effective. Moreover, it must have expression efficiency, extensibility, and scalability with the application and with the data. DS relevance and effectiveness are guaranteed if the DS components and relationships between these components are also relevant and effective. Expression

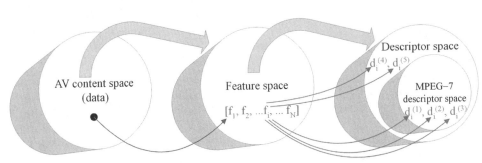

FIGURE 16 Relationship among data, features, and descriptors.

efficiency is guaranteed by obeying the parsimony principle, i.e., by employing the minimum number of DS components and relationships between these components. Finally, the DS should provide a multilevel representation of the data. Other MPEG-7 DS requirements are described in [30].

3.4.3 Description Definition Language

The description definition language (DDL) is the language used to specify the description schemes. MPEG-7 requires that the DDL be explicit by following an unambiguous grammar. Moreover, the DDL should have compositional capabilities, by allowing new DSs to be created and existing DSs to be extended. Most importantly, the DDL should be platform independent [57].

3.4.4 Coding of the Descriptions

A coded description is a representation of the description that allows efficient storage and transmission. MPEG-7 will standardize error resilient and low complexity methods for the efficient coding of the descriptions [57].

3.5 MPEG-7 Example: A Generic Visual Scene

In this section, we discuss how MPEG-7 will provide solutions to the problems associated with CBAM of visual data. Consider the audiovisual scene illustrated in Fig. 1. Suppose that we want to retrieve a picture that is similar to the one shown in the figure by submitting the same query "retrieve all the pictures containing fish" to two different CBAM systems: System A and System B. System A employs Fourier descriptors (coefficients of the Fourier transform of the fish boundary) for shape feature representation. Therefore, it will process the query by retrieving all the objects in the library having similar (according to a specific similarity measure) Fourier descriptors to those of the fish extracted from the query. System B uses geometric descriptors (e.g., area, perimeter) for shape feature representation, and it will then process the query by retrieving all the objects in the library having similar geometric descriptors to those extracted from the fish. The response of the systems A and B to the submitted query will clearly be different, even if both systems were to access the same digital library. This is due to the following two reasons. First, the query may be processed differently by these systems. For example, System A may accept sketch-based queries, whereas System B may accept picture-based queries. Second, even if the query were to be processed in an identical manner, the different shape feature representations and different similarity measures, would most definitely yield different results. This will likely pose problems for most of today's CBAM applications.

MPEG-7 addresses the above problems by providing a standardized description interface. That is, if System A and System B were MPEG-7 compliant, the shape feature representation would be the same in the sense that the two systems would use the same descriptors. Moreover, an MPEG-7 compliant system would achieve a better retrieval performance level than that of existing CBAM systems because of the following reasons. First, MPEG-7 would attach descriptors only to the relevant features. For instance, no descriptors would be attached to the texture feature for the black character shown in Fig. 1. Second, in an MPEG-7 description, the relevant features would be prioritized. For example, a higher importance level would be assigned to the shape descriptors than to the color descriptors for the fish. Finally, MPEG-7 would provide a hierarchical description of the audiovisual scene, as illustrated in Fig. 17. This would allow for coarse to fine representations of the audiovisual content and improve the description's accuracy.

4 Conclusions: Towards a Complete Multimedia Solution

In this chapter, we have presented a comprehensive technical description of the visual parts of the two emerging MPEG standards: MPEG-4 and MPEG-7. We showed, through examples, how these standards will enable many desired functionalities, such as efficient content-based representation, access, and manipulation of multimedia data, which are not addressed properly by today's multimedia standards.

MPEG-4 becomes an international standard in January 1999. A second version of MPEG-4, which will be backward compatible with the first version and will feature more functionalities and profiles, is expected to be completed by the end of 1999. The work of MPEG-7, however, is still in its infancy. In fact, the MPEG-7 call for proposals has just been issued (October 1998). MPEG-7 is expected to become an international standard in September 2001 [29].

MPEG-4 achieves high compression levels, making efficient the communication of multimedia content. Through its object-based representation and modeling tools (e.g., mesh, sprite), MPEG-4 allows us to combine graphics, text, and synthetic/natural objects in a single bit stream. MPEG-4 also features scalability and error resilience functionalities enabling efficient and robust transmission of multimedia data. MPEG-7 will build on MPEG-4, making use of the object-based representation and modeling tools, and providing complementary functionalities. MPEG-7 will facilitate, and even enable, the effective and efficient content-based access and manipulation of multimedia data by providing a standardized description interface.

A decoder that is compliant with both MPEG-4 and MPEG-7 will enable efficient and highly interactive multimedia applications. Consider our example of the visual scene shown in Fig. 1. While watching the TV, the user may want to search for "shirts" that have similar texture to the fish shown on the screen. Because of the object-based representation provided by MPEG-4, the "fish" object can be easily accessed by the user. Also, since MPEG-4 allows the embedding of user data in the bit stream, it is possible to attach MPEG-7 standardized texture descriptors to the corresponding object bit stream. Therefore, the user can access a database without performing expensive decoding,

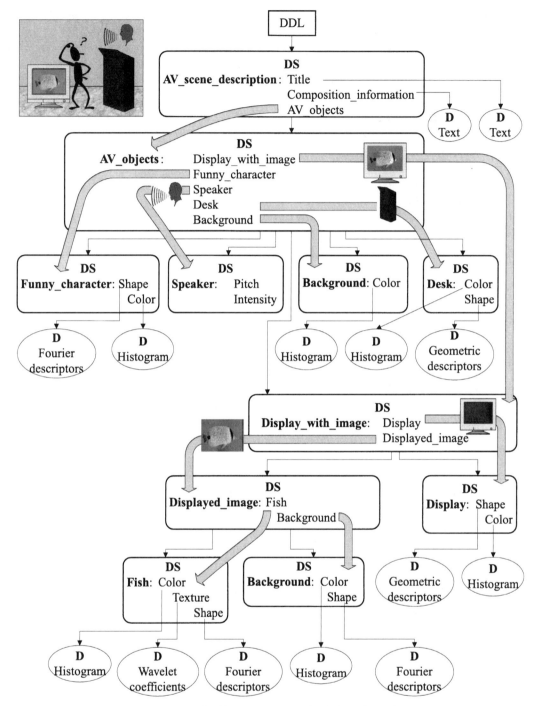

FIGURE 17   Example of description associated with the audiovisual scene.

segmentation, and feature extraction, which would have been required with other representations (e.g., JPEG, MPEG-1/ MPEG-2). The user may also want to search for video sequences that contain persons who are "walking". If the underlying bit stream were compliant with MPEG-2, the only way to achieve this would be to decode the bit stream, reconstruct the video sequences, perform spatiotemporal segmentation, and estimate the motion field corresponding to the person video object. On the other hand, the MPEG-4 mesh model can accurately represent continuous motion. Assuming MPEG-7 standardizes mesh motion, the corresponding descriptors can be used by the user to search for objects with similar motion trajectories. Another case is that in which the user wants to search for persons who are "smiling". MPEG-7 may standardize descriptors that are expressed in terms of MPEG-4s FAPs, described in the previous section. Since it is possible to tell the mood of the speaker (e.g., joyful, sad, angry) by the FAPs, the search for a "smiling" person can be easily performed, again without performing expensive processes, such as decoding, segmentation, and feature extraction. These expensive processes have to be performed only once at the encoder end, making MPEG-7/MPEG-4 compliant systems well suited for many applications.

Together, MPEG-4 and MPEG-7 will provide a complete multimedia system solution by allowing the efficient and effective representation, exchange, storage, access, and manipulation of multimedia data. They are expected to enable key technologies for the new generation multimedia applications, revolutionizing our multimedia world.

References

[1] ISO/IEC, "Information technology — coding of moving pictures and associated audio for digital storage media at up to about 1.5 mbits/s: video," 11172–2, 1993.

[2] ISO/IEC, "Information technology — generic coding of moving pictures and associated audio information: video," 13818–2, 1995.

[3] ITU-T, "Video coding for low bit rate communication," recommendation H.263, 1996.

[4] ITU-T, "Video coding for low bit rate communication," recommendation H.263, version 2, 1998.

[5] J. Osterman and A. Puri, "Natural and synthetic video in MPEG–4," in Proceedings of the IEEE International Conference on Acoustics, Speech, and Signal Processing (IEEE, New York, 1998), Vol. 5, pp. 3805–3808.

[6] ISO/IEC JTC1/SC29/WG11, "MPEG-4 overview," N2323, July 1998.

[7] MPEG-4 Systems Group, "Coding of audio-visual objects: systems," ISO/IEC JTC1/SC29/WG11 N2201, May, 1998.

[8] MPEG-4 Video Group, "Coding of audio-visual objects: video," ISO/IEC JTC1/SC29/WG11 N2202, March, 1998.

[9] MPEG-4 Audio Group, "Coding of audio-visual objects: audio," ISO/IEC JTC1/SC29/WG11 N2203, March, 1998.

[10] F. Pereira, "MPEG-4: a new challenge for the representation of audiovisual information," in Picture Coding Symposium '96 (Melbourne, Australia, 1996), pp. 7–16.

[11] T. Sikora, "MPEG-4 very low bit rate video," in Proceedings of the IEEE International Symposium on Circuits and Systems (IEEE, New York, 1997).

[12] ISO/IEC JTC1/SC29/WG11, "MPEG-4 systems FAQ, version 7.0a," N2527, October, 1998.

[13] ISO/IEC JTC1/SC29/WG11, "MPEG-4 video FAQ," N1713, April, 1997.

[14] ISO/IEC JTC1/SC29/WG11, "MPEG-4 requirements, version 9," N2456, October, 1998.

[15] ISO/IEC JTC1/SC29/WG11, "Description of MPEG-4," N1410, October, 1996.

[16] T. Sikora, "MPEG-4 video and its potential for future multimedia services," in Proceedings of the IEEE International Symposium on Circuits and Systems (IEEE, New York, 1997).

[17] MPEG-4 SNHC Group, "SNHC verification model 9.0," ISO/IEC JTC1/SC29/WG11 MPEG98/M4116, October, 1998.

[18] MPEG-4 Video Group, "MPEG-4 video verification model version 8.0," ISO/IEC JTC1/SC29/WG11 N1796, July, 1997.

[19] R. M. Haralick and L.H. Shapiro, "Image segmentation techniques," in Journal of Computer Vision, Graphics and Image Processing (Academic Press, 1985), pp. 100–132.

[20] A. K. Katsaggelos, L. P. Kondi, F. W. Meier, J. Ostermann, and G. M. Schuster, "MPEG-4 and rate-distortion-based shape-coding techniques," Proc. IEEE 86, 1029–1051 (1998).

[21] M. Lee, W. Chen, C. B. Lin, C. Gu, T. Markoc, S. Zabinsky, and R. Szeliski, "A layered video object coding system using sprite and affine motion model," in IEEE Trans. CSVT 7, 130–146 (1997).

[22] M. Tekalp, P. V. Beek, C. Toklu, and B. Gunsel, "Two-dimensional mesh-based visual-object representation for interactive synthetic/natural video," Proc. IEEE 86, 1126–1154 (1998).

[23] P. Kalra, A. Mangili, T. N. Magnenat, and D. Thalmann, "Simulation of facial muscle actions based on rational free form deformations," in Proc. Eurographics, pp. 59–69 (1992).

[24] P. Doenges, T. Capin, F. Lavagetto, J. Ostermann, I. S. Pandzic, and E. Petajan, "MPEG-4: Audio/video and synthetic graphics/audio for real-time, interactive media delivery," Image Commun. May, 433–463 (1997).

[25] B. G. Haskell, P. G. Howard, Y. A. Lecun, A. Puri, J. Ostermann, M. R. Civanlar, L. Rabiner, L. Bottou, and P. Haffner, "Image and video coding — emerging standards and beyond," Proc. CSVT 8, pp. 814–837 (1998).

[26] S. A. Martucci, I. Sodagar, T. Chiang, and Y. Zhang, "A zerotree wavelet video coder," in IEEE Trans. CSVT 7, 109–118 (1997).

[27] R. Talluri, "Error resilient video coding in the MPEG-4 standard," IEEE Commun. Mag. 26, pp. 112–119 (1998).

[28] Microsoft, "MPEG-4 video encoder/decoder," http://drogo.cselt. stet.it/ufv/leonardo/mpeg/public/mpeg-4_fcd/Visual/Natural, 1998.

[29] ISO/IEC JTC1/SC29/WG11, "MPEG-7: context and objectives," N2460, October, 1998.

[30] ISO/IEC JTC1/SC29/WG11, "MPEG-7: requirements document ver. 7," N2461, October, 1998.

[31] Y. Rui, T. S. Huang, and S. Mehrotra, "Content–based image retrieval with relevance feedback in Mars," in Proceedings of the International Conference on Image Processing (IEEE, Santa Barbara, California, 1997).

[32] Y. Kageyama and H. Saito, "Image retrieval system capable of learning the user's sensibility using neural networks," in Proceedings of the International Conference on Neural Networks (IEEE, Texas, US, 1997), pp. 1543–1567.

[33] University of Pisa, "Icon browser," http://www.cli.di.unipi.it/iconbrowser/, 1998.

[34] Excalibur Technologies Corporation, "Image surfer," http://www.interpix.com, 1998.

[35] Carnegie Mellon University, "Lycos media," http://www.lycos.com/picturethis/, 1998.

[36] Imagiware Inc., "Virtual image archive," http://www.imagiware.com/via/search.html/, 1998.

[37] Excalibur Technologies Corporation, "Yahoo image surfer," http://ipix.yahoo.com/, 1998.

[38] "Whoopie," http://www.whoopie.com, 1998.

[39] K. Messer and J. Kittler, "Using feature selection to aid an iconic search through an image database," in Proceedings of the IEEE International Conference on Acoustics, Speech, and Signal Processing (IEEE, New York, 1997), pp. 2605–2608.

[40] D. Androutsos, K. N. Plataniotis, and A. N. Venetsanopoulos, "Efficient image database filtering using color vector techniques," in

Proceedings of Canadian Conference on Electrical and Computer Engineering (Newfoundland, Canada, 1997), pp. 827–830.

[41] K. Liang and C. J. Kuo, "Progressive image indexing and retrieval based on embedded wavelet coding," in *Proceedings of the International Conference on Image Processing*, Santa Barbara, CA, October, 1997.

[42] M. Beatty and B. S. Manjunath, "Dimensionality reduction using multi-dimensional scaling for content-based retrieval," in *Proceedings of the International Conference on Image Processing*, Santa Barbara, CA, October, 1997.

[43] Y. Q. Chen, "Novel techniques for image texture classification," Ph.D. dissertation (Dept. of Electronics and Computer Science, U. of Southampton, UK, 1995).

[44] P. P. Raghu and B. Yegnanarayana, "Segmentation of Gabor-filtered textures using deterministic relaxation," *IEEE Trans. Image Process.* **5**, 1625–1636 (1996).

[45] R. M. Haralick, K. Shanmugam, and I. Dinstein, "Textural features for image classification," *IEEE Trans. Syst. Man Cybernet.* **SMC–3**, 610–621 (1973).

[46] A. K. Jain and K. Karu, "Learning texture discrimination masks," *IEEE Trans. Pattern Anal. Machine Intell.* **18**, 195–205 (1996).

[47] J. P. Eakins, J. M. Boardman, and M. E. Graham, "Similarity retrieval of trademark images," *IEEE Multimed.* April-June, 53–63 (1998).

[48] F. Dell'Acqua and P. Gamba, "Simplified modal analysis and search for reliable shape retrieval," *IEEE Trans. Circuits Syst. Video Technol.* **8**, pp. 654–666 (1998).

[49] X. Wan and C. J. Kuo, "A new approach to image retrieval with hierarchical color clustering," *IEEE Trans. Circuits Syst. Video Technol.* **8**, pp. 628–643 (1998).

[50] C. Y. Yee, K. Tan, T. S. Chua, and B. C. Ooi, "An empirical study of color-spatial retrieval techniques for large image databases," in *Proceedings of IEEE International Conference on Multimedia Computing and Systems* (IEEE, New York, 1998), pp. 218–221.

[51] Columbia University, "WebSeek — a content-based image and video search and catalog tool for the Web," http://www.ctr.columbia.edu/webseek/, 1998.

[52] E. Saber and A. M. Tekalp, "Integration of color, shape, and texture for image annotation and retrieval," in *Proceedings of the International Conference on Image Processing*, Lausanne, Switzerland, September 16–19, 1996.

[53] IBM, "Query by image content (QBIC) homepage," http://wwwqbic.almaden.ibm.com, 1998.

[54] W. Y. Ma and B. S. Manjunath, "NeTra: A toolbox for navigating large image databases," in *Proceedings of the International Conference on Image Processing*, Santa Barbara, CA, October 26–29, 1997.

[55] T. Chen and R. R. Rao, "Audio-visual interaction in multimedia communication," in *Proceedings of the IEEE International Conference on Acoustics, Speech, and Signal Processing* (IEEE, New York, 1997), pp. 179–182.

[56] ISO/IEC JTC1/SC29/WG11, "MPEG-7: proposal package description," N2464, October, 1998.

[57] ISO/IEC JTC1/SC29/WG11, "MPEG-7 Evaluation process document," N2463, October, 1998.

VII

Image and Video Acquisition

Image Scanning, Sampling, and Interpolation

Jan P. Allebach
Purdue University

1 Image Capture

Image capture takes us from the continuous-parameter real world in which we live to the discrete-parameter, amplitude quantized domain of the digital devices that comprise an electronic imaging system. The process of converting from a continuous-parameter image to one that is discrete parameter, i.e., consists of an array of numbers, is referred to as *sampling*. The meaning of the term *scanning* is somewhat less precise. Its common usage refers to the notion of sequential acquisition of data through some type of electromechanical motion. It is also used to refer to the process of converting a two-dimensional signal into a signal that is one dimensional. The process of quantizing an image that is continuous in amplitude to one that takes on values from a finite set is called *quantization*. Examples illustrating the effect of quantization may be found in Chapter 1.1.

1.1 Representations for the Sampled Image

Sampling a continuous-space image $g_c(x, y)$ yields a discrete-space image

$$g_d(m, n) = g_c(mX, nY), \qquad (1)$$

where the subscripts c and d denote, respectively, continuous

space and discrete space, and (X, Y) is the spacing between sample points, also called the pitch. However, it is also convenient to represent the sampling process by using the 2-D Dirac delta function $\delta(x, y)$. In particular, we have from the sifting property of the delta function that multiplication of $g_c(x, y)$ by a delta function centered at the fixed point (x_0, y_0) followed by integration will yield the sample value $g_c(x_0, y_0)$, i.e.,

$$g_c(x_0, y_0) = \iint g_c(x, y)\delta(x - x_0, y - y_0)\, dx\, dy, \qquad (2)$$

provided $g_c(x, y)$ is continuous at (x_0, y_0). It follows that

$$g_c(x, y)\delta(x - x_0, y - y_0) \equiv g_c(x_0, y_0)\delta(x - x_0, y - y_0); \qquad (3)$$

that is, multiplication of an impulse centered at (x_0, y_0) by the continuous-space image $g_c(x, y)$ is equivalent to multiplication of the impulse by the constant $g_c(x_0, y_0)$. It will also be useful to note from the sifting property that

$$g_c(x, y) * \delta(x - x_0, y - y_0) = g_c(x - x_0, y - y_0). \qquad (4)$$

That is, convolution of a continuous-space function with an impulse located at (x_0, y_0) shifts the function to (x_0, y_0).

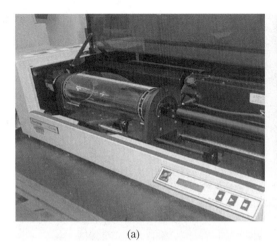 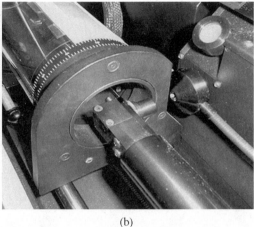

(a)   (b)

FIGURE 1   High-resolution drum scanner: (a) scanner with cover open, and (b) closeup view showing screw-mounted "C" carriage with light source on the inside arm, and detector optics on the outside arm. (See color section, p. C–32.)

To get all the samples of the image, we define the comb function

$$\text{comb}_{X,Y}(x, y) = \sum_m \sum_n \delta(x - mX, y - nY). \qquad (5)$$

Then we define the continuous-parameter sampled image, denoted with the subscript s, as

$$g_s(x, y) = g_c(x, y)\, \text{comb}_{X,Y}(x, y), \qquad (6)$$

$$= \sum_m \sum_n g_d(m, n)\delta(x - mX, y - nY). \qquad (7)$$

We see from Eq. (7) that the continuous- and discrete-space representations for the sampled image contain the same information about its sample values. In the sequel, we shall only use the subscripts c and d when necessary to provide additional clarity. In general, we can distinguish between functions that are continuous space and those that are discrete space on the basis of their arguments. We will usually denote continuous-space independent variables by (x, y) and discrete-space independent variables by (m, n).

1.2 Image Capture Technologies

There are two fundamental aspects of image capture. The first is the *raster* of points in two- or three-dimensional space where samples are taken. The second is the effect of the system *aperture*, which causes the data samples to consist of an average of the image or scene within a neighborhood of the nominal sampling point.

Devices for image capture may be divided into two classes, according to the mechanism by which the samples are acquired. The first class utilizes a *flying spot* mechanism for data acquisi-

tion. Examples of such mechanisms include an electron beam, as is used in an analog video camera, the electromechanical scan resulting from rotation of a drum and movement of a screw, as can be found in graphic arts drum scanners (see Fig. 1), diffractive optical beam formation, as is used in supermarket point-of-sale scanners, and phased array beam formation, as is used with radar. With all these systems, the spot trajectory and read times determine the sampling raster, whereas the aperture effects are governed by the shape of the illuminating and read spots and the dwell time, i.e., the time interval during which the read spot output signal is averaged to form a sample value. Although none of the examples cited above operate in this manner, flying spot scanners can also function in a passive mode. In this case, there is no write spot; and the read spot detects radiation emanating naturally from the scene. Air and spaceborne systems for remote sensing of the Earth's surface are examples of passively scanning systems.

The second class of image capture devices utilizes a *focal plane mosaic*, which consists of an array of detector sites. The scene is imaged onto the surface of the array; each detector integrates the radiation gathered from the active area of its surface. This gives rise to the aperture effect. The spatial arrangement of the detectors determines the sampling raster. Focal plane array technologies include charge coupled devices (CCDs), charge injection devices (CIDs), and CMOS devices. These technologies are widely used in digital still and video cameras. Some systems comprise a hybrid of the flying spot and focal plane mosaic architectures. The flatbed scanner, which uses a mechanical means to move a one-dimensional array of detectors across the surface of the document being scanned, is a good example.

1.3 General Model for the Image Capture Process

Despite the diversity of technologies and architectures for image capture devices, it is possible to cast the sampling process for all of these systems within a common framework. We will illustrate this fact for two examples. The first example is that of a flying spot scanner. Since this device acquires data *time sequentially*, i.e., one sample at a time, we can represent the scanned signal as a function of the single time parameter t. Accordingly, the operation of this device is described by

$$s(t) = \int_{-\infty}^{\infty} \int_{-\infty}^{\infty} p_i[\xi - x_s(t), \eta - y_s(t)]$$
$$\times p_r[\xi - x_s(t), \eta - y_s(t)] \, g(\xi, \eta) \, d\xi \, d\eta. \quad (8)$$

Here $s(t)$ is the continuous-time signal generated at the detector output, prior to A/D conversion; $p_i(x, y)$ and $p_r(x, y)$ are the illuminating and read spots, respectively; $[x_s(t), y_s(t)]$ is the trajectory of these spots across the image as a function of time; and $g(x, y)$ is the image to be sampled. What this equation shows is that at any time t, the detector output is given by an integral over the entire image $g(x, y)$, weighted by the spatially varying intensity of the illuminating spot and the spatially varying sensitivity of the detector (read spot), which are both centered at the trajectory coordinates $[x_s(t), y_s(t)]$ at that time. The final step in the sampling process is to sample the detector output at an appropriate set of times, yielding $s_d(k) = s(t_k)$, where again the subscript d denotes the fact that $s_d(k)$ is a discrete-time signal, and t_k is the sequence of sampling times, which are not necessarily uniformly spaced.

The scanning trajectory $[x_s(t), y_s(t)]$ and the set of sampling times t_k combine to determine the set of spatial points (x_k, y_k) at which samples are acquired. We shall represent each such sampling point by a 2-D Dirac delta function $\delta(x - x_k, y - y_k)$, and the entire set of sampling points by the sampling function

$$q(x, y) = \sum_k \delta(x - x_k, y - y_k). \quad (9)$$

Because the image is not time varying, the order in which the samples are acquired is immaterial to the characteristics of the sampled 2-D signal.

Since the illuminating and read spot functions have the same arguments, they may be combined as a single function $p(x, y) = p_i(-x, -y) p_r(-x, -y)$, which accounts for all aperture effects due to the flying spot scanning process. We have reflected the coordinates simply for mathematical convenience. In addition, the averaging effect of the aperture may be represented as a 2-D convolution of the continuous-parameter image $g(x, y)$ with the aperture function; so the continous-parameter representation of the sampled image is thus given by

$$g_s(x, y) = q(x, y)[p(x, y) * g(x, y)]. \quad (10)$$

With the appropriate choice of sampling times t_k for $s(t)$ in Eq. (8) and sampling points (x_k, y_k) for $q(x, y)$ in Eq. (10), these two representations for the sampled image are completely equivalent.

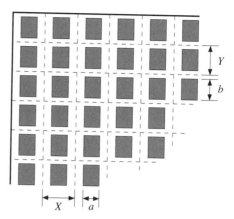

FIGURE 2   Focal plane array geometry.

The second example that we wish to consider is that of a 2-D focal plane array illustrated in Fig. 2. Here each sample is obtained by integrating over the active area of the corresponding detector; so we have

$$g_d(m, n) = \int_{mX - a/2}^{mX + a/2} \int_{nY - b/2}^{nY + b/2} g(\xi, \eta) \, d\xi \, d\eta, \quad (11)$$

where as before the spacing between sample points is $X \times Y$ and the size of the active area of each detector is $a \times b$. The averaging effect of the active area of the detector can again be accounted for by convolution with an appropriately chosen aperture function, in this case

$$p(x, y) = \text{rect}(x/a, y/b), \quad (12)$$

where $\text{rect}(x, y)$ is defined to be 1 if $|x| < 1/2$ and $|y| < 1/2$, and 0, otherwise. The sampling function is given by

$$q(x, y) = \text{comb}_{X, Y}(x, y). \quad (13)$$

With $p(x, y)$ given by Eq. (12) and $q(x, y)$ given by Eq. (13), Eq. (10) is completely equivalent to Eq. (11).

To summarize, the sampling process for a broad group of image capture devices may be modeled as a convolution with an appropriately chosen aperture function $p(x, y)$ followed by multiplication by an appropriate sampling function $q(x, y)$.

2 Fourier Analysis of Image Capture

Fourier analysis sheds a great deal of light on the effect of the sampling process. However, before we get to that, it will be helpful to first define the different spectral representations that we will be using.

2.1 Spectral Representations for Discrete- and Continuous-Space Signals

For continuous-space images, the appropriate spectral representation is the continuous-space Fourier transform (CSFT). The forward and inverse versions of this transform are given respectively by

$$G_c(u, v) = \iint g_c(x, y) e^{-i2\pi(ux + vy)} \, dx \, dy. \qquad (14)$$

$$g_c(x, y) = \iint G_c(u, v) e^{i2\pi(ux + vy)} \, du \, dv. \qquad (15)$$

The units of frequency for (u, v) are cycles/unit distance. For discrete-space images, we use the discrete-space Fourier transform (DSFT) defined as

$$G_d(U, V) = \sum_m \sum_n g_d(m, n) e^{-i2\pi(Um + Vn)}, \qquad (16)$$

$$g_d(m, n) = \int_{-0.5}^{0.5} \int_{-0.5}^{0.5} G_d(U, V) e^{i2\pi(Um + Vn)} \, dU \, dV. \qquad (17)$$

The units of frequency for (U, V) are cycles/pixel. Again, we shall use the subscripts c and d only where needed for clarity.

In Section 1.1, we defined both continous-parameter and discrete-parameter representations for the sampled signal. To examine the spectral form of continuous-parameter representation (6), we first note that

$$\text{comb}_{X,Y}(x, y) \overset{\text{CSFT}}{\longleftrightarrow} \frac{1}{XY} \text{comb}_{\frac{1}{X}, \frac{1}{Y}}(u, v). \qquad (18)$$

Then by the convolution theorem, we have that the CSFT of Eq. (6) is given by

$$G_s(u, v) = G_c(u, v) * \frac{1}{XY} \text{comb}_{\frac{1}{X}, \frac{1}{Y}}(u, v). \qquad (19)$$

It follows directly from the definition of comb function (5) and the convolution property of impulse (4) that

$$G_s(u, v) = \frac{1}{XY} \sum_k \sum_l G_c\left(u - \frac{k}{X}, v - \frac{l}{Y}\right). \qquad (20)$$

So sampling a continuous-space function on a lattice with interval (X, Y) causes the CSFT of that function to be replicated in the frequency domain on a lattice with interval $(1/X, 1/Y)$, and scaled overall by $1/XY$.

To relate this result to the DSFT of $g_d(m, n)$, we take the CSFT of Eq. (7) directly. Interchanging the summation over the terms in the comb function with the Fourier integral, and using sifting property (2), we obtain

$$G_s(u, v) = \sum_m \sum_n g_d(m, n) e^{-i2\pi(umX + vnY)}. \qquad (21)$$

Comparing this to Eq. (16), we see that Eq. (21) can be put in the form of the DSFT of $g_d(m, n)$ with an appropriate change of frequency variables. Thus

$$G_s(u, v) = G_d(u/\mathcal{U}, v/\mathcal{V}), \qquad (22)$$

where $\mathcal{U} = 1/X$ and $\mathcal{V} = 1/Y$ are the sampling frequencies in the horizontal and vertical directions in units of cycles/unit distance. Thus we see that there is a simple and direct relation between the CSFT of the continuous-space representation of the sampled image and the DSFT of the discrete-space representation of that image. Combining Eqs. (20) and (22), we obtain

$$G_d(U, V) = \mathcal{U}\mathcal{V} \sum_k \sum_l G_c((U - k)\mathcal{U}, (V - l)\mathcal{V}). \qquad (23)$$

2.2 The General Image Capture Model Revisited

We are now ready to examine our general model for image capture from a frequency domain perspective. Taking the CSFT of Eq. (10) and using the convolution and product theorems, we obtain

$$G_s(u, v) = Q(u, v) * [P(u, v)G(u, v)]. \qquad (24)$$

So we see that the spectrum of the sampled image is obtained by multiplying the spectrum of the continuous-space image by the CSFT $P(u, v)$ of the aperture function $p(x, y)$, and then convolving with the CSFT $Q(u, v)$ of the sampling function $q(x, y)$. Let us denote the effect of multiplication by the CSFT of the aperture with a tilde:

$$\tilde{G}(u, v) = P(u, v)G(u, v). \qquad (25)$$

For the special case where $q(x, y) = \text{comb}_{X,Y}(x, y)$, we then have from Eq. (20) that

$$G_s(u, v) = \mathcal{U}\mathcal{V} \sum_k \sum_l \tilde{G}(u - k\mathcal{U}, v - l\mathcal{V}). \qquad (26)$$

Figure 3 illustrates this result. Let us first assume that there is no aperture effect; so $P(u, v) \equiv 1$ and $\tilde{G}(u, v) \equiv G(u, v)$. We see that a sufficient condition for the spectral replications to not overlap is that

$$G(u, v) \neq 0 \quad \text{only if } |u| < \mathcal{U}/2 \text{ and } |v| < \mathcal{V}/2. \qquad (27)$$

This is referred to as the Nyquist condition. Since $1/X = \mathcal{U}$ and

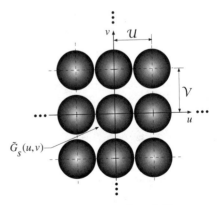

FIGURE 3   Spectrum of sampled image.

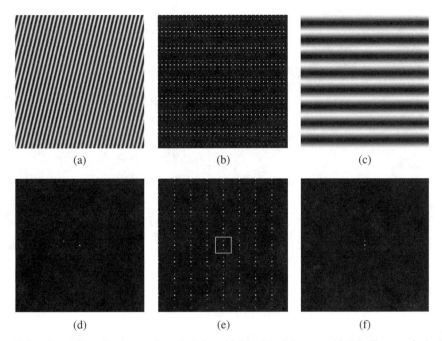

FIGURE 4   Effect of undersampling a 2-D sinusoid: (a) original sinewave with DC offset to make it nonnegative; (b) sampled sinewave (a); (c) reconstruction obtaining by band limiting (b) to the Nyquist limit; (d)–(f) spectra of (a)–(c), respectively. The square in (e) indicates frequencies below the Nyquist limit.

$1/Y = \mathcal{V}$, this condition has the interpretation that we must sample at least twice per cycle of the highest horizontal and vertical frequencies found in the image. Provided the Nyquist condition is satisfied, we see that $G(u, v)$ may be recovered from $G_s(u, v)$ by multiplication with a scaled 2-D rect function.

$$G(u, v) = \frac{1}{\mathcal{U}\mathcal{V}}\text{rect}(u/\mathcal{U}, v/\mathcal{V})G_s(u, v). \qquad (28)$$

Using the product theorem and the scaling property, we obtain in the spatial domain

$$g(x, y) = \text{sinc}(x/X, y/Y) * g_s(x, y), \qquad (29)$$

which using Eq. (7) can be expressed as

$$g(x, y) = \sum_m \sum_n g(mX, nY)\text{sinc}(x - mX, y - nY), \qquad (30)$$

where $\text{sinc}(x, y) \equiv \sin(\pi x)\sin(\pi y)/(\pi x)(\pi y)$. This is the Whitaker–Kotelnikov–Shannon sampling expansion, which shows that we can reconstruct an appropriately bandlimited image by interpolating between samples with a sinc function. This result is commonly known as the 2-D sampling theorem.

When the Nyquist condition is not satisfied, any frequency component in the continuous-parameter image that lies outside the region

$$\Omega_{\mathcal{U},\mathcal{V}}(u, v) = \{(u, v) : |u| < \mathcal{U}/2, |v| < \mathcal{V}/2\}, \qquad (31)$$

will fold back into $\Omega_{\mathcal{U},\mathcal{V}}$, thus mimicking a lower frequency. Figure 4 illustrates this for the case of a simple 2-D sinusoid. This phenomenon is known as *aliasing*. In images reconstructed from undersampled data, it manifests itself as moire patterns and staircasing or "jaggies" along straight edges.

Figure 5 illustrates the effect of undersampling a real image. At the top of Fig. 5(a), we see a jagged edge along the crest of the dune. In addition, close inspection of the ripples in the sand reveals what appear to be fine lines oriented at 90° to the ripples. Both these artifacts are due to undersampling. In Fig. 5(b), we see that the energy in the Fourier transform oriented along a fine line at about 75° from the positive U axis has folded back, creating short diagonal line segments in the second and fourth quadrants. This spectral component corresponds to the edge of the crest. In addition, the more diffuse cloud of energy oriented at 45° to the positive U axis has folded back, creating clouds in the upper left and lower right corners of the spectrum. This spectral component corresponds to the ripples in the sand.

The Nyquist condition may be stated more generally in a necessary and sufficient form: If and only if the support of $G(u, v)$ does not exceed an area of size $\mathcal{U}\mathcal{V}$, $g(x, y)$ may be reconstructed from its samples taken on a rectangular lattice at interval $(1/\mathcal{U}, 1/\mathcal{V})$. The interpolating function will be the inverse CSFT of the indicator function for the support region, scaled by $1/\mathcal{U}\mathcal{V}$.

Now let's consider the effect of the aperture indicated by Eq. (25). If the CSFT $P(u, v)$ of the aperture rolls off at

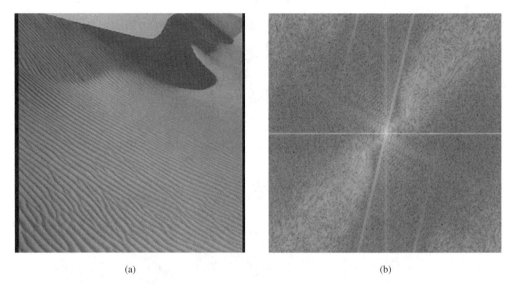

(a)   (b)

FIGURE 5 Effect of undersampling an image of a sand dune: (a) undersampled image, and (b) the magnitude of its Fourier transform.

frequencies outside $\Omega_{U,V}$, the aperture will attenuate any frequencies in the continuous-parameter image $g(x, y)$ outside $\Omega_{\mathcal{U},\mathcal{V}}$, thereby suppressing aliasing. This desirable effect is known as *prescan band limitation* or *antialiasing*. In contrast, if $P(u, v)$ rolls off, i.e., $|P(u, v)| < 1$, for frequencies $(u, v) \in \Omega_{\mathcal{U},\mathcal{V}}$, then the aperture will have the undesired effect of attenuating frequencies in $g(x, y)$ that are not undersampled. Typically, it is the higher frequencies in the image that are attenuated in this manner, resulting in an image that looks soft or slightly blurred after reconstruction using Eq. (28) or (30). Provided $|P(u, v)| > 0$ for all frequencies $(u, v) \in \Omega_{\mathcal{U},\mathcal{V}}$, this effect may be compensated by replacing Eq. (28) with

$$G(u, v) = \frac{1}{UV} \text{rect}(u/\mathcal{U}, v/\mathcal{V}) \left[P(u, v)\right]^{-1} G_s(u, v). \quad (32)$$

Of course, at frequencies within $\Omega_{\mathcal{U},\mathcal{V}}$ where $|P(u, v)|$ is small, this reconstruction procedure will amplify any noise present in the sampled data.

2.3 Sampling with Nonrectangular Lattices

We saw in the preceding section that sampling an image on a rectangular lattice with interval $X \times Y$ causes replication of the image spectrum on a reciprocal lattice that is also rectangular, and which has interval $1/X \times 1/Y$. To prevent aliasing, these replications must be spaced far enough apart to prevent overlap. For an image band limited to a circular band region with highest frequency W, we must have $1/X > 2W$ and $1/Y > 2W$. The minimum sampling density is given by

$$d_R = \frac{1}{XY} = 4W^2 \text{ samples/unit area.} \quad (33)$$

Figure 3 shows a situation in which the sampling in the vertical direction slightly exceeds the Nyquist rate. However, even if the

sampling were at the Nyquist rate, the spectral replications would not completely cover the frequency domain. This suggests that it may be possible to use a different lattice that will more tightly pack the spectra in the frequency domain, resulting in a spreading of samples on the reciprocal lattice in the spatial domain, and hence a lower sampling density.

It is well known that the lattice that most tightly packs circles is hexagonal. Figure 6 shows the corresponding spatial lattice. Each sample point has six equidistant neighbors, which are all separated from it by angles of $60°$. To determine the reciprocal lattice for this sampling structure, we represent it using two interlaced rectangular lattices with the same period, as indicated in Fig. 6; so

$$q(x, y) = \text{comb}_{X,Y}(x, y) + \text{comb}_{X,Y}(x - X/2, y - Y/2),$$

$$(34)$$

where $X = 1/(2W)$ and $Y = 1/(\sqrt{3}W)$. To determine the corresponding reciprocal lattice, we calculate the CSFT of Eq. (34),

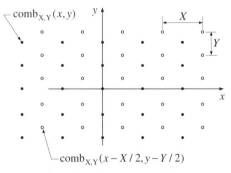

FIGURE 6 Hexagonal sampling lattice represented as superposition of two interleaved rectangular lattices.

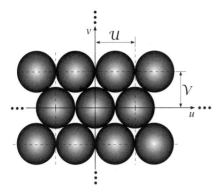

FIGURE 7   Spectrum of image sampled on a hexagonal lattice.

using the shifting property of the Fourier transform to yield

$$Q(u, v) = \frac{1}{XY}\text{comb}_{\frac{1}{X}, \frac{1}{Y}}(u, v)$$
$$+ \frac{1}{XY}\text{comb}_{\frac{1}{X}, \frac{1}{Y}}(u, v)e^{-i\pi(Xu+Yv)},$$
$$= \frac{1}{XY}\sum_k\sum_l\left(1 + e^{-i\pi(m+n)}\right)\delta\left(u - \frac{k}{X}, v - \frac{l}{Y}\right). \tag{35}$$

Because

$$1 + e^{-i\pi(m+n)} = \begin{cases} 2, & m+n \text{ even} \\ 0, & m+n \text{ odd} \end{cases}, \tag{36}$$

the reciprocal lattice is also hexagonal. So the spectrum

$$G_s(u, v) = \mathcal{U}\mathcal{V}\sum_k\sum_l\left(1 + e^{-i\pi(m+n)}\right)\tilde{G}\left(u - k\mathcal{U}, v - l\mathcal{V}\right) \tag{37}$$

of the sampled image appears as shown in Fig. 7. Here, $\mathcal{U} = 1/X$ and $\mathcal{V} = 1/Y$, as before. Now the sampling density is

$$d_H = \frac{2}{XY} = 4\sqrt{3}W^2. \tag{38}$$

The savings is $d_H/d_R = \sqrt{3}/2 = 0.866$, or 13.4%.

The hexagonal lattice is only one example of a nonrectangular lattice. Such lattices can be treated in a more general context of lattice theory. This framework is developed in Chapter 7.2.

3 Sampling Rate Conversion

In some instances, it is desirable to change the sampling rate of a digital image. This section addresses the procedures for doing this, and the effect of sampling rate changes.

3.1 Downsampling and Decimation

To decrease the sampling rate of a digital image $f(m, n)$ by integer factors of $C \times D$, we can *downsample* it by using

$$g_\downarrow(m, n) = f(Cm, Dn). \tag{39}$$

So we simply discard all but every Cth sample in the m direction and all but every Dth sample in the n direction. To understand the effect of this operation, we derive an expression for the DSFT of $g_\downarrow(m, n)$ in terms of that of $f(m, n)$. By definition (16),

$$G_\downarrow(U, V) = \sum_m\sum_n f(Cm, Dn)e^{-i2\pi(Um+Vn)}. \tag{40}$$

With a change of indices of summation, we can write

$$G_\downarrow(U, V) = \sum_m\sum_n s_C(m)s_D(n)f(m, n)e^{-i2\pi(Um/C+Vn/D)}, \tag{41}$$

where

$$s_C(m) = \begin{cases} 1, & m/C \text{ is an integer} \\ 0, & \text{else} \end{cases}. \tag{42}$$

Since $s_C(m)$ is periodic with period C, it can be expressed as a discrete Fourier series. Within one period, $s_C(m)$ consists of a single impulse; so the Fourier coefficients all have value unity. Thus we can write

$$s_C(m) = \frac{1}{C}\sum_{k=0}^{C-1}e^{-i2\pi(mk/C)}. \tag{43}$$

Substituting this into Eq. (41), and interchanging orders of summation, we obtain

$$G_\downarrow(U, V) = \frac{1}{CD}\sum_{k=0}^{C-1}\sum_{l=0}^{D-1}\sum_m\sum_n$$
$$\times f(m, n)e^{-i2\pi[(U-k)m/C+(V-l)n/D]}, \tag{44}$$

which may be recognized as

$$G_\downarrow(U, V) = \frac{1}{CD}\sum_{k=0}^{C-1}\sum_{l=0}^{D-1}F((U-k)/C, (V-l)/D). \tag{45}$$

So we see that downsampling the image causes the DSFT to be expanded by a factor of C in the U direction and a factor of D in the V direction. This is a consequence of the fact that the image has contracted by these same factors in the spatial domain. The DSFT $G(U, V)$ is comprised of a summation of CD replications of the expanded DSFT $F(U/C, V/D)$ shifted by unit intervals in both the U and V directions. Figure 8 illustrates the overall result. Here the downsampling has resulted in overlap of the spectral replications of $F(U/C, V/D)$, thus resulting in aliasing.

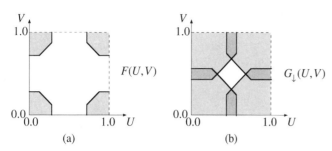

FIGURE 8   Effect of downsampling by factor of 2×2 on the DSFT of image: (a) before downsampling, (b) after downsampling.

The most important consequence of downsampling is the potential for additional aliasing, which will occur if $F(U, V) \neq 0$ for any $|U| \geq 1/(2C)$ or $|V| \geq 1/(2D)$. To prevent this, we can prefilter $f(m, n)$ prior to downsampling with a filter having frequency response

$$H(U, V) = CD \, \text{rect}(CU, DV). \tag{46}$$

The impulse response corresponding to this filter is

$$h(m, n) = \text{sinc}(m/C, n/D). \tag{47}$$

Because of the large negative sidelobes and slow roll-off of the filter, it can result in undesirable ringing at edges when it is truncated to finite extent. This is known as the *Gibb's phenomenon*. It can be avoided by tapering the filter with a window function. In practice, it is common to simply average the image samples within each $C \times D$ cell, or to use a Gaussian filter. The combination of a filter followed by a downsampler is called a *decimator*. Figure 9 shows the block diagram of such a system. Its net effect is

$$g(m, n) = \sum_k \sum_l f(k, l) h(Cm - k, Dn - l). \tag{48}$$

Here we have dropped the down-arrow subscript to denote the fact that we are not just downsampling, but rather are filtering first.

3.2 Upsampling and Interpolation

To understand how we increase the sampling rate of an image $f(m, n)$ by integer factors C in the m direction and D in the n direction, it is helpful to start with an *upsampler*, which inserts $C - 1$ zeros between each sample in the m direction and $D - 1$

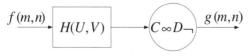

FIGURE 9   Decimator.

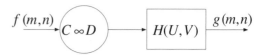

FIGURE 10   Interpolator.

zeros between each sample in the n direction:

$$g_\uparrow(m, n) = \begin{cases} f(m, n), & m/C \text{ and } n/D \text{ are integers} \\ 0, & \text{else} \end{cases}. \tag{49}$$

We again seek an expression for the DSFT of $g_\uparrow(m, n)$ in terms of that of $f(m, n)$. Applying the definition of the DSFT, we can write

$$G_\uparrow(U, V) = \sum_m \sum_n s_C(m) s_D(n) f(m/C, n/D) e^{-i2\pi(Um + Un)}, \tag{50}$$

which after a change of variable becomes

$$G_\uparrow(U, V) = \sum_m \sum_n f(m, n) e^{-i2\pi(UCm + UDn)}, \tag{51}$$

$$= F(CU, DV). \tag{52}$$

Thus upsampling contracts the spectrum by a factor of C in the U direction and D in the V direction. There is no aliasing, because no information has been lost. The DSFT $G_\uparrow(U, V)$ is periodic with period $1/C \times 1/D$. To generate an image that is oversampled by a factor of $C \times D$, we need to filter out all but the baseband replication, again using the ideal low-pass filter of Eq. (46). The combination of an upsampler followed by a filter is called an *interpolator*. Figure 10 shows the block diagram of such a system. Its net effect is given by

$$g(m, n) = \sum_k \sum_l f(k, l) h(m - Ck, n - Dl). \tag{53}$$

For the special case of the ideal low-pass filter,

$$g(m, n) = \sum_k \sum_l f(k, l) \text{sinc}(m/C - k, n/D - l). \tag{54}$$

In the frequency domain, the interpolator is described by

$$G(U, V) = H(U, V) F(CU, DV). \tag{55}$$

For reasons similar to those discussed in the context of decimation as well as undesirably large computational requirements, the sinc filter is not widely used for image interpolation. In the following section, we examine some alternative approaches.

4 Image Interpolation

Both decimation and interpolation are fundamental image processing operations. As the resolution of desktop printers grows, interpolation is increasingly needed to scale images up to the resolution of the printer prior to *halftoning*, which is discussed in detail in Chapter 8.1. In addition to the quality of the interpolated image, the effort required to compute it is a very important consideration in these applications. We will use the theory developed in the preceding section as the basis for describing a variety of methods that can be used for image interpolation. We shall assume that the image is to be enlarged by integer factors $C \times D$, where for convenience, we assume that both C and D are even. We begin with several methods that can be modeled as an upsampler followed by a linear filter, as shown in Fig. 10.

4.1 Linear Filtering Approaches

At the lowest level of computational complexity, we have *pixel replication*, also known as *nearest neighbor interpolation* or *zero-order interpolation*, which is widely used in many applications. In this case,

$$g(m, n) = f(\lfloor m/C \rceil, \lfloor n/D \rceil), \tag{56}$$

where $\lfloor \cdot \rceil$ denotes rounding to the nearest integer. The corresponding filter in the interpolator structure shown in Fig. 10 is given by

$$h_0(m, n) = \begin{cases} 1/(CD), & -C/2 \le m < C/2, \ -D/2 \le n < D/2 \\ 0, & \text{else} \end{cases}, \tag{57}$$

where the subscript denotes the order of the interpolation. Pixel replication yields images that appear blocky, as shown in Fig. 11. Looking at Eq. (55) and the frequency response

$$H_0(U, V) = \frac{e^{i\pi(U+V)}}{CD} \left[\frac{\sin(\pi U C)}{\sin(\pi U)} \right] \left[\frac{\sin(\pi V D)}{\sin(\pi V)} \right] \tag{58}$$

(a)   (b)

FIGURE 11 Interpolation by pixel replication: (a) original image, (b) image interpolated by $4\times$. (See color section, p. C–33.)

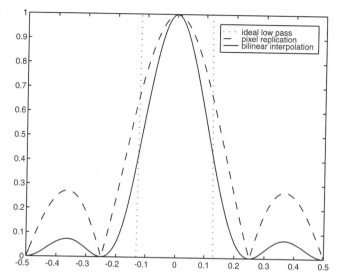

FIGURE 12 Magnitude of frequency response of linear filters for $4\times$ interpolation.

of the filter, we see that this is a consequence of the fact that the filter does not effectively block the replications of $F(CU, DV)$ outside the region $\Omega_{1/C, 1/D}(U, V)$, as shown in Fig. 12.

To obtain a smoother result, we can linearly interpolate between adjacent samples. The extension of this idea to 2-D, called *bilinear interpolation*, is described by

$$\begin{aligned} g(m, n) = {} & \alpha\beta f(\lfloor m/C \rfloor, \lfloor n/D \rfloor) \\ & + (1 - \alpha)\beta f(\lceil m/C \rceil, \lfloor n/D \rfloor) \\ & + \alpha(1 - \beta) f(\lfloor m/C \rfloor, \lceil n/D \rceil) \\ & + (1 - \alpha)(1 - \beta) f(\lceil m/C \rceil, \lceil n/D \rceil), \end{aligned} \tag{59}$$

where $\alpha = \lceil m/C \rceil - m/C$ and $\beta = \lceil n/D \rceil - n/D$. In this case,

$$\begin{aligned} h_1(m, n) &= \begin{cases} (1 - |m/C|)(1 - |n/D|), & |m| < C \text{ and } |n| < D \\ 0, & \text{else} \end{cases}, \\ &= h_0(m, n) * h_0(-m, -n). \end{aligned} \tag{60}$$

It follows directly from Eq. (60) that

$$H_1(U, V) = |H_0(U, V)|^2, \tag{61}$$

which provides better suppression of the nonbaseband replications of $F(U, V)$, as shown in Fig. 12. As can be seen in Fig. 13, bilinear interpolation yields an image that is free of the blockiness produced by pixel replication. However, the interpolated image has an overall appearance that is somewhat soft.

Both these strategies are examples of B-spline interpolation, which can be generalized to arbitrary order K. The corresponding frequency response is given by

$$H_K(U, V) = [H_0(U, V)]^K. \tag{62}$$

FIGURE 13   Interpolation by 4× by means of bilinear interpolation. (See color section, p. C–33.)

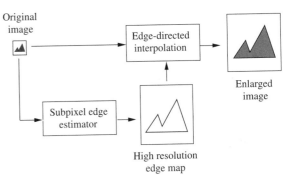

FIGURE 14   Framework for the edge-directed interpolation algorithm.

The choice $K = 3$ is popular, since it yields a good trade-off between smoothness of the interpolation and locality of dependence on the underlying data. For further discussion of image interpolation using splines, the reader is directed to [1] and [2]. The latter reference, in particular, discusses the design of an optimal prefilter for minimizing loss of information when splines are used for image reduction.

4.2 Model-Based Approaches

In many applications, spline interpolation does not yield images that are sufficiently sharp. This problem can be traced to the way in which edges and textures are rendered. In recent years, there has been a great deal of interest in techniques for improving the quality of interpolated images by basing the interpolation on some type of image model. With many of the algorithms, the fundamental idea is to identify edges and to avoid interpolating across them [3–16]. A few of these methods explicitly estimate high-resolution edge information from the low-resolution image, and use this information to control the interpolation [6,9,10]. These works use a variety of underlying interpolation methods: bilinear [7,9,10] cubic splines [6,8], directional filtering [3–5,8], and least-squares fit to a model [7,9].

Some of these methods are based on more general models that account for texture as well as edges. These approaches include Bayesian reconstruction based on a Markov random field model [11], a wavelet-transform based method [12], and a tree-based scheme [13]. In order to illustrate the kind of performance that can be achieved with methods of this type, we will briefly describe two approaches that have been reported in the literature, and show some experimental results.

4.2.1 Edge-Directed Interpolation

Figure 14 shows the framework within which the edge-directed interpolation algorithm operates. We will only sketch the highlights of the procedure here. For further details, the reader is directed to [10]. A subpixel edge estimation technique is used to generate a high-resolution edge map from the low-resolution image, and then the high-resolution edge map is used to guide the interpolation of the low-resolution image to the final high-resolution version. Figure 15 shows the structure of the edge-directed interpolation algorithm itself. It consists of two phases: rendering and data correction. Rendering is based on a modified form of bilinear interpolation of the low-resolution image data. An implicit assumption underlying bilinear interpolation is that the low-resolution data consists of point samples from the high-resolution image. However, most sensors generate low-resolution data by averaging the light incident at the focal plane over the unit cell corresponding to the low-resolution sampling lattice. We iteratively compensate for this effect by feeding the interpolated image back through the sensor model and using

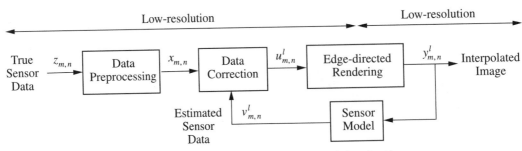

FIGURE 15   Structure of the edge-directed interpolation algorithm.

the disparity between the resulting estimated sensor data and the true sensor data to correct the mesh values on which the bilinear interpolation is based. Reference [11] also embodies a sensor model.

To estimate the subpixel edge map shown in Fig. 14, we filter the low-resolution image with a simple rectangular center-on-surround-off (COSO) filter with a constant positive center region embedded within a constant negative surround region. The relative heights are chosen to yield zero DC response. The filter coefficients are given by

$$h^{\text{COSO}}(m, n) = \begin{cases} h_c, & |m|, |n| \leq N_c \\ h_s, & N_c < |m| \leq N_s \text{ and } |n| \leq N_s \\ & N_c < |n| \leq N_s \text{ and } |m| \leq N_s \\ 0, & \text{otherwise} \end{cases} \quad (63)$$

This filter mimics the point-spread function for the Laplacian-of-Gaussian (LoG) given by [14]

$$h^{\text{LoG}}(m, n) = \frac{2}{\sigma^2}[1 - (m^2 + n^2)/2\sigma^2]e^{-(m^2+n^2)/2\sigma^2}. \quad (64)$$

as shown in Fig. 16. For a detailed treatment of the Laplacian-of-Gaussian filter and its use, the reader is directed to Chapter 4.11.

The COSO filter results in a good approximation to the edge map generated with a true LoG filter, but requires only nine additions/subtractions and two multiplies per output point when recursively implemented with row and column buffers. To determine the high-resolution edge map, we linearly interpolate the COSO filter output between points on the low-resolution lattice to estimate zero-crossing positions on the high-resolution lattice. Figure 17 shows a subpixel edge map estimated by using the COSO filter followed by piecewise linear interpolation, using the original low-resolution image shown in Fig. 11. The interpolation factor was 4×. For comparison, we show a subpixel edge map obtained by upsampling the low-resolution image,

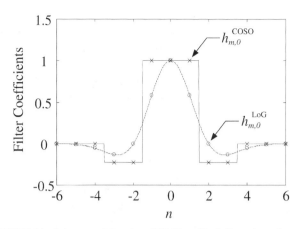

FIGURE 16 Point-spread function of COSO and LoG filters along the axes.

followed by filtering with a LoG filter, and detection of zero crossings. The COSO edge map does not contain the fine detail that can be seen in the LoG edge map. However, it does show the major edges corresponding to significant gray value changes in the original image.

Now let us turn our attention to Fig. 15. The essential feature of the rendering step is that we modify bilinear interpolation on a pixel-by-pixel basis to prevent interpolation across edges. To illustrate the approach, let's consider interpolation at the high-resolution pixel m in Fig. 18. We first determine whether or not any of the low-resolution corner pixels a, b, c, and d are separated from m by edges. For all those pixels that are, we compute replacement values according to a heuristic procedure that depends on the number and geometry of the pixels to be replaced. Figure 18(a) shows the situation in which a single corner pixel u_b is to replaced. In this case, we linearly interpolate to the midpoint i of the line $u_a - u_c$, and then extrapolate along the line $u_d - i$ to yield the replacement value $\bar{u}_b$. If two corner pixels are to be replaced, they can be either adjacent or not adjacent. Figure 18(b) shows the case in which two adjacent pixels u_a and

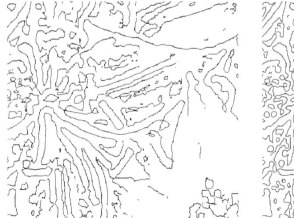 

FIGURE 17 High-resolution edge map interpolated by 4× using a rectangular COSO filter followed by piecewise linear interpolation of zero crossings (left) and high-resolution edge map interpolated by 4× using a LoG filter after upsampling by 4×.

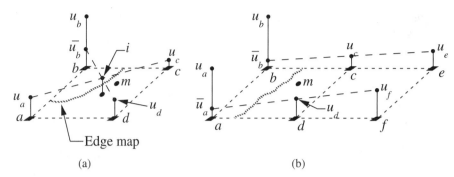

FIGURE 18 Computation of replacement values for the low-resolution corner pixels to be used when bilinearly interpolating the image value at high-resolution pixel m. The cases shown are (a) replacement of one pixel, and (b) replacement of two adjacent pixels.

u_b must be replaced. In this case, we check to see if any edges cross the lines $e - c$ and $f - d$. If none does, we linearly extrapolate along the lines $u_e - u_c$ and $u_f - u_d$ to generate the replacement values $\bar{u}_b$ and $\bar{u}_a$, respectively. If an edge crosses $e - c$, we simply let $\bar{u}_b = u_c$. The cases in which two nonadjacent pixels are to be replaced or in which three pixels are to be replaced are treated similarly. The final case to be considered is that which occurs when the pixel m to be interpolated is separated from all four corner pixels a, b, c, and d by edges. This case would only occur in regions of high spatial activity. In such areas, we assume that it is not possible to obtain a meaningful estimate of the high-resolution edge map from just the four low-resolution corner pixels; so the high-resolution image will be rendered with unmodified bilinear interpolation. It is interesting to note that the process of bilinear interpolation, except across edges, is very closely related to anisotropic diffusion, which is studied in detail in Chapter 4.12.

To describe the iterative correction procedure, we let l be the iteration index, and denote the true sensor data by $z(m, n)$, the preprocessed sensor data by $x(m, n)$, the corrected sensor data by $u^{(l)}(m, n)$, the edge-directed rendering step by the operator $\mathcal{R}$, the interpolated image by $y^{(l)}(m, n)$, the sensor model by the operator $\mathcal{S}$, and the estimated sensor data by $v^{(l)}(m, n)$. The sensor model $\mathcal{S}$ is a simple block average of the high-resolution pixels in the unit cell for each pixel in the low resolution lattice.

With this notation, we may formally describe the procedure depicted in Fig. 15 by the following equations:

$$y^{(l)}(m, n) = \mathcal{R}\big[u^{(l)}(m, n)\big], \tag{65}$$

$$v^{(l)}(m, n) = \mathcal{S}\big[y^{(l)}(m, n)\big], \tag{66}$$

$$u^{(l+1)}(m, n) = u^{(l)}(m, n) + \lambda\big(v^{(l)}(m, n) - x(m, n)\big), \tag{67}$$

where λ is a constant that controls the gain of the correction process. The iteration is started with the initial condition $u^0(m, n) = x(m, n)$. Equations (65–67) represent a classical successive approximation procedure [15]. We can think of $v^{(l)}(m, n) - x(m, n)$ as the closed loop error when an image

is interpolated and then decimated as it passes through the sensor model. If we have convergence in the iterative loop, i.e., if $u^{(l+1)}(m, n) = u^{(l)}(m, n)$, this implies that $v^{(l)}(m, n) = x(m, n)$. Hence the closed loop error is zero. Convergence of the iteration can be proved under mild restrictions on the location of edges [16].

Figure 19 shows the results of $4\times$ interpolation using the edge-directed interpolation algorithm after iterations 0 and 10. We see that edge-directed interpolation yields a much sharper result than bilinear interpolation. While some of the aliasing artifacts that occur with pixel replication can be seen in the edge-directed interpolation result, they are not nearly as prominent. The result after iteration 0 shows the effect of the edge-directed rendering alone, without data correction to account for the sensor model. While this image is sharper than that produced by bilinear interpolation, it lacks some of the crispness of the image resulting after 10 iterations of the algorithm.

4.2.2 Tree-Based Resolution Synthesis

Tree-based resolution synthesis (TBRS) [13] works by first performing a fast local classification of a window around the pixel being interpolated, and then applying an interpolation filter designed for the selected class, as illustrated in Fig. 20. The idea behind TBRS is to use a regression tree as a piecewise linear approximation to the conditional mean estimator of the high-resolution image given the low-resolution image. Intuitively, having the different regions of linear operation allows for separate filtering of distinct behaviors like edges of different orientation and smoother gradient transitions.

An overview of the TBRS algorithm appears in Fig. 21. Note that before TBRS can be executed, we must already have generated the parameters for the regression tree by training on sample images. This training procedure requires considerable computation, but it only has to be performed once. The resulting predictor may be used effectively on images that were not used in the training.

As illustrated in Fig. 20, we generate an $C \times C$ block of high-resolution pixels for every pixel in the low-resolution source

(a)   (b)

FIGURE 19 Image interpolated by 4× using edge-directed interpolation with (a) 0 and (b) 10 iterations. (See color section, p. C–34.)

image by filtering the corresponding $W \times W$ window of pixels in the low-resolution image, with the filter coefficients selected based on a classification. We have used $W = 5$. Thinking of the desired high-resolution pixels as a C^2-dimensional random vector X and the corresponding low-resolution pixels as the realization of a W^2-dimensional random vector Z, our approach is to use a regression tree that approximates the conditional mean estimator of $X \mid Z$, so that the vector $\hat{X}$ of interpolated pixels satisfies

$$\hat{X} = E[X \mid Z]. \tag{68}$$

It is well known that the conditional mean estimator minimizes the expected mean-squared error [17]. Here we will use capital letters to represent random quantities, and lowercase letters for their realizations. A closed-form expression for the true conditional mean estimator would be difficult to obtain for the present context. However, the regression tree T that we use provides a convenient and flexible piecewise linear approximation, with the M different linear regions being polygonal subsets which comprise a partition of the sample space $\mathcal{Z}$ of low-resolution vectors Z. These polygonal subsets, or classes, correspond to visually distinct behaviors like edges of different orientation.

With the main ideas in place, we return to Fig. 20 for a better look. To interpolate the shaded pixel in the low-resolution image, we first procure the vector z by stacking the pixels in the 5×5 window centered there. Then we obtain interpolated pixels as

$$\hat{x} = A_j z + \beta_j, \tag{69}$$

where A_j and β_j are respectively the $L^2 \times W^2$ matrix and L^2-dimensional vector comprising the interpolation filter for class j, and j is the index of the class obtained as

$$j = C_T(z), \tag{70}$$

where $C_T : \mathcal{Z} \to \{0, \ldots, M-1\}$ is a function that embodies the classifying action of T. To evaluate $C_T(z)$, we begin at the top and traverse down the tree T as illustrated in Fig. 22, making a decision to go right or left at each nonterminal node (circle), and taking the index j of the terminal node (square) that z lands in. Each decision has the form

$$e_m^t(z - U_m) \gtrless 0, \tag{71}$$

where m is the index of the node, e_m and U_m are W^2-dimensional

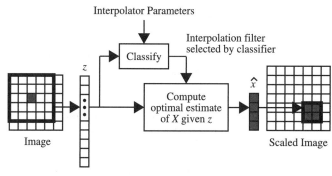

FIGURE 20 TBRS interpolation by a factor of 2.

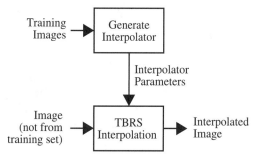

FIGURE 21 Overview of TBRS.

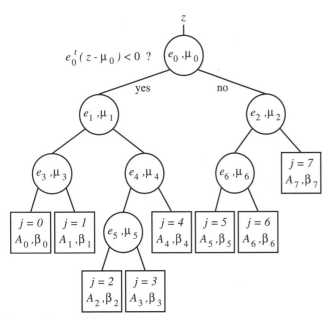

$e_0^t(z - \mu_0) < 0$?

FIGURE 22   Binary tree structure used in TBRS.

vectors, and a superscript t denotes taking the transpose. This decision determines whether z is on one side of a hyperplane or the other, with U_m being a point in the hyperplane and with e_m specifying its orientation. By convention, we go left if the quantity on the left-hand side is negative, and we go right otherwise.

In order to complete the design of the TBRS algorithm, we must obtain numerical values for the integer number $M \geq 1$ of terminal nodes in the tree; the decision rules $\{(e_m, U_m)\}_m = 0^M - 2$ for the nonterminal nodes (assuming that $M > 1$); and the interpolation filters $\{(A_m, \beta_m)\}_m = 0^M - 1$ for the terminal nodes. To compute these parameters, a training procedure is used, which is based on that given by Gelfand, Ravishankar, and Delp [18], suitably modified for the design of a regression tree

FIGURE 23   Interpolation by $4\times$ via TBRS. (See color section, p. C–34.)

rather than a classification tree. The training vector pairs are assumed to be independent realizations of (X, Z). Training vector pairs are extracted from low- and high-resolution renderings of the same image. For further details regarding the design process, the reader is directed to Ref. [13]. Figure 23 shows the flower image interpolated by $4\times$ using tree-based resolution synthesis. Comparing this image with those shown in Fig. 19, which were generated by means of edge-directed interpolation, we see that TBRS yields a higher quality than edge-directed interpolation after 0 iterations, and quality that is comparable to that of edge-directed interpolation after 10 iterations.

5 Conclusion

Most systems for image capture may be categorized into one of two classes: flying spot scanners and focal plane arrays. Both these classes of systems may be modeled as a convolution with an aperture function, followed by multiplication by a sampling function. In the frequency domain, the spectrum of the continuous-parameter image is multiplied by the Fourier transform of the aperture function, resulting in attenuation of the higher frequencies in the image. This modified spectrum is then replicated on a lattice of points that is reciprocal to the sampling lattice. If the sampling frequency is sufficiently high, the replications will not overlap; and the original image may be reconstructed from its samples. Otherwise, the overlap of the spectral replications with the baseband term may cause aliasing artifacts to appear in the image.

In many image processing applications, including printing of digital images, it is necessary to resize the image. This process may be analyzed within the framework of multirate signal processing. Decimation, which consists of low-pass filtering followed by downsampling, results in expansion and replication of the spectrum of the original digital image. Interpolation, which consists of upsampling following by low-pass filtering, causes the spectrum of the original digital image to contract; so it occupies only a portion of the baseband spectral region. With this approach, the interpolated image is a linear function of the sampled data. Linear interpolation may blur edges and fine detail in the image. A variety of nonlinear approaches have been proposed that yield improved rendering of edges and detail in the image.

References

[1] H. S. Hou and H. C. Andrews, "Cubic convolution interpolation for digital image processing," *IEEE Trans. Acoust. Speech Signal Process.* **26**, 508–517 (1978).

[2] M. Unser, A. Aldroubi, and M. Eden, "Enlargement or reduction of digital images with minimum loss of information," *IEEE Trans. Image Process.* **4**, 247–258 (1995).

[3] V. R. Algazi, G. E. Ford, and R. Potharlanka, "Directional interpolation of images based on visual properties and rank order filtering," presented at the 1991 International Conference on Acoustics, Speech, and Signal Processing, Toronto, CN, May 14–17, 1991.

[4] G. E. Ford, R. R. Estes, and H. Chen, "Space scale analysis for image sampling and interpolation," presented at the 1992 International Conference on Acoustics, Speech, and Signal Processing, San Francisco, CA, March 23–26, 1992.

[5] B. Ayazifar and J. S. Lim, "Pel-adaptive model-based interpolation of spatially subsampled images," presented at the 1992 International Conference on Acoustics, Speech, and Signal Processing, San Francisco, CA, March 23–26, 1992.

[6] K. Xue, A. Winans, and E. Walowit, "An edge-restricted spatial interpolation algorithm," *J. Electron. Imag.* **1**, 152–161 (1992).

[7] F. G. B. De Natale, G. S. Desoli, D. D. Guisto, and G. Vernazza, "A spline-like scheme for least-squares bilinear interpolation," presented at the 1993 International Conference on Acoustics, Speech, and Signal Processing, Minneapolis, MN, April 27–30, 1993.

[8] S. W. Lee and J. K. Paik, "Image interpolation using adaptive fast B-spline filtering," presented at the 1993 International Conference on Acoustics, Speech, and Signal Processing, Minneapolis, MN, April 27–30, 1993.

[9] K. Jensen and D. Anastassiou, "Subpixel edge localization and the interpolation of still images," *IEEE Trans. Image Process.* **4**, 285–295 (1995).

[10] J. P. Allebach and P. W. Wong, "Edge-directed interpolation," presented at the 1996 IEEE International Conference on Image Processing, Lausanne, Switzerland, September 16–19, 1996.

[11] R. R. Schultz and R. L. Stevenson, "A Bayesian approach to image expansion for improved definition," *IEEE Trans. Image Process.* **3**, 233–242 (1994).

[12] S. G. Chang, Z. Cvetkovic, and M. Vetterli, "Resolution enhancement of images using wavelet transform extrema extrapolation," *Proc. IEEE Int. Conf. Acoust. Speech Signal Process.* **4**, 2379–2382 (1995).

[13] C. B. Atkins, C. A. Bouman, and J. P. Allebach, "Tree-Based Resolution Synthesis," presented at the 1999 IS&T Image Processing, Image Quality, Image Capture Systems Conference, Savannah, GA, April 25–28, 1999.

[14] J. Canny, "A computational approach to edge detection," *IEEE Trans. Pattern Anal. Machine Intell.* **PAMI-8**, 679–698 (1986).

[15] R. W. Schafer, R M. Mersereau, and M. A. Richard, "Constrained iterative restoration algorithms," *Proc. IEEE* **69**, 432–450 (1981).

[16] P. W. Wong and J. P. Allebach, "Convergence of an iterative edge directed image interpolation algorithm," presented at the 1997 IEEE International Symposium on Circuits and Systems, Hong Kong, June 9–12, 1997.

[17] L. L. Sharf, *Statistical Signal Processing* (Addison-Wesley, Reading, MA, 1991).

[18] "An iterative growing and pruning algorithm for classification tree design," *IEEE Trans. Pattern Anal. Machine Intell.* **13**, 163–174 (1991).

7.2

Video Sampling and Interpolation

Eric Dubois
University of Ottawa

1 Introduction

This chapter is concerned with the sampled representation of time-varying imagery, often referred to as video. Time-varying imagery must be sampled in at least one dimension for the purposes of transmission, storage, processing, or display. Examples are one-dimensional temporal sampling in motion-picture film, two-dimensional vertical-temporal scanning in the case of analog television, and three-dimensional horizontal-vertical-temporal sampling in digital video. In some cases a single sampling structure is used throughout an entire video processing or communication system. This is the case in standard analog television broadcasting, in which the signal is acquired, transmitted, and displayed using the same scanning standard from end to end. However, it is becoming increasingly more common to have different sampling structures used in the acquisition, processing, transmission, and display components of the system. In addition, the number of different sampling structures in use throughout the world is increasing. Thus, sampling structure conversion for video systems is an important problem.

The initial acquisition and scanning is particularly critical because it determines what information is contained in the original data. The acquisition process can be modeled as an analog prefiltering followed by ideal sampling on a given sampling structure. The sampling structure determines the amount of spatiotemporal information that the sampled signal can carry, while the prefiltering serves to limit the amount of aliasing. At the final stage of the system, the desired display characteristics are closely related to the properties of the human visual system. The goal of the display is to convert the sampled signal to a continuous image presented to the viewer that approximates the original continuous scene as closely as possible. In particular, the effects caused by sampling should be attenuated sufficiently to be below the threshold of perceptibility.

This chapter has three main sections. First the sampling lattice, the basic tool in the analysis of spatiotemporal sampling, is introduced. The issues involved in the sampling and reconstruction of continuous time-varying imagery are then addressed. Finally, methods for the conversion of image sequences between different sampling structures are presented.

2 Spatiotemporal Sampling Structures

A continuous time-varying image $f_c(x, y, t)$ is a function of two spatial dimensions x and y and time t, usually observed in a rectangular spatial window $\mathcal{W}$ over some time interval $\mathcal{T}$. The spatiotemporal region $\mathcal{W} \times \mathcal{T}$ is denoted $\mathcal{W}_T$. The spatial window is of dimension pw $\times$ ph, where pw is the picture width and ph is the picture height. Since the absolute physical size of an image depends on the display device used, and the sampling density for a particular video signal may be variable, we choose to adopt the picture height ph as the basic unit of spatial distance, as is common in the broadcast video industry. The ratio pw/ph is called the aspect ratio, the most common values being 4/3 for standard TV and 16/9 for HDTV. The image f_c can be sampled in one, two, or three dimensions. It is almost always sampled in the temporal dimension at least, producing an *image sequence*.

An example of an image sampled only in the temporal dimension is motion picture film. Analog video is typically sampled in the vertical and temporal dimensions whereas digital video is sampled in all three dimensions. The subset of $\mathbb{R}^3$ on which the sampled image is defined is called the *sampling structure* Ψ; it is contained in $\mathcal{W}_T$.

The mathematical structure most useful in describing sampling of time-varying images is the *lattice*. A discussion of lattices from the point of view of video sampling can be found in [1] and [2]. Some of the main properties are presented here. A lattice Λ in D dimensions is a discrete set of points that can be expressed as the set of all linear combinations with integer coefficients of D linearly independent vectors in $\mathbb{R}^D$ (called basis vectors),

$$\Lambda = \{n_1 \boldsymbol{v}_1 + \cdots + n_D \boldsymbol{v}_D \mid n_i \in \mathbb{Z}\}, \qquad (1)$$

where $\mathbb{Z}$ is the set of integers. For our purposes, D will be one, two or three dimensions. The matrix $V = [\boldsymbol{v}_1 \mid \boldsymbol{v}_2 \mid \cdots \mid \boldsymbol{v}_D]$ whose columns are the basis vectors $\boldsymbol{v}_i$ is called a sampling matrix and we write $\Lambda = \mathrm{LAT}(V)$. The basis or sampling matrix for a given lattice is not unique however, since $\mathrm{LAT}(V) = \mathrm{LAT}(VE)$ where E is any unimodular ($|\det E| = 1$) integer matrix. Figure 1 shows an example of a lattice in two dimensions, with basis vectors $\boldsymbol{v}_1 = (2X, 0)$ and $\boldsymbol{v}_2 = (X, Y)$. The sampling matrix in this case is

$$V = \begin{bmatrix} 2X & X \\ 0 & Y \end{bmatrix}.$$

A *unit cell* of a lattice Λ is a set $\mathcal{P} \subset \mathbb{R}^D$ such that copies of $\mathcal{P}$ centered on each lattice point tile the whole space without overlap: $(\mathcal{P} + \boldsymbol{s}_1) \cap (\mathcal{P} + \boldsymbol{s}_2) = \emptyset$ for $\boldsymbol{s}_1, \boldsymbol{s}_2 \in \Lambda$, $\boldsymbol{s}_1 \neq \boldsymbol{s}_2$, and $\cup_{s \in \Lambda}(\mathcal{P} + \boldsymbol{s}) = \mathbb{R}^D$. The volume of a unit cell is $d(\Lambda) = |\det V|$, which is independent of the particular choice of sampling matrix. We can imagine that there is a region congruent to $\mathcal{P}$ of volume $d(\Lambda)$ associated with each sample in Λ, so that $d(\Lambda)$ is the reciprocal of the sampling density. The unit cell of a lattice is not unique. In Fig. 1, the shaded hexagonal region centered at the origin is a unit cell, of area $d(\Lambda) = 2XY$. The shaded parallelogram in the upper right is also a possible unit cell.

Most sampling structures of interest for time-varying imagery can be constructed using a lattice. In the case of 3-D sampling, the sampling structure can be the intersection of $\mathcal{W}_T$ with a lattice, or in a few cases, with the union of two or more shifted lattices. The latter case occurs relatively infrequently (although there are several practical situations where it is used), and so the discussion here is limited to sampling on lattices. The theory of sampling on the union of shifted lattices (cosets) can be found in [1]. In the case of one or two-dimensional (partial) sampling ($D = 1$ or 2), the sampling structure can be constructed as the Cartesian product of a D-dimensional lattice and a continuous $(3 - D)$ dimensional space. For one-dimensional sampling, the 1-D lattice is $\Lambda_t = \{nT \mid n \in \mathbb{Z}\}$, where T is the frame period. The sampling structure is then $\mathcal{W} \times \Lambda_t = \{(\boldsymbol{x}, t) \mid \boldsymbol{x} \in \mathcal{W}, t \in \Lambda_t\}$. For two-dimensional vertical-temporal sampling (scanning) using a 2-D lattice Λ_{yt}, the sampling structure is $\mathcal{W}_T \cap (\mathcal{H} \times \Lambda_{yt})$, where $\mathcal{H}$ is a one-dimensional subspace of $\mathbb{R}^3$ parallel to the scanning lines. In video systems, the scanning spot is moving down as it scans from left to right, and of course is moving forward in time. Thus $\mathcal{H}$ has both a vertical and temporal tilt, but this effect is minor and can usually be ignored; we assume that $\mathcal{H}$ is the line $y = 0$, $t = 0$. Most digital video signals are obtained by three-dimensional subsampling of signals that have initially been sampled with one or two-dimensional sampling as above. Although the sampling structure is space limited, the analysis is often simplified if the sampling structure is assumed to be of infinite extent, with the image either set to zero outside of $\mathcal{W}_T$ or replicated periodically.

Much insight into the effect of sampling time-varying images on a lattice can be achieved by studying the problem in the frequency domain. To do this, we introduce the Fourier transform for signals defined on different domains. For a continuous signal f_c the Fourier transform is given by

$$F_c(u, v, w) = \iiint f_c(x, y, t)$$
$$\times \exp[-j2\pi(ux + vy + wt)] \, \mathrm{d}x \, \mathrm{d}y \, \mathrm{d}t, \quad (2)$$

or, more compactly, setting $\boldsymbol{u} = (u, v, w)$ and $\boldsymbol{s} = (x, y, t)$, by

$$F_c(\boldsymbol{u}) = \int_{\mathcal{W}_T} f_c(\boldsymbol{s}) \exp(-j2\pi \boldsymbol{u} \cdot \boldsymbol{s}) \, \mathrm{d}\boldsymbol{s}, \quad \boldsymbol{u} \in \mathbb{R}^3. \quad (3)$$

The variables u and v are horizontal and vertical spatial frequencies in cycles/picture height (c/ph) and w is temporal frequency in Hz.

Similarly, a discrete signal $f(\boldsymbol{s})$, $\boldsymbol{s} \in \Lambda$ has a lattice Fourier transform (or discrete space-time Fourier transform)

$$F(\boldsymbol{u}) = \sum_{\boldsymbol{s} \in \Lambda} f(\boldsymbol{s}) \exp(-j2\pi \boldsymbol{u} \cdot \boldsymbol{s}), \quad \boldsymbol{u} \in \mathbb{R}^3. \quad (4)$$

With this nonnormalized definition, both $\boldsymbol{s}$ and $\boldsymbol{u}$ have the same units as in Eq. (3). As with the 1-D discrete-time Fourier transform, the lattice Fourier transform is periodic. If $\boldsymbol{k}$ is an element

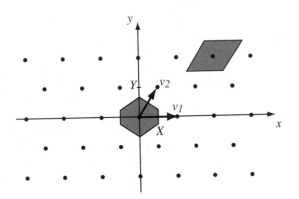

FIGURE 1   Example of a lattice in two dimensions with two possible unit cells.

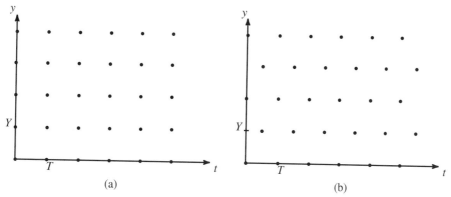

FIGURE 2   2-D vertical-temporal lattices: (a) rectangular lattice Λ_R; (b) hexagonal lattice Λ_H.

of $\mathbb{R}^3$ such that $k \cdot s \in \mathbb{Z}$ for all $s \in \Lambda$, then $F(u + k) = F(u)$. It can be shown that $\{k \mid k \cdot s \in \mathbb{Z} \text{ for all } s \in \Lambda\}$ is a lattice called the *reciprocal lattice* Λ^*, and that if V is a sampling matrix for Λ, then $\Lambda^* = \text{LAT}((V^T)^{-1})$. Thus $F(u)$ is completely specified by its values in a unit cell of Λ^*.

For partially sampled signals, a mixed Fourier transform is required. For the examples of temporal and vertical-temporal sampling mentioned previously, these Fourier transforms are

$$
F(u, w) = \int_{\mathcal{W}} \sum_n f(x, nT) \\
\times \exp[-j2\pi(u \cdot x + wnT)] \, dx, \quad (5)
$$

$$
F(u, v, w) = \int_{\mathcal{H}} \sum_{(y,t) \in \Lambda_{yt}} f(x, y, t) \\
\times \exp[-j2\pi(ux + vy + wt)] \, dx. \quad (6)
$$

These Fourier transforms are periodic in the temporal frequency domain (with periodicity $1/T$) and in the vertical-temporal frequency domain (with periodicity lattice Λ_{yt}^*), respectively.

The terminology is illustrated with two examples that will be discussed in more detail further on. Figure 2 shows two vertical-temporal sampling lattices: a rectangular lattice Λ_R in Fig. 2(a)

and a hexagonal lattice in Fig. 2(b). These correspond in video systems to progressive scanning and interlaced scanning, respectively. Possible sampling matrices for the two lattices are

$$
V_R = \begin{bmatrix} Y & 0 \\ 0 & T \end{bmatrix}, \quad V_H = \begin{bmatrix} Y & 0 \\ T/2 & T \end{bmatrix}. \quad (7)
$$

Both lattices have the same sampling density, with $d(\Lambda_R) = d(\Lambda_H) = YT$. Figure 3 shows the reciprocal lattices Λ_R^* and Λ_H^* with several possible unit cells.

3 Sampling and Reconstruction of Continuous Time-Varying Imagery

The process for sampling a time-varying image can be approximated by the system shown in Fig. 4. The light arriving on the sensor is collected and weighted in space and time by the sensor aperture $a(s)$ to give the output

$$
f_{ca}(s) = \int_{\mathbb{R}^3} f_c(s + s') a(s') \, ds', \quad (8)
$$

where it is assumed here that the sensor aperture is space and time invariant. The resulting signal $f_{ca}(s)$ is then sampled in an

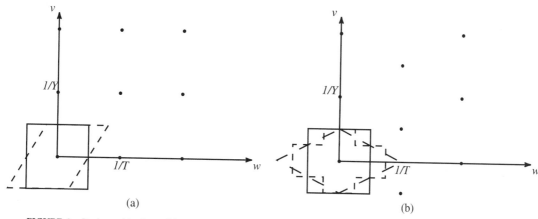

FIGURE 3   Reciprocal lattices of the 2-D vertical-temporal lattices of Fig. 2: (a) rectangular lattice Λ_R^*; (b) hexagonal lattice Λ_H^*.

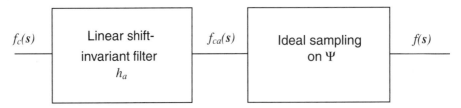

ideal fashion on the sampling structure Ψ:

$$f(s) = f_{ca}(s), \quad s \in \Psi. \tag{9}$$

By defining $h_a(s) = a(-s)$, we see that the aperture weighting is a linear filtering operation, i.e., the convolution of $f_c(s)$ with $h_a(s)$

$$f_{ca}(s) = \int_{\mathbb{R}^3} f_c(s - s') h_a(s') \, \mathrm{d}s'. \tag{10}$$

Thus, if $f_c(s)$ has a Fourier transform $F_c(u)$, then $F_{ca}(u) = F_c(u) H_a(u)$, where $H_a(u)$ is the Fourier transform of the aperture impulse response.

If the sampling structure is a lattice Λ, then the effect of sampling in the frequency domain is given by [1]

$$F(u) = \frac{1}{d(\Lambda)} \sum_{k \in \Lambda^*} F_{ca}(u + k). \tag{11}$$

In other words, the continuous signal spectrum F_{ca} is replicated on the points of the reciprocal lattice. The terms in the sum of Eq. (11) other than for $k = 0$ are referred to as *spectral repeats*. There are two main consequences of the sampling process. The first is that these spectral repeats, if not removed by the display/viewer system, may be visible in the form of flicker, line structure, or dot patterns. The second is that if the regions of support of $F_{ca}(u)$ and $F_{ca}(u + k)$ have nonzero intersection for some values $k \in \Lambda^*$, we have aliasing; a frequency u_a in this intersection can represent both the frequencies u_a and $u_a - k$ in the original signal. Thus, to avoid aliasing, the spectrum $F_{ca}(u)$ should be confined to a unit cell of Λ^*; this can be accomplished to some extent by the sampling aperture h_a. Aliasing is particularly problematic because once introduced it is difficult to remove, since there is more than one acceptable interpretation of the observed data. Aliasing is a familiar effect that tends to be localized to those regions of the image with high frequency details. It can be seen as moiré patterns in such periodic-like patterns as fishnets and venetian blinds, and as staircase-like effects on high-contrast oblique edges. The aliasing is particularly visible and annoying when these patterns are moving. Aliasing is controlled by selecting a sufficiently dense sampling structure and through the prefiltering effect of the sampling aperture.

If the support of $F_{ca}(u)$ is confined to a unit cell $\mathcal{P}^*$ of Λ^*, then it is possible to reconstruct f_{ca} exactly from the samples.

In this case, we have

$$F_{ca}(u) = \begin{cases} d(\Lambda) F(u) & \text{if } u \in \mathcal{P}^* \\ 0 & \text{if } u \notin \mathcal{P}^*, \end{cases} \tag{12}$$

and it follows that

$$f_{ca}(s) = \sum_{s' \in \Lambda} f(s') t(s - s') \tag{13}$$

where

$$t(s) = d(\Lambda) \int_{\mathcal{P}^*} \exp(j 2\pi u \cdot s) \, \mathrm{d}u \tag{14}$$

is the impulse response of an ideal low-pass filter (with sampled input and continuous output) having passband $\mathcal{P}^*$. This is the multidimensional version of the familiar Sampling Theorem.

In practical systems, the reconstruction is achieved by

$$\hat{f}_{ca}(s) = \sum_{s' \in \Lambda} f(s') d(s - s'), \tag{15}$$

where d is the display aperture, which generally bears little resemblance to the ideal $t(s)$ of Eq. (14). The display aperture is usually separable in space and time, $d(s) = d_s(x, y) d_t(t)$, where $d_s(x, y)$ may be Gaussian or rectangular, and $d_t(t)$ may be exponential or rectangular, depending on the type of display system. In fact, a large part of the reconstruction filtering is often left to the spatiotemporal response of the human visual system. The main requirement is that the first temporal frequency repeat at zero spatial frequency (at $1/T$ for progressive scanning and $2/T$ for interlaced scanning (Fig. 2)) be at least 50 Hz for large area flicker to be acceptably low.

If sampling is performed in only one or two dimensions, the spectrum is replicated in the corresponding frequency dimensions. For the two cases of temporal and vertical-temporal sampling, we obtain

$$F(u, w) = \frac{1}{T} \sum_{l=-\infty}^{\infty} F_{ca}\left(u, w + \frac{l}{T}\right) \tag{16}$$

$$F(u, v, w) = \frac{1}{d(\Lambda_{yt})} \sum_{k \in \Lambda_{yt}^*} F_{ca}(u, (v, w) + k). \tag{17}$$

Consider first the case of pure temporal sampling, as in motion-picture film. The main parameters in this case are the sampling period T and the temporal aperture. As shown in Eq. (16), the signal spectrum is replicated in temporal frequency at multiples of $1/T$. In analogy with one-dimensional signals, one might think that the time-varying image should be bandlimited in temporal frequency to $1/2T$ before sampling. However, this is not the case. To illustrate, consider the spectrum of an image undergoing translation with constant velocity $\boldsymbol{v}$. This can model the local behavior in a large class of time-varying imagery. The assumption implies that $f_c(\boldsymbol{x}, t) = f_{c0}(\boldsymbol{x} - \boldsymbol{v}t)$, where $f_{c0}(\boldsymbol{x}) = f_c(\boldsymbol{x}, 0)$. A straightforward analysis [3] shows that $F_c(\boldsymbol{u}, w) = F_{c0}(\boldsymbol{u})\delta(\boldsymbol{u} \cdot \boldsymbol{v} + w)$, where $\delta(\cdot)$ is the Dirac delta function. Thus, the spectrum of the time-varying image is not spread throughout spatiotemporal frequency space but rather it is concentrated around the plane $\boldsymbol{u} \cdot \boldsymbol{v} + w = 0$. When this translating image is sampled in the temporal dimension, these planes are parallel to each other and do not intersect, i.e., there is no aliasing, even if the temporal bandwidth far exceeds $1/2T$. This is most easily illustrated in two dimensions. Consider the case of vertical motion only. Figure 5 shows the vertical-temporal projection of the spectrum of the sampled image for different velocities $\boldsymbol{v}$. Assume that the image is vertically bandlimited to B c/ph. It follows that when the vertical velocity reaches $1/2TB$ picture heights per second (ph/s), the spectrum will extend out to the temporal frequency of $1/2T$ as shown in Fig. 5(b). At twice that velocity ($1/TB$), it would extend to a temporal fre-

quency of $1/T$, which might suggest severe aliasing. However, as seen in Fig. 5(c), there is no spectral overlap. To reconstruct the continuous signal correctly however, a vertical-temporal filtering adapted to the velocity is required. Bandlimiting the signal to a temporal frequency of $1/2T$ before sampling would effectively cut the vertical resolution in half for this velocity. Note that the velocities mentioned above are not really very high. To consider some typical numbers, if $T = 1/24$ s, as in film, and $B = 500$ c/ph (corresponding to 1000 scanning lines) the velocity $1/2TB$ is about $1/42$ ph/s. It should be noted that if the viewer is tracking the vertical movement, the spectrum of the image on the retina will be far less tilted, again arguing against sharp temporal bandlimiting. (This is in fact a kind of motion-compensated filtering by the visual system.) The temporal camera aperture can roughly be modeled as the integration of f_c for a period $T_a \leq T$. The choice of the value of the parameter T_a is a compromise between motion blur and signal-to-noise ratio.

Similar arguments can be made in the case of the two most popular vertical-temporal scanning structures, progressive scanning and interlaced scanning. In reference to Fig. 6, the vertical-temporal spectrum of a vertically translating image at the same three velocities (assuming that $1/Y = 2B$) is shown for these two scanning structures. For progressive scanning there continues to be no spectral overlap, whereas for interlaced scanning the spectral overlap can be severe at certain velocities [e.g., $1/TB$ as in

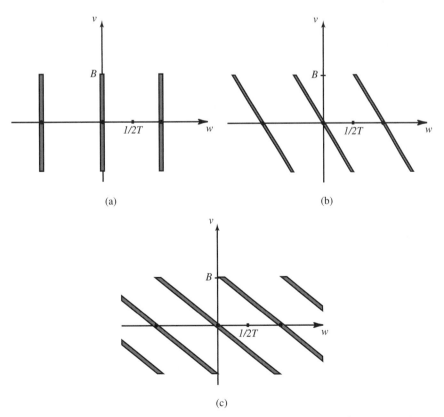

(a)

(b)

(c)

FIGURE 5 Vertical-temporal projection of the spectrum of temporally sampled time-varying image with vertical motion of velocity v: (a) $v = 0$, (b) $v = 1/2TB$, (c) $v = 1/TB$.

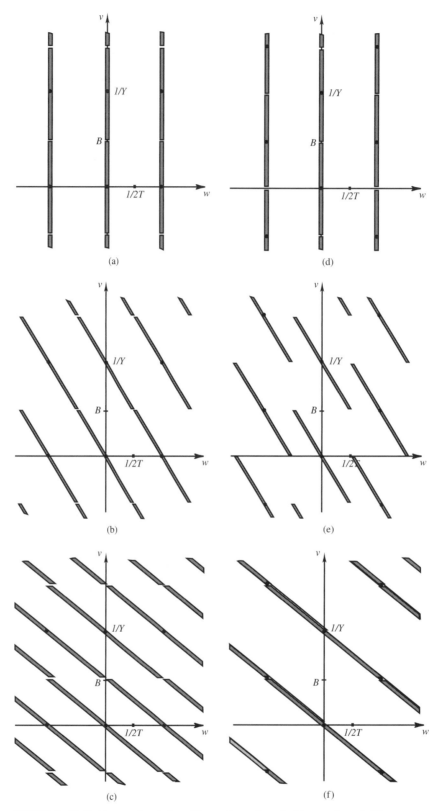

FIGURE 6 Vertical-temporal projection of the spectrum of a vertical-temporal sampled time-varying image with progressive and interlaced scanning: progressive, (a) $v = 0$, (b) $v = 1/2TB$, (c) $v = 1/TB$; interlaced, (d) $v = 0$, (e) $v = 1/2TB$, (f) $v = 1/TB$.

Fig. 6(f)]. This is a strong advantage for progressive scanning. Another disadvantage of interlaced scanning is that each field is spatially undersampled and pure spatial processing or interpolation is very difficult. An illustration in three dimensions of some of these ideas can be found in [4].

4 Sampling Structure Conversion

There are numerous spatiotemporal sampling structures used for the digital representation of time-varying imagery. However, the vast majority of those in use fall into one of two categories corresponding to progressive or interlaced scanning with aligned horizontal sampling. This corresponds to sampling matrices of the form

$$
\begin{bmatrix} X & 0 & 0 \\ 0 & Y & 0 \\ 0 & 0 & T \end{bmatrix} \quad \text{or} \quad \begin{bmatrix} X & 0 & 0 \\ 0 & Y & 0 \\ 0 & T/2 & T \end{bmatrix},
$$

respectively. Table 1 shows the parameters for a number of commonly used sampling structures covering a broad range of applications, from low-resolution QCIF used in videophone to HDTV and digitized IMAX film (the popular large-format film, about 70 mm by 52 mm, used by Imax Corporation). Note that of these, only HDTV and IMAX formats have $X = Y$ (i.e., square pixels). It is frequently required to convert a time-varying image sampled on one such structure to another. An input image sequence $f(\boldsymbol{x})$ sampled on lattice Λ_1 is to be converted to the output sequence $f_o(\boldsymbol{x})$ sampled on the lattice Λ_2.

Besides converting between different standards, sampling structure conversion can also be incorporated into the acquisition or display portions of an imaging system to compensate for the difficulty in performing adequate prefiltering with the camera aperture, or adequate postfiltering with the display aperture. Specifically, the time-varying image can initially be sampled at a higher density than required, using the camera aperture as prefilter, and then downsampled to the desired structure by using digital prefiltering, which offers much more flexibility. Similarly,

the image can be upsampled for the display device by using digital filtering, so that the subsequent display aperture has a less critical task to perform.

4.1 Frame-Rate Conversion

Consider first the case of pure frame-rate conversion. This applies when both the input and the output sampling structures are separable in space and time with the same spatial sampling structure, and where spatial aliasing is assumed to be negligible. The temporal sampling period is to be changed from T_1 to T_2. This situation corresponds to input and output sampling lattices

$$
\Lambda_1 = \begin{bmatrix} v_{11} & v_{12} & 0 \\ 0 & v_{22} & 0 \\ 0 & 0 & T_1 \end{bmatrix}, \quad \Lambda_2 = \begin{bmatrix} v_{11} & v_{12} & 0 \\ 0 & v_{22} & 0 \\ 0 & 0 & T_2 \end{bmatrix}. \tag{18}
$$

Pure Temporal Interpolation

The most straightforward approach is pure temporal interpolation, in which a temporal resampling is performed independently at each spatial location $\boldsymbol{x}$. A typical application for this is increasing the frame rate in motion-picture film from 24 frames/s to 48 or 60 frames/s, giving a significantly better motion rendition. With the use of linear filtering, the interpolated image sequence is given by

$$
f_o(\boldsymbol{x}, nT_2) = \sum_m f(\boldsymbol{x}, mT_1) h(nT_2 - mT_1). \tag{19}
$$

If the temporal spectrum of the underlying continuous time-varying image satisfies the Nyquist criterion, the output points can be computed by ideal sinc interpolation:

$$
h(t) = \frac{\sin(\pi t / T_1)}{\pi t / T_1}. \tag{20}
$$

However, aside from the fact that this filter is unrealizable, it is unlikely, and in fact undesirable according to the discussion of

TABLE 1 Parameters of several common scanning structures

System	X	Y	T	Structure	Aspect Ratio
QCIF	$\frac{1}{176} pw = \frac{1}{132} ph$	$\frac{1}{144} ph$	$\frac{1}{10}$ s	P	4:3
CIF	$\frac{1}{352} pw = \frac{1}{264} ph$	$\frac{1}{288} ph$	$\frac{1}{15}$ s	P	4:3
ITU-R-601 (30)	$\frac{1}{720} pw = \frac{1}{540} ph$	$\frac{1}{480} ph$	$\frac{1}{29.97}$ s	I	4:3
ITU-R-601 (25)	$\frac{1}{720} pw = \frac{1}{540} ph$	$\frac{1}{576} ph$	$\frac{1}{25}$ s	I	4:3
HDTV-P	$\frac{1}{1280} pw = \frac{1}{720} ph$	$\frac{1}{720} ph$	$\frac{1}{60}$ s	P	16:9
HDTV-I	$\frac{1}{1920} pw = \frac{1}{1080} ph$	$\frac{1}{1080} ph$	$\frac{1}{30}$ s	I	16:9
IMAX	$\frac{1}{4096} pw = \frac{1}{3002} ph$	$\frac{1}{3002} ph$	$\frac{1}{24}$ s	P	1.364

Section 3, that the temporal spectrum satisfy the Nyquist criterion. Thus high-order interpolation kernels that approximate Eq. (20) are not found to be useful and are rarely used. Instead, simple low-order interpolation kernels are frequently applied. Examples are zero-order and linear (straight-line) interpolation kernels given by

$$h(t) = \begin{cases} 1 & \text{if } 0 \le t \le T_1 \\ 0 & \text{otherwise} \end{cases}, \tag{21}$$

$$h(t) = \begin{cases} 1 - t/T_1 & \text{if } 0 \le t \le T_1 \\ 0 & \text{if } t > T_1 \\ h(-t) & \text{if } t < 0 \end{cases}, \tag{22}$$

respectively. Note that Eq. (22) defines a noncausal filter and that in practice a delay of T_1 must be introduced. Zero-order hold is also called frame repeat and is the method used in film projection to go from 24 to 48 frames/s. These simple interpolators work well if there is little or no motion, but as the amount of motion increases they will not adequately remove spectral repeats causing effects such as jerkiness, and they may also remove useful information, introducing blurring. The problems with pure temporal interpolation can easily be illustrated for the image corresponding to Fig. 5(c) for the case of doubling the frame rate, i.e., $T_2 = T_1/2$. Using a one-dimensional temporal low-pass filter with cutoff at about $1/2T_1$ removes the desired high vertical frequencies in the baseband signal above $B/2$ (motion blur) and leaves undesirable aliasing at high vertical frequencies, as shown in Fig. 7(a).

Motion-Compensated Interpolation

It is clear that to correctly deal with a situation such as in Fig. 4(c), it is necessary to adapt the interpolation to the local orientation of the spectrum, and thus to the velocity, as suggested in Fig. 7(b). This is called motion-compensated interpolation. An auxiliary motion analysis process determines information about local motion in the image and attempts to track the trajectory of scene points over time. Specifically, suppose we wish to estimate the signal value at position x at time nT_2 from neighboring frames at times mT_1. We can assume that the scene point imaged at position x at time nT_2 was imaged at position $c(mT_1; x, nT_2)$ at time mT_1 [5]. If we know c exactly, we can compute

$$f_o(x, nT_2) = \sum_m f(c(mT_1; x, nT_2), mT_1)h(nT_2 - mT_1). \tag{23}$$

Since we assume that $f(x, t)$ is very slowly varying along the motion trajectory, a simple filter such as the linear interpolator of Eq. (22) would probably do very well. Of course, we do not know $c(mT_1; x, nT_2)$, so we must estimate it. Furthermore, since the position $(c(mT_1; x, nT_2), mT_1)$ probably does not lie on the input lattice Λ_1, $f(c(mT_1; x, nT_2), mT_1)$ must be spatially interpolated from its neighbors. If spatial aliasing is low as we have assumed, this interpolation can be done well (see previous chapter).

If a two-point temporal interpolation is used, we only need to find the correspondence between the point at (x, nT_2) and points in the frames at times lT_1 and $(l+1)T_1$ where $lT_1 \le nT_2$ and $(l+1)T_1 > nT_2$. This is specified by the backward and forward displacements

$$d_b(x, nT_2) = x - c(lT_1; x, nT_2), \tag{24}$$

$$d_f(x, nT_2) = c((l+1)T_1; x, nT_2) - x, \tag{25}$$

respectively. The interpolated value is then given by

$$f_o(x, nT_2) = f(x - d_b(x, nT_2), lT_1)h(nT_2 - lT_1)$$

$$+ f(x + d_f(x, nT_2), (l+1)T_1)$$

$$\times h(nT_2 - (l+1)T_1). \tag{26}$$

There are a number of key design issues in this process. The main one relates to the complexity and precision of the motion estimator. Since the image at time nT_2 is not available, the trajectory must be estimated from the existing frames at times mT_1, and often just from lT_1 and $(l+1)T_1$ as defined above.

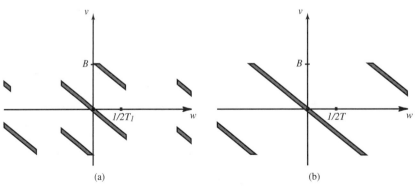

FIGURE 7 Frequency domain interpretation of 2:1 temporal interpolation of an image with vertical velocity $1/TB$: (a) pure temporal interpolation; (b) motion-compensated interpolation.

In the latter case, the forward and backward displacements will be collinear. We can assume that better motion estimators will lead to better motion-compensated interpolation. However, the tradeoff between complexity and performance must be optimized for each particular application. For example, block-based motion estimation (say one motion vector per 16×16 block) with accuracy rounded to the nearest pixel location will give very good results in large moving areas with moderate detail, giving significant overall improvement for most sequences. However, areas with complex motion and higher detail may continue to show quite visible artifacts, and more accurate motion estimates would be required to get good performance in these areas. Better motion estimates could be achieved with smaller blocks, parametric motion models, or dense motion estimates, for example. Motion estimation is treated in detail in Chapter 3.8. Some specific considerations related to estimating motion trajectories passing through points in between frames in the input sequence can be found in [5].

If the motion estimation method used sometimes yields unreliable motion vectors, it may be advantageous to be able to fall back to pure temporal interpolation. A test can be performed to determine whether pure temporal interpolation or motion-compensated interpolation is liable to yield better results, for example by comparing $|f(\boldsymbol{x}, (l+1) T_1) - f(\boldsymbol{x}, l T_1)|$ with $|f(\boldsymbol{x} + \boldsymbol{d}_f(\boldsymbol{x}, n T_2), (l+1) T_1) - f(\boldsymbol{x} - \boldsymbol{d}_b(\boldsymbol{x}, n T_2), l T_1)|$. Then the interpolated value can either be computed by the method suspected to be better, or by an appropriate weighted combination of the two.

Occlusions pose a particular problem, since the pixel to be interpolated may be visible only in the previous frame (newly covered area) or in the subsequent frame (newly exposed area). In particular, if $|f(\boldsymbol{x} + \boldsymbol{d}_f(\boldsymbol{x}, n T_2), (l+1) T_1) - f(\boldsymbol{x} - \boldsymbol{d}_b(\boldsymbol{x}, n T_2), l T_1)|$ is relatively large, this may signal that $\boldsymbol{x}$ lies in an occlusion area. In this case, we may wish to use zero-order hold interpolation based on either the frame at $l T_1$ or at $(l+1) T_1$, according to some local analysis. Figure 8 depicts the

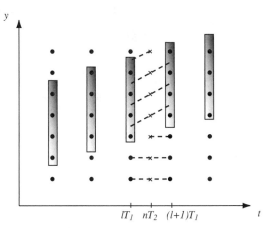

FIGURE 8 Example of motion-compensated temporal interpolation including occlusion handling.

motion-compensated interpolation of a frame midway between $l T_1$ and $(l+1) T_1$ including occlusion processing.

4.2 Spatiotemporal Sampling Structure Conversion

In this section, we consider the case in which both the spatial and the temporal sampling structures are changed, and when one or both of the input and output sampling structures is not separable in space and time (usually because of interlace). If the input sampling structure Λ_1 is separable in space and time, as in Eq. (18), and spatial aliasing is minimal, then the methods of the previous section can be combined with pure spatial interpolation. If we want to interpolate a sample at a time $m T_1$, we can use any suitable spatial interpolation. To interpolate at a sample at a time t that is not a multiple of T_1, the methods of the previous section can be applied.

The difficulties in spatiotemporal interpolation mainly arise when the input sampling structure Λ_1 is not separable in space and time, which is generally the case of interlace. This encompasses both interlaced-to-interlaced conversion, such as in conversion between NTSC and PAL television systems, and interlaced-to-progressive conversion (also called deinterlacing). The reason this introduces problems is that individual fields are undersampled, contrary to the assumption in all the previously discussed methods. Furthermore, as we have seen, there may also be significant aliasing in the spatiotemporal frequency domain as a result of vertical motion. Thus, a great deal of the research on spatiotemporal interpolation has been addressing these problems due to interlace, and a wide variety of techniques have been proposed, many of them very empirical in nature.

Deinterlacing

Deinterlacing generally refers to a 2:1 interpolation from an interlaced grid to a progressive grid with sampling lattices

$$
\begin{bmatrix} X & 0 & 0 \\ 0 & Y & 0 \\ 0 & T/2 & T \end{bmatrix}, \quad \begin{bmatrix} X & 0 & 0 \\ 0 & Y & 0 \\ 0 & 0 & T/2 \end{bmatrix},
$$

respectively (see Fig. 9). Both input and output lattices consist of fields at time instants $m T/2$. However, because each input field is vertically undersampled, spatial interpolation alone is inadequate. Similarly, because of possible spatiotemporal aliasing and difficulties with motion estimation, motion-compensated interpolation alone is inadequate. Thus, the most successful methods use a nonlinear combination of spatially and temporally interpolated values, according to local measures of which is most reliable. For example, in Fig. 9, sample A might best be reconstructed using spatial interpolation, sample B with pure temporal interpolation, and sample C with motion-compensated temporal interpolation. Another sample like D may be reconstructed by using a combination of spatial and motion-compensated temporal

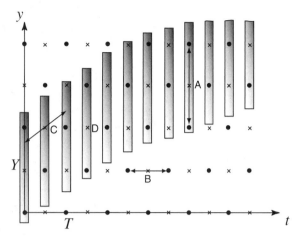

FIGURE 9 Input and output sampling structures for deinterlacing.

interpolation. See ref. [6] for a detailed presentation and discussion of a wide variety of deinterlacing methods. It is shown there that some adaptive motion-compensated methods can give reasonably good deinterlacing results on a wide variety of moving and fixed imagery.

5 Conclusion

This chapter has provided an overview of the basic theory related to sampling and interpolation of time-varying imagery. In contrast to other types of signals, it has been shown that it is *not* desirable to limit the spectrum of the continuous signal to a *fixed* three-dimensional frequency band prior to sampling, since this leads to excessive loss of spatial resolution. It is sufficient to ensure that the replicated spectra caused by sampling do not overlap. However, optimal reconstruction requires the use of motion-compensated temporal interpolation.

The interlaced scanning structure that is widely used in video systems has a fundamental problem whereby aliasing in the presence of vertical motion is inevitable. This makes operations such as motion estimation, coding, and so on more difficult to accomplish. Thus, it is likely that interlaced scanning will gradually disappear as camera technology improves and the full spatial resolution desired can be obtained with frame rates of 50–60 Hz and above.

Spatiotemporal interpolation will remain an important technology to convert between the wide variety of scanning standards in both new and archival material. Research will continue into robust, low-complexity methods for motion-compensated temporal interpolation that can be incorporated into any receiver.

Further Information

The classic paper on television scanning is ref. [7]. The use of lattices for the study of spatiotemporal sampling was introduced in [8]. A detailed study of camera and display aperture models for television can be found in [9]. Research papers on spatiotemporal interpolation can be found regularly in the *IEEE Transactions on Image Processing*, *IEEE Transactions on Circuits and Systems for Video Technology*, and *Signal Processing: Image Communication*. See ref. [10] for a special issue on motion estimation and compensation for standards conversion.

References

[1] E. Dubois, "The sampling and reconstruction of time-varying imagery with application in video systems," *Proc. IEEE* **73**, 502–522 (1985).

[2] T. Kalker, "On multidimensional sampling," in *The Digital Signal Processing Handbook*, V. Madisetti and D. Williams, eds. (CRC Press, Boca Raton, FL, 1998). Chap. 4, pp. 4-1–4-21.

[3] E. Dubois, "Motion-compensated filtering of time-varying images," *Multidimens. Syst. Signal Process.* **3**, 211–239 (1992).

[4] B. Girod and R. Thoma, "Motion-compensating field interpolation from interlaced and non-interlaced grids," in *Image Coding, Proc. SPIE* **594**, 186–193 (1985).

[5] E. Dubois and J. Konrad, "Estimation of 2-D motion fields from image sequences with application to motion-compensated processing," in *Motion Analysis and Image Sequence Processing*, M. Sezan and R. Lagendijk, eds. (Kluwer, Boston, MA, 1993). Ch. 3, pp. 53–87.

[6] G. de Haan, "Deinterlacing-an overview," *Proc. IEEE* **86**, 1839–1857 (1998).

[7] P. Mertz and F. Gray, "A theory of scanning and its relation to the characteristics of the transmitted signal in telephotography and television," *Bell Syst. Tech. J.* **13**, 464–515 (1934).

[8] D. Petersen and D. Middleton, "Sampling and reconstruction of wave-number-limited functions in 3-dimensional Euclidean spaces," *Inf. Contr.* **5**, 279–323 (1962).

[9] M. Isnardi, "Modeling the television process," Tech. Rep. 515, Research Lab. Electron. Massachusetts Institute of Technology, Cambridge, MA, May 1986.

[10] E. Dubois, G. de Haan, and T. Kurita, "Special issue on motion estimation and compensation technologies for standards conversion," *Signal Process. Image Commun.* **6**, no. 3, 189–280 (June 1994).

VIII

Image and Video Rendering and Assessment

Image Quantization, Halftoning, and Printing

Ping Wah Wong
Gainwise Limited

1 Introduction and Printing Technologies

The rapidly decreasing cost and increasing power of modern-day computing devices have contributed to the prevalent use of digital images in recent years. Image processing systems that have found applications in everyday life include digital photography, electronic publishing, on-line museum, electronic catlog shopping, home surveillance, image rich souvenirs, and so on. A typical end-to-end image processing system typically includes input devices, processing devices, and output devices. Output devices are very important in imaging systems because they allow the image data to be presented to the user in a human observable form. Popular imaging output devices include image printers, video displays, liquid crystal displays, projection devices, and many others.

In practical applications, quantities in digital images are usually quantized to 8 bits. Specifically, we use 8 bits to represent the image pixel intensity (gray level) for monochrome (gray-scale) images, and we use 24 bits to represent the combined red, green, and blue intensity levels for color images (i.e., 8 bits for each color plane). Hence we have 256 gray levels for monochrome images and 16 million colors for color images. It has been demonstrated in practice that such a representation is normally sufficient to be considered "continuous tone."

Many image output devices share characteristics that lead to two common problems requiring image processing solutions.

The first one is that many devices are only capable of rendering a small number of output levels. Examples include many printers that can only print at two levels (on–off) for each primary color, and liquid crystal displays that typically can display 16 gray levels. This means that an image must be converted from its "continuous-tone" representation to one that matches the characteristics of the printers or displays.

There are several printing technologies that are suitable for producing hardcopies of images. They include laser printing, inkjet printing, dye sublimation, thermal wax transfer, liquid toner laser transfer, offset printing, and so on. Among these, laser and inkjet printing are perhaps the most popular because of their relatively low cost, high performance, high quality, and excellent availability. Traditional laser and inkjet printers can only print at two levels for each primary color. To print a monochrome image, we first generate a halftone that consists of black and white pixels. Then we put black toners on the paper corresponding to the black pixel locations, and use the white paper background to represent the white pixels. This problem of rendering images in two levels is called *halftoning*.

In color printing, we generally use cyan (C), magenta (M), and yellow (Y) as the three primary colors. We mix dots of these three colors to give an impression of other colors. Putting all three colors at the same pixel location produces black output, which is called *composite black*. It is observed in practice that composite black is usually a bit dull in appearance. As a result, many color printers are designed to use four toners, namely, cyan, magenta, yellow, and black (K). For these four-toner printers, we have several ways to obtain black output; we can either mix CMY together, or we can use the black toner, or we can mix CMYK together in some proportion. Choices of these are generally dependent on the specific printer design.

Recently, both laser and inkjet printing technologies have advanced, and such printers that can print multiple (e.g., 4, 8, or 16) levels have appeared commercially. The corresponding problem to rendering at multiple levels is called *multitoning*.

Another problem is that many low-cost displays, while being able to handle "continuous tones," have limited-sized video buffers that necessitate only a subset of the 2^{24} colors to be used. As a result we need to choose the subset of colors (called a *color palette*) to be displayed, and perform the mapping from the original image to the chosen color palette. This problem is called *color quantization*.

Since quantization is a core component of both halftoning and color quantization, we first discuss in Section 2 the basics of scalar quantization. The design of optimum scalar quantizers is presented. In Section 3, we discuss the main approaches of halftoning, and show examples of various types of popular halftoning algorithms. The interplay between halftoning and compression is also considered. In Section 4, we discuss color quantization methods that enable images to be displayed in devices with a limited video buffer.

2 Scalar Quantization

2.1 Quantization Basics

The function of *scalar quantizers* is to convert a quantity to a finite precision representation (such as converting real numbers to integers in digital computing). We can represent scalar quantization as

$$y = Q(x),$$

where $Q(\cdot)$ is the quantization function. Mathematically, a K-level scalar quantizer is specified by the $K + 1$ *decision levels* $(d_0, d_1, \ldots, d_K)$ and the K *output levels* $(y_1, y_2, \ldots, y_K)$. Each region from d_{i-1} to d_i is referred to as a quantization bin. Specifically, we can specify a quantization function as

$$Q(u) = y_i \quad \text{if } d_{i-1} \le u < d_i; \qquad i = 1, 2, \ldots, K.$$

An example of the input–output characteristics of a scalar quantizer is shown in Fig. 1.

Define the quantizer error $e(u)$ by

$$e(u) = Q(u) - u.$$

This quantity is of crucial importance in describing the characteristics of a scalar quantizer as well as in many design problems

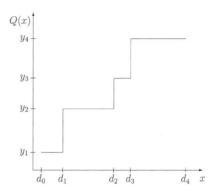

FIGURE 1   Input–output characteristics of a scalar quantizer.

associated with quantizers. Let $p_X(x)$ be the probability density function of a random variable X. We can write the expected nth power (the nth order moment) of the quantization error as

$$M_n = E(X^n) = \int e^n(x) p_X(x) \, dx$$
$$= \sum_{i=1}^{K} \int_{d_{i-1}}^{d_i} (y_i - x)^n p_X(x) \, dx. \tag{1}$$

Of particular importance in practice is the second-order moment, which is usually called the mean squared quantizer error (MSQE). The MSQE is often used in the design of optimum quantizers. In the next section, we will examine such a design.

2.2 Optimum Quantization

The problem of finding an optimum scalar quantizer was first solved by Lloyd in the mid-1950s [1]. Since Lloyd did not publish his results until 1982, the first optimum scalar quantizer design in the open literature was published by Max [2]. Hence, optimum scalar quantizers are usually referred to as Lloyd–Max quantizers.

If X has dynamic range (α, β), we set $d_0 = \alpha$ and $d_K = \beta$ in order to be able to generate an output for all possible values that X can take on. To determine the optimum quantizer, we first differentiate Eq. (1) for $n = 2$ with respect to d_i ($i = 1, 2, \ldots, K - 1$) to obtain

$$(y_i - d_i)^2 - (y_{i+1} - d_i)^2 = 0,$$

which is equivalent to

$$d_i = \frac{y_i + y_{i+1}}{2} \qquad i = 1, 2, \ldots, K - 1. \tag{2}$$

Equation (2) is a necessary condition for optimum scalar quantizers, namely, any decision level between two consecutive quantization bins is the midpoint of the two adjacent output levels. This is usually referred to as the *midpoint condition*.

Second, we differentiate Eq. (1) for $n = 2$ with respect to $y_i (i = 1, 2, \ldots, K)$ to obtain

$$\int_{d_{i-1}}^{d_i} (y_i - x) p_X(x) \, dx = 0, \qquad i = 1, 2, \ldots, K,$$

which gives

$$y_i = \frac{\int_{d_{i-1}}^{d_i} x p_X(x) \, dx}{\int_{d_{i-1}}^{d_i} p_X(x) \, dx}$$

$$= E[X | d_{i-1} \leq X < d_i] \qquad i = 1, 2, \ldots, K. \quad (3)$$

This is another necessary condition for optimum scalar quantizers. It says that the optimum output level at any quantization bin should be the conditional expectation (centroid) of X within that bin. This is the *centroid condition* for optimum quantizers.

The Lloyd algorithm for optimum quantizer design uses both the mid-point condition and the centroid condition. Essentially one starts with a set of quantizer boundaries, and use the centroid condition to find the optimum output levels based on the quantizer boundaries. From the new output levels, one uses the midpoint condition to determine a new set of quantizer boundaries. These two steps are repeated until convergence is achieved. The nice feature about the Lloyd algorithm is that it works either if one is given a probability density function, or if one is given a set of samples in a signal. In the latter case, one replaces probabilistic averages (expect values) by sample averages. Furthermore, the Lloyd algorithm always converges in practice. The Lloyd algorithm has also been extended for designing vector quantizers [3]. For further details on quantization, the reader is referred to the text [4].

2.3 Distortion Criteria and Image Quantization

There are three common problems that require quantization of either the image pixels values or some transformations of the pixels. They are halftoning, color quantization, and image compression. Image compression will be treated in a different chapter of this book. Here we only consider the problem of image quantization in the context of halftoning and color quantization.

Generally, we formulate the problem as follows: Given an image $f(n_1, n_2)$ of size N_1 by N_2, where each pixel takes on values in the range $\{0, 1, \ldots, \ldots, D - 1\}$, we want to quantize each image pixel to K output levels ($K < D$) so that a distortion criterion is minimized. As an example, we have $K = 2$ if we want to halftone the image so that it can be printed by using a binary laser or inkjet printer. In this case, the two output levels are 0 and D, respectively. Suppose we use the mean squared error criterion

$$D(x, y) = \sum_{n_1, n_2} (f(n_1, n_2) - b(n_1, n_2))^2;$$

FIGURE 2 Original boats image. (See color section, p. C–35.)

where $b(n_1, n_2)$ is the quantized output. Since the output levels are fixed, the mid-point condition in optimum scalar quantizers gives

$$b(n_1, n_2) = Q(f(n_1, n_2)) = \begin{cases} D & \text{if } f(n_1, n_2) \geq D/2 \\ 0 & \text{otherwise} \end{cases}.$$

This is constant thresholding of the image where the threshold is the midpoint of the dynamic range. Using the boats image in Fig. 2 as an example, we show in Fig. 3 the result of constant thresholding.

Although the quantizer is optimum with respect to the mean squared error distortion criterion, the resulting output quality is

FIGURE 3 Boats image thresholded to two levels, using a constant threshold for each primary color plane. (See color section, p. C–35.)

rather poor. This is because the mean squared error does not reflect the characteristics of human perception on images. A much better error criterion, often used in image processing applications, is the visual mean squared error criterion

$$\mathcal{E} = E((v(n_1, n_2) * (b(n_1, n_2) - f(n_1, n_2)))^2), \quad (4)$$

where $*$ denotes convolution, $E(\cdot)$ is the expectation (averaging) operator, and $v(n_1, n_2)$ is the impulse response of a filter that approximates the spectral response of the human visual system. Notice that a neighborhood operation is involved in the visual mean squared error due to the convolution operation. This is consistent with the notion of halftoning that the human perception system filters (mixes) the dots in halftone images to give an impression of graylevel.

3 Halftoning

Halftoning is a process where a continuous tone image of, say 256 levels, is converted into an output image of two levels so that the original and the halftone images appear similar when observed from a distance. The corresponding problem of generating output images of multiple levels is called multitoning. The main application of halftoning and multitoning is printing and in displaying images with limited bit-depth displays such as liquid crystal displays. Since the techniques for multitoning are straightforward extensions of halftoning (rendering at two levels), we will only focus on halftoning in the rest of this section.

The three popular halftoning techniques are ordered dithering, error diffusion, and optimization techniques. Generally, ordered dithering requires the lowest computational complexity, while optimization techniques are the most computationally intensive. In contrast, well-designed optimization techniques tend to generate halftones of the best quality. All three halftoning techniques will be described in this chapter.

Before we go into the specifics of halftoning algorithms, we comment that the shape of a basic output unit from a real-life printer is a very important factor to the ultimate quality of the output. Normally in image processing, we use rectangular grids and hence each pixel is considered to have a rectangular shape. In printing, the shapes of each output pixel (printing dot) is *not* rectangular in shape. Furthermore, the dot shapes between laser and inkjet printers have very different characteristics. Inkjet dots can be well represented by a round shape that has an area coverage larger than the rectangular pixel size. Halftoning considerations using this oversized round dot model can be found in [5]. The output dot shape of laser printers is round, but the intensity transition from the edge to the center of the dot is more gradual. If we cut a cross section across a laser printer dot and draw the intensity profile, we have a curve that resembles a bell shape. More detailed discussions of laser dot shapes can be found in [6].

For simplicity, we describe in the following sections halftoning algorithms for monochrome images (black and white halftoning). Halftoning of color images can be done on a plane-by-plane basis. We give examples of color halftones, using such an approach. Readers who are interested in plane-dependent halftoning, in which the color planes are halftoned jointly, are referred to the literature [7, 8].

3.1 Ordered Dithering

Consider an image $f(n_1, n_2)$ of size N_1 by N_2 with pixel values in the range $\{0, 1, \ldots, D-1\}$. We want to generate a halftone $b(n_1, n_2)$ by using ordered dithering. In this method, we use an array $a(n_1, n_2)$ of size M_1 by M_2, typically called a *dither matrix*, where the elements are threshold levels. Each element of the dither matrix is an integer that takes value in the range $\{0, 1, \ldots, D-1\}$. To generate the output, we perform an element-by-element thresholding of the image pixels, using the elements of the dither matrix as shown in Fig. 4.

In most applications, we have $M_1 \ll N_1$ and $M_2 \ll N_2$. That is, the size of the dither matrix is much smaller than the image size. Thus we periodically replicate $a(n_1, n_2)$ to give $\tilde{a}(n_1, n_2) = a(n_1 \bmod M_1, n_2 \bmod M_2)$ so that the entire image is covered. Then the halftoning operation can be represented mathematically as

$$b(n_1, n_2) = Q_{\tilde{a}(n_1, n_2)}(f(n_1, n_2))$$
$$= \begin{cases} 1 & \text{if } f(n_1, n_2) \geq a(n_1 \bmod M_1, n_2 \bmod M_2) \\ 0 & \text{otherwise} \end{cases}.$$

Here the output "1" represents a white pixel (do not print a dot) and "0" represents a black pixel (print a dot).

It is evident that the actual values of the dither matrix elements will have a great influence on the quality of the output halftone.

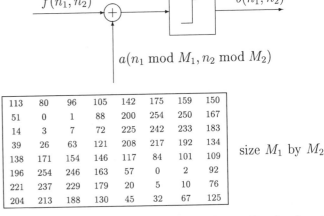

113	80	96	105	142	175	159	150
51	0	1	88	200	254	250	167
14	3	7	72	225	242	233	183
39	26	63	121	208	217	192	134
138	171	154	146	117	84	101	109
196	254	246	163	57	0	2	92
221	237	229	179	20	5	10	76
204	213	188	130	45	32	67	125

size M_1 by M_2

FIGURE 4 Pixel-by-pixel operation of ordered dithering. Here the values of the dither matrix elements correspond to images with an 8-bit intensity representation.

It can be easily verified that the following 8 by 8 matrix

$$\begin{bmatrix} 113 & 80 & 96 & 105 & 142 & 175 & 159 & 150 \\ 51 & 0 & 1 & 88 & 200 & 254 & 250 & 167 \\ 14 & 3 & 7 & 72 & 225 & 242 & 233 & 183 \\ 39 & 26 & 63 & 121 & 208 & 217 & 192 & 134 \\ 138 & 171 & 154 & 146 & 117 & 84 & 101 & 109 \\ 196 & 254 & 246 & 163 & 57 & 0 & 2 & 92 \\ 221 & 237 & 229 & 179 & 20 & 5 & 10 & 76 \\ 204 & 213 & 188 & 130 & 45 & 32 & 67 & 125 \end{bmatrix} \quad (5)$$

will give halftones where the dots tend to be clustered within 4 by 4 cells. In contrast, the matrix

$$\begin{bmatrix} 4 & 236 & 60 & 220 & 8 & 224 & 48 & 208 \\ 132 & 68 & 188 & 124 & 136 & 72 & 176 & 112 \\ 36 & 196 & 20 & 252 & 40 & 200 & 24 & 240 \\ 164 & 100 & 148 & 84 & 168 & 104 & 152 & 88 \\ 12 & 228 & 52 & 212 & 0 & 232 & 56 & 216 \\ 140 & 76 & 180 & 116 & 128 & 64 & 184 & 120 \\ 44 & 204 & 28 & 244 & 32 & 192 & 16 & 248 \\ 172 & 108 & 156 & 92 & 160 & 96 & 144 & 80 \end{bmatrix} \quad (6)$$

tends to generate halftones where the dots are dispersed. A very popular class of dispersed dither matrices is called Bayer dither [9], which has been shown to have certain optimality properties. Examples of clustered and dispersed halftones as well as their power spectra are shown in Figs. 5 and 6. It is evident from both the halftones and their spectra that these halftones exhibit a strong periodicity induced by the size of the dither matrices.

The major advantage of ordered dithering is its simplicity. To generate an output halftone pixel, only one thresholding or comparison operation is needed. As a result, ordered dithering is very popular in low-cost printing applications. Experimental evidence has shown that it is generally preferred to use dispersed halftones in inkjet printers for smoothness, and clustered dot halftones in laser printers for stability.

Traditionally, ordered dithering uses dither matrices of size between 8 by 8 to about 32 by 32. Recently, a class of ordered dither called *blue noise dithering* has been proposed that uses dither matrices of large sizes (e.g., 128 by 128 or larger) to generate halftones that are of higher quality than those of traditional ordered dithering with small dither matrices. These *blue noise dither matrices* are typically designed by using iterative methods [10].

3.2 Error Diffusion

Error diffusion [11] is an excellent method for generating high-quality halftones that are particularly suitable for low- to medium-resolution devices. A block diagram showing an error diffusion system is given in Fig. 7. It turns out [12] that error diffusion is the two-dimension equivalent of sigma-delta modulation (also called delta-sigma modulation) [13], which was developed for performing high-resolution analog-to-digital conversion by using a 1-bit quantizer embedded in a feedback loop.

To apply the error diffusion algorithm in producing a halftone $b(n_1, n_2)$, we need to scan the input image $f(n_1, n_2)$ in some fashion. One of the most popular strategies is raster scanning, in which we scan the image row by row from the top to the bottom, and within each row from the left to the right. At each pixel location, we perform the operations

$$u(n_1, n_2) = f(n_1, n_2) - \sum_{m_1, m_2} h(m_1, m_2) e(n_1 - m_1, n_2 - m_2), \quad (7)$$

$$b(n_1, n_2) = Q(u(n_1, n_2)) = \begin{cases} 1 & \text{if } u(n_1, n_2) \geq 0.5 \\ 0 & \text{otherwise} \end{cases}, \quad (8)$$

$$e(n_1, n_2) = b(n_1, n_2) - u(n_1, n_2) = Q(u(n_1, n_2)) - u(n_1, n_2). \quad (9)$$

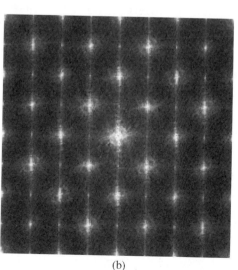

(a) (b)

FIGURE 5   Boats image: (a) halftoned with a clustered dot dither at 150 dots per in.; (b) the halftone power spectrum. (See color section, p. C–35.)

 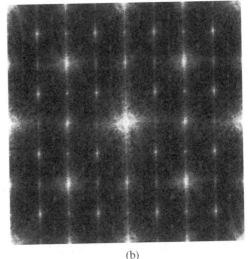

| (a) | (b) |

FIGURE 6   Boats image: (a) halftoned with a Bayer dither [9] at 150 dots per in.; (b) the halftone power spectrum. (See color section, p. C–36.)

In the signal processing literature, $u(n_1, n_2)$ is called either the *state variable* of the system or the *modified input*.

Notice that we need to buffer the binary quantizer error values $e(n_1, n_2)$ because of the convolution operation in Eq. (7). The extent of the buffer required is dependent on the support of the *error diffusion kernel* $h(m_1, m_2)$. The filter $h(m_1, m_2)$ generally has a low-pass characteristic, and the coefficients often satisfy

$$\sum_{(m_1, m_2)} h(m_1, m_2) = 1.$$

A very popular kernel suggested by Floyd and Steinberg [11] is shown in Fig. 8, with

$$h(0, 1) = 7/16, \quad h(1, -1) = 3/16, \quad h(1, 0) = 5/16,$$
$$h(1, 1) = 1/16.$$

This kernel consists of four coefficients, and hence the complexity of the error diffusion algorithm is quite mild. Two other popular error diffusion kernels are proposed by Jarvis, Judice, and Ninke [14] and by Stucki [15]. A Floyd–Steinberg error diffused halftone and its power spectrum is shown in Fig. 9.

Using Eqs. (7) and (9), we can write

$$b(n_1, n_2) = f(n_1, n_2) + \sum_{m_1, m_2} g(m_1, m_2) e(n_1 - m_1, n_2 - m_2),$$
$$\tag{10}$$

where

$$g(m_1, m_2) = \delta(m_1, m_2) - h(m_1, m_2),$$

and $\delta(m_1, m_2)$ is the Kronecker delta. Using the Floyd–Steinberg filter kernel as an example, we have

$$g(m_1, m_2) = \begin{cases} 1 & (m_1, m_2) = (0, 0) \\ -7/16 & (m_1, m_2) = (0, 1) \\ -3/16 & (m_1, m_2) = (1, -1) \\ -5/16 & (m_1, m_2) = (1, 0) \\ -1/16 & (m_1, m_2) = (1, 1) \\ 0 & \text{otherwise} \end{cases}.$$

A nice interpretation of Eq. (10) is that the output is the sum of the input and a filtered version of the quantizer noise.

To calculate the power spectral density of $b(n_1, n_2)$, we need to calculate its autocorrelation function. This calculation requires the autocorrelation function of $e(n_1, n_2)$ and the cross correlation function $E[f(n_1, n_2)e(n_1 + m_1, n_2 + m_2)]$, both as function of the statistical properties of the input image $f(n_1, n_2)$. This turns out to be a very difficult task because of the nonlinear nature of the system induced by the binary quantizer. A general solution for the two-dimensional case is still an open problem, although solutions to the one-dimensional case (sigma-delta modulation) have already been found [16, 17].

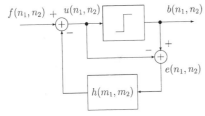

FIGURE 7   Block diagram of the error diffusion system.

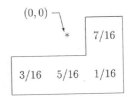

FIGURE 8   The Floyd–Steinberg error diffusion filter coefficients.

(a)

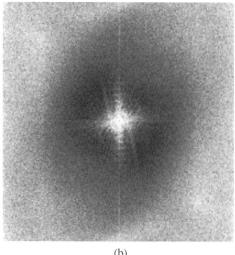
(b)

FIGURE 9 Boats image: (a) halftoned with a Floyd–Steinberg error diffusion [11] at 150 dots per in.; (b) the halftone power spectrum. (See color section, p. C–36.)

Despite the difficulty in an exact mathematical analysis of error diffusion, we can compute the spectrum of error diffused halftones empirically. An example of such a power spectrum is shown in Fig. 9. It is of interest to note that the noise of error diffusion is primarily in the high-frequency region, which agrees with the intuition that the quantization noise in a high-quality halftone should be mostly located in the high-frequency region.

Although Floyd–Steinberg error diffusion generally produces high-quality output, it also generates undesirable artifacts such as "worms" and undesirable patterns. Consequently there has been a large number of reports in the literature focusing on the improvement of error diffusion. These techniques can generally be categorized into two classes. The first class of techniques focuses on "breaking up" the "undesirable output patterns" of error diffusion. This includes using alternative scanning strategies [18,19] and injecting random noises into the error diffusion system [18,20].

Another class of techniques attempts to optimize the error diffusion filter with respect to a distortion criterion. To this end, one can use the frequency weighted mean squared error in Eq. (4). Substituting Eq. (10) into Eq. (4), we have

$$\mathcal{E} = E[(v(n_1, n_2) * g(n_1, n_2) * e(n_1, n_2))^2].$$

It is evident that $\mathcal{E}$ is a function of $h(n_1, n_2)$. Hence at least in principle, we can find an optimum error diffusion kernel $h(n_1, n_2)$ to minimize $\mathcal{E}$.

As discussed earlier, error diffusion is a highly nonlinear system where $e(n_1, n_2)$ depends on $f(n_1, n_2)$ and the structure of the feedback loop in a very complicated way. This makes the optimization problem very difficult. In the literature, there are reports in solving this optimization problem either by making certain assumptions on the quantizer noise [21], using the LMS algorithm in adaptive signal processing to solve the optimization problem [22], or using an iterative optimization approach [23].

3.3 Optimization Halftoning Techniques

Consider a distortion criterion $d(f(n_1, n_2), b(n_1, n_2))$ between a continuous tone image $f(n_1, n_2)$ and a halftone $b(n_1, n_2)$. Given a specific $f(n_1, n_2)$, we can find $b(n_1, n_2)$ that minimizes the average distortion $E[d(f(n_1, n_2), b(n_1, n_2))]$. There are many methods for finding a signal to minimize a certain cost function. Here, any specific method for minimizing $E[d(f(n_1, n_2), b(n_1, n_2))]$ will give rise to a halftoning algorithm.

Suppose we consider the frequency weighted squared error

$$d(f(n_1, n_2), b(n_1, n_2)) = (v(n_1, n_2) * (b(n_1, n_2) - x(n_1, n_2)))^2.$$

One way to find a halftone is to use a *greedy minimization* approach as follows: We scan the image $f(n_1, n_2)$ in raster scan fashion. At each pixel location (n_1, n_2), we determine the binary output pixel as

$$b(n_1, n_2) = \underset{b(n_1, n_2) \in \{0,1\}}{\arg \min} (v(n_1, n_2) * (b(n_1, n_2) - x(n_1, n_2)))^2.$$

Since there are only two possible values of $b(n_1, n_2)$ for each (n_1, n_2), we can compute $d(f(n_1, n_2), b(n_1, n_2))$ for both values, and then pick the one that results in a smaller distortion. As long as $v(n_1, n_2)$ has a causal support with respect to the scanning strategy, this minimization procedure can be performed very easily.

In general, greedy minimization does not generate output halftones of satisfactory quality. Since the greedy approach performs the minimization on a pixel by pixel basis with respect to a scanning strategy, there is no guarantee that it actually minimizes the overall average distortion. That is, any decision made by the greedy approach at a local pixel location will affect the local distortion of "future" pixels. Experimental results comparing greedy minimization and more sophisticated techniques indeed show that greedy minimization is far from optimal [24].

One method for generating halftones of excellent quality is the direct binary search (DBS) [25]. The DBS algorithm goes through the image in multiple passes. In the kth pass, one uses the output halftone of the previous pass $b_{k-1}(n_1, n_2)$ and the original continuous tone image $f(n_1, n_2)$ to obtain an improved output halftone $b_k(n_1, n_2)$. In the first pass, one needs an initial halftone $b_0(n_1, n_2)$ that can be generated, for example, by using greedy optimization. In each pass over the image, the DBS algorithm attempts to improve the quality of the output using the following strategy: At each pixel location (n_1, n_2), one considers possible improvement to the output halftone by flipping (inverting) the current halftone pixel or by swapping the current halftone pixel with one of its eight neighbors. The halftone pixel $b_k(n_1, n_2)$ at location (n_1, n_2) is chosen to be the one that results in the smallest distortion among the 10 possibilities (center pixel unchanged, center pixel inverted, center pixel swapped with one of its neighbors). Efficient methods for implementing the DBS algorithm have been suggested [25].

Other excellent methods for optimization based halftoning include tree coding [24], least-squares solution [26], combined halftoning compression [27], and many others. Figure 10 shows a halftone generated by using the tree-coding algorithm [24] and the corresponding halftone power spectrum.

3.4 Halftoning and Compression

The interplay between halftoning and compression is an interesting and important topic. This is important for both storage purposes (e.g., memory insider a printer) and for communications reasons (e.g., remote printing, fax). Consider a color page of U.S. letter size (8.5 in. by 11 in.). Assuming a print area of 8 by 10 in., a printer of 600 dots per in. resolution, and binary printing at 3 bits per pixel (i.e., two levels per color plane), we need 10.8 M bytes to store the halftone of just one page. It is obvious why compression is an important issue. There are two general categories of solutions related to compression and halftoning.

The first category of techniques attempt to directly compressed the halftones. Using the remote printing example, the continuous tone image is halftoned and then compressed at the source. The bit stream is transmitted. At the receiver, the bit stream is decoded and the resulting halftone is finally printed. Examples of work toward this area include lossless compression of halftones [28], lossy compression of halftones [29–31], joint halftone–compression design [27,32], and entropy-constrained halftoning [24,33].

Another category of solutions focuses on the optimization of continuous tone image compression taking into account the characteristics of halftoning. In the remote printing example, a continuous image is compressed at the source, and the compressed bit stream is transmitted. At the receiver, the bit stream is decompressed. In this case, the receiver has to halftone the decoded image before sending it for printing. Examples of work in this area include optimized subband coders [34] and the optimized JPEG coder [35], both designed with respect to the spectral characteristics of halftoning.

4 Color Quantization

Consider the problem of displaying a color image $f(n_1, n_2)$, where each pixel is represented by K bits (e.g., $K = 24$), i.e., assuming three primary color planes with $K/3$ bits precision

(a)

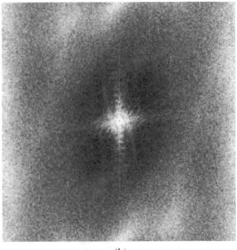

(b)

FIGURE 10   Boats image: (a) halftoned with a tree-coding algorithm [24] at 150 dots per in.; (b) the halftone power spectrum. (See color section, p. C–37.)

for each primary color. Suppose the display is only capable of handling L bits ($L < K$) or 2^L different colors. This restriction is usually due to a memory size constraint in the video display buffer. To display an image using such a device, we convert $f(n_1, n_2)$ to an image $c(n_1, n_2)$, where each pixel $c(n_1, n_2)$ is described by an L-bit index. In practice, we construct a look-up table that maps each index to an RGB color pixel value. The problem of color quantization is to design the color palette (hence the look-up table) and to determine how each pixel is to be mapped to the colors in the palette.

Although this problem may appear to be similar to halftoning, there are significant differences. First, the number of levels used in printing is small (usually between two and 16), whereas in color displays, we can usually have 256 different colors or more. Second, the colors (typically CMY or CMYK) of the inks and toners for color printing is fixed once a printer design is completed. In display, we can choose the optimum colors for displaying an image. This collection of colors is called a *color palette*. Third, the primary colors in printing are image independent. In contrast, the color palette in color quantization is often image dependent.

A very simple color quantization method is uniform quantization. For example, we can assign 3 bits each to the red and green channels, and 2 bits to the blue channel. We then use a uniform quantizer for each color channel. It is perhaps obvious that this method will generally give poor output results. In the rest of this section, we consider color quantization methods that can produce far better results.

4.1 Color Palette Design

There are many ways to design a color palette. The problem is similar to the optimum quantizer design problem. There are several quantization techniques, including sequential scalar quantization and vector quantization, that can be applied for designing

color palettes [36–38]. A very popular method for designing image dependent palettes is the median cut algorithm. It is a relatively simple algorithm both conceptually and computationally. At the same time, it produces very good results.

The median cut algorithm was originally reported in a thesis in 1980. It was subsequently published in a paper in 1982 [39]. The philosophy behind the algorithm is that each color within the color palette is used for representing approximately the same number of pixels in the original image $f(n_1, n_2)$.

Consider a 24-bit image $f(n_1, n_2)$. We first compute the color histogram of the image. For a 24-bit image, there can be a maximum of 16 million different colors. A histogram with that many entries could be difficult to handle. Hence Heckbert proposed to "pre-quantize" the original image into a 15-bit representation by using a 5-bit uniform quantizer for each color channel. Hence there are 32,768 colors after this step.

From the color histogram, one finds a rectangular box that tightly encloses the available colors within the image. The idea of the median cut algorithm is to partition the box recursively until we have 256 rectangular boxes. At each step of the algorithm, we choose the box that encloses the largest number of pixels to be partitioned. To perform the partition of a box, Heckbert proposed to make a cut along the coordinate that spans the largest range. The specific cut is made at the median point so that approximately equal number of pixels fall into each half of the subdivided boxes. This procedure is recursively applied until we have 256 boxes.

4.2 Error Diffusion Approach

Once a color palette is designed, the next step is to map each color pixel to one of the colors in the color palette. Nearest-neighbor mapping is a straightforward approach, in which we map each color pixel to the nearest color in the color palette.

(a)   (b)

FIGURE 11 Boats image, color quantized to 256 colors by using the median cut algorithm [39], with (a) nearest-neighbor mapping and (b) error diffusion. (See color section, p. C–37.)

This is equivalent to fixed vector quantization, using the color palette as the code book. It turns out that this approach generally produces output images where false contouring is quite visible, particularly when the size of the color palette is small (e.g., 256).

A better approach is to apply error diffusion for generating the output image. Here we use a system as shown in Fig. 7, except that we replace the 1-bit quantizer by a vector quantizer with the color palette as the code book. Experimental results have shown that this approach generates output images that are significantly better than nearest-neighbor mapping. Figures. 11(a) and 11(b) show color quantized images, using the median cut algorithm with nearest-neighbor mapping and error diffusion, respectively. False contouring is quite obvious with the nearest-neighbor mapping approach.

5 Conclusion

We have discussed in this chapter the basics of scalar quantization, halftoning, and color quantization. We reviewed some constraints imposed by printers and displays, which motivate the problems of halftoning and color quantization. The basics of optimum scalar quantization and its application to image quantization are presented. In halftoning we described and compared the common approaches of halftoning. Examples were given for several representative halftoning algorithms. We also described and showed examples of the median cut algorithm for color quantization. The readers who are interested in further probing into these areas are referred to the Reference Section and the references therein.

References

[1] S. P. Lloyd, "Least squares quantization in PCM," *IEEE Trans. Inf. Theory* **28**, 129–137 (1982).

[2] J. Max, "Quantizing for minimum distortion," *IRE Trans. Inf. Theory* **6**, 7–12 (1960).

[3] Y. Linde, A. Buzo, and R. M. Gray, "An algorithm for vector quantizer design," *IEEE Trans. Commun.* **COM-28**, 84–95 (1980).

[4] A. Gersho and R. M. Gray, *Vector Quantization and Signal Compression*. (Kluwer, Boston, 1992).

[5] T. N. Pappas, C. K. Dong, and D. L. Neuhoff, "Measurement of printer parameters for model-based halftoning," *J. Electron. Imag.* **2**, 193–204 (1993).

[6] Q. Lin and J. Wiseman, "Impact of electrophotographic printer dot modeling on halftone image quality," in *SID Digest of Technical Papers* (SID, Seattle, WA, 1993), pp. 147–150.

[7] J. R. Sullivan, R. L. Miller, and T. J. Wetzel, "Color digital halftoning with vector error diffusion." U.S. Patent 5070413, December 1991.

[8] Q. Lin and J. Allebach, "Color FM screen design using DBS algorithm," presented at the IS&T/SPIE Symposium on Electronic Imaging, San Jose, CA, January, 1998.

[9] B. E. Bayer, "An optimum method for two-level rendition of continuous-tone pictures," *Proc. IEEE Int. Conf. Commun.* 26.11–26.15 (1973).

[10] R. Ulichney, "The void-and-cluster method for dither array generation," in *Human Vision, Visual Processing, and Digital Display IV*, J. P. Allebach and B. E. Rogowitz, eds., *Proc. SPIE* **1913**, 332–343 (1993).

[11] R. Floyd and L. Steinberg, "An adaptive algorithm for spatial grey scale," in *SID International Symposium, Digest of Technical Papers* (SID, Seattle, WA, 1975), pp. 36–37.

[12] D. Anastassiou, "Error diffusion coding for A/D conversion," *IEEE Trans. Circuits Syst.* **CAS-36**, 1175–1186 (1989).

[13] H. Inose, Y. Yasuda, and J. Murakami, "A telemetering system by code modulation — Δ–Σ modulation," *IRE Trans. Space Electron. Telemetry* **SET-8**, 204–209 (1962).

[14] J. F. Jarvis, C. N. Judice, and W. H. Ninke, "A survey of techniques for the display of continuous tone pictures on bilevel displays," *Comput. Graphics Image Process.* 13–40 (1976).

[15] P. Stucki, "MECCA — a multiple-error correcting computation algorithm for bilevel image hardcopy reproduction," Tech. Rep. RZ1060, IBM Research Laboratory, Zurich, Switzerland, 1981.

[16] P. W. Wong and R. M. Gray, "Sigma-delta modulation with i.i.d. Gaussian inputs," *IEEE Trans. Inf. Theory* **IT-36**, 784–798 (1990).

[17] W. Chou, P. W. Wong, and R. M. Gray, "Multi-stage sigma-delta modulation," *IEEE Trans. Inf. Theory* **IT-35**, 784–796 (1989).

[18] R. A. Ulichney, "Dithering with blue noise," *Proc. IEEE* **76**, 56–79 (1988).

[19] I. H. Witten and M. Neal, "Using Peano curves for bilevel display of continuous-tone images," *Proc. IEEE CG&A*, 47–52 (May 1982).

[20] K. T. Knox and R. Eschbach, "Threshold modulation in error diffusion," *J. Electron. Imag.* **2**, 185–192 (1993).

[21] B. W. Kolpatzik and C. A. Bouman, "Optimized error diffusion for image display," *J. Electron. Imag.* **1**, 277–292 (1992).

[22] P. W. Wong, "Adaptive error diffusion and its application in multiresolution rendering," *IEEE Trans. Image Process.* **5**, 1184–1196 (1996).

[23] P. W. Wong and J. P. Allebach, "Optimum error diffusion kernel design," presented at the SPIE/IS&T Symposium on Electronic Imaging, San Jose, CA, January 1997.

[24] P. W. Wong, "Entropy constrained halftoning using multipath tree coding," *IEEE Trans. Image Process.* **6**, 1567–1579 (1997).

[25] T. J. Flohr, C. B. Atkins, and J. P. Allebach, "Can DBS ever be a practical halftoning technique?," in *SID Digest of Technical Papers* (SID, Seattle, WA, 1993), pp. 143–146.

[26] T. N. Pappas and D. L. Neuhoff, "Least-square model-based halftoning," in *Human Vision, Visual Processing, and Digital Display, III*, B. E. Rogowitz, ed., *Proc. SPIE* **1666**, 165–176 (1992).

[27] R. A. Vander Kam, P. A. Chou, and R. M. Gray, "Combined halftoning and entropy-constrained vector quantization," in *SID Digest of Technical Papers* (SID, Seattle, WA, 1993), pp. 223–226.

[28] ISO/IEC International Standard 11544, "Coded representation of bi-level and limited-bits-per-pixel grayscale and color images," February, 1993.

[29] S. Forchhammer and K. S. Jensen, "Data compression of scanned halftone images," *IEEE Trans. Commun.* **42**, 1881–1893 (1994).

[30] R. A. Vander Kam and R. M. Gray, "Lossy compression of clustered-dot halftones using subcell prediction," presented at the Data Compression Conference, Snowbird, UT, Month, 1995.

[31] M. Y. Ting and E. A. Riskin, "Error diffused image compression using a halftone-to-grayscale decoder and predictive pruned

tree-structured vector quantization," *IEEE Trans. Image Process.* **3**, 854–858 (1994).

[32] R. A. Vander Kam, P. A. Chou, E. A. Riskin, and R. M. Gray, "An algorithm for joint vector quantizer and halftoner design," presented at the ICASSP San Francisco, CA, March, 1992.

[33] P. W. Wong and H. Nguyen, "Entropy constrained error diffusion," in *SID '95 Digest of Technical Papers* (SID, Orlando, FL, 1995), pp. 905–908.

[34] D. L. Neuhoff and T. N. Pappas, "Perceptual coding of images for halftone display," *IEEE Trans. Image Process.* **3**, 341–354, (1994).

[35] R. A. Vander Kam, P. W. Wong, and R. M. Gray, "JPEG-compliant perceptual coding for a grayscale image printing pipeline," *IEEE Trans. Image Process* **8**, 1–14 (1999).

[36] M. T. Orchard and C. A. Bouman, "Color quantization of images," *IEEE Trans. Signal Process.* **39**, 2677–2690 (1991).

[37] R. Balasubramanian, C. A. Bouman, and J. P. Allebach, "Sequential scalar quantization of color images," *J. Electron. Imag.* **3**, 45–59 (1994).

[38] B. W. Kolpatzik and C. A. Bouman, "Optimized universal color palette design for error diffusion," *J. Electron. Imag.* **4**, 131–143 (1995).

[39] P. Heckbert, "Color image quantization for frame buffer display," *Comput. Graph.* **16**, 297–307 (1982).

Perceptual Criteria for Image Quality Evaluation

Thrasyvoulos N. Pappas
Northwestern University

Robert J. Safranek
Bell Laboratories, Lucent Technologies

1 Introduction

Recent advances in digital imaging technology, computational speed, storage capacity, and networking have resulted in the proliferation of digital images, both still and video. As the digital images are captured, stored, transmitted, and displayed in different devices, there is a need to maintain image quality. In this chapter we examine objective criteria for the evaluation of image quality that are based on models of visual perception.

An image is a two- (or three-) dimensional reproduction of the real world with an additional dimension for moving images. In this chapter we are only interested in images that are intended to be seen by humans. Such images could include a number of different imaging modalities, e.g., infrared, X-ray images, CAT scans, etc., as long as the ultimate destination is the human eye. The tasks could vary from looking at a photograph or watching a movie to reading text and trying to detect a target or a medical condition. However, the problem of image quality changes fundamentally when the ultimate user or interpreter of an image is not the human eye. For example, a typical quality metric for a range[1] image is a maximum (percent) deviation from the true value. In contrast, the human eye's sensitivity to luminance variations depends on a number of factors, including light level, spatial frequency, and signal content.

Even though we use the term image quality, we are primarily interested in image fidelity, i.e., how close an image is to a given original or reference image. It is very hard to develop objective metrics that evaluate image quality without a reference image, even though the human visual system is very good at doing that. Also, we are not considering image enhancement (see Chapters 3.1–3.4), which can improve the quality of an image by modifying it, e.g., by increasing the contrast or changing the colors.

In the following sections, we will examine objective criteria for image quality that are based on models of the human visual system (HVS). Mean squared error based metrics (such as peak signal-to-noise ratio) are still widely used for performance evaluation, and despite their well-known limitations, they can be quite helpful if used carefully. However, they fail when one compares different kinds of artifacts, e.g., artifacts of block-based versus subband or wavelet coders [1].

The perceptual metrics we will discuss were developed for different applications. Even though each metric was influenced by the particular application it was developed for, some of the metrics have general applicability. Such metrics are more elaborate and computationally intensive. We will take a closer look at the metrics developed by Daly [2], Lubin [3], and Teo and Heeger [4]. We also examine metrics that were designed for specific applications, e.g., compression, halftoning, printing, and displays. These metrics are simpler and computationally efficient. Our main focus in this second category will be on metrics for image compression.

Even though storage capacity and transmission bandwidth have been increasing, so have the demand for and the resolution of digital images. Thus, there is an ever increasing need for

---

[1]The values of a range image represent the distances from a point or planar surface to each point on the surface of an object or collection of objects.

image compression. In order to achieve high compression ratios, image compression techniques make use of the properties of the human visual system. The amount of compression that can be attained with lossless compression techniques is limited, typically to a factor of 2 (see Chapters 5.1 and 5.6). Most of the traditional lossy compression techniques make implicit use of the HVS characteristics to achieve much higher compression ratios (by a factor of 8 and higher) without significantly sacrificing image quality. (Chapters 5.2–5.5, 5.7, and 6.1–6.5 discuss lossy compression techniques for still image and video compression.) Recently, however, a number of algorithms have appeared that make use of explicit perceptual models [5–11]. The idea is to make the distortions introduced by the compression scheme invisible to the eye, i.e., perceptually lossless. Typically, the signal is analyzed into components (e.g., spatial and/or temporal subbands), and the role of the perceptual model is to provide the maximum amount of distortion that can be introduced to each component without resulting in any perceived distortion. This is usually referred to as the *just noticeable distortion* level, or JND. We will look closely at the metrics developed by Safranek and Johnston for subband coders [5], by Watson for DCT coders [6], and by Watson *et al.* for wavelet-based coders [12].

Most of the existing models for image quality and compression deal with the threshold of perception. In an increasing number of applications, however, there is a need to achieve very high compression ratios, and in such cases a certain amount of perceived distortion is unavoidable. In suprathreshold image compression, i.e., when the coding distortions exceed the threshold of visibility, there is a need to derive quantitative objective measures of perceived distortion. In general, it is easier to obtain models for the perceptually transparent case; it is much more difficult to quantify perceived distortion, especially across different types of artifacts.

The perceptual models we examine are based on properties of the visual system and measurements of the eye characteristics, e.g., the contrast sensitivity function, light adaptation, masking, etc. However, some models, especially those that are designed for specific applications, can be obtained empirically (experimentally). The disadvantage of such models is that it is very difficult to adapt them to different conditions.

All of the metrics we examine share a similar basic structure, which includes a calibration stage, linear filters tuned to different spatial frequencies and orientations, contrast sensitivity adjustments, and nonlinear mechanisms that account for masking. A final stage involves error pooling to obtain, either a single number that describes image quality, or a map of distortions or detection probabilities. Simpler (linear) models like the SQRI (square root integral) model [13] have been quite successful but are limited to specific applications. Similarly, simple linear models of the HVS have been very successful in image halftoning [14–18].

The advantage of objective quality metrics is that they are relatively easy to use. However, they are no substitute for subjective evaluations, which are accepted to be the most effective and reli- able, albeit quite cumbersome and expensive, way to assess image quality. A significant effort has been dedicated for the development of subjective tests for image quality. A considerable part of this effort has come from IPO, the Center for Research on User-System Interaction, in the Netherlands. Some recent contributions can be found in [19,20]. There has also been standards activity on subjective evaluation of image quality [21]. The study of the topic of subjective evaluation of image quality is beyond the scope of this chapter.

In the following sections, we review the fundamentals of human perception and image quality, and consider different metrics for image quality based on models of human perception.

2 Fundamentals of Human Perception and Image Quality

In this section we will introduce some of the concepts from the psychophysics of human perception that apply to image and video quality metrics. A more in depth coverage of this topic can be found in [22] and Chapter 1.3.

2.1 Psychophysics of Vision

The psychophysics of vision is a vast topic full of interesting experiments and illusions. The HVS models that are used for quality measurement are based on the lower order processing of the visual system, i.e., the modeling of the function of the optics, retina, lateral geniculate nucleus, and striate cortex. Higher level processing, such as attentive vision, Gestalt, and figure/ground effects are either too local in their effect or not understood well enough to be effectively utilized.

The approach taken by most visual quality models is to determine how the lower level physiology of the visual system limits visual sensitivity. In general these models incorporate four types of processes that introduce sensitivity variations: light level, spatial frequency, signal content, and in the case of video, temporal variation. These limits are then converted to *masking thresholds*, which determine the level of distortion to which an image can be exposed to before the alteration is apparent to a human observer. If the magnitude of the distortion is less than the masking threshold (also called the JND level), the original and distorted image will be indistinguishable.

Light level adaptation is caused primarily by the retina and is referred to as the *amplitude nonlinearity* of the visual system. Imperfect optics coupled with neural interactions produce a non-uniform frequency response that is called the *contrast sensitivity function*. Sensitivity variation due to signal content is due to postreceptor neural circuitry and gives rise to *masking* (also referred to as *texture masking*). Adaptation of neural state to an input signal gives rise to *temporal masking*. The next several sections will provide an introduction to each of these phenomena.

2.1.1 Amplitude Nonlinearity

It is well known that the perception of lightness is a nonlinear function of luminance. Consider the following experiment: create a series of images consisting of a background of uniform intensity, I, each with a square of a different intensity $I + \delta I$ inserted into its center. Show these to an observer in order of increasing δI. Ask the observer to determine the point at which he or she can first detect the square. Then, repeat this experiment for a large number of different values of background intensity. For a wide range of background intensities, the ratio of the threshold value δI divided by I is a constant. This equation,

$$\frac{\delta I}{I} = k, \tag{1}$$

is called *Weber's Law*. The value for k is roughly 0.33.

Most implementations of this aspect of visual sensitivity treat it as a point process using the value of a single pixel for the central square and the average of the surrounding pixels for the background. Various instantiations of the amplitude nonlinearity include [3, 23–26].

2.1.2 Contrast Sensitivity Function

The human visual system's contrast sensitivity function (also called the modulation transfer function) provides a characterization of its frequency response. The contrast sensitivity function can be thought of as a bandpass filter. There have been several different classes of experiments used to determine its characteristics which are described in detail in [22, Chapter 12] and Chapter 1.3.

One of these methods involves the measurement of visibility thresholds of sine-wave gratings in a manner analogous to the experiment described in the previous section. For a fixed frequency, a set of stimuli consisting of sine waves of varying amplitudes are constructed. These stimuli are presented to an observer and the detection threshold for that frequency is determined. This procedure is repeated for a large number of grating frequencies. The resulting curve is called the contrast sensitivity function and is illustrated in Fig. 1.

Note that these experiments used sine-wave gratings at a single orientation. To fully characterize the contrast sensitivity function, the experiments would have to be repeated with gratings at various orientations. This has been accomplished and the results show that the HVS is not perfectly isotropic. However, for the purposes of general quality measurement, it is close enough to isotropic that that assumption is normally used.

It should also be noted that the spatial frequencies are in units of cycles per degree of visual angle. This implies that the visibility of details at a particular frequency is a function of viewing distance. As an observer moves away from an image, a fixed size feature in the image takes up fewer degrees of visual angle. This action moves it to the right on the contrast sensitivity curve, possibly requiring it to have greater contrast to remain visible.

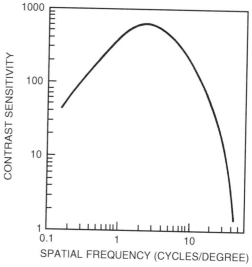

FIGURE 1 Spatial contrast sensitivity function. (Reprinted with permission from reference [63], page 269.)

In contrast, moving closer to an image can allow previously imperceivable details to rise above the visibility threshold. Given these observations, it is clear that the minimum viewing distance is where distortion is maximally detectable. Therefore quality metrics have to specify a minimum viewing distance and evaluate its distortion metric at that point. Several "standard" minimum viewing distances have been established for subjective quality measurement and have generally been used with objective models as well. These are six times image height for standard definition television and three times image height for high definition television.

2.1.3 Contrast Masking

In the previous two sections we have dealt with stimuli that are either constant or contain a single frequency. In general, this is not characteristic of natural scenes. They have a wide range of frequency content over many different scales. Consider for a moment the following thought experiment: consider two images, a constant intensity field and an image of a sand beach. Take a random noise process whose variance just exceeds the amplitude and contrast sensitivity thresholds for the flat field image. Add this noise field to both images. By definition, the noise will be detectable in the flat field image. However, it will not be detectable in the beach image. The presence of the multitude of frequency components in the beach image hides or *masks* the presence of the noise field.

Contrast masking refers to the reduction in visibility of one image component caused by the presence of another image component with similar spatial location and frequency content. As will be discussed further in a later section, the HVS can be thought of as a spatial frequency filter bank with octave spacing of subbands in radial frequency, and angular bands of

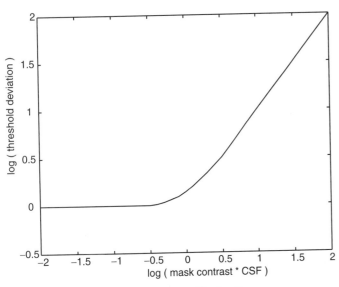

FIGURE 2 Contrast masking function.

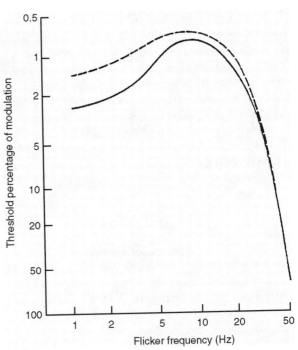

FIGURE 3 Temporal contrast sensitivity function [Kelly, 1969]. (Reprinted with permission from reference [22], page 406.)

roughly 30° spacing. The presence of a signal component in one of these subbands will raise the detection threshold for other signal components in the same subband [27–29] and Chapter 1.3.

The base masking threshold for each spatial frequency band is determined by a combination of the amplitude nonlinearity and the contrast sensitivity function. For the DC band (which corresponds to flat field images), this is the entire masking threshold. For the other bands, the amount by which the masking threshold is elevated is a nonlinear function of the energy in that band. Up to a point the masking threshold remains constant. Once that threshold is exceeded, the masking threshold rises with signal energy as shown in Fig. 2.

2.1.4 Temporal Masking

Temporal masking of time varying stimuli is extremely important in determining the quality of a video signal. This effect is complicated by the fact that the perception of a moving object depends heavily on whether or not the object is being tracked by the eye. Since the purpose of the metrics we will be discussing is to provide a quality measure for the *entire* image sequence, we will assume the worst case, i.e., all objects in a scene that can be tracked will be. There are two major forms of temporal masking that have been utilized in the literature: scene change and the temporal contrast sensitivity function.

In video, a *scene change* occurs when there is an abrupt change in the content of the entire image. This large change in the overall visual field induces a dramatic increase in the masking levels for period of up to 100 ms after the scene change [30–33].

The temporal contrast sensitivity function can be thought of as an extension to the spatial contrast sensitivity function. Experiments similar to those used to measure the spatial sensitivity function have been performed by Kelly [34–36]. In this case the stimulus was a uniform disk whose intensity was varied

sinusoidally in time. For a given temporal frequency, subjects were asked to adjust the amplitude of the sine wave so that it was just at the threshold of visibility. The results of these experiments are shown in Fig. 3. They show that distortion with a period of 4 to 8 Hz is most visible. Combining these results with those from the spatial contrast sensitivity function provides a means of visualizing the HVS spatiotemporal transfer function. This surface is illustrated in Fig. 4.

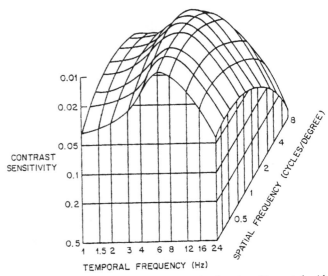

FIGURE 4 Spatiotemporal contrast sensitivity function. (Reprented with permission from reference [63], page 273.)

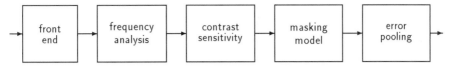

FIGURE 5 Perceptual metric.

3 Visual Models for Image Quality and Compression

In this section, we consider various metrics for image quality based on models of the human visual system. First, we will consider perceptual metrics that are intended for general applicability. Such metrics are quite elaborate, computationally intensive, and require careful calibration and parameter selection. Then, we will consider metrics that are designed specifically for image compression. Such metrics are simpler, more efficient computationally, and sometimes are designed with specific compression algorithms in mind. For example, they could apply to specific viewing conditions and display devices, and they may be intended for specific classes of coders (e.g., block-based or wavelet coders). Finally, we briefly consider perceptual metrics for other applications, e.g., halftoning, displays, and printing.

3.1 Basics of Perceptual Image Quality Metrics

The goal of a perceptual metric for image quality is to determine the differences between two images that are visible to the human visual system. Usually one of the images is the reference, which is considered to be "original," "perfect," or "uncorrupted." The second image has been modified or distorted in some sense. It is very difficult to evaluate the quality of an image without a reference. Thus, a more appropriate term would be image "fidelity" or "integrity," or alternatively, image "distortion" [37]. However, for historical reasons we adopt the term image quality.

In addition to the two digital images, an image quality metric requires a few other parameters, e.g., viewing distance, image size, display parameters, etc. The output is a number that represents the probability that a human eye can detect a difference in the two images or a number that quantifies the perceptual dissimilarity between the two images. Alternatively, the output of an image quality metric could be a map of detection probabilities or perceptual dissimilarity values.

The very first stage of an image quality metric is calibration.

3.1.1 Calibration

The array of numbers that represents an image may have come from a number of different devices and could have gone through different transformations (conversion to densities, gamma correction, etc.) before it is displayed for observation by a human eye. Many quality metrics require that the input image values be converted to physical luminances[2] before they enter the HVS

model. Alternatively, the quality metric could be designed for a specific set of conditions, and thus, all of the calibration could be incorporated in the model, but then the model would have to be rederived for each new set of conditions.

3.1.2 Registration

Registration, i.e., the point-by-point correspondence between two images, is necessary for any quality metric to make any sense. Otherwise, we could arbitrarily modify the value of a metric by arbitrarily shifting one of the images. The shift does not change the images but changes the value of the metric.

3.1.3 Display Model

An accurate model of the display device is an essential part of any image quality metric, e.g., [18], as the human visual system can only see what the display can reproduce. In some cases, when the perceptual model is obtained empirically, e.g., [5], the effects of the display are incorporated in the model. The obvious disadvantage of this approach is that when the display changes, a new set of the model parameters must be obtained; see, e.g., [8]. The study of display models is beyond the scope of this chapter.

3.2 General Models

A number of different quality metrics have been proposed that are based on models of the low-level processing of the HVS, such as the optics, the retina, the lateral geniculate nucleus, and the striate cortex [2]. Such metrics are quite general and are intended for a variety of image processing applications such as compression, halftoning, display evaluation, etc.

We discuss the following models in detail: the "visible differences predictor" by Daly [2], the model proposed by Lubin [3], and the "perceptual image distortion" metric by Teo and Heeger [4]. All such models have a similar basic structure, as shown in Fig. 5. The front end includes the calibration, display model, and registration.

3.2.1 Frequency Analysis

The frequency analysis comprises a hierarchy of filters that decompose the image into several components or channels (usually called subbands) with different spatial frequencies and orientations. Some examples of such decompositions are shown in Fig. 14 below. The range of each axis is from $-u_s/2$ to $u_s/2$ cycles per degree, where u_s is the sampling frequency. Fig. 14(a)

---

[2]In video practice, the term luminance is sometimes, incorrectly, used to denote a nonlinear transformation of luminance [38, p. 24].

shows the cortex transform that was proposed by Watson [39]. The cortex transform consists of two classes of filters applied sequentially that decompose the image, first into different radial frequency bands, and then into different orientation bands. A variation of the Cortex transform was used by Daly [2] and is shown in Fig. 14(b). One of the differences is that Daly used six orientation bands to better approximate the orientation selectivity of the human visual system. The radial filters of both decompositions have octave bandwidths.

A similar decomposition was used by Lubin [3]. He used the Laplacian pyramid of Burt and Adelson [40] to decompose the image into seven radial frequency bands, and the steerable filters of Freeman and Adelson [41] to decompose each pyramid level into four different orientations. Finally, Teo and Heeger [4] adopted the steerable pyramid transform by Simoncelli *et al.* [42], which also has octave radial bandwidths and six orientation bands. In an earlier paper [43] they considered a hexagonally sampled quadrature mirror filter transform, and concluded that its orientation selectivity was not adequate for matching the HVS data.

Both Daly and Lubin convert the filtered images to units of contrast. Daly [2] proposes two alternatives: local contrast, which uses the value of the baseband at any given location to divide the values of all the other bands, and global contrast, which divides all subbands by the average value of the input image. Lubin [3] converts to local contrast by dividing each point in each level of the Laplacian pyramid by the corresponding point obtained from the Gaussian pyramid two levels down in resolution. The conversion to contrast is necessary for general models to account for contrast variations that are not visible. As we will see below, this is not necessary for specific applications such as image compression, where the contrast of the original and coded image remains basically unchanged.

In the remainder of this paper, we will use $f(\mathbf{n})$ to denote the value (intensity, gray scale, etc.) of an image pixel at location $\mathbf{n}$. Usually the image pixels are arranged in a Cartesian grid and $\mathbf{n} = (n_1, n_2)$. The value of the kth image subband at location $\mathbf{n}$ will be denoted by $b(\mathbf{k}, \mathbf{n})$. The subband indexing $\mathbf{k} = (k_1, k_2)$ could be in Cartesian or polar or even scalar coordinates. The same notation will be used to denote the kth coefficient of the nth discrete cosine transform (DCT) block (both Cartesian coordinate systems). This notation underscores the similarity between the two transformations, even though we traditionally display the subband decomposition as a collection of subbands and the DCT as a collection of block transforms: a regrouping of coefficients in the blocks of the DCT results in a representation very similar to a subband decomposition. A more careful discussion of the relationship between the DCT, subband, and wavelet decompositions can be found in Chapter 5.4.

3.2.2 Baseline Contrast Sensitivity (Base Sensitivity)

The baseline contrast sensitivity determines the amount of energy in each subband that is required in order to detect the target

in an (arbitrary or) flat midgray image. As we discussed earlier, this is sometimes referred to as the JND. We will use $t_b(\mathbf{k})$ to denote the baseline sensitivity of the kth band or DCT coefficient. Note that the base sensitivity is independent of the location $\mathbf{n}$.

The base sensitivities can be obtained from the contrast sensitivity function (CSF), as in [3], or can be derived empirically and listed in a table, as in [5]. The base sensitivities are then adjusted to account for variations in luminance and texture masking to obtain the overall sensitivity $t(\mathbf{k}, \mathbf{n})$. An alternative approach, implemented in Daly's model [2], is to filter the image by the contrast sensitivity function before the frequency decomposition.

The key parameters for the contrast sensitivity are the viewing distance (in inches) and the resolution of the display device (in pixels per inch). Alternatively, one can specify the viewing distance in image heights and the image height in pixels (assuming the same horizontal and vertical display resolution). In either case, one must derive the "display visual resolution" in pixels per degree [12].

Since the contrast sensitivity function has a bandpass characteristic, e.g., see [2], if we assume a single fixed viewing distance, the metric may show a degradation in image quality as we move away from the image. To avoid this, one can assume a range of viewing distances [2], or a minimum viewing distance. This will result in a flattening of the CSF. This flattening is commonly assumed in image halftoning applications [14, 18] because it is the low-pass characteristic of the eye that is critical for halftoning.

3.2.3 Luminance Masking

Most of the perceptual models assume that the input image values have been converted to physical luminances. The human visual system's sensitivity to variations in luminance depends on (is a nonlinear function of) the local mean luminance. Some authors call this "light adaptation." Others prefer the term "luminance masking," e.g., [6], which groups it together with the other types of masking we will see below. It is called masking because the luminance of the original image signal masks the variations in the distortion signal. Thus, one approach to account for light adaptation or luminance masking is by including a modification of the base sensitivities. The luminance masking adjustment is a function of the local luminance (or gray level). A common simplification is to assume that it is independent of the subband index $\mathbf{k}$.

An alternative way to account for light adaptation is by including a nonlinear transformation before the frequency analysis. Commonly used transformations include conversion to density (log) and various power laws (e.g., cube root). Daly [2] includes a simple point-by-point amplitude nonlinearity that lies between these two. Daly even allows for an adaptation to absolute luminance levels. However, this is a second-order effect.

For simplicity, and since most image compression algorithms work on gray levels that are gamma-corrected[3] luminances (or sometimes image densities), many applied models are calibrated for gamma-corrected input images [12]. In a well-calibrated display system, the gray levels represent relatively uniform steps in perceived lightness and the luminance masking component of the perceptual model does not have a significant effect; i.e., it is relatively flat as in [8]. The disadvantage of this approach is that it is difficult to account for variations in gamma correction.

3.2.4 Contrast/Texture Masking

Masking is the reduction in the visibility of one image component (the target), which is due to the presence of another (the masker). It is strongest when both components have the same or similar frequency, orientation, and location. Sometimes the term *contrast masking* is used to denote the case in which both the target and the masker have the same frequency and orientation [43], and the term *texture masking* is used to refer to the more general case.

As we saw in Section 2, several experiments can be conducted to determine masking. For example, in the simplest case, both the signal and the mask may consist of single, possible different, frequencies. Alternatively, the two signals could be noise fields with different bandwidths, and in particular bandwidths that correspond to the components of the frequency decomposition of the visual model at hand.

Most of the metrics model contrast masking only. Watson [6] uses the following model for the contrast masking adjustment:

$$\tau_c(\mathbf{k}, \mathbf{n}) = \max\left\{1, |b(\mathbf{k}, \mathbf{n})|^{w_c(\mathbf{k})} t_l(\mathbf{k}, \mathbf{n})^{-w_c(\mathbf{k})}\right\} \quad (2)$$

where $t_l(\mathbf{k}, \mathbf{n})$ is the base sensitivity threshold adjusted for luminance masking and $w_c(\mathbf{k})$ is a number between 0 and 1. The overall sensitivity threshold $t(\mathbf{k}, \mathbf{n})$ is then equal to $\tau_c(\mathbf{k}, \mathbf{n})\, t_l(\mathbf{k}, \mathbf{n})$. A typical empirical value is $w_c(\mathbf{k}) = 0.7$ for $\mathbf{k} \neq 0$ and $w_c(\mathbf{0}) = 0$. Note that in the case of the DC coefficient, the contrast masking adjustment coincides with the luminance masking adjustment, which is applied separately. Lubin [3] uses a sigmoid[4] nonlinearity to account for contrast masking. Daly uses a similar nonlinearity for contrast masking [2]. He also allows for *mutual masking* which uses both the original and the distorted image to determine the degree of masking.

Safranek and Johnston [5] define as texture any deviation from a flat field within a subband and use the following texture masking adjustment:

$$\tau_t(\mathbf{k}, \mathbf{n}) = \max\left\{1, \left[\sum_{\mathbf{k}} w_{\mathrm{MTF}}(\mathbf{k}) e_t(\mathbf{k}, \mathbf{n})\right]^{w_t}\right\}, \quad (3)$$

where $e_t(\mathbf{k}, \mathbf{n})$ is the "texture energy" of subband $\mathbf{k}$ at location $\mathbf{n}$, $w_{\mathrm{MTF}}(\mathbf{k})$ is a weighting factor for subband $\mathbf{k}$ determined

empirically from the MTF of the HVS, and w_t is a constant equal to 0.15. The subband texture energy is given by

$$e_t(\mathbf{k}, \mathbf{n}) = \begin{cases} \text{local variance}_{\mathbf{m} \in \mathrm{N}(\mathbf{n})}(b(\mathbf{0}, \mathbf{m})), & \text{if } \mathbf{k} = 0 \\ b(\mathbf{k}, \mathbf{n})^2 & \text{otherwise} \end{cases}. \quad (4)$$

where $\mathrm{N}(\mathbf{n})$ is the neighborhood of the point $\mathbf{n}$ over which the variance is calculated. In the Safranek–Johnston model, the overall sensitivity threshold is the product of three terms,

$$t(\mathbf{k}, \mathbf{n}) = \tau_t(\mathbf{k}, \mathbf{n})\, \tau_l(\mathbf{k}, \mathbf{n})\, t_b(\mathbf{k}), \quad (5)$$

where $\tau_t(\mathbf{k}, \mathbf{n})$ is the texture masking adjustment, $\tau_l(\mathbf{k}, \mathbf{n})$ is the luminance masking adjustment, and $t_b(\mathbf{k})$ is the baseline sensitivity threshold. The Safranek–Johnston texture adjustment does not differentiate between random and structured texture. The contrast masking models used by Watson, Daly, and Lubin are more effective [44]. As we will see below, Teo and Heeger propose a more elaborate texture model that accounts for contrast masking as well as masking that occurs when the orientations of the target and the masker are different.

3.2.5 Error Pooling

The final step of an image quality metric is to combine the errors (normalized by the sensitivity thresholds, and therefore expressed in JNDs), which have been computed for each spatial frequency and orientation band and each spatial location, into a single number for each pixel of the image, or a single number for the whole image. Some metrics convert the JNDs to detection probabilities. Daly [2] converts to probabilities before error pooling and then uses probability summation to obtain a spatial map of detection probabilities for each point of the image. Lubin [3] converts to probabilities after error pooling. Watson [6] simply expresses the metric in JNDs.

An example of error pooling is the following Minkowski metric:

$$E(\mathbf{n}) = \frac{1}{M}\left\{\sum_{\mathbf{k}}\left|\frac{b(\mathbf{k}, \mathbf{n}) - \hat{b}(\mathbf{k}, \mathbf{n})}{t(\mathbf{k}, \mathbf{n})}\right|^Q\right\}^{1/Q}, \quad (6)$$

where $b_{\mathbf{k}}(\mathbf{n})$ and $\hat{b}_{\mathbf{k}}(\mathbf{n})$ are the $\mathbf{n}$th element of the $\mathbf{k}$th subband of the original and coded image, respectively, $t(\mathbf{k}, \mathbf{n})$ is the corresponding sensitivity threshold, and M is the total number of subbands. In this case, the errors are pooled across frequency to obtain a distortion measure (expressed in JNDs) for each spatial location. The value of Q varies from 2 (energy summation) to infinity (maximum error). Lubin [3] uses $Q = 2.4$. Teo and Heeger [4] use $Q = 2$. Watson [6] pools over DCT blocks first to obtain a "perceptual error matrix" and then pools across frequency to obtain a single valued perceptual error metric. He considers different values of Q, possibly different for each stage.

In order to be consistent with traditional error metrics, Van den Branden Lambrecht and Verscheure [45] suggest expressing

---

[3]Gamma correction is a nonlinear transformation (power law) of image luminance to compensate for the display characteristics, and at the same time, to obtain an efficient image representation [38, Chapter 6].

[4]A sigmoid function starts out flat; its slope increases to a maximum and then decreases back to zero, i.e., it changes curvature like the letter S.

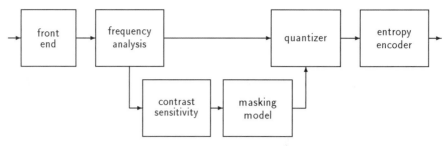

FIGURE 6   Perceptual coder.

the perceptual metric in terms of "visual decibels." They define the "masked peak signal-to-noise ratio (MPSNR)" as

$$\text{MPSNR} = 10 \log_{10} \frac{255^2}{E^2} \qquad (7)$$

where E is the value of the metric. In fact, Safranek and Johnston [5], even though they did not define a perceptual metric explicitly, also expressed the perceptual threshold of their coder as a PSNR in decibels.

3.2.6 Daly and Lubin Models

Daly's model [2] was developed for the evaluation of high-quality imaging systems. It is the most general and elaborate image quality metric. As we saw above, it accounts for variations in sensitivity due to light level, spatial frequency (CSF), and signal content (contrast masking).

Lubin's model [3] was developed for display evaluation. It is also quite general and elaborate and accounts for sensitivity variations due to spatial frequency and masking. In addition, it accounts for fixation depth and eccentricity of the images in the visual field.

We discussed most of the components of these models in the sections above. The main drawback of the metrics based on these models is that they are computationally intensive. Also, they are so elaborate and general that they are difficult to implement and to match to a given set of conditions.

3.2.7 Teo and Heeger Model

As we saw above, the Teo and Heeger metric [4] uses the steerable pyramid transform [42], which decomposes the image into several spatial frequency and orientation bands.[5] However, unlike the other two models we saw above, it does not attempt to separate the base sensitivity and the other masking effects. Instead, Teo and Heeger propose a *normalization model* that explains baseline contrast sensitivity, contrast masking, as well as masking that occurs when the orientations of the target and the

masker are different. The normalization model has the following form:

$$R(\mathbf{k}, \mathbf{n}, i) = R(\rho, \theta, \mathbf{n}, i) = \kappa_i \, \frac{[b(\rho, \theta, \mathbf{n})]^2}{\sum_\phi [b(\rho, \phi, \mathbf{n})]^2 + \sigma_i^2}, \quad (8)$$

where $R(\mathbf{k}, \mathbf{n}, i)$ is the normalized response of a sensor corresponding to the transform coefficient $b(\rho, \theta, \mathbf{n})$, $\mathbf{k} = (\rho, \theta)$ specifies the spatial frequency and orientation of the band, $\mathbf{n}$ specifies the location, and i specifies one of four different contrast discrimination bands characterized by different scaling and saturation constants, κ_i and σ_i^2, respectively. The scaling and saturation constants κ_i and σ_i^2 are chosen to fit the experimental data of Foley and Boynton. Thus, this model is tailored to a specific set of conditions, and would require a lot more work to be adapted to a new set of conditions.

3.3 Coder-Specific Models

We now look at models that have been developed specifically for image compression applications. While still based on the properties of the HVS, these models adopt the frequency decomposition of a given coder, which is chosen to provide high compression efficiency as well as computational efficiency. They are considerably simpler than the general models we saw above, as they only have to consider the properties of the HVS that are relevant for this application. For example, as we saw above, since the contrast of the original and coded image is basically the same, there is no need to account for local contrast variations. Another difference is that frequency decompositions used for compression are usually critically sampled, which means that the number of samples of the original image and the frequency decomposition is the same. Critical sampling is advantageous for compression, but it is not necessary for image quality metrics with general applicability.

The block diagram of a generic perceptually based coder is shown in Fig. 6. The frequency analysis decomposes the image into several components (subbands, wavelets, etc.), which are then quantized and entropy coded. The frequency analysis and entropy coding are virtually lossless; the only losses occur at the quantization step. The perceptual masking model is based on the frequency analysis and regulates the quantization parameters to minimize the visibility of the errors.

---

[5]A more detailed discussion of this model, with a different transform, can be found in [43].

The visual models can be incorporated in a compression scheme to minimize the visibility of the quantization errors, or they can be used independently to evaluate its performance. While the coder-specific image quality metrics are quite effective in predicting the performance of the given coder, some of them may not be as effective in predicting performance across different coders [1].

3.3.1 Safranek–Johnston Perceptual Image Coder and Metric

One of the first image coders to incorporate an elaborate perceptual model was the Safranek–Johnston perceptual subband image coder (PIC) [5]. It uses an empirically derived perceptual masking model that was obtained for a given CRT display and viewing conditions (six times image height). The PIC coder has the basic structure shown in Fig. 6. It uses a separable generalized quadrature mirror filter (GQMF) bank for subband analysis/synthesis. The baseband is coded with DPCM, while all other subbands are coded with PCM. All subbands use uniform quantizers with sophisticated entropy coding. The perceptual model specifies the amount of noise that can be added to each subband of a given image so that the difference between the output image and the original is just noticeable. The model contains the following components: the base sensitivity determines the noise sensitivity in each subband, given a flat midgray image, and was obtained by using subjective experiments. The results are listed in a table. The second component is a brightness adjustment. In general this would be a two-dimensional lookup table (for each subband and gray value). Safranek and Johnston made the reasonable simplification that the brightness adjustment is the same for all subbands. The final component is the texture masking adjustment described in the previous subsection.

In contrast to the general models we saw above, the PIC coder uses a decomposition into subbands of equal spacing and bandwidth as shown in Fig. 14(d). This decomposition is very efficient for compressing high-quality images with a lot of high-frequency content. It is also separable in Cartesian coordinates, which makes it a lot more efficient computationally than the decompositions used in the general models, which are separable in polar coordinates.

A simple metric based on the PIC coder can be defined as follows:

$$E = \left\{ \frac{1}{N} \sum_{\mathbf{n},\mathbf{k}} \left[\frac{b(\mathbf{k}, \mathbf{n}) - \hat{b}(\mathbf{k}, \mathbf{n})}{t(\mathbf{k}, \mathbf{n})} \right]^Q \right\}^{1/Q}, \qquad (9)$$

where $b_{\mathbf{k}}(\mathbf{n})$ and $\hat{b}_{\mathbf{k}}(\mathbf{n})$ are the $\mathbf{n}$th element of the $\mathbf{k}$th subband of the original and coded image, respectively, $t(\mathbf{k}, \mathbf{n})$ is the corresponding perceptual threshold, and N is the number of pixels in the image. A typical value for Q is 2. If the error pooling is done over the subband index $\mathbf{k}$ only, as in Eq. (6), we obtain a spatial map of perceptually weighted errors. This map is downsampled

FIGURE 7   Original 512×512 image.

by the number of subbands in each dimension. A full resolution map can also be obtained by doing the error pooling on the upsampled and filtered subbands.

Figures 7–13 demonstrate the performance of the PIC metric. Figure 7 shows an original 512×512 image. The gray-scale resolution is 8 bits/pixel. Figure 8 shows the image coded with

FIGURE 8   SPIHT coder at 0.52 bits/pixel; PSNR = 33.3 dB.

FIGURE 9   PIC coder at 0.52 bits/pixel; PSNR = 29.4 dB

FIGURE 10   PIC metric for SPIHT coder; perceptual PSNR = 46.8 dB.

the SPIHT coder [46] at 0.52 bits/pixel; the PSNR is 33.3 dB. Figure 9 shows the same image coded with the PIC coder [5] at the same rate. The PSNR is considerably lower at 29.4 dB. This is not surprising, as the SPIHT algorithm is designed to minimize the mean-squared error (MSE) and has no perceptual weighting. The PIC coder assumes a viewing distance of six image heights or 21 in. Depending on the quality of reproduction (which is not known at the time this chapter was written), at a close viewing distance, the reader may see ringing near the edges of the PIC image. In contrast, the SPIHT image has considerable blurring, especially on the wall near the left edge of the image. However, if the reader holds the image at the intended viewing distance (approximately at arm's length), the ringing disappears, and all that remains visible is the blurring of the SPIHT image. Figures 10 and 11 show the corresponding perceptual distortion maps provided by the PIC metric. The resolution is 128 × 128 and the distortion increases with pixel brightness. Observe that the distortion is considerably higher for the SPIHT image. In particular, the metric picks up the blurring on the wall on the left. The perceptual PSNR (pooled over the whole image) is 46.8 dB for the SPIHT image and 49.5 dB for the PIC image, in contrast to the PSNR values. Figure 12 shows the image coded with the standard JPEG algorithm at 0.52 bits/pixel, and Fig. 13 shows the PIC metric. The PSNR is 30.5 dB and the perceptual PSNR is 47.9 dB. At the intended viewing distance, the quality of the JPEG image is higher than the SPIHT image and worse than the PIC image as the metric indicates. Note that the quantization matrix provides some perceptual waiting, which explains why the SPIHT image is superior according to PSNR and inferior according to perceptual PSNR.

For the PIC coder and metric we used a 4 × 4 subband decomposition. We used the base sensitivity thresholds and texture masking adjustment but no brightness adjustment, i.e., we assumed that the gray levels represent relatively uniform steps in perceived lightness. We are hoping this turns out to be the case when this book is printed. Even though the metric is matched

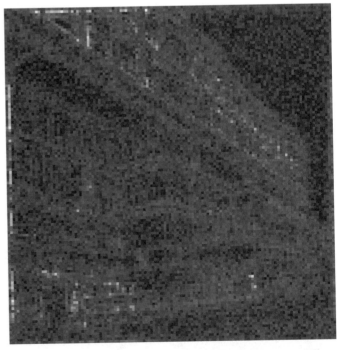

FIGURE 11   PIC metric for PIC coder; perceptual PSNR = 49.5 dB.

FIGURE 12   JPEG coder at 0.52 bits/pixel; PSNR = 30.5 dB.

to the PIC coder, we believe that the above examples illustrate the power of image quality metrics. In [1], Pappas *et al.* tested various metrics on different coders over many coding rates and several images, and found that the PIC metric provides better correlation with subjective evaluations than the MSE metric, the MTF weighted MSE, and Watson's DCT-based metric that

FIGURE 13   PIC metric for JPEG coder; perceptual PSNR = 47.9 dB.

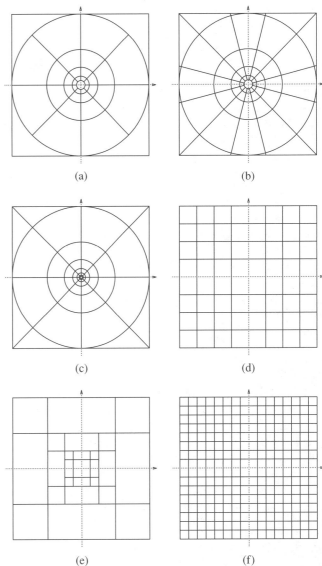

FIGURE 14   Decomposition of the frequency plane corresponding to various transforms: (a) cortex transform (Watson); (b) cortex transform (Daly); (c) lubin's transform; (d) subband transform; (e) wavelet transform; (f) DCT transform. The range of each axis is from $-u_s/2$ to $u_s/2$ cycles per degree, where u_s is the sampling frequency.

we discuss below.

3.3.2 Watson's DCT-Based Metric

Many current compression standards are based on a DCT decomposition. Watson [6] presented a model that computes the visibility thresholds for the DCT coefficients, and thus provides a metric for image quality. Watson's model was developed as a means to compute the perceptually optimal image dependent quantization matrix for DCT-based image coders like JPEG. It has also been used to further optimize JPEG-compatible coders [7, 9, 44]. The JPEG compression standard is discussed in Chapter 5.5.

The DCT decomposition is similar to the subband decomposition and is shown in Fig. 14(f). However, the filters are quite different and the coding artifacts that DCT-based coders produce are very different from those of subband coders. The DCT decomposition is separable and computationally efficient. Because of the popularity of DCT-based coders and computational efficiency of the DCT, we will give a more detailed overview of Watson's DCT-based perceptual model and its implementation, and how it can be used to obtain a metric of image quality.

The first step in the implementation is to convert the original reference and degraded images into a luminance/chrominance color space such as YC_rC_b. The luminance components are then partitioned into 8×8 pixel blocks and transformed to the frequency domain using the forward DCT. Next, the perceptual thresholds are computed from the DCT coefficients of the original image. These thresholds are computed for each block of data from the image. For each coefficient $b(\mathbf{k}, \mathbf{n})$, where $\mathbf{k}$ identifies the DCT coefficient and $\mathbf{n}$ denotes the block within the reference image, we will compute a threshold $t(\mathbf{k}, \mathbf{n})$, which accounts for the contrast sensitivity, luminance masking, and contrast masking.

The baseline contrast sensitivity thresholds $t_b(\mathbf{k})$ are determined by the Peterson, Ahumada, Watson method [47]. The threshold matrices for viewing distance equal to six image heights are provided in Table 1. Note that the quantization matrices can be obtained from the threshold matrices by multiplying by 2. These baseline thresholds are then modified to account, first for luminance masking, and then for contrast masking, in order to obtain the overall sensitivity thresholds.

Since luminance masking is a function of only the average value of a region, it depends only on the DC coefficient $b(\mathbf{0}, \mathbf{n})$ of each DCT block. The luminance-masked threshold is given by

$$t_l(\mathbf{k}, \mathbf{n}) = t_b(\mathbf{k}) \left(\frac{b(\mathbf{0}, \mathbf{n})}{\bar{b}(\mathbf{0})} \right)^{a_T}, \qquad (10)$$

where $\bar{b}(\mathbf{0})$ is the DC coefficient corresponding to average luminance of the display (1024 for an 8-bit image using a JPEG compliant DCT implementation) and a_T has a suggested value of 0.649. This parameter controls the amount of luminance masking that takes place. Setting it to zero turns off luminance masking. Note that since this is a power law, the effect of non-unity display Gamma can be accounted for by multiplying a_T by the gamma exponent.

The Watson model of contrast masking assumes that the visibility reduction is confined to each coefficient in each block. As we saw in Section 3.2.4, the contrast masking adjustment $\tau_c(\mathbf{k}, \mathbf{n})$ is a function of the coefficient $b(\mathbf{k}, \mathbf{n})$ and the luminance-masked threshold $t_l(\mathbf{k}, \mathbf{n})$,

$$\tau_c(\mathbf{k}, \mathbf{n}) = \max \left\{ 1, |b(\mathbf{k}, \mathbf{n})|^{w_c(\mathbf{k})} t_l(\mathbf{k}, \mathbf{n})^{-w_c(\mathbf{k})} \right\}, \qquad (11)$$

where $w_c(\mathbf{k})$ has values between 0 and 1. The exponent may be

TABLE 1 Perceptual threshold matrices for DCT-based coders for a viewing distance equal to six image heights

Channel thresholds							
Y							
5	3	4	7	11	16	24	34
3	4	4	6	8	12	18	25
4	4	8	9	11	15	20	28
7	6	9	14	16	20	26	33
11	8	11	16	26	28	34	42
16	12	15	20	28	41	46	54
24	18	20	26	34	46	63	71
34	25	28	33	42	54	71	95
C_r							
7	9	18	28	43	63	91	128
9	8	17	23	33	48	68	94
18	17	31	34	43	58	78	105
28	23	34	55	63	77	98	126
43	33	43	63	98	108	128	157
63	48	58	77	108	154	174	204
91	68	78	98	128	174	239	255
128	94	105	126	157	204	255	255
C_b							
14	18	46	71	109	161	232	255
18	17	43	59	85	122	173	240
46	43	80	87	111	148	200	255
71	59	87	142	160	196	249	255
109	85	111	160	251	255	255	255
161	122	148	196	255	255	255	255
232	173	200	249	255	255	255	255
255	240	255	255	255	255	255	255

different for each frequency, but it is typically set to a constant in the neighborhood of 0.7. If $w_c(\mathbf{k})$ is 0, no contrast masking occurs and the contrast masking adjustment is equal to 1. The overall sensitivity threshold is given by

$$t(\mathbf{k}, \mathbf{n}) = \tau_c(\mathbf{k}, \mathbf{n}) t_l(\mathbf{k}, \mathbf{n}). \qquad (12)$$

At this point, for each coefficient in each block we have a distortion visibility threshold. For each coefficient, we now need to determine the amount of distortion in terms of JND units. This is done by computing the error at each location (the difference between the DCT coefficients in the original and distorted images) weighted by the visibility threshold

$$d(\mathbf{k}, \mathbf{n}) = \frac{b(\mathbf{k}, \mathbf{n}) - \hat{b}(\mathbf{k}, \mathbf{n})}{t(\mathbf{k}, \mathbf{n})}, \qquad (13)$$

where $b(\mathbf{k}, \mathbf{n})$ and $\hat{b}(\mathbf{k}, \mathbf{n})$ are the reference and distorted images, respectively. Note that $d(\mathbf{k}, \mathbf{n}) < 1$ implies the distortion at that location is not visible, while $d(\mathbf{k}, \mathbf{n}) > 1$ implies the distortion is visible.

At this point, we have an array of distortion visibilities. These have to be combined into a single value denoting the quality of the image. This combination process is performed in two steps. First, error pooling is done spatially. Then the pools of spatial errors are pooled across frequency. Both pooling processes utilize the same probability summation framework.

$$p(\mathbf{k}) = \left\{ \sum_{\mathbf{n}} |d(\mathbf{k}, \mathbf{n})|^{Q_s} \right\}^{1/Q_s}. \qquad (14)$$

From psychophysical experiments, a value of 4 has been observed to be a good choice for Q_s.

The matrix $p(\mathbf{k})$ provides a measure of the degree of visibility of artifacts at each frequency. In order to determine a single valued metric for perceptual error, we need to pool these visibility measurements. This is accomplished by using a procedure similar to the spatial pooling.

$$P = \left\{ \sum_{\mathbf{k}} p(\mathbf{k})^{Q_f} \right\}^{1/Q_f}. \qquad (15)$$

Again, Q_f can have many values, depending on if the average or worst case error is more important. Low values emphasize average error, while setting Q_f to infinity reduces the summation to a maximum operator, thus emphasizing the worst case error.

An alternative means of representing image quality is to use a distortion map instead of a single number. A distortion map can be computed by not performing the spatial error pooling. By performing the frequency pooling on each block independently, and treating the result as an image, the distribution of error across the image is maintained. This approach has the advantage of providing a visual indication of quality. For example, using visibility maps it is easy to distinguish between an image where the peak distortion was confined to a single DCT block and one where that peak level appears in a large percentage of blocks, but both would have the same quality indicator if both the spatial and frequency pooling were performed.

Quality measures for the chrominance channels can be computed in a similar fashion, turning off the luminance and contrast masking portions of the model.

This metric has been shown to be very effective in predicting the performance of block-based coders. However, it is not as effective in predicting performance across different coders. In [1], Pappas *et al.* tested this and other metrics on different coders over many coding rates and several images, and they found that the metric predictions (they used $Q_f = Q_s = 2$) are not always consistent with subjective evaluations when comparing different coders. They found that this metric is strongly biased toward the JPEG algorithm. This is not surprising, as both the metric and JPEG are based on DCTs.

3.3.3 Watson's Wavelet Metric

Many recent coders are based on the discrete wavelet transform (DWT). Popular examples are Shapiro's EZW algorithm [48] and the SPIHT algorithm [46]. A more detailed description of wavelet-based compression algorithms can be found in Chapter 5.4. The DWT is a separable, hierarchical subband decom-

position. It has octave bandwidths in each dimension as shown in Fig. 14(e). It is also computationally efficient.

Watson *et al.* [12] measured the baseline sensitivity thresholds for the wavelet decomposition. They used the linear-phase 9/7 biorthogonal filters [49]. Note that the threshold values depend on the filter bank that is used for the wavelet decomposition. Even though [12] does not provide any detailed light adaptation and texture masking models, models such as those presented earlier in this chapter can be combined with the baseline sensitivity thresholds to obtain a perceptual image quality metric similar to the PIC and DCT-based metrics we discussed earlier.

3.3.4 Other Models

A number of other perceptual models that share many of the features of the ones we discussed above have been proposed. We mention some of them briefly. Silverstein and Klein [50] applied a DCT perceptual metric to a text-based scheme for image display. Westen *et al.* [51] proposed a metric based on a multiple channel HVS model. Horowitz and Neuhoff [52] developed a coder based on an image indistinguishability criterion based on a cortex transform similar to that of Watson [39]. Finally, some metrics have been designed to specific types of artifacts, e.g., blocking [53–55].

In all of the models in this section the threshold values were determined by subjective experiments. Such experiments are time consuming. In [56], Hahn and Mathews present an analytical method for estimating the baseline contrast sensitivity values as well as the luminance and contrast masking correction factors.

3.4 Other Applications

We briefly mention image quality models that have been developed for other specific applications. In such cases the models do not have to be as complicated and can be very successful.

A good example of such a model is Barten's SQRI (square root integral) model [13], which was developed for the evaluation of displays. It is a simple one-dimensional filter that has been very successful. However, it cannot be adapted to more general and complex applications.

An other application area where simple linear models of the HVS have proven to be quite successful is image halftoning [14–18]. Recently, Qian and Kimia have proposed more complicated multiscale models of the HVS for halftoning [57].

3.5 Video Quality Metrics

The basic principles we discussed above for still image metrics can be extended to video. The basic structure of the video quality metric is basically the same as that of Fig. 5. The frequency analysis decomposes the signal into channels with different spatial frequencies, orientations, and temporal frequencies. The baseline spatiotemporal contrast sensitivities are incorporated into the metric and modified by the luminance and contrast masking adjustments. Van den Branden Lambrecht and Verscheure [45] describe such a video quality metric. They use four spatial frequency bands with octave bandwidths, four orientation bands, and two temporal frequency bands. They compute a global metric that is a combination (Minkowski summation) over image blocks whose spatial dimensions depend on the focus of attention (size of the fovea) and whose temporal dimension depends on the persistence of images on the retina. They also segment the image into uniform regions, regions of texture, and regions of contours, and they compute distortion metrics for each region type. Lindh and Van den Branden Lambrecht [58] developed a similar metric that extends Teo and Heeger's still image metric [4].

These video quality metrics are examples of metrics of general applicability like the still metrics we saw above. Watson [59] proposed a coder-specific video quality metric that is based on the DCT. It is an extension of Watson's DCT-based still image metric [6] and is designed with computational efficiency in mind. For temporal filtering, the metric uses a first-order discrete IIR low-pass filter to minimize the number of frames that must be stored in memory. Watson *et al.* [60] evaluated the metric on DCT-based video coders and showed that the metric results correlate well with subjective evaluations.

Finally, Rohaly *et al.* [61] used a software version of a commercial video metric and compared various temporal pooling strategies. The metric is based on the Lubin and Bergen model [3]. The temporal pooling methods they used included an exponentially weighted Minkowski summation to model the recency effect described in [19, 62].

4 Conclusions

In this chapter we presented objective criteria for the evaluation of image quality that are based on models of visual perception. We considered two major classes of models for image quality: general models and models that were designed specifically for image compression applications. The general models are more elaborate and computationally intensive. The coder-specific models are simpler and computationally efficient. The coder-specific models have been shown to be quite effective when used to predict the performance of the given coder. However, the real challenging task is to predict the performance of coders with different structures, i.e., wavelet versus DCT-based coders. The general models are expected to be more effective in this case.

Most of the existing models for image quality and compression are designed for applications in which the distortions are near the threshold of perception. When the distortions exceed the threshold of visibility, it is considerably more difficult to derive quantitative measures of perceived distortion, especially across different types of artifacts. The models we discussed in this chapter can be used is such cases, but their predictions of subjective image quality may not be as accurate.

As the demand for high-quality still and video images increases, so does the need for their efficient storage and transmission. The existence of reliable and efficient metrics for image quality is critical in the development of algorithms that maximize the use of the available transmission bandwidth and storage capacity and produce the highest quality images. We have shown that an understanding of the properties of human visual system is critical in the development of such metrics. In recent years there has been a lot of activity in this field, and we have attempted to summarize it in this chapter. However, it is still a very active and exciting field. The basic principles and ideas we discussed in this chapter should be valuable tools in following its evolution.

References

[1] T. N. Pappas, T. A. Michel, and R. O. Hinds, "Supra-threshold perceptual image coding," in *Proc. ICIP-96* (1996), Vol. I, pp. 237–240.

[2] S. Daly, "The visible differences predictor: an algorithm for the assessment of image fidelity," in *Digital Images and Human Vision*, A. B. Watson, ed. (MIT, Cambridge, MA 1993), pp. 179–206.

[3] J. Lubin, "The use of psychophysical data and models in the analysis of display system performance," in *Digital Images and Human Vision*, A. B. Watson, ed. (MIT, Cambridge, MA, 1993), pp. 163–178.

[4] P. C. Teo and D. J. Heeger, "Perceptual image distortion," in *Proc. ICIP-94* (1994), vol. II, pp. 982–986.

[5] R. J. Safranek and J. D. Johnston, "A perceptually tuned subband image coder with image dependent quantization and post-quantization data compression," in *Proc. ICASSP-89* (1986), Vol. 3, pp. 1945–1948.

[6] A. B. Watson, "DCT quantization matrices visually optimized for individual images," in *Human Vision, Visual Processing, and Digital Display IV*, J. P. Allebach and B. E. Rogowitz, eds., *Proc. SPIE* **1913**, 202–216 (1993).

[7] R. J. Safranek, "A JPEG compliant encoder utilizing perceptually based quantization," in *Human Vision, Visual Processing and Digital Display V*, B. E. Rogowitz and J. P. Allebach, eds., *Proc. SPIE* **2179**, 117–126 (1994).

[8] D. L. Neuhoff and T. N. Pappas, "Perceptual coding of images for halftone display," *IEEE Trans. Image Process*, **3**, 341–354 (1994).

[9] R. Rosenholtz and A. B. Watson, "Perceptual adaptive JPEG coding," in *Proc. ICIP-96* (publisher, location, 1996) Vol. I, pp. 901–904.

[10] I. Höntsch and L. J. Karam, "Apic: adaptive perceptual image coding based on subband decomposition with locally adaptive perceptual weighting," in *Proc. ICIP-97* (1997), Vol. I, pp. 37–40.

[11] I. Höntsch, L. J. Karam, and R. J. Safranek, "A perceptually tuned embedded zerotree image coder," in *Proc. ICIP-97* (1997) Vol. I, pp. 41–44.

[12] A. B. Watson, G. Y. Yang, J. A. Solomon, and J. Villasenor, "Visibility of wavelet quantization noise," *IEEE Trans. Image Process.* **6** 1164–1175 (1997).

[13] P. G. J. Barten, "The SQRI method: a new method for the evaluation of visible resolution on a display," *Proc. Soc. Inf. Display* **28**, 253–262 (1987).

[14] J. Sullivan, L. Ray, and R. Miller, "Design of minimum visual modulation halftone patterns," *IEEE Trans. Syst. Man. Cybern.* **21**, 33–38 (1991).

[15] M. Analoui and J. P. Allebach, "Model based halftoning using direct binary search," in *Human Vision, Visual Processing, and Digital Display III*, B. E. Rogowitz, ed., *Proc. SPIE* **1666**, 96–108 (1992).

[16] J. B. Mulligan and A. J. Ahumada, Jr., "Principled halftoning based on models of human vision," in *Human Vision, Visual Processing, and Digital Display III*, (B. E. Rogowitz, ed., *Proc. SPIE* **1666**, 109–121 (1992).

[17] T. N. Pappas and D. L. Neuhoff, "Least-squares model-based halftoning," in *Human Vision, Visual Processing, and Digital Display III*, B. E. Rogowitz, ed., *Proc. SPIE* **1666**, 165–176 (1992).

[18] T. N. Pappas and D. L. Neuhoff, "Least-squares model-based halftoning," *IEEE Trans. Image Process.* **8**, 1102–1116 (1999).

[19] R. Hamberg and H. de Ridder, "Continuous assessment of time-varying image quality," in *Human Vision and Electronic Imaging II*, B. E. Rogowitz and T. N. Pappas, eds., *Proc. SPIE* **3016**, 248–259 (1997).

[20] H. de Ridder, "Psychophysical evaluation of image quality: from judgement to impression," in *Human Vision and Electronic Imaging III*, B. E. Rogowitz and T. N. Pappas, eds., *Proc. SPIE* **3299**, 252–263 (1998).

[21] "ITU/R Recommendation BT.500-7, 10/1995," Internet address http://www.itu.ch.

[22] T. N. Cornsweet, *Visual Perception*. (Academic, Orlands, 1970).

[23] C. F. Hall and E. L. Hall, "A nonlinear model for the spatial characteristics of the human visual system," *IEEE Trans. Syst. Man Cybern.* **SMC-7**, 162–170 (1977).

[24] T. J. Stockham, "Image processing in the context of a visual model," *Proc. IEEE* **60**, 828–842 (1972).

[25] J. L. Mannos and D. J. Sakrison, "The effects of a visual fidelity criterion on the encoding of images," *IEEE Trans. Inf. Theory* **IT-20**, 525–536 (1974).

[26] J. J. McCann, S. P. McKee, and T. H. Taylor, "Quantitative studies in the retinex theory," *Vision Res.* **16**, 445–458 (1976).

[27] J. G. Robson and N. Graham, "Probability summation and regional variation in contrast sensitivity across the visual field," *Vision Res.* **21**, 419–418 (1981).

[28] G. E. Legge and J. M. Foley, "Contrast masking in human vision," *J. Opt. Soc. Am.* **70**, 1458–1471 (1980).

[29] G. E. Legge, "A power law for contrast discrimination," *Vision Res.* **21**, 457–467 (1981).

[30] B. G. Breitmeyer, *Visual Masking: An Integrative Approach* (Oxford U. Press, New York, 1984).

[31] A. J. Seyler and Z. L. Budrikis, "Detail perception after scene change in television image presentations," *IEEE Trans. Inf. Theory* **IT-11**, 31–43 (1965).

[32] Y. Ninomiya, T. Fujio, and F. Namimoto, "Perception of impairment by bit reduction on cut-changes in television pictures (in japanese)," *Electric. Commun. Assoc. Essay Period.* **J62-B**.

[33] W. J. Tam, L. Stelmach, L. Wang, D. Lauzon, and P. Gray, "Visual masking at video scene cuts," in *Human Vision, Visual Processing, and Digital Display VI*, B. E. Rogowitz and J. P. Allebach, eds., *Proc. SPIE*, **2411**, 111–119 (1995).

[34] D. H. Kelly, "Visual response to time-dependent stimuli," *J. Opt. Soc. Am.* **51**, 422–429 (1961).

[35] D. H. Kelly, "Flicker fusion and harmonic analysis," *J. Opt. Soc. Am.* **51**, 917–918 (1961).

[36] D. H. Kelly, "Flickering patterns and lateral inhibition," *J. Opt. Soc. Am.* **59**, 1361–1370 (1969).

[37] D. A. Silverstein and J. E. Farrell, "The relationship between image fidelity and image quality," in *Proc. ICIP-96*, (1996), Vol. II, pp. 881–884.

[38] C. A. Poynton, *A Technical Introduction to Digital Video* (Wiley, New York, 1996).

[39] A. B. Watson, "The cortex transform: rapid computation of simulated neural images," *Comput. Vis. Graph. Image Process.* **39**, 311–327 (1987).

[40] P. J. Burt and E. H. Adelson, "The Laplacian pyramid as a compact image code," *IEEE Trans. Commun.*, **31**, 532–540 (1983).

[41] W. T. Freeman and E. H. Adelson, "The design and use of steerable filters," *IEEE Trans. Pattern Anal. Machine Intell.* **13**, 891–906 (1991).

[42] E. P. Simoncelli, W. T. Freeman, E. H. Adelson, and D. J. Heeger, "Shiftable multiscale transforms," *IEEE Trans. Inf. Theory* **38**, 587–607 (1992).

[43] P. C. Teo and D. J. Heeger, "Perceptual image distortion," in *Human Vision, Visual Processing, and Digital Display V*, B. E. Rogowitz and J. P. Allebach, eds., *Proc. SPIE* **2179**, 127–141 (1994).

[44] R. J. Safranek, "A comparison of the coding efficiency of perceptual models," *Human Vision, Visual Processing, and Digital Display VI*, B. E. Rogowitz and J. P. Allebach, eds., *Proc. SPIE* **2411**, 83–91 (1995).

[45] C. J. van den Branden Lambrecht and O. Verscheure, "Perceptual quality measure using a spatio-temporal model of the human visual system," in *Digital Video Compression: Algorithms and Technologies*, F. S. Vasudev Bhaskaran and S. Panchanathan, eds., *Proc. SPIE* **2668**, 450–461 (1996).

[46] A. Said and W. A. Pearlman, "A new fast and efficient image codec based on set partitioning in hierarchical trees," *IEEE Trans. Circuits Syst. Video Technol.* **6**, 243–250 (1996).

[47] H. A. Peterson, A. J. Ahumada, Jr., and A. B. Watson, "An improved detection model for DCT coefficient quantization," in *Human Vision, Visual Processing, and Digital Display IV*, J. P. Allebach and B. E. Rogowitz, eds., *Proc. SPIE* **1913**, 191–201 (1993).

[48] J. M. Shapiro, "Embedded image coding using zerotrees of wavelet coefficients," *IEEE Trans. Signal Process.* **SP-41**, 3445–3462, (1993).

[49] A. Cohen, I. Daubechies, and J. C. Feauveau, "Biorthogonal bases of compactly supported wavelets," *Commun. Pure Appl. Math.* **45**, 485–560, (1992).

[50] D. A. Silverstein and S. A. Klein, "A DCT image fidelity metric for application to a text-based scheme for image display," in *Human Vision, Visual Processing, and Digital Display IV*, J. P. Allebach and B. E. Rogowitz, eds., *Proc. SPIE* **1913**, 229–239 (1993).

[51] S. J. P. Westen, R. L. Lagendijk, and J. Biemond, "Perceptual image quality based on a multiple channel HVS model," in *Proc. ICASSP-95* (1995), Vol. 4, pp. 2351–2354.

[52] M. J. Horowitz and D. L. Neuhoff, "Image coding by perceptual pruning with a cortical snapshot indistinguishability criterion," in *Human Vision and Electronic Imaging III*, B. E. Rogowitz and T. N. Pappas, eds., *Proc. SPIE* **3299**, 330–339 (1998).

[53] C. Fenimore, B. Field, and C. V. Degrift, "Test patterns and quality metrics for digital video compression," in *Human Vision and Electronic Imaging II*, B. E. Rogowitz and T. N. Pappas, eds., *Proc. SPIE* **3016**, 269–276 (1997).

[54] J. M. Libert and C. Fenimore, "Visibility thresholds for compression-induced image blocking: measurement and models," in *Human Vision and Electronic Imaging IV*, B. E. Rogowitz and T. N. Pappas, eds., *Proc. SPIE* **3644**, 197–206 (1999).

[55] E. M. Yeh, A. C. Kokaram, and N. G. Kingsbury, "A perceptual distortion measure for edge-like artifacts in image sequences," in *Human Vision and Electronic Imaging III*, B. E. Rogowitz and T. N. Pappas, eds., *Proc. SPIE* **3299**, 160–172 (1998).

[56] P. J. Hahn and V. J. Mathews, "An analytical model of the perceptual threshold function for multichannel image compression," in *Proc. ICIP-98* (publisher, location, 1998), vol. III, pp. 404–408.

[57] W. Qian and B. Kimia, "On the perceptual notion of scale for halftone representations: nonlinear diffusion," in *Human Vision and Electronic Imaging*, B. E. Rogowitz and T. N. Pappas, eds., *Proc. SPIE* **3299**, 473–481 (1998).

[58] P. Lindh and C. J. van den Branden Lambrecht, "Efficient spatio-temporal decomposition for perceptual processing of video sequences," in *Proc. ICIP-96* (publisher, location, 1996) Vol. III, pp. 331–334.

[59] A. B. Watson, "Toward a perceptual video quality metric," in *Human Vision and Electronic Imaging III*, B. E. Rogowitz and T. N. Pappas, eds., *Proc. SPIE* **3299**, 139–147 (1998).

[60] A. B. Watson, J. Hu, J. F. McGowan, and J. B. Mulligan, "Design and performance of a digital video quality metric," in *Human Vision and Electronic Imaging IV*, B. E. Rogowitz and T. N. Pappas, eds., *Proc. SPIE* **3644**, 168–174 (1999).

[61] A. M. Rohaly, J. Lu, N. R. Franzen, and M. K. Ravel, "Comparison of temporal pooling methods for estimating the quality of complex video sequences," in *Human Vision and Electronic Imaging IV*, B. E. Rogowitz and T. N. Pappas, eds., *Proc. SPIE* **3644**, 218–225 (1999).

[62] D. Pearson, "Viewer response to time-varying video quality," in *Human Vision and Electronic Imaging III*, B. E. Rogowitz and T. N. Pappas, eds., *Proc. SPIE* **3299**, 16–25 (1998).

[63] A. N. Netravali and B. G. Haskell, *Digital Pictures: Representation and Compression* (Plenum Press, New York, 1988).

Image and Video Storage, Retrieval and Communication

9.1

Image and Video Indexing and Retrieval

Michael A. Smith
*Carnegie Mellon University
and AVA Media Systems*

Tsuhan Chen
Carnegie Mellon University

1 Introduction

Digital video and imagery is important for education, entertainment, and a host of multimedia applications. With increased computing power and electronic storage capacity, the potential for large digital video libraries is growing rapidly. The World Wide Web has seen an increased use of digital video, and digital video remains a key component of many educational and entertainment applications. For many users, the video of interest is not always a full-length film. Video libraries should provide informational access in the form of brief, content-specific segments as well as full-length videos. For a collection of images or video to be accessed, the embedded retrieval information must be easy to locate, manage and display.

As the size of accessible video collections grows to thousands of hours, potential viewers will need abstractions and technology that help them browse effectively and efficiently. Text-based search algorithms offer some assistance in finding specific segments among large video collections [16]. In most cases, these systems output many irrelevant video segments to ensure the retrieval of pertinent information. Intelligent indexing systems are essential for optimal retrieval of image and video data. In this chapter we explore the latest technology in image and video retrieval. We describe several methods for extracting features from image and video. These features are used to measure image and video similarity in multimedia databases. We also describe techniques for determining indexing and retrieving image and video data.

1.1 Video Terminology

There are many terms used to describe various attributes of video. This chapter uses common terminology from production standards, as well as image, audio, and video processing technology. To understand the notation, we define significant video terminology in Table 1 and Fig. 1.

1.2 Video Categories

Most commercial and academic research in image and video retrieval has focused on documentaries and broadcast news,

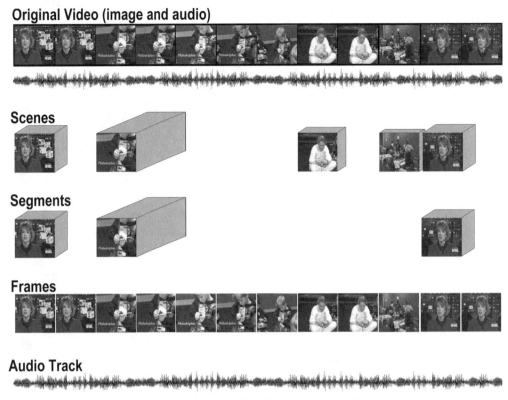

FIGURE 1   Video terminology. (See color section, p. C–38.)

although some experimentation has been devoted to sports, feature films, and stock footage. Documentary footage is very informative and follows many standard video production pro-

cedures. Access to public broadcast material is less stringent than private material since much of what is produced is educational, rather than commercial. There are three basic formats for documentary video: (1) factual, (2) historical, and (3) biographical. Broadcast news includes prerecorded and live footage from producers such as CNN Headline News, ABC, NBC, and CBS. Most live news segments present a variety of challenging editing styles and effects that are not commonly used in prerecorded segments. There are many different types of feature films. Most experiments test commercially successful films suitable for all audiences. For sports, most researchers examine video with recognizable formats and "plays" through the course of the event, such as football, basketball, baseball, soccer, and hockey.

1.3 Video Storage

In the analysis of digital video, compression schemes offer increased storage capacity and statistical image characteristics, such as filtering coefficients (discrete cosine transform, or DCT) and motion compensation data. One drawback to these compression schemes is loss in quality. Bit streams created by lossy compression schemes, however, typically preserve some statistical information of the original video in an explicit manner. For example, the DCT coefficients preserve colors, texture, and other spatial domain characteristics, and motion vectors preserve object motion, camera pan and zoom, and other temporal

TABLE 1   Significant video terminology

Term (medium)	Comments
Video (image/audio)	The term video will be used to represent a stream of images and audio. This term will never be used in the description of the image track without audio.
Scene (image)	A scene is a semantic unit of images. It is also used to describe change or borders of image content.
Frame (image)	Frames refer to the actual image portion of the video. A frame represents a single image in a video sequence.
Segment (image/audio)	A segment is a subset of a video that constitutes a single semantic unit. This unit may consist of several scenes and audio phrases.
Audio track (audio)	The audio track refers to the actual audio stream in the video.
Phrase (audio or text)	A phrase is a collection of words that make a semantic unit of audio or text. A phrase may be bound the range of a sentence, but this is not a necessary condition. A subregion of a sentence may also serve as a phrase.

characteristics. Lossless schemes, such as run length encoding (RLE) and Huffman coding, do not sacrifice quality but provide lower compression ratios. Furthermore, bit streams created by lossless algorithms do not explicitly contain any statistical information of the original video.

Many algorithms provide compression as high as 100 to 1 and often use DCT and motion compensation for compression. The parameters of the DCT may be used for video segmentation, while the motion compensation statistics may be used as a form of optical flow, as discussed in Section 3.

Storage capacity and data access are important in image and video retrieval systems. Data can be stored in a hard drive, CD-ROM, DVD, digital tape, or many other storage mediums. The time needed to access information on a particular medium will fluctuate, and so retrieval algorithms must be flexible. The storage requirements will also change depending on the size and quality of video. An hour of video at full SIF (352×240) constitutes approximately 640 Mbytes with the MPEG-1 compression standard.

1.4 Image and Video Features

A feature is defined as a descriptive parameter that is extracted from an image or video stream. Features may be used to interpret visual content, or as a measure for similarity in image and video databases. In this chapter, features are described in the following categories.

1. Statistical features: Features that are extracted from and image or video sequence without regard to content are described as statistical features. These include parameters derived from such algorithms as image difference and camera motion.
2. Compressed domain features: A feature that is extracted from a compressed image or video stream without regard to content is described as a compressed domain feature.
3. Content-based features: A feature that is derived for the purpose of describing the actual content in an image or video stream is a content-based feature.

In the sections that follow, we describe examples of each feature and potential applications in image and video databases.

2 Statistical Features

Certain features may be extracted from image or video without regard to content. These features include such analytical features as scene changes, motion flow and video structure in the image domain, and sound discrimination in the audio domain. In this section we describe techniques for image difference and motion analysis as statistical features. For more information on statistical image features, please see Chapters 4.7 and 4.8.

2.1 Image Difference

A difference measure between images serves as a feature to measure similarity. In the sections below, we describe two fundamental methods for image difference: absolute difference and histogram difference. The absolute difference requires less computation but is generally more susceptible to noise and other imaging artifacts, as described below.

2.1.1 Absolute Difference

The image difference of two images is defined as the sum of the absolute difference at each pixel. The first image I_t is analyzed with a second image, I_{t-T}, at a temporal distance T. The difference value is defined as

$$D(t) = \sum_{i=1}^{M} |I_{(t-T)}(i) - I_t(i)|,$$

where M is the resolution, or number of pixels in the image. This method for image difference is noisy and extremely sensitive to camera motion and image degradation. When applied to subregions of the image, $D(t)$ is less noisy and may be used as a more reliable parameter for image difference.

$$D_s(t) = \sum_{j=S}^{H/n} \sum_{i=S}^{W/n} |I_{(t-T)}(i, j) - I_t(i, j)|$$

Here $D_s(t)$ is the sum of the absolute difference in a subregion of the image, where S represents the starting position for a particular region and n^2, represents the number of subregions.

We may also apply some form of filtering to eliminate excess noise in the image and subsequent difference. For example, the image on the right in Fig. 2 represents the output of a Gaussian filter on the original image on the left. For more details on image enhancement, please refer to Chapter 3.

2.1.2 Histogram Difference

A histogram difference is less sensitive to subtle motion and is an effective measure for detecting similarity in images. By detecting significant changes in the weighted color histogram of two

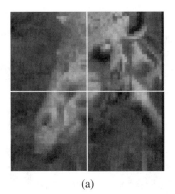 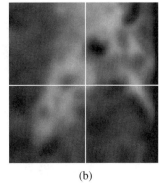

(a)   (b)

FIGURE 2   Image: (a) original; (b) filtered. (See color section, p. C–38.)

images, we form a more robust measure for image correspondence. The histogram difference may also be used in subregions to limit distortion that is due to noise and motion.

$$D_H(t) = \sum_{v=1}^{N} |H_{(t-1)}(v) - H_t(v)|.$$

The difference value, $D_H(t)$, will rise during scene changes, image noise, and camera or object motion. In the equation below, N represents the number of bins in the histogram, typically 256. Two adjacent images may be processed, although this algorithm is less sensitive to error when images separated by a spacing interval, D_i; D_i is typically of the order of 5–10 frames for video encoded at standard 30 fps. An empirical threshold may be set to detect values of $D_H(t)$ that correspond to scene changes. For inputs from multiple categories of video, an adaptive threshold for $D_H(t)$ should be used.

$$D_{H-R}(t) = \sum_{v=1}^{N} |H_{R(t-1)}(v) - H_{Rt}(v)|,$$

$$D_{H-G}(t) = \sum_{v=1}^{N} |H_{G(t-1)}(v) - H_{Gt}(v)|,$$

$$D_{H-B}(t) = \sum_{v=1}^{N} |H_{B(t-1)}(v) - H_{Bt}(v)|,$$

$$D_{H-RGB}(t) = \frac{\sum (D_{H-R}(t) + D_{H-G}(t) + D_{H-B}(t))}{3}$$

If the histogram is actually three separate sets for *RGB*, the difference may simply be summed. An alternative to summing the separate histograms is to convert the *RGB* histograms to a single color band, such as Munsell or LUV color. Color representations in digital imagery are discussed in Chapter 4.5.

2.2 Video Segmentation

An important application of image difference in video is the separation of visual scenes. A simple image difference represents one of the more common methods for detection of scene changes.

The difference measures, $D(t)$ and $D_H(t)$, may be used to determine the occurrence of a scene change. By monitoring the difference of two images over some time interval, one may set a threshold to detect significant differences or changes in scenery. This method provides a useful tool for detecting scene cuts, but is susceptible to errors during transitions. A block-based approach may be used to reduce errors in difference calculations. This method is still subject to errors when subtle object or camera motion occurs.

The most fundamental scene change is the video cut. For most cuts, the difference between image frames is so distinct that accurate detection is not difficult. Cuts between similar scenes, however, may be missed when only static properties such as image difference are used. Several research groups have developed working techniques for detecting scene changes through variations in image and histogram differencing. For more details on video segmentation technology, please see Chapter 4.9.

A histogram difference is less sensitive to subtle motion, and is an effective measure for detecting scene cuts and gradual transitions. By the detection of significant changes in the weighted color histogram of each successive frame, video sequences can be separated into scenes. This technique is simple and yet robust enough to maintain high levels of accuracy. An illustration of histogram based segmentation is shown in Fig. 3.

2.2.1 Scene Change Categories

There are a variety of complex scene changes used in video production, but the basic premise is a change in visual content. The video cut and other scene change procedures are discussed as follows.

1. Fast cut: A sequence of video cuts, each very short in duration, represents a fast cut. This technique heightens the sense of action or excitement. To detect a fast cut, we may look for a sequence of scene changes that are in close proximity.
2. Distance cut: A distance cut occurs when the camera cuts from one perspective of a scene to another some distance away. This shift in distance usually appears as a cut from a wide shot to a close-up shot, or vice versa.

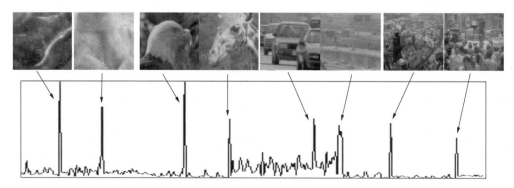

FIGURE 3 Histogram difference, $D_{H-RGB}(t)$, for scene segmentation.

3. Intercutting: When scenes change back and forth from one subject to another, we say the subjects are intercut. This concept is similar to the distance cut, but the images are separate and not inclusive of the same scenes. Intercutting is used to show a thought process between two or more subjects.

4. Dissolves and fades: Dynamic imaging effects are often used to change from one scene to another. A common effect in all types of video is the fade. A fade occurs when a scene changes over time from its original color scheme to a black background. This procedure is commonly used as a transition from one topic to another. Another dynamic effect is the dissolve. Similar to the fade, this effect occurs when a scene changes over time and morphs into a separate scene. This transition is less intrusive and is used when subtle change is needed.

5. Wipes and blends: These effects are most often used in news video. The actual format of each may change from one show to the next. A wipe usually consists of the last frame of a scene being folded like a page in a book. A blend may be shown as pieces of two separate scenes combined in some artistic manner. Like the fade and dissolve, wipes and blends are usually used to make a transition to a separate topic. In feature films a wipe is often used to convey a change in time or location.

2.2.2 Alternative Segmentation Technology

An alternative form of scene segmentation involves the use of traditional edge detection characteristics. Edges in images are useful information about the changes in background and object distribution between scenes. An effective algorithm for detecting cuts and gradual transitions was developed at Cornell University using edge detection technology [22]. For more details on edge detection technology, please see Chapters 4.10 and 4.11.

An analysis of the global motion of a video sequence may also be used to detect changes in scenery. For example, when the error in optical flow is high, this is usually attributed to its inability to track a majority of the motion vectors from one frame to the next. Such errors can be used to identify scene changes. A motion-controlled temporal filter may also be used to detect dissolves and fades, as well as separate video sequences that contain long pans. The use of motion as a statistical feature is discussed in the following section. The methods for scene segmentation described in this section may be used individually or combined for more robust segmentation.

2.3 Motion Analysis

Motion characteristics represent an important feature in video indexing. One aspect is based on interpreting camera motion [1, 17]. Many video scenes have dynamic camera effects but offer little in the description of a particular segment. Static scenes, such as interviews and still poses, contain essentially identical video

frames. Knowing the precise location of camera motion can also provide a method for video parsing. Rather than simply parse a video by scenes, one may also parse a video according to the type of motion. An important kind of video characterization is defined not just by the motion of the camera, but also by motion or action of the objects being viewed.

An analysis of optical flow can be used to detect camera and object motion. Most algorithms for computing optical flow require extensive computation, and more often, researchers are exploring methods to extract optical flow from video compressed with some form of motion compensation. Section 3 describes the benefits of using compressed video for optical flow and other image features.

Statistics from optical flow may also be used to detect scene changes. Optical flow is computed from one frame to the next. When the motion vectors for a frame are randomly distributed without coherency, this may suggest the presence of a scene change. In this sense, the quality of the camera motion estimate is used to segment video. Video segmentation algorithms often yield false scene changes in the presence of extreme camera or object motion. An analysis of optical flow quality may also be used to avoid false detection of scene changes.

Optical flow fields may be interpreted in many ways to estimate the characteristics of motion in video. Two such interpretations are the camera motion and object motion. For more details on motion analysis, please see Chapter 3.8.

2.3.1 Camera Motion

An affine model is used to approximate the flow patterns consistent with all types of camera motion.

$$u(x_i, y_i) = ax_i + by_i + c,$$
$$v(x_i, y_i) = dx_i + ey_i + f.$$

Affine parameters a, b, c, d, e, and f are calculated by minimizing the least-squares error of the motion vectors.

$$
\begin{bmatrix}
\sum x2 & \sum xy & \sum x & 0 & 0 & 0 \\
\sum xy & \sum x & \sum y & 0 & 0 & 0 \\
\sum x & \sum y & \sum N & 0 & 0 & 0 \\
0 & 0 & 0 & \sum x2 & \sum xy & \sum x \\
0 & 0 & 0 & \sum xy & \sum x2 & \sum y \\
0 & 0 & 0 & \sum x & \sum y & \sum N
\end{bmatrix}
\begin{bmatrix} a \\ b \\ c \\ d \\ e \\ f \end{bmatrix}
=
\begin{bmatrix}
\sum ux \\ \sum uy \\ \sum u \\ \sum vx \\ \sum vy \\ \sum v
\end{bmatrix}
$$

We also compute average flow $\bar{v}$ and $\bar{u}$. Where $\bar{v}$ and $\bar{u}$,

$$\bar{u} = \sum_{i=0}^{N} ax_i + by_i + c,$$

$$\bar{v} = \sum_{i=0}^{N} dx_i + ey_i + f.$$

Using the affine flow parameters and average flow, we classify the flow pattern. To determine if a pattern is a zoom, we first check if there is the convergence or divergence point (x_0, y_0), where $u(x_i, y_i) = 0$ and $v(x_i, y_i) = 0$. To solve for (x_0, y_0), the following relation must be true:

$$\begin{vmatrix} a & b \\ d & e \end{vmatrix} = 0.$$

If the above relation is true, and (x_0, y_0) is located inside the image, then it must represent the focus of expansion. If $\bar{v}$ and $\bar{u}$ are large, then this is the focus of the flow and camera is zooming. If (x_0, y_0) is outside the image or are large, then the camera is panning in the direction of the dominant vector.

If the above determinant is approximately 0, then (x_0, y_0) does not exist and the camera is panning or static. If $\bar{v}$ or $\bar{u}$ are large, the motion is panning in the direction of the dominant vector. Otherwise, there is no significant motion and the flow is static. We may eliminate fragmented motion by averaging the results over time in a window, W_A (typically 20 frames). Examples of the camera motion analysis results are shown in Fig. 4.

2.3.2 Object Motion

Object motion typically exhibits flow fields in specific regions of an image, whereas camera motion is characterized by flow throughout the entire image. The global distribution of motion vectors distinguishes between object and camera motion. The flow field is partitioned into a grid as shown in Fig. 5. If the average velocity for the vectors in a particular grid is high (typically >2.5 pixels), then that grid is designated as containing motion.

When the total number of connected motion grids, G_c,

$$G_c(i) = \begin{cases} 0, & (G_m(i-1) = 0, G_m(i+1) = 0, \ldots, M) \\ 1, & \text{otherwise} \end{cases}$$

is high (typically $G_c > 7$), the flow is some form of camera motion. Here $G_m(i)$ represents the status of motion grid at position i and M represents the number of neighbors. A motion grid should consist of at least a 4×4 array of motion vectors. If G_c is not high but greater than some small value (typically two grids), the motion is isolated in a small region of the image and the flow is probably caused by object motion. This result is averaged over a frame window of width W_A, just as with camera motion, but the number of object motion regions needed is typically of the order of 60%. This is 12 object motion frames for a typical W_A of 20 frames. Examples of the object motion analysis results are shown in Fig. 5.

2.4 Alternative Statistical Features

Texture: An analysis of image texture is useful in the discrimination of low-interest video from video containing complex features. A low-interest image may also contain uniform texture, as well as uniform color or low contrast. Perceptual features for individual video frames were computed by using common textual features, such as coarseness, contrast, directionality, and regularity. For more details on texture analysis, please see Chapters 4.7 and 4.9.

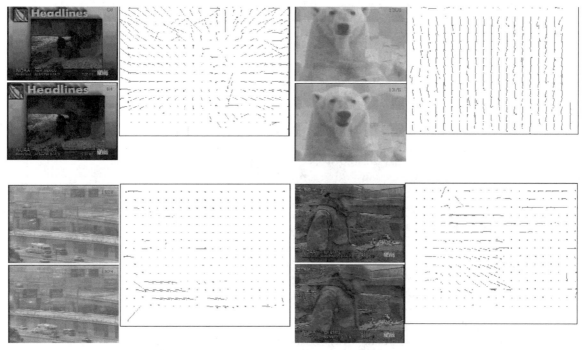

FIGURE 4   Optical flow fields for a pan (top right), zoom (top left), and object motion. (See color section, p. C–39.)

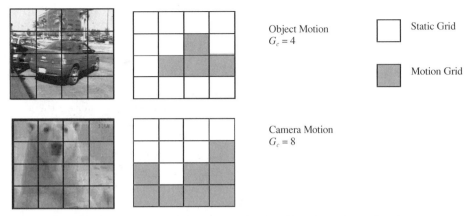

FIGURE 5   Camera and object motion detection. (See color section, p. C–39.)

Shape and position: The shape and appearance of objects may also be used as a feature for image correspondence. Color and texture properties will often change from one image to the next, making image difference and texture features less useful. An example of this is shown in Fig. 6, where the feature of interest is an anchorperson, but the color, texture and position of the subjects is different for each image. Commercial systems for shape-based image correspondence are discussed in Section 7. For more details on shape analysis, please see Chapter 4.6.

Audio features: In addition to image features, certain audio features may be extracted from video to assist in the retrieval task. Loud sounds, silence, and single-frequency sound markers may be detected analytically without actual knowledge of the audio content.

Loud sounds imply a heightened state of emotion in video, and they are easily detected by measuring a number of audio attributes, such as signal amplitude or power. Silent video may signify an area of less importance and can also be detected with straightforward analytical estimates. A video producer will often use single-frequency sound markers, typically a 1000 Hz tone, to mark a particular point in the beginning of a video. This tone may be detected to determine the exact point in which a video will start.

2.5 Hierarchical Video Structure

Most video is produced with a particular format and structure. This structure may be taken into consideration when analyzing particular video content. News segments are typically 30 min in duration and follow a rigid pattern from day to day. Commercials are also of fixed duration, making detection less difficult.

Another key element in video is the use of the black frame. In most broadcast video, a black frame is shown between a transition of two segments. In news broadcast this usually occurs between a story and a commercial. By detecting the location of black frames in video, a hierarchical structure may be created to determine transitions between segments. A black frame or any single intensity image may be detected by summing the total number of pixels in a particular color space, P_s.

$$P_s(t) = \sum_{i=1}^{M} \begin{cases} 0, & i > I_{\text{high}} \parallel i < I_{\text{low}} \\ 1, & \text{otherwise} \end{cases}.$$

In the detection of the black frame, I_{high}, the maximum allowable pixel intensity is of the order of 20% of the maximum color resolution (51 for a 256-bit image), and I_{low}, the minimum allowable pixel intensity, is 0. The separation of segments in video

FIGURE 6   Images with similar shapes (human face and torso). (See color section, p. C–40.)

is crucial in retrieval systems, where a user will most likely request a small segment of interest and not an entire full-length video. There are a number of ways to detect this feature in video, the simplest being to detect a high number of pixels in an image that are within a given tolerance of being a black pixel.

3 Compressed Domain Features

In typical applications of multimedia databases, the materials, especially the images and video, are often in a compressed format. Given large amounts of compressed materials (e.g., MPEG), how do we index and retrieve the content rapidly? To deal with these materials, we find a straightforward approach is to decompress all the data and utilize the same features as mentioned in previous section. Doing so, however, has some disadvantages. First, the decompression implies extra computation. Second, the process of decompression and recompression, often referred to as "recoding," results in a further loss of image quality. Finally, since the size of decompressed data is much larger than the compressed form, most hardware and CPU cycles are needed to process and store the data. The solution to these problems is to extract features directly from the compressed data. We call these the compressed-domain features, and these features can be useful for indexing and retrieval [1, 7, 19]. The question is how to explore unique information available in the compressed domain. We start by introducing a number of commonly used compressed-domain features.

The motion vectors that are available in all video data compressed by using standards such as H.261/H.263 and MPEG-1/2 are very useful. Analysis of motion vectors can be used to detect scene changes and other special effects such as dissolve, fade in, and fade out. For example, if the motion vectors for a frame are randomly distributed without coherency, that may suggest the presence of a scene change. Segmentation of a field of motion vectors into regions of similar vectors can be used to detect moving objects and track their positions. They can also be used to derive camera motion such as zoom and pan [1, 17]. Essentially, since motion vectors represent a low-resolution optical flow in the video, they can be used to extract all information that can be extracted by using the optical flow method.

The percentage of each type of block in a picture is also a good indicator of scene changes. For a P frame, a large percentage of intrablocks implies a lot of new information for the current frame that cannot be predicted from the previous frame. Therefore, such a P frame indicates the beginning of a new scene right after a scene change. For a B frame, the ratio between the number of forward predicted blocks and the number of backward predicted blocks can be used to conclude whether the scene change happens before this B frame or after this B frame. If the number of forward predicted blocks is larger than the number of backward predicted blocks, i.e., there is more correlation between the previous frame and the current B frame than there is between the current B frame and the following frames, then one can conclude that the scene change happens after the B frame. If the number of forward predicted blocks is smaller than the number of backward predicted blocks, then one can conclude that the scene change happens before the B frame.

The DCT provides a decomposition of the original image in the frequency domain. Therefore, DCT coefficients form a natural representation of texture in the original image. In addition to texture analysis, DCT coefficients can also be used to match images and to detect scene changes. If only the DC components are collected, we have a low-resolution representation of the original image, averaged over 8 × 8 blocks. This is very helpful because it means much less data to manipulate, and it is found that for some applications, DC components already contain sufficient information. For color analysis, usually only the DC components are used to estimate the color histogram. For scene change detection, usually only the DC components are used to compare the content in two consecutive frames.

Not only can information be extracted from the compressed data for indexing and retrieval; the parameters in the compression process that are not explicitly specified in the bit stream can be very useful as well. One example is the bit rate, i.e., the number of bits used for each picture. For intracoded video (i.e., no motion compensation), the number of bits per picture should remain roughly constant for a scene segment and should change when the scene changes. For example, a scene with simple color variation and texture requires fewer bits per picture compared with a scene that has detailed texture. For intercoding, the number of bits per picture is proportional to the action between the current picture and the previous picture. Therefore, if the number of bits for a certain picture is high, we can often conclude that there is a scene cut.

The compressed-domain approach does not solve all problems, though. The identification of useful features from compressed data is typically difficult because each compression technique poses additional constraints, e.g., nonlinear processing, rigid data structure syntax, and resolution reduction.

Another issue is that compressed-domain features depend on the underlying compression standard. For different compression standards, different feature extraction algorithms have to be developed. Ultimately, we would like to have new compression standards with maximal content accessibility. MPEG-4 and MPEG-7 already have considered this aspect. In particular, MPEG-7 is a standard that goes beyond the domain of "compression" and seeks efficient "representation" of image and video content. In conclusion, the compressed-domain approach provides significant advantages but also brings new challenges.

4 Content-Based Features

Section 2 described a number of image and video features that can be extracted by using well-known techniques in image processing. Section 3 described how many of these features are computed or approximated by using encoded parameters in image and video compression. Although in both cases there is considerable understanding of the structure of the video, the features in no way estimate the actual image or video content. In this section we describe several methods to approximate the actual content of an image or video. For many users, the query of interest is text based, and therefore the content is essential. The desired result has less to do with analytical features such as color or texture, and more with the actual objects within the image or video.

4.1 Object Detection

Identifying significant objects that appear in the video frames is one of the key components for video characterization. Several working systems have generated reasonable results for the detection of a particular object, such as human faces, text, or an automobile. These limited domain systems have much greater accuracy than do broad domain systems that attempt to identify any object in the image.

4.1.1 Human Subjects

The "talking head" image is common in interviews and news clips, and it illustrates a clear example of video production focusing on an individual of interest. A human interacting within an environment is also a common theme in video. The detection of a human subject is particularly important in the analysis of news footage. An anchorperson will often appear at the start and end of a news broadcast, which is useful for detecting segment boundaries. In sports, anchorpersons will often appear between plays or commercials.

The detection of humans in video is possible by using a number of algorithms. Figure 7 shows examples of faces detected using the neural network arbitration method [11]. Most techniques are dependent of scale and rely heavily on lighting conditions, limited occlusion, and limited facial rotation. For more details on face detection technology, please see Chapter 10.6.

4.1.2 Captions and Graphics

Text and graphics are used in a variety of ways to convey content to the viewer. They are most commonly used in broadcast news,

where information must be absorbed in a short time. Examples of text and graphics in video are discussed in the paragraphs that follow.

Video captions: Text in video provides significant information as to the content of a scene. For example, statistical numbers and titles are not usually spoken but are included in captions for viewer inspection. Moreover, this information does not always appear in closed captions, so detection in the image is crucial for identifying potentially important regions.

In news video, captions of the broadcasting company are often shown at low opacity as a watermark in a corner without obstructing the actual video. A ticker tape is widely used in broadcast news to display information such as the weather, sports scores, or the stock market. In some broadcast news, graphics such as weather forecasts are displayed in a ticker-tape format with the news logo in the lower right corner at full opacity. Captions that appear in the lower third portion of a frame are almost always used to describe a location, person of interest, title, or event in news video. In Figure 7, the anchorpersons location is listed.

Captions are used less frequently in video domains other than broadcast news. In sports, a score or some information about an ensuing play is often shown in a corner or border at low opacity. Captions are sometimes used in documentaries to describe a location, person of interest, title, or event. Almost all commercials use some form of captions to describe a product or institution, because their time is limited to only a few minutes.

For feature films a producer may use text at the beginning or end of a film for deliberate viewer comprehension, such as character listings or credits. A producer may also start a film with an introduction to the story being told. Throughout a film, captions may be used to convey a change in time or location, which would otherwise be difficult and time consuming for a video producer to create. A producer will seldom use fortuitous text in the actual video unless the wording is noticeable and easy to read in a short time. A typical text region can be characterized as a horizontal rectangular structure of clustered sharp edges, because characters usually form regions of high contrast against the background. By detecting these properties, we can extract potentially important regions from video frames that contain textual information. Most captions are high-contrast text, such as the black and white chyron commonly found in news video. Consistent detection of the same text region over a period of time is probable since text regions remain at an exact position for many video frames. This may also reduce the number of false detections that occur when text regions move or fade in and out between scenes.

Figure 8 illustrates the process of text detection, primarily regions of horizontal titles and captions. We first apply a global horizontal differential filter, F_{HD}, to the image.

$$F_{HD} = [-1/2 \quad 1 \quad 1/2]$$

FIGURE 7   Recognition of captions and faces [11]. (See color section, p. C–40.)

An appropriate binary threshold (Chapter 2) should be set for

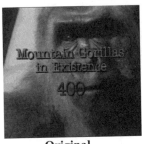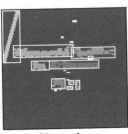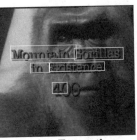

| **Original** | **HDF Filter** | **Clustering** | **Region Extraction** |

FIGURE 8   Text detection in video.

extraction of vertical edge features. A smoothing filter, F_S, is then used to eliminate extraneous fragments and to connect character sections that may have been detached.

$$F_S = \begin{bmatrix} 1/3 & 1/3 & 1/3 \end{bmatrix}$$

Individual regions must be identified though cluster detection (Chapter 2). A bounding box, B_B, should be computed for selection of text regions. We now select clusters with bounding regions that satisfy constraints in cluster size, C_S, cluster fill factor, C_{FF}, and horizontal-vertical aspect ratio.

$$C_s(n) = \sum_{i=1}^{P} C_n(i)$$

$$C_{FF}(n) = \frac{C_S(n)}{B_{B_{area}}(n)}$$

A sample set of parameters for the font size in Fig. 8 is listed as follows:

- cluster size >70 pixels
- cluster fill factor >0.45
- horizontal-vertical aspect ratio >0.75
- maximum cluster height $= 50$ pixels
- minimum cluster height $= 10$ pixels
- maximum cluster width $= 150$ pixels
- minimum cluster width $= 15$ pixels

A cluster's bounding region must have a small vertical-to-horizontal aspect ratio and must satisfy various limits in height and width. The fill factor of the region should be high to ensure dense clusters. The cluster size should also be relatively large to avoid small fragments. Other controlling parameters are listed below.

Finally, we examine the intensity histogram of each region to test for high contrast. This is because certain textures and shapes appear similar to text but exhibit low contrast when examined in a bounded region.

For some fonts a generic optical character recognition (OCR) package may accurately recognize video captions. For most OCR systems, the input is an individual character. This presents a problem in digital video since most of the characters experience

some degradation during recording, digitization and compression. For a simple font, we can search for blank spaces between characters and assume a fixed width for each letter [13].

Graphics: A graphic is usually a recognizable symbol, which may contain text. Graphic illustrations or symbolic logos are used to represent many institutions, locations, and organizations. They are used extensively in news video, where it is important to describe the subject matter as efficiently as possible. A logo representing the subject is often placed in a corner next to an anchorperson during dialogue. Detection of graphics is a useful method for finding changes in semantic content. In this sense, its appearance may serve as a scene break. Recognition of corner regions for graphics detection may be possible through an extension of the scene change technology. Histogram difference analysis, $D_{Hs}(t)$, of isolated image regions instead of the entire image can provide a simple method for detecting corner graphics. An example of a graphics logo detected with $D_{Hs}(t)$ is shown in Fig. 9. In this example, a change is detected in the upper corner, although no scene change is detected.

$$D_{Hs}(t) = \sum_{j=1}^{H/2} \sum_{i=W/2}^{W} |H_{t-T}(i, j) - H_t(i, j)|$$

4.1.3 Articulated Objects

A particular object is usually the emphasis of a query in image and video retrieval. Recognition of articulated objects poses a great challenge and represents a significant step in content-based

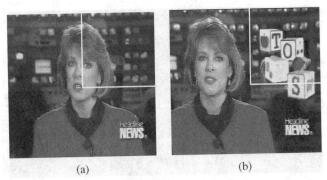

FIGURE 9   Graphics detection through subregion histogram differencing: (a) frame t; (b) frame $t + T$. (See color section, p. C–40.)

feature extraction. Many working systems have demonstrated an accurate recognition of animal objects, segmented objects, and rigid objects such as planes or automobiles.

The recognition of a single object is only one potential use of image-based recognition systems. Discrimination of synthetic and natural backgrounds or an animated or mechanical motion would yield a significant improvement content-based feature extraction. For more information on object recognition, please see Chapter 4.

4.2 Audio and Language

An important element in video indexing creation is the audio track. Audio is an enormous source for describing video content. Words specific to the actual content, or "keywords," can be extracted by using a number of language processing techniques [6, 16]. Keywords may be used to reduce indexing and provide abstraction for video sequences. There are many possibilities for language processing in video, but the audio track must first exist as an ASCII document, or speech recognition is necessary.

Audio segmentation is needed to distinguish spoken words from music, noise, and silence. Further analysis through speech recognition is necessary to align and translate these words into text. Audio selection is made on a frame-by-frame basis, so it is important to achieve the highest possible accuracy. At a sampling rate of 8 kHz, one frame corresponds to 267 samples of audio. Techniques in language understanding are used for selecting the most significant words and phrases.

In order to use the audio track, we must isolate each individual word. To transcribe the content of the video material, we recognize spoken words by using a speech recognition system. Speaker independent recognition systems have made great strides as of late and offer promise for application in video indexing [5]. Speech recognition works best when closed-captioned data are available. Captions usually occur in broadcast material, such as sitcoms, sports, and news. Documentaries and movies may not necessarily contain captions. Closed captions have become more common in video material throughout the United States since 1985, and most televisions provide standard caption display.

4.3 Rule-Based Features

The features described in previous sections may be used with rules that describe a particular type of video scene to create an additional set of content-based features [15]. By using examples from video production standards, we can identify a small set of heuristic rules. In some cases these rules involve the integration of image processing features with audio and language features. Below is a description of three rule-based features suitable for most types of video.

Introduction scenes: The scenes prior to the introduction of a person usually describe his or her accomplishments and often precede scenes with large views of the person's face. A person's name is generally spoken and then followed by supportive material. Afterward, the person's actual face is shown. If a scene contains a proper name, and a large human face is detected in the scenes that follow, we call this an introduction scene. Characterization of this type is useful when searching for a particular human subject, because identification is more reliable than using the image or audio features separately. Introduction scenes must meet the following criteria:

$$
\text{Scene}_{\text{introduction}}(i)
$$
$$
= \begin{cases} 1, & (\text{face}_i = \text{TRUE} \ \& \ \text{WORD}_i = \text{PROPER\_NAME}) \\ 0, & (\text{otherwise}) \end{cases}.
$$

Adjacent similar scenes: The color histogram difference measure gives us a simple routine for detecting similarity between scenes. Scenes between successive shots of a human face usually imply illustration of the subject. For example, a video producer will often interleave shots of research between shots of a scientist. Images that appear between two similar scenes that are less than T_{SS} seconds apart are characterized as an adjacent similar scene. Scene (i) is an adjacent similar scene if it meets the following criteria:

$$
\text{scene}_{\text{adjacent similar}}(i)
$$
$$
= \begin{cases} 1, & \text{scene}(i - T) = \text{scene}(i + T), \\ & |\text{scene}_{\text{start}}(i - T) - \text{scene}_{\text{start}}(i + T)| < T_{SS} \\ 0, & \text{otherwise} \end{cases}
$$

where T_{SS} is of the order of 10 s or less.

Short successive scenes: Short successive shots often introduce an important topic. By measuring the duration of each scene, S_D, we can detect these regions and identify short successive sequences. A set of scenes is short successive if a group of five or more scenes meet the following criteria:

$$
\text{scene}_{\text{short successive}}(i) =
$$
$$
\begin{cases} 1, & \begin{pmatrix} \text{scene}_{\text{duration}}(i - T) < S_D \ \& \ \text{scene}_{\text{duration}}(i + T) \\ \quad < S_D \ \& \\ \text{scene}_{\text{duration}}(i + 2T) < S_D \ \& \ \text{scene}_{\text{duration}}(i + 3T) \\ \quad < S_D \ldots \end{pmatrix} \\ 0, & \text{otherwise} \end{cases}
$$

where S_D for each scene is of the order of 3 s or less.

4.4 Embedded Video Features

A final solution for content-based feature extraction is the use of known procedures for creating video. Video production manuals provide insight into the procedures used during video editing and creation. There are many documents that describe the editing and production procedures for creating video segments, but one of the most recent is that by Pryluck [10].

One of the most common elements in video production is the ability to convey climax or suspense. Producers use a variety of different effects, ranging from camera positioning to lighting to special effects, to convey this mood to an audience. The detection of procedures such as these is beyond the realm of present image and language understanding technology. However, many of the important features described in Sections 2, 3, and 4 were derived from research in the video production industry.

Structural information as to the content of a video is a useful tool for indexing video. For example, the type of video being used (documentaries, news footage, movies and sports) and its duration may offer suggestions to assist in object recognition. In news footage, the anchorperson will generally appear in the same pose and background at different times. The exact locations of the anchorperson can then be used to delineate story breaks. In documentaries, a person of expertise will appear at various points throughout the story when topical changes take place. There are also many visual effects introduced during video editing and creation that may provide information for video content. For example, in documentaries the scenes prior to the introduction of a person usually describe his or her accomplishments and often precede scenes with large views of the person's face.

A producer will often create production notes that describe in detail action and scenery of a video, scene by scene. If a particular feature is needed for an application in image or video databases, the description may have already been documented during video production.

Another source of descriptive information may also embedded in the video stream in the form of timecode and geospatial

(GPS/GIS) data. These features are useful in indexing precise segments in video or a particular location in spatial coordinates. Aeronautic and automobile surveillance video will often contain GPS data that may be used as a source for indexing.

5 Retrieval Techniques

In Sections 2, 3 and 4, we described analytical and content-based features that can be extracted from image and video segments. This section describes techniques for establishing correspondence between these features.

5.1 Feature-Based Retrieval (Statistical and Compressed)

Correspondence between analytical features is established with the difference measures described in Sections 2 and 3. This is straightforward for image matching features, in which a match is based on the minimum absolute difference, $D(t)$, or histogram difference, $D_H(t)$. In the case of features based on motion, texture, and shape, the difference is based on the Euclidean distance between the parameters of the perspective feature. The difference measures may be applied to the entire image or a subregion of the image for better correspondence between objects in the image. Regardless of the granularity in applying difference measures, a key problem with color image matching is that similar colors do not necessarily provide similar content, as seen in Fig. 10.

FIGURE 10   Images with similar color. (See color section, p. C–41.)

5.1.1 Global and Subregion Image Matching

Image correspondence is important for identifying scenes that appear often in a video segment. The color histogram is not only useful for detecting scene changes, but also serves as an adequate method for image correspondence. A histogram from the first video frame of each scene is stored and compared with that of video frames in subsequent scenes. An analysis of the entire image requires less computation than subregion differencing, but the image match is less robust to foreground objects. Global image matching is particularly useful with images of uniform color and texture.

In news footage, an icon or logo is often used to symbolize the subject of the video. This icon is usually placed in the upper quarter of the image. Although the background of the image remains the same, changes in this icon represent changes in content. By applying histogram differencing to a small region in the image, we can detect changes in news icons. Processing subregions requires more computation, but the resulting image match is usually more robust. Objects that appear away from the background are usually easier to match with subregion differencing. Subregion differencing is also more affective with images of complex color and texture.

5.2 Content-Based Retrieval

The main problem in image in video indexing is that users query content and most systems only match statistical features such as color and texture. Figure 11 illustrates two images of essentially identical content, but that have almost no similarities in color, shape, texture, or motion. In this case, the motion of the players is similar, but the angle of camera will yield two separate forms of object motion from the original video sequence.

Content matching attempts to correlate actual objects with a given query. The user is not limited to selections based on similar color properties, but rather a collection based on content. In this form of matching, the query may be an image or text. The content features such as caption and face detection correspond to textual descriptions, so a query need not be an image.

TABLE 2 Potential query applications

Query type	Associated feature
Pans or zooms in video	Camera motion
Action or moving objects	Object motion
Important scenes	Short sequences, adjacent similar scenes, introduction scenes
Human subjects	Face detection, introduction scenes, video text detection
Video captions	Video text detection
Subject location	GPS, video text detection
Image scenery	Color difference, texture
Name or description	Audio and language analysis
Simple objects	Color difference
Segment boundaries	Face detection, scene changes, black frames

A number of content-based image and video systems are applicable to the features described in this chapter. In Table 2, we list several potential query applications associated with content-based and statistical features.

Several working systems have demonstrated the potential of content-based matching for identifying specific objects and stories. Three of the more interesting systems are discussed as follows.

Name-It is a system for matching a human face to a name in news video [12]. It approximates the likelihood of a particular face belonging to a name in close proximity within the transcript. Integrated language and image understanding technology make the automation of this system possible.

Spot-it is a topological system that attempts to identify known characteristics in news video for indexing and classification [8]. It has reasonable success in identifying common video themes such as interviews, group discussions, and conference room meetings.

Pictorial Transcripts a working system at AT&T Research Laboratories has shown promising results in video summarization when closed captions are used with statistical visual attributes [14]. CNN video is digitized and displayed in an HTML

FIGURE 11 Images with similar content. (See color section, p. C–41.)

environment with text for audio and a static image for every paragraph. More than 3000 hours of processed video can be searched and browsed.

5.3 Considerations in Multimedia Databases

The retrieval of an image or video segment is often limited in practical multimedia databases. There are many factors to consider when creating an image or video database, such as optimization for large databases, the type of query, the presentation of results, and the measure of success. Retrieval efficiency is an important concern for image and video databases. Flat file systems are sufficient when the size of a collection is moderate. However, a more robust solution is necessary when image and video libraries grow to several thousand units of data. Researchers have developed tree structure optimization systems that greatly reduce the search space by clustering image characteristics into small subsets for later retrieval [21]. For more details on image and video databases, please see Chapter 9.2.

5.3.1 Queries: Image or Text

For most image and video retrieval systems, the query is an image. When the comparison is based on analytical features, the results can often be ambiguous, as shown in Section 5.1. Content-based features provide a more accurate match to the given query, but the results are based on image processing technology that is only capable of recognizing a small number of objects.

Text queries eliminate ambiguity in the query and work only with content-based features. There is still a dependence on content-based feature extraction, but there is limited uncertainty in the query. This type of the query may also be used to match the title of the image or the transcript of the video.

5.3.2 Presentation of Results

The presentation of a query result is an important part of the image or visual query system. Presentations are usually visual and textural in layout. Textual presentations provide more specific information and are useful when presenting large collections of data. Visual, or iconic, presentations are more useful when the content of interest is easily recalled from imagery. This is quite often the case in stock footage video, where there is no audio to describe the content. Section 7 describes current working systems for presentation of image and video results.

5.3.3 Testing and Evaluation

In image and video databases, accuracy is based on the relevance of the output set of images or video to a particular query. A user defines the level of quality; therefore, the evaluation of an image or video retrieval system cannot be based on traditional analytical measures. The accuracy of these systems is purely subjective, which requires that some human intervention take place during evaluation.

User studies or some form of subjective rating are essential during the design and development of an image and video database system. Researchers have successfully demonstrated the utility of user studies in testing image and video retrieval applications [3, 4]. A subject is generally shown a query and asked to rank the resulting image or video segment's on a scale. For example, in a video database the user might be asked to rate the quality of selection on a scale ranging from "high relevance" to "low relevance." An example of a user-study interface for video retrieval is shown in Fig. 12.

6 Video Access and Browsing

With the size of video collections growing to thousands of hours, technology is needed to effectively browse segments in a short time without losing the content of the video. Simplistic browsing techniques, such as increasing playback speed and skipping video frames at fixed intervals, reduce video viewing time. However, increased video rates eliminate the majority of the audio information and distort much of the image information, and displaying video sections at fixed intervals merely gives a random estimate of the overall content. An ideal browser would display only the video pertaining to a segment's content, suppressing irrelevant data.

A multimedia abstraction ideally preserves and communicates the essential content of a video segment via a compact representation. Examples of multimedia abstractions include short text titles and single thumbnail images. Another commonly used abstraction presents an ordered set of representative, "thumbnail" images simultaneously on a computer screen. Image statistics, such as histogram analysis and texture, camera structure, and scene changes are the dominant factors in these systems. While these abstractions have proven useful in various contexts, their static nature ignores video's temporal dimension.

In addition, these abstractions often concentrate exclusively on the image content and neglect the audio information carried in a video segment. Preliminary investigations suggest that the opposite emphasis offers greater value. The video skim (see Fig. 13) is one of the first systems to integrate technology in image, language, and audio understanding for browsing and summarization [15]. Recently, researchers have proposed browsing representations based on information within the video. These systems rely on the motion in a scene and the placement of scenes breaks, but not on integrated image and language understanding.

Although the video skimming work represents the one of the first experiments in integrating language and image understanding, there are a number of efforts that combine language and image understanding as of late. The application of technology integration is different for these systems; however, they all demonstrate the advantages of using multiple modalities in video characterization and summarization. Examples of these systems are discussed as follows.

Browsing through clustering: this system was designed to cluster image regions for browsing digital video [20]. It uses many of the image statistics mentioned earlier, but it attempts to process scene transitions rather than just process individual frames.

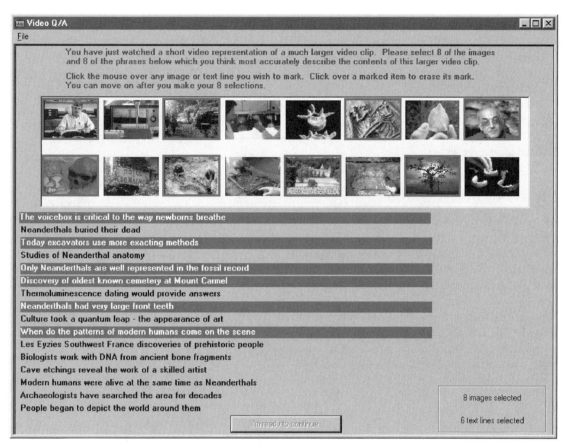

FIGURE 12 Video retrieval user-study interface [3]. (See color section, p. C–42.)

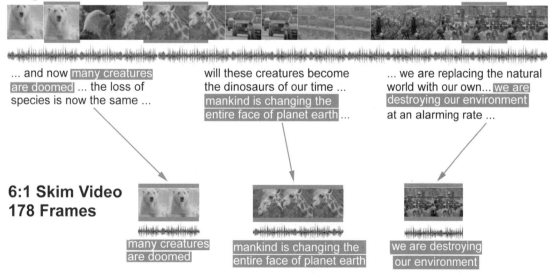

FIGURE 13 Illustration of video skimming. (See color section, p. C–43.)

High-rate keyframe browsing: the Digital Library Research Group at the University of Maryland, College Park, MD, has conducted a user study to test optimal frame rates for keyframe-based browsing [4]. They use many of the same image analysis techniques mentioned earlier to extract keyframes, and they quantify their research through studies of a video slide-show interface at various frame rates.

Video abstracts: the Movie Content Analysis (MoCA) group in Mannheim, Germany has created a system for movie abstraction based on the occurrence of image statistics and audio frequency analysis to detect dialogue scenes [9].

7 Case Study: Informedia and other Digital Library Projects

The Informedia digital video library project at Carnegie Mellon University has established a large, on-line digital video library by developing intelligent, automatic mechanisms to populate the library and allow for full-content and knowledge-based search and retrieval via desktop computer over local, metropolitan, and wide-area networks [18]. Initially, the library was populated with 1000 hours of raw and edited documentary and education videos drawn from video assets of WQED/Pittsburgh, Fairfax County (VA) Public Schools, and the Open University (U.K.). To assess the value of video reference libraries for enhanced learning at different ages, the library was deployed at a local area K–12 schools. Figure 14 shows an example of the Informedia interface, with poster frames, weighted query, and textual abstracts. The library's approach utilizes several techniques for content-based searching and video sequence retrieval. Content is conveyed in both the narrative (speech and language) and the image. The collaborative interaction of image, speech, and natural language understanding technology allows for successful population, segmentation, indexing, and search of diverse video collections with satisfactory recall and precision.

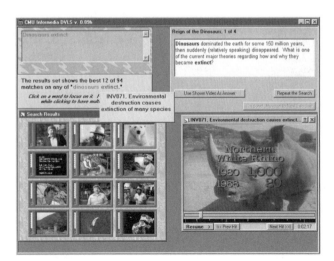

FIGURE 14   Informedia interface. (See color section, p. C–43.)

TABLE 3   Systems with unique characteristics

System	Description
U. C. Berkeley	Object extraction and recognition system. National Science Foundation Digital Library Initiative.
U. C. Santa Barbara	Image matching system based on region segmentation. National Science Foundation Digital Library Initiative.
IBM	One of the first well-known image matching systems, Query by Image Content, or QBIC. Fast image indexing through condensed hierarchical tree structure. Features based on color, texture, shape, and position.
VIRAGE	Image and video retrieval based on research at the University of California at San Diego. Features are based on color, texture, shape, position and language queries.
Media Site	Video, image and text indexing and retrieval for internet distribution. Primary technology is licensed from Carnegie Mellon University Informedia Project.
Excalibur	Video and text retrieval based on image matching and language queries.

There are many researchers working in the area of image matching. A few systems with unique characteristics are listed in Table 3.

8 The MPEG-7 Standard

As pointed out earlier in this chapter, instead of trying to extract relevant features, manually or automatically, from original or compressed video, a better approach for content retrieval should be to design a new standard in which such features, often referred to as metadata, are already available. MPEG-7, an ongoing effort by the Moving Picture Experts Group, is working toward this goal, i.e., the standardization of metadata for multimedia content indexing and retrieval.

MPEG-7 is an activity that is triggered by the growth of digital audiovisual information. The group strives to define a "multimedia content description interface" to standardize the description of various types of multimedia content, including still pictures, graphics, 3-D models, audio, speech, video, and composition information. It may also deal with special cases such as facial expressions and personal characteristics.

The goal of MPEG-7 is exactly the same as the focus of this chapter, i.e., to enable efficient search and retrieval of multimedia content. Once finalized, it will transform the text-based search and retrieval (e.g., keywords), as is done by most of the multimedia databases, into a content-based approach, e.g., using color, motion, or shape information. MPEG-7 can also be thought of as a solution to describing multimedia content. If one looks at PDF (portable document format) as a standard language to describe text and graphic documents, then MPEG-7 will be a standard description for all types of multimedia data, including audio, images, and video.

Compared with earlier MPEG standards, MPEG-7 possesses some essential differences. For example, MEPG-1, 2, and 4 all

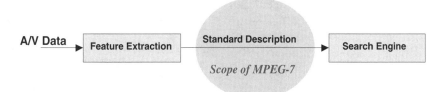

FIGURE 15 The scope of MPEG-7.

focus on the representation of audiovisual data, but MPEG-7 will focus on representing the metadata (information about data). MPEG-7, however, may utilize the results of previous MPEG standards, e.g., the shape information in MPEG-4 or the motion vector field in MPEG-1/2.

Figure 15 shows the scope of the MPEG-7 standard. Note that feature extraction is outside the scope of MPEG-7; so is the search engine. This is a result of one approach constantly taken by most of the standard activities, i.e., "to standardize the minimum." Therefore, the analysis (feature extraction) should not be standardized, so that after MPEG-7 is finalized, various analysis tools can still be further improved over time. This also leaves room for competition among vendors and researchers. This is similar to the fact that MPEG-1 does not specify motion estimation, and that MPEG-4 does not specify segmentation algorithms. Likewise, the query process (the search engine) should not be standardized. This allows the design of search engines and query languages to adapt to different application domains, and also leaves room for further improvement and competition. Summarizing, MPEG-7 takes the approach that it standardizes only what is necessary so that the description for the same content may adapt to different users and different application domains.

We now explain a few concepts of MPEG-7. One goal of MPEG-7 is to provide a standardized method of describing features of multimedia data. For images and video, colors or motion are example features that are desirable in many applications. MPEG-7 will define a certain set of descriptors to describe these features. For example, the color histogram can be a very suitable descriptor for color characteristics of an image, and motion vectors (as commonly available in compressed video bit streams) form a useful descriptor for motion characteristics of a video clip. MPEG-7 also uses the concept of description scheme (DS), which means a framework that defines the descriptors and their relationships. Hence, the descriptors are the basis of a DS.

Description then implies an instantiation of a DS. MPEG-7 not only to standardize the description; it also wants the description to be efficient. Therefore, MPEG-7 also considers compression techniques to turn descriptions into coded descriptions. Compression reduces the amount of data that need to be stored or processed. Finally, MPEG-7 will define a description definition language (DDL) that can be used to define, modify, or combine descriptors and description schemes. Summarizing, MPEG-7 will standardize a set of descriptors and DS's, a DDL, and methods for coding the descriptions.

The process to define MPEG-7 is similar to those of the previous MPEG standards. Since 1996, the group has been working on defining and refining the requirements of MPEG-7, i.e., what MPEG-7 should provide. The MPEG-7 process includes a competitive phase followed by a collaborative phase. During the competitive phase, a Call for Proposals is issued and participants respond by both submitting written proposals and demonstrating the proposed techniques. Proposals are then evaluated by experts to determine merit. During the collaborative phase, MPEG-7 will evolve as a series of experimentation models (XM), where each model outperforms the previous one. Eventually, MPEG-7 will turn into an international standard. Table 4 shows the timetable for MPEG-7 development. At the time of this writing, the group is going through the definition process of the first XM.

Once finalized, MPEG-7 has a large variety of applications, such as digital libraries, multimedia directory services, broadcast media selection, and multimedia authoring. Here are some examples. With MPEG-7, the user can draw a few lines on a screen to retrieve a set of images containing similar graphics. The user can also describe movements and relations between a number of objects to retrieve a list of video clips containing these objects with the described temporal and spatial relations. Also, for a given content, the user can describe actions and then get a list of scenarios where similar. Summarizing, here we presented an overview of recent MPEG-7 activities and their strong relationship with image and video indexing and retrieval. For more details on MPEG-4 and MPEG-7, please see Chapter 6.5.

TABLE 4 Timetable for MPEG-7

Development	Date
Call for test material	Mar. 1998
Call for proposals	Oct. 1998
Proposals due	Feb. 1999
1st experiment model (XM)	Mar. 1999
Working draft (WD)	Dec. 1999
Committee draft (CD)	Oct. 2000
Final committee draft (FCD)	Feb. 2001
Draft international standard (DIS)	July 2001
International standard (IS)	Sep. 2001

9 Conclusion

Image and video retrieval systems have been primarily based on the statistical analysis of a single image. With an increase in feature-based analysis and extraction, these systems are becoming usable and efficient in retrieving perceptual content. Powerful feature-based indexing and retrieval tools can be developed for image–video archives, complementing the traditional text-based techniques. There are no "best" features for "all" image domains. It's a matter of creating a good "solution" by using multiple features for a specific application. Performance evaluation of visual query is an important but unsolved issue.

References

[1] A. Akutsu and Y. Tonomura, "Video tomography: an efficient method for camerawork extraction and motion analysis," presented at ACM Multimedia '94, San Francisco, CA, October 15–20, 1994.

[2] F. Arman, A. Hsu, and M-Y. Chiu, "Image processing on encoded video sequences," *Multimedia Syst.* **1**, 211–219 (1994).

[3] M. G. Christel, D. B. Winkler, and C. R. Taylor, "Improving access to a digital video library," presented at the 6th IFIP Conference On Human-Computer Interaction, Sydney, Australia, July 14–18, 1997.

[4] L. Ding, et al., "Previewing video data: browsing key frames at high rates using a video slide show interface," presented at the International Symposium on Research Development and Practice in Digital Libraries, Tsukuba Science City, Japan, November 26–28, 1997.

[5] A. Hauptmann and M. Smith, "Text, speech, and vision for video segmentation," presented at the AAAI Fall Symposium on Computational Models for Integrating Language and Vision, Boston, Nov. 10–12, 1995.

[6] M. Mauldin, *Conceptual Information Retrieval: A Case Study in Adaptive Partial Parsing,* (Kluwer, Boston, MA, 1991).

[7] J. Meng and S.-F. Chang, "Tools for compressed-domain video indexing and editing," presented at the SPIE Conference on Storage and Retrieval for Image and Video Databases, San Jose, CA, February 1–2, 1996.

[8] Y. Nakamura and T. Kanade, "Semantic analysis for video contents extraction — spotting by association in news video," presented at the Fifth ACM International Multimedia Conference, Seattle, Nov. 9–13, 1997.

[9] S. Pfeiffer, R. Lienhart, S. Fischer, and W. Effelsberg, "Abstracting digital movies automatically," *J. Visual Commun. Image Rep.* **7**, 345–353 (1996).

[10] C. Pryluck, C. Teddlie, and R. Sands, "Meaning in film/video: order, time and ambiguity," *J. Broadcasting* **26**, 685–695 (1982).

[11] H. Rowley, S. Baluja, and T. Kanade, "Neural network-based face detection," *IEEE Trans. Pattern Anal. Machine Intell.* **20(1)**, Jan. 1998.

[12] S. Satoh, T. Kanade, and M. Smith, "NAME-IT: Association of face and name in video," presented at the meeting on Computer Vision and Pattern Recognition, San Juan, Puerto Rico, June 17–19, 1997.

[13] T. Sato, T. Kanade, E. Hughes, and M. Smith, "Video OCR for digital news archives," presented at the IEEE Workshop on Content-Based Access of Image and Video Databases (CAIVD'98), Bombay, India, Jan. 3, 1998.

[14] B. Shahraray and D. Gibbon, "Authoring of hypermedia documents of video programs," presented at the Third ACM Conference on Multimedia, San Francisco, CA, November 5–9, 1995.

[15] M. A. Smith and T. Kanade, "Video skimming and characterization through the combination of image and language understanding techniques," presented at the meeting on Computer Vision and Pattern Recognition, San Juan, Puerto Rico, June 17–19, 1997.

[16] "TREC 93," in D. Harmon, ed. *Proceedings of the Second Text Retrieval Conference* (ARPA/SISTO, location, 1993).

[17] Y. T. Tse and R. L. Baker, "Global zoom/pan estimation and compensation for video compression," *Proc. ICASSP*, 2725–2728 (1991).

[18] H. D. Wactlar, T. Kanade, M. A. Smith, and S. M. Stevens, "Intelligent access to digital video: informedia project," *IEEE Comput.* **29**, 46–52 (1996).

[19] H. Wang and S. F. Chang, "A highly efficient system for automatic face region detection in MPEG video sequences," *IEEE Trans. Circuits Syst. Video Technol.*, special issue on Multimedia Systems and Technologies, 1997.

[20] M. Yeung, B. Yeo, W. Wolf, and B. Liu, "Video browsing using clustering and scene transitions on compressed sequences," presented at the meeting of IS&T/SPIE Multimedia Computing and Networking, February 5–11, 1995.

[21] M. Flickner, et al., "Query by image content: The QBIC system," *IEEE Comput.* **28(9)**, 23–32 (September 1995).

[22] R. Zabih, J. Miller, and K. Mai, "A feature-based algorithm for detecting and classifying scene breaks," presented at the meeting of ACM International Conference on Multimedia, San Francisco, CA, November 5–9, 1995.

[23] H. J. Zhang, S. Tan, S. Smoliar, and G. Yihong, "Video parsing, retrieval and browsing: an integrated and content-based solution," presented at the ACM International Conference on Multimedia, San Francisco, CA, November 5–9, 1995.

9.2

Unified Framework for Video Browsing and Retrieval

Yong Rui
Microsoft Research

Thomas S. Huang
*University of Illinois at
Urbana-Champaign*

1 Introduction

Research on how to efficiently access video content has become increasingly active in the past few years [1–4]. Considerable progress has been made in video analysis, representation, browsing, and retrieval, which are the four fundamental bases for accessing video content. *Video analysis* deals with the *signal processing* part of the video system, including shot boundary detection, key frame extraction, etc. *Video representation* is concerned with the *structure* of the video. An example of a video representation is the tree structured key frame hierarchy [5, 3]. Built on top of the video representation, *video browsing* deals with how to use the representation structure to help viewers browse the video content. Finally, *video retrieval* is concerned with retrieving interesting video objects. The relationship among these four research areas is illustrated in Fig. 1.

So far, most of the research effort has gone into video analysis. Although it is the basis for all the other research activities, it is not the ultimate goal. Relatively less research exists on video representation, browsing, and retrieval. As seen in Fig. 1, video browsing and retrieval are on the very top of the diagram. They *directly* support users' access to the video content. For accessing a temporal medium, such as a video clip, browsing and retrieval are equally important. Browsing helps a user to quickly grasp the global picture of the data, whereas retrieval helps a user to find a specific query's results.

An analogy explains this argument. How does a reader efficiently access a 1000-page book's content? Without reading the whole book, the reader can first go to the book's Table of Contents (ToC), finding which chapters or sections suit his or her needs. If the reader has specific questions (queries) in mind, such as finding a term or a key word, he or she can go to the Index page and find the corresponding book sections containing that question. In short, the book's ToC helps a reader *browse*, and the book's index helps a reader *retrieve*. Both aspects are equally important in helping users access the book's content. For today's video data, unfortunately, we lack both the ToC and the Index. Techniques are urgently needed for automatically (or semiautomatically) constructing video ToCs and video Indexes to facilitate browsing and retrieval.

A great degree of power and flexibility can be achieved by simultaneously designing the video access components (ToC and Index) using a unified framework. For a long and continuous stream of data, such as video, a "back and forth" mechanism between browsing and retrieval is crucial.

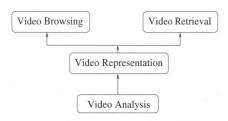

FIGURE 1   Relationships among the four research areas.

The goals of this chapter are to develop novel techniques for constructing both the video ToC and video Index as well as a method for integrating them into a unified framework. The rest of the chapter is organized as follows. In Section 2, important video terminologies are first introduced. We review video analysis, representation, browsing, and retrieval in Sections 2–5, respectively. In Section 6 we describe in detail a unified framework for video browsing and retrieval. Algorithms as well as experimental results on real-world video clips are presented. Conclusions and future research directions are summarized in Section 7.

2 Terminologies

Before we go into the details of the discussion, we find it beneficial to first introduce some important terminologies used in the digital video research field.

1. Video shot is a consecutive sequence of frames recorded from a single camera. It is the building block of video streams.
2. Key frame is the frame which represents the salient visual content of a shot. Depending on the complexity of the content of the shot, one or more key frames can be extracted.
3. Video scene is defined as a collection of semantically related and temporally adjacent shots, depicting and conveying a high-level concept or story. While shots are marked by physical boundaries, scenes are marked by semantic boundaries.[1]
4. Video group is an intermediate entity between the physical shots and semantic scenes and serves as the bridge between the two. Examples of groups are temporally adjacent shots [5] or visually similar shots [3].

In summary, the video data can be structured into a hierarchy consisting of five levels: video, scene, group, shot, and key frame, which increase in granularity from top to bottom [4] (see Fig. 2).

---

[1] Some of the early literature in video parsing misused the phrase *scene change detection* for *shot boundary detection*. To avoid any later confusion, we will use *shot boundary detection* to mean the detection of physical shot boundaries, and we will use *scene boundary detection* to mean the detection of semantic scene boundaries.

3 Video Analysis

As can be seen from Fig. 1, video analysis is the basis for later video processing. It includes *shot boundary detection* and *key frame extraction*.

3.1 Shot Boundary Detection

It is not efficient (sometimes not even possible) to process a video clip as a whole. It is beneficial to first decompose the video clip into shots and do signal processing at the shot level.

In general, automatic shot boundary detection techniques can be classified into five categories: *pixel based, statistics based, transform based, feature based*, and *histogram based*. Pixel-based approaches use pixelwise intensity difference to mark shot boundaries [1, 6]. However, they are highly sensitivity to noise. To overcome this problem, Kasturi and Jain propose to use intensity statistics (mean and standard deviation) as shot boundary detection measures [7]. Seeking to achieve faster processing, Arman *et al.* propose to use the compressed discrete cosine transform (DCT) coefficients (e.g., MPEG data) as the boundary measure [8]. Other transform-based shot boundary detection approaches make use of motion vectors, which are already embedded in the MPEG stream [9, 10]. Zabih *et al.* address the problem from another angle. Edge features are first extracted from each frame. Shot boundaries are then detected by finding sudden edge changes [11]. So far, the histogram-based approach is the most popular. Instead of using pixel intensities directly, the histogram-based approach uses histograms of the pixel intensities as the measure. Several researchers claim that it achieves a good tradeoff between accuracy and speed [1]. Representatives of this approach are [1, 12–15]. More recent work has been based on clustering and postfiltering [16], which achieves fairly high accuracy without producing many false positives. Two comprehensive comparisons of shot boundary detection techniques are [17, 18].

3.2 Key Frame Extraction

After the shot boundaries are detected, corresponding key frames can then be extracted. Simple approaches may just extract the first and last frames of each shot as the key frames [15]. More sophisticated key frame extraction techniques are based on visual content complexity indicators [19], shot activity indicators [20], and shot motion indicators [21].

4 Video Representation

Considering that each video frame is a two-dimensional (2-D) object and the temporal axis makes up the third dimension, a video stream spans a three-dimensional (3-D) space. Video representation is the *mapping* from the 3-D space to the 2-D

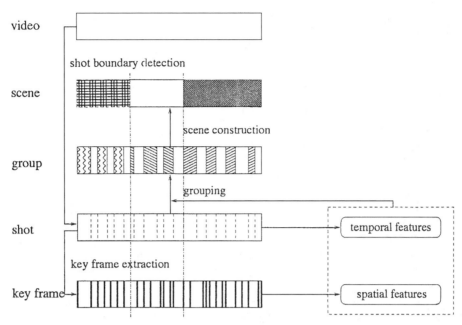

FIGURE 2   A hierarchical video representation.

view screen. Different mapping functions characterize different video representation techniques.

4.1 Sequential Key Frame Representation

After obtaining shots and key frames, an obvious and simple video representation is to sequentially lay out the key frames of the video, from top to bottom and from left to right. This simple technique works well when there are few key frames. When the video clip is long, this technique does not scale, since it does not capture the embedded information within the video clip, except for time.

4.2 Group-Based Representation

For a more meaningful video representation to be obtained when the video is long, related shots are merged into groups [3, 5]. In [5], Zhang *et al.* divide the entire video stream into multiple video segments, each of which contains an equal number of consecutive shots. Each segment is further divided into subsegments, thus constructing a tree structured video representation. In [3], Zhong *et al.* proposed a cluster-based video hierarchy, in which the shots are clustered based on their visual content. This method again constructs a tree structured video representation.

4.3 Scene-Based Representation

To provide the user with better access to the video, the construction of a video representation at the semantic level is needed [2, 4]. It is not uncommon for a modern movie to contain a few thousand shots and key frames. This is evidenced in [22] — there

are 300 shots in a 15-min video segment of the movie "Terminator 2 — Judgment Day," and the movie lasts 139 min. Because of the large number of key frames, a simple one-dimensional (1-D) sequential presentation of key frames for the underlying video (or even a tree structured layout at the group level) is almost meaningless. More importantly, people watch the video by its semantic scenes rather than the physical shots or key frames. While *shot* is the building block of a video, it *is scene* that conveys the semantic meaning of the video to the viewers. The discontinuity of shots is overwhelmed by the continuity of a scene [2]. Video ToC construction at the scene level is thus of fundamental importance to video browsing and retrieval. In [2], a scene transition graph (STG) of video representation is proposed and constructed. The video sequence is first segmented into shots. Shots are then clustered by using *time-constrained clustering*. The STG is then constructed based on the time flow of the clusters.

4.4 Video Mosaic Representation

Instead of representing the video structure based on the video-scene-group-shot-frame hierarchy as discussed above, this approach takes a different perspective [23]. The mixed information within a shot is decomposed into three components:

1. The Extended spatial information captures the appearance of the entire background imaged in the shot, and is represented in the form of a few mosaic images.
2. The Extended temporal information captures the motion of independently moving objects in the form of their trajectories.

3. The Geometric information captures the geometric transformations that are induced by the motion of the camera.

5 Video Browsing and Retrieval

These two functionalities are the ultimate goals of a video access system, and they are closely related to (and built on top of) video representations. The first three representation techniques discussed above are suitable for video browsing, while the last can be used in video retrieval.

5.1 Video Browsing

For "Sequential key frame representation," browsing is obviously sequential browsing, scanning from the top-left key frame to the bottom-right key frame.

For "Group-based representation," a hierarchical browsing is supported [3, 5]. At the coarse level, only the main themes are displayed. Once the user determines which theme he or she is interested in, the user can then go to the finer level of the theme. This refinement process can go on until the leaf level.

For the STG representation, a major characteristic is its indication of time flow embedded within the representation. By following the time flow, the viewer can browse through the video clip.

5.2 Video Retrieval

As discussed in Section 1, both the ToC and Index are equally important for accessing the video content. Unlike the other video representations, the mosaic representation is especially suitable

for video retrieval. Three components, moving objects, backgrounds, and camera motions, are perfect candidates for a video Index. After constructing such a video index, queries such as "find me a car moving like this," "find me a conference room having that environment," etc. can be effectively supported.

6 Proposed Framework

As we have reviewed in the previous sections, considerable progress has been made in each of the areas of video analysis, representation, browsing, and retrieval. However, so far, the interaction among these components is still limited and we still lack a unified framework to glue them together. This is especially crucial for video, given that the video medium is characteristically long and unstructured. In this section, we will explore the synergy between video browsing and retrieval.

6.1 Video Browsing

Among the many possible video representations, the "scene-based representation" is probably the most effective for meaningful video browsing [2, 4]. We have proposed a scene-based video ToC representation in [4]. In this representation, a video clip is structured into the scene-group-shot-frame hierarchy (see Fig. 2), which then serves as the basis for the ToC construction. This ToC frees the viewer from doing tedious "fast forward" and "rewind," and it provides the viewer with nonlinear access to the video content. Figures 3 and 4 illustrate the browsing process, enabled by the video ToC. Figure 3 shows a condensed ToC for a video clip, as we normally have in a long book. By looking at the representative frames and text annotation, the viewer can

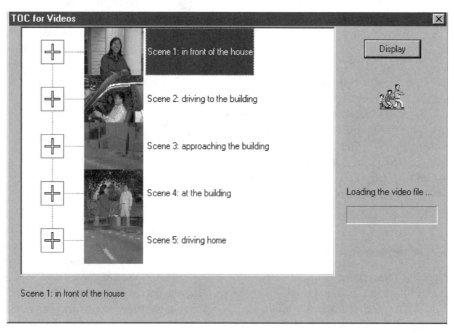

FIGURE 3   Condensed ToC. (See color section, p. C–44.)

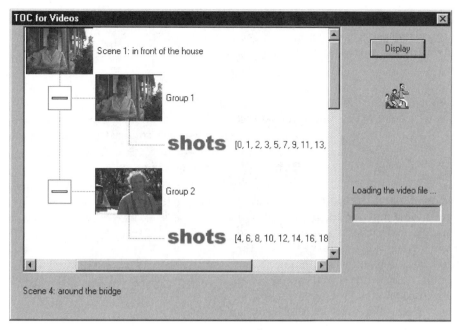

FIGURE 4   Expanded ToC. (See color section, p. C–44.)

determine which particular portion of the video clip he or she is interested in. Then, the viewer can further expand the ToC into more detailed levels, such as groups and shots. The expanded ToC is illustrated in Fig. 4. Clicking on the "Display" button will display the specific portion that is of interest to the viewer, without viewing the entire video.

The algorithm is described below. To learn details, interested readers are referred to [24].

[Main Procedure]

- Input: video shot sequence, $S = \{\text{shot } 0, \ldots, \text{shot } i\}$.
- Output: video structure in terms of scene, group, and shot.
- Procedure:
  1. Initialization: assign shot 0 to group 0 and scene 0; initialize the group counter numGroups $= 1$; initialize the scene counter numScenes $= 1$.
  2. If S is empty, quit; otherwise get the next shot. Denote this shot as shot i.
  3. Test if shot i can be merged to an existing group:
     (a) Compute the similarities between the current shot and existing groups: Call findGroupSim( ).
     (b) Find the maximum group similarity:

     $$\text{maxGroupSim}_i = \max_g \text{GroupSim}_{i,g},$$
     $$g = 1, \ldots, \text{numGroups},   (1)$$

     where $\text{GroupSim}_{i,g}$ is the similarity between shot i and group g. Let the group of the maximum similarity be group $g_{\max}$.
     (c) Test if this shot can be merged into an existing group:

If $\text{maxGroupSim}_i > \text{groupThreshold}$,   where groupThreshold is a predefined threshold:
  i. Merge shot i to group $g_{\max}$.
  ii. Update the video structure: call updateGroupScene( ).
  iii. Go to Step 2.
  otherwise:
  i. Create a new group containing a single shot i. Let this group be group j.
  ii. Set numGroups $=$ numGroups $+ 1$.
4. Test if shot i can be merged to an existing scene:
  (a) Calculate the similarities between the current shot i and existing scenes: call findSceneSim( ).
  (b) Find the maximum scene similarity:

  $$\text{maxSceneSim}_i = \max_s \text{SceneSim}_{i,s},$$
  $$s = 1, \ldots, \text{numScenes},   (2)$$

  where $\text{SceneSim}_{i,s}$ is the similarity between shot i and scene s. Let the scene of the maximum similarity be scene $s_{\max}$.
  (c) Test if shot i can be merged into an existing scene: If $\text{maxSceneSim}_i > \text{sceneThreshold}$, where sceneThreshold is a predefined threshold:
  i. Merge shot i to scene $s_{\max}$.
  ii. Update the video structure: call updateScene( ).
  otherwise:
  i. Create a new scene containing a single shot i and a single group j.
  ii. Set numScenes $=$ numScenes $+ 1$.
5. Go to Step 2.

[findGroupSim]

- Input: current shot and group structure.
- Output: similarity between current shot and existing groups.
- Procedure:
  1. Denote current shot as shot i.
  2. Calculate the similarities between shot i and existing groups:

$$\text{GroupSim}_{i,g} = \text{ShotSim}_{i,g_{\text{last}}}, \, g = 1, \ldots, \text{numGroups},\tag{3}$$

  where $\text{ShotSim}_{i,j}$ is the similarity between shots i and j; and g is the index for groups and g_{last} is the last (most recent) shot in group g. That is, the similarity between current shot and a group is the similarity between the current shot and the most recent shot in the group. The most recent shot is chosen to represent the whole group because all the shots in the same group are visually similar and the most recent shot has the largest *temporal attraction* to the current shot.
  3. Return.

[findSceneSim]

- Input: the current shot, group structure, and scene structure.
- Output: similarity between the current shot and existing scenes.
- Procedure:
  1. Denote the current shot as shot i.
  2. Calculate the similarity between shot i and existing scenes:

$$\text{SceneSim}_{i,s} = \frac{1}{\text{numGroups}_s} \sum_{g}^{\text{numGroups}_s} \text{GroupSim}_{i,g},\tag{4}$$

  where s is the index for scenes; numGroups_s is the number of groups in scene s; and $\text{GroupSim}_{i,g}$ is the similarity between current shot i and gth group in scene s.

That is, the similarity between the current shot and a scene is the average of similarities between the current shot and all the groups in the scene.

3. Return.

[updateGroupScene]

- Input: current shot, group structure, and scene structure.
- Output: an updated version of group structure and scene structure.
- Procedure:
  1. Denote current shot as shot i and the group having the largest similarity to shot i as group g_{max}. That is, shot i belongs to group g_{max}.
  2. Define two shots, top and bottom, where top is the secondmost recent shot in group g_{max} and bottom is the most recent shot in group g_{max} (i.e., current shot).
  3. For any group g, if any of its shots (shot g_j) satisfies the following condition

$$\text{top} < \text{shot } g_j < \text{bottom},\tag{5}$$

  merge the scene that group g belongs to into the scene that group g_{max} belongs to. That is, if a scene contains a shot that is interlaced with the current scene, merge the two scenes. This is illustrated in Fig. 5 (shot i = shot 4, $g_{\text{max}} = 0$, $g = 1$, top = shot 1, and bottom = shot 4).
  4. Return.

[updateScene]

- Input: current shot, group structure, and scene structure.
- Output: an updated version of scene structure.
- Procedure:
  1. Denote current shot as shot i and the scene having the largest similarity to shot i as scene s_{max}. That is, shot i belongs to scene s_{max}.
  2. Define two shots, top and bottom, where top is the secondmost recent shot in scene s_{max} and bottom is the current shot in scene s_{max} (i.e., current shot).

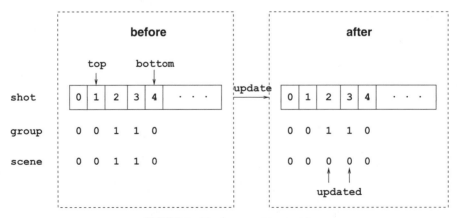

FIGURE 5   Merging scene 1 to scene 0.

TABLE 1 Scene structure construction results

Movie name	Frames	Shots	Groups	DS	FN	FP
Movie1	21717	133	16	5	0	0
Movie2	27951	186	25	7	0	1
Movie3	14293	86	12	6	1	1
Movie4	35817	195	28	10	1	2
Movie5	18362	77	10	6	0	0
Movie6	23260	390	79	24	1	10
Movie7	35154	329	46	14	1	2

3. For any scene s, if any of its shots (shot s_j) satisfies the following condition

$$\text{top} < \text{shot } s_j < \text{bottom}, \qquad (6)$$

merge scene s into scene $s_{\max}$. That is, if a scene contains a shot that is interlaced with the current scene, merge the two scenes.

4. Return.

Extensive experiments using real-world video clips have been carried out. The results are summarized in Table 1 [4], where DS (detected scene) denotes the number of scenes detected by the algorithm, FN (false negative) indicates the number of scenes missed by the algorithm; and FP (false positive) indicates the number of scenes detected by the algorithm, although they are considered scenes by humans.

Some observations can be summarized as follows.

1. The proposed scene construction approach achieves reasonably good results in most of the movie types.
2. The approach achieves better performance in "slow" movies than in "fast" movies. This follows since in the "fast" movies, the visual content is normally more complex and more difficult to capture. We are currently integrating closed-captioning information into the framework to enhance the accuracy of the scene structure construction.
3. The proposed approach seldom misses a scene boundary, but it tends to oversegment the video. That is, "false positives" outnumber "false negatives." This situation is expected for most of the automated video analysis approaches and has also been observed by other researchers [2, 22].

6.2 Video Retrieval

Video retrieval is concerned with how to return similar video clips (or scenes, shots, and frames) to a user given a video query. This is a little-explored research area. There are two major categories of existing work. One is to first extract key frames from the video data and then use image retrieval techniques to obtain the video data *indirectly*. Although easy to implement, it has the obvious problem of losing the temporal dimension. The other technique incorporates motion information (sometimes object tracking) into the retrieval process. Although this is a better technique, it requires the computationally expensive task of motion analysis. If object trajectories are to be supported, then this becomes more difficult.

Here we view video retrieval from a different angle. We seek to construct a video Index to suit various users' needs. However, constructing a video Index is far more complex than constructing an index for books. For books, the form of an index is fixed (e.g., key words). For videos, the viewer's interests may cover a wide range. Depending on his or her knowledge and profession, the viewer may be interested in semantic level labels (building, car, people), low-level visual features (color, texture, shape), or the camera motion effects (pan, zoom, rotation). In the system described here, we support all three Index categories:

- Visual Index
- Semantic Index
- Camera Motion Index

As a way to support semantic level and visual feature-based queries, frame clusters are first constructed to provide indexing. Our clustering algorithm is described as follows.

1. Feature extraction: color and texture features are extracted from each frame. The color feature is an 8×4 2-D color histogram in hue-saturation-value (HSV) color space. The V component is not used because of its sensitivity to lighting conditions. The H component is quantized finer than the S component because of the psychological observation that the human visual system is more sensitive to hue than to saturation. For texture features, the input image is fed into a wavelet filter bank and is then decomposed into decorrelated subbands. Each subband captures the feature of a given scale and orientation from the original image. Specifically, we decompose an image into three wavelet levels; thus there are 10 subbands. For each subband, the standard deviation of the wavelet coefficients is extracted. The 10 standard deviations are used as the texture representation for the image [25].

2. Global clustering: based on the features extracted from each frame, the entire video clip is grouped into clusters. A detailed description of the clustering process can be found in [19]. Note that each cluster can contain frames from multiple shots and each shot can contain multiple clusters. The cluster centroids are used as the visual Index and can be later labeled as a Semantic Index (see Section 6.3.). This procedure is illustrated in Fig. 6.

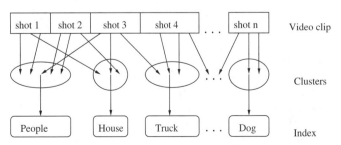

FIGURE 6 From video clip to cluster to Index.

After the above clustering process, the entire video clips are grouped into multiple clusters. Since color and texture features are used in the clustering process, all the entries in a given cluster are visually similar. Therefore these clusters naturally provide support for the visual queries.

In order to support semantic level queries, semantic labels have to be provided for each cluster. There are two techniques that have been developed in our research lab. One is based on the hidden Markov model (HMM), and the other is an annotation-based approach. Since the former approach also requires training samples, both approaches are semiautomatic. To learn details of the first approach, readers are referred to [16]. We will introduce the second approach here. Instead of attempting to attack the unsolved automatic image understanding problem, semiautomatic human assistance is used. We have built interactive tools to display each cluster centroid frame to a human user, who will label that frame. The label will then be *propagated* through the whole cluster. Since only the cluster centroid frame requires labeling, the interactive process is fast. For a 21,717 frame video clip (Movie1), ~20 min is needed. After this labeling process, the clusters can support both visual and semantic queries. The specific semantic labels for Movie1 are people, car, dog, tree, grass, road, building, house, etc.

To support camera motion queries, we have developed techniques to detect camera motion in the MPEG compressed domain [26]. The incoming MPEG stream does not have to be fully decompressed. The motion vectors in the bit stream form good estimates of camera motion effects. Hence, panning, zooming, and rotation effects can be efficiently detected [26].

6.3 Unified Framework for Browsing and Retrieval

Subsections 6.1 and 6.2 described video browsing and retrieval techniques separately. In this section, we integrate them into a unified framework to enable a user to go "back and forth" between browsing and retrieval. Going from the Index to the ToC, a user can get the *context* where the indexed entity is located. Going from the ToC to the Index, a user can *pin point* specific queries. Figure 7 illustrates the unified framework.

An essential part of the unified framework is the weighted links. The links can be established between Index entities and scenes, groups, shots, and key frames in the ToC structure. As a first step, in this paper we focus our attention on the links between Index entities and shots. Shots are the building blocks of the ToC. Other links are generalizable from the shot link.

To link shots and the *Visual Index*, we propose the following techniques. As we mentioned before, a cluster may contain frames from multiple shots. The frames from a particular shot form a subcluster. This subcluster's centroid is denoted as c_{sub}, and the centroid of the whole cluster is denoted as c. This is illustrated in Fig. 8.

Here c is a representative of the whole cluster (and thus the Visual Index) and c_{sub} is a representative of the frames from

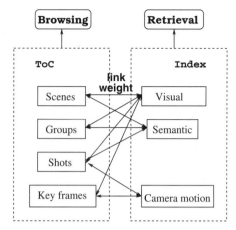

FIGURE 7   Unified framework.

a given shot in this cluster. We define the similarity between the cluster centroid and subcluster centroid as the link weight between Index entity c and that shot.

$$w_v(i, j) = \text{similarity}(c_{sub}, c_j), \qquad (7)$$

where i and j are the indices for shots and clusters, respectively, and $w_v(i, j)$ denotes the link weight between shot i and Visual Index cluster c_j.

After defining the link weights between shots and the Visual Index, and labeling each cluster, we can next establish the link weights between shots and the *Semantic Index*. Note that multiple clusters may share the same semantic label. The link weight between a shot and a Semantic Index is defined as

$$w_s(i, k) = \max_j(wv(i, j)), \qquad (8)$$

where k is the index for the Semantic Index entities, and j represents those clusters sharing the same semantic label k.

The link weight between shots and a *Camera Motion Index* (e.g., panning) is defined as

$$w_c(i, l) = \frac{n_i}{N_i}, \qquad (9)$$

where l is the index for the camera operation Index entities; n_i is

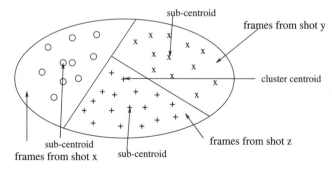

FIGURE 8   Subclusters.

FIGURE 9   Interface for going from the Semantic Index to the ToC. (See color section, p. C–45.)

the number of frames having that camera motion operation; and N_i is the number of frames in shot i.

Extensive tests have been carried out with real-world video clips. The video streams are MPEG compressed, with the digitization rate equal to 30 frames/s. Table 2 summarizes example results over the video clip Movie1. The first two rows are an example of going from the semantic Index (e.g., car) to the ToC (Fig. 9). The middle two rows are an example of going from the visual Index (e.g., Fig. 10) to the ToC (Fig. 11). The last two rows are going from the camera operation Index (panning) to the ToC.

By just looking at each isolated Index alone, a user usually cannot understand the context. By going from the Index to the ToC (as in Table 2), a user quickly learns when and under which circumstances (e.g., within a particular scene) that Index entity is happening. Table 2 summarizes how to go from the Index to

the ToC to find the *context*. We can also go from the ToC to the Index to *pin point* a specific Index. Table 3 summarizes which Index entities appeared in shot 33 of the video clip Movie1.

For a continuous and long medium such as video, a "back and forth" mechanism between browsing and retrieval is crucial. Video library users may have to browse the video first before

TABLE 2   From the semantic, visual, camera Index to the ToC

Shot id	0	2	10	12	14	31	33
W	0.958	0.963	0.919	0.960	0.957	0.954	0.920
shot id	16	18	20	22	24	26	28
Wv	0.922	0.877	0.920	0.909	0.894	0.901	0.907
shot id	0	1	2	3	4	5	6
Wc	0.74	0.03	0.28	0.17	0.06	0.23	0.09

FIGURE 10   Frame 2494 as a Visual Index.

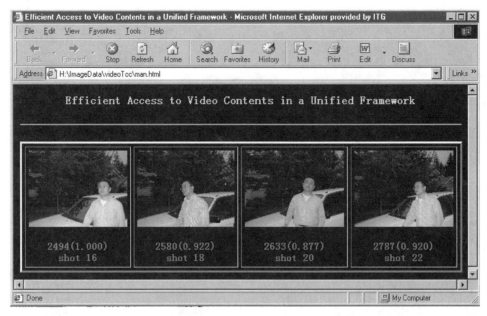

FIGURE 11   Interface for going from the Visual Index to the ToC. (See color section, p. C–45.)

they know what to retrieve. On the other hand, after retrieving some video objects, the users will be better able to browse the video in the correct direction. We have carried out extensive subjective tests employing users from various disciplines. Their feedback indicates that this unified framework greatly facilitated their access to video content, in both home entertainment and educational applications.

7 Conclusions and Promising Research Directions

This chapter reviewed and discussed recent research progress in video analysis, representation, browsing, and retrieval; it introduced the video ToC and the Index and presented techniques for constructing them; it proposed a unified framework for video browsing and retrieval; and it proposed techniques for establishing the link weights between the ToC and the Index. We should be aware that video is not just a visual medium. It contains text and audio information in addition to visual information and is thus "true" multimedia. Multimodel and multimedia processing is usually more reliable and robust than processing a single medium. We need to further extend our investigation to the integration of closed-captioning and audio track information into our algorithm to enhance the construction of ToCs, Indexes, and link weights.

TABLE 3   From the ToC (shots) to the index

Index	Fence	Mail box	Human hand	Mirror	Steer wheel
Weight	0.927	0.959	0.918	0.959	0.916

Acknowledgment

This work was supported in part by ARL Cooperative Agreement No. DAAL01-96-2-0003, and in part by a CSE Fellowship, College of Engineering, UIUC. The authors thank Sean X. Zhou and Roy R. Wang for their valuable discussions.

References

[1] H. Zhang, A. Kankanhalli, and S. W. Smoliar, "Automatic partitioning of full-motion video," *ACM Multimedia Sys.* **1**, 1–12 (1993).

[2] R. M. Bolle, B.-L. Yeo, and M. M. Yeung, "Video query: beyond the keywords," Tech. Rep., RC 200586 IBM Research, October 17, 1996.

[3] D. Zhong, H. Zhang, and S.-F. Chang, "Clustering methods for video browsing and annotation," *SPIE Conf. on Storage and Retrieval for Image and Video Databases*, San Jose. Feb 1996.

[4] Y. Rui, T. S. Huang, and S. Mehrotra, "Exploring video structures beyond the shots," in *Proceedings of the IEEE Conference on Multimedia Computing and Systems* (IEEE, New York, 1998).

[5] H. Zhang, S. W. Smoliar, and J. J. Wu, "Content-based video browsing tools," in *Proceedings of the IS&T/SPIE Conference on Multimedia Computing and Networking* (SPIE, New York, 1995).

[6] A. Hampapur, R. Jain, and T. Weymouth, "Digital video segmentation," in *Proceedings of the ACM Conference on Multimedia* (ACM, New York, 1994).

[7] R. Kasturi and R. Jain, "Dynamic vision," in *Proceedings of Computer Vision: Principles*, R. Kasturi and R. Jain, eds. (IEEE Computers Society Press, Washington, 1991).

[8] F. Arman, A. Hsu, and M.-Y. Chiu, "Feature management for large video databases," in *Storage and Retrieval for Image and Video Databases*, W. Niblack, ed., *Proc. SPIE* **1908**, –12 (1993).

[9] J. Meng, Y. Juan, and S.-F. Chang, "Scene change detection in a mpeg compressed video sequence," in *Digital Video Compression: Algorithms and Technologies*, A. A. Rodriguez, R. J. Safranek, and E. J. Delp, eds., *Proc. SPIE* **2419**, 1–12 (1995).

[10] B.-L. Yeo, "Efficient processing of compressed images and video," Ph.D. dissertation (Princeton University, Princeton, NJ, 1996).

[11] R. Zabih, J. Miller, and K. Mai, "A feature-based algorithm for detecting and classifying scene breaks," in *Proceedings of the ACM Conference on Multimedia* (ACM, New York 1995).

[12] A. Nagasaka and Y. Tanaka, "Automatic video indexing and full-video search for object appearances," in *Proceedings of Visual Database Systems II* Kruth E., Wegner L., eds. (Elsevier Science Publishers, 1992).

[13] D. Swanberg, C.-F. Shu, and R. Jain, "Knowledge guided parsing in video databases," in *Storage and Retrieval for Image and Video Databases*, W. Niblack, ed., *Proc. SPIE* **1908**, 22–24 (1993).

[14] H. Zhang and S. W. Smoliar, "Developing power tools for video indexing and retrieval," *SPIE Storage and Retrieval for Image and Video Databases II*, W. Niblack and R. C. Jain, eds., *Proc. SPIE* **2185**, 140–149 (1994).

[15] H. Zhang, C. Y. Low, S. W. Smoliar, and D. Zhong, "Video parsing, retrieval and browsing: An integrated and content-based solution," in *Proceedings of the ACM Conference on Multimedia* (ACM, New York, 1995).

[16] M. R. Naphade, R. Mehrotra, A. M. Ferman, T. S. Huang, and A. M. Tekalp, "A high performance algorithm for shot boundary detection using multiple cues," in *Proceedings of the IEEE International Conference on Image Processing* (IEEE, New York, 1998).

[17] J. S. Boreczky and L. A. Rowe, "Comparison of video shot boundary detection techniques," in *Storage and Retrieval for Image and Video Databases IV*, R. Joun, ed., 2670 170–179 (1996).

[18] R. M. Ford, C. Robson, D. Temple, and M. Gerlach, "Metrics for scene change detection in digital video sequences," in *Proceedings of the IEEE Conference on Multimedia Computing and Systems* (IEEE, New York, 1997).

[19] Y. Zhuang, Y. Rui, T. S. Huang, and S. Mehrotra, "Adaptive key frame extraction using unsupervised clustering," in *Proceedings of the IEEE International Conference on Image Processing* (IEEE, New York, 1998).

[20] P. O. Gresle and T. S. Huang, "Gisting of video documents: a key frames selection algorithm using relative activity measure," in *Proceedings of the 2nd International Conference on Visual Information Systems* (Knowledge Systems Institute, Skokie, IL 1997).

[21] W. Wolf, "Key frame selection by motion analysis," in *Proceedings of the IEEE International Conference on Acoustics, Speech, and Signal Processing* (IEEE, New York, 1996).

[22] M. Yeung, B.-L. Yeo, and B. Liu, "Extracting story units from long programs for video browsing and navigation," in *Proceedings of the IEEE Conference on Multimedia Computing and Systems* (IEEE, New York, 1996).

[23] M. Irani and P. Anandan, "Video indexing based on mosaic representations," *Proc. IEEE* **86**, 905–921 (1998).

[24] Y. Rui, T. S. Huang, and S. Mehrotra, "Constructing table-of-content for videos," *J. ACM/Springe, Mathmedias* Vol. 7, No. 5, 1999, 359–368.

[25] Y. Rui, T. S. Huang, and S. Mehrotra, "Content-based image retrieval with relevance feedback in MARS," in *Proceedings of the IEEE International Conference on Image Processing* (IEEE, New York, 1997).

[26] J. A. Schmidt, "Object and camera parameter estimation using mpeg motion vectors," M. S. thesis (University of Illinois at Urbana-Champaign, 1998).

9.3

Image and Video
Communication Networks

Dan Schonfeld
University of Illinois at Chicago

1 Introduction

Paul Baran from the RAND Corporation first proposed the notion of a distributed communication network in 1964. The aim of the proposal was to provide a communication network that could survive the impact of a nuclear war. This proposal employed a new approach to data communication based on packet switching.

The Department of Defense through the Advanced Research Projects Agency (ARPA) commissioned the ARPANET, later known as the Internet, in 1969. The ARPANET was initially an experimental communication network that consisted of four nodes: UCLA, UCSB, SRI, and the University of Utah.

The Internet grew very rapidly over the next two decades to encompass over 100,000 nodes by 1989, connecting research universities and government organizations around the world. Various protocols had been adopted to facilitate services such as remote connection, file transfer, electronic mail, and news distribution.

The proliferation of the Internet exploded over the past decade to over 10 million nodes since the release of the World Wide Web (WWW). Tim Berners-Lee proposed the WWW for the Corporation for Education and Research Networking (CERN) — the European center for nuclear research — in 1989. The Web grew out of a need for physics researchers from around the world to collaborate by using a large and dynamic collection of scientific documents.

Today the WWW provides a powerful framework for accessing linked documents throughout the Internet. The wealth of information available over the WWW has attracted the interest of commercial businesses and individual users alike. Its enormous popularity is enhanced by the graphical interfaces currently available for browsing multimedia information over the Internet.

The potential impact of multimedia information is currently restricted by the bandwidth of the existing communication networks. Recent proposals for the improvement of communication networks will be able to accommodate the data rates required for image and video information.

In the future, image and video communication networks will be used for a variety of applications such as videoconferencing, broadcast television, interactive television, video on demand (VoD), multimedia e-mail, telemedicine, and distance learning.

In this presentation, a broad overview of image and video communication networks is provided. The basic methods used for image and video communication are illustrated over a wide variety of communication networks: ATM, Internetworks, and Wireless networks. The efficient use of the various communication networks requires the transmission of image and video data in compressed form. A survey of the main image and video compression standards — JPEG and MPEG — is presented in Section 2.

The compressed image and video data are stored and transmitted in a standard format known as a compression stream. A discussion of the image and video compression stream standards is presented in Section 3. For brevity, this presentation will focus exclusively on the most popular current video compression standard: MPEG-2. A detailed presentation of the MPEG-2 compression stream standards — elementary stream, packetized elementary stream, program stream, and transport stream — is provided.

Initial efforts for image and video communication conducted over ATM networks are presented in Section 4. For brevity, this presentation will once again be restricted exclusively to the MPEG-2 compression standard. The mapping of the MPEG-2 transport stream to the ATM Application Layer (AAL) — AAL-1 and AAL-5 — is provided.

Current efforts are underway to expand the bandwidth of the Internet. For instance, the NSF has restructured its data networking architecture by providing the very high speed Backbone Network Service (vBNS). The vBNS Multicast Backbone (MBONE) network is intended to serve multicast real-time traffic such as audio and video communication over the Internet.

An overview of the existing protocols for image and video communication over the Internet is presented in Section 5. The standard protocol for the transport of real-time data is provided by the real-time transport protocol (RTP) presented in Section 5.1. For brevity, this presentation will once again focus exclusively on the MPEG-2 compression standard. Augmented to the RTP is the standard protocol for data delivery monitoring, as well as minimal control and identification capability, provided by the real-time transport control protocol (RTCP) presented in Section 5.2.

Preliminary plans have been in progress for image and video communication over wireless networks. A sketch of the proposed unified wideband wireless communication standard known as the International Mobile Telecommunications-2000 (IMT-2000) is discussed in Section 6.

Finally, a brief summary and discussion of the various methods used for image and video communication networks is presented.

2 Image and Video Compression Standards

Introduction

Numerous image and video compression standards have been proposed over the past decade by several international organizations.[1] In this section, a survey of the main image and video compression standards — JPEG and MPEG — is presented.[2]

2.1 JPEG: Joint Photographic Experts Group

The Joint Photographic Experts Group (JPEG) standard is used for the compression of continuous-tone still images. This compression standard is based on the Huffman and run-length encoding of the quantization of the discrete cosine transform (DCT) of image blocks. The widespread use of the JPEG standard is motivated by the fact that it consistently produces excellent perceptual picture quality at compression ratios in excess of 20:1.

A direct extension of the JPEG standard to video compression known as Motion JPEG (MJPEG) is obtained by the JPEG encoding of each individual picture in a video sequence. This approach is used when random access to each picture is essential, such as in video editing applications. The MJPEG compressed video yields data rates in the range of 8–10 Mbps.

For additional details about the lossy and lossless JPEG compression standard of continuous-tone still images, refer to Chapters 5.5 and 5.6, respectively.

2.2 MPEG-1: Motion Photographic Expert Group-1

The Motion Picture Expert Group (MPEG) proposals for compression of motion pictures have been adopted as the main video compression standards. Although the MPEG standards provide for both audio and video compression of motion pictures, our attention will be focused in this presentation exclusively on the video compression standards.

The goal of MPEG-1 was to produce VCR NTSC (352×240) quality video compression to be stored on CD-ROM (CD-I and CD-Video format) using a data rate of 1.2 Mbps. This approach is based on the arrangement of frame sequences into a group of pictures (GOP) consisting of four types of pictures: I picture (intra), P picture (predictive), B picture (bidirectional), and D picture (DC). I pictures are intraframe JPEG encoded pictures that are inserted at the beginning of the GOP. P and B pictures

---

[1] The organizations involved in the adoption of image and video compression standards include the International Standards Organization (ISO), the International Telecommunications Union (ITU), and the International Electrotechnical Commission (IEC).

[2] A closely related family of videoconferencing compression standards known as the H.26X Series — omitted from this presentation for brevity — is discussed in Chapter 6.1.

are interframe motion compensated JPEG encoded macroblock difference pictures that are interspersed throughout the GOP.[3] MPEG-1 restricts the GOP to sequences of 15 frames in progressive mode.

The system level of MPEG-1 provides for the integration and synchronization of the audio and video streams. This is accomplished by multiplexing and including time stamps in both the audio and video streams from a 90-kHz system clock. For additional information related to MPEG-1, refer to Chapter 6.4.

2.3 MPEG-2: Motion Photographic Expert Group-2

The aim of MPEG-2 was to produce broadcast-quality video compression and was expanded to support higher resolutions, including High Definition Television (HDTV).[4] The HDTV Grand Alliance standard has adopted the MPEG-2 video compression and transport stream standards in 1996.[5] MPEG-2 supports four resolution levels: low (352×240), main (720×480), high-1440 (1440×1152), and high (1920×1080). The MPEG-2 compressed video data rates are in the range of 3–100 Mbps.[6]

Although the principles used to encode MPEG-2 are very similar to MPEG-1, it provides much greater flexibility by offering several profiles that differ in the presence or absence of B pictures, chrominance resolution, and coded stream scalability.[7] MPEG-2 supports both progressive and interlaced modes.[8] Significant improvements have also been introduced in the MPEG-2 system level, as will be discussed in the following section. Additional details about MPEG-2 can also be found in Chapter 6.4.

2.4 MPEG-4: Motion Photographic Expert Group-4

The intention of MPEG-4 was to provide low bandwidth video compression at a data rate of 64 kbps that can be transmitted over a single N-ISDN B channel. This goal has evolved to the development of flexible scalable extendable interactive compression streams that can be used with any communication network for universal accessibility (e.g., Internet and wireless networks). MPEG-4 is a genuine multimedia compression standard that supports audio and video as well as synthetic and animated images, text, graphics, texture, and speech synthesis.

The foundation of MPEG-4 is on the hierarchical representation and composition of audio-visual objects (AVO). MPEG-4 provides a standard for the configuration, communication, and instantiation of classes of objects: The configuration phase determines the classes of objects required for processing the AVO by the decoder. The communication phase supplements existing classes of objects in the decoder. Finally, the instantiation phase sends the class descriptions to the decoder.

A video object at a given point in time is a video object plane (VOP). Each VOP is encoded separately according to its shape, motion, and texture. The shape encoding of a VOP provides a pixel map or a bitmap of the shape of the object. The motion and texture encoding of a VOP can be obtained in a manner similar to that used in MPEG-2. A multiplexer is used to integrate and synchronize the VOP data and composition information — position, orientation, and depth — as well as other data associated with the AVOs in a specified bit stream.

MPEG-4 provides universal accessibility supported by error robustness and resilience, especially in noisy environments at very low data rates (less than 64 kbps): bit-stream resynchronization, data recovery, and error concealment. These features are particularly important in mobile multimedia communication networks. For a thorough introduction to MPEG-4 refer to Chapter 6.5.

2.5 MPEG-7: Motion Photographic Expert Group-7

MPEG-7 — a recent initiative devoted to the standardization of the Multimedia Content Description Interface (MCDI) — is planned for completion by the year 2000. This standard will permit the description, identification, and access of audiovisual information from compressed multimedia databases. The search for audiovisual information will be retrieved by means of query material such as text, color, texture, shape, sketch, images, graphics, audio, and video, as well as spatial and temporal composition information. Although the MPEG-7 description can be attached to any multimedia representation, the standard will be based on MPEG-4. This standard will be used in applications such as medical imaging, home shopping, digital libraries, multimedia databases, and the Web. Additional information pertaining to MPEG-7 is also presented in Chapter 6.5.

3 Image and Video Compression Stream Standards

Introduction

The compressed image and video data are stored and transmitted in a standard format known as a compression stream. The discussion in this section will be restricted exclusively to the presentation of the video compression stream standards associated with the MPEG-2 systems layer: elementary stream (ES), packetized

---

[3] D pictures are used exclusively for low-resolution high-speed video scanning.

[4] The MPEG-3 video compression standard, which was originally intended for HDTV, was later cancelled.

[5] The HDTV Grand Alliance standard, however, has selected the Dolby Audio Coding 3 (AC-3) audio compression standard.

[6] The HDTV Grand Alliance standard video data rate is approximately 18.4 Mbps.

[7] The MPEG-2 video compression standard, however, does not support D pictures.

[8] The interlaced mode is compatible with the field format used in broadcast television interlaced scanning.

elementary stream (PES), program stream (PS), and transport stream (TS).

The MPEG-2 systems layer is responsible for the integration and synchronization of the ESs: audio and video streams, as well as an unlimited number of data and control streams that can be used for various applications such as subtitles in multiple languages. This is accomplished by first packetizing the ESs, thus forming the packetized elementary streams (PESs). These PESs contain time stamps from a system clock for synchronization.

The PESs are subsequently multiplexed to form a single output stream for transmission in one of two modes: PS and TS. The PS is provided for error-free environments such as storage in CD-ROM. It is used for multiplexing PESs that share a common time base, using long variable-length packets.[9] The TS is designed for noisy environments such as communication over ATM networks. This mode permits multiplexing streams (PESs and PSs) that do not necessarily share a common time base, using fixed-length (188 bytes) packets.

3.1 MPEG-2 Elementary Stream

As indicated earlier, MPEG-2 systems layer supports an unlimited number of ESs. Our focus is centered on the presentation of the ES format associated with the video stream. The structure of the video ES format is dictated by the nested MPEG-2 compression standard: video sequence, group of pictures (GOP), pictures, slices, and macroblocks. The video ES is defined as a collection of access units (pictures) from one source.

3.2 MPEG-2 Packetized Elementary Stream

The MPEG-2 systems layer packetizes all ESs — audio, video, data, and control streams — thus forming the PESs. Each PES is a variable-length packet with a variable format that corresponds to a single ES. The PES header contains time stamps to allow for synchronization by the decoder. Two different time stamps are used: the presentation time stamp (PTS) and the decoding time stamp (DTS). The PTS specifies the time at which the access unit should be removed from the decoder buffer and presented. The DTS represents the time at which the access unit must be decoded. The DTS is optional and is used only if the decoding time differs from the presentation time.[10]

3.3 MPEG-2 Program Stream

A PS multiplexes several PESs, which share a common time base, to form a single stream for transmission in error-free environments. The PS is intended for the storage and retrieval of programs from digital storage media such as CD-ROM. The PS uses relatively long variable-length packets. For a more detailed presentation of the MPEG-2 PS refer to [4].

3.4 MPEG-2 Transport Stream

A TS permits multiplexing streams (PESs and PSs) that do not necessarily share a common time base for transmission in noisy environments. The TS is designed for broadcasting over communication networks such as ATM networks. The TS uses small fixed-length packets (188 bytes) that make them more resilient to packet loss or damage during transmission. The TS provides the input to the transport layer in the OSI reference model.[11]

The TS packet is composed of a 4-byte header followed by 184 bytes shared between the variable-length optional adaptation field (AF) and the TS packet payload. The optional AF contains additional information that need not be included in every TS packet. One of the most important fields in the AF is the program clock reference (PCR). The PCR is a 42-bit field composed of a 9-bit segment incremented at 27 MHz as well as a 33-bit segment incremented at 90 kHz.[12] The PCR is used along with a voltage-controlled oscillator as a time reference for synchronization of the encoder and decoder clock.

A PES header must always follow the TS header and possible AF. The TS payload may consist of the PES packets or PSI. The PSI provides control and management information used to associate particular ESs with distinct programs. A program is once again defined as a collection of ESs that share a common time base. This is accomplished by means of a program description provided by a set of PSI associated signaling tables (ASTs): program association tables (PATs), program map tables (PMTs), network information tables (NITs), and conditional access tables (CATs). The PSI tables are sent periodically and carried in sections along with cyclic redundancy check (CRC) protection in the TS payload.

An example illustrating the formation of the TS packets is depicted in Fig. 1. The choice of the size of the fixed-length TS packets — 188 bytes — is motivated by the fact that the payload of the ATM Adaptation Layer-1 (AAL-1) cell is 47 bytes. Therefore, four AAL-1 cells can accommodate a single TS packet. A detailed discussion of the mapping of the TS packets to ATM networks is presented in the next section.

4 Image and Video ATM Networks

Introduction

Asynchronous transfer mode (ATM), also known as cell relay, is a method for information transmission in small fixed-size packets called cells based on asynchronous time-division multiplexing. ATM technology was proposed as the underlying foundation for the Broadband Integrated Services Digital Network (B-ISDN). B-ISDN is an ambitious very high data rate network that will replace the existing telephone system and all specialized networks

---

[9]The MPEG-2 PS is similar to the MPEG-1 systems stream.

[10]This is the situation for MPEG-2 video ES profiles that contain B pictures.

[11]The TS, however, is not considered as part of the transport layer.

[12]The 33-bit segment incremented at 90 kHz is compatible with the MPEG-1 system clock.

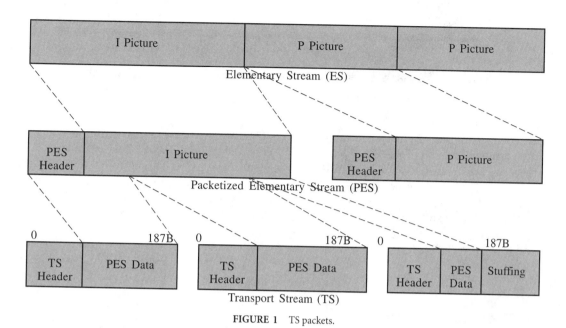

FIGURE 1 TS packets.

with a single integrated network for information transfer applications such as video on demand (VoD), broadcast television, and multimedia communication. These lofty goals not withstanding, ATM technology has found an important niche in providing the bandwidth required for the interconnection of existing local area networks (LAN); e.g., Ethernet.

The ATM cells are 53 bytes long, of which 5 bytes are devoted to the ATM header and the remaining 48 bytes are used for the payload. These small fixed-sized cells are ideally suited for the hardware implementation of the switching mechanism at very high data rates. The data rates envisioned for ATM are 155.5 Mbps (OC-3), 622 Mbps (OC-12), and 2.5 Gbps (OC-48).[13]

The B-ISDN ATM reference model consists of several layers: the physical layer, the ATM layer, the ATM Adaptation Layer (AAL), and the upper layers.[14] This layer can be further divided into the physical medium dependent (PMD) sublayer and the transmission convergence (TC) sublayer. The PMD sublayer provides an interface with the physical medium and is responsible for transmission and synchronization on the physical medium (e.g., SONET or SDH). The TC sublayer converts between the ATM cells and the frames — strings of bits — used by the PMD sublayer. ATM has been designed to be independent of the transmission medium. The data rates specified at the physical layer, however, require category 5 twisted pair or optical fibers.[15]

The ATM layer provides the specification of the cell format and cell transport. The header protocol defined in this layer provides generic flow control, virtual path and channel identification, payload type, cell loss priority, and header error checking. The ATM layer is a connection-oriented protocol that is based on the creation of end-to-end virtual circuits (channels). The ATM layer protocol is unreliable — acknowledgements are not provided — since it was designed for use of real-time traffic such as audio and video over fiber optic networks that are highly reliable. The ATM layer nonetheless provides quality of service (QoS) guarantees in the form of cell loss ratio, bounds on maximum cell transfer delay (MCTD), cell delay variation (CDV) — known also as delay jitter. This layer also guarantees the preservation of cell order along virtual circuits.

The structure of the AAL can be decomposed into the segmentation and reassembly sublayer (SAR) and the convergence sublayer (CS). The SAR sublayer converts between packets from the CS sublayer and the cells used by the ATM layer. The CS sublayer provides standard interface and service options to the various applications in the upper layers. This sublayer is also responsible for converting between the message or data streams from the applications and the packets used by the SAR sublayer. The CS sublayer is further divided into the common part convergence sublayer (CPCS) and the service specific convergence sublayer (SSCS).

Initially four service classes were defined for the AAL (Classes A–D). This classification has subsequently been modified by the characterization of four protocols: Class A is used to represent real-time (RT) constant bit-rate (CBR) connection-oriented (CO) services handled by AAL-1. This class includes applications such as circuit emulation for uncompressed audio and video transmission. Class B is used to define real-time (RT) variable bit-rate (VBR) CO services given by AAL-2. Among the

---

[13]The data rate of 155.5 Mbps was chosen to accommodate the transmission of HDTV and for compatibility with the Synchronous Optical Network (SONET). The higher data rates of 622 Mbps and 2.5 Gbps were chosen to accommodate four and 16 channels, respectively.

[14]Note that the B-ISDN ATM reference model layers do not map well into the OSI reference model layers.

[15]Existing twisted pair wiring cannot be used for B-ISDN ATM transmission for any substantial distances.

0	1	2	3	4	5	6	7b
CSI	SC			CRC			P

FIGURE 2 AAL-1 SAR-PDU header.

applications considered by this class are compressed audio and video transmission. Although the aim of the AAL-2 protocol is consistent with the focus of this presentation, we shall not discuss it in detail since the AAL-2 standard has not yet been defined. Classes C and D support nonreal-time (NRT) VBR services corresponding to AAL-3/4.[16] Class C is further restricted to NRT, VBR, connection-oriented services provided by AAL-5.[17] It is expected that this protocol will be used to transport IP packets and interface to ATM networks.

4.1 AAL-1: ATM Application Layer-1

The AAL-1 protocol is used for transmission of RT, CBR, connection-oriented traffic. This application requires transmission at constant rate, minimal delay, insignificant jitter, and low overhead.

Transmission using the AAL-1 protocol is in one of two modes: unstructured data transfer (UDT) and structured data transfer (SDT). The UDT mode is provided for data streams in which boundaries need not be preserved. The SDT mode is designed for messages where message boundaries must be preserved.

The CS sublayer detects lost and misinserted cells that occur due to undetected errors in the virtual path or channel identification. It also controls incoming traffic to ensure transmission at a constant rate. This sublayer also converts the input messages or streams into 46–47 byte segments to be used by the SAR sublayer.

The SAR sublayer has a 1-byte protocol header. The convergence sublayer indicator (CSI) of the odd-numbered cells forms a data stream that provides a 4-bit synchronous residual time stamp (SRTS) used for clock synchronization in SDT mode.[18] The timing information is essential for the synchronization of multiple media streams as well as for the prevention of buffer overflow and underflow in the decoder. The sequence count

(SC) is a modulo-8 counter used to detect missing or misinserted cells. The CSI and SC fields are protected by the cyclic redundancy check (CRC) field. An even parity (P) bit covering the protocol header affords additional protection of the CSI and SC fields. The AAL-1 SAR sublayer protocol header is depicted in Fig. 2. A corresponding glossary of the AAL-1 SAR sublayer protocol header is provided in Table 1.

An additional 1-byte pointer field is used on every even numbered cell in the STD mode.[19] The pointer field is a number in the range of 0–92 used to indicate the offset of the start of the next message, either in its own cell or the one following it in order to preserve message boundaries. This approach allows messages to be arbitrarily long and need not align on cell boundaries. In this presentation, however, we shall restrict ourselves to operation in the UDT mode for data streams in which boundaries need not be preserved and the pointer field will be omitted.

As we have already indicated, the MPEG-2 systems layer consists of 188-byte fixed-length TS packets. The CS sublayer directly segments each of the MPEG-2 TS packets into four 47-byte fixed-length AAL-1 SAR payloads. This approach is used when the cell loss ratio (CLR) that is provided by the ATM layer is satisfactory.

An alternative optional approach is used in noisy environments to improve reliability by the use of interleaved forward error correction (FEC); i.e., Reed-Solomon (128,124). The CS sublayer groups a sequence of 31 distinct 188-byte fixed-length MPEG-2 TS packets. This group is used to form a matrix written in standard format (row-by-row) of 47 rows and 124 bytes in each row. Four bytes of the FEC are appended to each row. The resulting matrix is composed of 47 rows and 128 bytes in each row. This matrix is forwarded to an interleaver that reads the matrix in transposed format (column by column) for transmission to the SAR sublayer. The interleaver ensures that a cell loss would be limited to the loss of a single byte in each row, which can be recovered by the FEC. A mild delay equivalent

---

[16] Classes C and D were originally used for the representation of NRT, VBR, CO, and connectionless services handled by AAL-3 and AAL-4, respectively. These protocols, however, were so similar — differing only in the presence or absence of a multiplexing header field — that they eventually decided to merge them into a single protocol provided by AAL-3/4.

[17] A new protocol AAL-5 — originally named simple efficient adaptation layer (SEAL) — was proposed by the computer industry as an alternative to the previously existing protocol AAL-3/4, which was presented by the telecommunications industry.

[18] The SRTS method encodes the frequency difference between the encoder clock and the network clock for synchronization of the encoder and receiver clock in the asynchronous service clock operation mode, despite the presence of delay jitter.

TABLE 1 AAL-1 SAR-PDU header glossary

Abbrev.	Function
CSI	convergence sublayer indicator
SC	sequence count
CRC	cyclic redundancy check
P	parity (even)

[19] The high-order bit of the pointer field is currently unspecified and reserved for future use.

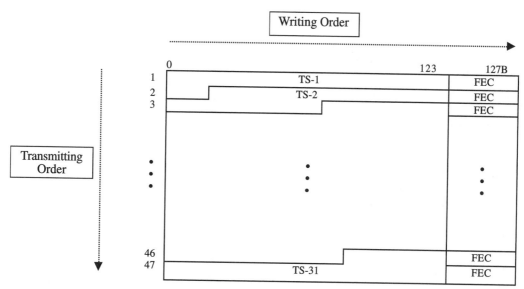

FIGURE 3   Interleaved transport stream (FEC).

to the processing of 128 cells is introduced by the matrix formation at the transmitter and the receiver. An illustration of the formation of the interleaved FEC TS packets is depicted in Fig. 3.

Whether the interleaved FEC of the TS packets is implemented or direct transmission of the TS packets is used, the AAL-1 SAR sublayer receives 47-byte fixed-length payloads that are appended by the 1-byte AAL-1 SAR protocol header to form 48-byte fixed-length packets. These packets serve as payloads of the ATM cells and are attached to the 5-byte ATM headers to comprise the 53-byte fixed-length ATM cells. An illustration of

the mapping of MPEG-2 systems layer TS packets into ATM cells using the AAL-1 protocol is depicted in Fig. 4.

4.2 AAL-5: ATM Application Layer 5

The AAL-5 protocol is used for NRT, VBR, CO, traffic. This protocol also offers the option of reliable and unreliable services.

The CS sublayer protocol is composed of a variable-length payload of length not to exceed 65,535 bytes and a variable-length trailer of length 8–55 bytes. The trailer consists of a padding (P)

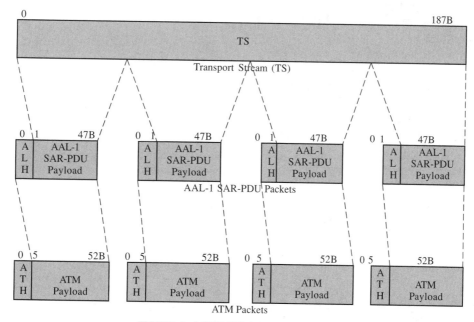

FIGURE 4   MPEG-2 TS AAL-1 PDU mapping.

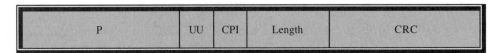

FIGURE 5　AAL-5 CPCS-PDU trailer.

field of length 0–47 bytes chosen to make the entire message — payload and trailer — be a multiple of 48 bytes. The user-to-user (UU) direct information transfer field is available for higher layer applications (e.g., multiplexing). The common part indicator (CPI) field designed for interpretation of the remaining fields in the CS protocol is currently not in use. The length field provides the length of the payload (not including the padding field). The standard 32-bit CRC field is used for error checking over the entire message — payload and trailer. This error checking capability allows for the detection of missing or misinserted cells without using sequence numbers. An illustration of the AAL-5 CPCS protocol trailer is depicted in Fig. 5. A corresponding glossary of the AAL-5 CPCS protocol trailer is provided by Table 2.

The SAR sublayer simply segments the message into 48-byte units and passes them to the ATM layer for transmission. It also informs the ATM layer that the ATM user-to-user (AAU) bit in the payload type indicator (PTI) field of the ATM cell header must be set on the last cell in order to preserve message boundaries.[20]

Encapsulation of a single MPEG-2 systems layer 188-byte fixed-length TS packet in one AAL-5 CPCS packet would introduce a significant amount of overhead because of the size of the AAL-5 CPCS trailer protocol. The transmission of a single TS packet using this approach to the implementation of the AAL-5 protocol would require five ATM cells in comparison to the four ATM cells needed with the AAL-1 protocol. More than one TS packets must be encapsulated in a single AAL-5 CPCS packet in order to reduce the overhead.

The encapsulation of more than one TS packets in a single AAL-5 CPCS packet is associated with an inherent packing jitter. This will manifest itself as delay variation in the decoder and may affect the quality of the systems clock recovered when one of the TS packets contains a PCR. For this problem to be alleviated,

the number of TS packets encapsulated in a single AAL-5 CPCS packet should be minimized.[21]

The preferred method adopted by the ATM Forum is based on the encapsulation of two MPEG-2 systems layer 188-byte TS packets in a single AAL-5 CPCS packet. The AAL-5 CPCS packet payload consequently occupies 376 bytes. The payload is appended to the 8-byte AAL-5 CPCS protocol trailer (no padding is required) to form a 384-byte AAL-5 CPCS packet. The AAL-5 CPCS packet is segmented into exactly eight 48-byte AAL-5 SAR packets, which serve as payloads of the ATM cells and are attached to the 5-byte ATM headers to comprise the 53-byte fixed-length ATM cells. An illustration of the mapping of two MPEG-2 systems layer TS packets into ATM cells using the AAL-5 protocol is depicted in Fig. 6.

The overhead requirements for the encapsulation of two TS packets in a single AAL-5 CPCS packet are identical to the overhead needed with the AAL-1 protocol — both approaches map two TS packets into eight ATM cells. This approach to the implementation of the AAL-5 protocol is currently the most popular method for mapping MPEG-2 systems layer TS packets into ATM cells.

5 Image and Video Internetworks

Introduction

A critical factor in our ability to provide worldwide multimedia communication is the expansion of the existing bandwidth of the Internet. The NSF has recently restructured its data networking architecture by providing the very high speed Backbone Network Service (vBNS). The vBNS currently employs ATM switches and OC-12c SONET fiber optic communications at data rates of 622 Mbps.

The vBNS Multicast Backbone (MBONE), a worldwide digital radio and television service on the Internet, was developed in 1992. MBONE is used to provide global digital multicast real-time audio and video broadcast via the Internet. The multicast process is intended to reduce the bandwidth consumption of the Internet.

MBONE is a virtual overlay network on top of the Internet. It consists of islands that support multicast traffic and tunnels that are used to propagate MBONE packets between these

TABLE 2　AAL-5 CPCS-PDU trailer glossary

Abbrev.	Function
P	padding
UU	user-to-user direct information transfer
CPI	common part indicator field
Length	length of payload
CRC	cyclic redundancy check

---

[20]Note that this approach is in violation of the principles of the open architecture protocol standards — the AAL layer should not invoke decisions regarding the bit pattern in the header of the ATM layer.

[21]An alternative solution to the packing jitter problem, known as PCR-aware packing, requires that TS packets containing a PCR appear in the last packet in the AAL-5 CPCS packet. This approach is rarely used because of the added hardware complexity in detecting TS packets with a PCR.

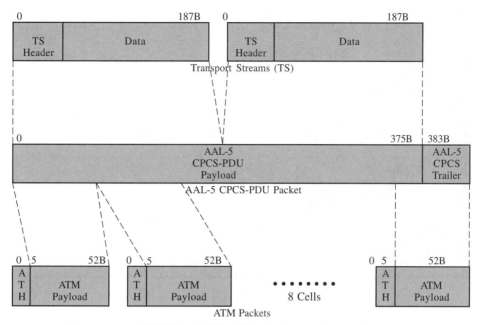

FIGURE 6 MPEG-2 TS AAL-5 PDU mapping.

islands. The islands are interconnected using mrouters (multicast routers), which are logically connected by tunnels.

The Multicast Internet Protocol (IP) was adopted as the standard protocol for multicast applications on the Internet. MBONE packets are transmitted as multicast IP packets between mrouters in different islands. Multicast IP packets are encapsulated within ordinary IP packets and regarded as standard unicast data by ordinary routers along a tunnel.

MBONE applications such as multimedia data broadcasting do not require reliable communication or flow control. These applications do require, however, real-time transmission over the Internet. The loss of an audio or video packet will not necessarily degrade the broadcast quality. Significant jitter delay, in contrast, cannot be tolerated. The user datagram protocol (UDP) — not the transmission control protocol (TCP) — is consequently used for transmission of multimedia traffic.

The UDP is an unreliable connectionless protocol for applications such as audio and video communications that require prompt delivery rather than accurate delivery and flow control. The UDP is restricted to an 8-byte header that contains the source and destination ports, the length of the packet, and an optional checksum over the entire packet.

In this section, an overview of the protocols used for image and video communications over the Internet is presented. For brevity, this presentation will focus exclusively on the MPEG-2 compression standard.

5.1 RTP: Real-Time Transport Protocol

The RTP provides end-to-end network transport functions for the transmission of real-time data such as audio or video over unicast or multicast services independent of the underlying network or transport protocols. Its functionality, however, is enhanced when run on top of the UDP. It is also assumed that resource reservation and quality of service have been provided by lower layer services (e.g., RSVP). The RTP protocol, however, does not assume nor provide guaranteed delivery or packet order preservation.

RTP services include time-stamp packet labeling for media stream synchronization, sequence numbering for packet loss detection, and packet source identification and tracing.

RTP is designed to be a flexible protocol that can be used to accommodate the detailed information required by particular applications. The RTP protocol is, therefore, deliberately incomplete and its full specification requires one or more companion documents: profile specification and payload format specification. The profile specification document defines a set of payload types and their mapping to payload formats. The payload format specification document defines the method by which particular payloads are carried.

The RTP protocol supports the use of intermediate system relays known as translators and mixers. Translators convert each incoming data stream from different sources separately. An example of a translator is used to provide access to an incoming audio or video packet stream beyond an application-level firewall. Mixers combine the incoming data streams from different sources to form a single stream. An example of a mixer is used to resynchronize an incoming audio or video packet stream from high-speed networks to a lower-bandwidth packet stream for communication across low-speed networks.

An illustration of the RTP packet header is depicted in Fig. 7. A corresponding glossary of the RTP packet header is provided in Table 3. The version number of the RTP is defined in the version (V) field. The version number of the current RTP is

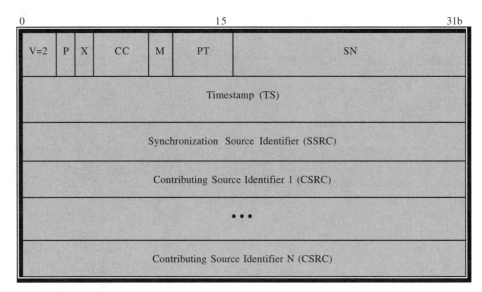

FIGURE 7 RTP packet header.

number 2.[22] A padding (P) bit is used to indicate if additional padding bytes, which are not part of the payload, have been appended at the end of the packet. The last byte of the padding field provides the length of the padding field. An extension (X) bit is used to indicate if the fixed header is followed by a header extension. The contributing source count (CC) provides the number (up to 15) of contributing source (CSRC) identifiers that follow the fixed header. A marker (M) bit is defined by a profile for various applications such as the marking of frame boundaries in the packet stream. The payload type (PT) field provides the format and interpretation of the payload. The mapping of the PT code to payload formats is specified by a profile. An incremental sequence number (SN) is used by the receiver to detect packet loss and restore packet sequence. The initial value of the SN is random in order to combat possible attacks on encryption. The time stamp provides the sampling instant of the first byte in the packet derived from a monotonically and linearly incrementing clock for synchronization and jitter delay estimation. The clock frequency is indicated by the profile or payload format specification. The initial value of the time stamp is once again random.

TABLE 3 RTP packet header glossary

Abbrev.	Function
V	version
P	padding
X	extension
CC	contibuting source count
M	marker
PT	payload type
SN	sequence number
TS	time stamp
SSRC	synchronization source identifier
CSRC	contributing source identifier

[22]Version numbers 0 and 1 have been used in previous versions of the RTP.

The synchronization source (SSRC) field is used to identify the source of a stream of packets from a synchronization source. A translator forwards the stream of packets while preserving the SSRC identifier. A mixer, on the other hand, becomes the new synchronization source and must therefore include its own SSRC identifier. The SSRC field is chosen randomly in order to prevent two synchronization sources from having the same SSRC identifier in the same session. A detection and collision resolution algorithm prevents the possibility that multiple sources will select the same identifier. The contributing source (CSRC) field designates the source of a stream of packets that has contributed to the combined stream, produced by a mixer, in the payload of this packet. The CSRC identifiers are inserted by the mixer and correspond to the SSRC identifiers of the contributing sources. As indicated earlier, the CC field provides the number (up to 15) of contributing sources.

Numerous options for the augmentation of the RTP protocol for various applications have been proposed. An important proposal for the generic forward error correction (FEC) data encapsulation in RTP packets has been presented in [2].

The most popular current video compression standards are based on MPEG. RTP payload encapsulation of MPEG data streams can be accomplished in one of two formats: systems stream — transport stream and program stream — as well as elementary stream. The format used for encapsulation of MPEG systems stream is designed for maximum interoperability with video communication network environments. The format used for the encapsulation of MPEG systems stream, however, provides greater compatibility with the Internet architecture including other RTP encapsulated media streams and current efforts in conference control.[23]

[23]RTP payload encapsulation of MPEG elementary stream format defers some of the issues addressed by the MPEG systems stream to other protocols proposed by the Internet community.

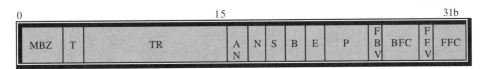

FIGURE 8 RTP MPEG ES video-specific header.

The RTP header for encapsulation of MPEG SS is set as follows: The payload type (PT) field should be assigned to correspond to the systems stream format in accordance with the RTP profile for audio and video conferences with minimal control [7]. The marker (M) bit is activated whenever the time stamp is discontinuous. The time-stamp field provides the target transmission time of the first byte in the packet derived from a 90 KHz clock reference, which is synchronized to the system stream PCR or system clock reference (SCR). This time-stamp is used to minimize network jitter delay and synchronize relative time drift between the sender and receiver. The RTP payload must contain an integral number of MPEG-2 transport stream packets — there are no restrictions imposed on MPEG-1 systems stream or MPEG-2 program stream packets.

The RTP header for encapsulation of MPEG ES is set as follows: the payload type (PT) field should once again be assigned to correspond to the elementary stream format in accordance with the RTP profile for audio and video conferences with minimal control [7]. The marker (M) bit is activated whenever the RTP packet contains an MPEG frame end code. The time-stamp field provides the presentation time of the subsequent MPEG picture derived from a 90-KHz clock reference, which is synchronized to the system stream program clock reference or system clock reference.

The RTP payload encapsulation of MPEG ES format requires that an MPEG ES video-specific header follow each RTP packet header. The MPEG ES video-specific header contains a must be zero (MBZ) field that is currently unused and must be set to zero. An indicator (T) bit is used to announce the presence of an MPEG-2 ES video-specific header extension following the MPEG ES video-specific header. The temporal reference (TR) field provides the temporal position of the current picture within the current group of pictures (GOP). The active N (AN) bit is used for error resilience and is activated when the following indicator (N) bit is active. The new picture header (N) bit is used to indicate parameter changes in the picture header information for MPEG-2 payloads.[24] A sequence header present (S) bit indicates the occurrence of an MPEG sequence header. A beginning of slice (B) bit indicates the presence of a slice start code at the beginning of the packet payload, possibly preceded by any combination of a video sequence header, group of pictures (GOP) header, and picture header. An end of slice (E) bit indicates that the last byte of the packet payload is the end of a slice. The picture type (PT) field specifies the picture type — I picture, P picture,

B picture, or D picture. The full pel backward vector (FBV), backward f code (BFC), full pel forward vector (FFV), and forward f code (FFC) fields are used to provide information necessary for determination of the motion vectors.[25] Figure 8 and Table 4 provide an illustration and corresponding glossary of the RTP MPEG ES video-specific header, respectively.

An illustration of the RTP MPEG-2 ES video-specific header extension is depicted in Fig. 9. A corresponding glossary used to summarize the function of the RTP MPEG-2 ES video-specific header extension is provided in Table 5. Particular attention should be paid to the composite display flag (D) bit, which indicates the presence of a composite display extension — a 32-bit extension that consists of 12 zeros followed by 20 bits of composite display information — following the MPEG-2 ES video-specific header extension. The extension (E) bit is used to indicate the presence of one or more optional extensions — quantization matrix extension, picture display extension, picture temporal scalable extension, picture spatial scalable extension, and copyright extension — following the MPEG-2 ES video-specific header extension as well as the composite display extension. The first byte of each of these extensions is a length (L) field that provides the number of 32-bit words used for the extension. The extensions are self-identifying since they must also include the extension start code (ESC) and the extension start code ID (ESCID). For additional information regarding the remaining fields in the MPEG-2 ES video-specific header extension refer to the MPEG-2 video compression standard.

TABLE 4 RTP MPEG ES video-specific header glossary

Abbrev.	Function
MBZ	must be zero
T	video-specific header extension
TR	temporal reference
AN	active N
N	new picture header
S	sequence header present
B	beginning of slice
E	end of slice
P	picture type
FBV	full pel backward vector
BFC	backward F code
FFV	full pel forward vector
FFC	forward F code

---

[24]The active N and new picture header indicator bits must be set to 0 for MPEG-1 payloads.

[25]Only the FFV and FFC fields are used for P pictures; none of these fields are used for I pictures and D pictures.

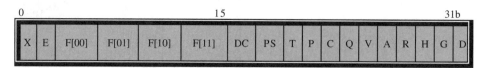

| 0 | | | | | 15 | | | | | | | | | | 31b |

| X | E | F[00] | F[01] | F[10] | F[11] | DC | PS | T | P | C | Q | V | A | R | H | G | D |

FIGURE 9   RTP MPEG-2 ES video-specific header extension.

The RTP payload encapsulation of MPEG ES format fragments the stream into packets such that the following headers must appear hierarchically at the beginning of a single payload of an RTP packet: MPEG video sequence header, MPEG GOP header, and MPEG picture header. The beginning of a slice — the fundamental unit of recovery — must be the first data (not including any MPEG ES headers) or must follow an integral number of slices in the payload of an RTP packet.

Efforts have also been devoted to the encapsulation of other video compression standards (e.g., Motion JPEG and MPEG-4).

5.2 RTCP: Real-Time Transport Control Protocol

The RTCP augments the RTP protocol to monitor the quality of service and data delivery monitoring as well as provide minimal control and identification capability over unicast or multicast services independent of the underlying network or transport protocols. The primary function of the RTCP protocol is to provide feedback on the quality of data distribution that can be used for flow and congestion control. The RTCP protocol is also used for the transmission of a persistent source identifier to monitor the participants and associate related multiple data streams from a particular participant. The RTCP packets are sent to all participants in order to estimate the rate at which control packets are sent. An optional function of the RTCP protocol can be used to convey minimal session control information.

The implementation of the RTCP protocol is based on the periodic transmission to all participants in the session of control information in several packet types summarized in Table 6. The sender report (SR) and receiver report (RR) provide reception quality feedback and are identical except for the additional sender information that is included for use by active senders. The SR or RR packets are issued depending on whether a site has sent any data packets during the interval since the last two reports were issued. The source description item (SDES) includes items such as the canonical end-point identifier (CNAME), user name (NAME), electronic mail address (EMAIL), phone number (PHONE), geographic user location (LOC), application or tool name (TOOL), notice/status (NOTE), and private extensions (PRIV). The end of participation (BYE) packet indicates that a source is no longer active. The application specific functions (APP) packet is intended for experimental use as new applications and features are developed.

RTCP packets are composed of an integral number of 32-bit structures and are, therefore, stackable; multiple RTCP packets may be concatenated to form compound RTCP packets. RTCP packets must be sent in compound packets containing at least two individual packets of which the first packet must always be a report packet. Should the number of sources for which reports are generated exceed 31 — the maximal number of sources that can be accommodated in a single report packet — additional RR packets must follow the original report packet. An SDES packet containing a CNAME item must also be included in each compound packet. Other RTCP packets may be included subject to bandwidth constraints and application requirements in any order, except that the BYE packet should be the last packet sent in a given session. These compound RTCP packets are forwarded to the payload of a single packet of a lower layer protocol (e.g., UDP).

An illustration of the RTCP SR packet is depicted in Fig. 10. A corresponding glossary of the RTCP SR packet is provided in Table 7. The RTCP SR and RR packets are composed of a header section, zero or more reception report blocks, and a possible profile-specific extension section. The SR packets also contain an additional sender information section.

TABLE 5   RTP MPEG-2 ES video-specific header extension glossary

Abbrev.	Function
X	unused (Zero)
E	extension
F[00]	forward horizontal F code
F[01]	forward vertical F code
F[10]	backward horizontal F code
F[11]	backward vertical F code
DC	intra DC Precision (intra macroblock DC difference value)
PS	picture structure (field/frame)
T	top field first (odd/even lines first)
P	frame predicted frame DCT
C	concealment motion vectors (I picture exit)
Q	Q-scale type (quantization table)
V	intra VLC format (Huffman code)
A	alternate scan (section/interlaced field breakup)
R	repeat first field
H	chroma 420 type (options also include 422 and 444)
G	progressive frame
D	composite display flag

TABLE 6   RTCP packet types

Abbrev.	Function
SR	sender report
RR	receiver report
SDES	source description item (e.g., CNAME)
BYE	end of participation indication
APP	application specific functions

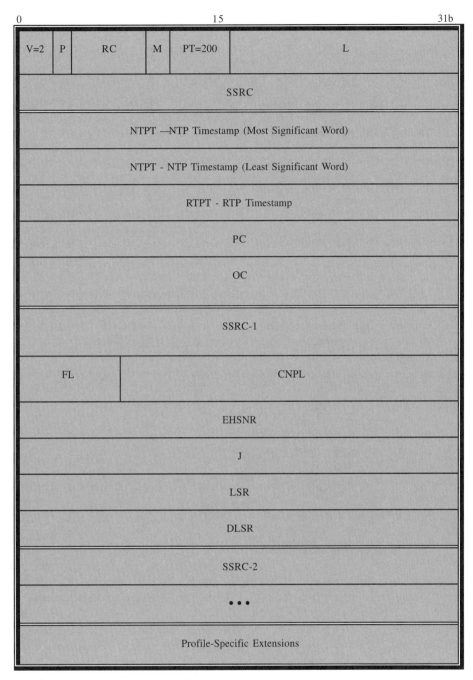

FIGURE 10 RTCP sender report packet.

The header section defines the version number of the RTCP protocol in the version (V) field. The version number of the current RTCP protocol is number 2 — the same as the version number of the RTP protocol. A padding (P) bit is used to indicate if additional padding bytes, which are not part of the control information, have been appended at the end of the packet. The last byte of the padding field provides the length of the padding field. In a compound RTCP packet, padding should only be required on the last individual packet. The reception report count (RC) field provides the number of reception report blocks contained in the packet. The packet type (PT) field contains the constant 200 and 201 to identify the packet as a sender report (SR) and receiver report (RR) RTCP packet, respectively. The length (L) field provides the number of 32-bit words of the entire RTCP packet — including the header and possible padding — minus one. The synchronization source (SSRC) field is used to identify the sender of the report packet.

The sender information section appears in the sender report packet exclusively and provides a summary of the data transmission from the sender. The network time protocol time stamp

TABLE 7 RTCP sender report packet glossary

Abbrev.	Function
V	version
P	padding
RC	reception report count
PT	packet type
L	length
SSRC	synchronization source identifier (sender)
NTPT	network time protocol time stamp
RTPT	real-time transport protocol time stamp
PC	packet count (sender)
OC	octet count (sender)
SSRC-N	synchronization source identifier-N
FL	fraction lost
CNPL	cumulative number of packets lost
EHSNR	extended highest sequence number received
J	interarrival jitter
LSR	last sender report time stamp
DLSR	delay since last sender report time stamp

(NTPT) indicates the wallclock time at that instant the report was sent.[26] This time stamp along with the time stamps generated by other reports is used to measure the round-trip propagation to the other receivers. The real-time protocol time stamp (RTPT) corresponds to the NTPT provided using the units and random offset used in the RTP data packets. This correspondence can be used for synchronization among sources whose NTP time stamps are synchronized. The packet count (PC) field indicates the total number of RTP data packets transmitted by the sender since the beginning of the session up until the generation of the SR packet. The octet count (OC) field represents the total number of bytes in the payload of the RTP data packets — excluding header and padding — transmitted by the sender since the beginning of the session up until the generation of the SR packet. This information can be used to estimate the average payload data rate.

All RTCP report packets must contain zero or more reception report blocks corresponding to the number of synchronization sources from which the receiver has received RTP data packets since the last report. These reception report blocks convey statistical data pertaining to the RTP data packets received from a particular synchronization source. The synchronization source (SSRC-N) field is used to identify the Nth synchronization source to which the statistical data in the Nth reception report block is attributed. The fraction lost (FL) field indicates the fraction of RTP data packets from the Nth synchronization source lost since the previous report was sent. This fraction is defined as the number of packets lost divided by the number of packets expected (NPE). The cumulative number of packets lost (CNPL) field provides the total number of RTP data packets from the Nth synchronization source lost since the beginning of

the session. The CNPL is defined as the number of packets expected (NPE) less the number of packets received. The extended highest sequence number received (EHSNR) field contains the highest sequence number of the RTP data packets received from the Nth synchronization source stored in the 16 least significant bits of the EHSNR field. In contrast, the extension of the sequence number provided by the corresponding count of sequence number cycles is maintained and stored in the 16 most significant bits of the EHSNR field. The EHSNR is also used to estimate the number of packets expected, which is defined as the last EHSNR less the initial sequence number received. The interarrival jitter (J) field provides an estimate of the statistical variance of the interarrival time of the RTP data packets from the Nth synchronization source. The interarrival jitter (J) is defined as the mean deviation of the interarrival time D between the packet spacing at the receiver compared with the sender for a pair of packets; i.e., $D(i, j) = (R(j) - R(i)) - (S(j) - S(i))$, where $S(i)$ and $R(i)$ are used to denote the RTP time stamp from the RTP data packet i and the time of arrival in RTP time-stamp units of RTP data packet i, respectively. The interarrival time D is equivalent to the difference in relative transit time for the two packets; i.e., $D(i, j) = (R(j) - S(j)) - (R(i) - S(i))$. An estimate of the interarrival jitter (J) is obtained by the first-order approximation of the mean deviation given by

$$ J = J + \frac{1}{16}[|D(i, i-1)| - J]. $$

The estimate of the interarrival jitter (J) is computed continuously as each RTP data packet is received from the Nth synchronization source and sampled whenever a report is issued. The last sender report time stamp (LSR) field provides the NPT time stamp (NTPT) received in the most recent RTCP sender report (SR) packet that arrived from the Nth synchronization source. The LSR field is confined to the middle 32 bits out of the 64-bit NTP time stamp (NTPT). The delay since last sender report (DLSR) expresses the delay between the time of the reception of the most recent RTCP sender report packet that arrived from the Nth synchronization source and sending the current reception report block. These measures can be used by the Nth synchronization source to estimate the round-trip propagation delay (RTPD) between the sender and the Nth synchronization source. The estimate of the RTPD obtained provided the time of arrival T of the reception report block from the sender is recorded at the Nth synchronization source is given by RTPD = T − LSR − DLSR.

6 Image and Video Wireless Networks

Wireless networks were until recently primarily devoted to paging as well as real-time speech communications. First generation wireless communication networks were analog systems. The most widely used analog wireless communication network

---

[26]The wallclock time (absolute time) represented with the Network Time Protocol time-stamp format is a 64-bit unsigned fixed-point number provided in seconds relative to 0h Universal Time Clock (UTC) on January 1, 1900.

is known as the Advanced Mobile Phone Service (AMPS).[27] The AMPS system is based on frequency-division multiple access (FDMA) and uses 832 30-kHz transmission channels in the range of 824–849 MHz and 832 30-kHz reception channels in the range of 869–894 MHz.

Second-generation wireless communication networks are digital systems based on two approaches: time-division multiple access (TDMA) and code-division multiple access (CDMA). Among the most common TDMA wireless communication networks are the IS-54 and IS-136, as well the Global Systems for Mobile communications (GSM). The IS-54 and IS-136 are dual mode (analog and digital) systems that are backward compatible with the AMPS system.[28] In IS-54 and IS-136, the same 30-kHz channels are used to accommodate three simultaneous users (six time slots) for transmission at data rates of approximately 8 kbps. The GSM system originated in Europe is, in contrast, a pure digital system based on both FDMA and TDMA. It consists of 50 200-kHz bands in the range of 900 MHz used to support eight separate connections (eight time slots) for transmission at data rates of 13 kbps.[29]

The second approach to digital wireless communication networks is based on CDMA. The origins of CDMA are based on spread-spectrum methods that date back to secure military communication applications during the Second World War.[30] The CDMA approach uses direct-sequence spread-spectrum (DSSS), which provides for the representation of individual bits by pseudo-random chip sequences. Each station is assigned a unique orthogonal pseudo-random chip sequence. The original bits are recovered by determining the correlation (inner product) of the received signal and the pseudo-random chip sequence corresponding to the desired station. The current CDMA wireless communication network is specified in IS-95.[31] In IS-95 the channel bandwidth of 1.25 MHz is used for transmission at data rates of 8 kbps or 13 Kbps.

Preliminary plans have proposed for the implementation of the third-generation wireless communication networks in the International Mobile Communications-2000 (IMT-2000). The motivation of IMT-2000 is to expand mobile communications to multimedia applications as well as to provide access to existing networks (e.g., ATM and Internet). This is accomplished by providing circuit and packet switched channel data connection as well as larger bandwidth used to support much higher data rates. The focus of IMT-2000 is on the integration of several technologies: CDMA-2000, Wideband CDMA (W-CDMA), and Universal Wireless Communications-136 (UWC-136).

The CDMA-2000 is designed to be a wideband synchronous intercell CDMA based network using the frequency-division duplex (FDD) mode and is backward compatible with the existing CDMA-One (IS-95). The CDMA-2000 channel bandwidth planned for the first phase of the implementation will be restricted to 1.25 MHz and 3.75 MHz for transmission at data rates of up to 1 Mbps. The CDMA-2000 channel bandwidth will be expanded during the second phase of the implementation to also include 7.5 MHz, 11.25 MHz, and 15 MHz for transmission that will support data rates that could possibly exceed 2.4 Mbps.

The W-CDMA is a wideband asynchronous intercell CDMA (with some TDMA options) based network that provides for both frequency-division duplex and time-division duplex (TDD) operations. The W-CDMA is backward compatible with the existing GSM. The W-CDMA channel bandwidth planned for the initial phase of the implementation is 5 MHz for transmission at data rates of up to 480 kbps. The W-CDMA channel bandwidth planned for a later phase of the implementation will reach 10 MHz and 20 MHz for transmission that will support data rates of up to 2 Mbps.

The UWC-136 is envisioned to be an asynchronous intercell TDMA based system that permits both frequency-division duplex and time-division duplex modes. The UWC-136 is backward compatible with the current IS-136 and provides possible harmonization with GSM. The UWC-136 is a unified representation of IS-136+ and IS-136 High Speed (IS-136 HS). The IS-136+ will rely on the currently available channel bandwidth of 30 kHz, for transmission at data rates of up to 64 Kbps. The IS-136 HS outdoor (mobile) channel bandwidth will be 200 kHz for transmission at data rates of up to 384 kbps, whereas the IS-136 HS indoor (immobile) channel bandwidth will be expanded to 1.6 MHz for transmission that will support data rates of up to 2 Mbps.

The larger bandwidth and significant increase in data rates supported by the various standards in IMT-2000 will facilitate image and video communication over wireless networks. Moreover, the packet switched channel data connection option provided by the various standards in IMT-2000 will allow for the implementation of many of the methods and protocols discussed in the previous sections over wireless communication networks (e.g., RTP).

7 Summary

In this presentation we have provided a broad overview of image and video communication networks. The fundamental image and video compression standards — JPEG and MPEG — were briefly discussed. The compression stream standards associated with the most popular video compression standard —MPEG-2 — were presented. These compression stream standards were subsequently mapped to various adaptation layers —AAL-1 and AAL-5 — of ATM communication networks. A comprehensive discussion of ATM communication networks must extend to

---

[27] The AMPS system is also known as TACS and MCS-L1 in England and Japan, respectively.

[28] The Japanese JDC system is also a dual mode (analog and digital) system that is backward compatible with the MCS-L1 analog system.

[29] The implementation of the GSM system in the range of 1.8 GHz is known as DCS-1800.

[30] In 1940, the actress Hedy Lamarr, at the age of 26, invented a form of spread spectrum, known as the frequency-hopping spread spectrum (FHSS).

[31] The IS-95 standard has recently been referred to as CDMA-One.

other image and video compression standards (e.g., MPEG-4). A broader topic addressing the issue of image and video communication over the Internet was discussed next.

Transport layer protocols — RTP and RTCP — that are essential for efficient and reliable image and video communication over the Internet were illustrated. Some complementary protocols in various stages of development were omitted for brevity. For instance, the resource reservation protocol (RSVP) is used to provide an integrated service resource reservation and quality of service control. Another example is the real-time streaming protocol used as an application level protocol that provides for the on-demand control over the delivery of real-time data. A more recent example is the advanced streaming format (ASF) used to provide interoperability through the standardization of a multimedia presentation file format. Other important developments in the effort to facilitate image and video communications over the Internet provided by various session layer protocols — session announcement protocol (SAP), session initiation protocol (SIP), and session description protocol (SDP) — were also omitted from this presentation. The final discussion pertained to the future implementation of image and video communications over wireless networks. The entirety of this presentation points to the imminent incorporation of a variety of multimedia applications into a seamless nested array of wireline and wireless communication networks.

References

[1] V. Garg and J. E. Wilkes, *Wireless and Personal Communication Systems* (Prentice-Hall, Englewood Cliffs, NJ, 1996).

[2] J. Rosenberg and H. Schulzrinne, Internet Draft draft-ietf-avt-fec-04, An RTP Payload Format for Generic Forward Error Correction, 1998.

[3] V. Kumar, *MBONE: Interactive Multimedia on the Internet* (New Riders, 1996).

[4] M. Orzessek and P. Sommer, *ATM & MPEG-2: Integrating Digital Video into Broadband Networks* (Hewlett-Packard Professional Books, location, 1998).

[5] T. S. Rappaport, *Wireless Communications: Principles & Practice* (Prentice-Hall and IEEE, New York, 1996).

[6] H. Schulzrinne, S. Casner, R. Frederick, and Y. Jacobson, "RTP: a transport protocol for real-time applications," RFC 1889, 1996.

[7] H. Schulzrinne, "RTP profile for audio and video conferences with minimal control," RFC 1890, 1996.

[8] D. Hoffman, G. Fernando, V. Goyal, and M. Civanlar, "RTP payload format for MPEG-1/MPEG-2 video," RFC 2250, 1998.

[9] L. Berc, W. Fenner, R. Frederick, S. McCanne, and P. Stewart, "RTP payload format for JPEG-compressed video," RFC 2435, 1998.

[10] W. Stallings, *Data and Computer Communications* (Prentice-Hall, Englewood Cliffs, NJ, 1997).

[11] R. Steinmetz and K. Nahrstedt, *Multimedia: Computing, Communications, and Applications* (Prentice-Hall, Englewood Cliffs, NJ, 1995).

[12] A. S. Tanenbaum, *Computer Networks* (Prentice-Hall, Englewood Cliffs, 1996).

[13] E. K. Wesel, *Wireless Multimedia Communications* (Addison-Wesley, Reading, MA, 1997).

[14] C.-H. Wu and J. D. Irwin, *Emerging Multimedia Computer Communication Technologies* (Prentice-Hall, Englewood Cliffs, NJ, 1998).

Image Watermarking for Copyright Protection and Authentication

George Voyatzis and
Ioannis Pitas
University of Thessaloniki

1 Introduction

The concepts of authenticity and copyright protection are of major importance in the framework of our information society. For example, TV channels usually place a small visible logo on the image corner (or a wider translucent logo) for copyright protection. In this way, unauthorized duplication is discouraged and the recipients can easily identify the video source. Official scripts are stamped or typed on watermarked papers for authenticity proof. Bank notes also use watermarks for the same purpose, which are very difficult to reproduce by conventional photocoping techniques. The above mentioned logos, patterns, and drawings are familiar examples of *visible watermarks*.

Nowadays, digital technology is rapidly replacing traditional techniques for information transmission, processing, and storage. Producers and customers find it very convenient to use digital images, video and audio, and multimedia products, and they are proving to be a revolutionary way for demonstrating information. A great number of tools and computer applications are available for producing and manipulating digital products.

However, at the same time, methods for piracy are becoming more powerful because duplications, forgery, and illegal retransmissions are easier than ever. Visible watermarks can also be applied to protect digital products in the traditional way. However, their contribution to copyright and authenticity protection is rather insufficient. Modern digital processing techniques can be used maliciously in order to remove or replace a visible watermark. In order to overcome such a problem, *invisible digital watermarks* or *invisible digital stamps* have been proposed [1–3]. A great number of various watermarking techniques have been presented in the literature. However, the problem of creating an efficient and robust watermarking system is still open.

In the following sections we present the basic concepts of invisible watermarking techniques applied on digital images. The presented watermark definitions, properties, and basic algorithms form a general watermarking framework. We note that invisible watermarks aim at protecting either authenticity (content verification) or copyright. Some watermark properties are common to both cases, but some others are not generic [4].

2 Piracy and Protection Schemes

Although we refer to digital images, most of the watermarking concepts are applicable to any type of multimedia information, including digital video, audio, documents, and computer graphics. Digital images are mostly delivered through network services or broadcasting. Figure 1 presents an outline of such a basic network-based distribution system. We adopt the following definitions:

1. A *provider* is the person or company that has the legal rights to distribute a digital image X_0 and to guarantee its authenticity.
2. A *customer* is the recipient of a distributed digital image X. He or she is also concerned about the authenticity of X.
3. A *pirate* is the person who receives an image X in some way and proceeds to one of the following actions:
   - copyright violation: he or she creates and resells product duplicates X_D without getting the proper rights from the copyright owner
   - intentional tampering: he or she modifies X for malicious reasons by extracting or inserting new features and, afterward, proceeds to the retransmission of the tampered (nonauthentic) image X_T

In the current multimedia and computer market, a potential deterrent of malicious modifications or duplications of digital images seems very difficult. Possible solutions against piracy include cryptography, digital signatures, and digital watermarks. Figure 2 illustrates these three solutions.

2.1 Private Public Key Cryptography

In this approach, the original data are encrypted by the providers, using a cryptographic algorithm and a private key. The users can decrypt the received data by using a decryption algorithm. A necessary condition for successful decryption is the possession of an associated public key [5]. Fast implementation of encryption–decryption algorithms is highly desirable. Furthermore, the increase of data size as a result of encryption should remain within reasonable limits. The key bit length should be sufficient for preventing an encryption break. The most significant weakness of such a method is that, once the digital data are decrypted, they are directly vulnerable to piracy,

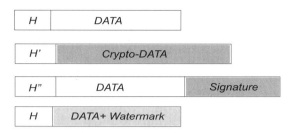

FIGURE 2   Typical data for encrypted, signed, and watermarked images.

because they are brought back to their original unprotected form.

2.2 Digital Signatures

Digital signatures have been proposed for content verification [5,6]. A digital signature is an encoded message that matches the content of a particular authentic digital image and is appended to the image data. Verification procedures are based on public algorithms and public keys. Any modification performed on the digital image data or on the signature causes verification failure. Generally, the signature size is proportional to the signed data size. Therefore, since usually images have a very large size, this scheme is not practical for their protection.

2.3 Digital Watermarks

Watermarking is related to *steganography*, which hides messages within other data for secret communication [7]. Invisible digital watermarks (or simply watermarks) are defined as small alterations of the image data. We can distinguish two watermarking schemes:

- private key watermarking for copyright protection

  1. Each provider possesses a unique private watermark key, K_{pr}.
  2. The provider alters the digital image data by using the private key and a *public* or *private* algorithm, thus producing the watermarked image, which is distributed to the customers.
  3. The provider can examine any accessible image and check for the existence of the watermark, by using a public and trustworthy detection algorithm and his or her personal private key.

- public key watermarking for content verification

  1. The provider possesses a unique private watermark key, K_{pr}, for watermark casting.
  2. Watermark casting should associate K_{pr} with a public key K_{pub}, which can demonstrate the watermark existence without disclosing the private key K_{pr}.
  3. The customer can use the particular public key K_{pub} and a public watermark detection algorithm in order to find out whether K_{pub} verifies the received digital data.

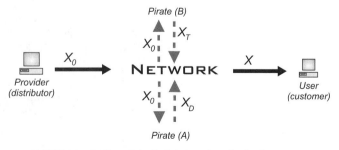

FIGURE 1   Outline of a basic digital product distribution system.

It is important to note that the private watermarking scheme aims to protect the provider, while the public scheme aims to protect the customer. The provider, who needs protection from copyright violations, is the only contributor to the first scheme. In this case a crucial point is a potential watermark removal by a pirate. Therefore, such watermarks should be very difficult to be removed by third parties. In the second scheme, the provider gives to the user the capability to verify the originality of the received data and, thus, to be protected from intentionally tampered copies. In this case, pirates do not aim at watermark removal but at reproducing the watermark in the tampered copy so as to create false authenticity proofs. We note that such watermarks are produced only by using the private key K_{pr} and should be easily destroyed when the image is modified.

Within the simplified distribution framework of Fig. 1, customers have no access to the original data. Watermarking does not affect the size of the data, as shown in Fig. 2. Although public key cryptography and public key digital signatures are feasible [5, 8], public key watermarking implementation seems to be a very difficult task. In the current stage, such watermarking is vulnerable to piracy [9]. Subsequently, the implementation of the second watermarking scheme is an open problem. However, private watermarking, which deals with fragile watermarks, can contribute to authenticity protection, e.g., in the following cases.

1. The provider exhibits his or her collection of images on an Internet server. Pirates or hackers may replace parts of the collection with nonauthentic images or may modify some of them. The provider is able at any time to examine the authenticity of the exhibited images by checking the existence of the particular watermark. When the watermark does not exist in an image, this image has been tampered with [10].
2. The provider disposes a securely accessible server, which can inform the customers about the authenticity of a questionable product through private key watermark detection.

Subsequently, we discuss private key watermarking, which has been extensively studied in the literature but is still a very hot research topic.

3 The Watermarking Framework

Watermarks can be described by digital signals defined as:

$$W = \{w(\mathbf{k}); \mid w(\mathbf{k}) \in U, \mathbf{k} \in \hat{W}^d\}, \tag{1}$$

where $\hat{W}^d$ denotes the watermark domain (grid) of dimension $d = 1, 2, 3$ for audio, still images, and video, respectively. The watermark data usually take values in the following sets:

$$U = \{0, 1\} \quad (\text{Binary [11, 12]})$$

$$U = \{-1, 1\} \quad (\text{Bipolar [13, 14]})$$

$$U = (-\alpha, \alpha) \subset I\!R \quad (\text{Gaussian noise [15, 16]}). \tag{2}$$

Sometimes, we call W "original watermark" in order to distinguish it from transformed watermarks ($W' = \mathcal{F}(W)$), which may be also used for watermark casting.

The watermarking framework can be defined as the *sixtuple* $(\mathbf{X}, \mathbf{W}, \mathbf{K}, \mathcal{G}, \mathcal{E}, \mathcal{D})$ related to the distribution system of Fig. 1.

1. $\mathbf{X}$ denotes the set of digital images X to be protected.
2. $\mathbf{W}$ is the set of the possible watermark signals defined by Eq. (1).
3. $\mathbf{K}$ is the set of watermark keys (ID numbers).
4. $\mathcal{G}$ denotes the algorithm that generates watermark signal (1) by using digital image (original or watermarked) and a key:

$$\mathcal{G} : \mathbf{X} \times \mathbf{K} \to \mathbf{W}, \quad W = \mathcal{G}(X, K). \tag{3}$$

The notation $\mathbf{A} \times \mathbf{B}$ means the Cartesian product of the sets $\mathbf{A}$ and $\mathbf{B}$.

5. $\mathcal{E}$ is the embedding algorithm that casts a watermark W in the original image X_0:

$$\mathcal{E} : \mathbf{X} \times \mathbf{W} \to \mathbf{X}, \quad X_w = \mathcal{E}(X_0, W) \tag{4}$$

X_w denotes the watermarked version of X_0.

6. Finally, $\mathcal{D}$ denotes the detection algorithm defined as follows:

$$\mathcal{D} : \mathbf{X} \times \mathbf{K} \to \{0, 1\} \tag{5}$$

$$\mathcal{D}(X, W) = \begin{cases} 1, & \text{if } W \text{ exists in } X \\ 0, & \text{otherwise} \end{cases}.$$

The overall watermark casting and detection procedures are formed by the pairs $(\mathcal{G}, \mathcal{E})$ and $(\mathcal{G}, \mathcal{D})$, respectively, and they are illustrated in Fig. 3.

The detection procedure may depend on the original image X_0 ($\mathcal{D} = \mathcal{D}(X, W; X_0)$). The use of the original in the watermark

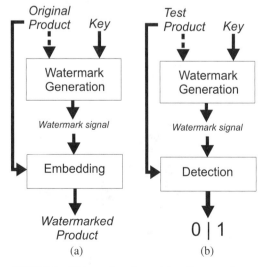

FIGURE 3 Watermark casting and detection procedures.

detection enhances the ability to develop more powerful and reliable techniques for countering attacks and, thus, improving detection performance. However, the use of the originals reduces significantly the capabilities of an automated watermark detector (AWD). An AWD is composed of a watermark detector $\mathcal{D}$ and a *network monitor* (e.g., a Web browser), which scans the accessible Web domains or monitors broadcasting, thus providing the detector with images to be examined. When the original image is required for the watermark detection, the AWD requires an additional efficient technique to search and localize the corresponding original in the provider's image archive.

4 Fundamental Properties and Demands

The watermarking framework, defined in the previous section, should be reliable, effective against malevolent attacks, and should not affect the perceived data quality. In order to satisfy these general demands as much as possible, the watermarking framework should obey basic rules. The *perceptual similarity* of products and *watermark equivalence* plays a central role in the watermarking framework.

4.1 Perceptual Similarity and Watermark Equivalence

Perceptual similarity: if $X, Y \in \mathbf{X}$, then the notation $X \sim Y$ denotes that the digital products X and Y seem perceptually the same. $X \nsim Y$ denotes that either X and Y are completely different products or Y shows significant perceived quality reduction with respect to X.

The capability of detector $\mathcal{D}$ to distinguish watermarks that are not exactly identical is generally limited. Two watermarks are

assumed different when possible detection of the first does not imply possible detection of the second. Thus, we introduce the following definition.

Watermark equivalence: We say that the watermark W_1 is equivalent to W_2 ($W_1 \simeq W_2$) when

$$\mathcal{D}(X, W_1) = 1 \;\Rightarrow\; \mathcal{D}(X, W_2) = 1.$$

Obviously, identical watermarks are equivalent but the inverse does not hold in general. Equivalent watermarks may differ significantly. When watermarking aims at copyright protection, perceptual similarity is associated with the commercial product value. When watermarking is performed for content verification, similarity is associated with content matching.

4.2 Basic Demands

Perceptual invisibility. The watermark embedding should not produce perceivable data alterations. X_w should not show any perceptual distortions that reduce the image quality with respect to the original data. This property implies that

$$X_0 \sim X_w.$$

Figure 4 demonstrates an 8-bit gray-scale original image and the corresponding watermarked one produced by the technique presented in [17]. The alterations on the watermarked image are unnoticed when they are displayed either on a computer screen or on printed copies. Therefore, the authenticity of the image is not affected by the watermark superposition, since the watermarked copy shows high quality and preserves content integrity. Furthermore, image quality (and therefore its commercial value) remains unaffected. Perceptual invisibility is usually

(a)   (b)

FIGURE 4   (a) Original and (b) watermarked image. Watermark alterations are almost invisible.

performed either by low-power alterations, which are not perceivable by the human eye, or by using visual (or audio) masking techniques [18, 19]. Visual masking can be used to render an image invisible, when embedded properly within another image.

Key uniqueness and adequacy. Different keys should not produce similar watermarks, i.e.,

$$K_1 \neq K_2 \implies W_1 \not\simeq W_2$$

for any product $X \in \mathbf{X}$ and $W_i = \mathcal{G}(X, K_i)$. This condition prevents possible conflicts between different providers who ask for unique watermarks. The key set $\mathbf{K}$ should be sufficiently large in order to supply all the providers with different keys and to hinder watermark key detection by trial and error procedures.

Product dependency. A provider may distribute a large amount of different images that generally consist of statistically independent data. When the same watermark data are embedded in each image, extraction of the watermark is possible by using statistical operations. For example, we consider a set of N 8-bit gray-scale images Y_n produced from the originals X_n by adding the same watermark alterations W. After averaging and for $N \to \infty$ we will get

$$\frac{1}{N} \sum_{n=1}^{N} Y_n = \frac{1}{N} \sum_{n=1}^{N} (X_n + W) = \bar{I} + W,$$

where $\bar{I}$ is a homogeneous image with approximately constant intensity. Therefore, when $\mathcal{G}$ is applied on different products with the same key, different watermarks should be produced, i.e., for any particular key $K \in \mathbf{K}$ and for any $X_1, X_2 \in \mathbf{X} : X_1 \not\simeq X_2 \implies W_1 \not\simeq W_2$, where $W_i = \mathcal{G}(X_i, K)$, so that such attacks fail. Another reason for using image-dependent watermarks is that a provider may give the customer both the original and the watermarked image, thus enabling him or her to subtract an image-independent watermark (if the embedding is simple, e.g., additive).

Reliable detection. In practice, the existence or not of a watermark in an image is indicated with a degree of certainty. The overall performance of the detector $\mathcal{D}$ should be characterized by a small error probability P_{err}. In particular, the realization of $\mathcal{D}$ may produce the following errors: Type I errors, in which the watermark is detected although it does not exist in the data (false positives); Type II errors, in which the watermark is not detected in the data although it does exist (false negative).

The above errors occur with specified probabilities of false alarm (P_{fa}) and rejection (P_{rej}), respectively, and the total probability error is

$$P_{\text{err}} = P_{\text{fa}} + P_{\text{rej}}. \qquad (6)$$

The certainty of a positive detection is $c = 1 - P_{\text{fa}}$ and the de-

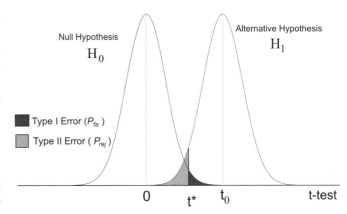

FIGURE 5 Detection by using a statistical test based on normal distributions.

tector output should be the following:

$$c \geq c_{\text{thres}} \implies \text{watermark exists.} \qquad (7)$$

The detection threshold c_{thres} is the *minimal certainty level* for establishing watermark existence in the test image. Hypothesis testing can be used for statistical certainty estimation and error manipulation [20]. Generally, when false positives become insignificant ($P_{\text{fa}} \to 0$) the probability to reject a watermark increases ($P_{\text{rej}} \to 1$), and vice versa. Figure 5 demonstrates typical detector output normal distributions, and the corresponding errors.

Computational efficiency. The watermarking algorithm should be efficiently implemented by hardware or software. Watermark casting is performed by applying the watermark generation and embedding only once for each image. However, the application of the overall detection procedure (browsing, watermark generation and detection) is frequently required. Subsequently, the development of a fast watermark detection algorithm is of great importance.

4.3 Necessary Conditions for Copyright Protection

Multiple watermarking. Watermarked images are like the original images with respect to their archival format and data range and size. Therefore, a watermarked image can be watermarked again without any technical restrictions. This feature is desirable in certain cases, e.g., for tracing the distribution channels when several resellers exist and are allowed to watermark the images. We consider the following multiply watermarked image:

$$X_{w_i} = \mathcal{E}(X_{w_{i-1}}, W_i), \quad i = 1, 2, \dots.$$

It is strongly recommended that the original watermark W_j, $j \leq i$, is still detectable in X_{w_i}:

$$\mathcal{D}(X_{w_i}, W_j) = 1, \quad \forall j \leq i \leq n,$$

where n is a sufficient number of coexisting watermarks such that $X_{w_n} \sim X_0$ and $X_{w_{n+1}} \not\sim X_0$. We remark also that

$$\mathcal{D}(X_{w_i}, W_j) = 0, \quad \forall j > i.$$

A pirate may embed his or her own watermark W_2 on an image X_{w_1} watermarked by the original owner with the watermark W_1. The pirate produces the product $X_{w_2} = \mathcal{E}(X_{w_1}, W_2)$. Both watermarks (the original and the piratical one) can be detected by using the corresponding unique key. However, the *true owner* can dispose of an image copy X_{w_1} that contains *only* his or her watermark W_1. In contrast, the pirate's copy X_{w_2} will always contain both watermarks.

Watermark validity and noninvertibility. Rightful ownership can be disputed and attacked when there is the possibility to produce *counterfeit* watermarks. Counterfeit watermark signals $\tilde{W} = \tilde{W}(\mathbf{k})$ are created by taking into account the features of a particular image X_w that is watermarked by the legal owner and the watermark generation method used in the public detector $\mathcal{D}$. The counterfeit watermark is never embedded, but it is designed in such a way so that it forces the detector to output a positive result for the particular image X, i.e., $\mathcal{D}(X_w, \tilde{W}) = 1$. In this case both the legal owner and the pirate can show the existence of their watermarks, W and $\tilde{W}$ respectively, in X_w. Since $\tilde{W}$ is formed by accounting the main features of X_w, generally, it can be detected in the original image X_0 as well. Subsequently, watermarking does not provide the legal owner with sufficient evidence to prove his or her ownership [21]. In order to overcome the above problem, only *valid* watermarks should be used in the watermarking scheme. A watermark signal $W \in \mathbf{W}$ is a valid watermark for a particular product $X \in \mathbf{X}$ if and only if it is associated with a key:

$$\exists K \in \mathbf{K} \text{ such that } \mathcal{G}(X, K) = W. \tag{8}$$

Watermark validity is effective when it is followed by *noninvertibility* of the watermark generation procedure: For any given image X^*, the watermark generation function $G_{X^*}(K) = \mathcal{G}(X^*, K)$ should not be invertible in the sense that, for a watermark $\tilde{W}$, it is infeasible to find a key $K^* \in \mathbf{K}$ that satisfies the relation $\mathcal{G}(X^*, K^*) \simeq \tilde{W}$.

Robustness to image modifications. A digital image can undergo many manipulations that deliberately (piratical attacks) or not (compression, filtering, etc.) affect the embedded watermark. Let X_0 be the original image and X_w be a watermarked version of it ($\mathcal{D}(X_w, W) = 1$). We denote by $\mathcal{M}$ an image processing operator that modifies somehow the digital images $X \in \mathbf{X}$. Robustness means that the watermark is still detected when the performed modifications preserve perceptual similarity:

$$\mathcal{D}(Y, W) = 1, \quad \forall Y \sim X_w \text{ and } Y = \mathcal{M}(X_w).$$

Image (or video) modifications usually include (but are not limited to):

- lossy compression up to a certain quality level that does not produce visible image degradations

- filtering for noise removal, enhancement for improving image quality, etc. Specialized filters for intentional watermark removal should be also accounted for as well
- geometric distortions (e.g., scaling, rotation, cropping, image/frame reflection, and line/column/frame extraction or insertion or their combinations)
- changes in presentation format, e.g., analog-to-digital or digital-to-analog conversion, printing, and rescanning.

4.4 Watermark Fragility and Content Verification

As in the case of copyright protection, efficiency for content verification demands watermarks that satisfy the basic demands discussed in Section 4.2. In this case, piracy is associated with forgery that aims to harm the credibility of the rightful providers or to distribute false information to the users. Pirates may tamper with and distribute an image that belongs to a rightful provider. They want to preserve the original watermark in the tampered copy. Furthermore, they may create their own images and put authenticity watermarks in them that belong to another rightful provider. Subsequently, protection against piracy requires secure and fragile watermarks.

Security against forgery. Determination and extraction of a watermark, without using the private key K_{pr}, and creation of forged authenticity proofs on other products should be impossible.

Watermark fragility. Any image modification that affects the original image content integrity should cause watermark distortions and, consequently, content verification failure.

High-performance protection demands that watermarks can reveal any slight image modifications. Watermarks based on the least significant bit (LSB) of the image data are very sensitive and fragile. However, such watermarks are not secure because a pirate can produce modifications leaving the LSB invariant. Although fragility is a basic watermark property, robustness may also be required in some special cases that include modifications that do not harm the original image content [22], e.g., High-quality compression and necessary insignificant modifications to incorporate the product in a multimedia environment. However, some researchers insist that no content modification should be allowed at all, since "minor" changes due, e.g., to compression render the image useless in legal terms. Generally, local image modifications affect image contents. Therefore, the watermark should be very sensitive to such modifications (e.g., object insertion or extraction in a photograph). It would be useful if the detection algorithm localizes the tampered regions, besides giving a negative authenticity answer.

5 Watermarking on the Spatial Image Domain

One of the first still image watermarking techniques was based on directly modifying the image pixel intensity [11, 13]. These techniques are applicable on 8-bit gray-scale or 24-bit color images. In the second case, the 8-bit luminance component is considered. In this section, we present the basic concepts of such watermarking and we demonstrate its capabilities to satisfy the demands of the general watermarking framework.

Subsequently, we consider the set $\mathbf{X}$ of gray-scale images of size $N \times M$, and an original image $X_0 \in \mathbf{X}$ is defined as

$$X_0 = \{x_0(n, m) \mid 0 \le n < N, \ 0 \le m < M\}. \quad (9)$$

5.1 Watermark Generation

The watermark set $\mathbf{W}$ is defined by the binary two-dimensional signals of size $N \times M$:

$$W = \{w(n, m) \in \{0, 1\} \mid 0 \le n < N, \ 0 \le m < M\}, \quad (10)$$

where the number P_0 of zeros is equal to the number P_1 of ones:

$$P_0 = P_1 = P = NM/2.$$

The number N_W of possible watermarks is very large and is estimated by the formula [11]

$$N_W = \binom{2P}{P} = \frac{(2P)!}{(P!)^2} \simeq \frac{2^{2P}}{\sqrt{\pi P}}. \quad (11)$$

However the choice of a watermark from $\mathbf{W}$ should not be arbitrary. Watermark validity requires well-defined key sets $\mathbf{K}$ and a noninvertible generation algorithm. An efficient way for producing the watermarks of Eq. (10) is to use a pseudo-random number generator (PNG), which provides an almost random binary sequence:

$$\mathrm{PNG}(n; S) = 0 \text{ or } 1, \quad n = 1, 2, 3, \ldots.$$

S is the seed of the PNG that coincides with the private watermark key K. The watermark generation procedure can be defined as $\mathcal{G} : \mathbf{K} \to \mathbf{W}$, and the produced watermarks are formed as follows:

$$w(n, m) = \mathrm{PNG}(k, K_{\mathrm{pr}}), \quad k = nM + m + 1. \quad (12)$$

Generally $P_0 - P_1 \ne 0$, e.g., when NM is an odd number. However, small deviations from zero do not affect the practical implementation of the algorithm. The seed bitlength determines the number of watermarks that can be produced. This number is generally less than the number estimated in Eq. (11). Generally, the inversion of $\mathcal{G}$ is very difficult. Furthermore, "key uniqueness" is not proven and we should account for the problem of equivalent watermarks described in Section 4.1.

5.2 Watermark Embedding

The embedding procedure $\mathcal{E}$ is based on intensity alterations that produce the watermarked image X_w:

$$X_w = \{x_w(n, m) \mid 0 \le n < N, \ 0 \le m < M\}. \quad (13)$$

The most straightforward embedding techniques are described by the following formulae:

$$x_w(n, m) = x_0(n, m) + \alpha w(n, m) \quad \text{(additive rule)}, \quad (14)$$

$$x_w(n, m) = (1 + \alpha w(n, m)) \, x_0(n, m) \quad \text{(multiplicative rule)}. \quad (15)$$

In the currently examined technique we perform additive watermark embedding [11, 13]:

$$x_w(n, m) = x_0(n, m) + \alpha(n, m),$$

$$\alpha(n, m) = \begin{cases} \delta, & \text{if } w(n, m) = 1 \\ -\delta, & \text{if } w(n, m) = 0 \end{cases}. \quad (16)$$

The positive parameter δ denotes alteration strength. In order to guarantee watermark invisibility, δ should be restricted by a maximum value $\delta_{\max}$, which depends on the image characteristics in the neighborhood of the particular pixel (n, m). The embedding procedure is exclusively responsible for watermark invisibility. It may include techniques based on the human visual system (HVS) (e.g., [12]) in order to get an estimate of $\delta_{\max}$. Generally, δ should be small at homogeneous image regions and large enough in highly textured image regions. In the following we consider a constant δ for simplicity.

5.3 Watermark Detection

Watermark detection is approached statistically. Watermark embedding produces a systematic intensity change, in the following two subsets:

$$I_+ = \{(n, m) \mid w(n, m) = 1\}, \quad (17)$$

$$I_- = \{(n, m) \mid w(n, m) = 0\}. \quad (18)$$

By considering the bipolar form of the $\{0, 1\}$-valued watermark, i.e., the signal $\tilde{W} = \{\tilde{w}(n, m) \in \{-1, 1\}\}$, and an image $X = \{x(n, m)\}$, we define the detection procedure $\mathcal{D}$ through the correlation

$$R(k) = \frac{2}{k} \sum_{n=0}^{N-1} \sum_{m=0}^{M-1} x(n, m) \, \tilde{w}(n, m), \quad (19)$$

where $k = nM + m + 1$. By taking into account that I_+ and I_- correspond to independent image samples we get the following expected values of R when it is applied on the images of Eqs. (9)

and (13) respectively:

$$\lim_{k \to \infty} R_0(k) = 0, \tag{20}$$

$$\lim_{k \to \infty} R_w(k) = 2\delta. \tag{21}$$

The above expected values provide a clear distinction between a watermarked and a nonwatermarked case. We note that the expected value of R for any image watermarked by a different key is zero:

$$\lim_{k \to \infty} \frac{2}{k} \sum_{n=0}^{N-1} \sum_{m=0}^{M-1} \tilde{w}_1(n, m) \, \tilde{w}_2(n, m) = 0,$$

where $\tilde{w}$ means bipolar watermark presentation.

Since k is limited by the number of total image pixels (NM), the expected values do not match exactly to the values R obtained at specific detection runs. At the limit $k = MN$ we get

$$\bar{R} = \frac{1}{P} \sum_{(n,m) \in I_+} x(n, m) - \frac{1}{P} \sum_{(n,m) \in I_-} x(n, m) = \mu_+ - \mu_-. \tag{22}$$

According to the central limit theorem, $\bar{R}$ follows a normal distribution $N(2\delta, \sigma_R^2)$ when the particular watermark is present in X and $N(0, \sigma_R^2)$ otherwise. The variance of R is estimated by the formula

$$\sigma_R^2 = s_+^2 + s_-^2 \approx 2s_X^2,$$

where s_+^2, s_-^2 and s_X^2 are the standard deviations of the subsets I_+, I_- and of the entire image X, respectively. The correlation output $\bar{R}$, calculated from Eq. (19) for a specific image X and a watermark W, belongs to the first or to the second distribution and, thus, indicates the absence or the presence of the watermark, respectively. However, the two distributions overlap as shown in Fig. 5. Therefore, the derivation of detector (5) should be based

on a threshold value R_{thres}, and it is formed as follows:

$$\mathcal{D}(X, W) = \begin{cases} 1, & \text{if } \bar{R} > R_{\text{thres}} \\ 0, & \text{otherwise} \end{cases}. \tag{23}$$

The choice of R_{thres} is associated with the false alarm and rejection probabilities, discussed in Section 4.2, and contributes to the overall detection error probability:

$$\begin{aligned} P_{\text{err}} &= P_{\text{fa}} + P_{\text{rej}} \tag{24} \\ &= \frac{1}{\sigma_R \sqrt{2\pi}} \int_{R_{\text{thres}}}^{\infty} \exp\left(-\frac{R^2}{2\sigma_R^2}\right) dR \\ &\quad + \frac{1}{\sigma_R \sqrt{2\pi}} \int_{-\infty}^{R_{\text{thres}}} \exp\left(-\frac{(R - 2\delta)^2}{2\sigma_R^2}\right) dR. \end{aligned}$$

Here P_{err} is minimized for $R_{\text{thres}} = \delta$ and decreases as the watermark strength δ increases. However, when δ is increased, perceptual distortions occur in the watermarked image.

Figure 6 demonstrates a sample of a "pseudo-random" watermark, and an 8-bit watermarked image of size 256×256, produced by using $\delta = 2$. Figure 7 shows the evolution of the correlation function $R(k)$ for the original and watermarked image, which converges to the expected values 0 and 2δ respectively for large k. The correlation $\bar{R}$ for 1,000 watermarks, produced with different keys, shows a major and well distinct peak that corresponds to the correct key.

5.4 Satisfaction of Basic Demands

The watermarks presented in the above section satisfy the basic demands of the watermarking framework under specific conditions.

1. Perceptual invisibility.   This demand is satisfied directly under the restrictions discussed in Section 5.2.

(a)   (b)

FIGURE 6   (a) A pseudo-random binary watermark; (b) the watermarked image.

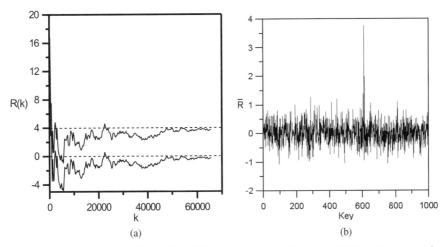

FIGURE 7 (a) Correlation $R(k)$ for the original and the watermarked image; (b) Detection output for 1,000 different keys. The correct key is $K = 611$.

2. Key uniqueness, adequacy, and noninvertibility. All the watermarks, generated by the chosen PNG, are valid since they correspond directly to the seed (key) of the PNG. The size of the key can be efficiently long for producing an enormous number of different watermarks. However, some watermarks might be equivalent. The number of nonequivalent watermarks is directly related to the choice of the threshold value in the detection procedure. In a set of $L \approx 1/P_{fa}$ keys, we expect to find one watermark that provides positive detection. Therefore, the number of nonequivalent watermarks is restricted approximately by the number $L = L(R_{thres})$. Key adequacy requires a very small false alarm detection probability. Invertibility of the procedure $\mathcal{G}$ requires invertibility of the PNG, which is extremely difficult. However, counterfeit watermarks can be derived by a trial and error procedure using about L different keys. This is an additional reason for operating at very small false alarm probability.

3. Reliable detection. The threshold R_{thres}, chosen by the provider, estimates the expected errors with good accuracy.

4. Computational efficiency. The computations for watermark generation, embedding, and detection are of rather small complexity.

5. Multiple watermarking. The embedding of a watermark W_2 on a watermarked image X_{W_1} does not significantly reduce the detection of the watermark W_1. This is a consequence of the statistical approach followed in the detection procedure [11].

6. Image dependency. The presented watermarking technique produces the same watermarks for all images and, subsequently, is directly vulnerable to the statistical attack mentioned in Section 4.2. Secure image-dependent watermarks can be produced by composing the PNG function with a function $F : \mathbf{X} \times \mathbf{K} \rightarrow \{0, 1\}$:

$$w(n, m) = \text{PNG}(k, K_{pr}) \otimes F(\hat{x}(n, m); K_{pr}),$$
$$k = nM + m + 1,$$

where $\hat{x}(n, m)$ denotes a robust image feature around the pixel (n, m).

7. Robustness and fragility. The watermarks of Eq. (10) are present on a watermarked image as a low-power white noise. Therefore, they are easily removed by low-pass filtering or JPEG compression. Also, correlation (19) demands "watermark synchronization." On one hand, when the watermarked image is resized, rotated, or cropped, the application of the detection procedure fails. On the other hand, the watermark fragility, under the above manipulations, is not proper for content verification. Local image modifications do not efficiently effect the detector output. In this case, the existence of watermark segments can be examined in particular image regions. Various optimizations that partially solve the above problems have been proposed, e.g., [12, 14]. Besides the watermarks of the form of Eq. (10), other watermark forms described by special constraints on the spatial domain components can be proven effective [17, 23].

6 Watermarking on Image Transform Domains

We mentioned that copyright protection requires watermarks that are robust to various attacks. Besides the spatial intensity image domain, discrete cosine transform (DCT) and discrete fourier transform (DFT) image domains are also convenient for watermarking. In this case, spread spectrum watermarks, embedded in a suitably chosen low-medium frequency range, provide increased security, invisibility, and robustness to lossy compression and certain geometrical modifications.

6.1 Watermarking in the DCT Domain

Spread spectrum watermarking in the image DCT domain has been proposed by Cox *et al.* [15]. Their scheme preserves image fidelity after proper alterations of the DCT coefficients. The detection procedure involves the use of the original image in order to overcome geometrical image modifications. A version of this technique, which bypasses the use of the original image in the detection procedure, has been proposed by Barni *et al.* [24].

We consider the one-dimensional (1-D) sequence of the DCT coefficients of an image X formed by a zig-zag ordering (see Chapter 5.5), denoted by Z, of the 2-D DCT domain:

$$\text{DCT}(X(n, m)) \circ Z \longrightarrow Y = \{y_1, y_2, y_3, \ldots\}.$$

The watermark signal is defined by a pseudo-random sequence of M real numbers that follows a normal distribution with zero mean and unit variance:

$$W = \{w_1, w_2, \ldots, w_M\}, \quad w_i \in (-d, d) \subset \mathbb{R}. \tag{25}$$

The watermark embedding takes place on a subset of the domain Y located in the medium frequency range in the interval $(L, M + L]$. The embedding is multiplicative:

$$y_i^{(w)} = y_i + \delta|y_i|w_{i-L}, \quad L < i \le M + L. \tag{26}$$

The watermarked image is obtained by applying the inverse transform:

$$X_w = Z^{-1} \circ \text{IDCT}(Y_w), \quad Y_w = \left\{y_1^{(w)}, y_2^{(w)}, \ldots\right\}. \tag{27}$$

Since alterations (26) may produce significant distortions in the watermarked image, *visual masking*, mentioned in Section 4.2, should be employed. In this way, the watermark casting is processed suitably in order to produce small changes in homogeneous image regions and higher ones in textured regions.

Detection is based on the correlation between the watermark W and a test image X^* with DCT coefficients $Y^* = \{y_1^*, y_2^*, \ldots\}$:

$$R = \frac{1}{M} \sum_{i=1}^{M} y_{L+i}^* w_i. \tag{28}$$

Similarly to correlation (22), R follows a normal distribution. In the absence of visual masking, the distribution has mean value and variance:

$$\mu_R = \begin{cases} \delta\mu_{|Y^*|}, & \text{if } Y^* = Y_w \\ 0, & \text{otherwise} \end{cases}, \quad \sigma_R^2 \simeq \frac{\sigma_{Y^*}^2}{M}. \tag{29}$$

The final decision about watermark existence requires to determine a threshold value R_{thres} as in definition (23). The total error is minimized for $R_{\text{thres}} = \delta\mu_{|Y^*|}/2$. Spread spectrum watermarks in the DCT domain show high resistance to modifications like JPEG compression, filtering, dithering, histogram equalization or stretching, and resizing. Also, internal cropping or replacement of some image objects preserves a significant part of the watermark power. However, such a watermark robustness is a disadvantage when content verification is desired. We should

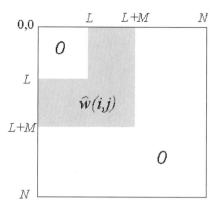

FIGURE 8   A 2-D watermark for embedding in the DCT image domain.

note that the DCT domain is not invariant under image rotation and, subsequently, watermark is not detected after such an attack. Cropped and resized image parts contain watermark information but watermark synchronization requires the knowledge of the size of the original image, which is generally not available.

The above technique can be implemented, in a similar manner as above, by using directly the 2-D DCT image domain $Y = \{y(i, j)\}$ and a watermark signal defined as

$$w_{i,j} = \begin{cases} \hat{w}(i, j), & \text{if } (i, j) \in U \\ 0, & \text{otherwise} \end{cases},$$

where $\hat{w}(i, j)$ follow the normal distribution $N(0, 1)$ and U is a subset of the frequency domain located in the medium band. An example is shown in Fig. 8. The altered coefficients are given by

$$y^{(w)}(i, j) = y(i, j) + \delta|y(i, j)| w(i, j), \quad \forall i, j$$

and provide the watermarked image after applying the inverse DCT. Figure 9 shows the watermarked image "Lena" (256×256) produced by using the watermark of Fig. 8 for $M = 45$ and various L, δ values. Similarly to Eq. (28), the watermark detection is defined by the correlation

$$R = \frac{1}{M'} \sum_i \sum_j y^*(i, j) w(i, j), \quad (i, j) \in U,$$

where M' is the number of elements of the subset U. Figure 10 shows the R values obtained for the image of Fig. 9(a) and for 1,000 different keys. The main peak P0 corresponds to the correct key. Peaks P1 and P2 indicate the R values calculated on JPEG versions of Fig. 9(a).

The parameters δ and L are essential for achieving watermark invisibility and robustness under lossy compression. By increasing δ (the strength of alterations), the detection performance also increases but, at the same time, the image fidelity is reduced and edge blurring and visible texture appear [Fig. 9(b)]. This is

FIGURE 9 Watermarked images for $M = 45$; (a) $\delta = 0.2$, $L = 128$; (b) $\delta = 1.0$, $L = 128$; (c) $\delta = 0.2$, $L = 10$; and (d) $\delta = 1.0$, $L = 128$, with visual masking.

also the case when embedding is applied in the low frequencies, i.e., when $L \rightarrow 0$ [Fig. 9(c)].

We can observe that quality degradation is most significant in the homogeneous image regions. In order to avoid undesirable effects, we may reduce the strength of alterations in homogeneous regions by using an image-dependent parameter and obtaining a new watermarked image $X'_w = \{x'_w(i, j)\}$ as follows:

$$x'_w(i, j) = x(i, j) + m(i, j)\,\Delta X(i, j), \qquad (30)$$

where is $\Delta X(i, j) = x_w(i, j) - x(i, j)$. The matrix $\mathbf{M} = \{m(i, j) \in I\!R\}$ is called "mask" and Eq. (30) is an example of *visual masking*. Barni *et al.* [24] proposed the mask $m(i, j) = |\text{var}(B_S(i, j))|$, where $B_S(i, j)$ is an image block of size $S \times S$ around the pixel (i, j) and $|.|$ denotes normalization to unity. For homogeneous image regions, $m(i, j) \ll 1$ and, subsequently, watermark alterations are filtered in such regions. The masked version of image 9(b) is shown in Fig. 9(d). We should remark that the above example of visual masking provides a correlation value

R less than the expected one given by Eq. (29). Detection, which is applied on the masked image, results the peak P3 in Fig. 10.

6.2 Watermarking Using Fourier–Mellin Transforms

Several geometrical image modification attacks can be countered if we use image domains that are invariant under rotation and scaling. Such domains can be derived by considering the 2-D DFT image transformation and *log polar* maps [25]. Ruanaidh and Pun [26] proposed a watermarking technique based on DFT amplitude spread spectrum modulation combined with a discrete Fourier–Mellin transformation. Let $A(k_1, k_2)$ denote the amplitude of the DFT transform of an image $X = \{x(n, m)\}$. We mention the following properties.

1. Scaling of the spatial domain of the image X by a factor ρ implies inverse scaling in the Fourier amplitude domain:

$$x(\rho n, \rho m) \longrightarrow \frac{1}{\rho} A\left(\frac{k_1}{\rho}, \frac{k_2}{\rho}\right). \qquad (31)$$

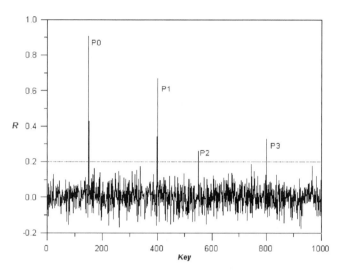

FIGURE 10   The detection response R for 1,000 different keys. The main peaks that exceed the dotted line (a possible threshold) correspond to correct positive detection on the original watermarked image (P0), on JPEG image versions (P1 and P2 for a compression ratio of 3:1 and 5:1, respectively), and on the visually masked image (P3).

2. Rotation of the image by an angle ϕ implies the same rotation in the amplitude domain:

$$x(n\cos\phi - m\sin\phi, n\sin\phi + m\cos\phi)$$

$$\longrightarrow A(k_1\cos\phi - k_2\sin\phi, k_1\sin\phi + k_2\cos\phi). \quad (32)$$

The log-polar mapping (LPM) is applied to provide a new coordinate system (μ, θ) given by equations

$$\mu = \ln\left(k_1^2 + k_2^2\right), \quad \theta = \tan^{-1}(k_2/k_1).$$

Let $\tilde{X}$ be a rotated and scaled version of an image X. Its DFT amplitude in the log polar coordinate system will be

$$\tilde{A}(\mu, \theta) = A(\mu + 2\rho, \theta + \phi).$$

The above relation means that image scaling and rotation is transformed to a translation of the DFT amplitude by a constant vector $(2\rho, \theta)$ in the log polar coordinate system. Such a translation can be eliminated by applying a new DFT transform on the above domain:

$$[DFT(\tilde{A}(\mu, \theta))] = [DFT(A(\mu, \theta))]$$

where [ ] denotes the amplitude domain of the transform. Therefore, we have the following RST (rotation, scaling, and translation) invariant domain:

$$DFT \circ LPM \circ DFT (X) = DFT \circ LPM \circ DFT (\tilde{X}). \quad (33)$$

The composition DFT $\circ$ LPM constitutes the discrete Fourier–Mellin transform. We remark that a suitable discrete space should

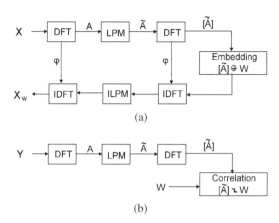

FIGURE 11   Schematic presentation of (a) the watermark embedding and (b) the detection procedure in the RST invariant domain. The Greek letter ϕ denotes the particular phase at each stage.

be used for the LPM transformation [25]. Figure 11 demonstrates the main steps for the watermark embedding and detection. The watermark generation and embedding should be performed for any one of the above amplitude domains. However, detection should be always applied in the RST invariant domain, Eq. (33), in order to compensate for scaling or rotation. However, simple combinations of cropping and scaling or the non-uniform scaling renders the watermark undetectable.

7 Conclusions

Digital watermarking is a new research topic. Important progress has occurred in the past years and many new techniques have been presented in the literature. Current watermarking research is mainly focused on watermark robustness issues for copyright protection. Can a watermark be robust to all processing technique attacks that preserve the perceived product quality? The answer may be "yes" for the currently known attacks. However, what will happen with future image processing attacks or lossy compression methods? For example, watermarking and compression are evolving techniques. A watermark may be robust under JPEG compression, but this may not be true for a more powerful technique that will possibly occur in the years to come. Once the watermarked product is out in public distribution, it is vulnerable to any future attack. Antiwatermarking techniques have been already developed based on miscellaneous processing methods [27].

References

[1] B. M. Macq and J. J. Quisquater, "Cryptology for digital TV broadcasting," *Proc. IEEE* **83**, 944–957 (1995).

[2] H. Berghel and L. Ó. Gorman, "Protecting ownership rights through digital watermarking," *IEEE Comput.* **29**, 101–103 (1996).

[3] A. Z. Tirkel R. G. Schyndel and C. F. Osborne, "A digital watermark," in *Proceedings of ICIP '94*, (IEEE, New York, 1994), Vol. II, pp. 86–90.

[4] F. Mintzer, G. W. Braudaway, and M. M. Yeung, "Effective and ineffective digital watermarks," in *Proceedings of ICIP '97* (IEEE, New York, 1997) Vol. III, pp. 9–12.

[5] D. R. Stinson, *Cryptography, Theory and Practice* (CRC Press, New York, 1995).

[6] G. L. Friedman, "The trustworthy digital camera: restoring gredibility to the photographic images," *IEEE Trans. Consumer Electron.* **39**, 905–910 (1993).

[7] N. F. Johnson and S. Jacodia, "Exploring steganography: seeing the unseen," *IEEE Comput.*, 26–34, (February 1998).

[8] T. ElGamal, "A public key cryptosystem and signature scheme base on discrete logarithms," *IEEE trans. Inf. Theory* **31**, 469–472 (1985).

[9] F. Hartung and B. Girod, "Fast public-key watermarking of compressed video," in *Proceedings of ICIP '97* (IEEE, New York, 1997), Vol. I, pp. 528–531.

[10] M. M. Yeung and F. Mintzer, "An invisible watermarking technique for image verification," in *Proceedings ICIP '97* (IEEE, New York, 1997), Vol. II, pp. 680–683.

[11] I. Pitas, "A method for signature casting on digital images, in *Proceedings of ICIP '96* (IEEE, New York, 1996), Vol. III, pp. 215–218.

[12] N. Nikolaidis and I. Pitas, "Robust image watermarking in the spatial domain," *Signal Process.* **66**, 385–403 (1998).

[13] W. Bender, D. Gruhl, N. Morimoto, and A. Lu, "Techniques for data hiding," *IBM Syst. J.* **35**, 313–335 (1996).

[14] A. Z. Tirkel, C. F. Osborne, and T. E. Hall, "Image and watermark registration," *Signal Process.*, **66**, 373–383 (1998).

[15] I. J. Cox, J. Kilian, F. T. Leighton, and T. Shamoon, "Secure spread spectrum watermarking for multimedia," *IEEE Trans. Image Process.* **6**, 1673–1687 (1997).

[16] M. D. Swanson, B. Zhu, A. H. Tewfik, and L. Boney, "Robust audio watermarking using perceptual masking," *Signal Process.* **66**, 337–355 (1998).

[17] G. Voyatzis and I. Pitas, "Digital image watermarking using mixing systems," *Comput. Graph.* **22**, 405–416 (1998).

[18] J. F. Delaigle, C. De Vleeschouwer, and B. Macq, "Watermarking algorithm based on a human visual model," *Signal Process.* **66**, 337–355 (1998).

[19] C. I. Podilchuk and W. Zeng, "Image adaptive watermarking using visual models," *IEEE J. Sel. Areas Commun.* **16**, 525–539 (1998).

[20] A. Papoulis, *Probability & Statistics*. (Prentice-Hall, Englewood Cliffs, NJ, 1991).

[21] S. Craver, N. Memon, B-L. Yeo, and M. Yeung, "Resolving rightful ownerships with invisible watermarking techniques: limitations, attacks and implications," *IEEE J. Sel. Areas Commun.* **16**, 573–586 (1998).

[22] B. Zhu, M. D. Swanson, and A. H. Tewfik, "Transparent robust authentication and distortion measurement technique for images," in *Proceedings of DSP '96*, (IEEE, New York, 1996), pp. 45–48.

[23] M. Kutter, F. Jordan, and F. Bossen, "Digital watermarking of color images using amplitude modulation," *J. Electron. Imag.* **7**, 326–332 (1998).

[24] M. Barni, F. Bartolini, V. Cappellini and A. Piva, "Robust audio watermarking using perceptual masking," *Signal Process.* **66**, 357–372 (1998).

[25] B. S. Reddy and B. N. Chatterji, "An FFT-based technique for translation, rotation, and scale invariant image registration," *IEEE Trans. Image Process.* **5**, 1266–1271 (1996).

[26] J. J. K. Ruanaidh and T. Pun, "Rotation, scale and translation invariant spead spectrum digital image watermarking," *Signal Process.* **66**, 303–317 (1998).

[27] F. Petitcolas, R. J. Anderson, and M. G. Kuhn, "Attacks on copyright marking systems," presented at the 2nd Workshop on Information Hiding, in Vol. 1525 of lecture notes in *Computer Science*, 218–238, Oregon, April, 1998.

Applications of Image Processing

10.1

Synthetic Aperture Radar Algorithms

Ron Goodman
ERIM International Inc.

Walter Carrara
Nonlinear Dynamics Inc.

1 Introduction

This chapter presents a sampling of key algorithms related to the generation and exploitation of fine resolution synthetic aperture radar (SAR) imagery. It emphasizes practical algorithms in common use by the SAR community. Based on function, these algorithms involve *image formation, image enhancement*, and *image exploitation*. Image formation transforms collected SAR data into a focused image. Image enhancement operates on the formed image to improve image quality and utility. Image exploitation refers to the extraction and use of information about the imaged scene.

Section 2 introduces the fundamental concepts that enable fine-resolution SAR imaging and reviews the characteristics of collected radar signal data and processed SAR imagery. These attributes determine the need for specific processing functions and the ability of a particular algorithm to perform such functions. Section 3 surveys leading SAR image formation algorithms and discusses the issues associated with their use. Section 4 introduces several enhancement algorithms for improving SAR image quality and utility. Section 5 samples image exploitation topics of current interest in the SAR community.

2 SAR Overview

Radar is an acronym for *r*adio *d*etection *a*nd *r*anging. In its simple form, radar *detects* the presence of a target by sensing energy that the target reflects back to the radar antenna. It *ranges* the target by measuring the time interval between transmitting a signal (for instance, in the form of a short pulse) and receiving a return (the backscattered signal) from the target. Radar is an active sensor that provides its own source of illumination. Radar operates at night without impact and through clouds or rain with only limited attenuation.

A radar image is a two-dimensional (2-D) map of the spatial variations in the radar backscatter coefficient (a measure of the strength of the signal returned to the radar sensor) of an illuminated scene. A scene includes targets, terrain, and other background. The image provides information regarding the position and strength of scatterers throughout the scene. While a common optical image preserves only amplitude, a radar image naturally contains phase and amplitude information. An optical sensor differentiates signals based on angle (in two dimensions) and makes no distinction based on range to various scene elements. An imaging radar naturally separates returns in range

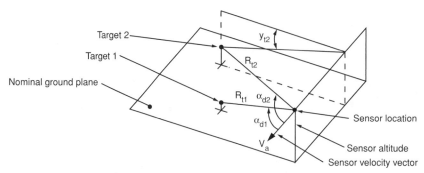

FIGURE 1   Resolution in range R_t and (Doppler) cone angle α_d.

and cone angle and does not differentiate signals based on depression (or elevation) angle. The (Doppler) cone angle α_d is the angle between the radar velocity vector V_a (indicating the direction of antenna motion) and the line-of-sight vector from the antenna to a particular scatterer. The depression angle ψ_t is the angle between the nominal ground plane and the projection of the line-of-sight vector onto a plane perpendicular to V_a. Figure 1 illustrates this range and angle differentiation by a SAR imaging system.

The ability to distinguish, or resolve, closely spaced features in the scene is an important measure of performance in an imaging system. In SAR imaging, it is common to define *resolution* as the −3-dB width of the system impulse response function with separate measures in each dimension of the image. The −3-dB width is the distance between two points, one on each side of the mainlobe peak, that are nearest to and one half the intensity of the peak.

The complex (phase and amplitude) nature of SAR imagery increases the ability of enhancement algorithms to improve the quality and interpretability of an image. It also increases the opportunity for image exploitation algorithms to derive additional information about an imaged scene. Traditional SAR provides 2-D scatterer location and resolution between scatterers in range and azimuth (or cross range). New applications extract 3-D information about the scene by using interferometric techniques applied to multiple images of a scene collected from similar viewing geometries.

SAR imaging involves the electromagnetic spectrum in the frequency bands encompassing VHF through K-band. Figure 2 relates these frequency bands to radio frequency and wavelength intervals. Various organizations throughout the world have successfully demonstrated and deployed SAR systems operating in most of these bands.

2.1 Image Resolution

Radar estimates the distance to a scatterer by measuring the time interval between transmitting a signal and receiving a return from the scatterer. Total time delay determines the distance to a scatterer; differential time delay separates scattering objects located at different distances from the radar sensor. The bandwidth B of the transmitted pulse limits time resolution to $1/B$ and corresponding range resolution ρ_r to

$$\rho_r \approx \frac{c}{2B}, \tag{1}$$

where c is the speed of light. To maintain a high average power at the large bandwidths required for fine resolution, it is common to transmit a longer pulse with linear frequency modulation (FM) rather than a shorter pulse at constant frequency. Pulse compression following reception of the linear FM pulses achieves a range resolution consistent with the transmitted bandwidth.

To generate a 2-D image, the radar separates returns arriving from the same distance based on differences in the angle of arrival. A real-beam radar achieves this angular resolution by scanning a narrow illuminating beam across the scene to provide azimuth samples *sequentially*. The angular resolution is comparable with the angular extent of the physical beam. A synthetic aperture radar generates an angular resolution much finer than its physical beamwidth. It transmits pulses from a series of locations as it moves along its path (the synthetic aperture) and processes the collection of returns to synthesize a much narrower beam. The image formation processor (IFP) adjusts the relative phase among the returns from successive pulses to remove the phase effects of the nominally quadratic range variation to scatterers within the scene. It coherently sums the returns (generally by means of a Fourier transform) to form the

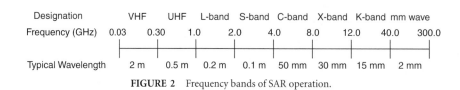

FIGURE 2   Frequency bands of SAR operation.

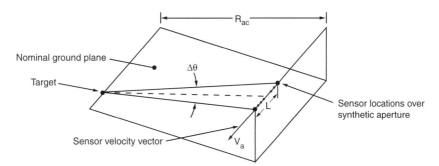

FIGURE 3   Synthetic aperture geometry.

synthetic beam and generate azimuth resolution cells. Signal processing provides azimuth samples *simultaneously* within a physical beamwidth.

The synthetic aperture concept is essential for achieving fine azimuth resolution when it is not practical to generate a sufficiently narrow real beam. The synthetic aperture provides an azimuth resolution capability ρ_a of

$$\rho_a \approx \frac{\lambda_c}{2\Delta\theta}. \tag{2}$$

Here, λ_c is the center wavelength of the transmitted signal and $\Delta\theta$ is the angular interval over which the processed data were collected. Azimuth resolution is proportional to range because $\Delta\theta$ decreases as distance to scatterers in the scene increases.

Figure 3 illustrates this synthetic aperture geometry. As an example, consider a SAR system that collects signals over a synthetic aperture distance L of 1 km with an antenna moving at velocity V_a of 100 m/s during a synthetic aperture time interval T_a of 10 s. At a minimum range R_{ac} of 20 km, the synthetic aperture angular interval $\Delta\theta$ is approximately 0.05 rad. With a transmitted bandwidth B of 500 MHz at a center wavelength of 0.03 m (X-band), these parameters offer azimuth resolution of 0.3 m and range resolution of 0.3 m.

2.2 Imaging Modes

Figure 4 illustrates two basic SAR data-collection modes. In the stripmap mode, the antenna footprint sweeps along a strip of terrain parallel to the sensor trajectory. Antenna pointing is fixed perpendicular to the flight line in a *broadside* collection, or pointed either ahead or behind the normal to the flight line in a *squinted* collection. The azimuth beamwidth of the antenna dictates the finest-achievable azimuth resolution by limiting the synthetic aperture, while the transmitted bandwidth sets the range resolution. The antenna elevation beamwidth determines the range extent (or *swath width*) of the imagery, while the length of the flight line controls the azimuth extent.

The stripmap mode naturally supports the coarser resolution, wide area coverage requirements of many natural resource and commercial remote-sensing applications. Most airborne SAR systems include a stripmap mode. Remote sensing from orbit generally involves a wide area coverage requirement that necessitates the stripmap mode.

In the spotlight mode, the antenna footprint continuously illuminates one area of terrain to collect data over a wide angular interval in order to improve azimuth resolution beyond that supported by the azimuth beamwidth of the antenna. The spotlight mode achieves this fine azimuth resolution at the cost of reduced image area. The angular interval over which the radar observes the scene determines the azimuth resolution. Antenna beamwidths in range and azimuth determine scene extent.

The spotlight mode naturally supports target detection and classification applications that emphasize fine resolution over a relatively small scene. While a fine-resolution capability is useful largely in military and intelligence missions, it also has value in various scientific and commercial applications.

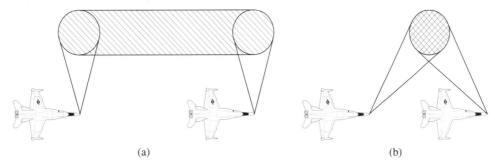

(a)   (b)

FIGURE 4   Basic SAR imaging modes: (a) stripmap mode for area search and mapping; (b) spotlight mode for fine resolution.

FIGURE 5   RADARSAT-1 C-band image of Ft. Irwin, CA. (Copyright Canadian Space Agency, 1998.)

2.3 Examples of SAR Imagery

The following examples indicate the diversity of imagery and applications available from SAR systems. They include stripmap and spotlight mode images in a variety of frequency bands. In each image, near range is at the top.

Figure 5 is a coarse resolution SAR image of Ft. Irwin, California, collected by the Canadian RADARSAT-1 satellite [1]. The RADARSAT-1 SAR operates at C-band (5.3 GHz) in the

stripmap mode with a variety of swath width and resolution options. The sensor collected this particular image at a resolution of 15.7 m in range and 8.9 m in azimuth. The processed image covers a ground area approximately 120 km in range by 100 km in azimuth, encompassing numerous large-scale geographic features including mountains, valleys, rivers, and lakes.

Figure 6(a) displays an X-band image of a region near Calleguas, California collected by the Interferometric SAR for Terrain Elevation (IFSARE) system [2]. The IFSARE system

(a)

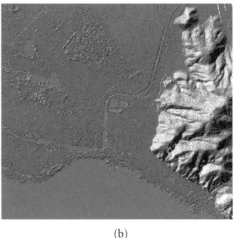

(b)

FIGURE 6   X-band image from the Interferometric SAR for Terrain Elevation system: (a) magnitude SAR image of Calleguas, CA; (b) corresponding elevation data displayed as a shaded relief map.

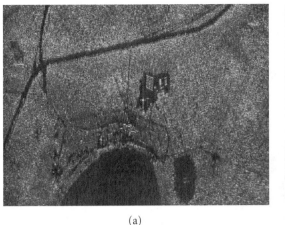

(a)   (b)

FIGURE 7 Stripmap mode VHF/UHF-band image of Northern Michigan tree stands: (a) forested area with several clearings and access roads; (b) close-up view of clearing.

is a dual-channel interferometric SAR built by ERIM International Incorporated and the Jet Propulsion Laboratory under the sponsorship of the Defense Advanced Research Projects Agency (DARPA). It simultaneously generates basic stripmap SAR images at two different depression angles and automatically produces terrain elevation maps from these images. The image in Fig. 6(a) is a composite image assembled from multiple strips. It covers a ground area of approximately 20 km by 20 km. The resolution of collected IFSARE imagery is 2.5 m in range by 0.8 m in azimuth. After several averaging operations (required to improve the fidelity of output digital terrain elevation data) and projection of the image into the nominal ground plane, the intrinsic resolution of the image in Fig. 6(a) is approximately 3.5 m in both range and azimuth. Figure 6(b) illustrates one way to visualize the corresponding terrain elevation. This type of presentation, known as a shaded relief map, uses a conventional linear mapping to represent the gradient of terrain elevation by assigning higher gray-scale values to steeper terrain slopes. The IFSARE system derives topographic data with a vertical accuracy of 1.5 m to 3.0 m depending on the collection altitude.

Figure 7(a) is a fine resolution VHF/UHF-band image of a forested region in northern Michigan with a spatial resolution of 0.33 m in range and 0.66 m in azimuth. This stripmap image originates from an ultrawideband SAR system that flies aboard a U. S. Navy P-3 aircraft. ERIM International designed and built this radar for DARPA in conjunction with the Naval Air Warfare Center (NAWC) for performing foliage penetration (FOPEN) and ground penetration (GPEN) experiments [3]. Figure 7(b) shows a close-up view of the clearing observed in Fig. 7(a). The numerous point like scatterers surrounding the clearing represent the radar signatures of individual tree trunks; a fraction of the incident radar energy has penetrated the forest canopy and returned to the sensor following double-bounce reflections between tree trunks and the ground.

The image of the Washington Monument in Fig. 8 originates from the ERIM International airborne Data Collection System

[4] operating at X-band in the spotlight mode. This 0.3-m resolution image illustrates the SAR phenomena of *layover* and *shadowing*. Layover occurs because scatterers near the top of the monument are closer to the SAR sensor and return echoes sooner than do scatterers at lower heights. Therefore, the system naturally positions higher scatterers on a vertical object at nearer ranges (toward the top of Fig. 8) than lower scatterers on the same object. As a result, vertical objects appear to lay over in a SAR image from far range to near range. Shadowing occurs in this example because the monument blocks the illumination of scatterers located behind it. Therefore, these scatterers can reflect no energy back to the sensor. The faint horizontal streaks observed throughout this image represent the radar signatures

FIGURE 8 Spotlight mode X-band image of the Washington Monument, collected by the Data Collection System.

FIGURE 9 Spotlight mode X-band image of the Pentagon building, collected by the Data Collection System.

of automobiles moving with various velocities during the synthetic aperture imaging time. Section 5.1 describes the image characteristics of moving targets.

The spotlight image of the Pentagon in Fig. 9 (from the Data Collection System) illustrates the extremely fine detail that a SAR can detect. Observable characteristics include low return areas, the wide dynamic range associated with SAR imaging, distinct shadows, and vehicles in the parking lots. Individual windowsills are responsible for the regular array of reflections observed along each ring of the Pentagon; as in the case of the Washington Monument, they exhibit considerable layover because of their vertical height. It is impressive to realize that SAR systems today are capable of generating such fine-resolution imagery in complete darkness during heavy rain from distances of many kilometers!

2.4 Characteristics of Signal Data

A SAR sensor transmits a sequence of pulses over time and receives a corresponding set of returns as it traverses its flight path. We visualize this sequence of returns as a 2-D signal, with one dimension being pulse number (or sensor position along the flight path) and the other being time delay (or round-trip range). Analogous to an optical signal reaching a lens, this 2-D radar signal possesses a quadratic phase pattern that the processor must match in order to compress the dispersed signal from each scatterer to a focused point or image of that scatterer. In a simple optical system, a spherical lens provides the required 2-D quadratic phase match to focus the incoming field and form an optical image. In a modern SAR imaging system,

a digital image formation algorithm generates and applies the required phase pattern. While the incoming SAR signal phase pattern is nominally quadratic in each coordinate, many variations and subtleties are present to challenge the IFP. For instance, the quadratic phase coefficient in the azimuth coordinate varies across the range swath. The quadratic phase in the range coordinate is a deterministic function of the linear FM rate of the transmitted radar pulses.

SAR signal data consist of a 2-D array of complex numbers. In the range dimension, these numbers result from analog-to-digital (A/D) conversion of the returns from each transmitted pulse. Each sample includes quantized amplitude and phase (or alternatively, in-phase and quadrature) components. In the azimuth dimension, samples correspond to transmitted pulses.

To alleviate high A/D sampling rates, most fine-resolution systems remove the quadratic phase associated with the incoming signals within each received pulse electronically in the receiver before storing the signals. This quadratic phase arises from the linear FM characteristic of the transmitted waveform. Thinking of the quadratic phase in range as a "chirping" signal with a linear variation in frequency over time, we refer to this electronic removal of the quadratic phase with the terminology *dechirp-on-receive* or *stretch processing*. Following range dechirp-on-receive, the frequency of the resulting intermediate frequency (IF) signal from each scatterer is proportional to the distance from the radar sensor to the scatterer. Figure 10 illustrates this process. Stretch processing is advantageous when the resulting IF signal has lower bandwidth than the RF bandwidth of the transmitted signal.

Similarly, it may be desirable to electronically remove the azimuth quadratic phase (or azimuth chirp) associated with a sequence of pulses in the receiver before storage and subsequent image formation processing. The quadratic phase characteristic in azimuth originates from the quadratic variation in range to each scatterer over the synthetic aperture interval. Processing such a *dechirped* signal in either dimension involves primarily a Fourier transform operation with preliminary phase adjustments to accommodate various secondary effects of the SAR data-collection modes and radar system peculiarities. If the radar receiver does not remove these quadratic phase effects, the image formation processor must remove them.

Requirements for a minimum number of range and azimuth samples arise from constraints on the maximum spacing between samples. These constraints are necessary to avoid the presence of energy in the desired image from undersampled signals originating from scatterers outside the scene. The number of complex samples in the range dimension must slightly exceed the number of range resolution cells represented by the range swath that is illuminated by the antenna elevation beam. Similarly, the number of complex samples in the azimuth dimension must exceed slightly the number of azimuth resolution cells represented by the azimuth extent illuminated by the azimuth antenna beam. In the spotlight mode, bandpass filtering in azimuth limits the azimuth scene size and reduces the number of data samples into the IFP.

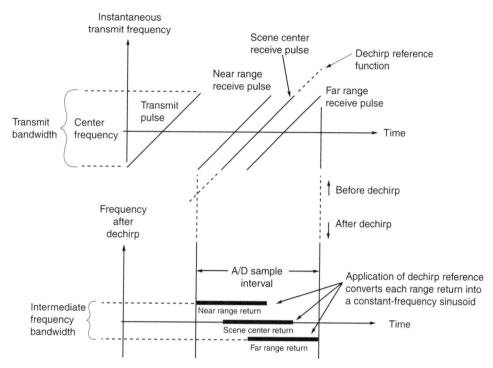

FIGURE 10 Effect of dechirp-on-receive operation on the time/frequency characteristics of radar signals.

Signal data include desired signals representing attributes of the scene being imaged, undesired phase effects related to transmitter and receiver properties or to the geometric realities of data collection, phase and amplitude noise from various sources, and ambiguous signals related to inadequate sampling density. Usually, the major error effect in SAR data is phase error in the azimuth dimension arising from uncertainty in the precise location of the radar antenna at the time of transmission and reception of each pulse. Without location accuracy of a small fraction of a wavelength, phase errors will exist across the azimuth signal aperture that degrade the quality of the SAR image. Other hardware and software sources of phase errors also are likely, even in a well-designed SAR system. Section 4.1 discusses autofocus algorithms to manage these error effects.

2.5 Characteristics of Image Data

SAR image data are a 2-D array of complex numbers with indices representing, for example, changing range and changing azimuth coordinates. Like signal data, each sample includes quantized amplitude and phase (or alternatively, in-phase and quadrature) components. Each element of the array represents an image pixel with amplitude related to the strength of the radar backscatter coefficient in the corresponding scene area. In general, the phase of an image pixel includes a deterministic component and a random component. The deterministic component is related to the distance between the corresponding scatterer and the radar sensor. The random component is related to the presence of *many* scattering centers in an area the size of a 2-D resolution cell in

most parts of the scene. Because of this random component, image phase generally is not useful when working with a single image. SAR interferometry, described in Section 5.2, surmounts this difficulty by controlling the data-collection environment adequately to achieve (and then cancel) the same random phase component in two images.

Characteristics of radar imagery include center frequency (for instance, X-band or L-band), polarization of transmit and receive antennas (for instance, horizontal or vertical and like or cross polarization), range and azimuth resolutions, and image display plane. Common choices for the image display plane are the nominal ground plane that includes the imaged terrain or the slant plane that contains the antenna velocity vector and the radar line-of-sight vector to scene center. Other attributes of SAR imagery include low return areas (shadows, roads, lakes, and other smooth surfaces), types of scatterers, range layover, targets moving during the data collection, multiple bounce (multipath) reflections, and coherent speckle patterns. Certain types of scatterers are common to manmade, metallic objects. These types include flat plates, cylinders, spheres, and dihedral and trihedral reflectors. Another type of scatterer is the distributed scatterer containing many scattering centers within the area of a resolution cell, such as a region covered by vegetation or a gravel-covered roof. Speckle refers to the characteristic nature of radar imagery of distributed scatterers to fluctuate randomly between high and low intensity. Such fluctuations about an average value appear throughout an otherwise uniform scene because the coherent summation of the echoes from the many scattering centers within each resolution cell yields a random value rather than the

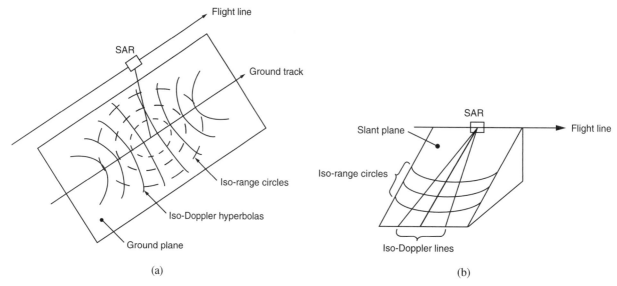

FIGURE 11   Intersection of range spheres with Doppler cones: (a) ground plane; (b) radar slant plane.

mean backscatter coefficient. Speckle is responsible for the mottled appearance of the grassy area surrounding the monument in Fig. 8.

The geometrical aspects of SAR image data naturally relate directly to scene geometry, data-collection geometry, and sensor parameters. Here we discuss the range and azimuth channels separately to describe these relationships.

Range refers to the distance R_t between the antenna phase center (APC) and a particular scatterer measured by the time delay ($t_d = 2R_t/c$) between transmission and reception of pulses. Spheres (indicating surfaces of constant range) centered at the APC will intersect a flat earth as circles centered at the radar nadir point. Figure 11(a) illustrates this geometric relationship. The illuminated parts of each of these circles appear as (straight) lines of constant range in a processed image.

Azimuth relates to angular location in terms of the Doppler cone angle, defined as the angle between the antenna velocity vector and the line of sight to a particular scatterer. A conical surface (indicating constant azimuth) with its vertex at the APC and its axis along the antenna velocity vector intersects a flat earth as a hyperbola. Figure 11(a) illustrates the shape of these intersections for a family of conical surfaces. The illuminated parts of each of these hyperbolas appear as (straight) lines of constant azimuth in a processed image.

While a conical surface and a spherical surface centered at the cone vertex intersect orthogonally in 3-D space, these circles of constant range and hyperbolas of constant Doppler on the flat earth generally are not orthogonal. As Fig. 11(b) illustrates, these intersections are orthogonal in the radar slant plane.

A typical set of image quality (IQ) parameters includes resolution, peak sidelobe levels, a measure of additive noise (primarily from thermal noise in the radar receiver), a measure of multiplicative noise, and geometric distortion. Resolution refers to the −3-dB width of the mainlobe of the system impulse response.

The sidelobe region is the area of the impulse response outside the mainlobe area. Peak sidelobe levels refer to the local peaks in intensity in the sidelobe region. Multiplicative noise refers to signal-dependent effects and includes digital quantization noise, energy in the sidelobes of the system impulse response, and energy from scatterers outside the scene that alias into the image as PRF (pulse repetition frequency) ambiguities. Geometric distortion involves a nonideal relationship between the image geometry and scene geometry, for instance, a square patch of terrain taking on a non-square shape in the image.

In practice, requirements on IQ parameters vary among task categories that include terrain imaging, target detection, target classification, and target identification. Each category indicates a different set of image quality, quantity, and timeliness requirements that a SAR system design and implementation must satisfy to perform that task acceptably [5].

3 Image Formation Algorithms

This section describes the principal image formation processing algorithms associated with operational spotlight and stripmap modes. We introduce this discussion with a short historical review of image formation processing of SAR data.

3.1 History of Image Formation Algorithms

The SAR sensor receives and processes analog electromagnetic signals in the form of time-varying voltages. While the modern digital signal processor requires that the receiver sample and quantize these analog signals, the first processor to successfully generate a fully focused SAR image operated on analog signals recorded in a 2-D format on a strip of photographic film. In this recording process, the signals returned from successively transmitted pulses were recorded side-by-side parallel to each other along the length of the film in a so-called *rectangular format*. The optical signal processor illuminated the signal film with a coherent (helium–neon) laser beam while an assortment of spherical, cylindrical, and conical lenses provided the needed quadratic focus to effect a Fourier transform operation. In a perspective analogous to optical imaging, the laser releases the radar wavefronts originating from the illuminated scene and stored in the photographic film while the lenses focus these wavefronts to form a 2-D image of the scene. Early digital signal processors performed essentially these same operations on the quantized signals, mimicking the rectangular format, the quadratic phase adjustments, and the Fourier transform operations inherent in the original optical processor.

Following these early processors, the SAR community has developed a succession of new approaches for processing SAR data in order to improve image quality, support different data-collection modes, and improve algorithm efficiency (particularly with respect to real-time and near-real-time imaging applications). Fortunately, the performance of digital signal processing (DSP) hardware has improved dramatically since the first digital SAR processors to keep pace with increasing processing demands of modern SAR sensors and associated algorithms.

3.2 Major Challenges in SAR Image Formation

The generation of high-quality SAR imagery requires that the IFP compensate a number of fundamental effects of radar system design, hardware implementation, and data-collection geometry. The more significant effects include scatterer motion through range and azimuth resolution cells, the presence of range curvature, effects of measured sensor motion, errors induced by nonideal sensor hardware components, errors induced by nonideal signal propagation, and errors caused by unmeasured sensor motion. Additional concerns involve computational complexity, quantity of digital data, and data rates.

Of these issues, motion through resolution cells (MTRC) and range curvature often present the greatest challenges to algorithm design. The remainder of this subsection defines and discusses these two challenges. Together with resolution requirements and scene size, they generally drive the choice of image formation algorithm. In addition, unmeasured sensor motion causes phase errors that often require the use of a procedure to detect, measure, and remove them. Section 4.1 discusses autofocus algorithms to address this need.

Over the synthetic aperture distance necessary to collect the data needed to form a single image, the changing position of the radar sensor causes changes in the instantaneous range and angle from the sensor to each scatterer in the scene being imaged. *Motion through resolution cells* refers to the existence of these changes. Because SAR uses the range and angle to a scatterer to position that scatterer properly within the image, the radar must estimate these changing quantities. For typical narrow-beam width sensors, the line-of-sight range to each scatterer is nominally a quadratic function of along-track sensor position. In a generic sense, this variation represents scatterer MTRC in range.

The drawings of imaging geometry in Fig. 12 help to relate MTRC to a change in range and define range curvature. In broadside stripmap imaging, the change in range to each scatterer is symmetrical about broadside and represents *range curvature*. In a squinted stripmap collection, the variation in range to each scatterer is not symmetrical over the synthetic aperture, but includes a large linear component. The SAR community refers to the linear component as *range walk* and the nonlinear (nominally quadratic) component as *range curvature*. Somewhat different terminology applies to the same effect in the arena of the fine-resolution spotlight mode, where all MTRC becomes *range curvature* regardless of whether the motion is linear or nonlinear. Figure 12 illustrates these effects for a stripmap collection (left side) and a spotlight collection (right side).

The key challenge in SAR image formation is the fact that range curvature varies with scatterer location within the imaged scene. The top right diagram in Fig. 12 suggests this variation. While it is easy to compensate range curvature for one scatterer, it can be difficult to compensate adequately and efficiently a different range curvature for each scatterer in the image.

For many systems having fine resolution or a wide swath width, this change in range curvature or *differential range curvature* (DRC) across the imaged swath can be large enough to challenge the approximations that most IFP algorithms use in their analytical basis for compensating MTRC. The consequences can include spatially variant phase errors that cause image defocus and geometric distortion.

3.3 Image Formation in the Stripmap Mode

In the *stripmap* mode, successively transmitted pulses interrogate the strip of terrain being imaged from successively increasing along-track positions as the antenna proceeds parallel to the strip. For image formation in the stripmap mode, we discuss range-Doppler processing, the range migration algorithm, and the chirp scaling algorithm.

Range-Doppler processing is the traditional approach for processing stripmap SAR data. It involves signal storage in a rectangular format analogous to the early optical stripmap processor described in Section 3.1. While many variations of this algorithm exist, the basic approach involves two common steps. First, the IFP compresses the signal data (pulses) in range. It then compresses the (synthetic aperture) data in azimuth to complete the imaging process. If range curvature is significant, the range-compressed track of each scatterer migrates through multiple range bins requiring use of a 2-D matched filter for azimuth compression. Otherwise, use of a 1-D matched filter is adequate. A range-Doppler processor usually implements the matched filter by means of the fast convolution algorithm involving a fast Fourier transform (FFT) followed by a complex multiply and an inverse FFT. The matched filter easily compensates the range

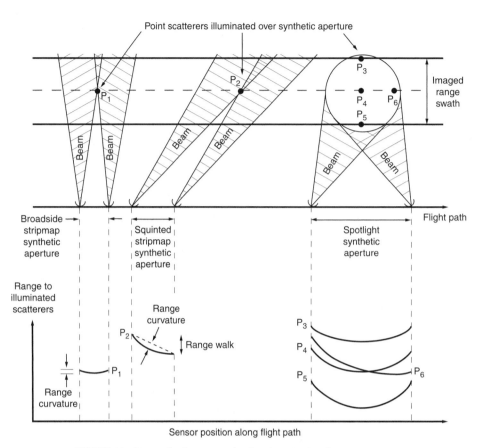

FIGURE 12   Data-collection geometries showing MTRC and range curvature.

curvature associated with scatterers at some reference range that is specified in the filter design.

A typical approach to accommodate DRC in range-Doppler processing divides the range swath being imaged into narrow subswaths. This division allows the use of the same matched filter for azimuth compression within each subswath tuned to its midrange, but a different matched filter from subswath to subswath. The IFP applies the same 2-D matched filter to all range bins within a specific subswath and accepts a gradual degradation in focus away from midrange. A common criterion allows $\pi/2$ rad of quadratic phase error and limits the maximum subswath width ΔR to

$$\Delta R \le \frac{8\rho_a^2}{\lambda_c} \qquad (3)$$

in order to avoid significant defocus [6]. As an example, an X-band ($\lambda_c = 0.03$ m) stripmap SAR with a 1-m azimuth resolution (requiring an azimuth beamwidth of 0.015 rad) corresponds to a range subswath width ΔR of 267 m.

A common implementation of the range-Doppler algorithm begins with an FFT of the azimuth chirped data in order to compensate directly for scatterer migration through range bins by means of a Doppler-dependent, 1-D digital interpolation in range. The idea is to straighten the curved trajectories that each

scatterer follows in range-Doppler (frequency) space by resampling the range compressed data. Figure 13 summarizes the steps in this process. This method is useful in processing medium-resolution and coarse-resolution SAR data but has difficulty with either fine-resolution data or data collected in a squinted geometry. While an additional processing stage can perform *secondary range compression* to partially overcome this difficulty, the range migration algorithm and the chirp scaling algorithm offer attractive alternatives for many applications.

The range migration algorithm (RMA) is a modern approach to stripmap SAR image formation [7]. As a key attribute, RMA provides a complete solution to the presence of range curvature and avoids any related geometric distortion or defocus. The RMA operates on input data after dechirp-on-receive (described in Section 2.4) in the receiver or subsequent range dechirp in the processor. It requires that the receiver preserve (or that the processor reapply) the natural azimuth chirp characteristics of the collected signals when compensating the received data for random sensor motion. We refer to this procedure of preserving the natural phase chirp in azimuth (common in conventional stripmap imaging) as *motion compensation to a line.*

Figure 14 illustrates the key steps in RMA processing. First, the RMA transforms the input signal data (already in the range frequency domain following the receiver dechirp-on-receive operation) into the 2-D spatial frequency (or *wavenumber*) domain by

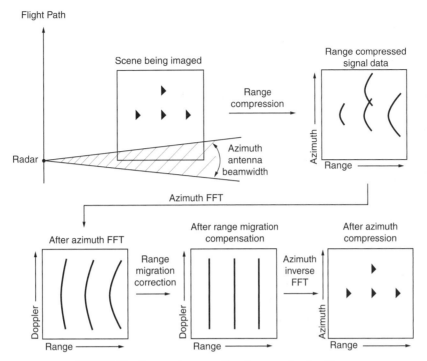

FIGURE 13   Key steps in a range-Doppler processing algorithm.

means of a 1-D along-track FFT. Operation in this 2-D wavenumber domain differentiates the RMA from range-Doppler algorithms. Next, a matched filter operation removes from all scatterers the along-track quadratic phase variation and range curvature associated with a scatterer located at swath center.

While this operation perfectly compensates the range curvature of scatterers located along swath center, it provides only partial compensation for scatterers at other ranges. In the next step, a 1-D coordinate transformation in the range frequency coordinate (known as the *Stolt interpolation*) removes the residual

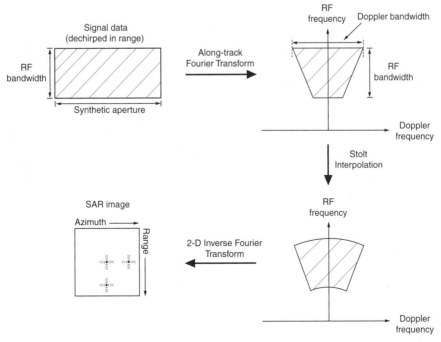

FIGURE 14   Key steps in RMA processing.

range curvature of all scatterers. Finally, a two-dimensional inverse FFT compresses the signal data in both range and azimuth to achieve the desired image.

The RMA outperforms other algorithms in situations in which differential range curvature is excessive. These situations are likely to occur in operations either at fine resolution, with a low center frequency, at a short standoff range, or with a large scene size. Thus, the RMA is a natural choice for processing fine-resolution stripmap imagery at VHF and UHF bands for FOPEN applications. With appropriate preprocessing of the signal data, the RMA can be a viable choice for spotlight processing applications as well [5].

The chirp scaling algorithm (CSA) requires SAR input data possessing chirp signal characteristics in both range and azimuth. Related to the RMA, the CSA requires only FFTs and complex multiplies to form a well-focused image of a large-scene; it requires no digital interpolations. This attribute often makes the CSA an efficient and practical alternative to the RMA.

The CSA avoids interpolation by approximating the Stolt transformation step of the RMA with a *chirp scaling* operation [8]. This operation applies a Doppler-dependent quadratic phase function to the range chirped data after an FFT of the azimuth chirped data. This process approximately equalizes DRC over the full swath width and permits partial range curvature compensation of all scatterers with a subsequent matched filtering step. With its efficiency and good focusing performance, the CSA and its various extensions have become standard image formation techniques for commercial and scientific orbital SAR systems that operate with coarse to medium resolutions over large swath widths.

3.4 Image Formation in the Spotlight Mode

In the spotlight mode, successively transmitted pulses interrogate the fixed scene being imaged at successively increasing cone angles as the antenna proceeds past the scene. This vision suggests the storage of collected pulses in a polar format for signal processing. In fact, the polar format algorithm (PFA) is the standard approach for image formation in the fine-resolution spotlight mode.

The PFA requires SAR signal data after dechirp in range. Such data occur naturally in systems employing dechirp-on-receive hardware. Unlike the range migration algorithm, the PFA requires that the receiver (or the IFP) remove the natural azimuth chirp characteristics of the collected signals. We refer to this procedure of removing the natural chirp when compensating the received data for random sensor motion as *motion compensation to a point*. This fixed reference point becomes scene center in the spotlight image.

The use of motion compensation to scene center completely removes the effect of MTRC from a scatterer at scene center and partially removes it from other scatterers. The PFA removes most of the remaining effects of MTRC by its choice of a data-storage format for signal processing. Using a 2-D interpolation, the al-

gorithm maps returns from successively transmitted pulses in an annular shape. It locates each return at a polar angle that tracks the increasing cone angle between the antenna velocity vector and its line-of-sight to scene center as the antenna proceeds past the scene. It locates the returns at a radial distance proportional to the radio frequency of the transmitted pulse. Figure 15 illustrates this data-storage format and its similarity to the data-collection geometry, particularly in terms of the Doppler cone angle α_d. The combination of motion compensation to a point and polar formatting leaves a small residual effect of MTRC that we call *range curvature phase error* in discussions of the PFA.

Range curvature phase error introduces geometric distortion in the image from residual linear phase effects and causes image defocus from quadratic and higher order phase effects. Based on sensor and data-collection parameters, these effects are deterministic and vary in severity over the scene. The digital processor is able to correct the geometric distortion by resampling the processed image to remove the deterministic distortion. The processor cannot easily remove the image defocus resulting from range curvature because the amount of defocus varies over the scene. Because the amount of defocus increases with distance from scene center, the usual method of dealing with it is simply to limit the processed scene to a size that keeps defocus to an acceptable level. A typical criterion allows $\pi/2$ rad of quadratic phase error. This criterion restricts the allowable scene radius r_o to

$$r_o \leq 2\rho_a \sqrt{\frac{R_{ac}}{\lambda_c}}, \tag{4}$$

where R_{ac} is the midaperture range between scene center and the SAR antenna [5]. As an example, a system design using $\lambda_c = 0.03$ m, $\rho_a = 0.3$ m, and $R_{ac} = 10$ km limits r_o to 346 m.

To process a larger scene, it is common to divide the scene into sections, process each section separately, and mosaic the sections together to yield an image of the entire illuminated scene. This subpatch processing approach can become inefficient because the IFP must process the collected signal data multiple times in order to produce the final output image. Amplitude and phase discontinuities are invariably present at section boundaries. Significant amplitude discontinuities affect image interpretability, while phase discontinuities impact utility in interferometry and other applications that exploit image phase.

The PFA requires a 2-D interpolation of digitized signal data to achieve the polar storage format. The IFP typically implements this 2-D interpolation separably in range and azimuth by means of two passes of 1-D finite impulse response filters [5].

The PFA is an important algorithm in fine-resolution SAR image formation because it removes a large component of MTRC in an efficient manner. In addition, the PFA is attractive because it can perform numerous secondary compensations along the way. These compensations include range and azimuth downsampling to reduce computational load, autofocus to remove

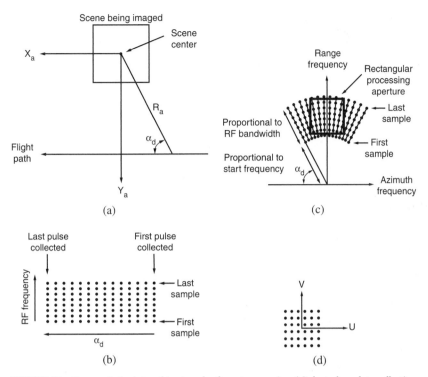

FIGURE 15 Geometrical relationships in polar format processing: (a) slant plane data-collection geometry; (b) signal data in rectangular format; (c) polar formatted signal data; (d) signal data after polar-to-rectangular interpolation.

unknown phase errors, and resampling to change the image display geometry. As a result, use of the PFA is common in many operational reconnaissance SAR systems.

4 Image Enhancement

The magnitude and phase of each image pixel can have significance in image exploitation. Additionally, the geometric relationship (mapping) between image pixel location and scatterer location in 3-D target space is an important aid in target detection, classification, and identification applications. It is the function of image enhancement algorithms to improve or accentuate these image characteristics for image understanding and information extraction.

The complex nature of the SAR image extends the capability of image enhancement algorithms to vary the quality and nature of the image. Important enhancement functions include autofocus, impulse response shaping, geometric distortion correction, intensity remapping, and noncoherent integration. Autofocus and distortion correction improve image quality by addressing deficiencies in the image formation process. Impulse response shaping and intensity remapping provide a capability to adjust image characteristics to match a specific application. Noncoherent integration smoothes speckle noise by noncoherently summing multiple images of the same scene collected at different frequencies or cone angles.

These image enhancement functions are standard considerations in SAR image improvement. In this section, we describe autofocus algorithms and impulse response shaping in detail and briefly discuss the remaining image enhancement functions.

4.1 Autofocus Algorithms

The synthetic aperture achieves fine cross-range resolution by adjusting the relative phase among signals received from various pulses and coherently summing them to achieve a focused image. A major source of uncertainty in the relative phase among these signals is the exact location of the radar antenna at the time of transmission and reception of each pulse. Location accuracy of a small fraction of a wavelength is necessary, perhaps to a few millimeters in the case of X-band operation at a 10-GHz center frequency. Without this location accuracy, phase errors will exist across the azimuth signal aperture and cause image distortion, defocus, and loss of contrast. Other hardware and software sources of phase error also are likely to be present, even in a well-designed system.

The high probability of significant phase error in the azimuth channel of a SAR system operating at fine resolution (typically better than 1-m azimuth resolution) necessitates the use of algorithms during or following image formation to measure and remove this phase error. We refer to the process that automatically estimates and compensates for phase error as *autofocus*. We describe two common autofocus algorithms in this chapter, the mapdrift algorithm and the phase gradient autofocus (PGA). The mapdrift algorithm is ideal for detecting and removing low-frequency phase error that causes image defocus. By low frequency, we mean phase error that varies slowly (for example, a quadratic or cubic variation) over the aperture. The PGA is an elegant algorithm designed to detect both low-frequency phase error and high-frequency phase error that varies rapidly over the aperture. High-frequency phase error primarily degrades image contrast.

Originating at Hughes Aircraft Corporation in the mid-1970s, the mapdrift algorithm became the first robust autofocus procedure to see widespread use in operational SAR systems. While mapdrift estimates quadratic and cubic phase errors best, it also extends to higher-frequency phase error [9]. With the aid of Fig. 16, we illustrate use of the mapdrift concept to detect and estimate an azimuth quadratic phase error with center-to-edge phase of Q over an aperture of length L. This error has the form $\exp(j2\pi k_q x^2)$, where x is the azimuth coordinate and k_q is the quadratic phase coefficient being measured. In its quadratic mode, mapdrift begins by dividing the signal data into two halves (or subapertures) in azimuth, each of length $L/2$. Mapdrift forms separate, but similar, images (or *maps*) from each subaperture. This process degrades the azimuth resolution of each map by a factor of 2 relative to the full-aperture image. Viewed separately over each subaperture, the original phase effect includes identical constant and quadratic components but a linear com-

ponent of opposite slope in each subaperture. Mapdrift exploits the fact that each subaperture possesses a different linear phase component. A measurement of the difference between the linear phase components over the two subapertures leads to an estimate of the original quadratic phase error over the full aperture. The constant phase component over each subaperture is inconsequential, while the quadratic phase component causes some defocus in the subaperture images that is not too troublesome.

By the Fourier shift theorem, a linear phase in the signal domain causes a proportional shift in the image domain. By estimating the shift (or *drift*) between the two similar maps, the mapdrift algorithm estimates the difference in the linear phase component between the two subapertures. This difference is directly proportional to Q. Most implementations of mapdrift measure the drift between maps by locating the peak of the cross-correlation of the intensity (magnitude squared) maps. After mapdrift estimates the error, a subsequent step removes the error from the full data aperture by multiplying the original signal by a complex exponential of unity magnitude and phase equal to the negative of the estimated error. Typical implementations improve algorithm performance by iterating the process after removing the current error estimate. Use of more than two subapertures to extend the algorithm to higher frequency phase error is rare because of the availability of more capable higher-order techniques, such as the PGA algorithm.

The PGA entered the SAR arena in 1989 as a method to estimate higher-order phase errors in complex SAR signal data [10, 11]. Unlike mapdrift, the PGA is a nonparametric technique in that it does not assume any particular functional model (for example, quadratic) for the phase error. The PGA follows an iterative procedure to estimate the derivative (or *phase gradient*) of a phase error in one dimension. The underlying idea is simple. The phase of the signal that results from isolating a dominant scatterer within an image and inverse Fourier transforming it in azimuth is a measure of the azimuth phase error in the signal data.

The PGA iteration cycle begins with a complex image that is focused in range but possibly blurred in azimuth by the phase error being estimated. The basic procedure isolates (by windowing) the image samples containing the azimuth impulse response of the dominant scatterer within each range bin and inverse Fourier transforms the windowed samples. The PGA implementation estimates the phase error in azimuth by measuring the change (or gradient) in phase between adjacent samples of the inverse transformed signal in each range bin, averaging these measurements over all range bins, and integrating the average. The algorithm then removes the estimated phase error from the original SAR data and proceeds with the next iteration. A number of techniques are available for selecting the initial window width. Typical implementations of the PGA decrease the window width following each iteration of the algorithm.

Figure 17 demonstrates use of the PGA to focus a 0.3-m resolution stripmap image of the University of Michigan engineering campus. The image in Fig. 17(a) contains a higher-order phase error in azimuth that seriously degrades image quality.

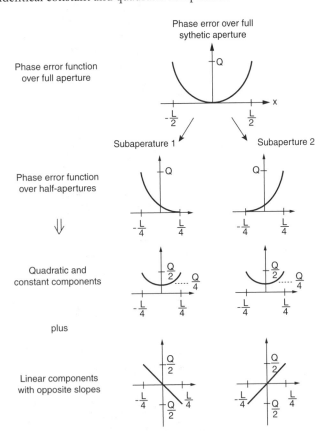

FIGURE 16 Subaperture phase characteristics in the mapdrift concept.

(a)   (b)

FIGURE 17   PGA algorithm example: (a) input image degraded with simulated phase errors; (b) output image after autofocus.

Figure 17(b) shows the focused image that results after three iterations of the PGA algorithm. This comparison illustrates the ability of the PGA to estimate higher-order phase errors accurately. While the presence of numerous dominant scatterers in this example eases the focusing task considerably, the PGA also exhibits robust performance against scenes without dominant scatterers.

4.2 Impulse Response Shaping

In the absence of errors, the impulse response of the SAR imaging system is the Fourier transform of the aperture weighting function. An unweighted (constant amplitude and phase) aperture yields a $\sin(x)/x$ impulse response. Control of the sidelobes of the impulse response is important in order to maintain image contrast and avoid interference with weaker nearby targets by a stronger scatterer. Conventional aperture weighting generally involves amplitude tapering at the data aperture edges to reduce their contribution to sidelobe energy. This type of weighting always widens the mainlobe as a consequence of reducing the energy in the sidelobes. Widening the mainlobe degrades resolution as measured by the −3-dB width of the impulse response function. Figure 18 compares the intensity impulse responses from an unweighted aperture and from −35-dB Taylor weighting, a popular choice for fine-resolution SAR imagery. With this weighting function, the first sidelobe is 35 dB below the mainlobe peak, compared with 13 dB without weighting. The weighted −3-dB mainlobe width is 1.3 times that in the unweighted case.

Dual apodization is a new approach to impulse response shaping for SAR imagery [12, 13]. In this approach, an algorithm generates two images from the same signal data, one using an unweighted aperture and one using heavy weighting that suppresses sidelobes and widens the mainlobe width. Logic within the algorithm compares the magnitude of the unweighted image with that resulting from the heavy weighting on a pixel-by-pixel

basis. This logic saves the minimum value at each pixel location to represent that pixel in the output image. In this way, dual apodization attempts to preserve both the narrow width of the unweighted aperture and the low sidelobe levels of the weighted aperture.

Our example of dual apodization compares the unweighted image with that resulting from half-cosine weighting, which we select specifically for use in a dual-apodization operation. Figure 19(a) illustrates the half-cosine weighting. Alone, half-cosine weighting is not useful because it greatly degrades the mainlobe of the impulse response. However, as a partner in dual apodization with the unweighted aperture, it performs adeptly to minimize sidelobes without increasing mainlobe width. Figures 19(b) and 19(c) show the weighted and unweighted impulse responses. Unlike many aperture weighting functions that do not significantly change the zero crossings of the impulse response function, half-cosine weighting does shift the zero crossings relative to those of the unweighted aperture. Figure 19(d) indicates the impulse response resulting from dual apodization. This result maintains the width of the unweighted aperture and the sidelobe levels of the half-cosine weighted aperture. Dual apodization with this pair of weightings requires that we multiply the magnitude of the weighted image by a factor of 2 before comparison to balance the reduction in amplitude from weighting. Figure 20 compares a SAR image containing a number of strong targets using an unweighted aperture and using this dual-apodization pairing.

Space variant apodization (SVA) is a step beyond dual apodization that uses logic queries regarding the phase and amplitude relationships among neighboring pixels to determine whether a particular pixel consists of primarily mainlobe energy, primarily sidelobe energy or a combination of the two [12, 13]. The logic directs the image enhancement algorithm to zero out the sidelobe pixels, maintain the mainlobe pixels, and suppress

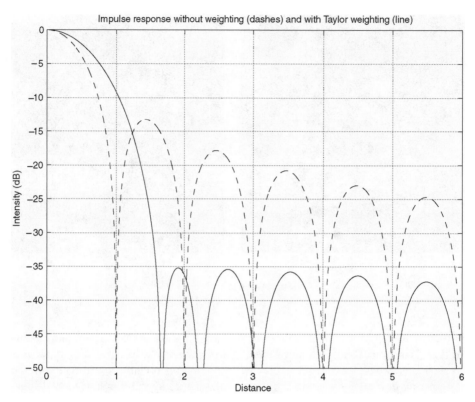

FIGURE 18   Effect of Taylor weighting on mainlobe width and sidelobe levels.

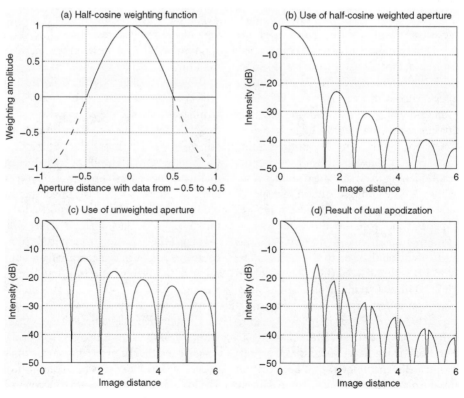

FIGURE 19   Impulse response comparison using dual apodization (half-cosine).

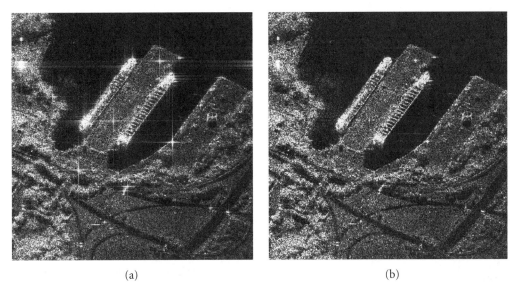

(a)   (b)

FIGURE 20 Image example using dual apodization: (a) original image with unweighted aperture; (b) image with dual apodization (half-cosine).

the pixels of mixed origin. The operation of SVA to zero out sidelobe pixels introduces some suppression of clutter patterns. Reference [5] supplements the original papers with heuristic explanations of SVA and SAR image examples.

4.3 Other Image Enhancement Functions

Other image improvement options include geometric distortion correction, intensity remapping, and noncoherent integration. Geometric distortion refers to the improper positioning of scatterers in the output image with respect to their true position when viewed in a properly scaled image display plane. Correction procedures remove the deterministic component of geometric distortion by resampling the digital SAR image from distorted pixel locations to undistorted locations. Intensity remapping refers to a (typically) nonlinear transformation between input pixel intensity values and output intensity. Such a remapping operation is particularly important when displaying SAR imagery in order to preserve the wide dynamic range inherent in the digital image data (typically 50 to 100 dB). Noncoherent integration refers to a process that detects the amplitude of SAR images (thereby eliminating the phase) and averages a number of these detected images taken at slightly different cone angles in order to reduce the variance of the characteristic speckle that naturally occurs in SAR images.

Geometric distortion arises largely from an inadequacy of the IFP algorithm to compensate for the geometrical relationships inherent in the range/angle imaging process. When necessary to satisfy image quality requirements, an image enhancement module after image formation compensates for deterministic distortion by interpolating between sample points of the original image to obtain samples on an undistorted output grid. This digital resampling operation (or interpolation) effectively unwarps the distorted image in order to reinstate geometric fidelity into the output image.

Intensity remapping is necessary and valuable because the wide dynamic range (defined as the ratio between the highest intensity scatterer present and system noise) inherent in radar imagery greatly exceeds that of common display media. It is often desirable to examine stronger targets in their natural background of terrain or in the presence of weaker targets. The common approach to remapping sets input pixels below a lower threshold level to zero, sets input pixels above an upper threshold level to that level, and maps pixels in between from input to output according to a prescribed (generally nonlinear) mapping rule. One popular remapping rule performs a linear mapping of image pixels having lower intensity and a logarithmic mapping of pixels having higher intensity. The output of this *linlog* mapping is typically an image with 8-bit samples that retains the proper linear relationship among the intensities of low-level scattering sources (such as terrain), yet compresses the wide dynamic range of the strongest scatterers (typically manmade, metallic objects).

Noncoherent integration (or multilook averaging) of fine-resolution radar images allows the generation of radar images with an almost optical-like appearance. This process smoothes out the pixel-to-pixel amplitude fluctuations (speckle noise) associated with a coherent imaging system. By including scatterers sensed at a multitude of cone angles, it adds detail to the target signature to enhance identification and provide a more literal image appearance. Figure 21 shows a fine-resolution SAR image of an automobile resulting from noncoherent summation of 36 images collected at unique cone angles.

5 Image Exploitation

The value of imagery is in its use. Information inherent in image data must be identified, accessed, quantified, often calibrated, and developed into a usable and observable form. Observation

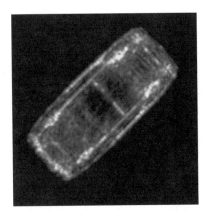

FIGURE 21 Use of noncoherent intergration to fill in the target signature.

may involve visual or numerical human study or automatic computer analysis.

An image naturally presents a spatial perspective to an observer with a magnitude presentation of specific features or characteristics. Beyond this presentation, SAR imagery offers additional information related to its coherent nature, with meaningful amplitude and phase associated with each pixel. This complex nature of SAR imagery represents special value when the image analyst can relate it to target or data-collection characteristics of value in specialized military or civilian applications.

Some examples of these special applications of SAR image data include moving target detection (possibly with tracking and focusing) using a single image, and digital terrain elevation data (DTED) extraction by means of interferometry using multiple images collected at different depression angles. We discuss these two applications in detail below.

Additional applications of a single SAR image include glint detection, automated road finding and following, and shadow exploitation. Glints (or specular flashes) refer to bright returns off the edges of linear surfaces, characteristic of manmade structures such as aircraft wings. Road finding and shadow detection naturally involve searches for low return areas in the image. Additional applications involving multiple images include target characterization using polarization diversity, and change detection using images of the same area collected at different times from a similar perspective. Differences in signatures from both terrain and cultural features as a function of the polarization characteristics of transmit and receive antennas support target classification and identification tasks. Change detection generally involves the subtraction of two detected images collected at different times. Image areas that are unchanged between collections will experience significant cancellation while features that have changed will not cancel, making the changes easier to identify.

5.1 Moving Target Detection

Target motion during the coherent aperture time used to generate azimuth resolution disturbs the pulse-to-pulse phase coherence required to produce an ideal impulse response function. The result is azimuth phase error in the signals received from moving target scatterers. In conventional SAR imagery, such phase error causes azimuth smearing of the moving target image. In the simple case of a target moving at constant velocity parallel to the antenna path (along-track velocity) or at constant acceleration toward the antenna (line-of-sight acceleration), the phase error is quadratic and the image smearing is proportional to the magnitude of the motion [5]. This image effect offers both a basis for detection of a moving target and a hope of refocusing the moving target image after image formation [14]. In the simple-motion case presented here, the image streak corresponding to a moving scatterer possesses a quadratic phase in the image deterministically related to the value of the target motion parameter and to the quadratic phase across the azimuth signal data. This quadratic phase characteristic of the streaks in the image offers an interesting approach to automatic detection and refocusing of moving targets in conventionally processed SAR images.

Equations relating target velocity to quadratic phase error in both domains and to streak length are well known [5]. A target moving with an along-track velocity V_{tat} parallel to the antenna velocity vector introduces a quadratic phase error across the azimuth signal data. The zero-to-peak size Q_{Vtat} of this phase effect is

$$Q_{Vtat} = \frac{\pi V_{tat} T_a}{2\rho_a S_{\alpha c}}. \tag{5}$$

Here, T_a is the azimuth aperture time and $S_{\alpha c}$ is the sine of the cone angle at aperture center.

Conventional image formation processing of the resulting signal data produces an azimuth streak in the image for each scattering center of the target. The length L_s of each streak is roughly

$$L_s = \frac{2 V_{tat} T_a}{S_{\alpha c}}. \tag{6}$$

Each image streak has a quadratic phase characteristic along its length of the same size but opposite sign as the phase effect in the signal data before the Fourier transform operation that produces the image. Figure 22 indicates these relationships. Line-of-sight target acceleration introduces a similar quadratic phase effect, while more complicated motions introduce higher order (for example, cubic, quartic and sinusoidal) phase effects.

A simple algorithm for automated detection of moving target streaks in conventional SAR imagery utilizes this low-frequency (largely quadratic) phase characteristic of the image streaks representing moving target scatterers. The procedure is to calculate the pixel-to-pixel change in phase in the azimuth direction along each range bin of the image. Normal stationary SAR image background areas including stronger extended targets such as trees and shrubbery vary almost randomly in phase from pixel to pixel while the streaks associated with moving scatterers vary more slowly and regularly in phase. This smooth phase derivative from azimuth pixel to azimuth pixel differentiates moving scatterers from stationary scatterers in a way easily detected by an automated process.

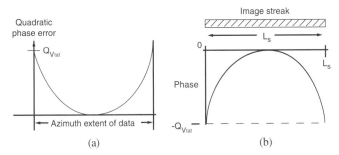

FIGURE 22 Characteristics of moving target signals: (a) phase associated with signal data; (b) phase along image data.

Figure 23(a) displays a 0.3-m resolution SAR image that includes a group of streaks associated with a defocused moving target. In this image, the horizontal coordinate is range and the vertical coordinate is azimuth. The moving target streaks are the brighter returns extending over much of the azimuth extent of the scene. The phase along each streak is largely quadratic. Figure 23(b) displays the azimuth derivative of the phase of this image from $-\pi$ change (dark) to $+\pi$ change (light). Various averaging, filtering, and thresholding operations in this phase derivative space will easily and automatically detect the moving target streak in the background. For instance, one simple approach detects areas where the second derivative of phase in azimuth is small.

A measure of L_s in Fig. 23, along with Eqs. (5) and (6), provides an estimate of the quadratic defocus parameter associated with this image. A moving target focus algorithm can make this estimate of defocus and apply a corrective phase adjustment to the original signal data to improve the focus of this moving target image. Ideally, this process generates a signature of the moving tar-

get identical to that of a similar stationary target. In reality, target motion is significantly more complex than that modeled here. In addition, the moving target streaks often do not stand out as well from the background as they do in this particular image. However, sophisticated implementations of this simple algorithm can provide reasonable detection performance, even for a relatively low ratio of target streak intensity to background intensity.

5.2 SAR Interferometry

SAR interferometry requires a comparison of two complex SAR images collected over the same Doppler cone angle interval but at different depression angles. This comparison provides an estimate of the depression angle from the sensor to each pixel in the image. Figure 24(a) illustrates an appropriate data-collection geometry using a vertical interferometer (second antenna directly below first antenna). Information on the depression angle from the sensor to each pixel in the image, along with the cone angle and range provided by a single SAR image, locates scatterers in three dimensions relative to the sensor location and velocity vector. With information about these sensor parameters, absolute height and horizontal position is available to generate a digital terrain elevation map. A natural product of SAR interferometry is a height contour map. Figure 6 presents example products from a modern interferometric SAR system. Major applications encompass both civilian and military activities.

We use the vertical interferometer in Fig. 24(a) to illustrate the geometrical basis for determining depression angle. The image from the first antenna locates the scatterer P_1 on the range-Doppler circle C_1 in a plane orthogonal to the sensor velocity vector. The image from the second antenna locates P_1 on the range-Doppler circle C_2. The point P_2 is the center of both

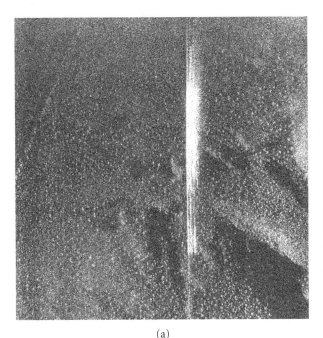

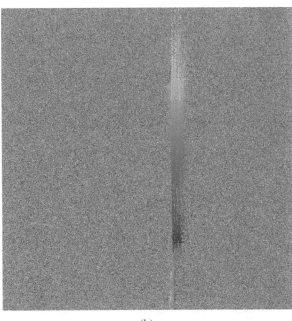

FIGURE 23 Example of moving target detection: (a) SAR image with moving target present; (b) phase derivative of image.

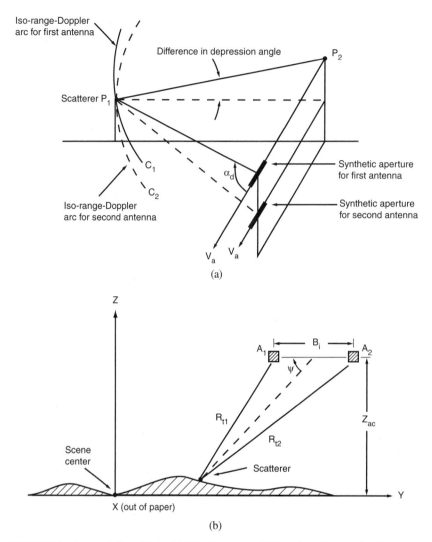

FIGURE 24 Geometrical models for SAR interferometry: (a) basis for estimating the depression angle; (b) model for interferometric analysis.

circles. In the absence of errors, the intersection of the two circles identifies the location of P_1.

The mathematical basis and sensitivity of SAR interferometry is readily available in the published literature [15–17]. To summarize the equations that characterize the interferometric SAR function, we use the horizontal interferometer illustrated in Fig. 24(b). The two antennas A_1 and A_2 are at the same height. They are separated by a rigid baseline of length B_i orthogonal to the flight line. Each antenna illuminates the same ground swath in a broadside imaging direction. The sensor travels in the **X** direction, ψ is the nominal depression angle from the interferometer to the scatterer relative to the horizontal baseline, and Z_{ac} is the height of the interferometer above the nominal ground plane XY.

Following image registration, multiplication of the first image by the complex conjugate of the second image yields the phase difference between corresponding pixels in the two images. For a particular scatterer, this phase difference is proportional to the difference in range to the scatterer from each antenna. This range difference, $R_{t1} - R_{t2}$ in Fig. 24(b), is adequate information to determine the depression angle to the scatterer. Without resolving the natural 2π ambiguity in the measurement of phase, this phase difference provides an estimate of only the difference in depression angle between the scatterers represented by image pixels rather than their absolute depression angle. The relationship between relative depression angle $\Delta\psi$ and the difference $\Delta\phi_{12}$ between pixels in this phase difference between images is

$$\Delta\psi = -\frac{\lambda_c \Delta\phi_{12}}{4\pi B_i \sin(\psi)}. \tag{7}$$

Two pixels with an interferometric phase difference $\Delta\phi_{12}$ differ in depression angle by $\Delta\psi$. A change in $\Delta\phi_{12}$ corresponds to a change in height Δh given by [5]:

$$\Delta h = -K_h \Delta\phi_{12}, \tag{8}$$

with

$$K_h = -\frac{\lambda_c Z_{ac} \cot(\psi)}{4\pi B_i \sin(\psi)}. \tag{9}$$

As an example, we consider an interferometer with horizontal baseline $B_i = 1$ m, center frequency $\lambda_c = 0.03$ m, operating at a depression angle $\psi = 30$ deg from a height $Z_{ac} = 4$ km. We have the coefficient $K_h = -33.1$ m/rad $= -0.58$ m/deg; thus 10 deg of interferometric phase difference corresponds to 5.8 m of height change.

6 Chapter Summary

Microwave imaging has been an attractive technology since its early roots in the World War II era, largely because of its potential for 24-hour remote surveillance in all weather conditions. In recent years, particularly with the advent of the synthetic aperture radar approach to realizing fine azimuth resolution, microwave imagery has come to represent a powerful remote sensing capability. With today's fine-resolution SAR techniques, the finest radar imagery begins to take on the appearance of optical imagery to which we are naturally accustomed. For many applications, the utility of SAR imagery greatly exceeds that of comparable optical imagery.

Four factors contribute significantly to this advanced state of radar imaging. First, advances in SAR sensor hardware technology (particularly with respect to resolving capability) provide the inherent information within the raw SAR data received by the radar sensor. Second, recent developments in image formation algorithms and computing systems provide the capability to generate a digital image in a computationally efficient manner that preserves the inherent information content of the raw radar signals. A combination of requirements on airborne SAR for finer and finer resolution in various military applications and requirements on orbital SAR for wide area coverage in natural resource and environmental applications provided the impetus for these developments. Third, improvements in image quality by means of state-of-the-art image enhancement algorithms extend the accessibility of information and emphasize that information of interest to the specialized user of SAR imagery. Autofocus and space-variant apodization exemplify these image quality improvements. Finally, an explosion in powerful exploitation techniques to extract information coded in the phase as well as the amplitude of SAR imagery multiplies the value of the radar imagery to the end user.

Acknowledgment

The image in Figure 5 is copyright Canadian Space Agency, 1998. All other SAR images are courtesy of ERIM International Incorporated.

References

[1] R. K. Raney, A. P. Luscombe, E. J. Langham, and S. Ahmed, "RADARSAT," *Proc. IEEE* **79**, 839–849 (1991).

[2] G. T. Sos, H. W. Klimach, and G. F. Adams, "High performance interferometric SAR description and capabilities," presented at the 10th Thematic Conference on Geologic Remote Sensing, San Antonio, TX, May 9–12, 1994.

[3] D. R. Sheen, S. J. Shackman, N. L. VandenBerg, D. L. Wiseman, L. P. Elenbogen, and R. F. Rawson, "The P-3 UltraWideband SAR: description and examples," presented at the 1996 IEEE National Radar Conference, Ann Arbor, MI, May 1996.

[4] M. A. DiMango, W. T. Hanna, and L. Andersen, "The Data Collection System (DCS) Airborne Platform," Record of the First International Airborne Remote Sensing Conference and Exhibition, Stasbourg, France, September 1994.

[5] W. G. Carrara, R. S. Goodman, and R. M. Majewski, *Spotlight Synthetic Aperture Radar: Signal Processing Algorithms* (Artech House, Boston, MA, 1995).

[6] J. C. Curlander and R. N. McDonough, *Synthetic Aperture Radar: Systems and Processing* (Wiley, New York, 1991).

[7] F. Rocca, C. Cafforio, and C. Prati, "Synthetic aperture radar: a new application for wave equation techniques," *Geophys. Prospect.* **37**, 809–830 (1989).

[8] R. K. Raney, H. Runge, R. Bamler, I. G. Cumming, and F. H. Wong, "Precision SAR processing using chirp scaling," *IEEE Trans. Geosci. Remote Sens.* **32**, 786–799 (1994).

[9] C. E. Mancill and J. M. Swiger, "A map drift autofocus technique for correcting higher order SAR phase errors (U)," presented at the 27th Annual Tri-Service Radar Symposium Record, Monterey, CA, June 23–25, 1981.

[10] P. Eichel, D. Ghiglia, and C. Jakowatz, Jr., "Speckle processing method for synthetic aperture radar phase correction," *Opt. Lett.* **14**, 1–3 (1989).

[11] D. E. Wahl, P. H. Eichel, D. C. Ghiglia, and C. V. Jakowatz, Jr., "Phase gradient autofocus — a robust tool for high resolution SAR phase correction," *IEEE Trans. Aerospace Electron. Syst.* **30**, 827–834 (1994).

[12] H. C. Stankwitz, R. J. Dallaire, and J. R. Fienup, "Spatially variant apodization for sidelobe control in SAR imagery," presented at the 1994 IEEE National Radar Conference, March 1994.

[13] H. C. Stankwitz, R. J. Dallaire, and J. R. Fienup, "Nonlinear apodization for sidelobe control in SAR imagery," *IEEE Trans. Aerospace Electron. Syst.* **31**, 267–279 (1995).

[14] S. Werness, W. Carrara, L. Joyce, and D. Franczak, "Moving target imaging algorithm for SAR data," *IEEE Trans. Aerospace Electron. Syst.* **26**, 57–67 (1990).

[15] E. Rodriguez and J. M. Martin, "Theory and design of interferometric synthetic aperture radars," *IEE Proc. F* **139**, 147–159 (1992).

[16] H. Zebker and R. M. Goldstein, "Topographic mapping from interferometric synthetic aperture radar observations," *J. Geophys. Res.* **91**, 4993–4999 (1986).

[17] H. Zebker and J. Villasensor, "Decorrelation in interferometric radar echoes," *IEEE Trans. Geosci. Remote Sens.* **30**, 950–959 (1992).

10.2

Computed Tomography

R. M. Leahy
*University of Southern
California*

R. Clackdoyle
University of Utah

1 Introduction

The term *tomography* refers to the general class of devices and procedures for producing two-dimensional (2-D) cross-sectional images of a three-dimensional (3-D) object. Tomographic systems make it possible to image the internal structure of objects in a noninvasive and nondestructive manner. By far the best known application is the computer assisted tomography (CAT or simply CT) scanner for X-ray imaging of the human body. Other medical imaging devices, including PET (positron emission tomography), SPECT (single photon emission computed tomography) and MRI (magnetic resonance imaging) systems, also make use of tomographic principles. Outside of the medical realm, tomography is used in diverse applications such as microscopy, nondestructive testing, radar imaging, geophysical imaging, and radio astronomy.

We will restrict our attention here to image reconstruction methods for X-ray CT, PET, and SPECT. In all three modalities the data can be modeled as a collection of line integrals of the unknown image. Many of the methods described here can also be applied to other tomographic problems. The reader should also refer to Chapter 3.6 for a more general treatment of image reconstruction in the context of ill-posed inverse problems.

We describe 2-D image reconstruction from parallel and fan-beam projections and 3-D reconstruction from sets of 2-D projections. *Analytic methods* derived from the relationships between functions and their line integrals are described in Sections 3–5. In Section 6 we describe the class of *iterative methods* that are based on a finite dimensional discretization of the problem. We will include key results and algorithms for a range of imaging geometries, including systems currently in development. References to the appropriate sources for a complete development are also included. Our objective is to convey the wide range of methods available for reconstruction from projections and to highlight some recent developments in what remains a highly active area of research.

2 Background

2.1 X-Ray Computed Tomography

In conventional X-ray radiography, a stationary source and planar detector are used to produce a 2-D projection image of the patient. The image has intensity proportional to the amount by which the X-rays are attenuated as they pass through the body, i.e., the 3-D spatial distribution of X-ray attenuation coefficients

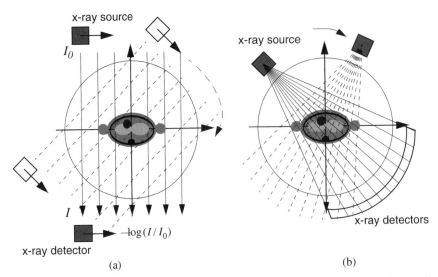

FIGURE 1   (a) Schematic representation of a first-generation CT scanner that uses translation and rotation of the source and a single detector to collect a complete set of 1-D parallel projections. (b) The current generation of CT scanners uses a fan X-ray beam and an array of detectors, which require rotation only.

is projected into a 2-D image. The resulting image provides important diagnostic information as a result of differences in the attenuation coefficients of bone, muscle, fat, and other tissues in the 40–120 keV range used in clinical radiography [1].

X-rays passing through an object experience exponential attenuation proportional to the linear attenuation coefficient of the object. The intensity of a collimated beam of monoenergetic X-radiation exiting a uniform block of material with linear attenuation coefficient μ and depth d is given by $I = I_0 e^{-\mu d}$, where I_0 is the intensity of the incident beam. For objects with spatially variant attenuation $\mu(z)$ along the path length z, this relationship generalizes to:

$$I = I_0 e^{-\int \mu(z)\,dz} \tag{1}$$

where $\int \mu(z)dz$ is a *line integral* through $\mu(z)$.

Let $\mu(x, y, z)$ represent the 3-D distribution of attenuation coefficients within the human body. Consider a simplified model of a radiography system that produces a broad parallel beam of X-rays passing through the patient in the z direction. An ideal 2-D detector array or film in the (x, y) plane would produce an image with intensity proportional to the negative logarithm of the attenuated X-ray beam, i.e., $-\log(I/I_0)$. The following projection image would then be formed at the ideal detector:

$$r(x, y) = \int \mu(x, y, z)\,dz \tag{2}$$

The utility of conventional radiography is limited because of the projection of 3-D anatomy into a 2-D image, causing certain structures to be obscured. For example, lung tumors, which have

a higher density than the surrounding normal tissue, may be obscured by a more dense rib that projects into the same area in the radiograph. Computed tomography systems overcome this problem by reconstructing 2-D cross sections of the 3-D attenuation coefficient distribution.

The concept of the line integral is common to the radiographic projection defined in Eq. (2) and to computed tomography. Consider the first clinical X-ray CT system for which the inventor, G. Hounsfield, received the 1979 Nobel prize in medicine (the prize was shared with mathematician A. Cormack) [2]. A collimated X-ray source and detector are translated on either side of the patient so that a single plane is illuminated, as illustrated in Fig. 1(a). After applying a logarithmic transformation, the detected X-ray measurements are a set of line integrals representing a 1-D parallel projection of the 2-D X-ray attenuation coefficient distribution in the illuminated plane. By rotating the source and detector around the patient, other 1-D projections can be measured in the same plane. The image can then be reconstructed from these *parallel-beam* projections using the methods described in Section 3.1.

One major limitation of the first-generation of CT systems was that the translation and rotation of the detectors was slow and a single scan would take several minutes. X-ray projection data can be collected far more quickly using the *fan-beam* X-ray source geometry employed in the current generation of CT scanners, as illustrated in Fig. 1(b). Since an array of detectors is used, the system can simultaneously collect data for all projection paths that pass through the current location of the X-ray source. In this case, the X-ray source need not be translated and a complete set of data is obtained through a single rotation of the source around the patient. Using this configuration, modern scanners can scan a single plane in less than 1s. Methods for reconstruction from fan-beam data are described in Section 3.2.

Recently developed spiral CT systems allow continuous acquisition of data as the patient bed is moved through the scanner [3]. The detector traces out a helical orbit with respect to the patient, allowing rapid collection of projections over a 3-D volume. These data require special reconstruction algorithms as described in Section 4.2. In an effort to simultaneously collect fully 3-D CT data, a number of systems have been developed that use a *cone-beam* of X-rays and a 2-D rather than 1-D array of detectors [3]. While cone-beam systems are rarely used in clinical CT, they play an important role in industrial applications. Methods for cone-beam reconstruction are described in Section 5.2.

The above descriptions can only be considered approximate because a number of factors complicate the X-ray CT problem. For example, the X-ray beam typically contains a broad spectrum of energies and therefore an energy dependence should be included in Eq. (1) [1]. The theoretical development of CT methods usually assumes a monoenergetic source. For broadband X-ray sources, the beam becomes "hardened" as it passes through the object; i.e., the lower energies are attenuated faster than the higher energies. This effect causes a beam hardening artifact in CT images that is reduced in practice by the use of a data calibration procedure [4].

In X-ray CT data the high photon flux produces relatively high signal-to-noise ratios. However, the data are corrupted by the detection of scattered X-rays that do not conform to the line integral model. Calibration procedures are required to compensate for this effect as well as for the effects of variable detector sensitivity. A final important factor in the acquisition of CT data is the issue of sampling. Each 1-D projection is undersampled by approximately a factor of 2 in terms of the attainable resolution as determined by detector size. This problem is dealt with in fan-beam systems by using fractional detector offsets or flying focal spot techniques [3].

2.2 Nuclear Imaging Using PET and SPECT

PET and SPECT are methods for producing images of the spatial distribution of biochemical tracers or "probes" that have been tagged with radioactive isotopes [1]. By tagging different molecules with positron or gamma-ray emitters, PET and SPECT can be used to reconstruct images of the spatial distribution of a wide range of biochemical probes. Typical applications include glucose metabolism and monoclonal antibody studies for cancer detection, imaging of cardiac function, imaging of blood flow and volume, and studies of neurochemistry using a range of neuroreceptors and transmitters [5, 6].

SPECT systems detect emissions by using a "gamma camera." This camera is a combination of a sodium iodide scintillation crystal and an array of photomultiplier tubes (PMTs). The PMTs measure the location on the camera surface at which each gamma ray photon is absorbed by the scintillator [1]. A mechanical collimator, consisting of a sheet of dense metal in which a large number of parallel holes have been drilled, is attached to the front of the camera as illustrated in Fig. 2(a). The collimated camera is only sensitive to gamma rays traveling in a direction parallel to the holes in the collimator. The total number of gamma rays detected at a given pixel in the camera will be approximately proportional to the total activity (or line integral) along the line that passes through the patient and is parallel to the holes in the collimator. Thus, when viewing a patient from a fixed camera position, we collect a 2-D projection image of the 3-D distribution of the tracer. By collecting data as the camera is rotated to multiple positions around the patient, we obtain *parallel-beam* projections for a contiguous set of parallel 2-D slices through the patient, as shown in Fig. 2(b). The distribution can be reconstructed slice by slice using the same parallel-beam reconstruction methods as are used for X-ray CT.

Other collection geometries can be realized by modifying the collimator design [7]. For imaging an organ such as the brain or

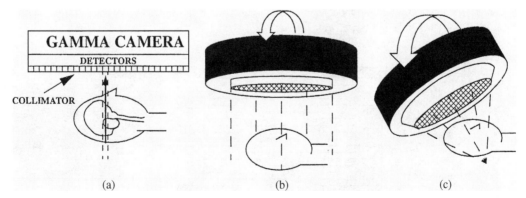

(a) (b) (c)

FIGURE 2 Schematic representation of a SPECT system: (a) cross-sectional view of a system with parallel hole collimator; gamma rays normally incident to the camera surface are detected, and others are stopped by the collimator so that the camera records parallel projections of the source distribution. (b) Rotation of the camera around the patient produces a complete set of parallel projections. (c) Different collimators can be used to collect converging or diverging fan- and cone-beam projections; shown is a converging cone-beam collimator.

heart, which is smaller than the surface area of the camera, improved sensitivity can be realized by using converging *fan-beam* or *cone-beam* collimators as illustrated in Fig. 2(c). Similarly, diverging collimators can be used for imaging larger objects. Images are reconstructed from these fan-beam and cone-beam data using the methods in Section 3.2 and Section 5.2, respectively. While the vast majority of SPECT systems use rotating planar gamma cameras, other systems have been constructed with a cylindrical scintillation detector that surrounds the patient. A rotating cylindrical collimator defines the projection geometry. Although the physical design of these cylindrical systems is quite different from that of the rotating camera, in most cases the reconstruction problem can still be reduced to one of the three basic forms: parallel, fan-beam, or cone-beam.

The physical basis for PET lies in the fact that a positron produced by a radioactive nucleus travels a very short distance and then annihilates with an electron to form a pair of high-energy (511 keV) photons [6]. The pair of photons travel in opposite directions along a straight line path. Detection of the positions at which the photon pair intersect a ring of detectors allows us to approximately define a line that contains the positron emitter, as illustrated in Fig. 3(a). The total number of photon pairs measured by a detector pair will be proportional to the total number of positron emissions along the line joining the detectors; i.e., the number of detected events between a detector pair is an approximate line integral of the tracer density.

A PET scanner requires one or more rings of photon detectors coupled to a timing circuit that detects coincident photon pairs by checking that both photons arrive at the detectors within a few nanoseconds of each other. PET detectors are usually constructed with a combination of scintillation crystals and PMTs. A unique aspect of PET is that the ring of detectors surrounding the subject allows simultaneous acquisition of a complete data set; no rotation of the detector system is required. A schematic view of two modern PET scanners is shown in Fig. 3. In the 2-D scanner, multiple rings of detectors surround the patient with dense material, or "septa," separating each ring. These septa stop photons traveling between rings so that coincidence events are collected only between pairs of detectors in a single ring. We refer to this configuration as a 2-D scanner because the data are separable and the image can be reconstructed as a series of 2-D sections. In contrast, the 3-D scanners have no septa so that coincidence photons can be detected between planes. In this case the reconstruction problem is not separable and must be treated directly in three dimensions.

PET data can be viewed as sets of approximate line integrals. In the 2-D mode the data are sets of *parallel-beam* projections and the image can be reconstructed by using methods equivalent to those in parallel-beam X-ray CT. In the 3-D case, the data are still line integrals but new algorithms are required to deal with the between-plane coincidences that represent *incomplete* projections through the patient. These methods are described in Sections 4 and 5.

As with X-ray CT, the line integral model is only approximate. Finite and spatially variant detector resolution is not accounted for in the line integral model and has a major impact on image quality [8]. The number of photons detected in PET and SPECT is relatively small so that photon-limited noise is also a factor limiting image quality. The data are further corrupted by additional noise that is produced by scattered photons. Also, in both PET and SPECT, the probability of detecting an emission is reduced by the relatively high probability of Compton scatter of photons before they reach the detector. These attenuation effects can be quantified by performing a separate "transmission" scan in which the scattering properties of the body are measured. This information must then be incorporated into the reconstruction algorithm [5, 6]. Although all of these effects can, to some degree, be compensated for within the framework of analytic reconstruction from line integrals, they are more readily and accurately dealt with by using the finite dimensional statistical formulations described in Section 6.

2.3 Mathematical Preliminaries

Since we deal with both 2-D and 3-D reconstruction problems here, we will use the following unified definition of the line

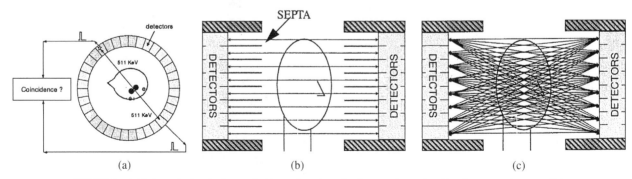

| (a) | (b) | (c) |

FIGURE 3   (a) Schematic showing how coincidence detection of a photon pair produced by electron-positron annihilation determines the line along which the positron was annihilated. (b) In 2-D systems septa between adjacent rings of detectors prevent coincidence detection between rings. (c) Removal of the septa produces a fully 3-D PET system in which cross-plane coincidences are collected and used to reconstruct the source distribution.

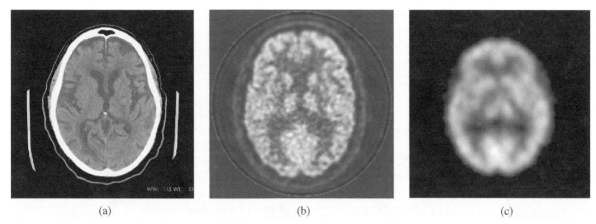

(a)   (b)   (c)

FIGURE 4 Examples of brain scans using: (a) X-ray CT, in which a nonlinear gray scale is used to enhance the contrast between soft tissue regions within the brain; (b) PET, showing an image of glucose metabolism obtained with an analog of glucose labelled with the positron emitting isotope, flourine-18; (c) SPECT, showing a brain perfusion scan using a technetium-99m ligand. (Courtesy of J. E. Bowsher, Duke University Medical Center.)

integrals of an image $f(\underline{x})$:

$$g(\underline{a}, \underline{\theta}) = \int_{-\infty}^{\infty} f(\underline{a} + t\underline{\theta})\, dt, \quad \|\underline{\theta}\| = 1. \qquad (3)$$

Here g is the integral of f over the line passing through $\underline{a}$ and oriented in the direction $\underline{\theta}$.

For parallel-beam data, each projection corresponds to a fixed $\underline{\theta}$ (the projection direction). To avoid redundant parameterization of line integrals we only consider those $\underline{a}$ perpendicular to $\underline{\theta}$ (i.e., $\underline{a} \cdot \underline{\theta} = 0$). We say a parallel projection $g(\cdot, \underline{\theta})$ is truncated if some nonzero line integrals are not measured. Generally, truncation occurs when a finite detector is too small to gather a complete projection of the object at some orientation $\underline{\theta}$.

For fan-beam and cone-beam systems, $\underline{a}$ is fixed for a single projection; $\underline{a}$ is the fan vertex or cone vertex, which in practice would be the position of the X-ray source or the focal point of a converging collimator. Again, truncation of a projection $g(\underline{a}, \cdot)$

refers to line integrals that are not available because of the limited extent of the detector.

2.4 Examples

We conclude this introductory section with examples in Figs. 4 and 5 of CT, PET, and SPECT images collected from the current generation of scanners. These images clearly reveal the differences between the high resolution, low noise images produced by X-ray CT scanners and the lower resolution and noisier images produced by the nuclear imaging instruments. These differences are primarily due to the photon flux in X-ray CT, which is many orders of magnitude higher than that for the individually detected photons in nuclear medicine imaging. In diagnostic imaging these modalities are highly complementary since X-ray CT reveals information about patient anatomy while PET and SPECT images contain functional information. For further insight into the ability of X-ray CT to produce high resolution

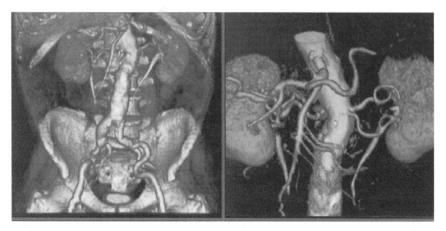

FIGURE 5 Volume rendering from a sequence of X-ray CT images, showing the abdominal cavity and kidneys (CT images courtesy of G. E. Medical Systems). (See color section, p. C–46.)

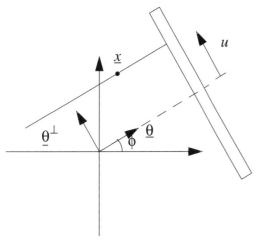

FIGURE 6   The coordinate system used to describe parallel-beam projection data.

anatomical images, we show a set of 3-D renderings from CT data in Fig. 5.

3 2-D Image Reconstruction

3.1 Fourier Space and Filtered Backprojection Methods for Parallel-Beam Projections

For 2-D parallel-beam projections, the general notation of Eq. (1) can be refined as illustrated in Fig. 6. We parameterize the direction of the rays using ϕ, so $\underline{\theta} = (\cos\phi, \sin\phi)$. For the position $\underline{a}$ perpendicular to $\underline{\theta}$, we write $\underline{a} = (-u\sin\phi, u\cos\phi) = u\underline{\theta}^{\perp}$, where u is the scalar coordinate indicating the distance

from the origin to the integration line, or equivalently, the projection element index for the ϕ projection. Since $\underline{\theta}$ depends only on ϕ, and $\underline{a}$ then depends on u, we simplify the notation by writing $g(u, \phi) = g(\underline{a}, \underline{\theta}) = \int f(\underline{a} + t\underline{\theta})\, dt$. For the parallel-beam case, the function g is the Radon transform of the image f [4].

Practical inversion methods can be developed by using the relationship between the Radon and Fourier transforms. The projection slice theorem is the basic result that is used in developing these methods [4]. This theorem states that the 1-D Fourier transform of the parallel projection at angle ϕ is equal to the 2-D image Fourier transform evaluated along the line through the origin in the direction $\phi + \pi/2$, i.e.,

$$G(U, \phi) = \int_{-\infty}^{\infty} g(u, \phi)e^{-juU}\, du = F(\underline{X})\Big|_{\underline{X}=U\underline{\theta}^{\perp}}$$
$$= F(-U\sin\phi,\, U\cos\phi), \tag{4}$$

where $F(\underline{X}) = F(X, Y)$ is the 2-D image Fourier transform

$$F(\underline{X}) = F(X, Y) = \int_{-\infty}^{\infty}\int_{-\infty}^{\infty} f(x, y)e^{-jxX}e^{-jyY}\, dx\, dy$$
$$= \iint_{R^2} f(\underline{x})e^{-j(\underline{X}\cdot\underline{x})}\, d\underline{x}. \tag{5}$$

This result, illustrated in Fig. 7, can be employed in a number of ways. The discrete Fourier transform (DFT, see Chapter 2.3) of the samples of each 1-D projection can be used to compute approximate values of the image Fourier transform. If the angular projection spacing is $\Delta\phi$, then the DFTs of all projections will produce samples of the 2-D image Fourier transform on a polar

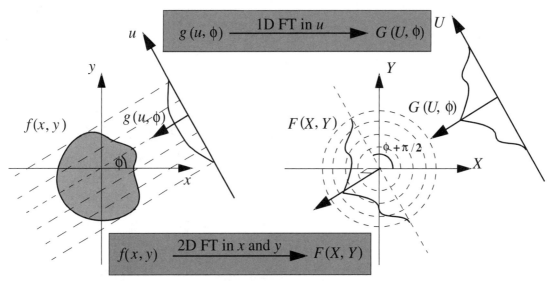

FIGURE 7   Illustration of the projection slice theorem. The 2-D image at the left is projected at angle ϕ to produce the 1-D projection $g(u, \phi)$. The 1-D Fourier transform, $G(U, \phi)$, of this projection is equal to the 2-D image Fourier transform, $F(X, Y)$, along the radial line at angle $\phi + \pi/2$.

sampling grid. The samples' loci lie at the intersections of radial lines, spaced by $\Delta\phi$, with circles of radii equal to integer multiples of the DFT frequency sampling interval. Once these samples are computed, the image can be reconstructed by first interpolating these values onto a regular Cartesian grid, and then applying an inverse 2-D DFT. Design of these Fourier reconstruction methods involves a tradeoff between computational complexity and accuracy of the interpolating function [4].

A more elegant solution can be found by reworking Eq. (4) into a spatial domain representation. It is then straightforward to show that the image can be recovered by using the following equations [9]:

$$f(\underline{x}) = \frac{1}{2} \int_0^{2\pi} \tilde{g}(u, \phi) \Big|_{u=u_{\underline{x},\phi}} d\phi, \qquad (6)$$

where

$$\tilde{g}(u, \phi) = \frac{1}{4\pi^2} \int_{-\infty}^{\infty} G(U, \phi)|U|e^{juU}\,dU \qquad (7)$$

and $u_{\underline{x},\phi} = \underline{x} \cdot \underline{\theta}^\perp$ is the u value of the parallel projection at angle ϕ of the point $\underline{x}$; see Fig. 6.

These two equations form the basis of the widely used filtered backprojection algorithm. Equation (7) is a linear shift-invariant filtering of the projection data with a filter with frequency response $H(U) = |U|$. The gain of this filter increases monotonically with frequency and is therefore unstable. However, by assuming that the data $g(u, \phi)$, and hence the corresponding image, are bandlimited to a maximum frequency $U = U_{max}$, we need only consider the finite bandwidth filter with impulse response:

$$h(u) = \int_{-U_{max}}^{U_{max}} |U|e^{juU}\,dU. \qquad (8)$$

The filtered projections $\tilde{g}(u, \phi)$ are found by convolving $g(u, \phi)$ with $h(u)$ scaled by $1/4\pi^2$. To reduce effects of noise in the data, the response of this filter can be tapered off at higher frequencies [4, 9].

The integrand $\tilde{g}(u_{\underline{x},\phi}, \phi)$ in Eq. (6) can be viewed as an image with constant values along lines in the ϕ direction that is formed by "backprojecting" the filtered projection at angle ϕ. Summing (or in the limit, integrating) these backprojected images for all ϕ produces the reconstructed image. Although this summation involves $\phi \in [0, 2\pi]$, in practice only 180° of projection measurements are collected because opposing parallel-beam projections contain identical information. In Eq. (6), the integration limits can be replaced with $\phi \in [0, \pi]$ and the factor of $\frac{1}{2}$ can be removed. This filtered backprojection method, or the modification described below for the fan-beam geometry, is the basis for image reconstruction in almost all commercially available computed tomography systems.

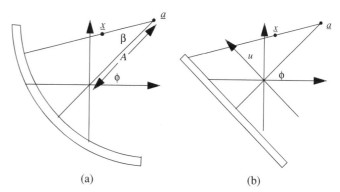

FIGURE 8 Illustration of the coordinate system for fan-beam tomography using (a) circular arc and (b) linear detector array arrangements.

3.2 Fan-Beam Filtered Backprojection

X-ray CT data can be collected more rapidly by using an array of detectors and a fan-beam X-ray source so that all elements in the array are simultaneously exposed to the X-rays. This arrangement gives rise to a natural fan-beam data collection geometry as illustrated in Fig. 1(b). The source and detector array are rotated around the patient and a set of fan-beam projections, $g(\underline{a}, \underline{\theta})$, are collected, where $\underline{a}$ represents the position of the source and $\underline{\theta}$ specifies the individual line integrals in the projections. For a radius of rotation A, we parametrize the motion of the source as $\underline{a} = (A\cos\phi, A\sin\phi)$. For the case of a circular arc of detectors whose center is the fan source and that rotates with the source, a particular detector element is conveniently specified by using the relative angle β as shown in Fig. 8(a). The fan-beam projection notation is then simplified to $g(\phi, \beta) = g(\underline{a}, \underline{\theta}) = \int f(\underline{a} + t\underline{\theta})\,dt$, where $\underline{\theta} = (-\cos(\phi - \beta), -\sin(\phi - \beta))$.

The projection data could be re-sorted into equivalent parallel projections and the above reconstruction methods applied. Fortuitously, this re-sorting is unnecessary. It can be shown [10] that reconstruction of the image can be performed by using a fan-beam version of the filtered backprojection method. Development of this inverse method involves substituting the fan-beam data in the parallel-beam formulas, (6) and (7), and applying a change of variables with the appropriate Jacobian. After some manipulation, the equations can be reduced to the form

$$f(\underline{x}) = \frac{1}{2} \int_0^{2\pi} \frac{1}{r^2} \tilde{g}(\phi, \beta) \Big|_{\beta=\beta_{\underline{x},\phi}} d\phi, \qquad (9)$$

where $r = \|\underline{x} - \underline{a}\|$ is the distance from the point $\underline{x}$ to the fan-beam source,

$$\tilde{g}(\phi, \beta) = \frac{1}{4\pi^2} \int_{-\gamma}^{\gamma} (A\cos\beta' g(\phi, \beta'))h(\sin(\beta - \beta'))\,d\beta', \qquad (10)$$

and γ is the maximum value of β required to ensure that data are not truncated. In Eq. (9), $\beta_{\underline{x},\phi} = \cos^{-1}((A^2 - (\underline{x} \cdot \underline{a}))/(rA))$ indicates the value in the ϕ projection for the line passing through the point $\underline{x}$.

As in the parallel-beam case, this reconstruction method involves a two-step procedure: filtering, in this case with a pre-weighting factor $A\cos\beta$, and backprojection. The backprojection for fan-beam data is performed along the paths converging at the location of the X-ray source and includes an inverse square-distance weighting factor. The filter $h(u)$ was given in Eq. (8) and, as before, can include a smoothing window tailored to the expected noise in the measured data.

In some fan-beam tomography applications the detector bank might be linear rather than curved. In principle, the same formula could be used by interpolating to obtain values sampled evenly in β. However, there is an alternative formula suitable for linear detectors. In this case we use u to indicate the projection line for a scaled version of the flat detector corresponding to a virtual flat detector passing through the origin, as shown in Fig. 8(b). The simplified notation is $g(\phi, u) = g(\underline{a}, \underline{\theta}) = \int f(\underline{a} + t\underline{\theta}) \, dt$, where $\underline{a} = (A\cos\phi, A\sin\phi)$ as before, and $\underline{\theta} = (u\sin\phi - A\cos\phi, -u\cos\phi - A\sin\phi)/\sqrt{u^2 + A^2}$.

The derivation of the fan-beam formula for linear detectors is virtually the same as for the curved detectors, and it results in equations of the form

$$f(\underline{x}) = \frac{1}{2} \int_0^{2\pi} \left(\frac{A^2 + u^2}{r^2} \tilde{g}(\underline{a}, u) \right) \Bigg|_{u = u_{\underline{x},\phi}} d\phi, \qquad (11)$$

$$\tilde{g}(\underline{a}, u) = \frac{1}{4\pi^2} \int_{-\infty}^{\infty} \left(\frac{A}{\sqrt{A^2 + u'^2}} g(\underline{a}, u') \right) h(u - u') \, du', \qquad (12)$$

where, as before, $r = \|\underline{x} - \underline{a}\|$ is the distance between $\underline{x}$ and the source point $\underline{a}$, and $u_{\underline{x},\phi} = A\tan\beta$ specifies the line passing through $\underline{x}$ in the ϕ projection. The limits of integration in the filtering step of Eq. (12) are replaced in practice with the finite range of u corresponding to nonzero values of the projection data $g(\phi, u)$.

The existence of a filtered backprojection algorithm for these two fan-beam geometries is quite fortuitous and does not occur for all detector sampling schemes. In fact these are two of only four sampling arrangements that have this convenient reconstruction form [11]. In general, the filtering step must be replaced with a more general linear operation on the weighted projection values, which results in a more computationally intensive algorithm.

For the fan-beam geometry, opposing projections do not contain the same information, although all line integrals are measured twice over the range of 2π measurements. The redundancy is interwoven in the projections. An angular range of $\pi + \gamma$ can be used with careful adjustments to Eqs. (11) and (12) to obtain a fast "short scan" reconstruction [4]. These short-scan modes

are used in clinical CT systems, including spiral CT systems as discussed in Section 4.2.

4 Extending 2-D Methods into Three Dimensions

4.1 Extracting 2-D Data from 3-D Data

A full 3-D image can be built up by repeatedly performing 2-D image reconstruction on a set of parallel contiguous slices. In X-ray CT, SPECT, and PET, this has been a standard method for volume tomographic reconstruction. Mathematically we use $f_z(\underline{x}) = f_z(x, y)$ to represent the z slice of $f(x, y, z)$, and $g_z(u, \underline{\theta})$ to represent the line integrals in this z slice. Reconstruction for each z is performed sequentially by using techniques described in Section 3.

More sophisticated methods of building 3-D tomographic images have been developed for a number of applications. For example, in spiral X-ray CT, the patient is moved continuously through the scanner so no fixed discrete set of tomographic slices is defined. In this case there is flexibility in choosing the slice spacing and the absolute slice location, but there is no slice position for which a complete set of projection data is measured. We describe image reconstruction for spiral CT in Section 4.2.

In a more general framework, we call an image reconstruction problem *fully 3-D* if the data cannot be separated into a set of parallel contiguous and independent 2-D slices. An example is 3-D PET, which allows measurement of oblique coincidence events and therefore must handle line integrals that cross multiple transverse planes, as shown in Fig. 3(c). Other examples of fully 3-D problems include cone-beam SPECT and cone-beam X-ray CT, where the diverging geometry of the rays precludes any sorting arrangement into parallel planes. Fully 3-D image reconstruction is described in more detail in Section 5, but a common feature of these methods is the heavy computational load associated with the 3-D backprojection step. Since 2-D reconstruction is generally very fast, a number of approaches reduce computation cost by converting a fully 3-D problem into a multislice 2-D problem. These *rebinning* procedures involve approximations that in some instances are very good, and significant improvements in image reconstruction time can be achieved with little resolution loss. One such example is the Fourier rebinning (FORE) method used in 3-D PET imaging, in which an order of magnitude improvement in computation time is achieved over the standard fully 3-D methods; the method is described in Section 4.3.

4.2 Spiral CT

In spiral CT a conventional fan-beam X-ray source and detector system rotates around the patient while the bed is translated along its long axis, as illustrated in Fig. 9. This supplementary

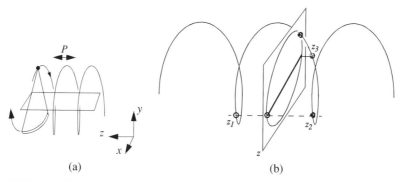

FIGURE 9 Illustration of spiral or helical CT geometry. (a) Relative to a stationary bed, the source and detector circle the patient in a helical fashion with pitch P; (b) to reconstruct cross section $f_z(x, y)$, missing projections are interpolated from neighboring points on the helix at which data were collected.

motion, although it complicates the image reconstruction algorithms and results in slightly blurred images, provides the capability to scan large regions of the patient in a single breath hold.

The helical motion is characterized by the pitch P, the amount of translation in the axial or z direction for a full rotation of the source and detector assembly. Therefore $\phi = 2\pi z/P$ and we can write $g(z, \beta) = g(\underline{a}, \underline{\theta}) = \int f(\underline{a} + t\underline{\theta})\,\mathrm{d}t$, which is similar to the fan-beam geometry of Section 3.2, with $\underline{a} = (A\cos\phi, A\sin\phi, z)$ and $\underline{\theta} = (-\cos(\phi - \beta), -\sin(\phi - \beta), 0)$. Note that ϕ now ranges from 0 to $2\pi n$ as z ranges from 0 to nP, where n is the number of turns of the helix. The usual method of reconstruction involves estimating a full set of fan-beam projections $g_z(\phi, \beta)$ for each transverse plane, using the available projections at other points on the helix. If the reconstruction on transverse plane z is required, the standard fan-beam CT algorithm is used:

$$f_z(\underline{x}) = \frac{1}{2}\int_0^{2\pi} \frac{1}{r^2}\tilde{g}_z(\phi, \beta)\bigg|_{\beta=\beta_{z,\underline{x},\phi}} \mathrm{d}\phi, \tag{13}$$

$$\tilde{g}_z(\phi, \beta) = \frac{1}{4\pi^2}\int_{-\gamma}^{\gamma} (A\cos\beta' g_z(\phi, \beta')) h(\sin(\beta - \beta'))\,\mathrm{d}\beta', \tag{14}$$

where $\beta_{z,\underline{x},\phi}$ indicates the relative projection angle β found by projecting in the z plane at angle ϕ through the point (x, y, z), and r is the distance (in the z plane) between the point (x, y, z) and the virtual source at angular position ϕ. In the simplest case, the z-plane projections $g_z(\phi, \beta)$ are estimated by a weighted sum of the measured projections at the same angular position above and below z on the helix:

$$g_z(\phi, \beta) \approx \frac{w_1 g(z_1, \beta) + w_2 g(z_2, \beta)}{w_1 + w_2} \tag{15}$$

for some suitable weights w_1 and w_2, and where, as illustrated

in Fig. 9, z_1 and z_2 lie within one pitch P of the reconstruction plane, z. Note that in Eq. (15), z_1/P differs from ϕ by some multiple of 2π, and similarly for z_2.

Various schemes for choosing the weights w_1 and w_2 exist [12]. Each weighting scheme establishes a tradeoff between increased image noise from unbalanced contributions, and inherent axial blurring artifacts from the geometric approximation of the estimation process. When the image noise is particularly low, a short-scan version of the fan-beam reconstruction algorithm might be used. This version reduces the range of contributing projections to $\pi + \gamma$ from 2π and correspondingly reduces the maximum distance required to estimate a projection $g_z(\phi, \beta)$. Even more elaborate estimation schemes exist, such as approximating $g_z(\phi, \beta)$ on a line-by-line basis. Figure 9(b) illustrates how in the short-scan mode, the line integral $g_z(\phi, \beta)$ could be estimated from a value in the z_3 projection rather than from the z_1 and z_2 projections.

The choice of pitch P represents a compromise between maximizing the axial coverage of the patient, and avoiding unacceptable artifacts from the geometric estimation. Generally the pitch is chosen between one and two times the thickness of the detector in the axial direction [12].

4.3 Rebinning Methods in 3-D PET

PET data are generally sorted into parallel-beam projections, $g(u, \phi)$, as described in Section 3.1. For a multiring 2-D PET scanner the data are usually processed slice by slice. With the use of z to denote the axis of the scanner, the data are reconstructed by using Eqs. (6) and (7) applied to each z slice as follows:

$$f_z(\underline{x}) = \frac{1}{2}\int_0^{2\pi} \tilde{g}_z(u, \phi)\bigg|_{u=u_{z,\underline{x},\phi}} \mathrm{d}\phi, \tag{16}$$

$$\tilde{g}_z(u, \phi) = \frac{1}{4\pi^2}\int_{-\infty}^{\infty} g_z(u', \phi) h(u - u')\,\mathrm{d}u, \tag{17}$$

with $u_{z,\underline{x},\phi} = (x, y, z) \cdot (-\sin \phi, \cos \phi, 0)$. The data $g_z(u, \phi)$ are found from sampled values of u and ϕ determined by the ring geometry: the radius R and the number of crystals (typically several hundreds). In Eqs. (16) and (17), z is usually chosen to match the center of each detector ring. In practice, 2-D scanners allow detection of coincidences between adjacent rings. Using the single-slice rebinning (SSRB) technique described below, slices midway between adjacent detector rings can also be reconstructed from 2-D scanner data.

Current commercial PET scanners usually consist of a few tens of detector rings and have supplementary 3-D capability to detect oblique photon pairs that strike detectors on different rings. These fully 3-D data require more advanced reconstruction techniques. The fully 3-D version of Eqs. (16) and (17) is given in Section 5. In this section we describe two popular rebinning methods in which the data are first processed to form independent 2-D projections $g_z(u, \phi)$. Equations (16) and (17) are then used for image reconstruction.

Let Δ denote the spacing between rings. Let $g_{l,m}(u, \phi)$ denote the resulting line integral, with endpoints on rings l and m, whose 2-D projection variables are (u, ϕ) when the line is projected onto the x–y plane, as shown in Fig. 10. In the SSRB method [13], all line-integral data are reassigned to the slice midway between the rings where the detection occurred. Thus

$$g_z(u, \phi) \approx \sum_{(l,m) \in \Omega_z} g_{l,m}(u, \phi), \quad \text{where}$$

$$\Omega_z = \left\{ (l, m): \frac{(l + m)\Delta}{2} = z, \right\}, \tag{18}$$

and reconstruction proceeds according to Eqs. (16) and (17).

A more sophisticated method, known as Fourier rebinning (FORE), [14] effectively performs the rebinning operation in the 2-D frequency domain. The rebinned data $g_z(u, \phi)$ are found by using the following transformations:

$$\tilde{g}_{l,m}(u, \phi) = g_{l,m}(u, \phi) \frac{2\sqrt{R^2 - u^2}}{\sqrt{4R^2 - 4u^2 + (l - m)^2 \Delta^2}}, \tag{19}$$

$$\tilde{G}_{l,m}(U, k_\phi) = \int_{-\infty}^{\infty} \int_{0}^{2\pi} \tilde{g}_{l,m}(u, \phi) e^{-j(Uu + k_\phi \phi)} \, d\phi \, du, \tag{20}$$

$$G_z(U, k_\phi) \approx \sum_{(l,m) \in \Omega_z} \tilde{G}_{l,m}(U, k_\phi), \quad \text{where}$$

$$\Omega_z = \left\{ l, m: \frac{(l + m)\Delta}{2} + \frac{(l - m)\Delta}{2} \frac{k_\phi}{U\sqrt{R^2 - U^2}} = z \right\}, \tag{21}$$

$$g_z(u, \phi) = \sum_{k_\phi} \int_{-\infty}^{\infty} G_z(U, k_\phi) e^{j(Uu + k_\phi \phi)} \, dU. \tag{22}$$

Both SSRB and FORE are approximate techniques, and the geometrical misplacement of the data can cause artifacts in the reconstructed images. However, FORE is far more accurate than

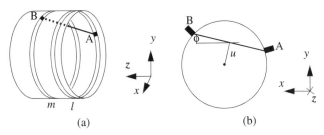

FIGURE 10 Oblique line integrals, along the path AB in (a), between different rings of detectors can be rebinned into equivalent in-plane data, either directly using SSRB or indirectly using FORE. (b) The relationship between the projected line integral path and the parameters (u, ϕ).

SSRB yet almost as fast computationally, when compared with the subsequent reconstruction time using Eqs. (16) and (17). In [14] a mathematically exact rebinning formula is presented, and it is shown that SSRB and FORE represent zeroth- and first-order versions of this formula. However, algorithms using the exact version are far less practical than SSRB or FORE.

5 3-D Image Reconstruction

5.1 Fully 3-D Reconstruction with Missing Data

In 3-D image reconstruction, parallel-beam projection data can be specified using $\underline{\theta}$, the direction of the line integrals, and two scalars (u, v) that indicate offsets in directions $\underline{\theta}^1$ and $\underline{\theta}^2$ perpendicular to $\underline{\theta}$. Therefore $g(u, v, \underline{\theta}) = g(\underline{a}, \underline{\theta}) = \int f(\underline{a} + t\underline{\theta}) \, dt$, where $\underline{a} = u\underline{\theta}^1 + v\underline{\theta}^2$, and $\{\underline{\theta}, \underline{\theta}^1, \underline{\theta}^2\}$ is an orthonormal system. Note that all vectors in this section are three dimensional.

Image reconstruction can be performed by using a 3-D version of the filtered backprojection formulas, Eqs. (6) and (7), given in Section 3.1:

$$f(\underline{x}) = \frac{1}{2} \int_\Omega \tilde{g}(u, v, \underline{\theta}) \Big|_{\substack{u = u_{\underline{x}, \theta} \\ v = v_{\underline{x}, \theta}}} \, d\underline{\theta} \tag{23}$$

$$\tilde{g}(u, v, \underline{\theta}) = \frac{1}{8\pi^3} \int_{-\infty}^{\infty} \int_{-\infty}^{\infty} G(U, V, \underline{\theta}) H_\Omega(U, V, \underline{\theta})$$
$$\times e^{j(uU + vV)} \, dU \, dV, \tag{24}$$

where, in the surface integral of Eq. (23), $d\underline{\theta}$ can be written as $\sin \theta \, d\theta \, d\phi$ for $\underline{\theta}$ having polar angle θ and azimuthal angle ϕ; and where $(u_{\underline{x}, \theta}, v_{\underline{x}, \theta}) = (\underline{x} \cdot \underline{\theta}^1, \underline{x} \cdot \underline{\theta}^2)$ represents the (u, v) coordinates of the line with orientation $\underline{\theta}$ passing through $\underline{x}$. The subset Ω of the unit sphere represents the measured directions $\{\underline{\theta}\}$ and must satisfy Orlov's condition for data completeness in order for Eqs. (23) and (24) to be valid. Orlov's condition requires that every great circle on the unit sphere intersect the region Ω. The tomographic reconstruction filter $H_\Omega(U, V, \underline{\theta})$ depends on the measured data set [15]. In the special case that Ω is the whole sphere S^2, then $H_\Omega(U, V, \underline{\theta}) = \sqrt{U^2 + V^2}$.

In 3-D PET imaging, data can be sorted according to the parameterization $g(u, v, \underline{\theta})$. The set of measured projections can be described by $\Omega_\Psi = \{\underline{\theta} = (\theta_x, \theta_y, \theta_z) : |\theta_z| \leq \sin\Psi\}$, where $\Psi = \text{atan}(L/(2R))$ represents the most oblique line integral possible for a scanner radius of R and axial extent L. Provided none of the projections are truncated, reconstruction can be performed according to Eqs. (23) and (24) by using the Colsher filter $H_{\Omega_\Psi}(U, V, \underline{\theta}) = H^\Psi(U, V, \underline{\theta})$ given by

$$H^\Psi(U, V, \underline{\theta}) =$$
$$\begin{cases} \sqrt{U^2 + V^2} & \text{if } \sqrt{U^2 + \theta_z^2 V^2} \\ & \quad < (\sin\Psi)\sqrt{U^2 + V^2} \\ \dfrac{\pi}{2} \dfrac{\sqrt{U^2+V^2}}{\sin^{-1}\left((\sin\Psi)\sqrt{U^2+V^2}/\sqrt{U^2+\theta_z^2 V^2}\right)} & \text{otherwise} \end{cases}.$$
$$(25)$$

In practice the object occupies most of the axial extent of the scanner so nearly all projections are truncated. However, there is always a subset $\Omega_{\Psi'}$ of the projections that are not truncated, and from these projections a reconstruction can be performed to obtain the image $f^{\Psi'}(\underline{x})$, using Eqs. (24) and (25) with $H^{\Psi'}(U, V, \underline{\theta})$. In the absence of noise, this reconstruction would be sufficient, but to include the partially measured projections a technique known as the "reprojection method" is used. All truncated projections are completed by estimating the missing line integrals based on the initial reconstruction $f^{\Psi'}(\underline{x})$. Then, in a second step, reconstruction from the entire data set is performed, using $H^\Psi(U, V, \underline{\theta})$ to obtain the final image $f^\Psi(\underline{x})$ [8,16].

5.2 Cone-Beam Tomography

For cone-beam projections it is convenient to use the general notation $g(\underline{a}, \underline{\theta})$, where $\underline{a}$ and $\underline{\theta}$ are three-dimensional vectors. For an X-ray system, the position of the source would be represented by $\underline{a}$ and the direction of individual line integrals obtained from that source would be indicated by $\underline{\theta}$. Because of difficulties obtaining sufficient tomographic data (see below), the source, detector, or both may follow elaborate trajectories in space relative to the object; therefore a general description of their orientations is required. For applications involving a planar detector, we replace $\underline{\theta}$ with (u, v), the coordinates on an imaginary detector centered at the origin and lying in the plane perpendicular to $\underline{a}$. The source point $\underline{a}$ is assumed never to lie on the scanner axis $\underline{e}_3$. The u axis lies in the detector plane in the direction $\underline{e}_3 \times \underline{a}$. The v direction is perpendicular to u and points in the same direction as $\underline{e}_3$, as shown in Fig. 11(b).

In the simplest applications, the detector and source rotate in a circle about the scanner axis $\underline{e}_3$. If the radius of rotation is A, the source trajectory is parameterized by $\phi \in [0, 2\pi]$ as $\underline{a} = (A\cos\phi, A\sin\phi, 0)$. In this case the v axis in the detector stays aligned with $\underline{e}_3$ and the u axis points in the tangent direction to the motion of the source. Physical detector measurements can easily be scaled to this virtual detector

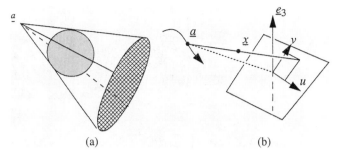

FIGURE 11 (a) 2-D planar projections of a 3-D object are collected in a cone-beam system as line integrals through the object from the cone vertex $\underline{a}$ to the detector array. (b) The coordinate system for the cone-beam geometry; the cone vertices $\underline{a}$ can follow an arbitrary trajectory provided Tuy's condition is satisfied.

system, just as for the fan-beam example of Section 3.2. Thus $g(\phi, u, v) = g(\underline{a}, \underline{\theta}) = \int f(\underline{a} + t\underline{\theta})\,dt$, where $\underline{\theta} = (-u\sin\phi - A\cos\phi, u\cos\phi - A\sin\phi, v)$.

The algorithm of Feldkamp *et al.* [17] is based on the fan-beam formula for flat detectors (see Section 3.2) and collapses to this formula in the central plane $z = 0$ where only fan-beam measurements are taken.

$$f_{\text{FDK}}(\underline{x}) = \frac{1}{2} \int_0^{2\pi} \left.\left(\frac{A^2 + u^2 + v^2}{r^2} \tilde{g}(\phi, u, v)\right)\right|_{\substack{u=u_{\underline{x},\phi} \\ v=v_{\underline{x},\phi}}} d\phi, \quad (26)$$

$$\tilde{g}(\phi, u, v) = \frac{1}{4\pi^2} \int_{-\infty}^{\infty} \left(\frac{A}{\sqrt{A^2 + u'^2 + v^2}} g(\phi, u', v)\right)$$
$$\times h(u - u')\,du'. \quad (27)$$

Similarly to Eq. (11), $r = \|\underline{x} - \underline{a}\|$ is the distance between $\underline{x}$ and the source position $\underline{a}$, and $(u_{\underline{x},\phi}, v_{\underline{x},\phi})$ are the coordinates on the detector of the cone-beam projection of $\underline{x}$; see Fig. 11.

Figure 12 shows two images of reconstructions from mathematically simulated data. Using a magnified gray scale to reveal the 1% contrast structures, the top images show both a high-quality reconstruction in the horizontal transverse slice at the level of the circular trajectory, and apparent decreased intensity on planes above and below this level. These artifacts are characteristic of the Feldkamp algorithm. The bottom images, showing reconstructions for the "disks" phantom, also exhibit cross-talk between transverse planes and some other less dramatic artifacts. The disks phantom is specifically designed to illustrate the difficulty in using cone-beam measurements for a circular trajectory. Frequencies along and near the scanner axis are not measured, and objects with high amplitudes in this direction produce poor reconstructions.

For the cone-beam configuration, requirements for a tomographically complete set of measurements are known as Tuy's condition. Tuy's condition is expressed in terms of a geometric relationship between the trajectory of the cone-beam vertex point (the source point) and the size and position of the object being scanned. Tuy's condition requires that every plane

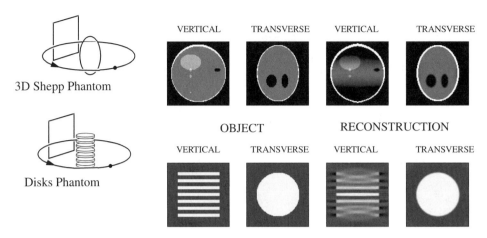

FIGURE 12   Example of cone-beam reconstructions from a circular orbit; the obvious artifacts are a result of the incompleteness in the data. Other trajectories, such as a helix or a circle plus line, give complete data and artifact-free reconstructions.

that cuts through the object must also contain some point of the vertex trajectory. Furthermore, it is assumed that the detector is large enough to measure the entire object at all positions of the trajectory, i.e., the projections should not be truncated. For the examples given in Fig. 12, the artifacts arose because the circular trajectory did not satisfy Tuy's condition (even though the projections were not truncated). In this sense the measurements were incomplete and artifacts were inevitable.

Analytic reconstruction methods for cone-beam configurations satisfying Tuy's completeness condition are generally based on a transform pair that plays a similar role to the Fourier transform in the projection slice theorem for classical parallel-beam tomography. A mathematical result from Grangeat [18] links the information in a single cone-beam projection to a subset of the transform domain, just as the Fourier slice theorem links a parallel projection to a certain subset of the Fourier domain. This relationship is defined through the " B transform," the derivative of the 3-D Radon transform:

$$Bf(s, \underline{\gamma}) = p(s, \underline{\gamma}) = \frac{1}{2\pi} \int_{R^3} f(\underline{x})\delta'(\underline{x} \cdot \underline{\gamma} - s)\, d\underline{x}, \quad (28)$$

$$B^{-1}p(\underline{x}) = f(\underline{x}) = \frac{1}{4\pi} \int_{S^2} \int_{R} p(s, \underline{\gamma})\delta'(s - \underline{x} \cdot \underline{\gamma})\, ds\, d\underline{\gamma} \quad (29)$$

where s is a scalar and $\|\underline{\gamma}\| = 1$. The symbol δ' represents the derivative of the Dirac delta function.

Grangeat's formula can be rewritten as

$$\frac{1}{4\pi} \int_{S^2} g(\underline{a}, \underline{\theta})\delta'(\underline{\theta} \cdot \underline{\gamma})\, d\underline{\theta} = p(s, \underline{\gamma})\Big|_{s=\underline{a}\cdot\underline{\gamma}}. \quad (30)$$

An analysis of Eq. (30) shows that if Tuy's condition is satisfied, then all values are available in the B domain representation of $f(\underline{x})$, namely $p(s, \underline{\gamma})$ [18].

Equations (29) and (30) form the basis for a reconstruction algorithm. All values in the B domain can be found from cone-beam projections, and $f(\underline{x})$ can be recovered from the inverse transform B^{-1}. Care must be taken to ensure that the B domain is sampled uniformly in s and $\underline{\gamma}$, and that if two different cone-beam projections provide the same value of $p(s, \underline{\gamma})$, then the contributions must be normalized. The method follows the concept of direct Fourier reconstruction described in Section 3.1.

A filtered backprojection type of formulation for cone-beam reconstruction is also possible [19]. If the trajectory is a piecewise smooth path, parameterized mathematically by $\phi \in \Phi \subset \mathbb{R}$, a reconstruction formula similar to filtered backprojection can be derived from Eqs. (29) and (30):

$$f(\underline{x}) = \frac{1}{2} \int_{\Phi} \frac{\tilde{g}(\underline{a}(\phi), \underline{\theta})}{r^2}\bigg|_{\underline{\theta}=\underline{\theta}_{\underline{x},\phi}}\, d\phi, \quad (31)$$

$$\tilde{g}(\underline{a}(\phi), \underline{\theta}) = \frac{-1}{8\pi^2} \int_{S^2} \left(\int_{S^2} g(\underline{a}(\phi), \underline{\theta}')\delta'(\underline{\theta}' \cdot \underline{\gamma})\, d\underline{\theta}' \right) \\ \times \delta'(\underline{\theta} \cdot \underline{\gamma}) M(\underline{\gamma}, \phi)|\underline{a}'(\phi) \cdot \underline{\gamma}|\, d\underline{\gamma}. \quad (32)$$

Here $r = \|\underline{x} - \underline{a}(\phi)\|$, $\underline{\theta}_{\underline{x},\phi} = (\underline{x} - \underline{a}(\phi))/\|\underline{x} - \underline{a}(\phi)\|$ is the line passing through $\underline{x}$ for the $\underline{a}(\phi)$ projection, and the function M must be chosen to normalize multiple contributions in the B domain [19]. The normalization condition is $1 = \sum_{k=1}^{n(\underline{\gamma},s)} M(\underline{\gamma}, \phi_k)$, where $n(\underline{\gamma}, s)$ is the number of vertices lying in the plane with unit normal $\underline{\gamma}$ and displacement s, and $\phi_1, \phi_2, \ldots, \phi_{n(\underline{\gamma},s)}$ indicate the vertex locations where the path $\underline{a}(\Phi)$ intersects the plane. By Tuy's condition, $n(\underline{\gamma}, s) > 0$ for $|s| < R$ for an object of radius R.

These equations must be tailored to the specific application. When the variables are changed to reflect the planar detector arrangement specified at the beginning of this subsection, the above equations resemble the Feldkamp algorithm with a much more complicated "filtering" step. To simplify notation, we write A for the varying distance $\|\underline{a}(\phi)\|$ of the vertex from the origin.

$$f(\underline{x}) = \frac{1}{2} \int_\Phi \left(\frac{A^2 + u^2 + v^2}{r^2} \tilde{g}(\underline{a}(\phi), u, v) \right) \Bigg|_{\substack{u = u_{\underline{x},\phi} \\ v = v_{\underline{x},\phi}}} d\phi, \tag{33}$$

$$\tilde{g}(\underline{a}(\phi), u, v) = \frac{-1}{4\pi^2} \int_0^\pi \left(\frac{d}{dt} T(\phi, \underline{\gamma}) \frac{\sqrt{A^2 + t^2}}{A^2} \left(\frac{d}{dt} \int_R \frac{A}{\sqrt{A^2 + u'^2 + v'^2}} g(\underline{a}, u', v')\, dl \right) \right) \Bigg|_{\substack{t = u \cos \mu \\ + v \sin \mu}} d\mu, \tag{34}$$

where, in the innermost integration, $(u', v') = (t \cos \mu - l \sin \mu,$ $t \sin \mu + l \cos \mu)$; the function $T(\phi, \underline{\gamma}) = |\underline{a}'(\phi) \cdot \underline{\gamma}| M(\underline{\gamma}, \phi)$ contains all the dependency on the particular trajectory. Note that in Eq. (34), $\underline{\gamma} = (A \cos \mu \underline{e}_u + A \sin \mu \underline{e}_v + t \underline{e}_w)/$ $(\sqrt{A^2 + t^2})$, where the detector coordinate axes are $\underline{e}_u$ and $\underline{e}_v$, and $\underline{e}_w = -\underline{a}(\phi)/A$.

Although these equations are only valid when the cone-beam configuration satisfies Tuy's condition, the algorithm of Eqs. (33) and (34) collapses to the Feldkamp algorithm when a circular trajectory is specified. This general algorithm has been refined and tailored for specific applications involving truncated projections. Practical methods have been published for the case of source trajectories containing a circle.

6 Iterative Reconstruction Methods

6.1 Finite Dimensional Formulations and ART

As noted above, the line-integral model on which all of the preceding methods are based is only approximate. Furthermore, there is no explicit modeling of noise in these approaches; noise in the data is typically reduced by tapering off the response of the projection filters before backprojection. In clinical X-ray CT, the beam is highly collimated, the detectors have low noise and have high resolution, and the number of photons per measurement is very large; consequently the line integral approximation is adequate to produce low noise images at submillimeter resolutions.

However, this may not be the case in industrial and other nonmedical applications, and these systems may benefit from more accurate modeling of the data and noise. In the case of PET and SPECT, the often low intrinsic resolution of detectors, depth dependent and geometric resolution losses, and the typically low photon count, can lead to rather poor resolution at acceptable noise levels. An alternative to the analytic approach is to use a finite dimensional model in which the detection system and the noise statistics can be modeled more accurately. Research in this area has lead to the development of a large class of reconstruction methods that often outperform the analytic methods.

We will assume that the image is adequately represented by a finite set of basis functions. While there has been some interest in alternative basis elements, almost all researchers currently use a cubic voxel basis function. Each voxel is an indicator function on a cubic region centered at one of the image sampling points in a regular 2-D or 3-D lattice. The image value at each voxel is proportional to the quantity being imaged integrated over the volume spanned by the voxel. For a unified treatment of 2-D and 3-D problems, a single index will be used to represent the lexicographically ordered elements of the image $f = \{f_1, f_2, \ldots, f_N\}$. Similarly, the elements of the measured projections will be represented in lexicographically ordered form as $y = \{y_1, y_2, \ldots, y_M\}$.

In X-ray CT we can model the attenuation of a finite width X-ray beam as the integral of the linear attenuation coefficient over the path (or strip) through which the beam passes. Thus the measurements can be written as

$$y_i = \iint_{x, y \in \text{strip}\{i\}} f(x, y)\, dx\, dy = \sum_{j=1}^N H_{ij} f_j, \tag{35}$$

where f_j is the attenuation coefficient at the jth voxel. The elements H_{ij} of the projection matrix, H, is equal to the area of intersection of the ith strip with the indicator function on the jth voxel, as illustrated in Fig. 13. Equation (35) represents a huge set of simultaneous linear equations, $y = Hf$, that can be solved to compute the CT image f. In principle the system can be solved

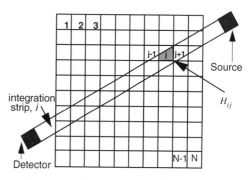

FIGURE 13 Illustration of the voxel-based finite-dimensional formulation used in iterative X-ray CT reconstruction. The matrix element H_{ij} gives the contribution of the jth voxel to the ith measurement and is proportional to the area of intersection of the voxel with the strip that joins the source and detector.

using standard methods. However, the size of these systems coupled with the special structure of H motivated research into more efficient specialized numerical procedures. These methods exploit the key property that H is very sparse; i.e., since the path along which each integration is performed intersects only a small fraction of the image pixels, most elements in the matrix are zero.

One algorithm that makes good use of the sparseness property is the algebraic reconstruction technique, or ART [20]. This method finds the solution to the set of equations in an iterative fashion through successive orthogonal projection of the current image estimate onto hyperplanes defined by each row of H. If this procedure converges, the solution will be a point where all of the hyperplanes intersect, i.e., a solution to Eq. (35). Let f^n represent the vector of image pixel values at the nth iteration, and let h_i^T represent the ith row of H. The ART method has the following form:

$$f^{n+1} = f^n + \left(\frac{y_i - h_i^T f^n}{h_i^T h_i} \right) h_i, \quad i = (n \bmod M) + 1. \quad (36)$$

ART can also be viewed in terms of the backprojection operator used in filtered backprojection: each iteration of Eq. (36) is equivalent to adding to the current image estimate f^n, the weighted backprojection of the error between the ith measured projection sample and the projection corresponding to f^n. ART will converge to a solution of Eq. (35) provided the system of equations is consistent. In the inconsistent case, the iterations will not converge to a single solution and the properties of the image at a particular stopping point will be dependent on the sequence in which the data are ordered. Many variations of the ART method can be found in the literature. These variations exhibit differences in convergence behavior, sensitivity to noise, and optimality properties [21].

6.2 Statistical Formulations

The ART method does not directly consider the presence of noise in the data. While acceptable in high signal-to-noise X-ray CT data, the low photon counting statistics found in PET and SPECT should be explicitly considered. The finite dimensional formulation in Section 6.1 can be extended to model both the physics of PET and SPECT detection and the statistical fluctuations caused by noise.

Rather than simply assume a strip integral model as in Eq. (35), we can instead use the matrix relating image and data to more exactly model the probability, P_{ij}, of detecting an emission from voxel site j at detector element i. To differentiate this probabilistic model from the strip integral one, we will denote the detection probability matrix by P. The elements of this matrix are dependent on the specific data acquisition geometry and many other factors including detector efficiency, attenuation effects within the subject, and the underlying physics of gamma-ray emission (for SPECT) and positron-electron annihilation (for

PET). See [22] and [23] for descriptions of the formation of these matrices.

In PET and SPECT the mean of the data can be estimated for a particular image as the linear transformation

$$E(y) = Pf, \quad (37)$$

where f represents the mean emission rates from each image voxel. In practice, these data are corrupted by additive background terms that are due to scatter and either "random coincidences" in PET [6] or background radiation in SPECT [7]. The methods described below can be modified relatively easily to include these factors, but these issues will not be addressed further here.

As mentioned in Section 2.2, PET and SPECT systems use an external radiation source to perform transmission measurements to determine the attenuation factors that must be included in the matrix P. These data represent line integrals of the patient's attenuation coefficient distribution at the energy of the transmission source. Just as in X-ray CT, it is possible to reconstruct attenuation images from these transmission measurements. As in Eq. (35), the image f would represent the map of attenuation coefficients and the elements of matrix H contain the areas of intersection of each projection path with each voxel. Let $E(y)$ represents the mean value of the transmission measurement. Assuming that the source intensity is a constant α, then we can model the mean of the transmission data as

$$E(y_i) = \alpha \exp \left\{ -\sum_j H_{ij} f_j \right\}. \quad (38)$$

Our emphasis in the following is the description of reconstruction methods for the emission problem, but we also indicate which methods can and cannot be applied to the transmission data.

For both emission and transmission measurements, the data can be modeled as collections of Poisson random variables, mean $E(y)$, with probability or likelihood

$$p(y/f) = \prod_{i=1}^{M} \frac{E(y_i)^{y_i} e^{-E(y_i)}}{y_i!}. \quad (39)$$

The physical model for the detection system is included in the likelihood function in the mapping from the image f to the mean of the detected events $E(y)$, using Eqs. (37) and (38) for the emission and transmission case, respectively. Using this basic model, we can develop estimators based on maximum likelihood (ML) or Bayesian image estimation principles.

6.3 Maximum Likelihood Methods

The maximum likelihood estimator is the image that maximizes the likelihood of Eq. (39) over the set of feasible images, $f \geq 0$. The EM (expectation maximization) algorithm can be applied to the emission CT problem, resulting in an iterative algorithm that has the elegant closed-form update equation [24]:

$$f_j^{n+1} = \frac{f_j^n}{\sum_i P_{ij}} \sum_i \frac{P_{ij} y_i}{\sum_l P_{il} f_l^n}. \qquad (40)$$

This algorithm has a number of interesting properties, including the fact that the solution is naturally constrained by the iteration to be nonnegative. Unfortunately, the method tends to exhibit very slow convergence and is often unstable at higher iterations. The variance problem is inherent in the ill-conditioned Fisher information matrix. This effect can be reduced using *ad hoc* stopping rules in which the iterations are terminated before convergence. An alternative approach to reducing variance is through penalized maximum likelihood or Bayesian methods as described in Section 6.4.

A number of modifications of the EM algorithm have been proposed to speed up convergence. Probably the most widely used of these is the ordered subsets EM (OSEM) algorithm, in which each iteration uses only a subset of the data [25]. Let $\{S_k\}$, $k = 1, \ldots, Q$ be a disjoint partition of the set $\{1, 2, \ldots, M\}$. Let n denote the number of complete cycles through the Q subsets, and define $f_j^{(n,0)} = f_j^{(n-1, Q)}$. Then one complete iteration of OSEM is given by

$$f_j^{(n,k)} = \frac{f_j^{(n,k-1)}}{\sum_{i \in S_k} P_{ij}} \sum_{i \in S_k} \frac{p_{ij} y_i}{\sum_l P_{il} f_l^{(n,k-1)}},$$
$$\text{for } j = 1, \ldots, N, k = 1, \ldots, Q. \qquad (41)$$

Typically, each subset will consist of a group of projections with the number of subsets equal to an integer fraction of the total number of projections. "Subset balance" is recommended [25], i.e., the subsets should be chosen so that an emission from each pixel has equal probability of being detected in each subset. The grouping of projections within subsets will alter both the convergence rate and the sequence of images generated. To avoid directional artifacts, the projections are usually chosen to have maximum separation in angle in each subset. In the early iterations OSEM produces remarkable improvements in convergence rates although subsequent iterations over the entire data is required for ultimate convergence to an ML solution.

The corresponding ML problem for Poisson distributed transmission data does not have a closed-form EM update. However, both emission and transmission ML problems can be solved effectively using standard gradient descent methods. In fact, it is easily shown that the EM algorithm for emission data can be written as a steepest descent algorithm with a diagonal preconditioner equal to the current image estimate. More powerful nonlinear optimization methods, and in particular the preconditioned conjugate gradient method, can produce far faster convergence than the original EM algorithm [26].

6.4 Bayesian Reconstruction Methods

As noted earlier, ML estimates of PET images exhibit a high variance as a result of ill-conditioning. Some form of regularization is required to produce acceptable images. Often regularization is accomplished simply by starting with a smooth initial estimate and terminating an ML search before convergence. Here we consider explicit regularization procedures in which a prior distribution is introduced through a Bayesian reformulation of the problem (see also Chapter 3.6). Some authors prefer to present these regularization procedures as penalized ML methods, but the differences are largely semantic.

By the introduction of random field models for the unknown image, Bayesian methods can address the ill-posedness inherent in PET image estimation. In an attempt to capture the locally structured properties of images, researchers in emission tomography, and many other image processing applications, have adopted Gibbs distributions as a suitable class of prior. The Markovian properties of these distributions make them both theoretically attractive as a formalism for describing empirical local image properties, as well as computationally appealing, since the local nature of their associated energy functions results in computationally efficient update strategies (see Chapter 4.3 for a description of Gibbs random field models for image processing). The majority of work using Gibbs distributions in tomographic applications involves relatively simple pairwise interaction models in which the Gibbs energy function is formed as a sum of potentials, each defined on neighboring pairs of pixels. These potential functions can be chosen to reflect the piecewise smooth property of many images. The existence of sharp intensity changes, corresponding to the edges of objects in the image, can be modeled by using more complex Gibbs priors. The Bayesian formulation also offers the potential for combining data from multiple modalities. For example, high-resolution anatomical X-ray CT or MR images can be used to improve the

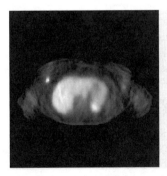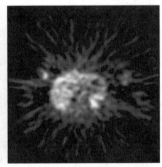

FIGURE 14 Example of a PET scan of metabolic activity using FDG, an F-18 tagged analog of glucose. This tracer is used in detection of malignant tumors. The image at left shows a slice through the chest of a patient with breast cancer; the tumor is visible in the bright region in the upper left region of the chest. This image was reconstructed using a Bayesian method similar to that in [23]. An analytic reconstruction method was used to form the image on the right from the same data.

quality of reconstructions from low-resolution PET or SPECT data [27]. See [28] for a recent review of statistical models and methods in PET.

Let $p(f)$ denote the Gibbs prior that captures the expected statistical characteristics of the image. The posterior probability for the image conditioned on the data is then given by Bayes theorem:

$$p(f \mid y) = \frac{p(y \mid f)\,p(f)}{p(y)}. \qquad (42)$$

Bayesian estimators in tomography are usually of the maximum *a posteriori* (MAP) type. The MAP solution is given by maximizing the posterior probability $p(f \mid y)$ with respect to f. For each data set, the denominator of the right-hand side of Eq. (42) is a constant so that the MAP solution can be found by maximizing the log of the numerator, i.e.,

$$\max_{f} \; \ln p(y \mid t) + \ln p(f). \qquad (43)$$

A large number of algorithms have been developed for computing the MAP solution. The EM algorithm, Eq. (40), can be extended to include a prior term — see, e.g., [27] — and hence maximize Eq. (43). This algorithm suffers from the same slow convergence problems as Eq. (40). Alternatively, Eq. (43) can be maximized by using standard nonlinear optimization algorithms such as the preconditioned conjugate gradient method [23] or coordinatewise optimization [29]. The specific algorithmic form is found by applying these standard methods to Eq. (43) after substituting both the log of the likelihood function, Eq. (39), in place of $\ln p(y \mid f)$ and the log of the Gibbs density in place of $\ln p(f)$. For compound Gibbs priors that involve line processes, mean-field annealing techniques can be combined with any of the above methods [27, 28].

7 Summary

We have summarized analytic and iterative approaches to 2-D and 3-D tomographic reconstruction for X-ray CT, PET, and SPECT. With the exception of the rebinning algorithms, which can be used in place of fully 3-D reconstruction methods, the choice of analytic reconstruction algorithm is determined primarily by the data collection geometry. In contrast, the iterative approaches (ART, ML, and MAP) can be applied to any collection geometry in PET and SPECT. Furthermore, after appropriate modifications to account for differences in the mapping from image to data, these methods are also applicable to transmission PET and SPECT data. X-ray CT data are not Poisson so that a different likelihood model is required if ML or MAP methods are to be used. The choice of approach for a particular problem should be determined by considering the factors limiting resolution and noise performance and weighing the relative importance of the computational cost of the algorithm and the desired and achievable image resolution and noise performance.

Image processing for computed tomography remains an active area of research. In large part, development is driven by the construction of new imaging systems, which are continuing to improve the resolution of these technologies. Carefully tailored reconstruction algorithms will help to realize the full potential of these new systems. In the realm of X-ray CT, new spiral and cone-beam systems are extending the capabilities of CT systems to allow fast volumetric imaging for medical and other applications. In PET and SPECT, recent developments are also aimed at achieving high-resolution volumetric imaging through combinations of new detector and collimator designs with fast, accurate reconstruction algorithms. In addition to advances resulting from new instrumentation developments, current areas of intense research activity include theoretical analysis of algorithm performance, combining accurate modeling with fast implementations of iterative methods, analytic methods that account for factors not included in the line integral model, and development of methods for fast dynamic volumetric (4-D) imaging.

References

[1] H. H. Barrett and W. Swindell, *Radiological Imaging* (Academic, New York, 1981), Vols. I and II.

[2] S. Webb, *From the Watching of Shadows: the Origins of Radiological Tomography* (Institute of Physics, London, 1990).

[3] H. P. Hiriyannaiah, "X-ray computed tomography," *IEEE Signal Process. Mag.* **14**, 42–59 (1997).

[4] A. C. Kak and M. Slaney, *Principles of Computerized Tomographic Imaging* (IEEE, New York, 1988).

[5] S. Webb, ed., *The Physics of Medical Imaging* (Institute of Physics, London, 1988).

[6] S. R. Cherry and M. E. Phelps, "Imaging brain function with positron emission tomography," in *Brain Mapping: The Methods*, A. W. Toga and J. C. Mazziotta, eds. (Academic, New York, 1996).

[7] G. Gullberg, G. Zeng, F. Datz, R. Christian, C. Tung, and H. Morgan, "Review of convergent beam tomography in single photon emission computed tomography," *Phys. Med. Biol.* **37**, 507–534 (1992).

[8] B. Bendriem and D. W. Townsend, eds, *The Theory and Practice of 3-D PET* (Kluwer, Boston, MA, 1998).

[9] L. A. Shepp and B. F. Logan, "The Fourier reconstruction of a head section," *IEEE Trans. Nucl. Sci.* **NS-21**, 21–33 (1974).

[10] B. K. Horn, "Fan-beam reconstruction methods," *Proc. IEEE* **67**, 1616–1623 (1979).

[11] G. Besson, "CT fan-beam parameterizations leading to shift-invariant filtering," *Inverse Problems* **12**, 815–833 (1996).

[12] C. Crawford and K. King, Computed tomography scanning with simultaneous patient translation, *Med. Phys.* **17**, 967–982 (1990).

[13] M. Daube-Witherspoon and G. Muehllehner, Treatment of axial data in three-dimensional PET, *J. Nucl. Med.* **28**, 1717–1724 (1987).

[14] M. Defrise, P. Kinahan, D. Townsend, C. Michel, M. Sibomana, and D. Newport, "Exact and approximate rebinning algorithms for 3-D PET data," *IEEE Trans. Med. Imag.* **16**, 145–158 (1997).

[15] J. Colsher, "Fully three dimensional positron emission tomography," *Phys. Med. Biol.* **25**, 103–115 (1980).

[16] P. Kinahan and L. Rogers, "Analytic 3-D image reconstruction

using all detected events," *IEEE Trans. Nucl. Sci.* **NS-36**, 964–968 (1996).

[17] L. Feldkamp, L. Davis, and J. Kress, "Practical cone-beam algorithm," *J. Opt. Soc. Am. A* **1**, 612–619 (1984).

[18] P. Grangeat, "Mathematical framework for cone-beam three-dimensional image reconstruction via the first derivative of the Radon transform," in G. Herman, A. Louis, and F. Natterer, eds., *Mathematical Methods in Tomography* (*Lecture Notes in Mathematics*, Vol. 1497, Springer, Berlin, 1991).

[19] M. Defrise and R. Clack, "A cone-beam reconstruction algorithm using shift-variant filtering and cone-beam backprojection," *IEEE Trans. Med. Imag.* **13**, 186–195 (1994).

[20] G. T. Herman, *Image Reconstruction From Projections: The Fundamentals of Computerized Tomography* (Academic, New York, 1980).

[21] Y. Censor, "Finite series-expansion reconstruction methods," *Proc. IEEE* **71**, 409–419 (1983).

[22] M. Smith, C. Floyd, R. Jaszczak, and E. Coleman, "Three-dimensional photon detection kernels and their application to SPECT reconstruction," *Phys. Med. Biol.* **37**, 605–622 (1992).

[23] J. Qi, R. Leahy, S. Cherry, A. Chatziioannou, and T. Farquhar, "High resolution 3-D Bayesian image reconstruction using the microPET small animal scanner," *Phys. Med. Biol.* **43**, 1001–1013 (1998).

[24] A. Shepp and Y. Vardi, "Maximum likelihood reconstruction for emission tomography," *IEEE Trans. Med. Imag.* **2**, 113–122 (1982).

[25] H. Hudson and R. Larkin, "Accelerated image reconstruction using ordered subsets of projection," *IEEE Trans. Med. Imag.* **13**, 601–609 (1994).

[26] L. Kaufman, "Maximum likelihood, least squares, and penalized least squares for PET," *IEEE Trans. Med. Imag.* **12**, 200–214 (1993).

[27] G. Gindi, M. Lee, A. Rangarajan, and I. G. Zubal, "Bayesian reconstruction of functional images using anatomical information as priors," *IEEE Trans. Med. Imag.* **12**, 670–680 (1993).

[28] R. Leahy and J. Qi, "Statistical approaches in quantitative PET," *Statist. Comput.* **10**, 147–165, 2000.

[29] J. Fessler, "Hybrid polynomial objective functions for tomographic image reconstruction from transmission scans," *IEEE Trans. Image. Process.* **4**, 1439–1450 (1995).

10.3

Cardiac Image Processing

Joseph M. Reinhardt
University of Iowa

William E. Higgins
Pennsylvania State University

1 Introduction

Heart disease continues to be the leading cause of death. Imaging techniques have long been used for assessing and treating cardiac disease [1, 2]. Among the imaging techniques employed are X-ray angiography, X-ray computed tomography (CT), ultrasonic imaging, magnetic-resonance (MR) imaging, positron emission tomography (PET), single-photon emission tomography (SPECT), and electrocardiography. These options span most of the common radiation types and have their respective strengths for assessing various disease conditions. Chapter 10.2 further discusses some relevant image-formation techniques, and references [1, 2] give a general discussion on cardiac image-formation techniques.

The heart is an organ that is constantly in motion. It receives deoxygenated blood from the body's organs via the venous circulation system (veins). It sends out oxygenated blood to the body via the arterial circulation system (arteries). The heart itself receives some of this blood via the coronary arterial network. Disease arises when the blood supply to the heart is interrupted or when the mechanics of the cardiac cycle change.

The available cardiac-imaging modalities produce a wide range of image data types for disease assessment: two-dimensional (2-D) projection images, reconstructed three-dimensional (3-D) images, 2-D slice images, true 3-D images, time sequences of 2-D and 3-D images, and sequences of 2-D interior-view (endoluminal) images. Each type of data introduces different processing issues. Fortunately, an extensive effort

has been made to devise computer-based techniques for managing these data and for extracting the useful information. This chapter focuses on techniques for processing cardiac images. Since a cardiac image is generally formed to diagnose a possible health problem, it is always essential that the physician have considerable control in managing the image data. Thus, visualization and manual data interaction play a major role in processing cardiac images. In general, the physician uses computer-based processing for guidance, not as the "final word." The various techniques for processing cardiac images can be broken down into four main classes:

1. Examination of the coronary arteries to find narrowed (stenosed) arteries.
2. Study of the heart's mechanics during the cardiac cycle.
3. Analysis of the temporal circulation of the blood through the heart.
4. Mapping of the electrical potentials on the heart's surfaces.

Subsequent sections of this chapter will focus on each of these four areas.

2 Coronary Artery Analysis

Perhaps the largest application of cardiac imaging is in the identification and localization of narrowed or blocked coronary arteries. Arteries become narrowed over time by means of a process known as coronary calcification ("hardening of the arteries"). If a major artery becomes completely blocked, this causes myocardial infarction ("heart attack"); the blood supply to the part of the heart provided by the blocked artery stops, causes tissue damage and, in many instances, death.

The region where an artery is narrowed or blocked is referred to as a stenosis. In the discussion to follow, the arteries will often be referred to as vessels. The inside of an artery is known as the lumen. The arterial network to the heart is often referred to as the coronary arterial tree. The major imaging modalities for examining the coronary arteries are X-ray angiography, CT imaging, and intravascular ultrasound. MR angiography, similar to X-ray angiography, is also possible. Digital image-processing techniques exist for all of these image types. As described further below, the primary aim of these methods is to provide human-independent aids for assessing the condition of the coronary arteries.

2.1 Single-Plane Angiography

Historically, angiographic imaging has been the standard for cardiovascular imaging. In angiography, a catheter is inserted into the body and positioned within the anatomical region under study. A contrast agent is injected through the catheter, and X-ray projection imaging is used to track the flow of contrast through the anatomy. An immediate problem with this imaging set-up is that 3-D anatomical information is mapped onto a 2-D plane. This results in information loss, structural overlap, and ambiguity.

Images may be obtained in a single plane or in two orthogonal planes (biplane angiography). Such images are referred to as angiograms. For coronary angiography, the contrast is used to highlight the coronary arteries. Figure 1 depicts a typical 2-D angiogram containing a stenosed artery.

The size of pixels in a digitized angiogram is of the order of 0.1 mm, permitting visualization of arteries around 1.0 mm in diameter. Sometimes separate angiograms can be collected before and after the contrast agent is introduced. Then, the no-contrast image is subtracted from the contrast-enhanced image to give an image that nominally contains only the enhanced coronary arteries. This procedure is referred to as digital subtraction angiography (DSA) [1, 2, 4].

For an X-ray coronary angiogram f, the value $f(x, y)$ represents the line integral of X-ray attenuation values of tissues situated along a ray L originating at the X-ray source and passing through the body to strike a detector at location (x, y):

$$f(x, y) = \int_{L_{x,y}} \mu(x, y, z)\, dz,$$

where L represents the ray (direction of X-ray) emanating from point (x, y) and $\mu(x, y, z)$ represents the attenuation coefficient of tissues. Encountered tissues can include muscle, fat, bone, blood, and contrast-agent enhanced blood. The value $f(x, y)$ tends to be darkest for rays passing through the contrast-enhanced arteries, since the contrast agent is radiodense (fully

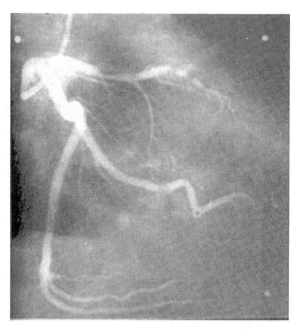

FIGURE 1 Typical 2-D angiogram. Image intensity is inverted to show the arteries as bright structures. The artery running horizontally near the top clearly shows a stenosis. From [3].

absorbs transmitted X-rays). Thus, the arteries of interest tend to appear dark in angiograms. The main image-processing problem is to locate the dark, narrow, branching structures — presumably this is the coronary arterial tree — and estimate the diameter or cross-sectional area along the extent of each identified branch. A stenosis is characterized by a local drop in vessel diameter or cross-sectional area.

Pappas proposed a complete mathematical model for structures contained in a 2-D angiogram [5]. In this model, a contrast-enhanced vessel is represented as a generalized cylinder having an elliptical cross-section; the 2-D projection of this representation can be captured by a function determined by two parameters. The background tissues (muscle, fat, etc.) are modeled by a low-order slowly varying polynomial, since such structures presumably arise from much bigger, and hence slowly varying functions. During the imaging process, unavoidable blurring occurs in the final image; this introduces another factor. Finally, a small noise component arises from digitization and attenuation artifacts. Thus, a point $f(x, y)$ on an angiogram can be modeled as

$$f(x, y) = \int_{L_{x,y}} (((v(x, y, z) + b(x, y, z)) * g(z)) + n(z))\, dz,$$
$$\tag{1}$$

where $v(x, y, z)$ represents a contrast-enhanced vessel, $b(x, y, z)$ represents the background, $g(z)$ is a Gaussian-smoothing function to account for image blurring, and $n(z)$ denotes the noise component. Pappas proposed a method in which parameters of this model can be estimated by using an iterative maximum-likelihood (ML) estimation technique. The procedure enables reasonable extraction of major arteries. Most importantly, it also provides estimates of vessel cross-sectional area profiles (a function showing the cross-sectional area measurement along the extent of a vessel); this permits identification of vessel stenoses.

Fleagle *et al.* proposed a fundamentally different approach for locating the coronary arteries and estimating vessel-diameter profiles [6]. Their study uses processing elements common to many other proposed approaches and contains many tests on real image data.

The first step of their approach requires a trained human observer to manually identify the centerline (central axis) of each artery of interest. The human uses a computer mouse to identify a few points that visually appear to approximately pass through the center of the vessel. Such manual intervention is common in many medical imaging procedures. These identified centerline points are then smoothed, using an averaging filter to give a complete centerline estimate. This step need not take more than 10 s per vessel. Next, two standard edge-detection operators — a Sobel operator and a Marr–Hildreth operator — are applied. A weighted sum of these output edge images is then computed. The composite edge image is then resampled along lines perpendicular to the centerline, at each point along the centerline. This produces a 2-D profile where the horizontal coordinate equals distance along the centerline and the vertical data correspond to the composite edge data. In effect, these resampled data represent a "straightened out" form of the artery. Next, this warped edge image is filtered, to reduce the effect of vessel border blurring, and a graph-search technique is applied to locate vessel borders. Finally, the detected borders are mapped back into the original space of the angiogram $f(x, y)$ to give the final vessel borders and diameters.

Sun *et al.* proposed a method especially suited for the insufficient resolution often inherent in digitized angiograms [7]. A human user first manually identifies the beginning and ending points of a vessel of interest. An adaptive tracking algorithm is then applied to identify the vessel's centerline. This centerline then serves as the axis traveled by a direction-sensitive low-pass filter. For each point along the centerline, angiographic data perpendicular to the centerline are retrieved and filtered. These new data are then filtered by a low-pass differentiator to identify vessel walls (outer borders). The differentiator acts as an edge detector. Figure 2 gives a typical output from this technique.

As an alternative to border-finding techniques, Klein *et al.* proposed a technique based on active contour analysis [8]. In their approach two direction-sensitive Gabor filters are applied to the original angiogram. These filtered images are then combined to form a composite energy-field image. The human operator then manually identifies several control points on this image to seed the contour finding process. Two B-spline curves, corresponding to the vessel borders, are then computed using an iterative dynamic-programming procedure. Figure 3 gives an example from the procedure.

2.2 Biplane Angiography and 3-D Reconstruction

Biplane angiography involves generating two 2-D angiograms at different viewing angles. Since the major coronary arteries are

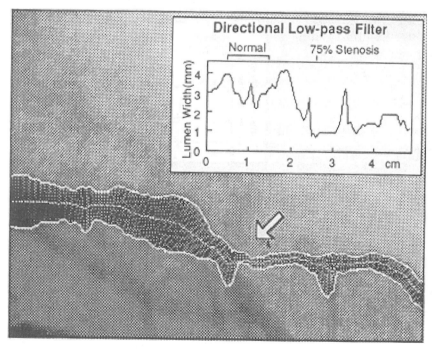

FIGURE 2   Example of an extracted artery and associated vessel-diameter (lumen-width) profile. The arrow points to the stenosis. From [7].

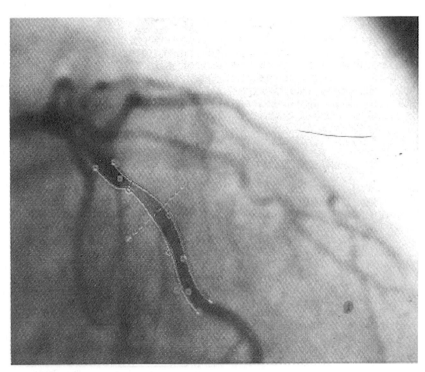

FIGURE 3   Result of an active-contour analysis applied to a selected artery in a typical 2-D angiogram. The green points are the manually identified control points. The red lines are the computed vessel wall borders. From [8]. (See color section, p. C–46.)

contrast enhanced, they can be readily identified and matched in the two given angiograms. This admits the possibility of 3-D reconstruction of the arterial tree. 3-D views provide many advantages over single 2-D views: (1) they provide unambiguous positional information, which is useful for catheter insertion and surgical procedures; (2) they enable true vessel cross-sectional area calculations; and (3) they are useful for monitoring the absolute motion of the myocardium. Biplane angiography is essentially a form of stereo imaging, but the term "biplane" has evolved in the medical community. Many computer-based approaches have been proposed for 3-D reconstruction of the arterial tree from a set of biplane angiograms [3, 4, 9, 10].

Parker *et al.* proposed a procedure in which the user first manually identified the axes of the arterial tree in each given angiogram [9]. Next, a dynamic search, employing vessel edge information, improves the manually identified axes. A least-squares-based point-matching algorithm then correlates points from the two skeletons to build the final 3-D reconstructed tree. The point-matching algorithm takes into account manually identified key points, the sparseness of the 3-D data, and the known geometry between the two given angiograms.

Kitamura *et al.* proposed a two-stage 3-D reconstruction technique [4]. First, the skeleton (central axes) and artery boundaries are computed for each 2-D angiogram. Next, a correspondence technique is applied to build a 3-D reconstructed artery model and skeleton.

Stage 1 employs the same generalized cylinder model (1) as used by Pappas [5]. Figure 4 shows a portion of this model and its relationship to each of the two known angiograms. Kitamura *et al.* allow the user to manually set parameters for all artery end points. Thus, all preidentified parts of the arterial tree are estimated. A nonlinear least-squares technique is used to estimate the model parameters. A few arteries can be situated parallel to the transmitted X-rays; these ill-defined portions of branches must be manually preidentified. Stage-2 reconstruction requires the user to manually identify bifurcation points (where a mother artery forms two smaller daughter branches) and stenotic points (where a stenosis occurs). These identified points then enable an automatic correspondence calculation of all skeleton points for the two reconstructed trees. This is done by backprojecting the points from the two trees into 3-D space, as depicted in Fig. 4. Since the structure of the 3-D tree is known from the manually identified points, the resulting correspondence is straightforward. The final output is a 3-D reconstructed tree and associated cross-sectional areas. The approach of Wahle *et al.* is similar [3]; Fig. 5 gives a typical result.

Note that the anatomy of the coronary arterial tree is well known. Also, the imaging geometry is known. This admits the possibility of using a knowledge-based system for reconstructing the 3-D arterial tree. Recently, Liu and Sun proposed such a method that is fully automatic [10].

2.3 X-ray CT Imaging

Recently, ultrafast high-resolution X-ray CT has emerged as a true 3-D cardiac imaging technique. CT can give detailed

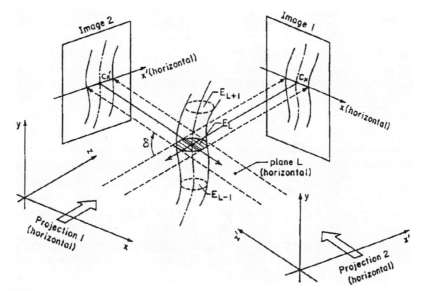

FIGURE 4 Geometry for reconstructing the 3-D arterial tree from two biplane images. The artery is modeled as a generalized cylinder having elliptical 2-D cross-sections. These cross-sections project as weighted line segments onto the two known 2-D angiograms (Images 1 and 2). From [4].

information on the 3-D geometry and function of the heart. Because of the heart motion during the cardiac cycle, high-speed scanning combined with electrocardiogram (ECG)-gated image acquisition is required to obtain high-resolution images. Over the past 20 years, cardiac imaging has been performed on scanners such as the experimental Dynamic Spatial Reconstructor [11, 12], electron beam CT (EBCT) scanner [13], and the newer spiral (helical) CT scanners [14]. CT can provide a stack of 2-D cross-sectional images to form a high-resolution 3-D image. Thus, true 3-D anatomic information is possible in a CT image, without the 2-D projection artifacts of angiograms that cause structural ambiguities.

Once again, to image the coronary arteries, generally one must inject a contrast agent into the patient prior to scanning.

Fortunately, the contrast can be injected intravenously, requiring significantly less invasion. Such an image is referred to as a 3-D coronary angiogram. Recently, the EBCT scanner has received considerable attention for use as an early screening device for coronary artery disease [13].

A complete system for 3-D coronary angiographic analysis has been devised by Higgins *et al.* [12]. The first component of the system automatically processes the 3-D angiogram to produce a complete 3-D coronary arterial tree. It also outputs vessel cross-sectional area information and vessel branching relationships. The second component of the system permits the user to visualize the analysis results.

The first processing component uses true, automatic, 3-D, digital image-processing operations. The raw 3-D angiogram

(a)

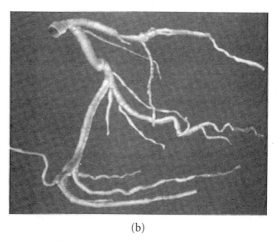

(b)

FIGURE 5 Extracted 3-D tree using the method of Wahle *et al.*: (a) the angiogram with superimposed tree (angiogram is the same as Fig. 1); (b) the reconstructed rendered 3-D tree. From [3].

undergoes 3-D nonlinear filtering to reduce image noise and sharpen the thin, bright coronary arteries. Next, a 3-D seeded region growing approach is applied to segment the raw 3-D coronary arterial tree. Cavity filling and other shape-based image-processing operations, based on 3-D mathematical morphology, are next applied to clean up the raw segmentation. Next, a 3-D skeletonization technique is applied to generate the raw central axes of the major tree branches. (Chapter 2.2 discusses mathematical morphology and skeletonization techniques.) This skeleton undergoes pruning to remove distracting short branches. Finally, the skeleton is converted into a series of piecewise linear line segments to give the final tree. Vessel cross-sectional area measurements and other quantities are also computed for the tree.

After automatic analysis, the second system component, a visualization tool, provides 2-D projection views, 3-D rendered views, along-axis cross-sectional views, and plots of cross-sectional area profiles. Figure 6 provides an overview of this visual tool. The various tools clearly show evidence of a stenosis. As shown in this example and in Section 2.4, 3-D imaging applications routinely require visualization tools to give adequate means for assessing the image data beyond simple 2-D image planes.

High-resolution CT imaging techniques are also in use to image the microvasculature. These so-called micro-CT scanners give a voxel resolution of the order of 0.001 mm (10 μm). Micro-CT scanners are being used to track the anatomical changes in genetically engineered mice to determine the long-term impact of various genes on disease states.

2.4 Intravascular Ultrasound Imaging

Standard coronary angiography does not give reliable information on the cross-sectional structure of arteries. This makes it difficult to accurately assess the buildup of plaque along the artery walls. Intravascular ultrasound (IVUS) imaging has emerged as a complementary technique for providing such cross-sectional data [15]. To perform IVUS, one inserts a catheter equipped with an ultrasonic transducer into a vessel of interest. As the catheter is maneuvered through the vessel, real-time cross-sectional images are generated along the vessel's extent. IVUS, however, does not provide positional information for the device. However, when IVUS is used in conjunction with biplane angiography, precise positional information can be computed. Thus, true 3-D information, as well as local detailed cross-sectional information, can be collected. This admits the possibility of using sophisticated viewing tools drawing upon the virtual reality modeling language (VRML). See Fig. 7 for an example.

For this view to be produced, standard biplane analysis, similar to that described in Section 2.2, must first be performed on a given pair of biplane angiograms. Next, the 3-D position of the IVUS probe, as given by its spatial location and rotation, must be computed from the given sequence of IVUS cross-sectional images. This positional information can then be easily correlated to the biplane information.

3 Analysis of Cardiac Mechanics

Imaging can be used to make a clinically meaningful assessment of heart structure and function. The human heart consists of four chambers separated by four valves. The right atrium receives deoxygenated blood from the venous circulation and delivers it to the right ventricle. The right ventricle is a low-pressure pump that moves the blood through the pulmonary artery into the lungs for gas exchange. The left atrium receives the oxygenated blood from the lungs and empties it into the left ventricle (LV). The LV is a high-pressure pump that distributes the oxygenated blood to the rest of the body. The heart muscle, called the myocardium, receives blood via the coronary arteries. During the diastolic phase of the heart cycle, the LV chamber fills will blood from the left atrium. At the end of the diastolic phase (end diastole) the LV chamber is at its maximum volume. During the systolic phase of the heart cycle, the LV chamber pumps blood to the systemic circulation. At the end of the systole phase (end systole), the LV chamber is at its minimum volume. The cycle of diastole–systole repeats for each cardiac cycle.

Cardiac imaging can be used to qualitatively assess heart morphology, for example, by checking for a four-chambered heart with properly functioning heart valves. More quantitatively, parameters such as chamber volumes and myocardial muscle mass can be estimated from either 2-D or 3-D imaging modalities. If 3-D images are available, it is possible to construct a 3-D surface model of the inner and outer myocardium walls. If 3-D images are available at multiple time points (a 4-D image sequence), the 3-D model can be animated to show wall motion and estimate wall thickening, velocity, and myocardial strain.

From an image engineering perspective, cardiac imaging provides a number of unique challenges. Since the heart is a dynamic organ that dramatically changes size and shape across the cardiac cycle (~1 s), image acquisition times must be short or the cardiac structures will be blurred as a result of the heart motion. Good spatial resolution is required to accurately image the complex heart anatomy. Adjacent structures, including the chest wall, ribs, and lungs, all contribute to the difficulties associated with obtaining high quality cardiac images. A variety of image processing techniques, ranging from simple edge detection to sophisticated 3-D shape models, have been developed for cardiac image analysis. Once the cardiac anatomy has been segmented in the image data, measurements such as heart chamber volume, ejection fraction, and muscle mass can be computed.

3.1 Chamber Analysis

The LV chamber is the high-pressure heart pump that moves oxygenated blood from the heart to other parts of the body. An assessment of LV geometry and function can provide information on overall cardiac health. Many cardiac image acquisition protocols and image analysis techniques have been developed specifically for imaging the LV chamber to estimate chamber volume. There are two specific points in the cardiac cycle that

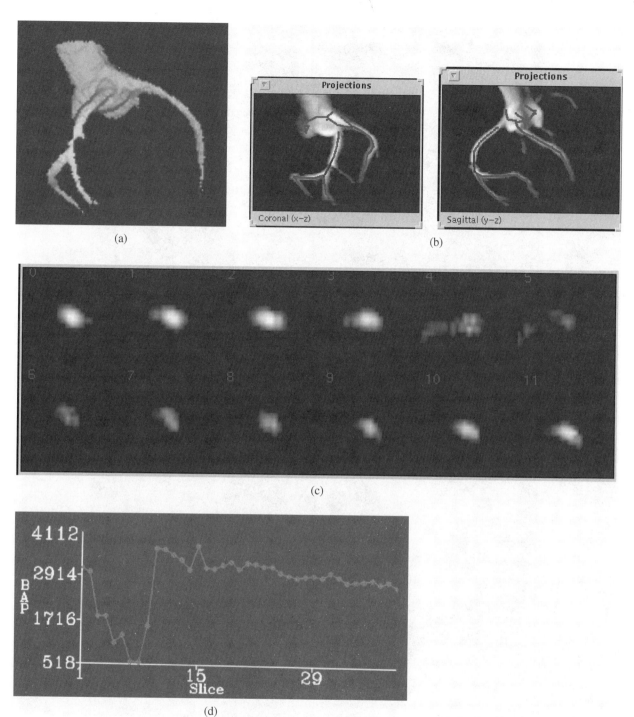

FIGURE 6 Composite view of a visual tool for assessing a 3-D angiogram [12]. (a) Volume-rendered version of the extracted 3-D arterial tree. (b) 2-D coronal (*x-z*) and sagittal (*y-z*) maximum-intensity projection images, with extracted arterial axes superimposed; red lines are extracted axes, green squares are bifurcation points, and the blue line is a selected artery segmented highlighted below. (c) Series of local 2-D cross-sectional images along a stenosed branch; these views lie orthogonal to the automatically defined axis through this branch. (d) Cross-sectional area plot along the stenosed branch. (See color section, p. C–47.)

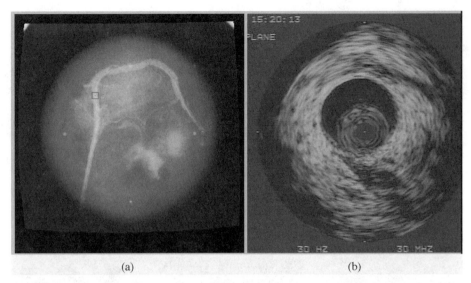

(a)   (b)

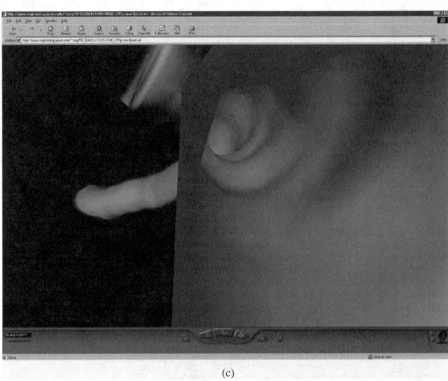

(c)

FIGURE 7   Views from a VRML-based IVUS system. (a) 2-D angiogram; the square indicates interior arterial viewing site of interest. (b) Corresponding cross-sectional IVUS frame of the arterial lumen. (c) 3-D surface rendering of the artery surface. (Courtesy of Dr. Milan Sonka, University of Iowa.) (See color section, p. C–48.)

are of particular interest: the end of the LV filling phase (end diastole), when the LV chamber is a maximum volume; and the end of the LV pumping phase (end systole), when the LV chamber is at minimum volume. Let V_{ES} and V_{ED} represent the end systolic and end diastolic chamber volumes. Then the total cardiac stroke volume $SV = V_{ED} - V_{ES}$, and the cardiac ejection fraction is $EF = SV/V_{ED}$. Both of these parameters can be used as indices of cardiac efficiency [16].

3.1.1 Angiography

Both single and biplane angiography can be used to study the heart chambers [2, 16]. For this analysis, sometimes called ventriculography, the imaging planes are typically oriented so that one image is acquired on a coronal projection (called the anterior–posterior, or A–P plane), and the other image is acquired on a sagittal projection (called the lateral, or LAT plane).

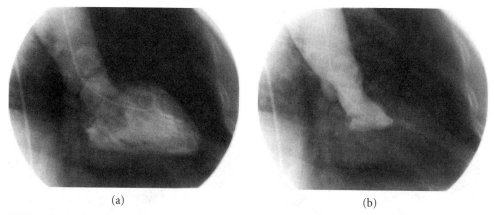

(a)   (b)

FIGURE 8 Angiographic images showing the contrast-enhanced LV chamber: (a) end diastole; (b) end systole. (Courtesy of the Adult Cardiac Catheterization Laboratory, University of Iowa.)

A radio-opaque liquid is injected into the LV chamber to increase contrast between the LV chamber and surrounding myocardium. Images may be acquired at a single time point (e.g., end diastole or end systole), or a sequence of images can be acquired over time showing the changing geometry of the chamber across the entire heart cycle (cineangiography). An angiogram showing the contrast-enhanced LV chamber is shown in Fig. 8.

Once the projection images have been obtained, the boundary between the LV chamber and myocardium must be identified. The LV boundary can be manually traced by using a computer mouse or trackball, and for single plane or biplane images acquired at only a small number of time points, this manual processing may be acceptable. However, for routine clinical use and in the case of cineangiography, manual analysis is prohibitively time consuming. Computer-assisted processing can reduce the time required to analyze the data, and it can eliminate human error and variability.

Automatic processing can be technically challenging because the LV chamber will be overlapped by nearby structures (ribs, vessels, catheter, etc.) in the projection image. Some of this overlap problem can be addressed by using DSA. With DSA, a reference image (or image sequence) is acquired just prior to contrast injection. This reference can be digitally subtracted from the contrast-enhanced image to remove the background and overlapping adjacent structures. Care must be taken to gather both the reference image and contrast-enhanced image at exactly the same point in the cardiac and respiratory cycle [2]. Other challenges to automatic LV border detection include variations in the image intensity in the chamber that are due to inhomogeneous mixing of the contrast material, geometric distortions and non-linearities in the imaging system, and random image acquisition noise [2].

Automatic and semiautomatic LV border detection methods typically identify the border as the local maximum of the image intensity gradient. Semiautomatic approaches rely on the operator to identify a starting pixel on the LV border; automatic methods may try to detect the starting location based on the image gradient. In either case, the border is tracked around the

LV by computing the image gradient magnitude and direction. Constraints are used to construct a contour that approximately follows the gradient maximum around the border and encloses the LV with a closed, smooth curve [2].

Once the LV boundary has been identified, measurements can be made to assess heart function. The first step in making these measurements is to estimate the LV chamber volume from the boundaries identified on the projection images. Methods for computing chamber volume assume a 3-D ellipsoidal chamber shape and require estimates of the minor and major axes of the ellipsoid from the projection images [16]. For single plane angiography, the ellipsoid is assumed to be rotationally symmetric about the long axis [2, 16].

3.1.2 Computed Tomography

X-ray CT imaging can also be used for heart chamber analysis. CT imaging is able to take advantage of injected contrast agents to better delineate the interface between the myocardium and the chamber. A typical cardiac CT image with a contrast-enhanced LV chamber is shown in Fig. 9.

The cardiac chambers can be segmented by identifying the chamber boundaries on 2-D CT slices, or more specialized 3-D processing can be used. Once the chamber boundaries have been identified, the total chamber volume can be obtained by integrating over all the 2-D cross-sectional areas across slices and scaling by the CT slice thickness. Because both the inner (endocardial) and outer (epicardial) borders of the myocardium can be identified on the CT images, parameters such as myocardial wall thickness and myocardial muscle volume can be computed [16].

Much of the early cardiac CT image analysis was performed manually. Manual image analysis requires that a skilled operator trace region boundaries on 2-D slices projected on a computer screen. This processing is extremely time consuming and prone to human error and variability. A number of methods for automatic and semiautomatic LV chamber segmentation in CT images have been developed [17–19]. Higgins *et al.* developed a semiautomatic method that relies on some manual slice-by-slice

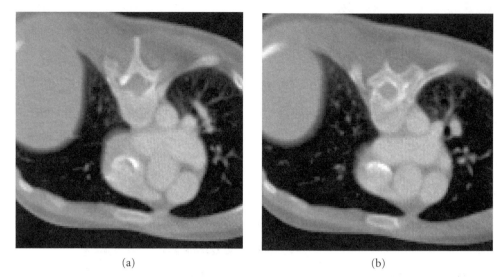

(a)   (b)

FIGURE 9   CT cross-sectional images of a human thorax obtained by using an electron beam CT scanner. Images show heart (oval-shaped gray region near the center of the images, with several bright ovals inside), lungs (dark regions on either side of heart), and vertebrae (bright regions at top middle). The heart region contains the myocardium (medium gray) and contrast-enhanced heart chambers (bright gray). (Images provided by Dr. Eric A. Hoffman, University of Iowa).

editing to guide automatic gray-scale and shape-based processing; their method shows good LV chamber volume correlation with manual analyses [17]. A popular LV chamber segmentation approach is to use deformable 2-D contours or 3-D surfaces attracted to the gradient maxima. Staib and Duncan used a 3-D surface model of the LV to segment the chamber from CT data [18]. Their method is initialized by configuring the model to an average chamber shape, and then deforming the model based on local gradient information. Related work from the same group uses a 3-D shape model and combined gray-scale region statistics with edge information for robust LV chamber segmentation [19].

Figure 10 shows a surface-rendered view of a canine heart from a DSR data set. This figure was created by manually tracing region boundaries on the image, and then shading surface pixels based on the angle between the viewing position and the local surface normal. The image clearly shows the four-chambered heart, the valves, and the myocardium. Figure 11 shows how the time series of images can be processed to yield data about cardiac function. The figure shows the LV chamber volume as

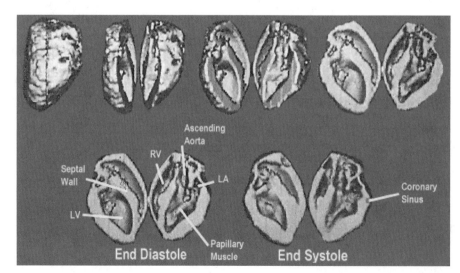

FIGURE 10   3-D surface-rendered heart image. The top row shows the computer-generated "dissection" of the 3-D heart volume; the bottom row has partially labeled heart anatomy. LA is left atrium and RV is right ventricle. From [20]. (See color section, p. C–49.)

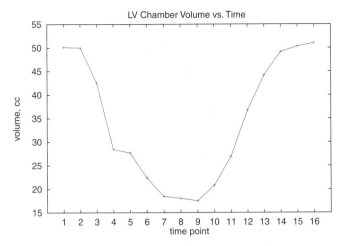

FIGURE 11 Plot of canine LV chamber volume vs. time. The image data set consists of 16 3-D images gathered over one heart cycle, using the DSR. LV volume was computed for each image after segmenting the chamber, using the method of [17].

a function of time across the heart cycle. In this case the LV chamber volume was computed by identifying the pixels within the LV, using the semiautomatic method described in [17]. The peak of the curve in Fig. 11 occurs at end diastole; the minimum of the curve occurs at end systole.

3.1.3 Echocardiography

Echocardiography uses ultrasound energy to image the heart [21]. The ultrasound energy (in the form of either a longitudinal or transverse wave) is applied to the body through a transducer with piezoelectric transmit and receive ultrasound crystals. As the ultrasound wave propagates through the body, some energy is reflected when the wavefront encounters a change in acoustic impedance (caused by a change in tissue type). The ultrasound receiver detects this return signal and uses it to form the image. Because ultrasound imaging does not use ionizing radiation to construct the image, ultrasound exams can be repeated many times without worries of cumulative radiation exposure. Ultrasound systems are often inexpensive, portable, and easy to operate, and as a result, exams are often performed at the bedside or in an examination room.

Common cardiac ultrasound imaging applications use energy in the range of 1 to ~25 MHz, although higher frequencies may be used for IVUS imaging. Most clinical ultrasound scanners can acquire B-mode (brightness) images, M-mode (motion) images, and Doppler (velocity) images. In B-mode imaging, a 2-D sector scan is used to create an image where pixel brightness in the image is proportional to the strength of the received echo signal. Several B-mode images may be obtained at different orientations to approximate volume imaging. In M-mode imaging, a 2-D image is formed where one image axis is the distance from the transducer and the other axis is the time. As with the B mode, pixel intensities in the M-mode image are proportional to the strength of the received echo signal. M-mode images can be used

to track myocardial wall and valve motion. Doppler imaging uses the frequency shift in the received signal to estimate the velocity of ultrasound scatterers. Doppler imaging can be used to measure wall and valve motion, and to assess blood flow through the arteries and heart. New 3-D ultrasound scanners have been introduced. These scanners use an electronically steered 2-D phased array transducer to acquire a volumetric data set. The 3-D ultrasound scanners can acquire data sets at near video rates (10–20 3-D images per second).

As shown in Fig. 12, ultrasound images are often considerably noisier and lower in resolution than images obtained using X-ray or magnetic resonance imaging. They present a number of interesting image processing challenges.

Much work in cardiac ultrasound image processing has been focused on edge detection in 2-D B-mode images to eliminate the need for manual tracing of the endocardial and epicardial borders [2]. The first step in the processing is often some preprocessing filtering, such as a median filter, to reduce noise in the image. Preprocessing is followed by an edge detection step, with a 2-D operator such as the Sobel or Prewitt edge detection mask, to identify strong edges in the image. Finally, the strong edges are linked together to form a closed boundary around the ventricle. This automatic 2-D processing shows good correlation with contours manually traced by a human [2]. After identifying the ventricle on each slice of a 3-D stack of B-mode images, a 3-D surface can be reconstructed and visualized.

Deformable contour models have also been successfully applied to the segmentation of LV chamber borders in echocardiographic images [22]. Another approach for LV chamber detection in echocardiography has focused on using optimization algorithms to identify likely border pixels. For these approaches, the image is processed with an edge detection operator to compute the edge strength at each pixel. The edge strength at each pixel is converted to a cost value, where the cost assigned to a pixel is inversely proportional to the likelihood that the pixel lies on the true LV border. Graph searching or dynamic programming is used to find a minimum-cost path through the image, corresponding to the most likely location of the LV chamber border. More sophisticated optimization methods can incorporate *a priori* anatomic information into the cost computation. Related work on MR images has focused on extending these 2-D optimal border detection algorithms to three dimensions [23].

3.1.4 Magnetic Resonance Imaging

Magnetic resonance (MR) imaging uses RF magnetic fields to construct tomographic images based on the principle of nuclear magnetic resonance (NMR) [21]. The pixel values in MR images are a function of the chemical configuration of the tissue under study. For most imaging protocols the pixel values are proportional to the density of hydrogen nuclei within a region of interest, although new imaging techniques are being developed to measure blood flow and other physiologic parameters. Diagnostic MR imaging uses nonionizing radiation, so exams

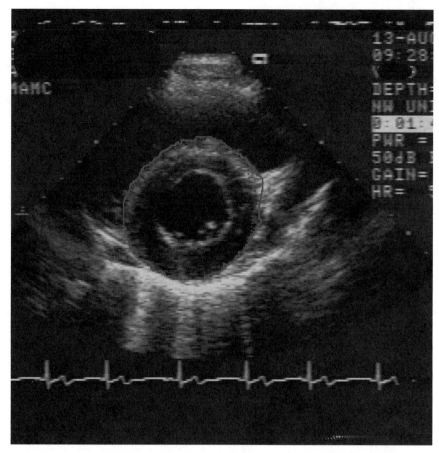

FIGURE 12 Ultrasound B-mode image of a human heart. The LV chamber is a center of image. The red border is the automatically detected epicardial border from 3-D graph search; the green border is the manually traced border. (Figure courtesy of Dr. Edwin L. Dove, University of Iowa and Dr. David D. McPherson, Northwestern University.) (See color section, p. C–49.)

can be repeated without the dangers associated with cumulative radiation exposure. Because the magnetic fields are electrically controlled, MR imaging is capable of gathering planar images at arbitrary orientations. New, faster MR scanners are being developed especially for cardiovascular applications. Because of differences in their magnetic resonance, there is natural contrast between the myocardium and the blood pool. MR contrast agents are now available to further enhance cardiac and vascular imaging. Many of the same techniques used in echocardiography and CT are applicable to cardiac MR image analysis; for example, 2-D and 3-D border detection algorithms based on optimal graph searches have been applied to LV chamber segmentation in MR images [23].

3.2 Myocardial Wall Motion

If a time series of images showing the heart chamber motion is available, information such as regional chamber wall velocity, myocardial thickening, and muscle strain can be computed. This analysis requires that the LV boundary be determined at each time point in the image sequence. After the LV boundary has been determined, motion estimation requires that the point-to-point correspondences be determined between the LV border pixels in images acquired at different times. For this difficult problem, algorithms based on optical flow [24] and shape-constrained minimum-energy deformations [25] have been successfully applied to CT and echocardiographic images.

One of the most dramatic recent advances in cardiac imaging has been the development of noninvasive techniques to "tag" specific regions of tissue within the body [26]. These tagging techniques, all based on MR imaging, use a presaturation RF pulse to temporarily change the magnetic characteristics of the nuclei in the tagged region just prior to image acquisition. The tagged region will have a greatly attenuated NMR response signal compared with the untagged tissue. Because the tags are associated with a particular spatial region of tissue, if the tissue moves, the tags move as well. Thus, by acquiring a sequence of images across time, the local displacement of the tissue can be determined by tracking the tags. One common cardiac tagging technique is called spatial modulation of magnetization (SPAMM) [26]. SPAMM tags are often applied as grid lines, as illustrated in Fig. 13.

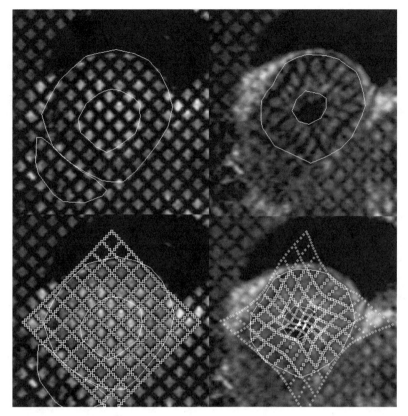

FIGURE 13 MR image showing SPAMM tag lines. The top left shows the initial tag line configuration (manually traced contours show chamber borders), and the top right is after the heart has changed shape. Tag lines have deformed to provide an indication of myocardial deformation. The bottom left and right show detected tag lines. From [27].

The two major image analysis problems in SPAMM imaging are the detection of the tag points and tracking and registering the tag points as the tissue deforms. Young *et al.* used a mesh of snakes to detect the tag lines in SPAMM images and tracked the tag lines and their intersection points between images [27]. The deformation information in [27] was used to drive a finite element model of the myocardium. Park *et al.* analyzed the dynamic LV chamber by using 3-D deformable models. The models were parameterized by functions representing the local LV surface shape and deformation parameters. Their approach gave estimates of LV radial contraction, longitudinal contraction, and twisting. Amini used B-spline snakes to detect the tag lines. The B splines were part of a thin-plate myocardial model that could be used to estimate myocardial deformation (compression, torsion, etc.) and strain at sample points between the tag line intersections [28]. Figure 14 shows a 3-D myocardial wall model computed by tracking SPAMM tag line motion during the heart cycle.

4 Myocardial Blood Flow (Perfusion)

Coronary angiography can be used to evaluate the structure of the coronary artery tree and to detect and quantify arterial stenoses. However, the precise linkage between coronary artery stenoses and blood flow (perfusion) to the myocardium is unclear [30]. Angiographic imaging is also limited by the spatial resolution of the imaging system. The largest coronary arteries

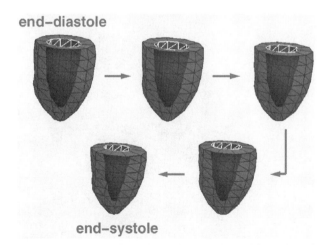

FIGURE 14 3-D myocardial wall model derived from deformable surface tracking SPAMM tag lines. The model shows inner and outer borders of the myocardium. Also shown is the evolution of myocardial wall and LV chamber shape from end diastole to end systole. (Figure courtesy of Dr. Jinah Park, University of Pennsylvania.) (See color section, p. C–50.)

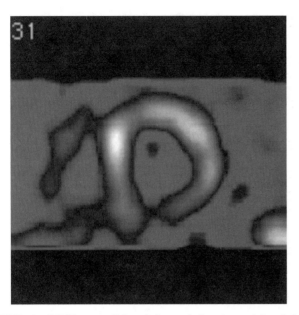

FIGURE 15 SPECT myocardial perfusion analysis, using an injected thallium-201 tracer. Shown is a cross-sectional view of the myocardium (LV chamber is the cavity at center of the image), with pixel intensity proportional to myocardial blood flow distribution. (Image courtesy of Dr. Richard Hichwa, PET Imaging Center, University of Iowa.) (See color section, p. C–50.)

are easily identified and analyzed. However, the vast network of smaller arteries that actually deliver blood to the myocardium remain mostly undetectable on the images. In this section we describe imaging modalities capable of directly assessing myocardial perfusion. The primary use of these techniques is to detect perfusion flow deficits beyond an arterial stenosis.

Positron emission tomography (PET) and single photon emission computed tomography (SPECT) both use intravenously injected radiopharmacueticals to track the flow of blood into the myocardial tissue. An image is formed where the pixels in the image represent the spatial distribution of the radiopharmacuetical. An example SPECT myocardial perfusion image is shown in Fig. 15.

Both echocardiographic and MR imaging can also be used to assess myocardial blood flow. In both cases, a contrast agent is used to increase the signal response from the blood. In echocardiography, small microbubbles (of the order of 5 μm in diameter) are injected into the bloodstream [30]. Bubbles this small can move through the pulmonary circulation and travel to the myocardium through the coronary arteries. The large difference in acoustic impedance between the blood and the bubbles results in a dramatic increase in the echo signal back from the perfused myocardium. For MR imaging, new injectable MR contrast agents have been developed to serve a similar purpose.

An interesting image processing challenge related to perfusion imaging is the problem of registering the functional (blood flow) images to structural (anatomic) images obtained with other modalities [31]. This structure–function matching typically uses anatomic landmarks, external fiducial markers, or both to find an affine transformation to align the two image data sets. The results can be visualized by combining the images so that a thresholded blood flow image is overlaid in pseudo-color on the anatomic image.

5 Electrocardiography

The constant muscular contractions of the heart during the cardiac cycle are triggered by regular electrical impulses originating from the heart's sinoatrial node (the heart's "pacemaker"). These impulses conduct throughout the heart, causing the movement of the heart's muscle. Certain diseases can produce iregularities in this activity; if it is sufficiently interrupted, it can cause death. This electrical activity can be recorded and monitored as an electrocardiogram (ECG).

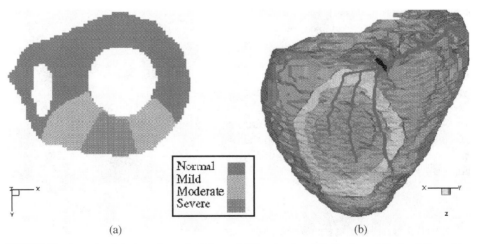

                    (a)   (b)

FIGURE 16 Example of a body-surface potential map. (a) Mapping for a 2-D slice through the heart; the cavities correspond to the ventricles. (b) 3-D surface-rendered view of the same map. The color-coding indicates the degree of myocardial ischemia (reduction in blood flow). The red lines on the 3-D view indicate a stenosed arterial region that brought about the ischemia. From [32]. (See color section, p. C–51.)

Through the techniques of electrocardiographic imaging, ECG data can be mapped into a 2-D or 3-D image [32]. These so-called body-surface potential maps are constructed by simultaneously recording and assembling a series of ECGs. Such image data can be used to visualize and evaluate various disease states, such as myocardial ischemia, in which the blood flow is reduced to a portion of the myocardium. Angiographic and CT imaging cannot provide such data. Body-surface potential maps also permit the study of ventricular fibrillation, a condition when the heart is excited by chaotic — and potentially lethal — electrical impulses.

Standard analytical methods from electromagnetics, such as the application of Green's theorem to compute the electric field distributions within the heart volume, are applied to evaluate such image data. Figure 16 gives an example.

6 Summary and View of the Future

Cardiovascular imaging is a major focus of modern healthcare. Many modalities are available for cardiac imaging. The image processing challenges include the development of robust image segmentation algorithms to minimize routine manual image analysis, methods for accurate measurement of clinically relevant parameters, techniques for visualizing and modeling these complex multidimensional data sets, and tools for using the image information to guide surgical interventions.

As technology continues to advance, scanner hardware and imaging software will continue to improve as well. Faster scanners with higher-resolution detectors will improve image quality. Researchers will continue the move toward scanning systems that provide true 3-D and 4-D image acquisition. From the image processing perspective, there will be a need to quickly process large multidimensional data sets, and to provide easy-to-use tools to inspect and visualize the results.

The interest in cardiovascular imaging is evidenced by the large number of journals, conferences, and workshops devoted to this area of research. From the engineering perspective, the *IEEE Transactions on Medical Imaging, IEEE Transactions on Biomedical Engineering IEEE Transactions on Image Processing*, and the *IEEE Engineering in Medicine and Biology Society Magazine* carry articles related to cardiac imaging. Conferences such as Computers in Cardiology, the IEEE International Conference on Image Processing, the SPIE Conference on Medical Imaging, and the IEEE Engineering in Medicine and Biology Annual Meeting are good sources for the most recent advances in cardiac imaging and image processing.

Acknowledgments

The following people have contributed figures and comments to this chapter: Drs. Edwin Dove, Richard Hichwa, Eric Hoffman, Charles Rossen, and Milan Sonka, of the University of Iowa; Dr. David McPherson of Northwestern University; and Dr. Jinah Park of the University of Pennsylvania.

References

[1] M. L. Marcus, H. R. Schelbert, D. J. Skorton, and G. L. Wolf, *Cardiac Imaging* (Saunders, Philadelphia, 1991).

[2] S. M. Collins and D. J. Skorton, eds., *Cardiac Imaging and Image Processing* (McGraw-Hill, New York, 1986).

[3] A. Wahle, E. Wellnhofer, I. Mugaragu, H. U. Sauer, H. Oswald, and E. Fleck, "Assessment of diffuse coronary artery disease by quantitative analysis of coronary morphology based upon 3-D reconstruction from biplane angiograms," *IEEE Trans. Med. Imag.* **14**, 230–241 (1995).

[4] K. Kitamura, J. B. Tobis, and J. Sklansky, "Estimating the 3-D skeletons and transverse areas of coronary arteries from biplane angiograms," *IEEE Trans. Med. Imag.* **7**, 173–187 (1988).

[5] T. N. Pappas and J. S. Lim, "A new method for estimation of coronary artery dimensions in angiograms," *IEEE Trans. Acoust. Speech Signal Process* **36**, 1501–1513 (1988).

[6] S. R. Fleagle, M. R. Johnson, C. J. Wilbricht, D. J. Skorton, R. F. Wilson, C. W. White, and M. L. Marcus, "Automated analysis of coronary arterial morphology in cineangiograms: geometric and physiologic validation in humans," *IEEE Trans. Med. Imag.* **8**, 387–400 (1989).

[7] Y. Sun, R. J. Lucariello, and S. A. Chiaramida, "Directional low-pass filtering for improved accuracy and reproducibility of stenosis quantification in coronary arteriograms," *IEEE Trans. Med. Imag.* **14**, 242–248 (1995).

[8] A. K. Klein, F. Lee, and A. A. Amini, "Quantitative coronary angiography with deformable spline models," *IEEE Trans. Med. Imag.* **16**, 468–482 (1997).

[9] D. L. Parker, D. L. Pope, R. van Bree, and H. W. Marshall, "Three-dimensional reconstruction of moving arterial beds from digital subtraction angiography," *Comput. Biomed. Res.* **20**, 166–185 (1987).

[10] I. Liu and Y. Sun, "Fully automated reconstruction of three-dimensional vascular tree structures from two orthogonal views using computational algorithms and production rules," *Opt. Eng.* **31**, 2197–2207 (1992).

[11] M. Block, Y. H. Liu, D. Harris, R. A. Robb, and E. L. Ritman, "Quantitative analysis of a vascular tree model with the dynamic spatial reconstructor," *J. Comp. Assist. Tomogr.* **8**, 390–400 (1984).

[12] W. E. Higgins, R. A. Karwoski, W. J. T. Spyra, and E. L. Ritman, "System for analyzing true three-dimensional angiograms," *IEEE Trans. Med. Imag.* **15**, 377–385 (1996).

[13] J. A. Rumberger, B. J. Rensing, J. E. Reed, E. L. Ritman, and P. H. Sheedy, "Noninvasive coronary angiography using electron beam computed tomography," in *Medical Imaging '96: Phys. Funct. Multidimensional Images*, E. A. Hoffman, ed., *Proc. SPIE* **2709**, 93–106 (1996).

[14] J. P. Heiken, J. A. Brink, and M. W. Vannier, "Spiral (helical) CT," *Radiology* **189**, 647–656 (1993).

[15] G. P. M. Prause, S. C. DeJong, C. R. McKay, and M. Sonka, "Towards a geometrically correct 3-D reconstruction of tortuous coronary arteries based on biplane angiography and intravascular ultrasound," *Int. J. Cardiac Imag.* **13**, 451–462 (1997).

[16] K. B. Chandran, *Cardiovascular Biomechanics* (New York, U. Press, New York, 1992).

[17] W. E. Higgins, N. Chung, and E. L. Ritman, "Extraction of the left-ventricular chamber from 3-D CT images of the heart," *IEEE Trans. Med. Imag.* **9**, 384–395 (1990).

[18] L. H. Staib and J. S. Duncan, "Model-based deformable surface finding for medical images," *IEEE Trans. Med. Imag.* **15**, 720–731 (1996).

[19] A. Chakraborty, L. H. Staib, and J. S. Duncan, "Deformable boundary finding in medical images by integrating gradient and region information," *IEEE Trans. Med. Imag.* **15**, 859–870 (1996).

[20] J. K. Udupa and G. T. Herman, eds., *3-D Imaging in Medicine* (CRC Press, Boca Raton, FL, 1991).

[21] K. K. Shung, M. B. Smith, and B. M. W. Tsui, *Principles of Medical Imaging* (Academic, San Diego, CA, 1992).

[22] V. Chalana, D. T. Linker, D. R. Haynor, and Y. Kim, "A multiple active contour model for cardiac boundary detection on echocardiographic sequences," *IEEE Trans. Med. Imag.* **15**, 290–298 (1996).

[23] D. R. Thedens, D. J. Skorton, and S. R. Fleagle, "Methods of graph searching for border detection in image sequences with applications to cardiac magnetic resonance imaging," *IEEE Trans. Med. Imag.* **14**, 42–55 (1995).

[24] S. Song and R. Leahy, "Computation of 3-D velocity fields from 3-D cine CT images," *IEEE Trans. Med. Imag.* **10**, 295–306 (1991).

[25] J. C. McEachen, II and J. S. Duncan, "Shape-based tracking of left ventricular wall motion," *IEEE Trans. Med. Imag.* **16**, 270–283 (1997).

[26] L. Axel and L. Dougherty, "Heart wall motion: improved method of spatial modulation of magnetization for MR imaging," *Radiology* **172**, 349–350 (1989).

[27] A. A. Young, D. L. Kraitchman, L. Dougherty, and L. Axel, "Tracking and finite element analysis of stripe deformation in magnetic resonance tagging," *IEEE Trans. Med. Imag.* **14**, 413–421 (1995).

[28] A. A. Amini, Y. Chen, R. W. Curwen, and J. Sun, "Coupled B-snake grids and constrained thin-plate splines for the analysis of 2-D tissue deformations from tagged MRI," *IEEE Trans. Med. Imag.* **17**, 344–356 (1998).

[29] J. Park, D. Metaxas, A. Young, and L. Axel, "Deformable models with parameter functions for cardiac motion analysis from tagged MRI data," *IEEE Trans. Med. Imag.* **15**, 278–289 (1996).

[30] J. H. C. Reiber and E. E. van der Wall, eds., *Cardiovascular Imaging* (Kluwer, Dordrecht, The Netherlands, 1996).

[31] M. C. Gilardi, G. Rizzo, A. Savi, and F. Fazio, "Registration of multimodality biomedical images of the heart," *Q. J. Nucl. Med.* **40**, 142–150 (1996).

[32] R. M. Gulrajani, "The forward and inverse problems of electrocardiography," *IEEE Eng. Med. Biol. Mag.* **17**, 84–101 (1998).

10.4

Computer Aided Detection for Screening Mammography

Michael D. Heath and
Kevin W. Bowyer
University of South Florida

Abstract

Breast cancer is the second leading cause of death for women in the U.S. Screening asymptomatic women is the most effective method for early detection of this disease. Despite its proven effectiveness, screening still misses about 20% of cancers and is the reason for an estimated 536,100[1] negative biopsies in 1998. Several studies have shown that double reading of mammograms (by a second radiologist) improves the accuracy of mammogram interpretation. The desire to use computers in place of the second radiologist, or as a prescreener to separate out clearly normal mammograms, are motivations for computer-aided detection research.

1 Introduction

Screening mammography is the X-ray examination of the breasts to check for cancer in asymptomatic women. Its goal is to identify breast cancer in an early stage of growth, before it becomes palpable or metastasizes (spreads to other parts of the body). Screening with mammography, accompanied by clinical exams and routine breast self-examination, currently provides the best means for early detection and survival of breast cancer.

The effectiveness of screening mammography can be measured by the reduction of mortality from breast cancer. Combined data from several randomized controlled trials showed the mortality rate from breast cancer was reduced with breast cancer screening. The size of the reduction was related to the age of women entering screening trials [1]. Women aged 40–49 showed a 17% reduction 15 years after starting screening, and women aged 50–69 showed a reduction in mortality of 25–30% 10–12 years after beginning screening. In women over age 70, there was insufficient information on the effectiveness of mammography because of the small numbers of women that started screening at this age. In addition to the reduction in mortality, early detection of breast cancer can provide a benefit through less invasive treatments.

Despite the observed reduction in mortality through screening, there is still room for improvement. A recent retrospective

---

[1]This figure of 536,100 assumes that 178,700 breast cancers were found in 1998 with a true positive biopsy rate of 25%.

805

study of women screened over a 10-year period, with a median of four mammogram exams and five clinical breast exams, showed that 23.8% of women had at least one false positive[2] mammogram, 13.4% had at least one false positive breast examination, and 31.7% had at least one false positive result for either test.

Another improvement in screening could be achieved by finding cancers that are missed by current screening programs. The FDA has estimated that "for every 80 cancers currently detected through routine mammogram screening of healthy women, an estimated 20 additional cancers are missed and not found until later. Of those missed, about half have cancerous features that are simply overlooked; the other half have cancerous features but look benign" [3]. It is estimated that 10–25% of palpable cancers are not visible on mammograms [4].

To improve screening with mammography, one must improve both the sensitivity (find a higher fraction of cancers) and specificity (obtain a higher fraction of malignancies in reported abnormal findings) or improve one without changing the other. This can be done by improving the quality of mammograms and the accuracy in interpreting them. Another potential improvement is to find invasive cancers earlier when they are smaller.

The image quality of mammograms has gone through significant technological improvements over the past 20 years (e.g., dedicated mammography X-ray equipment, optimized film screen combinations, automatic exposure control, improved quality control, and improved film processing), and it is reasonable to believe that digital mammography may lead to further improvements (mainly by increasing contrast and lowering the noise).

Interpretation accuracy can be improved by double reading mammograms by a second radiologist [5–7]. Similar results may be achieved when computers, programmed to detect suspicious regions in mammograms, direct radiologists' attention to them [8]. Another possibility is to apply artificial intelligence methods to help classify the suspicious features as being malignant or benign.

This chapter will introduce the reader to the mammographic exam, describe the digitization of mammograms, discuss issues in preprocessing the digitized images, introduce algorithms for detecting cancers, and describe methods of measuring their performance. In short, we will summarize the technical background for the engineer to work in the area of computer-aided detection (CAD) in mammography. Of course, collaboration with radiologists specializing in mammography is also an important component of research in this area.

---

[2]A false positive screening result was defined as "mammograms or clinical breast examinations that were interpreted as indeterminate, aroused a suspicion of cancer, or prompted recommendations for additional work up in women whom breast cancer was not diagnosed within the next year" [2].

2 Mammographic Screening Exam

2.1 Breast Positioning and Compression

A mammographic screening exam involves obtaining one or two images of each breast. A single view exam can be performed by using a mediolateral oblique (MLO) projection, and a two-view exam can be done by using a craniocaudal (CC) projection with the MLO projection. Figure 1 illustrates the four images collected in a typical two-view exam. Single-view exams are common in countries in Europe, while two-view exams are standard practice in the United States.

A mediolateral oblique mammogram images the most breast tissue of any single view. It is imaged at an angle approximately 45° to vertical with X-rays entering the patient anteriorly and medially (from the upper center of the body). The inferolateral (the lower outside) aspect of the breast is positioned near the film holder. A craniocaudal view is imaged vertically with X-rays entering the breast from the top. Medial tissue is better visualized in the CC projection than in the MLO projection.

In both projections, the breast is positioned by lifting it up and out such that the nipple is in profile. Compression is applied to the breast to improve image quality and to reduce the dose to the patient by helping to separate overlapping structures, reducing geometric unsharpness, reducing motion unsharpness, obtaining more uniform tissue thickness, and reducing scattered radiation.

The number of views taken of each breast and the exact positioning will depend on the patient. Examples of these include the following: (1) two films may be required to show all of the tissue in an MLO view of a large breast and (2) the angle used in positioning the patient for an MLO view may vary from 40–60° from vertical for large breasted women or from 60–70° from vertical for small breasted women [9]. The amount of compression used also varies between patients. It can be 10 cm or more with a median value of 4.5–5.5 cm, depending on the population [10].

2.2 Film Labeling

Mammograms are initially labeled with radiopaque markers that indicate (1) right or left laterality (R/L) and projection (CC or MLO); (2) patient identification; (3) the date of the exam; (4) technologist identification; and (5) the name of the facility. This information is recorded on the film when the image is acquired so it is permanently recorded on the mammogram. The label is placed on top of the film holder near the axillary portion of the breast.

Additional sticker labels may be attached to the film. These may include, but are not limited to, the following information: (1) patient name, age, sex, and social security number; (2) date of study; and (3) technical factors such as the mAs, kVp, compression force, compressed breast thickness, and angle for MLO views.

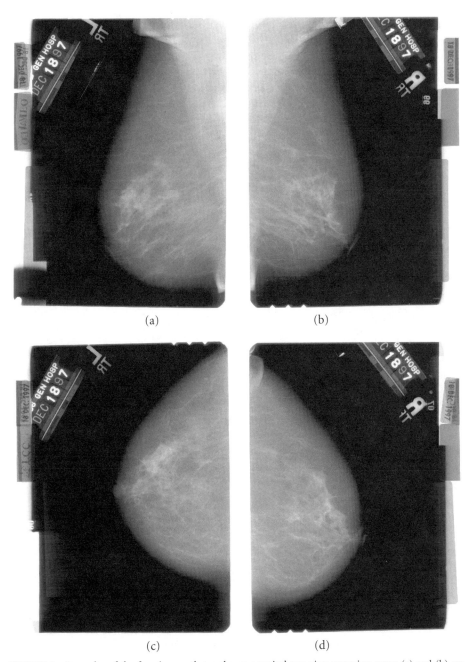

(a) (b)

(c) (d)

FIGURE 1 Examples of the four images that make up a typical two-view screening exam; (a) and (b) are mediolateral oblique projections and (c) and (d) are craniocaudal projections. Note the lettering used to record the date of the exam (December 18, 1997), the institution (Massachusetts General Hospital), the breast laterality (R or L), and the technician's initials (T. R.). A patient identification number appeared below the date but was covered before digitizing the film. The label is found near the axillary tail of the breast (by the armpit). Stickers with the view, date, and patient data can also be seen on the images.

3 Recording the Image

The image quality of mammograms must be very high for them to be useful. This is because the mammogram must accurately record small, low-contrast features that are critical to the detection of cancers such as those with microcalcifications or those for which the margin characteristics of a mass must be determined. The need for high quality mammograms was stated by the "Mammography Quality Standards Act of 1992" (PL 102–539, Oct. 27, 1992) 106 United States Statutes at Large, pp. 3547–3562, which requires the use of dedicated mammography equipment and certification of mammography facilities.

Many technical factors interact with one another and must be balanced to achieve high-quality mammograms. For example, the focal spot must be small enough to image small breast structures. Smaller focal spots reduce the X-ray intensity and increase the exposure time. This in turn may lead to a reduction of image quality from motion-induced blur during the longer exposure. Another tradeoff involves the use of an antiscatter grid placed between the patient and the film. As the name implies, this device reduces the amount of scattered radiation to improve image contrast, but it does so at the cost of reducing the exposure to the film. Subsequently, the amount of X-ray exposure must be increased to overcome the exposure reduction to the film, thereby increasing the radiation dose to the patient.

The following subsections will introduce film screen mammography, which is in common use today, and will describe the digitization of mammographic films for computer analysis. Direct digital systems, which are nearing deployment, will also be introduced.

3.1 Film Screen Mammography

In a film screen mammography system, the image is captured, stored, and displayed by using photographic film. By itself, film has a poor sensitivity to X-rays. To compensate for this, a sheet of phosphorescent material that converts X-rays to visible light is placed tightly against the film. This "screen" substantially reduces the X-ray dose to the patient but does so at the cost of blurring the image somewhat.

In film screen mammography systems, high spatial resolution may be achieved, but the quality of the image is limited by film granularity (noise), non-uniform contrast with relative exposure, and the blur introduced by the phosphor screen. The degradation in image quality from these sources (noise, contrast, and blur) reduces the interpretability of the mammogram in ways that are not well expressed as a single numeric spatial resolution limit.

3.2 Film Digitization

Image digitization is the process of converting the image stored on a physical medium (film) into an electronic image. Scanners or digitizers do this by dividing the image up into tiny picture elements (pixels) and assigning a number that corresponds to the average transmission or optical density in each area. This process involves illuminating the film with a known light intensity, and measuring the amount of light transmitted by each point (small area) of the film.

Scanners have physical limitations that introduce noise and artifacts. The quality of a scanner can be expressed by the following four primary performance criteria. "The *spatial resolution* measures the ability of the scanner to distinguish fine spatial structure in the film image. The *photometric accuracy* measures the uncertainty in the density values produced by the instrument.

The *scanning speed* measures the rate at which the instrument scans images. *Image artifacts* are errors in the density values that are not random. Artifacts usually fall outside the stated error bounds for the instrument and may be correlated over many pixels." [11].

Several of these performance criteria can be quantified, and this may be useful in comparing and contrasting scanners. However, it is important to understand that a scanner is but one component of a larger system and that the degree to which the entire system meets its goals is the best measure of performance.

Pixel values have no explicit relation to film density other than increasing or decreasing with density. Digital images obtained from the same radiograph by two scanners may be very different. To compensate for these differences, a normalization procedure may be applied to the images to remap the pixels values to a common measurement such as optical density.

There are no specifications on the required spatial resolution or photometric accuracy in scanning mammograms. One study by Chan *et al.* [12] showed that the accuracy of an algorithm for detecting microcalcifications decreased in performance with increasing sampling distance (35–140 µm). At the time of this writing, the preferred sampling resolution for digitizing mammograms is around 50 µm with a photometric digitization of 12–16 bits over an optical density range of roughly 0 to 3.5.

3.3 Direct Digital Mammography

Direct digital mammography replaces the film exposure and processing by directly digitizing the X-ray signal. The design of such a system involves a separate design of the image detector, storage, and display subsystems. This allows separate optimization of each, which in turn should produce a system with better overall performance.

Various configurations for the acquisition of digital mammograms have been proposed [13]. The principle differences between them are the scanning method (point, line and slot, and area systems) and X-ray detection method (indirect phosphor conversion or direct X-ray to electrical charge conversion). The digitization resolution of systems under development is 100 µm or better.

The development of detector technologies for digital mammography is well underway and will likely produce a practical system. The ultimate success of a digital mammography system will, however, rely on several factors, including the detection, storage, processing and display of digital mammograms. In summary, direct digital mammography will provide radiologists the control to visualize more detail in mammograms, but it must do so in a time-efficient manner. Computer-assisted detection may be of great importance here by serving to direct the physician to particular regions to examine in detail. Direct digital acquisition will provide data in the format necessary for CAD technology to be applied.

4 Image Preprocessing

Preprocessing digital mammographic images is a useful step before any interpretation of the image is performed. This preprocessing involves correcting artifacts introduced by the scanner and mammography equipment to better relate pixel values to the transmission of the breast. Depending on the magnitude of the artifacts, and the detection algorithm to be used, preprocessing can have a range of effects on the ultimate success of the algorithm. In situations in which an algorithm performs well on some images, preprocessing may still be useful when the algorithm is to be applied to images from different digitizers or mammographic equipment.

4.1 CCD Non-Uniformity

As discussed previously, the digitization process requires measuring the average density or exposure of tiny regions at regular intervals. The individual CCD detectors for most high speed devices are either arranged in a line that is swept across the field, or in a full two-dimensional lattice used in full-field digital mammography.

Each detector may have a slightly different sensitivity to light. The effect of this across the image is the addition of a noise pattern. In a linear scanning device, this noise will appear as stripes oriented in the direction of the linear scanning device with time. In a full-field digital device, this may appear as any pattern.

To some degree, this noise pattern artifact can be reduced by estimating the relative sensitivity function for each detector and then simulating an image that would have been obtained if each detector had the same sensitivity.

Measurements of light of two different intensities by each CCD can be used to estimate the relative sensitivity function of each

detector. In a film scanner these can be obtained by scanning a film with two uniform regions of different optical density such that all CCD elements are used to scan each region. In a full-field digital system, images can be recorded of a uniform object at two different exposures. Several repetitions may be recorded at each exposure.

For each CCD element, $CCD(k)$ where $1 \leq k \leq K$, the average of the high-intensity measurements is $CCD_{hi}(k)$ and the average of the low-intensity measurements for the same CCD is denoted $CCD_{low}(k)$. The average of all $CCD_{hi}(k)$ and all $CCD_{low}(k)$ values is $\overline{CCD}_{hi}$ and $\overline{CCD}_{low}$, respectively. A corrected value can be calculated for any pixel $f(n_1, n_2)$ that was recorded by CCD element k using Eq. (1):

$$\hat{f}(n_1, n_2) = CCD_o(k) + CCD_g(k) \times f(n_1, n_2) \qquad (1)$$

where the coefficients $CCD_o(k)$ and $CCD_g(k)$ are defined by Eqs. (2) and (3).

$$CCD_g(k) = \frac{\overline{CCD}_{hi} - \overline{CCD}_{low}}{CCD_{hi}(k) - CCD_{low}(k)}, \qquad (2)$$

$$CCD_o(k) = \overline{CCD}_{hi} - CCD_g(k) \times CCD_{hi}(k). \qquad (3)$$

Figure 2 illustrates the CCD non-uniformity in an image obtained with a HOWTEK 960 film digitizer and the removal of this artifact by processing with the algorithm described above.

4.2 Calibration to Film Density

Calibrating a digitized film screen mammogram to optical density is one way to normalize images digitized on different scanners. Algorithms written to detect abnormalities in mammograms calibrated to film density will be more generally applicable

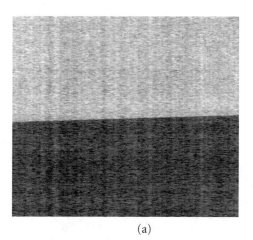

| (a) | (b) |

FIGURE 2   Non-uniformity correction applied to an image scanned with a HOWTEK 960 film digitizer; (a) is a subsection of a step wedge calibration film that shows artifacts introduced by non-uniform sensitivity of CCD elements in the digitizer, and (b) illustrates the image resulting from applying the correction algorithm described in Section 4.1. Note that the vertical lines have been removed by this processing.

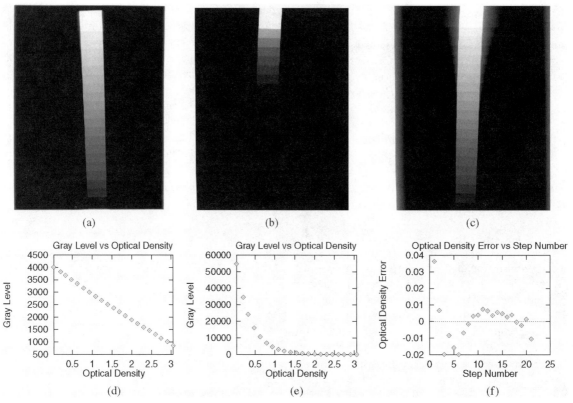

FIGURE 3 (a) and (b) Step wedge film scanned on both HOWTEK and DBA scanners, respectively. Plots of the average gray level vs. optical density in (d) and (e) show that the HOWTEK scanner has a linear response with density and the DBA scanner has an exponential response with density. The residuals from fitting a polynomial to the DBA data, using a regression model $OD = a_0 + a_1 * \log_{10}(GL) + a_2 * \log_{10}^2(GL)$, are plotted in (f). Image (c) illustrates the result of applying this model to the DBA image in (b) and then linearly scaling it for display. Note how much more similar the gray levels are in images (a) and (c) than in (a) and (b). The bleeding of the brighter steps in (c) is an artifact introduced by the DBA scanner.

to mammograms digitized on different scanners or at different times.

To calibrate an image to film density, one can scan a film that has regions of uniform optical density. If the optical density of each patch is not known, it can be measured with a spot densitometer operated in the transmission mode. The average digital count of each patch in the scanned image can be calculated and then plotted as a function of the known optical density. The optical density of any pixel value can be estimated by using an equation resulting from a regression analysis. Figure 3 illustrates how images produced by scanning the same film on two scanners are very different. After calibration, the images look more similar.

4.3 Calibration to Relative Exposure

The relationship between optical density and exposure is not linear over a broad range in film. This relationship is most easily expressed as a plot of the characteristic curve of a film (density vs. log exposure). Figure 4 shows the characteristic curve of Kodak MIN-R 2000 film. The shape of this plot is largely due to the

film and processing chemicals and conditions used, but it is influenced by many factors including, for example, the amount of time between exposure and development.[3]

An effect of this non-linear relationship of optical density and log exposure is that the contrast changes with exposure. A gamma plot shows the contrast (change in optical density) as a function of optical density. Figure 4 illustrates this for Kodak MIN-R 2000 film. Inspection of this plot shows that the contrast is reduced at both low and high optical densities (low and high relative exposures).

Since the film and development affect the optical density, we want to back out their effects to more accurately measure the relative transmission of the breast. To do this, we must image an artificial object (called a phantom) on the mammographic unit. This phantom may be made by stacking uniform thickness material of the same X-ray transmission in a stair-step fashion to achieve a variety of thicknesses that can all be imaged on the same film. Once the film is developed, digitized and calibrated to optical density, the average value of each constant thickness

---

[3]A delay of 4 h between exposure and processing can reduce the film speed by 10% [14].

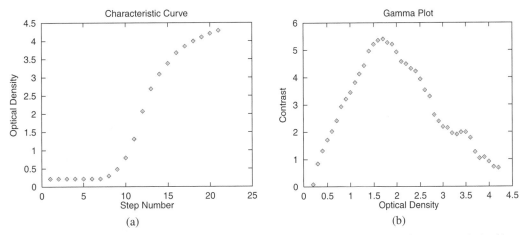

FIGURE 4 (a) Illustration of the nonlinear relationship between optical density and relative log exposure obtained by measuring the optical densities from KODAK MIN-R 2000 film exposed with a step pattern. (b) Shown is the contrast (change in optical density) as a function of optical density. The loss of contrast is evident at both low and high optical densities.

patch can be calculated and plotted against the corresponding thickness. The equation produced by application of regression analysis can be used to produce an image in which the contrast is nearly uniform with exposure.

4.4 Noise Equalization

Another approach to normalizing the mammograms is to remap the pixel values to equalize the noise [15]. This procedure will map the image to an isoprecision scale, which is a scale in which the noise level does not change with intensity. The advantage of this approach is that differences in pixel values are differences in a constant signal-to-noise scale, so detection thresholds should be uniform over gray level.

This mapping can be found by recording a number of uniform samples at different exposure levels and estimating the noise by measuring the variance in each sample. When it is not possible to perform these measurements on the mammographic system, a different approach can be taken where the high-frequency noise characteristics are estimated from a mammogram. Assuming that there are more pixels in homogeneous regions than there are near region boundaries, the conditional probability distribution of the noise can be estimated as a function of gray level. This can be done by computing a histogram of contrast for each intensity value in the image $k = 1, \ldots, K$, hist($f(n_1, n_2), c(n_1, n_2)$), where $c(n_1, n_2)$ is the local contrast,

$$c(n_1, n_2) = f(n_1, n_2) - \frac{1}{N} \sum_{j \in \delta_{n_1, n_2}} f(j), \qquad (4)$$

with δ_{n_1, n_2} specifying a square neighborhood centered at position n_1, n_2 of size N. Assuming the noise process is symmetric and the relationship between the pixel value and the X-ray exposure is approximately linear, the mean of each sample probability density function $g(c \mid k)$ should be zero. The standard deviation

$\hat{s}_c(k)$ of the contrast distribution for each intensity level k can be estimated from the histogram. The scale transform $L(k)$ that rescales pixel values to a scale with uniform noise level can then be calculated by numerically solving

$$\frac{\partial L(k)}{\partial k} = \frac{S_r}{\hat{s}_c(k)}, \qquad (5)$$

where the constant S_r is a free parameter that represents the noise level on the transformed scale. The equation can be numerically solved to create a look-up table $L(k)$ from the array $\hat{s}_c(k)$ by computing a normalized cumulative sum of $1/\hat{s}_c(k)$ from 1 to k. Figure 5 shows examples of the steps described above applied to create an isoprecision remapping look-up table. In practice, the intensities can be placed in bins that increase exponentially in width to obtain histograms for which it is easier to measure the standard deviation.

5 Abnormal Mammographic Findings

Masses and calcifications are the most common abnormalities on mammograms. A mass is a space-occupying lesion seen in at least two mammographic projections. A calcification is a deposit of calcium salt in a tissue. Both can be associated with either malignant or benign abnormalities, and can have a variety of visual appearances. To aid in standardized reporting, the American College of Radiology, in cooperation with the National Cancer Institute, the Centers for Disease Control, the Food and Drug Administration, the American Medical Association, the American College of Surgeons, and the College of American Pathologists, formulated the Breast Imaging Reporting and Data System (BI-RADS) [16]. The lexicon used for describing mammographic abnormalities is organized by mass and calcifications. Masses are described by their geometry, border characteristics, and density. Calcifications are described by their size

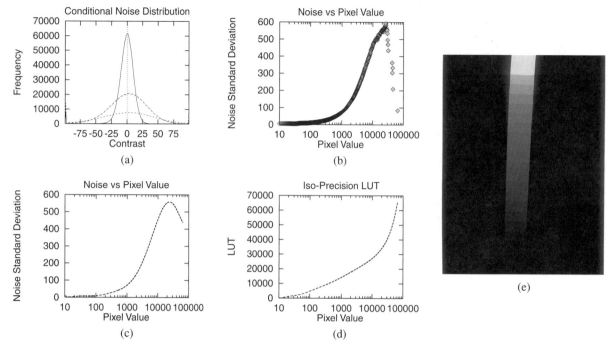

FIGURE 5 Plots showing steps of a noise equalization process. First, the histogram of the contrast is calculated for each gray level in a set of mammograms; (a) illustrates this for three gray levels. The standard deviation of each plot is then estimated; (b) shows this for each gray level. Note that the standard deviation is underestimated for high valued pixels because they occur with low frequency. A polynomial is then fit to the data in (b), where several of the highest points were dropped because of their poor estimation. A plot of the polynomial is shown in (c). A look-up table (LUT) is constructed from (c), where the derivative of the LUT at each point is the inverse of (c) at that point. The LUT was then normalized to the range 0–65535 and then plotted in (d); (e) shows the result of applying this LUT to the step wedge image scanned on the DBA scanner, i.e., (b) from Fig. 3.

morphology and distribution. Several books provide example illustrations and descriptions of abnormalities found in mammograms [17–20].

5.1 Masses

The shape of a mass can be round, oval, lobular, or irregular, and its margins may be circumscribed, microlobulated, obscured, indistinct, or spiculated. Both the shape and margins are indicators of the likelihood of malignancy with round and oval shapes and circumscribed margins having a lower likelihood of malignancy. Several examples of masses, described with the BI-RADS lexicon, are shown in image subsections in Fig. 6. Another secondary sign of cancer is architectural distortion of the normal breast structure with no visible mass.

5.2 Calcifications

Calcifications are described by their type, which refers to their size and shape. Typically benign types include skin, vascular, coarse, large rodlike, round, spherical or lucent centered, eggshell, milk of calcium, suture, dystrophic, or punctate. Amorphous or indistinct types are of more intermediate concern. Pleomorphic or heterogeneous and fine branching calcification types indicate a higher probability of malignancy.

The type is modified by keywords that indicate the distribution of the calcifications. The distribution can be clustered, linear, segmental regional, or diffuse (scattered). Regional and diffuse distributions are more likely to be benign. Figure 7 shows several examples of calcifications in different distributions.

6 Cancer Detection

6.1 Breast Segmentation

Breast segmentation is usually the first step applied in most cancer detection algorithms. The reason for this is that the breast region can be segmented quickly and the results can be used to limit the search area for the more computationally intensive abnormality detection algorithms.

An adequate segmentation of the breast tissue can often be achieved by thresholding the image and finding the largest region of pixels above the threshold. Histograms of mammograms have a characteristic large peak in a low-valued bin as a result of the large area of the background. Automated location of this peak can be accomplished by finding the maximum valued bin in the histogram. This bin value will be a typical background pixel value. Searching the histogram for the upper end of this peak will reveal a threshold value that should segment the breast. This can be

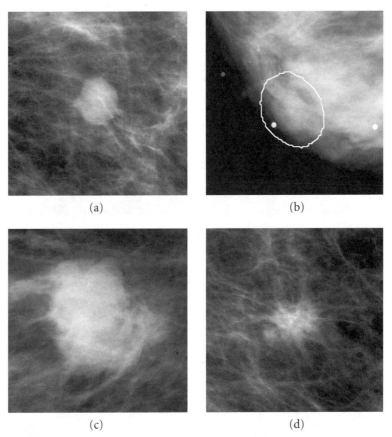

FIGURE 6 Examples of several types of masses: (a) a circumscribed oval mass; (b) an oval mass with obscured margins outlined for illustration; (c) a lobulated mass with microlobulated margins; (d) an irregular mass with spiculated margins.

accomplished by automatically searching the histogram for the lowest value bin position that has a higher value than the typical background bin value and is a local minimum in the histogram. There may be regions other than the breast, such as labels, which contain pixels above the threshold. Since these regions are generally smaller than the breast, keeping only the largest region should yield an adequate segmentation of the breast.

Some problems may be encountered with this approach when the label partially overlaps the breast tissue or when the intensifying screen does not cover the entire film. More sophisticated segmentation procedures may have to be developed to contend with these problems.

6.2 Mass Detection

The reliable detection of masses is a difficult problem because of their nonspecific appearance.[4] Masses can be many shapes and sizes, and may not even be directly visible, as in the case of architectural distortions.

---

[4]For the very reason that masses are difficult to detect, CAD may have its largest impact on breast cancer mortality if reliable detection methods can be found.

Common approaches to mass detection search for a "bright region" in a single image [21, 22], a region that is brighter in the image of one breast than the corresponding region in the contralateral breast [23, 24] or a mass that is spiculated with radial lines emanating from the center [25].

Most of these methods consist of computing several features (properties) for each pixel in the image and applying a classification procedure to decide which pixels are part of a mass. The features may include the average brightness, the direction of the gradient, the difference in brightness between corresponding positions in each breast, a measure of the distribution of the directions of the gradient, or any other feature thought to have a different value for pixels in a mass than for those not in a mass.

One method for detecting spiculated lesions [25] uses a binary decision tree classifier to assign a "suspiciousness" probability to each pixel in the breast. This probability of suspiciousness image is then blurred and thresholded to yield a map of suspicious areas in the mammogram.

Five features are precomputed for each pixel in the breast region and are then used by the binary decision tree classifier. This classifier uses a set of rules, that, when applied to the feature vector associated with a pixel, determines a category or class label for the pixel. The rules are automatically generated by training

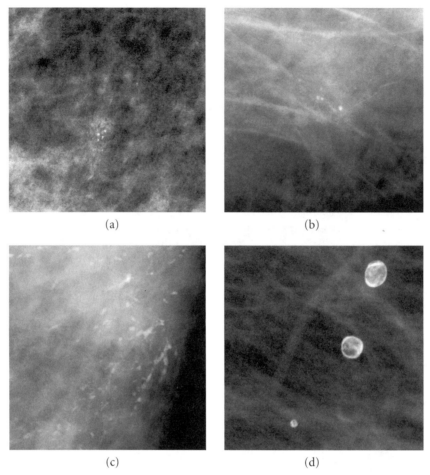

(a)   (b)

(c)   (d)

FIGURE 7 Examples of several types of calcifications: (a) a cluster of pleomorphic calcifications; (b) a cluster of punctate calcifications; (c) a regional distribution of fine linear branching calcifications; (d) an example of three lucent centered calcifications.

the classifier with features from mammograms that have spiculated lesions with known locations. The training produces a set of rules that can be represented graphically by a tree. When data are classified, it begins at the root and takes the path specified by the result of the first rule. This continues with subsequent rules until the last rule has been applied. At this point a leaf of the tree has been reached and the pixel is assigned a probability of suspiciousness associated with that leaf.

The five feature vectors include four Laws texture energy features and one novel feature, named ALOE for analysis of local oriented edges. Each of the Laws texture energy features is obtained by convolving the mammogram with two one-dimensional kernels and then computing the sum of the absolute values of the filtered pixel values in a local window. The equations for the four feature images $F1$, $F2$, $F3$, and $F4$ are provided in Eqs. 6–9. The kernel A is a 15×15 matrix[5] containing elements

that all have the value 1.0.

$$F1 = ABS(F * L5 * E5^t) * A, \tag{6}$$

$$F2 = ABS(F * E5 * S5^t) * A, \tag{7}$$

$$F3 = ABS(F * L5 * S5^t) * A, \tag{8}$$

$$F4 = ABS(F * R5 * R5^t) * A, \tag{9}$$

$$L5 = \begin{pmatrix} 1.0 & 4.0 & 6.0 & 4.0 & 1.0 \end{pmatrix},$$
$$E5 = \begin{pmatrix} -1.0 & -2.0 & 0.0 & 2.0 & 1.0 \end{pmatrix},$$
$$S5 = \begin{pmatrix} -1.0 & 0.0 & 2.0 & 0.0 & -1.0 \end{pmatrix},$$
$$R5 = \begin{pmatrix} 1.0 & -4.0 & 6.0 & -4.0 & 1.0 \end{pmatrix}.$$

The ALOE feature is computed as the standard deviation of a histogram of edge orientations (quantized to 180 discrete values) in a 4×4 cm window of the image, centered at each pixel. The edge orientation, ϕ, is computed at each site n_1, n_2 by

[5]This assumes that the mammogram being processed is sampled at or resampled to 280 μm.

using Eq. 10.

$$\phi_{n_1, n_2} = \arctan(sx(n_1, n_2), sy(n_1, n_2)), \qquad (10)$$

where

$$sx = f * \begin{pmatrix} -1 & 0 & 1 \\ -2 & 0 & 2 \\ -1 & 0 & 1 \end{pmatrix},$$

$$sy = f * \begin{pmatrix} -1 & -2 & -1 \\ 0 & 0 & 0 \\ 1 & 2 & 1 \end{pmatrix}.$$

The decision tree classifier is trained by using a set of images for which the positions of spiculated lesions are known. A random sample of pixels in spiculated lesions, and pixels outside of spiculated lesions is input to a decision tree classifier[6] operated in the training mode, and the rules are automatically generated. All pixels from the training images are then passed through the tree and the fraction of suspicious pixels is computed for each leaf of the tree. This fraction serves as the probability of suspiciousness for all pixels classified in this leaf.

Mammograms that are to have spiculated lesions detected can then be processed by computing the five feature values for each pixel, classifying each pixel with the decision tree, assigning the probability of suspiciousness to each pixel, convolving this image with a 7.5×7.5 mm kernel filled with equal weights that sum to 1.0, and thresholding the convolved image at a value of 0.5. This produces a binary image in which probabilities above 0.5 are assigned a value of 255 and probabilities smaller than or equal to 0.5 are assigned the value zero. All pixels with the value 255 are part of regions that are suspicious as being part of spiculated lesions.

Figure 8 illustrates the application of this algorithm to a mammogram. The figure shows the original image with overlayed ground truth indicating the position of a spiculated lesion, the five feature images, the blurred probability image and a thresholded probability image. One true positive (correct detection) and two false positive (incorrect detections) can be seen in the figure.

6.3 Calcification Detection

The detection of microcalcifications is an important component of CAD. Calcifications are small densities that appear as bright spots on mammograms, as illustrated in Fig. 7. Calcification detection is generally regarded as a much easier problem than the detection of masses as a result of their more specific appearance. Their visual detection may be difficult without the aid of a magnifying glass. Computer-aided detection of microcalcifications has been an intense area of research. Many approaches

[6]One decision tree classifier program that can be used for this purpose is C4.5 [26].

have been taken to the problem and impressive results have been reported [15, 27].

One straightforward approach that applies standard image processing steps will be described and illustrated here. This is basically the approach used in [28], written for application on images digitized at a 100 μm. This method can be easily implemented and used as a baseline for comparisons with other algorithms.

The first step in the algorithm is to create an image in which calcifications are enhanced for easier identification and detection. This is done by subtracting a signal-suppressed image from a signal-enhanced image. The signal enhanced image, g, is obtained by convolving the original mammogram with a small kernel. The signal suppressed image is obtained by using a median filter (described in Chapter 3.2). This nonlinear filter selects the median value of the 49 intensity values in a 7×7 window centered at each pixel in the original mammogram. If calcifications are present in the original mammogram, they will appear as bright dots in the difference image, h.

$$g = f * \frac{1}{7} \begin{pmatrix} 0.75 & 0.75 & 0.75 \\ 0.75 & 1.00 & 0.75 \\ 0.75 & 0.75 & 0.75 \end{pmatrix},$$

$$h(n_1, n_2) = g(n_1, n_2) - \text{median}\{f(p, q)\}, \qquad (11)$$

where $n_1 - 3 \le p \le n_1 + 3$ and $n_2 - 3 \le q \le n_2 + 3$.

Calcifications can be segmented in the difference image by thresholding. One method for selecting the threshold value is to calculate the cumulative histogram of the enhanced image, h, for all pixels in the segmented breast, and automatically searching the cumulative histogram for the lowest numbered bin where the count exceeds a large percentage (e.g., 99.995%) of the total number of pixels. Once a binary image is obtained from thresholding, individual calcifications can be labeled by using a connected component labeling algorithm (described in Chapter 2.2). The result of the connected component algorithm will be an image in which all of the pixels in each separate connected group of pixels have the same value and this value will not be shared by any other pixels in the image.

The next step is to remove any connected group of pixels that has less than two or three pixels in it. This can be done by computing the histogram of the connected component image and setting any bin value that is less than 3 to zero. The histogram can then be applied as a look-up table to the image to remap pixel values of small components to the background value of zero.

The final step in the process is to find "clusters" of calcifications, where a cluster is defined to be more than three calcifications (i.e., connected regions) in a 1- to 1.5-cm diameter circle. Figure 9 illustrates the results of this method.

Additional processing can be applied to improve the accuracy of calcification detection algorithms. The usual approach for this is to measure features such as the size and shape of individual calcifications and the geometry of a group of calcifications and

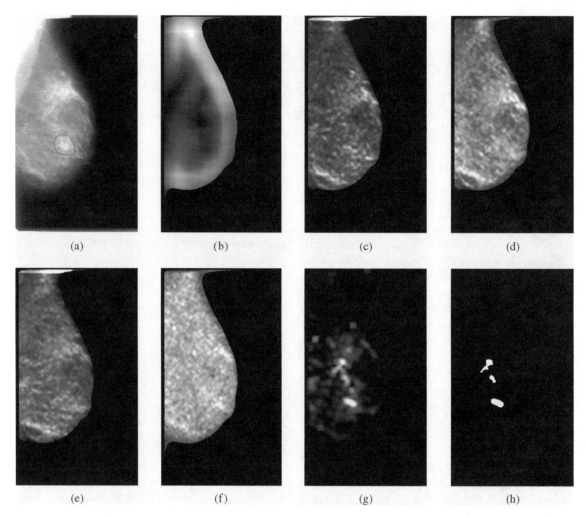

FIGURE 8   Illustration of a spiculated lesion detection algorithm, showing the original mammogram with the ground truth marking the location of a spiculated lesion in red in (a), the ALOE feature image in (b), and the Laws feature images $L5 * E5^t$, $E5 * S5^t$, $L5 * S5^t$, and $R5 * R5^t$ in (c)–(f). The blurred probability image is shown in (g). Image (h) is a thresholded version of (g); it shows three detected spiculated lesions. The bottom one is a correct detection corresponding the the ground truth outline in (a), and the other two white blobs are false detections. (See color section, p. C–51.)

then train a classifier to differentiate false detections from true detections of clusters of calcifications. Many features commonly used for this are described in [29].

7 Performance Assessment

Thorough assessment of the performance of a CAD algorithm is critical. Many algorithms for detecting and classifying both masses and microcalcifications have been designed and implemented, but few have undergone rigorous testing with large databases of proven cases, and even fewer have been evaluated in a clinical setting [30, 31].

It is usually desirable to first evaluate an algorithm using retrospective testing on a database of previously diagnosed cases with radiologist specified ground truth. Several publicly available databases [32–35] simplify this task because high-quality images with ground truth are available either for free or for a minimal charge. If testing is done properly according to standard train, test, and evaluation procedures, it may even be possible to estimate the relative merit of competing algorithms.

The largest publicly available mammography database is the Digital Database for Screening Mammography (DDSM) at the University of South Florida [36]. It currently contains 2620 cases. Each case contains all four images from the mammographic exam. Both cancer and benign cases include ground truth markings. Keyword descriptions of abnormalities are specified using the BI-RADS lexicon. Additional data with each case include radiologist-assigned values for the ACR breast density rating, the ACR assessment code, and a subtlety rating for the case, on a 1–5 scale. All of the images in this chapter were selected from this database.

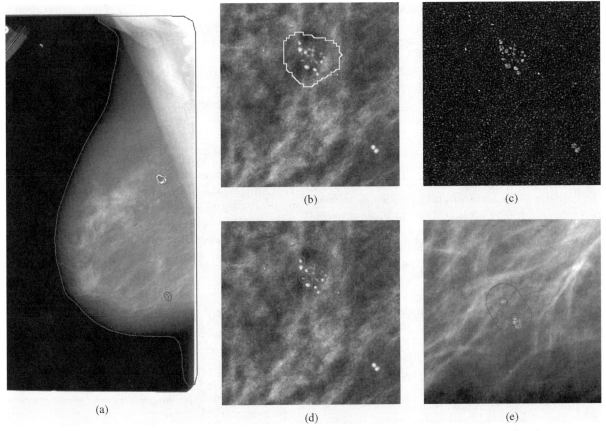

FIGURE 9 Illustration of a calcification detection algorithm, showing one true positive and one false positive detection of a malignant cluster of pleomorphic calcifications. (a) An overview of a segmented breast with one ground truth region (white) and two detections (green and red). The border of the segmented breast is shown in purple. (b) A closeup of the cluster of calcifications with ground truth overlaid in white. (c) The result of enhancing the calcifications in the image using the algorithm described in Section 6.3. (d) The result of thresholding the enhanced mammogram, labeling individual calcifications, finding a cluster group of more than three calcifications linked by intercalcification distances of <4 mm. Individual calcifications in the group are circled in green and the cluster is marked with a green border. (e) A false detection of a group of calcifications. (See color section, p. C–52.)

Once testing has been done on the computer, and satisfactory results are obtained, it is necessary to evaluate a CAD algorithm in a clinical setting with radiologists using the system. A high-performance algorithm for prompting a radiologist's attention to suspicious regions must still be shown to result in an improvement in the radiologist's interpretation of the case.

7.1 Computer Analysis of Algorithm Performance

The performance of an algorithm for detecting suspicious regions in mammograms can be calculated from a set of digital mammograms when ground truth markings of the abnormalities are available. Many decisions must be made in the course of evaluating the performance. The selection of cases, the training procedure used, and the method of scoring detections can all dramatically affect the measured performance.

The subtlety of lesions of the same type will vary in mammograms. For example tumors may have a variety of size, contrast, margin definition, and similarity to normal parenchymal tissue, and calcifications may vary in size, number, contrast, and the number of noninteresting calcifications (e.g., vascular calcifications) may vary per mammogram. Thus, some mammograms will have lesions that are easier to detect than others, and the mix of mammograms will affect the overall performance. The number of normal mammograms included in the evaluation will also affect the performance. Since cancers are only found in a small percentage of mammograms, the measured performance on a set of images that all contain cancers will not reflect the performance of the algorithm when run on consecutive cases in a screening program.

Mammography databases contain a large number of cases of a variety of cancers (e.g., calcifications and masses with varied visual appearances). Some bias toward easier cases may be present in the cases in the database, because low-quality mammograms may be excluded or because more interesting or difficult cases may not have been available when the database was created. Additional bias will be introduced when a subset of mammograms is selected for evaluating an algorithm designed to detect a particular subtype of abnormalities. For example, should cases with calcifications inside a mass or cases with calcifications not distributed in clusters be used when testing an algorithm for detecting clustered calcifications? Such restrictions on the selection of cases will reduce the size of the dataset and reduce the generality of the estimated performance.

To avoid the possibility of overtraining an algorithm and obtaining artificially high-performance scores, no data in the test set should be allowed to influence the algorithm or its parameter settings. An algorithm must be fixed before training or testing is done. In training, the algorithm is tuned to the data by running it on a fixed subset of the data in an iterative fashion, measuring the performance of the algorithm at each step and adjusting parameter values to obtain the best performance. Once this is completed, the algorithm is run on the remaining cases (a disjoint subset of the cases) and the performance is evaluated by using the ground truth. There are two common procedures for selecting the subset to use in training and testing. The first method is to randomly divide the cases in half. Training and testing is done twice, first training on one half of the data and then testing on the other half and then reversing the process. The other method is to train on all of the cases except one, and then evaluate the performance on the remaining case. This is repeated many times such that each case is left out one time.

The method of scoring detections as correct or incorrect must be fixed, and listed with any performance result. In a prompting system, a correct detection may be assumed when a prompt is generated inside a ground truth region. Prompts that are outside all ground truth regions are false positive detections and the first prompt inside a ground truth region is a true positive. Additional prompts in the same ground truth region are not scored because only one detection of an abnormality is productive. With this measure it is clear that the method of marking the ground truth will affect the performance of an algorithm. If ground truth is specified as a circle around an abnormality, the area of the marking will be larger than if free-form markings are used. Also, if the ground truth represents all regions that initially looked suspicious, the performance of an algorithm will measure higher than when only cancers are marked.

Another decision to make in measuring the performance of an algorithm is how the fraction of true positive and false positive detections are calculated. When an algorithm prompts three regions in an MLO mammogram and three regions in a CC mammogram of the same breast, and of these six prompts one falls on a cancer, is the average true positive rate 1 or 0.5? Either could be correct, as long as the method of calculation is consistent and clearly stated.

The preferred method for showing the results of a detection algorithm in mammography is through a free-response receiver operating characteristic (FROC) plot. This is a plot of the average fraction of correct detections (TP/(TP + FN)) versus the average number of false detections per image obtained on a set of images. Displaying results in this form shows the performance

of a detection algorithm at a range of possible operating points. Typically, operating a detection algorithm at a point where more correct detections are made will lead to more false detections as well. Comparing FROC curves generated from two algorithms allows a quick comparison of the algorithms at any of a range of possible operating points. Excellent coverage of ROC analysis can be found in [37] and [38].

7.2 Testing in a Clinical Setting

Analysis of a computer algorithm alone is not sufficient for obtaining approval to use CAD in a clinical setting. The evaluation must consider the effects that the computer-generated information ultimately has on patient care [39]. In the United States, the Food and Drug Administration (FDA) has the authority to ensure safe and effective devices. This authority includes the approval of computerized medical image analysis and computer-aided detection.

To evaluate a CAD prompting system, one may want to demonstrate that the sensitivity and specificity in finding breast cancer are improved when the system is used. Measuring this directly would be very difficult and time consuming because of the low incidence of cancers in screening exams. A more practical approach may be to break the problem down into two parts. First, one could determine whether or not the biopsy rate for a particular radiologist increases compared to his or her prior biopsy rate. Second, one could demonstrate whether or not a system is able to prompt regions on previous exams where a cancer was not found until the next exam at that location. This could demonstrate the the system is capable of detecting cancers missed by radiologists. Clinical studies of this type were performed to demonstrate the safety and effectiveness of a commercial CAD system (the ImageChecker M1000 System, R2 Technologies Inc., Los Altos, CA). In June 1998 this system was approved for clinical use by the FDA [3].

8 Summary

Mammography has proven to be an effective tool for the early detection of breast cancer, and when used in a screening program has been shown to decrease the mortality rate. This being stated, there is room for improvement by either finding cancers earlier or by decreasing the number of false positive mammograms. Initial experience with CAD in mammography has shown potential, and this technology is at a stage of transition to commercial application. It will take some time to determine the true effect CAD has on reducing the mortality rate from breast cancer.

Acknowledgments

This work was supported in part by a NASA Florida Space Grant Consortium graduate fellowship. The Digital Database

10.5

Fingerprint Classification
and Matching

Anil Jain
Michigan State University

Sharath Pankanti
*IBM T. J. Watson Research
Center*

1 Introduction

The problem of resolving the identity of a person can be categorized into two fundamentally distinct types of problems with different inherent complexities [1]: (1) verification and (2) recognition. Verification (authentication) refers to the problem of confirming or denying a person's claimed identity (Am I who I claim I am?). Recognition (Who am I?) refers to the problem of establishing a subject's identity.[1] A reliable personal identification is critical in many daily transactions. For example, access control to physical facilities and computer privileges are becoming increasingly important to prevent their abuse. There is an increasing interest in inexpensive and reliable personal identification in many emerging civilian, commercial, and financial applications.

Typically, a person could be identified based on (1) a person's possession ("something that you possess"), e.g., permit physical access to a building to all persons whose identities could be authenticated by possession of a key; (2) a person's knowledge of a piece of information ("something that you know"), e.g., permit log-in access to a system to a person who knows the user i.d.

[1] Often, recognition is also referred to as identification.

and a password associated with it. Another approach to positive identification is based on identifying physical characteristics of the person. The characteristics could be either a person's physiological traits, e.g., fingerprints, hand geometry, etc., or his or her behavioral characteristics, e.g., voice and signature. This method of identification of a person based on his or her physiological or behavioral characteristics is called *biometrics*. Since the biological characteristics can not be forgotten (like passwords) and can not be easily shared or misplaced (like keys), they are generally considered to be a more reliable approach to solving the personal identification problem.

2 Emerging Applications

The accurate identification of a person could deter crime and fraud, streamline business processes, and save critical resources. Here are a few mind-boggling numbers: about one billion dollars in welfare benefits in the United States are annually claimed by "double dipping" welfare recipients with fraudulent multiple identities [33]. MasterCard estimates the credit card fraud at $450 million per annum, which includes charges made on lost and stolen credit cards: unobtrusive positive personal identification of the legitimate ownership of a credit card at the

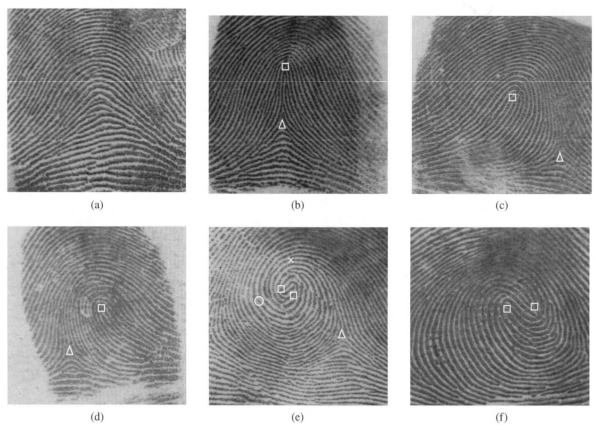

FIGURE 1   Fingerprints and a fingerprint classification schema involving six categories: (a) arch, (b) tented arch, (c) right loop, (d) left loop, (e) whorl, and (f) twin loop. Critical points in a fingerprint, called core and delta, are marked as squares and triangles. Note that an arch does not have a delta or a core. One of the two deltas in (e) and both the deltas in (f) are not imaged. A sample minutiae ridge ending (○) and ridge bifurcation (×) is illustrated in (e). Each image is 512 × 512 with 256 grey levels and is scanned at 512 dpi resolution. All features points were manually extracted by one of the authors.

point of sale would greatly reduce the credit card fraud. About 1 billion dollars worth of cellular telephone calls are made by cellular bandwidth thieves — many of these calls are made from stolen PINS or cellular telephones. Again, an identification of the legitimate ownership of the cellular telephones would prevent cellular telephone thieves from stealing the bandwidth. A reliable method of authenticating the legitimate owner of an ATM card would greatly reduce ATM-related fraud, worth approximately $3 billion annually [6]. A positive method of identifying the rightful check payee would also reduce billions of dollars that are misappropriated through fraudulent encashment of checks each year. A method of positive authentication of each system log-in would eliminate illegal break-ins into traditionally secure (even federal government) computers. The United States Immigration and Naturalization service stipulates that it could each day detect or deter about 3,000 illegal immigrants crossing the Mexican border without delaying legitimate persons entering the United States if it had a quick way of establishing positive personal identification.

High-speed computer networks offer interesting opportunities for electronic commerce and electronic purse applications. The accurate authentication of identities over networks is ex-

pected to become one of the important application of biometric-based authentication.

Miniaturization and mass-scale production of relatively inexpensive biometric sensors (e.g., solid-state fingerprint sensors) will facilitate the use of biometric-based authentication in asset protection.

3 Fingerprint as a Biometric

A smoothly flowing pattern formed by alternating crests (ridges) and troughs (valleys) on the palmar aspect of hand is called a palmprint. Formation of a palmprint depends on the initial conditions of the embryonic mesoderm from which they develop. The pattern on pulp of each terminal phalanx (of a finger) is considered as an individual pattern and is commonly referred to as a *fingerprint* (see Fig. 1). A fingerprint is believed to be unique to each person (and each finger).[2] Fingerprints of even identical twins are different.

[2]There is some anecdotal evidence that a fingerprint expert once found two (possibly latent) fingerprints belonging to two distinct individuals having 10 identical minutiae.

Fingerprints are one of the most mature biometric technologies and are considered legitimate proofs of evidence in courts of law all over the world. Fingerprints are, therefore, used in forensic divisions worldwide for criminal investigations. More recently, an increasing number of civilian and commercial applications are either using or actively considering the use of fingerprint-based identification because of a better understanding of fingerprints as well as a better demonstrated matching performance than any other existing biometric technology.

4 History of Fingerprints

Humans have used fingerprints for personal identification for a very long time [23]. Modern fingerprint matching techniques were initiated in the late 16th century [7]. Henry Fauld, in 1880, first scientifically suggested the individuality and uniqueness of fingerprints. At the same time, Herschel asserted that he had practiced fingerprint identification for about 20 years [23]. This discovery established the foundation of modern fingerprint identification. In the late 19th century, Sir Francis Galton conducted an extensive study of fingerprints [23]. He introduced the minutiae features for single fingerprint classification in 1888. The discovery of the uniqueness of fingerprints caused an immediate decline in the prevalent use of anthropometric methods of identification and led to the adoption of fingerprints as a more efficient method of identification [29]. An important advance in fingerprint identification was made in 1899 by Edward Henry, who (actually his two assistants from India) established the famous "Henry system" of fingerprint classification [7, 23] — an elaborate method of indexing fingerprints very much tuned to facilitating the human experts performing (manual) fingerprint identification. In the early 20th century, fingerprint identification was formally accepted as a valid personal identification method by law enforcement agencies and became a standard procedure in forensics [23]. Fingerprint identification agencies were set up worldwide, and criminal fingerprint databases were established [23]. With the advent of livescan fingerprinting and the availability of cheap fingerprint sensors, fingerprints are increasingly used in government and commercial applications for positive person identification.

5 System Architecture

The architecture of a fingerprint-based automatic identity authentication system is shown in Fig. 2. It consists of four components: (1) user interface, (2) system database, (3) enrollment module, and (4) authentication module. The user interface provides mechanisms for a user to indicate his or her identity and input his or her fingerprints into the system. The system database consists of a collection of records, each of which corresponds to an authorized person that has access to the system. Each record contains the following fields, which are used for authentication purpose: (1) user name of the person, (2) minutiae template(s)

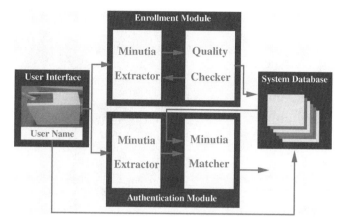

FIGURE 2 Architecture of an automatic identity authentication system. © IEEE.

of the person's fingerprint(s), and (3) other information (e.g., specific user privileges).

The task of enrollment module is to enroll persons and their fingerprints into the system database. When the fingerprint images and the user name of a person to be enrolled are fed to the enrollment module, a minutiae extraction algorithm is first applied to the fingerprint images and the minutiae patterns are extracted. A quality checking algorithm is used to ensure that the records in the system database only consist of fingerprints of good quality, in which a significant number (default value is 25) of genuine minutiae may be detected. If a fingerprint image is of poor quality, it is enhanced to improve the clarity of ridge/valley structures and mask out all the regions that cannot be reliably recovered. The enhanced fingerprint image is fed to the minutiae extractor again.

The task of authentication module is to authenticate the identity of the person who intends to access the system. The person to be authenticated indicates his or her identity and places his or her finger on the fingerprint scanner; a digital image of this fingerprint is captured; minutiae pattern is extracted from the captured fingerprint image and fed to a matching algorithm, which matches it against the person's minutiae templates stored in the system database to establish the identity.

6 Fingerprint Sensing

There are two primary methods of capturing a fingerprint image: inked (off line) and live scan (inkless) (see Fig. 3). An inked fingerprint image is typically acquired in the following way: a trained professional[3] obtains an impression of an inked finger on a paper and the impression is then scanned with a flat bed document scanner. The live-scan fingerprint is a collective term for a fingerprint image directly obtained from the finger without

---

[3] Possibly for reasons of expediency, MasterCard sends fingerprint kits to their credit card customers. The kits are used by the customers themselves to create an inked fingerprint impression to be used for enrollment.

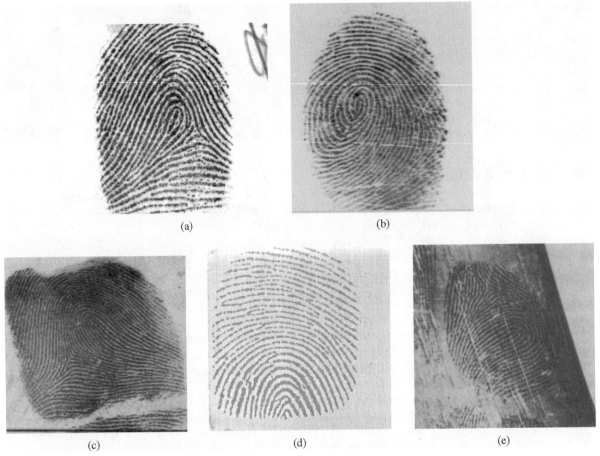

FIGURE 3   Fingerprint sensing: (a) an inked fingerprint image could be captured from the inked impression of a finger;
(b) a live-scan fingerprint is directly imaged from a live finger based on the optical total internal reflection principle;
(c) rolled fingerprints are images depicting the nail-to-nail area of a finger; (d) fingerprints captured with solid-state
sensors show a smaller area of finger than a typical fingerprint dab captured with optical scanners; (e) a latent fingerprint
refers to partial print typically lifted from the scene of a crime.

the intermediate step of getting an impression on a paper. The
acquisition of inked fingerprints is cumbersome; in the context
of an identity authentication system, it is both infeasible and
socially unacceptable. The most popular technology to obtain
a live-scan fingerprint image is based on the optical frustrated
total internal reflection (FTIR) concept [22]. When a finger is
placed on one side of a glass platen (prism), ridges of the finger
are in contact with the platen, while the valleys of the finger are
not in contact with the platen. The rest of the imaging system
essentially consists of an assembly of an LED light source and a
CCD placed on the other side of the glass platen. The laser light
source illuminates the glass at a certain angle and the camera
is placed such that it can capture the laser light reflected from
the glass. The light that is incident upon the platen at the glass
surface touched by the ridges is randomly scattered, while the
light that is incident upon the glass surface corresponding to
valleys suffers total internal reflection. Consequently, portions
of the image formed on the imaging plane of the CCD corresponding to ridges are dark, and those corresponding to valleys
are bright. More recently, capacitance-based solid-state live-scan
fingerprint sensors have been gaining popularity since they are

very small in size and hold the promise of becoming inexpensive in the near future. A capacitance-based fingerprint sensor
essentially consists of an array of electrodes. The fingerprint skin
acts as the other electrode, thereby forming a miniature capaci-
tor. The capacitance from the ridges is higher than that from the
valleys. This differential capacitance is the basis of operation of
a capacitance-based solid-state sensor [34].

7 Fingerprint Representation

Fingerprint representations are of two types: local and global.
Major representations of the local information in fingerprints
are based on the entire image, finger ridges, pores on the ridges,
or salient features derived from the ridges. Representations predominantly based on ridge endings or bifurcations (collectively
known as minutiae; see Fig. 4) are the most common, primarily
because of the following reasons: (1) minutiae capture much of
the individual information, (2) minutiae-based representations
are storage efficient, and (3) minutiae detection is relatively robust to various sources of fingerprint degradation. Typically,

Ridge Ending   **Ridge Bifurcation**

FIGURE 4   Ridge ending and ridge bifurcation. © IEEE.

minutiae-based representations rely on locations of the minutiae and the directions of ridges at the minutiae location. Fingerprint classification identifies the typical global representations of fingerprints and is the topic of Section 10. Some global representations include information about locations of critical points (e.g., core and delta) in a fingerprint.

8 Feature Extraction

A feature extractor finds the ridge endings and ridge bifurcations from the input fingerprint images. If ridges can be perfectly located in an input fingerprint image, then minutiae extraction is just a trivial task of extracting singular points in a thinned ridge map. However, in practice, it is not always possible to obtain a perfect ridge map. The performance of currently available minutiae extraction algorithms depends heavily on the quality of the input fingerprint images. As a result of a number of factors (aberrant formations of epidermal ridges of fingerprints, postnatal marks, occupational marks, problems with acquisition devices, etc.), fingerprint images may not always have well-defined ridge structures.

A reliable minutiae extraction algorithm is critical to the performance of an automatic identity authentication system using fingerprints. The overall flowchart of a typical algorithm [18,28] is depicted in Fig. 5. It mainly consists of three components: (1) orientation field estimation, (2) ridge extraction, and (3) minutiae extraction and postprocessing.

1. Orientation estimation : The orientation field of a fingerprint image represents the directionality of ridges in the

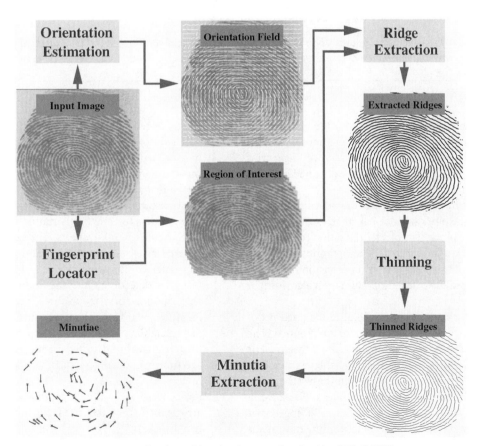

FIGURE 5   Flowchart of the minutiae extraction algorithm [18]. © IEEE.

(a) *Divide the input fingerprint image into blocks of size $W \times W$.*

(b) *Compute the gradients G_x and G_y at each pixel in each block [4].*

(c) *Estimate the local orientation at each pixel (i, j) using the following equations [28]:*

$$V_x(i,j) = \sum_{u=i-\frac{W}{2}}^{i+\frac{W}{2}} \sum_{v=j-\frac{W}{2}}^{j+\frac{W}{2}} 2G_x(u,v)G_y(u,v), \tag{1}$$

$$V_y(i,j) = \sum_{u=i-\frac{W}{2}}^{i+\frac{W}{2}} \sum_{v=j-\frac{W}{2}}^{j+\frac{W}{2}} (G_x^2(u,v) - G_y^2(u,v)), \tag{2}$$

$$\theta(i,j) = \frac{1}{2}tan^{-1}(\frac{V_x(i,j)}{V_y(i,j)}), \tag{3}$$

where W is the size of the local window; G_x and G_y are the gradient magnitudes in x and y directions, respectively.

(d) *Compute the* consistency level *of the orientation field in the local neighborhood of a block (i, j) with the following formula:*

$$C(i,j) = \frac{1}{N}\sqrt{\sum_{(i',j')\in D} |\theta(i',j') - \theta(i,j)|^2}, \tag{4}$$

$$|\theta' - \theta| = \begin{cases} d & if\,(d = (\theta' - \theta + 360)\,mod\,360) < 180, \\ d - 180 & otherwise, \end{cases} \tag{5}$$

where D represents the local neighborhood around the block (i, j) (in our system, the size of D is 5×5); N is the number of blocks within D; $\theta(i', j')$ and $\theta(i, j)$ are local ridge orientations at blocks (i', j') and (i, j), respectively.

(e) *If the* consistency level *(Eq.(4)) is above a certain threshold T_c, then the local orientations around this region are re-estimated at a lower resolution level until $C(i, j)$ is below a certain level.*

FIGURE 6   Hierarchical orientation field estimation algorithm. © IEEE.

fingerprint image. It plays a very important role in fingerprint image analysis. A number of methods have been proposed to estimate the orientation field of fingerprint images [22]. Fingerprint image is typically divided into a number of nonoverlapping blocks (e.g., 32×32 pixels), and an orientation representative of the ridges in the block is assigned to the block based on an analysis of gray-scale gradients in the block. The block orientation could be determined from the pixel gradient orientations based on, say, averaging [22], voting [25], or optimization [28]. We have summarized the orientation estimation algorithm in Fig. 6.

2. Segmentation: it is important to localize the portions of fingerprint image depicting the finger (foreground). The simplest approaches segment the foreground by global or adaptive thresholding. A novel and reliable approach to segmentation by Ratha *et al.* [28] exploits the fact that there is significant difference in the magnitudes of variance in the gray levels along and across the flow of a fingerprint ridge. Typically, block size for variance computation spans 1–2 interridge distance.

3. Ridge detection: The approaches to ridge detection use either simple or adaptive thresholding. These approaches may not work for noisy and low-contrast portions of the image. An important property of the ridges in a fingerprint image is that the gray-level values on ridges attain their local maxima along a direction normal to the local ridge orientation [18, 28]. Pixels can be identified to be ridge pixels based on this property. The extracted ridges may be thinned or cleaned by using standard thinning [26] and connected component algorithms [27].

4. Minutiae detection: Once the thinned ridge map is available, the ridge pixels with three ridge pixel neighbors are identified as ridge bifurcations, and those with one ridge pixel neighbor identified as ridge endings. However, all the minutiae thus detected are not genuine because of image processing artifacts and the noise in the fingerprint image.

5. Postprocessing: In this stage, typically, genuine minutiae are gleaned from the extracted minutiae using a number of heuristics. For instance, too many minutiae in a small neighborhood may indicate noise, and they could be

discarded. Very close ridge endings oriented antiparallel to each other may indicate spurious minutia generated by a break in the ridge, caused by either a poor contrast or a cut in the finger. Two very closely located bifurcations sharing a common short ridge often suggest extraneous minutia generated by bridging of adjacent ridges as a result of dirt or image processing artifacts.

9 Fingerprint Enhancement

The performance of a fingerprint image matching algorithm relies critically on the quality of the input fingerprint images. In practice, a significant percentage of acquired fingerprint images (approximately 10% according to our experience) is of poor quality. The ridge structures in poor-quality fingerprint images are not always well defined, and hence they cannot be correctly detected. This leads to the following problems: (1) a significant number of spurious minutiae may be created, (2) a large percentage of genuine minutiae may be ignored, and (3) large errors in minutiae localization (position and orientation) may be introduced. In order to ensure that the performance of the minutiae extraction algorithm will be robust with respect to the quality of fingerprint images, an enhancement algorithm that can improve the clarity of the ridge structures is necessary.

Typically, fingerprint enhancement approaches [5, 9, 14, 20] employ frequency domain techniques [9,10,20] and are computationally demanding. In a small local neighborhood, the ridges and furrows approximately form a two-dimensional sinusoidal wave along the direction orthogonal to local ridge orientation. Thus, the ridges and furrows in a small local neighborhood have well-defined local frequency and local orientation properties. The common approaches employ bandpass filters that model the frequency domain characteristics of a good-quality fingerprint image. The poor-quality fingerprint image is processed by using the filter to block the extraneous *noise* and pass the fingerprint *signal*. Some methods may estimate the orientation or frequency of ridge in each block in the fingerprint image and adaptively tune the filter characteristics to match the ridge characteristics.

One typical variation of this theme segments the image into nonoverlapping square blocks of widths larger than the average interridge distance. With the use of a bank of directional bandpass filters, each filter is matched to a predetermined model of generic fingerprint ridges flowing in a certain direction; the filter generating a strong response indicates the dominant direction of the ridge flow in the finger in the given block. The resulting orientation information is more accurate, leading to more reliable features. A single block direction can never truly represent the directions of the ridges in the block and may consequently introduce filter artifacts.

For instance, one common directional filter used for fingerprint enhancement is a Gabor filter [17]. Gabor filters have both frequency-selective and orientation-selective properties and have optimal joint resolution in both spatial and frequency domains. The even-symmetric Gabor filter has the general form [17]

$$h(x, y) = \exp\left\{-\frac{1}{2}\left[\frac{x^2}{\delta_x^2} + \frac{y^2}{\delta_y^2}\right]\right\}\cos(2\pi u_0 x), \qquad (6)$$

where u_0 is the frequency of a sinusoidal plane wave along the x axis, and δ_x and δ_y are the space constants of the Gaussian envelope along x and y axes, respectively. Gabor filters with arbitrary orientation can be obtained by a rotation of the x–y coordinate system. The modulation transfer function (MTF) of Gabor filter can be represented as

$$H(u, v) = 2\pi\delta_x\delta_y\left(\exp\left\{-\frac{1}{2}\left[\frac{(u - u_0)^2}{\delta_u^2} + \frac{v^2}{\delta_v^2}\right]\right\}\right.$$
$$\left. + \exp\left\{-\frac{1}{2}\left[\frac{(u + u_0)^2}{\delta_u^2} + \frac{v^2}{\delta_v^2}\right]\right\}\right), \qquad (7)$$

where $\delta_u = 1/2\pi\delta_x$ and $\delta_v = 1/2\pi\delta_y$. Figure 7 shows an

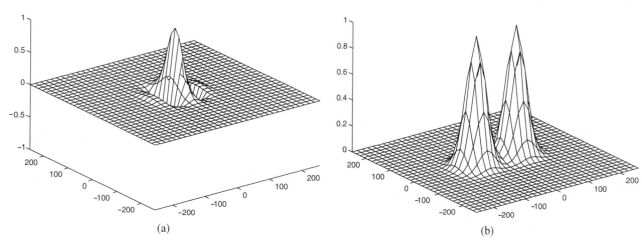

(a)　　　　　　　　　　　　　　　　(b)

FIGURE 7 An even-symmetric Gabor filter: (a) Gabor filter tuned to 60 cycles/width and 0° orientation; (b) corresponding MTF.

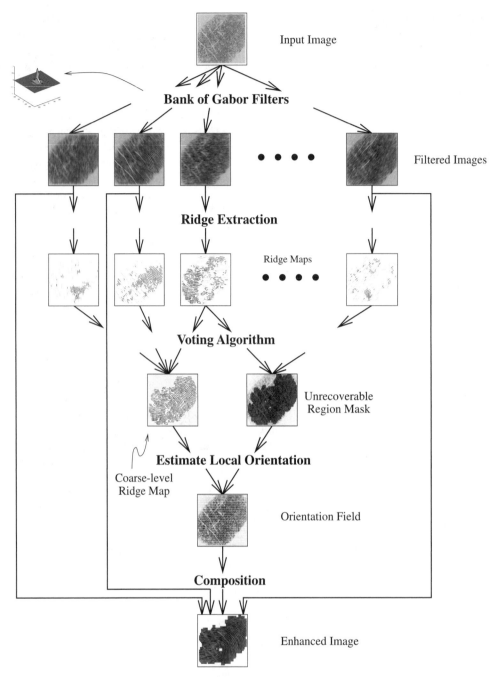

FIGURE 8   Fingerprint enhancement algorithm [11].

even-symmetric Gabor filter and its MTF. Typically, in a 500 dpi, 512×512 fingerprint image, a Gabor filter with $u_0 = 60$ cycles per image width (height), the radial bandwidth of 2.5 octaves, and orientation θ models the fingerprint ridges flowing in the direction $\theta + \pi/2$.

We summarize a novel approach to fingerprint enhancement proposed by Hong *et al.* [11] (see Fig. 8). It decomposes the given fingerprint image into several component images by using a bank of directional Gabor bandpass filters and extracts ridges from each of the filtered bandpass images by using a typical feature extraction algorithm [18]. By integrating informa-

tion from the sets of ridges extracted from filtered images, the enhancement algorithm infers the region of fingerprint where there is sufficient information to be considered for enhancement (recoverable region) and estimates a coarse-level ridge map for the recoverable region. The information integration is based on the observation that genuine ridges in a region evoke a strong response in the feature images extracted from the filters oriented in the direction parallel to the ridge direction in that region, and at most a weak response in feature images extracted from the filters oriented in the direction orthogonal to the ridge direction in that region. The coarse ridge map thus generated consists of the

 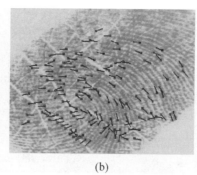 

(a)   (b)   (c)

FIGURE 9   Fingerprint enhancement results: (a) a poor-quality fingerprint; (b) minutiae extracted without image enhancement; (c) minutiae extracted after image enhancement [11]. (See color section, p. C–53.)

ridges extracted from each filtered image that are mutually consistent, and portions of the image where the ridge information is consistent across the filtered images constitute the *recoverable* region. The orientation field estimated from the coarse ridge map is more reliable than the orientation estimation from the input fingerprint image.

After the orientation field is obtained, the fingerprint image can then be adaptively enhanced by using the local orientation information. Let $f_i(x, y)(i = 0, 1, 2, 3, 4, 5, 6, 7)$ denote the gray-level value at pixel (x, y) of the filtered image corresponding to the orientation θ_i, $\theta_i = i * 22.5°$. The gray-level value at pixel (x, y) of the enhanced image can be interpolated according to the following formula:

$$f_{enh}(x, y) = a(x, y) f_{p(x,y)}(x, y) + (1 - a(x, y)) f_{q(x,y)}(x, y),$$
(8)

where $p(x, y) = \lfloor \frac{\theta(x,y)}{22.5} \rfloor$, $q(x, y) = \lceil \frac{\theta(x,y)}{22.5} \rceil \mod 8$, $a(x, y) = \frac{\theta(x,y) - p(x,y)}{22.5}$, and $\theta(x, y)$ represents the value of local orientation field at pixel (x, y). The major reason that we interpolate the enhanced image directly from the limited number of filtered images is that the filtered images are already available and the above interpolation is computationally efficient.

An example illustrating the results of minutiae extraction algorithm on a noisy input image and its enhanced counterpart is shown in Fig. 9. The improvement in performance caused by image enhancement was evaluated by using the fingerprint matcher described in Section 11. Figure 10 shows an improvement in accuracy of the matcher with and without image enhancement on the MSU database consisting of 700 fingerprint images of 70 individuals (10 fingerprints per finger per individual).

10 Fingerprint Classification

The fingerprints have been traditionally classified into categories based on information in the global patterns of ridges. In large-scale fingerprint identification systems, elaborate methods of manual fingerprint classification systems were developed to index individuals into bins based on classification of their finger-

prints; these methods of binning eliminate the need to match an input fingerprint(s) to the entire fingerprint database in identification applications and significantly reduce the computing requirements [8, 19].

Efforts in automatic fingerprint classification have been exclusively directed at replicating the manual fingerprint classification system. Figure 1 shows one prevalent manual fingerprint classification scheme that has been the focus of many automatic fingerprint classification efforts. It is important to note that the distribution of fingers into the six classes (shown in Fig. 1) is highly skewed. A fingerprint classification system should be invariant to rotation, translation, and elastic distortion of the frictional skin. In addition, often a significant part of the finger may not be imaged (e.g., dabs frequently miss deltas) and the classification methods requiring information from the entire fingerprint may be too restrictive for many applications.

A number of approaches to fingerprint classification have been developed. Some of the earliest approaches did not make use of the rich information in the ridge structures and exclusively depended on the orientation field information. Although fingerprint landmarks provide very effective fingerprint class clues, methods relying on the fingerprint landmarks alone may not be very successful because of a lack of availability of such information in many fingerprint images and because of the difficulty in extracting the landmark information from the noisy fingerprint images. As a result, the most successful approaches have to (1) supplement the orientation field information with ridge information; (2) use fingerprint landmark information when available but devise alternative schemes when such information cannot be extracted from the input fingerprint images; and (3) use reliable structural/syntactic pattern recognition methods in addition to statistical methods.

We summarize a method of classification [12] that takes into consideration the above-mentioned design criteria that has been tested on a large database of realistic fingerprints to classify fingers into five major categories: right loop, left loop, arch, tented arch, and whorl.[4]

---

[4]Other types of prints, e.g., twin loop, are not considered here but, in principle, could be lumped into the "other" or "reject" category.

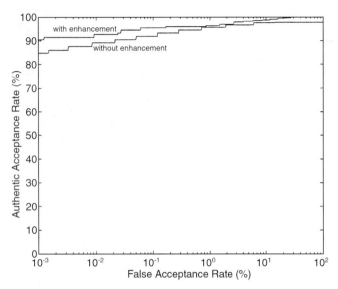

FIGURE 10   Performance of the fingerprint enhancement algorithm.

The orientation field determined from the input image may not be very accurate and the extracted ridges may contain many artifacts and, therefore, cannot be directly used for fingerprint classification. A ridge verification stage assesses the reliability of the extracted ridges based upon the length of each connected ridge segment and its alignment with other adjacent ridges. Parallel adjacent subsegments typically indicate a good quality fingerprint region; the ridge/orientation estimates in these regions are used to refine the estimates in the orientation field/ridge map.

1. Singular points: the Poincaré index [22] on the orientation field is used to determine the number of delta (N_D) and core (N_C) points in the fingerprint. A digital closed curve, Ψ, about 25 pixels long, around each pixel is used to compute the Poincaré index as defined here:

$$\text{Poincaré}(i, j) = \frac{1}{2\pi} \sum_{k=0}^{N_\Psi} \Delta(k),$$

   where

$$\Delta(k) = \begin{cases} \delta(k), & \text{if } |\delta(k)| < \pi/2 \\ \pi + \delta(k), & \text{if } \delta(k) \le -\pi/2, \\ \pi - \delta(k), & \text{otherwise} \end{cases}$$

$$\delta(k) = \mathcal{O}'(\Psi_x(i'), \Psi_y(i')) - \mathcal{O}'(\Psi_x(i), \Psi_y(i)),$$

$$i' = (i + 1) \bmod N_\Psi,$$

   $\mathcal{O}$ is the orientation field, and $\Psi_x(i)$ and $\Psi_y(i)$ denote coordinates of the ith point on the arc length parameterized closed curve Ψ.

2. Symmetry: the feature extraction stage also estimates an axis locally symmetric to the ridge structures at the core and computes (1) α, the angle between the symmetry axis

and the line segment joining core and delta, (2) β, the average angle difference between the ridge orientation and the orientation of the line segment joining the core and delta, and (3) γ, the number of ridges crossing the line segment joining core and delta. The relative position, R, of the delta with respect to symmetry axis is determined as follows: $R = 1$ if the delta is on the right side of symmetry axis; $R = 0$, otherwise.

3. Ridge structure: the classifier not only uses the orientation information but also utilizes the structural information in the extracted ridges. This feature summarizes the overall nature of the ridge flow in the fingerprint. In particular, it classifies each ridge of the fingerprint into three categories:

   - nonrecurring ridges: the ridges that do not curve very much
   - Type-1 recurring ridges: ridges that curve approximately π
   - Type-2 fully recurring ridges: ridge that curve by more than π

The classification algorithm summarized here (see Fig. 11) essentially devises a sequence of tests for determining the class of a fingerprint and conducts simpler tests earlier in the decision tree. For instance, two core points are typically detected for a whorl (see Fig. 11), which is an easier condition to verify than detecting the number of Type-2 recurring ridges. Another highlight of the algorithm is that if it does not detect the salient characteristics of any category from features detected in a fingerprint, it recomputes the features with a different preprocessing method. For instance, in the current implementation, the differential preprocessing consists of a different method/scale of smoothing. As can be observed from the flowchart, the algorithm detects (1) whorls based upon detection of either two core points or a sufficient number of Type-2 recurring ridges; (2) arch based upon the inability to detect either delta or core points; (3) left (right) loops based on the characteristic tilt of the symmetric axis, detection of a core point, and detection of either a delta point or a sufficient number of Type-1 recurring curves; and (4) tented arch based on relatively upright symmetric axis, detection of a core point, and detection of either a delta point or a sufficient number of Type-1 recurring curves.

Table 1 shows the results of the fingerprint classification algorithm on the NIST-4 database, which contains 4,000 images (image size is 512×480) taken from 2,000 different fingers, two images per finger. Five fingerprint classes are defined: (1) arch, (2) tented arch, (3) left loop, (4) right loop, and (5) whorl. Fingerprints in this database are uniformly distributed among these five classes (800 per class). The five-class error rate in classifying these 4,000 fingerprints is 12.5%. The confusion matrix is given in Table 1; numbers shown in bold font are correct classifications. Since a number of fingerprints in the NIST-4 database are labeled as belonging to possibly two different classes, each row of the confusion matrix in Table 1 does not sum up to 800. For the five-class problem, most of the classification errors are due

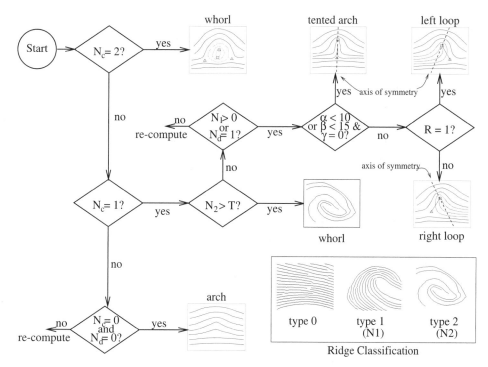

FIGURE 11 Flowchart of fingerprint classification algorithm. The inset also illustrates ridge classification [12]. The re-compute option involves starting the classification algorithm with a different preprocessing (e.g., smoothing) of the image.

TABLE 1 Five-class classification results on the NIST-4 database

True Class	Assigned Class				
	A	T	L	R	W
A	**885**	13	10	11	0
T	179	**384**	54	14	5
L	31	27	**755**	3	20
R	30	47	3	**717**	16
W	6	1	15	15	**759**

Note: A, arch; T, tented arch; L, left loop; R, right loop; W, whorl.

to misclassifying a tented arch as an arch. By combining these two arch categories into a single class, the error rate drops from 12.5% to 7.7%. Besides the tented arch/arch errors, the other errors mainly come from misclassifications between arch/tented arch and loops and are due to poor image quality.

11 Fingerprint Matching

Given two (input and template) sets of features originating from two fingerprints, the objective of the feature matching system is to determine whether or not the prints represent the same finger. Fingerprint matching has been approached from several different strategies, like image-based [2], ridge pattern-based, and point (minutiae) pattern-based fingerprint representations. There also exist graph-based schemes [15, 16, 30] for

fingerprint matching. Image-based matching may not tolerate large amounts of nonlinear distortion in the fingerprint ridge structures. Matchers critically relying on extraction of ridges, or their connectivity information may display drastic performance degradation with a deterioration in the quality of the input fingerprints. We, therefore, believe that the point pattern matching (minutiae matching) approach facilitates the design of a robust, simple, and fast verification algorithm while maintaining a small template size.

The matching phase typically defines the similarity (distance) metric between two fingerprint representations and determines whether a given pair of representations is captured from the same finger (mated pair) based on whether this quantified (dis)similarity is greater (less) than a certain (predetermined) threshold. The similarity metric is based on the concept of correspondence in minutiae-based matching. A minutia in the input fingerprint and a minutia in the template fingerprint are said to be corresponding if they represent the identical minutia scanned from the same finger.

Before the fingerprint representations could be matched, most minutia-based matchers first transform (*register*) the input and template fingerprint features into a common frame of reference. The registration essentially involves alignment based on rotation/translation and may optionally include scaling. The parameters of alignment are typically estimated either from (1) singular points in the fingerprints, e.g., core and delta locations; (2) pose clustering based on minutia distribution [28]; or (3) any other landmark features. For example, Jain *et al.* [18]

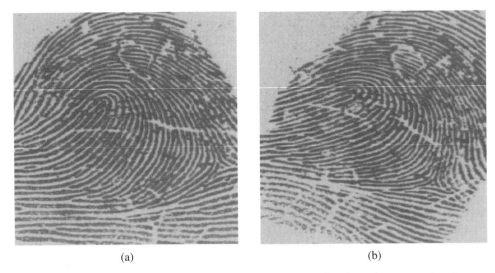

| (a) | (b) |

FIGURE 12 Two different fingerprint impressions of the same finger. In order to know the correspondence between the minutiae of these two fingerprint images, all the minutiae must be precisely localized and the deformation must be recovered. © IEEE.

use a rotation/translation estimation method based on properties of ridge segment associated with ridge ending minutiae.[5]

There are two major challenges involved in determinating the correspondence between two aligned fingerprint representations (see Fig. 12). (1) dirt or leftover smudges on the sensing device and the presence of scratches or cuts on the finger either introduce spurious minutiae or obliterate the genuine minutiae; (2) variations in the area of finger being imaged and its pressure on the sensing device affect the number of genuine minutiae captured and introduce displacements of the minutiae from their "true" locations as a result of elastic distortion of the fingerprint skin. Consequently, a fingerprint matcher should not only assume that the input fingerprint is a transformed template fingerprint by a similarity transformation (rotation, translation, and scale), but it should also tolerate both spurious minutiae as well as missing genuine minutiae and accommodate perturbations of minutiae from their true locations. Figure 13 illustrates a typical situation of aligned ridge structures of mated pairs. Note that the best alignment in one part (top left) of the image may result in a large magnitude of displacements between the corresponding minutiae in other regions (bottom right). In addition, observe that the distortion is nonlinear: given the amount of distortions at two arbitrary locations on the finger, it is not possible to predict the distortions at all the intervening points on the line joining the two points.

The adaptive elastic string matching algorithm [18] summarized in this chapter uses three attributes of the aligned minutiae for matching: its distance from the reference minutiae (*radius*), the angle subtended to the reference minutiae (*radial angle*), and the local direction of the associated ridge (*minutiae direc-*

tion). The algorithm initiates the matching by first representing the aligned input (template) minutiae as an input (template) minutiae string. The string representation is obtained by imposing a linear ordering based on radial angles and radii. The resulting input and template minutiae strings are matched using an inexact string matching algorithm to establish the correspondence.

The inexact string matching algorithm essentially transforms (*edits*) the input string to the template string, and the number of edit operations is considered as a metric of the (dis)similarity

FIGURE 13 Aligned ridge structures of mated pairs. Note that the best alignment in one part (midleft) of the image results in a large displacements between the corresponding minutiae in the other regions (bottom right). © IEEE. (See color section, p. C–53.)

[5]The input and template minutiae used for the alignment will be referred to as reference minutiae below.

TABLE 2 False acceptance and false reject rates on two data sets with different threshold values © IEEE

Threshold Value	False Accept. Rate (MSU) (%)	False Reject Rate (MSU) (%)	False Accept. Rate (NIST 9) (%)	False Reject Rate (NIST 9) (%)
7	0.07	7.1	0.073	12.4
8	0.02	9.4	0.023	14.6
9	0.01	12.5	0.012	16.9
10	0	14.3	0.003	19.5

between the strings. While permitted edit operators model the impression variations in a representation of a finger (deletion of the genuine minutiae, insertion of spurious minutiae, and perturbation of the minutiae), the penalty associated with each edit operator models the likelihood of that edit. The sum of penalties of all the edits (*edit distance*) defines the similarity between the input and template minutiae strings. Among several possible sets of edits that permit the transformation of the input minutiae string into the reference minutiae string, the string matching algorithm chooses the transform associated with the minimum cost based on dynamic programming.

The algorithm tentatively considers a candidate (aligned) input and a candidate template minutia in the input and template minutiae string to be a mismatch if their attributes are not within a tolerance window (see Fig. 14) and penalizes them for deletion/insertion edit. If the attributes are within the tolerance window, the amount of penalty associated with the tentative match is proportional to the disparity in the values of the attributes in the minutiae. The algorithm accommodates for the elastic distortion by adaptively adjusting the parameters of the tolerance window based on the most recent successful tentative match. The tentative matches (and correspondences) are accepted if the edit distance for those correspondences is smaller than any other correspondences.

Figure 15 shows the results of applying the matching algorithm to an input and a template minutiae set pair. The outcome

TABLE 3 Average CPU time for minutiae extraction and matching on a Sun ULTRA 1 workstation © IEEE

Minutiae Extraction (s.)	Minutiae Matching (s.)	Total (s.)
1.1	0.3	1.4

of the matching process is defined by a matching score. The matching score is determined from the number of mated minutia from the correspondences associated with the minimum cost of matching input and template minutiae string. The raw matching score is normalized by the total number of minutia in the input and template fingerprint representations and is used for deciding whether input and template fingerprints are mates. The higher the normalized score, the larger the likelihood that the test and template fingerprints are the scans of the same finger.

The results of a performance evaluation of the fingerprint matching algorithm are illustrated in Fig. 16 for 1,350 fingerprint images in NIST 9 database [31] and in Fig. 10 for 700 images of 70 individuals from the MSU database. Some sample points on the receiver operating characteristics curve are tabulated in Table 2.

In order for an automatic identity authentication system to be acceptable in practice, the response time of the system has to be within a few seconds. Table 3 shows that our implemented system does meet the practical response time requirement.

12 Summary and Future Prospects

With recent advances in fingerprint sensing technology and improvements in the accuracy and matching speed of the fingerprint matching algorithms, automatic personal identification based on the fingerprint is becoming an attractive alternative or complement to the traditional methods of identification. We have provided an overview of the fingerprint-based identification and summarized algorithms for fingerprint feature extraction, enhancement, matching, and classification. We have also presented a performance evaluation of these algorithms.

The critical factor for the widespread use of fingerprints is in meeting the performance (e.g., matching speed and accuracy) standards demanded by emerging civilian identification applications. Unlike an identification based on passwords or tokens, performance of the fingerprint-based identification is not perfect. There will be a growing demand for faster and more accurate

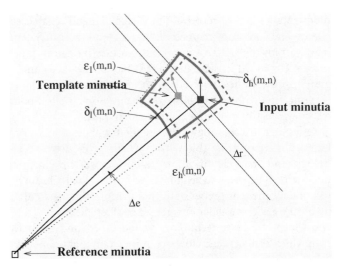

FIGURE 14 Bounding box and its adjustment. © IEEE.

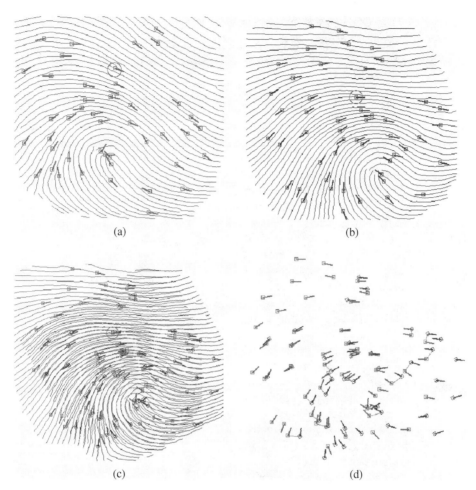

(a)   (b)

(c)   (d)

FIGURE 15   Results of applying the matching algorithm to an input minutiae set and a template: (a) input minutiae set; (b) template minutiae set; (c) alignment result based on the minutiae marked with green circles; (d) matching result, where template minutiae and their correspondences are connected by green lines. © IEEE. (See color section, p. C–54.)

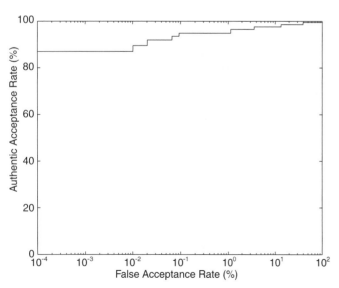

FIGURE 16   Receiver operating characteristic curve for NIST 9 (CD No. 1). © IEEE.

fingerprint matching algorithms that can (particularly) handle poor-quality images. Some of the emerging applications (e.g., fingerprint-based smartcards) will also benefit from a compact representation of a fingerprint. The design of highly reliable, accurate, and foolproof biometrics-based identification systems may warrant the effective integration of discriminatory information contained in several different biometrics or technologies [13]. The issues involved in integrating fingerprint-based identification with other biometric or nonbiometric technologies may constitute an important research topic.

As biometric technology matures, there will be an increasing interaction among the (biometric) market, (biometric) technology, and the (identification) applications. The emerging interaction is expected to be influenced by the added value of the technology, the sensitivities of the population, and the credibility of the service provider. It is too early to predict where, how, and which biometric technology would evolve and be mated with which applications. However, it is certain that biometrics-based identification will have a profound influence on the way we conduct our daily business. It is also certain that, as the most mature and well-understood biometric, fingerprints will remain an integral part of the preferred biometric-based identification solutions in the years to come.

References

[1] A. K. Jain, R. Bolle, and S. Pankanti, eds., *Biometrics: Personal Identification in Networked Society* (Kluwer, Boston, MA, 1999).

[2] R. Bahuguna, "Fingerprint verification using hologram matched filterings," presented at the Biometric Consortium Eighth Meeting, San Jose, CA, June 11–12, 1996.

[3] G. T. Candela, P. J. Grother, C. I. Watson, R. A. Wilkinson, and C. L. Wilson, "PCASYS: a pattern-level classification automation system for fingerprints," NIST Tech. Rep. NISTIR 5647, August, 1995.

[4] J. Canny, "A computational approach to edge detection," *IEEE Trans. PAMI* **8**, 679–698 (1986).

[5] L. Coetzee and E. C. Botha, "Fingerprint recognition in low quality images," *Pattern Recog.* **26**, 1441–1460 (1993).

[6] L. Lange and G. Leopold, "Digital identification: it's now at our fingertips," EEtimes at http://techweb.cmp.com/eet/823/, March 24, Vol. 946, 1997.

[7] Federal Bureau of Investigation, *The Science of Fingerprints: Classification and Uses* (U. S. GPO, Washington, D. C., 1984).

[8] R. Germain, A. Califano, and S. Colville, "Fingerprint matching using transformation parameter clustering," *IEEE Comput. Sci. Eng.* **4**, 42–49 (1997).

[9] L. O'Gorman and J. V. Nickerson, "An Approach to fingerprint filter design," *Pattern Recog.* **22**, 29–38 (1989).

[10] L. Hong, A. K. Jain, S. Pankanti, and R. Bolle, "Fingerprint enhancement," in *Proceedings of the IEEE Workshop on Applications of Computer Vision* (IEEE, New York, 1996), pp. 202–207.

[11] L. Hong, "Automatic personal identification using fingerprints," Ph.D. dissertation (Michigan State University, 1998).

[12] L. Hong and A. K. Jain, "Classification of fingerprint images," MSU Tech. Rep. MSUCPS:TR98-18, June, 1998.

[13] L. Hong and A. K. Jain, "Integrating faces and fingerprints," *IEEE Trans. Pattern Anal. Machine Intell.* **20**, 1295–1307 (1998).

[14] D. C. Douglas Hung, "Enhancement and feature purification of fingerprint images," *Pattern Recog.* **26**, 1661–1671 (1993).

[15] A. K. Hrechak and J. A. McHugh, "Automated fingerprint recognition using structural matching," *Pattern Recog.* **23**, (1990).

[16] D. K. Isenor and S. G. Zaky, "Fingerprint identification using graph matching," *Pattern Recog.* **19**, pp. 113–122 (1986).

[17] A. K. Jain and F. Farrokhnia, "Unsupervised texture segmentation using Gabor filters," *Pattern Recog.* **24**, 1167–1186 (1991).

[18] A. Jain, L. Hong, S. Pankanti, and R. Bolle, On-line identity-authentication system using fingerprints, *Proc. IEEE* (Special Issue on Automated Biometrics) **85**, 1365–1388 (1997).

[19] A. K. Jain, S. Prabhakar, and L. Hong, "A multichannel approach to fingerprint classification," presented at the Indian Conference on Computer Vision, Graphics, and Image Processing (ICVGIP '98), New Delhi, India, December 21–23, 1998.

[20] T. Kamei and M. Mizoguchi, "Image filter design for fingerprint enhancement," in *Proceedings of ISCV '95* Coral Gables, FL, 1995, pp. 109–114.

[21] K. Karu and A. K. Jain, "Fingerprint classification," *Pattern Recog.* **29**, 389–404 (1996).

[22] M. Kawagoe and A. Tojo, "Fingerprint pattern classification," *Pattern Recog.* **17**, 295–303 (1984).

[23] H. C. Lee and R. E. Gaensslen, *Advances in Fingerprint Technology* (Elsevier, New York, 1991).

[24] D. Maio, D. Maltoni, "Direct gray-scale minutiae detection in fingerprints," *IEEE Trans. Pattern Anal. Machine Intell.* **19**, 27–40 (1997).

[25] B. M. Mehtre and B. Chatterjee, "Segmentation of fingerprint images — a composite method," *Pattern Recog.* **22**, 381–385 (1989).

[26] N. J. Naccache and R. Shinghal, "An investigation into the skeletonization approach of Hilditch," *Pattern Recog. J.* **17**, 279–284 (1984).

[27] T. Pavlidis, *Algorithms for Graphics and Image Processing* (Computer Science Press, Rockville, MD, 1982).

[28] N. Ratha, K. Karu, S. Chen and A. K. Jain, "A real-time matching system for large fingerprint database," *IEEE Trans. Pattern Anal. Machine Intell.* **18**, 799–813 (1996).

[29] H. T. F. Rhodes, *Alphonse Bertillon: Father of Scientific Detection* (Abelard-Schuman, New York, 1956).

[30] M. K. Sparrow and P. J. Sparrow, "A topological approach to the matching of single fingerprints: development of algorithms for use of rolled impressions," National Bureau of Standards Tech. Rep., Gaithersburg, MD, May, 1985.

[31] C. I. Watson, *NIST Special Database 9, Mated Fingerprint Card Pairs* (National Institute of Standards and Technology, May, 1993).

[32] C. L. Wilson, G. T. Candela and C. I. Watson, "Neural-network fingerprint classification," *J. Artif. Neural Networks* **1**, 203–228 (1994).

[33] J. D. Woodward, "Biometrics: privacy's foe or privacy's friend?," *Proc. IEEE* (Special Issue on Automated Biometrics) **85**, 1480–1492 (1997).

[34] N. D. Young, G. Harkin, R. M. Bunn, D. J. McCulloch, R. W. Wilks, and A. G. Knapp, "Novel fingerprint scanning arrays using polysilicon TFT's on glass and polymer substrates," *IEEE Electron Device Lett.* **18**, 19–20 (1997).

10.6

Probabilistic, View-Based, and Modular Models for Human Face Recognition

Baback Moghaddam
and Alex Pentland

*Massachusetts Institute
of Technology*

1 Introduction

Developing a computational model for the recognition of natural objects such as human faces is quite difficult, because they are complex, multidimensional, and meaningful visual stimuli. They are a natural class of objects, and they stand in stark contrast to sine wave gratings and other artificial stimuli used in human and computer vision research. Thus, unlike most image processing functions, for which we may construct detailed mathematical models, recognition of natural objects such as human faces is a high-level task in which computational approaches must rely on features and structures learned from examples by statistical modeling.

The general approach we have developed is to attempt to describe the *range* of two-dimensional (2-D) appearances of objects to be recognized. To obtain such an "appearance-based" representation, one must first transform the image into a low-dimensional coordinate system that preserves the general perceptual quality of the target object's image. The necessity for such a transformation is to address the "curse of dimensionality": the raw image data have so many degrees of freedom that it

would require millions of examples to learn the range of appearances directly. Once a low-dimensional representation of the target class (face, eye, hand, car, etc.) has been obtained, then standard methods can be used to learn the range of appearance that the target exhibits in the new, low-dimensional coordinate system.

1.1 Appearance-Based Detection and Recognition

What do we mean by "the range of appearances of the human face"? The range of appearances is precisely the *probability density function (PDF)* of the image data for the target class. For instance, given several examples of a target class Ω in such a low-dimensional representation, it is straightforward to model the probability distribution function $P(\mathbf{x} \mid \Omega)$ of its image-level features $\mathbf{x}$ as a mixture of Gaussian distributions, thus obtaining a low-dimensional, parametric *appearance model* for the target class [21]. Once the target classes' *probability distribution function* has been learned, we can use Bayes' rule to perform maximum *a postieriori* (MAP) detection and recognition.

The result is a very simple, representation of the target class's appearance, which can be used to detect occurances of the class, to compactly describe its appearance, and to efficiently compare different examples from the same class. We have shown that this method is very powerful for the detection and recognition of human faces, hands, and facial expressions [22]. Other researchers have used extensions of this basic method to recognize industrial objects and household items [24].

The use of parametric appearance models to characterize the PDF of an object's appearance in the image is related to the simpler idea of a view-based representation [31, 38]. As originally developed, the idea of view-based recognition was to accurately describe the spatial structure of the target object by interpolating between previously seen views. However, in order to describe natural objects such as faces or hands, which display a wide range of structural and nonrigid variation, one must extend the notion of "view" to include characterizing the range of geometric and feature variation, as well as the likelihood associated with such variation. That is, one must use an appearance-based approach such as the one described here, instead of the simpler idea of a view-based approach.

We typically use the Karhunen–Loève transform (KLT), also called the principal components analysis (PCA), as the dimensionality-reducing preprocessing transform. This is because the KLT is known to provide an optimally compact linear basis (with respect to the RMS error) for a given class of signal. This transform has also been used by pioneers in face recognition research [1, 16]. However, our use of the same preprocessing step leads to a false impression of similarity. Previous researchers have used the KLT as a feature extraction step, which is followed by a simple classification algorithm. In contrast, we are using the KLT to facilitate learning of the range of appearances (the PDF), which we then use to make MAP estimates for target detection and recognition.

In this chapter we will first address the problem of *detecting* faces and facial features, and describe how our method of learning the range of facial/feature appearances is used to accomplish this first, all-important step. We will then describe how to the same models of facial appearance can be used for facial recognition, and we report on the robustness and accuracy of the combined detection/recognition method. Finally, we will discuss how recognition is extended to included variation in head orientation, and how facial features can be usefully included in the face recognition proccess.

2 Visual Attention and Object Detection

Visual attention is the process of restricting higher-level processing to a subset of the visual field, referred to as the *focus of attention* (FOA). The critical component of visual attention is the *selection* of the FOA. In humans this process is not based purely on bottom-up processing and is in fact goal driven. The

measure of interest or *saliency* is thus defined by the demands of the particular visual task.

Palmer [26] has suggested that visual attention is the process of locating the object of interest and placing it in a *canonical* (or object-centered) reference frame suitable for recognition (or template matching). We have developed a computational technique for automatic object recognition, which is in accordance with Palmer's model of visual attention. The system uses a probabilistic formulation for the estimation of the position and scale of the object in the visual field and remaps the FOA to an object-centered reference frame, which is subsequently used for recognition and verification.

At a simple level the underlying mechanism of attention during a visual search task can be based on a spatiotopic *saliency* map $S(i, j)$, which is a function of the image information $I(x, y)$ in a local region R:

$$S(i, j) = f\left[\{I(i + r, j + c) : (r, c) \in R\}\right] \qquad (1)$$

For example, saliency maps have been constructed that employ spatiotemporal changes as cues for foveation [2] or other low-level image features such as local symmetry for detection of interest points [33]. However, bottom-up techniques based on low-level features lack *context* with respect to high-level visual tasks such as object recognition. In a recognition task, the selection of the FOA is driven by higher-level goals and therefore requires internal representations of an object's appearance and a means of comparing candidate objects in the FOA to the stored object models.

Specifically, in an object-based visual search the saliency map is a function of the degree of match between a candidate object in a local image region and an internal model of the target object. In view-based recognition (as opposed to 3-D geometric or invariant-based recognition), the saliency can be formulated in terms of visual similarity by using a variety of metrics, ranging from simple template matching scores to more sophisticated measures using, for example, robust statistics for image correlation [6]. In this chapter, however, we are primarily interested in saliency maps that have a *probabilistic* interpretation as object-class membership functions or *likelihoods*. These likelihood functions are learned by applying density estimation techniques in complementary subspaces obtained by an eigenvector decomposition. Our approach to this learning problem is *view-based*, that is, the learning and modeling of the visual appearance of the object from a (suitably normalized and preprocessed) set of training imagery. Figure 1 shows examples of the automatic selection of FOA for the detection of human faces. In each case, the target object's probability distribution was *learned* from training views and then was subsequently used in computing likelihoods for detection. The face representation is based on the visual appearance (normalized gray-scale image). The maximum likelihood (ML) estimates of position and scale are shown in the figure by the cross-hairs and bounding box, respectively.

| (a) | (b) | (c) | (d) |

FIGURE 1   (a) Input image, (b) face detection, (c) face centering, (d) facial feature detection.

2.1 Object Detection

The standard detection paradigm in image processing is that of normalized correlation or template matching (see Chapter 3.1 of this *Handbook*). However, this approach is only optimal in the simplistic case of a *deterministic* signal embedded in additive white Gaussian noise. When we begin to consider a target *class* detection problem — e.g, finding a generic human face or a human hand in a scene — we must incorporate the underlying probability distribution of the object. Subspace methods and eigenspace decompositions are particularly well suited to such a task since they provide a compact and *parametric* description of the object's appearance and also automatically identify the *degrees of freedom* of the underlying statistical variability.

In particular, the eigenspace formulation leads to a powerful alternative to standard detection techniques such as template matching or normalized correlation. The reconstruction error (or residual) of the eigenspace decomposition (referred to as the "distance from face space" in the context of the work with "eigenfaces" [37]) is an effective indicator of similarity. The residual error is easily computed by using the projection coefficients and the original signal energy. This detection strategy is equivalent to matching with a linear combination of *eigentemplates* and allows for a greater range of distortions in the input signal (including lighting, and moderate rotation and scale). In a statistical signal detection framework, the use of eigentemplates has been shown to yield superior performance in comparison with standard matched filtering [17, 27].

In [28] we used this formulation for a modular eigenspace representation of facial features where the corresponding residual — referred to as distance-from-*feature*-space, or DFFS — was used for localization and detection. Given an input image, a saliency map was constructed by computing the DFFS at each pixel. When using *M* eigenvectors, this requires *M* convolutions (which can be efficiently computed using an FFT) plus an additional local energy computation. The global minimum of this distance map was then selected as the best estimate of the location of the target.

In this chapter we will show that the DFFS can be interpreted as an estimate of a marginal component of the probability density of the object and that a complete estimate must also incorporate a second marginal density based on a complementary "distance *in* feature space" (DIFS). Using our estimates of the object densities, we formulate the problem of target detection from the point of view of a ML estimation problem. Specifically, given the visual field, we estimate the position (and scale) of the image region that is most representative of the target of interest. Computationally this is achieved by sliding an *m*-by-*n* observation window throughout the image and at each location computing the *likelihood* that the local subimage **x** is an instance of the target class Ω, i.e., $P(\mathbf{x} \mid \Omega)$. After this probability map is computed, we select the location corresponding to the highest likelihood as our ML estimate of the target location. Note that the likelihood map can be evaluated over the entire parameter space affecting the object's appearance, which can include transformations such as scale and rotation.

3 Eigenspace Methods for Visual Modeling

In recent years, computer vision research has witnessed a growing interest in eigenvector analysis and subspace decomposition methods. In particular, eigenvector decomposition has been shown to be an effective tool for solving problems that use high-dimensional representations of phenomena that are intrinsically low-dimensional. This general analysis framework lends itself to several closely related formulations in object modeling and recognition that employ the *principal modes* or characteristic *degrees of freedom* for description. The identification and parametric representation of a system in terms of these principal modes is at the core of recent advances in physically based modeling [29], correspondence and matching [34], and parametric descriptions of shape [9].

Eigenvector-based methods also form the basis for data analysis techniques in pattern recognition and statistics, where they are used to extract low-dimensional subspaces comprising statistically uncorrelated variables that tend to simplify tasks such as classification. The KLT [18] and PCA [13] are examples of eigenvector-based techniques that are commonly used for dimensionality reduction and feature extraction in pattern recognition.

In computer vision, eigenvector analysis of *imagery* has been used for the characterization of human faces [15] and automatic face recognition using "eigenfaces" [27, 37]. More recently, a principal components analysis of imagery has also been applied for robust target detection [8, 27], nonlinear image interpolation [5], visual learning for object recognition [24, 40], and visual servoing for robotics [25].

3.1 Probabilistic Eigenspaces

However, these authors (with the exception of [27]) have used eigenvector analysis primarily as a dimensionality reduction technique for subsequent modeling, interpolation, or classification. In contrast, our methods use an eigenspace decomposition as an integral part of an efficient technique for probability density estimation of high-dimensional data.

Our learning method estimates the *complete* probability distribution of the object's appearance by using an eigenvector decomposition of the sample covariance matrix of a set of training views. The desired target density is decomposed into two components: the density in the principal subspace (containing the traditionally defined principal components) and its orthogonal complement (which is usually discarded in standard PCA). We have derived the form for an optimal density estimate for the case of Gaussian data and a near-optimal estimator for arbitrarily complex distributions in terms of a mixture-of-Gaussians density model [22].

We note that this learning method differs from *supervised* visual learning with function approximation networks [32] in which a hypersurface representation of an input/output map is automatically learned from a set of training examples. Instead, we use a probabilistic formulation, which combines the two standard paradigms of *unsupervised* learning — PCA and density estimation — to arrive at a computationally feasible estimate of the class conditional density function which is then used for maximum likelihood detection of faces and facial features, as well as Bayesian modeling for recognition.

The key to our approach to automatic visual learning is density estimation. However, instead of applying estimation techniques directly to the original high-dimensional space of the imagery, we use an eigenspace decomposition to yield a computationally feasible estimate. Specifically, the eigenspace analysis is applied to a set of training views of the object in order to identify a principal subspace that captures the *intrinsic* dimensionality of the data. The component of the complete density in this lower-dimensional subspace is then estimated by using a suitable parametric form. In addition, we implicitly model the component of the distribution in the *orthogonal* subspace. The complete density estimate can be efficiently computed from the lower-dimensional principal components. Our density estimate is shown to be *optimal* in the case of Gaussian-distributed training data.

3.1.1 Principal Component Imagery

Given a training set of *m*-by-*n* images $\{I^t\}_{t=1}^{N_T}$, we can form a training set of vectors $\{\mathbf{x}^t\}$, where $\mathbf{x} \in \mathcal{R}^{N=mn}$, by lexicographic ordering of the pixel elements of each image I^t. The basis functions for the KLT [18] are obtained by solving the eigenvalue

problem,

$$\Lambda = \Phi^T \Sigma \Phi, \qquad (2)$$

where Σ is the covariance matrix, Φ is the eigenvector matrix of Σ, and Λ is the corresponding diagonal matrix of eigenvalues. The unitary matrix Φ defines a coordinate transform (rotation) that *decorrelates* the data and makes explicit the *invariant subspaces* of the matrix operator Σ. In PCA, a partial KLT is performed to identify the largest-eigenvalue eigenvectors and obtain a principal component feature vector $\mathbf{y} = \Phi_M^T \tilde{\mathbf{x}}$, where $\tilde{\mathbf{x}} = \mathbf{x} - \bar{\mathbf{x}}$ is the mean-normalized image vector and Φ_M is a submatrix of Φ containing the principal eigenvectors. PCA can be seen as a linear transformation $\mathbf{y} = \mathcal{T}(\mathbf{x}) : \mathcal{R}^N \to \mathcal{R}^M$, which extracts a lower-dimensional subspace of the KL basis corresponding to the maximal eigenvalues. These principal components preserve the major linear correlations in the data and discard the minor ones.[1]

By ranking the eigenvectors of the KL expansion with respect to their eigenvalues and selecting the first M principal components, we form an orthogonal decomposition of the vector space $\mathcal{R}^N$ into two mutually exclusive and complementary subspaces: the principal subspace (or feature space) $F = \{\Phi_i\}_{i=1}^M$ containing the principal components, and its orthogonal complement $\bar{F} = \{\Phi_i\}_{i=M+1}^N$. This orthogonal decomposition is illustrated in Fig. 2(a), where we have a prototypical example of a distribution that is embedded entirely in F. In practice there is always a signal component, in $\bar{F}$ because of the minor statistical variabilities in the data or simply because of the observation noise that affects every element of $\mathbf{x}$.

In a partial KL expansion, the residual reconstruction error is defined as

$$\epsilon^2(\mathbf{x}) = \sum_{i=M+1}^N y_i^2 = \|\tilde{\mathbf{x}}\|^2 - \sum_{i=1}^M y_i^2 \qquad (3)$$

and can be easily computed from the first M principal components and the L_2 norm of the mean-normalized image $\tilde{\mathbf{x}}$. Consequently the L_2 norm of every element $\mathbf{x} \in \mathcal{R}^N$ can be decomposed in terms of its projections in these two subspaces. We refer to the component in the orthogonal subspace $\bar{F}$ as the DFFS, which is a simple Euclidean distance and is equivalent to the residual error $\epsilon^2(\mathbf{x})$ in Eq. (3). The component of $\mathbf{x}$ that lies *in* the feature space F is referred to as the DIFS, but it is generally not a distance-based norm, but can be interpreted in terms of the probability distribution of y in F.

3.1.2 Density Estimation in Eigenspace

One difficulty with probabilistic visual modeling is that the intensity or intensity difference vectors are very high dimensional,

---

[1]In practice, the number of training images N_T is far less than the dimensionality of the imagery N; consequently, the covariance matrix Σ is singular. However, the first $M < N_T$ eigenvectors can always be computed (estimated) from N_t samples by using, e.g., a singular value decomposition [12].

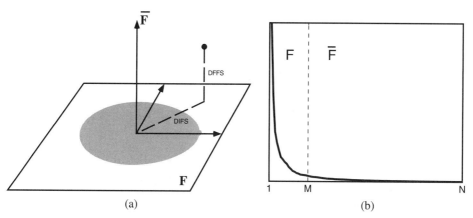

FIGURE 2   (a) Decomposition into the principal subspace F and its orthogonal complement $\bar{F}$ for a Gaussian density; (b) a typical eigenvalue spectrum and its division into the two orthogonal subspaces.

with $\Delta \in \mathcal{R}^N$ and $N = O(10^4)$. Therefore we typically lack sufficient independent training observations to compute reliable second-order statistics for the likelihood densities (i.e., singular covariance matrices will result). Even if we were able to estimate these statistics, the computational cost of evaluating the likelihoods is formidable. Furthermore, this computation would be highly inefficient since the *intrinsic* dimensionality or major degrees of freedom of the data vectors of each class is likely to be significantly smaller than N.

Recently, an efficient density estimation method was proposed by Moghaddam and Pentland [22], which divides the vector space $\mathcal{R}^N$ into two complementary subspaces by using an eigenspace decomposition. This method relies on a PCA [13] to form a low-dimensional estimate of the complete likelihood, which can be evaluated by using only the first M principal components, where $M \ll N$. This decomposition is illustrated in Fig. 2, which shows an orthogonal decomposition of the vector space $\mathcal{R}^N$ into two mutually exclusive subspaces: the principal subspace F containing the first M principal components and its orthogonal complement $\bar{F}$, which contains the residual of the expansion. The component in the orthogonal subspace $\bar{F}$ is the so-called DFFS, a Euclidean distance equivalent to the PCA residual error. The component of Δ that lies *in* the feature space F is referred to as the DIFS and is a *Mahalanobis* distance for Gaussian densities.

As shown in [22], the complete likelihood estimate can be written as the product of two independent marginal Gaussian densities:

$$\hat{P}(\vec{x}\,|\,\Omega) = \left[\frac{\exp\left(-\frac{1}{2}\sum_{i=1}^{M}\frac{y_i^2}{\lambda_i}\right)}{(2\pi)^{M/2}\prod_{i=1}^{M}\lambda_i^{1/2}}\right] \times \left[\frac{\exp\left(-\frac{\epsilon^2(\vec{x})}{2\rho}\right)}{(2\pi\rho)^{(N-M)/2}}\right], \quad (4)$$

$$= P_F(\vec{x}\,|\,\Omega)\,\hat{P}_{\bar{F}}(\vec{x}\,|\,\Omega),$$

where $P_F(\vec{x}\,|\,\Omega)$ is the true marginal density in F, $\hat{P}_{\bar{F}}(\vec{x}\,|\,\Omega)$ is the estimated marginal density in the orthogonal complement $\bar{F}$, y_i are the principal components, and $\epsilon^2(\vec{x})$ is the residual (or

DFFS). The optimal value for the weighting parameter ρ is then found to be simply the average of the $\bar{F}$ eigenvalues

$$\rho = \frac{1}{N-M}\sum_{i=M+1}^{N}\lambda_i. \quad (5)$$

We note that in actual practice, the majority of the $\bar{F}$ eigenvalues are unknown but *can* be estimated, for example, by fitting a nonlinear function to the available portion of the eigenvalue spectrum and estimating the average of the eigenvalues beyond the principal subspace.

3.2 Maximum Likelihood Detection

The density estimate $\hat{P}(\mathbf{x}\,|\,\Omega)$ can be used to compute a local measure of target saliency at each spatial position (i, j) in an input image based on the vector $\mathbf{x}$ obtained by the lexicographic ordering of the pixel values in a local neighborhood R:

$$S(i, j; \Omega) = \hat{P}(\mathbf{x}\,|\,\Omega), \quad \mathbf{x} = \downarrow [\{I(i+r, j+c) : (r, c) \in R\}], \quad (6)$$

where $\downarrow [\bullet]$ is the operator that converts a subimage into a vector by raster scanning the image elements into the vector. The ML estimate of position of the target Ω is then given by finding the position (i, j) that maximizes $S(i, j; \Omega)$, e.g.,

$$(i, j)^{\mathrm{ML}} = S(i, j; \Omega). \quad (7)$$

An example of a saliency may for ML detection is shown in Fig. 3. This ML formulation can be extended to estimate object scale with *multiscale* saliency maps. The likelihood computation is performed (in parallel) on linearly scaled versions of the input image $I^{(\sigma)}$ corresponding to a predetermined set of scales $\{\sigma_1, \sigma_2, \ldots, \sigma_n\}$:

$$S(i, j, k; \Omega) = \hat{P}(\downarrow \{I^{(\sigma_k)}(\sigma_k i + r, \sigma_k j + c) : (r, c) \in R\}\,|\,\Omega), \quad (8)$$

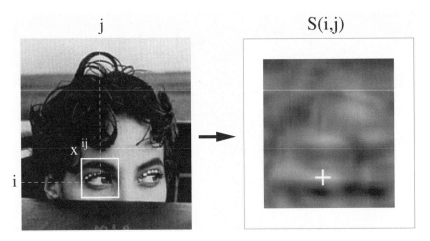

FIGURE 3 Target saliency map $S(i, j)$, showing the probability of a left eye pattern over the input image.

where the ML estimate of the spatial and scale indices is defined by

$$(i, j, k)^{\mathrm{ML}} = S(i, j, k; \Omega). \qquad (9)$$

One important factor of variability in the appearance of the object in gray-scale imagery is that of lighting and contrast. However, one can normalize for global illumination changes (as well as the linear response characteristics of the CCD camera) by normalizing each subimage **x** by its mean and standard deviation. This lighting normalization is performed both during training (density estimation) and also in the operational mode (e.g., in detection).

This maximum likelihood detection framework can be viewed as a Bayesian formulation of some neural network approaches to target detection. Perhaps the most closely related is the neural network face detector of Sung and Poggio [35], which is essentially a trainable nonlinear binary pattern classifier. They too learn the distribution of the object class with a mixture-of-Gaussians model (using an elliptical k-means algorithm instead of EM). Instead of likelihoods, however, input patterns are represented by a set of distances to each mixture component (similar to a combination of the DIFS and DFFS), thus forming a feature vector indicative of the overall class membership. In addition, Sung and Poggio explicitly model the "not-class" by learning the distribution of nearby *nonface* patterns. The set of distances to both classes are then used to train a neural network to discriminate between face and nonface patterns (similar to computing a likelihood ratio in MAP).

4 Bayesian Model of Facial Similarity

Current approaches to image matching for visual object recognition and image database retrieval often make use of simple image similarity metrics such as Euclidean distance or normalized correlation, which correspond to a standard template-matching approach to recognition. For example, in its simplest form, the similarity measure $S(I_1, I_2)$ between two images I_1 and I_2 can be set to be inversely proportional to the norm $\| I_1 - I_2 \|$. Such a simple formulation suffers from a major drawback: it does not exploit knowledge of which type of variations are critical (as opposed to incidental) in expressing similarity. In this chapter, we formulate a *probabilistic* similarity measure which is based on the probability that the image intensity differences, denoted by $\Delta = I_1 - I_2$, are characteristic of typical variations in appearance of the *same* object. For example, for purposes of face recognition, we can define two classes of facial image variations: *intrapersonal* variations Ω_I (corresponding, for example, to different facial expressions of the *same* individual) and *extrapersonal* variations Ω_E (corresponding to variations between *different* individuals). Our similarity measure is then expressed in terms of the probability

$$S(I_1, I_2) = P(\Delta \in \Omega_I) = P(\Omega_I \mid \Delta), \qquad (10)$$

where $P(\Omega_I \mid \Delta)$ is the *a posteriori* probability given by Bayes rule, using estimates of the likelihoods $P(\Delta \mid \Omega_I)$ and $P(\Delta \mid \Omega_E)$, which are derived from training data by using an efficient subspace method for density estimation of high-dimensional data [22]. This Bayesian (MAP) approach can also be viewed as a generalized nonlinear extension of linear discriminant analysis (LDA) [11, 36] or "FisherFace" techniques [3] for face recognition. Moreover, our nonlinear generalization has distinct computational/storage advantages over these linear methods for large databases.

4.1 Analysis of Intensity Differences

We now consider the problem of characterizing the type of differences that occur when matching two images in a face recognition task. We define two distinct and mutually exclusive classes: Ω_I, representing *intrapersonal* variations between multiple images of the same individual (e.g., with different expressions and lighting conditions); and Ω_E, representing *extrapersonal* variations that result when matching two different individuals. We will assume that both classes are Gaussian distributed and seek to obtain estimates of the likelihood functions $P(\Delta \mid \Omega_I)$ and $P(\Delta \mid \Omega_E)$ for a given intensity difference $\Delta = I_1 - I_2$.

Given these likelihoods, we can define the similarity score $S(I_1, I_2)$ between a pair of images directly in terms of the intrapersonal *a posteriori* probability as given by Bayes rule:

$$
\begin{aligned}
S &= P(\Omega_I \mid \Delta) \\
&= \frac{P(\Delta \mid \Omega_I)\,P(\Omega_I)}{P(\Delta \mid \Omega_I)\,P(\Omega_I) + P(\Delta \mid \Omega_E)\,P(\Omega_E)},
\end{aligned} \tag{11}
$$

where the priors $P(\Omega)$ can be set to reflect specific operating conditions (e.g., number of test images vs. the size of the database) or other sources of *a priori* knowledge regarding the two images being matched. Additionally, this particular Bayesian formulation casts the standard face recognition task (essentially an M-ary classification problem for M individuals) into a *binary* pattern classification problem with Ω_I and Ω_E. This much simpler problem is then solved by using the maximum *a posteriori* rule; i.e., two images are determined to belong to the same individual if $P(\Omega_I \mid \Delta) > P(\Omega_E \mid \Delta)$, or equivalently, if $S(I_1, I_2) > 1/2$.

We note that the Bayesian classification of identity outlined above is perhaps closely related to human "categorical perception," and the *a posteriori* probability itself is perhaps a more meaningful and accurate computational model of *perceptual similarity* as judged by humans.

5 Face Detection and Recognition

In this section we will present several examples of our face detection and recognition systems, including the following: ML detection of faces and facial features (e.g., eyes) used for facial

alignment, recognition using "eigenfaces" on large databases, and recognition using the Bayesian similarity measure.

Over the years, various strategies for facial feature detection have been proposed, ranging from edge map projections [14], to more recent techniques that use generalized symmetry operators [33] and multilayer perceptrons [39]. In any robust face processing system, this task is critically important since a face must be first geometrically normalized by aligning its features with those of a stored model before recognition can be attempted.

The eigentemplate approach to the detection of facial features in "mugshots" was proposed in [27], where the DFFS metric was shown to be superior to standard template matching for target detection. The detection task was the estimation of the position of facial features (the left and right eyes, the tip of the nose, and the center of the mouth) in frontal view photographs of faces at fixed scale. Figure 4 shows examples of facial feature training templates and the resulting detections on the MIT Media Laboratory's database of 7,562 "mugshots."

We have compared the detection performance of three different detectors on approximately 7,000 test images from this database: a sum-of-square-differences (SSD) detector based on the average facial feature (in this case the left eye), an eigentemplate or DFFS detector, and a ML detector based on $S(i, j; \Omega)$ as defined in Section 3.1.2. Figure 5(a) shows the *receiver operating characteristic* (ROC) curves for these detectors, obtained by varying the detection threshold independently for each detector. The DFFS and ML detectors were computed based on a five-dimensional principal subspace. Since the projection coefficients were unimodal, a Gaussian distribution was used in modeling the true distribution for the ML detector as in Section 3.1.2. Note that the ML detector exhibits the best detection versus false-alarm tradeoff and yields the highest detection rate (95%). Indeed, at the *same* detection rate the ML detector has a false-alarm rate that is nearly 2 orders of magnitude lower than the SSD.

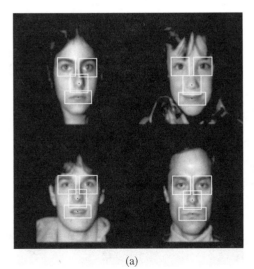
(a)

(b)

FIGURE 4   (a) Examples of facial feature training templates and (b) the resulting typical detections.

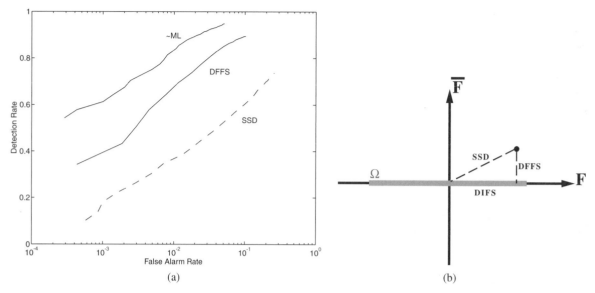

FIGURE 5   (a) Detection performance of an SSD, DFFS, and a ML detector; (b) geometric interpretation of the detectors.

Figure 5(b) provides the geometric intuition regarding the operation of these detectors. The SSD detector's threshold is based on the *radial* distance between the average template (the origin of this space) and the input pattern. This leads to hyperspherical detection regions about the origin. In contrast, the DFFS detector measures the orthogonal distance to F, thus forming planar acceptance regions about F. Consequently, to accept valid object patterns in Ω that are very different from the mean, the SSD detector must operate with high thresholds, which results in many false alarms. However, the DFFS detector cannot discriminate between the object class Ω and non-Ω patterns in F. The solution is provided by the ML detector, which incorporates both the $\bar{F}$-space component (DFFS) and the F-space likelihood (DIFS). The probabilistic interpretation of Fig. 5(b) is as follows: SSD assumes a *single* prototype (the mean) in additive white Gaussian noise, whereas the DFFS assumes a *uniform* density in F. The ML detector, in contrast, uses the complete probability density for detection.

We have incorporated and tested the multiscale version of the ML detection technique in a face detection task. This multiscale head finder was tested on the ARPA FERET database, where in 97% of 2,000 images the face, eyes, nose, and mouth were correctly detected and localized to within one pixel error. Figure 6 shows examples of the ML estimate of the position and scale on these images. The multiscale saliency maps $S(i, j, k; \Omega)$ were computed based on the likelihood estimate $\hat{P}(\mathbf{x} \mid \Omega)$ in a ten-dimensional principal subspace using a Gaussian model (Section 3.1.2). Note that this detector is able to localize the position and scale of the head despite variations in hair style and hair color, as well as the presence of sunglasses. Illumination invariance was obtained by normalizing the input subimage $\mathbf{x}$ to a zero-mean unit-norm vector.

5.1 Using ML Detection for Attention and Alignment

We have also used the multiscale version of the ML detector as the *attentional* component of an automatic system for recognition and model-based coding of faces. The block diagram of this

FIGURE 6   Examples of multiscale face detection.

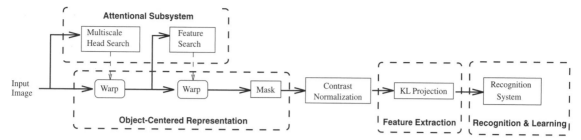

FIGURE 7 The face processing system.

system is shown in Fig. 7, which consists of a two-stage object detection and alignment stage, a contrast normalization stage, and a feature extraction stage whose output is used for both recognition and coding. Figure 8 illustrates the operation of the detection and alignment stage on a natural test image containing a human face. The function of the face finder is to locate regions in the image that have a high likelihood of containing a face.

The first step in this process is illustrated in Fig. 8(b), where the ML estimate of the position and scale of the face is indicated by the cross-hairs and bounding box. Once these regions have been identified, the estimated scale and position are used to normalize for translation and scale, yielding a standard "head-in-the-box" format image [Fig. 8(c)]. A second feature detection stage operates at this fixed scale to estimate the position of four facial features: the left and right eyes, the tip of the nose, and the center of the mouth [Fig. 8(d)]. Once the facial features have been detected, the face image is warped to align the geometry and shape of the face with that of a canonical model. Then the facial region is extracted (by applying a fixed mask) and subsequently normalized for contrast. The geometrically aligned and normalized image [shown in Fig. 9(a)] is then projected onto a custom set of eigenfaces to obtain a feature vector, which is then used for recognition purposes as well as facial image coding.

Figure 9 shows the normalized facial image extracted from Fig. 8(d), its reconstruction with a 100-dimensional eigenspace representation (requiring only 85 bytes to encode), and a comparable nonparametric reconstruction obtained with a standard transform-coding approach for image compression (requiring 530 bytes to encode). This example illustrates that the eigenface representation used for recognition is also an effective *model-based* representation for data compression.

To test our Bayesian recognition strategy, we used a collection of images from the FERET face database. This collection of images consists of hard recognition cases that have proven difficult for all face recognition algorithms previously tested on the FERET database. The difficulty posed by this dataset appears to stem from the fact that the images were taken at different times, at different locations, and under different imaging conditions. The set of images consists of pairs of frontal views (FA/FB) and are divided into two subsets: the "gallery" (training set) and the "probes" (testing set). The gallery images consisted of 74 pairs of images (two per individual), and the probe set consisted of 38 pairs of images, corresponding to a subset of the gallery members. The probe and gallery datasets were captured a week apart and exhibit differences in clothing, hair, and lighting.

Before we can apply our matching technique, we need to perform an affine alignment of these facial images. For this purpose we have used an automatic face-processing system, which extracts faces from the input image and normalizes for translation, scale as well as slight rotations (both in plane and out of plane). This is achieved by using the maximum-likelihood detection and alignment method described earlier, which is summarized in Fig. 7. All the faces in our experiments were geometrically aligned and normalized in this manner prior to further analysis.

5.1.1 Comparison with Eigenface Matching

As a baseline comparison, we first used an eigenface matching technique for recognition [37]. The normalized images from the gallery and the probe sets were projected onto a 100-dimensional eigenspace, and a nearest-neighbor rule based on a Euclidean distance measure was used to match each probe image to a gallery

    (a)   (b)   (c)   (d)

FIGURE 8 (a) Original image, (b) position and scale estimate, (c) normalized head image, (d) position of facial features.

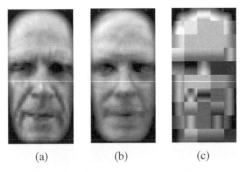

FIGURE 9 (a) Aligned face, (b) eigenspace reconstruction (85 bytes), (c) JPEG reconstruction (530 bytes).

image. We note that this method corresponds to a generalized template-matching method that uses a Euclidean norm type of similarity $S(I_1, I_2)$, which is restricted to the principal component subspace of the data. We note that these eigenfaces represent the principal components of an entirely different set of images; i.e., none of the individuals in the gallery or probe sets were used in obtaining these eigenvectors. In other words, neither the gallery nor the probe sets were part of the "training set." The rank-1 recognition rate obtained with this method was found to be 84% (64 correct matches out of 76), and the correct match was always in the top 10 nearest neighbors. Note that this performance is better than or similar to recognition rates obtained by any algorithm tested on this database, and that it is lower (by about 10%) than the typical rates that we have obtained with the FERET database [20]. We attribute this lower performance to the fact that these images were selected to be particularly challenging. In fact, using an eigenface method to match the first views of the 76 individuals in the gallery to their second views, we obtain a higher recognition rate of 89% (68 out of 76), suggesting that the gallery images represent a less challenging dataset since these images were taken at the same time and under identical lighting conditions.

5.1.2 Intrapersonal- and Extrapersonal-Based Matching

For our probabilistic algorithm, we first gathered training data by computing the intensity differences for a training subset of 74 intrapersonal differences (by matching the two views of every individual in the gallery) and a random subset of 296 extrapersonal differences (by matching images of *different* individuals in the gallery), corresponding to the classes Ω_I and Ω_E, respectively.

It is interesting to consider how these two classes are distributed; for example, are they linearly separable or embedded distributions? One simple method of visualizing this is to plot their mutual principal components, i.e., perform PCA on the *combined* dataset and project each vector onto the principal eigenvectors. Such a visualization is shown in Fig. 10(a), which is a 3-D scatter plot of the first three principal components. This plot shows what appears to be two completely enmeshed distributions, both having near-zero means and differing primarily in the amount of scatter, with Ω_I displaying smaller intensity differences as expected. It therefore appears that one cannot reliably distinguish low-amplitude extrapersonal differences (of which there are many) from intrapersonal ones.

However, direct visual interpretation of Fig. 10(a) is very misleading, since we are essentially dealing with low-dimensional (or "flattened") hyperellipsoids that are intersecting near the origin of a very high-dimensional space. The key distinguishing factor between the two distributions is their relative orientation. Fortunately, we can easily determine this relative orientation by performing a separate PCA on each class and computing the dot product of their respective first eigenvectors. This analysis yields the cosine of the angle between the major axes of the two hyperellipsoids, which was found to be 124°, implying that the orientation of the two hyperellipsoids is quite different. Figure 10(b) is a schematic illustration of the geometry of this configuration, where the hyperellipsoids have been drawn to approximate scale by using the corresponding eigenvalues.

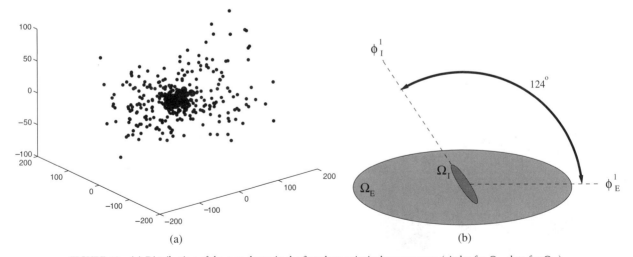

FIGURE 10 (a) Distribution of the two classes in the first three principal components (circles for Ω_I, dots for Ω_E) and (b) schematic representation of the two distributions showing the orientation difference between the corresponding principal eigenvectors.

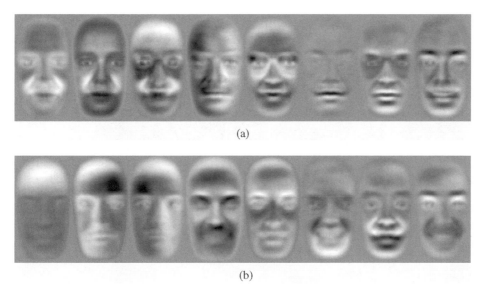

FIGURE 11 "Dual" eigenfaces: (a) intrapersonal, (b) extrapersonal.

5.1.3 Dual Eigenfaces

We note that the two mutually exclusive classes Ω_I and Ω_E correspond to a "dual" set of eigenfaces as shown in Fig. 11. Note that the intrapersonal variations shown in Fig. 11(a) represent subtle variations caused mostly by expression changes (and lighting), whereas the extrapersonal variations in Fig. 11(b) are more representative of general eigenfaces that code variations such as hair color, facial hair, and glasses. This suggests the basic intuition that intensity differences of the extrapersonal type span a larger vector space similar to the volume of face space spanned by standard eigenfaces, whereas the *intrapersonal* eigenspace corresponds to a more tightly constrained subspace. It is the representation of this intrapersonal subspace that is the critical part of formulating a probabilistic measure of facial similarity. In fact, our experiments with a larger set of FERET images have shown that this intrapersonal eigenspace alone is sufficient for a simplified *maximum likelihood* measure of similarity (see Section 5.1.4).

Finally, we note that since these classes are not linearly separable, simple linear discriminant techniques (e.g., using hyperplanes) cannot be used with any degree of reliability. The proper decision surface is inherently nonlinear (quadratic, in fact, under the Gaussian assumption) and is best defined in terms of the *a posteriori* probabilities — i.e., by the equality $P(\Omega_I \mid \Delta) = P(\Omega_E \mid \Delta)$. Fortunately, the optimal discriminant surface is automatically implemented when invoking a MAP classification rule.

Having analyzed the geometry of the two distributions, we then computed the likelihood estimates $P(\Delta \mid \Omega_I)$ and $P(\Delta \mid \Omega_E)$ by using the PCA-based method outlined in Section 3.1.2. We selected principal subspace dimensions of $M_I = 10$ and $M_E = 30$ for Ω_I and Ω_E, respectively. These density estimates were then used with a default setting of equal priors, $P(\Omega_I) = P(\Omega_E)$, to evaluate the *a posteriori* intrapersonal prob-

ability $P(\Omega_I \mid \Delta)$ for matching probe images to those in the gallery.

Therefore, for each probe image we computed probe-to-gallery differences and sorted the matching order, this time using the *a posteriori* probability $P(\Omega_I \mid \Delta)$ as the similarity measure. This probabilistic ranking yielded an improved rank-1 recognition rate of 89.5%. Furthermore, out of the 608 extrapersonal warps performed in this recognition experiment, only 2% (11) were misclassified as being intrapersonal — i.e., with $P(\Omega_I \mid \Delta) > P(\Omega_E \mid \Delta)$.

5.1.4 The 1996 FERET Competition Results

The interpersonal/extrapersonal approach to recognition has produced a significant improvement over the accuracy we obtained by using a standard eigenface nearest-neighbor matching rule. The probabilistic similarity measure was used in the September 1996 FERET competition (with subspace dimensionalities of $M_I = M_E = 125$) and was found to be the top-performing system by a typical margin of 10–20% over the other competing algorithms [30]; see Fig. 12(a). Figure 12(b) shows the performance comparison between standard eigenfaces and the Bayesian method from this test. Note the 10% gain in performance afforded by the new Bayesian similarity measure. Thus we note that the new probabilistic similarity measure has effectively *halved* the error rate of eigenface matching.

We have recently experimented with a more simplified probabilistic similarity measure, which uses only the *intrapersonal* eigenfaces with the intensity difference Δ to formulate a ML matching technique, using

$$S' = P(\Delta \mid \Omega_I) \qquad (12)$$

instead of the MAP approach defined by Eq. (11). Although this

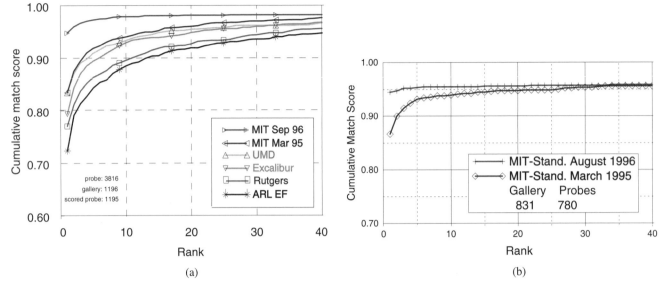

FIGURE 12 (a) Cumulative recognition rates for frontal FA/FB views for the competing algorithms in the FERET 1996 test. The top curve (labeled "MIT Sep 96") corresponds to our Bayesian matching technique. Note that second placed is standard eigenface matching (labeled "MIT Mar 95"). (b) Cumulative recognition rates for frontal FA/FB views with standard eigenface matching and the newer Bayesian similarity metric.

simplified measure has not yet been officially FERET tested, our own experiments with a database of size 2000 have shown that using S' instead of S results in only a minor (2%) deficit in the recognition rate while cutting the computational cost by a factor of 2 (requiring a single eigenspace projection as opposed to two).

6 View-Based Face Recognition

The problem of face recognition under general viewing conditions (change in pose) can also be approached by using an eigenspace formulation. There are essentially two ways of approaching this problem by using an eigenspace framework. Given N individuals under M different views, one can do recognition and pose estimation in a universal eigenspace computed from the combination of NM images. In this way a single "parametric eigenspace" will encode both identity as well as pose. Such an approach, for example, has recently been used by Murase and Nayar [24] for general 3-D object recognition.

Alternatively, given N individuals under M different views, we can build a "view-based" set of M distinct eigenspaces, each capturing the variation of the N individuals in a common view. The view-based eigenspace is essentially an extension of the eigenface technique to multiple sets of eigenvectors, one for each combination of scale and orientation. One can view this architecture as a set of parallel "observers," each trying to explain the image data with their set of eigenvectors, see also Darrell and Pentland [10]. In this view-based, multiple-observer approach, the first step is to determine the location and orientation of the target object by selecting the eigenspace that best describes the input

image. This can be accomplished by calculating the likelihood estimate, using each viewspace's eigenvectors, and then selecting the maximum.

The main advantage of the parametric eigenspace method is its simplicity. The encoding of an input image using n eigenvectors requires only n projections. In the view-based method, M different sets of n projections are required, one for each view. However, this does not imply that a factor of M times more computation is necessarily required. By progressively calculating the eigenvector coefficients while pruning alternative viewspaces, one can greatly reduce the cost of using M eigenspaces.

The key difference between the view-based and parametric representations can be understood by considering the geometry of facespace. In the high-dimensional vector space of an input image, multiple-orientation training images are represented by a set of M distinct regions, each defined by the scatter of N individuals. Multiple views of a face form nonconvex (yet connected) regions in image space [4]. Therefore, the resulting ensemble is a highly complex and nonseparable manifold.

The parametric eigenspace attempts to describe this ensemble with a projection onto a single low-dimensional linear subspace (corresponding to the first n eigenvectors of the NM training images). In contrast, the view-based approach corresponds to M independent subspaces, each describing a particular region of the facespace (corresponding to a particular view of a face). The relevant analogy here is that of modeling a complex distribution by a single cluster model or by the union of several component clusters. Naturally, the latter (view-based) representation can yield a more accurate representation of the underlying geometry.

This difference in representation becomes evident when considering the quality of reconstructed images using the two

FIGURE 13 Some of the images used to test the accuracy of face recognition, despite wide variations in head orientation. The average recognition accuracy was 92%; the orientation error had a standard deviation of 15°.

different methods. Figure 14 below compares reconstructions obtained with the two methods when trained on images of faces at multiple orientations. In Fig. 14(a) top row, we see first an image in the training set, followed by reconstructions of this image using first the parametric eigenspace and then the view-based eigenspace. Note that in the parametric reconstruction neither the pose nor the identity of the individual is adequately captured. The view-based reconstruction, in contrast, provides a much better characterization of the object. Similarly, in Fig. 14(a) bottom row, we see a novel view (+68°) with respect to the training set (−90° to +45°). Here, both reconstructions correspond to the nearest view in the training set (+45°), but the view-based reconstruction is seen to be more representative of the individual's identity. Although the quality of the reconstruction is not a direct indicator of the recognition power, from an information-theoretic point of view the multiple eigenspace representation is a more accurate representation of the signal content.

We have evaluated the view-based approach with data similar to that shown in Fig. 13. These data consist of 189 images,

made up of nine views of 21 people. The nine views of each person were evenly spaced from −90° to +90° along the horizontal plane. In the first series of experiments the *interpolation* performance was tested by training on a subset of the available views {±90°, ±45°, 0°} and testing on the intermediate views {±68°, ±23°}. A 90% average recognition rate was obtained. A second series of experiments tested the *extrapolation* performance by training on a range of views (e.g., −90° to +45°) and testing on novel views outside the training range (e.g., +68° and +90°). For testing views separated by ±23° from the training range, the average recognition rates were 83%. For ±45° testing views, the average recognition rates were 50%; see [28] for further details.

7 Modular Descriptions for Recognition

The eigenface recognition method is easily extended to facial features as shown in Fig. 15(a). Eye-movement studies indicate that

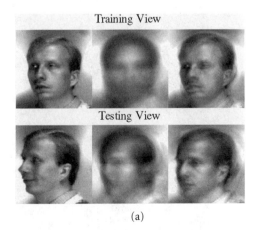

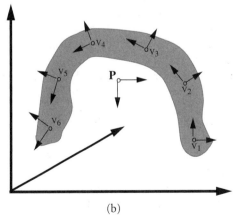

(a)   (b)

FIGURE 14 (a) Parametric vs. view-based eigenspace reconstructions for a training view and a novel testing view. The input image is shown in the left column. The middle and right columns correspond to the parametric and view-based reconstructions, respectively. All reconstructions were computed using the first 10 eigenvectors. (b) Schematic representation of the two approaches.

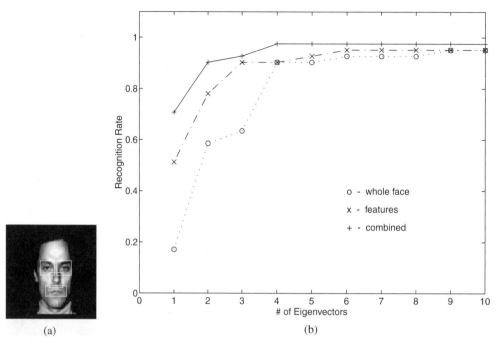

(a)   (b)

FIGURE 15 (a) Facial eigenfeature regions; (b) recognition rates for eigenfaces, eigenfeatures, and the combined modular representation.

these particular facial features represent important landmarks for fixation, especially in an attentive discrimination task [41]. This leads to an improvement in recognition performance by incorporating an additional layer of description in terms of facial features. This can be viewed as either a modular or layered representation of a face, where a coarse (low-resolution) description of the whole head is augmented by additional (higher-resolution) details in terms of salient facial features.

The utility of this layered representation (eigenface plus eigenfeatures) was tested on a small subset of our large face database. We selected a representative sample of 45 individuals with two views per person, corresponding to different facial expressions (neutral versus smiling). These set of images was partitioned into a training set (neutral) and a testing set (smiling). Since the difference between these particular facial expressions is primarily articulated in the mouth, this feature was discarded for recognition purposes.

Figure 15(b) shows the recognition rates as a function of the number of eigenvectors for eigenface-only, eigenfeature-only, and the combined representation. What is surprising is that (for this small dataset at least) the eigenfeatures alone were sufficient in achieving an (asymptotic) recognition rate of 95% (equal to that of the eigenfaces). More surprising, perhaps, is the observation that in the lower dimensions of eigenspace, eigenfeatures outperformed the eigenface recognition. Finally, by using the combined representation, we gain a slight improvement in the asymptotic recognition rate (98%). A similar effect was reported by Brunelli and Poggio [7], in which the cumulative normalized correlation scores of templates for the face, eyes, nose, and mouth showed improved performance over the face-only templates.

A potential advantage of the eigenfeature layer is the ability to overcome the shortcomings of the standard eigenface method. A pure eigenface recognition system can be fooled by gross variations in the input image (hats, beards, etc.). Figure 16(a) shows the additional testing views of three individuals in the above dataset of 45. These test images are indicative of the type of variations that can lead to false matches: a hand near the face, a painted face, and a beard. Figure 16(b) shows the nearest matches found based on standard eigenface matching. None of the three matches

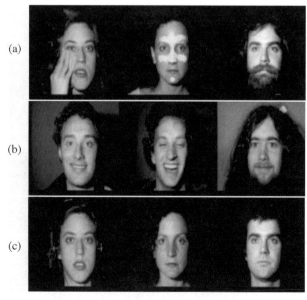

FIGURE 16 (a) Test views, (b) eigenface matches, (c) eigenfeature matches.

correspond to the correct individual. In contrast, Fig. 16(c) shows the nearest matches based on the eyes and nose, and it results in a correct identification in each case. This simple example illustrates the potential advantage of a modular representation in disambiguating low-confidence eigenface matches.

8 Discussion

In this chapter we have described an eigenspace density estimation technique for unsupervised visual learning that exploits the *intrinsic* low-dimensionality of the training imagery to form a computationally simple estimator for the complete likelihood function of the object. Our estimator is based on a subspace decomposition and can be evaluated by using only the *M*-dimensional principal component vector. In contrast to previous work on learning and characterization — which uses PCA primarily for dimensionality reduction or feature extraction — our method uses the eigenspace decomposition as an integral part of estimating *complete* density functions in high-dimensional image spaces.

These density estimates were then used in a maximum likelihood formulation for target detection. The multiscale version of this detection strategy was demonstrated in applications in which it functioned as an attentional subsystem for object recognition. The performance was found to be superior to existing detection techniques in experiments with large numbers of test data. We have also shown that the same representation can be extended to multiple head poses, to incorporate edge or texture features, and to utilize facial features such as eye or nose shape. Each of these extensions has provided additional robustness and generality to the core idea of detection and recognition using probabilistic appearance models.

References

[1] H. Abdi, "Generalized approaches for connectionist auto-associative memories: interpretation, implication, and illustration for face processing," in *AI and Cognitive Science* (Manchester University Press, Manchester, 1988), pp. 151–164.

[2] C. H. Anderson, P. J. burt, and G. S. Van der Wall," Change detection and tracking using pyramid transform techniques," in *Intelligent Robots and Computer Vision IV*, D. P. Casasent, ed., Proc. SPIE **579**, 72–78 (1985).

[3] V. I. Belhumeur, J. P. Hespanha, and D. J. Kriegman, "Eigenfaces vs. fisherfaces: recognition using class specific linear projection," *IEEE Trans. Pattern Anal. Machine Intell.* **19**, 711–720 (1997).

[4] M. Bichsel and A. Pentland, "Human face recognition and the face image set's topology," *CVGIP: Image Understand.* **59**, 254–261 (1994).

[5] C. Bregler and S. M. Omohundro, "Surface learning with applications to lip reading," in G. Tesauro, J. D. Cowan, and J. Alspector, eds., *Advances in Neural Information Processing Systems 6* (Morgan Kaufman, San Fransisco, CA, 1994), pp. 43–50.

[6] R. Brunelli and S. Messelodi, "Robust estimation of correlation:

[7] R. Brunelli and T. Poggio, Face recognition: Features vs. templates. *IEEE Trans. on Pattern Analysis and Machine Intelligence*, 15(10), 1993.

[8] M. C. Burl, U. M. Fayyad, P. Perona, P. Smyth, and M. P. Burl, "Automating the hunt for volcanos on venus," in *Proceedings of the IEEE Conference on Computer Vision & Pattern Recognition* (IEEE, New York, 1994).

[9] T. F. Cootes and C. J. Taylor, "Active shape models: smart snakes," in *Proceedings of the British Machine Vision Conference* (Springer-Verlag, New York, 1992), pp. 9–18.

[10] T. Darrell and A. Pentland, "Space-time gestures," in *Proceedings of the IEEE Conference on Computer Vision & Pattern Recognition* (IEEE, New York, 1993).

[11] K. Etemad and R. Chellappa, "Discriminant analysis for recognition of human faces," in *Proceedings of the International Conference on Acoustics, Speech and Signal Processing* (IEEE, New York, 1996), pp. 2148–2151.

[12] G. H. Golub and C. F. Van Loan, *Matrix Computations* (Johns Hopkins U. Press, Baltimore, MD, 1989).

[13] I. T. Jolliffe, *Principal Component Analysis* (Springer-Verlag, New York, 1986).

[14] T. Kanade, "Picture processing by computer complex and recognition of human faces," Tech. Rep., Kyoto University, Dept. of Information Science, 1973.

[15] M. Kirby and L. Sirovich, "Application of the Karhunen-Loeve procedure for the characterization of human faces," *IEEE Trans. Pattern Anal. Machine Intell.* **12**, x–x (1990).

[16] T. Kohonen, *Self Organization and Associative Memory* (Springer-Verlag, Berlin, 1989).

[17] B. Kumar, D. Casasent, and H. Murakami, "Principal component imagery for statistical pattern recognition correlators," *Opt. Eng.* **21**, x–x (1982).

[18] M. M. Loeve, *Probability Theory* (Van Nostrand, Princeton, NJ, 1955).

[19] B. Moghaddam, T. Jebara, and A. Pentland, "Efficient MAP/ML similarity matching for face recognition," presented at the International Conference on Pattern Recognition, Brisbane, Australia, August x–x, 1998.

[20] B. Moghaddam and A. Pentland, "Face recognition using view-based and modular eigenspaces," in *Automatic Systems for the Identification and Inspection of Humans*, R. J. Mammone and J. D. Murley, eds., *Proc. SPIE* **2277**, x–x (1994).

[21] B. Moghaddam and A. Pentland, "Probabilistic visual learning for object detection," in *IEEE Proceedings of the Fifth International Conference on Computer Vision (ICCV'95)* (IEEE, New York, 1995).

[22] B. Moghaddam and A. Pentland, "Probabilistic visual learning for object representation," *IEEE Trans. Pattern Anal. Machine Intell.* **19**, 696–710 (1997).

[23] B. Moghaddam, W. Wahid, and A. Pentland, "Beyond eigenfaces: probabilistic matching for face recognition," in *Proceedings of the IEEE International Conference on Automatic Face and Gesture Recognition* (IEEE, New York, 1998).

[24] H. Murase and S. K. Nayar, "Visual learning and recognition of 3-D objects from appearance," *Int'l J. Comput. Vision* **14**, x–x (1995).

[25] S. K. Nayar, H. Murase, and S. A. Nene, "General learning algorithm for robot vision," *Neural and Stochastic Methods in Image*

[6] an application to computer vision," Tech. Rep. 9310-015, IRST, October, 1993.

and Signal Processing, III, S.-S. Chen, ed., *Proc. SPIE* **2304**, x–x (1994).

[26] S. E. Palmer, *The Psychology of Perceptual Organization: A transformational Approach* (Academic, New York, 1983).

[27] A. Pentland, B. Moghaddam, and T. Starner, "View-based and modular eigenspaces for face recognition, in *Proceedings of the IEEE Conference on Computer Vision and Pattern Recognition* (IEEE, New York, 1994).

[28] A. Pentland, R. Picard, and S. Sclaroff, "Photobook: tools for content-based manipulation of image databases," *Storage and Retrieval for Image and Video Databases II*, W. Niblack and R. C. Jain, eds., *Proc. SPIE* **2185**, x–x (1994).

[29] A. Pentland and S. Sclaroff, "Closed-form solutions for physically based shape modeling and recovery," *IEEE Trans. Patteern Anal. Machine Intell.* **13**, 715–729 (1991).

[30] P. J. Phillips, H. Moon, P. Rauss, and S. Rizvi, "The FERET evaluation methodology for face-recognition algorithms," in *IEEE Proceedings of Computer Vision and Pattern Recognition* (IEEE, New York, 1997), pp. 137–143.

[31] T. Poggio and S. Edelman, "A network that learns to recognize three-dimensional objects," *Nature* **343**, x–x (1990).

[32] T. Poggio and F. Girosi, "Networks for approximation and learning," *Proc. IEEE* **78**, 1481–1497 (1990).

[33] D. Reisfeld, H. Wolfson, and Y. Yeshurun, "Detection of interest points using symmetry," presented at the International Conference on Computer Vision, Osaka, Japan, month x–x, 1990.

[34] S. Sclaroff and A. Pentland, "Modal matching for correspondence and recognition," *IEEE Trans. Pattern Anal. Machine Intell.* **17**, 545–561 (1995).

[35] K. K. Sung and T. Poggio, "Example-based learning for view-based human face detection," presented at the Image Understanding Workshop, Monterey, CA, November x–x, 1994.

[36] D. Swets and J. Weng, "Using discriminant eigenfeatures for image retrieval," *IEEE Trans. Pattern Anal. Machine Intell.* **18**, 831–836 (1996).

[37] M. Turk and A. Pentland, "Eigenfaces for recognition," *J. Cognitive Neurosci.* **3**, x–x (1991).

[38] S. Ullman and R. Basri, "Recognition by linear combinations of models," *IEEE Trans. Pattern Anal. Machine Intell.* **13**, 992–1006 (1991).

[39] J. M. Vincent, J. B. Waite, and D. J. Myers, "Automatic location of visual features by a system of multilayered perceptrons," *IEEE Proc.* **139**, x–x (1992).

[40] J. J. Weng, "On comprehensive visual learning," presented at the NSF/ARPA Workshop on Performance vs. Methodology in Computer Vision, Seattle, WA, June x–x, 1994.

[41] A. L. Yarbus, *Eye Movement and Vision* (Plenum, New York, 1967), B. Haigh (trans.).

10.7

Confocal Microscopy

Fatima A. Merchant
Perceptive Scientific Instruments

Keith A. Bartels
Southwest Research Institute

Alan C. Bovik and
Kenneth R. Diller
The University of Texas at Austin

1 Introduction

Confocal microscopes have been built and used in research laboratories since the early 1980s and have been commercially available for only the last few years. The concept of the confocal microscope, however, is over 40 years old. In 1957, Marvin Minsky [1] applied for a patent on the confocal idea. At that time, Minsky demonstrated great insight into the power of the confocal microscope. He realized that the design of the confocal microscope would give increased resolution and increased depth discrimination ability over conventional microscopes. Independently, in Czechoslovakia, M. Petrán and M. Hadravsky [2] developed the idea for the tandem scanning optical microscope (a form of the confocal microscope) in the mid-1960s. However, it was not until the 1980s that the confocal microscope became a useful tool in the scientific community. At the time the confocal scope was introduced, the electron microscope was receiving a great deal of attention as it was becoming commercially available. Meanwhile, the confocal microscope required a very high intensity light source, and thus its commercialization was delayed until the emergence of affordable lasers in the technological market. Finally, without the aid of high-speed data processing equipment and large computer memories, taking advantage of the three-dimensional (3-D) capabilities of the confocal microscope was not practical. Visualization of the data was also not feasible without high-powered computers and advanced computer graphics techniques.

Since the early 1980s, research and application of confocal microscopy has grown substantially. A great deal of research has now been done in understanding the imaging properties of the confocal microscope. Moreover, confocal microscopes of different varieties are now commercially available from several quality manufacturers.

2 Image Formation in Confocal Microscopy

There are several different designs of the confocal microscope. Each of these designs is based on the same underlying physical principles. First these underlying principles will be discussed, and then some of the specific designs will be briefly described.

The confocal microscope has three important features that make it advantageous over a conventional light microscope. First, the lateral resolution can be as great as one and a half times that of a conventional microscope. Second, and most importantly, the confocal microscope has the ability to remove out-of-focus information and thus produce an image of a very thin "section" of a specimen. Third, because of the absence of out-of-focus information, much higher contrast images are obtained.

A schematic representation of a reflectance (dark field) or fluorescence type confocal microscope is shown in Fig. 1. The illumination pinhole produces a point source from which the

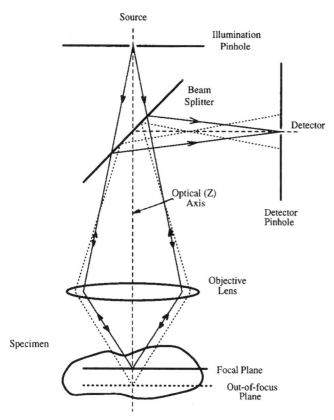

FIGURE 1 Diagram of a confocal microscope. The dashed lines represent light rays from an out-of-focus plane within the specimen; these rays are blocked by the imaging pinhole and do not reach the detector.

light ray originates. The ray passes through the beam splitter and down to the objective lens, where it is focused to a point spot inside of the specimen on the focal plane. If the ray reflects off a point in the focal plane, it will take the same path back up through the objective and pass, via the beam splitter, through the imaging pinhole and to the detector. If the ray instead reflects off a point that is out of the focal plane, the ray will take a new path back through the objective lens and will be blocked by the imaging pinhole from reaching the detector. From this simple explanation it is seen that only the focal plane is imaged. This analysis was purely in terms of geometrical optics. However, since the resolution of a high-quality microscope is diffraction limited, a diffraction analysis is needed to compare the resolutions of the conventional and the confocal microscope.

2.1 Lateral Resolution

First the lateral resolution of the microscope will be considered. The lateral resolution refers to the resolution in the focal plane of the microscope. The point spread function (PSF) of a circular converging lens is well known to be the Airy disk [3]. The Airy disk is defined in terms of $J_1(v)$, the Bessel function of order 1. The PSF is defined as the square of the modulus of the amplitude

point spread function, $h(v)$, which has the form

$$h(v) = \left[\frac{2 J_1(v)}{v} \right]. \tag{1}$$

The independent variable v, known as the optical distance, is defined in terms of r, the radial distance from the optical axis in the focal plane:

$$v = \frac{2\pi}{\lambda} \left(\frac{a}{f} \right) r, \tag{2}$$

where a is the radius of the lens, λ is the wavelength of the light, and f is the focal length of the lens.

Light is most often detected on an intensity basis. Sheppard and Wilson [4] give the following formulas for calculating the distribution of intensity, $I(x, y)$, for the coherent conventional microscope, the incoherent conventional microscope, and the confocal microscope in terms of the amplitude point spread function. A *coherent microscope* is a microscope in which the illumination source is coherent light. Likewise, an *incoherent microscope* has an incoherent illumination source. Letting $t(x, y)$ be the object *amplitude* transmittance, for the conventional coherent microscope the intensity is

$$I_{cc} = |t * h|^2, \tag{3}$$

for the conventional incoherent microscope the intensity is

$$I_{ci} = |t|^2 * |h|^2, \tag{4}$$

and for the confocal microscope the intensity is

$$I_c = |t * h^2|^2. \tag{5}$$

From a quick examination of these equations, it may not be obvious that the resolution of the confocal microscope is superior. The responses of each type of microscope to a point object are $I_{cc} = |h|^2$, $I_{ci} = |h|^2$, and $I_c = |h|^4$. These responses, with h as defined in Eq. (1), are plotted in one dimension in Fig. 2. In both cases of the conventional microscope, the PSFs are identical and equal to the Airy disk. The confocal PSF is equal to the square of the Airy disk and hence is substantially narrower and has very weak sidelobes. Because of the different imaging properties of the microscopes, the width of the PSF is not a sufficient means by which to describe resolution. Using the width of the PSF, one might conclude that the coherent and incoherent conventional microscopes have the same resolution. This, as is shown below, is not the case. The resolution of the incoherent microscope is in fact greater than that of the coherent microscope.

The resolution of an optical system is often given in terms of its *two-point resolution*. The two-point resolution is defined as the closest distance between two point objects such that each

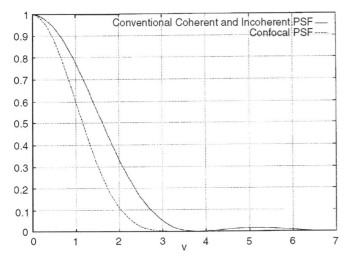

FIGURE 2 Plots of the PSFs for the conventional coherent and incoherent microscopes ($|h|^2$), and the confocal microscope ($|h|^4$).

object can just be resolved. This is a somewhat loose definition, since one must explain what is meant by *just* resolved. The Rayleigh criterion is often used to define the two-point resolution. The Rayleigh criterion (somewhat arbitrarily) states that the two points are just resolved when the center of the Airy disk generated by one point coincides with the first zero of the Airy disk generated by the second point. The Rayleigh distances for the coherent and incoherent conventional microscope are given in [3] as 0.77λ/N.A. and 0.6λ/N.A., respectively, where N.A. represents the numerical aperture of the objective lens. The numerical aperture is computed as $n \sin \theta$, where n is the index of refraction of the immersion medium and θ is the half-angle of the cone of

light that exits the objective. For the confocal microscope, the Rayleigh distance is given in [5] as 0.56λ/N.A.

Figure 3 shows the one-dimensional response to two point objects separated by the Rayleigh distance for the conventional incoherent microscope. The point objects are shown with reduced amplitude on the plot for reference purposes. From Fig. 3, it is evident that the conventional coherent microscope cannot resolve the two point objects. The two points appear as a single large point. The superior resolution of the confocal microscope is demonstrated from this simulation.

2.2 Depth Resolution and Optical Sectioning

The confocal microscope's most important property is its ability to discriminate depth. It is easy to show by the conservation of energy that the conventional microscope has no depth discrimination ability. Consider the conventional detector setup in Fig. 4. The output of the large area detector is the integral of the intensity of the image formed by the lens. When a point object is in focus (at A), the Airy disk is formed on the detector. If the point object is moved out of the focal plane (at B), a pattern of greater spatial extent is formed on the detector (a mathematical description of the out-of-focus PSF is given in [6]). By the conservation of light energy, the integral of these two intensity patterns must be equal and hence the detector output is the same for the in-focus and out-of-focus objects.

In the case of the confocal microscope, the pinhole aperture blocks the light from the extended size of the defocused point object's image. Early work by Born and Wolf [3] gave a description of the defocused light amplitude along the optical axis of such a lens system. Wilson *et al.* [5, 6], have adapted this analysis

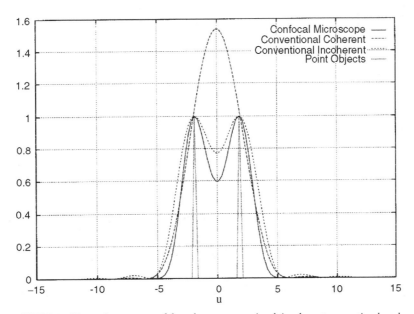

FIGURE 3 Two-point response of the coherent conventional, incoherent conventional, and confocal microscopes. The object points are spaced apart by one Rayleigh distance of the conventional incoherent system.

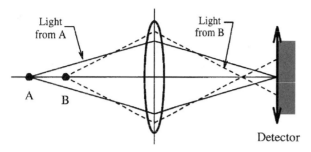

FIGURE 4 In the conventional microscope, the detector output for an in-focus and an out-of-focus point is the same.

to the confocal microscope. An optical distance along the optical axis of the microscope is defined by

$$u = \frac{8\pi}{\lambda} \sin^2(\alpha/2)z, \qquad (6)$$

where z is distance along the optical (z) axis, and $\sin \alpha$ is the numerical aperture of the objective. With this definition, the intensity along the optical axis is given by

$$I_u = \left[\frac{(\sin u/2)}{(u/2)} \right]^2 \qquad (7)$$

Experimental verification of Eq. (7) has been performed by sectioning through a highly planar mirror [7–9]. Figure 5 shows a plot of $I(u)$ versus u. The resolution of the z-axis sectioning is most often given as the full width at the half-intensity point. A plot of the z-sectioning width as a function of numerical aperture is given in [7]. A typical example is for an air objective with a N.A. of 0.8, the z-sectioning width is approximately 0.8 μm. For an oil immersion objective with N.A. equal to 1.4, the z-sectioning is approximately 0.25 μm.

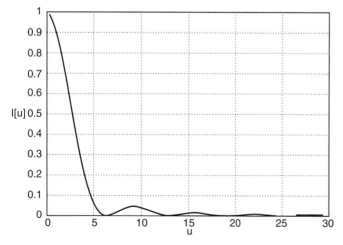

FIGURE 5 Plot of $I(u)$ vs. u, showing the optical sectioning along the optical axis of a confocal microscope.

3 Confocal Fluorescence Microscopy

The analysis presented herein has assumed that the radiation emitted from the specimen is of the same wavelength as the radiation incident on the specimen. This is true for reflectance and transmission confocal microscopy, but not for fluorescence confocal microscopy. In fluorescence confocal microscopy, the image formation no longer takes the form of Eq. (5), but rather of

$$I_c = |t * h(u, v)h(u/\beta, v/\beta)|^2, \qquad (8)$$

where β is the ratio of the fluorescent wavelength (λ_2) to the incident wavelength (λ_1), i.e., $\beta = \lambda_2/\lambda_1$, and u and v are rectangular distances in the focal plane. Considering, as before, the case of the circular pupil function

$$I(v) = \left[4\frac{J_1(v)}{v} \frac{J_1(v/\beta)}{v/\beta} \right]^2 \qquad (9)$$

as the lateral PSF in the focal plane. Obviously, if $\beta = 1$ the PSF of the reflection (and transmission) confocal microscope is obtained. As $\beta \rightarrow \infty$, the PSF of the conventional (nonconfocal) microscope is obtained. In practice, β will be generally less than 2. A detailed analysis of a confocal microscope in fluorescence mode is given in [7, 10].

4 Further Considerations

In all of the analyses presented here, it is assumed that the pinhole apertures are infinitely small. In practice, the pinhole apertures are of finite radius. In [9], Wilson presents theoretical and experimental results of the effects of various finite pinhole sizes. As one would expect, the resolution in both the axial and transverse directions is degraded by a larger pinhole. Also in [9], Wilson discusses the use of slit, rather than circular, apertures at the detector. The slit detector allows more light to reach the detector than the circular aperture with a compromise of sectioning ability. Wilson has also shown that using an annular rather than a circular lens pupil can increase the resolution of the confocal microscope at the expense of higher sidelobes in the point spread function [5, 7].

5 Types of Confocal Microscopes

Confocal microscopes are categorized into two major types, depending on the instrument design employed to achieve imaging. One type of confocal microscope scans the specimen by either moving the stage or the beam of light, whereas the second type employs both a stationary stage and light source.

5.1 Scanning Confocal Microscope

The *scanning* confocal microscope is by far the most popular on the market today, and it employs a laser source for specimen scanning. If a laser is not used, then a very high power light source is needed to get sufficient illumination through the source and detector pinhole apertures.

There are two practical methods for the raster scanning of a specimen. One method is to use a mechanical scanning microscope stage. With a scanning stage, the laser beam is kept stationary while the specimen is raster scanned through the beam. The other method is to keep the specimen still and scan the laser in a raster fashion over the specimen. There are, of course, advantages to using either of these scanning methods.

There are two qualities that makes scanning the specimen relative to the stationary laser attractive. First, the field of view is not limited by the optics, but by the range of the mechanical scanners. Therefore, very large areas of a specimen can be imaged. A second important advantage of scanning the specimen is that only a very narrow optical path is necessary in the design of the optics. This means that aberrations in the images due to imperfections in the lenses will be less of a problem. A disadvantage of this type of scanning is that image formation is very slow.

The main advantage of scanning the laser instead of the specimen is that the imaging speed is greatly increased. A mobile mirror can be used to scan the laser, in which case an image of 512×512 pixels can be obtained in ~ 1 s. A newer technology of laser scanning confocal microscopes uses acousto-optical deflection devices that can scan out an image at speeds up to TV frame rates. The problem with these acousto-optic scanners is that they are highly nonlinear and special care must be taken in order to obtain distortion-free images.

5.2 Tandem Scanning Optical Microscope

The tandem scanning optical microscope (TSOM) was patented in Czechoslovakia in the mid-1960s by M. Petrán and M. Hadravsky. The main advantage of the TSOM over the scanning confocal microscope is that images are formed in real time (at video frame rates or greater). Figure 6 shows a simple diagram of the tandem scanning optical microscope. The most important feature of the TSOM is the Nipkow disk. The holes in the Nipkow disk are placed such that when the disk is spun, a sampled scan of the specimen is produced. Referring to Fig. 6, the source light enters a pinhole on the Nipkow disk and is focused onto the specimen through the objective lens. The light reflected off of the specimen goes back up through the objective and up through a corresponding pinhole on the opposite side of the Nipkow disk. The light exiting from the eyepiece can be viewed by the operator, captured on video, or digitized and sent to a computer. In early TSOMs, sunlight was used as the illuminating source. Today, though, an arc or filament lamp is generally used. Figure 6 shows the path of a single ray through the system, but it should be noted that several such rays are focused on the specimen at any given instant of time.

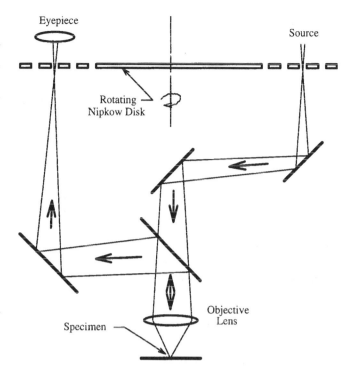

FIGURE 6 Diagram of the tandem scanning optical microscope.

Kino *et al.* [10] altered the above design so that the light enters and exits through the same pinhole. With this design, smaller pinholes can be used since mechanical alignment of the optics is not as difficult. Smaller pinholes, of course, are desirable since the depth of the in-focus plane is directly related to the pinhole size. Kino *et al.* were able to construct a Nipkow disk with 200,000 pinholes, 20 μm in diameter each, that spun at 2000 RPM. This gave them a frame rate of 700 frames/s with 5000 lines/image.

The TSOM does have certain drawbacks. Because the total area of the pinholes on the Nipkow disk must be negligible (less than 1%) with respect to the total area of the disk [7], the intensity of the light actually reaching the specimen is a very small fraction of that of the source. Depending on the specimen, the amount of light reflected may not be detectable. Another disadvantage of the TSOM is that it is mechanically more complex than the scanning confocal microscopes. Very precise adjustment is needed to keep the tiny pinholes in the rapidly spinning Nipkow disk aligned.

6 Biological Applications of Confocal Microscopy

Confocal microscopy is widely used in a variety of fields, including materials science, geology, metrology, forensics, and biology. The enhanced imaging capability of the confocal microscope has resulted in its increased application in the field of biomedical sciences. In general, there is considerable interest in the biological sciences to study and analyze the 3-D structure of cells and tissues. Confocal imaging is a high-resolution microscopy technique that provides both fine structural details and 3-D information without the need to physically slice the specimen into thin sections. In the area of biological imaging, confocal microscopy has been extensively used and has led to increase our understanding of the cell's 3-D structure, as well as its physiology and motility.

Recent technical advances have made 3-D imaging more accessible to researchers, and the collection of 3-D data sets is now routine in several biomedical laboratories. With the dramatic improvements in computing technology, the visualization of 3-D data is no longer a daunting task. Several software packages for 3-D visualization, both commercial and freeware (http://www.cs.ubc.ca/spider/ladic/software.html) are now readily available. These packages include special rendering algorithms that allow (1) the visualization of 3-D structures from several viewing angles, (2) the analysis of surface features, (3) the generation of profiles across the surface, and through the 3-D volume, and (4) the production of animations, anaglyphs (red-green images) and stereo image pairs. Several books and articles have been written covering the different visualization and reconstruction techniques for 3-D data [11–13]. However, little work has yet been done in the quantitative assessment of 3-D confocal microscope images. Moreover, the current emphasis in biology is now on engineering quantification and quantitative analysis of information, so that observations can be integrated and their significance understood. Information regarding the topological properties of structure such as the number of objects and their spatial localization per unit volume, or the connectivity of networks cannot be made by using single two-dimensional images. Such quantitative measurements have to be made in 3-D, using volume data sets. In the following sections, we will present some of the digital image processing methods that may be implemented to obtain quantitative information from 3-D confocal microscope images of biological specimens.

6.1 Quantitative Analysis of 3-D Confocal Microscope Images

Three-dimensional data obtained from confocal microscopes comprise a series of optical sections, referred to as the *z* series. The optical sections are obtained at fixed intervals at successively higher or lower focal planes along the *z* axis. Each two-dimensional (2-D) image is called an "optical slice," and all the slices together comprise a volume data set. Building up the *z* series in depth allows the 3-D structure to be reconstructed.

Most of the image processing algorithms for 2-D images discussed in the preceding chapters can be easily extended into three dimensions. Quantitative measurements in 3-D involve the identification, classification, and tracing of voxels that are connected to each other throughout the volume data set. For the volume data sets, 3-D image measurements are generally performed by using two different approaches, either independently, or in conjunction with each other. The first approach involves performing image processing operations on the individual optical sections (2-D) of the *z* series, and then generating a new (processed) 3-D image set to make measurements. The second approach is to perform image processing by using the voxel (volume element), which is the 3-D analog of pixel (the unit of brightness in two dimensions). In this case, cubic voxel arrays are employed to perform operations such as kernel multiplication, template matching, and others using the 3-D neighborhood of voxels. In either case, quantitative measurements have to be made on the volume data set to determine the 3-D relationship of connecting voxels. A summary of the different image processing algorithms for 2-D images, which can be applied to the individual slices of a 3-D data set without compromising the 3-D measurements, is discussed by Russ in [14]. Certain operations such as skeletonization, however, cannot be applied to single optical slices, and they have to be performed in three dimensions, using voxel arrays to maintain the true connectivity of 3-D structure. See [15, 16] for a discussion.

In the following sections, we will use examples to demonstrate the application of image processing algorithms to perform quantitative measurements at both the cellular and tissue level in biological specimens. It will be evident from the examples presented that each volume data set requires a specific set of image processing operations, depending on the image parameters to be measured. There are no generic image processing algorithms that can be used to make 3-D measurements, so in most cases it is necessary to customize a set of image analysis operations for a particular data set.

6.2 Cells and Tissues

Confocal fluorescence microscopy is increasingly used to study dynamic changes in the physiology of living cell's and tissues, and to determine the spatial relationships between fluorescently labeled features in fixed specimens. Live cell imaging is used to determine cell and tissue viability, and to study dynamic processes such as membrane fusion and fission, calcium-ion fluxes, volumetric transitions, and FRAP (fluorescence recovery after photobleaching). Similarly, immunofluorescence imaging is used determine cellular localization of organelles, cytoskeletal elements, and macromolecules such as proteins, RNA, and DNA. We present examples demonstrating the use of image analysis for confocal microscope images to estimate viability, determine the spatial distribution of cellular components, and to track volume and shape changes in cells and tissues.

6.2.1 Viability Measurements

Fluorescence methods employing fluorescent dyes specifically designed for assaying vital cell functions are now routinely used in biological research. Propidium iodide (PI) is one such dye that is highly impermeant to membranes, and it stains only cells that are dead or have injured cell membranes. Similarly, acridine orange (AO) is a weakly basic dye that concentrates in acidic organelles in living plant and animal cells, and it is used to assess cell viability. Dead cells are stained red with the PI dye, while the live cells are stained green with AO. Laser scanning confocal microscopy (LSCM) allows the reconstruction of the 3-D morphology of both the viable and dead cells. Digital image processing algorithms can then be implemented to obtain an estimate of the proportion of viable and dead cells throughout the islet volume as described below.

Figures 7(a) and 7(b) present series of optical sections that were obtained through an individual islet, at two different excitation wavelengths, 488 nm and 514 nm, for viable and damaged tissue, respectively. We implemented image analysis algorithms consisting of template masking, binarization, and median filtering (Chapter 2.2) to estimate viability, as described next. The first step involved the processing of each 2-D (512×512) image in the sequence of N sections. Template masking was applied to perform object isolation, in which the domain of interest (islet) was separated from the background region. The template mask is a binary image in which the mask area has an intensity of 1 and the background has an intensity of 0. Point wise multiplication of this mask with the individual serial optical sections isolates the islet cross-sections, since the intensity of the background is forced to zero. The advantage of masking, especially in the case of biological samples, is that the processed images are free of background noise and other extraneous data (i.e., surrounding regions of varying intensity that may occur as a result of the presence of exocrine tissue or impurities in the culture media). The masked images were then binarized by using gray-level thresholding operations (discussed earlier in Chapter 2.2). For 3-D (volume) data sets, it is critical to choose a threshold that produces a binary image retaining most of the relevant information for the entire sequence of images. The result of the gray-level thresholding operation is a binary image with each pixel value greater than or equal to the threshold set to 255 and

the remaining pixels values set to 0. Binary median filtering was then applied to smooth the binary image. The algorithm to perform median filtering on binary images in the neighborhood (eight-connected) of a pixel counts the incidence of (255 and 0) values of the pixels and its neighbors, determines the majority, and assigns this value to the pixel. The function of median filtering is to smooth the image by eliminating isolated intensity spikes. Following these preprocessing steps on each 2-D optical section, the 3-D data set was then used to determine the total number fluorescently stained voxels (dead/live) present in the islet. The total number of pixels at an intensity of 255 (indicating the local presence of the fluorescent stain) was recorded for each cross-section of the live and dead cell data sets. The sum of the total pixels for N sections was computed, and the ratio of the sum of the live tissue to that of the dead was determined. This technique was successfully applied to investigate the effect of varying cooling rates on the survival of cryopreserved pancreatic islets [17]. These image processing algorithms can be easily applied to determine the viability in various cells and tissues that have been labeled with vital fluorescent dyes.

6.2.2 Quantification of Spatial Localization and Distribution

In order to take full advantage of the 3-D data available by means of confocal microscopy, it is imperative to quantitatively analyze and interpret the volume data sets. An application where such quantification is most beneficial constitutes the spatial localization and distribution of objects within the 3-D data set. This is particularly applicable to biological specimens, because the exact location or distribution of cellular components (e.g., organelles or proteins) within cells is often desired. We will present an example, each for living and fixed cells wherein a 3-D quantitative

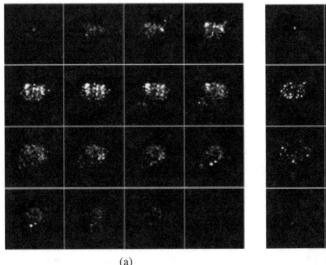 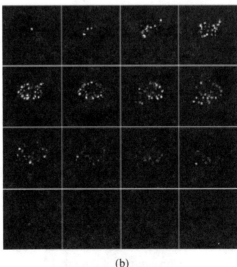

(a)　　　　　　　　　　　　　(b)

FIGURE 7　Series of 14 optical sections through an islet: (a) viable cells imaged at 488 nm; (b) dead cells imaged at 514 nm (reproduced with permission from [17]).

analysis is required, to estimate the distribution of damage within cells, and to determine the cellular localization of a protein; respectively.

Frequently, the elements of interest are represented by either individual voxels (indicating the presence of fluorescently labeled elements) or clusters of connected voxels. It is typically required to determine the location and frequency of occurrence of these objects within the volume data set. There are three steps involved in performing a spatial distribution analysis: (1) identify objects, (2) determine their local position relative to the 3-D imaged volume, and (3) determine the frequency of their occurrence within the 3-D volume. The first step involves identifying the elements of interest in the data sets, whose spatial distribution is desired. If individual voxels are to be analyzed, there is no special processing that has to be performed. However, if the objects of interest consist of clusters of connected voxels, an image processing algorithm called *region labeling* or *blob coloring* (Chapter 2.2) is implemented to identify and isolate these objects.

3-D Region Labeling. Each image element in 3-D is a voxel, and each voxel has 26 neighboring voxels; eight voxels, one at each corner, 12 voxels, one at each edge, and six voxels, one at each surface. A 3-D region array may then be defined wherein a similar value (region number/unique color) is assigned for each nonzero voxel in the image depending on its connectivity. The connectivity of a voxel is tested based on a predetermined neighborhood so that all voxels belonging to the same connected region may have the same region number. The size of the neighborhood is chosen depending on image parameters, and the size of the features of interest. Each region or blob is identified by its unique color, and hence the procedure is called blob coloring [18]. For example, the volume data set presented in Fig. 7(b) was analyzed by region labeling to identify and isolate the dead nuclei within the islet volume. The connectivity of voxels was tested by using a ten-connected neighborhood. Since the diameter of each nucleus is ~7–9 μm and the serial sectioning was performed at a z interval of ~2–5 μm, it was necessary to use only the six surface voxels and four edge voxels for comparison. This decision was made because the use of the voxels at the remaining eight edges and the 12 corners produced artificially connected regions extending from the first to the last section in the 3-D image. These artificial regions were larger in size and did not compare with the typical size of a nucleus. An algorithm for 3-D blob coloring was implemented, to first scan the data set and check for connectedness so that pixels belonging to the same eight-connected region in the X–Y plane had the same color for each nonzero pixel. The remaining two surface neighbors in the Z direction were then checked for connectedness so that voxels belonging to the same two-connected (voxels in the previous and following z sections) region had the same color for each non-zero voxel. The final results of this procedure thus contained information on the connectedness of voxels in the 3-D image. All voxels belonging to the same ten-connected re-

gion were assigned the same color. A threshold was set for the size of each region. Only regions containing more than ten voxels were counted; the rest were assumed to be noise and neglected.

Once the elements of interest [individual voxels/connected voxels (objects)] have been identified, the second step is to determine their spatial location or position within the volume data set. On one hand, for individual voxels, the spatial coordinates along the x, y, and z axes are used to represent position. The position of objects, on the other hand, can be represented in terms of its centroid.

Determination of Centroid. The centroid of an object may be defined as the center of mass of an object of the same shape with constant mass per unit area. The center of mass is in turn defined as that point where all the mass of the object could be concentrated without changing the first moment of the object about any axis [19]. In the 3-D case the moments about the X, Y, and Z axes are:

$$X_c \iint_I f(x, y, z) \mathrm{d}x \, \mathrm{d}y \, \mathrm{d}z = \iint_I x \, f(x, y, z) \mathrm{d}x \, \mathrm{d}y \, \mathrm{d}z,$$

$$Y_c \iint_I f(x, y, z) \mathrm{d}x \, \mathrm{d}y \, \mathrm{d}z = \iint_I y \, f(x, y, z) \mathrm{d}x \, \mathrm{d}y \, \mathrm{d}z,$$

$$Z_c \iint_I f(x, y, z) \mathrm{d}x \, \mathrm{d}y \, \mathrm{d}z = \iint_I z \, f(x, y, z) \mathrm{d}x \, \mathrm{d}y \, \mathrm{d}z,$$

$$\tag{10}$$

where (X_c, Y_c, Z_c) is the position of the center of mass. The expressions appearing on the left of these equations are the total mass, with integration over the entire image I. For discrete binary images the integrals become sums, thus the center of mass for 3-D binary images can be computed using the following:

$$X_c = \frac{\sum_i \sum_j \sum_k i f(i, j, k)}{\sum_i \sum_j \sum_k f(i, j, k)},$$

$$Y_c = \frac{\sum_i \sum_j \sum_k j f(i, j, k)}{\sum_i \sum_j \sum_k f(i, j, k)}, \tag{11}$$

$$Z_c = \frac{\sum_i \sum_j \sum_k k f(i, j, k)}{\sum_i \sum_j \sum_k f(i, j, k)},$$

where $f(i, j, k)$ is the value of the 3-D binary image (i.e., the intensity) at the point in the ith row, jth column and kth section of the 3-D image, i.e., at voxel (i, j, k). Intensities are assumed to be analogous to mass so that zero intensities represented zero mass. The above expressions were used to determine the centroid of the 3-D islet volume shown in Fig. 7(b), and the centroid of each damaged nuclei isolated using the region labeling technique. Thus, the spatial position of each damaged nuclei within the islet was determined.

It should be noted here that the position of the individual voxels defined by the (x, y, z) spatial coordinates, or that of objects in terms of the centroid, represent their "global" location with respect to the entire 3-D data set. In order to determine the spatial distribution locally, it is necessary to estimate their position relative to some specific feature in the imaged volume. For example, Fig. 7(b) presents a z series or volume data set of the damaged nuclei within an islet. The specific feature of interest (or image volume) in this case comprises the islet. The spatial position of the nuclei when expressed only in terms of the centroid then represents their "global" position within the z series. In order to establish their distribution locally within the imaged islet, it is necessary to determine their location in terms of some feature specific to the islet. Thus, the final stage of a spatial distribution analysis is to determine the frequency and location of the objects with reference to the imaged volume. For cellular structures, this can be accomplished by estimating a 3-D surface that encloses the imaged volume. In the islet example, the local distribution of the damaged nuclei can then be described relative to the surface of the islet within which they lie. A technique to estimate the 3-D surface of spherical objects is described as follows.

Estimation of 3-D surface. Superquadrics are a family of parametric shapes that are used as primitives for shape representation in computer graphics and computer vision. An advantage of using these geometric modeling primitives is that they allow complex solids and surfaces to be constructed and altered easily from a few interactive parameters. Superquadric solids are based on the parametric forms of quadric surfaces such as the superellipse or superhyperbola, in which each trigonometric function is raised to an exponent. The spherical product of pairs of such curves produces a uniform mathematical representation for the superquadric. This function is referred to as the inside–outside function of the superquadric or the cost function. The cost function represents the surface of the superquadric that divides the 3-D space into three distinct regions: inside, outside, and surface boundary.

Model recovery may be implemented by using 3-D data points as input. The cost function is defined such that its value depends on the distance of points from the model's surface and on the overall size of the model. A least-squares minimization method is used to recover model parameters, with initial estimates for minimization obtained from the rough position, orientation, and size of the object. During minimization, all the model parameters are iteratively adjusted to recover the model surface, such that most of the input 3-D data points lie close to the surface. To summarize, a superquadric surface is defined by a single analytic function that is differentiable everywhere, and can be used to model a large set of structures like spheres, cylinders, parallelepipeds, and shapes in between. Further, superquadrics with parametric deformations can be implemented to include tapering, bending, and cavity deformation [20].

We will demonstrate the use of superellipsoids to estimate the 3-D bounding surface of pancreatic islets. In the example presented, our aim was to approximate a smooth surface to define the shape of islets, and parametric deformations were not implemented. A 3-D surface for pancreatic islets was estimated by formulating a least-squares minimization of the superquadric

cost function with the imaged 3-D data points as input [21]. The inside–outside cost function, $F(x, y, z)$, of a superquadric surface is defined by the following equation:

$$F(x, y, z) = \left(\left(\left(\frac{x}{a_1} \right)^{2/\epsilon_2} + \left(\frac{y}{a_2} \right)^{2/\epsilon_2} \right)^{\epsilon_2/\epsilon_1} + \left(\frac{z}{a_3} \right)^{2/\epsilon_1} \right)^{\epsilon_1}, \tag{12}$$

where x, y, and z are the position coordinates in 3-D; a_1, a_2, a_3 define the superquadric size; and ϵ_1 and ϵ_2 are the shape parameters.

The input 3-D points were initially translated and rotated to the center of the world coordinate system (denoted by the subscript W) and the superquadric cost function in the general position was defined as follows [21]:

$$F(x_W, y_W, z_W)$$
$$= F(x_W, y_W, z_W : a_1, a_2, a_3, \epsilon_1, \epsilon_2, \phi, \theta, \psi, c_1, c_2, c_3) \tag{13}$$

where a_1, a_2, a_3, ϵ_1 and ϵ_2 are as described earlier; ϕ, θ, ψ represent orientation; and c_1, c_2, c_3 define the position in space of the islet centroid. To recover a 3-D surface it was necessary to vary the above 11 parameters to define a set of values such that most of the outermost 3-D input data points will lie on or close to the surface. The orientation parameters ϕ, θ, ψ were neglected in accordance with the rationale of Solina and Bajcsy [20], for the analysis of bloblike objects. Only the size and the shape parameters were varied, and the cost function was minimized by using the Levenberg–Marquardt method [22]. Further, since multiple sets of parameter values can produce identical shapes, typically certain severe constraints are essential to obtain an unique solution. However, since the recovered 3-D surface was used only to represent space occupancy or shape, such ambiguities did not impose a problem [20]. The initial estimates for the size parameters were obtained from the input data points, whereas the shape parameters were initially set to 1. The final parameter values for the 3-D surface were determined based on the criterion that the computed surface would enclose >90% of the 3-D input data points. Figure 8 presents a graph of an estimated superquadric surface illustrating the imaged tissue voxels enclosed within or lying on the 3-D surface along with the outlying tissue voxels. The estimated surface was then used as a local reference boundary, relative to which the spatial distribution of individual voxels or objects within the islet was determined.

Localization and Distribution. The spatial localization of an element in 3-D space can be estimated by describing its position with reference to a morphological feature, such as an enclosing surface. This information can then be organized into groups to determine the distribution of elements by computing the frequency of elements that occur at similar spatial positions. In the example presented, the 3-D spatial distribution of tissue was determined by identifying each voxel (viable and damaged) and computing its relative location in the islet. The spatial location

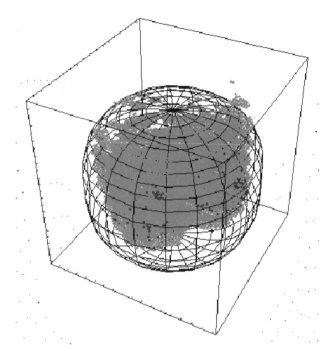

FIGURE 8 Graph of an estimated 3-D superquadric surface illustrating the viable (green) and dead (red) tissue voxels enclosed within or lying on the 3-D surface along with a few outlying voxels (reproduced with permission from [17]).

of tissue within the islet was measured by computing the normalized distance of each voxel from the recovered superquadic surface, as described below.

After the surface model was identified, the distance of each viable or damaged image voxel from the centroid of the 3-D islet volume was obtained. The distance was then normalized with respect to the length of a vector containing the voxel and extending from the centroid to its intersection with the estimated superquadric 3-D surface. Defining the origin O to be fixed at the centroid, and ρ_c to be the length of the vector originating at O, passing through a voxel P, and terminating at the point of intersection with the superquadric surface, S, we then have the coordinates of voxel S as (ρ_c, θ, ϕ). Voxels P and S have similar θ and ϕ values and different ρ values. Thus, ρ_c is easily obtained from

$$\rho_c = \frac{1.0}{\sqrt{\left(\left(\left(\frac{\sin\phi\cos\theta}{a_1}\right)^{2/\epsilon_2} + \left(\frac{\sin\phi\sin\theta}{a_2}\right)^{2/\epsilon_2}\right)^{\epsilon_2/\epsilon_1} + \left(\frac{\cos\phi}{a_3}\right)^{2/\epsilon_1}\right)^{\epsilon_1}}},$$

(14)

where the parameters a_1, a_2, a_3, ϵ_1 and ϵ_2 were estimated by means of the nonlinear least-squares minimization of the superquadric cost function.

After ρ_c was obtained, the normalized distance of voxel P from the centroid was computed as ρ/ρ_c. All voxels inside the estimated 3-D surface had a normalized distance value less than 1,

and surface voxels had a value of 1. Thus the "local" spatial location of each voxel within the islet volume was determined.

For estimating the spatial distribution, each tissue voxel was then assigned to a regional group as a function of its computed normalized distance from the centroid. Thereby 10 serial annular shells were obtained, each having a normalized shell width of 0.1. Thus, the spatial distribution of viable and damaged tissue was computed in the form of a histogram, i.e., the number of voxels were determined for each shell depending upon the normalized distance from the centroid. This technique was used to determine the 3-D nature of cryopreservation induced injury in pancreatic islets, and the information was used to obtain a better understanding of the fundamental phenomena underlying the mechanisms of freeze-thaw induced injury [17]. A similar analysis was implemented to determine the spatial distribution of a bacterial protein in mouse fibroblasts cells, fluorescently labeled by using indirect immunofluorescence methods [23].

These methods may be easily extended to other applications, biologically oriented or otherwise, to determine the spatial distribution of 3-D data.

6.2.3 Dynamic Volumetric Transitions and Shape Analysis

The confocal microscope has the ability to acquire 3-D images of an object that is moving or changing shape. A complete volumetric image of an object can be acquired at discrete time instances. By acquiring a sequence of images this way, the time dimension is added to the collected data, and a 4-D data set is produced. The addition of the time dimension makes analyses of the data even more difficult, and manual techniques become nearly impossible. Some of the volumetric morphological techniques described in the previous sections can be easily extended into the time domain. Quantities such as the total volume, surface area, or centroid of an object can be measured over time by simply computing these quantities for each time sample. Simple extensions into the time domain such as this cannot give a detailed picture of how a nonrigid object has changed shape from one time frame to the next. The most difficult analysis is to determine where each portion of an object undergoing nonrigid deformations has moved from one frame to the next.

An overview of a technique that produces detailed localized information on nonrigid object motion is presented. The technique is described in detail in [24, 25]. The technique works by initially defining a material coordinate system for the specimen in the initial frame and computing the deformations of that coordinate system over time. It assumes that the 3-D frames are sampled at a sufficiently fast rate so that displacements are relatively small between image frames.

Let $f_i(x, y, z)$ represent the 3-D image sequence in which each 3-D frame was sampled at time t_i where i is an integer. The material coordinate system which is "attached" to the object changing shape is given by (u_1, u_2, u_3). The function that defines the location and deformation of the material coordinate system within the fixed (x, y, z) coordinate system is defined as $\alpha(u_1, u_2, u_3) = (x, y, z)$. To define the position of the material coordinate system at a particular time t_i, the subscript i is added, giving $\alpha_i(u_1, u_2, u_3) = (x_i, y_i, z_i)$. The deformation of the material coordinate system between times t_{i-1} and t_i is given by the function Δ_i, i.e., $\alpha_i = \alpha_{i-1} + \Delta_i$.

The goal of the shape-change technique is to find the functions Δ_i given the original image sequence and the initially defined material coordinates $\alpha_0(u_1, u_2, u_3)$. The functions are found by minimizing the following functional using the calculus of variations [26]:

$$E(\Delta_i) = \lambda P(\Delta_i) + S(\Delta_i), \tag{15}$$

where E is a nonnegative functional that is a measure of the shape-change smoothness, S, and the penalty functional P that measures how much the brightness of each material coordinate changes as a result of a given deformation Δ_i. The parameter λ is a positive real number that weights the tradeoff between the fidelity to the data given by P and the shape-change limit imposed by S. Specifically, the brightness continuity constraint is given by

$$P(\Delta_i) = \int_{u_1} \int_{u_2} \int_{u_3} [\, f_i(\alpha_{i-1} + \Delta_i) - f_{i-1}(\alpha_{i-1})]^2 \, du_1 \, du_2 \, du_3. \tag{16}$$

The shape-change constraint is given by

$$S(\Delta_i) = \int_{u_1} \int_{u_2} \int_{u_3} (g_i - g_{i-1})^2 du_1 \, du_2 \, du_3, \tag{17}$$

where g_i is a 3×3 matrix and function of (u_1, u_2, u_3) called the *first fundamental form* [27] of the material coordinate system. The first fundamental form is a differential geometric property of the coordinate system which completely defines the shape of the coordinate system up to a rigid motion in (x, y, z) space.

The formulation of the shape-change technique is similar to the well-known *optical flow* algorithm presented in [19], except that in this case the smoothness constraint is based on the actual shape of the object rather than simple derivatives of the image. Also, this formulation is presented in three dimensions and produces a model of the shape change for an entire image sequence.

The solution of Eq. (15) requires solution of 3 coupled, nonlinear partial differential equations. A finite difference approach can be used to solve the equations. The resulting solution depends highly on the selection of the parameter λ in Eq. (15). Selection of λ is generally done by trial and error. Once an appropriate value for λ is found, however, it can generally be held constant throughout solution for the entire image sequence. Figure 9 shows the result of running the shape-change algorithm on human pancreatic islets undergoing dynamic volumetric changes in response to osmotic changes caused by the presence of a cryoprotective additive (dimethyl sulfoxide) [28].

6.3 Microvascular Networks

Microvascular research is another area in biology that employs various imaging methods to study the dynamics of blood flow, and vascular morphology. One of the problems associated with evaluating microvascular networks relates to the measurement of the tortuous paths followed by blood vessels in thick tissue samples. It is difficult to acquire this information by means of conventional light/fluorescence microscopy without having to physically section the specimen under investigation. The use of

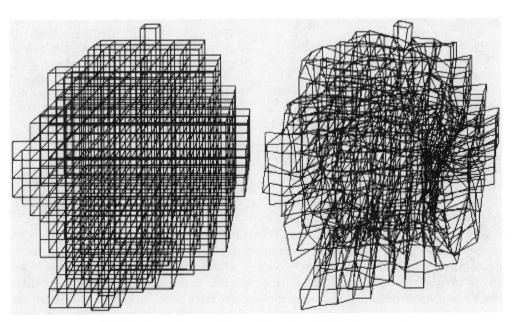

FIGURE 9 Shape-change analysis in a human islet subjected to osmotic stress.

confocal microscopy overcomes this problem by providing, in three dimensions, additional spatial information related to the vascular morphology. However, this now presents the issue of allowing quantitative measurements to be made in the 3-D space. In the past, even with 2-D data, morphometric evaluation of blood vessel density and diameter has involved manual counting and estimation procedures. There is considerable ambiguity involved in the manual measurement of vessel diameters. Estimating the location of vessel boundaries within the image of microvessels presents a difficult problem. Manual counting of blood vessels is often tedious and time consuming, and the error in measurements typically increases with time. The problem is only compounded in 3-D space. Hence, it is necessary to develop computer algorithms to automate the quantitative measurements, thus providing an efficient alternative for measurements of the vascular morphology. We present an example in which digital imaging was used to measure the angiogenesis and revascularization processes occurring in rat pancreatic islets transplanted at the renal subcapsular site [29]. Confocal microscopy was employed to image the 3-D morphology of the microvasculature, and image processing algorithms were used to analyze the geometry of the neovasculature. Vascular morphology was estimated in terms of 3-D vessel lengths, branching angles, and diameters, whereas vascular density was measured in terms of vessel to tissue area (2-D) and volume (3-D) ratios. The image processing algorithms employed are described in the following sections. It should be noted that the methodology described here is suited for microvascular networks wherein the vessel lengths are perpendicular to the optical axis. For vascular networks, where the vessel direction is parallel to the optical axis so that only vessel cross-sections (circular or elliptical) are known by the 3-D image, different image processing algorithms are needed [30].

6.3.1 Data Acquisition and 3-D Representation

The revascularization of pancreatic islet grafts transplanted at the renal subcapsular site in rats was evaluated experimentally by means of intravital LSCM of the blood vessels [29]. Three-dimensional imaging of the contrast-enhanced microcirculation (5% fluorescein labeled dextran) was performed to obtain serial optical cross-sections through the neovascular bed at defined z increments. In this example, the acquisition of the optical sections was influenced by the curvilinear surface of the kidney. During optical sectioning of the graft microvasculature, images were captured along an inclined plane rather then vertically through the area being sectioned. This occurred as adjacent areas on the surface of the kidney came into focus during optical sectioning. This effect is demonstrated in Fig. 10, which presents the results of a computerized 3-D reconstruction performed on 25 optical sections (z interval of 5 μm) obtained through the vascular bed of an islet graft. As seen in Fig. 10, the curvaceous shape of the kidney is easily distinguished in the 3-D reconstruction. Thus, in order to evaluate the 3-D vascular morphology, a 2-D image was projected from the 3-D reconstruction. The compos-

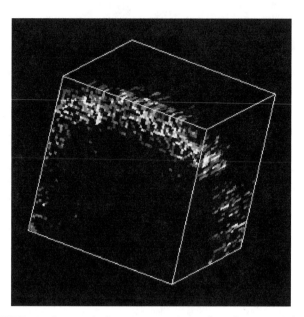

FIGURE 10 Computerized 3-D reconstruction performed on 25 optical sections (z interval of 5 μm) obtained through the vascular bed of the kidney. As seen, the curvaceous shape of the kidney is easily distinguished in the 3-D reconstruction.

ite 2-D image representing the 3-D morphology was obtained by projection of the individual sections occurring at varying depths (along the z plane) onto the x–y plane. As shown in Fig. 11, the resulting image consisted of blood vessels that were contiguous in the third dimension. All the morphological measurements were performed on the composite image.

6.3.2 Determination Of Vascular Density

The measurement of the vascular density included a combination of the gray-level thresholding, binarization, and median filtering algorithms described in the preceding chapters. Binary images were initially generated by image segmentation, using gray-level thresholding. Two-dimensional images of similar spatial resolutions were then smoothed with a 3×3 or 5×5 median filter (Chapter 3.2). The total of the number of pixels at 255 was used as an estimate of the vessel area, and the remaining pixels represented the tissue area. The vessel to tissue area ratios were then computed for each section (areas) or for an entire sequence of sections (volumes).

6.3.3 Determination of Vascular Morphology

Vascular morphology was determined in terms of 3-D vessel lengths, vessel diameter, and tortuosity index as described below.

Unbiased Estimation of Vessel Length. Composite images of the projected microvasculature were segmented by using gray-level thresholding to extract the blood vessels from the background. The segmented image was then used to obtain a skeleton of the vascular network by means of a thinning operation [31]. The skeletonization algorithm obtains the skeletons from binary images by thinning regions, i.e., by progressively eliminating

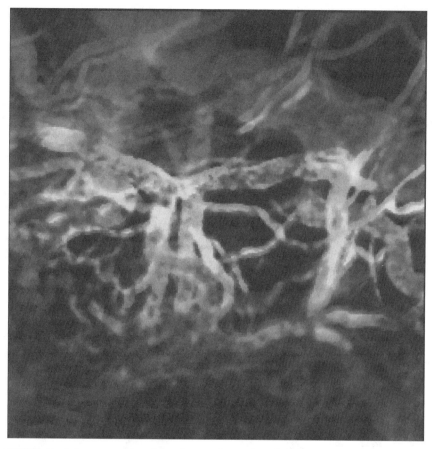

FIGURE 11   3-D representation of the microvasculature of an islet graft at the renal subcapsular site. The image is color coded to denote depth. The vessels appearing in the lower portion (blue) are at a depth of 30 μm, whereas those in the middle and upper portions of the image are at a depth of ~85 (green) and 135 μm (violet), respectively (reproduced with permission from [29]). (See color section, p. C–55.)

border pixels that do not break the connectivity of the neighboring (eight-connected) pixels, thus preserving the shape of the original region. The skeletonized image was labeled by using the procedure of region labeling and chain coding. The region labeling procedure was implemented with a eight-connected neighborhood for identifying connected pixels. It was used to identify and isolate the different blood vessel skeletons and to determine the length of each segment. Further, the chain coding operation was applied to identify nodes and label vessel segments. The labeled image was scanned to isolate the nodes, by checking for connectivity in the eight-connected neighborhood. Pixels with only one neighbor were assigned as the terminating nodes. Those having greater than two neighbors were classified as junction nodes with two, three, or four branches, depending on the connectivity of pixels. The labeled image was pruned to remove isolated short segments without affecting the connectivity of the vascular network. The vessel length was determined as the sum of the total number of pixels in each labeled segment. This approach introduces some systematic bias, because the projection of the 3-D data onto a 2-D composite results in the lost of some information. An unbiased estimation of the 3-D vessel lengths

was implemented by applying a modification of the technique described by Gokhale [32] and Cruz–Orive and Howard [33]. This technique eliminated the error introduced in the measurement of the vessel lengths caused by the bias generated during the vertical projection of volume data sets.

Gokhale [32] and Cruz-Orive *et al.* [33] have addressed the issue of estimating the 3-D lengths of curves using stereological techniques. These studies describe a method to obtain an unbiased estimate the 3-D length of linear features from "total vertical projections," obtained by rotating the curve about a fixed axis and projecting it onto a fixed vertical plane. The length of linear structures is measured for each of the vertical projections. The final estimate of the 3-D length is then obtained as the maximum of the different projected lengths. This technique was adapted for our application and implemented as follows. The 3-D reconstruction (Fig. 10) was rotated about a fixed axis (y axis) in varying amounts, and the vertical projections were performed to obtain the composite image for each orientation.

The 3-D rotations were implemented by means of 3-D transformations represented by 3×3 matrices using nonhomogenous coordinates. A right-handed 3-D coordinate system was implemented. By convention, positive rotations in the right-handed system are such that, when looking from a positive axis toward the origin, a 90° counterclockwise rotation transforms one positive axis into another. Thus, for a rotation of the x axis the direction of positive rotation is y to z, for a rotation of the y axis the direction of positive rotation is z to x, and for a rotation of the z axis the direction of positive rotation is x to y. The z axis (optical axis) was fixed as the vertical axis. The y axis was fixed as the axis about which the 3-D rotations were performed, and the vertical projections were obtained in the x–y plane. The 3-D morphology of the microvascular bed, i.e., the blood vessels, were projected onto the fixed plane (x–y plane) in a systematic set of directions between $0°$ and $180°$, about the y axis as shown in Fig. 12.

The 3×3 matrix representation of the 3-D rotation at angle θ about the y axis is

$$R_y(\theta) = \begin{bmatrix} \cos\theta & 0 & -\sin\theta \\ 0 & 1 & 0 \\ -\sin\theta & 0 & \cos\theta \end{bmatrix}. \quad (18)$$

Thus, the geometrical transformation of the 3-D volume is

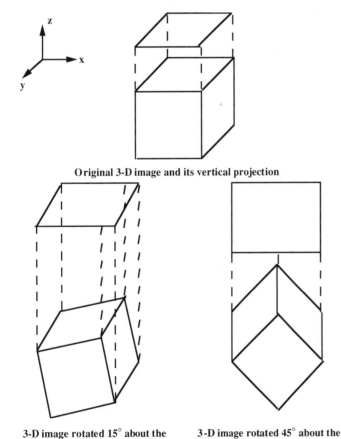

Original 3-D image and its vertical projection

3-D image rotated 15° about the y-axis and its vertical projection   **3-D image rotated 45° about the y-axis and its vertical projection**

FIGURE 12   3-D solid rotated about the y axis and its vertical projections in the x–y plane.

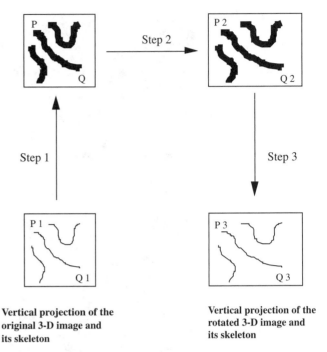

Vertical projection of the original 3-D image and its skeleton

Vertical projection of the rotated 3-D image and its skeleton

FIGURE 13   Correspondence of vessel segments in the various projected images by means of mapping and 3-D transformations.

computed as follows:

$$\begin{bmatrix} x' \\ y' \\ z' \end{bmatrix} = \begin{bmatrix} \cos\theta & 0 & -\sin\theta \\ 0 & 1 & 0 \\ -\sin\theta & 0 & \cos\theta \end{bmatrix} \times \begin{bmatrix} x \\ y \\ z \end{bmatrix}, \quad (19)$$

where x', y', and z' are the transformed coordinates, x, y, and z are the original coordinates of the reconstructed 3-D image, and θ is the angle of rotation about the y axis. The projected length of individual vessel segments may vary in the different projections obtained. Vessel segments were uniquely labeled in each of the composites at different orientations, and the connecting node junctions identified. The unbiased 3-D lengths were determined as the maximum of the projected lengths estimated for the various rotations. In order to achieve this, the individual vessel segments at the different rotations have to be matched. The problem involves the registration of each individual vessel segment as it changes in its projected orientation. It was resolved by performing a combination of mapping and inverse mapping transformations. For example, as shown in Fig. 13, an unbiased estimate of the 3-D length of segment PQ may be determined as the maximum of the length of the projected segments P1Q1 and P3Q3. It is essential that P1Q1 and P3Q3 are matched as projections of the same vessel segment. This was achieved in three steps. The first step was to match the points P1 and Q1 in the skeletonized image to the points P and Q in the binary image. This was achieved by a simple mapping of points because the skeleton P1Q1 maps onto the centerline of the binary segment PQ. In step 2, the points P and Q were mapped onto the points P2 and Q2 by performing the required transformation to rotate the 3-D image. Finally, the binary segment P2Q2 was mapped on to its skeleton P3Q3. Thus a correspondence was established between the two projected lengths P1Q1 and P3Q3. The maximum value of these lengths was a measure of the unbiased length. The tortuosity index was then defined as the ratio of the length of a straight line vector between two points to the length of the vessel segment between the same points. An index of 1 represents a straight vessel, and <1.0 represents a curvilinear or tortuous vessel.

Automated Estimation of the Vessel Diameter. In order to automate the vascular diameter measurements, a technique employing linear rotating structuring elements (ROSE) described by Thackray and Nelson [34], was implemented. In this method, various linear structuring elements/templates of known orientations were constructed to represent shapes frequently occurring in the images [35]. The template was then passed over the labeled image until a match was obtained. At this step, a path was identified through a matched point on the skeleton such that its direction was along the normal to the edge of the vessel (in the image or x–y plane) at the corresponding point in the segmented (binary) image. The diameter was then measured by traversing the two sides from the corresponding point in the segmented image along the defined path until an intensity change occurred from white (255) to black (0). The total distance traversed on both sides was then used as the diameter estimate at that point. The diameter measurements were obtained by starting at a point a few pixels (determined as 20% of the total number of pixels in the vessel) from the nodes in order to avoid any erroneous measurements caused by the presence of overlapping blood vessels. Further, since each blood vessel segment was labeled individually, the diameter measurements were obtained at various intervals along each segment at points where a template match was found, and the average was determined to obtain an estimate of the vessel diameter along that length.

Determination of Contiguous Vessel Segments. The 3-D lengths of the blood vessels were then obtained by identifying all vessel segments that were contiguous in depth. The continuity of vessel segments was determined by two parameters, namely the vessel diameters and the branching and junction angles. The diameters of the various vessel segments meeting at each junction node were examined, and vessel segments having similar diameters and a common junction node were identified as contiguous vessels. Further, at junction nodes of three or more vessel segments, the branching angles were measured with the following:

$$\tan \phi_{12} = \left| \frac{m_1 - m_2}{1 + m_1 m_2} \right|, \tag{20}$$

where m_1 and m_2 are the slopes of two vessel segments that are at angle ϕ_{12} to each other.

Two vessel segments were considered to be vessel branches when their junction angle was $<90°$, and their diameters were different. A vessel segment was identified as a parent vessel if its junction angle with other vessels was $>90°$. Vessels segments were identified as crossing or overlapping vessels when their junction angles were $\sim90°$.

We applied these algorithms to assess and compare the microvasculature of cultured and cryopreserved islets transplanted at the renal subcapsular site in rats [36]. These algorithms may be employed to estimate the morphology of various other vascular networks, including tumor microvasculature, angiograms of patients evaluated for heart disease, and the retinal microvasculature.

7 Conclusion

The past 5 years have seen a virtual explosion in the application of confocal microscopy to biological specimens. There is no doubt that the need for quantification of 3-D biological data will steadily grow. Digital image processing can provide numerical data to quantify and substantiate biological processes. Most often, digital analysis algorithms have to be customized to meet the requirements of the application. We have presented several examples to demonstrate the application of image processing algorithms for analyzing confocal microscope images of biological specimens. The methodology developed here would be applicable to the general problem of 3-D image analysis in both cellular and network structures.

References

[1] M. Minsky, "Microscopy apparatus," U. S. Patent No. 3,013,467, December 19, 1961.

[2] M. Petrán and M. Hadravsky "Method and arrangement for improving the resolving power and contrast," U. S. Patent No. 3,517,980, June 30, 1970.

[3] M. Born and E. Wolf, *Principles of Optics* (Pergamon, New York, 1970).

[4] C. J. R. Sheppard and T. Wilson, "Image formation in confocal scanning microscopes," *Optik*, **55**, 331–342 (1980).

[5] T. Wilson and C. J. R. Sheppard, *Theory and Practice of Scanning Optical Microscopy* (Academic, London, 1984).

[6] M. Petrán, A. Boyde, and M. Hadravsky, *Confocal Microscopy* (Academic, London, 1990). Chap. 9, pp. 245–283.

[7] T. Wilson, ed., *Confocal Microscopy* (Academic, London, 1990).

[8] G. J. Brackenhoff, H. T. M. Van Der Voort, E. A. Van Spronsen, and N. Nanninga, "Three-dimensional imaging by confocal scanning fluorescence microscopy," *Ann. N.Y. Acad. Sci.* **483**, 405–415 (1986).

[9] T. Wilson, *The Handbook of Biological Confocal Microscopy* (IMR Press, Madison, WI, 1989). Chap. 11, pp. 99–112.

[10] G. S. Kino and G. Q. Xiao, *Confocal Microscopy* (Academic, London, 1990). Chap. 14, pp. 361–387.

[11] A. Kriete. *Visualization in Biomedical Microscopies: 3D Imaging and Computer Applications* (VCH Publishers, Weinheim, 1992).

[12] E. M. Johnson, and J. J. Capowski, "Principles of reconstruction and 3D display of serial sections using a computer," in *The Microcomputer in Cell and Neurobiology Research*, R. R. Mize, ed. (Elsevier, New York, 1985), pp. 249–263.

[13] J. K. Stevens, L. R. Mills, and J. E. Trogadis, *Three-Dimensional Confocal Microscopy: Volume Investigation of Biological Systems* (Academic, New York, 1994).

[14] J. C. Russ, *The Image Processing Handbook* (CRC Press, Boca Raton, FL, 1994). Chap. 10, pp. 589–659.

[15] K. J. Halford and K. Preston, "3-D skeletonization of elongated solids," *Comput. Vis. Graphics Image Process.* **27**, 78–91 (1984).

[16] S. Lobregt, P. W. Verbeek, and F. C. A. Groen, "Three-dimensional skeletonization: principle and algorithm," *IEEE Trans. PAMI* **2**, 75–77 (1980).

[17] F. A. Merchant, K. R. Diller, S. J. Aggarwal, and A. C. Bovik, "Viability analysis of cryopreserved rat pancreatic islets using laser scanning confocal microscopy," *Cryobiology* **33**, 236–252 (1996).

[18] N. H. Kim, A. B. Wysocki, A. C. Bovik, and K. R. Diller, "A microcomputer based vision system for area measurements," *Computat. Biol. Med.* **17**, 173–183 (1987).

[19] B. K. P. Horn, "Binary images geometric properties," *Robot Vision.* (MIT Press, Cambridge, MA, 1986), pp. 46–64.

[20] F. Solina and R. Bajcsy, "Recovery of parametric models from range images: the case for superquadrics with global deformations," *IEEE Trans. PAMI* **12**, 131–147 (1990).

[21] T. E. Boult, and A. D. Gross, "Recovery of superquadrics from depth information," in *Proceeding of the Spatial Reasoning Multi-Sensor Fusion Workshop* (1987), pp. 128–137.

[22] L. E. Scales, *Introduction to Non-Linear Optimization* (Springer, New York, 1985).

[23] F. A. Merchant, H. Bayley, and M. Toner, "Analysis of spatial and dynamic interaction of an engineered pore forming protein with cell membranes," in press.

[24] K. A. Bartels, A. C. Bovik, S. J. Aggarwal, and K. R. Diller, "The analysis of biological shape changes from multi-dimensional dynamic images," *J. Comput. Med. Imag. Graphics* **17**, pp. 89–99 (1993).

[25] K. A. Bartels, C. E. Griffin, and A. C. Bovik, "Spatio-temporal tracking of material shape change via multi-dimensional splines,"

[26] R. Weinstock, *Calculus of Variations with Applications to Physics and Engineering* (Dover, New York, 1974).

[27] M. P. Do Carmo, *Differential Geometry of Curves and Surfaces* (Prentice-Hall, Englewood Cliffs, NJ, 1976).

[28] F. A. Merchant, S. J. Aggarwal, K. R. Diller, K. A. Bartels, and A. C. Bovik, "Analysis of volumetric changes in rat pancreatic islets under osmotic stress using laser scanning confocal microscopy," *Biomed. Scie. Instrumen.* **29**, 111–119 (1993).

[29] F. A. Merchant, S. J. Aggarwal, K. R. Diller, and A.C. Bovik, "In-vivo analysis of angiogenesis and revascularization of transplanted pancreatic islets using confocal microscopy," *J. Microsc.* **176**, 262–275 (1994).

[30] W. E. Higgins, C. Morice, and E. L. Ritman, "Shape based interpolation of tree-like structures in 3-D images," *IEEE Trans. Med. Imag.* **12**, 439–450 (1993).

[31] T. Pavlidis, "A thinning algorithm for discrete binary images," Comput. Graphics Image Process. **13**, 142–157 (1980).

[32] A. M. Gokhale, "Unbiased estimation of curve length in 3-D using vertical slices," *J. Microsc.* **159**, 133–141 (1990).

[33] L. M. Cruz-Orive and C. V. Howard, "Estimating the length of a bounded curve in three dimensions using total vertical projections," *J. Microsc.* **163**, 101–113 (1990).

[34] B. D. Thackray, and A. C. Nelson, "Semi-automatic segmentation of vascular network images using a rotating structuring element (ROSE) with mathematical morphology and dual feature thresholding," *IEEE Trans. Med. Imag.* **12**, 385–392 (1993).

[35] F. A. Merchant, S. J. Aggarwal, K. R. Diller, and A. C. Bovik, "Semi-automatic morphological measurements of 2-D and 3-D microvascular images," *Proc. IEEE Int. Conf. Image Process.* **1**, 416–420 (1994).

[36] F. A. Merchant, K. R. Diller, S. J. Aggarwal, and A. C. Bovik, "Angiogenesis in cultured and cryopreserved pancreatic islet grafts," *Transplantation* **63**, 1652–1660 (1997).

in *IEEE Workshop on Biomedical Image Analysis* (IEEE, New York, 1994).

10.8

Bayesian Automated Target Recognition

Anuj Srivastava
Florida State University

Michael I. Miller
The Johns Hopkins University

Ulf Grenander
Brown University

1 Introduction

When human beings look at camera images of known objects, such as a table, a chair, or a car, we recognize them immediately. For example, the top left panel in Fig. 1 shows a picture in which we can easily identify a car. Even if the pictures are corrupted or noisy, or the objects are partially obscured by other objects, we can still recognize the car. This observation points to an important fact: *the human visual recognition system is an awesome system with extraordinary processing power.* Can we design an automated system, equipped with cameras, computers, databases and algorithms, to achieve a similar performance in object recognition? The answer so far has been no! In this chapter we analyze this issue in the context of a very specific problem in automated image analysis, called automated target recognition (ATR). By restricting ourselves to ATR we can utilize the additional contextual information available in designing ATR algorithms. In a general ATR situation, a number of remote sensors (cameras, radars, ladars, etc.) observe a scene containing a number of dynamic or stationary targets; a more detailed introduction can be found in [2]. These sensors produce observations, in the form of images or signals, which are then analyzed by computer algorithms to *detect, track,* and *recognize* the targets of interest in that scene. Our goal is to derive ATR algorithms and analyze them for their performance. Our approach relies on two main building blocks: (i) efficient mathematical representations of the scenes containing targets, and (ii) efficient algorithms for infer-

ences on these representation spaces. This article describes these two steps to ATR.

One fundamental issue in automated target recognition is the following. Consider a normal hand-held camera taking pictures of a car. Depending on the relative orientation between the camera and the car, and the distance between them, the car appears vastly different in different pictures. The possible variability in relative orientation, also called the *pose*, causes a tremendous variability in the profiles of the targets as seen by a camera, or a sensor in general. This fact underlines one difficulty in the design of a completely automated algorithm of target recognition: how to mathematically model the variability in the sensor outputs caused by the variability in target pose? The task is further complicated by relative motion between the sensors and the targets, imperfections in sensor operations, and the presence of structured clutter in the scene, which often obscures the targets.

We will utilize elements of *deformable template theory* to mathematically model the variations in target pose. For each possible object, we define a template (using CAD models and other descriptors) of standard size, pose, and location. All occurrences of a target in a scene can then be represented by scaling, rotating and translating its template appropriately. All possible scales, rotations, and translations form sets that have interesting geometrical properties. As described later, they have a *group* structure. In short, these transformations are utilized to transform the templates to match the occurrence of targets in a scene. The objects and the scenes containing them are three dimensional

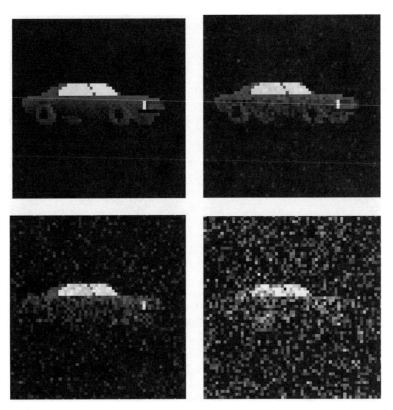

FIGURE 1 Synthetic images of a toy model of a car; the pictures become noisier from top left to right bottom.

even though our observations of them are one or two dimensional. Using the physics of the sensor operation, we will derive operators that transform three-dimensional scenes into sensor outputs, thus mathematically modeling the sensor operation. These operators can be deterministic or random with known probability distributions.

In view of several competing ATR approaches presented in recent years, it becomes important to develop a coherent framework for performance analysis. This analysis should include both prognostics (e.g., the best performance that can be achieved irrespective of the algorithm) and diagnostics (e.g., the performance analysis of a given algorithm). Several authors have presented metrics for ATR performance analyses, although in limited frameworks [16–20]. A detailed review of current ATR approaches is also presented in a recent report [3], in the context of synthetic aperture radar (SAR) ATR. One advantage of the Bayesian framework is that it provides metrics and bounds for comparing algorithmic performance, both between the algorithms and with the best that can be achieved.

Section 2 introduces the deformable template approach to representing the target variabilities, Section 3 defines statistical models for some commonly used sensors. Section 4 sets up a Bayesian framework to solve pose and location estimation, and target recognition problems. Section 5 defines and computes minimum mean square error (MMSE) estimates for the target pose and location, and Section 6 summarizes the procedure for target recognition.

2 Target Representations

Representation is an essential element of image understanding and target recognition. The generation of efficient models, for representations of target shapes, supporting recognition invariant to orientations and locations is crucial. Targets are observed at arbitrary positions and orientations, in highly variable environments. The variability in target pose, with respect to the sensor, is important because at different orientations the targets appear very different. Even the same target can appear completely different at two different orientations. Because of the nonlinear relationship between target orientation and image pixel values, the orientation parameter has to be modeled explicitly and estimated for target recognition. The task is complicated by relative motion between the sensors and the targets, imperfections in sensor operations, and the presence of clutter elements in the scene. Furthermore, different sensors capture widely different aspects of the target. A video captures the visible light reflection, radar captures the electromagnetic scattering, forward-looking infrared (FLIR) captures the thermodynamic profile, and so on. For these widely varying sensor outputs, what should be chosen to represent the targets?

An emerging paradigm for target representations is the **deformable template** theory. In this approach the starting point is to select a standard template for each of the targets and then define a family of transformations to account for the variability associated with target occurrences.

FIGURE 2 Templates for various targets.

1. Templates: start by defining a set of target labels:

$$\mathcal{A} = \{\text{airplane, chair, car, lamp,}$$

$$\text{table, jeep, truck, tank,} \ldots\}.$$

Each $\alpha \in \mathcal{A}$ denotes a particular target. For each $\alpha \in \mathcal{A}$, we define I^α to be a template associated with that target. It includes all the physical attributes of the target that are reflected in the sensor output, including shape, size, material, surface reflectivity, and thermal profile. Clearly, the constituents of I^α depend upon the sensor(s) being used. For a visible spectrum video camera, I^α may consist of a finite element description of its surface, surface texture, and the colors. Shown in Fig. 2 are three-dimensional renderings of sample target templates. In this case each template consists of a set of polygonal patches covering the surface, the material description (texture and reflectivity), and surface colors.

2. Transformations: the targets when they appear in a scene do so at arbitrary positions, orientations, light conditions, and thermal profiles. The next issue is to account for this variability by defining a family of transformations, on the templates, to generate all possible occurrences of the targets. To understand the basic idea, consider this simple example from high-school geometry. We define two triangles to be *similar* if they have equal corresponding angles, for example the two triangles shown in Fig. 3. If we rotate, translate, and (uniformly) scale the left triangle appropriately, we will obtain the right triangle and vice versa. The transformation that takes one triangle to another is called the *similarity transformation*. The set of all possible similarity transformations, call it S, forms a *group*. A group is a set endowed with a group operation (denoted here by $\circ$, often called the product) such that for any two elements in the group their product also lies in the group. Additionally, there exists an identity element, e, such that its product with any element of the group does not change that element; please refer to [12] for more details. As an example, $I\!R^n$ is a group with vector-addition

as the group operation and zero vector as the identity element. Similarly, the set of $n \times n$ nonsingular matrices is a group with matrix multiplication as the group operation and the identity matrix as the identity element. The group structure is instrumental in defining compositions of the transformations: one transformation (s_1) applied after another transformation (s_2) has the equivalent effect of a third transformation (s_3) applied alone. The third transformation is a product of the first two; $s_3 = s_2 \circ s_1$.

Now we extend the same idea to more complicated objects and seek groups that model their variations. We need groups to rigidly rotate and translate three-dimensional objects. Let O be a 3×3 matrix such that $OO^\dagger =$ identity ($\dagger$ denotes matrix transpose) and the determinant of O is 1. Then, for any point $x \in I\!R^3$ on an object, Ox is just a rotated version of x. O is called a rotation matrix, and the set of all such rotation matrices is denoted by $SO(3)$, the *special orthogonal group* in three dimensions. $SO(3)$ is a group with matrix multiplication as the group operation and a 3×3 identity matrix as the identity element. If we fix an axis of rotation, as is the case for ground-based objects, then there is only one rotational freedom left. This rotation is modeled by 2×2 rotation matrices, and their set is denoted by $SO(2)$. For translations, if we translate an object by a vector $p \in I\!R^3$, then each point x on the object becomes $x + p$. The set

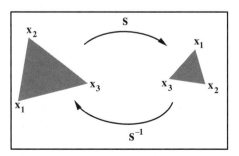

FIGURE 3 Two similar triangles in Euclidean geometry.

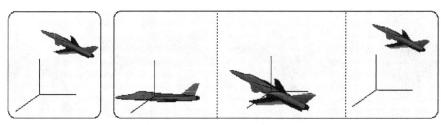

FIGURE 4 Left, an airplane at an arbitrary orientation and position; right, the airplane template rotated and translated from an initial pose and location to match the pose and location on the left.

of all possible translations in three dimensions is the whole of $\mathbb{R}^3$. Similarly, if the translations are restricted to ground, then $\mathbb{R}^2$ is the *translation group*. More generally, in n-dimensional spaces, $SO(n)$ is the rotation group and $\mathbb{R}^n$ is the translation group. To accomplish both rotation and translation, we utilize a combination of $SO(n)$ and $\mathbb{R}^n$. Let U be a $(n+1) \times (n+1)$ matrix such that

$$\begin{bmatrix} O & p \\ 0 & 1 \end{bmatrix}, \quad \text{where } O \in SO(n), \ p \in \mathbb{R}^n.$$

For a vector $x \in \mathbb{R}^n$, define an augmented $(n+1)$ vector $x_1 = \begin{bmatrix} x \\ 1 \end{bmatrix}$. Then, the first n entries of the vector Ux_1 represent a rotated and translated version of x. The set of all such matrices U is denoted by $SE(n)$, the *special Euclidean group*. $SE(n)$ is a group with matrix multiplication as the group operation and the $(n+1) \times (n+1)$ identity matrix as its identity element.

Depending on the specific problem, the group of transformations S can be $\mathbb{R}^n$, $SO(n)$, $SE(n)$, or Cartesian products of them. For an element $s \in S$, let sI^α denote the target template I^α transformed by the element s. For example if $S = SE(3)$, then sI^{air} is the airplane template rotated and translated according to s, as shown in Fig. 4. The set of all possible transformations of a target α is given by

$$\mathcal{O}_\alpha = \{sI^\alpha, s \in S\}.$$

$\mathcal{O}_\alpha$ is called an *orbit* associated with the target α. Then, S is said to *act on* $\mathcal{O}_\alpha$ (on the left) because it satisfies the following two conditions:

(a) If e is the identity element of S, then

$$eI^\alpha = I^\alpha, \quad \text{for all } \alpha \in \mathcal{A}.$$

(b) If $s_1, s_2 \in S$, then

$$s_2(s_1 I^\alpha) = (s_1 \circ s_2) I^\alpha, \quad \text{for all } \alpha \in \mathcal{A}.$$

The strength of a deformable template approach comes from the fact that all targets' occurrences can be modeled by using appropriate transformations on appropriate templates. Therefore, given an observed image of a target, the task reduces to finding the template and the transformation that fits that image best.

It must be noted that the variability in targets is not caused only by arbitrary orientations, and positions. There are other factors such as light conditions, targets' surface temperatures, texture variations, and their operational status. These factors can also be incorporated through more general transformations that are much higher dimensional than rigid rotation and translation. As an example, the thermodynamic variability in target surfaces as observed by FLIR cameras is modeled and estimated as a high-dimensional scalar field in [14, 15].

3 Sensor Modeling

So far we have considered three-dimensional target templates and a set of transformations on them to describe their occurrences in arbitrary scenes. The observations are, however, in general restricted to one- or two-dimensional arrays of numbers as generated by the sensors. Therefore, for a better understanding of images we have to build detailed models for these sensors. In these models the physics of sensor operation plays an important role because different sensors may produce very different pictures of the same scene. Microwave radars generate very different "pictures" of the target than second-generation FLIR cameras or video cameras.

In most sensors, imaging is essentially a projective mechanism operating by accumulating responses from the scene elements that project to the same pixel in the image. Mathematically, we will model the mechanism that maps the scene to some observation space $\mathcal{I}^{\mathcal{D}}$. In most cases $\mathcal{I}^{\mathcal{D}} = \mathbb{R}^d$ or $\mathbb{C}^d$ for some fixed number d. This mechanism can either be deterministic or random and constitutes a mapping T by which a transformed target, sI^α, appears to the observer as an image $I^D \in \mathcal{I}^{\mathcal{D}}$. In addition to T, a sensor may also generate random noise image, w, which is assumed to be additive. Then, the observation is modeled by

$$I^D = TsI^\alpha + w \in \mathcal{I}^{\mathcal{D}}. \tag{1}$$

In the ATR context, we must abstract this T in some generality to accommodate various sensors. The particular transformation T and the noise properties are determined by the sensor. For example, in case of an infrared camera, TsI^α is the mean field of a Poisson process for which the additive noise is not appropriate; see, for example, the discussion in [10]. It must be noted that accurate analytical expressions for T may not be available in all situations, but very often a high-quality simulation experiment (using special hardware) can be used to sample T at some predefined target orientations. For modeling radar returns, the XPATCH simulator has been widely used, whereas for FLIR cameras, PRISM is used. Visible spectrum images can be simulated on high-performance silicon graphics machines.

I^D may have multiple components corresponding to multiple sensors observing the scene simultaneously: $I^D \equiv (I_1^D, I_2^D, \ldots)$. Since the images are random, they are characterized by means of a statistical transition law, called the *likelihood function* $P(\cdot|\cdot)$: $\mathcal{I}^D \times (S, \mathcal{A}) \to I\!R_+$, summarizing completely the mapping from the target α at transformation s to the output I^D. Some of the sensors used frequently in ATR applications are as follows.

1. *Video imager*: A video sensor provides two-dimensional high-resolution real-valued images of rigid targets sampled

on a lattice of certain size, $I^D \equiv \{I^D(y), y \in Y \equiv \{1, 2, \ldots, \}^2, I^D(y) \in I\!R\}$. The images are assumed to result from an orthographic or a perspective projection of a three-dimensional surface intensity on to the camera focal plane, as shown in Fig. 5. Figures 5(a) and (c) depict an orthographic projection scheme utilized in pose estimation, when the target position is assumed to be known. Figures 5(b) and (d) illustrate the perspective projection system utilized when both the target pose and location are unknown. Figures 5(c) and (d) show TsI^{tank} for orthographic and perspective systems, respectively. It is assumed that the reflected light intensity is high so that $I^D \equiv \{I^D(y), y \in Y\}$ is taken to be a Gaussian random field, with the mean field given by TsI^α. Shown in Fig. 6(a) is an example of a simulated noisy video image of a truck.

2. *High range resolution radar*: A high range resolution (HRR) radar provides one-dimensional range profiles of rigid targets;

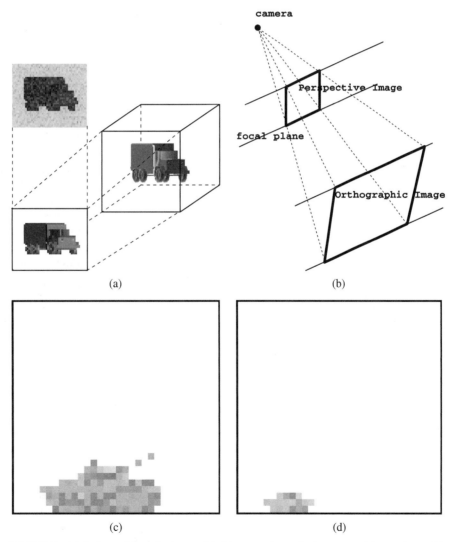

(a) (b)

(c) (d)

FIGURE 5 (a) Orthographic projection model; (b) perspective projection system; (c) an orthographic image; (d) a perspective image.

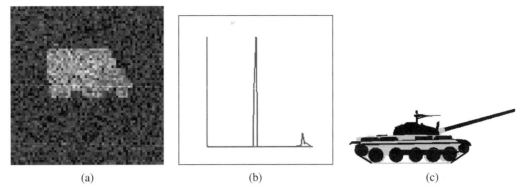

(a) (b) (c)

FIGURE 6 Simulated sample images obtained from different sensors: (a) video imagery (SGI), (b) high range resolution range imagery (XPATCH), and (c) FLIR imagery (PRISM).

see Jacobs *et al.* [6]. The transmitted electromagnetic pulses directed at a target are received back at the receiver, at times proportional to the distance traveled, representing the superposition of the echoes from all the reflectors in a bin along the range direction. The received signal is processed by a matched filter to generate a one-dimensional magnitude profile versus range, $I^D \equiv \{I^D(y), y = 1, 2, \ldots, I^D(y) \in IR\}$. The middle panel of Fig. 6 shows a range profile of a T62 tank at certain orientation, for a carrier frequency in the millimeter-wave region.

3. *Forward-looking infrared*: A second-generation FLIR camera captures the thermodynamical profile of a target body by means of CCD detectors; see Snyder *et al.* [7,10]. The measured data $I^D \equiv \{I^D(y), y \in \{1, 2, \ldots, \}^2, I^D(y) \in IR\}$ are each assumed to be Poisson with means given by the corresponding pixels of the perspective projection of the target's three-dimensional thermodynamic state. Figure 6(c) shows a tank's thermal profile, which when projected and blurred by the point-spread function of the camera, provides an infrared image.

4 Bayesian Framework

To analyze observed images and to set up estimation problems, we utilize the classical Bayesian framework. Similar to the optimal conditional mean estimators and their covariances as derived in the Kalman filtering, we will seek optimal estimators for ATR transformation groups.

A probability density function is often defined as the derivative of a probability distribution function. For probabilities on IR^n, this derivative is with respect to the infinitesimal volume element in $IR^n : dx = dx_1 dx_2, \ldots, dx_n$. On $SO(n)$, the volume element, has a different form since $SO(n)$ is not a vector space. The derivatives of functions are evaluated with respect to an infinitesimal volume element, which we will denote by $\gamma(dO)$; please refer to [1] for a description of this volume element, also called the Haar measure. The product of the volume elements on $SO(n)$ and IR^n provides a volume element on $SE(n)$. Note that just like $\int_{IR^n} f(x)\, dx$, the integration of a function on any set is defined with respect to the volume element of that set.

Now to model the uncertainty in associating an observed image to a particular template (indexed by α) and a particular transformation (denoted by s), we derive a posterior density on these unknowns. The posterior density is the product of the prior probability density on the unknowns and the likelihood of the data according to

$$P(s, \alpha \mid I^D) = \frac{1}{P(I^D)} P(s, \alpha) P(I^D \mid s, \alpha), \quad s \in S, \ \alpha \in \mathcal{A}.$$

The prior density $P(s, \alpha)$ incorporates our prior knowledge on finding a target α, at the pose and location dictated by the transformation s, in the scene. For example, in case of moving targets, the knowledge of target location may imply a higher probability of there being a future target presence in certain areas and low probability in others. The likelihood function $P(I^D \mid s, \alpha)$ quantifies the probability that a target α at the pose and location resulting from the transformation s will give rise to the observed image I^D. It is derived from the physical characteristics of the sensor map T and the statistics of the sensor noise. As an example, for the video sensor described earlier, the likelihood function takes the form

$$P(I^D \mid s, \alpha) = \frac{1}{(2\pi\sigma^2)^{d/2}} \exp\left(\frac{-1}{2\sigma^2} \| I^D - TsI^\alpha \|^2\right).$$

The resulting posterior includes all the information we have for target recognition.

Having obtained the posterior density, we will generate the classical estimators such as maximum a posteriori probability (MAP), MMSE, minimum absolute error (MAE), and entropy-based estimators. Following the classical Kalman filtering framework, we will seek MMSE estimators for the transformation, s, and a MAP estimation for the target type, α. Along with the estimators, we will also compute quantities that represent errors in estimation and impose a lower bound on these errors. First we construct MMSE estimators on the transformation groups $SO(n)$ and $SE(n)$, and then we seek a MAP estimator for α.

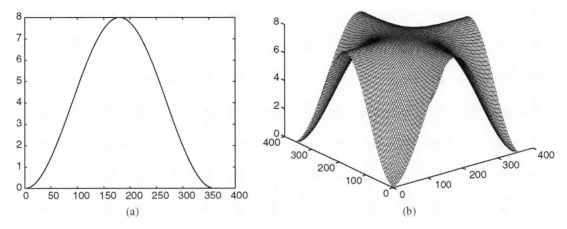

FIGURE 7   Plots of the Hilbert–Schmidt distance between orientations in (a) $SO(2)$ and (b) $SO(3)$ as a function of the Euler rotation angles.

5 Pose-Location Estimation and Performance

In view of the nonlinear relationship between the template transformations and the observed images, we have to estimate the transformations explicitly. As a first step we restrict our task to estimating the appropriate transformation from a given image I^D and for a fixed target hypothesis α. This α can be any index from our label set $\mathcal{A}$. Given the pose and location estimates for each α, we will seek the α that best matches the observed image in the next section.

A major difficulty being faced in the ATR setting is that, unlike the more classical estimation results, the parameter spaces of the scene of targets are groups with a group operation that is not necessarily addition. These groups can have a curved geometry, meaning that they may not be vector spaces. As an example, if we add two rotation matrices then the result is not a rotation matrix, so $SO(n)$ is said to have a curved geometry. In addition, the sensor outputs result from a sequence of nonlinear transformations on the scenes, including projective transformation, occlusion, ray tracing, etc. Therefore, we cannot inherit the more direct results obtained for estimators in Euclidean spaces associated with additive Gaussian channels.

To illustrate, we isolate and focus on estimating the target orientation, represented by special orthogonal group $SO(n)$, since it is a group with curved geometry. We will define an optimal estimator, called the *Hilbert–Schmidt estimator*, which is the MMSE estimator under a chosen norm. It is shown that the error associated with this estimator provides a lower bound on the error associated with any estimator. To establish a notion of the error on $SO(n)$, we have to define a function that computes a distance between any two points on $SO(n)$. We do so by considering the elements of $SO(n)$ as $n \times n$ matrices. For $O \in SO(n)$ and $O = \{O_{ij} : i, j = 1, 2, \ldots, n\}$, define the Hilbert–Schmidt (HS) norm given by $\|O\| = \sqrt{(\sum_{i,j=1}^{n} O_{ij}^2)}$. As a result of orthogonality, $\|O\|^2 = n$ and $\|O_1 - O_2\|^2 = 2[n - \text{trace}(O_1 O_2^\dagger)]$;

$\dagger$ denotes the matrix transpose. For $n = 3$, expressing the rotation $O_1 O_2^\dagger$ in terms of the usual Euler angles ϕ, ψ, θ (pitch, roll, and yaw), the norm becomes

$$\|O_1 - O_2\|^2 = (1 + \cos\theta)\cos(\phi + \psi) + \cos\theta.$$

Shown in Fig. 7 are the illustrations of this distance as a function of the angles parameterizing the orientations. Figure 7(a) plots the distance $4 - 4\cos(\theta)$ as a function of the rotation angle θ for $SO(2)$; Fig. 7(b) shows the distance $\|O_1 - O_2\|^2$ as a function of the angles θ and $(\phi + \psi)$ for $SO(3)$. It must be noted that other distance functions on $SO(n)$ (such as the Riemann distance) can also be used, as long as the resulting estimator is computationally tractable.

5.1 MMSE Estimator

Given an observed image I^D and the target type α, the MMSE estimate of the target orientation, also called the Hilbert–Schmidt estimate, is defined as follows.

Definition 1. *A Hilbert–Schmidt estimate (HSE) is given by* $\hat{s} : \mathcal{I}^D \to SO(n)$, *such that*

$$\hat{O}(I^D) = \operatorname*{argmin}_{O \in SO(n)} \int_{SO(n)} \|O - O'\|^2 P(O' \mid I^D) \gamma(dO'), \quad (2)$$

where γ is the infinitesimal volume element on $SO(n)$.

The HSE can be interpreted as a MMSE estimator when the error is computed by using the HS norm. As the norm squared $\|\cdot\|^2$ is continuous in its entries and $SO(n)$ is compact, the minimizer is attained in $SO(n)$, and hence the estimator is well defined. It should be noted that if the minimum is attained at multiple points, all these points are MMSE estimates, i.e., the estimator is then set valued. Instead of choosing the MMSE criterion, other error functions (such as absolute difference, step function, etc.) can also be used, resulting in the corresponding estimators.

Because of the geometric properties of rotation matrices, the evaluation of the HSE simplifies to the following form:

$$\hat{O}(I^D) = \underset{O \in SO(n)}{\mathrm{argmax}} \; \mathrm{trace}(OA^\dagger) \tag{3}$$

$$= \begin{cases} UV^\dagger, & \text{if determinant}(A) \geq 0 \\[4pt] ULV^\dagger, & L = \begin{bmatrix} 1 & 0 & \dots & 0 \\ 0 & 1 & \dots & 0 \\ \vdots & & & \\ 0 & 0 & \dots & -1 \end{bmatrix}, \\ & \qquad\qquad \text{if determinant}(A) < 0, \end{cases} \tag{4}$$

where

$$A = \int_{SO(n)} O P(O \mid I^D) \gamma(\mathrm{d}O), \tag{5}$$

and where $A = U \Sigma V^\dagger$ is the standard singular value decomposition of A, as described in [22]. The matrices U, Σ, V are arranged such that the singular values occur in decreasing order along the diagonal of Σ. Equation (5) can be interpreted as element-by-element integration in $\mathbb{R}^{n^2}$ with non-zero contributions only from the rotation matrices. This integral can be computed by using one of several numerical integration techniques: a Monte Carlo sampling technique is presented in [11], and the trapezoidal integration is utilized in [4] to compute $\hat{O}$, the orientation estimate.

5.2 Lower Bound on Expected Error

The next issue is to define a quantity that can be used to assess any given estimator in terms of its expected estimation errors. For example, in the case of Euclidean parameters, Cramer–Rao lower bounds are often used to establish the optimum performance and the estimators are judged through these comparisons. In the context of orientation estimation in ATR, we will derive Hilbert–Schmidt bounds, which provide a way of comparing different algorithms. The Hilbert–Schmidt bound (HSB) is defined to be the minimum error attainable when the error is specified using the HS norm.

Definition 2. *Define the HSB as the quantity $\int_{\mathcal{I}^D} \varrho(I^D) P(I^D) \, \mathrm{d}I^D$, where $\mathrm{d}I^D$ is the base measure on $\mathcal{I}^D$, and*

$$\varrho(I^D) = \int_{SO(n)} \|\hat{s} - s'\|^2 P(s' \mid I^D) \gamma(\mathrm{d}s'). \tag{6}$$

The importance of the HSB stems from the fact that for any estimator $\tilde{s} : \mathcal{I}^D \to SO(n)$,

$$E\|\tilde{O} - O\|^2 \geq E\|\hat{O} - O\|^2 \equiv \text{HSB}, \tag{7}$$

where $\hat{O}$ is the HSE as defined earlier. The expectation is over both: the randomness in the data and the randomness in the

unknown parameters according to

$$E\|\tilde{O} - O\|^2 = \int_{\mathcal{I}^D} \left(\int_{SO(n)} \|\tilde{O}(I^D) - O'\|^2 P(O' \mid I^D) \gamma(\mathrm{d}O') \right) \times P(I^D) \, \mathrm{d}I^D.$$

For the proof of this result please refer to [4]. Because of the structure of $SO(n)$, the HSB takes the form

$$\int_{\mathcal{I}^D} \varrho(I^D) p(I^D) \, \mathrm{d}I^D,$$

where $\rho(I^D) = 2(n - \mathrm{trace}(A^\dagger \hat{O}))$ for A as defined in Eq. (5). We shall say that the HSE is efficient in the sense that it has HS efficiency $= 1$, with the HS efficiency of an arbitrary estimator $\tilde{s} : \mathcal{I}^D \to SO(n)$ defined as the ratio

$$\text{efficiency}(\hat{O}) = \frac{E\|\hat{O} - O\|^2}{E\|\tilde{O} - O\|^2}.$$

Shown in Fig. 8(a) is a plot of the HSB for estimating the truck orientation, in $SO(2)$, as a function of the noise standard deviation, σ. To avoid some symmetry issues (please refer to [21] for a discussion on symmetry issues), this bound is computed by considering only the half-circle. The zero expected error implies perfect orientation estimation; the maximal expected error of 1.45 implies completely unreliable estimation of the truck orientation. Superimposed on the error plot are three x's, corresponding to three noise levels. The three truck images in Figs. 8(b)–8(d) are samples at the noise levels corresponding to the x's. Figure 8(b) corresponds to low-noise resulting in a perfect pose estimation; however, notice the rapid increase in the estimation error as the noise level increases.

To explain the performance curves, at a given noise level, say at noise standard deviation 0.4, the HSB value of the video sensor is 1.0; i.e., the minimum expected error in estimating truck orientation in this environment is 1.0. Also, for a noise level <0.2, the HSB $\simeq 0$ and for deviation >0.6, the error is maximal. Errorless estimation is, thus, possible in the case of the video sensor for a noise level ≤ 0.2, and reasonable estimates (HSB of 0–1.2) are possible for deviations in the range $[0.2, 0.6]$; beyond that the data are too noisy to provide any information for inference on target orientation. To illustrate the significance of the HSB $= 1.0$, consider Fig. 9. Figure 9(f) shows a noisy image of the truck at the noise level corresponding to HSB $= 1.0$. At this particular noise level, the estimation is degraded to such a point that, on average, the estimates span a 1.0 HSB unit around the mean. Four sample orientations, all within the 1.0 HSB unit of the orientation shown in Fig. 9(a), are shown in the other panels. Naturally, the target geometry should determine the bound associated with pose estimation by a given sensor suite, as is depicted in Fig. 10. Shown here are HSB curves for two different targets: tank and truck, when imaged by a video camera. This curve shows that, in low-noise situations, the tank orientation estimates are better than the truck estimates by the

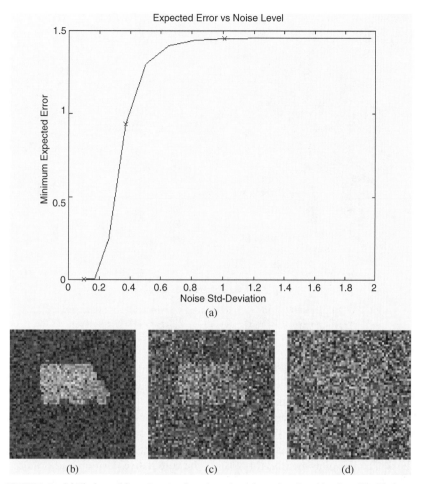

FIGURE 8 (a) The bound for estimating the orientation of a truck, using video data. (b)–(d) show sample images of the truck at three noise levels consistent with the x's in (a).

FIGURE 9 (a) Orientation with a 1.0 HSB unit; (b)–(d) show four different truck orientations within HSB = 1.0 of the orientation in (a); (f) shows the associated imagery with this uncertainty level.

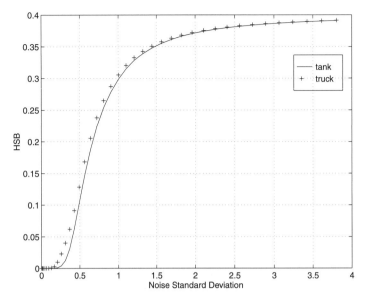

FIGURE 10   Panel shows the variation of the HSB with noise for two targets: truck and tank. For low-noise levels the bounds are different; at higher noise levels the performance is identical.

video sensor, whereas at higher noise levels, the performance is similar.

Sensor fusion occurs automatically in this setting. The increased number of data observations I_1^D, I_2^D, ... increases the accuracy of the estimator. Figure 11(a) shows plots of the HSB on expected error versus noise level in estimating tank orientation for the two sensors: the broken line plots the HSB for HRR, the solid line shows the HSB for video, and the x's display the HSB for the joint case. Figure 11(b) shows the HSB curves for the tank pose estimation by three individual sensors: the solid line for FLIR, the broken line for HRR, and the dotted line for video. The HSB for the joint case is shown by the crosses. Notice that

since the information is being optimally fused in the Bayesian setting, the joint curves always deliver a higher accuracy for the estimation.

For joint estimation of target pose and location, the transformation s is an element of $SE(n)$. As described in [8] both HSE and HSB simply extend from $SO(n)$ to $SE(n)$. To illustrate the cumulative position and orientation estimation bound, we have utilized a dataset involving real FLIR images of a tank, mounted on a pedestal and imaged at 120 different orientations. (This dataset is obtained courtesy of Dr. Richard Sims at Army Missile Command). Shown in Fig. 12 are six sample images from this dataset. Shown in Fig. 13 is the variation of the cumulative

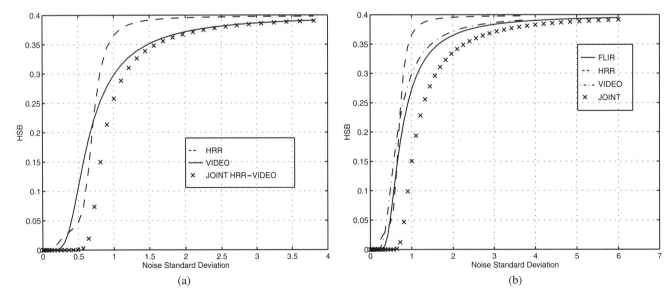

FIGURE 11   (a) the plots of the HSB on expected error versus noise level in estimating tank orientation for the two sensors; (b) the HSB curves for the tank pose estimation by three individual sensors.

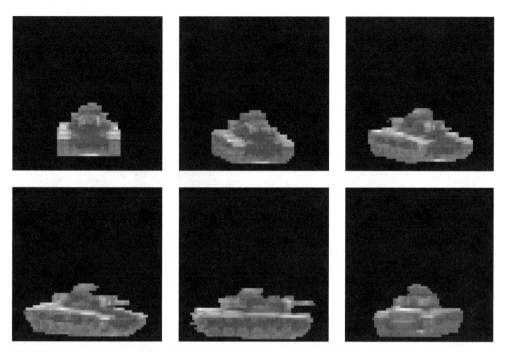

FIGURE 12 Sample images from a dataset of real FLIR images of a tank (data courtesy of Dr. Richard Sims of AMCOM). The images are downscaled to 64 × 64 for the results described in this paper.

position and orientation error (on $SE(2)$) versus the sensor noise. This error bound can be utilized to analyze multisensor, multitarget situations.

6 Target Recognition and Performance

Having established a framework for target orientation and location estimation, we now focus on the main task: finding the index α that best matches a given image I^D. As described earlier, in a Bayesian framework the estimated target type is given by

the index with maximum a posteriori probability. It becomes an M-ary hypothesis test. That is,

$$\hat{\alpha} = \underset{\alpha \in \mathcal{A}}{\operatorname{argmax}}\, P(\alpha \mid I^D), \tag{8}$$

where the posterior is calculated by using the Bayes' rule,

$$P(\alpha \mid I^D) = \frac{P(I^D \mid \alpha)\, P(\alpha)}{P(I^D)}.$$

The term $P(I^D \mid \alpha)$ is the likelihood of observing I^D given that the true target is α and can be evaluated as the integration over all transformations

$$P(I^D \mid \alpha) = \int_S P(I^D \mid s, \alpha)\, P(s \mid \alpha)\, \mathrm{d}s.$$

In the context of selecting α and ATR, s can be considered as a nuisance parameter. This important integral governs the relationship between target recognition (selecting α) and the pose-location estimation (estimating s). It is intuitively clear that recognition and pose estimation are inherently linked; accuracy of target recognition is directly determined by the accuracy of pose estimation.

In most practical situations, the integrand is too complicated to be computed analytically, and one of several approximations, numerical and analytical, can be used. To illustrate some of these methods we simplify to *binary* target recognition. That is, given an observed image, our task is to select one of the two targets: α_0 or α_1. For binary decision and $S = SO(n)$, the nuisance integral can be evaluated by using one of the following three methods.

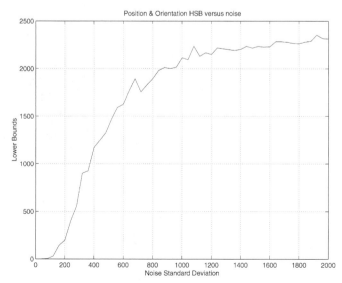

FIGURE 13 HSB for joint pose-location estimation on $SE(2)$.

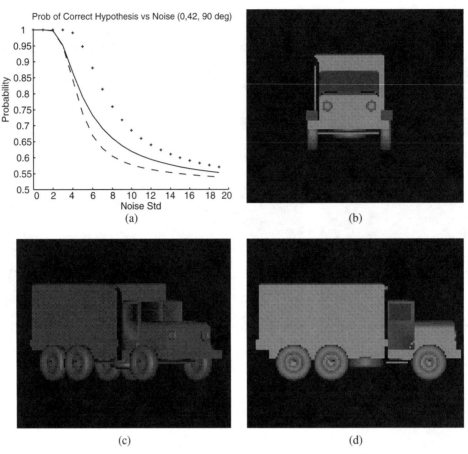

FIGURE 14 The probability of a correct hypothesis in binary Bayesian identification plotted against increasing noise for three underlying orientations: the crosses for 90, the solid line for 42 and the broken line for 0 deg. (b)–(d) show the three underlying truck orientations: (b), 0, (c), 42, and (d), 90.

1. Quadrature integration: since $SO(n)$ is compact, one can compute the integral approximately by evaluating the integrand at some sampled points and using one of the many established formulas (trapezoidal, Simpson's, Gauss-quadrature). As an example, for ground targets ($n = 2$), we have evaluated the integral by using the trapezoidal rule and performed hypothesis selection for target recognition. Shown in Fig. 14 are the results from binary recognition for $\alpha_0 = $ truck and $\alpha_1 = $ tank. A video image was simulated for α_0 at some orientation s_0 with respect to the sensor, the integral was computed for that image, and a decision is made following Bayes' selection. Plotted in Fig. 14(a) are the probabilities of selecting the correct target, α_0, studied against the sensor noise for three different target orientations. Notice that when the target is broadside with most of the pixels in the image, the probability of recognizing it is the highest.

2. Generalized likelihood ratio: in this procedure the integral value is approximated by the maximum value of the integrand as a function of the integration variable [13]. The test is given by

$$\frac{\max_{s \in S} P(I^D \mid s, \alpha_1)}{\max_{s \in S} P(I^D \mid s, \alpha_0)} \underset{\alpha_0}{\overset{\alpha_1}{\lessgtr}} \frac{P(\alpha_0)}{P(\alpha_1)} .$$

In other words, the maximum likelihood estimation of s is cal-

culated for both of the hypotheses, and the ratio of maximum likelihoods compared to the ratio of prior probabilities decides the hypothesis selection.

3. Asymptotics: to obtain analytical expressions, which are often more useful than the numerical approximations, asymptotic approximations using Laplace's method ([23]) can be derived. The basic approach is to assume a very large signal to noise ratio, either through large sample size or small sensor noise, and approximate the integrand using normal approximation of the integrand [5]. This result is then used in computing the likelihood ratio and, furthermore, the probability of error in the hypothesis selection. The error probability decreases exponentially with the decrease in the sensor noise, with the rate depending on the accuracy in pose estimation. This highlights the relevance of transformation estimation accuracy in hypothesis testing. A more accurate pose estimator can lead to a better recognition system.

7 Discussion

In this paper we have described a model-based Bayesian approach to automated target recognition. Models for targets are

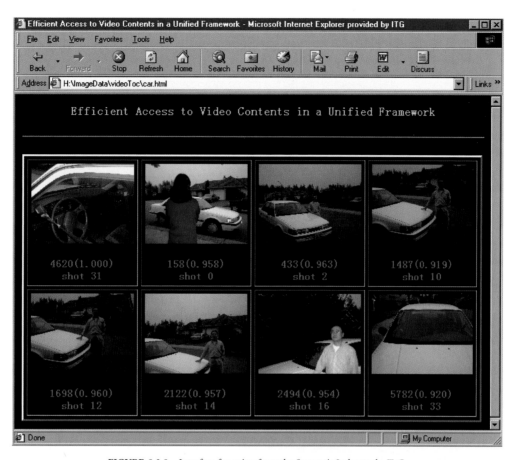

FIGURE 9.2.9   Interface for going from the Semantic Index to the ToC.

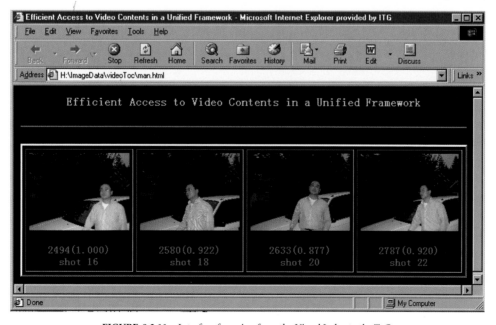

FIGURE 9.2.11   Interface for going from the Visual Index to the ToC.

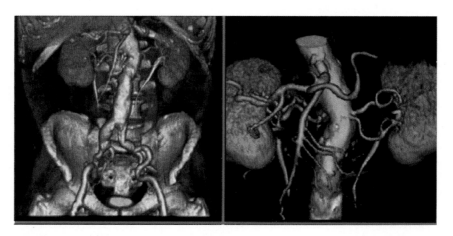

FIGURE 10.2.5 Volume rendering from a sequence of X-ray CT images, showing the abdominal cavity and kidneys (CT images courtesy of G. E. Medical Systems).

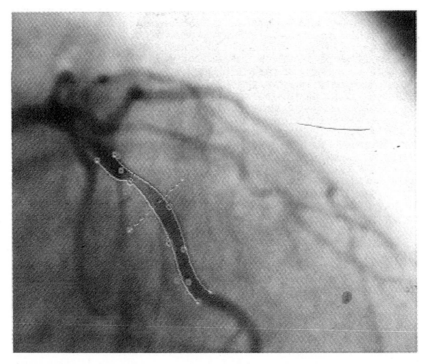

FIGURE 10.3.3 Result of an active-contour analysis applied to a selected artery in a typical 2-D angiogram. The green points are the manually identified control points. The red lines are the computed vessel wall borders. From [8].

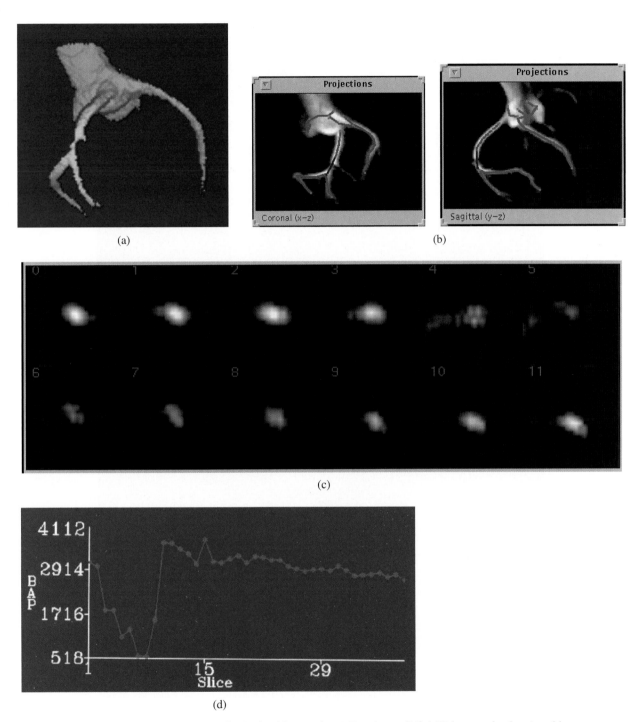

(a)

(b)

(c)

(d)

FIGURE 10.3.6 Composite view of a visual tool for assessing a 3-D angiogram [12]. (a) Volume-rendered version of the extracted 3-D arterial tree. (b) 2-D coronal (x-z) and sagittal (y-z) maximum-intensity projection images, with extracted arterial axes superimposed; red lines are extracted axes, green squares are bifurcation points, and the blue line is a selected artery segmented highlighted below. (c) Series of local 2-D cross-sectional images along a stenosed branch; these views lie orthogonal to the automatically defined axis through this branch. (d) Cross-sectional area plot along the stenosed branch.

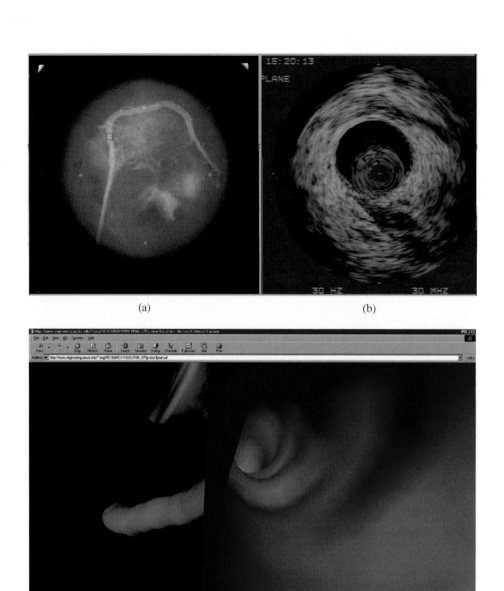

(a)

(b)

(c)

FIGURE 10.3.7 Views from a VRML-based IVUS system. (a) 2-D angiogram; the square indicates interior arterial viewing site of interest. (b) Corresponding cross-sectional IVUS frame of the arterial lumen. (c) 3-D surface rendering of the artery surface. (Courtesy of Dr. Milan Sonka, University of Iowa.)

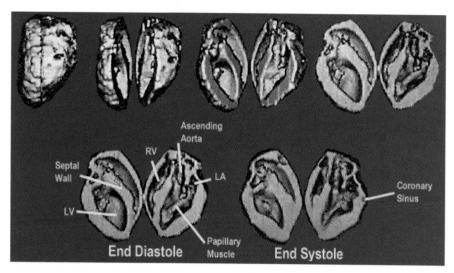

FIGURE 10.3.10 3-D surface-rendered heart image. The top row shows the computer-generated "dissection" of the 3-D heart volume; the bottom row has partially labeled heart anatomy. LA is left atrium and RV is right ventricle. From [20].

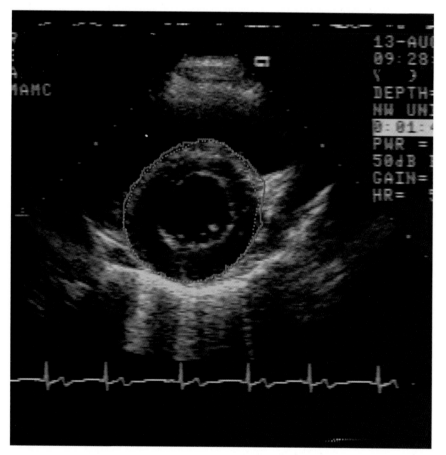

FIGURE 10.3.12 Ultrasound B-mode image of a human heart. The LV chamber is a center of image. The red border is the automatically detected epicardial border from 3-D graph search; the green border is the manually traced border. (Figure courtesy of Dr. Edwin L. Dove, University of Iowa and Dr. David D. McPherson, Northwestern University.)

end–diastole

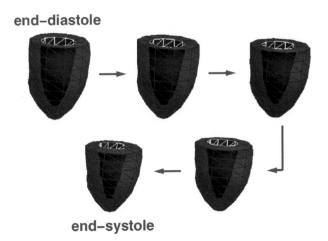

end–systole

FIGURE 10.3.14   3-D myocardial wall model derived from deformable surface tracking SPAMM tag lines. The model shows inner and outer borders of the myocardium. Also shown is the evolution of myocardial wall and LV chamber shape from end diastole to end systole. (Figure courtesy of Dr. Jinah Park, University of Pennsylvania).

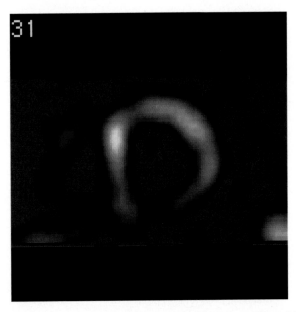

FIGURE 10.3.15   SPECT myocardial perfusion analysis, using an injected thallium-201 tracer. Shown is a cross-sectional view of the myocardium (LV chamber is the cavity at center of the image), with pixel intensity proportional to myocardial blood flow distribution. (Image courtesy of Dr. Richard Hichwa, PET Imaging Center, University of Iowa).

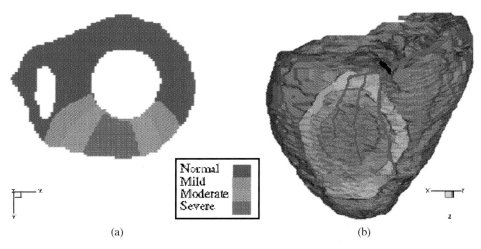

(a)   (b)

FIGURE 10.3.16 Example of a body-surface potential map. (a) Mapping for a 2-D slice through the heart; the cavities correspond to the ventricles. (b) 3-D surface-rendered view of the same map. The color-coding indicates the degree of myocardial ischemia (reduction in blood flow). The red lines on the 3-D view indicate a stenosed arterial region that brought about the ischemia. From [32].

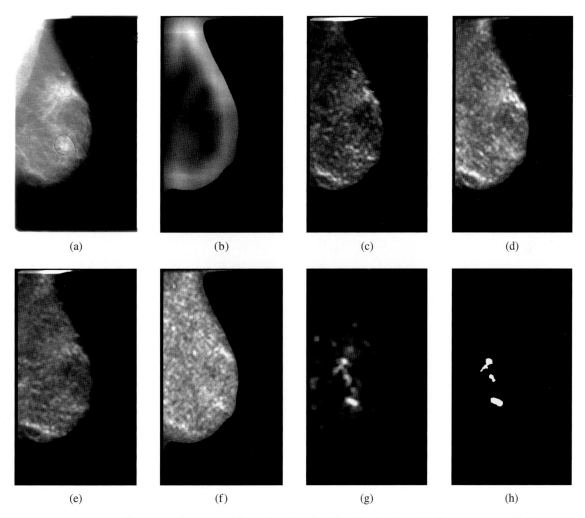

(a)   (b)   (c)   (d)

(e)   (f)   (g)   (h)

FIGURE 10.4.8 Illustration of a spiculated lesion detection algorithm, showing the original mammogram with the ground truth marking the location of a spiculated lesion in red in (a), the ALOE feature image in (b), and the Laws feature images $L5 * E5^t$, $E5 * S5^t$, $L5 * S5^t$, and $R5 * R5^t$ in (c)–(f). The blurred probability image is shown in (g). Image (h) is a thresholded version of (g); it shows three detected spiculated lesions. The bottom one is a correct detection corresponding the the ground truth outline in (a), and the other two white blobs are false detections.

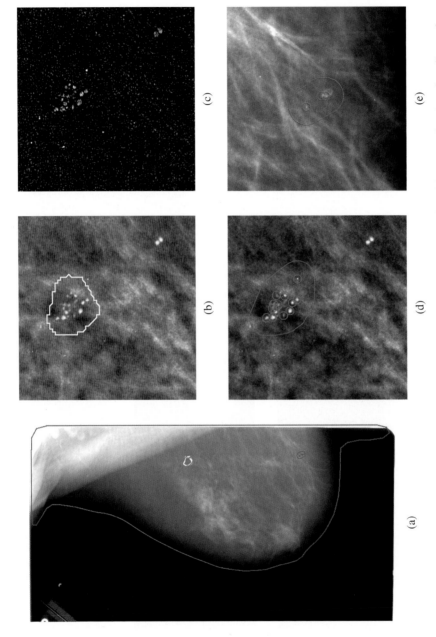

FIGURE 10.4.9 Illustration of a calcification detection algorithm, showing one true positive and one false positive detection of a malignant cluster of pleomorphic calcifications. (a) An overview of a segmented breast with one ground truth region (white) and two detections (green and red). The border of the segmented breast is shown in purple. (b) A closeup of the cluster of calcifications with ground truth overlaid in white. (c) The result of enhancing the calcifications in the image using the algorithm described in Section 6.3. (d) The result of thresholding the enhanced mammogram, labeling individual calcifications, finding a cluster group of more than three calcifications linked by intercalcification distances of <4 mm. Individual calcifications in the group are circled in green and the cluster is marked with a green border. (e) A false detection of a group of calcifications.

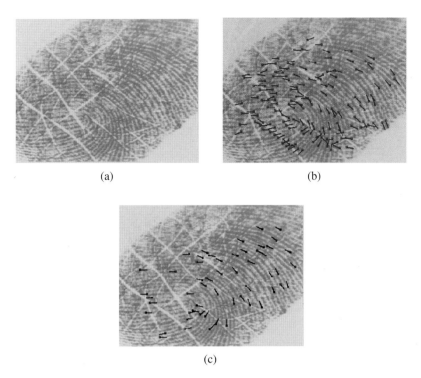

(a)   (b)

(c)

FIGURE 10.5.9 Fingerprint enhancement results: (a) a poor-quality fingerprint; (b) minutiae extracted without image enhancement; (c) minutiae extracted after image enhancement [11].

FIGURE 10.5.13 Aligned ridge structures of mated pairs. Note that the best alignment in one part (midleft) of the image results in a large displacements between the corresponding minutiae in the other regions (bottom right). © IEEE.

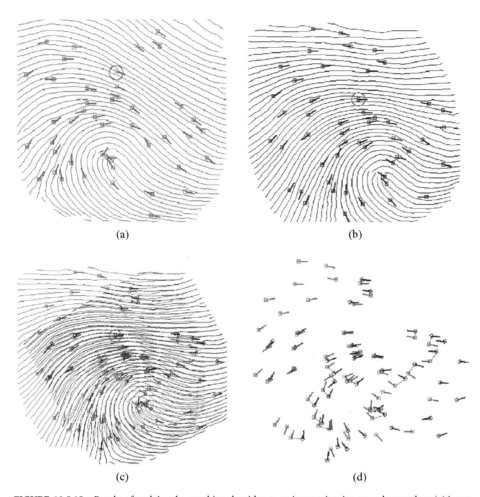

(a)

(b)

(c)

(d)

FIGURE 10.5.15 Results of applying the matching algorithm to an input minutiae set and a template: (a) input minutiae set; (b) template minutiae set; (c) alignment result based on the minutiae marked with green circles; (d) matching result, where template minutiae and their correspondences are connected by green lines. © IEEE.

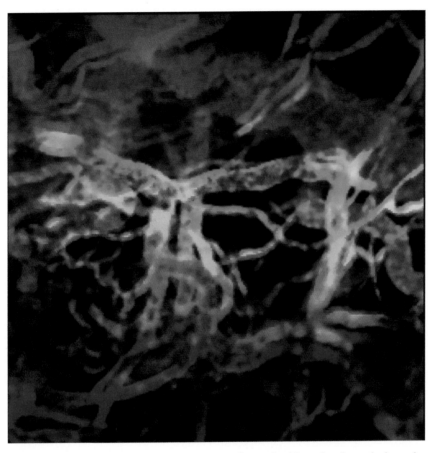

FIGURE 10.7.11 3-D representation of the microvasculature of an islet graft at the renal subcapsular site. The image is color coded to denote depth. The vessels appearing in the lower portion (blue) are at a depth of 30 μm, whereas those in the middle and upper portions of the image are at a depth of ~85 (green) and 135 μm (violet), respectively (reproduced with permission from [29]).

developed by using a deformable template approach in which each target occurrence in a given scene is modeled by using a template and a transformation. The transformations associated with ATR form groups and have curved geometry. Utilizing the Hilbert–Schmidt norm, we have defined a MMSE estimator for pose and pose/location estimates, and also lower bounded the expected squared error for any estimator. Pose-location estimates are incorporated in target recognition, which is performed using Bayesian hypothesis selection. The posterior calculation includes an integration over the nuisance parameters, and several methods are presented to perform this numerically. The asymptotic technique leads to an analytic expression for the performance analysis by providing the probability of errors in recognition.

Among the remaining challenges in developing a general ATR system has to be developing reasonable clutter models. Any element of the scene that is not a target of interest and influences the observed images can be called clutter. If the cluttered is so structured that it appears like a target, then it can severely affect the ATR performance. Statistical models are being developed to tackle this issue.

Acknowledgment

We gratefully acknowledge contributions from our colleagues, Matt Cooper and Marc Loizeaux, in generating some implementation results presented here. This research was supported by grants ARO CIS DAA-H04-95-1-0494, ONR MURI N00014-98-1-0606, ARO MURI DAA-H04-96-1-0445, NSF-9871196, ARO DAA-G55-98-1-0102, and ARO DAAD19-99-1-0267.

References

[1] W. M. Boothby, *An Introduction to Differential Manifolds and Riemannian Geometry* (Academic, New York, 1986).

[2] D. E. Dudgen and R. T. Lacoss, "An overview of automatic target recognition," *MIT Lincoln Lab. J.* **6**, 3–10 (1993).

[3] D. E. Dudgeon, "Atr performance modeling and estimation," MIT Lincoln Lab. Tech. Rep. 1051, (MIT, Lexington, MA, 1998).

[4] U. Grenander, M. I. Miller, and A. Srivastava, "Hilbert-Schmidt lower bounds for estimators on matrix lie groups for ATR," *IEEE Trans. Pattern Anal. Machine Intell.* **20**, 790–802 (1998).

[5] U. Grenander, A. Srivastava, and M. I. Miller, "Asymptotic performance analysis of Bayesian object recognition," *IEEE Trans. Inf. Theory* **x**, (in review) (1998).

[6] Steve P. Jacobs, "The utility of HRR radar data in automatic target recognition," *Electron. Signals Syst. Res. Lab. Mono.* (1994).

[7] A. D. Lanterman, M. I. Miller, and D. L. Snyder, "Implementa-

tion of jump-diffusion processes for understanding flir scenes," in *Automatic Object Recognition V*, F. A. Sadjadi, ed., *Proc. SPIE* **2485**, 309–320 (1995).

[8] M. Loizeaux, A. Srivastava, and M. I. Miller, "Pose/location estimation of ground targets," in *Signal Processing, Sensor Fusion, and Target Recognition*, I. Kadar, ed., *Proc. SPIE* **3720**, 140–151 (1999).

[9] D. Marr, *VISION: A Computational Investigation into the Human Representation and Processing of Visual Information* (W. H. Freeman, New York, 1982).

[10] D. L. Snyder, A. M. Hammoud, and R. L. White, "Image recovery from data acquired with a charge-coupled-device camera," *J. Opt. Soc. Am. A* **10**, 1014–1023 (1993).

[11] A. Srivastava, M. Miller, and U. Grenander, *Ergodic Algorithms on Special Euclidean Groups for ATR*. Systems and Control in the Twenty-First Century: Progress in Systems and Control, Vol. 22 (Birkhauser, 1997).

[12] M. Steinberger, *Algebra* (PWS, Boston, 1994).

[13] H. L. Van Trees, *Detection, Estimation, and Modulation Theory*, Vol. I (Wiley, New York, 1971).

[14] M. L. Cooper and M. I. Miller, "Information measures for object recognition," *Algorithms for Synthetic Aperture Radar Imagery V*, E. G. Zelnio, Ed., *Proc. SPIE* **3370**, 637–645 (1998).

[15] M. Cooper, U. Grenander M. Miller, A. Srivastava, Accommodating geometric and thermodynamic variability for forward-looking infrared sensors Proc. SPIE Vol. 3070, p. 162–172, *Algorithms for Synthetic Aperture Radar Imagery, IV*, E. G. Zelnio, Ed. (1997).

[16] J. K. Aggarwal and S. Shah, "Object recognition and performance bounds," *Image Analysis and Processing*, 9th International Conference, ICIAP Proc. **2**, 722–794, **1**, 343–360 (1997).

[17] F. D. Garber and E. G. Zelnio, "On simple estimates of ATR performance, and initial comparisons for a small data set." In *Algorithms for Synthetic Aperture Radar Imagery IV*, E. G. Zelnio, Ed., *Proc. SPIE* **3070**, 150–161 (1997).

[18] M. Lindenbaum, "Bounds on shape recognition performance," *IEEE Trans. Pattern Anal. Machine Intell.* **17**, 666–680 (1995).

[19] L. M. Novak, G. R. Benitz, G. J. Owirka, and L. A. Bessette, "ATR performance using enhanced resolution SAR," in *Algorithms for Synthetic Aperture Radar Imagery III*, E. G. Zelnio, R. J. Douglass, Eds., *Proc. SPIE* **2757**, 332–337 (1996).

[20] A. C. Williams and B. N. Clark, "Evaluation of SAR ATR," in *Single Processing, Sensor Fusion, and Target Recognition V*, I. Kadar, V. Zibby, Eds., *Proc. SPIE* **2755**, 36–45 (1996).

[21] A. Srivastava and U. Grenander, "Metrics for target recognition," in *Applications of Artificial Neural Networks in Image Processing III*, N. M. Nosrabadi, A. K. Katsaggelos, Eds., *Proc. SPIE*, **3307**, 29–37 (1998).

[22] Gene H. Golub and C. F. Vanloan, *Matrix Computations* (Johns Hopkins U. Press, Baltimore, MD, 1989).

[23] G. Polya and G. Szego, *Problems and Theorems in Analysis*, Translated by D. Aeppli (Springer-Verlag, New York, 1976).

Index

883